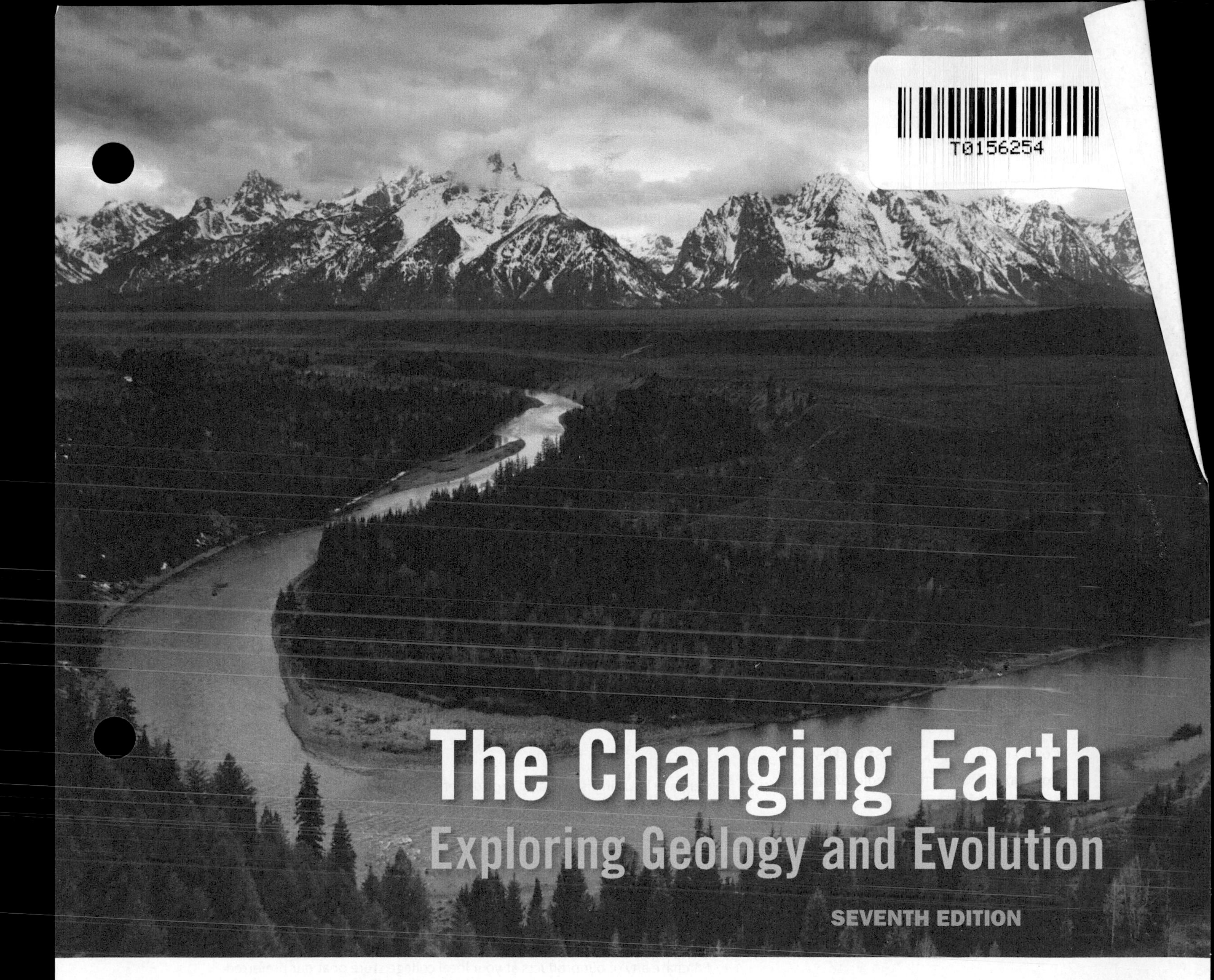

The Changing Earth

Exploring Geology and Evolution

SEVENTH EDITION

James S. Monroe
Professor Emeritus
Central Michigan University

Reed Wicander
Central Michigan University

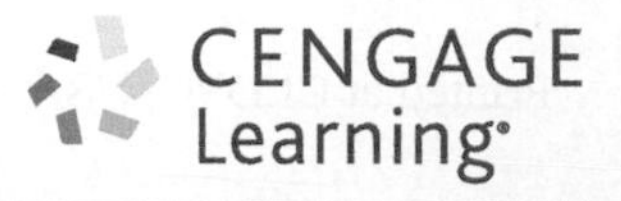

Australia • Brazil • Mexico • Singapore • United Kingdom • United States

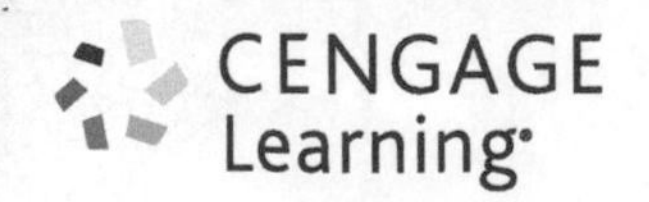

***The Changing Earth: Exploring Geology and Evolution*, Seventh Edition**
James S. Monroe and Reed Wicander

Product Team Manager: Yolanda Cossio

Senior Product Manager: Aileen Berg

Content Developer: Jake Warde

Product Assistant: Victor Luu

Content Coordinators: Shannon Holt and Kellie Petruzzelli

Media Developer: Stefanie Beeck

Marketing Manager: Lindsy Lettre

Senior Content Project Manager: Carol Samet

Senior Art Director: Pam Galbreath

Manufacturing Planner: Rebecca Cross

Senior Rights Acquisitions Specialist: Dean Dauphinais

Production Service: Chris Schoedel and Andrea Clemente, Cenveo Publisher Services

Photo Researcher: Megan Lessard, PreMediaGlobal

Text Researcher: Nirmala Arumugam, PreMediaGlobal

Copy Editor: Cenveo Publisher Services.

Cover and Text Designer: Lisa Buckley

Cover Image: © Michael Medford/National Geographic Stock

Compositor: Cenveo Publisher Services

Library of Congress Control Number: 2013940238

ISBN-13: 978-1-285-73341-8
ISBN-10: 1-285-73341-X

Loose-leaf Edition:
ISBN-13: 978-1-305-25860-0
ISBN-10: 1-305-25860-6

Cengage Learning
200 First Stamford Place, 4th Floor
Stamford, CT 06902
USA

Cengage Learning is a leading provider of customized learning solutions with office locations around the globe, including Singapore, the United Kingdom, Australia, Mexico, Brazil, and Japan. Locate your local office at **www.cengage.com/global**.

Cengage Learning products are represented in Canada by Nelson Education, Ltd.

To learn more about Cengage Learning Solutions, visit **www.cengage.com.**

Purchase any of our products at your local college store or at our preferred online store **www.cengagebrain.com.**

About the cover image:

The Snake River and Teton Range, Wyoming The Snake River, in the foreground, meanders through Bridger-Teton National Forest, Wyoming. A large point bar (see Chapter 12) can be clearly seen in the lower left corner. The snow-covered Teton Range, including the Grand Teton (the highest peak visible) of the Rocky Mountains is dramatically set off in the background with several easily identified glacial features such as U-shaped valleys and horns (see Chapter 14). Bridger-Teton National Forest, located in western Wyoming, consists of 14,000 km^2 of forest and is part of the 81,000 km^2 Greater Yellowstone Ecosystem, one of few remaining sizeable, northern temperate zone ecosystems.

Printed at CLDPC, USA, 02-23

Brief Contents

Contents

Understanding Earth: A Dynamic and Evolving Planet 3

Plate Tectonics: A Unifying Theory 27

Minerals—The Building Blocks of Rocks 61

4

5

Weathering, Erosion, and Soil 137

Sediment and Sedimentary Rocks 159

Metamorphism and Metamorphic Rocks 181

Earthquakes and Earth's Interior 201

10

11

12

13

14

15

Oceans, Shorelines, and Shoreline Processes 385

Geologic Time: Concepts and Principles 413

Organic Evolution—The Theory and Its Supporting Evidence 447

Precambrian Earth and Life History 471

Paleozoic Earth History 501

Paleozoic Life History 539

Mesozoic Earth and Life History 575

Cenozoic Earth and Life History 613

Geology in Perspective 655

APPENDICES

Preface

Earth is a dynamic planet that has changed continuously during its 4.6 billion years of existence. The size, shape, and geographic distribution of the continents and ocean basins have changed through time, as have the atmosphere and biota. As scientists and concerned citizens, we have become increasingly aware of how fragile our planet is and, more importantly, how interdependent all of its various systems and subsystems are.

We also have learned that we cannot continually pollute our environment and that our natural resources are limited and, in most cases, nonrenewable. Furthermore, we are coming to realize how central geology is to our everyday lives. For example, on March 11, 2011, a magnitude 9.0 earthquake struck Japan, killing more than 20,000 people and generating a tsunami that wreaked destruction along the coast of northeastern Japan, as well as damaging three nuclear power plants and causing radioactive leakage in one of them. A major oil spill in the Gulf of Mexico in 2010 resulted in much ecological damage along the shorelines of the Gulf Coast of the United States, as well as to the biota in the Gulf of Mexico. And, finally, Hurricane Sandy caused tremendous damage and major flooding along the Eastern Seaboard of the United States in October 2012.

All of these events point out how much geology affects our lives, as well as the global economy. For these and other reasons, geology is one of the most important college or university courses that a student can take.

The seventh edition of *The Changing Earth: Exploring Geology and Evolution* is designed to be an introductory course in geology that can serve both majors and nonmajors in geology and the Earth sciences. One of the problems with any introductory science course is that students are overwhelmed by the amount of material that must be learned. Furthermore, most of the material does not seem to be linked by any unifying theme and does not always appear to be relevant to their lives. This book, however, is written to address that problem in that it shows, in its easy-to-read style, that geology is an exciting and ever-changing science in which new discoveries and insights are continually being made.

The goals of this book are to provide students with a basic understanding of geology and its processes and, most importantly, with an understanding of how geology relates to the human experience—that is, how geology affects not only individuals, but society in general. It is also our intent to present the geologic and biologic history of Earth, not as a set of encyclopedic facts to memorize, but rather as a continuum of interrelated events reflecting the underlying geologic and biologic principles and processes that have shaped our planet and life upon it.

Instead of emphasizing individual, and seemingly unrelated, events, we seek to understand the underlying causes of why things happened the way they did and how all of Earth's systems and subsystems are interrelated. Using this approach, students will gain a better understanding of how everything fits together, and why geology is such an important course.

With these goals in mind, we introduce the major themes of the book in the first chapter to provide students with an overview of the subject and to enable them to see how the various systems, subsystems, and cycles of Earth are interrelated. We then cover the unifying theme of geology—plate tectonics—in the second chapter. Plate tectonic theory is central to the study of geology because it links together many aspects of geology. It is a theme that is woven throughout this edition.

The economic and environmental aspects of geology are emphasized throughout the book rather than treating these topics in separate chapters. In this way, students can see, through topical and interesting examples, how geology affects our lives. Climate change is an especially relevant and important topic that currently is in the news and being discussed and debated by scientists, politicians, and citizens alike. Because of its importance, we introduce the topic in the first chapter and integrate it throughout the book as it relates to the various topics covered. Geology is unique in that it can provide the perspective of geologic time in this important debate as to what, and the possible degree to which, humans have contributed to climate change.

Another topic that has been in the news of late is hydraulic fracturing, popularly called "fracking." This controversial method of releasing oil and gas from nearly impermeable shales, is both an environmental and energy issue that elicits strong feelings from both its proponents and opponents. Because of the importance of this topic and its cross-disciplinary nature, we cover it in several chapters.

Features in the Seventh Edition

Just as Earth is dynamic and evolving, so too is *The Changing Earth: Exploring Geology and Evolution*. The seventh edition has undergone significant rewriting and updating, resulting in a volume that is still easy to read and contains a high level of current information. Drawing on the comments and suggestions of reviewers and users of the sixth edition, we have retained those features that were both relevant and popular in the sixth edition as well as incorporated several new features into this edition.

- Many new, bold, and dramatic photos open each chapter, a number of which are recent geologic events, which add relevancy to the text and emphasize the theme of how geology relates to humans.
- Chapter content has been extensively updated and rewritten to (1) help clarify concepts, (2) emphasize underlying processes, and (3) make the material more exploratory.
- *Have You Ever Wondered?* questions follow the chapter *Outline* at the beginning of each chapter. These intriguing questions are designed to spark student interest about what is covered in the chapter and motivate them to find the answers by reading the chapter.
- The *Connection Link* boxed features have been retained. These help students see the big picture of how Earth systems are interrelated and connected. In this edition, a green icon in the text refers students to the *Connection Link* box, which then refers the reader to other locations in the book where more information can be found and connections can be made with other related important topics.
- Many of the popular *Geo-Focus* and *Geo-Insight* features contain either new topics or have been updated, with an emphasis on environmental and economic topics.
- *Geo-Impact* boxed features now replace the former *What Would You Do?* These boxed features continue to encourage students to think critically about what they are learning by asking open-ended questions related to the chapter material. These features also emphasize current issues related to natural resources and the environment, and many of them have an accompanying photo to better illustrate the topic at hand.
- *Critical Thinking Questions* are part of many of the figures. These questions are designed to encourage active student learning, guide observational skill development, and deepen understanding of geologic processes.
- The *Summary* is now called *Key Concepts Review* to emphasize the important concepts covered in the chapter.
- The previous format of 10 multiple-choice questions and 10 short-answer questions in the *Review Questions* section at the end of each chapter has been reduced to five multiple-choice and five short answer questions. A number of the questions have been rewritten or are new, and answers to all of the multiple-choice questions, as well as several of the short-answer questions, are provided at the back of the book.
- The last short-answer question is now titled *Creative Thinking Visual Question* and challenges students to describe the geologic process being depicted, engage in quantitative solutions, or address an issue using the information provided in the photo or graphic. These images and questions were chosen to encourage students to develop strong observational and critical thinking skills.
- A new feature, *Global GeoScience Watch,* directs students to the Cengage Learning website, which is a one-stop site for studying Earth and the environment. The website is interactive and current, and allows users to navigate issue, country, organization, and *Global GeoScience Watch*–specific portals. A question relating to the chapter material is asked that requires students to search the website for the answer.

It is our strong belief that the rewriting and updating done in the text, as well as the new features introduced, significantly improve the seventh edition of *The Changing Earth: Exploring Geology and Evolution*. We think that these changes and the new enhancements make this textbook easier to read and comprehend, as well as making it a more effective teaching tool that engages students in the learning process, thereby fostering a better understanding of the material and how it relates to Earth in the 21st century.

Text Organization

Plate tectonic theory is the unifying theme of geology and also for this book. This theory has revolutionized geology because it provides a global perspective of Earth and allows geologists to treat many seemingly unrelated geologic phenomena as part of a total planetary system.

A second, and equally important, theme is that Earth is a complex, dynamic planet that has changed continuously since its origins some 4.6 billion years ago. We can better understand this complexity by using a systems approach to the study of Earth and emphasizing this concept throughout the book.

We therefore have organized *The Changing Earth: Exploring Geology and Evolution,* seventh edition, into the following informal categories:

- Chapter 1 is an introduction to geology and Earth systems, geology's relevance to the human experience, and the debate about climate change and humans' possible role and effect, as well as the origin of the solar system and Earth's place in it.
- Chapter 2 deals with plate tectonics in detail and sets the stage for its integration throughout the rest of the book. Particular emphasis is placed on the evidence substantiating plate tectonic theory, why this theory is one of the cornerstones of geology, and why plate tectonic theory serves as a unifying paradigm in explaining many apparently unrelated geologic phenomena.
- Chapters 3–8 examine Earth's materials (minerals, and igneous, sedimentary, and metamorphic rocks) and the geologic processes accounting for them, including the role of plate tectonics in their origin and distribution.
- Chapters 9–10 deal with the related topics of Earth's interior, earthquakes, and deformation and mountain building.

- Chapters 11–16 cover Earth's surface processes, including such features as mass wasting, running water, groundwater, glaciers and glaciation, the work of wind and deserts, and shorelines and shoreline processes.
- Chapter 17 discusses geologic time, and Chapter 18 explores fossils and evolution.
- Chapters 19–23 constitute our chronological treatment of the geologic and biologic history of Earth.
- Chapter 24 summarizes and synthesizes the concepts, themes, and major topics covered in this book.

Of particular assistance to students are the end-of-chapter summary tables found in Chapters 20–22. These tables are designed to give an overall perspective of the geologic and biologic events that occurred during the particular time interval covered in that chapter and to show how the events are interrelated.

We have found that presenting the material in the order discussed above allows for an integration of the major themes of this book, as well as an emphasis on the underlying principles of geology and how they relate to the human experience and in deciphering Earth's history. We also know, however, that many professors prefer an entirely different order of topics, depending on the emphasis in their course. Therefore, we have written this book so that instructors can present the chapters in whatever order that suits the needs of a particular course.

Chapter Organization

All chapters have the same organizational format as follows:

- Each chapter opens with a dramatic photograph, many of which are new, followed by an *Outline* of the topics covered and a series of questions under the title *Have You Ever Wondered?* that are designed to pique student interest in the topics covered in that chapter.
- An *Introduction* follows that is intended to stimulate interest in the chapter and show how the chapter material fits into the larger geologic perspective. A number of the *Introductions* have been rewritten and updated in this edition.
- The text is written in a clear, informal style, making it easy for students to comprehend.
- Within each chapter are several *Connection Link* boxes that refer the reader to other locations in the book where more information can be found, and connections are made to important concepts and topics.
- Numerous *Critical Thinking Questions* are found associated with selected figures in each chapter. These are designed to encourage active student learning and deepen understanding of geologic processes.
- Numerous color diagrams and photographs complement the text and provide a visual representation of the concepts and information presented.
- Each chapter contains at least one *Geo-Focus* or *Geo-Insight* feature that presents a brief discussion or visual representation of an interesting aspect of geology or geologic research. Several of these features are new to this edition and emphasize economic and environmental issues.
- At least one *Geo-Impact* feature per chapter encourages students to engage in critical thinking by solving hypothetical problems or issues that are related to the chapter material.
- Topics related to environmental issues, such as climate change and hydraulic fracturing, are discussed throughout the text. Integrating economic geology and environmental issues with the chapter material helps students relate the importance and relevance of geology to their lives.
- The end-of-chapter materials begin with *Key Concepts Review*, which summarizes the important concepts covered in the chapter.
- The *Important Terms*, which are printed in boldface type in the chapter text, are listed at the end of each chapter for easy review, along with the page numbers on which they are first defined.
- The *Review Questions* are another important feature of this book and include multiple-choice questions with answers as well as short-answer questions, some of which have the answers provided at the end of the book. Many new questions have been added to each chapter of the seventh edition.
- The *Global GeoScience Watch* directs students to the Cengage Learning website, which is interactive, current, and allows users to navigate topics related to issue, country, and professional organizations.
- A full *Glossary* of important terms appears at the end of the text.

Ancillary Materials

For Instructors

We are pleased to offer a full suite of text and multimedia products to accompany the seventh edition of *The Changing Earth: Exploring Geology and Evolution*.

The *Earth Science CourseMate* features a rich array of study tools and learning resources for your students. This text-specific companion website includes quizzing, flashcards, and other web-based activities that will help students explore the concepts presented in the text.

The *Instructor Companion Website* contains everything you need for your course in one place! This collection of book-specific features and class tools in available online via www.cengage.com/login. Access and download PowerPoint presentations, images, instructor's manual, videos, and more.

The *Online Instructor's Manual* contains resources designed to streamline your course preparation. The Instructor's Manual includes Chapter Outlines, Learning Objectives, Chapter Summaries, Enrichment Topics, Common Misconceptions, Lecture Suggestions, "Consider This" questions, Important Terms, and Weblinks/Videos suggestions. This guide is available on the Instructor Companion Website.

New to this edition is *Cengage Learning Testing Powered by Cognero*, a flexible, online system that allows you to

- author, edit, and manage test bank content from multiple Cengage Learning solutions;
- create multiple test versions in an instant; and
- deliver tests from your Learning Management System (LMS), your classroom, or wherever you want.

Instructors can start right away. *Cengage Learning Testing Powered by Cognero* works on any operating system or browser; no special installations or downloads are needed. You'll be able to create tests from school, home, the coffee shop—anywhere with Internet access. You'll also find the following features:

- Simplicity at every step: A desktop-inspired interface features drop-down menus and familiar, intuitive tools that take you through content creation and management with ease.
- Full-featured test generator: Create ideal assessments with your choice of 15 question types, including true–false, multiple choice, opinion or Likert scale, and essay.
- Multilanguage support, an equation editor, and unlimited metadata: Ensure your tests are complete and compliant.
- Cross-compatible capability: Import and export content into other systems.

Global Geoscience Watch is an online resource center that provides access to a rich array of media resources to help you keep up with current events and highlight the concepts taught in class, as well as show the human impact on our planet. The following features are included:

- Articles
- Case studies
- Podcasts
- Videos
- World map (searchable by topic and country)
- Citation tools
- Sharing options (e-mail, Twitter, Facebook, etc.)
- Topic browsing and advanced searching
- And more!

For Students

The Changing Earth features *Earth Science CourseMate*, which helps you make the grade. *Earth Science CourseMate* includes an interactive e-book, with highlighting, note taking, and search capabilities as well as the following interactive learning tools:

- Quizzes
- Flashcards
- Video exercises
- Animations
- And more!

Go to login.cengagebrain.com to access these resources (ISBN-13: 9781285776255).

Virtual Field Trips in Geology are concept-based modules that teach you geology by using famous locations throughout the United States. Grand Canyon, Arches, and Hawaii Volcanoes National Parks are included, as well as many others. Designed to be used as homework assignments or lab work, the modules use a rich array of multimedia to demonstrate concepts. High-definition videos, images, animations, quizzes, and Google Earth activities work together in *Virtual Field Trips* to bring the concepts to life.

Global Geoscience Watch, updated several times a day, is a focused portal into GREENR—our Global Reference on the Environment, Energy, and Natural Resources—an ideal one-stop site for current events and research projects for all things geoscience. Divided into the four key course areas (geography, geology, meteorology, and oceanography), you can easily find the most relevant information for the course you are taking.

You will have access to the latest information from trusted academic journals, news outlets, and magazines. You also will receive access to statistics, primary sources, case studies, podcasts, and much more (ISBN-13: 9781111429065).

Acknowledgments

As the authors, we are, of course, responsible for the organization, style, and accuracy of the text, and any mistakes, omissions, or errors are our responsibility. The finished product is the culmination of many years of work during which we received numerous comments and advice from many geologists who reviewed all or parts of the text for the first five editions. They are as follows: Kenneth Beem, Montgomery College; David Berry, California State Polytechnic University, Pomona; Wesley A. Brown, Stephen F. Austin State University; Patricia J. Bush, Dèlgado Community College; Paul J. Bybee, Utah Valley State College; Brian Campbell, Southwestern Oklahoma State University; Deborah Caskey, El Paso Community College; Renee M. Clary, University of Louisiana at Lafayette; Michael Conway, Arizona Western College; David Cordero, Lower Columbia College; William C. Cornell, University of Texas at El Paso; Kathleen Devaney, El Paso Community College; Richard Diecchio, George Mason University; Robert Ewing, Portland Community College; David J. Fitzgerald, St. Mary's University; Yongli Gao, East Tennessee State University; Susan

Grandy, St. Clair County Community College; Ken Griffin, Tarrant County College; Dann M. Halverson, University of Southwestern Louisiana; Kristi Higginbotham, San Jacinto College; Ray Kenny, New Mexico Highlands University; Gary L. Kinsland, University of Louisiana at Lafayette; Jorg Maletz, University at Buffalo-SUNY; James MacDonald, Florida Gulf Coast University; Kevin McCartney, University of Maine-Presque Isle; Lynn Milwood, Mountain View College; Bob Mims, Richland College; Michael O'Donnell, Lord Fairfax Community College; Clair Ossian, Tarrant County College; Joseph Sarnecki, Macomb Community College; Roger Steinberg, Del Mar College; Michelle Stoklosa, Boise State University; Glenn B. Stracher, East Georgia College; Azam M. Tabrizi, Tidewater Community College; Thomas J. Weiland, Georgia Southwestern State University; Monte D. Wilson, Boise State University; and Guy Worthey, St. Ambrose University.

We especially wish to express our sincere appreciation to the reviewers of the sixth edition who made numerous helpful and useful comments that led to many improvements and new features seen in this seventh edition. They are as follows: Joel S. Aquino, Gainesville State College; Elaine K. Alexander Fagner, McLennan Community College; Brian Lock, University of Louisiana; and Krista Syrup, Moraine Valley Community College.

We are also grateful for the generosity of the various agencies and individuals from many countries that provided photographs.

Special thanks must go to Aileen Berg at Cengage Learning, who initiated this seventh edition, and to our content developer, Jake Warde, who not only kept us on task but also superbly edited and managed the content for this edition, as well as provided a fresh perspective on this edition and gave excellent suggestions throughout the continuing gestation of this new edition. We thank Andrea Clemente and Chris Schoedel, project managers at Cenveo Publisher Services, for their excellent work in overseeing this edition, as well as Radhey Balabh, compositor, and also part of the Cenveo Publisher Services team. We are indebted to the copyeditor Maureen O'Driscoll for her attention to detail and consistency throughout the book production process. We would also like to thank Pamela Galbreath and Lisa Buckley for the fresh design. We thank Parvinder Sethi for his help in locating appropriate photographs. We would also like to recognize Carol Samet, Cengage Learning production project manager; Stefanie Beeck, media developer; Shannon Holt and Kellie Petruzzelli, content coordinators; Alexandria Brady, product development manager, for developing the media program; and Janet del Mundo, senior market development manager. As always, our families were very patient and encouraging when much of our spare time and energy were devoted to this book. We again thank them for their continued support and understanding

James S. Monroe
Reed Wicander

The Changing Earth

True color satellite image of Asia (partly in shadow), the Arctic ice cap, and the Sun. In this book, we examine Earth as a system of interconnected components that interact with each other. The atmosphere, biosphere, hydrosphere, and lithosphere are four of Earth's major subsystems that are visible in this image. The complex interactions among these subsystems, as well as Earth's interior, results in a dynamically changing planet.

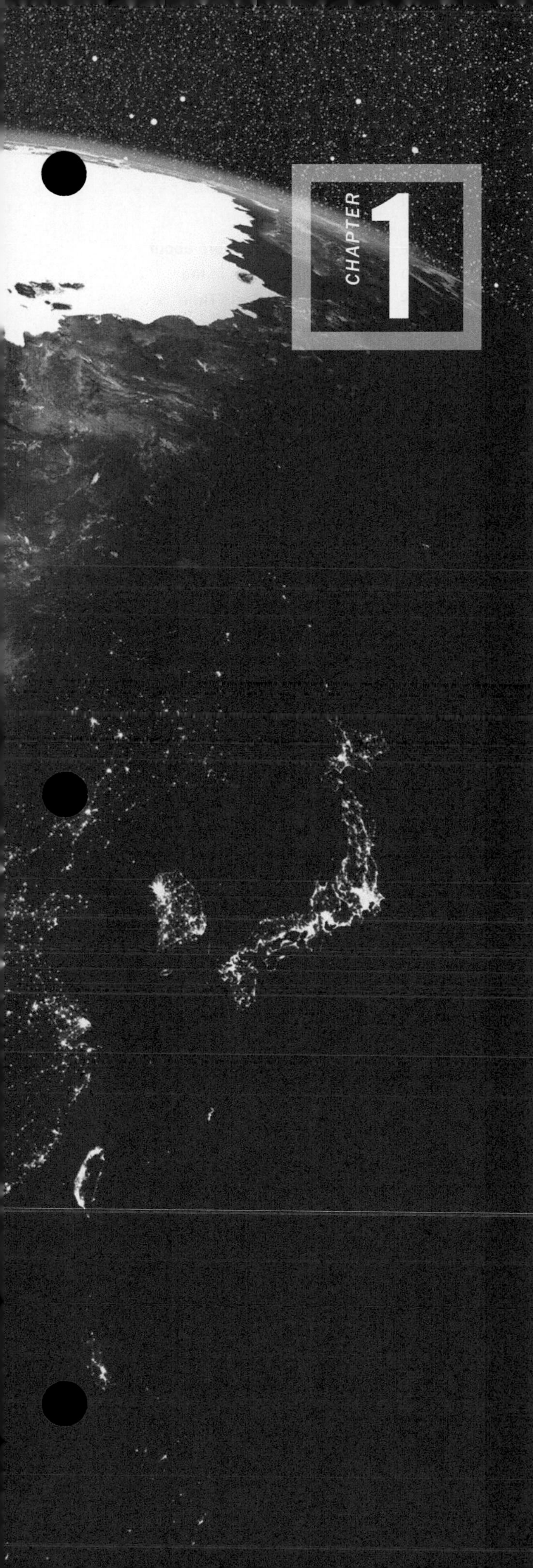

Planet Observer/Universal Images Group/Getty Images

Understanding Earth

A Dynamic and Evolving Planet

OUTLINE

HAVE YOU EVER WONDERED?

- How all of Earth's different components are interconnected and how the interactions among them are what make Earth a dynamic and ever-changing planet?
- Why a theory is not an unsubstantiated wild guess at explaining various natural phenomena?
- How geology relates to the human experience and affects our everyday lives?
- How environmental issues such as overpopulation, climate change, and rising sea level, to name a few, directly affect you?
- How the universe, solar system, and the planets came into being and evolved to what we now see when we look beyond our own planet Earth?
- What Earth is composed of and how its history has changed through time?
- Why you should study geology and the benefits that you will derive from a better understanding of this most important science?

1.1 Introduction

A major benefit of the space age has been the ability to look back from space and view our planet in its entirety. Every astronaut has remarked in one way or another on how Earth stands out as an inviting oasis in the otherwise black void of space (see the chapter opening photo). We are able to see not only the beauty of our planet but also its fragility and how humans are affecting the environment. And lastly, even though we did not witness it firsthand, we can still read the story of Earth's long and turbulent 4.6-billion-year history by deciphering the clues found in the geologic record, that is, the evidence for prehistoric physical and biological events that are preserved in rocks.

A major theme of this book is that Earth is a complex, dynamic planet that has changed continuously since its origin some 4.6 billion years ago. These changes and the present-day features we observe result from the interactions among Earth's internal and external systems, subsystems, and cycles. By viewing Earth as a whole—that is, thinking of it as a system—we not only see how its various components are interconnected, but can also better appreciate its complex and dynamic nature.

The system concept makes it easier for us to study a multifaceted subject such as Earth, because it divides the whole into smaller components that we can easily understand, without losing sight of how the separate components fit together as a whole. In the same way, you can think of this book as a large, panoramic landscape painting. Each chapter fills in the details of the landscape, thereby enhancing the overall enjoyment and understanding of the entire painting.

A **system** is a combination of related parts that interact in an organized manner. An automobile is a good example of a system. Its various components or subsystems, such as the engine, transmission, steering, and brakes, are all interconnected in such a way that a change in any one of them affects the others.

We can examine Earth in the same way we view an automobile—that is, as a system of interconnected components that interact and affect one another in many ways. The principal subsystems of Earth are the *atmosphere, biosphere, hydrosphere, lithosphere, mantle*, and *core* (Figure 1.1). The complex interactions among these subsystems result in a dynamically changing planet in which matter and energy are continuously recycled into different forms. For example, the movement of plates has profoundly affected the formation and evolution of its surface features and the distribution of mineral resources, as well as atmospheric and oceanic circulation patterns, which in turn, have affected global climate changes. Examined in this manner, the continuous evolution of Earth and its life is not a series of isolated and unrelated events, but rather it is a dynamic interplay among its various subsystems.

We also must not forget that humans are part of the Earth system, and our activities can produce changes with potentially wide-ranging consequences. When people discuss and debate such environmental issues as pollution and global warming, it is important to remember that these are not isolated issues, but rather they are part of the larger Earth system. Furthermore, Earth goes through much longer time cycles than humans are used to. Although global warming may have deleterious short-term effects on Earth's biota, climate change is part of long-term cycles that have resulted in large-scale periods of soaring global temperature and numerous episodes of glaciation.

 **ConnectionLink**

You can learn more about climate change in the section Geologic Time and Climate Change in Chapter 17.

As you study the various topics covered in this book, keep in mind the themes discussed in this chapter and how, like the parts of a system, they are interrelated. By relating each chapter's topic to its place in the entire Earth system, you will gain a greater appreciation of why geology is so integral to our lives.

1.2 What Is Geology?

Geology, from the Greek *geo* and *logos*, is defined as the study of Earth but now must also include the study of the planets and moons in our solar system. The discipline of geology is generally divided into two broad areas—physical geology and historical geology. *Physical geology* is the study of Earth materials, such as minerals and rocks, as well as the processes operating within Earth and on its surface. *Historical geology* examines the origin and evolution of Earth, its continents, oceans, atmosphere, and life.

Although the discipline of geology is broad and subdivided into numerous fields or specialties, nearly every aspect of geology has some economic or environmental relevance. For example, many geologists are involved in exploration for mineral and energy resources, using their specialized knowledge to locate the natural resources on which our industrial society is based. As the demand for these nonrenewable resources increases, geologists are applying the basic principles of geology in increasingly sophisticated ways, thereby enabling them to focus on areas that have the greatest chance for economic success.

Other geologists are using their expertise to address various environmental and societal issues. Not only is finding adequate sources of groundwater for the ever-burgeoning needs of communities and industries important but so too is the monitoring and prevention of surface and groundwater pollution, and when necessary, its cleanup. Geologic engineers help find safe locations for dams, waste-disposal sites, and power plants, as well as designing earthquake-resistant structures.

Geologists are increasingly asked to make short- and long-range predictions about earthquakes and volcanic eruptions and the potential destruction that may result. In fact, geologists are now more involved than ever in working with various government agencies and civil defense planners to ensure that contingency plans are in place and timely warnings are given for areas potentially affected by natural disasters, such as a tsunami.

 **ConnectionLink**

Learn more about the connection between plate boundaries, earthquakes, and tsunami in Chapter 2, Geo-Insight 2.1.

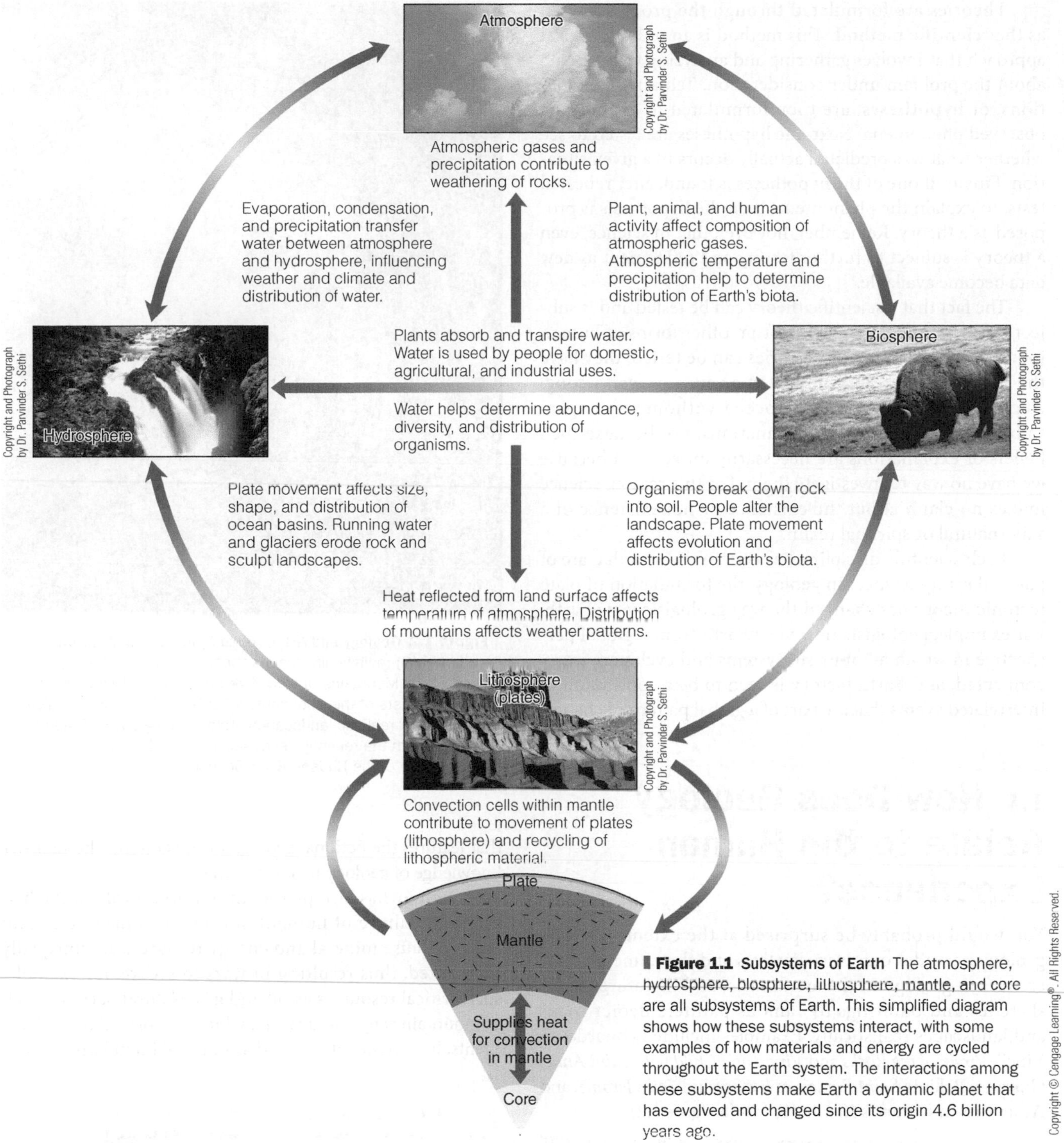
Atmosphere
Copyright and Photograph by Dr. Parvinder S. Sethi
Atmospheric gases and precipitation contribute to weathering of rocks.
Evaporation, condensation, and precipitation transfer water between atmosphere and hydrosphere, influencing weather and climate and distribution of water.
Plant, animal, and human activity affect composition of atmospheric gases. Atmospheric temperature and precipitation help to determine distribution of Earth's biota.
Hydrosphere
Copyright and Photograph by Dr. Parvinder S. Sethi
Plants absorb and transpire water. Water is used by people for domestic, agricultural, and industrial uses.
Water helps determine abundance, diversity, and distribution of organisms.
Biosphere
Copyright and Photograph by Dr. Parvinder S. Sethi
Plate movement affects size, shape, and distribution of ocean basins. Running water and glaciers erode rock and sculpt landscapes.
Organisms break down rock into soil. People alter the landscape. Plate movement affects evolution and distribution of Earth's biota.
Heat reflected from land surface affects temperature of atmosphere. Distribution of mountains affects weather patterns.
Lithosphere (plates)
Copyright and Photograph by Dr. Parvinder S. Sethi
Convection cells within mantle contribute to movement of plates (lithosphere) and recycling of lithospheric material.
Plate
Mantle
Supplies heat for convection in mantle
Core

Figure 1.1 **Subsystems of Earth** The atmosphere, hydrosphere, biosphere, lithosphere, mantle, and core are all subsystems of Earth. This simplified diagram shows how these subsystems interact, with some examples of how materials and energy are cycled throughout the Earth system. The interactions among these subsystems make Earth a dynamic planet that has evolved and changed since its origin 4.6 billion years ago.

1.3 Geology and the Formulation of Theories

The term **theory** has various meanings and is frequently misunderstood and consequently misused. In colloquial usage, it means a speculative or conjectural view of something—hence, the widespread belief that scientific theories are little more than unsubstantiated wild guesses. In scientific usage, however, a theory is a coherent explanation for one or more related natural phenomena supported by a large body of objective evidence. From a theory, scientists derive predictive statements that can be tested by observations and/or experiments so that their validity can be assessed. The law of universal gravitation is an example of a theory that describes the attraction between masses (an apple and Earth in the popularized account of Newton and his discovery).

Theories are formulated through the process known as the **scientific method**. This method is an orderly, logical approach that involves gathering and analyzing facts or data about the problem under consideration. Tentative explanations, or **hypotheses**, are then formulated to explain the observed phenomena. Next, the hypotheses are tested to see whether what was predicted actually occurs in a given situation. Finally, if one of the hypotheses is found, after repeated tests, to explain the phenomena, then the hypothesis is proposed as a theory. Remember, however, that in science, even a theory is subject to further testing and refinement as new data become available.

The fact that a scientific theory can be tested and is subject to such testing separates it from other forms of human inquiry. Because scientific theories can be tested, they have the potential for being supported or even proven wrong. Accordingly, science must proceed without any appeal to beliefs or supernatural explanations, not because such beliefs or explanations are necessarily untrue, but because we have no way to investigate them. For this reason, science makes no claim about the existence or nonexistence of a supernatural or spiritual realm.

Each scientific discipline has certain theories that are of particular importance. In geology, the formulation of plate tectonic theory has changed the way geologists view Earth. For example, geologists now view Earth from a global perspective in which all of its subsystems and cycles are interconnected, and Earth history is seen to be a continuum of interrelated events that are part of a global pattern of change.

Figure 1.2 **Geology and Art** *Kindred Spirits* by Asher Brown Durand (1849) realistically depicts the layered rocks along gorges in the Catskill Mountains of New York State. Durand was one of numerous artists of the 19th-century Hudson River School, which was known for realistic landscapes. This painting shows Durand conversing with the recently deceased Thomas Cole, the original founding force of the Hudson River School.

1.4 How Does Geology Relate to the Human Experience?

You would probably be surprised at the extent to which geology pervades our everyday lives and the numerous references to geology in the arts, music, and literature. Many sketches and paintings by famous painters depict rocks and landscapes realistically. Examples include Leonardo da Vinci's *Virgin of the Rocks* and *Virgin and Child with Saint Anne*, Giovanni Bellini's *Saint Francis in Ecstasy* and *Saint Jerome*, and Asher Brown Durand's *Kindred Spirits* (Figure 1.2).

In the field of music, Ferde Grofé's *Grand Canyon Suite* was no doubt inspired by the grandeur and timelessness of Arizona's Grand Canyon and its vast rock exposures. The rocks on the Island of Staffa in the Inner Hebrides provided the inspiration for Felix Mendelssohn's famous *Hebrides Overture*.

References to geology abound in *The German Legends of the Brothers Grimm*. Jules Verne's novel *Journey to the Center of the Earth* describes an expedition into Earth's interior. The poem *Ozymandias* by Percy B. Shelley deals on one level with the fact that nothing lasts forever and even solid rock eventually disintegrates under the ravages of time and weathering. There is even a series of mystery books by Sarah Andrews that features the fictional geologist Em Hansen, who uses her knowledge of geology to solve crimes.

Geology has also played an important role in the history and culture of humankind. Empires have risen and fallen because mineral and energy resources are unequally distributed, thus resulting in wars to secure, for example, such critical resources as oil and gas. Natural barriers, such as mountain ranges and rivers, which are formed by geologic agents, have frequently served as political boundaries.

1.5 How Does Geology Affect Our Everyday Lives?

The most obvious connection between geology and our everyday lives is made when such natural disasters as volcanic eruptions, earthquakes, landslides, tsunami, and floods strike. Although we cannot prevent most of these natural disasters from happening, the more we learn about what causes them, the better we will be able to predict and mitigate the severity of their impact.

Less apparent, but equally significant, are the connections between geology and economic, social, and political issues. Although most readers of this book will not become professional geologists, everyone should have a basic understanding of the geologic processes that ultimately affect all of us. It is possible that at some time in the future you may become involved in geologic decisions as a member of a planning board or as a property owner with mineral rights. A perfect example is the current controversy regarding the use of hydraulic fracturing to extract oil and gas, and the environmental impact that results.

With government playing a greater role in environmental issues and regulations, members of Congress have increased the number of staff devoted to studying geology-related topics, such as climate change, energy policy, and environmental protection. It is important, therefore, to have a basic understanding of not only geology but also science, in general, to better understand these major issues that affect all of us.

ConnectionLink

You can learn more about the controversial issue of hydraulic fracturing in Chapter 13, Geo-Focus 13.1.

If such topics as nonrenewable energy resources, waste disposal, and pollution seem too far removed or too complex to be fully appreciated, consider for a moment just how dependent we are on geology in our daily routines (Figure 1.3).

Much of the electricity for our appliances comes from the burning of coal, oil, natural gas, or uranium consumed in nuclear-generating plants. Geologists locate the coal, petroleum, natural gas, and uranium. The copper and other metal wires through which electricity travels are manufactured from materials found as the result of mineral exploration. The concrete foundation (concrete is a mixture of clay, sand or gravel, and limestone), drywall (made largely from the mineral gypsum), and windows (the mineral quartz is the principal ingredient in the manufacture of glass) of the buildings we live and work in owe their very existence to geologic resources.

The car or public transportation we use to go to work is powered and lubricated by some type of petroleum by-product and is constructed of metal alloys and plastics. And the roads or rails we ride over come from geologic materials, such as gravel, asphalt, concrete, or steel. All of these items are the result of processing geologic resources.

As individuals and societies, we enjoy a standard of living that obviously directly depends on the consumption of geologic materials. We therefore need to be aware of how our use and misuse of geologic resources may affect the environment and thus develop policies that not only encourage management of our natural resources but also allow for continuing economic development among all the world's nations.

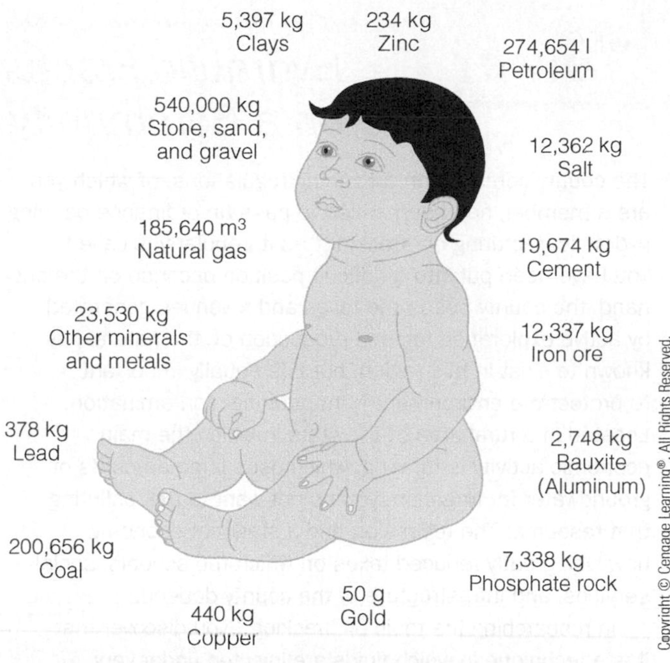

Figure 1.3 Lifetime Mineral Usage According to the Minerals Education Coalition, the average American born in 2012 has a life expectancy of 78.7 years and will need 1,350,000 kg of minerals, metals, and fuels to sustain his or her standard of living over a lifetime. That is an average of 17,154 kg of mineral and energy resources per year for every man, woman, and child in the United States. Data from the Minerals Education Coalition, The Society for Mining, Metallurgy and Exploration Foundation. http://www.mineralseducationcoalition.org/.

Critical Thinking Question Every year the life expectancy of the average American increases as well as our usage of minerals, metals, and fuels needed over a lifetime to maintain our standard of living. Is this increase sustainable, and is there anything that can be done to balance the depletion of natural resources but still maintain a high standard of living? How does our increasing consumption of these natural resources impact the rest of the world's population?

1.6 Global Geologic and Environmental Issues Facing Humankind

Most scientists would argue that overpopulation is the greatest environmental problem facing the world today. The world's population was slightly more than 7 billion at the end of 2012, and projections indicate that this number will reach between 8.0 and 10.5 billion people by 2050. Although this may not seem to be a geologic problem, remember that these people must be fed, housed, and clothed, and all with a minimal impact on the environment. Much of this population growth will be in areas that are already at risk from such hazards as earthquakes, tsunami, volcanic eruptions, and floods. Adequate water supplies must be found and protected from pollution. Additional energy resources will be needed to help fuel the economies of nations with ever-increasing populations. New techniques must be developed to reduce the use of our

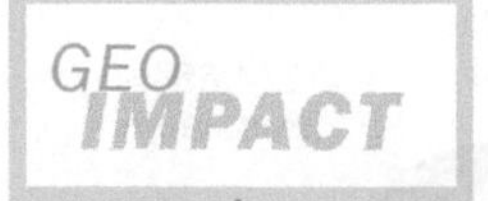

Hydraulic Fracturing and Its Impact on a Community

The county commission for zoning regulations, of which you are a member, has been asked to pass an ordinance banning hydraulic fracturing or "fracking" as it popularly is called. You have been put into a difficult position because on the one hand, the county needs the taxes and revenues generated by active exploration for, and production of, the natural gas known to exist in this region, but it is equally important to protect the environment from potential contamination. Located in a rural area of the state in which the main economic activity is farming, which uses large amounts of groundwater for irrigation, you do not want to risk polluting that resource. The recession and a stagnant economy, however, greatly reduced taxes on which the schools, social services, and infrastructure of the county depend.

In researching the topic of "fracking," you discover that it is a technique in which fluids are injected under very high pressure into organic-rich shales to create a network of fractures into which trapped natural gas can flow and ultimately be recovered. Proponents of "fracking" are quick to point out the economic benefits, whereby formerly inaccessible hydrocarbons, usually natural gas, can now be extracted.

The opponents of "fracking," however, point to the possible problems of environmental contamination, especially to groundwater reservoirs, which can be polluted by the migration of hydraulic fluids and other chemicals.

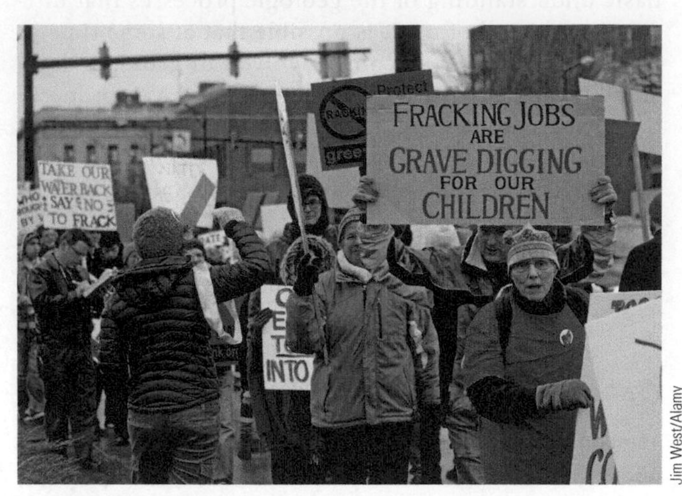

Jim West/Alamy

Protestors at an anti-fracking rally.

At the various public hearings held, there has been passionate debate about this issue by both proponents and opponents of "fracking." As these hearings will soon be coming to a close, you will be asked to vote on whether to allow or ban "fracking" in your county. How will you vote and what will be your reasons for that vote?

---

dwindling nonrenewable resource base and to increase our recycling efforts so that we can decrease our dependence on new sources of these materials.

The problems of overpopulation and how it affects the global ecosystem vary from country to country. For many poor and nonindustrial countries, the problem is too many people and not enough food. For the more developed and industrial countries, the problem is too many people rapidly depleting both the nonrenewable and renewable natural resource base. And in the most industrially developed countries, it is people producing more pollutants than the environment can safely recycle on a human time scale. The common thread tying these varied situations together is an environmental imbalance created by a human population that is exceeding Earth's short-term carrying capacity.

Other global issues in the news all the time are the greenhouse effect, global warming, and climate change. These topics affect not only all of us but also the planet we live on. But just how will global warming and the resultant climate change personally affect you? Should you really be concerned about them? After all, you do have to worry about exams, graduation, and employment, let alone the everyday issues we all must deal with, not to mention your personal life. Yet, part of the college experience is examining and debating the "big picture" and issues facing society today. So what about global warming and you?

The relationship between the greenhouse effect and global warming is an excellent example of how Earth's various subsystems are interrelated. As a by-product of respiration and the burning of organic material, carbon dioxide is a component of the global ecosystem and constantly is being recycled as part of the carbon cycle. The concern in recent years over the increase in atmospheric carbon dioxide levels is related to its role in the greenhouse effect.

The recycling of carbon dioxide between Earth's crust and atmosphere is an important climate regulator because carbon dioxide and other gases, such as methane, nitrous oxide, chlorofluorocarbons, and water vapor, allow sunlight to pass through them, but they trap the heat reflected back from Earth's surface. This retention of heat is called the *greenhouse effect.* It results in an increase in the temperature of Earth's surface and, more importantly, its atmosphere, thus producing global warming (■ Figure 1.4). The issue is not whether we have a greenhouse effect, because we do, but rather the degree to

Figure 1.4 The Greenhouse Effect and Global Warming

a Short-wavelength radiation from the Sun that is not reflected back into space penetrates the atmosphere and warms Earth's surface.

b Earth's surface radiates heat in the form of long-wavelength radiation back into the atmosphere, where some of it escapes into space. The rest is absorbed by greenhouse gases and water vapor and reradiated back toward Earth.

c Increased concentrations of greenhouse gases trap more heat near Earth's surface, causing a general increase in surface and atmospheric temperatures, which leads to global warming.

which human activity, such as the burning of fossil fuels, is increasing the greenhouse effect and thus contributing to global warming.

Because of the increase in human-produced greenhouse gases during the past 200 years, many scientists are concerned that a global warming trend has already begun and will result in severe global climatic shifts. Presently, most climate researchers use a range of scenarios for greenhouse gas emissions when predicting future warming rates. Climate model simulations published in the *2007 Fourth Intergovernmental Panel on Climate Change* show a predicted increase in global average temperature from 2000 to 2100 of 1–3°C under the best conditions, to a 2.5–6.5°C rise under "business-as-usual" conditions. These predicted increases in temperatures are based on various scenarios that explore different global development pathways. The fifth assessment by the Intergovenmental Panel on Climate Change was issued in 2013 and addressed the question of how much global warming actually results from increased carbon dioxide production.

Regardless of which scenario is followed, the global temperature change will be uneven, with the greatest warming occurring in the higher latitudes of the Northern Hemisphere. As a consequence of this warming, rainfall patterns will shift dramatically. This realignment will have a major effect on the largest grain-producing areas of the world, such as the United States Midwest. Drier and hotter conditions will intensify the severity and frequency of droughts, leading to more crop failures and higher food prices. With such shifts in climate, Earth's deserts may expand, with a resulting decrease in valuable crop and grazing land.

As climates change, diseases such as malaria are easily spreading into areas of warmer, wetter climates. Disease-carrying mosquitoes are expanding their range as climate changes allow them to survive in formerly inhospitable regions.

ConnectionLink

To learn more about desertification, which is the expansion of deserts into formerly productive lands, go to the Introduction in Chapter 15.

Higher temperature will also affect regional water supplies, creating potential water crises in the western United States within the next 20 years, as well as in other areas such as Peru and western China. Just as many regions will experience longer and hotter summers, other areas will suffer from intense and increased rainfall, which will result in severe flooding and landslides.

Moreover, continued global warming will result in a rise in mean sea level, as ice caps and glaciers melt and contribute their water to the world's oceans (Figure 1.5). It is predicted that at the current rate of glacial melting, sea level will rise 21 cm by around 2050, thus increasing the number of people at risk from flooding in coastal areas by approximately 20 million.

We would be remiss, however, if we did not point out that not all scientists are convinced that the global warming trend is the direct result of increased human activity related to industrialization. In fact, there has been much heated debate concerning the data and statistics used in the various models that are then used to make climate change predictions. These scientists indicate that although the level of

Figure 1.5 Rising Sea Level Rising sea level due to climate change threatens many island nations such as Tuvalu, a chain of low-lying islands and atolls located in the South Pacific Ocean approximately 1,000 km north of Fiji. This aerial view of Funafuti Island shows how vulnerable these islands, with a population of approximately 4,500 people, are to a slight rise in sea level, which would easily submerge them.

Ashley Cooper/Corbis

greenhouse gases has increased, we are still uncertain about their rate of generation and rate of removal, and whether the rise in global temperatures during the past century resulted from normal climatic variations through time or from human activity. Furthermore, they conclude that even if a general global warming occurs during the next hundred years, it is not certain that the dire predictions made by some scientists will come true.

Earth, as we know, is a remarkably complex system, with many feedback mechanisms and interconnections throughout its various subsystems and cycles. It is very difficult to predict with any certainty, all of the various consequences that global warming would have for atmospheric and oceanic circulation patterns and its ultimate effect on Earth's biota. It is, however, important to remember that although everyone is vulnerable to weather-related disasters, large-scale changes brought about by climate change will affect people in poor countries much more than those in the more industrial countries. Whether these climate changes are part of a natural global cycle taking place over thousands or hundreds of thousands of years—that is, on a geological time scale—or are driven, in part, by human activities, is immaterial. The bottom line is that we already are, or eventually will be, affected in some way, be it economic or social.

1.7 Origin of the Universe and Solar System, and Earth's Place in Them

How did the universe begin? What has been its history? What is its eventual fate, or is it infinite? These are just some of the basic questions people have asked and wondered about since they first looked into the nighttime sky and saw the vastness of the universe beyond Earth.

Origin of the Universe—Did It Begin with a Big Bang?

Most scientists think that the universe originated about 14 billion years ago in what is popularly called the **Big Bang**. The Big Bang is a model for the evolution of the universe in which a dense, hot state, was followed by expansion, cooling, and a less dense state.

According to modern *cosmology* (the study of the origin, evolution, and nature of the universe), the universe has no edge and therefore no center. Thus, when the universe began, all matter and energy were compressed into an infinitely small high-temperature and high-density state in which both time and space were set at zero. Therefore, there is no "before the Big Bang," only what occurred after it. As demonstrated by Einstein's Theory of Relativity, space and time are unalterably linked to form a space–time continuum—that is, without space, there can be no time.

How do we know that the Big Bang took place approximately 14 billion years ago? Why couldn't the universe have always existed as we know it today? Two fundamental phenomena indicate that the Big Bang occurred: (1) the universe is expanding, and (2) it is permeated by background radiation.

When astronomers look beyond our own solar system, they observe that everywhere in the universe galaxies are moving away from each other at tremendous speeds. Edwin Hubble first recognized this phenomenon in 1929. By measuring the optical spectra of distant galaxies, Hubble noted that the velocity at which a galaxy moves away from Earth increases proportionally to its distance from Earth. He observed that the spectral lines (wavelengths of light) of the galaxies are shifted toward the red end of the spectrum; that is, the lines are shifted toward longer wavelengths. Galaxies receding from each other at tremendous speeds would produce such a redshift. This is an example of the *Doppler effect*, which is a change in the frequency of a sound, light, or other wave caused by movement of the wave's source relative to the observer (Figure 1.6).

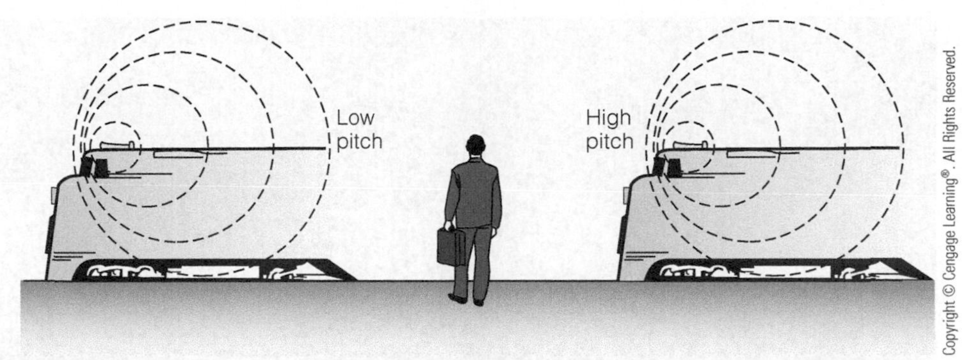

Figure 1.6 The Doppler Effect One way to understand the Doppler effect is by analogy to the sound of a passing train's whistle. As the train approaches, the sound waves are compressed slightly, so that an individual hears a shorter-wavelength, higher-pitched sound. As the train passes, and recedes from the individual, the sound waves are slightly expanded, and a longer-wavelength, lower-pitched sound is heard.

An easy way to envision how velocity increases with increasing distance is by reference to the popular analogy of a rising loaf of raisin bread, in which the raisins are uniformly distributed throughout the loaf (Figure 1.7). As the dough rises, the raisins are uniformly pushed away from each other at velocities directly proportional to the distance between any two raisins. The farther away a given raisin is to begin with, the farther it must move to maintain the regular spacing during the expansion, and hence the greater its velocity must be.

In the same way that raisins move apart in a rising loaf of bread, galaxies are receding from each other at a rate proportional to the distance between them, which is exactly what astronomers see when they observe the universe. By measuring this expansion rate, astronomers can calculate how long ago the galaxies were all together at a single point, which turns out to be about 14 billion years, the currently accepted age of the universe.

In 1965, Arno Penzias and Robert Wilson of Bell Telephone Laboratories made the second important observation that provided evidence of the Big Bang. They discovered that there is a pervasive background radiation of 2.7 Kelvin (K) above absolute zero (absolute zero equals −273°C; 2.7 K = −270.3°C) everywhere in the universe. This background radiation is thought to be the fading afterglow of the Big Bang.

Currently, cosmologists cannot say what it was like at time zero of the Big Bang, because they do not understand the physics of matter and energy under such extreme conditions. It is thought that during the first second following the Big Bang, the four basic forces—(1) *gravity* (the attraction of one body toward another), (2) *electromagnetic force* (combines electricity and magnetism into one force and binds atoms into molecules), (3) *strong nuclear force* (binds protons and neutrons together), and (4) *weak nuclear force* (responsible for the breakdown of an atom's nucleus, producing radioactive decay)—separated and the universe experienced enormous expansion.

As the universe continued expanding and cooling, stars and galaxies began to form and the chemical makeup of the universe changed. Initially, the universe was 100% hydrogen and helium, whereas today it is 98% hydrogen and helium and 2% all other elements by weight.

How did such a change in the universe's composition occur? Throughout their life cycle, stars undergo many nuclear reactions in which lighter elements are converted into heavier elements by nuclear fusion. When a star dies, often explosively, the heavier elements that were formed in its core are returned to interstellar space and are available for inclusion in new stars. In this way, the composition of the universe gradually is enhanced by heavier elements.

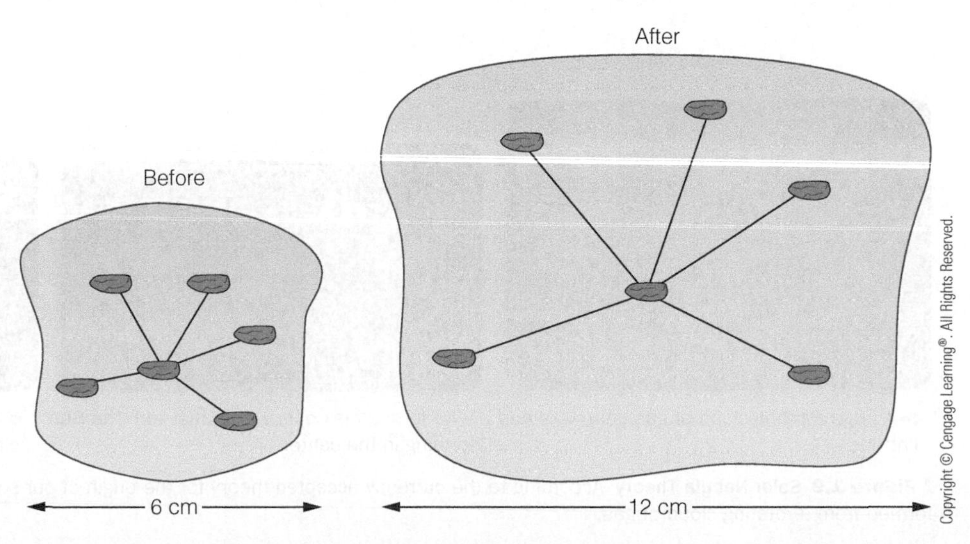

Figure 1.7 The Expanding Universe The motion of raisins in a rising loaf of raisin bread illustrates the relationship that exists between distance and speed and is analogous to an expanding universe. In this diagram, adjacent raisins are located 2 cm apart before the loaf rises. After one hour, any raisin is now 4 cm away from its nearest neighbor and 8 cm away from the next raisin over, and so on. Therefore, from the perspective of any raisin, its nearest neighbor has moved away from it at a speed of 2 cm per hour, and the next raisin over has moved away from it at a speed of 4 cm per hour. In the same way that raisins move apart in a rising loaf of bread, galaxies are receding from each other at a rate proportional to the distance between them.

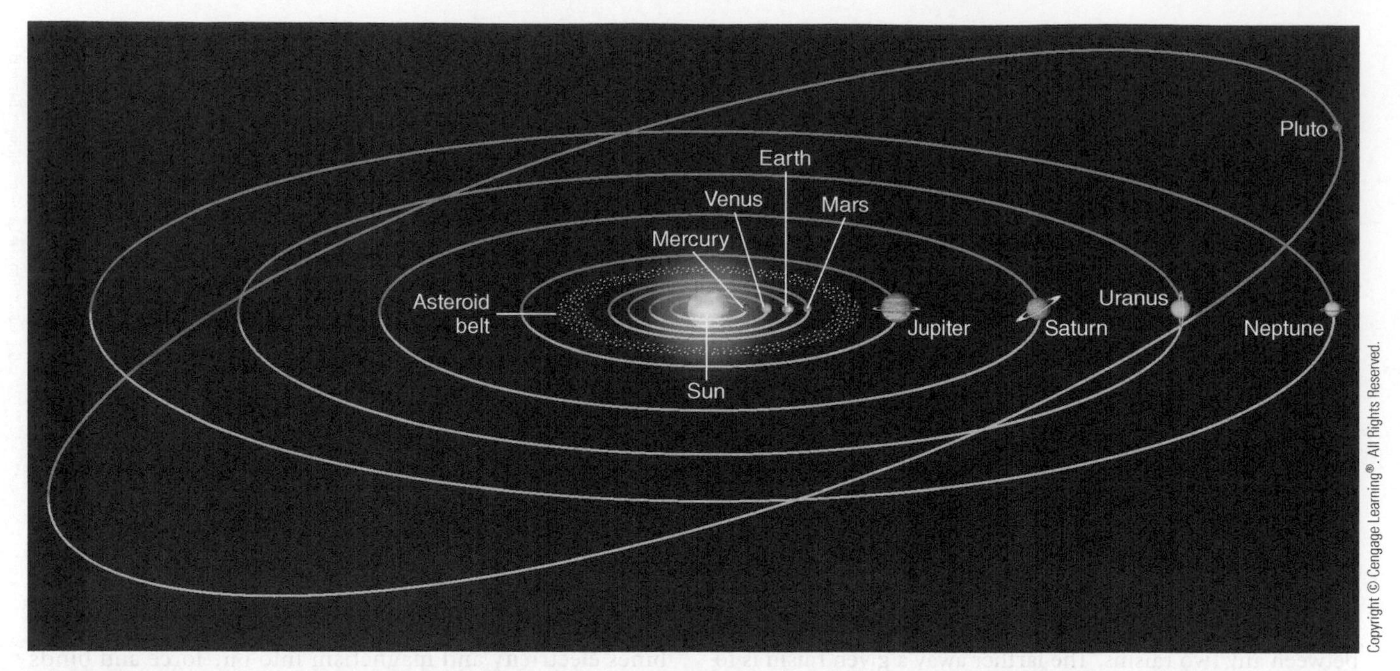

Figure 1.8 Diagrammatic Representation of the Solar System This representation of the solar system shows the planets and the dwarf planet Pluto, and their orbits around the Sun. Pluto orbits among the icy debris of the Kuiper Belt, a disc-shaped region beyond Neptune, in which it and four other dwarf planets are known.

Our Solar System—Its Origin and Evolution

Our solar system, which is part of the Milky Way galaxy, consists of the Sun; eight planets; five known dwarf planets (including Pluto); dozens of moons or satellites; a tremendous number of asteroids, most of which orbit in a zone between Mars and Jupiter; the Kuiper Belt—a disc-shaped region beyond Neptune, where such icy worlds as Pluto reside; and the even more distant Oort Cloud, which is thought to be the source of many of the millions of comets that orbit our Sun (Figure 1.8). Any theory formulated to explain the origin and evolution of our solar system must, therefore, take into account all of its various features and characteristics.

Many scientific theories for the origin of the solar system have been proposed, modified, and discarded since the French scientist and philosopher René Descartes first proposed, in 1644, that the solar system formed from a gigantic whirlpool within a universal fluid. Today, the **solar nebula theory** for the formation of the solar system not only best explains the features of the solar system but also provides a logical explanation for its evolutionary history (Figure 1.9).

According to the solar nebula theory, the condensation and subsequent collapse of interstellar material in a spiral arm of the Milky Way galaxy resulted in a counterclockwise-rotating disk of gases and small grains. About 90% of the material was concentrated in the central part of the disk, thus forming an embryonic Sun, around which swirled a rotating cloud of material called a *solar nebula*.

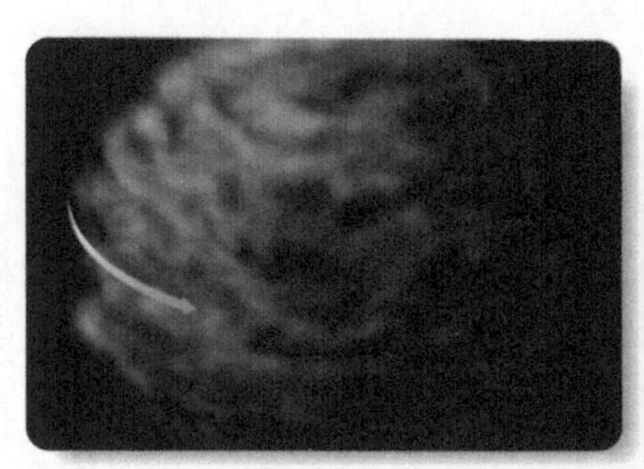

a A huge rotating cloud of gas contracts and flattens

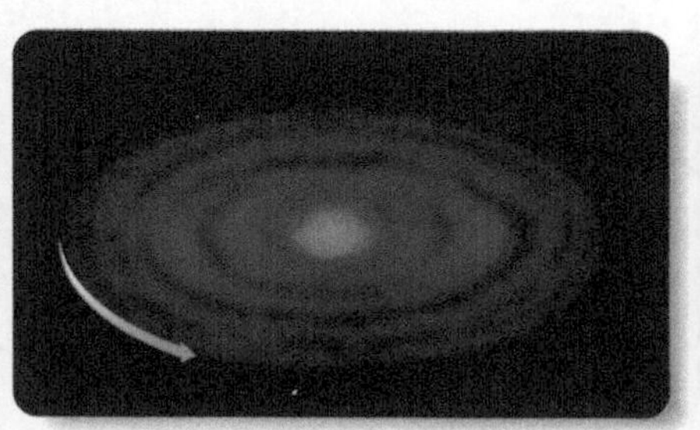

b to form a disk of gas and dust with the Sun forming in the center

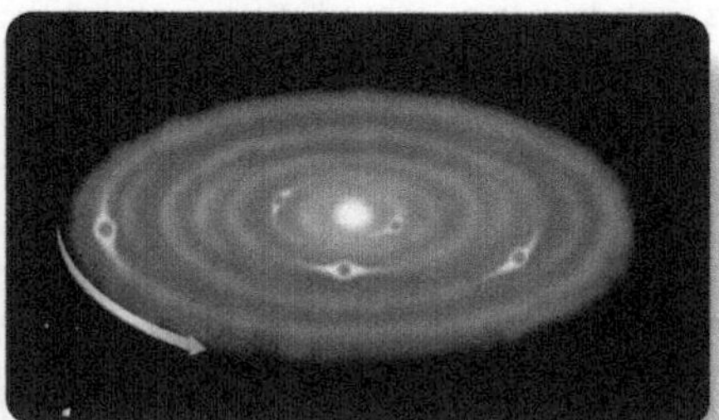

c and eddies gathering up material to form planets.

Figure 1.9 Solar Nebula Theory According to the currently accepted theory for the origin of our solar system, the planets and the Sun formed from a rotating cloud of gas.

Within this solar nebula were localized eddies in which gases and solid particles condensed. During the condensation process, gaseous, liquid, and solid particles began to accrete into ever-larger masses called *planetesimals*, which collided and grew in size and mass until they eventually became planets.

The composition and evolutionary history of the planets are a consequence, in part, of their distance from the Sun. The **terrestrial planets**—Mercury, Venus, Earth, and Mars—so named because they are similar to *terra*, Latin for "earth," are all small and composed of rock and metallic elements that condensed at the high temperatures of the inner nebula (Geo-Insight 1.1). The **Jovian planets**—Jupiter, Saturn, Uranus, and Neptune—so named because they resemble Jupiter (the Roman god was also called *Jove*), all have small rocky cores compared with their overall size and are composed mostly of hydrogen, helium, ammonia, and methane, which condense at low temperatures.

While the planets were accreting, material that had been pulled into the center of the nebula also condensed, collapsed, and was heated to several million degrees by gravitational compression. The result was the birth of a star, our Sun.

During the early accretionary phase of the solar system's history, collisions between various bodies were common, as indicated by the numerous craters on many planets and moons. Asteroids probably formed as planetesimals in a localized eddy between what eventually became Mars and Jupiter in much the same way that other planetesimals formed the terrestrial planets. The tremendous gravitational field of Jupiter, however, prevented this material from ever accreting into a planet. Comets, which are interplanetary bodies composed of loosely bound rocky and icy materials, are thought to have condensed beyond the orbit of Neptune.

The solar nebula theory for the formation of the solar system thus accounts for most of the characteristics of the planets and their moons, the differences in composition between the terrestrial and Jovian planets, and the presence of the asteroid belt. On the basis of the available data, the solar nebula theory best explains the features of the solar system and provides a logical explanation for its evolutionary history.

Recall from our earlier discussion that theories are formulated on the basis of the scientific method, that is, the gathering and analysis of facts and observations relevant to the problem—in this case, the formation of the solar system. A hypothesis is then proposed and tested to account for the observed phenomenon, and if the hypothesis explains the phenomenon, then a theory is proposed. This is exactly how the solar nebula theory came to explain the formation of the solar system. New observations or data, however, may show that the theory is incomplete or does not explain some newly discovered aspect of the solar system. In that case, the theory will have to be modified to incorporate and explain the new observations or data. This is how science works.

Earth—Its Place in Our Solar System

Some 4.6 billion years ago, various planetesimals in our solar system gathered enough material together to form Earth and the other planets. Scientists think that this early Earth was probably cool, of generally uniform composition and density throughout, and composed mostly of silicates (compounds consisting of silicon and oxygen), iron and magnesium oxides, and smaller amounts of all the other chemical elements (Figure 1.10a). Subsequently, when the combination of meteorite impacts, gravitational compression, and heat from radioactive decay increased the temperature of Earth enough to melt iron and nickel, this homogeneous composition disappeared (Figure 1.10b) and was replaced by a series of concentric layers of differing composition and density, resulting in a differentiated planet (Figure 1.10c).

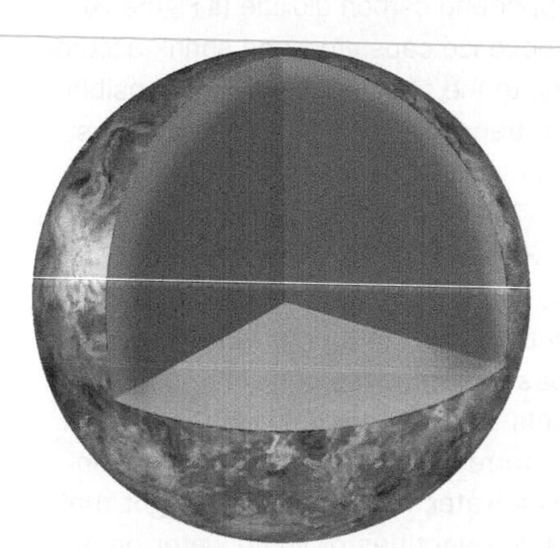

a Early Earth probably had a uniform composition and density throughout.

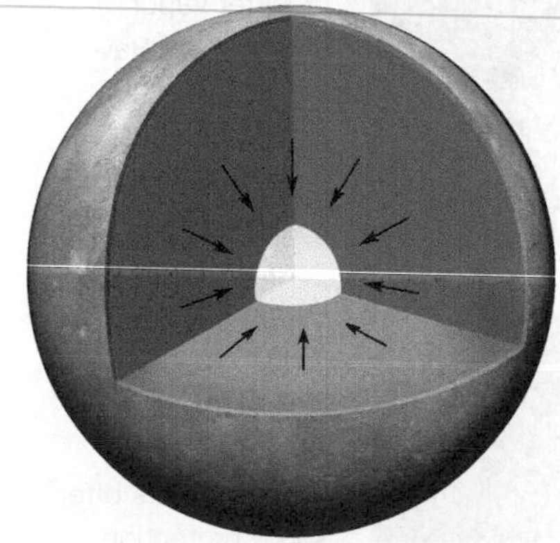

b The temperature of early Earth reached the melting point of iron and nickel, which, being denser than silicate minerals, settled to Earth's center. At the same time, the lighter silicates flowed upward to form the mantle and the crust.

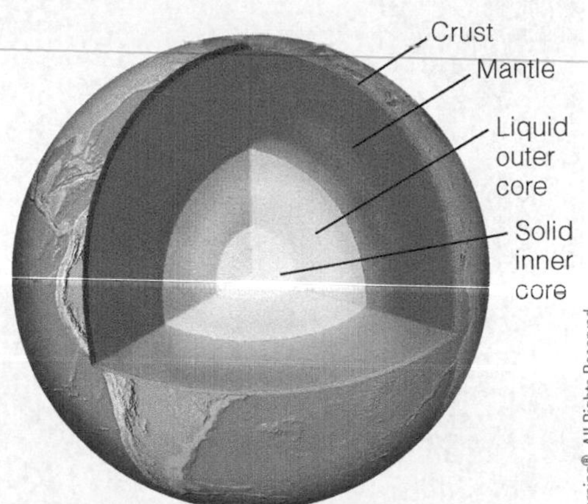

c In this way, a differentiated Earth formed, consisting of a dense iron–nickel core, an iron-rich silicate mantle, and a silicate crust with continents and ocean basis.

Figure 1.10 Homogeneous Accretion Theory for the Formation of a Differentiated Earth

GEO INSIGHT 1.1 Mars—The "Red Planet"

What do many people think of when they think of Mars? The possibility of life on the planet, spectacular photos of its surface, and perhaps even the prospect that humans will someday visit it. Mars holds a special place in the collective psyche of Earth's people. Named after *Mars*, the Roman god of war, it is the fourth planet from the Sun, and is one of the four terrestrial planets (▮ Figure 1). It is popularly called the "Red Planet" because its surface is composed of iron-rich minerals that have oxidized, giving it a reddish color. Dust from these iron oxide minerals are frequently blown into the atmosphere by storms, lending a reddish hue to the atmosphere as well.

There have been numerous missions to Mars, beginning with the flyby in 1965 of *Mariner 4*, and culminating with the successful landing on August 6, 2012, of the Mars rover *Curiosity*. All of these missions were designed to provide a greater understanding of the geologic history of the planet, and whether conditions are, or ever were, suitable for hosting life.

Generally speaking, scientists know that Mars has a thin atmosphere, distinct seasons, and little water. Its southern hemisphere is heavily cratered like the surfaces of Mercury and the Moon. The northern hemisphere has large, smooth plains, fewer craters, and evidence of extensive volcanism. The largest volcano in the solar system is found in the northern hemisphere. Huge canyons, the largest of which would stretch from San Francisco to New York, also appear in the northern half of the planet (▮ Figure 2). Circling Mars are two small potato-shaped moons, Phobos and Deimos, that may be captured asteroids.

▲ **2.** The largest canyon known in the solar system, Valles Marineris, can be clearly seen in the center of this image, while three of the planet's volcanoes are visible on the left side of the image.

The thin atmosphere of Mars, which is about 1% the thickness of Earth's atmosphere, consists of about 95% carbon dioxide, 3% nitrogen, 1.6% argon, and traces of water vapor and oxygen. Having lost its global magnetic field around 3.5 billion years ago, Mars does not shield its surface from radiation in the same way Earth does. Its atmosphere, however, does offer some protection against incoming radiation, and based on initial measurements taken by *Curiosity*, scientists think that current surface radiation levels would not be lethal to humans.

▲ **1.** A variety of geologic structures, as well as the northern polar ice cap and water-ice clouds, are visible in this full-disk image of Mars.

Mars has an axial tilt similar to Earth, and thus has seasons, albeit longer than Earth's seasons, because it takes longer to complete its orbit around the Sun. Mars has two permanent polar ice caps composed primarily of water vapor and carbon dioxide (▮ Figure 3). These ice caps grow and shrink according to the season and are responsible for transporting large amounts of dust and water vapor into the atmosphere.

The cold surface temperatures (−87°C to −5°C) and thin atmosphere currently preclude the presence of liquid water on the surface of Mars. Based on images of numerous channels and canyons that appear similar to terrestrial features formed by running water, however, it is thought that large quantities of liquid water once flowed freely on Mars' surface and produced the many landforms that on Earth result from running water

3. Martian polar ice caps

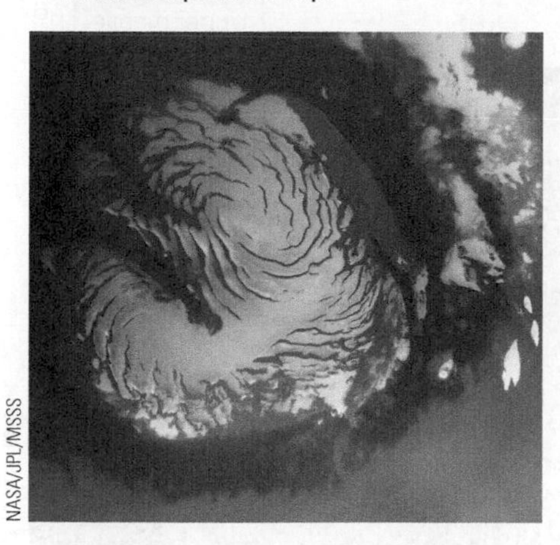

NASA/JPL/MSSS

a Martian north polar ice cap on March 13, 1999. The north polar ice cap is approximately 1,100 km across. The light-toned surfaces surrounding the ice are residual water ice that remains through the summer, whereas the nearly circular band of dark material is mostly sand dunes that are formed and shaped by blowing winds.

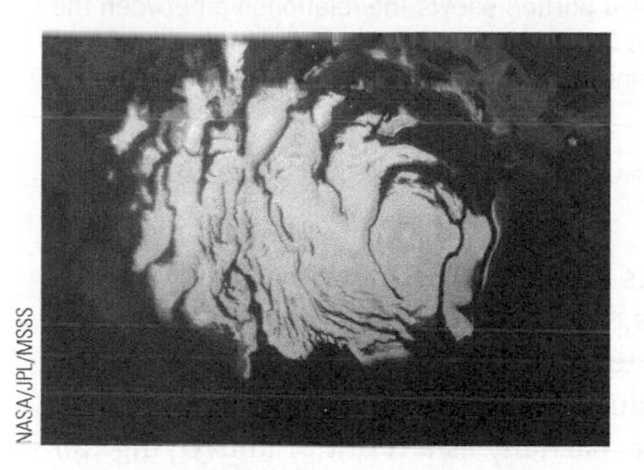

NASA/JPL/MSSS

b Martian south polar ice cap on April 17, 2000. The long dimension of the south polar ice cap is about 420 km. Because Mars experiences seasons, the polar ice caps expand and contract, reaching a minimum size in summer, as shown here, and expanding as the area covered by carbon dioxide frost, seen here as white, begins to grow.

(▮ Figure 4). Such features strongly suggest Mars once had a warmer and wetter climate. Furthermore, the occurrence of hematite and goethite (both of which are iron oxide minerals), as well as gypsum, along with discovery, by the rover *Opportunity* in 2004, of the mineral jarosite, which only forms in the presence of acidic water, clearly demonstrates that water once existed on Mars. Where this water came from, how long it lasted, and where it went, are still questions perplexing scientists as they unravel the history of the "Red Planet."

Just as with the other terrestrial planets, Mars shows evidence of an early period of meteorite bombardment. Thousands of craters can be seen in its southern hemisphere, as well as evidence of an enormous impact basin in the northern hemisphere. The northern hemisphere also has numerous features that indicate an extensive early period of volcanism. These include the shield volcano, Olympus Mons, the solar systems' largest known volcano (▮ Figure 5), lava flows, and uplifted regions thought to have resulted from mantle convection. In addition to volcanic features, Mars displays abundant evidence of tensional tectonics, including numerous faults and large fault-produced valley structures. Whereas Mars was once tectonically active, no evidence indicates that plate tectonics comparable to that on Earth has ever occurred there.

Perhaps the most intriguing question of all about Mars is whether there was ever life on the planet. Most scientists would agree that having liquid water increases a planet's ability to harbor life. There is a strong consensus in the scientific community that Mars once did have considerable surface waters, making it significantly more habitable to life than it is today. Although there are tantalizing hints that Mars could have supported life at one time, whether living organisms ever existed on Mars remains unknown. It is hoped that information collected by the rover *Curiosity* will begin to provide some definitive answers as to whether Mars currently, or ever, possessed an environment suitable for life.

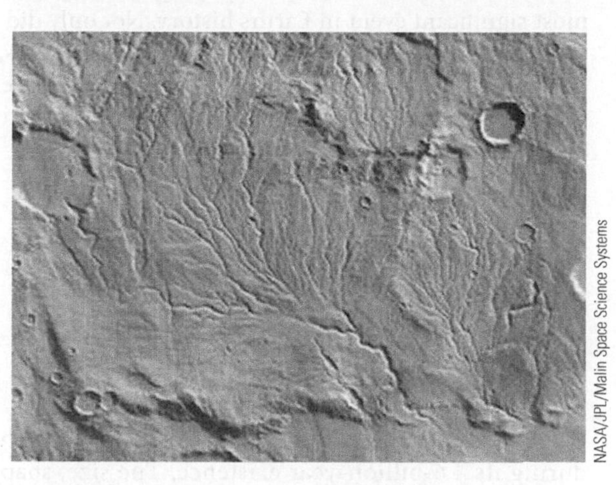

NASA/JPL/Malin Space Science Systems

4. Landforms produced by running water (dendritic stream beds) and sedimentary outcrops deposited by running water.

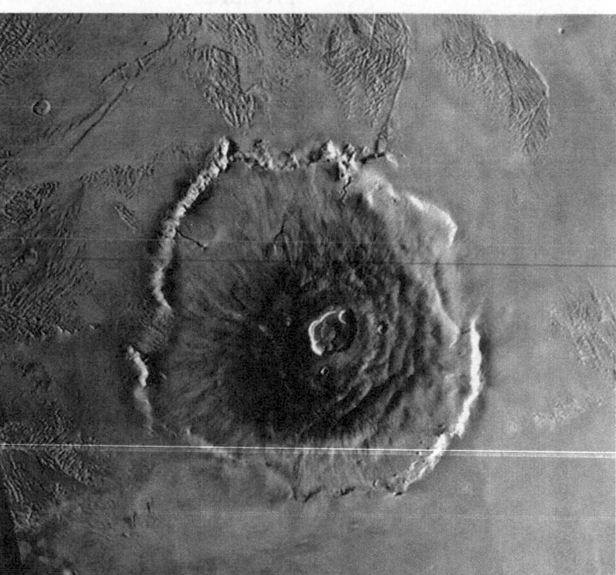

NSSDC Photo Gallery

5. A vertical view of Olympus Mons, a shield volcano and the largest volcano in our solar system. The edge of the Olympus Mons caldera is marked by a cliff several kilometers high rather than a moat as in Mauna Loa, Earth's largest shield volcano.

This differentiation into a layered planet is probably the most significant event in Earth's history. Not only did it lead to the formation of a crust and eventually continents, but it also was probably responsible for the emission of gases from the interior that eventually led to the origin of the oceans and atmosphere.

1.8 Why Earth Is a Dynamic and Evolving Planet

Earth is a dynamic planet that has continuously changed during its 4.6-billion-year existence. The size, shape, and geographic distribution of continents and ocean basins have changed throughout time; the composition of the atmosphere has evolved; and life-forms existing today differ from those that lived during the past. Mountains and hills have been worn away by erosion, and the forces of wind, water, and ice have sculpted a diversity of landscapes. Volcanic eruptions and earthquakes reveal an active interior, and folded and fractured rocks are testimony to the tremendous power of Earth's internal forces.

Earth consists of three concentric layers: the core, the mantle, and the crust (Figure 1.11). This orderly division results from density differences between the layers as a function of variations in composition, temperature, and pressure.

> **ConnectionLink**
> You can learn more about Earth's core, mantle, and crust, how they were discovered, and how we know their composition and physical properties in Chapter 9.

The **core** has a calculated density of 10–13 grams per cubic centimeter (g/cm^3) and occupies about 16% of Earth's total volume. Seismic (earthquake) data indicate that the core consists of a small, solid inner region and a larger, apparently liquid, outer portion. Both are thought to consist mostly of iron and a small amount of nickel.

The **mantle** surrounds the core and accounts for about 83% of Earth's volume. It is less dense than the core (3.3–5.7 g/cm^3) and is thought to be composed mostly of *peridotite*, a dark, dense igneous rock containing abundant iron and magnesium. The mantle can be divided into three distinct zones based on physical characteristics. The lower mantle is solid and forms most of the volume of Earth's interior.

The **asthenosphere** surrounds the lower mantle. It has the same composition as the lower mantle but behaves plastically and flows slowly. Partial melting within the asthenosphere generates *magma* (molten material), some of which rises to the surface because it is less dense than the rock

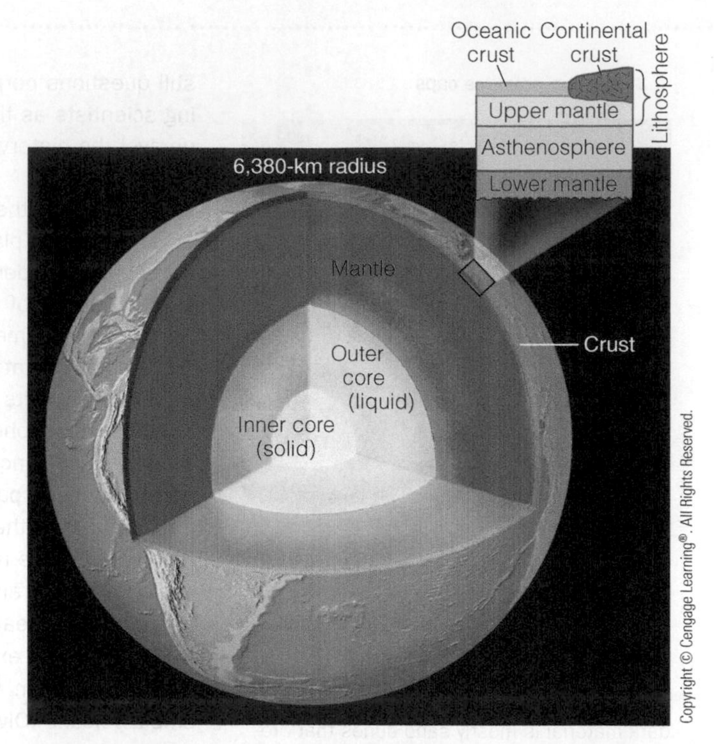

Figure 1.11 Cross Section of Earth Illustrating the Core, Mantle, and Crust The enlarged portion shows the relationship between the lithosphere (composed of the continental crust, oceanic crust, and solid upper mantle) and the underlying asthenosphere and lower mantle.

from which it was derived. The upper mantle surrounds the asthenosphere. The solid upper mantle and the overlying crust constitute the **lithosphere**, which is broken into numerous individual pieces called **plates** that move over the asthenosphere, partially as a result of underlying *convection cells* (Figure 1.12). Interactions of these plates are responsible for such phenomena as earthquakes and volcanic eruptions and for the formation of mountain ranges and ocean basins.

The **crust**, Earth's outermost layer, consists of two types. *Continental crust* is thick (20–90 km), has an average density of 2.7 g/cm^3, and contains considerable silicon and aluminum. *Oceanic crust* is thin (5–10 km), denser than continental crust (3.0 g/cm^3), and is composed of the dark igneous rocks *basalt* and *gabbro*.

Plate Tectonic Theory

The recognition that the lithosphere is divided into rigid plates that move over the asthenosphere (Figure 1.13) forms the foundation of **plate tectonic theory**, a unifying theory of geology holding that large segments of Earth's outer part (lithospheric plates) move relative to one another (discussed in greater detail in Chapter 2). Zones of volcanic activity, earthquakes, or both mark most

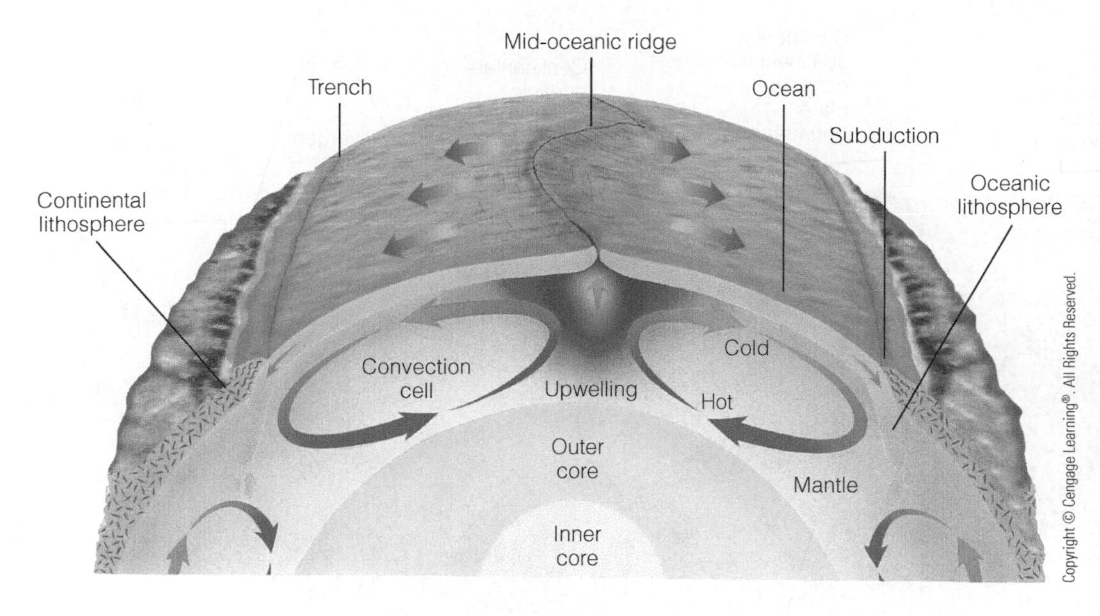

Figure 1.12 Movement of Earth's Plates Earth's plates are thought to move partially as a result of underlying mantle convection cells in which warm material from deep within Earth rises toward the surface, cools, and then, upon losing heat, descends back into the interior, as shown in this diagrammatic cross section.

plate boundaries. Along these boundaries, plates separate (diverge), collide (converge), or slide sideways past each other (Figure 1.14).

The acceptance of plate tectonic theory is recognized as a major milestone in the geologic sciences, comparable to the revolution that Darwin's theory of evolution caused in biology. Plate tectonics has provided a framework for interpreting the composition, structure, and internal processes of Earth on a global scale. It has led to the realization that the continents and ocean basins are part of a lithosphere–atmosphere–hydrosphere system that evolved together with Earth's interior.

A revolutionary concept when it was proposed in the 1960s, plate tectonic theory has had far-reaching consequences in all fields of geology because it provides the basis for relating many seemingly unrelated phenomena, such as the formation and occurrence of Earth's natural resources, as well as the distribution and evolution of the world's biota.

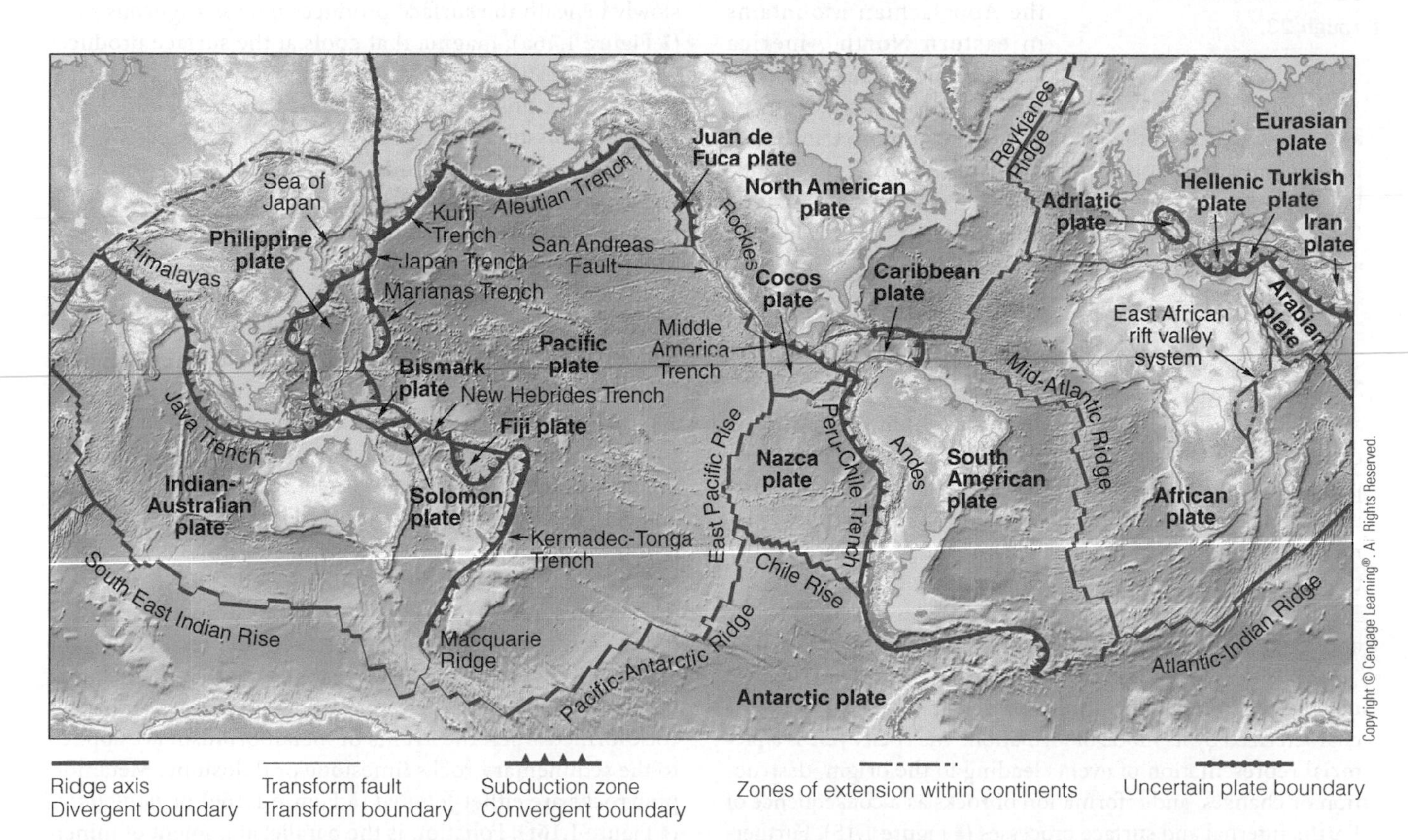

Figure 1.13 Earth's Plates Earth's lithosphere is divided into rigid plates of various sizes that move over the asthenosphere.

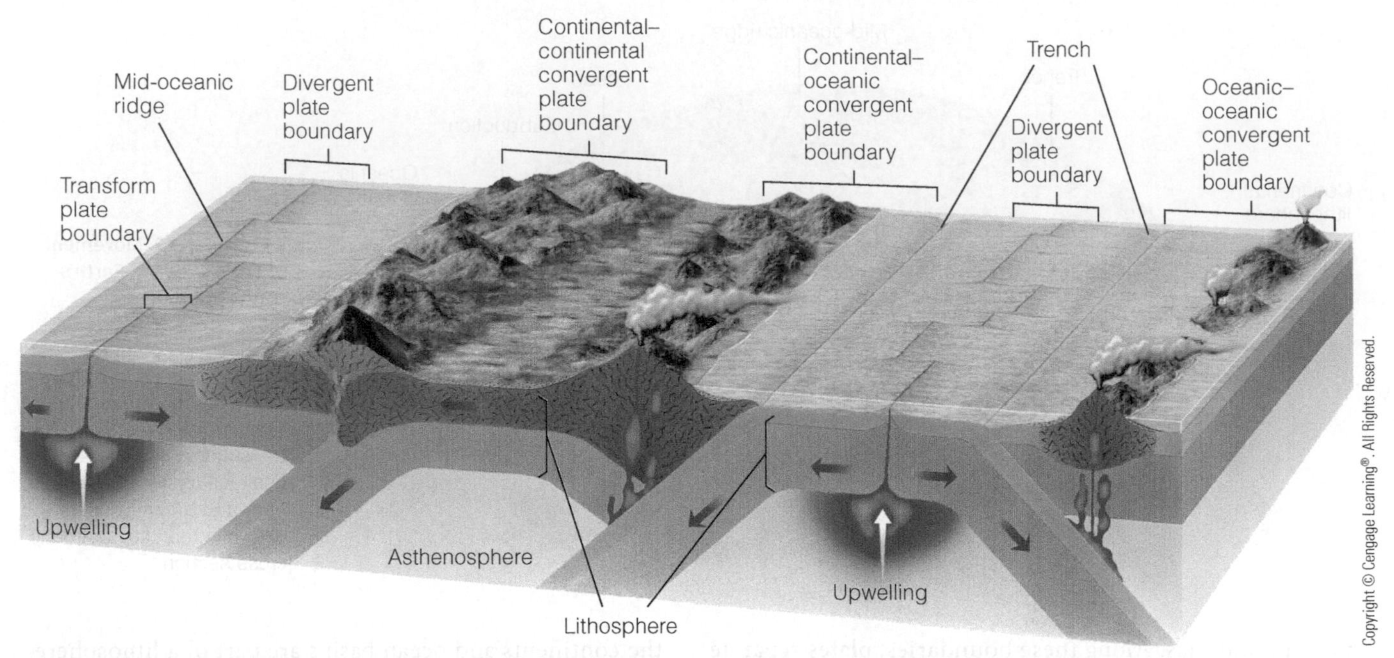

Figure 1.14 Relationship Between Lithosphere, Asthenosphere, and Plate Boundaries An idealized block diagram illustrating the relationship between the lithosphere and the underlying asthenosphere and the three principal types of plate boundaries: divergent, convergent, and transform.

ConnectionLink
You can find more information on Earth's history in Chapters 19 through 23.

Furthermore, the impact of plate tectonic theory has been particularly notable in the interpretation of Earth's history. For example, the Appalachian Mountains in eastern North America and the mountain ranges of Greenland, Scotland, Norway, and Sweden are not the result of unrelated mountain-building episodes, but, rather, they are part of a larger mountain-building event that involved the closing of an ancient Atlantic Ocean and the formation of the supercontinent Pangaea approximately 251 million years ago.

1.9 The Rock Cycle

A **rock** is an aggregate of **minerals**, which are naturally occurring, inorganic, crystalline solids that have definite physical and chemical properties. Minerals are composed of elements, such as oxygen, silicon, and aluminum, and elements are made up of atoms, the smallest particles of matter that retain the characteristics of an element. Approximately 3,800 minerals have been identified and described, but only about a dozen make up the bulk of the rocks in Earth's crust.

Geologists recognize three major groups of rocks—*igneous, sedimentary,* and *metamorphic*—each of which is characterized by its mode of formation. The **rock cycle** is a pictorial representation of events leading to the origin, destruction or changes, and reformation of rocks as a consequence of Earth's internal and surface processes (Figure 1.15). Furthermore, it shows that the three major rock groups are interrelated; that is, any rock type can be derived from the others.

Igneous rocks result when magma or lava crystallizes, or when volcanic ejecta (pyroclastic materials), such as ash, accumulate and consolidate. As magma cools, minerals crystallize, and the resulting rock is characterized by a texture of interlocking mineral grains. Magma that cools slowly beneath the surface produces *intrusive igneous rocks* (Figure 1.16a); magma that cools at the surface produces *extrusive igneous rocks* (Figure 1.16b).

Rocks exposed at Earth's surface are broken into particles and dissolved by various weathering processes. The particles and dissolved materials may be transported by wind, water, or ice and eventually deposited as *sediment.* This sediment may then be compacted or cemented (lithified) into sedimentary rock.

Sedimentary rocks are composed of sediment and form in one of three ways: (1) consolidation of mineral or rock fragments, (2) precipitation of mineral matter from solution, or (3) compaction of plant or animal remains (Figure 1.16c, 1.16d). Because sedimentary rocks form at or near Earth's surface, geologists can infer some information about the environment in which they were deposited, the transporting agent, and perhaps even something about the source from which the sediments were derived (see Chapter 7). Accordingly, sedimentary rocks are especially useful for interpreting Earth history.

Metamorphic rocks result from the alteration of other rocks, usually beneath the surface, by heat, pressure, and the chemical activity of fluids. For example, marble—a rock preferred by many sculptors and builders—is a metamorphic rock formed when the agents of metamorphism are applied to the sedimentary rocks limestone or dolostone. Metamorphic rocks are either *foliated* (Figure 1.16e) or *nonfoliated* (Figure 1.16f). Foliation is the parallel alignment of minerals caused by the application of pressure, and gives the rock a layered or banded appearance.

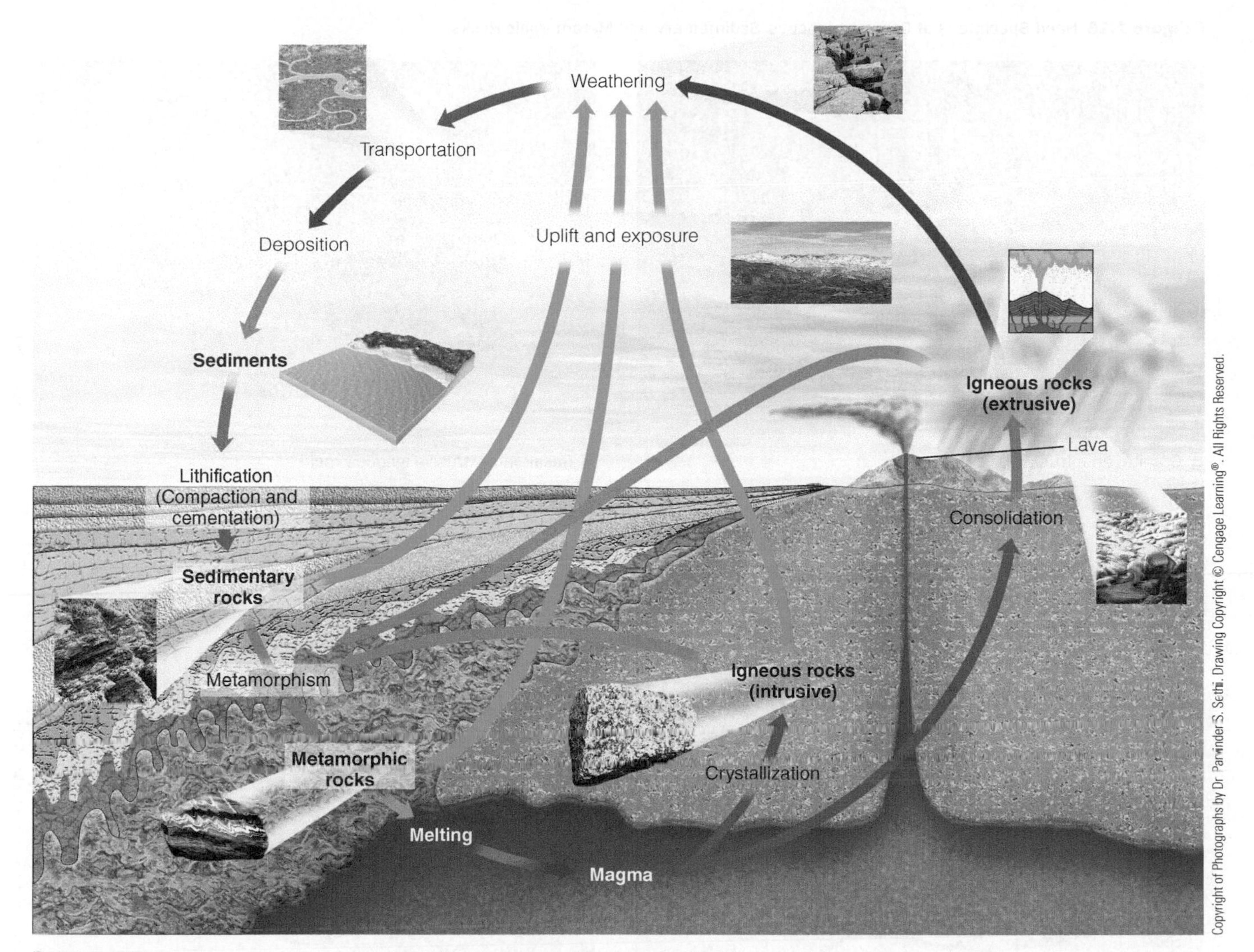

Figure 1.15 The Rock Cycle This cycle shows the interrelationships between Earth's internal and external processes and how the three major rock groups are related. An ideal cycle includes the events on the outer margin of the cycle, but interruptions, indicated by internal arrows, are common.

Critical Thinking Question What types of geologic events would cause interruptions to the idealized cycle shown on the margin of the rock cycle? What do you think is more common and why—completion of an idealized cycle, or one in which there are interruptions?

How Are the Rock Cycle and Plate Tectonics Related?

Interactions between plates determine, to some extent, which of the three rock groups will form (Figure 1.17). For example, when plates converge, the heat and pressure generated along the plate boundary may lead to igneous activity and metamorphism within the descending oceanic plate, thus producing various igneous and metamorphic rocks.

Some of the sediments and sedimentary rocks on the descending plate are melted, whereas other sediments and sedimentary rocks along the boundary of the nondescending plate are metamorphosed by the heat and pressure produced along the convergent plate boundary. Later, the mountain range or chain of volcanic islands formed along this plate boundary will be weathered and eroded, and the new sediments will be transported to the ocean, where they will be deposited, to begin yet another cycle.

The connection between the rock cycle and plate tectonics is just one of many examples of how Earth's various subsystems and cycles are interrelated. Heating within Earth's interior results in convection cells that power the movement of plates and magma, which forms intrusive and extrusive igneous rocks. Movement along plate boundaries may result in volcanic activity, earthquakes, and, in some cases, mountain building. The interaction among the atmosphere, hydrosphere, and biosphere contributes to the weathering of rocks exposed on Earth's surface. Plates descending into Earth's interior are subjected to increasing heat and pressure, which may lead to metamorphism, as well as the generation of magma and yet another recycling of materials.

Figure 1.16 Hand Specimens of Common Igneous, Sedimentary, and Metamorphic Rocks

a Granite, an intrusive igneous rock.

b **Basalt**, an extrusive igneous rock.

c Conglomerate, a sedimentary rock formed by the consolidation of rounded rock fragments.

d **Limestone**, a sedimentary rock formed by the extraction of mineral matter from seawater by organisms or by the inorganic precipitation of the mineral calcite from seawater.

e Gneiss, a foliated metamorphic rock.

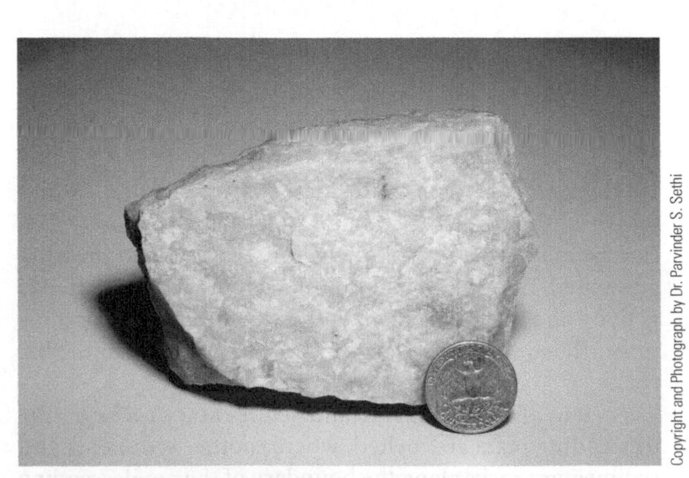

f **Quartzite**, a nonfoliated metamorphic rock.

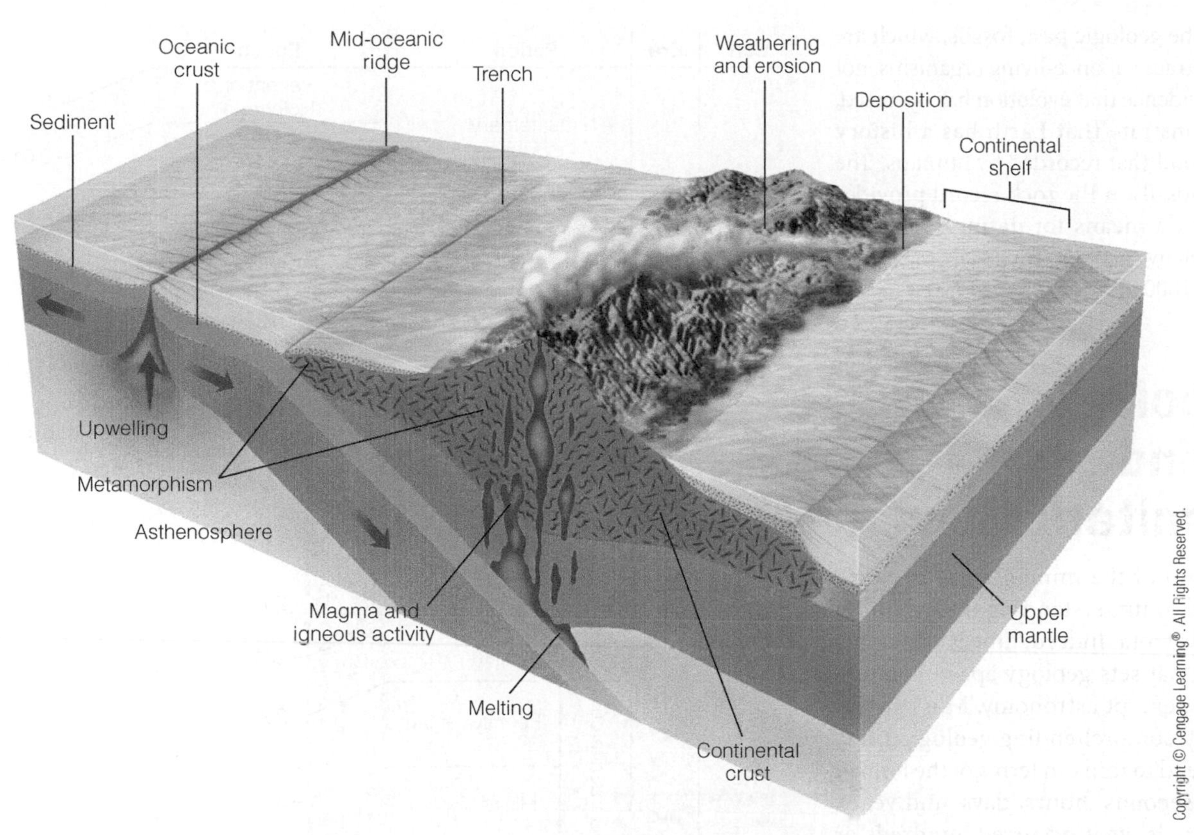

Figure 1.17 Plate Tectonics and the Rock Cycle Plate movement provides the driving mechanism that recycles Earth materials. This block diagram shows how the three major rock groups—igneous, sedimentary, and metamorphic—are recycled through both the continental and oceanic regions. Subducting plates are partially melted to produce magma, which rises and either crystallizes beneath Earth's surface as intrusive igneous rock, or spills out on the surface, solidifying as extrusive igneous rock. Rocks exposed at the surface are weathered and eroded to produce sediments that are transported and eventually lithified into sedimentary rocks. Metamorphic rocks result from pressure generated along converging plates or adjacent to rising magma.

1.10 Organic Evolution and the History of Life

Plate tectonic theory provides us with a model for understanding the internal workings of Earth and their effect on Earth's surface. The theory of **organic evolution** (whose central thesis is that all present-day organisms are related and that they have descended with modifications from organisms that lived in the past) provides the conceptual framework for understanding the history of life. Together, the theories of plate tectonics and organic evolution have changed the way we view our planet, and we should not be surprised at the intimate association between them. Although the relationship between plate tectonic processes and the evolution of life is incredibly complex, paleontological data provide indisputable evidence of the influence of plate movement on the distribution of organisms.

The publication in 1859 of Darwin's *On the Origin of Species by Means of Natural Selection* revolutionized biology and marked the beginning of modern evolutionary biology. Upon its publication, most naturalists recognized that evolution provided a unifying theory that explained an otherwise-encyclopedic collection of biologic facts.

ConnectionLink

You can learn more about Charles Darwin, the theory of evolution, and its supporting evidence in Chapter 18.

When Darwin proposed his theory of organic evolution, he cited a wealth of supporting evidence, including the way organisms are classified, embryology, comparative anatomy, the geographic distribution of organisms, and, to a limited extent, the fossil record. Furthermore, Darwin proposed that *natural selection*, which results in the survival to reproductive age of those organisms best adapted to their environment, is the mechanism that best accounts for evolution.

ConnectionLink

You can learn how the various principles of relative dating and the discovery of radioactivity played an important role in the construction of the geologic time scale in Chapter 17.

Perhaps the most compelling evidence in favor of evolution can be found in the fossil record. Just as the geologic record allows geologists to interpret physical events and conditions in the geologic past, **fossils**, which are the remains or traces of once-living organisms, not only provide evidence that evolution has occurred, but also demonstrate that Earth has a history extending beyond that recorded by humans. The succession of fossils in the rock record provides geologists with a means for dating rocks and allowed for a relative geologic time scale to be constructed in the 1800s.

1.11 Geologic Time and Uniformitarianism

An appreciation of the immensity of geologic time is central to understanding the evolution of Earth and its biota. Indeed, time is one of the main aspects that sets geology apart from the other sciences, except astronomy. Most people have difficulty comprehending geologic time, because they tend to think in terms of the human perspective—seconds, hours, days, and years. Ancient history is what occurred hundreds or even thousands of years ago. When geologists talk of ancient geologic history, however, they are referring to events that happened hundreds of millions or even billions of years ago.

It is important to remember that Earth goes through cycles of much longer duration than the human perspective of time. Although some of these cycles, such as global warming and cooling, may have disastrous effects on the human species, they are, nonetheless, part of the larger cycle of global change that has, for example, resulted in numerous glacial advances and retreats during the past 2.6 million years. Because of their perspective on time and the interrelationships of Earth's various systems and cycles, geologists are in a unique position to make valuable contributions to such current environmental debates as global warming and sea-level changes.

The **geologic time scale** subdivides geologic time into a hierarchy of increasingly shorter time intervals, each of which has a specific name and duration. During the 19th-century, many geologists pieced together information from numerous rock exposures to construct a chronology, or time scale, based on changes in Earth's biota through time. Subsequently, with the discovery of radioactivity in 1895 and the development of various radiometric-dating techniques, geologists have been able to assign numerical ages (also known as *absolute ages*) in years to the subdivisions of the geologic time scale (▌Figure 1.18).

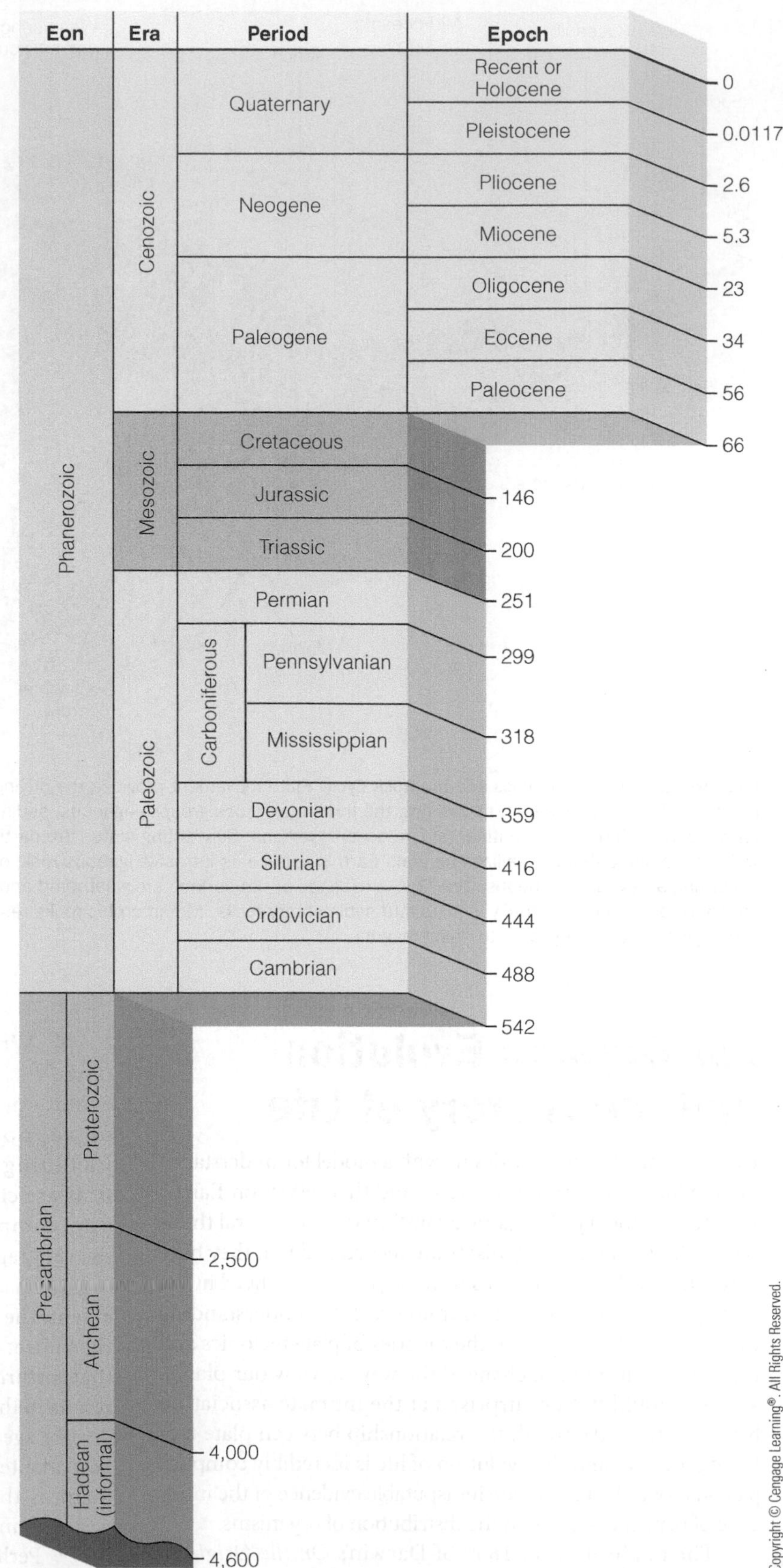

▌**Figure 1.18 The Geologic Time Scale** The numbers to the right of the columns are the ages in millions of years before the present. Dates are from the 2009 International Stratigraphic Chart, © 2009 by the International Commission on Stratigraphy.

One of the cornerstones of geology is the **principle of uniformitarianism**, which is based on the premise that present-day processes have operated throughout geologic time. Therefore, to understand and interpret geologic events from evidence preserved in rocks, we must first understand present-day processes and their results. In fact, uniformitarianism fits completely with the system approach that we are following for the study of Earth.

Uniformitarianism is a powerful principle that allows us to use present-day processes as the basis for interpreting the past and for predicting potential future events. We should keep in mind, however, that uniformitarianism does not exclude sudden or catastrophic events, such as volcanic eruptions, earthquakes, tsunami, landslides, or floods.

What uniformitarianism does mean is that even though the rates and intensities of geologic processes have varied during the past, the physical and chemical laws of nature have remained the same. Although Earth is in a dynamic state of change and has been ever since it formed, the processes that shaped it during the past are the same ones operating and modifying it today.

1.12 How Does the Study of Geology Benefit Us?

The most meaningful lesson to learn from the study of geology is that Earth is an extremely complex planet in which interactions are taking place between its various subsystems and have been for the past 4.6 billion years. If we want to ensure the survival of the human species, we must first understand how the various subsystems work and interact with each other and, more importantly, how our actions affect the delicate balance between these systems. We can do this, in part, by studying what has happened in the past, particularly on the global scale, and using that information to try to predict how our actions might affect the balance between Earth's various subsystems now and in the future.

The study of geology also goes beyond learning numerous facts about Earth. In reality, we do not just study geology—we *live* it. Geology is an integral part of our lives. Our standard of living depends directly on our consumption of natural resources, most of which formed millions and billions of years ago. However, the way that we consume natural resources and interact with the environment—as individuals and as a society—also determines our ability to pass on this standard of living to the next generation.

As you study the various subjects covered in this book, keep in mind the themes and topics discussed in this chapter and how, like the parts of a system, they are interrelated and responsible for the events that occurred during Earth's 4.6-billion-year history. View each chapter's subject matter in the context of how it fits into the whole Earth system, and remember that Earth's history is a continuum resulting from the interaction among its various subsystems. By relating each chapter's topic to its place in the Earth system, you will gain a greater appreciation of why geology is so integral to our lives.

Key Concepts Review

- Earth can be viewed as a system of interconnected components that interact with and affect each other. The principal subsystems of Earth are the atmosphere, hydrosphere, biosphere, lithosphere, mantle, and core.
- Earth is a continually changing and dynamic planet because of the interactions among its various subsystems and cycles.
- Geology, the study of Earth, is divided into (1) physical geology, which is the study of Earth materials and the processes that operate both within Earth and on its surface; and (2) historical geology, which examines the origin and evolution of Earth, its continents, oceans, atmosphere, and life.
- The scientific method is an orderly, logical approach that involves gathering and analyzing facts about a particular phenomenon, formulating hypotheses to explain the phenomenon, testing the hypotheses, and finally proposing a theory. A theory is a testable explanation for some natural phenomenon that has a large body of supporting evidence.
- Geology is not only part of the human experience, examples of which can be found in art, music, and literature, but it also affects our daily lives as individuals, societies, and nation-states. A basic understanding of geology, and science in general, is critical for dealing with and finding solutions to the many environmental problems and issues facing humankind.
- The universe began, in what is popularly called the Big Bang, approximately 14 billion years ago. Astronomers have deduced this age by observing that celestial objects are moving away from each other in an ever-expanding universe. Furthermore, the universe has a pervasive background radiation of 2.7 K above absolute zero (2.7 K = –270.3°C), which is thought to be the faint afterglow of the Big Bang.
- About 4.6 billion years ago, our solar system formed from a rotating cloud of interstellar matter. As this cloud condensed, it eventually collapsed under the influence of gravity and flattened into a counterclockwise-rotating disk. Within this rotating disk, the Sun, planets, and moons formed from the turbulent eddies of nebular gases and solids.
- Earth formed from a swirling eddy of nebular material 4.6 billion years ago, accreting as a solid body and soon thereafter differentiating into a layered planet.
- Earth's outermost layer is the crust, which is divided into continental and oceanic portions. The crust and

underlying solid upper mantle, together known as the lithosphere, overlie the asthenosphere, a zone that behaves plastically and flows slowly. The asthenosphere is underlain by the solid lower mantle. Earth's core consists of an outer liquid portion and an inner solid portion.

- The lithosphere is divided into a series of plates that diverge, converge, and slide sideways past one another.
- Plate tectonic theory provides a unifying explanation for many geologic features and events. The interaction between plates is responsible for volcanic eruptions, earthquakes, the formation of mountain ranges and ocean basins, and the recycling of rock materials.
- The three major rock groups are igneous, sedimentary, and metamorphic. Igneous rocks result from the crystallization of magma or the consolidation of volcanic ejecta. Sedimentary rocks are typically formed by the consolidation of rock fragments, precipitation of mineral matter from solution, or compaction of plant or animal remains. Metamorphic rocks result from the alteration of other rocks, usually beneath Earth's surface, by heat, pressure, and chemically active fluids.
- The rock cycle illustrates the interactions between Earth's internal and external processes and how the three rock groups are interrelated.
- The central thesis of the theory of organic evolution is that all living organisms evolved (descended with modifications) from organisms that existed in the past.
- Time sets geology apart from the other sciences except astronomy, and an appreciation of the immensity of geologic time is central to understanding Earth's evolution. The geologic time scale is the calendar geologists use to date past events.
- The principle of uniformitarianism is basic to the interpretation of Earth history. This principle holds that the laws of nature have been constant through time and that the same processes operating today have operated in the past, although not necessarily at the same rates.
- Geology is an integral part of our lives. Our standard of living depends directly on our consumption of natural resources, most of which formed millions and billions of years ago.

Important Terms

asthenosphere (p. 16)
Big Bang (p. 10)
core (p. 16)
crust (p. 16)
fossil (p. 22)
geologic time scale (p. 22)
geology (p. 4)
hypothesis (p. 6)
igneous rock (p. 18)
Jovian planet (p. 13)
lithosphere (p. 16)
mantle (p. 16)
metamorphic rock (p. 18)
mineral (p. 18)
organic evolution (p. 21)
plates (p. 16)
plate tectonic theory (p. 16)
principle of uniformitarianism (p. 23)
rock (p. 18)
rock cycle (p. 18)
scientific method (p. 6)
sedimentary rock (p. 18)
solar nebula theory (p. 12)
system (p. 4)
terrestrial planet (p. 13)
theory (p. 5)

Review Questions

1. The movement of plates is thought to result from
   a. ____ density differences between the inner and outer core.
   b. ____ rotation of the mantle around the core.
   c. ____ gravitational forces.
   d. ____ the Coriolis effect.
   e. ____ convection cells.
2. Which of the following statements about a scientific theory is *not* true?
   a. ____ It is an explanation for some natural phenomenon.
   b. ____ Predictive statements can be derived from it.
   c. ____ It is a conjecture or guess.
   d. ____ It has a large body of supporting evidence.
   e. ____ It is testable.
3. The study of the origin and evolution of Earth, its continents, oceans, atmosphere, and life is
   a. ____ meteorology.
   b. ____ cosmology.
   c. ____ historical geology.
   d. ____ geomorphology.
   e. ____ physical geology.
4. A combination of related parts interacting in an organized fashion is
   a. ____ a cycle.
   b. ____ a theory.
   c. ____ uniformitarianism.
   d. ____ a hypothesis.
   e. ____ a system.

5. That present-day processes have operated throughout geologic time is the premise for the principle of
   a. _____ organic evolution.
   b. _____ plate tectonics.
   c. _____ uniformitarianism.
   d. _____ geologic time.
   e. _____ scientific deduction.
6. Why is plate tectonic theory a unifying theory of geology, and how does it fit into a systems approach to the study of Earth?
7. How does the solar nebula theory account for the formation of our solar system, its features, and evolutionary history?
8. Why do most scientists think overpopulation is the greatest environmental problem facing the world today? Do you agree? If not, what do you think is the greatest threat to our existence as a species?
9. Using plate movement as the driving mechanism of the rock cycle, explain how the three rock groups are related and how each rock group can be converted into a different rock group.
10. **Creative Thinking Visual Question:** One of the major environmental issues facing humankind today is global warming and the effect, if any, human activity has had on raising global average temperatures. Proponents on both sides of the debate point to increasing mean global temperature changes from 1880 to 2008, which is a historical time interval (▌Figure 1). On the other hand, mean global temperature and precipitation have changed, in some cases significantly, when viewed from a geologic time perspective (▌Figure 2).

    Discuss why an understanding of geologic time is crucial to the global warming debate and its consequences, and whether it is possible to extrapolate changes based on a historical time frame to one of a geologic time frame.

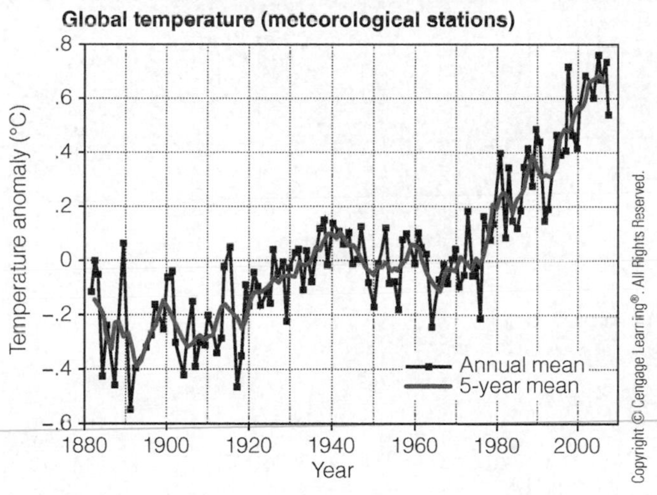

▌**Figure 1** **Mean Global Temperature Changes from 1880 to 2008** The zero line represents the average from 1951 to 1980. The plus and minus values represent deviations (in °C) from the average for the annual mean (in black) and 5-year mean (in red). Note the dramatic increase in temperature from approximately 1970 to 2008.

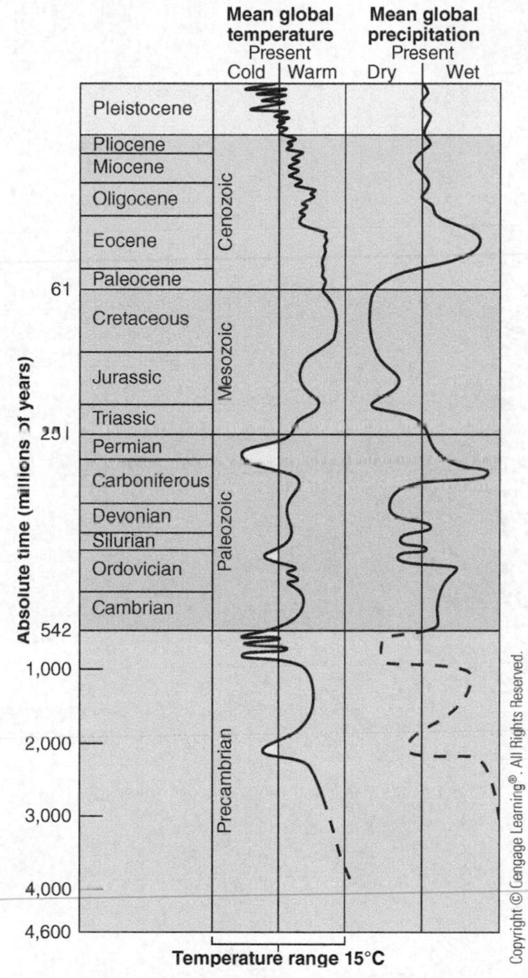

▌**Figure 2** **Mean Global Temperature and Precipitation Changes Throughout Earth History** There have been several episodes of global cooling throughout Earth's history, notably at the end of the Precambrian (Proterozoic Eon), the end of the Ordovician Period, the Late Carboniferous to Permian periods, and a series of glacial and interglacial episodes during the Pleistocene Epoch.

Global GeoScience Watch One of the themes in Chapter 1 is the connection between geology and our everyday lives, especially our dependence on natural resources. Within the GREENR database, search "mining" to be taken to this portal, and then click on "VIEW FULL OVERVIEW." Do you think this article presents a balanced coverage of the topic? Summarize the main points of the article in terms of the benefits of mining as well as some of the environmental results of mining and quarrying operations. Next, click on "View All" next to the Academic Journals heading. Using the titles of the first 20 articles listed, what would you say is the general theme of the articles? Include this observation in your summary.

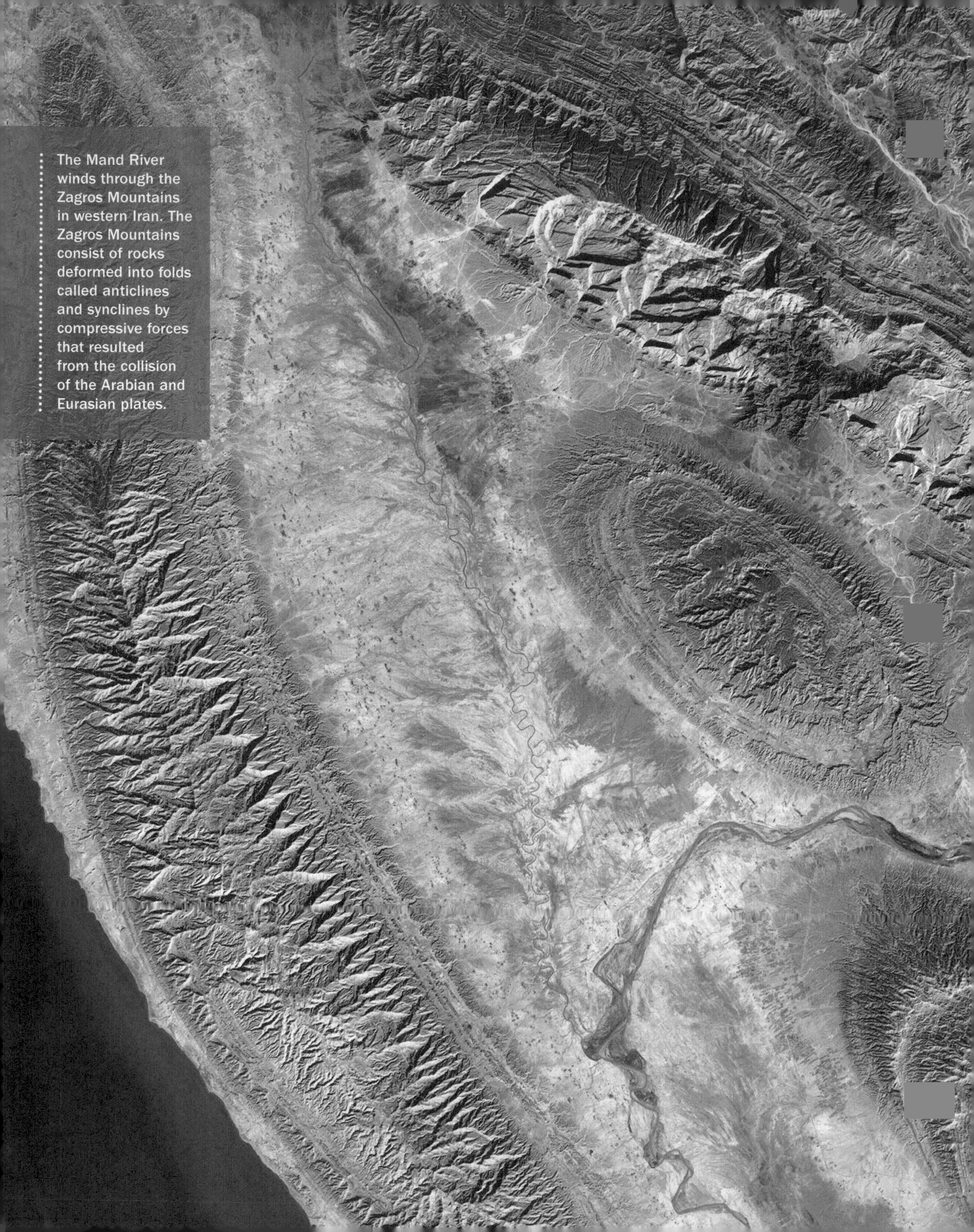

The Mand River winds through the Zagros Mountains in western Iran. The Zagros Mountains consist of rocks deformed into folds called anticlines and synclines by compressive forces that resulted from the collision of the Arabian and Eurasian plates.

USGS EROS Data Center/Landsat/NASA

Plate Tectonics: A Unifying Theory

OUTLINE

HAVE YOU EVER WONDERED?

- Whether Earth looked different in the past than it does today in terms of the position of continents and ocean basins?
- Why such natural disasters as earthquakes, volcanic eruptions, and tsunami seem to occur repeatedly in the same areas of the world?
- If there is a unifying theory in geology, similar to the theory of evolution in biology?
- Why the Middle East has so much oil?
- Why natural resources are seemingly distributed unevenly throughout the world?
- Why the present distribution of plants and animals is not random?

2.1 Introduction

Imagine it is the day after Christmas, December 26, 2004, and you are vacationing on a beautiful beach in Thailand. You look up from the book you are reading to see the sea suddenly retreat from the shoreline. Within minutes of this unusual event, a powerful tsunami will sweep over your resort and everything in its path for several kilometers inland. In the next few hours, the coasts of Indonesia, Sri Lanka, India, Thailand, Somalia, Myanmar, Malaysia, and the Maldives will be inundated by the deadliest tsunami in history. More than 230,000 people will die, and the region will incur billions of dollars in damage.

Now go forward to October 25, 2010, when Mount Merapi in Central Java, Indonesia, began a series of violent eruptions that lasted into late November. Huge quantities of ash, lava, and ash flows were emitted during these eruptions, and although warnings were issued, at least 350 people were killed by the eruptions, and more than 350,000 were evacuated from areas around the volcano.

Almost five months later, on March 11, 2011, a catastrophic 9.0-magnitude earthquake struck Japan, killing more than 20,000 people, and leaving thousands injured and homeless. This devastating earthquake also generated a tsunami that sent walls of water crashing into the northeastern shores of the island, causing further damage and casualties.

What do these three recent tragic events have in common? They are part of the dynamic interactions involving Earth's plates. When two plates come together, one plate is pushed or pulled under the other plate, triggering large earthquakes such as the recent ones that shook Haiti in 2010, and New Zealand and Japan in 2011. If conditions are right, earthquakes also can produce a tsunami such as the one in Indonesia in 2004 and in Japan in 2011.

As the descending plate moves downward and is assimilated into Earth's interior, magma is generated. Being less dense than the surrounding material, the magma rises toward the surface, where it may erupt as a volcano. It therefore should not be surprising that the distribution of volcanoes and earthquakes closely follows plate boundaries.

If you are like most people, you probably have only a vague notion of what plate tectonic theory is. Yet plate tectonics affects all of us. Volcanic eruptions, earthquakes, and tsunami are the result of interactions between plates. Global weather patterns and oceanic currents are caused, in part, by the configuration of the continents and ocean basins. The formation and distribution of many natural resources are related to plate movement and, thus, have an impact on the economic well-being and political decisions of nations. It is therefore important to understand this unifying theory, not only because it affects us as individuals and as citizens of nation-states, but also because it ties together many aspects of the geology you will be studying.

2.2 Early Ideas About Continental Drift

The idea that Earth's past geography was different from todays is not new. The earliest maps showing the east coast of South America and the west coast of Africa probably provided people with the first evidence that continents may have once been joined and then broken apart and moved to their present positions. As far back as 1620, Sir Francis Bacon commented on the similarity of the shorelines of western Africa and eastern South America. However, he did not make the connection that the Old and New Worlds might once have been joined.

During the late 19th century, the Austrian geologist Edward Suess noted the similarities between the Late Paleozoic plant fossils of India, Australia, South Africa, and South America, as well as evidence of glaciation in the rock sequences of these continents. The plant fossils make up a unique flora that occurs in the coal layers just above the glacial deposits of these southern continents. This flora is very different from the contemporaneous coal swamp flora of the northern continents and is collectively known as the ***Glossopteris* flora** after its most conspicuous genus (▌Figure 2.1).

Suess also proposed the name *Gondwanaland* (or **Gondwana** as we will use here) for a supercontinent composed of the aforementioned southern continents. Abundant fossils of the *Glossopteris* flora are found in coal beds in Gondwana, a province in India. Suess thought these southern continents were at one time connected by land bridges over which plants and animals migrated. Thus, in his view, the similarities of fossils on these continents were due to the appearance and disappearance of the connecting land bridges.

▌**Figure 2.1 Fossil *Glossopteris* Leaves** Plant fossils, such as these *Glossopteris* leaves from the Upper Permian Dunedoo Formation in Australia, are found on all five of the Gondwana continents. Their presence on continents with widely varying climates today is evidence that the continents were at one time connected. The distribution of the plants at that time was in the same climatic latitudinal belt.

Figure 2.2 Alfred Wegener Alfred Wegener, a German meteorologist, proposed the continental drift hypothesis in 1912 based on a tremendous amount of geologic, paleontologic, and climatologic evidence. He is shown here waiting out the Arctic winter in an expedition hut in Greenland.

Alfred Wegener and the Continental Drift Hypothesis

Alfred Wegener, a German meteorologist (Figure 2.2), is generally credited with developing the hypothesis of **continental drift**. In his monumental book, *The Origin of Continents and Oceans* (first published in 1915), Wegener proposed that all landmasses were originally united in a single supercontinent that he named **Pangaea**, from the Greek meaning "all land." Wegener portrayed his grand concept of continental movement in a series of maps showing the breakup of Pangaea and the movement of the various continents to their present-day locations. Wegener amassed a tremendous amount of geologic, paleontologic, and climatologic evidence in support of continental drift; however, initial reaction of scientists to his then-heretical ideas can best be described as mixed.

Nevertheless, Alexander du Toit, a South African geologist, and one of Wegener's more ardent supporters, further developed Wegener's arguments and gathered more geologic and paleontologic evidence in support of his continental drift hypothesis. In 1937, du Toit published *Our Wandering Continents*, in which he contrasted the glacial deposits of Gondwana with coal deposits of the same age found in the continents of the Northern Hemisphere. To resolve this apparent climatologic paradox, du Toit moved the Gondwana continents to the South Pole and brought the northern continents together such that the coal deposits were located at the equator. He named this northern landmass **Laurasia**. It consisted of present-day North America, Greenland, Europe, and Asia (except for India, which was part of Gondwana).

2.3 What Is the Evidence for Continental Drift?

What then was the evidence Wegener, du Toit, and others used to support the hypothesis of continental drift? It included the fit of the shorelines of continents, the appearance of the same rock sequences and mountain ranges of the same age on continents now widely separated, the matching of glacial deposits and paleoclimatic zones, and the similarities of many extinct plant and animal groups whose fossil remains are found today on widely separated continents. Wegener and his supporters argued that this vast amount of evidence from a variety of sources surely indicated that the continents must have been close together in the past.

Continental Fit

Wegener, like some before him, was impressed by the close resemblance between the coastlines of continents on opposite sides of the Atlantic Ocean, particularly South America and Africa. He cited these similarities as partial evidence that the continents were at one time joined together as a supercontinent that subsequently split apart. His critics pointed out, however, that the configuration of coastlines results from erosional and depositional processes and therefore is continuously being modified.

A more realistic approach is to fit the continents together along the continental slope where erosion would be minimal. Sir Edward Bullard, an English geophysicist, and two associates showed in 1965 that the best fit between continents occurs at a depth of about 2,000 m, confirming the close fit between continents when they are reassembled to form Pangaea (Figure 2.3).

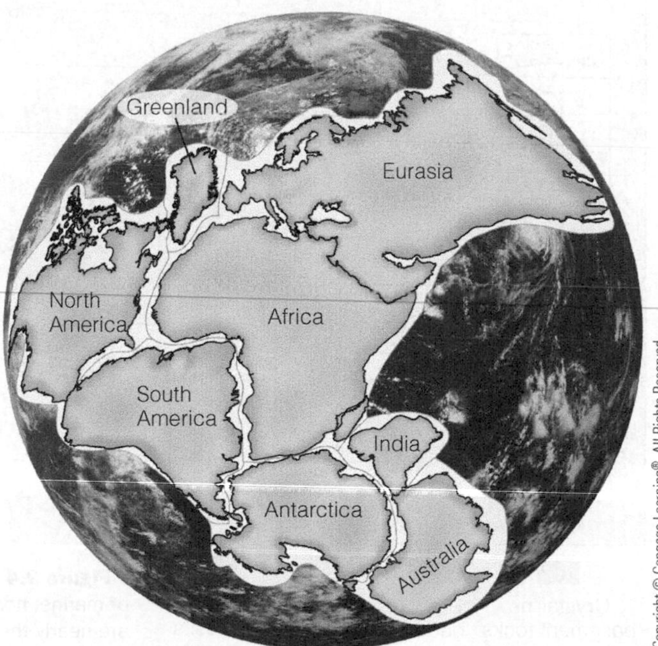

Figure 2.3 Continental Fit When continents are placed together based on their outlines, the best fit is not along their present-day coastlines but, rather, along the continental slope at a depth of about 2,000 m.

Critical Thinking Question Why is the best fit along the continental slope and not along the current coastline?

Similarity of Rock Sequences and Mountain Ranges

If the continents were at one time joined, then the rocks and mountain ranges of the same age in adjoining locations on the opposite continents should closely match. Such is the case for the Gondwana continents (▮ Figure 2.4). Marine, nonmarine, and glacial rock sequences of Pennsylvanian to Jurassic age are almost identical on all five Gondwana continents, strongly indicating that they were joined at one time.

Furthermore, the trends of several major mountain ranges also support the hypothesis of continental drift. These mountain ranges seemingly end at the coastline of one continent only to apparently continue on another continent across the ocean. The folded Appalachian Mountains of North America, for example, trend northeastward through the eastern United States and Canada and terminate abruptly at the Newfoundland coastline. Mountain ranges of the same age and deformational style are found in eastern Greenland, Ireland, Great Britain, and Norway. When the continents are positioned next to each other as they were during the Paleozoic Era, they form an essentially continuous mountain range.

ConnectionLink

To learn more about how Pangaea formed and how its origin accounted for Paleozoic mountain building, see Chapter 20.

Glacial Evidence

During the Late Paleozoic Era, massive glaciers covered large continental areas of the Southern Hemisphere. Evidence for this glaciation includes layers of *till* (sediments deposited by glaciers) and glacial *striations* (scratch marks) in the bedrock beneath the till (▮ Figure 2.5). Fossils and sedimentary rocks of the same age from the Northern Hemisphere, however, give no indication of glaciation. Fossil plants found in coals indicate that the Northern Hemisphere had a tropical climate during the time that the Southern Hemisphere was glaciated.

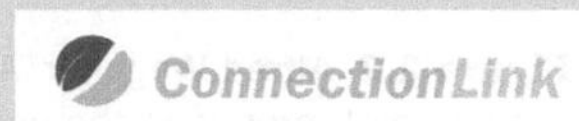

ConnectionLink

To better understand continental glaciers and the erosional and depositional landforms they create, see Chapter 14.

All of the Gondwana continents except Antarctica are currently located near the equator in subtropical to

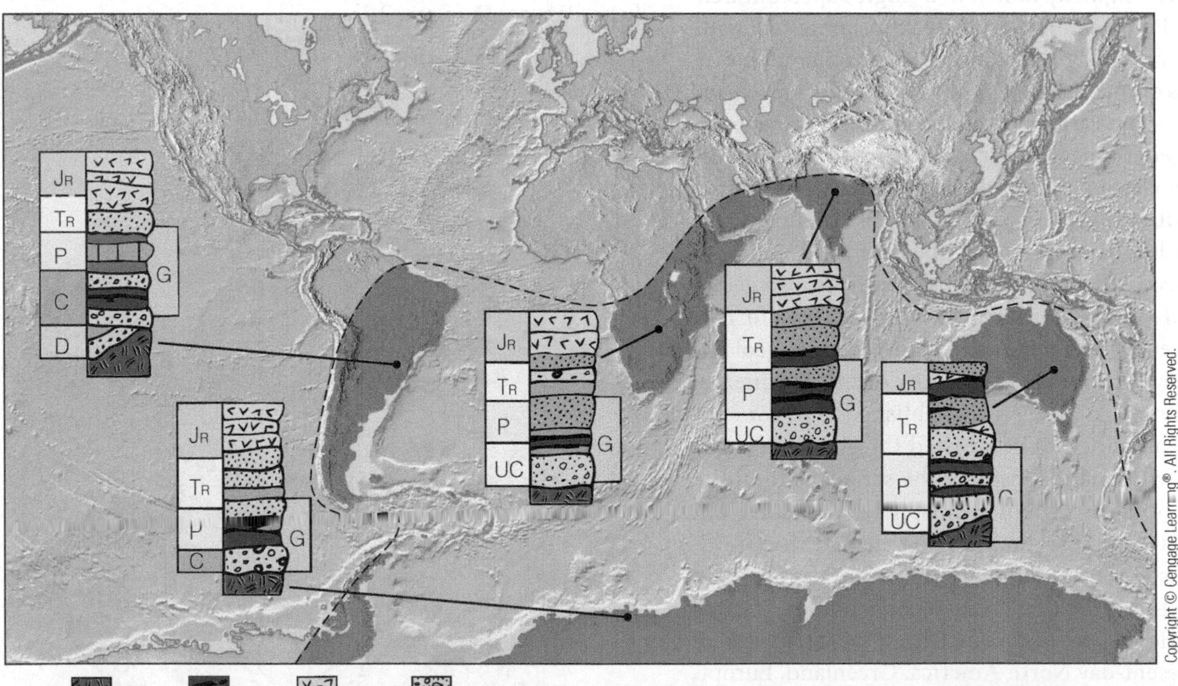

▮ **Figure 2.4 Similarity of Rock Sequences on the Gondwana Continents** Sequences of marine, nonmarine, and glacial rocks of Pennsylvanian (UC) to Jurassic (JR) age are nearly the same on all five Gondwana continents (South America, Africa, India, Australia, and Antarctica). These continents are widely separated today and have different environments and climates ranging from tropical to polar. Thus, the rocks forming on each continent are very different. When the continents were all joined in the past, however, the environments of adjacent continents were similar, and the rocks forming in those areas were similar. The range indicated by G in each column is the age range of the *Glossopteris* flora.

Figure 2.5 Glacial Evidence Indicating Continental Drift

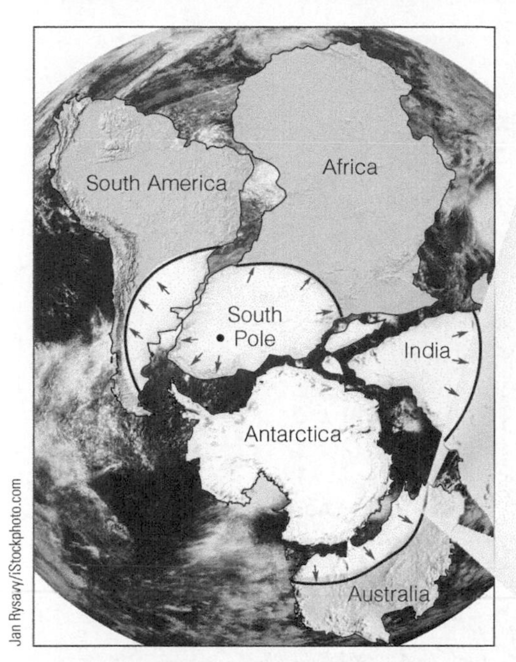

a When the Gondwana continents are placed together so that South Africa is located at the South Pole, the glacial movements indicated by striations (red arrows) found on rock outcrops on each continent make sense. In this situation, the glacier (white area) is located in a polar climate and has moved radially outward from its thick central area toward its periphery.

b Glacial striations (scratch marks) on an outcrop of Permian-age bedrock exposed at Hallet's Cove, Australia, indicate the general direction of glacial movement more than 200 million years ago. As a glacier moves over a continent's surface, it grinds and scratches the underlying rock. The glacial striations that are preserved on a rock's surface thus provide evidence of the direction (red arrows) the glacier moved at that time.

tropical climates. Mapping of glacial striations in bedrock in Australia, India, and South America indicates that the glaciers moved from the areas of the present-day oceans onto land. Yet, this would be highly unlikely because large continental glaciers (such as occurred on the Gondwana continents during the Late Paleozoic Era) flow outward from their central area of accumulation toward the sea.

Therefore, if the continents did not move during the past, one would have to explain how glaciers moved from the oceans onto land and how large-scale continental glaciers formed near the equator. But, if the continents are reassembled as a single landmass with South Africa located at the South Pole, then the direction of movement of Late Paleozoic continental glaciers makes sense (Figure 2.5a). Furthermore, this geographic arrangement places the northern continents nearer the tropics, which is consistent with the fossil and climatologic evidence from Laurasia.

Fossil Evidence

Some of the most compelling evidence for continental drift comes from the fossil record (▌Figure 2.6). For example, fossils of the *Glossopteris* flora (which include the seed fern *Glossopteris*, as well as many other distinctive and easily identifiable plants) are found in equivalent Pennsylvanian- and Permian-age coal deposits on all five Gondwana continents. The present-day climates of South America, Africa, India, Australia, and Antarctica range from tropical to polar, however, and are much too diverse to support the type of plants comprising the *Glossopteris* flora. Wegener therefore reasoned that these continents must once have been joined so that these widely separated localities were all in the same latitudinal belt (Figure 2.6).

The fossil remains of animals also provide strong evidence for continental drift. One of the best examples is *Mesosaurus*, a freshwater reptile whose fossils are found in Permian-age rocks in certain regions of Brazil and South Africa and nowhere else in the world (Figure 2.6). Because the physiologies of freshwater and marine animals are completely different, it is hard to imagine how a freshwater reptile could have swum across the Atlantic Ocean and found a freshwater environment nearly identical to its former habitat. Moreover, if *Mesosaurus* could have swum across the ocean, its fossil remains should be widely dispersed. It is more logical to assume that *Mesosaurus* lived in lakes in what were once adjacent areas of South America and Africa when it was united into a single continent. Discoveries of fossils from additional land-dwelling animals on these and other Gondwana continents, further solidify the argument that these landmasses were at one time in proximity.

Notwithstanding all of the empirical evidence presented by Wegener and later by du Toit and others, most geologists simply refused to entertain the idea that continents might have moved in the past. The geologists were not necessarily being obstinate about accepting new ideas; rather, they found the evidence for continental drift inadequate and unconvincing. In part, this was because no one could provide a suitable mechanism to explain how continents could move over Earth's surface.

Interest in continental drift waned until studies of Earth's magnetic field and oceanographic research, conducted during the 1950s, showed that the present-day ocean basins are not as old as continents but are geologically young features that resulted from the breakup of Pangaea.

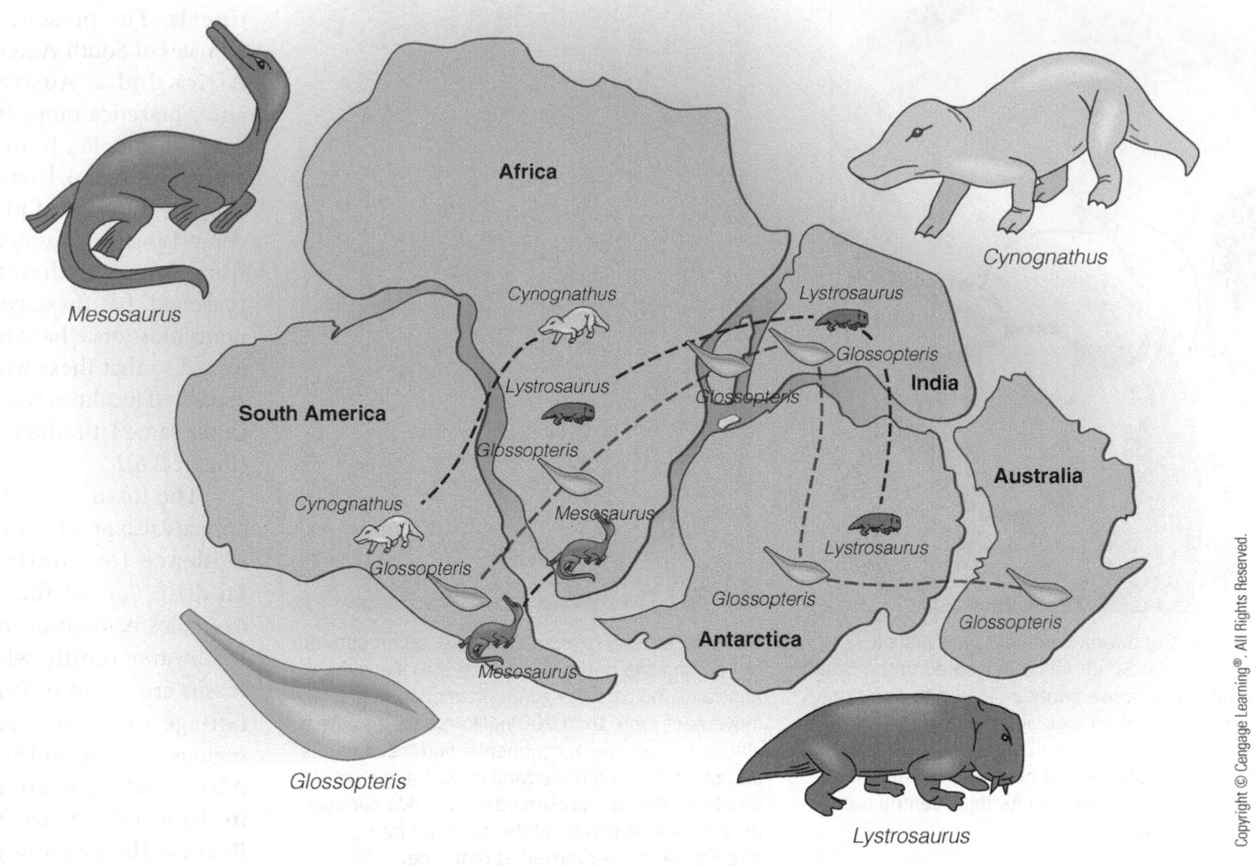

Figure 2.6 Fossil Evidence Supporting Continental Drift Some of the plants and animals whose fossils are found today on the widely separated continents of South America, Africa, India, Australia, and Antarctica. During the Late Paleozoic Era, these continents were joined to form Gondwana, the southern landmass of Pangaea. Plants of the *Glossopteris* flora are found on all five continents, which today have widely different climates; however, during the Pennsylvanian and Permian periods, they were all located in the same general climatic belt. *Mesosaurus* is a freshwater reptile whose fossils are found only in similar nonmarine Permian-age rocks in Brazil and South Africa. *Cynognathus* and *Lystrosaurus* are land reptiles that lived during the Early Triassic Period. Fossils of *Cynognathus* are found in South America and Africa, whereas fossils of *Lystrosaurus* have been recovered from Africa, India, and Antarctica. It is hard to imagine how a freshwater reptile and land-dwelling reptiles could have swum across the wide oceans that presently separate these continents. It is more logical to assume that the continents were once connected.

2.4 Features of the Seafloor

At this point, it is useful to discuss some of the various features of Earth's seafloor. Many of the topographic features found on the seafloor and along the continental margins are the manifestations of Earth's internal processes and activity taking place along plate margins. Thus, it is important to know how these features relate to plate tectonic theory.

Most people think of continents as land areas outlined by the oceans, but the true geologic margin of a continent—where granitic continental crust changes to basalt and gabbro oceanic crust—is below sea level. A **continental margin** is made up of a gently sloping continental shelf; a more steeply inclined continental slope; and, in some cases, a deeper, gently sloping continental rise (Figure 2.7). Thus, the continental margins extend to increasingly greater depths until they merge with the deep seafloor. Continental crust changes to oceanic crust somewhere beneath the continental rise, so part of the continental slope and the continental rise actually rest on oceanic crust.

The Continental Shelf, Slope, and Rise

As one proceeds seaward from the shoreline across the continental margin, the first area encountered is the gently sloping **continental shelf** lying between the shore and the more steeply dipping continental slope (Figure 2.7). The width of the continental shelf varies considerably, ranging from a few tens of meters to more than 1,000 km; the shelf terminates where the inclination of the seafloor increases abruptly from 1 degree or less to several degrees.

The seaward margin of the continental shelf is marked by the *shelf–slope break* (at an average depth of 135 m), where

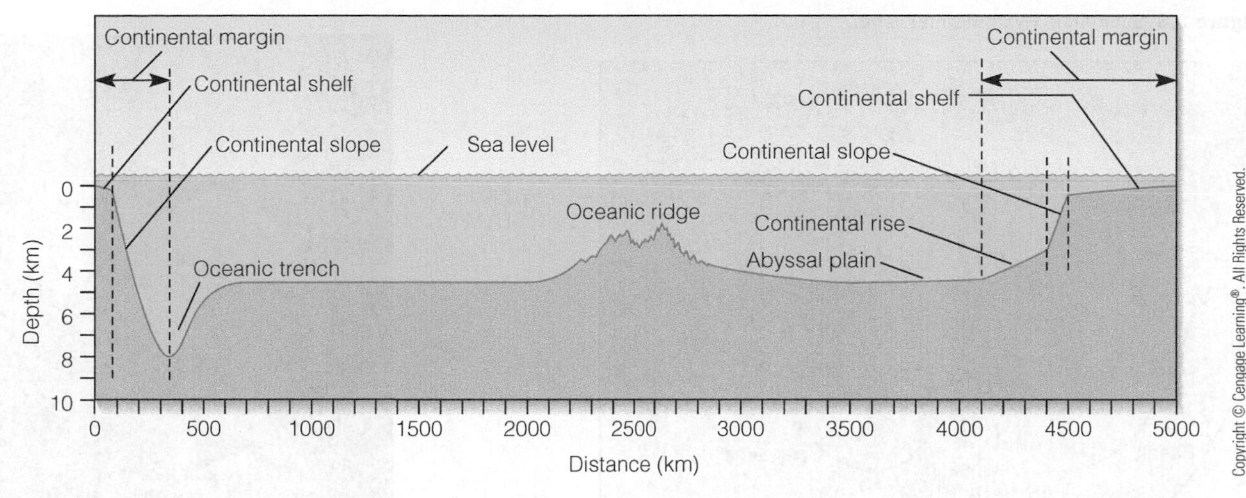

Figure 2.7 Features of Continental Margins A generalized profile showing features of the continental margins. The vertical dimensions of the features in this profile are greatly exaggerated, because the vertical and horizontal scales differ.

the more steeply inclined **continental slope** begins (Figure 2.7). In most areas around the margins of the Atlantic, the continental slope merges with a more gently sloping **continental rise**. This rise is absent around the margins of the Pacific, where continental slopes descend directly into an oceanic trench (Figure 2.7)

Abyssal Plains, Oceanic Ridges, Submarine Hydrothermal Vents, and Oceanic Trenches

Beyond the continental rises are **abyssal plains**—flat surfaces covering vast areas of the seafloor. In some areas, they are interrupted by peaks rising more than 1 km, but abyssal plains are nevertheless the flattest, most featureless areas on Earth (Figure 2.7). Their flatness is a result of sediment deposition covering the usually rugged topography of the seafloor.

A renewed interest in oceanographic research led to extensive mapping of the ocean basins during the 1960s. Such mapping revealed an **oceanic ridge** system more than 65,000 km long, constituting the most extensive mountain range in the world (Figure 2.7). This system runs from the Arctic Ocean through the middle of the Atlantic and curves around South Africa, where the Indian Ridge continues into the Indian Ocean; the Atlantic–Pacific Ridge extends eastward, and a branch of it, the East Pacific Rise, trends northeast until it reaches the Gulf of California. Perhaps the best-known part of the ridge system is the Mid-Atlantic Ridge, which divides the Atlantic Ocean basin into two nearly equal parts (see Figure 1.13).

Oceanic ridges are composed almost entirely of the igneous rocks basalt and gabbro and possess features produced by tensional forces. Thus, they are the sites where new oceanic crust is generated and plates move away from each other along divergent plate boundaries.

First seen on the seafloor in 1979, **submarine hydrothermal vents** are found at or near spreading ridges. Here, cold seawater seeps through oceanic crust, is heated by the hot rocks at depth, and then rises and discharges into the seawater as plumes of hot water with temperatures as high as 400°C. Many of the plumes are black because dissolved minerals give them the appearance of black smoke—hence the name *black smoker* (▌Figure 2.8).

Submarine hydrothermal vents are interesting from the biologic, geologic, and economic points of view. Near the vents live communities of organisms, such as bacteria, crabs, mussels, starfish, and tube worms—many of which have never been seen before. No sunlight is available, so these organisms depend on bacteria that oxidize sulfur compounds for their ultimate source of nutrients. The vents are also interesting because of their economic potential. The heated seawater reacts with oceanic crust, transforming it into a metal-rich solution that discharges into seawater and cools, precipitating iron, copper, and zinc sulfides and other minerals (Figure 2.8a). A chimney-like vent forms that eventually collapses and forms a mound of sediments rich in the elements just mentioned (Figure 2.8b).

Oceanic trenches are long, steep-sided depressions on the seafloor near convergent plate boundaries and constitute no more than 2% of the seafloor (Figure 2.7). It is here, however, that oceanic lithosphere is consumed by subduction; that is, oceanic lithosphere plunges into Earth's interior along convergent plate boundaries (see Figure 1.14). The greatest oceanic depths are found in trenches; the Challenger Deep of the Marianas Trench in the Pacific is more than 11,000 m deep.

Seamounts, Guyots, and Aseismic Ridges

Except for the abyssal plains, the seafloor is not a flat, featureless expanse. In fact, a large number of volcanic hills, seamounts, and guyots rise above the seafloor in all

Figure 2.8 Submarine Hydrothermal Vents

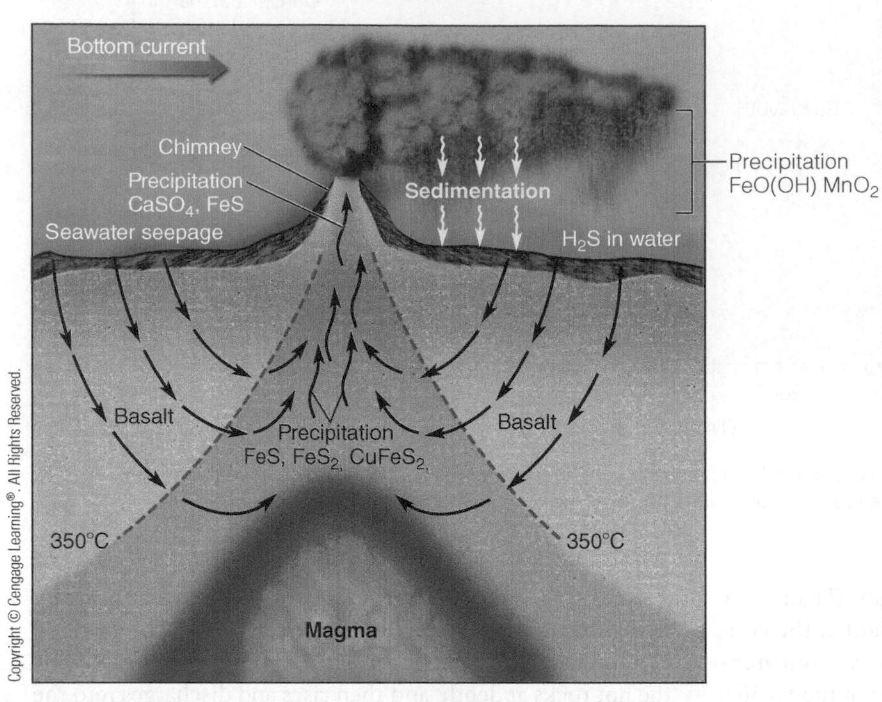

a Cross section showing the origin of a submarine hydrothermal vent called a black smoker.

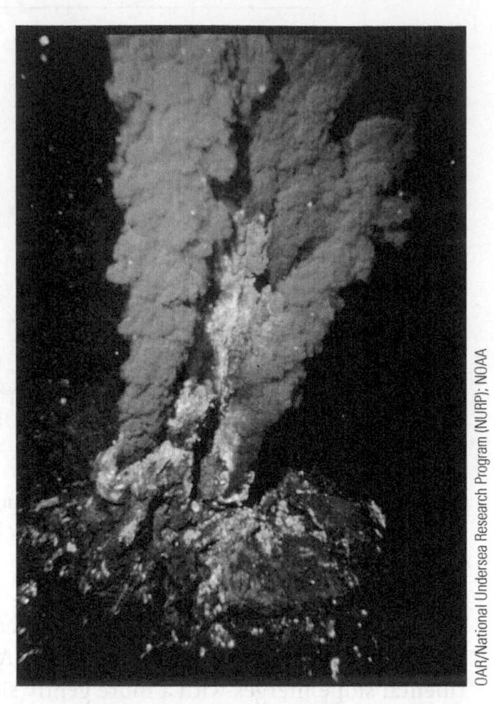

b This black smoker on the East Pacific Rise is at a depth of 2,800 m. The plume of "black smoke" is heated seawater with dissolved minerals.

ocean basins and are particularly abundant in the Pacific. All are of volcanic origin and differ mostly in size. *Seamounts* rise more than 1 km above the seafloor, and if flat-topped, they are called *guyots* (Figure 2.9). Guyots are volcanoes that originally extended above sea level. However, as the plate upon which they were located continued to move, they were carried away from a spreading ridge, and as the oceanic crust cooled, it descended to greater depths. Thus, what was once an island slowly sank beneath the sea, and as it did, wave erosion produced the typical flat-topped appearance of a guyot (Figure 2.9).

Other common features in the ocean basins are long, narrow ridges and broad, plateau-like features rising as much as 2 to 3 km above the surrounding seafloor. These *aseismic ridges* are so called because they lack seismic (earthquake) activity (see Hot Spots and Mantle Plumes). A few of these ridges are probably small fragments separated from continents during rifting and are referred to as *microcontinents*. Avalonia is a good example of a Paleozoic microcontinent (see Figure 20.2b).

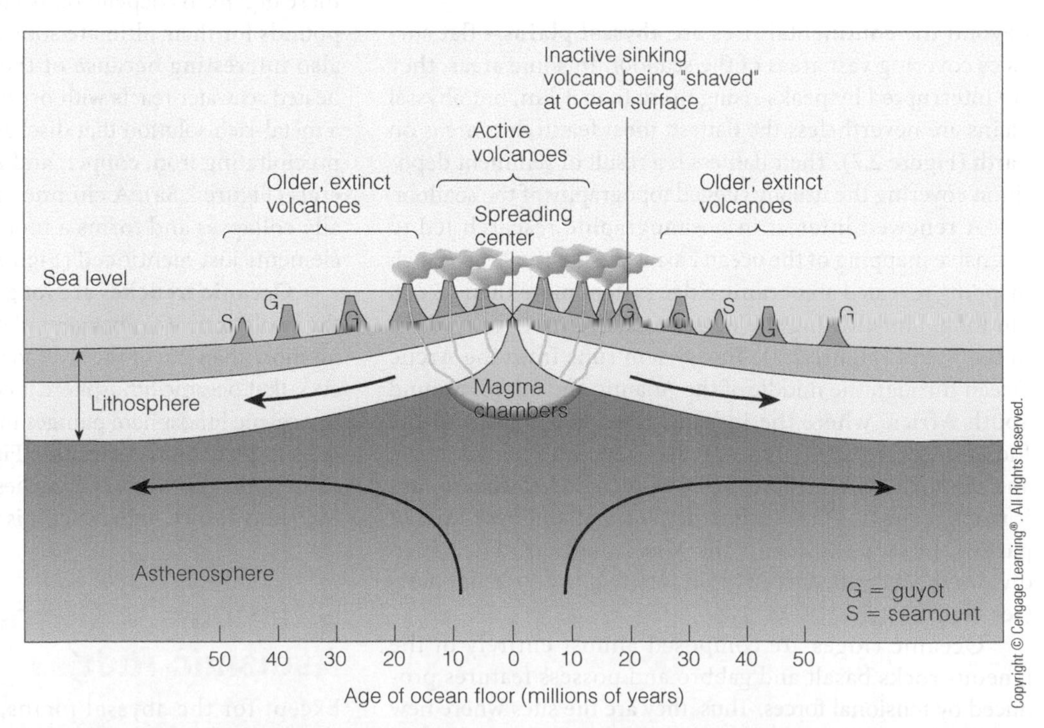

Figure 2.9 The Origin of Seamounts and Guyots As the plate on which a volcano rests moves into greater water depths, the submerged volcanic island is called a *seamount*. Those that are flat-topped are called *guyots*.

Continental Margins

Although we will be discussing convergent and divergent plate boundaries later in this chapter, here is a good place to cover the two types of continental margins associated with the respective aforementioned plate boundaries.

Active continental margins develop at the leading edge of a continental plate where oceanic lithosphere is subducted. The western margin of South America is a good example of where an oceanic plate is subducted beneath the continent, resulting in seismic activity, a geologically young mountain range (Andes Mountains), and active volcanism (▮ Figure 2.10). In addition, the continental shelf is narrow, and the continental slope descends directly into an oceanic trench, so sediment is dumped into the trench and no continental rise develops (Figure 2.7, left side).

The western margin of North America is also considered an active continental margin, although much of it is now bounded by transform faults (see the section on transform boundaries later in this chapter) rather than a subduction zone. However, plate convergence and subduction continue in the Pacific Northwest along the continental margins of northern California, Oregon, and Washington.

The continental margins of eastern North America and South America differ considerably from their western margins. For one thing, they possess broad continental shelves, as well as a continental slope and rise, with abyssal plains adjacent to the rises (Figure 2.7, right side). Furthermore, these **passive continental margins** are within a plate rather than at a plate boundary, and they lack the volcanic and seismic activity found at active continental margins (Figure 2.10). Nevertheless, earthquakes do take place there occasionally, such as the magnitude 5.8 earthquake that struck the east coast of the United States on August 23, 2011, and caused slight damage to the Washington National Cathedral.

2.5 Earth's Magnetic Field

Having just discussed those features of the seafloor that are related to plate movement, we now turn our attention to the phenomenon of magnetism and its role in the formulation of plate tectonic theory.

Magnetism is a physical phenomenon resulting from the spin of electrons in some solids—particularly those of iron—and moving electricity. A **magnetic field** is an area in which magnetic substances such as iron are affected by lines of magnetic force emanating from a magnet (▮ Figure 2.11). The magnetic field shown in Figure 2.11 is *dipolar*, meaning that it possesses two unlike magnetic poles referred to as the north and south poles.

Although Earth's interior is too hot for a permanent magnet to exist, it is thought that thermal and compositional convection within the liquid metallic outer core, coupled with Earth's rotation, produce complex electrical currents (known as a *self-exciting dynamo*) that, in turn, generate the magnetic field. A useful analogy is to think of Earth as a giant dipole magnet in which the magnetic poles are in proximity to the geographic pole (▮ Figure 2.12). This arrangement means that the strength of the magnetic field is not constant but varies.

Notice in Figure 2.12 that the lines of magnetic force around Earth parallel its surface only near the equator, just as the iron filings do around a bar magnet (Figure 2.11). As the lines of force approach the poles, they are oriented at increasingly larger angles with respect to the surface, and the strength of the magnetic field increases; it is strongest at the poles and weakest at the equator.

Another important aspect of the magnetic field is that the magnetic poles, where the lines of force leave and enter Earth, do not coincide with the geographic

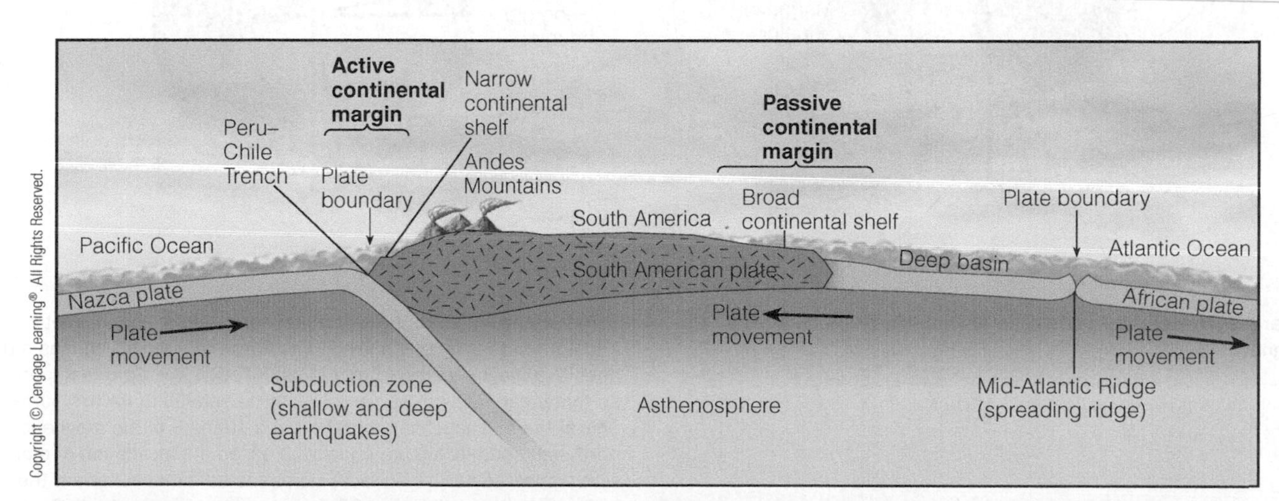

▮ **Figure 2.10 Active and Passive Continental Margins** Active and passive continental margins along the west and east coasts of South America. Notice that the passive margins are much wider than active margins. Seafloor sediment is not shown.

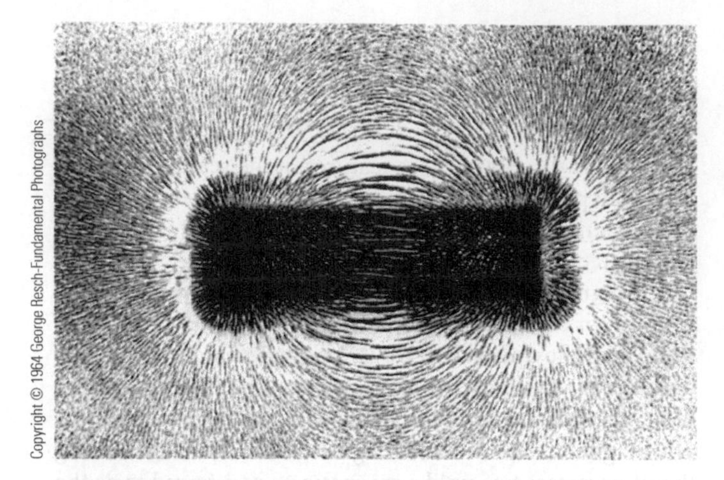

Figure 2.11 **Magnetic Field** Iron filings align along the lines of magnetic force radiating from a bar magnet.

(rotational) poles. Currently, an 11.5-degree angle exists between the two (Figure 2.12). Studies of Earth's magnetic field show that the locations of the magnetic poles vary slightly over time but that they still correspond closely, on average, with the locations of the geographic poles.

2.6 Paleomagnetism and Polar Wandering

Interest in continental drift revived during the 1950s as a result of evidence from paleomagnetic studies, a relatively new discipline at the time. **Paleomagnetism** is the remanent magnetism in ancient rocks recording the direction and intensity of Earth's magnetic poles at the time of the rock's formation.

When magma cools, the magnetic iron-bearing minerals align themselves with Earth's magnetic field, recording both its direction and strength. The temperature at which iron-bearing minerals gain their magnetization is called the **Curie point**. As long as the rock is not subsequently heated above the Curie point, it will preserve that remanent magnetism. Thus, an ancient lava flow provides a record of the orientation and strength of Earth's magnetic field at the time the lava flow cooled.

As paleomagnetic research progressed during the 1950s, some unexpected results emerged. When geologists measured the paleomagnetism of geologically recent rocks from different continents, they found that it was generally consistent with Earth's current magnetic field.

Figure 2.12 **Earth's Magnetic Field**

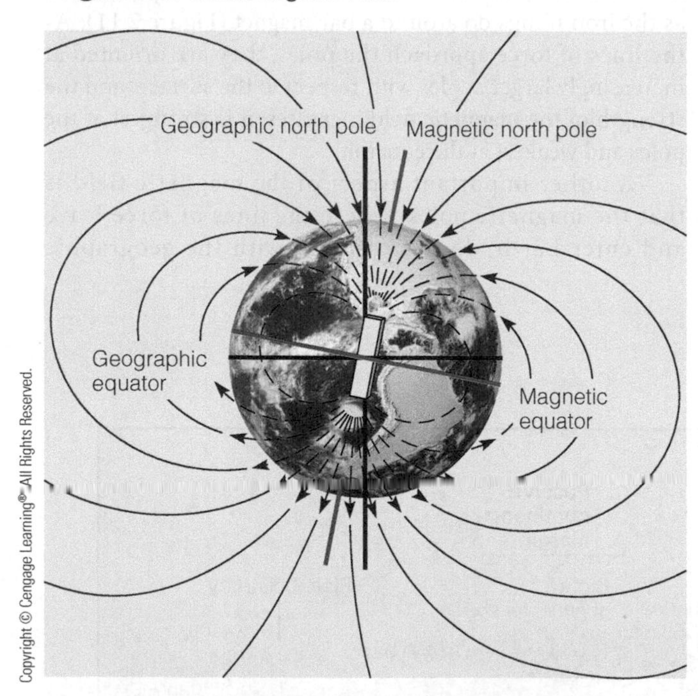

a Earth's magnetic field has lines of force like those of a bar magnet.

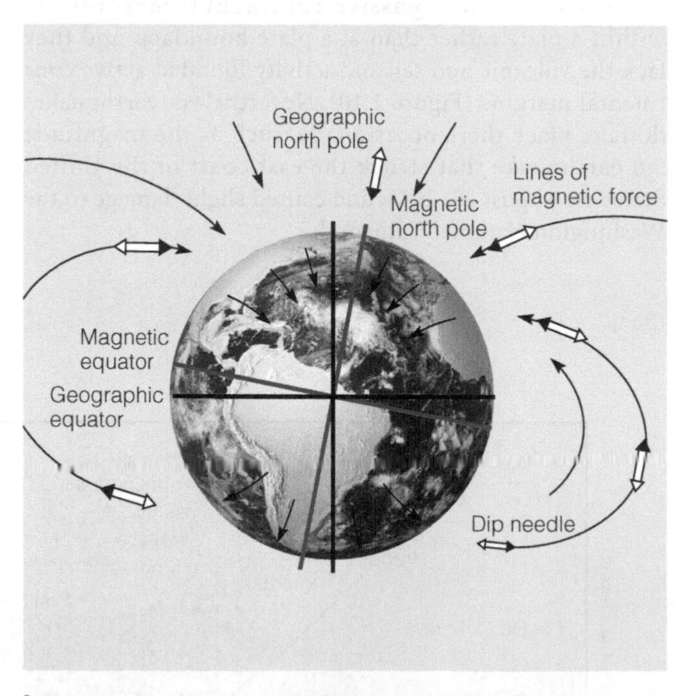

b The strength of the magnetic field changes from the magnetic equator to the magnetic poles. This change in strength causes a dip needle (a magnetic needle that is balanced on the tip of a support so that it can freely move vertically) to be parallel to Earth's surface only at the magnetic equator, where the strength of the magnetic north and south poles are equally balanced. Its inclination or dip, with respect to Earth's surface, increases as it moves toward the magnetic poles until it is at 90 degrees, or perpendicular to Earth's surface at the magnetic poles.

The paleomagnetism of ancient rocks, though, showed different orientations. For example, paleomagnetic studies of Silurian lava flows in North America indicated that the north magnetic pole was located in the western Pacific Ocean at that time, whereas the paleomagnetic evidence from Permian lava flows pointed to yet another location in Asia. When plotted on a map, the paleomagnetic readings of numerous lava flows from all ages in North America trace the apparent movement of the magnetic pole (called *polar wandering*) through time (❚ Figure 2.13).

Furthermore, analysis of lava flows from all continents indicated that each continent seemingly had its own series of magnetic poles! How could this be? Does it really mean there were different north magnetic poles for each continent?

The best explanation for such data is that the magnetic poles have remained near their present locations at the geographic north and south poles and the continents have moved. When the continental margins are fit together so that the paleomagnetic data point to only one magnetic pole, we find, just as Wegener did, that the rock sequences and glacial deposits match and that the fossil evidence is consistent with the reconstructed paleogeography.

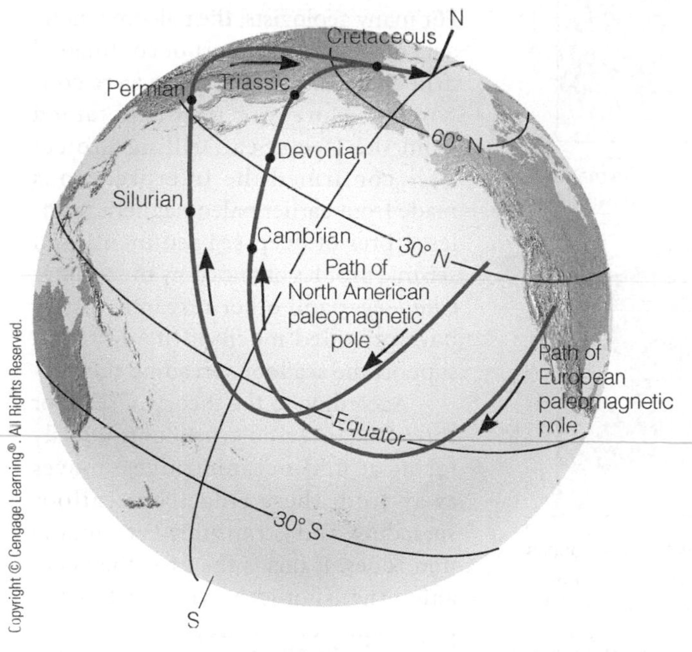

❚ Figure 2.13 Polar Wandering The apparent paths of polar wandering for North America and Europe. The apparent location of the north magnetic pole is shown for different periods on each continent's polar wandering path. Because Earth has only one magnetic pole, the paleomagnetic readings taken on different continents for the same time in the past should all point to the same location if the continents have not moved. However, the north magnetic pole has different locations for the same time in the past when measured on different continents, indicating multiple north magnetic poles. The logical explanation for this dilemma is that the magnetic north pole has remained at the same approximate geographic location during the past, and the continents have moved.

2.7 Magnetic Reversals and Seafloor Spreading

Geologists refer to Earth's present magnetic field as being normal—that is, with the north and south magnetic poles located approximately at the north and south geographic poles. At various times in the geologic past, however, Earth's magnetic field has completely reversed. The magnetic north and south poles have switched positions so that the magnetic north pole became the magnetic south pole, and the magnetic south pole became the magnetic north pole. During such a reversal, the magnetic field weakens until it temporarily disappears. When the magnetic field returns, the magnetic poles have reversed their position. The existence of such **magnetic reversals** was discovered by dating and determining the orientation of the remanent magnetism in lava flows on land (❚ Figure 2.14). Although the cause of magnetic reversals is still uncertain, their occurrence in the geologic record is well documented.

As a result of oceanographic research conducted during the 1950s, Harry Hess of Princeton University proposed, in a 1962 landmark paper, the theory of **seafloor spreading** to account for continental movement. He suggested that continents do not move through oceanic crust as do ships plowing through sea ice, but, rather, that the continents and oceanic crust move together as a single unit. Thus, the theory of seafloor spreading answered a major objection of the opponents of continental drift—namely, how could continents move through oceanic crust? The answer is that they do not. The fact is, the continents move with the oceanic crust as part of a lithospheric system.

As a mechanism to drive this system, Hess revived the idea (first proposed in the late 1920s by the British geologist Arthur Holmes) of a heat transfer system—or **thermal convection cells**—within the mantle to move the plates. According to Hess, hot magma rises from the mantle, intrudes along fractures defining oceanic ridges, and thus forms new crust. Cold crust is subducted back into the mantle at oceanic trenches, where it is heated and recycled, thus completing a thermal convection cell (see Figure 1.12).

How could Hess's hypothesis be confirmed? Magnetic surveys of the oceanic crust revealed a pattern of striped **magnetic anomalies** (deviations from the average strength of Earth's present-day magnetic field) in the rocks that are both parallel to and symmetric around the oceanic ridges (❚ Figure 2.15). A positive magnetic anomaly results when Earth's magnetic field at the time of oceanic crust formation along an oceanic ridge summit was the same as today, thus yielding a stronger than normal (positive) magnetic signal. A negative magnetic anomaly results when Earth's magnetic field at the time of oceanic crust formation was reversed, therefore yielding a weaker than normal (negative) magnetic signal.

Thus, as new oceanic crust forms at oceanic ridge summits and records Earth's magnetic field at the time, the previously formed crust moves laterally away from

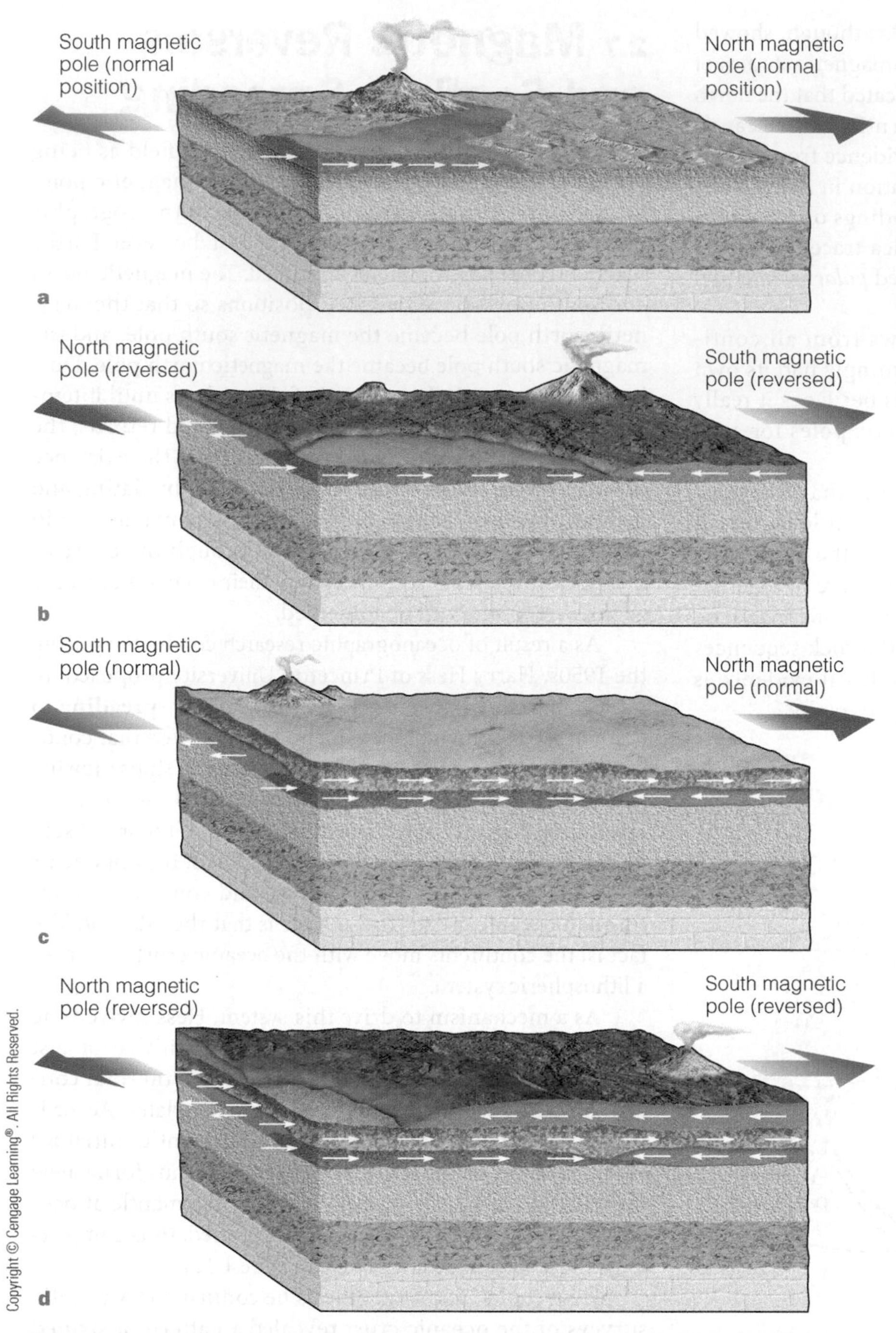

Figure 2.14 Magnetic Reversals During the time period shown (a–d), volcanic eruptions produced a succession of overlapping lava flows. At the time of these volcanic eruptions, Earth's magnetic field completely reversed—that is, the magnetic north pole moved to the geographic south pole, and the magnetic south pole moved to the geographic north pole. Thus, the end of the needle of a magnetic compass that today would point to the North Pole would point to the South Pole if the magnetic field should again reverse. We know that Earth's magnetic field has reversed numerous times in the past because when lava flows cool below the Curie point, magnetic minerals within the flow orient themselves parallel to the magnetic field at the time. They thus record whether the magnetic field was normal or reversed at that time. The white arrows in this diagram show the direction of the north magnetic pole for each individual lava flow, thus confirming that Earth's magnetic field has reversed during the past.

the ridge. These magnetic stripes therefore represent times of normal and reversed polarity at oceanic ridges (where upwelling magma forms new oceanic crust) and conclusively confirm Hess's theory of seafloor spreading.

One of the consequences of the seafloor spreading theory is its confirmation that ocean basins are geologically young features whose openings and closings are partially responsible for continental movement (Figure 2.16). Radiometric dating reveals that the oldest oceanic crust is somewhat younger than 180 million years old, whereas the oldest continental crust is about 4 billion years old. Although geologists do not universally accept the idea of thermal convection cells as the sole driving mechanism for plate movement, most accept that plates are created at oceanic ridges and destroyed at deep-sea trenches (Figure 2.10), regardless of the driving mechanism involved.

Deep-Sea Drilling and the Confirmation of Seafloor Spreading

For many geologists, the paleomagnetic data amassed in support of continental drift and seafloor spreading were convincing. Moreover, results obtained from the Deep-Sea Drilling Project later confirmed the interpretations made from earlier paleomagnetic studies. Cores of deep-sea sediments and seismic profiles obtained by the *Glomar Challenger* and other research vessels have provided much of the data that support the seafloor spreading theory.

According to the theory of seafloor spreading, oceanic crust continuously forms at mid-oceanic ridges, moves away from these ridges by seafloor spreading, and is consumed at subduction zones. If this is the case, then oceanic crust should be youngest at the ridges and become progressively older with increasing distance away from them. Furthermore, the age of the oceanic crust should be symmetrically distributed about the ridges. As we have just noted, paleomagnetic data confirm these statements. In addition, fossils from sediments overlying the oceanic crust and radiometric dating of rocks found on oceanic islands both substantiate this predicted age distribution.

Sediments in the open ocean accumulate, on average, at a rate of less than 0.3 cm in 1,000 years. If the ocean basins

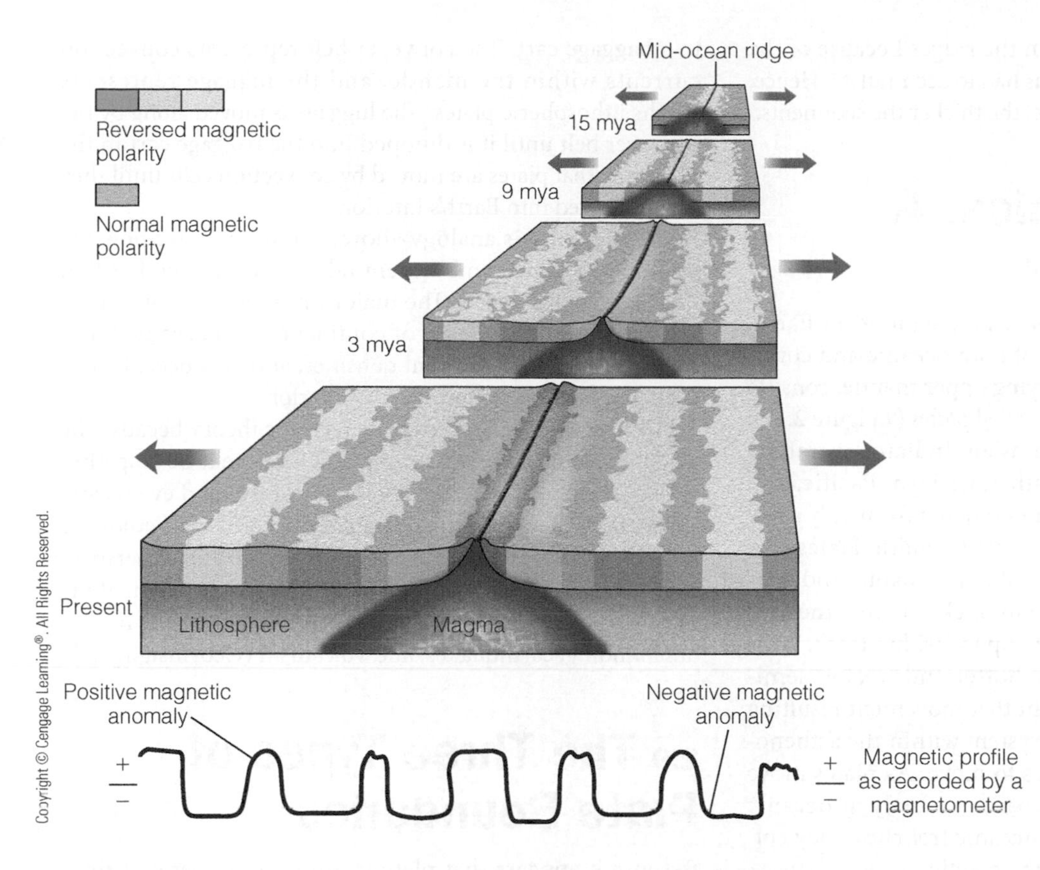

Figure 2.15 Magnetic Anomalies and Seafloor Spreading The sequence of magnetic anomalies preserved within the oceanic crust is both parallel to and symmetric around oceanic ridges. Basaltic lava intruding into an oceanic ridge today and spreading laterally away from the ridge records Earth's current magnetic field or polarity (considered by convention to be normal). Basaltic intrusions 3, 9, and 15 million years ago record Earth's reversed magnetic field at those times. This schematic diagram shows how the solidified basalt moves away from the oceanic ridge (or spreading ridge), carrying with it the magnetic anomalies that are preserved in the oceanic crust. Magnetic anomalies are magnetic readings that are either higher (positive magnetic anomalies) or lower (negative magnetic anomalies) than Earth's current magnetic field strength. The magnetic anomalies are recorded by a magnetometer, which measures the strength of the magnetic field.

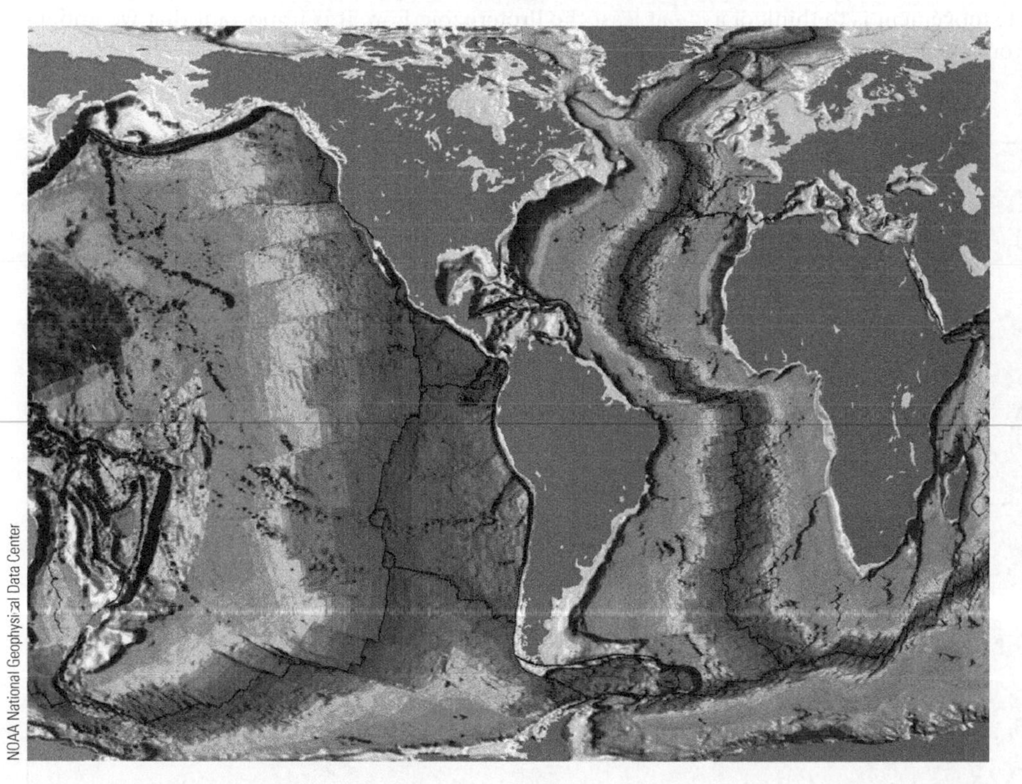

Figure 2.16 Age of the World's Ocean Basins The age of the world's ocean basins has been determined from magnetic anomalies preserved in oceanic crust. The red colors adjacent to the oceanic ridges are the youngest oceanic crust. Moving laterally away from the ridges, the red colors grade to yellow at 48 million years ago, to green at 68 million years ago, and to dark blue some 155 million years ago. The darkest blue color is adjacent to the continental margins and is just somewhat less than 180 million years old.

Critical Thinking Question How does the age of the oceanic crust confirm the theory of seafloor spreading?

were as old as the continents, we would expect deep-sea sediments to be several kilometers thick. However, data from numerous drill holes indicate that deep-sea sediments are, at most, only a few hundred meters thick and are thin or absent at oceanic ridges.

Their near-absence at the ridges should come as no surprise because these are the areas where new crust is continuously produced by volcanism and seafloor spreading. Accordingly, sediments have had little time to accumulate at or very close to spreading ridges. However, their thickness increases with distance away from the ridges because of the longer amount of time sediment has had to accumulate. Hence, the farther away one is from a ridge, the thicker the sediments.

2.8 Plate Tectonics: A Unifying Theory

Plate tectonic theory is based on a simple model of Earth. The rigid lithosphere, composed of both oceanic and continental crust, as well as the underlying upper mantle, consists of numerous variable-sized pieces called *plates* (❚ Figure 2.17). There are seven major plates (Eurasian, Indian-Australian, Antarctic, North American, South American, Pacific, and African) and numerous smaller ones ranging from only a few tens to several hundreds of kilometers in width. Plates also vary in thickness; those composed of upper mantle and continental crust are as much as 250 km thick, whereas those of upper mantle and oceanic crust are up to 100 km thick.

The lithosphere overlies the hotter and weaker semiplastic asthenosphere. It is thought that movement resulting from some type of heat-transfer system within the asthenosphere causes the overlying plates to move. As plates move over the asthenosphere, they separate, mostly at oceanic ridges; in other areas, such as at oceanic trenches, they collide and are subducted back into the mantle.

An easy way to visualize plate movement is to think of a conveyer belt moving luggage from an airplane's cargo hold to a baggage cart. The conveyer belt represents convection currents within the mantle, and the luggage represents Earth's lithospheric plates. The luggage is moved along by the conveyer belt until it is dumped into the baggage cart in the same way that plates are moved by convection cells until they are subducted into Earth's interior.

Although this analogy allows you to visualize how the mechanism of plate movement takes place, remember that this analogy is limited. The major limitation is that, unlike the luggage, plates consist of continental and oceanic lithosphere, which have different densities, and only oceanic lithosphere is subducted into Earth's interior.

Most geologists accept plate tectonic theory because the evidence for it is overwhelming and because it ties together many seemingly unrelated geologic features and events and shows how they are interrelated. Consequently, geologists now view many geologic processes from the global perspective of plate tectonic theory in which plate interaction along plate margins is responsible for such phenomena as mountain building, earthquakes, and volcanism (Geo-Insight 2.1).

2.9 The Three Types of Plate Boundaries

Because it appears that plate tectonics has operated since at least the Proterozoic Eon, it is important that we understand how plates move and interact with each other and how

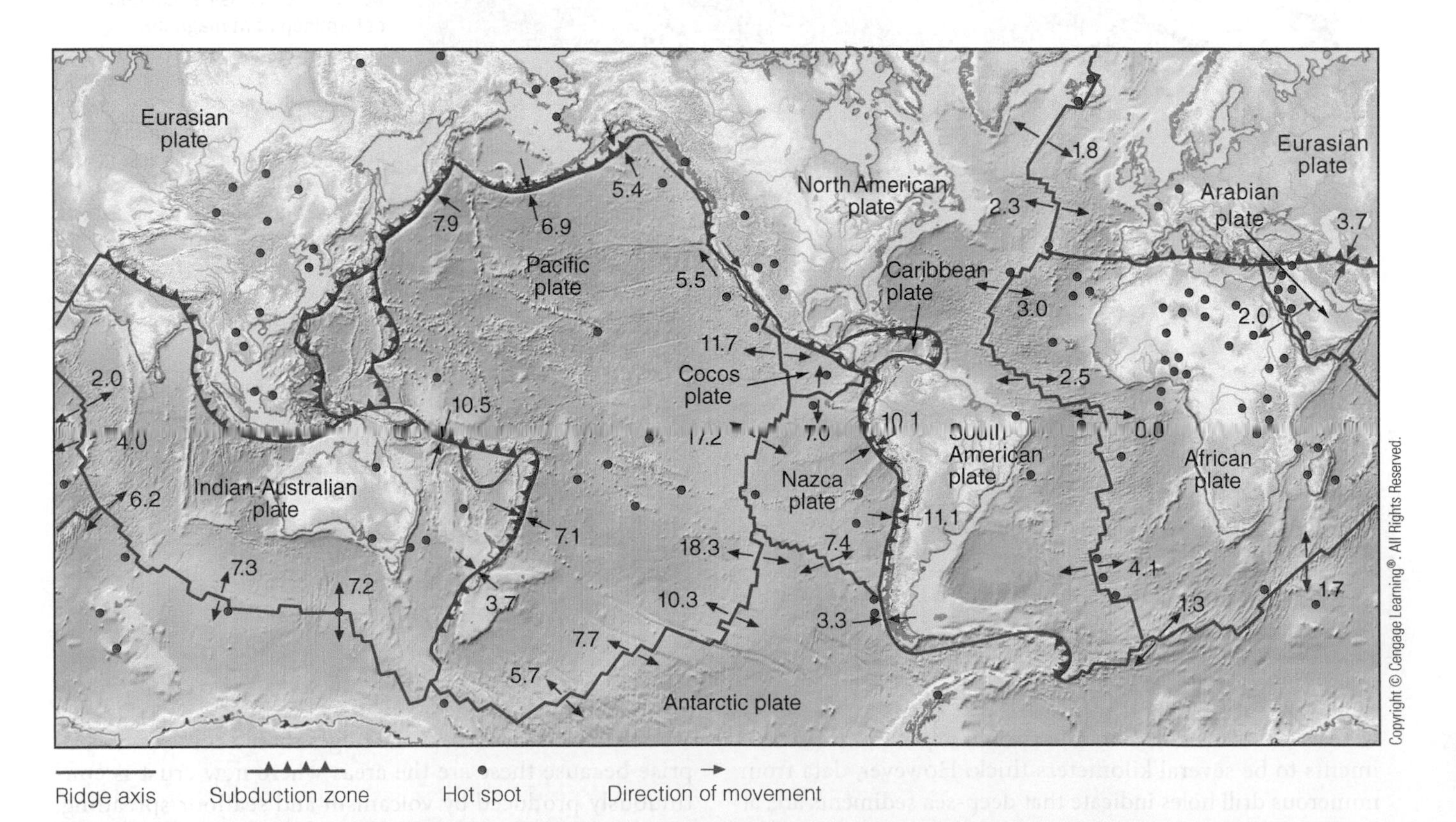

❚ **Figure 2.17 Earth's Plates** A world map showing Earth's plates, their boundaries, their relative motion, average rates of movement in centimeters per year, and hot spots.

The Hazards of Living Near a Convergent Plate Boundary

Volcanic eruptions, earthquakes, and tsunami are all manifestations associated with convergent plate boundaries. As such, it is important to know the risks involved in living near an active plate margin. One such area is the Pacific Northwest of the United States. Stretching from Lassen Peak in northern California through Oregon, Washington, and into British Columbia, Canada, the Cascade Range is the result of subduction of the Juan de Fuca plate beneath the North American plate. With a number of cities located near the coast, and also in proximity to some of the Cascade Range volcanoes, it is important to know the extent and severity of past eruptions, earthquakes, and tsunami that have occurred in this area. What type of evidence can geologists look for to determine past volcanic eruptions, earthquakes, and tsunami? Can this information be used in predicting and planning for future activity associated with the Juan de Fuca–North American convergent plate boundary? To what extent should frequency of eruptions and seismic activity in the past play in determining future land use planning?

ancient plate boundaries are recognized. After all, the movement of plates has profoundly affected the geologic and biologic history of this planet.

Geologists recognize three types of plate boundaries: *divergent*, *convergent*, and *transform* (Table 2.1). Along these boundaries, new plates are formed, are consumed, or slide laterally past one another. Interaction of plates at their boundaries accounts for most of Earth's volcanic eruptions and earthquakes, as well as the formation and evolution of its mountain systems.

Divergent Boundaries

Divergent plate boundaries, or *spreading ridges*, occur where plates are separating and new oceanic lithosphere is forming. Divergent boundaries are places where the crust is extended, thinned, and fractured as magma, derived from the partial melting of the mantle, rises to the surface. The magma is almost entirely basaltic and intrudes into vertical fractures to form dikes and pillow lava flows (see Figure 5.6). As successive injections of magma cool and solidify, they form new oceanic crust and record the intensity and orientation of Earth's magnetic field (Figure 2.15).

Divergent boundaries most commonly occur along the crests of oceanic ridges—for example, the Mid-Atlantic Ridge. Oceanic ridges are thus characterized by rugged topography with high relief resulting from displacement of rocks along large fractures, shallow-depth earthquakes, high heat flow, and basaltic flows or pillow lavas.

Divergent boundaries are also present under continents during the early stages of continental breakup. When magma wells up beneath a continent, the crust is initially elevated, stretched, and thinned, producing fractures, faults, rift valleys, and volcanic activity (▮ Figure 2.18a). As magma intrudes into faults and fractures, it solidifies or flows out onto the surface as lava flows; the latter often covering the rift valley floor (▮ Figure 2.18b). The East African Rift Valley is an excellent example of continental breakup at this stage (▮ Figure 2.19).

As spreading proceeds, some rift valleys continue to lengthen and deepen until the continental crust eventually breaks and a narrow linear sea is formed, separating two continental blocks (▮ Figure 2.18c). The Red Sea, separating the Arabian Peninsula from Africa (Figure 2.19), and the Gulf of California, which separates Baja California from mainland Mexico, are good examples of this more advanced stage of rifting.

TABLE 2.1 Types of Plate Boundaries

Type	Example	Landforms	Volcanism
Divergent			
Oceanic	Mid-Atlantic Ridge	Mid-oceanic ridge with axial rift valley	Basalt
Continental	East African Rift Valley	Rift valley	Basalt and rhyolite, no andesite
Convergent			
Oceanic-oceanic	Aleutian Islands	Volcanic island arc,offshore oceanic trench	Andesite
Oceanic-continental	Andes	Offshore oceanic trench, volcanic mountain chain, mountain belt	Andesite
Continental-continental	Himalayas	Mountain belt	Minor
Transform	San Andreas Fault	Fault valley	Minor

GEO INSIGHT 2.1

Plate Boundaries, Earthquakes, and Tsunami

In the afternoon of January 12, 2010, a magnitude 7.0 earthquake struck the island nation of Haiti. According to official estimates, 222,570 people died, at least 300,000 were injured, and more than 285,000 residences and businesses were destroyed or severely damaged (▮ Figure 1). Widespread devastation occurred in the capital of Port-au-Prince and elsewhere throughout the region, exacerbated by an almost total collapse of the vital infrastructure needed to respond to such a disaster, including medical, transportation, and communications systems.

A little more than a year later, on February 23, 2011, a 6.3-magnitude earthquake struck Christchurch, New Zealand. In addition to the tremendous damage to buildings and infrastructure, 181 people were killed in this earthquake, making it the second deadliest earthquake to strike New Zealand (▮ Figure 2).

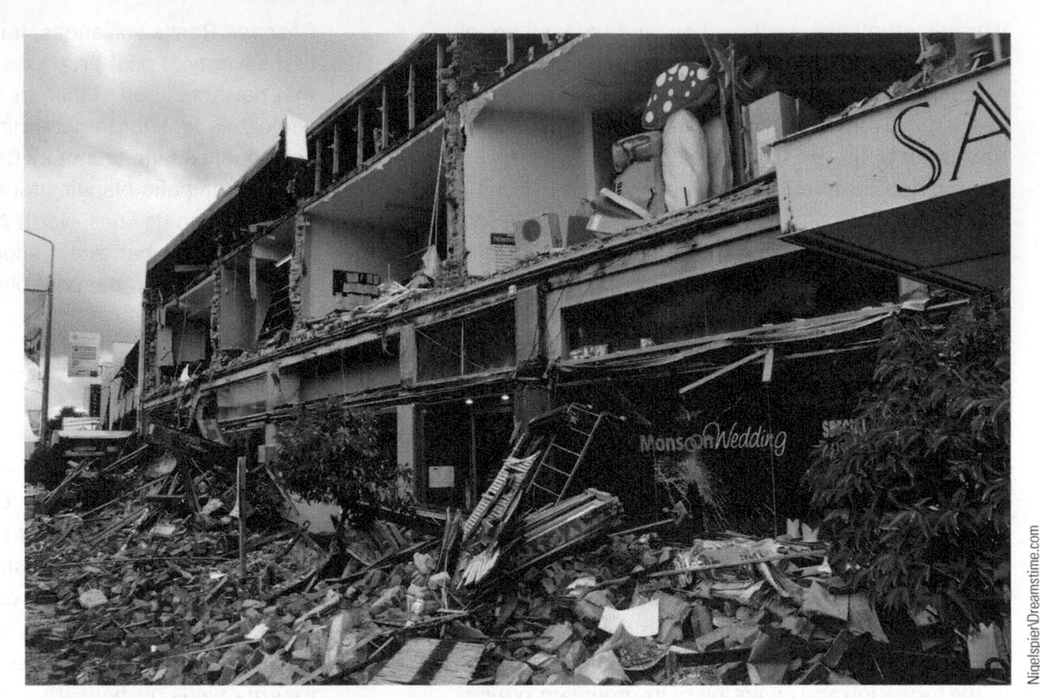

Nigelspier\Dreamstime.com

▲ **2.** Some of the damage done to buildings by the violent ground shaking experienced in the heart of Christchurch, New Zealand, as a result of the 6.3-magnitude earthquake that struck on February 23, 2011.

KOZAK NICK/SIPA/AP Images

▲ **1.** People walking over the rubble of destroyed buildings in downtown Port-au-Prince, Haiti, following the 7.0-magnitude earthquake that struck the island nation on January 12, 2010, and left more than 222,500 dead.

Less than a month following the New Zealand earthquake, a 9.0-magnitude earthquake and tsunami struck Japan on March 11, 2011, causing more than 20,000 deaths and tremendous property damage, including severe damage to a nuclear power plant in the northeastern part of the island. Within minutes after the earthquake, walls

KOZAK NICK/SIPA/AP Images

3. Aerial view of Minamisanriku, Miyagi Prefecture in northeastern Japan, showing houses clogged with debris and a large boat that had been swept inland as a result of the tsunami.

of water, some as high as 37 m, inundated low-lying areas along the Japanese coast and extending as far as 10 km inland, sweeping aside boats, vehicles, and structures, as if they were toys (■ Figure 3).

What do these horrific events share, besides the terrible loss of lives and massive property damage? They are all associated with Earth's plates and the interactions taking place along their boundaries. For example, the Haiti earthquake in 2010 resulted from movement along a strike-slip fault zone (a strike-slip fault is one involving horizontal movement of rock) that is part of the boundary separating the southwestward-moving North American plate, from the eastward-moving Caribbean plate to the south. This earthquake also struck a small fishing village and swept at least three people out to sea.

The New Zealand earthquake also resulted from movement along a strike-slip fault, but in this case, there was an additional component of vertical movement. Some geologists think that this particular earthquake was actually an aftershock of the 7.1-magnitude Canterbury earthquake that struck the south island of New Zealand on September 4, 2010. Although both earthquakes occurred on previously unknown faults, the faults and subsequent earthquakes are both associated with the regional plate boundary deformation taking place between the Pacific and Australian plates in the central South Island of New Zealand (Figure 2.17).

Lastly, the 2011 Japanese earthquake and resulting tsunami were caused by the Pacific plate subducting beneath the North American plate in the region of northern Japan. Here, the Pacific plate is moving approximately westward with respect to the North American and Eurasian plates and plunges beneath Japan along the Japan Trench, which marks the surface expression of an oceanic-oceanic convergent plate boundary (Figure 2.20a).

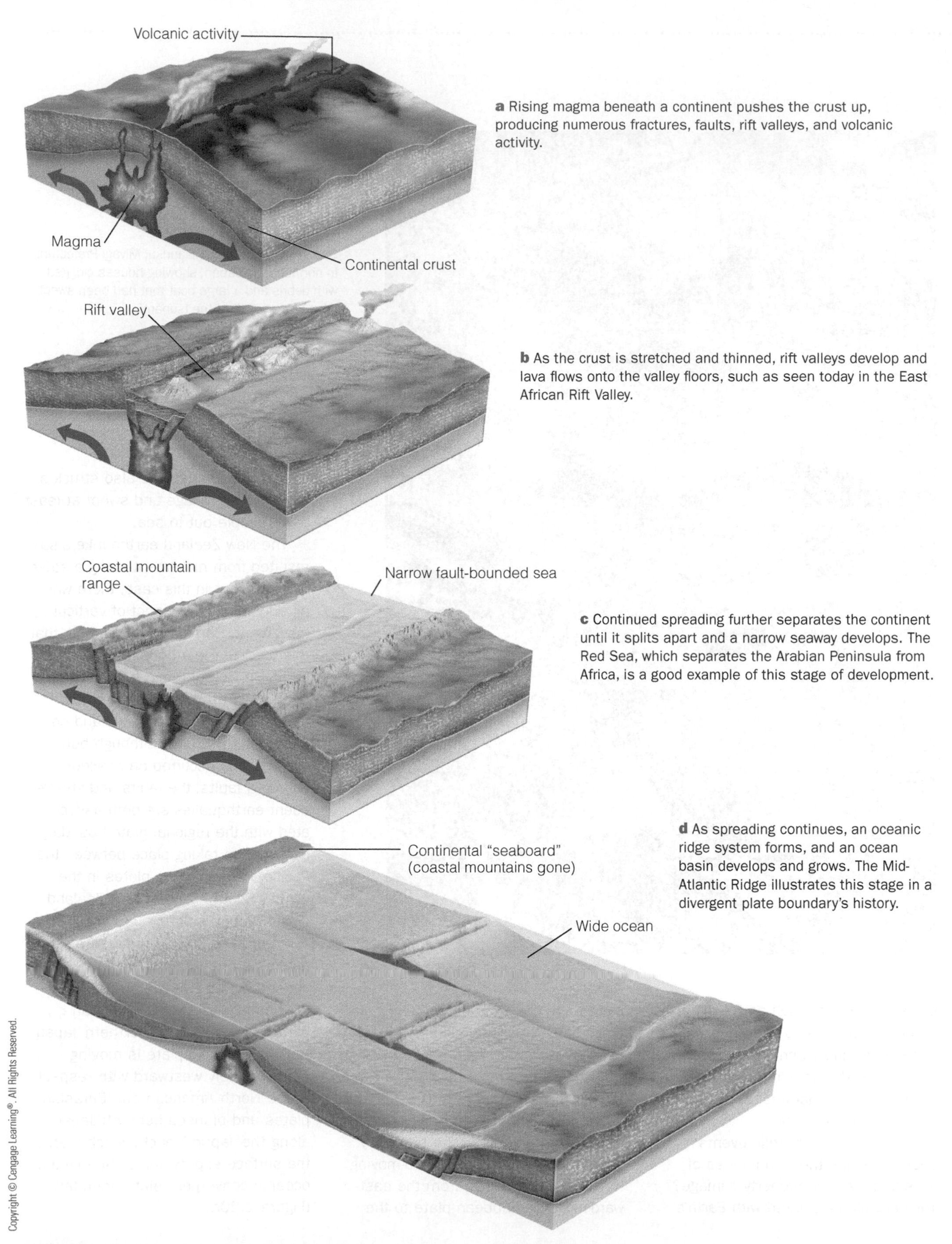

a Rising magma beneath a continent pushes the crust up, producing numerous fractures, faults, rift valleys, and volcanic activity.

b As the crust is stretched and thinned, rift valleys develop and lava flows onto the valley floors, such as seen today in the East African Rift Valley.

c Continued spreading further separates the continent until it splits apart and a narrow seaway develops. The Red Sea, which separates the Arabian Peninsula from Africa, is a good example of this stage of development.

d As spreading continues, an oceanic ridge system forms, and an ocean basin develops and grows. The Mid-Atlantic Ridge illustrates this stage in a divergent plate boundary's history.

Figure 2.18 History of a Divergent Plate Boundary

As a newly created narrow sea continues to enlarge, it may eventually become an expansive ocean basin, such as the Atlantic Ocean basin is today, separating North and South America from Europe and Africa by thousands of kilometers (Figure 2.18d). The Mid-Atlantic Ridge is the boundary between these diverging plates; the American plates are moving westward, and the Eurasian and African plates are moving eastward.

Extending outward from the eastern coasts of North and South America, as well as the western coasts of Europe and Africa, are broad continental shelves, continental slopes, and rises. These features, discussed earlier, are referred to as passive continental margins (Figure 2.10). And, although they are found within a plate, rather than at a plate boundary, they result from the breakup of a continent and the subsequent movement away from a divergent plate boundary (Figure 2.18).

An Example of Ancient Rifting What features in the geologic record can geologists use to recognize ancient rifting? Associated with regions of continental rifting are faults, dikes (vertical intrusive igneous bodies), sills (horizontal intrusive igneous bodies), lava flows, and thick sedimentary sequences within rift valleys, all features that are preserved in the geologic record. The Triassic fault basins of the eastern United States are a good example of ancient continental rifting (see Figure 22.7). These fault basins mark the zone of rifting that occurred when North America split apart from Africa. The basins contain thousands of meters of continental sediment and are riddled with dikes and sills (see Chapter 22).

Pillow lavas, in association with deep-sea sediment, are also evidence of ancient rifting. The presence of pillow lavas marks the formation of a spreading ridge in a narrow linear sea (Figures 2.18c and 2.19). A narrow linear sea forms when the continental crust in the rift valley finally breaks apart, and the area is flooded with seawater. Magma, intruding into the sea along this newly formed spreading ridge, solidifies as pillow lavas, which are preserved in the geologic record, along with the sediment being deposited on them.

ConnectionLink
To learn more about the breakup of Pangaea and how it affected the eastern margin of North America during the Mesozoic Era, see Chapter 22.

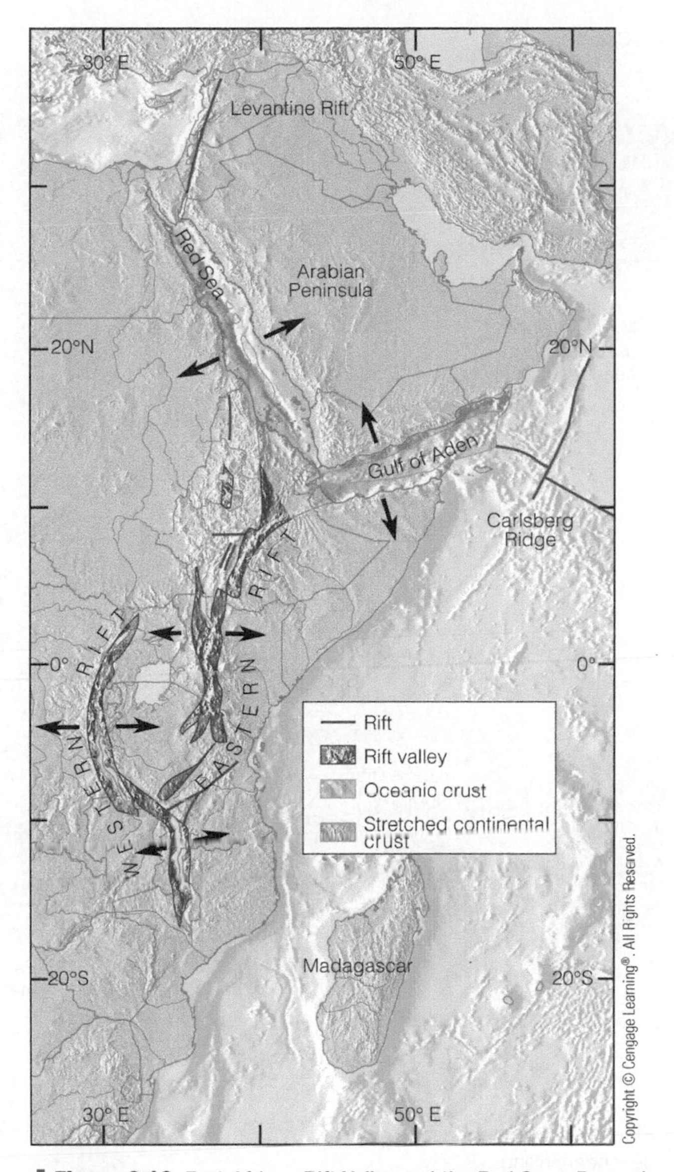

Figure 2.19 East African Rift Valley and the Red Sea—Present-Day Examples of Divergent Plate Boundaries The East African Rift Valley and the Red Sea represent different stages in the history of a divergent plate boundary. The East African Rift Valley is being formed by the separation of eastern Africa from the rest of the continent along a divergent plate boundary. The Red Sea represents a more advanced stage of rifting, in which two continental blocks (Africa and the Arabian Peninsula) are separated by a narrow sea.

Convergent Boundaries

Whereas new crust forms at divergent plate boundaries, older crust must be destroyed and recycled in order for the entire surface area of Earth to remain the same. Otherwise, we would have an expanding Earth. Such plate destruction occurs along **convergent plate boundaries** (Figure 2.20), where two plates collide and the leading edge of one plate is subducted beneath the margin of the other plate and eventually incorporated into the asthenosphere. A dipping plane of earthquake foci, called a *Benioff zone*, defines subduction zones (see Figure 9.5). Most of these planes dip from oceanic trenches beneath adjacent island arcs or continents, marking the surface of slippage between the converging plates.

Deformation, volcanism, mountain building, metamorphism, earthquake activity, and deposits of valuable minerals characterize convergent boundaries. Three types of convergent plate boundaries are recognized: *oceanic–oceanic, oceanic–continental,* and *continental–continental.*

ConnectionLink
More information about the relationship between plate tectonics and mountain building can be found in Chapter 10.

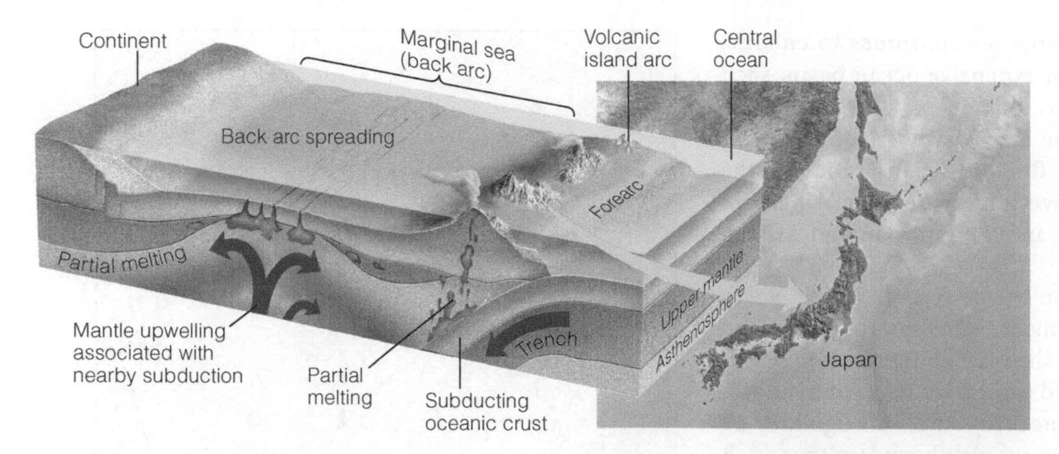

a Oceanic-oceanic plate boundary. An oceanic trench forms where one oceanic plate is subducted beneath another. On the nonsubducted plate, a volcanic island arc forms from the rising magma generated from the subducting plate. The Japanese Islands are a volcanic island arc resulting from the subduction of one oceanic plate beneath another oceanic plate.

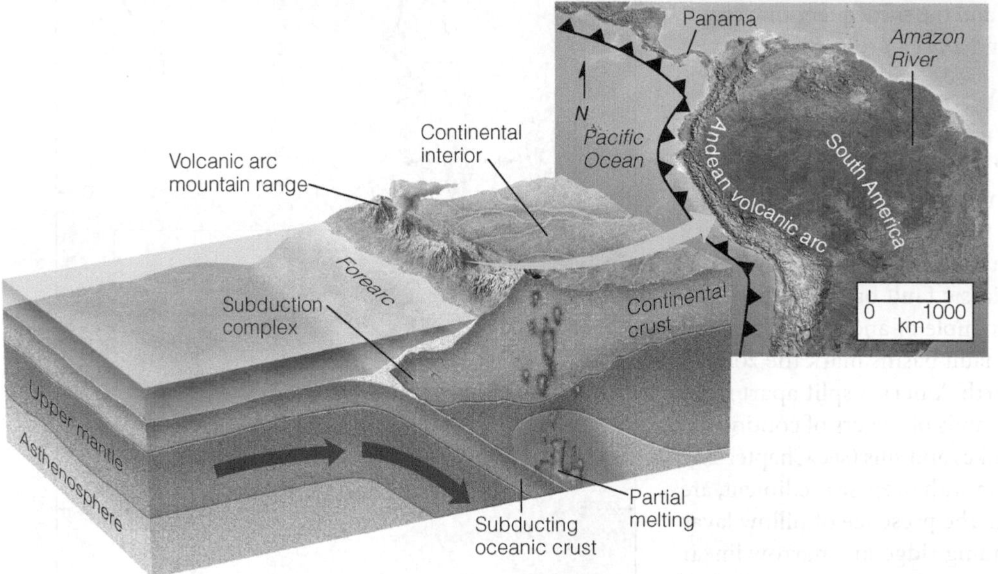

b Oceanic-continental plate boundary. When an oceanic plate is subducted beneath a continental plate, an andesitic volcanic mountain range is formed on the continental plate as a result of rising magma. The Andes Mountains in Peru are one of the best examples of continuing mountain building along an oceanic–continental plate boundary.

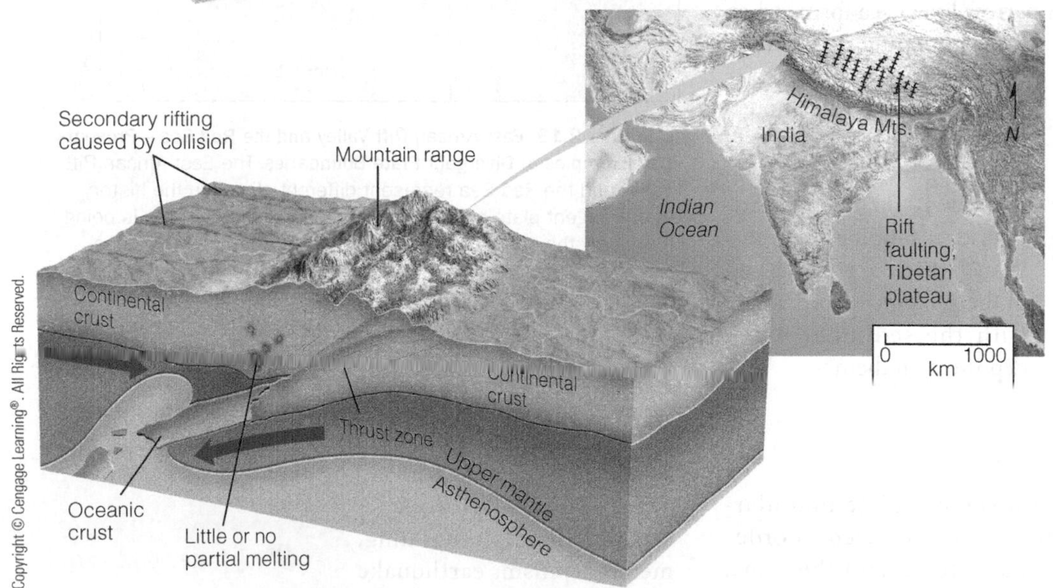

c Continental-continental plate boundary. When two continental plates converge, neither is subducted because of their great thickness and low and equal densities. As the two continental plates collide, a mountain range is formed in the interior of a new and larger continent. The Himalayas in central Asia resulted from the collision between India and Asia approximately 40 to 50 million years ago.

Figure 2.20 Three Types of Convergent Plate Boundaries

Oceanic–Oceanic Boundaries When two oceanic plates converge, one is subducted beneath the other along an **oceanic–oceanic plate boundary** (Figure 2.20a). The subducting plate bends downward to form the outer wall of an oceanic trench. A subduction complex, composed of wedge-shaped slices of highly folded and faulted marine sediments and oceanic lithosphere scraped off the descending plate, forms along the inner wall of the oceanic trench. As the subducting plate descends into the mantle, it is heated and partially melted, generating magma commonly of andesitic composition (see Chapter 4). This magma is less dense than the surrounding mantle rocks and rises to the surface of the nonsubducted plate to form a curved chain of volcanic islands called a *volcanic island arc* (any plane intersecting a sphere makes an arc). This arc is nearly parallel to the oceanic trench and is separated from it by a distance of up to several hundred kilometers—the distance depends on the angle of dip of the subducting plate (Figure 2.20a).

In those areas where the rate of subduction is faster than the forward movement of the overriding plate, the lithosphere on the landward side of the volcanic island arc may be subjected to tensional stress and may be stretched and thinned, resulting in the formation of a *back-arc basin*. This back-arc basin may grow by spreading if magma breaks through the thin crust and forms new oceanic crust (Figure 2.20a). A good example of a back-arc basin associated with an oceanic–oceanic plate boundary is the Sea of Japan between the Asian continent and the islands of Japan.

Most present-day active volcanic island arcs are in the Pacific Ocean basin and include the Aleutian Islands, the Kermadec–Tonga arc, and the Japanese (Figure 2.20a) and Philippine Islands. The Scotia and Antillean (Caribbean) island arcs are in the Atlantic Ocean basin.

Oceanic–Continental Boundaries When an oceanic and a continental plate converge, the denser oceanic plate is subducted under the continental plate along an **oceanic–continental plate boundary** (Figure 2.20b). Just as at oceanic–oceanic plate boundaries, the descending oceanic plate forms the outer wall of an oceanic trench.

The magma generated by subduction rises beneath the continent and either crystallizes as large intrusive bodies, called *plutons*, before reaching the surface or erupts at the surface to produce a chain of andesitic volcanoes, also called a *volcanic arc*. An excellent example of an oceanic–continental plate boundary is the Pacific Coast of South America where the oceanic Nazca plate is currently being subducted under South America (Figure 2.20b; see also Chapter 10). The Peru–Chile Trench marks the site of subduction, and the Andes Mountains are the resulting volcanic mountain chain on the nonsubducting plate.

Just as there are passive continental margins, there are also active continental margins (Figure 2.10). The aforementioned oceanic–continental plate boundary between the west coast of the South American plate and the eastern side of the oceanic Nazca plate is an excellent example of an active continental margin (Figure 2.20b). Here the continental shelf is narrow, and the continental slope descends directly into the Peru–Chile Trench, so sediment is dumped into the trench and no continental rise develops, such as along a passive continental margin (Figure 2.10).

Continental–Continental Boundaries Two continents approaching each other are initially separated by an ocean floor that is being subducted under one continent. The edge of that continent displays the features characteristic of oceanic–continental convergence along an active continental plate margin. As the ocean floor continues to be subducted, the two continents come closer together until they eventually collide. Because continental lithosphere, which consists of continental crust and the upper mantle, is less dense than oceanic lithosphere (oceanic crust and upper mantle), it cannot sink into the asthenosphere. Although one continent may partially slide under the other, it cannot be pulled or pushed down into a subduction zone (Figure 2.20c).

When two continents collide, they are welded together along a zone marking the former site of subduction. At this **continental–continental plate boundary**, an interior mountain belt is formed consisting of deformed sediments and sedimentary rocks, igneous intrusions, metamorphic rocks, and fragments of oceanic crust. In addition, the entire region is subjected to numerous earthquakes. The Himalayas in central Asia, the world's youngest and highest mountain system, resulted from the collision between India and Asia that began 40 to 50 million years ago and is still continuing (Figure 2.20c; see Chapter 10).

Recognizing Ancient Convergent Plate Boundaries How can former subduction zones be recognized in the geologic record? Igneous rocks provide one such clue. The magma erupted at the surface, forming island arc volcanoes and continental volcanoes, is of andesitic composition. Another clue is the zone of intensely deformed rocks between the deep-sea trench where subduction is taking place and the area of igneous activity. Here, sediments and submarine rocks are folded, faulted, and metamorphosed into a chaotic mixture of rocks termed a *mélange*.

During subduction, pieces of oceanic lithosphere are sometimes incorporated into the mélange and accreted onto the edge of the continent. Such slices of oceanic crust and upper mantle are called *ophiolites* (▌Figure 2.21). Detailed

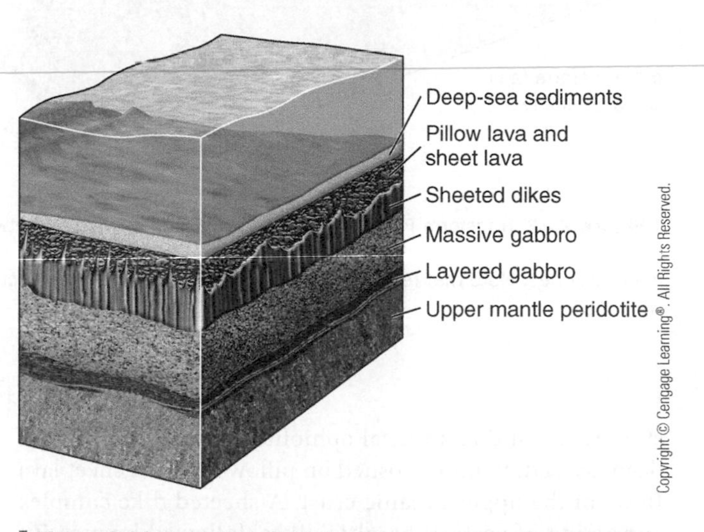

▌**Figure 2.21 Ophiolites** Ophiolites are sequences of rock on land consisting of deep-sea sediments, oceanic crust, and upper mantle. Ophiolites are one feature used to recognize ancient convergent plate boundaries.

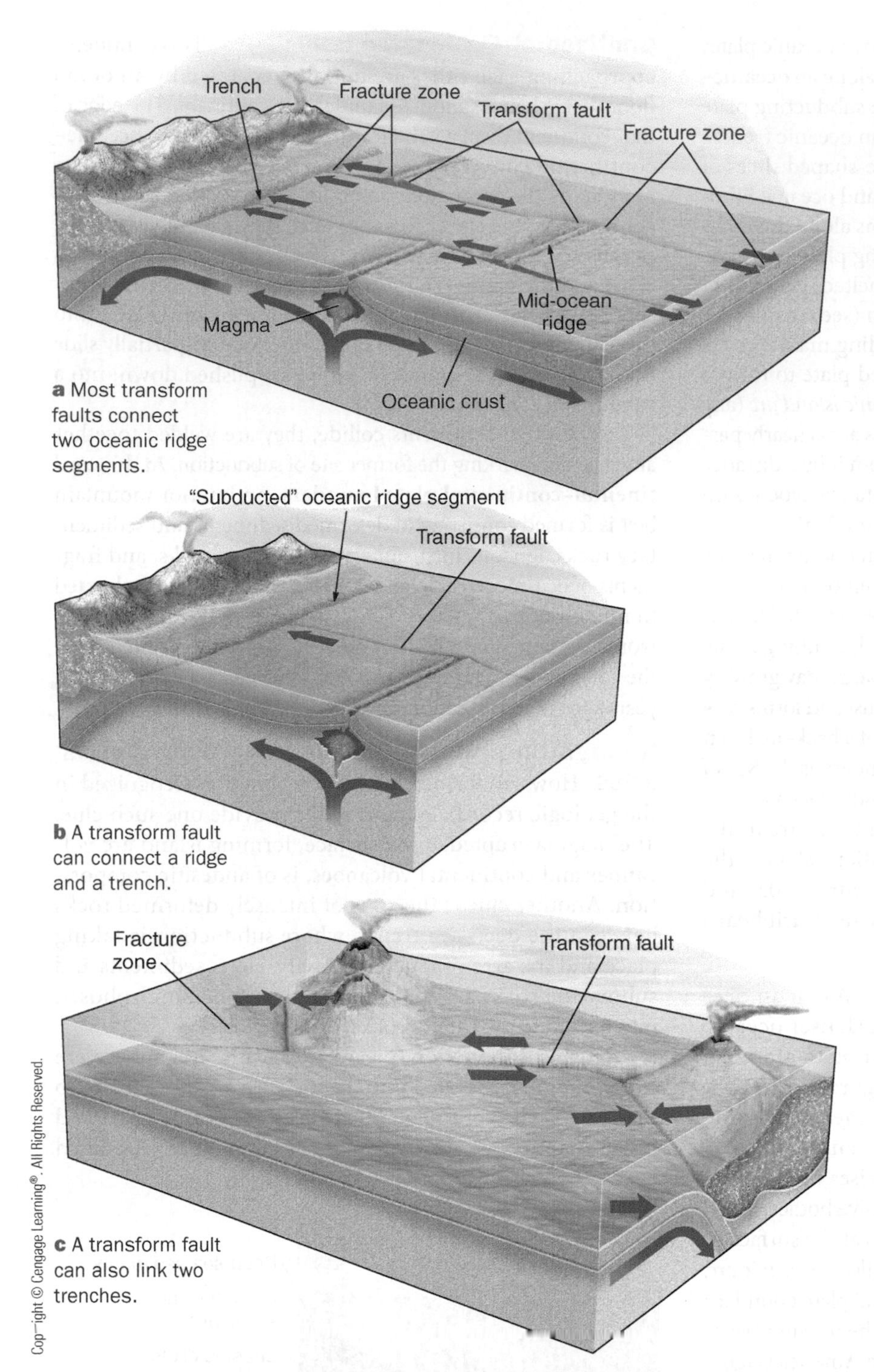

a Most transform faults connect two oceanic ridge segments.

b A transform fault can connect a ridge and a trench.

c A transform fault can also link two trenches.

▮ Figure 2.22 Transform Plate Boundaries Horizontal movement between plates occurs along transform faults. Extensions of transform faults on the seafloor form fracture zones. Note that relative motion between the plates only occurs between the two ridges.

studies reveal that an ideal ophiolite consists of a layer of deep-sea sediments deposited on pillow lava and sheet lava flows of the upper oceanic crust. A sheeted dike complex consisting of vertical basaltic dikes, followed by massive gabbro (a dark intrusive igneous rock), and layered gabbro, form the rest of the oceanic crust. Beneath the gabbro is peridotite (a dark intrusive igneous rock composed of the mineral olivine), which probably represents the upper mantle.

This peridotite is sometimes altered by metamorphism to a greenish rock known as serpentinite. Thus, a complete ophiolite consists of deep-sea sedimentary rocks underlain by rocks of the oceanic crust and upper mantle. The presence of ophiolites in an outcrop or drilling core is a key indicator of plate convergence along a subduction zone.

Elongated belts of folded and faulted marine sedimentary rocks, andesites, and ophiolites are found in the Appalachians, Alps, Himalayas, and Andes. The combination of such features is significant evidence that these mountain ranges resulted from deformation along convergent plate boundaries.

Transform Boundaries

The third type of plate boundary is a **transform plate boundary,** which mostly occur along fractures in the seafloor, known as *transform faults*, where plates slide laterally past one another roughly parallel to the direction of plate movement. Although lithosphere is neither created nor destroyed along a transform boundary, the movement between plates results in a zone of intensely shattered rock and numerous shallow-depth earthquakes.

Transform faults "transform" or change one type of motion between plates into another type of motion. Most commonly, transform faults connect two oceanic ridge segments; however, they can also connect ridges to trenches and trenches to trenches (▮ Figure 2.22). Although the majority of transform faults are in oceanic crust and are marked by distinct fracture zones, they may also extend into continents.

One of the best-known transform faults is the San Andreas Fault in California. It separates the Pacific plate from the North American plate and connects spreading ridges in the Gulf of California with the Juan de Fuca and Pacific plates off the coast of northern California

ConnectionLink
Learn more about the San Andreas Fault in Geo-Insight 9.1 in Chapter 9.

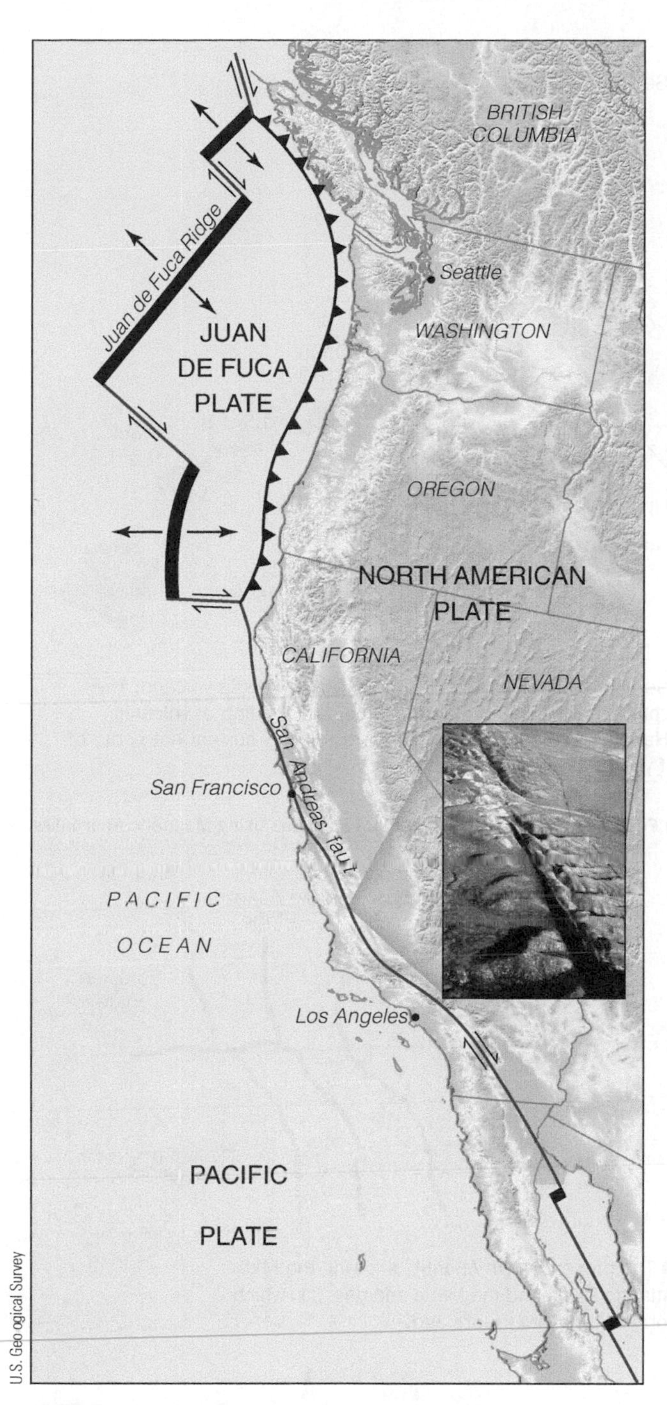

Figure 2.23 The San Andreas Fault—A Transform Plate Boundary The San Andreas Fault is a transform fault separating the Pacific plate from the North American plate. It connects the spreading ridges in the Gulf of California with the Juan de Fuca and Pacific plates off the coast of northern California. Movement along the San Andreas Fault has caused numerous earthquakes. The insert photograph shows a segment of the San Andreas Fault as it cuts through the Carrizo Plain, California.

Critical Thinking Question If the movement along the San Andreas Fault, which separates the Pacific plate from the North American plate, averages 5.5 cm per year, how long will it take before Los Angeles is opposite San Francisco?

(Figure 2.23). Many of the earthquakes that affect California are the result of movement along this fault (see Chapter 9).

Unfortunately, transform faults generally do not leave any characteristic or diagnostic features except for the obvious displacement of the rocks with which they are associated. This displacement is usually large, on the order of tens to hundreds of kilometers. Such large displacements in ancient rocks can sometimes be related to transform fault systems.

2.10 Hot Spots and Mantle Plumes

Before leaving the topic of plate boundaries, we should mention an intraplate feature found beneath both oceanic and continental plates. A **hot spot** (Figure 2.17) is the location on Earth's surface where a stationary column of magma, originating deep within the mantle (*mantle plume*), has slowly risen to the surface and formed a volcano. Because the mantle plumes apparently remain stationary (although some evidence suggests that they might not) within the mantle while the plates move over them, the resulting hot spots leave a trail of extinct and progressively older volcanoes (aseismic ridges) that record the movement of the plate.

One of the best examples of aseismic ridges and hot spots is the Emperor Seamount–Hawaiian Island chain (Figure 2.24). This chain of islands and seamounts extends from the island of Hawaii to the Aleutian Trench off Alaska, a distance of some 6,000 km, and consists of more than 80 volcanic structures.

Currently, the only active volcanoes in this island chain are on the island of Hawaii and the Loihi Seamount. The rest of the islands are extinct volcanic structures that become progressively older toward the north and northwest. This means that the Emperor Seamount–Hawaiian Island chain records the direction that the Pacific plate traveled as it moved over an apparently stationary mantle plume. In this case, the Pacific plate first moved in a north-northwesterly direction and then, as indicated by the sharp bend in the chain, changed to a west-northwesterly direction approximately 43 million years ago. The reason that the Pacific plate changed directions is not known, but the shift might be related to the collision of India with the Asian continent at about the same time (see Figure 10.21).

Mantle plumes and hot spots help geologists explain some of the geologic activity occurring within plates as opposed to activity occurring at or near plate boundaries. In addition, if mantle plumes are essentially fixed with respect to earth's rotational axis, they can be used to determine not only the direction of plate movement but also the rate of movement.

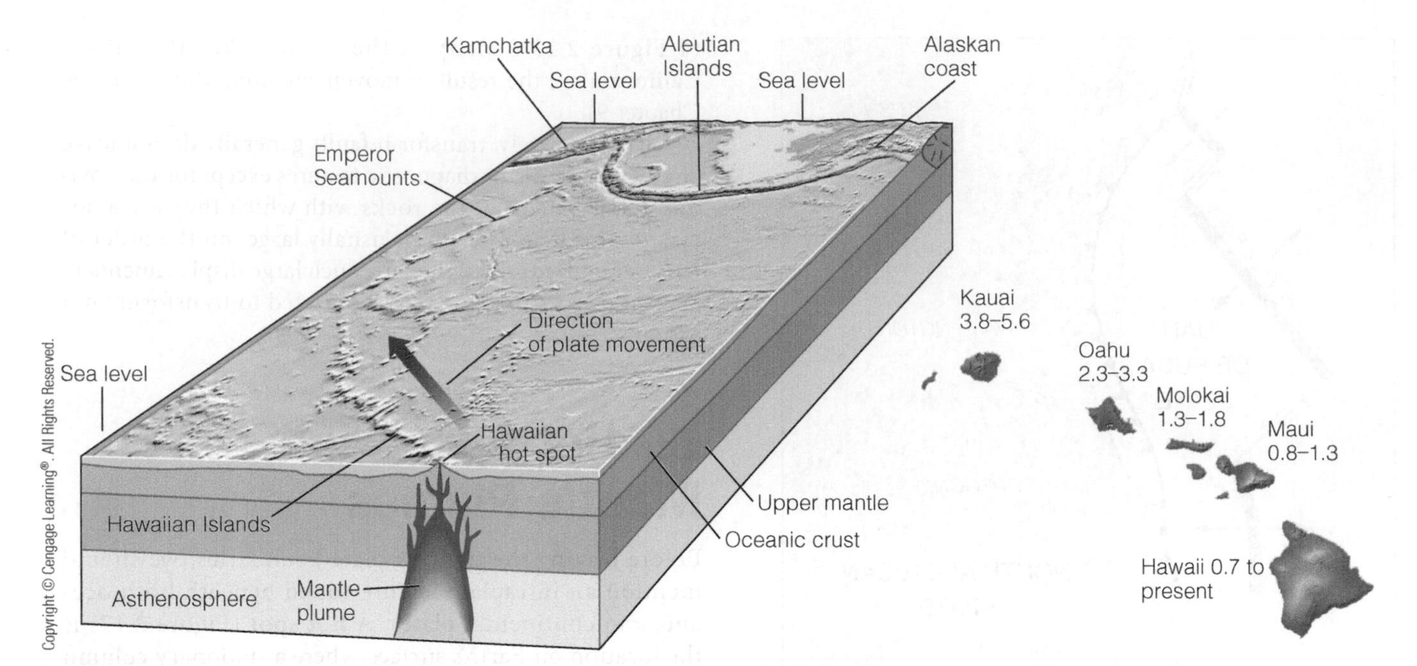

Figure 2.24 **Hot Spots** A hot spot is the location where a stationary mantle plume has risen to the surface and formed a volcano. The Emperor Seamount–Hawaiian Island chain formed as a result of the Pacific plate moving over a mantle plume, and the line of volcanic islands in this chain traces the direction of plate movement. The island of Hawaii and the Loihi Seamount are the only current hot spots of this island chain. The numbers indicate the age of the islands in millions of years.

2.11 Plate Movement and Motion

How fast and in what direction are Earth's plates moving? Do they all move at the same rate? Rates of plate movement can be calculated in several ways. The least accurate method is to determine the age of the sediments immediately above any portion of the oceanic crust and then divide the distance from the spreading ridge by that age. Such calculations give an average rate of movement.

A more accurate method of determining both the average rate of movement and relative motion is by dating the magnetic anomalies in the crust of the seafloor. The distance from an oceanic ridge axis to any magnetic anomaly indicates the width of new seafloor that formed during that time interval. For example, if the distance between the present-day Mid-Atlantic Ridge and anomaly 31 is 2,010 km, and anomaly 31 formed 67 million years ago (Figure 2.25), then the average rate of movement during the past 67 million years has been 3 cm per year (2,010 km, which equals 201 million cm divided by 67 million years; 201,000,000 cm/67,000,000 years = 3 cm/year). Thus, for a given interval of time, the wider the strip of seafloor, the faster the plate has moved. In this way, not only can the present average rate of movement and relative motion be determined (Figure 2.17), but also the average rate of movement in the past can be calculated by dividing the distance between anomalies by the amount of time elapsed between anomalies.

Figure 2.25 **Reconstructing Plate Positions Using Magnetic Anomalies**

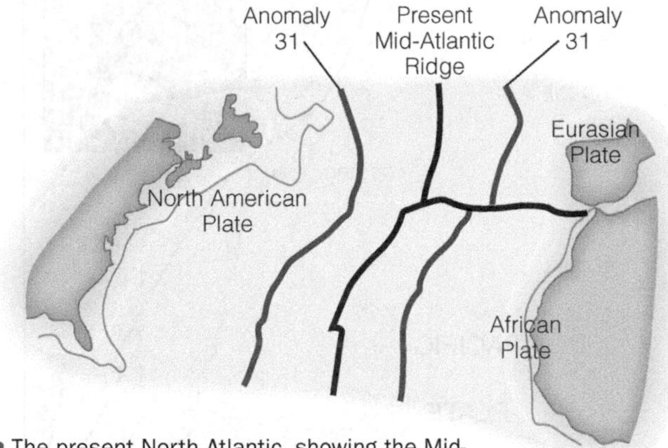

a The present North Atlantic, showing the Mid-Atlantic Ridge and magnetic anomaly 31, which formed 67 million years ago.

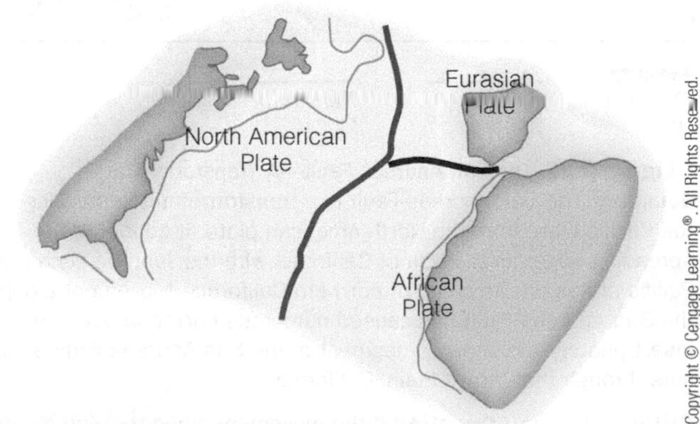

b The Atlantic 67 million years ago. Anomaly 31 marks the plate boundary 67 million years ago. By moving the anomalies back together, along with the plates they are on, we can reconstruct the former positions of the continents.

Geologists use magnetic anomalies not only to calculate the average rate of plate movement but also to determine plate positions at various times in the past. Because magnetic anomalies are parallel and symmetric with respect to spreading ridges, all one must do to determine the position of continents when particular anomalies formed is to move the anomalies back to the spreading ridge, which will also move the continents with them (Figure 2.25). Because subduction destroys oceanic crust and the magnetic record that it carries, we have an excellent record of plate movements since the breakup of Pangaea, but not as good an understanding of plate movement before that time.

The average rate of movement, as well as the relative motion between any two plates, can be determined by satellite-laser ranging techniques. Laser beams from a station on one plate are bounced off a satellite (in geosynchronous orbit) and returned to a station on a different plate. As the plates move away from each other, the laser beam takes more time to go from the sending station to the stationary satellite and back to the receiving station. This difference in elapsed time is used to calculate the rate of movement and the relative motion between plates.

2.12 The Driving Mechanism of Plate Tectonics

A major obstacle to the acceptance of the continental drift hypothesis was the lack of a driving mechanism to explain continental movement. When it was shown that continents and ocean floors moved together, not separately, and that new crust is formed at spreading ridges by rising magma, most geologists accepted some type of convective heat system (convection cells) as the basic process responsible for plate motion. However, the question of what exactly drives the plates remains.

Most of the heat from Earth's interior results from the decay of radioactive elements, such as uranium (see Chapter 17), in the core and lower mantle. The most efficient way for this heat to escape Earth's interior is through some type of slow convection system. Heat from the core, supplemented by heat generated from radioactive decay, thus drives large mantle convection cells (Figure 2.26). In this manner, hot rock from the interior rises toward the surface, loses heat to the overlying lithosphere, becomes denser as it cools, and then sinks back into the interior where it is heated, and the process repeats itself. This type of convective heat system is analogous to a pot of stew cooking on a stove (Figure 2.27).

In this mantle convection cell model, spreading ridges mark the ascending limbs of adjacent convection cells, and trenches are present where convection cells descend back into Earth's interior. The convection cells therefore determine the location of spreading ridges and trenches, with the lithosphere lying above the thermal convection cells. Thus, each plate corresponds to a single convection cell that moves as a result of the convective movement of the cell itself (Figure 2.26).

> **ConnectionLink**
> More can be learned about Earth's interior in Chapter 9.

Although most geologists agree that Earth's internal heat plays an important role in plate movement, two other processes, referred to as "slab-pull" and "ridge-push," might also help facilitate the movement of plates (Figure 2.28).

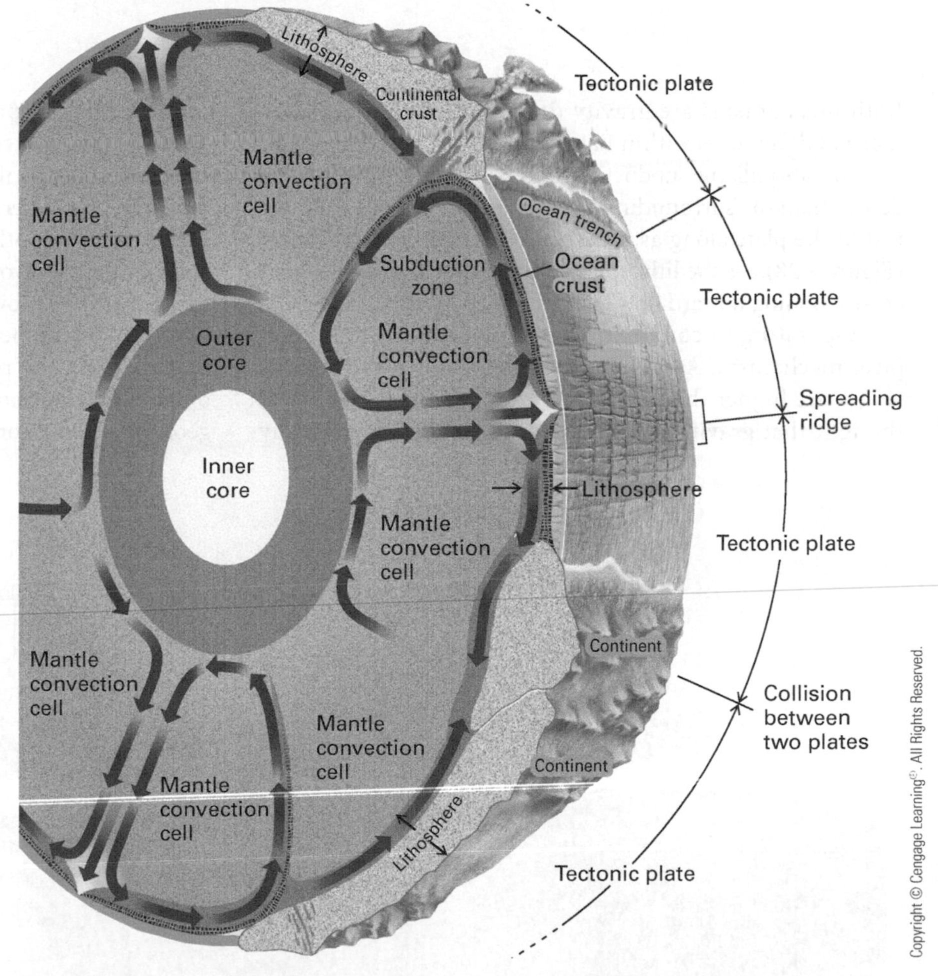

Figure 2.26 Thermal Convection Cells as the Driving Force of Plate Movement A cutaway view of Earth shows that the lithosphere glides horizontally across the asthenosphere. Heat from the core, supplemented by heat produced from radioactive decay, drives huge mantle convection cells that move the lithosphere. Spreading ridges mark the location of ascending limbs of the convection cells, and oceanic trenches are the surface expression of subduction, where plates descend into Earth's interior.

Roman Sigaev/iStockphoto

Figure 2.27 Convection in a Pot of Stew Heat from the stove is applied to the base of the stew pot, causing the stew to heat up. As heat rises through the stew, pieces of the stew are carried to the surface, where the heat is dissipated, the pieces of stew cool, and then sink back to the bottom of the pot. The bubbling seen at the surface of the stew is the result of convection cells churning the stew. In the same manner, heat from the decay of radioactive elements produces convection cells within Earth's interior.

Both mechanisms are gravity driven, but still depend on thermal differences within Earth.

In slab-pull, the subducting cold slab of lithosphere, being denser than the surrounding warmer asthenosphere, pulls the rest of the plate along as it descends into the asthenosphere (Figure 2.28). As the lithosphere moves downward, there is a corresponding upward flow back into the spreading ridge.

Operating in conjunction with slab-pull is the ridge-push mechanism. As a result of rising magma, the oceanic ridges are higher than the surrounding oceanic crust. It is thought that gravity pushes the oceanic lithosphere away from the higher spreading ridges and toward the trenches, where it is subducted back into Earth's interior (Figure 2.28).

Currently, geologists are fairly certain that some type of convective system is involved in plate movement, but the extent to which other mechanisms, such as slab-pull and ridge-push, are involved is still unresolved. However, the fact that plates have moved in the past and are still moving today has been proven beyond a doubt. And although a comprehensive theory of plate movement has not yet been developed, more and more of the pieces are falling into place as geologists learn more about Earth's interior.

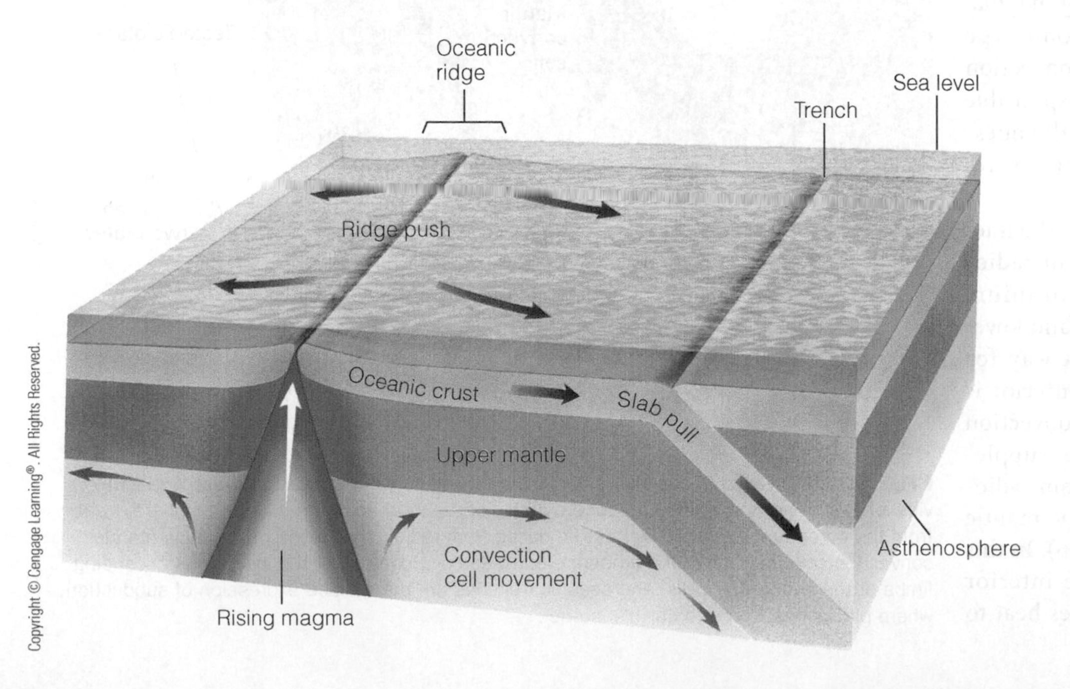

Figure 2.28 Plate Movement Resulting from Gravity-Driven Mechanisms Plate movement is also thought to occur, at least partially, from gravity-driven "slab-pull" or "ridge-push" mechanisms. In slab-pull, the edge of the subducting plate descends into Earth's interior, and the rest of the plate is pulled downward. In ridge-push, rising magma pushes the oceanic ridges higher than the rest of the oceanic crust. Gravity thus pushes the oceanic lithosphere away from the ridges and toward the trenches.

Molybdenum Mining and Economic, Environmental, and Political Concerns

Molybdenum is an element that is important in many industrial metallurgical applications. It has an extremely high melting point, making it useful as an alloy for high-strength and high-temperature steels, as well as in other products requiring resistance to very high temperatures. Like many metals, it is mined from porphyry-style deposits, which typically form in high-temperature igneous environments, such as those associated with convergent plate boundaries. Although the United States is a leading producer of molybdenum, one of the world's largest undeveloped molybdenum deposits is beneath Mount Hope, approximately 50 km northwest of Eureka, Nevada.

To mine this ore deposit, General Moly, Inc., a Colorado-based mining company, has recently received the necessary permits from the Bureau of Land Management and the Nevada Division of Environmental Protection to proceed with its Mount Hope Project. Plans call for blasting off the top of Mount Hope to reach the molybdenum-bearing ore. Part of the financing of this $1.3 billion project will come from the Sichuan Hanlong Group, a Chinese company that will ultimately own a 30% stake in General Moly, but will also buy most of its molybdenum output.

Numerous economic, environmental, and political factors come into play when dealing with metallic ores critical to an industrial society. Discuss some of the factors that you think might have been involved in the decision to go ahead and exploit this important ore deposit.

Mount Lewis Field Office/U.S. Department of the Interior Bureau of Land Management

The location and a view of Mount Hope, near Eureka, Nevada, where one of the world's largest undeveloped molybdenum-bearing ore deposits will soon be mined.

2.13 Plate Tectonics and the Distribution of Natural Resources

In addition to being responsible for the major features of Earth's crust and influencing the distribution and evolution of the world's biota, plate movement also affects the formation and distribution of some natural resources. Consequently, geologists are using plate tectonic theory in their search for petroleum and mineral deposits and in explaining the occurrence of these natural resources. It is becoming increasingly clear that if we are to keep up with the continuing demands of a global industrial society, the application of plate tectonic theory to the origin and distribution of natural resources is essential.

Petroleum

Although significant concentrations of petroleum occur in many areas of the world, more than 50% of all proven reserves are in the Persian Gulf region. It should therefore not be surprising that many of the conflicts in the Middle East have shared as their underlying cause the desire to control these vast amounts of petroleum. Most people, however, are not aware of *why* there is so much oil in this region of the world. The answer lies in the paleogeography and plate movement of this region during the Mesozoic and Cenozoic eras.

During the Mesozoic Era, and particularly the Cretaceous Period when most of the petroleum formed, the Persian Gulf area was a broad marine shelf extending eastward from Africa. This passive continental margin lay near the equator, where countless microorganisms lived in the surface waters. The remains of these organisms accumulated with the bottom sediments and were buried, beginning the long, complex process of petroleum generation and the formation of source beds in which petroleum forms.

As a result of rifting in the Red Sea and the Gulf of Aden during the Cenozoic Era, the Arabian plate is moving northeast away from Africa and subducting beneath Iran (Figure 2.19). During the early stages of collision between Arabia and Iran, as the sediments of the passive continental margin were initially subducted, heating broke down the organic molecules and led to the formation of petroleum.

The continued subduction and collision with Iran folded the rocks, creating traps for petroleum to accumulate, so much so that the vast area south of the collision zone is now a major oil-producing region (see chapter opening photo). Elsewhere in the world, plate tectonics is also responsible for concentrations of petroleum.

Connection Link

To learn more about petroleum and natural gas, go to the Important Resources in Sedimentary Rocks section in Chapter 7.

Mineral Deposits

Many metallic mineral deposits such as copper, gold, lead, silver, tin, and zinc are related to igneous and associated hydrothermal (hot water) activity. So it is not surprising that a close relationship exists between plate boundaries and the occurrence of these valuable deposits.

The magma generated by partial melting of a subducting plate rises toward the surface, and as it cools, it precipitates and concentrates various metallic ores. Many of the world's major metallic ore deposits are associated with convergent plate boundaries, including those in the Andes of South America and the Coast Ranges and Rockies of North America, Japan, the Philippines, Russia, and a zone extending from the eastern Mediterranean region to Pakistan. In addition, the majority of the world's gold is associated with deposits located at ancient convergent plate boundaries in such areas as Canada, Alaska, California, Venezuela, Brazil, Russia, southern India, and Western Australia.

The copper deposits of western North and South America are an excellent example of the relationship between convergent plate boundaries and the distribution, concentration, and exploitation of valuable metallic ores (Figure 2.29a). The world's largest copper deposits are found along this belt. The majority of the copper deposits in the Andes and the southwestern United States were formed less than 60 million years ago when oceanic plates were subducted under the North and South American plates. The rising magma and associated hydrothermal fluids carried minute amounts of copper, which were originally widely disseminated but eventually became concentrated in the cracks and fractures of the surrounding andesites. These low-grade copper deposits contain from 0.2 to 2% copper and are extracted from large open-pit mines (Figure 2.29b).

Divergent plate boundaries also yield valuable ore resources. For example, the island of Cyprus in the Mediterranean is rich in copper and has been supplying all or part of the world's needs for the past 3,000 years. The concentration of copper on Cyprus formed as a result of precipitation adjacent to hydrothermal vents along a divergent plate boundary (Figure 2.8). This deposit was brought to the surface when the copper-rich seafloor collided with the European plate, warping the seafloor and forming Cyprus.

Figure 2.29 Copper Deposits and Convergent Plate Boundaries

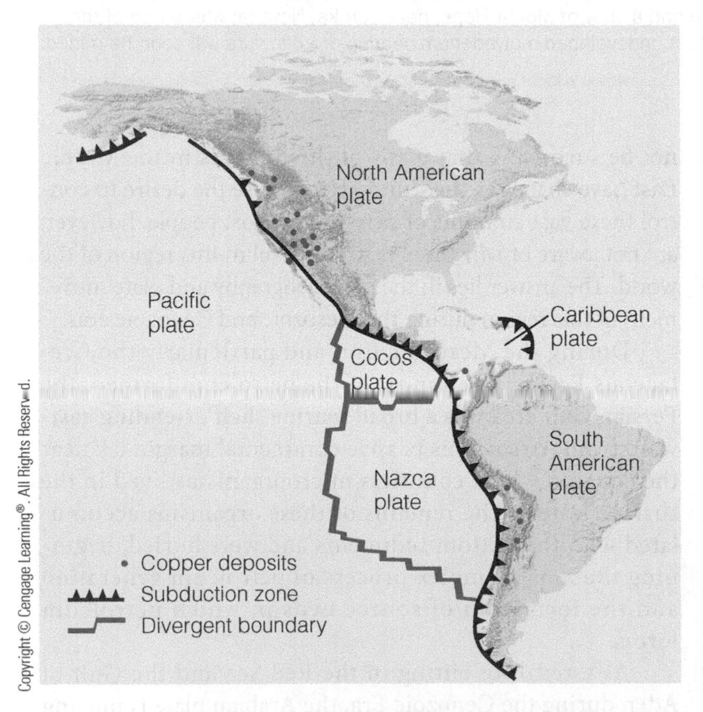

a Valuable copper deposits are located along the west coasts of North and South America in association with convergent plate boundaries. Through time, the rising magma and associated hydrothermal activity resulting from subduction carried small amounts of copper that then became trapped and concentrated in the surrounding rocks.

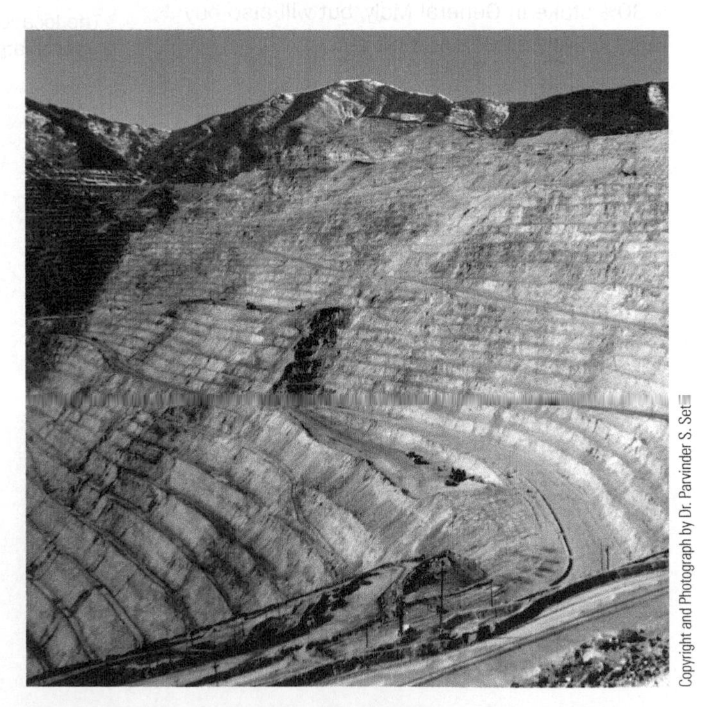

b Bingham Copper Mine, near Salt Lake City, Utah, is a huge open-pit copper mine with reserves estimated at 1.7 billion tons. More than 400,000 tons of rock are removed for processing each day. Note the small specks toward the middle of the photograph that are the 12-foot-high dump trucks.

2.14 Plate Tectonics and the Distribution of Life

Plate tectonic theory is as revolutionary and far-reaching in its implications for geology as the theory of evolution was for biology when it was proposed. Interestingly, it was the fossil evidence that convinced Wegener, Suess, and du Toit, as well as many other geologists, of the correctness of continental drift. Together, the theories of plate tectonics and evolution have changed the way we view our planet, and we should not be surprised at the intimate association between them. Although the relationship between plate tectonic processes and the evolution of life is incredibly complex, paleontological data provide convincing evidence of the influence of plate movement on the distribution of organisms.

The present distribution of plants and animals is not random, but it is controlled mostly by climate and geographic barriers. The world's biota occupy *biotic provinces*, which are regions characterized by a distinctive assemblage of plants and animals. Organisms within a province have similar ecological requirements, and the boundaries separating provinces are

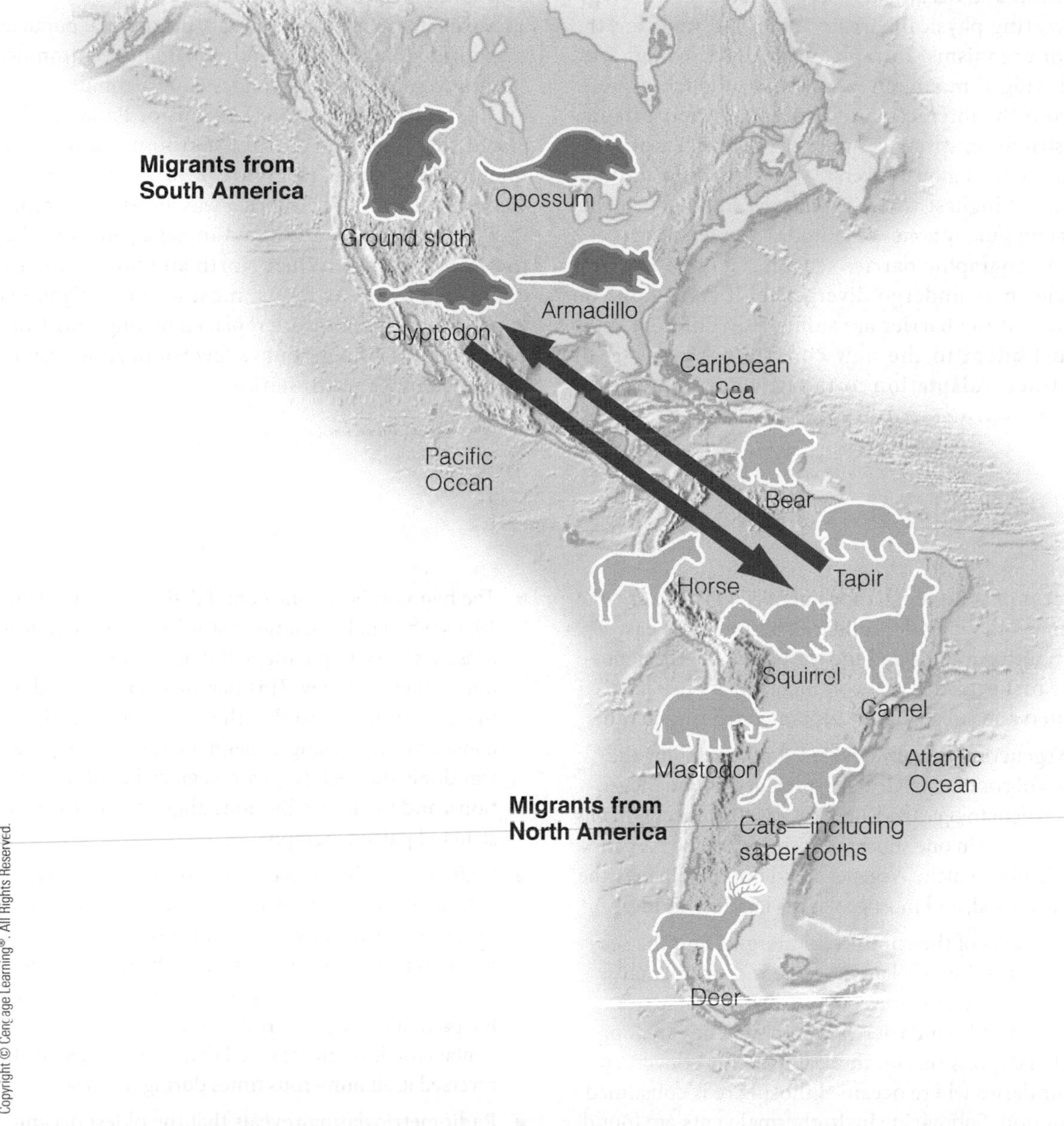

Figure 2.30 Plate Tectonics and the Distribution of Organisms The Isthmus of Panama forms a barrier that divides a once-uniform fauna of molluscs that inhabited the shallow seas of both the Pacific Ocean and Caribbean Sea. Its creation also formed a land corridor in which migration between the two continents took place. Prior to the formation of this isthmus, South America was isolated from all other landmasses during much of the Cenozoic, and its mammal fauna consisted of marsupials (pouched mammals) and placentals that lived nowhere else. When the Isthmus of Panama formed during the Late Pliocene, many placental mammals migrated south, resulting in numerous South American mammals becoming extinct. A few South American mammals migrated north and successfully occupied North America.

Critical Thinking Question Why is the mammalian fauna of Australia so different from elsewhere?

therefore natural ecological breaks. Climatic or geographic barriers are the most common province boundaries, and these are mostly controlled by plate movement.

The complex interaction between wind and ocean currents has a strong influence on the world's climates. Wind and ocean currents, in turn, are thus strongly influenced by the number, distribution, topography, and orientation of continents. For example, the southern Andes Mountains in South America act as an effective barrier to moist, easterly blowing Pacific winds, resulting in a desert east of the southern Andes that is virtually uninhabitable.

The distribution of continents and ocean basins not only influences wind and ocean currents but also affects provinciality by creating physical barriers to, or pathways for, the migration of organisms. Intraplate volcanoes, island arcs, mid-oceanic ridges, mountain ranges, and subduction zones all result from the interaction of plates, and their orientation and distribution strongly influence the number of provinces and hence total global diversity. Thus, provinciality and diversity will be highest where numerous small continents are spread across many zones of latitude.

When a geographic barrier separates a once-uniform fauna, species may undergo divergence. If conditions on opposite sides of the barrier are sufficiently different, then species must adapt to the new conditions, migrate, or become extinct. Adaptation to the new environment by various species may involve enough change that new species eventually evolve.

The marine invertebrates found on opposite sides of the Isthmus of Panama provide an excellent example of divergence caused by the formation of a geographic barrier. Prior to the rise of this land connection between North and South America, a homogeneous population of bottom-dwelling invertebrates inhabited the shallow seas of the area. After the formation of the Isthmus of Panama by subduction of the Pacific plate approximately 5 million years ago, the original population was thus divided. In response to the changing environment, new species evolved on opposite sides of the isthmus.

You can learn more about divergent, convergent, and parallel evolution in Chapter 18.

The formation of the Isthmus of Panama also influenced the evolution of the North and South American mammalian faunas (Figure 2.30). During most of the Cenozoic Era, South America was an island continent, and its mammalian fauna evolved in isolation from the rest of the world's faunas. When North and South America were connected by the Isthmus, most of the indigenous South American mammals were replaced by migrants from North America. Surprisingly, only a few South American mammal groups migrated northward.

Key Concepts Review

- The concept of continental movement is not new. The earliest maps showing the similarity between the east coast of South America and the west coast of Africa provided the first evidence that continents may once have been united and subsequently separated from each other.
- Alfred Wegener is generally credited with developing the hypothesis of continental drift. He provided abundant geologic and paleontologic evidence to show that the continents were once united in one supercontinent, which he named Pangaea. Unfortunately, Wegener could not explain how the continents moved, and most geologists ignored his ideas.
- Various features of the continental margins and the seafloor are a reflection of plate movement. Continental margins are active or passive, depending on their relationship to plate boundaries. Oceanic trenches are long, steep-sided depressions on the seafloor near convergent plate boundaries where oceanic lithosphere is consumed by subduction. Submarine hydrothermal vents are found at or near spreading ridges and are associated with divergent plate boundaries.
- The hypothesis of continental drift was revived during the 1950s when paleomagnetic studies of rocks indicated the presence of multiple magnetic north poles instead of just one as there is today. This paradox was resolved by moving the continents so that the paleomagnetic data became consistent with a single magnetic north pole. When this was done, the rock sequences, glacial deposits and striations, and fossil distributions aligned with the reconstructed paleogeography.
- Seafloor spreading was confirmed by the discovery of magnetic anomalies in the oceanic crust that were both parallel to and symmetric around oceanic ridges, indicating that new oceanic crust must have formed as the seafloor was spreading. The pattern of oceanic magnetic anomalies matched the pattern of magnetic reversals already known from continental lava flows and showed that Earth's magnetic field has reversed itself numerous times during the past.
- Radiometric dating reveals that the oldest oceanic crust is less than 180 million years old, whereas the oldest continental crust is approximately 4 billion years old. Fossil

evidence and the thickness of sediments overlying the oceanic crust further support and confirm that ocean basins are recent geologic features.

- Plate tectonic theory became widely accepted by the 1970s because the evidence overwhelmingly supports it and because it provides geologists with a powerful theory for explaining such phenomena as volcanism, earthquake activity, mountain building, global climatic changes, the distribution of the world's biota, and the distribution of many mineral resources.
- Geologists recognize three types of plate boundaries: divergent boundaries, where plates move away from each other; convergent boundaries, where two plates collide; and transform boundaries, where two plates slide past each other.
- Ancient plate boundaries can be recognized by their associated rock assemblages and geologic structures. For divergent boundaries, these may include rift valleys with thick sedimentary sequences and numerous dikes and sills. For convergent boundaries, ophiolites and andesitic rocks are two characteristic features. Transform plate boundaries generally do not leave any characteristic or diagnostic features in the geologic record.
- The average rate of movement and relative motion of the plates can be calculated in several ways. The results of these different methods all agree and indicate that the plates move at different average velocities.
- The absolute motion of plates can be determined by the movement of plates over mantle plumes. A mantle plume is an apparently stationary column of magma that rises to the surface from deep within the mantle and forms either a subsurface mushroom-shaped plume head, or erupts at the surface as a volcano.
- Although a comprehensive theory of plate movement has yet to be developed, geologists think that some type of convective system is involved in plate movement.
- A close relationship exists between the formation of petroleum, as well as some mineral deposits, and plate boundaries. Furthermore, the formation and distribution of many natural resources are related to plate movements.
- The relationship between plate tectonic processes and the evolution of life is complex. The distribution of plants and animals is not random, but rather is controlled mostly by climate and geographic barriers, which, in turn, are influenced, to a great extent, by the movement of plates.

Important Terms

abyssal plain (p. 33)
active continental margin (p. 35)
continental–continental plate boundary (p. 47)
continental drift (p. 28)
continental margin (p. 32)
continental rise (p. 32)
continental shelf (p. 32)
continental slope (p. 32)
convergent plate boundary (p. 45)
Curie point (p. 36)
divergent plate boundary (p. 41)
Glossopteris flora (p. 28)
Gondwana (p. 28)
hot spot (p. 49)
Laurasia (p. 29)
magnetic anomaly (p. 37)
magnetic field (p. 35)
magnetic reversal (p. 37)
magnetism (p. 35)
oceanic–continental plate boundary (p. 47)
oceanic–oceanic plate boundary (p. 47)
oceanic ridge (p. 33)
oceanic trench (p. 33)
paleomagnetism (p. 36)
Pangaea (p. 29)
passive continental margin (p. 35)
plate tectonic theory (p. 40)
seafloor spreading (p. 37)
submarine hydrothermal vent (p. 33)
thermal convection cell (p. 37)
transform fault (p. 48)
transform plate boundary (p. 48)

Review Questions

1. Magnetic surveys of the ocean basins indicate that
   a. ___ the oceanic crust is youngest adjacent to mid-oceanic ridges.
   b. ___ the oceanic crust is oldest adjacent to mid-oceanic ridges.
   c. ___ the oceanic crust is youngest adjacent to the continents.
   d. ___ the oceanic crust is the same age everywhere.
   e. ___ answers b and c.
2. The most common biotic province boundaries are
   a. ___ geographic barriers.
   b. ___ biologic barriers.
   c. ___ climatic barriers.
   d. ___ answers a and b.
   e. ___ answers a and c.
3. The man credited with developing the continental drift hypothesis is
   a. ___ Wilson.
   b. ___ Wegener.
   c. ___ Hess.
   d. ___ du Toit.
   e. ___ Vine.
4. The driving mechanism of plate movement is thought to be largely the result of
   a. ___ isostasy.
   b. ___ Earth's rotation.
   c. ___ thermal convection cells.
   d. ___ magnetism.
   e. ___ polar wandering.
5. Along what type of boundary does subduction occur?
   a. ___ divergent.
   b. ___ transform.
   c. ___ convergent.
   d. ___ answers a and b.
   e. ___ answers a and c.
6. What evidence convinced Wegener and others that continents must have moved in the past and at one time formed a supercontinent?
7. Why is some type of thermal convection system thought to be the major force driving plate movement? How have slab-pull and ridge-push, both mainly gravity driven, modified a purely thermal convection model for plate movement?
8. In addition to the volcanic eruptions and earthquakes associated with convergent and divergent plate boundaries, why are these boundaries also associated with the formation and accumulation of various metallic ore deposits?
9. Plate tectonic theory builds on the continental drift hypothesis and the theory of seafloor spreading. As such, it is a unifying theory of geology. Explain why it is a unifying theory.

10. **Creative Thinking Visual Question:** Using the ages (the numbers represent ages in millions of years) for each of the Hawaiian Islands, as well as the scale given in the figure on the next page, calculate the average rate of movement per year for the Pacific plate since each island formed (▌Figure 1). Is the average rate of movement the same for each island? Would you expect it to be? Explain why it may not be and why there are different ages for some of the islands.

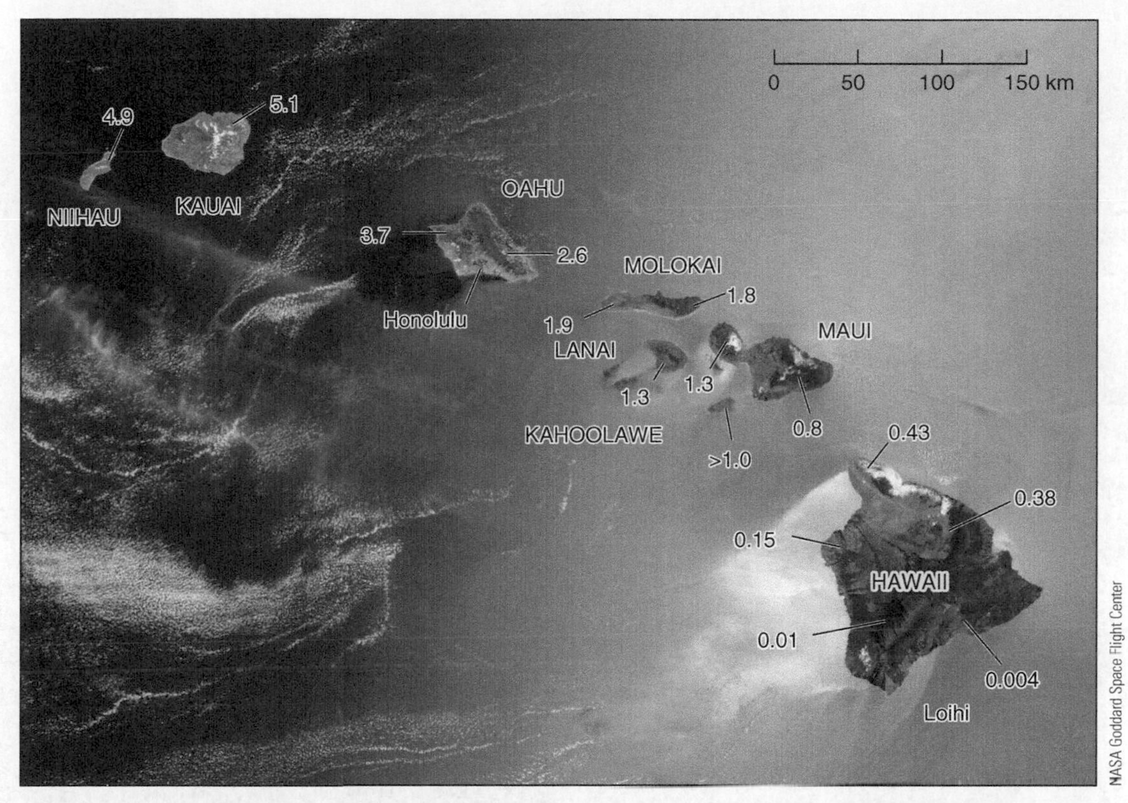

Figure 1 Hawaiian Islands Image of the Hawaiian islands with the age of each island in millions of years.

Global GeoScience Watch Plate tectonic theory is the unifying theory of geology because it ties together many seemingly unrelated geologic features and events. Plate interaction along plate boundaries results in earthquakes, which can cause tsunami. Within the GREENR database, search "tsunami warning systems" and then click on "View All" next to the News heading. Of the first 20 articles, where have most tsunami recently occurred? How does a tsunami early warning system work, and how can it save lives? Write a short report answering these questions.

This blue quartz is agate that formed when minerals filled a hollow inside a host rock. Agate is often found with concentric bands that more or less parallel the interior of the hollow. The mass of colorless crystals that surround the agate consist of the mineral quartz.

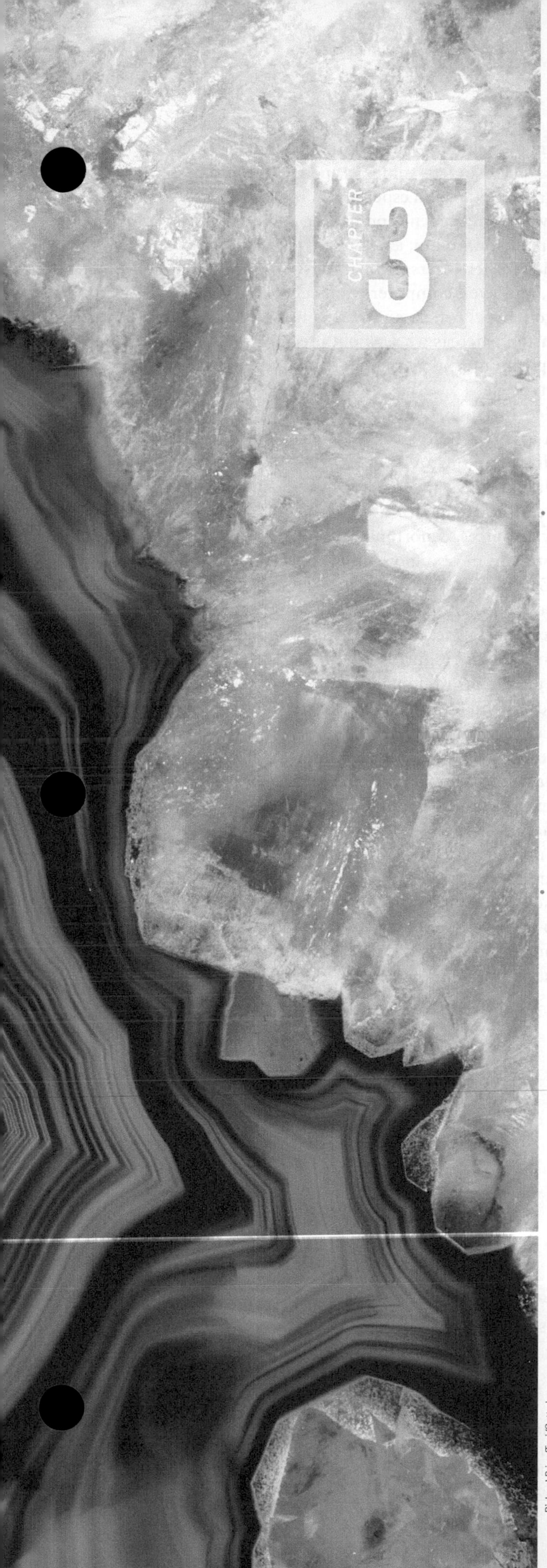

Richard Price/Taxi/Getty Images

Minerals—The Building Blocks of Rocks

OUTLINE

HAVE YOU EVER WONDERED?

- What minerals are and how they form?
- How to distinguish minerals from other naturally occurring substances?
- Why so few minerals are abundant and yet there are so many of them?
- What criteria geologists use to identify and classify minerals?
- Which minerals are valued as gemstones?
- What kinds of minerals are important to industrial societies?

3.1 Introduction

The term *mineral* usually brings to mind dietary substances such as calcium, potassium, and magnesium that are necessary for good health. Geologists, however, define a **mineral** as an inorganic, naturally occurring, crystalline solid, with a narrowly defined chemical composition, and characteristic physical properties (▮ Figure 3.1a). The definition is lengthy, but the only part that is likely unfamiliar is *crystalline solid.* A solid is any substance that retains its shape and volume unless deformed by a force, but a crystalline solid is one in which its constituent atoms are arranged in a specific three-dimensional framework. Ice is a mineral because it is inorganic, that is, it is not composed of the complex carbon-based molecules of organisms, and it is certainly naturally occurring. Chemically it consists of hydrogen and oxygen atoms (H_2O), and it has distinguishing features, such as color, hardness, and density. And ice is a crystalline solid, because its hydrogen and oxygen atoms are arranged in an orderly fashion, as opposed to the atoms in glass that have no such arrangement.

According to a publication by the American Geological Institute titled, *Minerals: The Foundations of Society,* "Materials from the Earth have played a crucial role in the development of civilization since the dawn of modern humans 100,000 years ago." In fact, minerals and rocks play an essential role in all societies, and especially in industrial ones. Our economic well-being depends on finding and extracting ores of iron, copper, aluminum, and tin, as well as minerals and rocks needed for fertilizers, animal feed supplements, glass, wallboard, and even computers and cell phones. Mineral additives to paint, lipstick, and nail polish give them their resinous luster, whereas we seek out precious metals (gold, silver, and platinum) and gemstones (diamond, emerald, etc.) for industrial uses as well as for decorative purposes and jewelry (Geo-Insight 3.1 and Figure 3.1a). Additionally, we depend on energy resources such as petroleum, natural gas, coal, and uranium.

The United States and Canada owe much of their economic success and standard of living to abundant mineral and energy resources. Unfortunately, resources are finite and now both nations must import some essential commodities, although Canada is more self-reliant than the United States. In any case, both nations must import aluminum ore and manganese, which is necessary in steel production. Indeed, the consumption of resources by the industrial nations as well as the uneven distribution of resources make it necessary for them to depend on foreign sources for many important commodities.

One reason to study minerals is their economic importance. In fact, the distribution of resources has important implications for foreign policy decisions and economic ties between nations. For students, though, it is important to know that minerals are the building blocks of rocks, so **rocks** are made up of one or more minerals. Granite, for example, is composed of specific amounts of the minerals quartz, potassium feldspars, and plagioclase feldspars, plus a few others in minor quantities (▮ Figure 3.1b). The only necessary component of limestone, however, is calcite (see Chapter 7).

▮ **Figure 3.1** Minerals and Rocks

Critical Thinking Question You must explain to an interested audience the distinction between minerals and rocks. How would you do so, and can you think of any analogies that might clarify the points that you make?

Los Angeles County Museum specimen © Harold and Erica Van Pelt

a A spectacular specimen of tourmaline (the elongate minerals) and quartz (colorless) from the Himalaya Mine in San Diego County, California. Notice the change in color along the lengths of the the tourmaline crystals.

Copyright and Photograph by Dr. Parvinder S. Sethi

b Minerals in Granite The igneous rock granite (see Chapter 4) is made up of mostly three minerals—quartz, potassium feldspar, and plagioclase feldspar—but it may also contain small amounts of biotite, muscovite, and hornblende.

Minerals and the Economy

The United States Geological Survey (USGS) maintains a site titled *Mineral Commodity Summaries 2012*. What are some of the commodities that the United States relies on entirely from foreign sources? Also, what are the most valuable nonfuel mineral commodities in your state?

Some minerals and rocks are eagerly sought out by collectors because they are attractive or valuable (see the chapter opening photo).

From our discussion so far, we have a formal definition of the term *mineral,* and we know that rocks are composed of minerals. Now let's delve deeper into what minerals are composed of by considering matter, elements, atoms, and bonding and compounds.

3.2 Matter—What Is It?

Anything that has mass and occupies space is matter, and accordingly includes water, all organisms, and the atmosphere, as well as minerals and rocks. Physicists recognize four states of matter; *plasma* (composed of ionized gases as in the Sun and stars), *liquids, gases,* and *solids.* Liquids and gases are important in our understanding of several surface processes such as running water and wind, but here our main concern is solids, because, by definition, minerals are solids. That is, they are rigid substances that resist changes in shape or volume.

Atoms and Elements

Matter is made up of chemical **elements,** which, in turn, are composed of **atoms,** the smallest units of matter that retain the characteristics of a particular element (▮ Figure 3.2). That is, elements cannot be changed into different substances except through radioactive decay (discussed in Chapter 17). Thus, an element is made up of atoms, all of which have the same properties. Scientists have discovered 92 naturally occurring elements, some of which are listed in Figure 3.2. All naturally occurring elements have a name and a symbol—for example, oxygen (O), aluminum (Al), and potassium (K) (▮ Figure 3.3).

At the center of an atom is a **nucleus** made up of one or more particles known as **protons,** which have a positive electrical charge, and **neutrons,** which are electrically neutral. The nucleus is only about 1/100,000 of the diameter of an atom, yet it contains virtually all of the atom's mass. **Electrons,** particles with a negative electrical charge, orbit rapidly around the nucleus at specific distances in one or more **electron shells.** The electrons control how an atom interacts with other atoms, but the nucleus determines how many electrons an atom has, because the positively charged protons attract and hold the negatively charged electrons in their orbits.

The number of protons in an atom's nucleus determines its identity and **atomic number.** Hydrogen (H), for instance, has one proton in its nucleus and thus has an atomic number of 1. The nuclei of helium (He) atoms possess 2 protons, whereas those of carbon (C) have 6, and uranium (U) 92, so their atomic numbers are 2, 6, and 92, respectively. Atoms also have an **atomic mass number,** which is the sum of protons and neutrons in the nucleus (electrons contribute negligible mass to atoms). However, atoms of the same chemical element might have different atomic mass numbers because

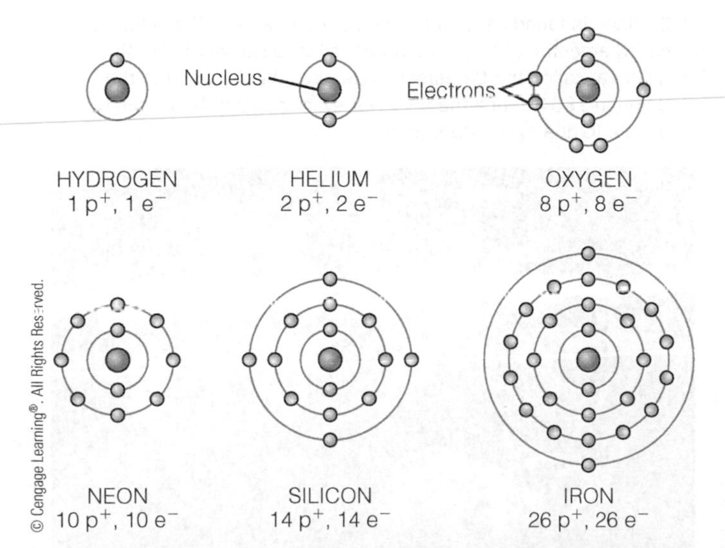

			Distribution of Electrons			
Element	Symbol	Atomic Number	First Shell	Second Shell	Third Shell	Fourth Shell
Hydrogen	H	1	1	—	—	—
Helium	He	2	2	—	—	—
Carbon	C	6	2	4	—	—
Oxygen	O	8	2	6	—	—
Neon	Ne	10	2	8		—
Sodium	Na	11	2	8	1	—
Magnesium	Mg	12	2	8	2	—
Aluminum	Al	13	2	8	3	—
Silicon	Si	14	2	8	4	—
Phosphorus	P	15	2	8	5	—
Sulfur	S	16	2	8	6	—
Chlorine	Cl	17	2	8	7	—
Potassium	K	19	2	8	8	1
Calcium	Ca	20	2	8	8	2
Iron	Fe	26	2	8	14	2

▮ **Figure 3.2 Shell Models for Common Atoms** The shell model for several atoms and their electron configurations. A blue circle represents the nucleus of each atom, but remember that atomic nuclei are made up of protons and neutrons, as shown in Figure 3.4.

GEO INSIGHT 3.1 The Precious Metals and Gemstones

The precious metals include gold, silver, and platinum. The discovery of gold in California in 1848 sparked the California Gold Rush (1849–1853) during which $200 million in gold was recovered. However, the first gold rush in the United States took place in 1829 in Georgia and lasted until the 1840s. The precious metals are symbols of wealth and are used for jewelry as are gemstones, which is any mineral or rock or rock-like substance prized for decorative stones and jewelry. The precious gemstones include diamond, emerald (green beryl), and ruby and sapphire, which are red and blue varieties of corundum, respectively. Many others are semiprecious gemstones and include turquoise, amber, several types of quartz, opal, pearls, and so on.

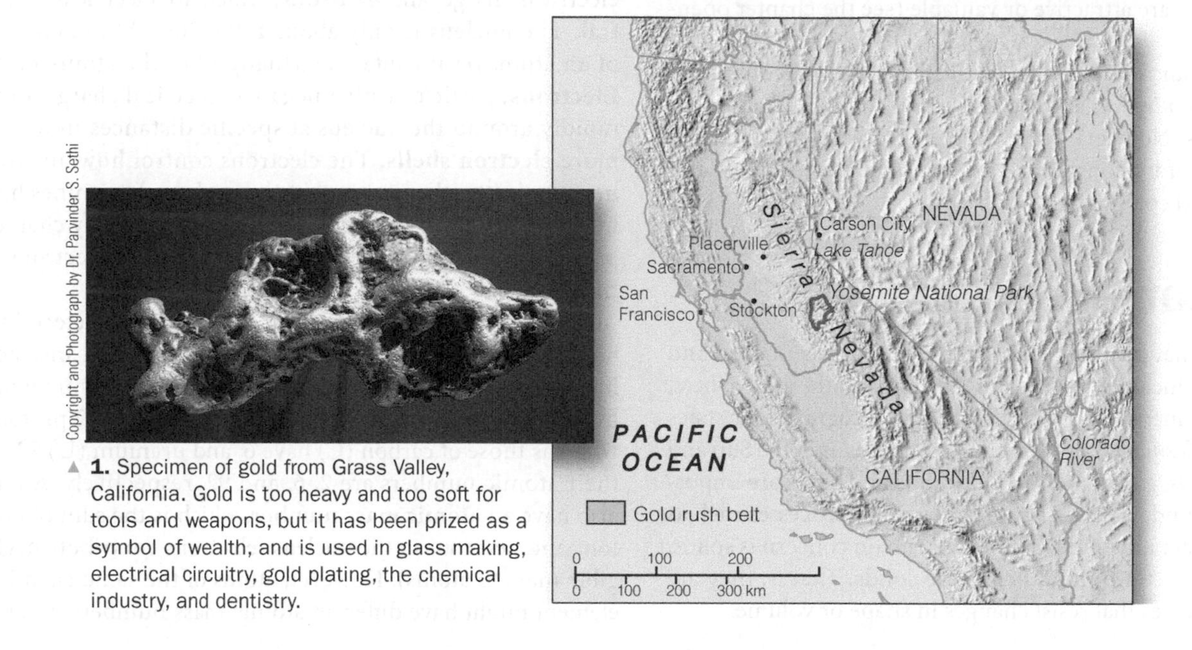

▲ **1.** Specimen of gold from Grass Valley, California. Gold is too heavy and too soft for tools and weapons, but it has been prized as a symbol of wealth, and is used in glass making, electrical circuitry, gold plating, the chemical industry, and dentistry.

▼ **2.** A miner pans for gold (foreground) by swirling water, sand, and gravel in a broad, shallow pan. The heavier gold sinks to the bottom. At the far left, a miner washes sediment in a cradle. As in panning, the cradle separates the heavier gold from other materials.

▼ **3.** Silver is found as a native element and as a sulfide mineral called argentite (Ag_2S). The largest silver discovery in North America, called the Comstock Lode, brought Nevada into the Union in 1864 during the Civil War even though it had too few people to qualify for statehood.

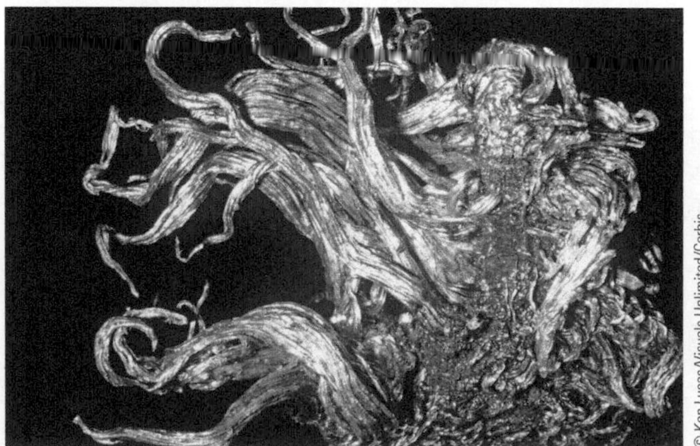

Bettmann/CORBIS

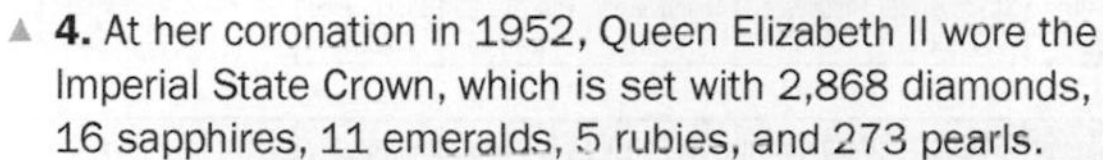

▲ **4.** At her coronation in 1952, Queen Elizabeth II wore the Imperial State Crown, which is set with 2,868 diamonds, 16 sapphires, 11 emeralds, 5 rubies, and 273 pearls.

Copyright and Photograph by Dr. Parvinder S. Sethi

◄ **5.** The pendant in this necklace is the Victoria Transvaal Diamond in the Smithsonian Institute in Washington, D.C. The pendant is 67.89 carats and was cut from a larger stone found in 1951. A total of 106 other diamonds are also in the necklace.

Sue Monroe

◄ **6.** These emeralds, a green variety of the mineral beryl, are on display at the Natural History Museum in Vienna, Austria. The specimen measures about 17 cm high.

▼ **7.** Pearls are semiprecious stones. These pearls, valued at about $13,000, are on display at Maui Pearls on the island of Roratonga, which is part of the Cook Islands in the South Pacific.

Nevada Wier/Encyclopedia/Corbis

◄ **8.** Turquoise is a sky blue, blue green, or light green semiprecious gemstone used for jewelry and as a decorative stone.

Sue Monroe

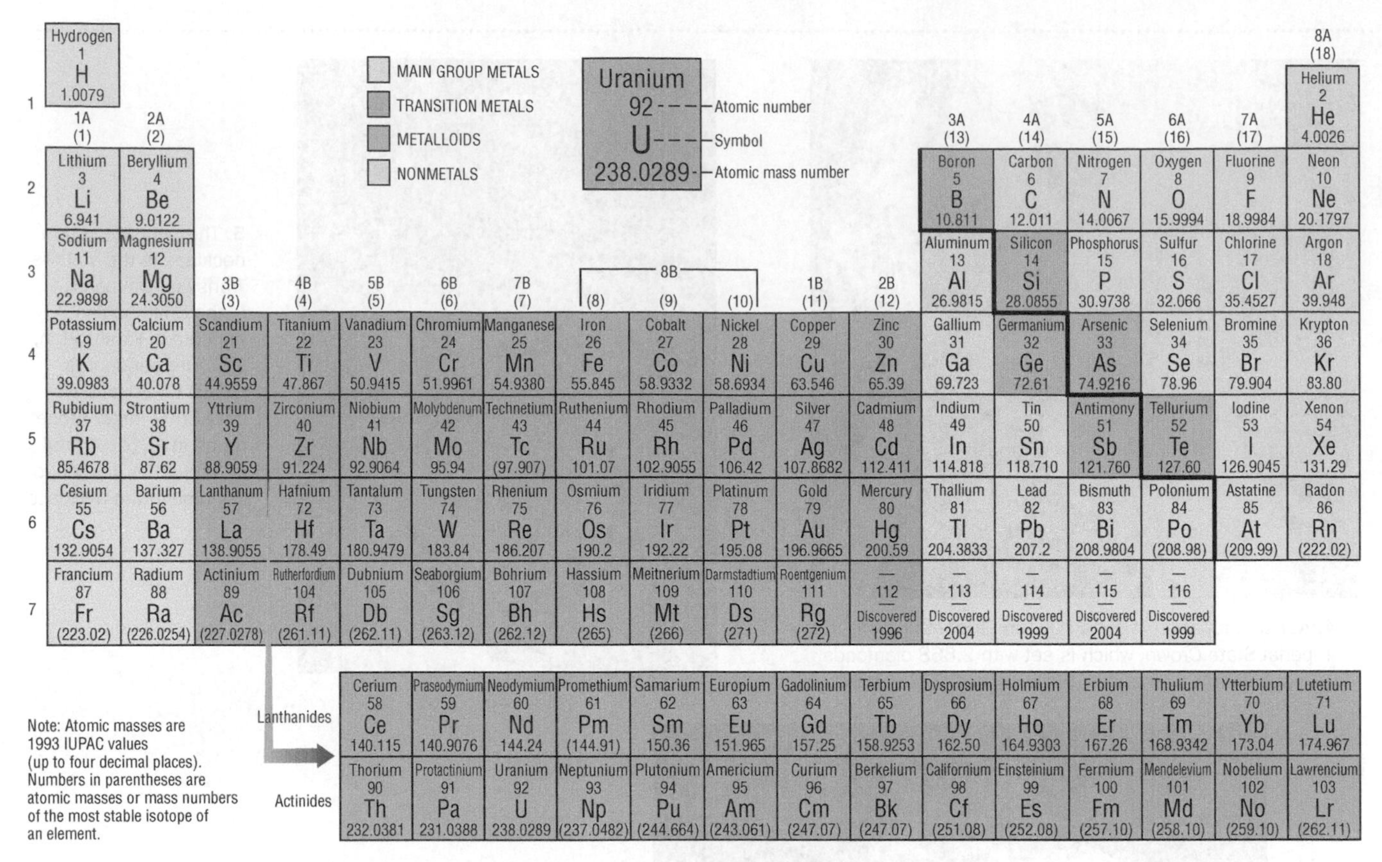

Figure 3.3 The Periodic Table of Elements Only about a dozen elements are common in minerals and rocks, but many uncommon ones are important sources of natural resources. For example, lead (Pb) is not found in many minerals, but it is present in the mineral galena, the main ore of lead. Silicon (Si) and oxygen (O), in contrast, are important elements in most of the minerals in Earth's crust.

the number of neutrons can vary. All carbon (C) atoms have six protons—otherwise they would not be carbon—but the number of neutrons can be 6, 7, or 8. Thus, we recognize three types or isotopes of carbon (Figure 3.4), each with a different atomic mass number.

> **ConnectionLink**
>
> **You can learn more about the role of radioactive isotopes in absolute or numerical dating in Chapter 17.**

The isotopes of carbon, or those of any other element, behave the same chemically; carbon 12 and carbon 14 are both present in carbon dioxide (CO_2), for example. However, some isotopes are radioactive, meaning that they spontaneously decay or change to other elements. Carbon 14 is radioactive, whereas both carbon 12 and carbon 13 are not. Radioactive isotopes are important for determining the absolute or numerical ages of rocks.

Bonding and Compounds

Interactions among electrons around atoms can result in two or more atoms joining together, a process known as **bonding.** If atoms of two or more elements bond, the resulting substance is a **compound.** Gaseous oxygen consists of only oxygen atoms and is thus an element, whereas the mineral quartz, consisting of silicon and oxygen atoms, is a compound.

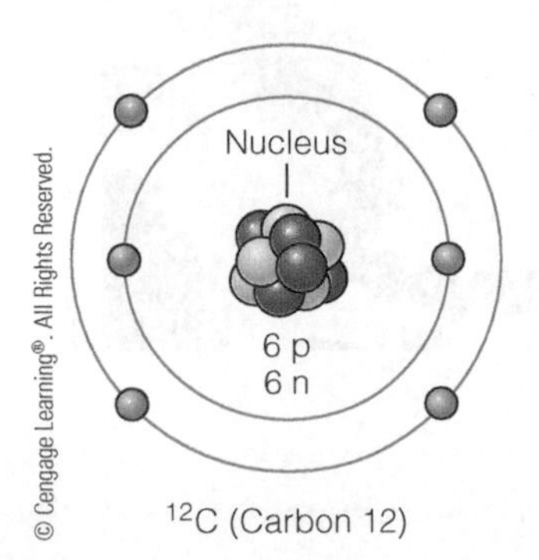

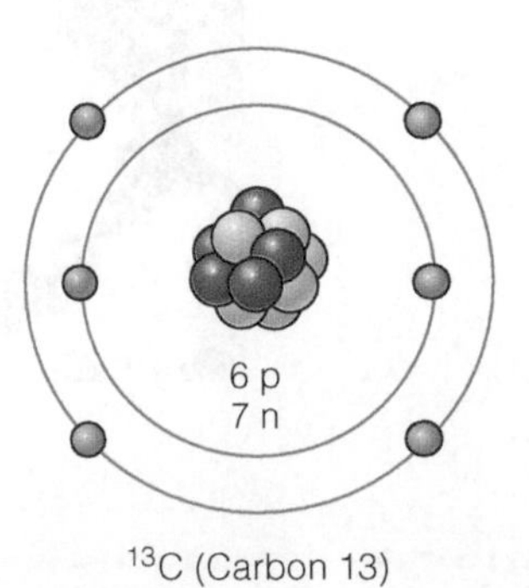

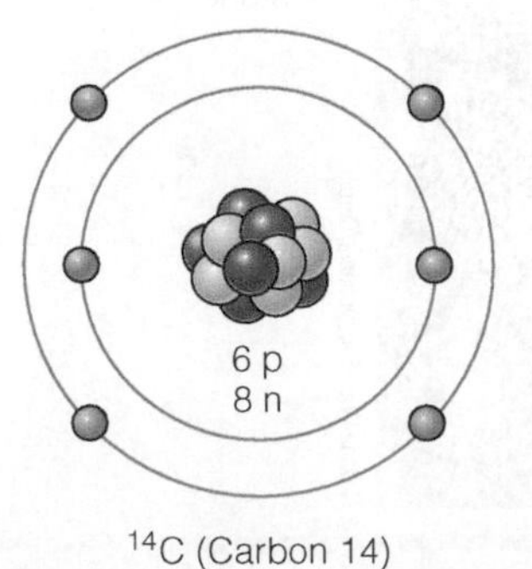

Figure 3.4 Carbon Isotopes Schematic representation of the isotopes of carbon. Carbon has an atomic number of 6 and an atomic mass number of 12, 13, or 14, depending on the number of neutrons in its nucleus.

Most minerals are compounds, although gold, platinum, and several others are important exceptions.

To understand bonding, it is necessary to delve deeper into the structure of atoms. Recall that negatively charged electrons orbit the nuclei of atoms in electron shells. With the exception of hydrogen, which has only one proton and one electron, the innermost electron shell of an atom contains only two electrons. The other shells contain various numbers of electrons, but the outermost shell never has more than eight (Figure 3.2), and it is these that are usually involved in chemical bonding.

Two types of chemical bonds, *ionic* and *covalent,* are particularly important in minerals, and many minerals contain both types of bonds. Two other types of chemical bonds, *metallic* and *van der Waals,* are much less common, but are very important in determining the properties of some useful minerals.

Ionic Bonding Notice in Figure 3.2 that most atoms have fewer than eight electrons in their outermost electron shell. However, some elements, including neon and argon, have complete outer shells with eight electrons; because of this electron configuration, these elements, known as the *noble gases,* do not react readily with other elements to form compounds. Interactions among atoms tend to produce electron configurations similar to those of the noble gases. That is, atoms interact so that their outermost electron shell is filled with eight electrons, unless the first shell (with two electrons) is also the outermost electron shell, as in helium.

One way to attain the noble gas configuration is by the transfer of one or more electrons from one atom to another. Common salt is composed of the elements sodium (Na) and chlorine (Cl); each element is poisonous, but when combined chemically they form the compound sodium chloride (NaCl), the mineral halite. Notice in ▮ Figure 3.5a that sodium has 11 protons and 11 electrons; thus the positive electrical charges of the protons are exactly balanced by the negative charges of the electrons, and the atom is electrically neutral. Likewise, chlorine with 17 protons and 17 electrons is electrically neutral (Figure 3.5a). However, neither sodium nor chlorine has eight electrons in its outermost electron shell; sodium has only one, whereas chlorine has seven. To attain a stable configuration, sodium loses the electron in its outermost electron shell, leaving its next shell with eight electrons as the outermost one (Figure 3.5a). Sodium now has one fewer electron (negative charge) than it has protons (positive charge), so it is an electrically charged **ion** and is symbolized Na^+.

The electron lost by sodium is transferred to the outermost electron shell of chlorine, which had seven electrons to begin with. The addition of one more electron gives chlorine an outermost electron shell of eight electrons, the configuration of a noble gas. But its total number of electrons is now 18, which exceeds by 1 the number of protons. Accordingly, chlorine also becomes an ion, but it is negatively charged (Cl^-). An **ionic bond** forms between sodium and chlorine because of the attractive force between the positively charged sodium ion and the negatively charged chlorine ion (Figure 3.5a).

In ionic compounds, such as sodium chloride (the mineral halite), the ions are arranged in a three-dimensional framework that results in overall electrical neutrality. In halite, sodium ions are bonded to chlorine ions on all sides, and chlorine ions are surrounded by sodium ions (Figure 3.5b).

Covalent Bonding **Covalent bonds** form between atoms when their electron shells overlap and they share electrons. For example, atoms of the same element, such as carbon, cannot bond by transferring electrons from one atom to another. Carbon (C), has four electrons in its outermost electron shell (▮ Figure 3.6a). If these four electrons were transferred to another carbon atom, the atom receiving the electrons would have the noble gas configuration of eight electrons in its outermost electron shell, but the atom contributing the electrons would not.

In such situations, adjacent atoms share electrons by overlapping their electron shells. A carbon atom in diamond shares all four of its outermost electrons with a neighbor to produce a stable noble gas configuration (Figure 3.6a). Diamond and graphite, both crystalline forms of carbon, have covalent bonds, but in diamond the atoms form a three-dimensional framework, whereas in graphite the carbon atoms form sheets that are weakly bonded together (Figures 3.6b and c).

Covalent bonds are not restricted to substances composed of atoms of a single kind. Among the most common minerals, the silicates (discussed later in this chapter), the element silicon forms partly covalent and partly ionic bonds with oxygen.

Metallic and van der Waals Bonds *Metallic bonding* involves an extreme type of electron sharing. The electrons of the outermost electron shell of metals such as gold, silver, and copper readily move about from one atom to another. This electron mobility accounts for the metallic luster of metals (their appearance in reflected light), their good electrical and thermal conductivity, and their ability to be easily reshaped. Only a few minerals possess metallic bonds, but those that do are very useful; copper, for example, is used for electrical wiring because of its high electrical conductivity.

Some electrically neutral atoms and molecules* have no electrons available for ionic, covalent, or metallic bonding. They nevertheless have a weak attractive force between them, called a *van der Waals* or *residual bond,* when in proximity. The carbon atoms in the mineral graphite are covalently bonded to form sheets, but the sheets are weakly held together by van der Waals bonds (Figure 3.6c). This type of bonding makes graphite useful for pencil lead—when a pencil point is moved across a piece of paper, small pieces of graphite flake off along the planes held together by van der Waals bonds and adhere to the paper.

*A molecule is the smallest unit of a substance that has the properties of that substance. A water molecule (H_2O), for example, possesses two hydrogen atoms and one oxygen atom.

Figure 3.5 Ionic Bonding and the Origin of the Mineral Halite (NaCl)

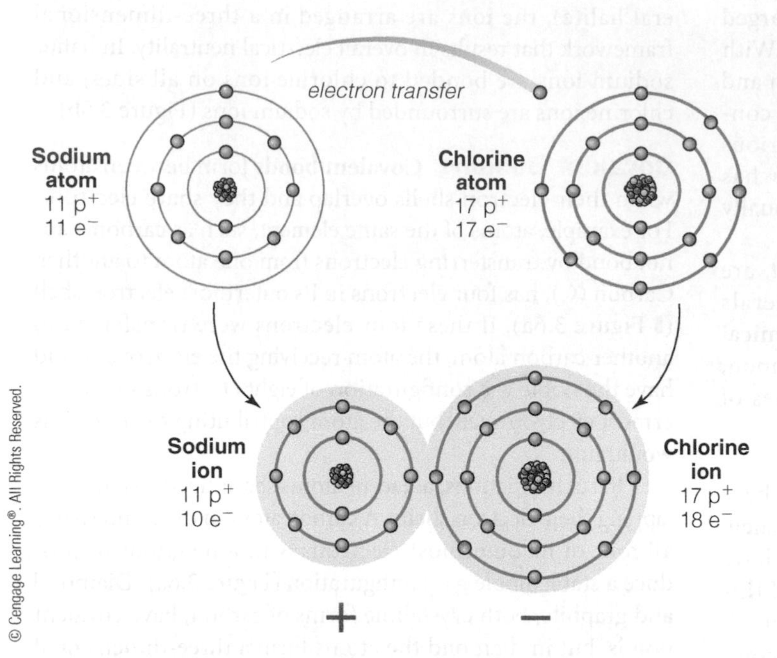

a Transfer of the electron in the outermost shell of sodium to the outermost shell of chlorine. After electron transfer, the sodium and chlorine atoms are positively and negatively charged ions, respectively.

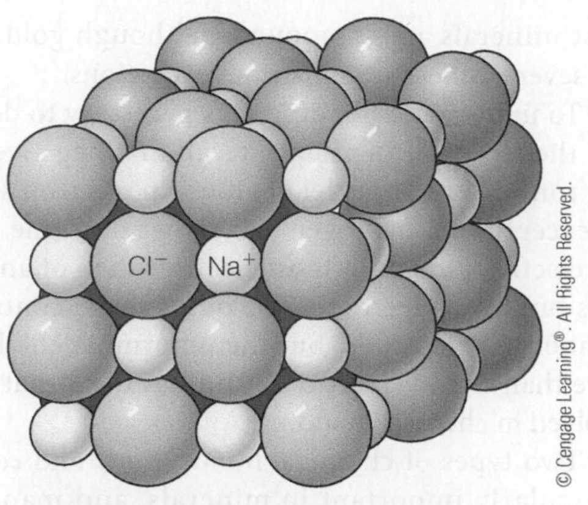

b This diagram shows the relative sizes of the sodium and chlorine atoms and their locations in a crystal of halite.

c Tiny crystals of halite.

Figure 3.6 Covalent Bonds

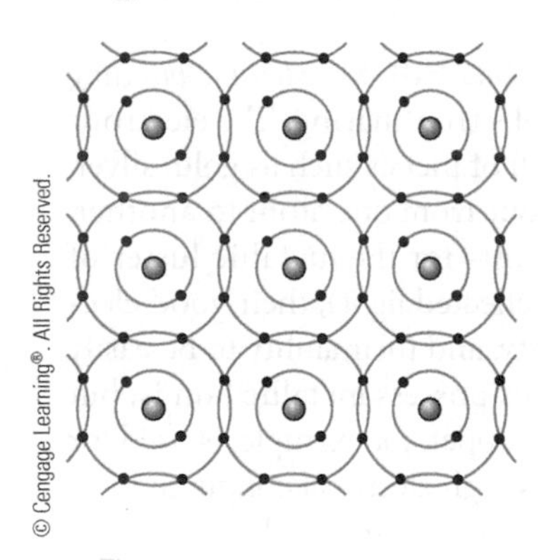

a The orbits in the outermost electron shell overlap, so electrons are shared in diamond.

b Covalent bonding of carbon atoms in diamond forms a three-dimensional framework.

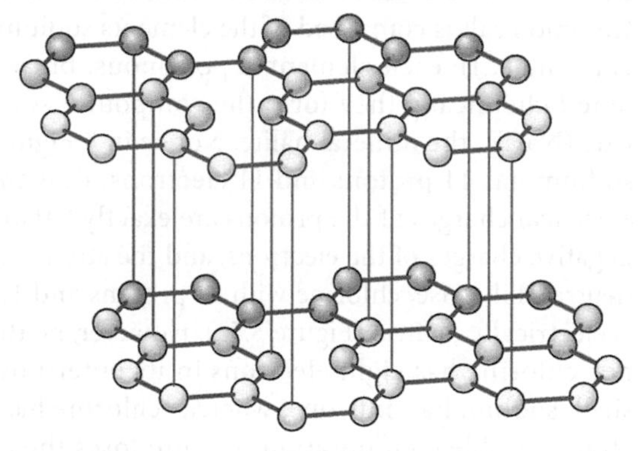

c Covalent bonds in graphite form strong sheets, but the van der Waals bonds between sheets are weak.

3.3 Explore the World of Minerals

We defined a *mineral* as a naturally occurring, inorganic crystalline solid with a narrowly defined chemical composition and characteristic physical properties. In the following sections, we will examine each part of this formal definition.

Naturally Occurring Inorganic Substances

The criterion *naturally occurring* excludes from minerals all manufactured substances such as synthetic diamonds and rubies. This criterion is particularly important to those who buy and sell gemstones, most of which are minerals, because some human-made substances are very difficult to distinguish from natural gem minerals.

Figure 3.7 A Variety of Mineral Crystal Shapes

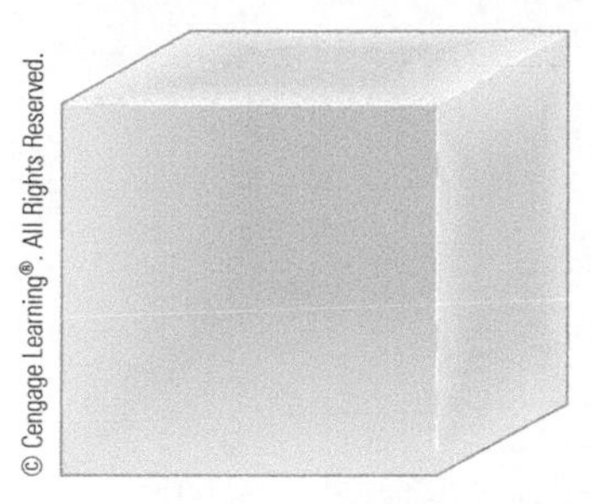

a Cubic crystals are typical of the minerals halite and galena.

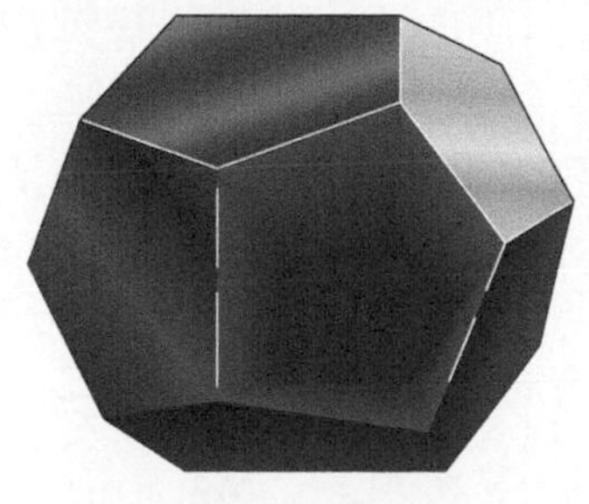

b Pyritohedron crystals such as those of pyrite have 12 sides.

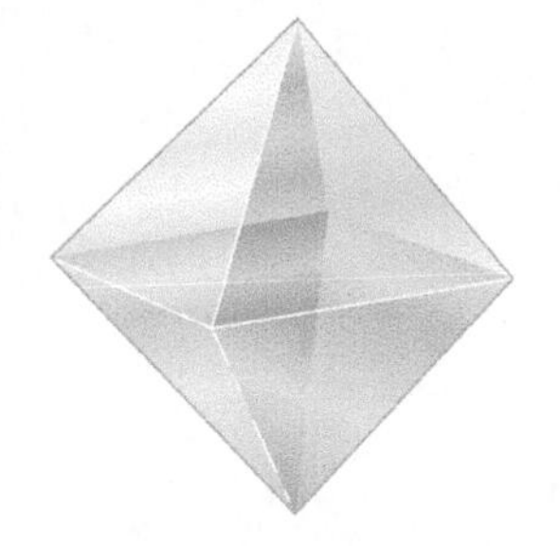

c Diamond has octahedral, or eight-sided, crystals.

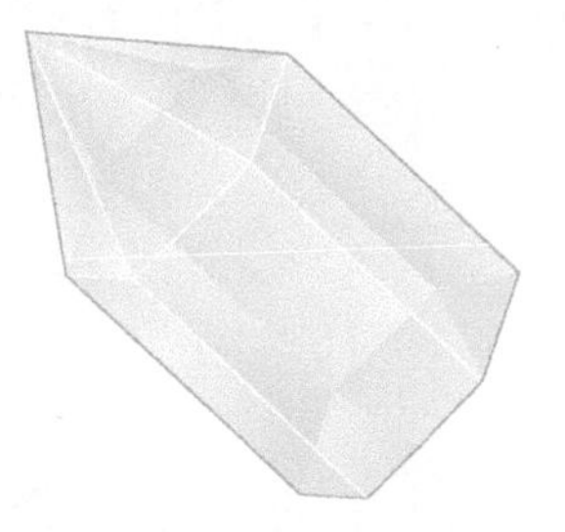

d A prism terminated by a pyramid is found in quartz.

Some geologists think the term *inorganic* in the mineral definition is unnecessary, but it does remind us that animal matter and vegetable matter are not minerals. Nevertheless, some organisms, including corals, clams, and a number of other animals and some plants, construct their shells of calcium carbonate ($CaCO_3$), which is either the mineral aragonite or calcite, or their shells are made of silicon dioxide (SiO_2) as in quartz.

Mineral Crystals

By definition, minerals are **crystalline solids**, in which the constituent atoms are arranged in a regular, three-dimensional framework (Figure 3.5b). Under ideal conditions, such as in a cavity, these crystalline solids can grow and form perfect **crystals** that possess planar surfaces (crystal faces), sharp corners, and straight edges (▌Figure 3.7). In other words, the regular geometric shape of a well-formed mineral crystal is the exterior manifestation of an ordered internal atomic arrangement. Not all rigid substances are crystalline solids; natural and manufactured glass lack the ordered arrangement of atoms and are said to be *amorphous*, meaning "without form." Minerals are crystalline solids, but crystalline solids do not always yield well-formed crystals. The reason is that when crystals form, they may grow in proximity and form an interlocking mosaic in which individual crystals are not apparent or easily discerned (▌Figure 3.8).

So how do we know that the mineral in Figure 3.8b is actually crystalline? X-ray beams and light transmitted through mineral crystals or crystalline solids behave in a predictable manner, which provides compelling evidence for an internal orderly structure. Another way to determine that minerals with no obvious crystals are crystalline is by their *cleavage*, the property of breaking or splitting repeatedly along smooth, closely spaced planes. Not all minerals have cleavage planes, but many do, and such regularity certainly indicates that splitting is controlled by internal structure.

As early as 1669, the Danish scientist Nicholas Steno determined that the angles of intersection of equivalent crystal faces on different specimens of quartz are identical. Since then, this *constancy of interfacial angles* has been demonstrated for many other minerals, regardless of their size, shape, age, or geographic occurrence (▌Figure 3.9). Steno postulated that mineral crystals are made up of very small, identical building blocks, and that the arrangement of these building blocks determines the external form of mineral crystals, a proposal that has since been verified.

Figure 3.8 Quartz and the Constancy of Interfacial Angles

Sue Monroe

a Well-shaped crystal of smoky quartz.

Harry Taylor/Dorling Kindersley/Getty Images

b Specimen of rose quartz in which no obvious crystals can be discerned.

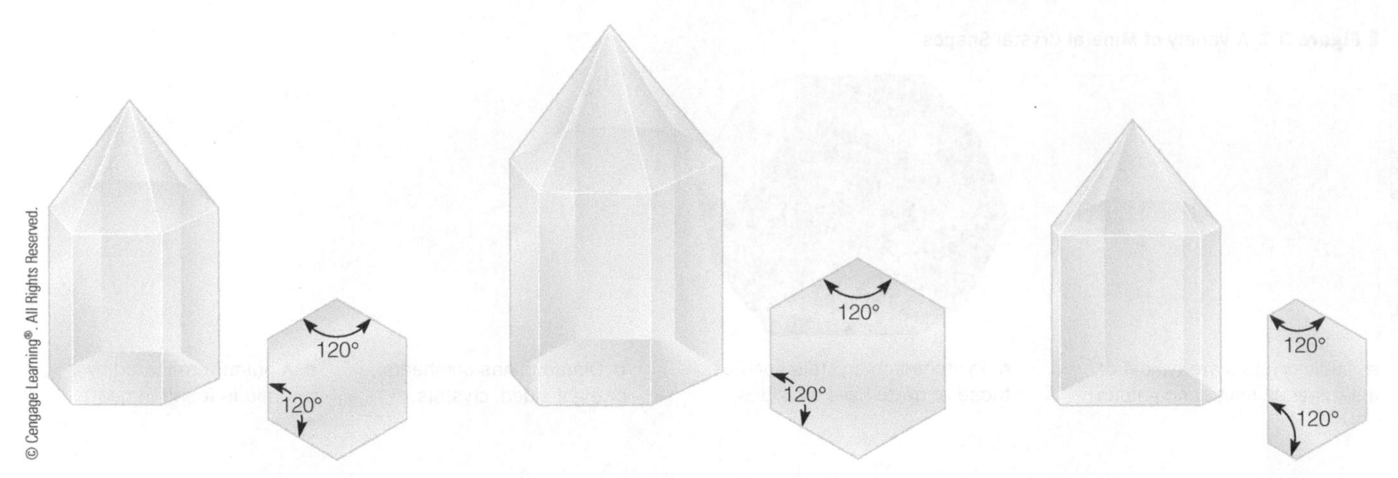

Figure 3.9 Constancy of Interfacial Angles Side views and cross sections of quartz crystals showing the constancy of interfacial angles. A well-shaped crystal (left), a larger well-shaped crystal (middle), and a poorly shaped crystal (right). The angles formed between equivalent crystal faces on different specimens of the same mineral are the same regardless of size, shape, age, or geographic occurrence of the specimens.

Chemical Composition of Minerals

A mineral's chemical composition is shown by a formula, which is a shorthand way of indicating the numbers of atoms of different elements. Quartz consists of one silicon (Si) atom for every two oxygen (O) atoms and thus has the formula SiO_2; the subscript number indicates the number of atoms. Orthoclase is composed of one potassium, one aluminum, three silicon, and eight oxygen atoms, so its formula is $KAlSi_3O_8$. Some minerals known as *native elements* consist of a single element and include silver (Ag), platinum (Pt), gold (Au), and graphite and diamond, both of which are composed of carbon (C).

For many minerals, the chemical composition does not vary. Quartz is made up only of silicon and oxygen (SiO_2), and halite contains only sodium and chlorine (NaCl). However, some minerals have a range of compositions because one element can substitute for another if the atoms of two or more elements are nearly the same size and the same charge (Figure 3.10).

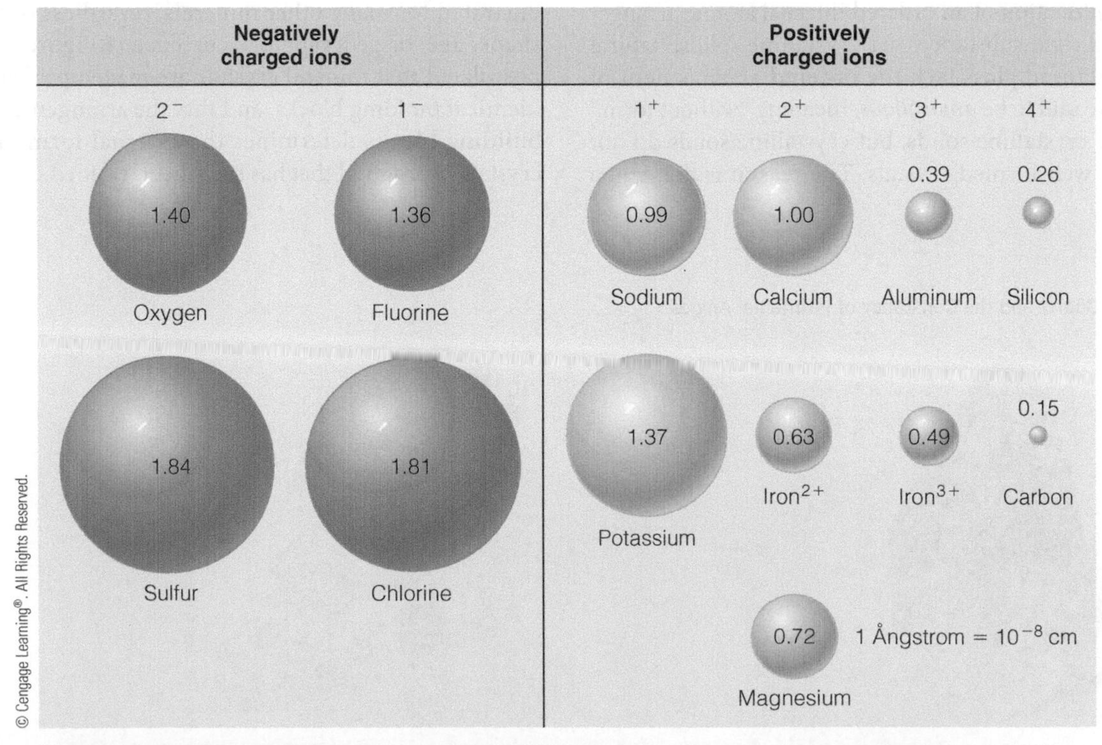

Figure 3.10 Sizes and Charges of Ions Electrical charges and relative sizes of ions common in minerals. The numbers within the ions are the radii shown in Ångstrom units.

Iron and magnesium meet these criteria and can substitute for one another. The chemical formula for olivine is $(Mg,Fe)_2SiO_4$, meaning that in addition to silicon and oxygen, it may contain magnesium, iron, or both. A number of other minerals also have ranges of composition, so they are actually mineral groups with several members. Thus, the definition of the term *mineral* has the phrase *narrowly defined chemical composition.*

Physical Properties of Minerals

The last criterion in our definition of a mineral, *characteristic physical properties,* refers to such properties as hardness, color, and crystal form. These properties are controlled by composition and structure. We will have more to say about the physical properties of minerals in the section titled Mineral Identification.

3.4 Mineral Groups Recognized by Geologists

According to the Mineralogical Society of America, geologists have identified and described about 3,800 minerals, but only a few—perhaps two dozen—are common. One might think that an extremely large number of minerals could form from 92 naturally occurring elements; however, the bulk of Earth's crust is made up of only eight chemical elements, and even among these, silicon and oxygen are by far the most common. In fact, most common minerals in Earth's crust consist of silicon, oxygen, and one or more of the elements in ❚ Figure 3.11.

Geologists recognize mineral classes or groups, each with members that share the same negatively charged ion or ion group (Table 3.1). We have mentioned that ions are atoms that have either a positive or negative electrical charge resulting from the loss or gain of electrons in their outermost shell. In addition to ions, some minerals contain tightly bonded, complex groups of different atoms known as *radicals* that act as single units. A good example is the carbonate radical, consisting of a carbon atom bonded to three oxygen atoms and thus having the formula CO_3 and a -2 electrical charge. Other common radicals and their charges are sulfate (SO_4, -2), hydroxyl (OH, -1), and silicate (SiO_4, -4) (❚ Figure 3.12).

Silicate Minerals

Because silicon and oxygen are so abundant in Earth's crust, it is not surprising that many minerals contain these elements. A combination of silicon and oxygen is known as **silica,** and minerals that contain silica are **silicates.** Quartz (SiO_2) is pure silica because it is composed entirely of silicon and oxygen. But most silicates have one or more additional elements, as in orthoclase ($KAlSi_3O_8$) and olivine [$(Mg,Fe)_2SiO_4$]. Silicate minerals include about one-third of all known minerals, but their abundance is even more impressive when one considers that they make up perhaps 90% of Earth's crust.

The basic building block of all silicate minerals is the **silica tetrahedron,** consisting of one silicon atom and four oxygen atoms (❚ Figure 3.13a). These atoms are arranged so that the four oxygen atoms surround a silicon atom, which occupies the space between the oxygen atoms, thus forming a four-faced pyramidal structure (❚ Figure 3.13b). The silicon atom has a positive charge of 4, and each of the four oxygen atoms has a negative charge of 2, resulting in a radical with a total negative charge of 4 $(SiO_4)^{-4}$.

Because the silica tetrahedron has a negative charge, it does not exist in nature as an isolated ion group; rather, it combines with positively charged ions or shares its oxygen atoms with other silica tetrahedra. In the simplest silicate minerals, the silica tetrahedra exist as single units bonded to positively charged ions. In minerals that contain isolated tetrahedra, the

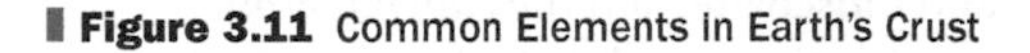

❚ **Figure 3.11** Common Elements in Earth's Crust

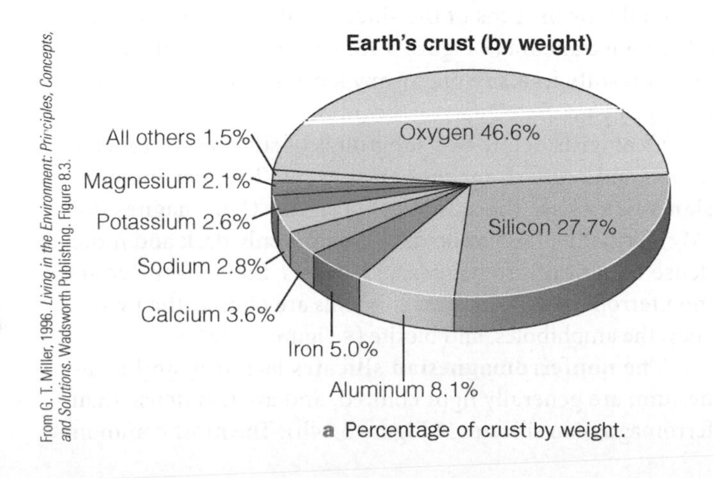

a Percentage of crust by weight.

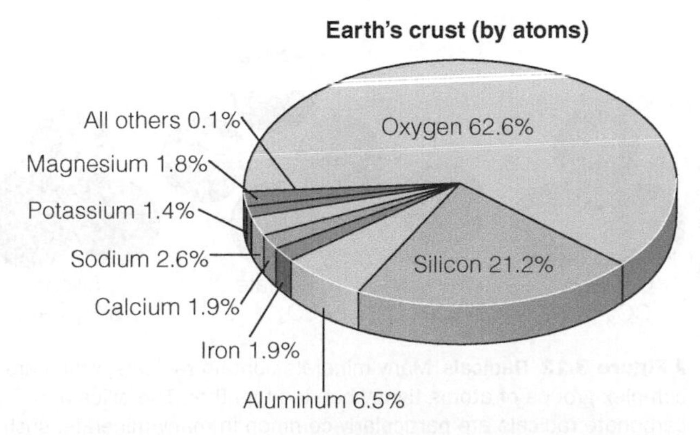

b Percentage of crust by atoms.

From G. T. Miller, 1996. *Living in the Environment: Principles, Concepts, and Solutions.* Wadsworth Publishing. Figure 8.3.

TABLE 3.1 Mineral Groups Recognized by Geologists

Mineral Group	Negatively Charged Ion or Radical	Examples	Composition
Carbonate	$(CO_3)^{-2}$	Calcite	$CaCO_3$
		Dolomite	$CaMg(CO_3)_2$
Halide	Cl^{-1}, F^{-1}	Halite	$NaCl$
		Fluorite	CaF_2
Hydroxide	$(OH)^{-1}$	Brucite	$Mg(OH)_2$
Native element	—	Gold	Au
		Silver	Ag*
		Diamond	C
Phosphate	$(PO_4)^{-3}$	Apatite	$Ca_5(PO_4)_3(F,Cl)$
Oxide	O^{-2}	Hematite	Fe_2O_3
		Magnetite	Fe_3O_4
Silicate	$(SiO_4)^{-4}$	Quartz	SiO_2
		Potassium feldspar	$KAlSi_3O_8$
		Olivine	$(Mg,Fe)_2SiO_4$
Sulfate	$(SO_4)^{-2}$	Anhydrite	$CaSO_4$
		Gypsum	$CaSO_4 \cdot 2H_2O$
Sulfide	S^{-2}	Galena	PbS
		Pyrite	FeS_2
		Argentite	Ag_2S*

*Note that silver is found as both a native element and a sulfide mineral.

silicon-to-oxygen ratio is 1:4, and the negative charge of the silica ion is balanced by positive ions (Figure 3.13c). Olivine [$(Mg,Fe)_2SiO_4$], for example, has either two magnesium (Mg^{+2}) ions, two iron (Fe^{+2}) ions, or one of each to offset the -4 charge of the silica ion.

Silica tetrahedra may also join together to form chains of indefinite length (Figure 3.13d). Single chains, as in the pyroxene minerals, form when each tetrahedron shares two of its oxygens with an adjacent tetrahedron, resulting in a silicon-to-oxygen ratio of 1:3. Enstatite, a pyroxene-group mineral, reflects this ratio in its chemical formula $MgSiO_3$. Individual chains, however, possess a net -2 electrical charge, so they are balanced by positive ions, such as Mg^{+2}, that link parallel chains together (Figure 3.13d).

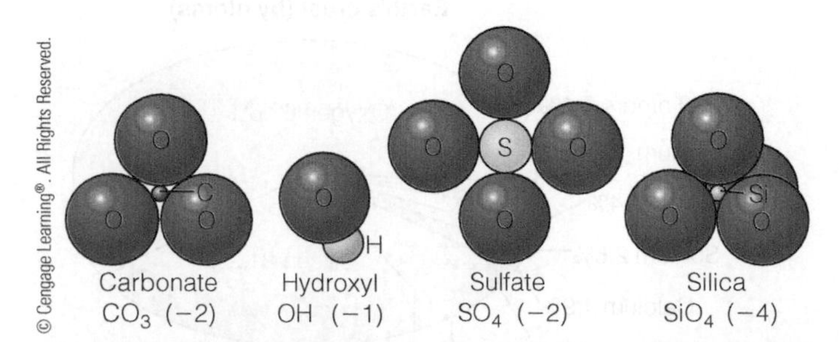

Figure 3.12 Radicals Many minerals contain radicals, which are complex groups of atoms tightly bonded together. The silica and carbonate radicals are particularly common in many minerals, such as quartz (SiO_2) and calcite ($CaCO_3$).

The amphibole group of minerals is characterized by a double-chain structure in which alternate tetrahedra in two parallel rows are cross-linked (Figure 3.13d). The formation of double chains results in a silicon-to-oxygen ratio of 4:11, so each double chain possesses a -6 electrical charge. Mg^{+2}, Fe^{+2}, and Al^{+3} are usually involved in linking the double chains together.

In sheet-structure silicates, three oxygens of each tetrahedron are shared by adjacent tetrahedra (Figure 3.13e). Such structures result in continuous sheets of silica tetrahedra with silicon-to-oxygen ratios of 2:5. Continuous sheets also possess a negative electrical charge satisfied by positive ions located between the sheets. This particular structure accounts for the characteristic sheet structure of the *micas,* such as biotite and muscovite, and the *clay minerals.*

Three-dimensional networks of silica tetrahedra form when all four oxygens of the silica tetrahedra are shared by adjacent tetrahedra (Figure 3.13f). Such sharing of oxygen atoms results in a silicon-to-oxygen ratio of 1:2, which is electrically neutral. Quartz is a common framework silicate.

Geologists define two subgroups of silicates: ferromagnesian and nonferromagnesian silicates. The **ferromagnesian silicates** are those that contain iron (Fe), magnesium (Mg), or both. These minerals are commonly dark and more dense than nonferromagnesian silicates. Some of the common ferromagnesian silicate minerals are olivine, the pyroxenes, the amphiboles, and biotite (Figure 3.14a).

The **nonferromagnesian silicates** lack iron and magnesium, are generally light colored, and are less dense than ferromagnesian silicates (Figure 3.14b). The most common

Figure 3.13 The Silica Tetrahedron and Silicate Materials

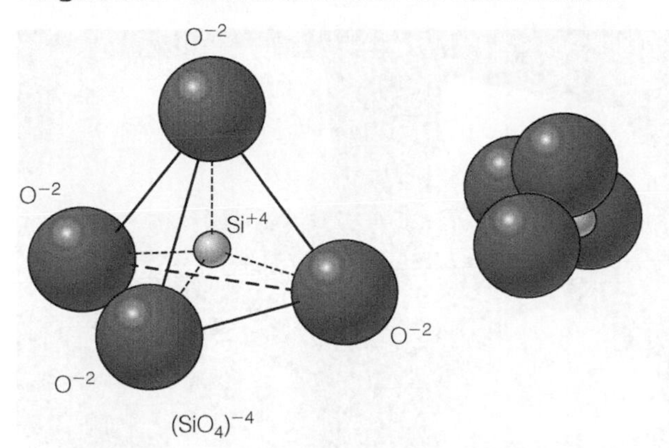

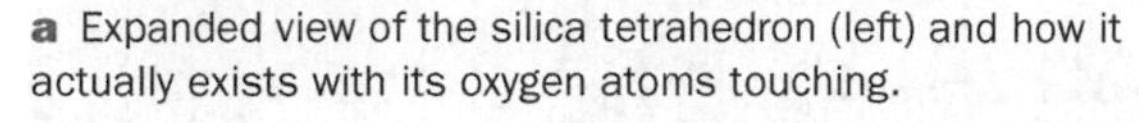

a Expanded view of the silica tetrahedron (left) and how it actually exists with its oxygen atoms touching.

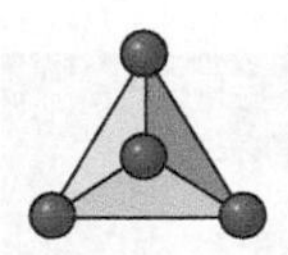

b View of the silica tetrahedron from above. Only the oxygen atoms are visible.

			Formula of negatively charged ion group		Example
c	Isolated tetrahedra		$(SiO_4)^{-4}$	No oxygen atoms shared	Olivine
d	Continuous chains of tetrahedra	Single chain	$(SiO_3)^{-2}$	Each tetrahedra shares two oxygen atoms with adjacent tetrahedra	Pyroxene group (augite)
		Double chain	$(Si_4O_{11})^{-6}$	Single chains linked by sharing oxygen atoms	Amphibole group (hornblende)
e	Continuous sheets		$(Si_4O_{10})^{-4}$	Three oxygen atoms shared with adjacent tetrahedra	Micas (muscovite)
f	Three-dimensional networks	SiO_2	$(SiO_2)^0$	All four oxygen atoms in tetrahedra shared	Quartz Potassium feldspars Plagioclase feldspars

c–f Structures of the common silicate minerals shown by various arrangements of the silica tetrahedra.

minerals in Earth's crust are nonferromagnesian silicates known as *feldspars*. Feldspar is a general name, however, and includes two distinct groups, each of which consists of several species. The *potassium feldspars* are represented by microcline and orthoclase ($KAlSi_3O_8$). The second group of feldspars, the *plagioclase feldspars,* range from calcium-rich ($CaAl_2Si_2O_8$) to sodium-rich ($NaAlSi_3O_8$) varieties.

Quartz (SiO_2) is another common nonferromagnesian silicate. It is a framework silicate that can usually be recognized by its glassy appearance and hardness. Another fairly common nonferromagnesian silicate is muscovite, which is a mica (Figure 3.14b).

Carbonate Minerals

Carbonate minerals, are those that contain the negatively charged carbonate radical $(CO_3)^{-2}$ and include calcium carbonate ($CaCO_3$) as the minerals *aragonite* or *calcite* (Figure 3.15a). Aragonite is unstable and commonly changes to calcite, the main constituent of the sedimentary rock *limestone*. Several other carbonate minerals are

To learn more about limestone and dolostone, see Chapter 7.

Figure 3.14 Common Rock-Forming Silicate Minerals

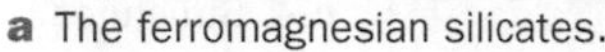
a The ferromagnesian silicates.

b The nonferromagnesian silicates.

known, but only one of these need concern us: *Dolomite* [$CaMg(CO_3)_2$] forms by the chemical alteration of calcite by the addition of magnesium. Sedimentary rock composed of the mineral dolomite is *dolostone.*

Other Mineral Groups

Even though minerals from the other groups in Table 3.1 are less common, many are found in rocks in small quantities and others are important resources. In the oxides, an element combines with oxygen as in hematite (Fe_2O_3) and magnetite (Fe_3O_4). Rocks with high concentrations of these minerals in the Lake Superior region of Canada and the United States are sources of iron ores for the manufacture of steel. The related hydroxides form mostly by the chemical alteration of other minerals.

We have noted that the *native elements* are minerals composed of a single element, such as diamond and graphite (C) and the precious metals gold (Au), silver (Ag), and platinum (Pt). Some elements, such as silver and copper, are found both as native elements and as compounds and are thus also included in other mineral groups; argentite (Ag_2S), a silver sulfide, is an example (Table 3.1).

Several minerals and rocks that contain the phosphate radical $(PO_4)^{-3}$ are important sources of phosphorus for fertilizers. The sulfides, such as galena (PbS), the ore of lead, have a positively charged ion combined with sulfur (S^{-2}) (Figure 3.15b),

Figure 3.15 Representative Specimens from Four Mineral Groups

Critical Thinking Question Which of these minerals has a metallic luster and nonmetallic luster?

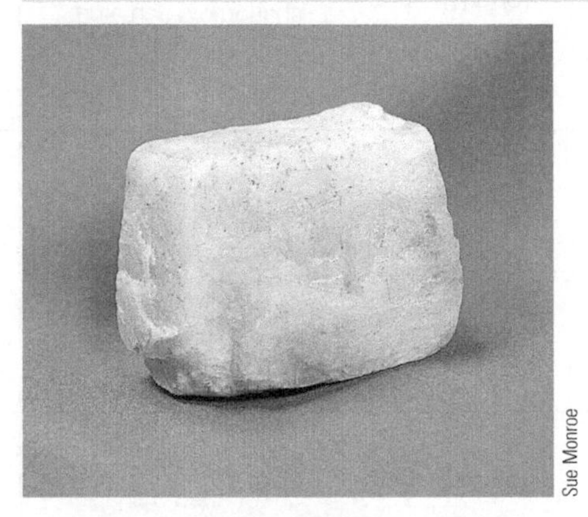

a Calcite ($CaCO_3$) is the most common carbonate mineral.

b The sulfide mineral galena (PbS) is the ore of lead.

c Gypsum ($CaSO_4{\cdot}2H_2O$) is a common sulfate mineral.

d Halite (NaCl) is a good example of a halide mineral.

whereas the sulfates have an element combined with the complex radical $(SO_4)^{-2}$, as in gypsum $(CaSO_4 \cdot 2H_2O)$ (▮ Figure 3.15c). The halides contain the halogen elements, fluorine (F^{-1}) and chlorine (Cl^{-1}); examples are halite (NaCl) (▮ Figure 3.15d) and fluorite (CaF_2).

3.5 Mineral Identification

Many physical properties of minerals are remarkably constant for a given mineral species, but some, especially color, may vary (Figure 3.8). Although professional geologists use sophisticated techniques to study and identify minerals, you can identify most common minerals by using the physical properties described next.

Luster and Color

Luster (not to be confused with *color*) is the quality and intensity of light reflected from a mineral's surface. Geologists define two main types of luster: *metallic,* having the appearance of a metal, and *nonmetallic.* Notice that of the four minerals shown in Figure 3.15, only galena has a metallic luster. Among the several types of nonmetallic luster are glassy or vitreous (as in quartz), dull or earthy, waxy, greasy, and brilliant (as in diamond).

Beginning geology students are distressed by the fact that the color of some minerals varies, making the most obvious physical property of little use for identifying some minerals. In any case, we can make some helpful generalizations about color. Ferromagnesian silicates are typically black, brown, or dark green, although olivine is olive green (Figure 3.14a). Nonferromagnesian silicates, on the other hand, vary in color, but are rarely very dark. White, cream, colorless, and shades of pink and pale green are typical (Figure 3.14b).

Another helpful generalization is that the color of minerals with a metallic luster is more consistent than is the color of nonmetallic minerals. For example, galena is always lead-gray (Figure 3.15b) and pyrite is invariably brassy yellow. In contrast, quartz, a nonmetallic mineral, may be colorless, smoky brown to almost black, rose, yellow-brown, milky white, blue, or violet to purple (Figure 3.8).

Crystal Form

As we noted, many mineral specimens do not show the perfect crystal form typical of that mineral species. Nevertheless, some minerals do commonly occur as crystals. For example, 12-sided crystals of garnet are common, as are 6- and 12-sided crystals of pyrite. Minerals that grow in cavities or are precipitated from circulating hot water (hydrothermal solutions) in cracks and crevices in rocks also commonly occur as crystals (▮ Figure 3.16) and, under ideal conditions, some minerals grow to incredible sizes (see Chapter 13 opening photo).

Crystal form is a useful characteristic for mineral identification, but a number of minerals have the same crystal form. Pyrite (FeS_2), galena (PbS), and halite (NaCl) all occur as cubic crystals, but they can be easily identified by other properties, such as color, luster, hardness, and density.

Cleavage and Fracture

Not all minerals possess **cleavage,** but those that do break, or split, along smooth planes of weakness determined by the strength of their chemical bonds. Cleavage is characterized in terms of quality (perfect, good, poor), direction, and angles of intersection of cleavage planes. Biotite, a common ferromagnesian silicate, has perfect cleavage

Sue Monroe

▮ **Figure 3.16** Crystals of Smoky Quartz The cluster of well formed quartz crystals measures about 15 cm across. Much of the quartz you will ever see is colorless or milky white, but this variety, called smoky quartz, is brown to black. Regardless of its color, all quartz is composed of silicon and oxygen (SiO_2). Most mineral crystals are not very large, but under some conditions they can grow to several meters long (see Chapter 13).

Critical Thinking Question Refer to the rock cycle in Figure 1.15 and explain at what points in the cycle you would expect minerals to form.

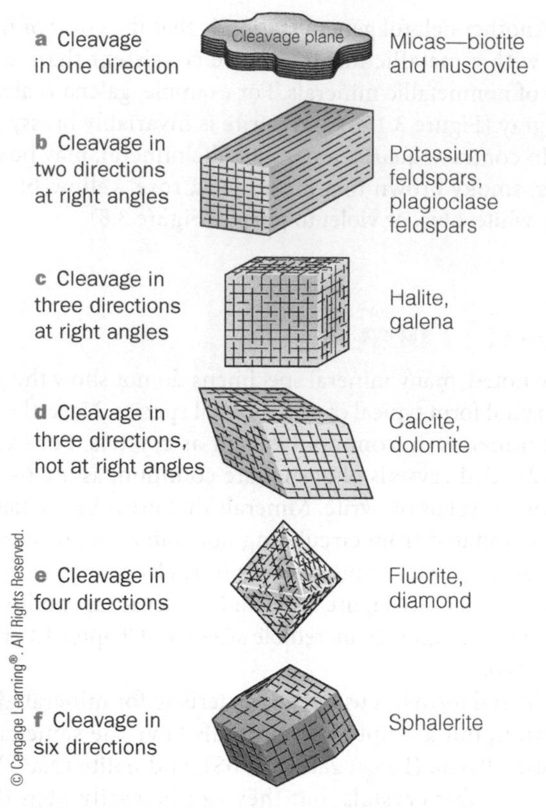

Figure 3.17 Several Types of Mineral Cleavage

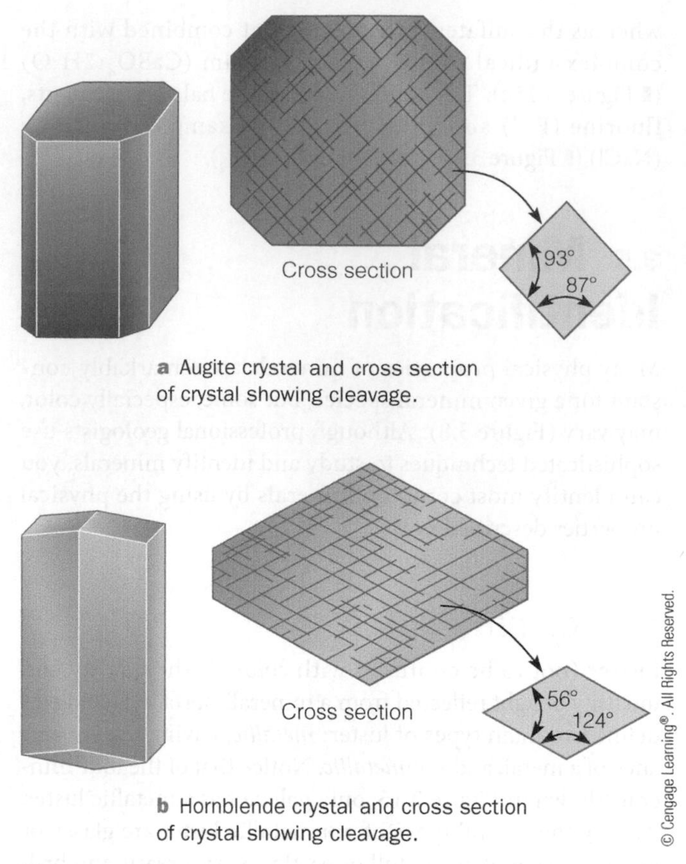

Figure 3.18 Cleavages in Augite and Hornblende

in one direction (Figure 3.17a). Biotite is a sheet silicate with the sheets of silica tetrahedra weakly bonded to one another by iron and magnesium ions.

Feldspars possess two directions of cleavage that intersect at right angles (Figure 3.17b), and the mineral halite has three directions of cleavage, all of which intersect at right angles (Figure 3.17c). Calcite also possesses three directions of cleavage, but none of the intersection angles is a right angle, so cleavage fragments of calcite are rhombohedrons (Figure 3.17d). Minerals with four directions of cleavage include fluorite and diamond (Figure 3.17e). Ironically, diamond, the hardest mineral, can be cleaved easily. A few minerals, such as sphalerite, an ore of zinc, have six directions of cleavage (Figure 3.17f).

Cleavage is an important diagnostic property of minerals, and recognizing it is essential in distinguishing between some minerals. The pyroxene mineral augite and the amphibole mineral hornblende, for example, look much alike: Both are dark green to black, have the same hardness, and possess two directions of cleavage. But the cleavage planes of augite intersect at about 90 degrees, whereas the cleavage planes of hornblende intersect at angles of 56 degrees and 124 degrees (Figure 3.18).

In contrast to cleavage, *fracture* is mineral breakage along irregular surfaces. Any mineral will fracture if enough force is applied, but the fracture surfaces are uneven or conchoidal (curved) rather than smooth.

Hardness

An Austrian geologist, Friedrich Mohs (1773–1839), devised a relative hardness scale for 10 minerals. He arbitrarily assigned a hardness value of 10 to diamond, the hardest mineral known, and lesser values to the other minerals. Relative hardness is easily determined by the use of Mohs hardness scale (Table 3.2). Quartz will scratch fluorite but cannot be scratched by fluorite, gypsum can be scratched by a fingernail, and so on. So **hardness** is defined as a mineral's resistance to abrasion and is controlled mostly by internal structure. For example, both graphite and diamond are composed of carbon, but the former has a hardness of 1 to 2, whereas the latter has a hardness of 10.

Specific Gravity (Density)

Specific gravity and density are two separate concepts, but here we will use them more or less as synonyms. A mineral's **specific gravity** is the ratio of its weight to the weight of an equal volume of pure water at 4°C. Thus, a mineral with a specific gravity of 3.0 is three times as heavy as water. **Density,** in contrast, is a mineral's mass (weight) per unit of volume expressed in grams per cubic centimeter. So the specific gravity of galena (Figure 3.15b) is 7.58 and its density is 7.58 g/cm^3. In most instances, we will refer to a mineral's density, and in some of the following chapters, we will mention the density of various rocks.

TABLE 3.2 Mohs Hardness Scale

Hardness	Mineral	Hardness of Some Common Objects
10	Diamond	
9	Corundum	
8	Topaz	
7	Quartz	
		Steel file (6½)
6	Orthoclase	
		Glass (5½–6)
5	Apatite	
4	Fluorite	
3	Calcite	Copper penny (3)
		Fingernail (2½)
2	Gypsum	
1	Talc	

Structure and composition control a mineral's specific gravity and density. Because ferromagnesian silicates contain iron, magnesium, or both, they tend to be denser than nonferromagnesian silicates. In general, the metallic minerals, such as galena and hematite, are denser than nonmetals. Pure gold with a density of 19.3 g/cm^3 is about 2.5 times as dense as galena, "the ore of lead." Diamond and graphite, both of which are composed of carbon (C), illustrate how structure controls specific gravity or density. The specific gravity of diamond is 3.5, whereas that of graphite varies from 2.09 to 2.33.

Other Useful Mineral Properties

Talc has a distinctive soapy feel, graphite writes on paper, halite tastes salty, and magnetite is magnetic. Calcite possesses the property of *double refraction,* meaning that an object when viewed through a transparent piece of calcite will have a double image. Some sheet silicates are plastic and, when bent into a new shape, will retain that shape; others are flexible and, if bent, will return to their original position when the forces that bent them are removed.

A simple chemical test to identify the minerals calcite and dolomite involves applying a drop of dilute hydrochloric acid to the mineral specimen. If the mineral is calcite, it will react vigorously with the acid and release carbon dioxide, which causes the acid to bubble or effervesce. Dolomite, in contrast, will not react with hydrochloric acid unless it is powdered.

3.6 Rock-Forming Minerals

You probably have some idea of what the term *rock* means, but specifically it refers to a solid aggregate of minerals, although it also includes masses of mineral-like matter as in the natural glass obsidian, and masses of altered organic matter as in coal. Neither obsidian nor coal is made up of minerals, but they are found in or associated with other rocks that are composed of minerals, hence their inclusion with rocks. Many minerals are found in rocks, but only a few, designated **rock-forming minerals**, are sufficiently common for rock identification and classification (Table 3.3, and

TABLE 3.3 Important Rock-Forming Minerals

Mineral	Primary Occurrence
Ferromagnesian silicates	
Olivine	Igneous and metamorphic rocks
Pyroxene group Augite most common	Igneous and metamorphic rocks
Amphibole group Hornblende most common	Igneous and metamorphic rocks
Biotite	All rock types
Nonferromagnesian silicates	
Quartz	All rock types
Potassium feldspar group Orthoclase, microcline	All rock types
Plagioclase feldspar group	All rock types
Muscovite	All rock types
Clay mineral group	Soils, sedimentary rocks, and some metamorphic rocks
Carbonates	
Calcite	Sedimentary rocks
Dolomite	Sedimentary rocks
Sulfates	
Anhydrite	Sedimentary rocks
Gypsum	Sedimentary rocks
Halides	
Halite	Sedimentary rocks

Figure 3.19 Rock-Forming Minerals in the Igneous Rock Basalt In Figure 3.1b, we showed the rock-forming minerals in granite, but in that rock the minerals are easily seen with the unaided eye. For basalt, though, the minerals are too small to be seen without magnification.

J.D. Griggs/U.S. Geological Survey

a Basalt forms when lava cools and crystallizes thereby forming minerals.

Sue Monroe

b A specimen of basalt measuring about 4 cm across.

G.J. Taylor

c Magnified view of basalt showing its minerals, none of which exceed about 0.5 mm long. The grey minerals are plagioclase feldspar, the bluish green to green minerals near the top are olivine, and the rest is mostly calcium-rich pyroxene.

Figures 3.1b and 3.19). Others known as *accessory minerals* are commonly present but in such small quantities that they can be disregarded.

We already noted that silicate minerals make up perhaps 90% of Earth's crust, so it follows that most rocks are composed of silicates. Indeed, feldspar minerals (potassium feldspars and plagioclase feldspars) alone make up more than 60% of the crust, whereas about 12% consists of quartz. So, even though there are about 2,500 known silicate minerals, only a very few of them are particularly common in rocks.

The most common nonsilicate rock-forming minerals are the carbonate minerals calcite ($CaCO_3$) and dolomite [$CaMg(CO_3)_2$], which are main components of the sedimentary rocks limestone and dolostone, respectively. By far the most common rock-forming minerals that you will ever encounter are silicate and carbonate minerals, but a few others are common enough in some localities to qualify. For instance, among the sulfates and halides, gypsum ($CaSO_4 \cdot 2H_2O$) in rock gypsum and halite (NaCl) in rock salt meet these criteria (Table 3.3).

ConnectionLink

To learn more about calcite, gypsum, and halite in sedimentary rocks, see Chapter 7.

3.7 How Do Minerals Form?

Thus far, we have discussed the composition, structure, and physical properties of minerals but have not addressed how they originate. One phenomenon that accounts for the origin of minerals is the cooling of molten rock material known as *magma* (magma that reaches the surface is called *lava*) (Figure 3.19).

As magma or lava cools, minerals crystallize and grow, thereby determining the mineral composition of various igneous rocks, such as basalt (dominated by ferromagnesian silicates), and granite (dominated by nonferromagnesian silicates). Hot-water solutions derived from magma commonly invade cracks and crevasses in adjacent rocks, and from these solutions, several minerals crystallize, some of economic importance. Minerals also originate when water in hot springs cools (see Chapter 13) and when hot, mineral-rich water discharges onto the seafloor at hot springs known as hydrothermal vents (see Figure 2.8).

Dissolved materials in seawater, more rarely in lake water, combine to form minerals such as halite (NaCl), gypsum ($CaSO_4 \cdot 2H_2O$), and several others when the water evaporates. Aragonite and calcite, both varieties of calcium carbonate ($CaCO_3$), might also form from evaporating water, but most originate when organisms such as clams, oysters, corals, and floating microorganisms use this compound to construct their shells. A few plants and animals use silicon dioxide

(SiO_2) for their skeletons, which accumulate as mineral matter on the seafloor when the organisms die (see Chapter 16).

Some clay minerals form when chemical processes compositionally and structurally alter other minerals (see Chapter 6), and others originate when rocks are changed during metamorphism (see Chapter 8). In fact, the agents that cause metamorphism—heat, pressure, and chemically active fluids—are responsible for the origin of many minerals. A few minerals even originate when gases such as hydrogen sulfide (H_2S) and sulfur dioxide (SO_2) react at volcanic vents to produce sulfur.

3.8 Economic Geology

Any aspect of geology concerned with the search for minerals and rocks of economic value is called **economic geology** and might involve exploration for iron ore deposits, petroleum and natural gas, and minerals needed for glass and wallboard. In short, economic geology is concerned with finding natural resources. Geologists at the United States Geological Survey define a **resource** as a naturally occurring concentration of solid, liquid, or gaseous material in or on Earth's crust in such form and amount that economic extraction of a commodity is currently or potentially feasible. In fact, we refer to *metallic resources* (ores of iron, tin, copper, zinc, etc.), *nonmetallic resources,* also called *industrial minerals* (sand and gravel, crushed stone, sulfur, salt, etc.) (Geo-Focus 3.1), and *energy resources* (petroleum, natural gas, coal, and uranium).

All of these are examples of resources, but we must make a distinction between a resource, the total amount of a commodity whether discovered or not—and a **reserve,** which is that part of the resource base that can be economically recovered. In principle, this distinction is simple, but in practice it depends on several factors, not all of which remain constant. Market price is the most obvious variable; consider that the price of gold is now more than $1,400 per ounce, making it profitable to open mines that were closed when the price was lower. Location may play a role, too. Sand and gravel are essential to the construction industry, but supplies must be nearby because shipping them very far is too expensive.

In addition to the better known mineral and rock commodities, economic geology is also concerned with many industrial minerals such as pure quartz sand for glass; gypsum for wallboard; clay minerals for ceramics and paper; feldspars for use in porcelain, ceramics, and enamel; as well as phosphate-rich rocks needed for fertilizers and animal feed supplements; and many others. In 2012, the extraction of nonfuel natural resources in the United States totaled more than $76.5 billion (Table 3.4).

Access to natural resources is essential to all societies but especially to industrial societies with their high standards of living. However, most natural resources are *nonrenewable,* meaning that a limited supply exists and they cannot be replenished by natural processes as fast as they are depleted. For some essential resources, the United States is totally dependent on imports, some from politically unstable parts of the world (■ Figure 3.20). During 2011, the United States imported 75% of the cobalt it needed for gas-turbine aircraft engines, magnets, and corrosion-resistant alloys. Manganese is necessary for making steel, and all that is used is imported. Which brings us to the concept of *strategic minerals,* which are those minerals not produced in sufficient quantities to meet our needs for national defense, industry, and other needs during a national emergency. Accordingly, large quantities of bauxite (aluminum ore), manganese, chromium, cobalt, and many others are stockpiled.

As of May 2012, 51.5% of all petroleum used in this country was imported, more than half of it from Venezuela, Canada, and Mexico, although about 22% came from nations in the

TABLE 3.4 The 10 Leading U.S. States in Nonfuel Mineral Production for 2011

State	$	Most Important Commodities in Order of Value
1. Nevada	10,400,000,000	Gold, copper, silver, lime, sand and gravel
2. Arizona	8,250,000,000	Copper, molybdenum, sand and gravel, silver, cement
3. Minnesota	5,120,000,000	Iron ore, sand and gravel, stone,[1] lime
4. Utah	4,570,000,000	Copper, molybdenum, gold, potash, magnesium
5. Alaska	3,790,000,000	Zinc, gold, silver, lead, sand and gravel
6. Florida	3,270,000,000	Phosphate rock, stone, cement, sand and gravel
7. California	2,870,000,000	Boron minerals, sand and gravel, cement, gold, stone
8. Texas	2,810,000,000	Cement, stone, salt, sand and gravel
9. Michigan	2,470,000,000	Iron ore, cement, sand and gravel, salt, stone
10. Missouri	2,140,000,000	Cement, stone, lead, lime, sand and gravel
All states	74,000,000,000[2]	Stone, sand and gravel, gold, copper, cement, phosphate rock, iron ore

[1]Most stone is crushed stone for construction. Dimension stone cut or shaped for specific purposes constitutes a minor part of all stone mined.
[2]Slightly more than 50% of the total value of mineral production was from metals.
Source: USGS, Mineral Commodity Summaries 2012.

GEO FOCUS 3.1 The Industrial Minerals

Industrial societies depend on finding, extracting, processing, and using many minerals and rocks, not the least of which are ores of metals, such as iron, lead, tin, and zinc, as well as minerals and/or rocks used as energy sources. Here we are concerned with the *industrial minerals* that are necessary for some aspect of industrial production, but more specifically they are minerals that are extracted for their commercial value with the exception of those used to produce metals or energy. Accordingly, uranium-rich minerals and hematite, an ore of iron, are not included among the industrial minerals. Many minerals meet the criteria that define an industrial mineral, but here we discuss only a few of them—quartz, gypsum, muscovite, and clay minerals.

One of the most common minerals in Earth's crust is quartz (SiO_2) (Figure 3.14b). Indeed, most sand on beaches, in desert dunes, and in stream and river channels is made up of quartz, although there are a few notable exceptions. Quartz is essential in the production of glass, and optical instruments, and it is also used for abrasives, and some varieties of quartz are semiprecious gemstones. For example, amethyst (purple quartz), rose quartz, smoky quartz (Figure 3.8), and quartz in geodes (Figure 1) are used for decorative "stones" and jewelry, and rhinestones are simply small colorless quartz crystals, although most of it is now artificially produced.

Your watch likely has a thin wafer of a quartz crystal that expands and contracts about 100,000 times per second in response to an electrical current from the watch's battery. This so-called piezoelectric effect is essential in instruments that measure time, pressure, and acceleration. Quartz clocks were first developed in 1928, and now quartz watches and clocks are commonplace, although the quartz wafer in timepieces is now artificially synthesized.

As noted previously, gypsum ($CaSO_4 \cdot 2H_2O$) is one of the evaporite minerals because it forms when minerals precipitate from evaporating saline waters in the seas or some lakes (Figure 3.15c). The United States Geological Survey reports that in 2011, domestic production of gypsum totaled nearly 9.4 million metric tons valued at about $65.9 million. Nevertheless, 13% of the gypsum consumed in this country is imported, mostly from Canada, Mexico, and Venezuela. About 90% of all gypsum used in the United States goes into the manufacture of wallboard and plaster products, with lesser amounts used in cement and agriculture.

What makes the paint on cars and appliances so lustrous? Why are lipstick, eyeliner, and glitter so attractive? The remarkable substance in these commodities is the mineral muscovite (Figure 2a), which is one of the micas, minerals with a sheet structure that have similar physical properties. There are more than three dozen varieties of micas, but only a few are common in rocks, especially biotite (black mica) and muscovite (colorless, white, or pale red or green mica). Biotite has no commercial value, although geologists use it in potassium-argon dating to assign numerical ages to rocks (see Chapter 17). Muscovite, on the other hand, has several commercial uses.

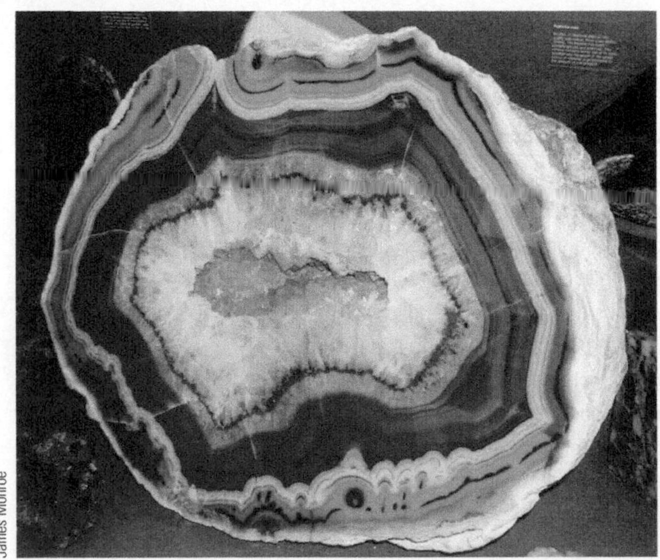

James Monroe

Figure 1 This object is a geode, which forms when minerals grow along the margins of a cavity, partially filling it. In this geode, the outermost layers are thin layers of agate, which is composed of tiny crystals. Next inward in toward the geode's center are quartz crystals that point inward. Because geodes are so attractive, almost any rock shop and many other souvenir shops have geodes that have been sawed in half like this one.

When ground up dry, muscovite loses much of its luster but retains its platy nature and is ideal for wallboard joint compound and as an additive to paint. It is important in wallboard joint compound because it makes the compound smoother and easier to work with and it prevents cracking. In addition, dry ground muscovite is used in plastics, roofing, rubber, and welding rods. If muscovite is ground up wet, it retains its sparkling shine and is used in many cosmetics, including body powder, eye shadow, lipstick, blush, and nail polish (▌Figure 2b). The brilliant, resinous sheen of some paints on automobiles and their changing colors depending on viewing angles depends on muscovite additives to paint (▌Figure 2c).

In Chapter 7 we note that the term *clay* has two meanings: clay sized refers to sedimentary particles measuring less than 1/256 mm across; and clay mineral refers to a group of sheet silicates with similar properties. There are several types of clay minerals, some of which are important geologically as well as commercially. For instance, clay minerals are the main components of some sedimentary rocks (see Chapter 7), they are important in soils (see Chapter 6), and they facilitate some landslides that move on slippery layers of clay (see Chapter 11).

We give little thought to clay, but it is nevertheless an important mineral commodity. You might remember playing with modeling clay, and perhaps you know that it is used in pottery and porcelain. It is also used as a filler in paints and in plastics and rubber. Although processed wood is the main component of paper, the glossy paper in magazines is manufactured with a clay additive. One type of clay is used as drilling mud in drilling oil wells (see Chapter 7), and historically clay has been used in one form or another in construction as in adobe and bricks.

▌**Figure 2** Muscovite

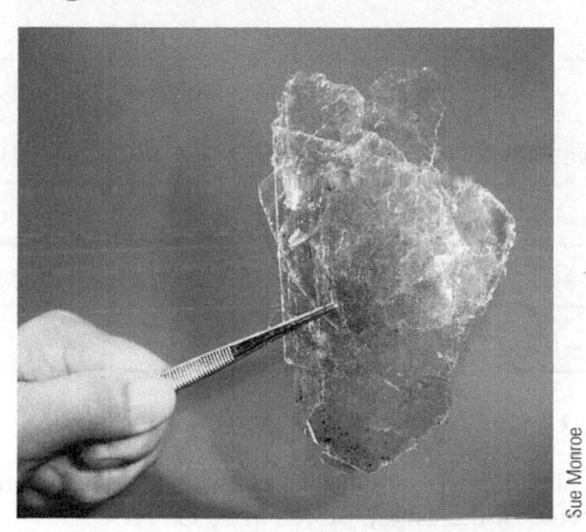

Sue Monroe

a Muscovite mica is colorless, white, pale red, or pale green. It is a sheet silicate that has many industrial uses in paint wallboard compound, eyeliner, lipstick, and nail polish.

Laurence Monneret/The Image Bank/Getty Images

b Muscovite is used in several kinds of makeup as well as glitter.

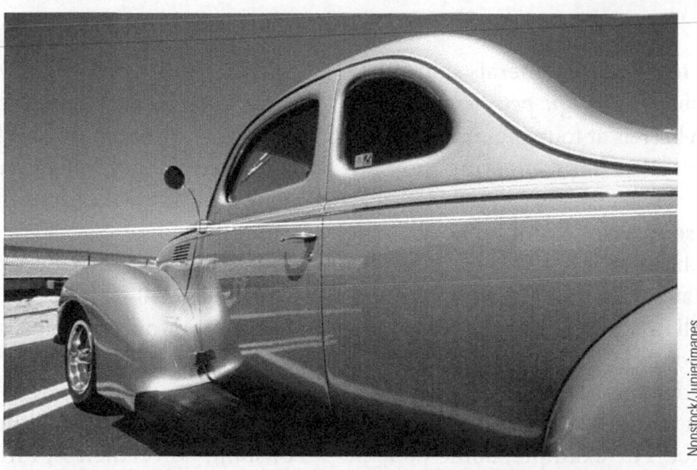

Nonstock/Jupiterimages

c Muscovite in paint on automobiles and appliances gives them their lustrous sheen.

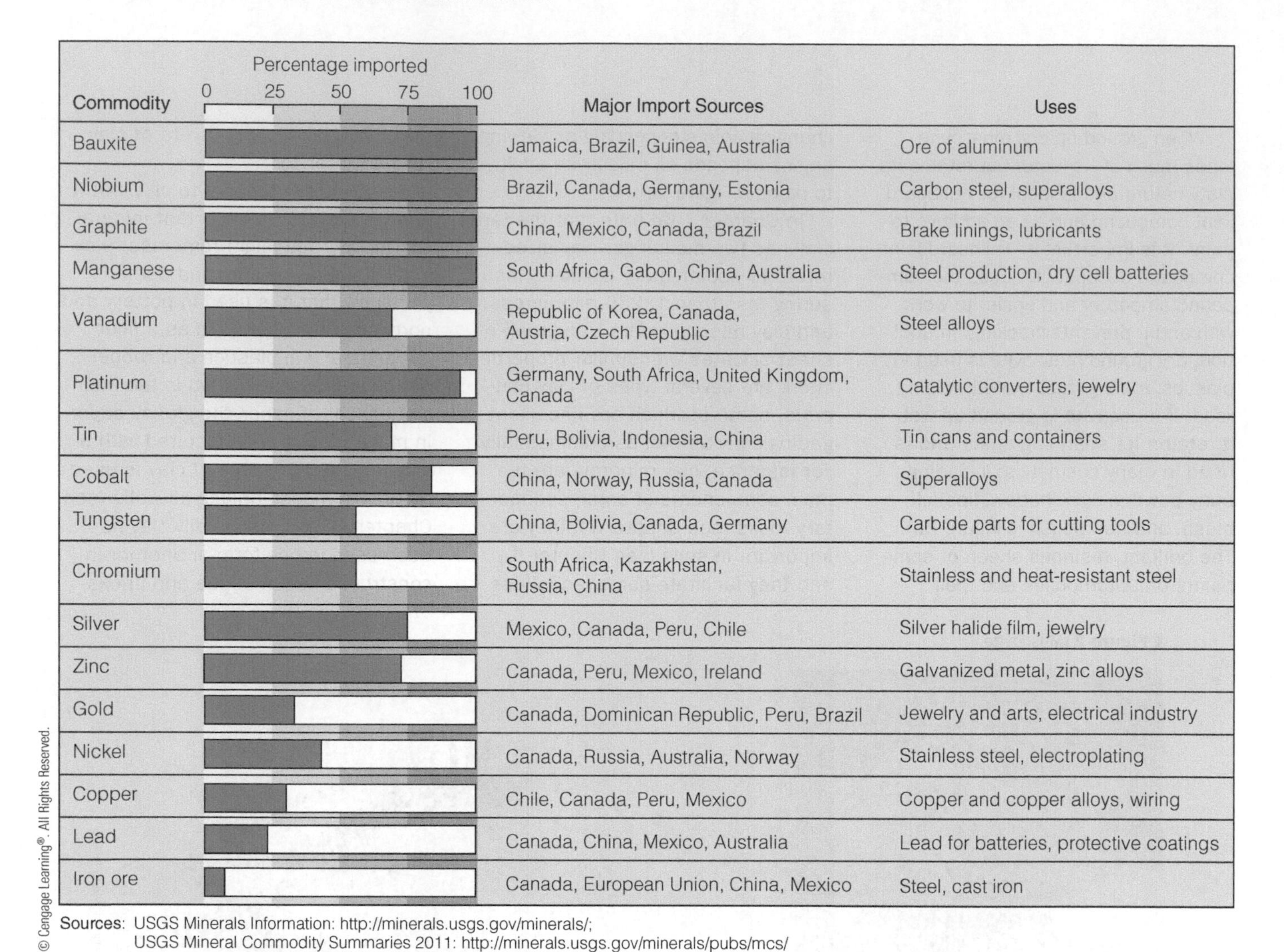

Commodity	Percentage imported	Major Import Sources	Uses
Bauxite		Jamaica, Brazil, Guinea, Australia	Ore of aluminum
Niobium		Brazil, Canada, Germany, Estonia	Carbon steel, superalloys
Graphite		China, Mexico, Canada, Brazil	Brake linings, lubricants
Manganese		South Africa, Gabon, China, Australia	Steel production, dry cell batteries
Vanadium		Republic of Korea, Canada, Austria, Czech Republic	Steel alloys
Platinum		Germany, South Africa, United Kingdom, Canada	Catalytic converters, jewelry
Tin		Peru, Bolivia, Indonesia, China	Tin cans and containers
Cobalt		China, Norway, Russia, Canada	Superalloys
Tungsten		China, Bolivia, Canada, Germany	Carbide parts for cutting tools
Chromium		South Africa, Kazakhstan, Russia, China	Stainless and heat-resistant steel
Silver		Mexico, Canada, Peru, Chile	Silver halide film, jewelry
Zinc		Canada, Peru, Mexico, Ireland	Galvanized metal, zinc alloys
Gold		Canada, Dominican Republic, Peru, Brazil	Jewelry and arts, electrical industry
Nickel		Canada, Russia, Australia, Norway	Stainless steel, electroplating
Copper		Chile, Canada, Peru, Mexico	Copper and copper alloys, wiring
Lead		Canada, China, Mexico, Australia	Lead for batteries, protective coatings
Iron ore		Canada, European Union, China, Mexico	Steel, cast iron

Sources: USGS Minerals Information: http://minerals.usgs.gov/minerals/;
USGS Mineral Commodity Summaries 2011: http://minerals.usgs.gov/minerals/pubs/mcs/

Figure 3.20 Mineral Commodities The dependence of the United States on imports of various mineral commodities is apparent from this chart. The lengths of the green bars correspond to the amounts of the resources imported by the United States in 2011.

Critical Thinking Question Why do you think that the United States and Canada import bauxite (the ore of aluminum) given that both countries have vast exposures of igneous rocks and clay-rich rock from which we can extract aluminum? What factors(s) may change the status of these rocks so that we do extract aluminum from them?

Middle East. So, in addition to strategic minerals stockpiles, the United States has established a *strategic petroleum reserve* with a capacity of 727 million barrels at four sites in Louisiana and Texas. Canada is more self-reliant, meeting most of its domestic mineral and energy needs, but nevertheless it is partly dependent on foreign sources for aluminum ore, chromium, manganese, and phosphate.

Given that the United States uses such large quantities of resources and must import all or some of them, what can be done? Of course one partial solution is to use resources more efficiently and to recycle. Indeed, much of all aluminum is recycled. Another approach is to more fully exploit those resources we already extract. For example, some copper mining companies with newly developed technology have plans to excavate even deeper into Earth and to explore nearby areas for more recoverable resources. And there are alternative sources of some resources. We currently derive iron mostly from deposits on the continents, but there are iron deposits, and other valuable minerals, on the seafloor adjacent to hydrothermal (hot water) vents (see Chapter 2). Whether we can feasibly mine these resources in the near future is another matter.

It is difficult to answer how long our current inventory of reserves will last, because estimates depend on several variables. For one thing it is not easy to determine how much of a resource is potentially available, and such estimates are commonly based on current rates of production and current rates of consumption, neither of which are likely to remain constant. Nevertheless, estimates are made. For example, a recent article in the *Wall Street Journal* (June 5, 2012) gives 18.9 years for gold, 136 years for copper, and 590 years for iron.

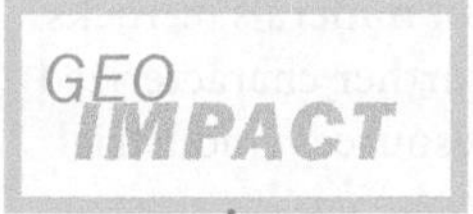

Investing in Natural Resources

Some reputable businesspeople tell you of opportunities to invest in natural resources. Two ventures look promising: a gold mine and a sand and gravel pit. Given that gold sells for more than $1,400 per ounce (as of August 2012), whereas sand and gravel are worth a few dollars per ton, would it be more prudent to invest in the gold mine? Explain not only how market price would influence your decision but also what other factors you might need to consider.

Sand and gravel pit in California.

To ensure continued supplies of mineral and energy resources, scientists, government agencies, and executives in business and industry continually assess the status of resources in view of changing economic and political conditions and changes in science and technology. The United States Geological Survey keeps detailed statistical records of mine production, imports, and exports, and regularly publishes reports on the status of numerous resources in its *Mineral Commodity Summaries.* Similar reports also appear regularly in the *Canadian Minerals Yearbook.*

Key Concepts Review

- Matter is composed of chemical elements, each of which consists of atoms. Protons and neutrons are present in an atom's nucleus, and electrons orbit around the nucleus in electron shells.
- The number of protons in an atom's nucleus determines its atomic number. The atomic mass number is the number of protons plus neutrons in the nucleus.
- Bonding results when atoms join with other atoms; different elements bond to form compounds. With few exceptions, minerals are compounds.
- Ionic and covalent bonds are most common in minerals, but metallic and van der Waals bonds are found in some.
- Minerals are crystalline solids, which means that they possess an ordered internal arrangement of atoms.
- Mineral composition is indicated by a chemical formula, such as SiO_2 for quartz.
- Some minerals have a range of compositions because some elements substitute for one another if their atoms are about the same size and have the same electrical charge.

- About 3,800 minerals are known, and most of them are silicates. The two types of silicates are ferromagnesian and nonferromagnesian.
- In addition to silicates, geologists recognize carbonates, native elements, hydroxides, oxides, phosphates, halides, sulfates, and sulfides.
- Structure and composition control the physical properties of minerals, such as luster, crystal form, hardness, color, cleavage, fracture, and specific gravity.
- Several processes account for the origin of minerals, including cooling magma and lava, weathering, evaporation of seawater, metamorphism, and organisms using dissolved substances in seawater to build their shells.
- A few minerals, designated rock-forming minerals, are common enough in rocks to be essential in their identification and classification. Most rock-forming minerals are silicates, but some carbonates are also common.
- Many resources are concentrations of minerals or rocks of economic importance. They are further characterized as metallic resources, nonmetallic resources (industrial minerals), and energy resources.
- Reserves are that part of the resource base that can be extracted profitably. Distinguishing a resource from a reserve depends on market price, labor costs, geographic location, and developments in science and technology.
- The United States must import many resources to maintain its industrial capacity. Canada is more self-reliant, but it too must import some commodities.

Important Terms

atom (p. 63)
atomic mass number (p. 63)
atomic number (p. 63)
bonding (p. 66)
carbonate mineral (p. 73)
cleavage (p. 75)
compound (p. 66)
covalent bond (p. 67)
crystal (p. 69)
crystalline solid (p. 69)
density (p. 76)
economic geology (p. 79)
electron (p. 63)
electron shell (p. 63)
element (p. 63)
ferromagnesian silicate (p. 72)
hardness (p. 76)
ion (p. 67)
ionic bond (p. 67)
luster (p. 75)
mineral (p. 62)
neutron (p. 63)
nonferromagnesian silicate (p. 72)
nucleus (p. 63)
proton (p. 63)
reserve (p. 79)
resource (p. 79)
rock (p. 62)
rock-forming mineral (p. 77)
silica (p. 71)
silica tetrahedron (p. 71)
silicate (p. 71)
specific gravity (p. 76)

Review Questions

1. By far the most common minerals in Earth's crust are ________ minerals.
   a. ____ silicate
   b. ____ carbonate
   c. ____ oxide
   d. ____ sulfide
   e. ____ sulfate
2. In ionic bonding, electrons
   a. ____ are shared by adjacent atoms.
   b. ____ freely migrate from atom to atom.
   c. ____ change from negative to positive electrical charge.
   d. ____ are transferred from one atom to another.
   e. ____ double in number and size.
3. A naturally occurring solid with its atoms arranged in a specific three-dimensional framework is said to be
   a. ____ amorphous.
   b. ____ crystalline.
   c. ____ covalent.
   d. ____ siliceous.
   e. ____ electronic.
4. Resources are characterized as metallic resources, nonmetallic resources, and ________ resources.
   a. ____ oxide
   b. ____ carbonate
   c. ____ commercial
   d. ____ lithologic
   e. ____ energy

5. The atomic mass number of an atom having 10 protons, 16 neutrons, and 10 electrons is
   a. _____ 22.
   b. _____ 26.
   c. _____ 32.
   d. _____ 38.
   e. _____ 48.
6. What accounts for the fact that some minerals, such as plagioclase feldspars, have a range of chemical compositions rather than one specific composition?
7. One part of the definition of the term *mineral* is that minerals are crystalline solids. What are crystalline solids and how do they differ from noncrystalline solids?
8. What factors determine whether a mineral or rock commodity is a resources or a reserve?
9. Although about 3,800 minerals have been named and described, only a few are very common. Why and which ones?

10. **Creative Thinking Visual Question:** If diamond is perfectly cleaved, it would yield geometric figures like those in ▌Figure 1 but this mineral is fluorite. From the image alone you should be able to tell that these specimens are not diamond. How? Look up the other properties of diamond and fluorite in Appendix C and see what other mineral properties are used to differentiate one from the other.

Sue Monroe

▌**Figure 1** Fluorite This mineral has four directions of cleavage as in Figure 3.17e.

Global GeoScience Watch Within the GREENR database, search "strategic minerals" and under "Reference" open the article titled "Strategic Minerals." What are some minerals for which the United States has a sufficient supply to meet demand? What are several of the minerals the United States needs but does not have in sufficient quantity and where do we import them from? List some of the minerals we import from friendly, stable nations as opposed to those from less friendly, less stable, and less dependable nations.

Stone Mountain, North Carolina, is a body of granite composed of quartz, potassium feldspar, plagioclase feldspar, biotite, and muscovite. It is in Stone Mountain State Park and is made up of several bodies of granite that were intruded into Earth's crust when plates collided during the Devonian Period (359 to 415 years ago). When this area was uplifted and the overlying rocks eroded, Stone Mountain was revealed in its present form. It stands more than 200 m above the surrounding valley.

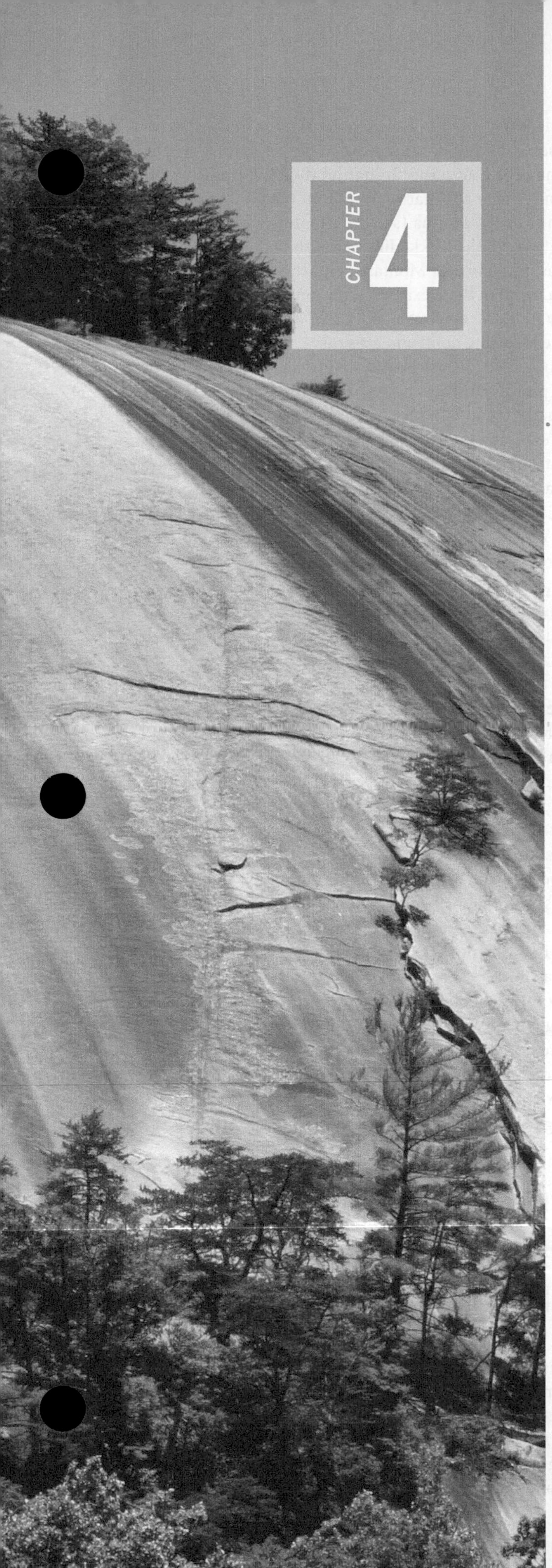

Pat & Chuck Blackley/Alamy

Igneous Rocks and Plutons

OUTLINE

HAVE YOU EVER WONDERED?

- Why geologists make a distinction between magma and lava?
- How hot lava is and how fast it flows?
- How and why igneous rocks differ from other types of rocks?
- Whether plate tectonics plays a role in the origin of magma and lava?
- How the vast bodies of granite in many parts of North America originated?
- What natural resources are found in or near igneous rocks?

4.1 Introduction

From our discussions so far you know that the term *rock* applies to aggregates of one or more minerals (see Figures 3.1b and 3.19), mineral-like matter as in obsidian (natural glass), and coal (altered organic matter). We also noted in Chapter 1 that the three groups of rocks are igneous, sedimentary, and metamorphic (see Figure 1.15). Here we are concerned only with **igneous rocks,** which are rocks made up of minerals that crystallized from molten rock matter called **magma,** if below Earth's surface, and **lava,** if at the surface (▌Figure 4.1). So, magma and lava are the same material, although the gas content of lava is less because of decreased pressure. Also included among the igneous rocks are those made up of particulate matter known as **pyroclastic materials** ejected from volcanoes during explosive eruptions (Figure 4.1).

Although you are no doubt familiar with volcanoes, which are simply hills or mountains where magma, pyroclastic materials, and gases reach the surface, you probably have little knowledge about how or where magma is generated, or how it rises toward the surface. In fact, most magma cools and crystallizes underground, a phenomenon known as *intrusive igneous activity,* thereby forming several types of igneous rock bodies known as *plutons,* named for Pluto, the Roman god of the underworld. Granite and similar rocks are common in the plutons at Stone Mountain, North Carolina (see the chapter opening photo), as well as those in the Black Hills of South Dakota, the Rocky Mountains of Idaho and Montana, and the Sierra Nevada of California. In all of these examples, plutons formed far underground by intrusive igneous activity and were later exposed at the surface following uplift and erosion.

ConnectionLink
You can find more information on volcanoes in Chapter 5.

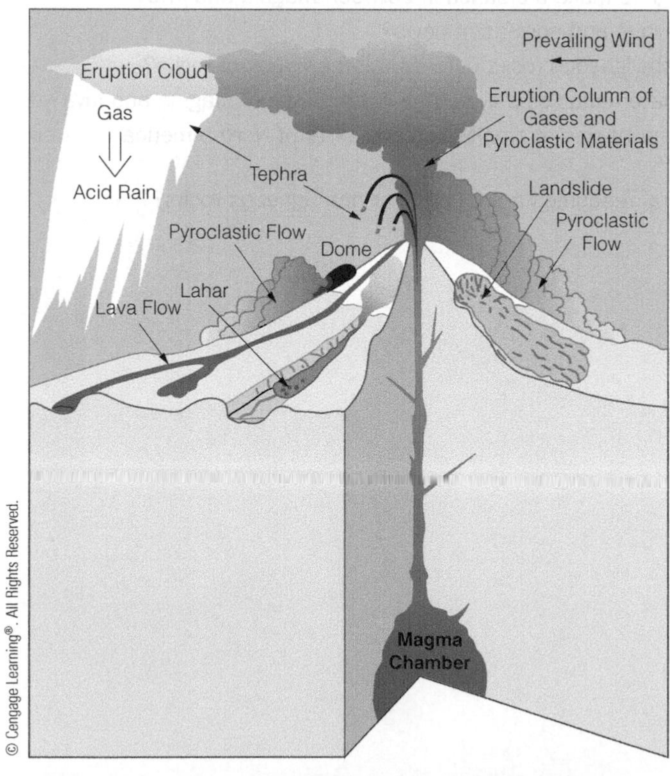

▌**Figure 4.1 Magma, Lava, and Pyroclastic Materials** Magma (i.e., molten rock below the surface) may cool and crystallize within the crust, but some of it is erupted at the surface as lava flows and pyroclastic materials. The reservoir below the surface where magma accumulates is a magma chamber. Notice in this illustration that igneous activity also releases gases into the atmosphere.

Although we discuss intrusive igneous activity and the origin of plutons in this chapter and volcanism (extrusive igneous activity) in Chapter 5, both processes are related. Indeed, the same kinds of magmas are involved, but magma varies in its mobility, so only some reaches the surface as lava or pyroclastic materials. In addition, plutons lie below areas of volcanism and are the sources of the lava and pyroclastic materials erupted at volcanoes or along fissures, a topic we cover more fully in Chapter 5.

One reason to study igneous rocks and plutons is that they make up part of the continents and all of the oceanic crust, which forms continuously at divergent plate boundaries. As a matter of fact, most plutons and volcanoes are found at or near divergent and convergent plate boundaries, so their presence in the geologic record is one criterion for identifying ancient plate boundaries. Another reason to study these topics is their economic importance, especially the origin of the larger plutons. When these huge masses of magma began to cool and crystallize to form minerals, fluids emanating from them follow cracks and crevasses in adjacent rocks where mineral resources such as copper may form.

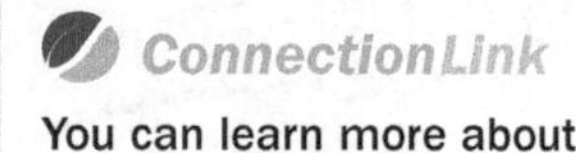
You can learn more about the origin of oceanic crust in Chapter 10.

Part of this chapter is devoted to simply describing and classifying igneous rocks. However, descriptions and classifications are not ends in themselves but simply ways of organizing and conveying information. For example, to anyone familiar with geology, the terms granite and basalt immediately bring to mind sequences of events in Earth history.

4.2 The Properties and Behavior of Magma and Lava

Magma is less dense than the rock from which it formed, so it tends to rise toward the surface, but much of it cools and crystallizes far underground. All igneous rocks ultimately come from magma, but two separate processes account for them. **Plutonic rocks,** also called **intrusive igneous rock,** result from magma cooling and crystallizing below the surface—that is, from magma intruded into Earth's crust. In contrast,

TABLE 4.1 The Most Common Types of Magmas and Their Characteristics

Type of Magma	Silica Content (%)	Sodium, Potassium, and Aluminum	Calcium, Iron, and Magnesium
Ultramafic	<45	↓	Increase
Mafic	45–52	↓	↑
Intermediate	53–65	↓	↑
Felsic	>65	Increase	↑

volcanic rocks, also known as **extrusive igneous rocks,** form when lava cools, or when pyroclastic materials become consolidated, both of which are extruded onto the surface.

Figure 4.2 How Hot Is Lava? A United States Geological Survey geologist uses a thermocouple to measure the temperature of a lava flow in Hawaii.

Composition of Magma

By far the most abundant minerals in Earth's crust are silicates such as feldspars, quartz, and several ferromagnesian silicates, all made up of silicon and oxygen, and other elements shown in Figure 3.11. As a result, melting of the crust yields mostly silica-rich magmas that also contain substantial aluminum, calcium, sodium, iron, magnesium, and potassium, and several other elements in lesser quantities. Another source of magma is Earth's upper mantle, which is composed of rocks that contain mostly ferromagnesian silicates. Thus, magma from this source contains comparatively less silicon and oxygen (silica) and more iron and magnesium.

With few exceptions, the primary constituent of magma is silica, which varies enough to distinguish magmas called felsic, intermediate, mafic, and ultramafic* (Table 4.1). **Felsic magma,** with more than 65% silica, is silica-rich and contains considerable sodium, potassium, and aluminum but little calcium, iron, and magnesium. As you would expect, **intermediate magma** (53%–65% silica) has a composition between felsic and mafic magma (Table 4.1). **Mafic magma** has 45%–52% silica, but it contains proportionately more calcium, iron, and magnesium than does felsic magma. Any magma with less than 45% silica is termed **ultramafic,** which cools to form rocks made up mostly of ferromagnesian silicates.

How Hot Are Magma and Lava?

Erupting lava has a temperature in the range of 700° to 1,200°C, although a temperature of 1,350°C was recorded above a lava lake in Hawaii where volcanic gases reacted with the atmosphere. Magma must be even hotter than lava, but no direct measurements of magma temperatures have ever been made.

Most lava temperatures are taken at volcanoes that show little or no explosive activity, so our best information comes from mafic lava flows such as those in Hawaii (Figure 4.2). In contrast, eruptions of felsic lava are not as common, and the volcanoes that these flows issue from tend to be explosive and thus cannot be approached safely. Nevertheless, the temperatures of some bulbous masses of felsic lava in lava domes have been measured at a distance with an optical pyrometer. The surfaces of these lava domes are as hot as 900°C, but their interiors must surely be even hotter.

When Mount St. Helens erupted in 1980 in Washington State, it ejected felsic magma as particulate matter in pyroclastic flows. Two weeks later, these flows had temperatures between 300° and 420°C, and a steam explosion took place more than a year later when water encountered some of the still-hot pyroclastic materials.

The reason that lava and magma retain heat so well is because rock conducts heat so poorly. Accordingly, the interiors of thick lava flows and pyroclastic flow deposits may remain hot for months or years. In 1959, lava filled a crater to 85 m deep in Hawaii, but when the lava was drilled in 1988, it still had not completely solidified near its base. Plutons, depending on their size and depth, may not cool completely for thousands to millions of years.

Viscosity—Resistance to Flow

Viscosity, or resistance to flow, is a property of all liquids. Water's viscosity is very low so it is highly fluid and flows readily. For other liquids, viscosity is so high that they flow much more slowly. Good examples are cold motor oil and syrup, both of which are quite viscous and thus flow only with difficulty. But when these liquids are heated, their viscosity is much lower and they flow more easily; that is, they become more fluid with increasing temperature. Accordingly, you might suspect that temperature controls the viscosity of magma and lava, and this inference is partly correct. We can generalize and say that the hotter the magma or lava, the more readily it moves, but we must qualify this statement by noting that temperature is not the only control of viscosity.

*Lava from some volcanoes in Africa cools to form carbonitite, an igneous rock with at least 50% carbonate minerals, mostly calcite and dolomite.

▌Figure 4.3 Viscosity of Magma and Lava Temperature is an important control on viscosity, but so is composition. Mafic lava tends to be fluid, whereas felsic lava is much more viscous.

Critical Thinking Question Which one of these volcanoes would you expect to erupt explosively? Explain.

a A mafic lava flow in 1984 on Mauna Loa Volcano in Hawaii. These flows move rapidly and form thin layers.

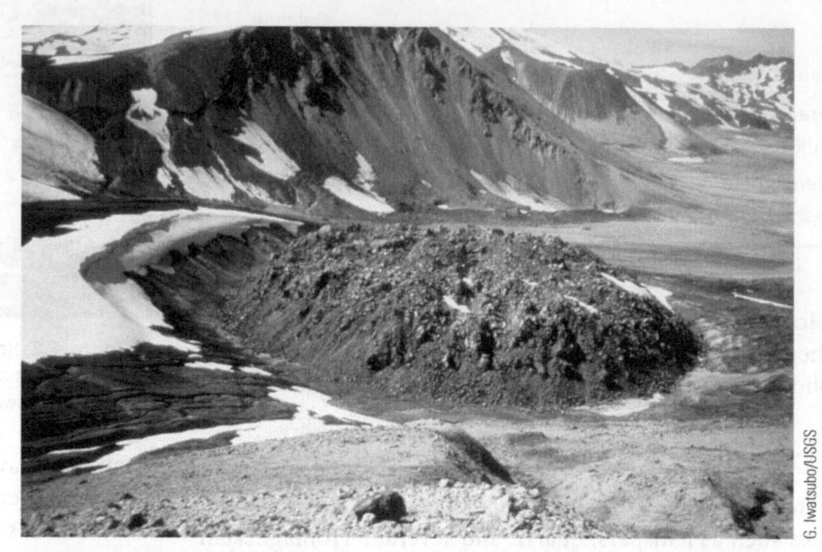

b The Novarupta lava dome in Katmai National Park in Alaska. The lava is felsic and viscous, so it was extruded as a bulbous mass. This image was taken in 1987.

Silica content also strongly controls viscosity. With increasing silica content, numerous networks of silica tetrahedra form and retard flow because for flow to take place, the strong bonds of the networks must be ruptured. Mafic magma and lava with 45%–52% silica have fewer silica tetrahedra networks and, as a result, are more mobile than felsic magma and lava flows (▌Figure 4.3). One mafic flow in 1783 in Iceland flowed nearly 80 km, and geologists traced ancient flows in Washington State for more than 500 km. Felsic magma, because of its higher viscosity, does not reach the surface as often as mafic magma. And when felsic lava flows do occur, they tend to be slow moving and thick, and move only short distances. A thick, pasty lava flow that erupted in 1915 from Lassen Peak in California flowed only about 300 m before it ceased moving.

Temperature and silica content are important controls on the viscosity of magma and lava, but other factors include gases, mostly water vapor and CO_2, as well as the presence of mineral crystals and friction from the surface over which lava flows. Lava with a high content of dissolved gases flows more readily than one with a lesser amount of gases, whereas lava that has many crystals or that flows over a rough surface tends to be more viscous.

4.3 How Does Magma Originate and Change?

Most people are unaware of how and where magma originates, how it rises, and how it might change. Indeed, many believe the misconception that lava comes from a continuous layer of molten rock beneath the crust or that it comes from Earth's molten outer core.

First, let us address how magma originates. The atoms in a solid are in constant motion, so when a solid is heated, the energy of motion eventually exceeds the binding forces and the solid melts. We are all familiar with this phenomenon, and we are also aware that not all solids melt at the same temperature. Once magma forms, it tends to rise because as rocks become hotter, they expand and their density decreases.

Magma may come from 100 to 300 km deep, but most forms at much shallower depths in the upper mantle or lower crust and accumulates in reservoirs known as **magma chambers** (Figure 4.1). Beneath spreading ridges, where the crust is thin, magma chambers exist at a depth of only a few kilometers, but along convergent plate boundaries, they are usually a few tens of kilometers deep. The volume of a magma chamber ranges from a few to many hundreds of cubic kilometers of molten rock within the otherwise-solid lithosphere. Some simply cools and crystallizes within Earth's crust, thus accounting for the origin of plutons, whereas some rises to the surface and is erupted as lava flows or pyroclastic materials.

Bowen's Reaction Series

During the early 1900s, N. L. Bowen knew that some minerals occurred together in igneous rocks but others did not. Based on his observations and laboratory experiments with cooling molten material, Bowen determined a predictable sequence in which specific minerals crystallized, rather than all crystallizing simultaneously. Furthermore, he proposed how intermediate and felsic magma could be derived from mafic magma. **Bowen's reaction series,** as it is now called,

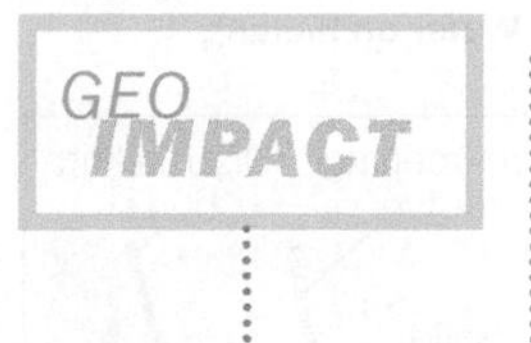

Teaching About Magma and Lava

You are a high school teacher interested in developing experiments to show your students that (1) composition and temperature affect the viscosity of a lava flow and (2) when magma or lava cools, some minerals crystallize before others. Describe the experiments you might devise to illustrate these points.

actually consists of two branches: a *discontinuous branch* and a *continuous branch* (▌Figure 4.4). As magma cools, minerals crystallize along both branches simultaneously, but for convenience, we will discuss them separately.

In the discontinuous branch, which contains only ferromagnesian silicates, one mineral changes to another over specific temperature ranges (Figure 4.4). As the temperature decreases, a temperature range is reached in which a given mineral begins to crystallize. A previously formed mineral reacts with the remaining liquid magma (the melt) so that it forms the next mineral in the sequence. For instance, olivine [$(Mg,Fe)_2SiO_4$] is the first ferromagnesian silicate to crystallize. As the magma continues to cool, it reaches the temperature range at which pyroxene is stable; a reaction occurs between the olivine and the remaining melt, and pyroxene forms.

With continued cooling, a reaction takes place between pyroxene and the melt, and the pyroxene structure is rearranged to form amphibole. Further cooling causes a reaction between the amphibole and the melt, and its structure is rearranged so that the sheet structure of biotite mica forms.

Although the reactions just described tend to convert one mineral to the next in the series, the reactions are not always complete. Olivine, for example, might have a rim of pyroxene, indicating an incomplete reaction. If magma cools rapidly enough, the early-formed minerals do not have time to react with the melt, and thus all the ferromagnesian silicates in the discontinuous branch can be in one rock. In any case, by the time biotite has crystallized, essentially all the magnesium and iron present in the original magma have been used up.

Plagioclase feldspars, which are nonferromagnesian silicates, are the only minerals in the continuous branch of Bowen's reaction series (Figure 4.4). Calcium-rich plagioclase crystallizes first, and as the magma continues to cool, it reacts with the melt, and plagioclase containing proportionately more sodium crystallizes until all of the calcium and sodium are used up. In many cases, cooling is too rapid for a complete transformation from calcium-rich to sodium-rich plagioclase to take place. Plagioclase forming under these conditions is *zoned,* meaning that it has a calcium-rich core surrounded by zones progressively richer in sodium.

As minerals crystallize simultaneously along the two branches of Bowen's reaction series, iron and magnesium are depleted because they are used in ferromagnesian silicates, whereas calcium and sodium are used up in plagioclase feldspars. At this point, any leftover magma is enriched in potassium, aluminum, and silicon, which combine to form orthoclase ($KAlSi_3O_8$), a potassium feldspar, and if water pressure is high, the sheet silicate muscovite forms. Any remaining magma is enriched in silicon and oxygen (silica) and forms the mineral quartz (SiO_2). The crystallization of orthoclase and quartz is not a true reaction series because they form independently rather than by a reaction of orthoclase with the melt.

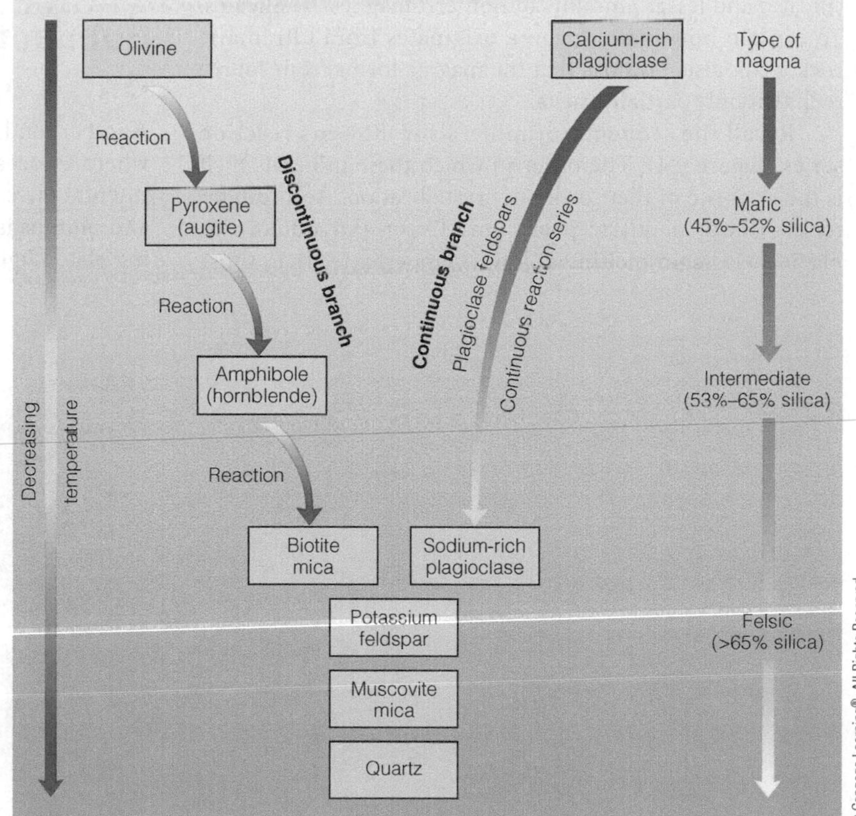

▌**Figure 4.4 Bowen's Reaction Series** Bowen's reaction series consists of a discontinuous branch along which a succession of ferromagnesian silicates crystallize as the magma's temperature decreases, and a continuous branch along which plagioclase feldspars with increasing amounts of sodium crystallize. Notice also that the composition of the initial mafic magma changes as crystallization takes place along the two branches.

The Origin of Magma at Spreading Ridges

One fundamental observation regarding the origin of magma is that Earth's temperature increases with depth. Known as the *geothermal gradient,* this temperature increase averages about 25°C/km. Accordingly, rocks at depth are hot but remain solid because their melting temperature rises with increasing pressure (Figure 4.5). However, beneath spreading ridges, the temperature locally exceeds the melting temperature, at least in part because pressure decreases. That is, plate separation at ridges probably causes a decrease in pressure on the already hot rocks at depth, thus initiating melting (Figure 4.6). In addition, the presence of water decreases the melting temperature beneath spreading ridges because water aids thermal energy in breaking the chemical bonds in minerals.

ConnectionLink

You can find more information about Earth's geothermal gradient in Chapter 9.

Magma formed beneath spreading ridges is invariably mafic (45%–52% silica). However, the upper mantle rocks from which this magma is derived are characterized as ultramafic (<45% silica), consisting mostly of ferromagnesian silicates and lesser amounts of nonferromagnesian silicates. To explain how mafic magma originates from ultramafic rock, geologists propose that the magma forms from source rock that only partially melts.

Recall the sequence of minerals in Bowen's reaction series (Figure 4.4). The order in which these minerals melt is the opposite of their order of crystallization. Accordingly, rocks made up of quartz, potassium feldspar, and sodium-rich plagioclase begin melting at lower temperatures than those composed of ferromagnesian silicates and the calcic varieties of plagioclase. So when ultramafic rock starts to melt, the minerals richest in silica melt first, followed by those containing less silica. Therefore, if melting is not complete, mafic magma containing proportionately more silica than the source rock results.

Figure 4.5 The Effects of Pressure and Water on Melting

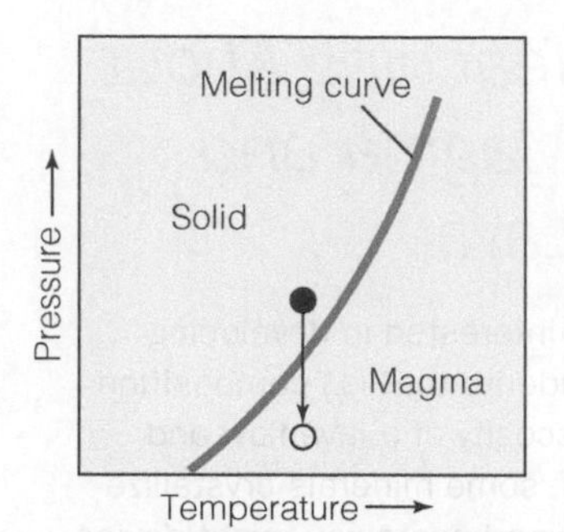

a As pressure decreases, even when temperature remains constant, melting takes place. The black circle represents rock at high temperature. The same rock (open circle) melts at lower pressure.

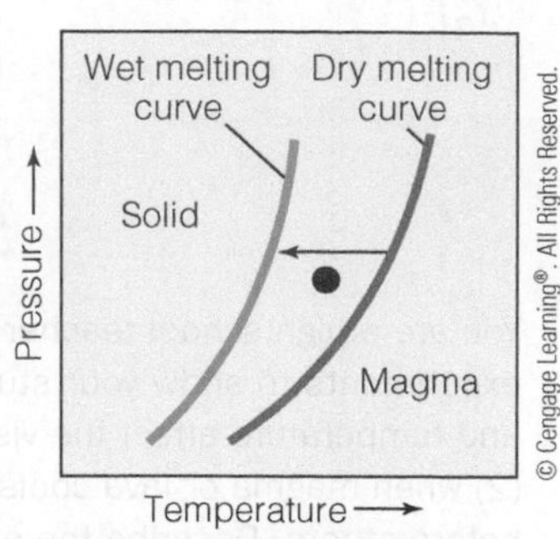

b If water is present, the melting curve shifts to the left because water provides an additional agent to break chemical bonds. Accordingly, rocks melt at a lower temperature (green melting curve) if water is present.

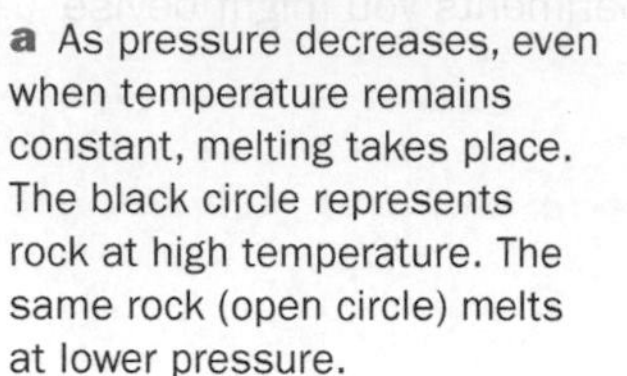

Subduction Zones and the Origin of Magma

Another fundamental observation regarding magma is that, where an oceanic plate is subducted beneath either a continental plate or another oceanic plate, a belt of volcanoes and plutons is found near the leading edge of the overriding plate (Figure 4.6). It would seem, then, that subduction

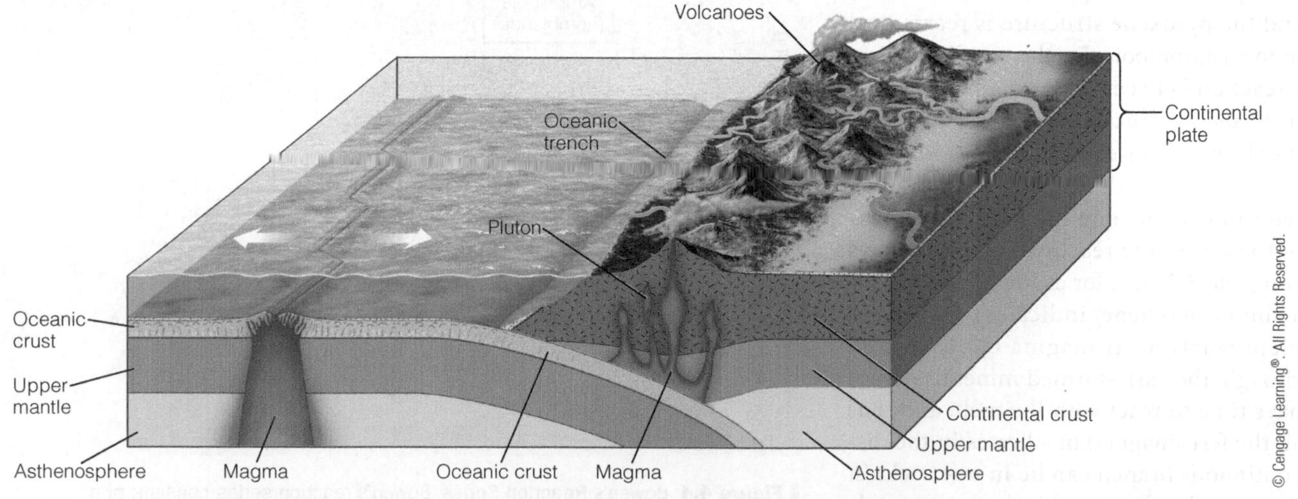

Figure 4.6 The Origin of Magma Magma forms beneath spreading ridges, because as plates separate, pressure is reduced on the hot rocks and partial melting of the upper mantle begins. Invariably, the magma formed is mafic. Magma also forms at subduction zones where water from the subducted plate aids partial melting of the upper mantle. This magma is also mafic, but as it rises, melting of the lower crust makes it more felsic.

Figure 4.7 Mantle Plume and Hot Spot

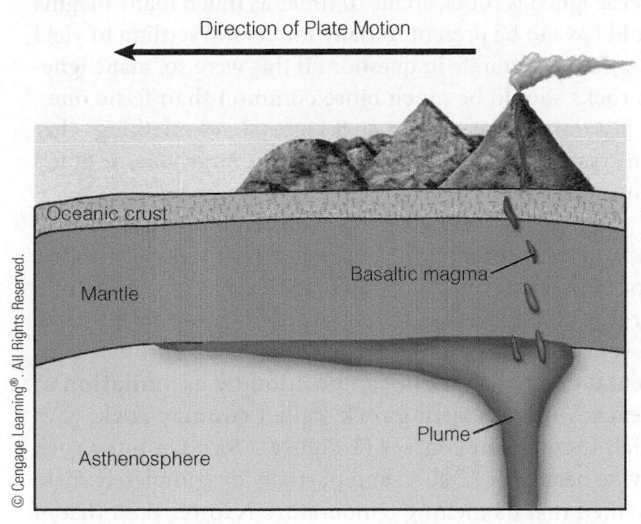

a A mantle plume beneath oceanic crust with a hot spot. Rising magma forms a series of volcanoes that become younger in the direction of plate movement.

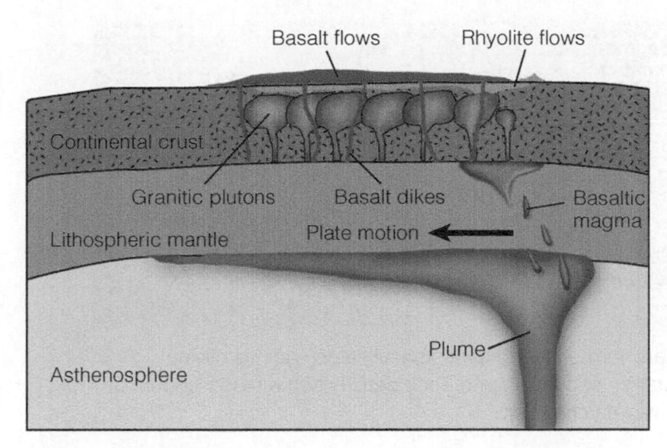

b A mantle plume with an overlying hot spot yields flood basalts, and some of the continental crust melts to form felsic magma.

and the origin of magma must be related in some way, and indeed they are. Furthermore, magma at these convergent plate boundaries is mostly intermediate (53%–65% silica) or felsic (>65% silica).

Once again, geologists invoke the phenomenon of partial melting to explain the origin and composition of magma at subduction zones. As a subducted plate descends toward the asthenosphere, it eventually reaches a depth at which the temperature is high enough to initiate partial melting. In addition, the oceanic crust descends to a depth at which dewatering of hydrous minerals takes place, and as the water rises into the overlying mantle, it enhances melting and magma forms (Figure 4.5).

Recall that partial melting of ultramafic rock at spreading ridges yields mafic magma. Similarly, partial melting of mafic rocks of the oceanic crust yields intermediate (53%–65% silica) and felsic (>65% silica) magmas, both of which are richer in silica than the source rock. Moreover, some of the silica-rich sediments and sedimentary rocks of continental margins are probably carried downward with the subducted plate and contribute their silica to the magma. Also, mafic magma rising through the lower continental crust must be contaminated with silica-rich materials, which changes its composition.

Hot Spots and the Origin of Magma

Most volcanism occurs at divergent and convergent plate boundaries, but there are some chains of volcanic outpourings in the ocean basins and on continents that are not near either of these boundaries. The Emperor Seamount–Hawaiian Islands, for instance, form a chain of volcanic islands 6,000 km long, and the volcanic rocks become progressively older toward the northwest (see Figure 2.24). In 1963, Canadian geologist J. Tuzo Wilson proposed that the Hawaiian Islands and other areas showing similar trends lay above a **hot spot** over which a plate moves, thereby yielding a succession of volcanoes (Figure 4.7a).

Many geologists now think that hot-spot volcanism results from a rising **mantle plume,** a cylindrical plume of hot mantle rock that rises from perhaps near the core–mantle boundary. As it rises toward the surface, the pressure decreases on the hot rock and melting begins, thus yielding magma. Hot-spot volcanism may also account for vast flat-lying areas of overlapping lava flows or what geologists call *flood basalts* (Figure 4.7b). Figure 2.17 shows the locations of many hot spots, but of particular interest to us is the Yellowstone hot spot in Wyoming.

ConnectionLink

For more on the Yellowstone region and the Yellowstone hot spot, see Chapter 23.

Compositional Changes in Magma

Obviously the composition of any magma depends on what was melted in the first place. Should mafic rock melt completely, the resulting mama would also be mafic. However, as we discussed previously, partial melting yields magma that differs from its parent rock. In any case, once magma forms, it may change by **crystal settling,** which involves gravitational settling of minerals as they crystallize (Figure 4.8). A good example is olivine, the first ferromagnesian silicate in the discontinuous branch of Bowen's reaction series,

▌Figure 4.8 **Crystal Settling Is a Process That Changes the Composition of Magma**

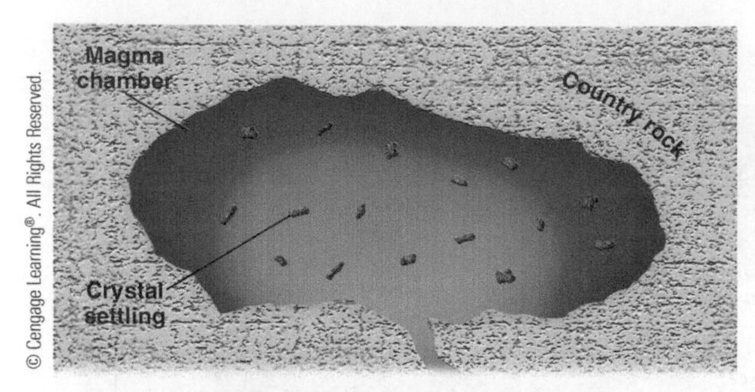

a Early formed ferromagnesian silicates such as olivine crystallize and because of their density settle to the bottom of the magma chamber.

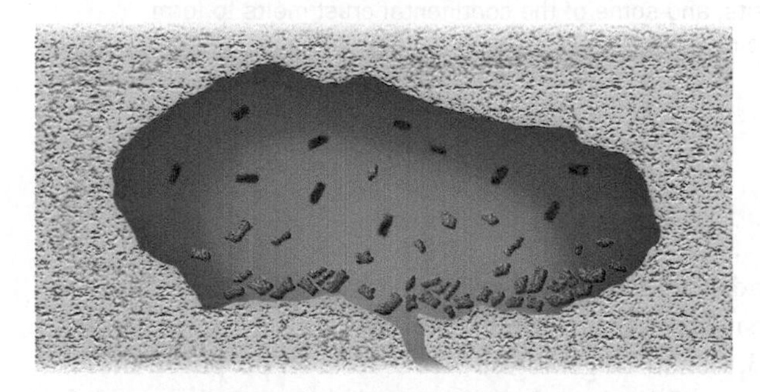

b Ferromagnesian silicates continue to form and settle.

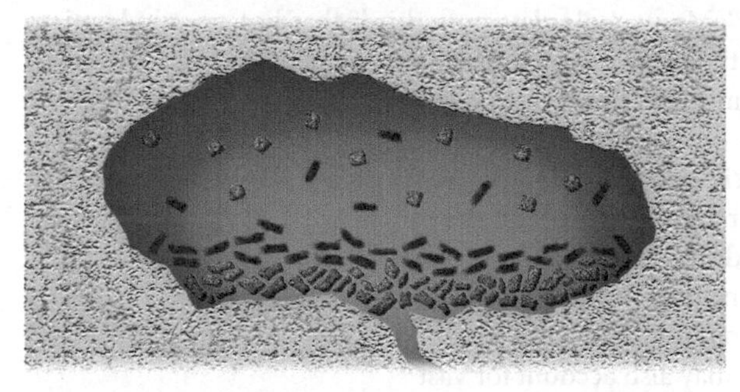

c The remaining melt becomes richer in silicon, sodium, and potassium because much of the iron and magnesium originally present is now in the ferromagensian minerals that settled.

which has a density greater than the remaining melt and tends to sink. As a result, the remaining melt becomes richer in silica, sodium, and potassium, because much of the iron and magnesium were removed as olivine and perhaps pyroxene crystallized. In other words, the remaining melt becomes more felsic.

In some thick, sheetlike plutons called *sills,* the first-formed ferromagnesian silicates are in fact concentrated in their lower parts, thus making their upper parts less mafic. But even so, very little felsic magma forms by this process.

Calculations show, that to yield a given volume of granite (a felsic igneous rock), about 10 times as much mafic magma would have to be present initially for crystal settling to yield the volume of granite in question. If this were so, mafic igneous rocks should be much more common than felsic ones, but just the opposite is true, so it seems that something other than crystal settling must account for the large volume of felsic magma.

Once again, geologists refer to partial melting to solve this apparent dilemma. Remember that partial melting of oceanic crust and silica-rich sediments of continental margins yields magma richer in silica (more felsic) than the source rock. In addition, magma rising through continental crust changes in composition by **assimilation** as it reacts with preexisting rock, called **country rock,** with which it comes in contact (▌Figure 4.9a). Country rock may be heated to 1,300°C and partially or completely melt, provided that its melting temperature is lower than that of the magma. Because assimilated rocks rarely have the same composition as the magma, the composition of the magma changes.

The fact that assimilation takes place is clearly indicated by *inclusions,* which are incompletely melted pieces of country rock that are fairly common in many igneous rocks (▌Figure 4.9b). As magma rises, it forces its way into cracks and crevasses and wedges loose pieces of country rock. Assimilation certainly takes place, but its effect on the bulk composition of magma must be slight. The reason is that the heat for melting must come from the magma, which has the effect of cooling the magma, so only a limited amount of country rock can be assimilated.

Neither crystal settling nor assimilation can produce a significant amount of felsic magma from a mafic one. But both processes, if operating concurrently, can bring about greater changes than either process acting alone. Some geologists think that this is one way that intermediate magma forms where oceanic lithosphere is subducted beneath continental lithosphere.

A single volcano can erupt lavas of different composition, indicating that magmas of differing composition are present. It seems likely that some of these magmas would come into contact and mix with one another. If this is the case, we would expect that the composition of the magma resulting from **magma mixing** would be a modified version of the parent magmas. Suppose rising mafic magma mixes with felsic magma of about the same volume (Figure 4.9a). The resulting "new" magma would have a more intermediate composition. To account for the fact that the eruptions of Eyjafjallajökull in Iceland changed from mafic to more felsic from March to April 2010, geologists propose either that magma mixing occurred or that the later eruptions tapped into a different magma chamber.

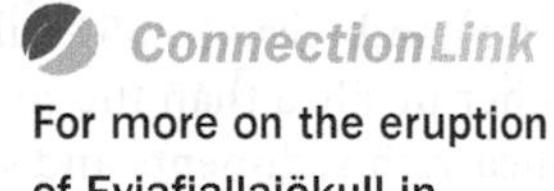

For more on the eruption of Eyjafjallajökull in Iceland, see Chapter 5.

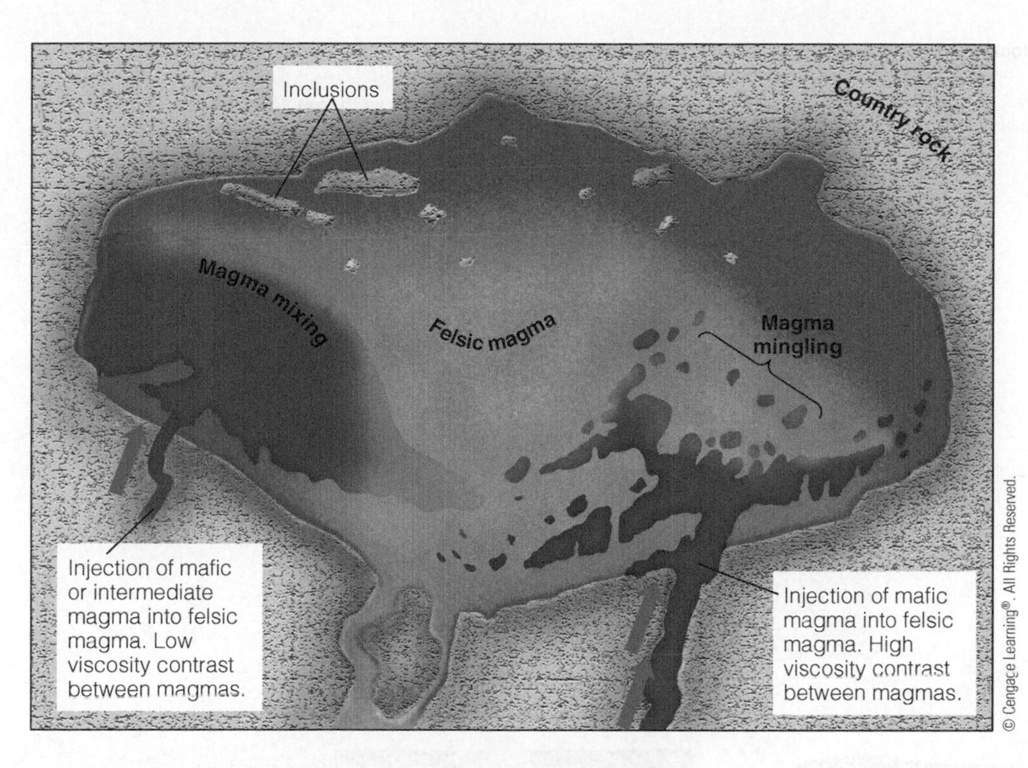

Figure 4.9 Assimilation and Magma Mixing

a Fragments of rock dislodged by rising magma may melt and become incorporated into the magma, a process called assimilation, or they may remain as inclusions. Magma mixing is shown where mafic magma is injected into felsic magma.

b Dark inclusions in granitic rocks in the Sierra Nevada in California.

4.4 Igneous Rocks—Their Characteristics and Classification

We have already defined *plutonic* or *intrusive igneous rocks* and *volcanic* or *extrusive igneous rocks*. Here we will have much more to say about the texture, composition, and classification of these rocks, which constitute one of the three major rock groups depicted in the rock cycle (see Figure 1.15).

Igneous Rock Textures

Texture refers to the size, shape, and arrangement of the minerals that make up igneous rocks. Size is the most important because mineral crystal size is related to the cooling history of magma or lava and usually indicates whether an igneous rock is volcanic or plutonic. The atoms in magma and lava are in constant motion, but when cooling begins, some atoms bond to form small mineral nuclei. As other atoms in the liquid chemically bond to these nuclei, they do so in an orderly geometric arrangement and the nuclei grow into crystalline *mineral grains,* the individual particles that make up igneous rocks.

If cooling takes place rapidly as in lava flows, the rate at which mineral nuclei form exceeds the rate of growth and an aggregate of many small mineral grains forms. The result is a fine-grained or **aphanitic texture,** in which individual minerals are too small to be seen without magnification (❚ Figure 4.10a). If cooling is slow, the rate of growth exceeds the rate of nuclei formation, and large mineral grains form, thus yielding a coarse-grained or **phaneritic texture,** in which minerals are clearly visible (❚ Figure 4.10b). Aphanitic textures usually indicate an extrusive origin, whereas rocks with phaneritic textures are commonly intrusive. However, shallow plutons might have an aphanitic texture, and the rocks that form in the interiors of thick lava flows might be phaneritic.

Another common igneous texture is one termed **porphyritic,** in which minerals of markedly different size are present in the same rock. The larger minerals are *phenocrysts* and the smaller ones collectively make up the *groundmass,* which is simply the grains between phenocrysts (❚ Figure 4.10c). The only requirement for a porphyritic texture is that the phenocrysts be considerably larger than the minerals in the groundmass, which may be either aphanitic or phaneritic. Igneous rocks with porphyritic textures are designated *porphyry,* as in basalt porphyry. These rocks have more complex cooling histories than those with aphanitic or phaneritic textures and might

Figure 4.10 Textures of Igneous Rocks

a Rapid cooling, as in lava flows, results in many small minerals and an aphanitic (fine-grained) texture.
b Slower cooling in plutons yields a phaneritic (coarse-grained) texture.
c These porphyritic textures indicate a complex cooling history.
d Obsidian has a glassy texture because magma cooled too quickly for mineral crystals to form.
e Gases expand in lava to yield a vesicular texture.
f Microscopic view of a rock with a fragmental texture. The colorless, angular particles of volcanic glass measure up to 2 mm.

involve, for example, magma partly cooling beneath the surface followed by its eruption and rapid cooling at the surface.

Lava may cool so rapidly that its constituent atoms do not have time to become arranged in the ordered, three-dimensional frameworks of minerals. As a consequence, *natural glass,* such as *obsidian,* forms (▌Figure 4.10d). Even though obsidian with its glassy texture is not composed of minerals, geologists nevertheless classify it as an igneous rock.

Some magmas contain large amounts of water vapor and other gases. These gases may be trapped in cooling lava where they form numerous small holes or cavities known as **vesicles;** rocks with many vesicles are termed *vesicular,* as in vesicular basalt (▌Figure 4.10e).

A **pyroclastic (fragmental) texture** characterizes igneous rocks formed by explosive volcanic activity (▌Figure 4.10f). For example, ash discharged high into the atmosphere eventually settles to the surface where it accumulates; if consolidated, it forms pyroclastic igneous rock.

Composition of Igneous Rocks

Most igneous rocks, like the magma from which they originate, are mafic (45%–52% silica), intermediate (53%–65% silica), or felsic (>65% silica). A few are *ultramafic* (<45% silica), but these are probably derived from mafic magma by a process we discuss later. The parent magma plays an important role in determining the mineral composition of igneous rocks, yet it is possible for the same magma to yield a variety of igneous rocks because its composition can change as a result of the sequence in which minerals crystallize, or by crystal settling, assimilation, and magma mixing (Figures 4.4, 4.8, and 4.9).

Classifying Igneous Rocks

Recall from the introduction that classification of igneous rocks is not an end in itself but simply a way of organizing and conveying information. Geologists use texture and composition to classify igneous rocks, although a few are classified mostly by texture. Notice in ▌Figure 4.11 that all rocks except peridotite are in pairs; the members of a pair have the same composition but different textures. Basalt and gabbro, andesite and diorite, and rhyolite and granite are compositional (mineralogical) pairs, but basalt, andesite, and rhyolite are aphanitic and most commonly volcanic, whereas gabbro, diorite, and granite are phaneritic and mostly plutonic. The volcanic and plutonic members of each pair can usually be distinguished by texture, but remember that rocks in some shallow plutons may be

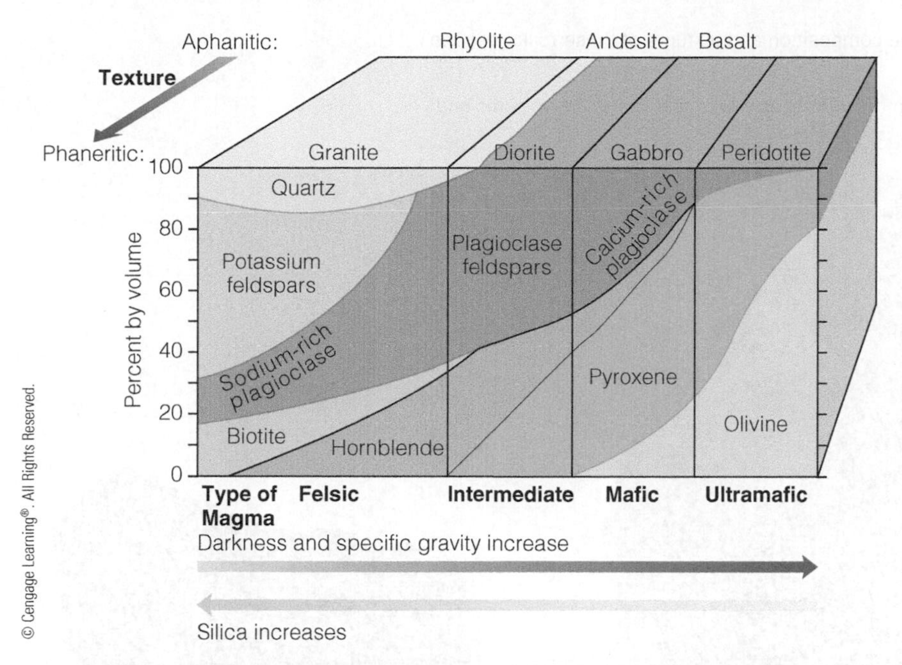

Figure 4.11 Classification of Igneous Rocks This diagram shows the percentages of minerals, as well as the textures of common igneous rocks. For example, an aphanitic (fine-grained) rock of mostly calcium-rich plagioclase and pyroxene is basalt.

Critical Thinking Question Of the magmas that crystallized to form rhyolite, andesite, and basalt, which one would have been the most viscous? How do you know?

aphanitic and rocks that formed in thick lava flows may be phaneritic. In other words, all these rocks exist in a textural continuum.

The igneous rocks in Figure 4.11 are also differentiated by composition—that is, by their mineral content. Reading across the chart from rhyolite to andesite to basalt, for example, we see that the proportions of nonferromagnesian and ferromagnesian silicates change. The differences in composition, however, are gradual along a compositional continuum. In other words, there are rocks with compositions that correspond to the lines between granite and diorite, basalt and andesite, and so on. For example, Lassen Peak in California, which erupted from 1914 through 1917, is made up mostly of *dacite,* a rock with a composition between andesite and rhyolite.

Ultramafic Rocks Ultramafic rocks (<45% silica) are composed mostly of ferromagnesian silicates. *Peridotite* contains olivine, lesser amounts of pyroxene, and usually a little plagioclase feldspar (Figure 4.11 and Figure 4.12), and *pyroxenite* is composed predominately of pyroxene. Because these minerals are dark, the rocks are black or green. Peridotite is a likely candidate for the rock that makes up the upper mantle (see Chapter 9). Ultramafic rocks in Earth's crust probably originate by concentration of the early-formed ferromagnesian minerals that separated from mafic magmas.

ConnectionLink

For more about Earth's mantle and its composition, see Chapter 9.

Ultramafic lava flows, called *komatiites,* are known in rocks older than 2.5 billion years but are rare or absent in younger ones. The reason is that to erupt, ultramafic lava must have a near-surface temperature of about 1,600°C; the surface temperatures of present-day mafic lava flows are rarely more than 1,200°C. During early Earth history, though, more radioactive decay heated the mantle to as much as 300°C hotter than now, and ultramafic lavas could erupt onto the surface. Because the amount of heat has decreased over time, Earth has cooled, and eruptions of ultramafic lava flows are rare.

Basalt-Gabbro *Basalt* and *gabbro* are the aphanitic and phaneritic rocks that crystallize from mafic magma (45%–52% silica) (Figure 4.13a, b). Both have the same composition—mostly calcium-rich plagioclase and pyroxene, with smaller amounts of olivine and amphibole (Figure 4.11). Because they contain a large proportion of ferromagnesian silicates, basalt and gabbro are dark; those that are porphyritic contain calcium plagioclase or olivine phenocrysts.

Extensive basalt lava flows cover vast areas in Washington, Oregon, Idaho, and northern California (see Chapters 5 and 23). Oceanic islands such as Iceland, the Galápagos, the Azores, and the Hawaiian Islands are composed mostly of basalt, and basalt makes up the upper part of the oceanic crust.

Gabbro is much less common than basalt, at least in the continental crust or where it can be easily observed. Small intrusive bodies of gabbro are present in the continental

Sue Monroe

Figure 4.12 Peridotite This specimen of the ultramafic rock peridotite is made up mostly of olivine. Notice in Figure 4.11 that peridotite is the only phaneritic rock that does not have an aphanitic counterpart. Peridotite is rare at Earth's surface, but is very likely the rock that makes up the mantle.

Figure 4.13 The Common Igneous Rocks For the composition and textures of these rocks see the classification of igneous rocks in Figure 4.11.

Critical Thinking Question How can you tell from these images which specimens are volcanic and which are plutonic?

Sue Monroe

a Basalt.

Sue Monroe

b Gabbro.

Sue Monroe

c Andesite.

Sue Monroe

d Diorite.

Sue Monroe

e Rhyolite.

Sue Monroe

f Granite.

crust, but intermediate to felsic intrusive rocks are much more common. However, the lower part of the oceanic crust is composed of gabbro.

Andesite-Diorite Intermediate-composition magma (53%–65% silica) crystallizes to form *andesite* and *diorite*, which are compositionally equivalent fine- and coarse-grained igneous rocks (▌Figure 4.13c, d). Andesite and diorite are composed predominately of plagioclase feldspar, with the typical ferromagnesian component being amphibole or biotite (Figure 4.11). Andesite is medium to dark gray, but diorite has a salt-and-pepper appearance because of its white to light gray plagioclase and dark ferromagnesian silicates.

Andesite is a common extrusive igneous rock formed from lava erupted in volcanic chains at convergent plate boundaries. The volcanoes of the Andes Mountains of South America and the Cascade Range in western North America are composed, in part, of andesite. Intrusive bodies of diorite are fairly common in the continental crust.

ConnectionLink

For more about the volcanoes of the Cascade Range, see Chapter 23.

Rhyolite-Granite *Rhyolite* and *granite* crystallize from felsic magma (>65% silica) and are therefore silica rich rocks (▌Figure 4.13e, f). They consist mostly of potassium feldspar, sodium-rich plagioclase, and quartz, with perhaps some biotite and, rarely, amphibole (Figure 4.11). Because nonferromagnesian silicates predominate, rhyolite and granite are typically light colored. Rhyolite is fine grained, although most often it contains phenocrysts of potassium feldspar or quartz, and granite is coarse-grained. Granite porphyry is also fairly common.

Rhyolite lava flows are much less common than andesite and basalt flows. Recall that one control of magma viscosity is silica content. Thus, if felsic magma rises to the surface, it begins to cool, the pressure on it decreases, and gases are released explosively, usually yielding rhyolitic pyroclastic materials. The rhyolitic lava flows that do occur are thick and highly viscous and move only short distances.

Granite is a coarsely crystalline igneous rock with a composition corresponding to that of the field shown in Figure 4.11. Strictly speaking, not all rocks in this field are granites. For example, a rock with a composition close to the line separating granite and diorite is called *granodiorite*. To avoid the confusion that might result from introducing more rock names, we will follow the practice of referring to rocks to the left of the granite-diorite line in Figure 4.11 as *granitic*.

Granitic rocks are by far the most common plutonic rocks, although they are restricted to the continents (Geo-Focus 4.1). Most granitic rocks were intruded at or near convergent plate margins during mountain-building episodes. When these mountainous regions are uplifted and eroded, the vast bodies of granitic rocks forming their cores are exposed.

Pegmatite The term *pegmatite* refers to an igneous rock with a particular texture rather than a specific composition, but most pegmatites are composed mostly of quartz, potassium feldspar, and sodium-rich plagioclase, thus corresponding closely to granite. A few pegmatites are mafic or intermediate in composition and are appropriately called *gabbro* and *diorite pegmatites*. The most remarkable feature of pegmatites is the size of their minerals, which measure at least 1 cm across, and in some pegmatites they measure tens of centimeters or meters (▌Figure 4.14). Many pegmatites are adjacent to large granite plutons and are composed of minerals that formed from the water-rich magma that remained after most of the granite crystallized.

▌**Figure 4.14** Pegmatite

Courtesy of Steve Stahl

a This pegmatite, the light-colored rock, is exposed in the Black Hills of South Dakota.

Copyright and Photograph by Dr. Parvinder S. Sethi

b Close-up view of a specimen from a pegmatite with minerals measuring 8 to 10 cm across.

GEO FOCUS 4.1 Granite—Common, Attractive, and Useful

Granite and closely related rocks are common in the large batholiths of the world. Indeed, if you live in the Northeast you are probably aware of New Hampshire's Old Man of the Mountain, a rugged granite outcrop that collapsed in 2003. Students in Georgia and North Carolina have likely visited Stone Mountain in their respective states (see chapter opening photo), which are huge masses of granite. Mount Rushmore in South Dakota is another mass of granite with the carved images of Presidents Washington, Jefferson, Lincoln, and Theodore Roosevelt. The Sierra Nevada batholith, mostly in California, is a huge composite body of granitic rocks with the highest peak in the continental United States (▮ Figure 1).

Today, no one doubts that granite and related rocks formed when magma cooled beneath the surface, but this has not always been the case. In 1787, a German professor of mineralogy, Abraham Gottlob Werner (1749–1817), proposed that all rocks, including basalt and granite, had precipitated in an orderly sequence from a primeval, worldwide ocean. Werner further claimed that metamorphic rocks and granite were the oldest rocks on Earth. His concept, called Neptunism, was popular for many years, but it eventually became apparent that basalt cooled from lava, whereas granite cooled from magma as

Sue Monroe

▮ Figure 1 Granite in the Sierra Nevada View of the Sierra Nevada taken west of Lone Pine, California. The rocks in this view are part of the Sierra Nevada batholith, a huge mass of granite and related rocks made up of many intrusive bodies. The high peak toward the right is Mount Whitney, which at 4,421 m is the highest peak in the continental United States.

When felsic magma cools and forms granite, the remaining water-rich magma has properties that differ from the magma from which it separated. It has a lower density and viscosity and invades cracks in the nearby rocks where minerals crystallize. This water-rich magma also contains elements that rarely enter into the common minerals that form granite. Pegmatites that are essentially very coarsely crystalline granite are simple pegmatites, whereas those with minerals containing elements such as lithium, beryllium, cesium, boron, and several others are complex pegmatites. Some complex pegmatites contain 300 mineral species, a few of which are important economically. In addition, several gem minerals such as emerald and aquamarine, both of which are varieties of the silicate mineral beryl, and tourmaline are found in some pegmatites.

The formation and growth of mineral-crystal nuclei in pegmatites are similar to those processes in other magmas, but with one critical difference: the water-rich magma from which pegmatites crystallize inhibits the formation of nuclei. However, some nuclei do form, and because the appropriate ions in the liquid can move easily and attach themselves to a growing crystal, individual minerals may grow very large.

Other Igneous Rocks Geologists classify the igneous rocks in Figure 4.11 by texture and composition, but a few others are identified primarily by their textures (▮ Figure 4.15).

we have outlined. Furthermore, rock type is not an indication of its age as Werner proposed.

Granitic rocks are quarried, cut and polished, and otherwise shaped to meet specific purposes. Any stone used in this fashion is called *dimension stone*, and in the case of granitic rocks, they may be used for facing stones on buildings, walkways (▮ Figure 2), tombstones, mantelpieces, countertops, and monuments. A 23-m-high, red granite obelisk weighing 250 metric tons was carved from a single piece of granite at Aswan, Egypt, about 3,300 years ago and transported nearly 200 km to its present location at Luxor, Egypt.

Plymouth Rock at Plymouth, Massachusetts, has great symbolic value, but otherwise it is a rather ordinary rock (▮ Figure 3). Actually, it is a rather small stone from the 600-million-year-old Dedham Granodiorite, which was transported by a glacier to its present location. As its name indicates, granodiorite has a composition between granite and diorite (Figure 4.11). Legend has it that Plymouth Rock is the landing place where the Pilgrims first set foot in the New World in 1620. In fact, the Pilgrims first landed at Provincetown on Cape Cod and later went to the Plymouth area, but even then they probably landed north of Plymouth.

The term *granite* has a specific meaning in geology, but in commercial use the term is applied to several rocks, such as basalt and gabbro, as in black granite. True granite is in fact used for many purposes, although many granite-like substances are manufactured. This *cultured granite*, as it is called, is made by chemically bonding material and molding them into various shapes, which cannot be done with natural stones.

James S. Monroe

▮ **Figure 2 Granite Dimension Stone** Any rock that is shaped for a specific purpose is called dimension stone. The granite used here for the walkway, steps, and facing for the plant enclosure is at the Mena House Oberia, a luxury hotel near Cairo, Egypt.

Mark E. Gibson/Encyclopedia/Corbis

▮ **Figure 3** Plymouth Rock is the first piece of land on which the Pilgrims supposedly set foot when they arrived in Massachusetts in 1620. The rock first became a patriotic icon during the Revolutionary War.

Much of the fragmental material erupted by volcanoes is *ash*, a designation for pyroclastic materials measuring less than 2 mm, most of which consists of pieces of minerals or shards of volcanic glass (Figure 4.10f). The consolidation of ash forms the pyroclastic rock *tuff* (▮ Figure 4.16a). Most tuff is silica-rich and light colored and is appropriately called *rhyolite tuff.* Some ash flows are so hot that as they come to rest, the ash particles fuse together and form a *welded tuff.* Consolidated deposits of larger pyroclastic materials, such as cinders, blocks, and bombs, are *volcanic breccia* (Figure 4.15).

Both *obsidian* and *pumice* are varieties of volcanic glass (▮ Figure 4.16b, c). Obsidian may be black, dark gray, red, or brown, depending on the presence of iron. Obsidian breaks with the conchoidal (smoothly curved) fracture typical of glass. Analyses of many samples indicate that most obsidian has a high silica content and is compositionally similar to rhyolite.

 Pumice is a variety of volcanic glass containing numerous vesicles that develop when gas escapes through lava and forms a froth (Figure 4.16c). If pumice falls into water, it can be carried great distances because it is so porous and not very permeable so it floats. Another vesicular rock is *scoria.* It is more crystalline and denser than pumice, but it has more vesicles than solid rock (▮ Figure 4.16d).

ConnectionLink

To learn more about porosity and permeability, see Chapter 13.

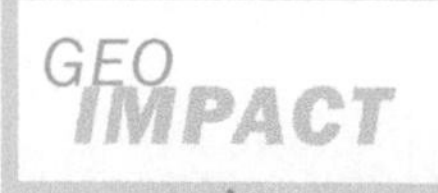

Classification and History of Igneous Rocks

As the only member of your community with any geology background you are considered the local expert on minerals and rocks. Suppose one of your friends brings you a rock specimen with the following features—composition: mostly potassium feldspar and plagioclase feldspar with about 10% quartz and minor amounts of biotite; texture: minerals average 3 mm across, but several potassium-feldspars are up to 5 cm. Give the specimen a rock name and tell your friend as much as you can about the rock's history.

	Composition	Felsic ←→ Mafic	
Texture	Vesicular	Pumice	Scoria
	Glassy	Obsidian	
	Pyroclastic or fragmental	← Volcanic Breccia → Tuff/welded tuff	

Figure 4.15 Texture Classification Classification of igneous rocks for which texture is the main consideration. Composition is shown, but it is not essential for naming these rocks.

Figure 4.16 Examples of Igneous Rocks Classified Primarily by Their Texture

Critical Thinking Question Why do pumice and scoria have so many vesicles but obsidian has none?

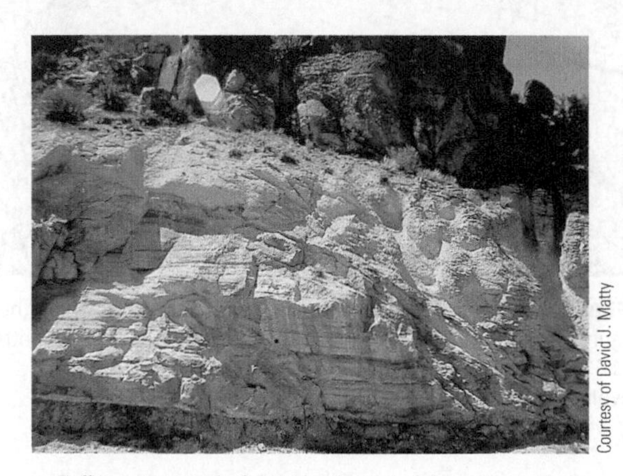

Courtesy of David J. Matty

a Tuff is composed of pyroclastic materials such as those in Figure 4.10f.

Sue Monroe

b The natural glass obsidian.

Sue Monroe

c Pumice is glassy and extremely vesicular.

Sue Monroe

d Scoria is also vesicular, but it is darker, heavier, and more crystalline than pumice.

4.5 Intrusive Igneous Bodies—Plutons

Unlike volcanism and the origin of volcanic rocks, we can study intrusive igneous bodies, collectively called **plutons,** only indirectly because intrusive rocks form when magma cools and crystallizes within Earth's crust. Furthermore, geologists cannot duplicate the conditions under which intrusive rocks form except in small-scale laboratory experiments.

Geologists recognize several types of plutons based on their geometry (three-dimensional shape) and relationships to the country rocks. In terms of their geometry, plutons are tabular, cylindrical, or irregular (massive). Furthermore, they may be **concordant,** meaning they have boundaries that parallel the layering in the country rock, or **discordant,** with boundaries that cut across the country rock's layering (❚ Figure 4.17).

Dikes, Sills, and Laccoliths

Dikes and **sills** are tabular or sheetlike igneous bodies that differ only in that dikes are discordant and sills are concordant (❚ Figure 4.18). Dikes are quite common and range from a few centimeters to more than 100 m thick. Invariably, they are intruded into preexisting fractures or where fluid pressure is great enough for them to form their own fractures as they move upward into county rock. Erosion of the Hawaiian volcanoes exposed dikes in fracture zones, and the Columbia River basalts in Washington State issued from long fissures in which magma solidified to form dikes. Dikes underlie the Eldgja fissure and Laki fissure in Iceland, both of which have been the sites of voluminous eruptions in 934 and 1783–1784, respectively.

Sills are tabular just as dikes are, but they are concordant (Figure 4.18b). Many sills are a meter or less thick, although some are much thicker. A well-known sill in the United States is the Palisades sill that forms the Palisades along the west side of the Hudson River in New York and New Jersey (see Figure 22.7d). This 300-m-thick sill is exposed for 60 km along the river. Most sills were intruded into sedimentary rocks, but eroded volcanoes also reveal that sills are injected into piles of volcanic rocks. In fact, some of the inflation of a volcano that precedes an eruption may be caused by the injection of sills. In contrast to dikes, which follow zones of weakness, sills are intruded between layers in country rock when the fluid pressure is great enough for the magma to lift the overlying rocks.

Under some circumstances, a sill inflates and causes the overlying rocks to bow upward, forming an igneous body called a **laccolith.** A laccolith has a flat floor and is domed in its central part, giving it a mushroom-like geometry (Figure 4.17). Like sills, laccoliths are rather shallow intrusions that lift the overlying rocks. Well-known laccoliths in the United States are in the Henry Mountains of southeastern Utah, and several buttes in Montana are eroded laccoliths (❚ Figure 4.19).

Volcanic Pipes and Necks

Volcanoes have a cylindrical conduit known as a **volcanic pipe** that connects to an underlying magma chamber. Magma rises through this structure; however, when a volcano

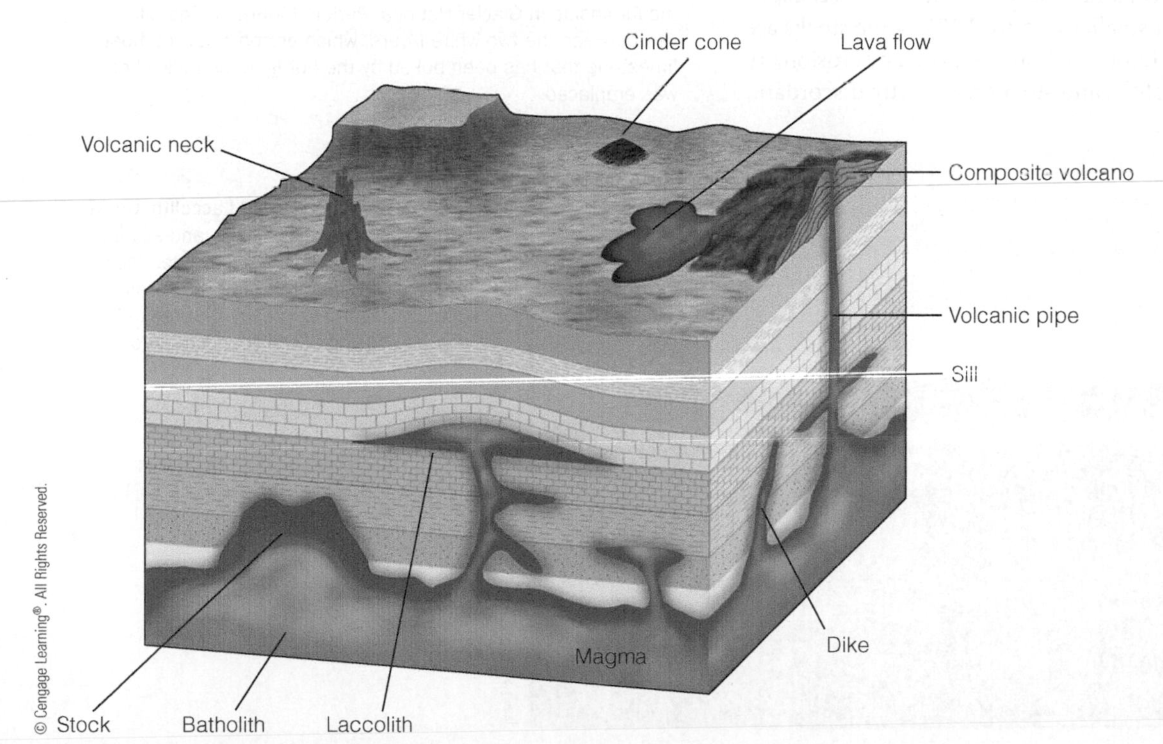

❚ Figure 4.17 Block Diagram Showing Various Plutons Some plutons cut across the layering in country rock and are discordant, whereas others parallel the layering and are concordant.

Figure 4.18 Dike and Sill Dikes and sills are made up of igneous rocks emplaced in country rock. Dikes cut across the layering in the country rock and are thus discordant, whereas sills parallel the layering in the country rock and are concordant. Both dikes and sills have a sheetlike geometry, but in these views we can see them only in two dimensions.

a This dike is made up of dark colored igneous rock that cuts across the layering in metamorphic rocks in Death Valley, California.

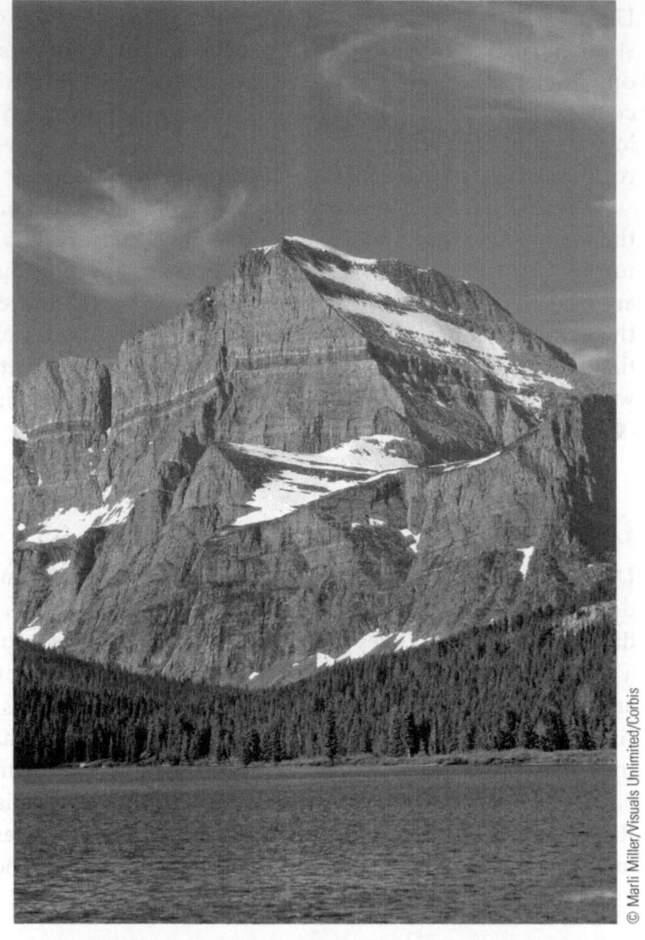

b The Purcell Sill is the black band of igneous rock on this mountainside in Glacier National Park in Montana. The sill lies between the two white layers, which are parts of the host limestone that has been baked by the hot igneous rock when it was emplaced.

ceases to erupt, its slopes are attacked by weathering and erosion, but the magma in the pipe is commonly more resistant to erosion and is left as a remnant called a **volcanic neck** (Figures 4.20 and 4.21). Several volcanic necks are found in the southwestern United States, especially in Arizona and New Mexico, and others are recognized elsewhere.

Batholiths and Stocks

A **batholith,** the largest of all plutons, must have at least 100 km^2 of surface area, and most are far larger. A **stock,** in contrast, is similar but smaller (Figure 4.17). Some stocks are simply parts of large plutons that once exposed by erosion are batholiths. Both batholiths and stocks are mostly discordant,

Figure 4.19 An Eroded Laccolith Crown Butte near Simms, Montana, and several other buttes in the area are eroded laccoliths that were intruded into the crust during an episode of igneous activity that took place between 70 and 80 million years ago.

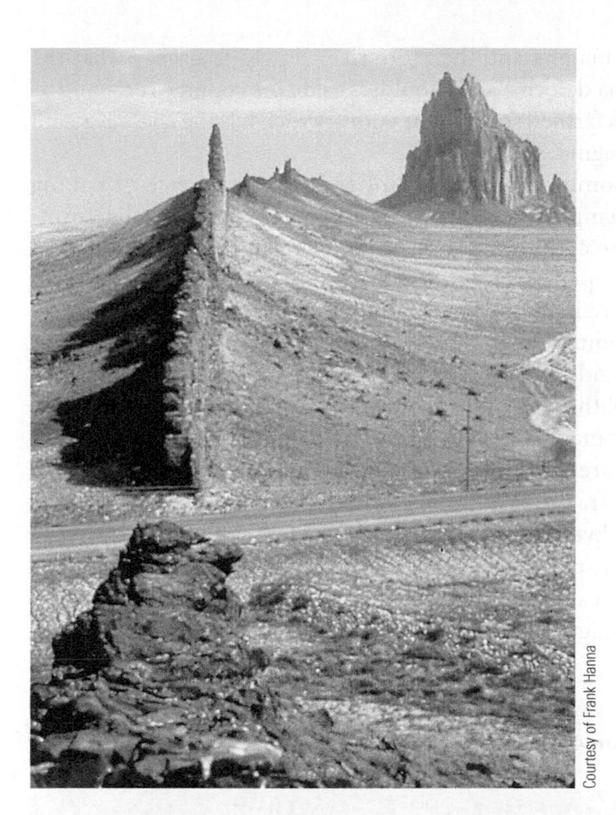

Figure 4.20 Volcanic Neck Shiprock, a volcanic neck in northwestern New Mexico, rises nearly 550 m above the surrounding plain. One of the dikes radiating from Shiprock is in the foreground.

Critical Thinking Question How can you account for the fact that the dike in the foreground stands above the surface like a wall?

although locally they may be concordant, and batholiths, especially, consist of multiple intrusions. In other words, a batholith is a large composite body produced by repeated, voluminous intrusions of magma in the same region. The coastal batholith of Peru, for instance, was emplaced during a period of 60–70 million years and is made up of as many as 800 individual plutons.

The igneous rocks that make up batholiths are mostly granitic, although diorite may also be present. Batholiths and stocks are emplaced mostly near convergent plate boundaries during episodes of mountain building. One example is the Sierra Nevada batholith of California, which formed over millions of years. Other large batholiths in North America include the Idaho batholith, the Boulder batholith in Montana, and the Coast Range batholith in British Columbia, Canada (Geo-Focus 4.1).

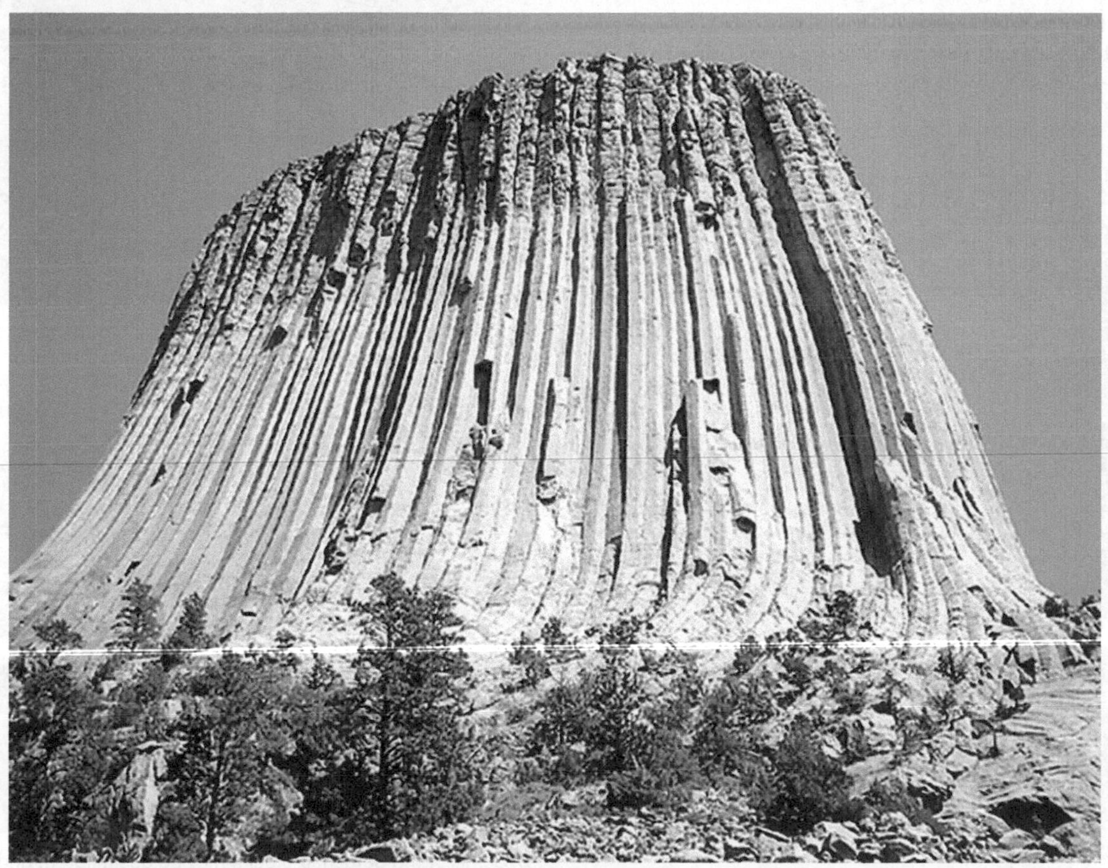

Figure 4.21 Devils Tower, Wyoming Devils Tower National Monument was the first area in the United States so designated. The tower rises more than 260 m above its base and it can be seen from 48 km away. The vertical lines result from intersections of fractures called columnar joints (see Chapter 5), but according to Cheyenne legend, a gigantic grizzly bear made the scratches. Traditionally, Devils Tower was thought to be a volcanic neck, but it now seems more likely that it is an eroded laccolith.

Mineral resources are found in rocks of batholiths and stocks and in the adjacent country rocks. Near Salt Lake City, Utah, copper is mined from the mineralized rocks of the Bingham stock, a composite pluton composed of granite and granite porphyry. Granitic rocks also are the primary source of gold, which forms from mineral-rich solutions moving through cracks and fractures of the igneous body.

4.6 The Origin of Batholiths

Geologists realized long ago that the origin of batholiths posed a space problem. What happened to the rock that was once in the space now occupied by a batholith? One solution to the space problem is that these large igneous bodies melted their way into the crust. In other words, they simply assimilated the country rock as they moved upward (Figure 4.9a). The presence of inclusions, especially near the tops of some plutons, indicates that assimilation does occur. Nevertheless, as we noted, assimilation is a limited process because magma cools as country rock is assimilated. Calculations indicate that far too little heat is available in magma to assimilate the huge quantities of country rock necessary to make room for a batholith.

Geologists now generally agree that batholiths were emplaced by *forceful injection* as magma moved upward (▌Figure 4.22). Recall that granite is derived from viscous felsic magma and therefore rises slowly. It appears that the magma deforms and shoulders aside the country rock, and as it rises farther, some of the country rock fills the space beneath the magma.

Some batholiths do indeed show evidence of having been emplaced forcefully by shouldering aside and deforming the country rock. This mechanism probably occurs in the deeper parts of the crust where temperature and pressure are high and the country rocks are easily deformed in the manner described. At shallower depths, the crust is more rigid and tends to deform by fracturing. In this environment, batholiths may move upward by **stoping,** a process in which rising magma detaches and engulfs pieces of country rock (▌Figure 4.23). According to this concept, magma moves up along fractures and the planes separating layers of country rock. Eventually, pieces of country rock detach and settle into the magma. No new room is created during stoping; the magma simply fills the space formerly occupied by country rock (Figure 4.23).

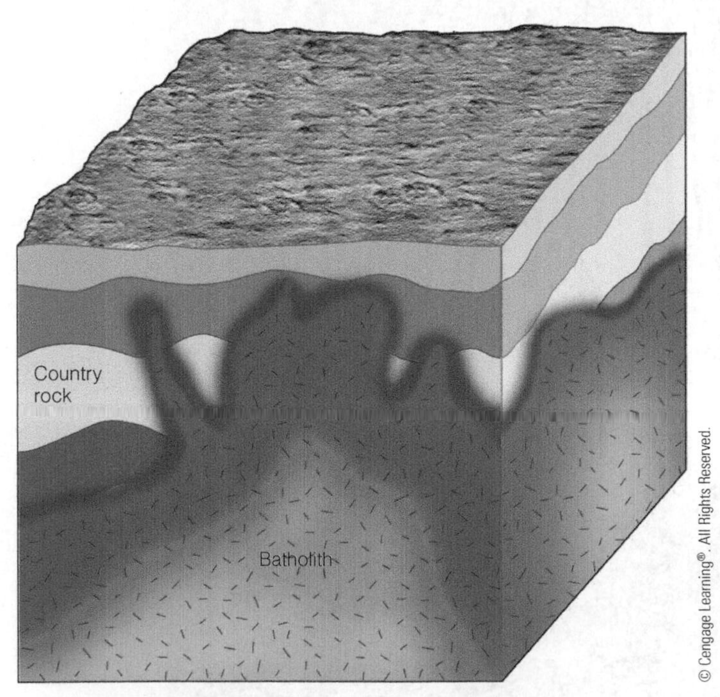

▌**Figure 4.22 Emplacement of a Hypothetical Batholith** As the magma rises, it shoulders aside and deforms the country rock.

▌**Figure 4.23 Emplacement of a Batholith by Stoping**

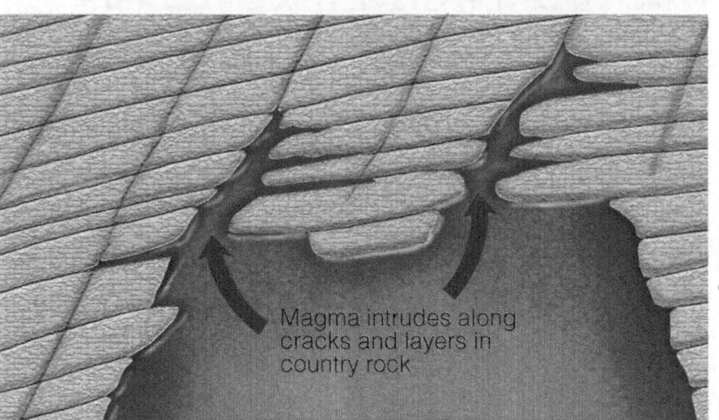

a Stoping takes place when magma rises into the crust by detaching and engulfing pieces of country rock.

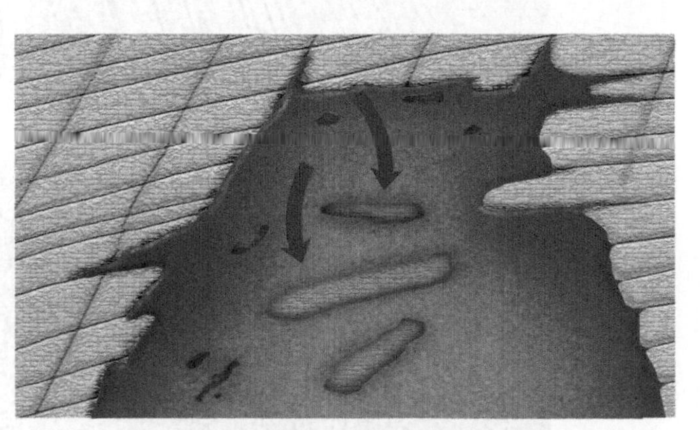

b Some of the detached blocks may be assimilated, and some may remain as inclusions (Figure 4.9).

Key Concepts Review

- Magma is the term for molten rock below Earth's surface, whereas the same material at the surface is lava.
- Silica content is what distinguishes ultramafic (<45% silica), mafic (45%–52% silica), intermediate (53%–65% silica), and felsic (>65% silica) magmas.
- Magma and lava viscosity depends mostly on temperature and composition: the higher the temperature, the lower the viscosity; the more silica, the greater the viscosity.
- Minerals crystallize from magma and lava when small crystal nuclei form and grow.
- Rapid cooling accounts for the aphanitic textures of volcanic rocks, whereas comparatively slow cooling yields the phaneritic textures of plutonic rocks. Igneous rocks with markedly different-sized minerals are porphyritic.
- Igneous rock composition is determined mostly by the composition of the parent magma, but magma composition can change so that the same magma may yield more than one kind of igneous rock.
- According to Bowens reaction series, cooling mafic magma yields a sequence of minerals, each of which is stable within specific temperature ranges. Only ferromagnesian silicates are found in the discontinuous branch of Bowen's reaction series. The continuous branch yields only plagioclase feldspars that become increasingly enriched with sodium as cooling occurs.
- A chemical change in magma may take place as early ferromagnesian silicates form and, because of their density, settle in the magma.
- Compositional changes also take place in magma when it assimilates country rock or one magma mixes with another.
- Geologists recognize two broad categories of igneous rocks: volcanic or extrusive and plutonic or intrusive.
- Texture and composition are the criteria used to classify igneous rocks, although a few are defined mostly by texture.
- Crystallization from water-rich magma results in very large minerals that form rocks known as pegmatite. Most pegmatite has an overall composition similar to granite.
- Intrusive igneous bodies known as plutons vary in their geometry and their relationship to country rock: some are concordant, whereas others are discordant.
- The largest plutons, known as batholiths, consist of multiple intrusions of magma during long periods of time.
- Most plutons, including batholiths, are found at or near divergent and convergent plate boundaries.

Important Terms

aphanitic texture (p. 95)
assimilation (p. 94)
batholith (p. 104)
Bowen's reaction series (p. 90)
concordant (p. 103)
country rock (p. 94)
crystal settling (p. 93)
dike (p. 103)
discordant (p. 103)
felsic magma (p. 89)
hot spot (p. 93)
igneous rock (p. 88)
intermediate magma (p. 89)
laccolith (p. 103)
lava (p. 88)
mafic magma (p. 89)
magma (p. 88)
magma chamber (p. 90)
magma mixing (p. 94)
mantle plume (p. 93)
phaneritic texture (p. 95)
pluton (p. 103)
plutonic (intrusive igneous) rock (p. 88)
porphyritic texture (p. 95)
pyroclastic (fragmental) texture (p. 96)
pyroclastic materials (p. 88)
sill (p. 103)
stock (p. 104)
stoping (p. 106)
ultramafic magma (p. 89)
vesicle (p. 96)
viscosity (p. 89)
volcanic (extrusive igneous) rock (p. 88)
volcanic neck (p. 104)
volcanic pipe (p. 103)

Review Questions

1. A laccolith is what kind of pluton?
   a. _____ disordant/massive.
   b. _____ concordant/porphyritic.
   c. _____ felsic/gabbroic.
   d. _____ concordant/mushroom-shaped.
   e. _____ ultramafic/viscous.
2. Which pair of the following igneous rocks has the same mineral composition?
   a. _____ andesite-diorite.
   b. _____ granite-basalt.
   c. _____ tuff-gabbro.
   d. _____ obsidian-aphaneitic.
   e. _____ peridotite-rhyolite.
3. Which one of the following statements about batholiths is correct?
   a. _____ They are composed of a single, huge mass of magma.
   b. _____ They consist primarily of basalt and gabbro.
   c. _____ They form at divergent plate boundaries.
   d. _____ They are mostly concordant.
   e. _____ They are composed mostly of granitic rocks.
4. The phenomenon by which pieces of country rock are detached and engulfed by rising magma is
   a. _____ crystal inversion.
   b. _____ magma mixing.
   c. _____ recapitulation.
   d. _____ stoping.
   e. _____ ultramafic accumulation.
5. The order in which minerals crystallized from mafic magma is called
   a. _____ crystal settling.
   b. _____ Bowen's reaction series.
   c. _____ Richard's assimilation index.
   d. _____ plutonic mixing.
   e. _____ crustal differentiation.
6. How do assimilation and stoping contribute to the emplacement of batholiths in the continental crust?
7. What are the controls on the viscosity of a lava flow?
8. Two phaneritic rocks have the following compositions: Specimen 1: 5% biotite, 20% sodium-rich plagioclase, 65% potassium feldspar, and 10% quartz; Specimen 2: 10% olivine, 55% pyroxene, 5% hornblende, and 30% calcium-rich plagioclase. Use Figure 4.11 and classify these rocks. Which one would you expect to be the darkest and densest? Explain.
9. How do aphanitic and phaneritic textures form, and what do they tell you about rocks with these textures?

10. **Creative Thinking Visual Question:** Figure 1 on the next page shows an imaginary landscape made up of sandstone (SS) and the following igneous rocks: basalt (B), diorite (D), granite (GR), and pegmatite (PG).

    What type of igneous body is made up of gabbro?
    Is the diorite older or younger than the sandstone, and is the granite older or younger than the diorite? How do you know?

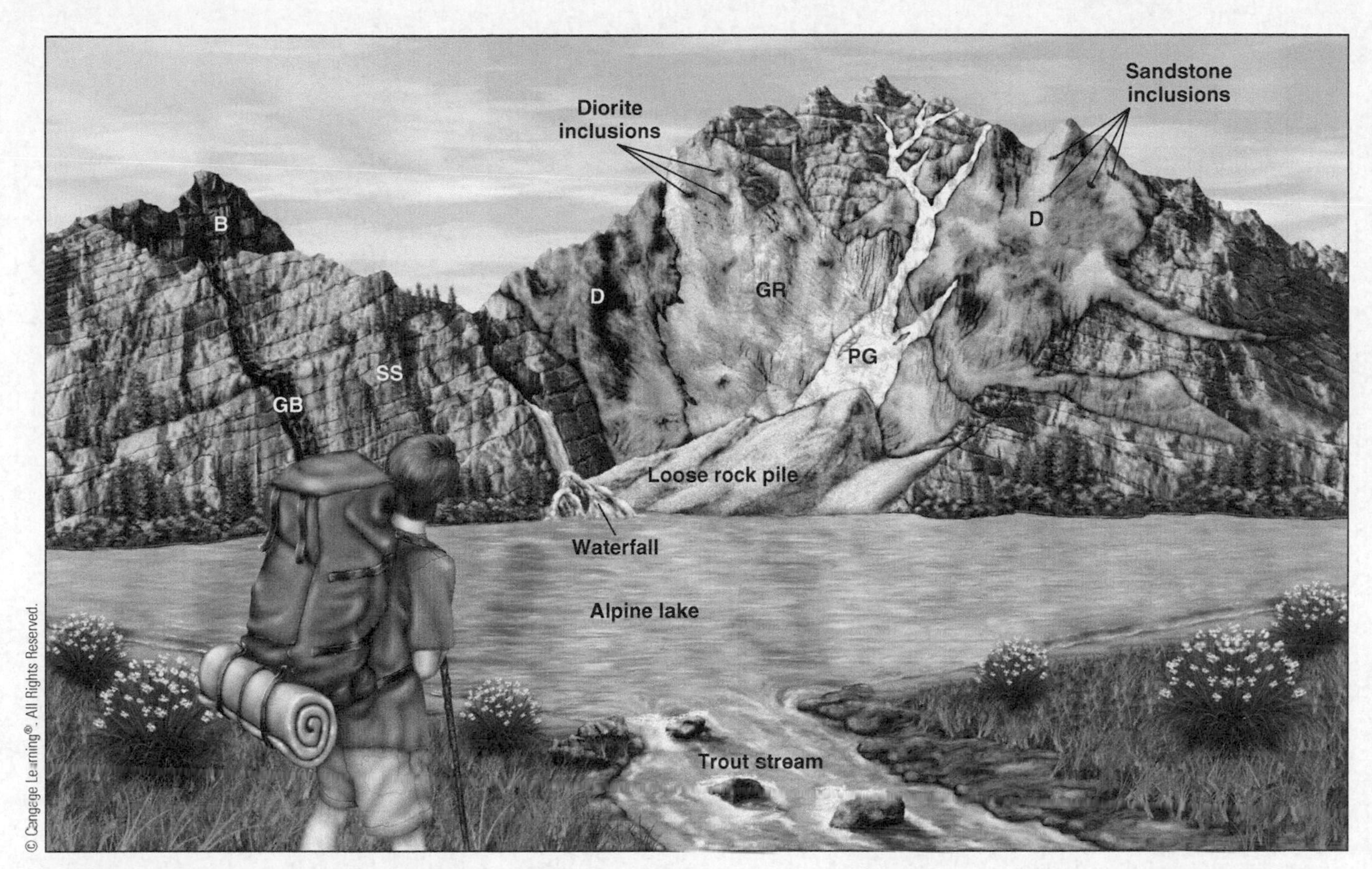

Figure 1 Imaginary landscape showing the relationships between several bodies of igneous rocks and sedimentary rocks.

Global GeoScience Watch Within the GREENR database, search for the term "magma chamber." In the "Magazines" section, open the article titled "Yellowstone Rising: Magma Floods into Chamber Beneath Park." Where does the heat come from that accounts for the hydrothermal activity in that area? How deep is the magma chamber beneath Yellowstone National Park, and how large an area does it affect?

Anak Krakatau, which means “daughter of Krakatau,” is a small volcano in the Sundra Strait between Java and Sumatra that is the remnant of a much larger volcano known as Krakatau. In 1883, Krakatau erupted with such explosive force that most of the island disappeared and the volcano collapsed in on itself forming a huge depression known as a caldera. Renewed volcanism built up Anak Krakatau which rose from the ocean in 1927. It is shown here erupting on November 3, 2010.

CHAPTER 5

Volcanoes and Volcanism

OUTLINE

HAVE YOU EVER WONDERED?

- What besides lava is emitted from erupting volcanoes?
- How and why volcanoes differ from one another?
- What the most dangerous types of volcanoes are?
- Whether volcanoes are really as dangerous as depicted in movies and television shows?
- Where the active volcanoes are in the United States?
- Whether there is a relationship between plate boundaries and volcanoes?
- How large volcanic eruptions are and whether they can be forecast with any certainty?

Stephen Belcher/Foto Natura/Minden Pictures/Getty Images

5.1 Introduction

Volcanic eruptions are certainly one of the most awe-inspiring geologic phenomena, especially when glowing streams of lava pour from volcanic vents or when volcanoes erupt explosively and eject pyroclastic materials high into the atmosphere. What better topic for a disaster movie, some of which show lava flows as a great danger to people. Lava flows do in fact destroy buildings and cover what may have been productive land, but overall, they are the least dangerous manifestation of volcanic eruptions. In 2002, though, lava flows in Goma, Democratic Republic of the Congo, killed 147 people when they caused gasoline storage tanks to explode.

It is difficult to get a precise number of *active volcanoes,* that is, volcanoes that have erupted during historic time, but a common estimate is about 550, 10 to 20 of which are erupting at any one time. We also know of several active submarine volcanoes, but there are certainly many more. Most volcanic eruptions are minor and go unreported in the popular press unless they occur in populated areas or have tragic consequences. Recall the April 2010 eruptions of Eyjafjallajökull in Iceland. The eruptions were not very large, but they took place beneath 200 m of glacial ice resulting in steam explosions that ejected huge amounts of volcanic ash into the atmosphere, disrupting air traffic over the North Atlantic for several days (❚ Figure 5.1).

In addition to active volcanoes, Earth has many *dormant volcanoes,* meaning that they have not erupted in historic time but may do so again. The distinction between active and dormant is not easy to make, because some volcanoes that have not erupted for centuries become active again. For instance, in 1991, Mount Pinatubo in the Philippines erupted after lying dormant for 600 years, and similarly Mount Sinabung in Indonesia began erupting in August 2010 after a hiatus of 400 years. Mount Vesuvius in Italy had not erupted in human memory until its AD 79 eruptions destroyed the towns of Herculaneum, Pompeii, and Stabiae (❚ Figure 5.2).

Thousands of volcanoes have not erupted during historic time and show no signs of doing so in the future so they are *extinct* or *inactive volcanoes.* The Sutter Buttes in California last erupted about 1.5 million years ago but now show no activity and will probably not erupt again. Nevertheless, whether a volcano is truly extinct is not easy to determine, because some volcanic centers have long lifetimes with tens of thousands of years of little or no activity following by renewed eruptions. The Yellowstone caldera in Yellowstone National Park, Wyoming, probably falls into this category; its last eruption was 70,000 years ago, but ongoing seismic

bt3/ZUMA Press/ Newscom

❚ **Figure 5.1** Eyjafjallajökull in Iceland erupting a huge cloud of volcanic ash and steam on April 17, 2010. The eruption, although not very large, took place beneath a glacier that added to the cloud over the volcano. Air traffic over the North Atlantic was disrupted for several days.

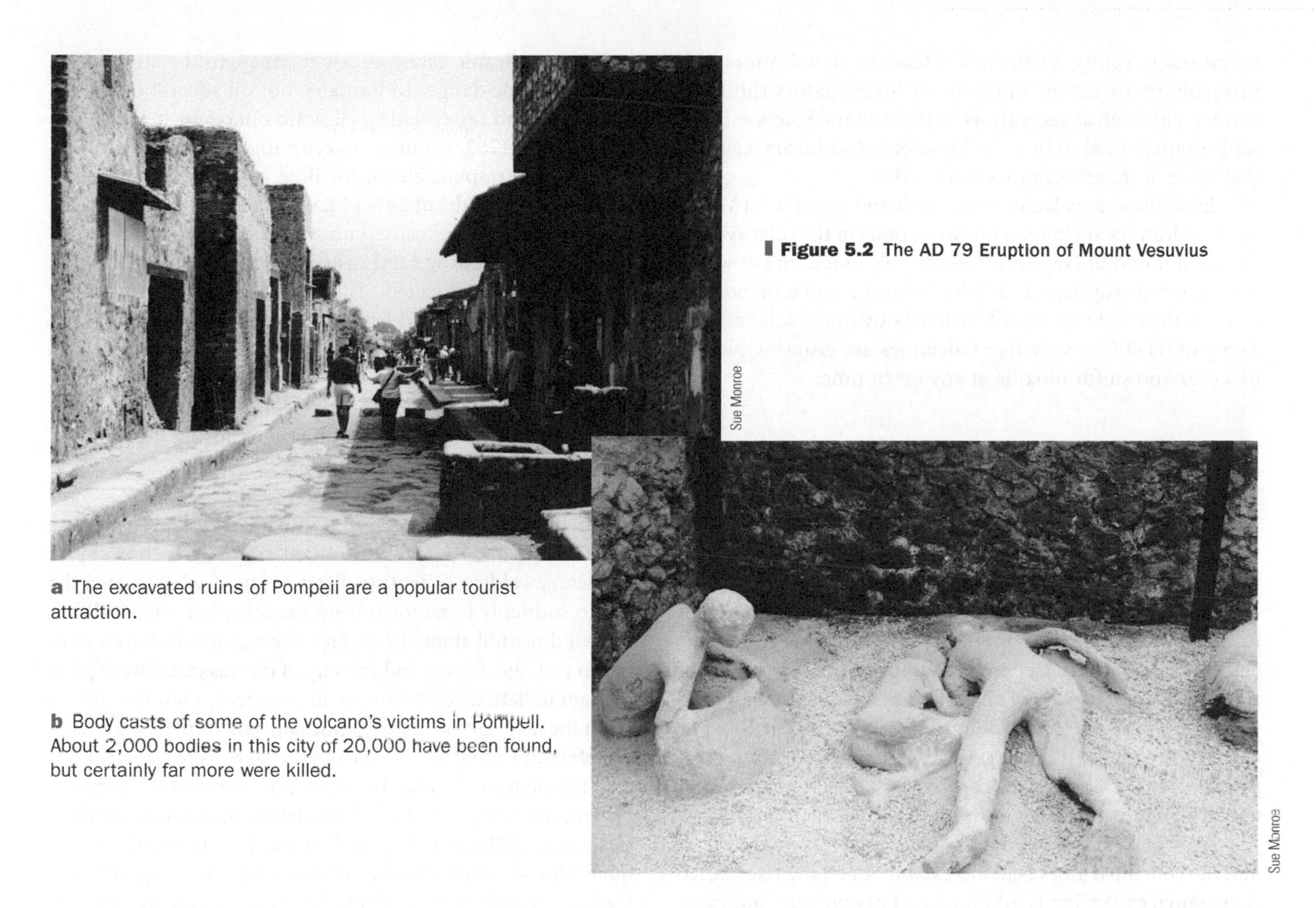

Figure 5.2 The AD 79 Eruption of Mount Vesuvius

a The excavated ruins of Pompeii are a popular tourist attraction.

b Body casts of some of the volcano's victims in Pompeii. About 2,000 bodies in this city of 20,000 have been found, but certainly far more were killed.

Sue Monroe

Sue Monroe

activity and hydrothermal (hot water) activity indicate that it is not extinct.

Erupting volcanoes do cause injuries and fatalities and considerable damage, but when considered in the context of Earth history, volcanism is a constructive process. Volcanic emissions during early Earth history played an important role in the origin and evolution of the atmosphere and hydrosphere. The continents are composed in part of volcanic and plutonic rocks, and oceanic crust is produced continuously by extrusive and intrusive igneous activity at divergent plate boundaries. Volcanic islands such as the Hawaiian Islands, Iceland, the Azores, and many others owe their existence, and in some cases, their continuing evolution, to volcanic eruptions. In addition, in areas such as Indonesia weathering of lava flows, pyroclastic materials, and volcanic mudflows converts them to productive soils.

One reason to study volcanism is that eruptions of lava and pyroclastic materials are responsible for some igneous rocks, specifically those that geologists call *volcanic* rocks or *extrusive igneous rocks*. Recall from Chapter 4 that *intrusive igneous rocks* or *plutonic rocks* result when magma cools and crystallizes within Earth's crust, and also that extrusive and intrusive igneous activity are related phenomena. Another reason to study this topic is that it illustrates the complex interactions among Earth's systems. The eruption of gases and pyroclastic materials has an immediate and profound effect on the atmosphere and hydrosphere, and biosphere, at least in the area of an eruption. And in some cases, the effects of eruptions are worldwide, as they were following the eruptions of Tambora in 1815, Krakatau in 1883, and Mount Pinatubo in 1991.

5.2 Volcanoes and Volcanism

A **volcano** is a hill or mountain that forms around a vent where lava, pyroclastic materials, and gases erupt, so it is a landform—a feature on Earth's surface. The term **volcanism** refers to all processes related to the rise and discharge of magma and gases at the surface or into the atmosphere. Thus, volcanism accounts for the origin of all volcanic (extrusive igneous) rocks, such as basalt, tuff, and obsidian, as well as volcanoes, but not all volcanism results in the origin of volcanoes. Some eruptions take place along fissures and build up basalt plateaus (discussed later in this chapter).

All of the other terrestrial planets (Mercury, Venus, Mars) and Earth's Moon had many active volcanoes during their early history, but now the only one with possible active volcanoes is Venus. Venus has at least 1,600 volcanoes, and probably many more, which most investigators think are extinct, although as recently as 2010, Suzanne Smerkar of the Jet Propulsion Laboratory in Pasadena, California, claimed that three of these volcanoes were active.

In addition to volcanism on Earth and possibly on Venus, active volcanoes are known on two moons in the solar system. Triton, a moon of Neptune, has ice volcanoes, one of which was captured erupting in 1989 by *Voyager 2*. Jupiter's moon Io is by far the most volcanically active body in the solar system. Many of its 400 or so active volcanoes are erupting plumes of sulfur and sulfur dioxide at any given time.

Volcanic Gases

Present-day volcanoes emit several gases, but in most cases water vapor (H_2O) is the most common followed by carbon dioxide (CO_2), sulfur dioxide (SO_2), and some others in minor quantities. In areas of recent or ongoing volcanism, one cannot help but notice the rotten-egg odor of hydrogen sulfide (H_2S) (▌Figure 5.3a). Depending on the volcano, water vapor may make up 37%–97% of all gases, but the ones that are dangerous to humans, livestock, and plants are sulfur dioxide, carbon dioxide, and hydrogen fluoride. Indeed, some people think that chlorine gas from volcanoes is responsible for ozone depletion in Earth's atmosphere (Geo-Focus 5.1).

As magma rises toward the surface, the pressure is reduced and the contained gases begin to expand. In highly viscous, felsic magma, expansion is inhibited and gas pressure increases. Eventually, the pressure may become great enough to cause an explosion and produce pyroclastic materials. In contrast, low-viscosity mafic magma allows gases to expand and escape easily, so mafic magma usually erupts rather quietly.

Most volcanic gases quickly dissipate in the atmosphere and pose little danger to humans, but on several occasions they have had far-reaching climatic effects or have caused fatalities. In 1783, sulfur gases erupting from Laki fissure in Iceland were responsible for the Blue Haze famine, crop failures, and the deaths of 24% of Iceland's population. In 1816, a persistent "dry fog" caused unusually cold spring and summer weather in Europe and eastern North America. In North America, 1816 was called "The Year Without a Summer" or "Eighteen Hundred and Froze to Death." Killing frosts during the summer resulted in crop failures and food shortages. This particularly cold summer was attributed to the huge eruption of Tambora in 1815 in Indonesia, although an eruption of Mayon volcano in the Philippines during the previous year also probably contributed.

In 1986, in the African nation of Cameroon, 1,746 people died when a cloud of carbon dioxide engulfed them. The gas accumulated in the waters of Lake Nyos, which occupies a volcanic caldera. Scientists disagree about what caused the gas to suddenly burst forth from the lake, but once it did, it flowed downhill along the surface because it was denser than air. In fact, the density and velocity of the gas cloud were great enough to flatten vegetation, including trees, a few kilometers from the lake. Unfortunately, thousands of animals and many people, some as far as 23 km from the lake, were asphyxiated.

Residents of the island of Hawaii have coined the term *vog* for volcanic smog. Kilauea volcano has been erupting continuously since 1983, releasing small amounts of lava and copious quantities of carbon dioxide and sulfur dioxide (▌Figure 5.3b). Carbon dioxide is no problem because it dissipates quickly in the atmosphere, but sulfur dioxide produces a haze and the unpleasant odor of sulfur. Vog probably poses little or no health risk for tourists, but a long-term threat exists for residents of the west side of the island where vog is most common.

▌Figure 5.3 Volcano Gases

Critical Thinking Question Are volcanic gases dangerous to humans?

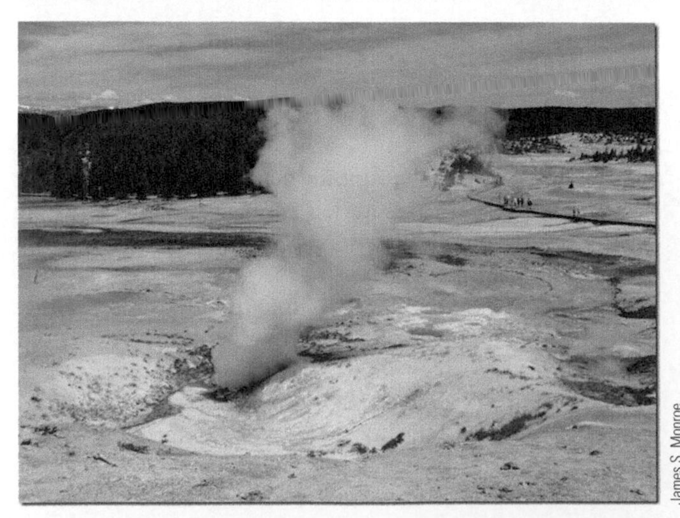

James S. Monroe

a This fumerole called the Black Growler is in Norris Geyser Basin in Yellowstone National Park, Wyoming.

James S. Monroe

b Gases Billowing from Halema 'uma'u Crater on Kilauea Volcano in Hawaii Volcanic gases are mostly water vapor, but carbon dioxide, sulfur dioxide, and several others are emitted, too. Sulfur dioxide produces a haze and the unpleasant odor of sulfur, or what people in Hawaii call *vog*.

Lava Flows

Movies and television shows depict lava flows as a great danger to humans, but only rarely do they cause fatalities. The reason is that most lava flows do not flow very rapidly, and because they are fluid, they follow low areas. So once a flow begins moving, determining the path it will take is fairly easy, and anyone in areas likely to be affected can be evacuated. From April 1990 to January 1991, lava flows covered Kalapana, Hawaii, and in the process destroyed 180 homes, highways, and points of archaeological interest. Civil Defense authorities working with geologists at the Hawaiian Volcano Observatory made important decisions regarding evacuations and road closures, and as a result, there were no injuries or fatalities.

Even low-viscosity flows, such as those in Hawaii, do not move very rapidly. They can move much more rapidly, though, when their margins cool to form a channel, and especially when insulated on all sides as in a **lava tube,** where geologists have recorded speeds of more than 50 km/hr. A lava tube forms within a flow when its margins and upper surface solidify, thereby forming a conduit through which lava can move rapidly and for great distances (❚ Figure 5.4a). When an eruption ceases, the tube drains, leaving an empty tunnel-like structure. Part of the roof of a lava tube may collapse to form a *skylight* through which an active flow can be observed, or an inactive lava tube can be accessed (❚ Figure 5.4b).

Geologists in Hawaii characterize basalt lava flows as pahoehoe or aa, although these are terms that are now also used elsewhere. **Pahoehoe** (pronounced *pay-hoy-hoy*) has a smooth ropy surface much like taffy (❚ Figure 5.5a). The surface of an **aa** (pronounced *ah-ah*) flow consists of rough, jagged, angular blocks and fragments (❚ Figure 5.5b). Pahoehoe flows are hotter and thinner than aa flows; indeed, aa flows are viscous enough to break into blocks and move forward as a wall of rubble. A pahoehoe flow may change along its length to aa as its viscosity increases, partly because it cools, but aa does not change to pahoehoe.

Much of the igneous rock in the upper oceanic crust is a distinctive type made up of bulbous masses of basalt that resemble pillows, hence the name **pillow lava.** Geologists realized long ago that pillow lava forms when lava is rapidly chilled underwater, but its formation was not observed until 1971. Divers near Hawaii saw pillows form when a blob of lava broke through the crust of an underwater lava flow and cooled very quickly, forming a pillow- or loaf-shaped structure with a glassy exterior. The remaining fluid lava inside then broke though the crust of an already-formed pillow and formed another pillow, and so on (❚ Figure 5.6).

Mafic lava flows, and some intermediate ones, as well as some rocks in dikes, sills, and volcanic necks, show a pattern of columns bounded by fractures, or what geologists call **columnar joints.** For columnar joints to form, a lava flow must cease moving, and cool and contract, thereby setting up forces that cause fractures, known as *joints,* to open. On a lava flow's surface, the fractures are commonly polygonal (often six-sided) cracks that extend downward, thus outlining columns with their long axes perpendicular to the cooling surface (❚ Figure 5.7 and see Figure 4.21).

Pyroclastic Materials

In addition to lava flows, erupting volcanoes eject pyroclastic materials (❚ Figure 5.8), especially **volcanic ash,** a designation for pyroclastic particles that measure less than

❚ **Figure 5.4** Lava Tubes When the margins and top of a lava flow cool and become solid, the lava in the interior of the flow continues to move leaving an open space called a lava tube.

USGS/Hawaii Volcanoes National Park

a An active lava tube in Hawaii. Part of the tube's roof has collapsed, forming a skylight.

Sue Monroe

b Jot Dean Cave is a lava tube on Medicine Lake Volcano in northeastern California. Notice that part of the lava tube's roof has collapsed, thus allowing access. This and several other lava tubes in the area are called ice caves, because ice that forms during the winter persists throughout the summer.

GEO FOCUS 5.1 Do Volcanic Gases Cause Ozone Depletion?

Earth supports life because of its distance from the Sun and the fact that it has abundant liquid water and an oxygen-rich atmosphere. Furthermore, an ozone layer (O_3) in the stratosphere (10–48 km above the surface) protects Earth because it blocks out most of the harmful ultraviolet radiation that bombards our planet (▮ Figure 1). During the early 1980s, scientists discovered an *ozone hole* over Antarctica that has continued to grow. In fact, depletion of the ozone layer is now also recognized over the Arctic region and elsewhere. Any depletion in ozone levels is viewed with alarm because it would allow more dangerous radiation to reach the surface, increasing the risk of skin cancer, among other effects.

This discovery unleashed a public debate about the primary cause of ozone depletion and how best to combat the problem. Scientists proposed that one cause of ozone depletion is chlorofluorocarbons (CFCs), which are used in various consumer products—for instance, in aerosol cans. According to this theory, CFCs rise into the upper atmosphere where reactions with ultraviolet radiation liberate chlorine, which in turn reacts with and depletes ozone (▮ Figure 2). As a result of this view, an international agreement called the Montreal Protocol was reached in 1983, limiting the production of CFCs, along with other ozone-depleting substances.

However, during the 1990s this view was challenged by some radio talk show hosts as well as a few government officials. They proposed an alternative idea that ozone depletion occurred because of natural causes rather than commercial products such as CFCs. They pointed out that volcanoes release copious quantities of hydrogen chloride (HCl) gas that rises into the stratosphere and that could be responsible for ozone depletion. Furthermore, they claimed that because CFCs are heavier than air they would not rise into the stratosphere.

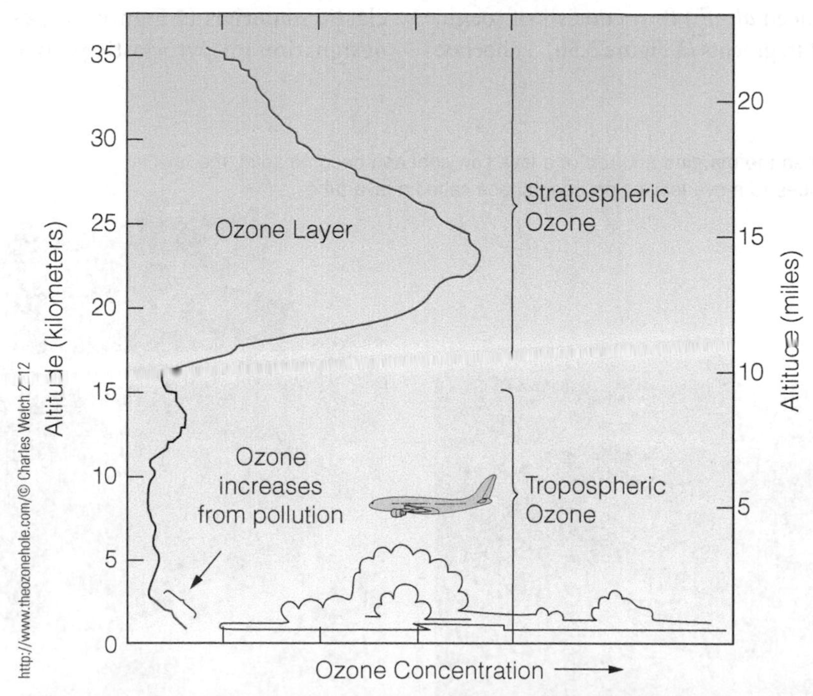

▮ **Figure 1** Most of the ozone in Earth's atmosphere is in the stratosphere where it filters out most of the incoming ultraviolet radiation.

It is true that volcanoes release HCl gas as well as several other gases, some of which are quite dangerous—recall the cloud of CO_2 gas in Cameroon that killed 1,746 people and the Blue Haze Famine in Iceland. However, most eruptions are too weak to inject gases of any kind high into the stratosphere. Even when it is released, HCl gas from volcanoes is very soluble and quickly removed from the atmosphere by rain and even by steam (water vapor) from the same eruption that released HCl gas in the first place. Measurements of chlorine concentrations in the stratosphere show that only temporary increases occur following huge eruptions. For example, the largest volcanic outburst since 1912, the eruption of Mount Pinatubo in the Philippines in 1991, caused little increase in upper atmosphere chlorine. The impact of volcanic eruptions is certainly not enough to cause the average rate of ozone depletion taking place each year.

It is true that CFCs are heavier than air, but this does not mean that they cannot rise into the stratosphere. Earth's surface heats differentially, meaning that more heat may be absorbed in one area than in an adjacent one. The heated air above a warmer area becomes less dense, rises by convection, and carries with it CFCs and other substances that are actually denser than air. Once in the stratosphere, ultraviolet radiation, which is usually absorbed by ozone, breaks up CFC molecules and releases chlorine that reacts with ozone. Indeed, a single chlorine atom can destroy 100,000 ozone molecules (Figure 2). In contrast to the HCl gas produced by volcanoes, CFCs are absolutely insoluble; it is the fact that they are inert that made them so desirable for various uses. Because a CFC molecule can last for decades, any increase in CFCs is a long-term threat to the ozone layer.

Another indication that CFCs are responsible for the Antarctic ozone hole is that the rate of ozone depletion has slowed since the implementation of the Montreal Protocol. A sound understanding of the science behind these atmospheric processes helped world leaders act quickly to address this issue. Now scientists hope that with continued compliance with the protocol, the ozone layer will recover by the middle of this century.

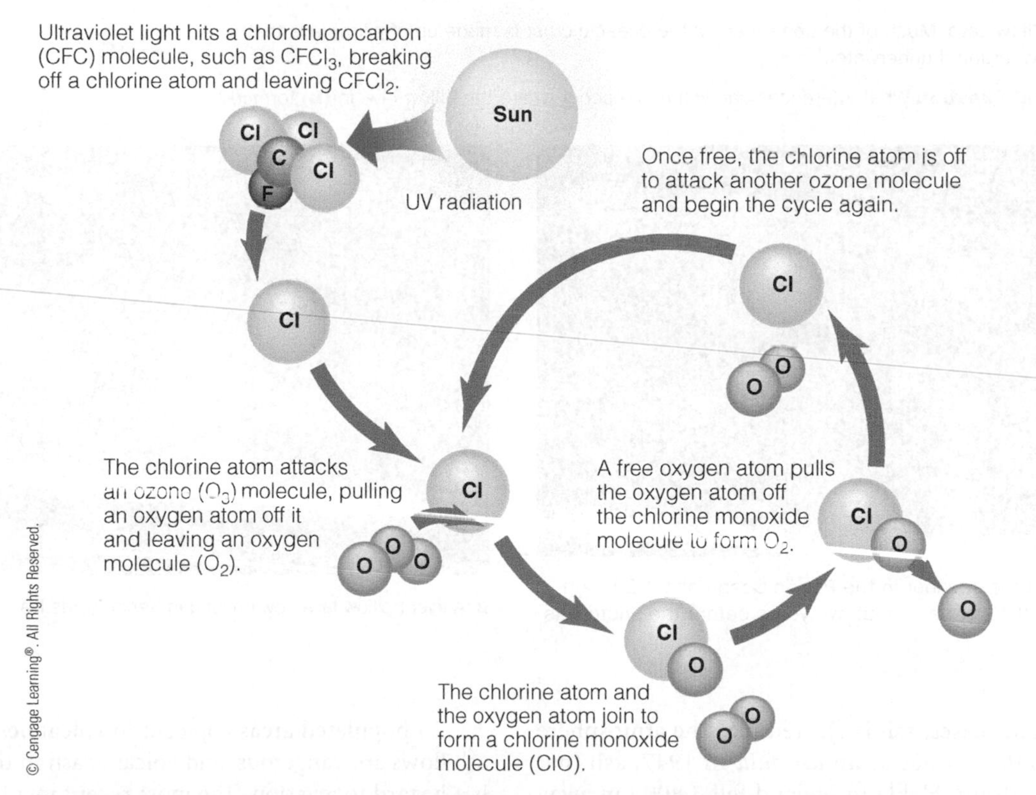

Figure 2 Ozone is destroyed by chlorofluorocarbons (CFCs). Chlorine atoms are continuously regenerated, so one chlorine atom can destroy many ozone molecules.

J.D. Griggs/USGS

a An excellent example of the taffylike appearance of pahoehoe.

Figure 5.5 Pahoehoe and Aa Lava Flows Pahoehoe and aa were named for lava flows in Hawaii, but the same kinds of flows are found in many other areas.

Critical Thinking Question What controls whether a lava flow is aa or pahoehoe?

Robert Tilling/USGS

b An aa lava flow advances over an older pahoehoe flow. Notice the rubbly nature of the aa flow.

Figure 5.6 Pillow Lava Much of the upper part of the oceanic crust is made up of pillow lava that formed when lava erupted underwater.

Critical Thinking Question What inferences can you make about where the pillow lava in **(b)** formed?

NOAA

a Pillow lava on the seafloor in the Pacific Ocean about 240 km west of Oregon that formed about five years before the photo was taken.

James S. Monroe

b Ancient pillow lava now on land in Kenai Fjords National Park, Alaska.

2.0 mm. In some cases, ash is ejected into the atmosphere and settles to the surface as an *ash fall.* In 1947, ash that erupted from Mount Hekla in Iceland fell 3,800 km away on Helsinki, Finland. In contrast to an ash fall, an *ash flow* is a cloud of ash and gas that flows along or close to the land surface. Ash flows can move faster than 100 km/hr, and some cover vast areas.

In populated areas adjacent to volcanoes, ash falls and ash flows are dangerous, and volcanic ash in the atmosphere is a hazard to aviation. The most recent incident took place in April 2010 when an eruption in Iceland disrupted air traffic across the North Atlantic (Figure 5.1a). Since 1980, approximately 80 aircraft have been damaged when they encountered clouds of volcanic ash. The most serious incident took place

Figure 5.7 Columnar Jointing Columnar jointing is seen mostly in mafic lava flows and related intrusive rocks.

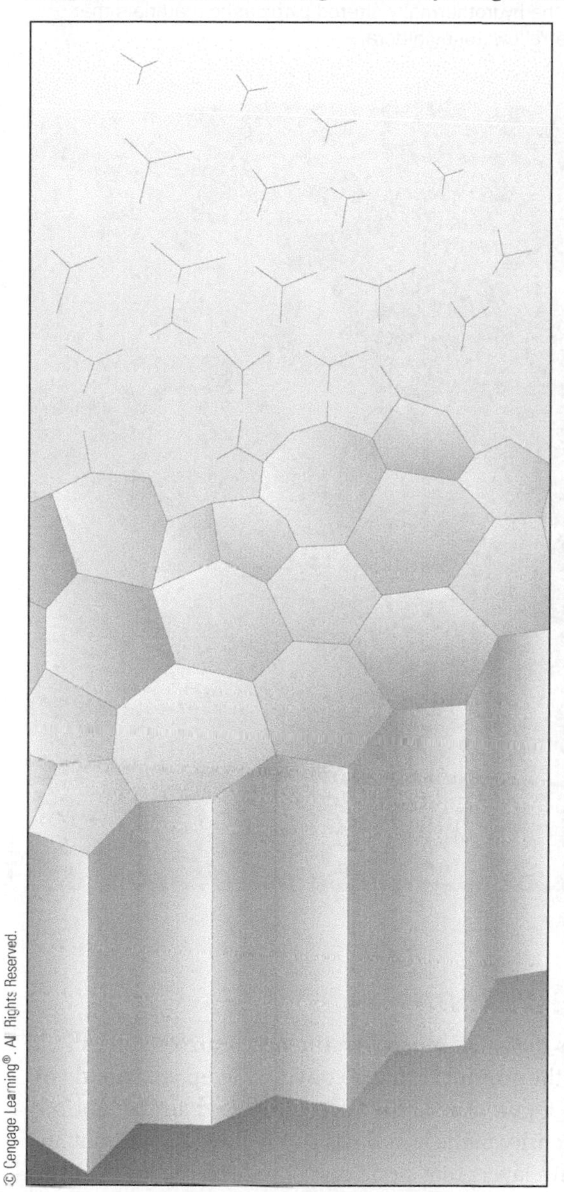

a As lava cools and contracts, three-pronged cracks form that grow and intersect to form four- to seven-sided columns, most of which are six-sided.

Sue Monroe

b Columnar joints in a basalt lava flow in Washington State.

James S. Monroe

c Surface view of columns at Devils Postpile National Monument, California. The straight lines and polished surface resulted from abrasion by a sediment-laden glacier that moved over this surface.

in 1989 when ash from Redoubt volcano in Alaska caused all four jet engines to fail on KLM Flight 867. The plane, carrying 231 passengers, nearly crashed when it fell more than 3 km before the crew could restart the engines. The plane landed safely in Anchorage, Alaska, but it required $80 million in repairs.

In addition to volcanic ash, volcanoes erupt *lapilli,* consisting of pyroclastic materials that measure from 2 to 64 mm, and *blocks* and *bombs,* both larger than 64 mm (Figure 5.8a). Bombs have a twisted, streamlined shape, which indicates that they were erupted as globs of magma that cooled and solidified during their flight through the air. Blocks, in contrast, are angular pieces of rock ripped from a volcanic conduit or pieces of a solidified crust of a lava flow. Because of their size, lapilli, bombs, and blocks are confined to the immediate area of an eruption.

5.3 Types of Volcanoes

Although volcanoes vary in size and shape, all have a conduit or conduits leading to a magma chamber beneath the surface (see Figure 4.1). Vulcan, the Roman deity of fire, was the inspiration for calling these mountains volcanoes, and because of their danger and obvious connection to Earth's interior, they have been held in awe by many cultures.

Figure 5.8 Pyroclastic Materials

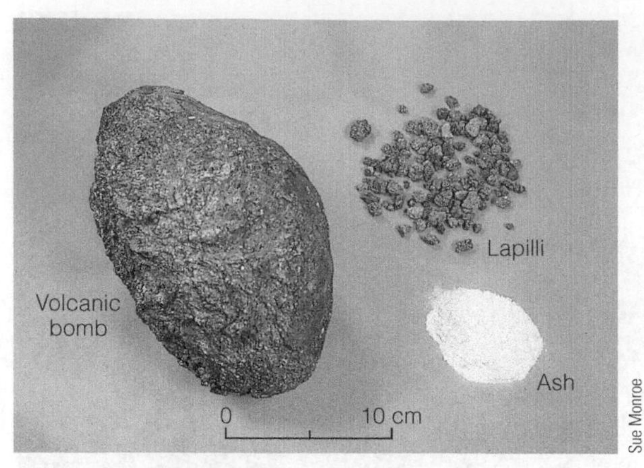

a Pyroclastic materials are all particles ejected from volcanoes, especially during explosive eruptions. The volcanic bomb is elongate, because it was molten when it descended through the air. The lapilli was collected at a small volcano in Oregon, whereas the ash came from the 1980 eruption of Mount St. Helens in Washington.

b Tuff: The walls of the Grand Canyon of the Yellowstone River are made up of the hydrothermally altered pyroclastic materials that partly fill the Yellowstone caldera.

In Hawaiian legends, the volcano goddess Pele resides in the crater of Kilauea on Hawaii. During one of her frequent rages, Pele causes earthquakes and lava flows, and she may hurl flaming boulders at those who offend her. Native Americans in the Pacific Northwest tell of a titanic battle between the volcano gods Skel and Llao to account for huge eruptions that took place about 7,700 years ago in Oregon and California.

Most volcanoes have a circular depression known as a **crater** at their summit, or on their flanks, that forms by explosions or collapse. Most craters are less than 1 km across, whereas much larger rimmed depressions on volcanoes are **calderas.** In fact, some volcanoes have a summit crater within a caldera. Calderas are huge structures that form following voluminous eruptions during which part of a magma chamber drains and the mountain's summit collapses into the vacated space below. An excellent example is misnamed Crater Lake in Oregon (Figure 5.9). Crater Lake is actually a steep-rimmed caldera that formed about 7,700 years ago in the manner just described; it is more than 1,200 m deep and measures 9.7 by 6.5 km. As impressive as Crater Lake is, it is not nearly as large as some other calderas, such as the Toba caldera in Sumatra, which is 100 km long and 30 km wide.

Geologists recognize several major types of volcanoes, but one must realize that each volcano is unique in its history of eruptions and development. For instance, the frequency of eruptions varies considerably; the Hawaiian volcanoes and Mount Etna on Sicily have erupted repeatedly, whereas Pinatubo in the Philippines erupted in 1991 for the first time in 600 years. Furthermore, some volcanoes are complex and have the characteristics of more than one type of volcano.

Shield Volcanoes

Volcanoes that look much like the outer surface of a shield laying on the ground with the convex side up are **shield volcanoes** (Figure 5.10). They are composed almost entirely of low-viscosity mafic lava flows so the flows spread out and formed thin layers that slope only 2 to 10 degrees. Erupting shield volcanoes, sometimes called *Hawaiian-type eruptions,* are rather quiet compared with eruptions of many other volcanoes, particularly those at convergent plate boundaries. Magma rises to the surface and issues as lava flows that pose little danger to humans. However, jets of incandescent lava may be forcefully but not explosively ejected as lava fountains, some up to 400 m high, when magma reaches the surface and gases expand (Figure 5.10c). These lava fountains contribute some pyroclastic materials to shield volcanoes, but they are made up mostly of layers of basalt.

Although eruptions of shield volcanoes tend to be rather quiet, some of the Hawaiian volcanoes have, on occasion, produced sizable explosions when groundwater instantly vaporizes as it comes in contact with magma. One such explosion in 1790 killed about 80 warriors in a party headed by Chief Keoua, who was leading them across the summit of Kilauea volcano.

Kilauea volcano is impressive because it has been erupting continuously since January 3, 1983, making it the longest

Figure 5.9 The Origin of Crater Lake, Oregon Remember, Crater Lake is actually a caldera that formed by partial draining of a magma chamber.

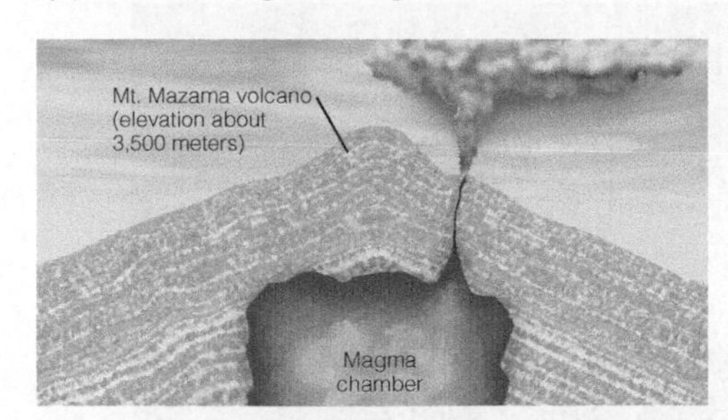

a Eruption begins as huge quantities of ash are ejected from the volcano.

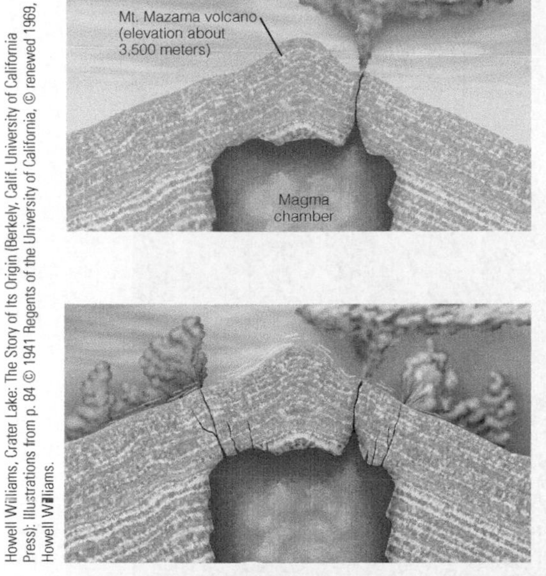

b The eruption continues as more ash and pumice are ejected into the air and pyroclastic flows move down the flanks of the mountain.

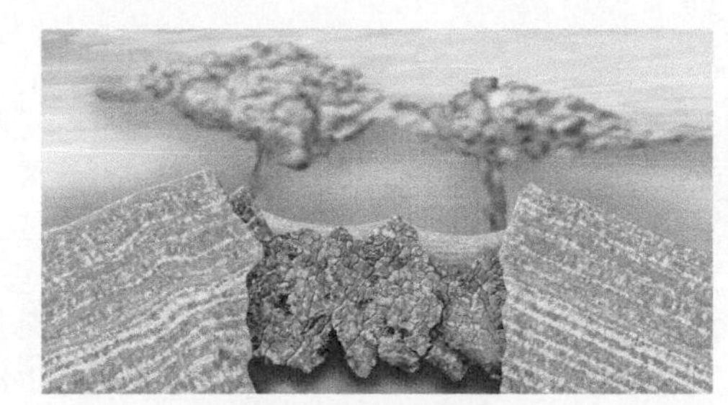

c The collapse of the summit into the partially drained magma chamber forms a huge caldera.

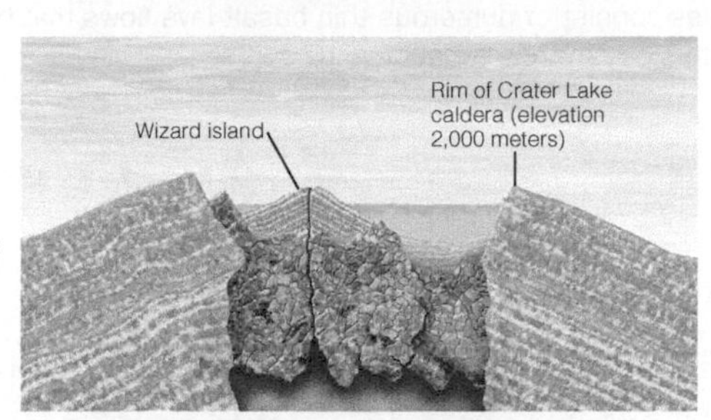

d Postcaldera eruptions partly cover the caldera floor, and the small cinder cone called Wizard Island forms.

e View from the rim of Crater Lake showing Wizard Island. The lake is 594 m deep, making it the second deepest in North America.

recorded eruption. During these 30 years, more than 2.5 km^3 of molten rock has flowed out at the surface, much of it reaching the sea and forming 2.2 km^2 of new land on the island of Hawaii.

Shield volcanoes are most common in the ocean basins, but some are also present on the continents—in California and Oregon in North America and in East Africa, for instance. The island of Hawaii is made up of five huge shield volcanoes, two of which, Kilauea and Mauna Loa, are active much of the time. Mauna Loa is nearly 100 km across its base and stands more than 9.5 km above the surrounding seafloor. It has a volume estimated at 50,000 km^3, making it the world's largest volcano in volume.

Mauna Loa is certainly a large volcano, but the record for size goes to Olympus Mons on Mars, which is by far the largest volcano of any kind known in the solar system. It stands 25 km high, about three times as high as Mount Everest, Earth's highest peak, and measures more than 500 km across. Olympus Mons probably attained its great size because Mars lacks any plate tectonic activity, so prolonged eruptions in the same place simply built up this huge mountain. Scientists have found no evidence that Olympus Mons is active.

ConnectionLink

For more information on Olympus Mons and Mars, see Chapter 2.

Figure 5.10 Shield Volcanoes

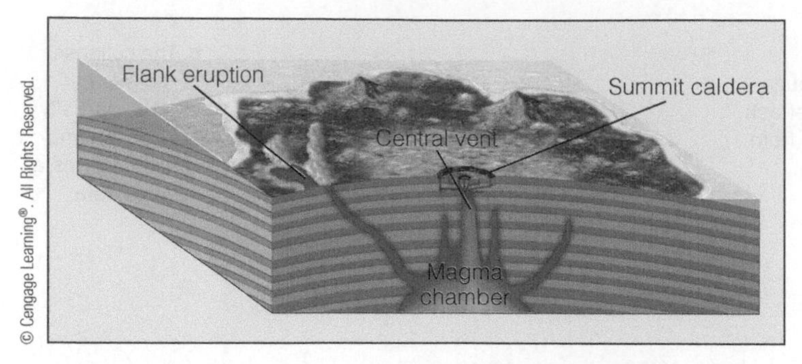

a Shield volcanoes consist of numerous thin basalt lava flows that build up mountains with slopes rarely exceeding 10 degrees.

b Profile of Mauna Loa in Hawaii. Mauna Loa is one of five huge shield volcanoes that make up the island of Hawaii.

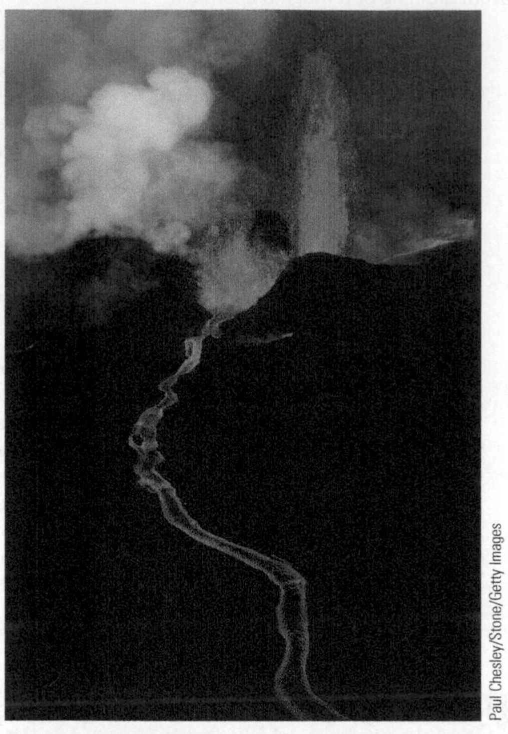

c Shield volcanoes are made up mostly of basalt lava flows, but some pyroclastic materials are erupted at lava fountains.

Cinder Cones

Small, steep-sided **cinder cones** made up of particles resembling cinders form when pyroclastic materials accumulate around a vent from which they erupted (Figure 5.11). Cinder cones are small, rarely exceeding 400 m high, with slope angles up to 33 degrees, depending on the angle that can be maintained by the angular pyroclastic materials. Many have a large, bowl-shaped crater, and if they issue any lava flows, they usually break through the base or lower flanks of the mountains. Although all cinder cones are conical, their symmetry varies from those that are almost perfectly symmetrical to those that formed when prevailing winds caused pyroclastic materials to build up higher on the downwind side of the vent.

Many cinder cones form on the flanks or within the calderas of larger volcanoes and represent the final stages of activity, particularly in areas of basaltic volcanism. Wizard Island in Crater Lake, Oregon, is a small cinder cone that formed after the summit of Mount Mazama collapsed to form a caldera (Figure 5.9e). Cinder cones are common in the southern Rocky Mountain states, particularly New Mexico and Arizona, and many others are in California, Oregon, Washington, and Hawaii.

In 1973, on the Icelandic island of Heimaey, the town of Vestmannaeyjar was threatened by a new cinder cone. The initial eruption began on January 23, and within two days, a cinder cone, later named Eldfell, rose to about 100 m above the surrounding area (Figure 5.11b). Pyroclastic materials from the volcano buried parts of the town, and by February, a massive aa lava flow was advancing toward the town. The flow's leading edge ranged from 10 to 20 m thick, and its central part was as much as 100 m thick. The residents of Vestmannaeyjar sprayed the leading edge of the flow with seawater in an effort to divert it from the town. The flow was in fact diverted, but how effective the efforts of the townspeople were is not clear. They may have been simply lucky.

Composite Volcanoes (Stratovolcanoes)

Pyroclastic layers, as well as lava flows, both of intermediate composition, are found in **composite volcanoes,** which are also called *stratovolcanoes* (Figure 5.12). As the lava flows cool, they typically form andesite; recall that intermediate lava flows are more viscous than mafic ones. Geologists use the term **lahar** for volcanic mudflows, which are also common on composite volcanoes. A lahar may form when rain falls on unconsolidated pyroclastic materials and creates a muddy slurry that moves downslope (Figure 5.13). On November 13, 1985, a minor eruption of Nevado del Ruiz

U.S. Geological Survey

Figure 5.11 Cinder Cones Cinder cones are small, steep-sided volcanoes made up of pyroclastic materials that resemble cinders.

© Solarfilma/GeoScience Features

b Eldfell, a cinder cone in Iceland. Eldfell began erupting in 1973 and in two days grew to 100 m high. The steam visible on the left side of the image resulted from aa lava flowing into the sea. Another cinder cone, known as Helgafel, is also visible.

a This 400-m-high cinder cone named Paricutín formed in a short time in Mexico in 1943, when pyroclastic materials began to erupt in a farmer's field. Lava flows from the volcano covered two nearby villages, but all activity ceased by 1952.

Figure 5.12 Composite Volcanoes

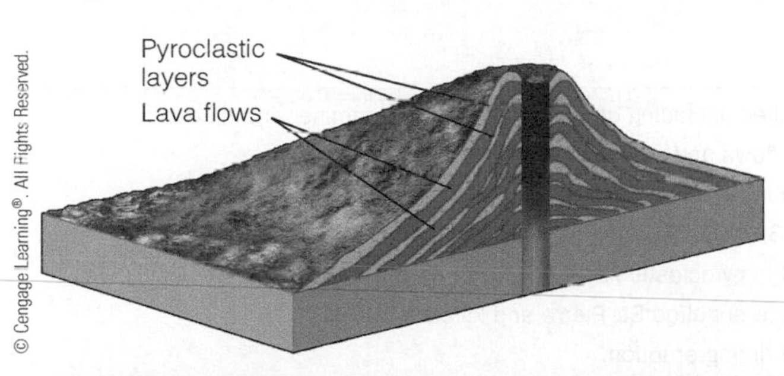

a Composite volcanoes, also called stratovolcanoes, are made up mostly of lava flows and pyroclastic materials of intermediate composition, although mudflows (lahars) are also common.

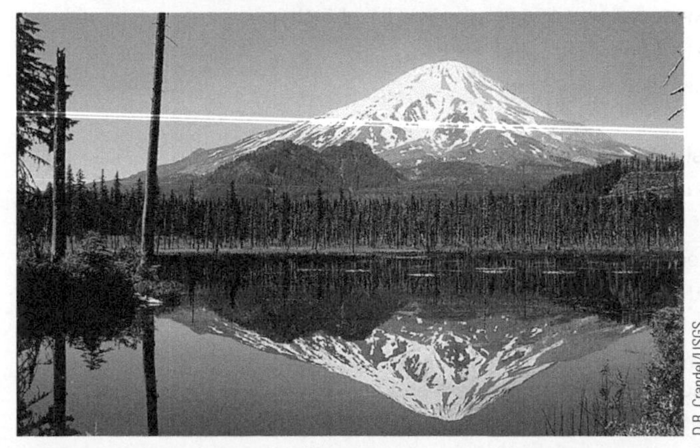

D.R. Crandel/USGS

b Mount St. Helens in Washington State as it appeared from the east in 1978.

R. Solkowski

c Mayon volcano in the Philippines, a nearly symmetrical composite volcano that erupted 13 times during the 1900s and also in 2009 and 2010.

David Johnson/USGS

Figure 5.13 **Lahar** Lahars or volcanic mudflows are common on composite volcanoes. Homes partly buried by a lahar on June 15, 1991, following the eruption of Mount Pinatubo in the Philippines.

in Colombia melted snow and ice on the volcano, causing lahars that killed approximately 23,000 people (Table 5.1).

Composite volcanoes differ from shield volcanoes and cinder cones in composition and in their overall shape. Remember that shield volcanoes have very low slopes, whereas cinder cones are small, steep-sided, conical mountains. In contrast, composite volcanoes are steep-sided near their summits, perhaps as much as 30 degrees, but the slope decreases toward the base, where it may be no more than 5 degrees (Figure 5.12). Mayon volcano in the Philippines is one of the most nearly symmetrical composite volcanoes anywhere (Figure 5.12c). It has erupted 48 times during the last 400 years, most recently in 2010.

When most people think of volcanoes, they picture the graceful profiles of composite volcanoes, which are the typical large volcanoes found on the continents and island arcs. And some of these volcanoes are indeed large; Mount Shasta in northern California is made up of about 350 km^3 of material and measures 20 km across its base. Remember, though, that Mauna Loa in Hawaii has an estimated volume of 50,000 km^3.

Lava Domes

Most volcanoes are classified as shield volcanoes, cinder cones, or composite volcanoes. Less common are **lava domes,** also known as *volcanic domes* and *plug domes,* which are steep-sided, bulbous mountains that form when viscous felsic magma, and occasionally intermediate magma, is forced toward the surface (Figure 5.14). Because felsic magma is so viscous, it moves upward very slowly and only when the pressure from below is great.

Beginning in 1980, a number of lava domes were emplaced in the crater of Mount St. Helens in Washington, most of which were destroyed during subsequent eruptions. Since 1983, Mount St. Helens has been characterized by sporadic dome growth and renewed eruptions in 2004.

TABLE 5.1 **Some Notable Volcanic Eruptions**

Date Deaths	Volcano	Deaths
Apr 10, 1815	Tambora, Indonesia	117,000 killed, including deaths from eruption, famine, and disease.
Oct 8, 1822	Galunggung, Java	Pyroclastic flows and mudflows killed 4,011.
Mar 2, 1856	Awu, Indonesia	2,806 died in pyroclastic flows.
Aug 27, 1883	Krakatau, Indonesia	More than 36,000 died, most killed by tsunami.
June 7, 1892	Awu, Indonesia	1,532 died in pyroclastic flows.
May 8, 1902	Mount Pelée, Martinique	Nuée ardente engulfed St. Pierre and killed 28,000.
Oct 24, 1902	Santa Maria, Guatemala	5,000 died during eruption.
May 19, 1919	Kelut, Java	Mudflows devastated 104 villages and killed 5,110.
Jan 21, 1951	Lamington, New Guinea	Pyroclastic flows killed 2,942.
Mar 17, 1963	Agung, Indonesia	1,148 perished during eruption.
May 18, 1980	Mount St. Helens, Washington	63 killed, 600 km^2 of forest devastated.
Mar 28, 1982	El Chichon, Mexico	Pyroclastic flows killed 1,877.
Nov 13, 1985	Nevado del Ruiz, Colombia	Minor eruption triggered mudflows that killed 23,000.
Aug 21, 1986	Oku volcanic field, Cameroon	Cloud of CO_2 released from Lake Nyos killed 1,746.
June 15, 1991	Mount Pinatubo, Philippines	281 killed during eruption, 83 died in later mudflows, 358 died of illness.
July 1999	Soufriére Hills, Montserrat	19 killed, 12,000 evacuated.
Jan 17, 2002	Nyiragongo, Zaire	Lava flow killed 147 in Goma.
Aug 16, 2006	Tungurahua, Ecuador	Explosive eruption killed 7; continuously active since 1999, periodic evacuations and several villages destroyed.
Apr 2010	Eyjafjallajökull, Iceland	No fatalities, but disrupted air traffic over the North Atlantic for several days.
Oct–Nov 2010	Mount Merapi, Indonesia	353 died in pyroclastic flows; 350,000 displaced.
June 2011	Nabro Volcano, Eritrea-Ethiopia	31 killed by volcanic ash on Ethiopian side of volcano.

Figure 5.14 Lava Domes Lava domes are bulbous masses of magma that are emplaced in the craters of composite volcanoes or stand alone as irregularly shaped mountains flanked by debris shed from the dome.

Critical Thinking Question Why do you think lava domes are so dangerous?

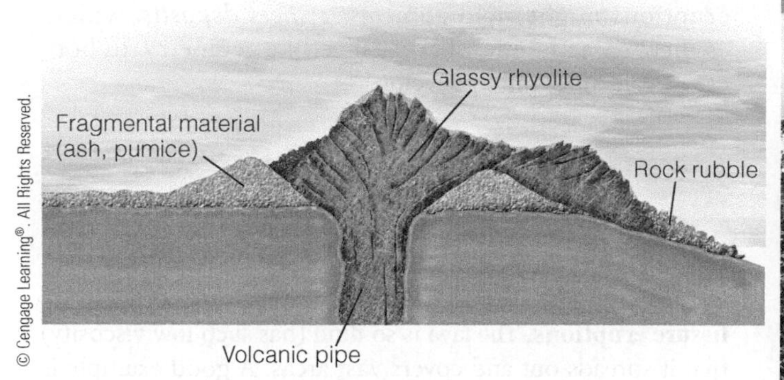

a Diagram of a mass of viscous magma forming a lava dome.

Sue Monroe

b The volcano in the distance is Lassen Peak in Lassen Volcanic National Park in California. The dark masses of rock cooled from the viscous magma that was injected in this area. Lassen Peak erupted from 1914 to 1917.

Lava dome eruptions are some of the most violent and destructive. In 1902, viscous magma accumulated beneath the summit of Mount Pelée on the island of Martinique. Eventually, the pressure increased until the side of the mountain blew out in a tremendous explosion, ejecting a mobile, dense cloud of pyroclastic materials and a glowing cloud of gases and dust called a **nuée ardente** (French for "glowing cloud"). The pyroclastic flow followed a valley to the sea, but the nuée ardente jumped a ridge and engulfed the city of St. Pierre (Figure 5.15).

A tremendous blast hit St. Pierre, leveling buildings; hurling boulders, trees, and pieces of masonry down the

Figure 5.15 Nuée Ardente

© Bettman/CORBIS

a St. Pierre, Martinique, after it was destroyed by a nuée ardente from Mount Pelée in 1902. Only two of the city's 28,000 inhabitants survived.

USGS

b An April 1986 pyroclastic flow rushing down Augustine volcano in Alaska. This flow is similar to the one that wiped out St. Pierre.

streets; and moving a 3-ton statue 16 m. Accompanying the blast was a swirling cloud of incandescent ash and gases with an internal temperature of 700°C that incinerated everything in its path. The nuée ardente passed through St. Pierre in two or three minutes, only to be followed by a firestorm as combustible materials burned and casks of rum exploded. But by then most of the 28,000 residents of the city were already dead. In fact, in the area covered by the nuée ardente, only two survived!* One survivor was on the outer edge of the nuée ardente, but even there, he was terribly burned and his family and neighbors were all killed. The other survivor, a stevedore incarcerated the night before for disorderly conduct, was in a windowless cell partly below ground level. He remained in his cell badly burned for four days after the eruption until rescue workers heard his cries for help. He later became an attraction in the Barnum and Bailey Circus where he was advertised as "The only living object that survived in the 'Silent City of Death' where 40,000 beings were suffocated, burned or buried by one belching blast of Mont Pelée's terrible volcanic eruption."**

Supervolcano Eruptions

Geologists have no formal definition for *supervolcano eruptions,* but we can take it to mean an explosive eruption of hundreds of cubic kilometers of pyroclastic materials and the origin of a huge caldera. No supervolcano eruptions have occurred in historic times, but geologists know of several that took place during the past 2 million years—Long Valley in eastern California, Toba in Indonesia, and Taupo in New Zealand, for example.

On three occasions, supervolcano eruptions followed the accumulation of rhyolitic magma beneath Yellowstone National Park, which is mostly in Wyoming, each yielding a widespread blanket of volcanic ash and pumice and gigantic calderas. We can summarize Yellowstone's volcanic history by noting that supervolcano eruptions took place 2 million years ago, 1.3 million years ago, and 600,000 years ago. Then, between 150,000 and 75,000 years ago, an additional 1,000 km^3 of pyroclastic materials were erupted within the Yellowstone caldera (Figure 5.8b).

Geologists think that these huge eruptions were caused by a rising *mantle plume,* a cylindrical mass of magma probably of rhyolitic composition. Because this type of magma is viscous, it triggers explosive eruptions when it nears the surface. Personnel from the United States Geological Survey (USGS) and the University of Utah continue to monitor the Yellowstone area for any signs of renewed activity.

*Although reports commonly claim that only two people survived the eruption, at least 69 and possibly as many as 111 people survived beyond the extreme margins of the nuée ardente and on ships in the harbor. Many, however, were badly injured.

**Quoted from A. Scharth, *Vulcan's Fury: Man Against the Volcano* (New Haven, CT: Yale University Press, 1999), p. 177.

5.4 Other Volcanic Landforms

During *fissure eruptions,* fluid lava pours out and simply builds up rather flat-lying areas, whereas huge explosive eruptions might yield *pyroclastic sheet deposits,* which, as their name implies, have a sheetlike geometry. In both cases, volcanoes fail to develop.

Fissure Eruptions and Basalt Plateaus

Rather than erupting from central vents, the lava flows making up **basalt plateaus** issue from long cracks or fissures during **fissure eruptions.** The lava is so fluid (has such low viscosity) that it spreads out and covers vast areas. A good example is the Columbia River basalt in eastern Washington and parts of Oregon and Idaho. This huge accumulation of 17- to 6-million-year-old overlapping lava flows covers about 164,000 km^2 (Figure 5.16a and b), and has an aggregate thickness of more than 1,000 m. And some individual flows are enormous—the Roza flow advanced along a front about 100 km wide and covered 40,000 km^2.

ConnectionLink
You can find more information about the Columbia River basalts in Chapter 23.

Similar accumulations of vast, overlapping lava flows are also found in the Snake River Plain in Idaho (Figure 5.16a and c). These flows are 5.0 to 1.6 million years old, and they represent a style of eruption between fissure eruptions and those of shield volcanoes. In fact, there are small, low shields, as well as fissure flows, in the Snake River Plain.

Currently, fissure eruptions occur only in Iceland. Iceland has a number of volcanoes, but the bulk of the island is composed of basalt lava flows that issued from fissures. In fact, about half of the lava erupted during historic time in Iceland came from two fissure eruptions, one in AD 930 and the other in 1783. The 1783 eruption from Laki fissure, which is more than 30 km long, accounted for lava that covered 560 km^2 and, in one place, filled a valley to a depth of about 200 m.

Pyroclastic Sheet Deposits

Geologists have long been aware of vast areas covered by felsic volcanic rocks a few meters to hundreds of meters thick. It seemed improbable that these could be vast lava flows, but it seemed equally unlikely that they were ash fall deposits. Based on observations of historic pyroclastic flows, such as the nuée ardente erupted by Mount Pelée in 1902, it seems that these ancient rocks originated as pyroclastic flows—hence the name **pyroclastic sheet deposits.**

Pyroclastic sheet deposits cover far greater areas than any observed during historic time, however, and apparently erupted from long fissures rather than from a central vent.

The pyroclastic materials of many of these flows were so hot that they fused together to form *welded tuff.*

Geologists now think that major pyroclastic flows issue from fissures formed during the origin of calderas. For instance, pyroclastic flows erupted during the formation of a large caldera now occupied by Crater Lake in Oregon (Figure 5.9) and in the Yellowstone caldera in Wyoming.

Similarly, the Bishop Tuff of eastern California erupted shortly before the formation of the Long Valley caldera. Interestingly, earthquake activity in the Long Valley caldera and nearby areas beginning in 1978 may indicate that magma is moving upward beneath part of the caldera. Thus, the possibility of future eruptions in that area cannot be discounted.

5.5 Volcano Belts

Most of the world's active volcanoes are in well-defined zones or belts rather than randomly distributed. The **circum-Pacific belt,** popularly called the Ring of Fire, has more than 60% of all active volcanoes. It includes volcanoes in the Andes of South America; the volcanoes of Central America, Mexico, and the Cascade Range of North America; and the Alaskan volcanoes as well as those in Japan, the Philippines, Indonesia, and New Zealand (▮ Figure 5.17). Also in the circum-Pacific belt are the southernmost active volcanoes at Mount Erebus that erupted in Antarctica during 2011 and a large caldera at Deception Island that erupted most recently during 1970.

The second area of active volcanism is the **Mediterranean belt** (Figure 5.17). About 20% of all active volcanism takes

Classification and History of Igneous Rocks

You are an enthusiast of natural history and would like to share your interests with your family. Accordingly, you plan a vacation to see some of the volcanic features in United States national parks and monuments. Let's assume your planned route will take you through Wyoming, Idaho, Washington, Oregon, and California. What specific areas might you visit, and what kinds of volcanic features would you see in these areas? What other parts of the United States might you visit in the future to see additional evidence for volcanism?

▮ **Figure 5.16** Basalt Plateaus Basalt plateaus are vast areas of overlapping lava flows that issued from long fissures. Fissure eruptions take place today in Iceland; however, in the past, they formed basalt plateaus in various areas.

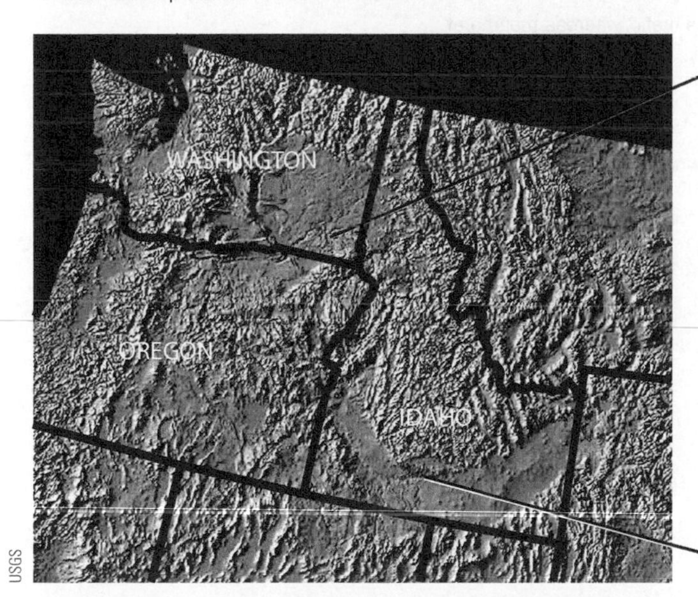

USGS

a Relief map of the northwestern United States showing the locations of the Columbia River basalt and the Snake River Plain.

James S. Monroe

b Lava flows of the Columbia River basalt in Washington.

Sue Monroe

c Basalt lava flows of the Snake River Plain at Malad Gorge State Park, Idaho.

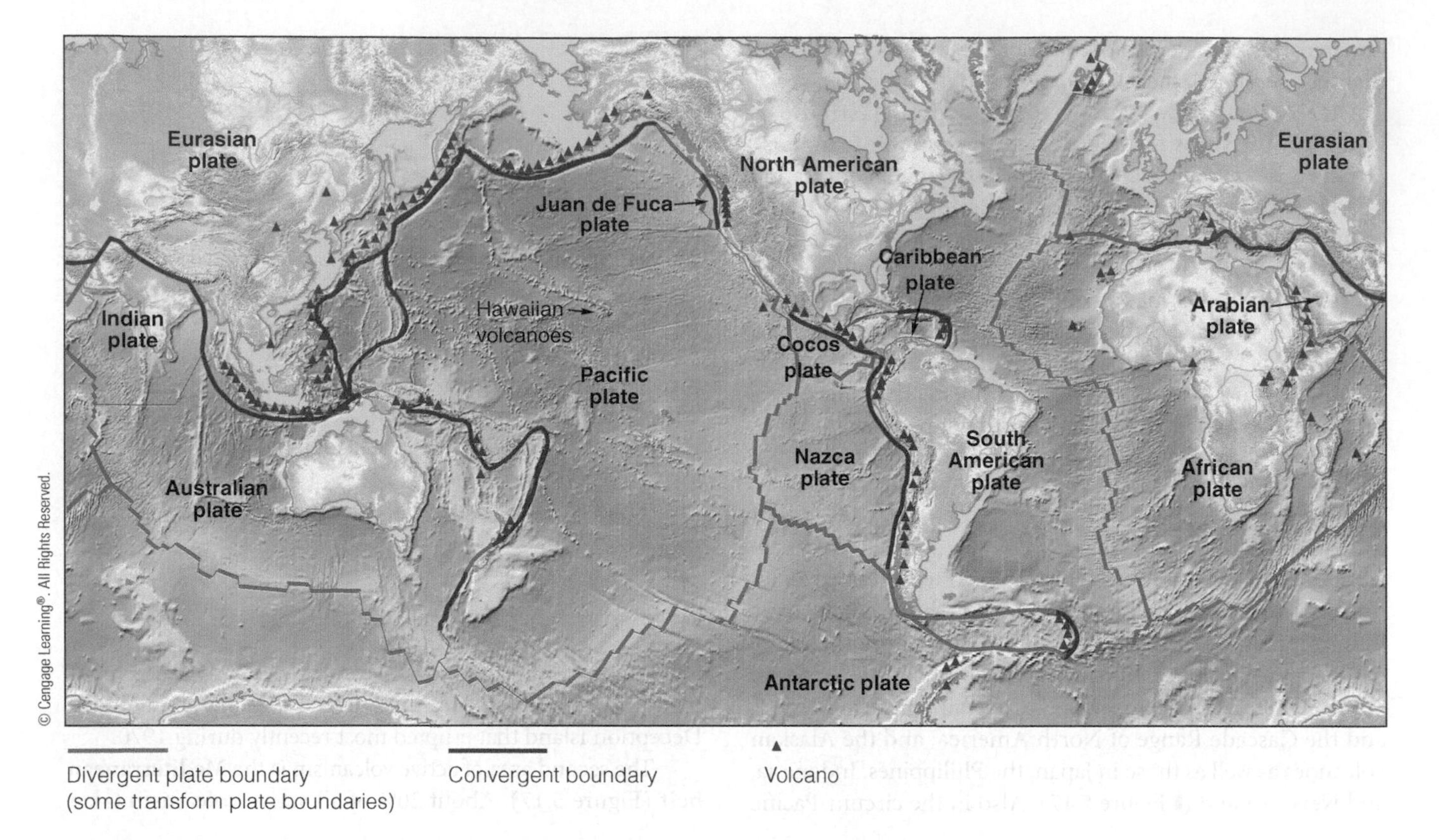

Figure 5.17 **Volcanoes of the World** Most volcanoes are at or near convergent and divergent plate boundaries. The two major volcano belts are the circum-Pacific belt, commonly known as the Ring of Fire, with about 60% of all active volcanoes, and the Mediterranean belt, with 20% of active volcanoes. Most of the rest lie near the mid-oceanic ridges.

Critical Thinking Question Why is magma at divergent plate boundaries mafic, whereas magma at convergent plate boundaries is intermediate or felsic?

place in this belt, where the famous Italian volcanoes such as Mounts Etna and Vesuvius and the Greek volcano Santorini are found. Mount Etna has issued lava flows 190 times since 1500 BC, when activity was first recorded. A particularly violent eruption of Santorini in 1390 BC might be the basis for the myth about the lost continent of Atlantis, and, in AD 79, an eruption of Mount Vesuvius destroyed Pompeii and other nearby cities (Figure 5.2).

Nearly all the remaining active volcanoes are at or near mid-oceanic ridges or the extensions of these ridges onto land (Figure 5.17). These include the East Pacific Rise and the longest of all mid-oceanic ridges, the Mid-Atlantic Ridge. The latter is located near the center of the Atlantic Ocean basin, accounting for the volcanism in Iceland and elsewhere. It continues around the southern tip of Africa, where it connects with the Indian Ridge. Branches of the Indian Ridge extend into the Red Sea and East Africa, where such volcanoes as Kilamanjaro in Tanzania, Nyiragongo in Zaire, and Erta Ale in Ethiopia with its continuously active lava lake are found.

Ridge volcanism also occurred in 1980 and 2009 when Axial Volcano (also called Axial Seamount) erupted on the Juan de Fuca Ridge about 180 km west of Oregon. Even more recently, a new island appeared in the Red Sea as a result of volcanism on the Red Sea rift.

5.6 North America's Active Volcanoes

Part of the circum-Pacific belt includes volcanoes in the Pacific Northwest, as well as those in Alaska, both of which are at convergent plate boundaries. Of the 80 or so potentially active volcanoes in Alaska, at least half have erupted since 1760. The other active North American volcanoes are in the Cascade Range in the Pacific Northwest where the Juan de Fuca plate is subducted beneath North America. Many of these volcanoes have been historically active, although since 1900, only Lassen Peak in California and Mount St. Helens in Washington have erupted.

Alaska's Volcanoes

Many of the volcanoes in mainland Alaska and in the Aleutian Islands (Figure 5.17) are composite volcanoes, some with huge calderas. Mount Spurr has erupted explosively at least 35 times during the past 5,000 years, but its eruptions pale by comparison with that of Novarupta in 1912. Its defining event was the June 1912 eruption, the largest in the world since the late 1800s. At least 15 km^3 of mostly pyroclastic materials erupted during about 60 hours.

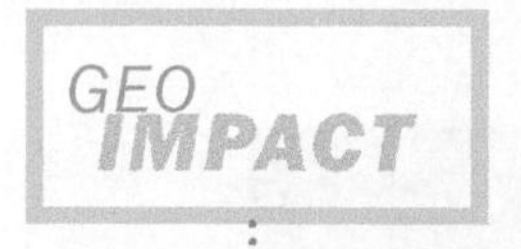

Cascade Range Volcanoes

No one doubts that some of the Cascade Range volcanoes will erupt again, but we do not know when or how large these eruptions will be. A job transfer takes you to a community in Oregon that has several nearby large volcanoes. You have some concerns about future eruptions. What kinds of information would you seek before buying a home in this area? In addition, as a concerned citizen, can you make any suggestions about what should be done in case of a large eruption?

When the eruption was over, 120 km^2 of land was buried beneath pyroclastic deposits as deep as 213 m. In fact, the deposits filled the Valley of Ten Thousand Smokes—so named because of the hundreds of fumaroles where gases vented through the hot deposits for as long as 15 years following the eruption. Fortunately, the eruption took place in a remote area so there were no injuries or fatalities, but enough ash, gases, and pumice were ejected that, for several days, the sky was darkened over much of the Northern Hemisphere.

By the time you read this chapter, several more volcanoes in Alaska will have erupted as the Pacific plate moves relentlessly northward only to be subducted at the Aleutian Trench. The Alaska Volcanoes Observatory in Anchorage, Alaska, continues to monitor these volcanoes and issue warnings about potential eruptions.

The Cascade Range

The **Cascade Range** (▌Figure 5.18a) stretches from Lassen Peak in northern California north through Oregon and Washington into British Columbia, Canada. Most of the large volcanoes in the range are composite volcanoes, but Lassen Peak in California is the world's largest lava dome (▌Figure 5.18b). Actually, it is a rather small volcano that developed 27,000 years ago on the flank of a much larger, deeply eroded composite volcano. It erupted from 1914 to 1917, but has since been quiet except for ongoing hydrothermal activity. Two large shield volcanoes lie just to the east of the main Cascade Range volcanoes—Medicine Lake Volcano in California and Newberry Volcano in Oregon.

What was once a nearly symmetrical composite volcano changed markedly on May 6, 1980, when Mount St. Helens in Washington erupted explosively, killing 57 people and leveling some 600 km^2 of forest (▌Figure 5.18c). A huge lateral blast caused much of the damage and fatalities, but snow and ice on the volcano melted and pyroclastic materials displaced water in lakes and rivers, causing lahars and extensive flooding.

Mount St. Helens's renewed activity, beginning in late September 2004, has resulted in dome growth and small steam and ash explosions. Scientists at the Cascades Volcano Observatory in Vancouver, Washington, issued a low-level alert for an eruption and continue to monitor the volcano.

Several other Cascade Range volcanoes will almost certainly erupt again. Mount Hood in Oregon, one of the most symmetrical volcanoes in the range, last erupted in 1865, but should a large eruption take place, it lies only about 50 km from Portland, an urban area with more than 2 million people (▌Figure 5.18d). However, the most dangerous is probably Mount Rainier in Washington. The greatest danger from Mount Rainier is volcanic mudflows or huge debris flows. Of the 60 large flows that have occurred during the last 100,000 years, the largest, consisting of 4 km^3 of debris, covered an area now occupied by more than 120,000 people. Indeed, in August 2001, a sizable debris flow took place on the south side of the mountain, but it caused no injuries or fatalities. No one knows when the next flow will take place, but one community has taken the threat seriously enough to formulate an emergency evacuation plan. Unfortunately, the residents would have only one or two hours to evacuate.

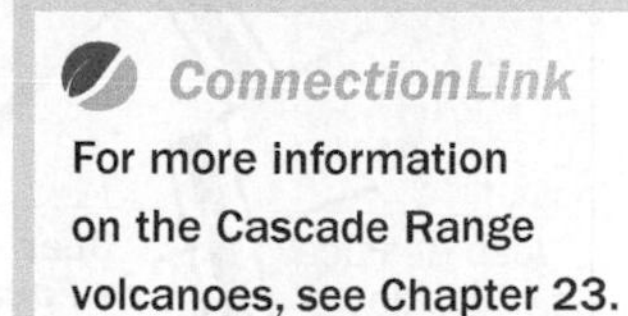

For more information on the Cascade Range volcanoes, see Chapter 23.

5.7 Plate Tectonics, Volcanoes, and Plutons

In Chapter 4, we discussed the origin and evolution of magma and concluded that (1) mafic magma is generated beneath spreading ridges, and (2) intermediate or felsic magma forms where an oceanic plate is subducted beneath another oceanic plate or a continental plate. Accordingly, with few exceptions, volcanism and emplacement of plutons take place at or near divergent and convergent plate boundaries.

Igneous Activity at Divergent Plate Boundaries

Much of the mafic magma that originates at spreading ridges is emplaced as vertical dikes and gabbro plutons, thus composing the lower part of the oceanic crust. However, some rises to the surface and issues forth as submarine lava flows and pillow lava (Figure 5.6), which constitutes the upper part of the oceanic crust. Much of this volcanism goes undetected, but researchers in submersibles have seen the results of recent eruptions.

Mafic lava is very fluid, allowing gases to escape easily, and at great depth in the oceans, the water pressure is so great that explosive volcanism is prevented. In short, pyroclastic materials are rare to absent unless, of course, a volcanic center builds up above sea level. Even if this occurs, however, the mafic magma is so fluid that it forms the gently sloping layers found on shield volcanoes.

Excellent examples of divergent plate boundary volcanism are found along the Mid-Atlantic Ridge, particularly

Figure 5.18 The Cascade Range of the Pacific Northwest

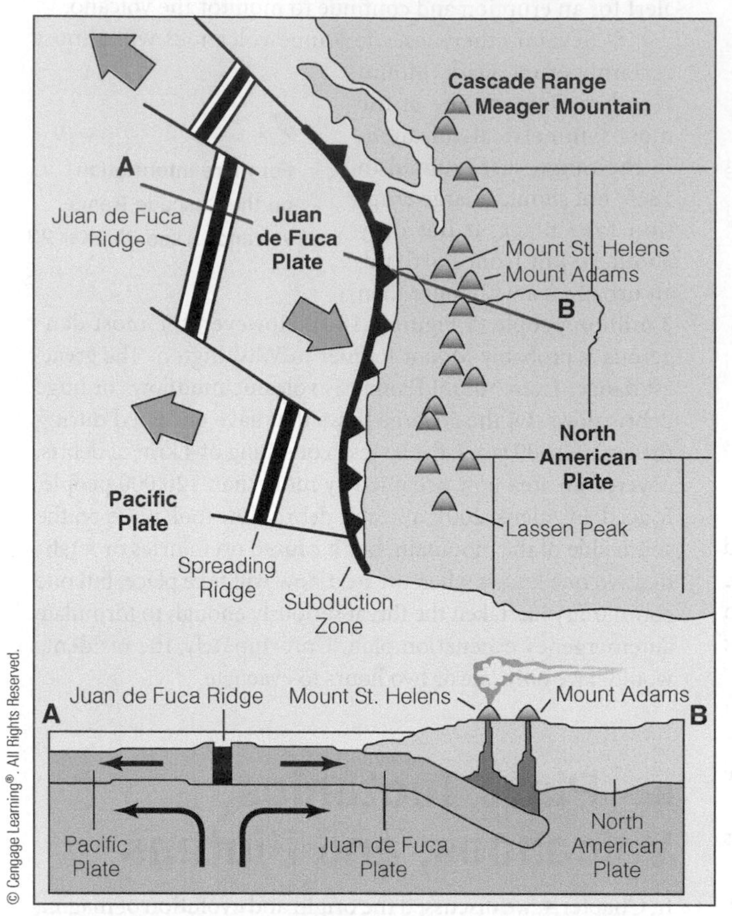

a Plate tectonic setting for the Pacific Northwest. Suduction of the Juan de Fuca plate accounts for ongoing volcanism in the region.

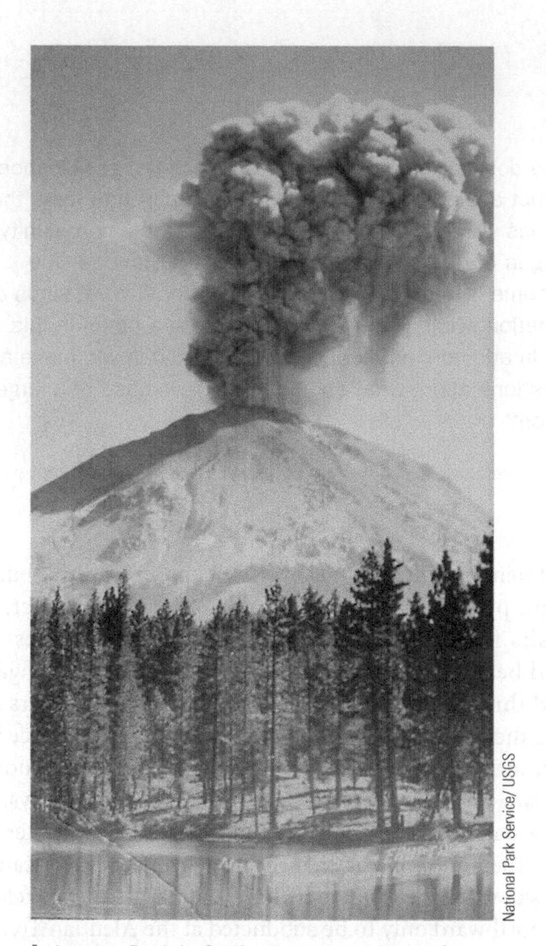

National Park Service/ USGS

b Lassen Peak in California erupted from 1914 to 1917. This eruption took place in 1915.

Courtesy of Keith Ronnholm

c Mount St. Helens, Washington. The lateral blast on May 18, 1980, took place when a bulge on the volcano's north face collapsed, reducing the pressure on gas-charged magma.

Sue Monroe

d Mount Hood in Oregon as seen from Portland, a city of more than two million people.

where it rises above sea level as in Iceland (Figure 5.17). In November 1963, a new volcanic island, later named Surtsey, rose from the sea south of Iceland. The East Pacific Rise and the Indian Ridge are areas of similar volcanism. A divergent plate boundary is also present in Africa as the East African Rift system, which is well known for its volcanoes.

Igneous Activity at Convergent Plate Boundaries

Most of the large active volcanoes in the circum-Pacific and Mediterranean belts are composite volcanoes near the leading edges of overriding plates at convergent plate boundaries.

The overriding plate, with its chain of volcanoes, may be oceanic as in the case of the Aleutian Islands, or it may be continental as is, for instance, the South American plate with its chain of volcanoes along its western margin (Figure 5.17).

As we noted, these volcanoes at convergent plate boundaries consist mostly of lava flows and pyroclastic materials of intermediate to felsic composition. Remember that when mafic oceanic crust partially melts, some of the magma generated is emplaced near plate boundaries as plutons and some is erupted to build up composite volcanoes. More viscous magmas, usually of felsic composition, are emplaced as lava domes, thus accounting for the explosive eruptions that typically occur at convergent plate boundaries.

Good examples of volcanism at convergent plate boundaries are the explosive eruptions of Mount Pinatubo and Mayon volcano in the Philippines; both are near a plate boundary beneath which an oceanic plate is subducted. Mount St. Helens, Washington, is similarly situated, but it is on a continental rather than an oceanic plate.

Intraplate Volcanism

Mauna Loa and Kilauea on the island of Hawaii and Loihi just 32 km to the south are within the interior of a rigid plate far from any divergent or convergent plate boundary (Figure 5.17). The magma is derived from the upper mantle, as it is at spreading ridges, and accordingly is mafic, so it builds up shield volcanoes. Loihi is particularly interesting because it represents an early stage in the origin of a new Hawaiian island. It is a submarine volcano that rises more than 3,000 m above the adjacent seafloor, but its summit is still about 940 m below sea level.

Even though the Hawaiian volcanoes are not at or near a divergent or convergent plate boundary, their evolution is nevertheless related to plate movements. Notice in Figure 2.24 that the ages of the rocks that make up the Hawaiian islands increase toward the northwest. Kauai formed 5.6–3.8 million years ago, whereas Hawaii began forming less than 1 million years ago, and Loihi began to form even more recently. The islands have formed in succession as the Pacific plate moves continuously over a hot spot now beneath Hawaii and just to the south at Loihi.

5.8 Volcanic Hazards, Volcano Monitoring, and Forecasting Eruptions

Undoubtedly you suspect that living near an active volcano poses some risk, and of course this assessment is correct. But what exactly are volcanic hazards, is there any way to anticipate eruptions, and what can we do to minimize the dangers of eruptions? We have already mentioned that lava flows, with few exceptions, pose little threat to humans although they may destroy property. Lava flows, nuée ardentes, and volcanic gases are threats during an eruption (■ Figures 5.19 and 5.20); however, lahars and landslides may take place even when no eruption has occurred for a long time. Certainly, the most vulnerable areas in the United States are Alaska, Hawaii, California, Oregon, and Washington, but some other parts of the West might also experience renewed volcanism.

■ **Figure 5.19 Volcanic Hazards** A volcanic hazard is any manifestation of volcanism that poses a threat, including lava flows and, more importantly, volcanic gas, ash, and lahars.

a This sign at Mammoth Mountain volcano in California warns of the potential danger of CO_2 gas, which has killed 170 acres of trees.

b When Mount Pinatubo in the Philippines erupted on June 15, 1991, this huge cloud of ash and steam formed over the volcano.

Figure 5.20 Kalapana, Hawaii Lava flows from Kilauea Volcano entered Kalapana in 1986 but caused little damage. However, in 1990, the flows moved through the town destroying houses and businesses. These images show the progressive burial of Walter's Kalapana Store and Drive Inn.

a April 23, 1990

U.S. Geological Survey

b June 6, 1990

c June 13, 1990

How Large Is an Eruption, and How Long Do Eruptions Last?

The most widely used indication of the size of a volcanic eruption is the **volcanic explosivity index (VEI)** (Figure 5.21). The VEI ranges from 0 (gentle) to 8 (cataclysmic) and is based on several aspects of an eruption, such as the volume of material explosively ejected and the height of the eruption plume. However, the volume of lava, fatalities, and property damage are not considered. For instance, the 1985 eruption of Nevado del Ruiz in Colombia killed 23,000 people, yet has a VEI value of only 3. In contrast, the huge eruption (VEI = 6) of Novarupta in Alaska in 1912 caused no fatalities or injuries. Since AD 1500, only the 1815 eruption of Tambora had a value of 7; it was both large and deadly (Table 5.1). Of the several eruptions of Eyjafjallajökull in Iceland during April 2010, the largest one had a VEI of no more than 4, but because of its location and the fact that it erupted beneath glacial ice, it caused disruption of air traffic over the North Atlantic.

The duration of eruptions varies considerably. Fully 42% of about 3,300 historic eruptions lasted less than one month. About 33% erupted for one to six months, but some 16 volcanoes have been active more or less continuously for more than 20 years. Stromboli and Mount Etna in Italy and Erta Ale in Ethiopia are good examples. For some explosive volcanoes, the time from the onset of their eruptions to the climactic event is weeks or months. A case in point is the colossal explosive eruption of Mount St. Helens on May 18, 1980, that occurred two months after eruptive activity began (Figure 5.18c). Unfortunately, many volcanoes give little or no warning of such large-scale events; of 252 explosive eruptions, 42% erupted most violently during their first day of activity.

Is It Possible to Forecast Eruptions?

Only a few of Earth's potentially dangerous volcanoes are monitored, including some in Japan, Italy, Russia, New Zealand, and the United States. Volcano monitoring involves recording and analyzing physical and chemical changes at volcanoes (Figure 5.22). Tiltmeters detect changes in the slopes of a volcano as it inflates when magma rises beneath it, and a geodimeter uses a laser beam to measure horizontal distances, which change as a volcano inflates. Geologists also monitor changes in gas emissions, groundwater level and temperature, hot springs activity, and local magnetic and electrical fields. Even the accumulating snow and ice, if any, are evaluated to anticipate hazards from floods should an eruption take place.

Of critical importance in volcano monitoring and warning of an imminent eruption is the detection of **volcanic tremor,** continuous ground motion that lasts for minutes to hours as opposed to the sudden, sharp jolts produced by most earthquakes. Volcanic tremor, also known as *harmonic tremor,* indicates that magma is moving beneath the surface.

To more fully anticipate the future activity of a volcano, its eruptive history must be known. Accordingly,

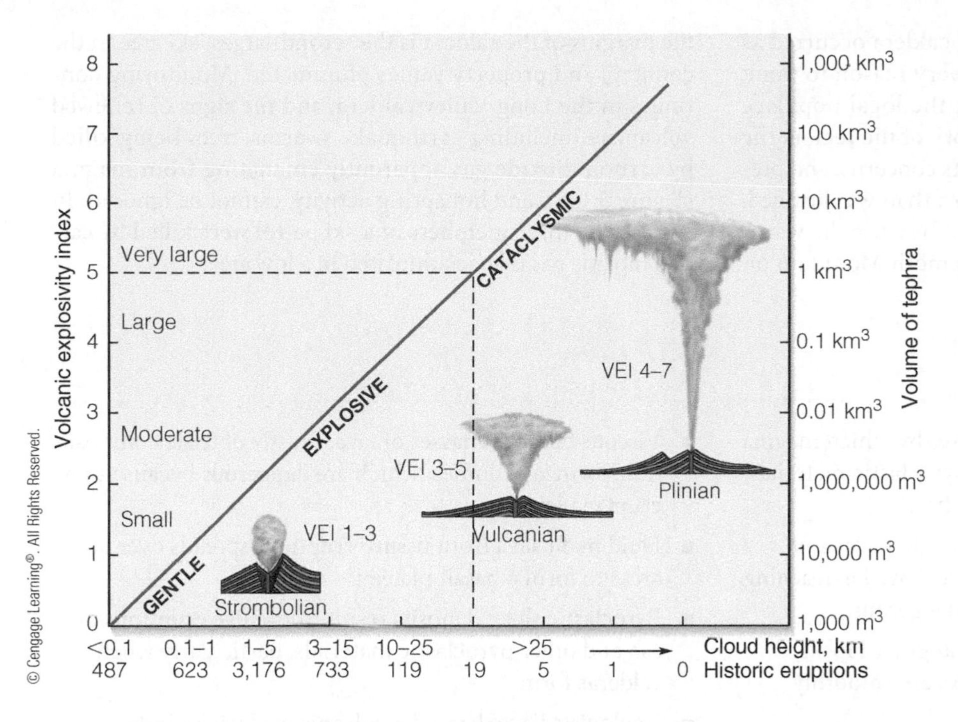

Figure 5.21 The Volcanic Explosivity Index In this example, an eruption with a VEI of 5 has an eruption cloud up to 25 km high and ejects at least 1 km^3 of tephra, a collective term for all pyroclastic materials. Geologists characterize eruptions as Hawaiian (nonexplosive), Strombolian, Vulcanian, and Plinian.

Critical Thinking Question Why are the numbers of fatalities and property damage not used in assigning a VEI value to an eruption?

geologists study the record of past eruptions preserved in rocks. Detailed studies before 1980 indicated that Mount St. Helens, Washington, had erupted explosively 14 or 15 times during the past 4,500 years, so geologists concluded that it was one of the most likely Cascade Range volcanoes to erupt again. In fact, maps they prepared showing areas in which damage from an eruption could be expected were helpful in determining which areas should have restricted access and evacuations once an eruption did take place.

Geologists successfully gave timely warnings of impending eruptions of Mount St. Helens in Washington and Mount Pinatubo in the Philippines, but in both cases, the climactic eruptions were preceded by eruptive activity of lesser intensity. In some cases, however, the warning signs are much more subtle and difficult to interpret. Numerous small earthquakes and other warning signs indicated to USGS geologists that magma was moving beneath the surface of the Long Valley caldera in eastern California, so in 1987, they issued a low-level warning, and then nothing happened.

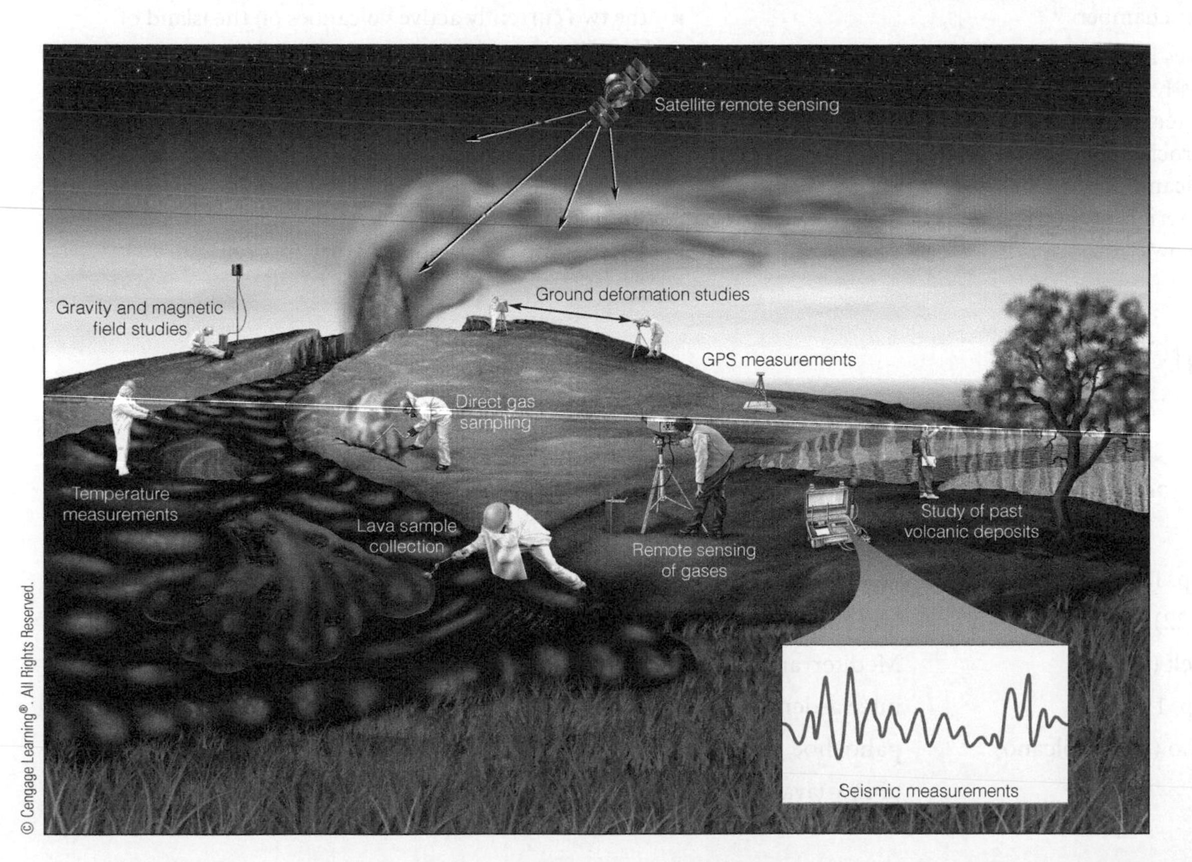

Figure 5.22 Volcanic Monitoring Some important techniques used to monitor volcanoes.

Volcanic activity in the Long Valley caldera occurred as recently as 250 years ago, and there is every reason to think that it will occur again. Unfortunately, the local populace was largely unaware of the geologic history of the region, the USGS did a poor job in communicating its concerns, and premature news releases caused more concern than was justified. In any case, local residents were outraged because the warnings caused a decrease in tourism (Mammoth Mountain on the margins of the caldera is the second-largest ski area in the country) and property values plummeted. Monitoring continues in the Long Valley caldera, and the signs of renewed volcanism, including earthquake swarms, trees being killed by carbon dioxide gas apparently emanating from magma (Figure 5.19a), and hot spring activity, cannot be ignored. In April 2006, three members of a ski patrol were killed by carbon dioxide gas that accumulated in a low area.

Key Concepts Review

- Volcanism encompasses those processes by which magma rises to the surface as lava flows and pyroclastic materials, and gases are released into the atmosphere.
- Gases make up only a few percent by weight of magma. Most is water vapor, but sulfur gases may have far-reaching climatic effects, and carbon dioxide is dangerous.
- Aa lava flows have surfaces of jagged, angular blocks, whereas the surfaces of pahoehoe flows are smoothly wrinkled.
- Other features of lava flows are lava tubes and columnar joints. Lava erupted under water typically forms bulbous masses known as pillow lava.
- Volcanoes are found in various shapes and sizes, but all form where lava and pyroclastic materials are erupted from a vent.
- The summits of volcanoes have either a crater or a much larger caldera. Calderas form following voluminous eruptions, and the volcanic peak collapses into a partially drained magma chamber.
- Shield volcanoes have low, rounded profiles and are composed mostly of mafic flows that cool and form basalt. Small, steep-sided cinder cones form around a vent where pyroclastic materials erupt and accumulate. Composite volcanoes are made up of lava flows and pyroclastic materials of intermediate composition and volcanic mudflows.
- Viscous bulbous masses of lava, mostly of felsic composition, form lava domes, which are dangerous because they erupt explosively.
- Fluid mafic lava from fissure eruptions spreads over large areas to form a basalt plateau.
- Pyroclastic sheet deposits result from huge eruptions of ash and other pyroclastic materials, particularly when calderas form.
- Geologists have devised a volcanic explosivity index (VEI) to give a semiquantitative measure of the size of an eruption. Volume of material erupted and height of the eruption plume are criteria used to determine the VEI; fatalities and property damage are not considered.
- Approximately 80% of all volcanic eruptions take place in the circum-Pacific and the Mediterranean belts, mostly at convergent plate boundaries. Most of the rest of the eruptions occur along mid-oceanic ridges or their extensions onto land.
- The two currently active volcanoes on the island of Hawaii and one just to the south lie above a hot spot over which the Pacific plate moves.
- To effectively monitor volcanoes, geologists evaluate several physical and chemical aspects of volcanic regions. Of particular importance in monitoring volcanoes and forecasting eruptions is detecting volcanic tremor and determining the eruptive history of a volcano.

Important Terms

aa (p. 115)
basalt plateau (p. 126)
caldera (p. 120)
Cascade Range (p. 129)
cinder cone (p. 122)
circum-Pacific belt (p. 127)
columnar joint (p. 115)
composite volcano (stratovolcano) (p. 122)
crater (p. 120)
fissure eruption (p. 126)
lahar (p. 122)
lava dome (p. 124)
lava tube (p. 115)
Mediterranean belt (p. 127)
nuée ardente (p. 125)
pahoehoe (p. 115)
pillow lava (p. 115)
pyroclastic sheet deposit (p. 126)
shield volcano (p. 120)
volcanic ash (p. 115)
volcanic explosivity index (VEI) (p. 132)
volcanic tremor (p. 132)
volcanism (p. 113)
volcano (p. 113)

Review Questions

1. Which one of the following statements is correct?
   a. _____ Most volcanism takes place in continental interiors.
   b. _____ Lava flows on composite volcanoes are predominantly intermediate in composition.
   c. _____ Volcanism in the Cascade Range takes place at a divergent plate boundary.
   d. _____ The VEI rating for an eruption depends on the number of fatalities and injuries.
   e. _____ Earth's Moon has many active volcanoes.
2. The shaking that occurs when magma moves beneath the surface is called
   a. _____ columnar vibrations.
   b. _____ basalt accumulation.
   c. _____ volcanic tremor.
   d. _____ cratering.
   e. _____ fissure eruption.
3. An incandescent cloud of gas and particles erupted by a volcano is a
   a. _____ spatter cone.
   b. _____ lapelli.
   c. _____ pressure ridge.
   d. _____ nuée ardent.
   e. _____ welded tuff.
4. Water-saturated flows of volcanic debris are called _________ and are common on _________ volcanoes.
   a. _____ fissure eruption/cinder cone.
   b. _____ pillow lava/submarine.
   c. _____ lahars/composite.
   d. _____ lava flows/dome.
   e. _____ volcanic bombs/shield.
5. The most common gas emitted by volcanoes is
   a. _____ hydrogen sulfide.
   b. _____ fluorine.
   c. _____ methane.
   d. _____ pahoehoe.
   e. _____ water vapor.
6. What kinds of data do geologists evaluate when they monitor volcanoes and warn of impending erutions?
7. Why are eruptions of mafic magma rather quiet, whereas those of felsic magma are commonly explosive?
8. What criteria do geologists use to assign a volcanic explosivity index (VEI) value to an eruption?
9. Suppose you find rock exposures on land made up of pillow lava overlain by deep-sea sedimentary rocks. Where and how did the pillow lava form, and what type of rock would you expect to find beneath the pillow lava?

10. **Creative Thinking Visual Question:** Identify the type of volcano and the kind of lava flow shown in this image (▌Figure 1).

Sue Monroe

▌**Figure 1** Volcanic features on Medicine Lake Volcano in California.

Global GeoScience Watch Search "volcanoes" in the GREENR database, and then search within these results (on the left-hand side) "USGS News: August." Click on the article titled "USGS News: August Science Picks." Scan down to the subhead titled "Explosive Eruption of Kasatochi, Cleveland, and Okmok Volcanoes in Alaska." When did these volcanoes erupt, what kinds of hazards do they pose to travel in this area, and how do scientists track the progress of these eruptions? How does plate tectonics account for the location of these volcanoes?

Weathering and erosion has yielded this badlands topography in Badlands National Park in South Dakota. Notice the sharp, angular slopes and ridges and numerous ravines that are typical of this kind of topography. In addition to their scenery, these rocks and those underlying them are noted for their numerous fossil mammals. The rocks in this view were originally deposited as sand in stream channels and mud on floodplains.

CHAPTER 6

Weathering, Erosion, and Soil

OUTLINE

HAVE YOU EVER WONDERED?

- How weathering brings about changes in Earth materials?
- What physical processes break rocks into smaller pieces?
- How chemical reactions decompose rocks?
- What factors control the rate at which Earth materials are altered?
- What soil is, how it forms, and why it varies in fertility?
- Why soil loses its fertility?
- How alteration of rocks is responsible for some important natural resources?

6.1 Introduction

Weathering includes several physical and chemical processes that alter surface and near-surface rocks and minerals, whereas **erosion** involves the removal of weathered materials, by running water, wind, glaciers, and waves from the area where weathering takes place. Actually, weathering alters rocks and minerals so that they are more nearly in equilibrium with a new set of environmental conditions. For instance, many igneous and metamorphic rocks form within Earth's crust where pressure and temperature are high and little or no water or oxygen is present. These same rocks at the surface, though, are exposed to low pressure and temperature, water, the atmosphere, and the activities of organisms (▌Figure 6.1). Thus, interactions of Earth materials with the hydrosphere, atmosphere, and biosphere bring about changes as they break down physically (*disintegrate*) and change chemically (*decompose*).

Weathering is such a pervasive phenomenon that it is easy to overlook. Nevertheless, it takes place continuously at variable rates on all surface and near-surface rocks and minerals, including rocklike substances used in construction. Roadways and runways, bricks, and concrete in sidewalks, foundations, and bridges change with time as they are relentlessly attacked by the elements. Obviously, weathering of sidewalks, foundations, and other structures requires costly repairs or replacement.

Weathering and erosion are partly responsible for some of the magnificent landscapes that we enjoy visiting. The beautiful scenery of Bryce Canyon National Park in Utah, Badlands National Park in South Dakota, and Acadia National Park in Maine are only a few examples of landscapes partly fashioned by weathering and erosion (see chapter opening photo and ▌Figure 6.2). Mountains form mostly at convergent plate boundaries by deformation of rocks as well as metamorphism and emplacement of plutons. Nevertheless, their present-day topographic expression is also related to ongoing weathering and erosion, especially erosion by mass wasting (gravity-driven processes), running water, and, in some cases, glaciers.

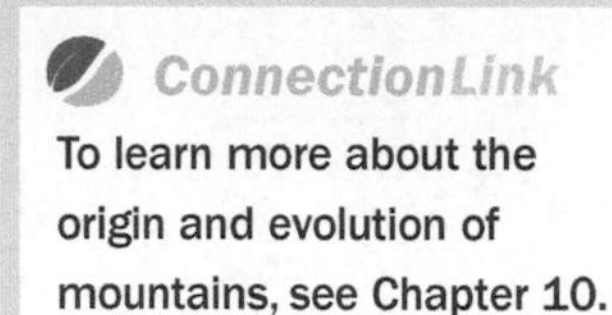

▌**Figure 6.1** Weathering of Granite

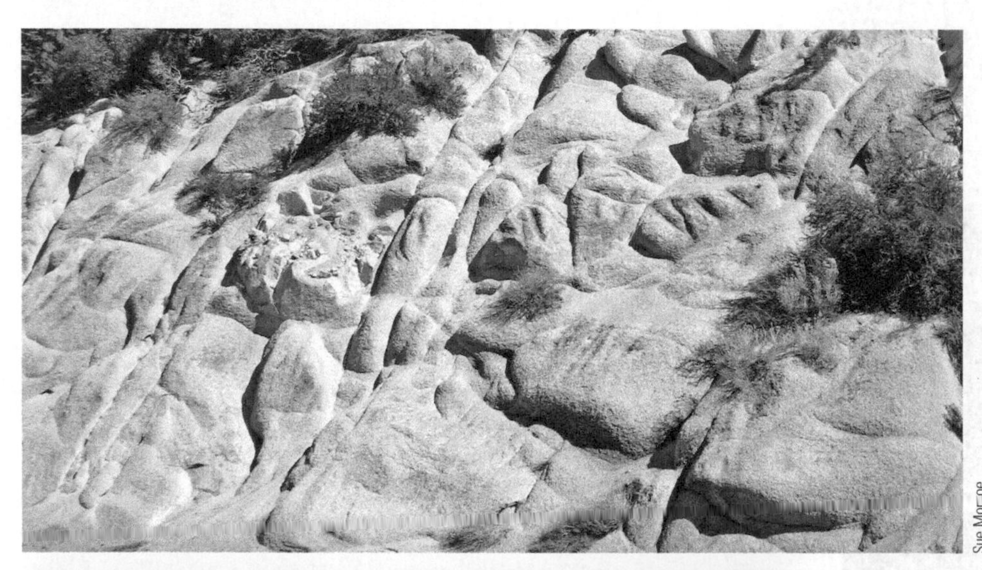

a This exposure of granite has been so thoroughly altered by weathering that only spherical masses of the original rock are visible.

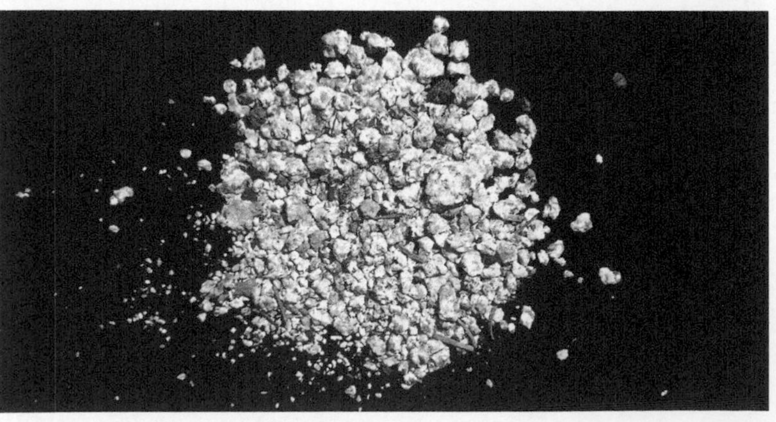

b Close-up view of the weathered granite. Mechanical weathering has predominated, so the particles are mostly small pieces of granite and minerals such as quartz and feldspars.

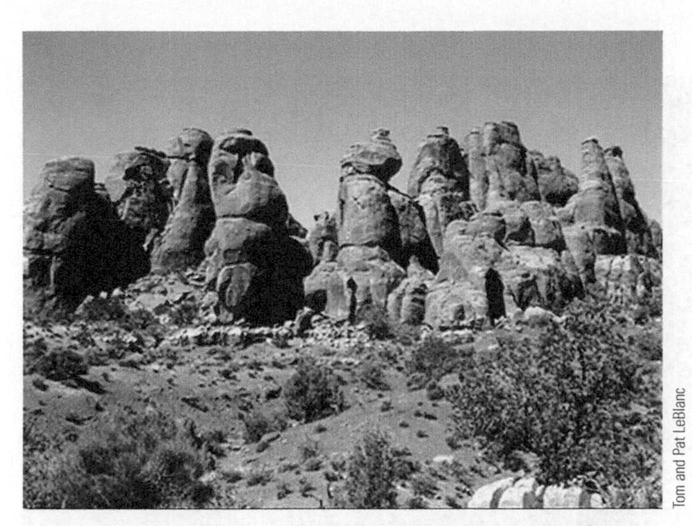

Tom and Pat LeBlanc

Figure 6.2 Differential Weathering and Erosion Rocks are not compositionally and physically homogeneous, so irregular surfaces develop as they weather and erode such as this one in Arches National Park, in Utah.

From the academic point of view, weathering is an important group of processes that again illustrates the interactions among Earth's systems. But there are other good reasons to study weathering. One reason is so that we can develop construction materials that are more resistant to physical and chemical changes, or develop more effective methods to protect them from the elements. Another reason is that weathering is an essential part of the *rock cycle* (see Figure 1.15). When **parent material**—that is, rocks and minerals exposed to weathering—breaks down into smaller pieces (Figure 6.1b) or perhaps dissolves, this weathered material may be eroded from the weathering site and transported elsewhere, by running water or wind, and deposited as *sediment,* the raw materials for sedimentary rocks.

ConnectionLink

You can find more information about sediments and sedimentary rocks in Chapter 7.

In addition to providing the raw materials for sedimentary rocks, weathering is responsible for the origin of *soils.* Needless to say, we depend on soils directly or indirectly for our existence. In some areas, however, erosion takes place faster than soil-forming processes operate, thereby decreasing the amount of productive soil. Of course, erosion is an ongoing natural process, so some soil losses are expected as a normal consequence of the evolution of the land. Unfortunately, some human activities compound the problem, leading to less fertile soil and reduced agricultural production.

Another reason to study weathering is that it accounts for the origin or concentration of some natural resources—the ore of aluminum, for instance. Weathering is also responsible for many clay deposits that are used in ceramics and the manufacture of paper, and weathering coupled with erosion and deposition yields deposits of tin, gold, and diamonds.

6.2 Alteration of Minerals and Rocks

Weathering is a surface or near-surface process, but the rocks it acts on are not structurally and compositionally homogeneous throughout, which accounts for **differential weathering.** That is, weathering takes place at different rates even in the same area and on the same rocks, so it commonly results in irregular surfaces. Differential weathering and *differential erosion*—that is, variable rates of erosion—combine to yield some unusual and even bizarre features, such as hoodoos, spires, and arches (Figure 6.2).

Geologists characterize weathering as *mechanical* and *chemical,* both of which proceed simultaneously on parent material as well as on materials in transport and those deposited as sediment. In short, all surface or near-surface materials weather, although one type of weathering may predominate depending on such variables as climate and rock type.

6.3 Mechanical Weathering—Disaggregation of Earth Materials

Mechanical weathering takes place when physical forces break Earth materials into smaller pieces that retain the composition of the parent material. Granite, for instance, might mechanically weather and yield smaller pieces of granite or individual grains of quartz, potassium feldspars, plagioclase feldspars, and biotite (Figure 6.1b). Several physical processes account for mechanical weathering.

Frost Action

Frost action involving water repeatedly freezing and thawing in cracks and pores in rocks is particularly effective where temperatures often fluctuate above and below freezing. Frost action is effective because water expands by about 9% when it freezes, thus exerting great force on the walls of a crack, widening and extending it by *frost wedging* (Figure 6.3a). Repeated freezing and thawing dislodge angular pieces of rock from the parent material that tumble downslope and accumulate as **talus** (Figure 6.3b). Frost action is most effective in high mountains, even during the summer months, but it has little or no effect where the temperature rarely drops below freezing or where Earth materials are permanently frozen.

Figure 6.3 Frost Wedging

a Frost wedging takes place when water seeps into cracks and expands as it freezes. Angular pieces of rock are pried loose by repeated freezing and thawing.

b Frost wedging and other mechanical weathering processes produced this talus accumulation at the base of this bluff along the Henrys Fork of the Snake River in Idaho.

Pressure Release

Some rocks form at great depth and are stable under tremendous pressure. Granite crystallizes far below the surface, so when it is uplifted and eroded, its contained energy is released by outward expansion, a phenomenon known as **pressure release.** The outward expansion results in the origin of fractures called **sheet joints** (Figure 6.4) that more or less parallel the exposed rock surface. Sheet-joint-bounded slabs of rock slip or slide off the parent rock, leaving large, rounded masses known as **exfoliation domes** (Figure 6.5).

For more information on joints, see Chapter 10.

That solid rock expands and produces fractures might be counterintuitive, but it is a well-known phenomenon. In deep mines, masses of rock detach from the sides of the excavation, often explosively. These *rock bursts* and less violent *popping* pose a danger to mine workers, and in South Africa they are responsible for approximately 20 deaths per year. In some quarries for building stone, excavations to only 7–8 m exposed rocks in which sheet joints formed, in some cases with enough force to throw quarrying machines weighing more than a ton from their tracks.

Thermal Expansion and Contraction

During **thermal expansion and contraction,** the volume of rocks changes as they heat up and then cool down. The temperature may vary as much as 30°C a day in a desert, and rock—being a poor conductor of heat—heats and expands on its outside more than its inside. Even dark minerals absorb heat faster than light-colored ones, so differential expansion takes place between minerals. Surface expansion might generate enough stress to cause fracturing, but experiments in which rocks are heated and cooled repeatedly to simulate years of such activity indicate that thermal expansion and contraction are of minor importance in mechanical weathering.

Growth of Salt Crystals

The formation of salt crystals can exert enough force to widen cracks and dislodge particles in porous, granular rocks such as sandstone. And even in rocks with an interlocking mosaic of crystals, such as granite, **salt crystal growth** pries loose individual minerals. It takes place mostly in hot, arid regions but also probably affects rocks in some coastal areas.

Activities of Organisms

Animals and plants also participate in the mechanical alteration of rocks (Figure 6.6a). Burrowing animals, such as worms, reptiles, rodents, termites, and ants, constantly mix soil and sediment particles and bring material from depth to the surface where further weathering occurs. The roots of plants, especially large bushes and trees, wedge themselves into cracks in rocks and further widen them (Figure 6.6b).

Figure 6.4 Sheet Joints

Critical Thinking Question A highway is visible in the foreground in both **(a)** and **(b)**. Why do you think these sheet joints might be a problem for highway maintenance crews?

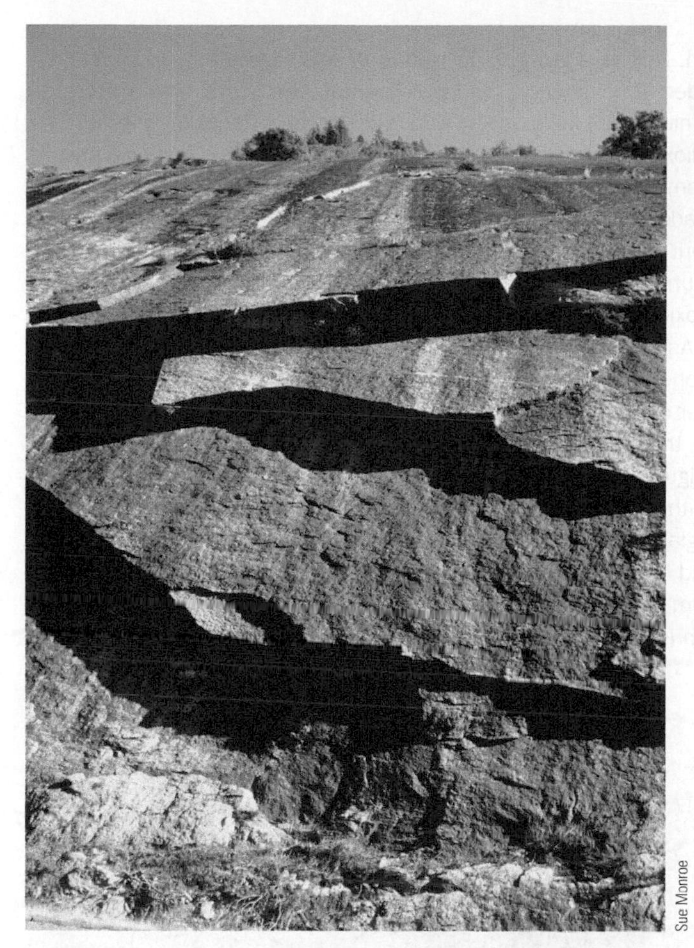

Sue Monroe

a Slabs of granitic rock bounded by sheet joints in the Sierra Nevada of California. The slabs are inclined downward toward the roadway visible at the lower left.

Sue Monroe

b Notice in this image that the sheet-joint bounded slabs have started moving down-slope toward the road.

James S. Monroe

Figure 6.5 Exfoliation Domes North Dome and Basket Dome are two of many exfoliation domes in Yosemite National Park, California. In the distance you can see several other exfoliation domes.

6.4 Chemical Weathering—Decomposition of Earth Materials

Chemical weathering decomposes rocks and minerals by chemical alteration of the parent material. In contrast to mechanical weathering, chemical weathering changes the composition of weathered materials. For example, several clay minerals (sheet silicates) form by the chemical and structural alteration of other minerals, such as potassium feldspars and plagioclase feldspars, both of which are framework silicates. Other minerals completely decompose during chemical weathering, but some chemically stable minerals are simply liberated from the parent material.

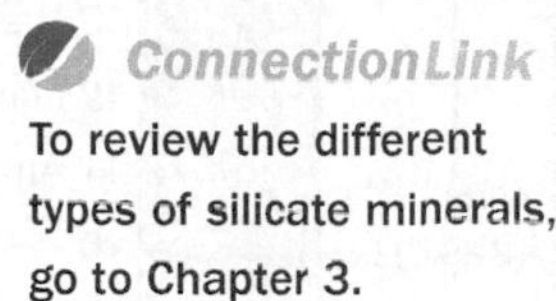

ConnectionLink

To review the different types of silicate minerals, go to Chapter 3.

Important agents of chemical weathering include atmospheric gases, especially oxygen, water, and acids (Geo-Focus 6.1). Organisms also play an important role. Rocks with lichens composite organisms made up of fungi and algae) on their surfaces undergo more rapid chemical alteration than lichen-free rocks (Figure 6.6a). In addition, plants remove ions from soil water and reduce the chemical stability of soil minerals, and plant roots release organic acids.

Solution

When **solution** takes place, the ions of a substance separate in a liquid, and the solid substance dissolves. Water is a remarkable solvent because its molecules have an asymmetric shape,

GEO FOCUS 6.1 Industrialization and Acid Rain

One result of industrialization is atmospheric pollution, which causes smog, possible disruption of the ozone layer, global warming, and acid rain. Acidity, a measure of hydrogen ion concentration, is measured on the pH scale ((▌Figure 1). A pH value of 7 is neutral, whereas acidic conditions correspond to values less than 7, and values greater than 7 denote alkaline, or basic, conditions. Normal rain has a pH value of about 5.6, making it slightly acidic, but acid rain has a pH of less than 5.0. In addition, some areas experience acid snow and even acid fog with a pH as low as 1.7.

Several natural processes, including soil bacteria metabolism and volcanism, release gases into the atmosphere that contribute to acid rain. Human activities also produce added atmospheric stress, especially burning fossil fuels that release carbon dioxide and nitrogen oxide from internal combustion engines. Both of these gases add to acid rain, but the greatest culprit is sulfur dioxide released mostly by burning coal that contains sulfur that oxidizes to form sulfur dioxide (SO_2). As sulfur dioxide rises into the atmosphere, it reacts with oxygen and water droplets to form sulfuric acid (H_2SO_4), the main component of acid rain (▌Figure 2).

Robert Angus Smith first recognized acid rain in England in 1872, but not until 1961 did it become an environmental concern when scientists realized that acid rain is corrosive and irritating, kills vegetation, and has a detrimental effect on surface waters. Since then, the effects of acid rain are apparent in Europe (especially in eastern Europe) and the eastern part of North America, where the problem has been getting worse for the past three decades.

The areas affected by acid rain invariably lie downwind from plants that emit sulfur gases, but the effects of acid rain in these areas may be modified by the local geology. For instance, if the area is underlain by limestone or alkaline soils, acid rain tends to be neutralized, but granite has little or no modifying effect. Small lakes lose their ability to neutralize acid rain and become more and more acidic until

▌Figure 1 Acid Rain

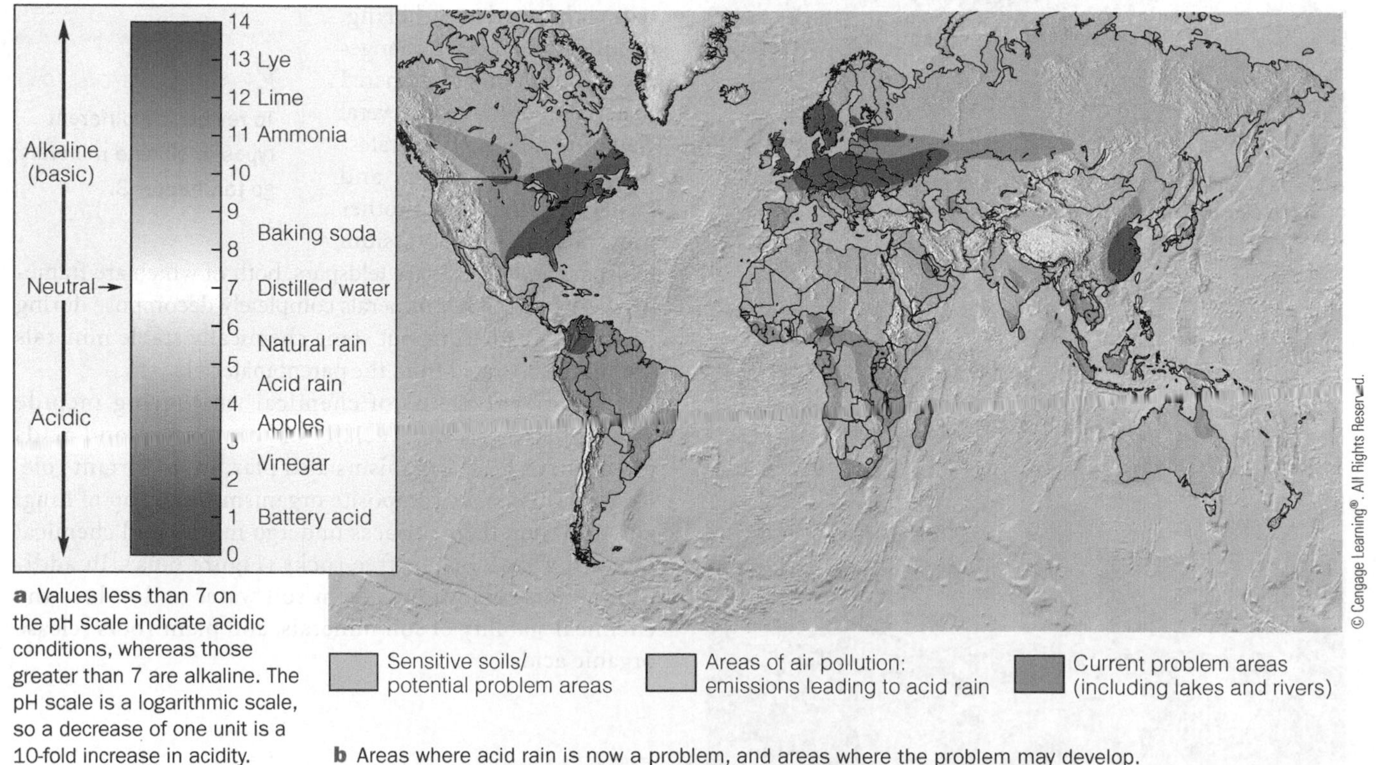

a Values less than 7 on the pH scale indicate acidic conditions, whereas those greater than 7 are alkaline. The pH scale is a logarithmic scale, so a decrease of one unit is a 10-fold increase in acidity.

b Areas where acid rain is now a problem, and areas where the problem may develop.

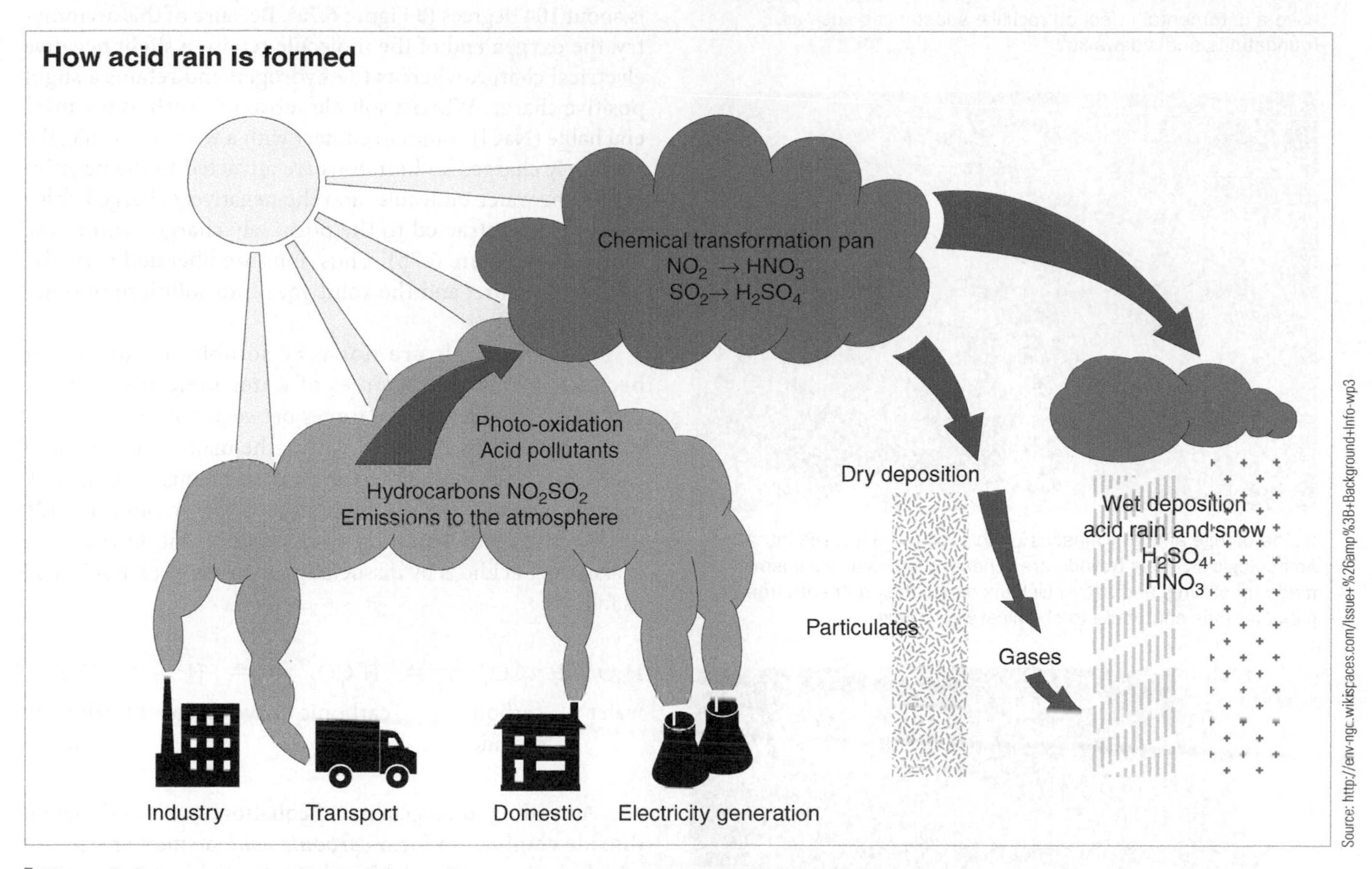

Figure 2 Sources of sulfur dioxide (SO_2) and its reaction to form sulfuric acid (H_2SO_4). About half of the acidic gases and particles fall to Earth in dry deposition, whereas wet deposition applies to acid rain, snow, or fog.

various types of organisms disappear, and in some cases all life-forms eventually die.

Acid rain also causes increased chemical weathering of limestone and marble and, to a lesser degree, sandstone. The effects are especially evident on buildings, monuments, and tombstones, as in Gettysburg National Military Park in Pennsylvania.

The devastation caused by sulfur gases on vegetation near coal-burning plants is apparent, and many forests in the eastern United States show signs of stress that cannot be attributed to other causes (see Figure 24.2).

Millions of tons of sulfur dioxide are released yearly into the atmosphere in the United States. Power plants built before 1975 have no emission controls, but the problems they pose must be addressed if emissions are to be reduced to an acceptable level. The most effective way to reduce emissions from these older plants is with flue-gas desulfurization, a process that removes up to 90% of the sulfur dioxide from exhaust gases.

Flue-gas desulfurization has some drawbacks. One is that some plants are simply too old to be profitably upgraded. Other problems include disposal of sulfur wastes, the lack of control on nitrogen gas emissions, and reduced efficiency of the power plant, which must burn more coal to make up the difference.

Other ways to control emissions are burning low-sulfur coal, fluidized bed combustion, and conservation of electricity. Natural gas contains practically no sulfur, but converting to this alternative energy source would require the installation of expensive new furnaces in existing plants.

Acid rain, like global warming, is a worldwide problem that knows no national boundaries. Wind may blow pollutants from the source in one country to another where the effects are felt. For instance, much of the acid rain in eastern Canada actually comes from sources in the United States.

Figure 6.6 Organisms and Weathering

Critical Thinking Question Does vegetation, especially trees, have a detrimental effect on rocklike substances such as foundations and sidewalks?

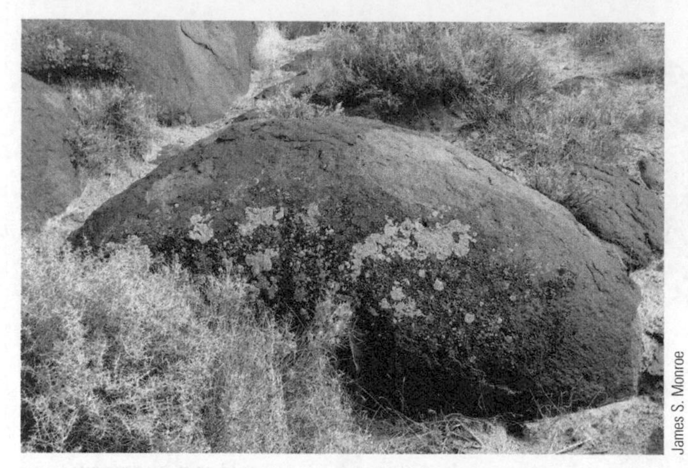

James S. Monroe

a The orange and gray masses on this rock at Grimes Point Archeological Site in Nevada are lichens, composite organisms made up of fungi and algae. Lichens derive their nutrients from the rock and thus contribute to chemical weathering.

James S. Monroe

b These trees in the Black Hills of South Dakota contribute to mechanical weathering as they grow in cracks in the rocks, thereby breaking the parent material into smaller pieces.

consisting of one oxygen atom with two hydrogen atoms arranged so that the angle between the two hydrogen atoms is about 104 degrees (Figure 6.7a). Because of this asymmetry, the oxygen end of the molecule retains a slight negative electrical charge, whereas the hydrogen end retains a slight positive charge. When a soluble substance such as the mineral halite (NaCl) comes in contact with a water molecule, the positively charged sodium ions are attracted to the negative end of the water molecule, and the negatively charged chloride ions are attracted to the positively charged end of the molecule (Figure 6.7b). Thus, ions are liberated from the crystal structure, and the solid goes into solution; in other words, it dissolves.

Most minerals are not very soluble in pure water because the attractive forces of water molecules are not sufficient to overcome the forces between particles in minerals. The mineral calcite ($CaCO_3$), the major constituent of the sedimentary rock limestone and the metamorphic rock marble, is practically insoluble in pure water, but it rapidly dissolves if a small amount of acid is present. One way to make water acidic is by dissociating the ions of carbonic acid as follows:

$$\underset{\text{water}}{H_2O} + \underset{\text{carbon dioxide}}{CO_2} \rightleftharpoons \underset{\text{carbonic acid}}{H_2CO_3} \rightleftharpoons \underset{\text{hydrogen ion}}{H^+} + \underset{\text{bicarbonate ion}}{HCO_3^-}$$

According to this chemical equation, water and carbon dioxide combine to form *carbonic acid,* a small amount of which dissociates to yield hydrogen and bicarbonate ions. The concentration of hydrogen ions determines the acidity of a solution; the more hydrogen ions present, the stronger the acid (Geo-Focus 6.1).

Carbon dioxide from several sources may combine with water and react to form acid solutions. The atmosphere is mostly nitrogen and oxygen, but approximately 0.03% is carbon dioxide, causing rain to be slightly acidic. Decaying organic matter and the respiration of organisms produce carbon dioxide in soils, so groundwater is also usually slightly acidic. Climate also affects the acidity, with arid regions tending to have alkaline groundwater (that is, it has a low concentration of hydrogen ions).

Whatever the source of carbon dioxide, once an acidic solution is present, calcite rapidly dissolves according to the reaction.

$$\underset{\text{calcite}}{CaCO_3} + \underset{\text{water}}{H_2O} + \underset{\text{carbon dioxide}}{CO_2} \rightleftharpoons \underset{\text{calcium ion}}{Ca^{++}} + \underset{\text{bicarbonate ion}}{2HCO_3^-}$$

Oxidation

The term **oxidation** has a variety of meanings for chemists, but in chemical weathering, it refers to reactions with oxygen to form an oxide (one or more metallic elements combined with oxygen) or, if water is present, a hydroxide (a metallic

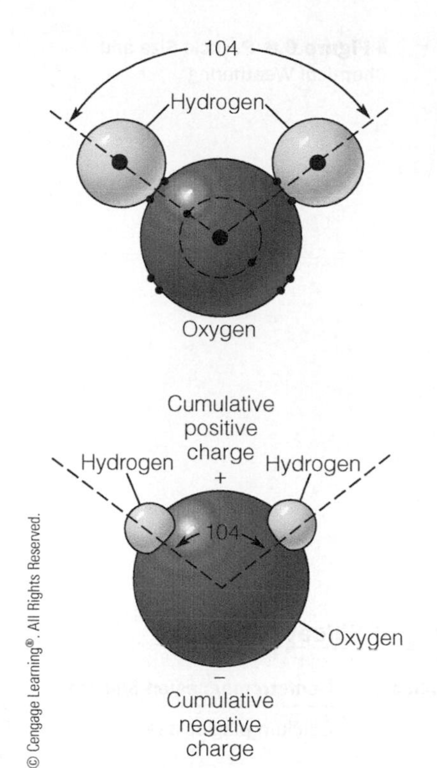

Figure 6.7 The Solution of Halite

Critical Thinking Question Which do you think would dissolve faster, rock salt or table salt? Explain.

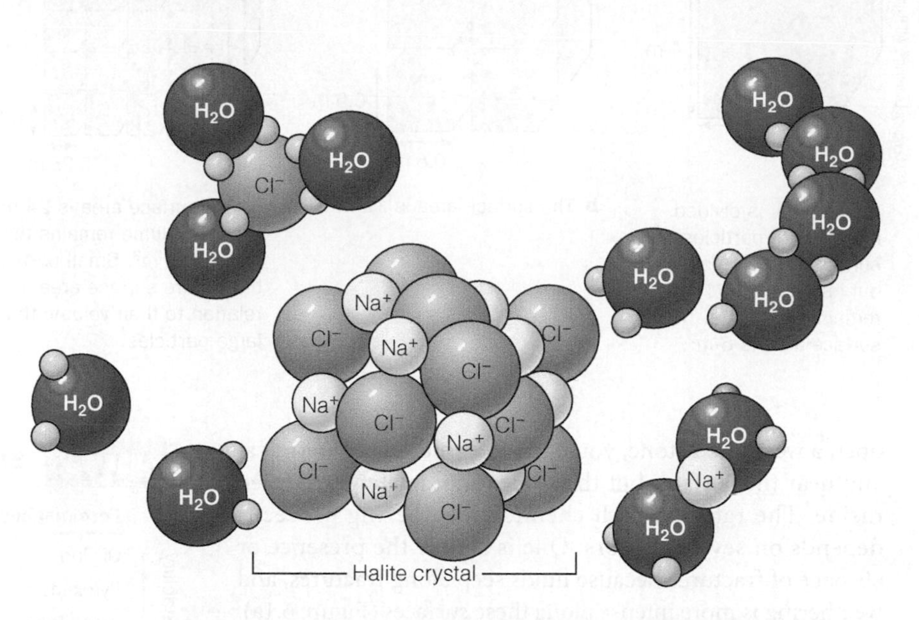

a The structure of a water molecule. The asymmetric arrangement of hydrogen atoms causes the molecule to have a slight positive electrical charge at its hydrogen end and a slight negative charge at its oxygen end.

b Solution of sodium chloride (NaCl), the mineral halite, in water. Note that the sodium atoms are attracted to the oxygen end of a water molecule, whereas chloride ions are attracted to the hydrogen end of the molecule.

element or radical combined with OH^-). For example, iron rusts when it combines with oxygen to form the iron oxide hematite:

$$\underset{\text{iron}}{4Fe} + \underset{\text{oxygen}}{3O_2} \rightarrow \underset{\text{iron oxide (hematite)}}{2Fe_2O_3}$$

Atmospheric oxygen is abundantly available for oxidation reactions, but oxidation is generally a slow process unless water is present. Thus, most oxidation is carried out by oxygen dissolved in water.

Oxidation is important in the alteration of ferromagnesian silicates, such as olivine, pyroxenes, amphiboles, and biotite. Iron in these minerals combines with oxygen to form the reddish iron oxide hematite (Fe_2O_3) or the yellowish or brown hydroxide limonite [$FeO(OH) \cdot nH_2O$]. The yellow, brown, and red colors of many soils and sedimentary rocks are caused by the presence of small amounts of hematite or limonite.

Hydrolysis

The chemical reaction between the hydrogen (H^+) ions and hydroxyl (OH^-) ions of water and a mineral's ions is known as **hydrolysis.** In hydrolysis, hydrogen ions actually replace positive ions in minerals. Such replacement changes the composition of minerals and liberates iron that then may be oxidized.

The chemical alteration of the potassium feldspar orthoclase provides a good example of hydrolysis. All feldspars are framework silicates, but when altered, they yield soluble salts and clay minerals, such as kaolinite, which are sheet silicates. The chemical weathering of orthoclase by hydrolysis occurs as follows:

$$\underset{\text{orthoclase}}{2KAlSi_3O_8} + \underset{\text{hydrogen ion}}{2H^+} + \underset{\text{bicarbonate ion}}{2HCO_3^-} + \underset{\text{water}}{H_2O} \rightarrow$$

$$\underset{\text{clay (kaolinite)}}{Al_2Si_2O_5(OH)_4} + \underset{\text{potassium ion}}{2K^+} + \underset{\text{bicarbonate ion}}{2HCO_3^-} + \underset{\text{silica}}{4SiO_2}$$

In this reaction, hydrogen ions attack the ions in the orthoclase structure, and some liberated ions are incorporated in a developing clay mineral. The potassium and bicarbonate ions go into solution and combine to form a soluble salt. On the right side of the equation is excess silica that would not fit into the crystal structure of the clay mineral.

The Rate of Chemical Weathering Chemical weathering operates on the surfaces of particles, so it alters rocks and minerals from the outside inward. In fact, if you break

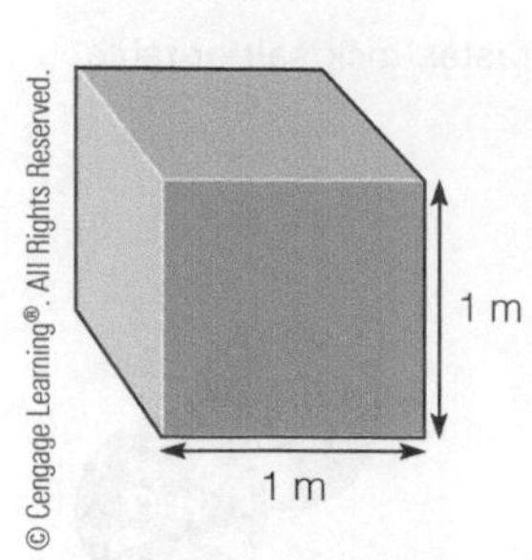

a As a rock is divided into smaller particles, its surface area increases, but its volume of 1 m^3 remains the same. The surface area is 6 m^2.

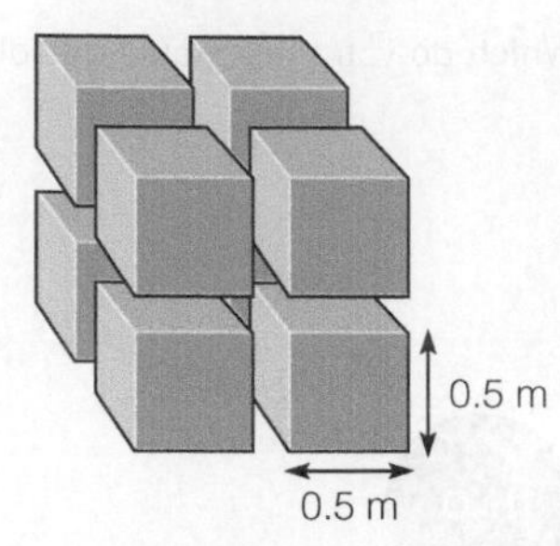

b The surface area is 12 m^2.

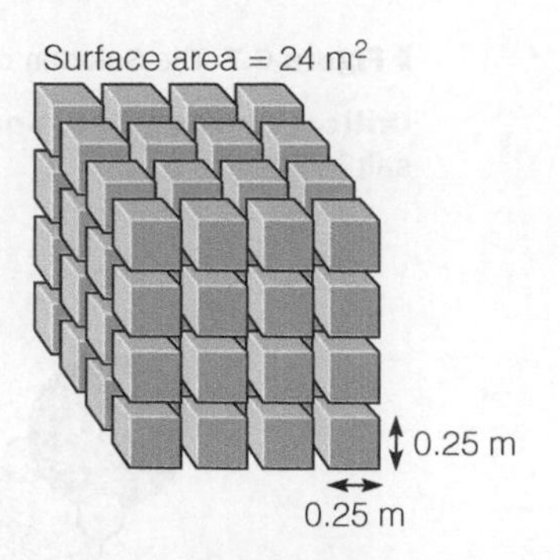

c The surface area is 24 m^2, but the volume remains the same at 1 m^3. Small particles have more surface area in relation to their volume than do large particles.

Figure 6.8 Particle Size and Chemical Weathering

open a weathered stone, you will see a rind of weathering at and near the surface, but the stone is completely unaltered inside. The rate at which chemical weathering proceeds depends on several factors. One is simply the presence or absence of fractures because fluids seep along fractures, and weathering is more intense along these surfaces (Figure 6.1a). The other factors that control chemical weathering include particle size, climate, and parent material.

Because chemical weathering affects particle surfaces, the greater the surface area, the more effective the weathering. It is important to realize that small particles have larger surface areas compared with their volume than do large particles. Notice in Figure 6.8 that a block measuring 1 m on a side has a total surface area of 6 m^2, but when the block is broken into particles measuring 0.5 m on a side, the total surface area increases to 12 m^2. And if these particles are all reduced to 0.25 m on a side, the total surface area increases to 24 m^2. Note that although the surface area in this example increases, the total volume remains the same at 1 m^3.

We can conclude that mechanical weathering contributes to chemical weathering by yielding smaller particles with greater surface area compared with their volume. Actually, your own experiences with particle size verify our contention about surface area and volume. Because of its very small particle size, powdered sugar gives an intense burst of sweetness as the tiny pieces dissolve rapidly, but otherwise it is the same as the granular sugar we use on our cereal or in our coffee.

It is not surprising that chemical weathering is more effective in the tropics than in arid and arctic regions because temperatures and rainfall are high and evaporation rates are low. In addition, vegetation and animal life are much more abundant. Consequently, the effects of weathering extend to depths of several tens of meters, but they extend only centimeters to a few meters deep in arid and arctic regions.

Some rocks are more resistant to chemical alteration than others, so parent material is another control on the rate of chemical weathering. The metamorphic rock quartzite is an extremely stable substance that alters slowly compared with most other rock types. In contrast, basalt,

TABLE 6.1 Stability of Silicate Minerals

Increasing Stability ↓	Ferromagnesian Silicates	Nonferromagnesian Silicates
	Olivine	Calcium plagioclase
	Pyroxene	
	Amphibole	Sodium plagioclase
	Biotite	Potassium feldspar
		Muscovite
		Quartz

which contains large amounts of calcium-rich plagioclase and pyroxene minerals, decomposes rapidly because these minerals are chemically unstable. In fact, the stability of common minerals is just the opposite of their order of crystallization in Bowen's reaction series (Table 6.1 and see Figure 4.4). The minerals that form last in this series are more stable, whereas those that form early are easily altered because they are most out of equilibrium with their conditions of formation.

One manifestation of chemical weathering is **spheroidal weathering** (Figure 6.9). In spheroidal weathering, a stone, even one that is rectangular to begin with, weathers to form a more spherical shape because that is the most stable shape it can assume. The reason is that on a rectangular stone, the corners are attacked by weathering from three sides, and the edges are attacked from two sides, but the flat surfaces weather more or less uniformly (Figure 6.9). Consequently, the corners and edges are altered more rapidly, the material sloughs off, a more spherical shape develops, and all surfaces weather at the same rate.

6.5 Soil and Its Origin

Much of Earth's land surface is covered by a layer of **regolith** consisting of sediment, pyroclastic materials, and the residue formed in place by weathering. Part of the regolith that

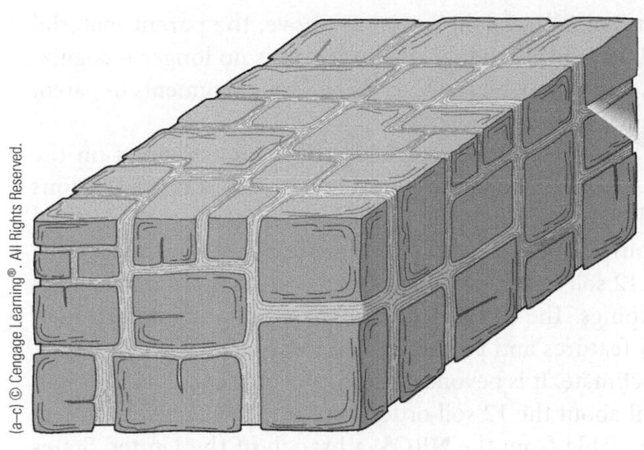

a The rectangular blocks outlined by fractures are attacked by chemical weathering processes.

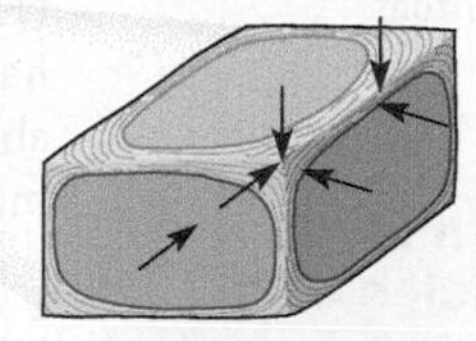

b Corners and edges weather most rapidly.

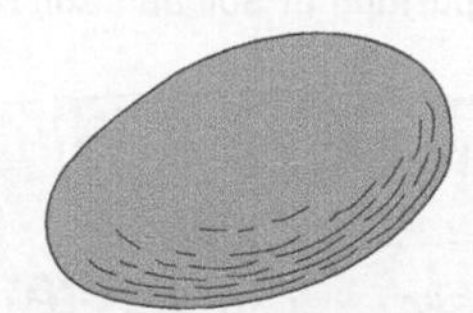

c When the blocks are weathered so that they are nearly spherical, their surfaces weather evenly and no further change in shape takes place.

James S. Monroe

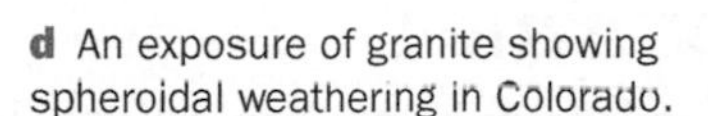

d An exposure of granite showing spheroidal weathering in Colorado.

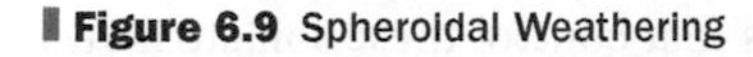

Figure 6.9 Spheroidal Weathering

Critical Thinking Question If you placed some rectangular ice cubes in the sun, how do you think their shape would change as they melted?

contains air, water, and organic matter and supports vegetation is **soil.** Obviously plants grow in soil from which they receive most of their nutrients and water. All land-dwelling animals depend directly or indirectly on soils for their nutrients.

A good soil for farming and gardening is made up of about 45% solid particles derived by weathering of parent material, and most of the rest of its volume is voids filled with air and/or water (Figure 6.10a). Another important constituent of soil is *humus,* which is carbon that forms by bacterial decay of organic matter and is very resistant to further decay. Even fertile soils may have as little as 5% humus, but it is essential as a source of plant nutrients and it enhances a soil's capacity to retain moisture.

The solid particles in soils may be sand- and silt-sized mineral grains, such as quartz, feldspars, and others, which hold soil particles apart and allow oxygen and water to circulate more freely. Clay minerals are also important, because they supply nutrients to plants and aid in soil–water retention. Should excess clay be present, though, a soil drains poorly and is sticky when wet and hard when dry.

We characterize soils as *residual* or *transported* depending on whether they formed in place or the materials composing them were transported from the weathering site. For example, if a body of granite weathers and the residue that accumulates over the granite is converted to soil, the soil so formed is residual. In contrast, transported soil forms if this same weathering residue is transported elsewhere, deposited, and then converted to soil.

The Soil Profile

Observed in vertical cross section, soil is made up of distinct layers or **soil horizons** that differ in texture, structure, composition, and color (Figure 6.10b). From the surface downward, the soil horizons are designed O, A, E, B, and C, although the boundaries between horizons are transitional, and in some cases horizon E is not present.

Horizon O is only a few centimeters thick and is composed of organic matter (Figure 6.10b). Plant remains in various states of decomposition are clearly visible in the upper part of this horizon, but its lower part consist of humus. Indeed, the upper and lower parts of horizon O are sometimes referred to as O1 and O2, respectively.

Horizon A, also called *topsoil,* has more organic matter than the horizons below and is characterized by intense biological activity because plant roots, fungi, bacteria, and worms are abundant (Figure 6.10b). In fact, the earthy aroma of freshly plowed soils comes from threadlike soil bacteria. In soils developed over a long time, horizon A is composed mostly of clays and chemically stable minerals such as quartz. Because soil formation starts at the surface and works downward, horizon A has been altered longest and is the most changed from the parent material than the horizons below.

Below horizon A in some soils is a pale layer with little carbon from which much of the small particles have been removed. This horizon E, as it is called, is present in older, more mature soils, and results from *eluviation,* a process of

Figure 6.10 The Composition of Soil and Soil Horizons

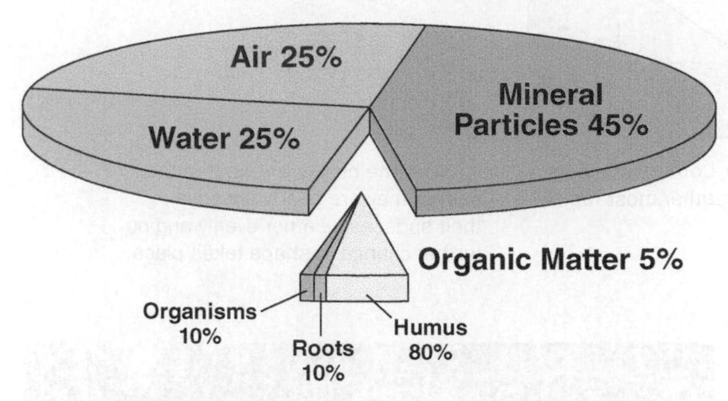

a Soils are made up mostly of minerals and rock fragments derived by weathering, air, water, and organic matter. Most of the organic matter is humus.

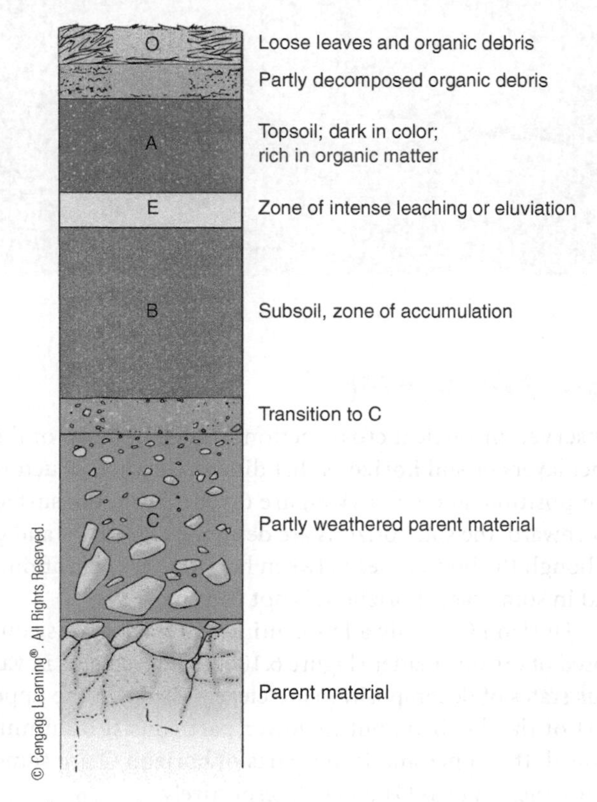

b The soil horizons in a fully developed soil. Horizon O is only a few centimeters thick, but it has been exaggerated here to show some detail.

leaching of minerals by downward moving soil water. Some of this material is then deposited in the horizon below.

Horizon B, or *subsoil,* has fewer organisms and less organic matter than horizon A (Figure 6.10b). This horizon is also called the *zone of accumulation* because soluble materials leached from above accumulate as irregular masses. If horizon A is eroded leaving horizon B exposed, plants do not grow as well, and if it is clayey, it is harder when dry and stickier when wet than the other soil horizons.

Partially altered bedrock grading down into unaltered bedrock with little organic matter characterizes horizon C (Figure 6.10b). In the horizons above, the parent material has been so thoroughly altered that it is no longer recognizable, but in horizon C, minerals and rock fragments of parent material are easy to identify.

Soils are subdivided, classified, and mapped on the basis of the development and composition of the various soil horizons. In fact, the Soil Survey Division of the Natural Resources Conservation Service (the NRCS) divides soils into 12 soil orders, and further subdivides them into smaller groupings. The 12 soil orders are based on the interaction of such features and processes as parent material, vegetation, and climate. It is beyond the scope of this chapter to go into detail about the 12 soil orders, but information about them is available from the NRCS, a branch of the United States Department of Agriculture.

Factors That Control Soil Formation

All soils form by mechanical and chemical weathering, but they differ in color, texture, thickness, and fertility. Thus, we are interested in the factors that control these soil attributes as well as the locations of various soils and how rapidly soil-forming processes operate. Climate, parent material, organic activity, relief, and time are the critical factors.

Climate and Soil Soil scientists know that climate is the most important factor influencing soil type and thickness (▌Figure 6.11). Soils that form in rather humid regions also have much of their soluble minerals leached out and horizon A may be gray, but more commonly it is black because of abundant organic matter.

Soils that form in semiarid to arid regions have much less organic matter and more unstable minerals in horizon A because so little water is available to leach them out (▌Figure 6.12). However, soluble minerals such as calcite ($CaCO_3$) is taken into solution and precipitated in horizon B as irregular masses of *caliche* (▌Figure 6.13a). Precipitation of sodium salts in some arid-region soils where soil water evaporates yields *alkali soils* that are so alkaline that they support little or no vegetation (▌Figure 6.13b).

In the tropics where chemical weathering is intense and leaching of most minerals is complete, a soil termed **laterite** forms (▌Figure 6.14a). These red soils extend to depths of many meters and are composed largely of aluminum hydroxides, iron oxides, and clay minerals. Laterite supports lush vegetation but is not very fertile because most plant nutrients have been leached out; the vegetation depends mostly on the surface layer of organic matter. In fact, when laterite is cleared of its vegetation and planted for crops, it can sustain farming for only a few years until the soil is depleted, in which case native farmers simply clear another area and repeat the process (▌Figure 6.14b, c).

Parent Material The same rock type can yield different soils in different climatic regimes and, in the same climatic regime, the same soils can develop on different rock types.

Figure 6.11 Climate and Soil Formation Generalized diagram showing soil formation as a function of the relationships between climate and vegetation, which alter parent material over time.

Critical Thinking Question Why does soil extend to much greater depths in the tropics compared to deserts?

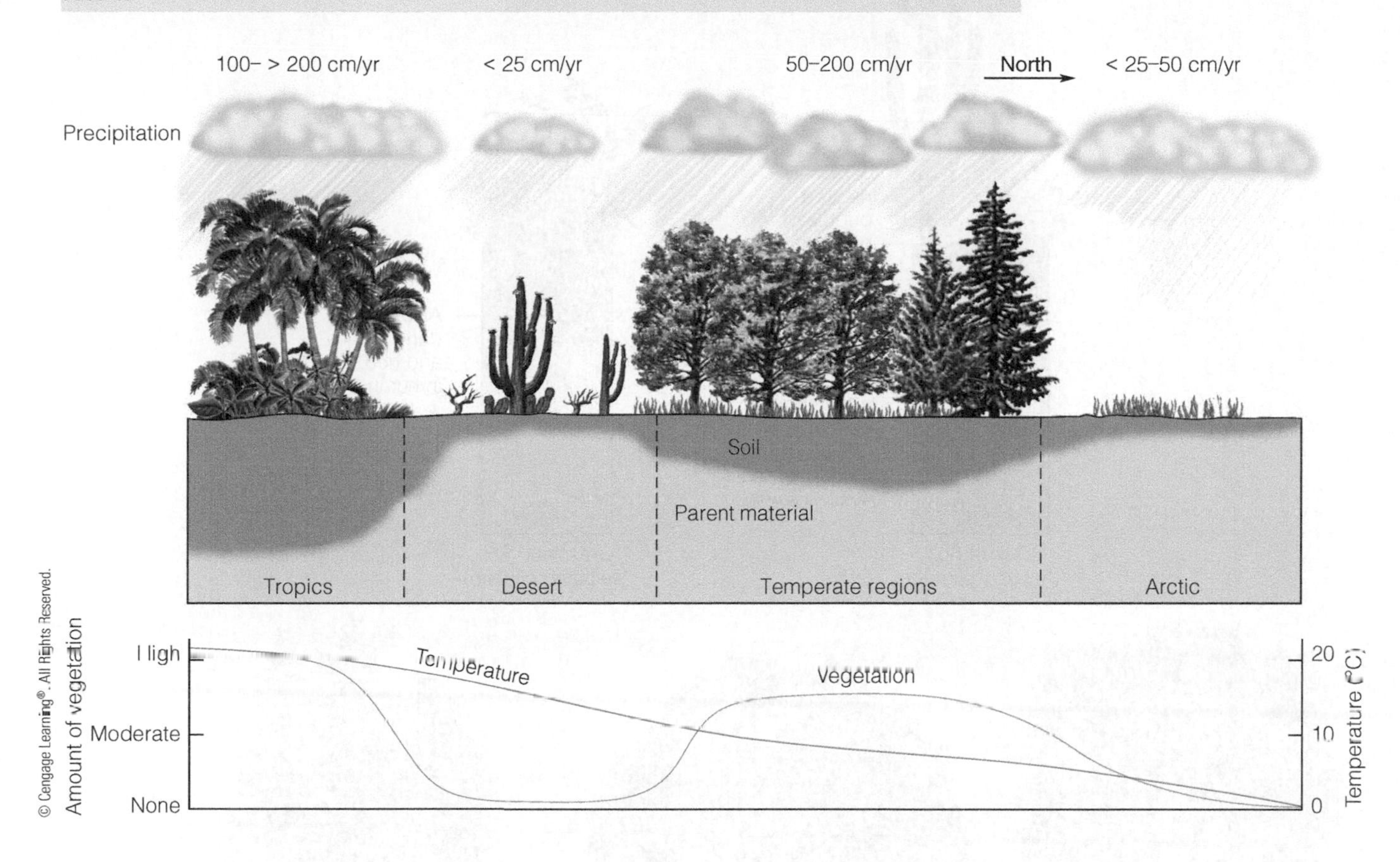

Thus, climate is more important than parent material in determining the type of soil. Nevertheless, rock type does exert some control. For example, the metamorphic rock quartzite will have a thin soil over it because it is chemically stable, whereas an adjacent body of granite will have a much deeper soil.

Organic Activity Soils depend on organisms for their fertility, and in return they provide a suitable habitat for many organisms. Earthworms, ants, sowbugs, termites, centipedes, millipedes, and nematodes, along with fungi, algae, and single-celled organisms, make their homes in soil. All contribute to soil formation and provide humus when they die and decompose by bacterial action.

Much of the humus in soils comes from grasses or leaf litter that microorganisms decompose to obtain food. In so doing, they break down organic compounds in plants and release nutrients back into the soil. In addition, organic acids from decaying soil organisms are important in further weathering of parent materials and soil particles.

Burrowing animals constantly churn and mix soils, and their burrows provide avenues for gases and water. Soil organisms, especially some types of bacteria, are extremely important in changing atmospheric nitrogen into a form of soil nitrogen suitable for use by plants.

Relief and Slope The difference in elevation between high and low points in a region is called *relief*. And because climate is such an important factor in soil formation and climate changes with elevation, areas with considerable relief have different soils in mountains and adjacent lowlands. *Slope,* another important control, influences soil formation in two ways. One is simply *slope angle;* steep slopes have little or no soil because weathered materials are eroded faster than soil-forming processes can operate. The other factor is *slope direction.* In the Northern Hemisphere, north-facing slopes receive less sunlight than south-facing slopes and have cooler internal temperatures, support different vegetation, and if in a cold climate, remain snow covered or frozen longer.

Time How much time is needed to develop a centimeter of soil or a fully developed soil a meter or so deep? We cannot give a definite answer because weathering proceeds at vastly different rates depending on climate and parent material, but an overall average might be about 2.5 cm per century. Nevertheless, a lava flow a few centuries old in Hawaii may have a well-developed soil on it, whereas a flow the same age in Iceland will have considerably less soil. Given the same climatic conditions, soil develops faster on unconsolidated sediment than it does on bedrock.

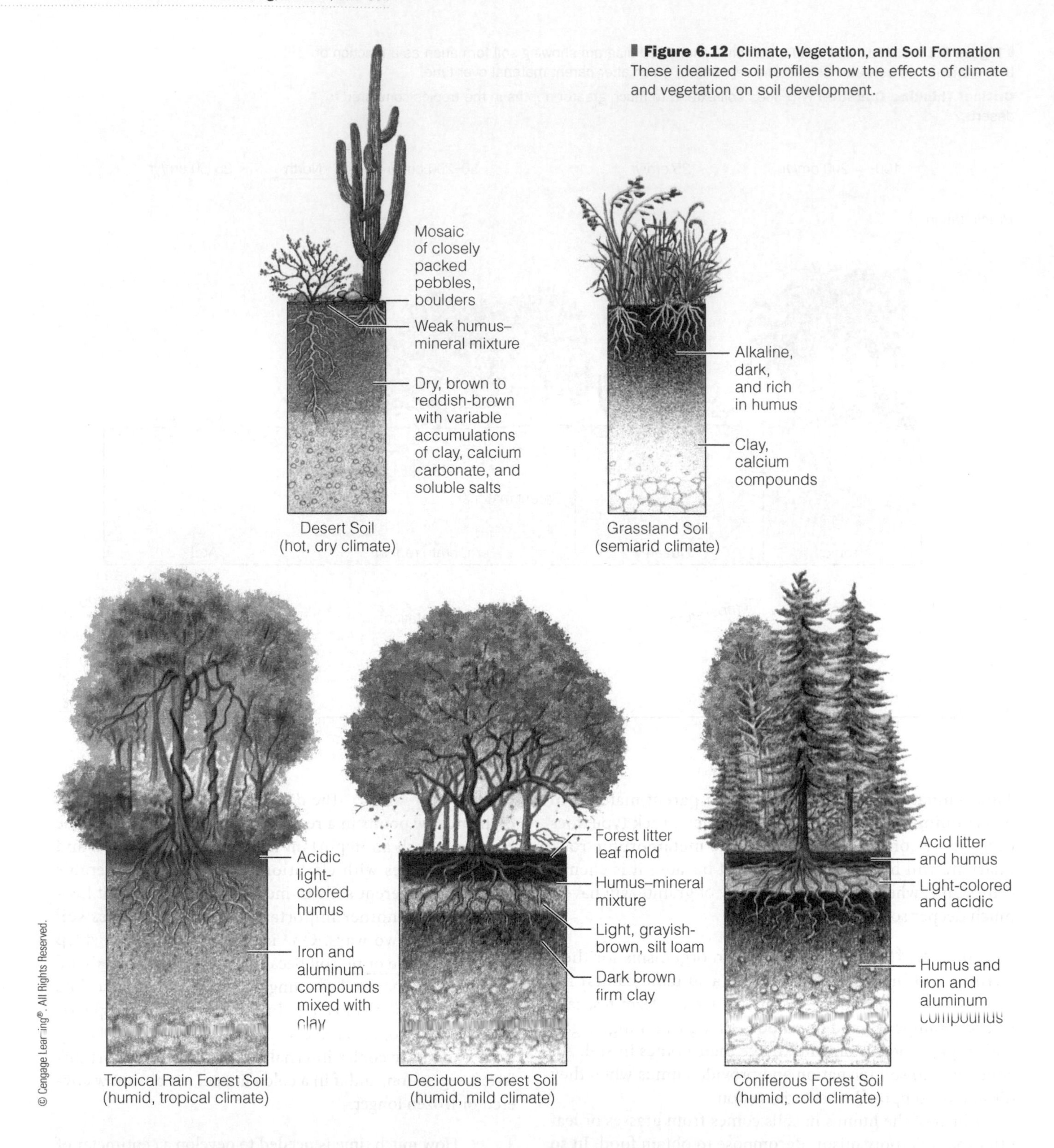

Figure 6.12 Climate, Vegetation, and Soil Formation These idealized soil profiles show the effects of climate and vegetation on soil development.

6.6 Expansive Soils and Soil Degradation

As you might imagine, soils that expand and contract pose problems for homeowners, developers, and engineers. And given the fact that our survival depends on soils, any loss of soil fertility, contamination of soils, or loss of soils to erosion is viewed with alarm.

Expansive Soils

Some soils contain clay minerals that increase in volume when wet and shrink when they dry out. Any soil that expands by 6% or more is an **expansive soil** (also called swelling soil) (Figure 6.15). When soil expands and contracts, overlying structures are first uplifted and then subside, thus experiencing forces that usually are not equally applied. About $6 billion in damage to foundations, roadways,

Figure 6.13 Features of Soils in Arid Regions

Sue Monroe

a This boulder has been turned over to show the scaley white material known as *caliche,* which formed on its underside. Irregular masses of caliche are common in horizon B of arid region soils.

Sue Monroe

b An alkali soil near Fallon, Nevada. The white material is sodium carbonate or potassium carbonate. Notice that only a few hardy plants grow in this soil.

Figure 6.14 Laterite and Slash-and-Burn Agriculture

Critical Thinking Question Laterite supports lush vegetation but is not very good for agriculture. Why?

Walt Anderson/Visuals Unlimited/Getty Images

a Laterite, shown here in Madagascar, is a deep-red soil that forms in response to intense chemical weathering in the tropics.

b Indigenous people in some rain forest areas clear and burn the vegetation from a small area where they plant and harvest crops for 2 to 5 years.

c Then the soil fertility is depleted and the farmers move on and repeat the process. An abandoned plot takes 10 to 30 years to completely revegetate and for soil fertility to be restored.

▮ Figure 6.15 The Distribution of Expansive Soils The areas with the most expansive soils are shown in red, followed by the areas in blue. The areas not colored do have some expansive soils; they are not extensive enough to be shown at this scale.

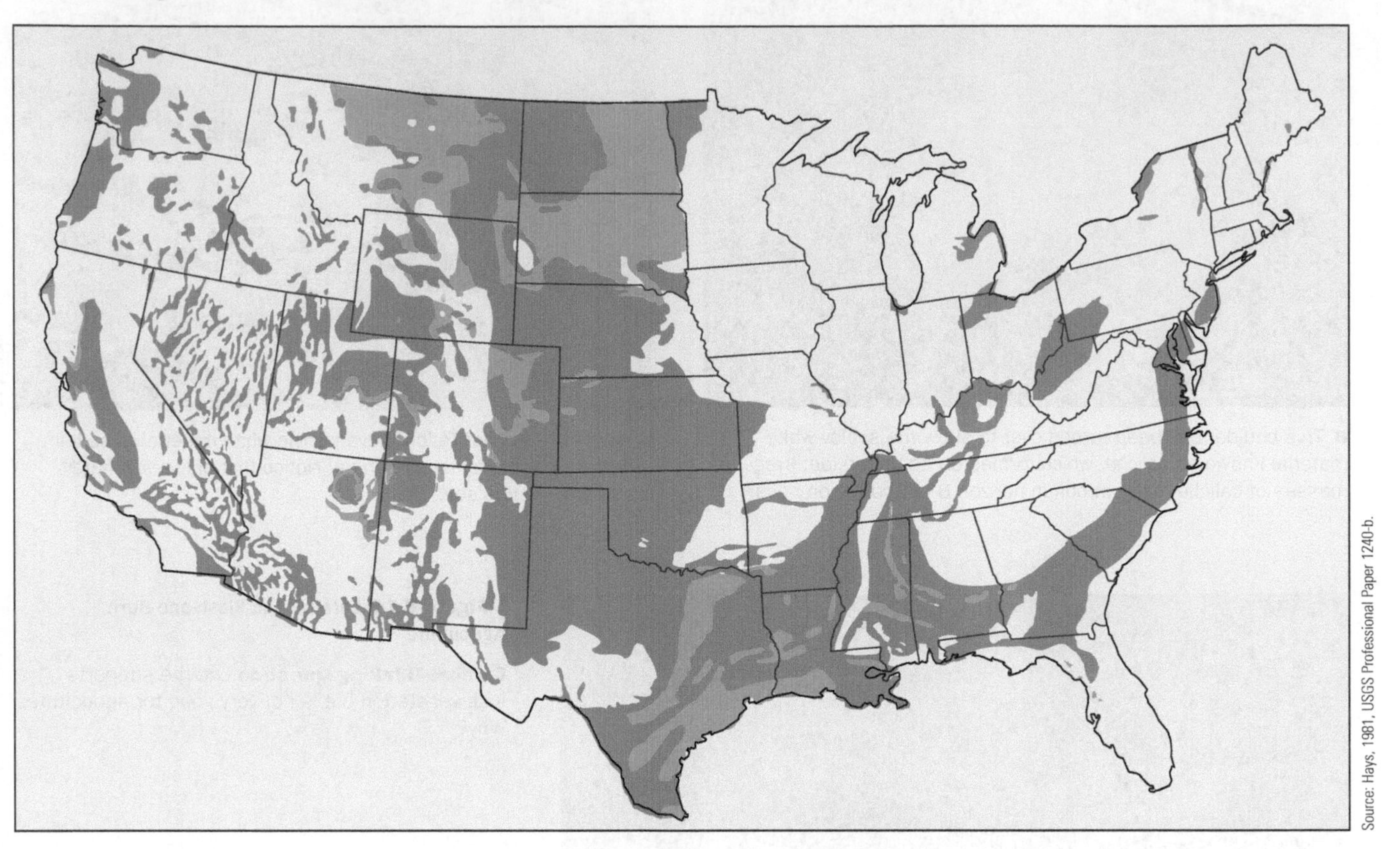

Source: Hays, 1981, USGS Professional Paper 1240-b.

sidewalks, and other structures takes place each year in the United States.

Avoiding areas of expansive soils is the best way to prevent damages, but what can be done for structures already sited on these soils? In some cases, the soil can be removed, or mixed with chemicals that inhibit expansion, or covered by a layer of nonexpansive fill, all expensive but perhaps necessary remedies. Also, expansive soils should be kept as dry as possible, and specialized construction methods may be used such as placing structures on piers or reinforced foundations.

Soil Degradation

In the context of geologic time, soils form rapidly; however, from the human perspective, the processes are so slow that soil is a nonrenewable resource. So any soil losses that exceed the rate of soil formation and any reduction in soil fertility or production are causes for concern. Any process that removes soil or makes it less productive is defined as **soil degradation,** a serious problem in many parts of the world that includes erosion, chemical deterioration, and physical changes.

Erosion, an ongoing natural process, is usually slow enough for soil formation to keep pace, but unfortunately, some human practices add to the problem. Removing natural vegetation by plowing, overgrazing, overexploitation for fire wood, and deforestation all contribute to erosion by wind and running water. The Dust Bowl that developed in several Great Plains states during the 1930s is a poignant example of just how effective wind erosion is on soil pulverized and exposed by plowing (▮ Figure 6.16).

Running water removes soil by *sheet erosion,* which involves the removal of thin layers more or less evenly over a broad, sloping surface. *Rill erosion,* in contrast, takes place when running water scours small, troughlike channels. Channels shallow enough to be eliminated by plowing are *rills* (▮ Figure 6.17a), but those too deep (about 30 cm) to be plowed over are *gullies* (▮ Figure 6.17b). Where gullying is extensive, croplands can no longer be tilled and must be abandoned.

Soil undergoes chemical deterioration when its nutrients are depleted and its productivity decreases. Loss of soil nutrients is most notable in many of the populous developing nations where soils are overused to maintain high levels of agricultural productivity. Chemical deterioration is also caused by insufficient use of fertilizers and by clearing soils of their natural vegetation.

Other types of chemical deterioration are pollution and *salinization,* which occurs when the concentration of salts increases in a soil, making it unfit for agriculture. Improper disposal of domestic and industrial wastes, oil and chemical spills, and the concentration of insecticides and pesticides in soils all cause pollution.

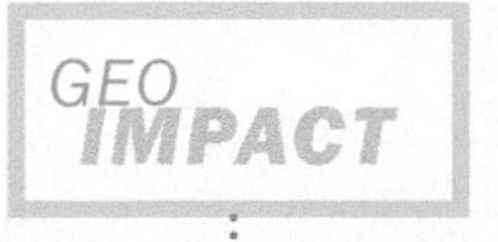

Expansive Soils

You have inherited a piece of property ideally located for everything you consider important. Unfortunately, as you prepare to have a house built, your contractor tells you that the soil is rich in clay that expands when wet and contracts when it dries. You nevertheless go ahead with construction but must now decide what measures to take to prevent damage to the structure. Make several proposals that might solve this problem. Which one or ones do you think would be the most cost effective?

Colorado Geological Survey

Uneven expansion of soils caused this road in Colorado to undulate.

Soil deteriorates physically when it is compacted by the weight of heavy machinery and livestock, especially cattle. Compacted soils are more costly to plow, and plants have a more difficult time emerging from them. Furthermore, water does not readily infiltrate, so more runoff occurs, which in turn accelerates the rate of water erosion.

In North America, the rich prairie soils of the midwestern United States and the Great Plains of the United States and Canada are suffering soil degradation. Problems experienced in the past have stimulated the development of methods to minimize soil erosion on agricultural lands. Crop rotation, contour plowing, and strip cropping (**▌** Figure 6.18),

▌ Figure 6.16 The Dust Bowl of the 1930s was a time of drought, dust storms resulting from wind erosion, and economic hardship.

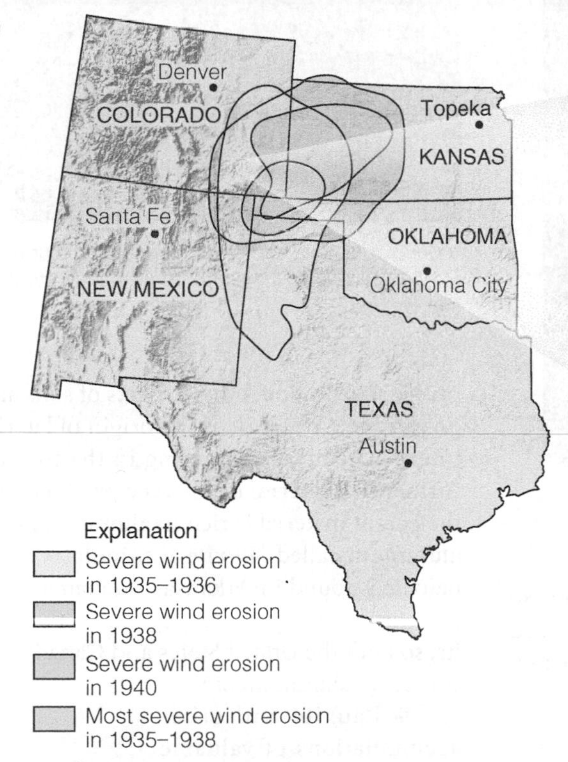

a Map of the Dust Bowl of the 1930s.

Source: "Map: Extent of Area Subject to Severe Wind Erosion," p. 30, from *Dust Bowl: The Southern Plains in the 1930s,* by Donald Worster. Copyright © 1979, 1982 by Oxford University Press, Inc. Used by permission of Oxford University Press, Inc.

The Granger Collection, NYC

b This huge dust storm was photographed at Lamar, Colorado, in 1934.

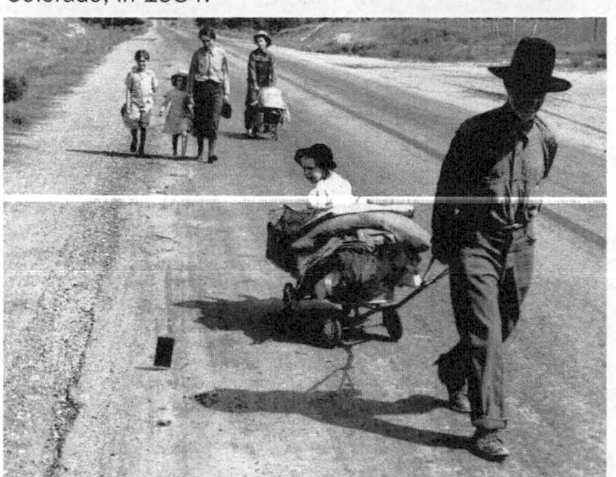
Dorothea Lange/CORBIS

c By the mid-1930s, tens of thousands of people were on relief, homeless, or leaving the Dust Bowl. In 1939, Dorothea Lange photographed this homeless family of seven in Pittsburg County, Oklahoma

Figure 6.17 Soil Degradation Resulting from Erosion

James S. Monroe

a Rill erosion in a field in Michigan during a rainstorm. The rill was later plowed over.

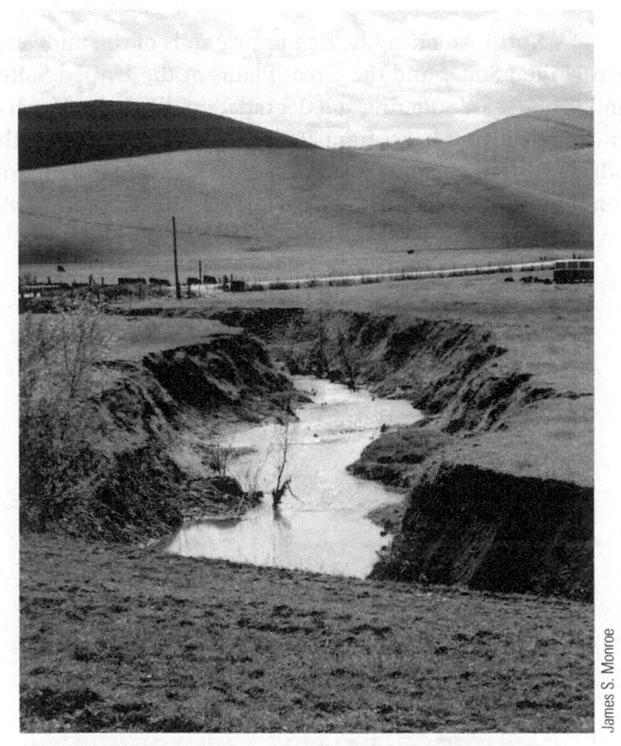

James S. Monroe

b This large gully cuts across a farmer's field in Glenn County, California. Notice the cattle in the field for scale.

Figure 6.18 Soil Conservation Practices

James S. Monroe

a An example of strip cropping in northern Montana which involves alternating row crops with other crops.

JERRY IRWIN/Photo Researchers/Getty Images

b Contour plowing and strip cropping. Contour plowing involves plowing parallel to the contours of the land to inhibit runoff and soil erosion.

and the construction of terraces have all proved helpful. So has no-till planting, in which the residue from the harvested crop is left on the ground to protect the surface from the ravages of wind and water.

6.7 Weathering and Resources

Soils are certainly one of our most precious natural resources. Indeed, if it were not for soils, food production on Earth would be vastly different and capable of supporting far fewer people. In addition, other aspects of soils are important economically. We discussed the origin of laterite in response to intense chemical weathering in the tropics, and we noted further that laterite is not very productive (Figure 6.14). If the parent material is rich in aluminum, however, the ore of aluminum called *bauxite* accumulates in horizon B. Some bauxite is found in Arkansas, Alabama, and Georgia, but at present, it is cheaper to import rather than mine these deposits, so both the United States and Canada depend on foreign sources of aluminum ore.

Bauxite and other accumulations of valuable minerals by the selective removal of soluble substances during chemical weathering are known as *residual concentrations.* Certainly, bauxite is a good example, but other deposits that formed in a

 ConnectionLink

For more about sedimentary iron deposits, see Chapter 7 and Chapter 19.

similar manner are those rich in iron, manganese, clays, nickel, phosphate, tin, diamonds, and gold. Some of the sedimentary iron deposits in the Lake Superior region of the United States and Canada were enriched by chemical weathering when soluble parts of the deposits were carried away. Some kaolinite deposits in the southern United States formed when chemical weathering altered feldspars in pegmatites or as residual concentrations of clay-rich limestones and dolostones. Kaolinite is a clay mineral used in the manufacture of paper and ceramics.

A *gossan* is a yellow to red deposit made up mostly of hydrated iron oxides that formed by oxidation and leaching of sulfide minerals such as pyrite (FeS_2). The dissolution of pyrite and other sulfides forms sulfuric acid, which causes other metallic minerals to dissolve, and these tend to be carried down toward the groundwater table, where the descending solutions form minerals containing copper, lead, and zinc. Gossans have been mined for iron, but they are far more important as indicators of underlying ore deposits.

Key Concepts Review

- Mechanical and chemical weathering disintegrate and decompose parent material so that it is more nearly in equilibrium with new physical and chemical conditions.
- The products of weathering include rock fragments and minerals liberated from parent material as well as soluble compounds and ions in solution.
- Weathering yields materials that may become soil or sedimentary rock.
- Mechanical weathering processes include frost action, pressure release, thermal expansion and contraction, salt crystal growth, and the activities of organisms. The particles yielded retain the composition of the parent material.
- Chemical weathering by solution, hydrolysis, and oxidation results in a chemical change in parent material and proceeds most rapidly in hot, wet environments.
- Mechanical weathering contributes to chemical weathering by breaking parent material into smaller pieces, thereby exposing more surface area.
- Soils possess horizons designated, in descending order, as O, A, E, B, and C, which differ from one another in texture, composition, structure, and color.
- The important factors controlling soil formation are climate, parent material, organic activity, relief and slope, and time.
- Soils in humid regions are darker and more fertile than those of semiarid regions. Laterite is soil that forms in the tropics where chemical weathering is intense.
- Soil degradation results from erosion as well as from physical and chemical deterioration. Human activities, such as construction, agriculture, deforestation, waste disposal, and chemical spills, contribute to soil degradation.
- Chemical weathering is responsible for the origin of some mineral deposits, such as residual concentrations of iron, lead, manganese, and clay.

Important Terms

chemical weathering (p. 141)
differential weathering (p. 139)
erosion (p. 138)
exfoliation dome (p. 140)
expansive soil (p. 150)
frost action (p. 139)
hydrolysis (p. 145)
laterite (p. 148)
mechanical weathering (p. 139)
oxidation (p. 144)
parent material (p. 139)
pressure release (p. 140)
regolith (p. 146)
salt crystal growth (p. 140)
sheet joins (p. 140)
soil (p. 147)
soil degradation (p. 152)
soil horizon (p. 147)
solution (p. 141)
spheroidal weathering (p. 146)
talus (p. 139)
thermal expansion and contraction (p. 140)
weathering (p. 138)

Review Questions

1. The unconsolidated sediment and pyroclasic materials as well as the debris derived by weathering that cover much of the land surface is called
   a. ____ soil degradation.
   b. ____ regolith.
   c. ____ caliche.
   d. ____ exfoliation.
   e. ____ humus.
2. Which one of the following processes is responsible for exfoliation domes?
   a. ____ pressure release.
   b. ____ oxidation.
   c. ____ lateritization.
   d. ____ solution.
   e. ____ intensification.
3. The loose, angular debris derived by weathering that accumulates at the base of a slope is called
   a. ____ horizon B.
   b. ____ alkali soil.
   c. ____ expansive soil.
   d. ____ bauxite.
   e. ____ talus.
4. Horizon C differs from other soil horizons in that it
   a. ____ is the most fertile.
   b. ____ has been weathered the most.
   c. ____ is made up of sodium sulfate.
   d. ____ grades downward into parent material.
   e. ____ contains the most humus.
5. Which one of the following is a soil conservation practice?
   a. ____ frost heaving.
   b. ____ contour plowing.
   c. ____ slash-and-burn agriculture.
   d. ____ crop depletion.
   e. ____ weathering prevention.
6. Draw, label, and describe the horizons found in a typical soil.
7. How do parent material, particle size, and climate control the rate of chemical weathering?
8. How does mechanical weathering differ from and contribute to chemical weathering?
9. What are expansive soils and why are they a problem?

10. **Creative Thinking Visual Question:** These images (❚ Figure 1) show exposures of granite. Explain which weathering phenomena account for their present surface expression.

❚ **Figure 1** Exposures of granite modified by weathering.

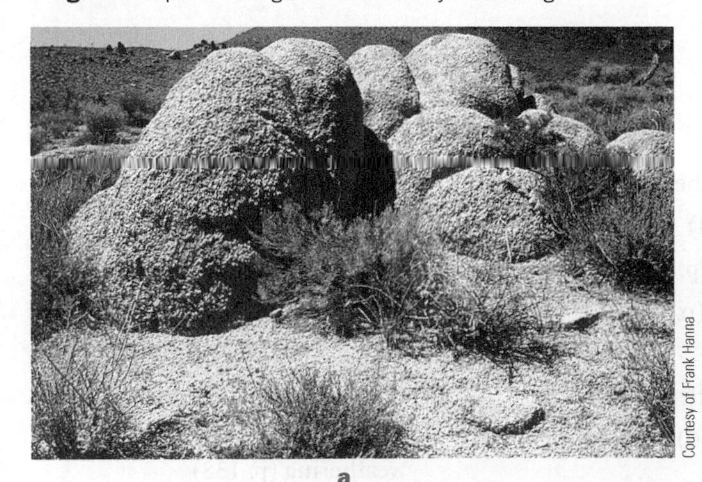

Courtesy of Frank Hanna

a

James S. Monroe

b

Global GeoScience Watch Search for "rocks and bacteria" in the GREENR database, and then scroll down to the "Magazines" section and click on "Attack of the Rock-eating Microbes! Some Bacteria Break Down Minerals, While Others Make Them." Explain how bacteria are involved in the origin or alteration of rocks and minerals. In addition, how might microbes be responsible for some environmental problems at abandoned mines?

View of Mount Haynes in Yellowstone National Park, Wyoming. This mountain is made up of the Madison Group, which in turn consists of two formations of mostly limestone with many fossils of ocean-dwelling animals. The Madison Group is about 520 m thick and was deposited between 325 and 360 million years ago (the Mississippian Period). The Madison River is in the foreground.

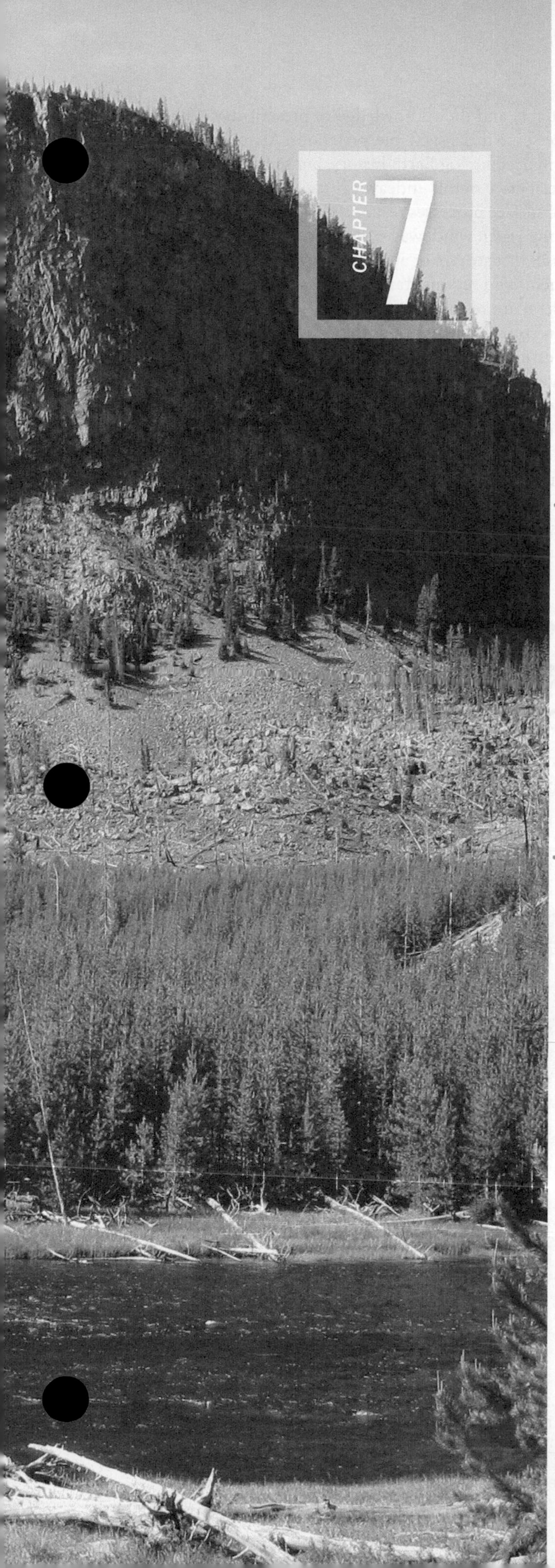

Sediment and Sedimentary Rocks

OUTLINE

HAVE YOU EVER WONDERED?

- Where gravel, sand, and mud come from?
- How sediments such as gravel, sand, and mud become sedimentary rocks?
- What kinds of sedimentary rocks come from seawater and lake water?
- Which organisms are important in the origin of sedimentary rocks?
- What features in rocks geologists use to determine Earth history?

James S. Monroe

7.1 Introduction

One definition of the term **sediment** is any solid that settles to the bottom of a liquid, but in geology we are interested in sediment that forms in two ways: First, solid particles derived by weathering of preexisting rocks is called detritus or detrital sediment; and, second, solid particles that precipitate from solution or that are extracted from solution by organisms is chemical sediment. Examples of detrital sediment include small pieces of weathered rock (the parent material) and individual mineral grains (see Figure 6.1). Chemical sediment may form by inorganic precipitation of minerals from solution, such as evaporating seawater, or it may form as an accumulation of shells or pieces of shells of organisms.

Regardless of how sediment originates, it is a loose aggregate of solids as in beach sand, sand in a river channel, or mud in a lake. If these loose solids are somehow bound together to form a solid aggregate, they become **sedimentary rocks,** which by definition are composed of sediment. Some examples of sedimentary rocks that you are likely familiar with are sandstone and limestone.

Earth's crust is made up mostly of *crystalline* rock, a term that applies loosely to metamorphic rocks and igneous rocks with the exception of those composed of pyroclastic materials. Sediment and sedimentary rocks constitute perhaps 5% of the crust, but they are the most commonly encountered Earth materials, because they cover about two-thirds of the continents and most of the seafloor, except spreading ridges. Their near absence at spreading ridges should not be surprising, though, because new oceanic crust forms at these ridges and has had little time for sediment to accumulate, whereas the seafloor distant from ridges has had sediment deposited on it for a long time.

The Mississippian-age rocks in the chapter opening photo were deposited in a shallow marine environment, whereas the Permian- to Jurassic-age sedimentary rocks in ❚ Figure 7.1 were deposited in several environments, including tidal flats, streams, lakes, swamps, and sand dunes. No one witnessed these events, so geologists must use the principle of uniformitarianism (see Chapter 1) and their knowledge of present-day depositional processes to make such interpretations. All rock types are important in deciphering Earth history, but sedimentary rocks have a special place in this endeavor because (1) they preserve evidence of surface processes responsible for sediment deposition (running water, wind, waves, and glaciers), and (2) many contain fossils that are rare or absent in other types of rocks.

Sediments and sedimentary rocks are important because many of them are resources themselves or they are the host rocks for resources such as oil and natural gas. Sand and gravel are essential to the construction industry, particularly as aggregate in concrete; rock gypsum is used in wallboard; quartz is used for abrasives and glass; and clay minerals are used in ceramics. Most coal is burned at power plants to generate electricity, and phosphorous-rich sedimentary rocks are used for fertilizers and animal feed supplements. Another sedimentary rock called *banded iron formation* is the main source of the world's iron ore, which is needed to manufacture steel (see Chapter 19).

7.2 Sediment Sources, Transport, and Deposition

Weathering, erosion, and sediment transport are fundamental aspects of the rock cycle and so is deposition of sediment (see Figures 1.15 and ❚ 7.2). As we noted in Chapter 6,

❚ **Figure 7.1 Sedimentary rocks in Monument Valley Navajo Tribal Park in Arizona** These rocks preserve evidence that show they were deposited on tidal flats and in streams, lakes, swamps, and sand dunes.

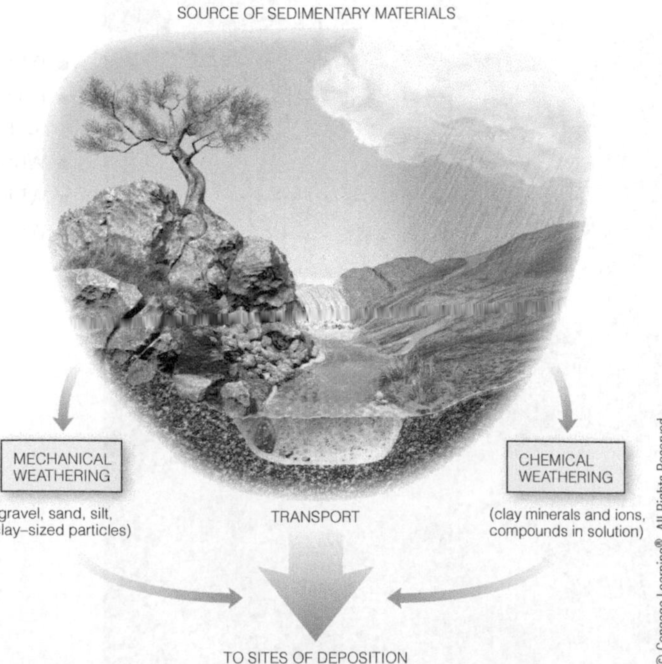

❚ **Figure 7.2 Sources of Sediments** Whether derived from preexisting rocks by mechanical or chemical weathering, solid particles and ions and compounds in solution are transported and deposited elsewhere. If they are lithified, they become detrital and chemical sedimentary rocks.

some sediment is further modified to form soil, but here we are interested in the sediment that is converted to sedimentary rocks. By far the most effective geologic agent that erodes sediment from its weathering site and transports it elsewhere is running water, but waves, wind, and glaciers are also important. Glaciers are moving solids and thus can transport sediment of any size, but wind transports only sand and smaller particles. Along shorelines, waves and marine currents accomplish much of the work of sediment transport and deposition.

One important criterion for classifying detrital sediment and sedimentary rocks is particle size. Particles described as *gravel* measures more than 2 mm; *sand* measures between 1/16 mm and 2 mm; *silt* is composed of particles 1/256–1/16 mm across; and, finally, *clay* measures less than 1/256 mm. The terms gravel, sand, and silt are size designations only and do not imply anything about composition. Most gravel is made up of rock fragments—that is, small pieces of granite, basalt, or any other rock type, but sand- and silt-size particles are mostly single-mineral grains. Clay is also a size designation, but it also applies to certain sheet silicates called *clay minerals*. However, most clay minerals are also clay size. The common name for mixtures of silt and clay is *mud*.

Remember that detrital sediment is solid particles derived from other rocks. Chemical sediment, in contrast, comes from chemicals that were derived from other rocks that were then extracted from solution by inorganic chemical processes, such as evaporating seawater, or by the activities of organisms. Clams, oysters, corals, and some plants make their skeletons of minerals, especially aragonite or calcite ($CaCO_3$) or silica (SiO_2). In any case, minerals form that may be converted into sedimentary rock.

All detrital sediment is transported some distance from its source, but chemical sediment forms in the area where it is deposited. During transport of detrital sediment, sand and gravel particles collide, and *abrasion* wears away the sharp corners and edges, causing **rounding** (▌Figure 7.3a). Transport also results in **sorting,** which refers to the size distribution of particles in sediment or sedimentary rocks that are poorly sorted if a wide range of size is present and well sorted if all the particles are about the same size (▌Figure 7.3b, c). Rounding and sorting may seem unimportant, but both influence how readily groundwater, petroleum, and gas move through sediments and sedimentary rocks, which is essential to our efforts to recover these materials. They are also useful for determining how sediment deposition occurred, a topic covered more fully in a later section.

Regardless of how detrital sediment is transported or how chemical sediment forms, both are eventually deposited in some geographic area known as a **depositional environment** (▌Figure 7.4). Deposition may take place in a stream channel or on its floodplain, in a lake, on a beach, or on the deep seafloor where physical and biological processes impart distinctive characteristics to the accumulating sediment. The three broad depositional settings geologists recognize are continental (on the land), transitional (on or near seashores), and marine (in the seas), each with several specific depositional environments (Figure 7.4).

7.3 How Does Sediment Become Sedimentary Rock?

We mentioned that deposits of detrital sediment are made up of loose aggregates of solid particles, such as sand or gravel in a stream channel or mud in a lake. The process whereby these aggregates of particles are converted into sedimentary rock is called **lithification,** which involves compaction, cementation, or both (▌Figure 7.5).

To illustrate the relative importance of compaction and cementation, consider a detrital deposit made up of mud and another composed of sand. In both cases, the sediment consists of solid particles and *pore spaces,* the voids between particles. These deposits are subjected to **compaction** from

▌Figure 7.3 Rounding and Sorting in Sediments

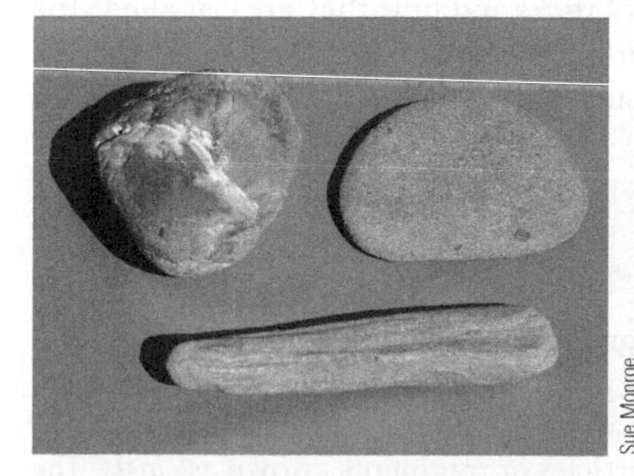

Sue Monroe

a Beginning students mistake rounding to mean ball-shaped or spherical. These three stones are all rounded, but only the one at the upper left is spherical.

James S. Monroe

b Deposit of moderately sorted, well rounded gravel.

James S. Monroe

c Angular, poorly sorted gravel. Note the quarter for scale.

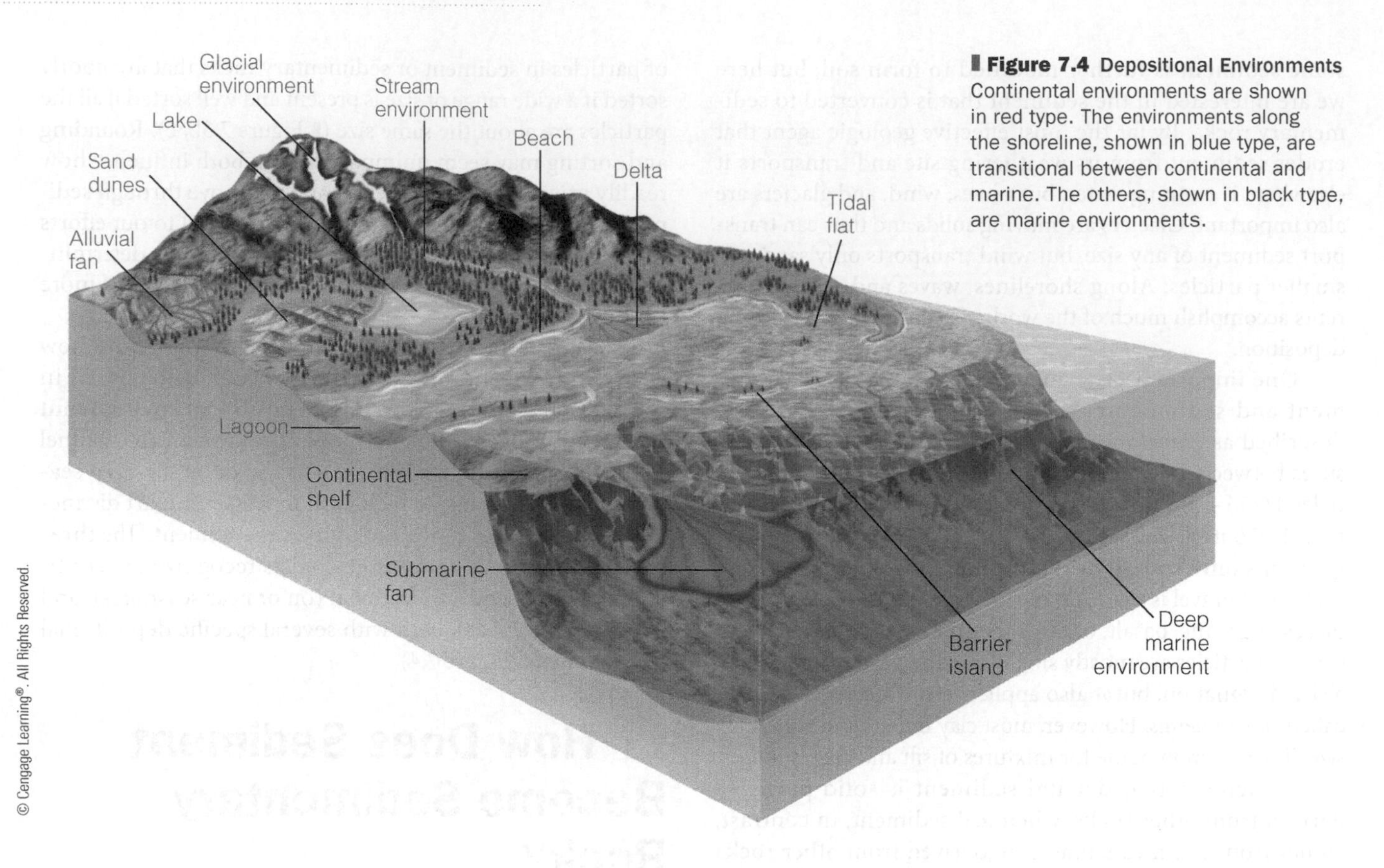

Figure 7.4 Depositional Environments Continental environments are shown in red type. The environments along the shoreline, shown in blue type, are transitional between continental and marine. The others, shown in black type, are marine environments.

their own weight and the weight of any additional sediment deposited on top of them, thereby reducing the amount of pore space and the volume of the deposit. Our hypothetical mud deposit may have 80% water-filled pore space, but after compaction, its volume is reduced by as much as 40% (Figure 7.5a). The sand deposit, with as much as 50% pore space, is also compacted, but far less than the mud deposit, so that the grains fit more tightly together (Figure 7.5a).

Compaction alone is sufficient for lithification of mud, but for sand and gravel, **cementation** involving the precipitation of minerals in pore spaces is also necessary (Figure 7.5b). The two most common chemical cements are calcium carbonate ($CaCO_3$) and silicon dioxide (SiO_2), but iron oxide and hydroxide cement, such as hematite (Fe_2O_3) and limonite [$FeO(OH) \cdot nH_2O$], are found in some sedimentary rocks. Recall that calcium carbonate readily dissolves in water that contains a small amount of carbonic acid, and chemical weathering of feldspars and other minerals yields silica in solution. Cementation takes place when minerals precipitate in the pore spaces of sediment from circulating water, thereby binding the particles together. Iron oxide and hydroxide cements account for the red, yellow, and brown sedimentary rocks found in many areas (Figure 7.1).

We have explained lithification of detrital sediments, but we have not yet considered this process in chemical sediments. By far the most common chemical sediments are calcium carbonate mud and sand- and gravel-sized accumulations of calcium carbonate grains, such as shells and shell fragments. Compaction and cementation also take place in these sediments, converting them into types of limestone, but compaction is less effective because cementation takes place soon after deposition. In any case, the cement is calcium carbonate derived by partial solution of some of the particles in the deposit.

7.4 Types of Sedimentary Rocks

Thus far, we have considered the origin of sediment, its transport, deposition, and lithification. We now turn to the types of sedimentary rocks and how they are classified. The two broad classes or types of sedimentary rocks are *detrital* and *chemical,* although the latter has a subcategory known as *biochemical.*

Detrital Sedimentary Rocks

Detrital sedimentary rocks are made up of gravel-, sand-, silt-, and clay-sized particles derived from parent material. Furthermore, they all have a *clastic texture,* meaning they are composed of particles or fragments known as *clasts.* The several varieties of detrital rocks are classified by the size of their constituent particles, although composition is used to modify some rock names (Table 7.1).

Figure 7.5 Lithification of Detrital Sediment

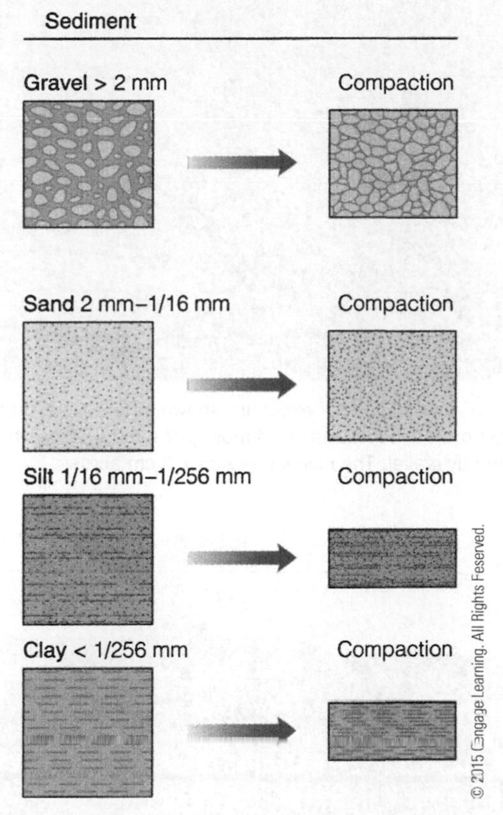

a Compaction of gravel, sand, silt, and clay. Notice that little compaction takes place in gravel and sand, but the volume of silt and clay deposits is considerably reduced by compaction.

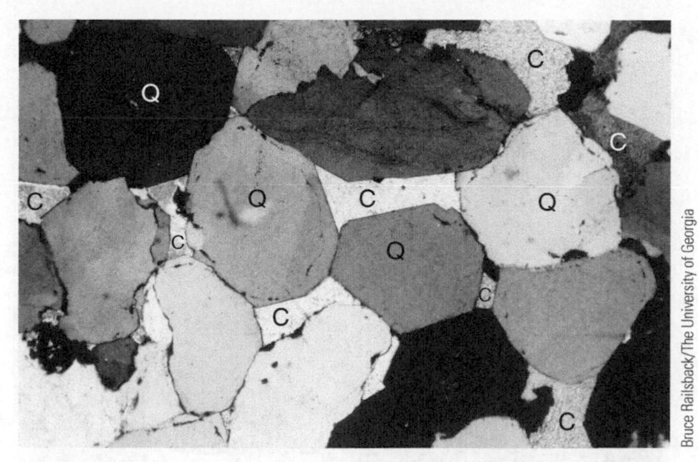

b Cementation of sand to form sandstone. This sandstone is made up mostly of quartz (Q) and the cement is calcite (C).

Both *conglomerate* and *sedimentary breccia* are composed of gravel-sized particles (▮ Figure 7.6a–c), but conglomerate has rounded gravel, whereas sedimentary breccia has angular gravel. Conglomerate is common, but sedimentary breccia is not because gravel becomes rounded very quickly during transport. Thus, if you encounter sedimentary breccia, you can conclude that its angular gravel has experienced little transport, probably less than a kilometer. Considerable energy is needed to transport gravel, so conglomerate is usually found in high-energy environments, such as stream channels and beaches.

Sand is a size designation for particles between 1/16 and 2 mm, so any mineral or rock fragment in that size range can be in *sandstone.* Geologists recognize varieties of sandstone based on mineral content (Table 7.1). *Quartz sandstone* is the most common and, as the name implies, is made up mostly of quartz sand. Another variety of sandstone called *arkose* contains at least 25% feldspar minerals. Sandstone is found in many depositional environments, including stream channels, sand dunes, beaches, barrier islands, deltas, and the continental shelf.

Mudrock is a general term that encompasses all detrital sedimentary rocks composed of silt- and clay-sized particles (Table 7.1). Varieties include *siltstone* (mostly silt-sized particles), *mudstone* (a mixture of silt and clay), and *claystone* (primarily clay-sized particles) (▮ Figure 7.6d). Some mudstones and claystones are designated *shale* if they are fissile, meaning that they break along closely spaced parallel planes. Even weak currents transport silt- and clay-sized particles, and deposition takes place only where currents and fluid turbulence are minimal, as in the quiet offshore waters of lakes or in lagoons.

TABLE 7.1 Classification of Detrital Sedimentary Rocks

Sediment Name and Size	Composition	Other Features	Rock Name	
Gravel (>2 mm)	Variable; rock fragments and quartz common	Rounded gravel	Conglomerate	
		Angular gravel	Sedimentary breccia	
Sand (1/16–2 mm)	Mostly quartz	Feels like sandpaper	Quartz sandstone	
	>25% feldspar	Feldspars obvious	Arkose	
Silt (1/256–1/16 mm)	Variable, but quartz, and	Mostly silt	Siltstone	Mudrocks
and	clay common	Silt-clay mixture	Mudstone*	Mudrocks
Clay (<1/256) mm	Clay minerals	Mostly clay	Claystone*	Mudrocks

*If mudstone and claystone split along closely spaced planes, they are called shale.

Figure 7.6 Detrital Sedimentary Rocks

James S. Monroe

a Exposure of sandstone/conglomerate in the Garden of the Gods, at Colorado Springs, Colorado.

James S. Monroe

b Close-up view of the rock exposure shown in **(a)**. Note that the upper part of this rock is made up mostly of sand whereas the lower part is mostly gravel. The nickel measures 2 cm across.

Sue Monroe

c Sedimentary breccia in Death Valley, California. Notice the angular, gravel-sized particles. The largest clast is about 12 cm across.

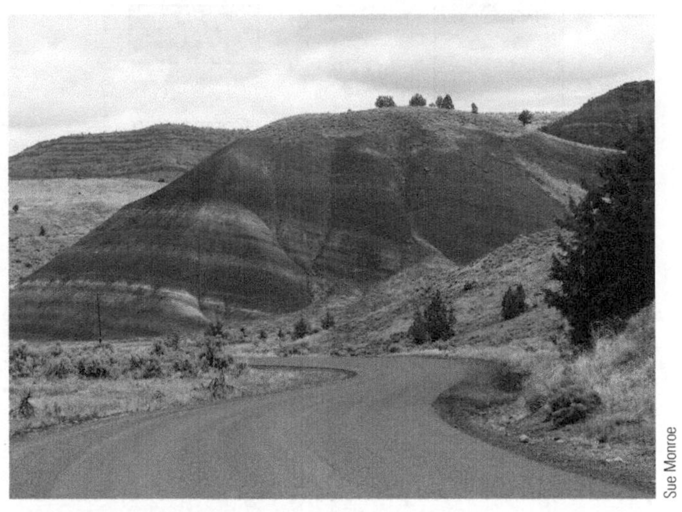

Sue Monroe

d These mudrocks at John Day Fossils Beds National Monument in Oregon are made up of silt and clay.

Chemical and Biochemical Sedimentary Rocks

During chemical weathering several compounds and ions go into solution and provide the raw materials for chemical **and biochemical sedimentary rocks.** For example, seawater contains silica (SiO_2), calcium (Ca), carbonate (CO_3), sulfate (SO_4), potassium (K), sodium (Na), and chlorine (Cl), as well as many other substances that under certain conditions are extracted from the water to form minerals that make up **chemical sedimentary rocks.** Organisms play an important role in the origin of some of these rocks, which are designated **biochemical sedimentary rocks** (Table 7.2). Some chemical sedimentary rocks have a crystalline texture, meaning that they are made up of a mosaic of interlocking mineral crystals as in rock salt. Others have a clastic texture as in some limestones composed of fragmented shells.

By far the most common chemical sedimentary rocks are **carbonate rocks,** so-called because they contain the carbonate radical $(CO_3)^{-2}$. Several rocks meet this criterion, but only two are common: limestone, composed of calcite ($CaCO_3$), and dolostone, composed of dolomite [$CaMg(CO_3)_2$] (Table 7.2). The origin of limestone is fairly straightforward. Recall from Chapter 6 that calcite in the presence of acidic groundwater rapidly goes into solution, but the chemical reaction leading to solution is reversible, so under some conditions calcite can precipitate from solution. Thus, some limestone forms by inorganic chemical precipitation from seawater or, more rarely, lake water.

Most limestone is biochemical because organisms are so important in its origin. Indeed, the skeletons of sea-dwelling animals are common in many varieties of limestone (▮ Figure 7.7a). Coquina is a type of limestone composed almost entirely of fragmented seashells (▮ Figure 7.7b), and chalk is a soft type of limestone that consists of microscopic shells. One distinctive variety of limestone contains small spherical grains called *ooids* that have a small nucleus around which concentric layers of calcite precipitate. Lithified deposits of ooids form *oolitic limestone* (▮ Figure 7.7c).

TABLE 7.2 Classification of Chemical and Biochemical Sedimentary Rocks

Texture	Composition	Rock Name	
Chemical Sedimentary Rocks			
Varies	Calcite ($CaCO_3$)	Limestone	Carbonate rocks
Varies	Dolomite [$CaMg(CO_3)_2$]	Dolostone	Carbonate rocks
Crystalline	Gypsum ($CaSO_4 \cdot 2H_2O$)	Rock gypsum	Evaporites
Crystalline	Halite (NaCl)	Rock salt	Evaporites
Biochemical Sedimentary Rocks			
Clastic	Calcite ($CaCO_3$) shells	Limestone (various types, such as chalk and coquina)	
Usually crystalline	Altered microscopic shells of SiO_2	Chert (various color varieties)	
	Carbon from altered land plants	Coal (lignite, bituminous, anthracite)	

Figure 7.7 Chemical Sedimentary Rocks—Limestone

Critical Thinking Question Coquina is made up of fragmented seashells, so would you expect it to accumulate on a Florida beach, on the deep seafloor, in a lake, or on a river's floodplain?

Sue Monroe

a Limestone with numerous fossil shells is called fossiliferous limestone.

Sue Monroe

b Coquina is limestone made up entirely of broken shells.

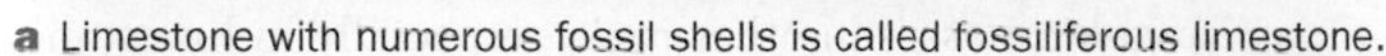

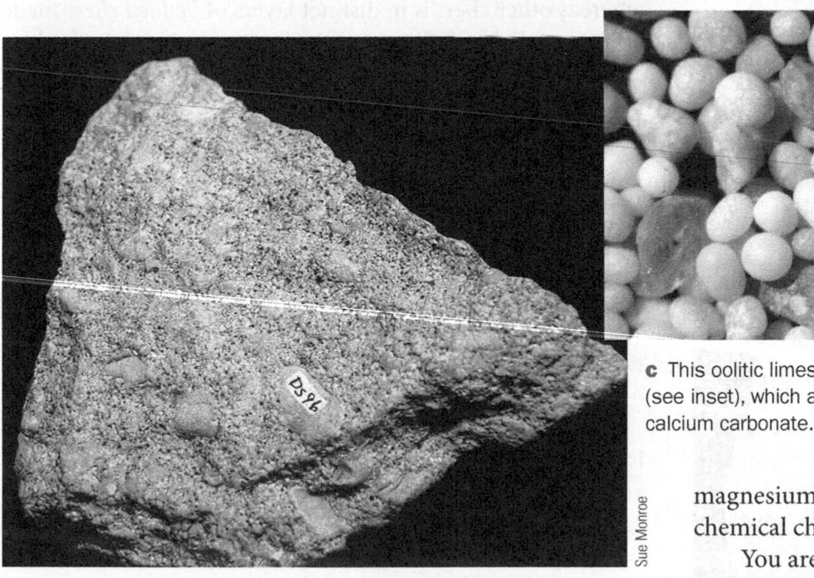

Sue Monroe

Stan Celestian/ Glendale Community College

c This oolitic limestone is made up partly of ooids (see inset), which are rather spherical grains of calcium carbonate.

Dolostone is similar to limestone, but it forms mostly by the alteration of limestone when magnesium replaces some of the calcium in calcite, thereby converting it to dolomite. This process may occur in a lagoon where evaporation of seawater takes place, enriching the remaining seawater in magnesium, which permeates limestone and brings about a chemical change.

You are undoubtedly aware that seawater is salty; that is, it contains sodium and chlorine and several other compounds and ions in solution. If you were to take a glass of seawater and let it evaporate completely, you would find a layer of minerals on the bottom of the glass. Obviously evaporation is involved in the origin of these minerals and their corresponding rocks, which are collectively called **evaporites** (Table 7.2). The most familiar evaporites are *rock salt,* composed of the mineral halite (NaCl), and *rock gypsum,* made up of the mineral gypsum ($CaSO_4 \cdot 2H_2O$) (▌Figure 7.8a, b). Compared with sandstone, mudrocks, and limestone, evaporites are not very common, and yet they are significant deposits in such areas as Michigan, Ohio, New York, and the Gulf Coast region.

Chert is hard rock consisting of microscopic crystals of silica (SiO_2) (Table 7.2, ▌Figure 7.9). Perhaps you have heard of *flint,* which is simply chert colored black by inclusions of organic matter, or *jasper,* which is red or brown chert because of its iron oxide content. Because chert is hard and lacks cleavage, it can be shaped to form sharp cutting edges for tools, spear points, and arrowheads. Some chert is found as irregular masses in other rocks, especially in limestone, whereas other chert is in distinct layers of *bedded chert* made up of tiny shells of silica-secreting organisms and thus is biochemical (Figure 7.9).

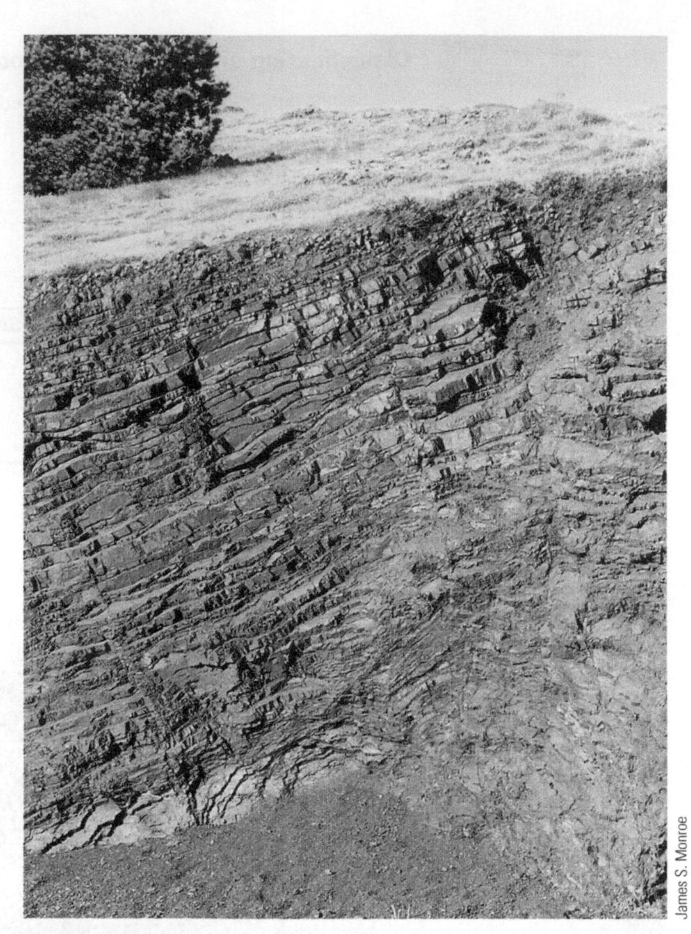

James S. Monroe

▌**Figure 7.9 Chert** Bedded chert exposed in Marin County, California. Most of the layers are about 5 cm thick.

▌**Figure 7.8 Chemical Sedimentary Rocks—Rock Salt and Rock Gypsum**

Critical Thinking Question Do you know of any commercial uses of rock salt and rock gypsum?

Sue Monroe

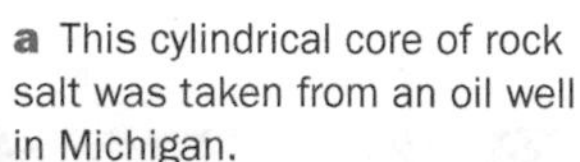

a This cylindrical core of rock salt was taken from an oil well in Michigan.

Sue Monroe

b Rock gypsum. When deeply buried, gypsum ($CaSO_4 \cdot 2H_2O$) loses its water and is converted to anhydrite ($CaSO_4$).

We mentioned that *coal* consists of altered organic matter but is nevertheless a biochemical sedimentary rock (Table 7.2, ▌Figure 7.10). It forms from vegetation that accumulates in bogs and swamps where the water is deficient in oxygen. Bacteria that decompose vegetation can live without oxygen, but their wastes must be oxidized, and because little or no oxygen is present, wastes build up and kill the bacteria. Decomposition ceases, and the vegetation forms organic muck, which if buried and compressed becomes *peat,* the first step in forming coal (Figure 7.10a).

Where peat is abundant, as in Ireland and Scotland, it is used for fuel, but if it is altered further by deeper burial, and especially if it is also heated, it becomes dull-black coal called *lignite* (Figure 7.10b). During the change from peat to lignite, the volatile elements are driven off, increasing the amount of carbon; peat has about 50% carbon, whereas lignite has about 70%. Further changes yield *bituminous coal,* with about

Connection Link

You can find more information on anthracite in Chapter 8.

Figure 7.10 The Origin of Coal

a Peat is partly decomposed plant material. It represents the first stage in the origin of coal.

b Lignite is a dull variety of coal in which plant remains are still visible.

c Bituminous coal is shinier and darker than lignite, and only rarely are plant remains visible.

80% carbon, which is dense, black, and so thoroughly altered that plant remains are rarely seen (Figure 7.10c). The highest grade of coal is *anthracite,* a metamorphic type of coal with up to 98% carbon.

7.5 Sedimentary Facies

Long ago, geologists realized that when they traced a layer of sediment or sedimentary rock laterally, the layers changed in composition, texture, or both. These changes resulted from the simultaneous operation of different processes in adjacent depositional environments. For example, sand may be deposited in a high-energy nearshore marine environment, whereas mud and carbonate sediments accumulate simultaneously in the laterally adjacent low-energy offshore environments (Figure 7.11a). Deposition in each environment produces **sedimentary facies,** bodies of sediment each possessing distinctive physical, chemical, and biological attributes. Figure 7.11a illustrates three sedimentary facies: a sand facies, a mud facies, and a carbonate facies. If these sediments become lithified, they are sandstone, mudstone (or shale), and limestone facies, respectively.

Many sedimentary rocks in the interiors of continents show clear evidence of deposition in marine environments. The rock layers in Figure 7.12, for example, consist of a sandstone facies that was deposited in a nearshore marine environment overlain by shale and limestone facies deposited in offshore environments. Geologists explain this vertical sequence of facies by deposition during a time when sea level rose with respect to the continents. As sea level rises, the shoreline moves inland, giving rise to a **marine transgression** (Figure 7.11a), and the depositional environments parallel to the shoreline migrate landward. As a result, offshore facies are superimposed over nearshore facies, thus accounting for the vertical succession of sedimentary facies. Even though the nearshore environment is long and narrow at any particular time, deposition takes place continuously as the environment migrates landward. The sand deposit may be tens to hundreds of meters thick but have horizontal dimensions of length and width measured in hundreds of kilometers.

The opposite of a marine transgression is a **marine regression** (Figure 7.11b). If sea level falls with respect to a continent, the shoreline and environments that parallel the shoreline move seaward. The vertical sequence produced by a marine regression has facies of the nearshore environment superposed over facies of offshore environments. It is important to understand marine transgressions and regressions and the sequences of facies resulting from these events, because we discuss them in some detail in the chapters on Paleozoic and Mesozoic geologic history (see Chapters 20 and 22).

7.6 Sedimentary Rocks—The Archives of Earth History

No one was present when ancient sediments were deposited, so geologists must evaluate those aspects of sedimentary rocks that allow them to make inferences about the

Figure 7.11 Marine Transgressions and Regressions

a Three Stages of Marine Transgression

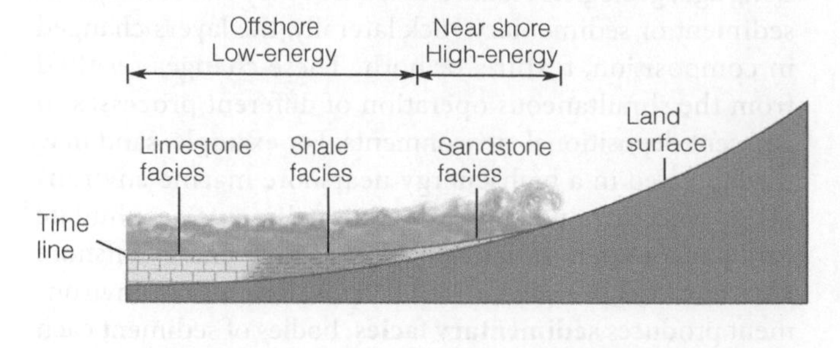

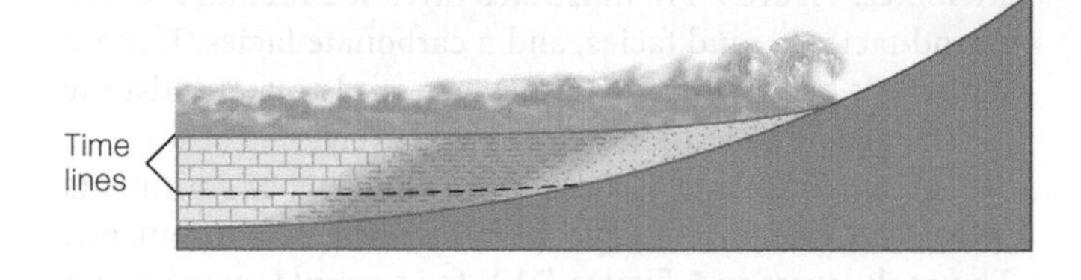

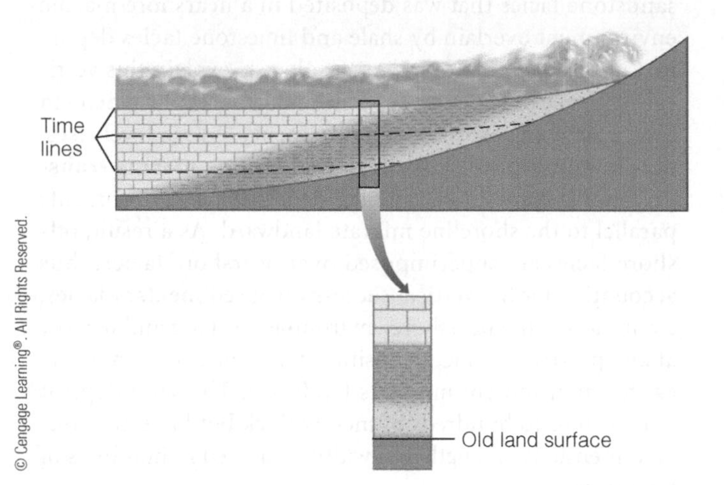

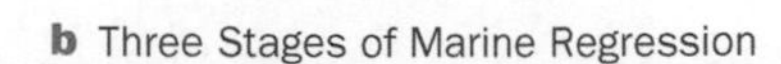

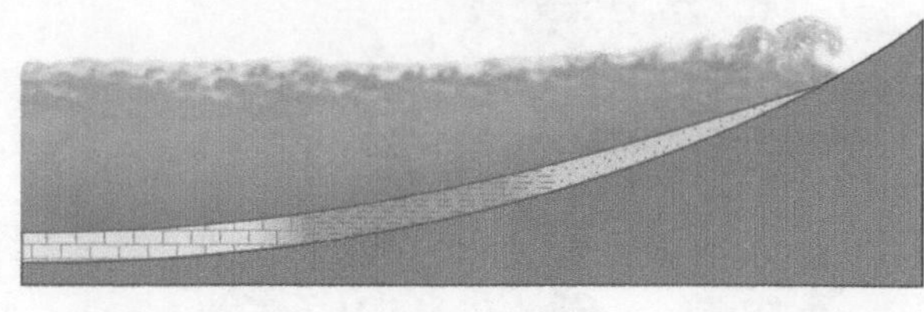

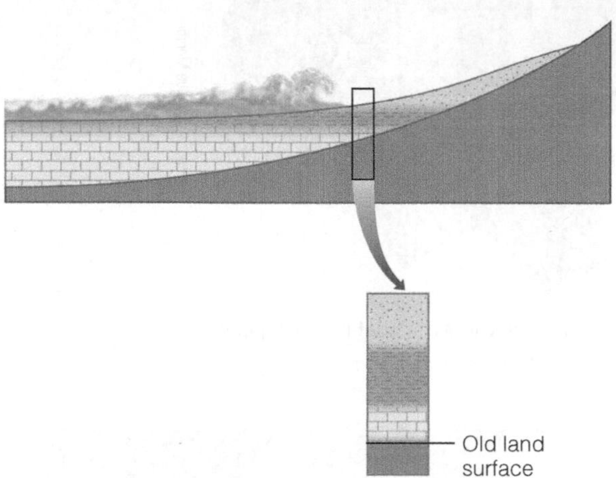

original depositional environment. And making such determinations is of more than academic interest. For instance, barrier island sand deposits make good reservoirs for hydrocarbons, so knowing the environment of deposition and the geometry of these deposits is helpful in exploration for resources.

Sedimentary textures such as sorting and rounding give clues to depositional processes. The sand in wind-blown dunes tends to be well sorted and well rounded, but poor sorting is typical of glacial deposits. The geometry or three-dimensional shape is another important aspect

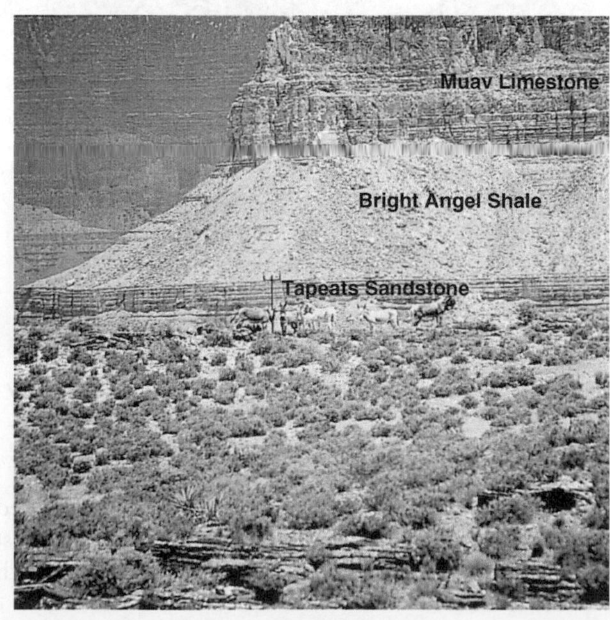

Figure 7.12 Sedimentary Rocks in the Grand Canyon of Arizona
View of the Tapeats Sandstone, Bright Angel Shale, and Muav Limestone. The three rock layers are about 300 m thick, although part of the Tapeats Sandstone is buried beneath debris. The rocks were deposited in the sea during a marine transgression. Also notice the layered aspect of these rocks, or what geologists call bedding or stratification.

Figure 7.13 Cross-Bedding Cross-bedding forms when individual beds or strata are deposited at an angle to the surface upon which they accumulate.

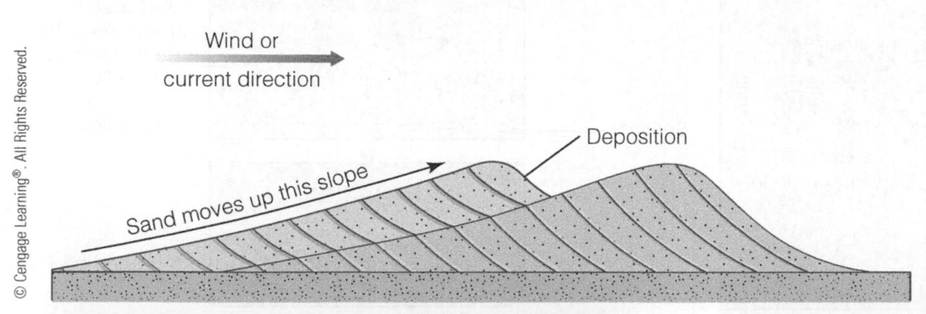

a Origin of cross-bedding by deposition on the sloping surface of a desert dune. Cross-bedding also forms in dunelike structures in stream and river channels and on the continental shelf.

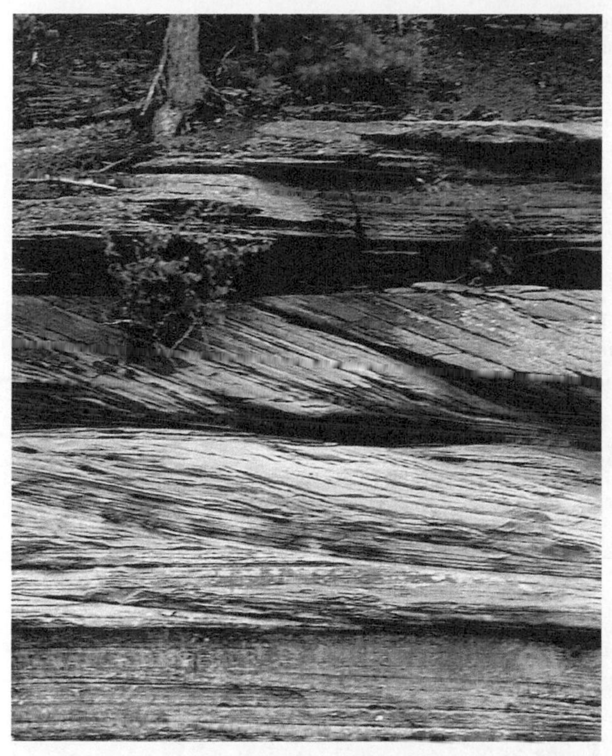

Sue Monroe

b Horizontal bedding and cross-bedding in sandstone at Wisconsin Dells, Wisconsin. About 10 m of strata are visible in this image.

James S. Monroe

c Outcrop of cross-bedding at Rock City, Kansas. These cross-beds are about 45 cm high.

of sedimentary rock bodies. Marine transgressions and regressions yield sediment bodies with a blanket or sheet-like geometry, but sand deposits in stream channels are long and narrow and are described as having a shoestring geometry. Sedimentary textures and geometry alone are usually insufficient to determine depositional environment, but when considered with other sedimentary rock properties, especially *sedimentary structures* and *fossils,* they enable geologists to reliably determine the history of a deposit.

Sedimentary Structures

Physical and biological processes operating in depositional environments are responsible for a variety of features known as **sedimentary structures.** One of the most common is distinct layers known as **strata** or **beds** (Figure 7.12), with individual layers of less than a millimeter up to many meters thick. These strata or beds are separated from one another by surfaces above and below which the rocks differ in composition, texture, color, or a combination of features. Layering of some kind is present in almost all sedimentary rocks; however, a few, such as limestone that formed in coral reefs, lack this feature.

Many sedimentary rocks possess **cross-bedding,** in which layers are arranged at an angle to the surface on which they are deposited (▮ Figure 7.13). Cross-beds are found in many depositional environments, such as sand dunes in deserts and along shorelines, as well as in stream-channel deposits and shallow marine sediments. Invariably, cross-beds result from transport and deposition by wind or water currents, and the cross-beds are inclined downward in the same direction that the current flowed. Thus, ancient deposits with cross-beds inclined down toward the south, for example, indicate that the currents responsible for them flowed from north to south.

Some sedimentary rock layers show an upward decrease in grain size called **graded bedding.** Most graded bedding is deposited by *turbidity currents,* which are underwater flows of sediment and water that have a greater density then sediment-free water. Because of its greater density, a turbidity current flows downslope along the bottom until it reaches the relatively flat seafloor or lakefloor, where it slows and begins depositing large particles followed by progressively smaller ones (▮ Figure 7.14).

Figure 7.14 Turbidity Currents and the Origin of Graded Bedding

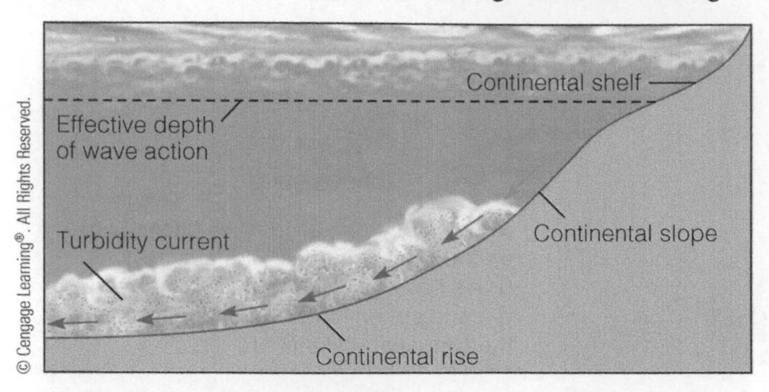

a A turbidity current flows downslope along the seafloor (or a lake bottom) because it is denser than sediment-free water.

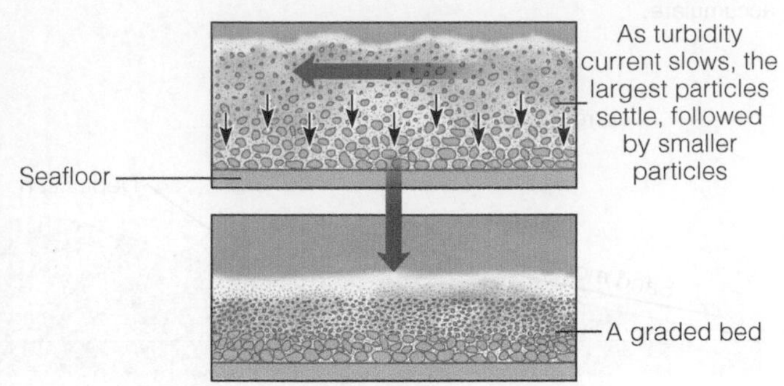

b The flow slows and deposits progressively smaller particles, thus forming a graded bed.

Figure 7.15 Current and Wave-Formed Ripple Marks

Critical Thinking Question How can you determine whether the flow that formed the ripples shown in part **(b)** was from right to left or from left to right?

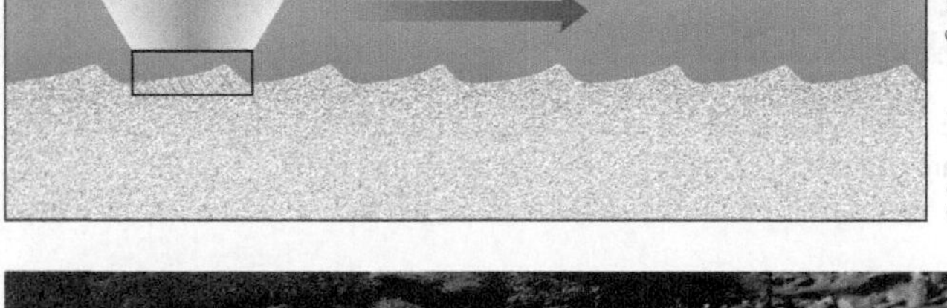

a Current ripple marks form in response to currents that flow in one direction, as in a stream. The enlargement shows the cross-beds in an individual ripple.

c The to-and-fro motion of waves in shallow water yields wave-formed ripple marks.

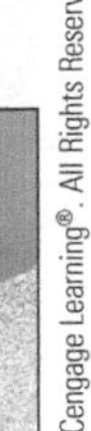

b Current ripple marks that formed in a stream channel.

d Wave-formed ripple marks in sand in shallow seawater.

Small-scale alternating ridges and troughs called **ripple marks** are common on the surfaces of some beds, especially on layers of sand (Figure 7.15). *Current ripple marks* are asymmetrical in profile and are invariably generated by currents, either water or wind, that flow in one direction (Figure 7.15a, b). Because of their asymmetry with the gentle slope upstream and their steep slope downstream, current ripple marks are good indicators of ancient current directions. For instance, should you observe ancient current ripple marks with their steep downstream side on the north, you can be sure the original currents flowed from south to north.

Wave-formed ripple marks, on the other hand, tend to have symmetrical profiles, and, as their name implies, are generated by the to-and-fro motion of waves (Figure 7.15c, d). Obviously, they cannot be used to determine ancient current directions, but they most commonly form in the shallow waters near seashores or lakeshores.

When clay-rich sediment dries, it shrinks and develops intersecting fractures called **mud cracks** (Figure 7.16).

Figure 7.16 Mud Cracks Mud cracks form in clay-rich sediments when they dry and contract.

Critical Thinking Question What inferences can you make about the environment in which the mud in the image below was deposited.

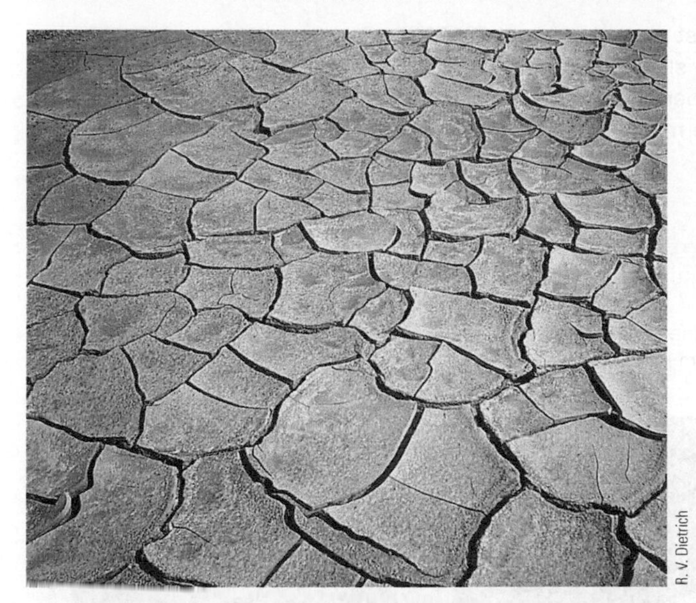
R. V. Dietrich

a Mud cracks in a present-day environment.

Jeff Mayo/Geophoto Publishing

b Ancient mud cracks in Glacier National Park in Montana. Note that the cracks have been filled in with sediment.

Mud cracks in ancient sedimentary rocks indicate that the sediment was deposited in an environment where periodic drying took place, such as on a river floodplain, near a lakeshore, or where muddy deposits are exposed along seacoasts at low tide.

Fossils—Remains and Traces of Ancient Life

Fossils, the remains or traces of ancient organisms, are interesting as evidence of prehistoric life (Geo-Insight 7.1 and Figure 7.17), and are also important for determining

Figure 7.17 Fossils

Sue Monroe

a Skull of *Allosaurus* on display at the Natural History Museum in Vienna, Austria.

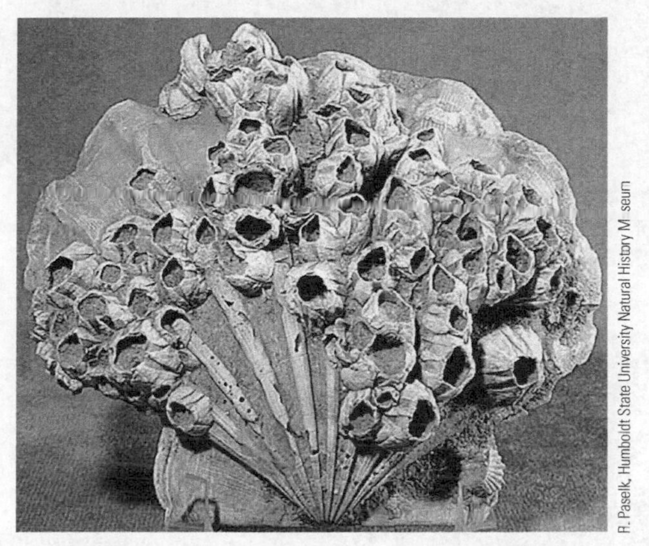
R. Paselk, Humboldt State University Natural History Museum

b Fossil bivalve encrusted with barnacles, from the Miocene of Virginia.

depositional environments. Most people are familiar with fossils of dinosaurs and some other land-dwelling animals but are unaware that fossils of invertebrates—animals lacking a segmented vertebral column, such as corals, clams, oysters, and a variety of microorganisms—are much more useful because they are so common.

Clams with heavily constructed shells typically live in shallow turbulent seawater, whereas organisms living in low-energy environments commonly have thin, fragile shells. Marine organisms that carry on photosynthesis are restricted to the zone of sunlight penetration, which is usually less than 200 m. The amount of sediment is also a limiting factor; for example, many corals live in shallow, clear seawater because suspended sediment clogs their respiratory and food-gathering organs, and some have photosynthesizing algae living in their tissues.

Microfossils are particularly useful for environmental interpretations because hundreds or even thousands are recovered from small rock samples. When drilling for oil,

GEO INSIGHT 7.1 Fossilization

Although the chance that any one organism will be preserved in the fossil record is slight, fossils are nevertheless common because so many billions of organisms lived during the past several hundred million years. Hard skeletal parts of organisms living where burial was likely and rapid are most common.

▼ **1.** ► **2.** The bones of this dinosaur and the shells of these marine animals called ammonites have had minerals added to their pores, making them more durable.

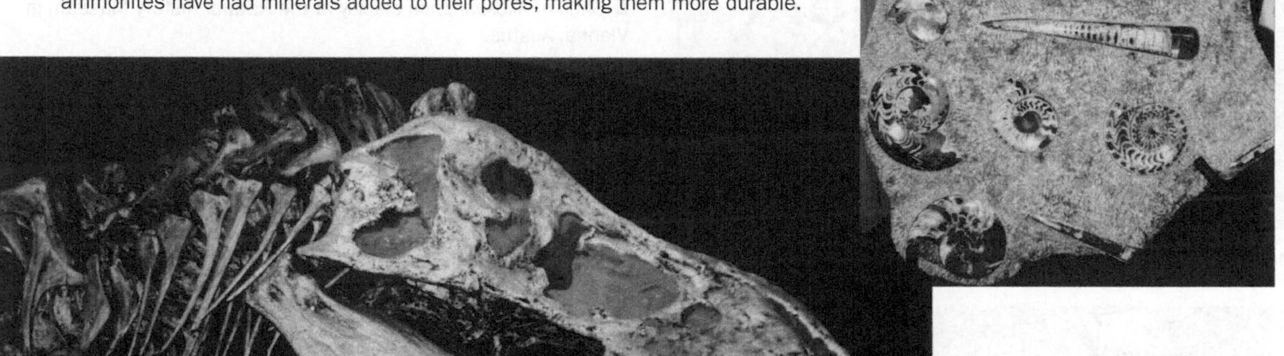

Sue Monroe

Sue Monroe

Sue Monroe

▲ **3.** A coprolite (fossilized feces) from a carnivorous mammal. This coprolite is about 5.5 cm long and is 40 million years old.

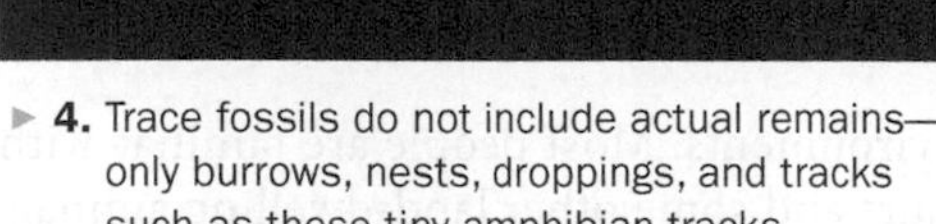

► **4.** Trace fossils do not include actual remains—only burrows, nests, droppings, and tracks such as these tiny amphibian tracks.

Sue Monroe

Sue Monroe

◄ **5.** This object looks like a clam, but it is simply sediment that filled a space formed when a clam shell dissolved.

▼ **6.** Fossil insect replaced by silicon dioxide (SiO_2).

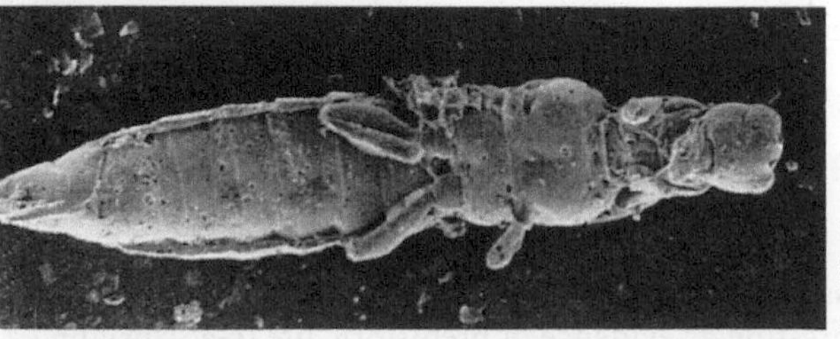

Reed Wicander

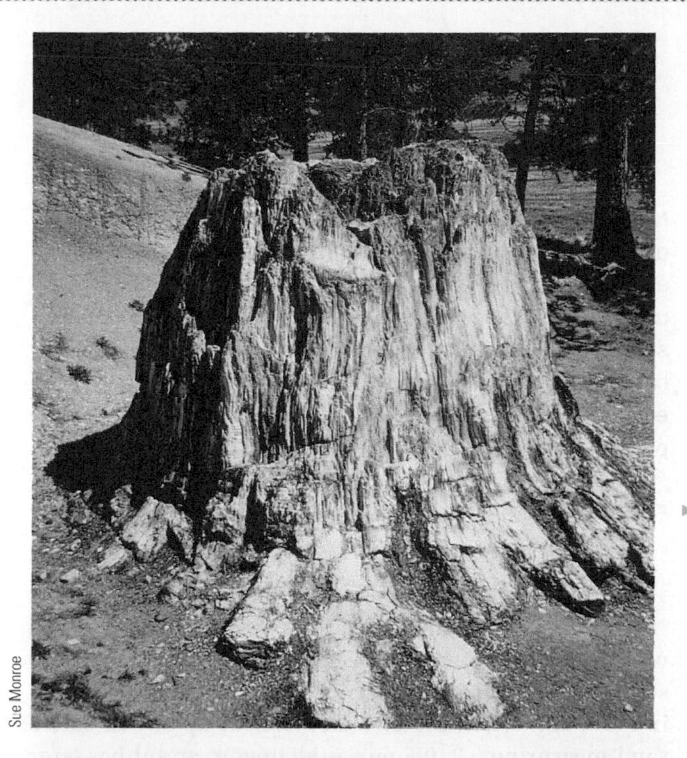

Sue Monroe

◄ **7.** Fossilized tree stump at Florissant Fossil Beds National Monument, Colorado. The woody tissue has been replaced by silicon dioxide. Mudflows 3–6 m deep buried the lower parts of many trees at this site.

Sue Monroe

► **8.** A palm frond and an insect preserved by carbonization.

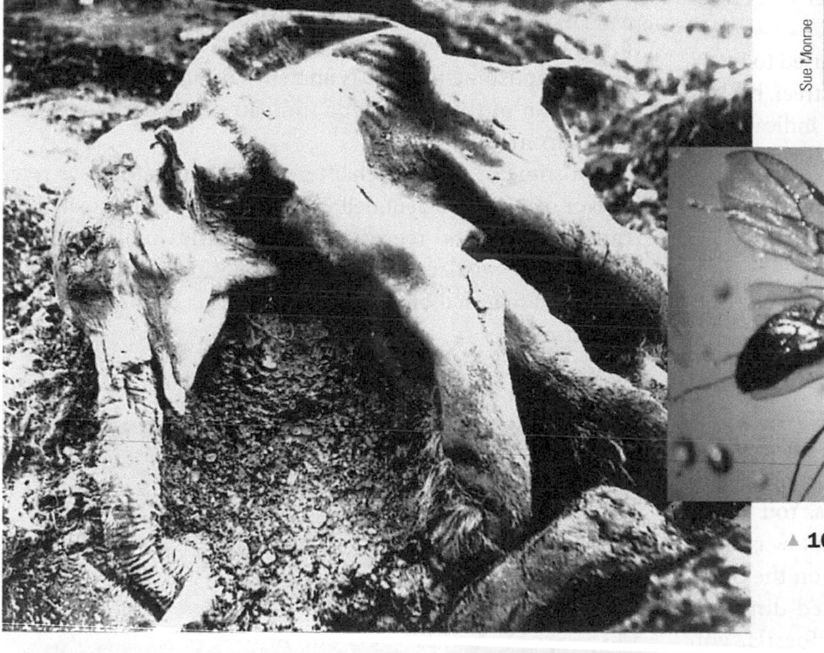

Sovfoto/V.Khristoforov

▲ **9.** This six- or seven-month-old frozen baby mammoth was found in Siberia in 1977. It measures 1.15 m long and 1 m high. Most of its hair has fallen out except around the feet. The carcass is about 40,000 years old.

John Koivula/Science Source

▲ **10.** Insects preserved in amber, which is hardened resin secreted by coniferous trees.

Sue Monroe

◄ **11.** This insect from the La Brea Tar Pits in Los Angeles is just one of hundreds of species of animals found in the asphalt-like substance in the pits.

ConnectionLink
You can find more information on using fossils for correlation in Chapter 17, and the use of fossils as evidence for organic evolution in Chapter 18.

geologists recover small rock chips called *well cuttings* that may contain numerous fossils of tiny organisms. These fossils are routinely used to determine depositional environments and to match up or correlate rocks of the same relative age in different areas. In addition, fossils provide some of the evidence for organic evolution and many states have a designated state fossil (Geo-Focus 7.1).

Determining the Environment of Deposition

What kinds of evidence would allow you to determine how a layer of sandstone was deposited? Certainly you would consider texture—that is, rounding and sorting—and also the kinds of sedimentary structures and fossils, if any. You might also compare features in the sandstone with those seen in sand deposits forming today. But are you justified in using present-day processes and deposits to infer what happened when no human observers were present?

Actually you are familiar with the reasoning used to interpret events you did not witness. Skid marks on a street, broken glass, and a damaged power pole almost certainly indicate that a vehicle hit the pole. If you see a badly burned, shattered tree in the forest, you could conclude that a bomb damaged it, but in the absence of any bomb fragments or residue, you would no doubt decide that the tree was hit by lightning. Geologists use exactly the same kind of reasoning—that is, their understanding of natural processes—when they evaluate evidence preserved in sedimentary rocks, when the rocks formed is irrelevant. Geologists simply rely on the principle of uniformitarianism in making these interpretations (see Chapter 1).

So what about that sandstone we mentioned at the beginning of this section? Suppose that it has symmetrical ripples and fossils of marine-dwelling organisms. You would no doubt conclude that it was deposited in a shallow marine environment. If, on the other hand, it contained dinosaur and land-plant fossils, you would probably conclude that it was deposited near a lakeshore. Many other features in sedimentary rocks are used in a similar fashion. Ooids (Figure 7.6c) form today in shallow marine environments with vigorous currents, and we have every reason to think that ancient ones formed in the same way. Glacial deposits are typically poorly sorted, show little stratification, and have other features that indicate glacial transport and deposition.

ConnectionLink
You can find more information on glacial transport and deposition in Chapter 14.

Sedimentary Rocks and Earth History

No one was present millions of years ago to record data about the climate, the fauna and flora, the geography, and geologic processes. So how is it possible to decipher unobserved past events? In other words, what features in sedimentary rocks would you look for to determine what happened during the distant past? Can you think of any economic reasons to interpret Earth history from the record preserved in rocks?

The evidence indicates that the rocks in the lower part of the Grand Canyon (Figure 7.12) were deposited in shallow seas during a marine transgression. The Navajo Sandstone in the southwestern United States is made up of well-sorted sand measuring 0.2–0.5 mm in diameter, and it has large cross-beds and the footprints of land-dwelling animals (▮ Figure 7.18). Long ago, geologists concluded that the Navajo Sandstone formed as ancient sand dunes. In fact, the inclination of the cross-beds indicates that the wind blew mostly from the northeast.

In some of the later chapters on Earth history, we refer to river deposits, ancient deltas, carbonate shelf deposits, and transgressions and regressions. We cannot include all the supporting evidence for these interpretations, but we can say that they are based on the kinds of criteria discussed in this chapter.

James S. Monroe

▮ **Figure 7.18** **Interpretation of Ancient Sedimentary Rocks**
The Jurassic-age Navajo Sandstone in Zion National Park, Utah, is a wind-blown sand deposit. Notice the well developed cross-bedding in this image which shows more than 200 m of strata.

GEO FOCUS 7.1 State Fossils

Most U.S. states and territories have a state gemstone, mineral, and rock, and 42 states have a designated state fossil. State fossils include plants (Maine), petrified wood (Arizona), invertebrate animals (Kentucky and New York), dinosaurs (Colorado and Montana) or dinosaur tracks (Connecticut and Massachusetts), and several mammals (Alaska, California, and West Virginia). Even Washington, D.C., has a fossil dinosaur called *Capitolosaurus*, although it has never been formally described.

All state fossils are interesting in their own right, but perhaps one of the most notable is the fossil whale *Basilosaurus* that is the state fossil for Alabama and Mississippi, though it was first discovered in Louisiana in Cenozoic-age rocks dating from 34 to 40 million years ago. This ancient whale was about 18 m long, and even though it is truly a whale, it differed in proportions and other features from whales today. For example, it was rather long and slender and had a small head; whales today are much heavier bodied and have very large heads.

The discovery of a nearly complete skeleton of *Basilosaurus* in the 1840s by Albert Koch and his subsequent reconstruction of what he called a huge sea serpent was remarkable. Albert Koch was born in 1804 in Germany, immigrated to the United States when he was 22, and eventually took up residence in St. Louis, Missouri, where he owned a small museum. Koch was certainly a keen observer of geologic phenomena and is now credited with discovering some information about Earth history, but he was also a hoaxer.

It all began when Koch recovered the remains of a large mastodon and reassembled it with extra vertebra and wooden blocks to make it look even larger than a normal mastodon; his restoration yielded an animal 10 m long and 4.6 m tall. He also placed the mastodon's tusks at a peculiar angle, giving his creature an even more bizarre appearance (Figure 1). In 1842, he sold his creation, called *Missourium*, to the British Museum, where scientists removed the extra bones and blocks and placed the tusks in the correct positon, thus making it a normal-size mastodon.

Koch attracted visitors to his museum with unusual displays, and the more unusual the better, so when he heard rumors of people finding sea serpent bones in Alabama, he went there and eventually collected a nearly complete skeleton of *Basilosaurus*, or what was originally called *Zeuglodon*. In any case, the fossils he collected were real enough, but in his restoration Koch used vertebrae from five animals, thereby making his sea serpent 35 m long, which is nearly twice the actual length of the animal (Figure 2). He displayed his "sea serpent" in the United States and Europe, for a fee of course, and made a considerable amount of money. Scientists were well aware that Koch's restorations were hoaxes, so, ironically, when he reported legitimate fossil evidence, no one took him seriously.

There have been several other hoaxes, as well as honest mistakes, in the history of paleontology, both before and after Koch built his sea serpent. Most of these fooled no one, at least for very long, although one perpetrated in 1913 called *Piltdown man* was not revealed as a hoax by scientists for several decades.

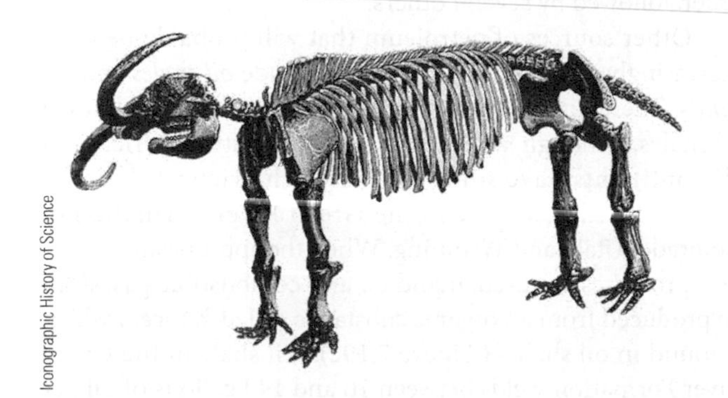

Iconographic History of Science

Figure 1 Koch's mastodon, known as *Missourium*, was a restoration of a real fossil, but he added extra vertebrae and wood blocks to make the animal larger than it actually was. Also, note the odd position of the tusks.

© Classic Image/Alamy

Figure 2 This "sea serpent" on display in New York was constructed by Albert Koch from the bones of several fossil whales.

7.7 Sedimentary Rocks and Economic Geology

Sand and gravel are essential to the construction industry; pure clay deposits are used for ceramics, and limestone is used in the manufacture of cement and in blast furnaces where iron ore is refined to make steel. Evaporites are the source of table salt as well as chemical compounds, and rock gypsum is used to manufacture wallboard. Phosphate-bearing sedimentary rock is used in fertilizers and animal feed supplements.

Some valuable sedimentary deposits are found in streams and on beaches where minerals were concentrated during transport and deposition. These *placer deposits* are surface accumulations resulting from the separation and concentration of materials of greater density from those of lesser density. Much of the gold recovered during the initial stages of the California gold rush (1849–1853) was mined from placer deposits, and placers of other minerals such as diamonds and tin are important.

Coal

Historically, most coal mined in the United States has been bituminous coal from the Appalachian region that formed in coastal swamps during the Pennsylvanian Period (299–318 million years ago). Huge lignite and subbituminous coal deposits in the western United States are becoming increasingly important. During 2011, more than 1 billion metric tons of coal were mined in this country, more than 62% of it from mines in Wyoming, West Virginia, and Kentucky.

You can find more information on Pennsylvanian-age coal deposits in Chapter 20.

Anthracite coal is especially desirable because it burns more efficiently than other types of coal. Unfortunately, it is the least common variety, so most coal used for heating buildings and generating electricity is bituminous (Figure 7.10c). *Coke*, a hard, gray substance consisting of the fused ash of bituminous coal, is used in blast furnaces where steel is produced. Synthetic oil and gas and a number of other products are also made from bituminous coal and lignite.

Petroleum and Natural Gas

Petroleum and natural gas are *hydrocarbons,* meaning that they are composed of hydrogen and carbon. The remains of microscopic organisms settle to the seafloor, or lakefloor in some cases, where little oxygen is present to decompose them. If buried beneath layers of sediment, they are heated and transformed into petroleum and natural gas. The rock in which hydrocarbons form is known as *source rock,* but for hydrocarbons to accumulate in economic quantities, they must migrate from the source rock into some kind of *reservoir rock.* And finally, the reservoir rock must have an overlying, nearly impervious *cap rock;* otherwise, the hydrocarbons would eventually reach the surface and escape (Figure 7.19a, b). Effective reservoir rocks must have appreciable pore space and good permeability, the capacity to transmit fluids; otherwise, hydrocarbons cannot be extracted from them in reasonable quantities.

You can learn more about porosity and permeability of sediments and rocks in Chapter 13.

Many hydrocarbon reservoirs are nearshore marine sandstones with nearby fine-grained, organic-rich source rocks. Such oil and gas traps are called *stratigraphic traps* because they owe their existence to variations in the strata (Figure 7.19a). Indeed, some of the oil in the Persian Gulf region and Michigan is trapped in ancient reefs that are also good stratigraphic traps. *Structural traps* result when rocks are deformed by folding, fracturing, or both. In sedimentary rocks that have been deformed into a series of folds, hydrocarbons migrate to the high parts of these structures. Displacement of rocks along faults (fractures along which movement has occurred) also yields traps for hydrocarbons (Figure 7.19b).

We noted in Chapter 3 that the United States uses approximately 19 million barrels of oil per day, 51.5% of which is imported, even though the United States is the world's third largest oil producer. In addition, the United States has established a strategic petroleum reserve with a capacity of 727 million barrels. Much of the United States domestic production comes from offshore wells, especially in the Gulf of Mexico, whereas the leading oil-producing states are Texas, Alaska, California, and North Dakota, in that order, followed by several others.

Other sources of petroleum that will probably become increasingly important in the future include *oil shales* and *tar sands*. The United States has about two-thirds of all known oil shales, although large deposits are in South America, and all continents have some oil shale. The richest deposits in the United States are in the Green River Formation of Colorado, Utah, and Wyoming. When the appropriate extraction processes are used, liquid oil and combustible gases can be produced from an organic substance called *kerogen,* which is found in oil shale (Figure 7.19c). Oil shale in the Green River Formation yields between 10 and 140 gallons of oil per ton of rock processed, and the total amount of oil recoverable with present processes is estimated at 80 billion barrels. Currently, no oil is produced from oil shale in the United States because conventional drilling and pumping are less expensive.

Figure 7.19 Oil and Natural Gas The arrows in **(a)** and **(b)** indicate the direction of migration of hydrocarbons.

a Two examples of stratigraphic traps: one in sand within shale and the other in a buried reef.

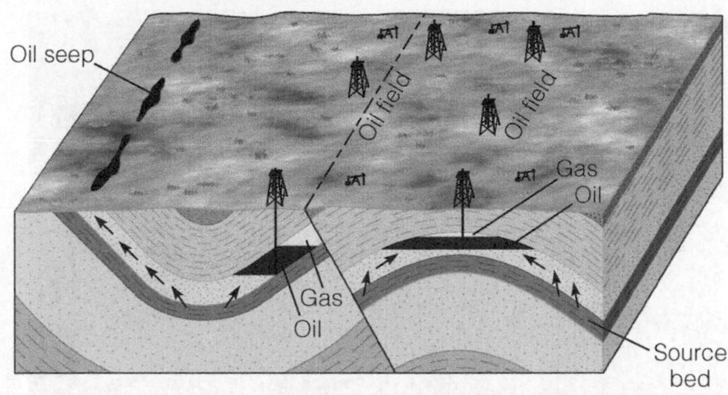

b Two examples of structural traps: one formed by folding and the other by faulting.

c The sedimentary rock oil shale (left) and oil extracted from it. The United States has vast oil shale deposits.

Tar sand is a type of sandstone in which viscous, asphalt-like hydrocarbons fill the pore spaces. This substance is the sticky residue of once-liquid petroleum from which the volatile constituents have been lost. Liquid petroleum can be recovered from tar sand, but for this to happen, large quantities of rock must be mined and processed. Because the United States has few tar-sand deposits, it cannot look to this source as a significant future energy resource. The Athabaska tar sands in Alberta, Canada, however, are one of the largest deposits of this type. These deposits are currently being mined, and it is estimated that they contain several hundred billion barrels of recoverable petroleum.

Uranium

Most of the uranium used in nuclear reactors in North America comes from the complex potassium-, uranium-, and vanadium-bearing mineral *carnotite,* which is found in some sedimentary rocks. Some uranium is also derived from *uraninite* (UO_2), a uranium oxide in granitic rocks and hydrothermal veins. Uraninite is easily oxidized and dissolved in groundwater, transported elsewhere, and chemically reduced and precipitated in the presence of organic matter.

The richest uranium ores in the United States are widespread in the Colorado Plateau area of Colorado and adjoining parts of Wyoming, Utah, Arizona, and New Mexico. These ores, consisting of fairly pure masses and encrustations of carnotite, are associated with plant remains in sandstones that formed in ancient stream channels. Although most of these ores are associated with fragmentary plant remains, some petrified trees also contain large quantities of uranium.

Large reserves of low-grade uranium ore also are found in the Chattanooga Shale. The uranium is finely disseminated in this black, organic-rich mudrock that underlies large parts of several states, including Illinois, Indiana, Ohio, Kentucky, and Tennessee. Canada is the world's largest producer and exporter of uranium.

Banded Iron Formation

The chemical sedimentary rock known as *banded iron formation* consists of alternating thin layers of chert and iron minerals, mostly the iron oxides hematite and magnetite (Figure 7.20). Banded iron formations are present on all continents and account for most of the iron ore mined in the world today. Vast banded iron formations are present in the Lake Superior region of the United States and Canada and in the Labrador Trough of eastern Canada.

ConnectionLink

You can find more information about banded iron formation in Chapter 19.

Figure 7.20 Banded Iron Formation Is an Important Sedimentary Rock Because It Is the Main Source of Iron Ore

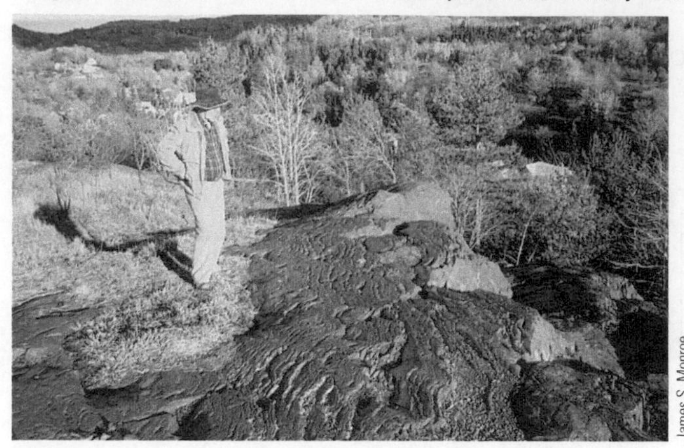

James S. Monroe

a The banded iron formation at Ishpeming, Michigan, consists of alternating layers of red chert and silver-colored iron minerals.

Sue Monroe

b Much of the iron ore in the Great Lakes region is processed and shaped into pellets about 1 cm across and then either shipped directly to steel mills or stored for later shipment.

Key Concepts Review

- Detrital sediment consists of weathered solid particles, whereas chemical sediment consists of minerals extracted from solution by inorganic chemical processes and the activities of organisms.
- Sedimentary particles are designated in order of decreasing size as gravel, sand, silt, and clay.
- During transport, sedimentary particles are rounded and sorted.
- Major depositional settings are continental, transitional, and marine, each of which includes several specific depositional environments.
- Lithification takes place when sediments are compacted and cemented and converted into sedimentary rock. Silica and calcium carbonate are the most common chemical cements.
- Sedimentary rocks are classified as detrital or chemical:
  a. Detrital sedimentary rocks consist of particles (gravel, sand, silt, and clay) derived from preexisting rocks.
  b. Chemical sedimentary rocks are derived from substances in solution by inorganic chemical processes or the activities of organisms. A subcategory called biochemical sedimentary rocks is recognized.
- Carbonate rocks contain minerals with the carbonate radical $(CO_3)^{-2}$ as in limestone and dolostone.
- Evaporites include rock salt and rock gypsum, both of which form by inorganic precipitation of minerals from evaporating water.
- Coal is a type of biochemical sedimentary rock composed of the altered remains of land plants.
- Sedimentary facies are bodies of sediment or sedimentary rock that are recognizably different from adjacent sediments or rocks.
- Vertical sequences of rocks with offshore facies overlying nearshore facies form when sea level rises with respect to the land, causing a marine transgression. A rise in the land relative to sea level causes a marine regression, which results in nearshore facies overlying offshore facies.
- Sedimentary structures, such as bedding, cross-bedding, and ripple marks, help geologists determine ancient current directions and depositional environments.
- Fossils provide the only record of prehistoric life and are useful for correlation and environmental interpretations.
- Depositional environments of ancient sedimentary rocks are determined by studying all aspects of the rocks and making comparisons with present-day sediments deposited by known processes.
- Many sediments and sedimentary rocks, including sand, gravel, evaporates, coal, and banded iron formations, are important natural resources. Most oil and natural gas are found in sedimentary rocks.

Important Terms

biochemical sedimentary rock (p. 164)
carbonate rock (p. 164)
cementation (p. 162)
chemical sedimentary rock (p. 164)
compaction (p. 161)
cross-bedding (p. 169)
depositional environment (p. 161)
detrital sedimentary rock (p. 162)
evaporite (p. 165)
fossil (p. 171)
graded bedding (p. 169)
lithification (p. 161)

marine regression (p. 167)
marine transgression (p. 167)
mud crack (p. 170)
ripple mark (p. 170)
rounding (p. 161)
sediment (p. 160)
sedimentary facies (p. 167)
sedimentary rock (p. 160)
sedimentary structure (p. 169)
sorting (p. 161)
strata (bed) (p. 169)

Review Questions

1. Graded bedding commonly results from deposition by
   a. ____ turbidity currents.
   b. ____ wind.
   c. ____ landslides.
   d. ____ glaciers.
   e. ____ mud settling in a lake.
2. A type of sandstone that contains at least 25% feldspar minerals is called
   a. ____ sedimentary breccia.
   b. ____ chert.
   c. ____ anthracite.
   d. ____ claystone.
   e. ____ arkose.
3. Cross-bedding preserved in sedimentary rocks is a good indication of
   a. ____ the intensity of organic activity.
   b. ____ how old the rocks are.
   c. ____ the amount of silica cement.
   d. ____ ancient current directions.
   e. ____ a marine regression.
4. A vertical sequence of sedimentary rocks in which marine offshore facies overly nearshore facies results from
   a. ____ granitization.
   b. ____ deposition in a river system.
   c. ____ changing limestone to dolostone.
   d. ____ a marine transgression.
   e. ____ turbidity current deposition.
5. Traps for petroleum and natural gas that from by folding and fracturing of rocks are called ________ traps.
   a. ____ structural.
   b. ____ bedding.
   c. ____ transgressive.
   d. ____ compaction.
   e. ____ lithified.
6. Under what conditions are evaporates deposited, and what are the two most common types of evaporite rocks?
7. How is it possible to determine how ancient sedimentary rocks were deposited given that no human observers were present to witness their deposition?
8. Describe how deposits of mud and sand are lithified.
9. The United States uses about 1,060,000,000 metric tons of coal per year from a recoverable reserve base of 237,295,000,000 metric tons. How long will this coal last at the current rate of consumption? Do you see any problems with projections like this?

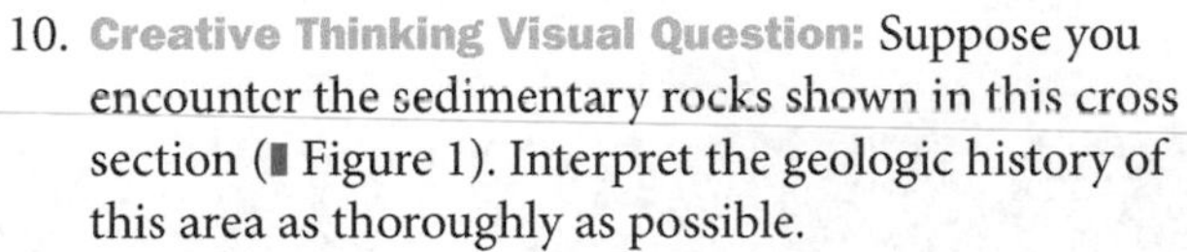

10. **Creative Thinking Visual Question:** Suppose you encounter the sedimentary rocks shown in this cross section (Figure 1). Interpret the geologic history of this area as thoroughly as possible.

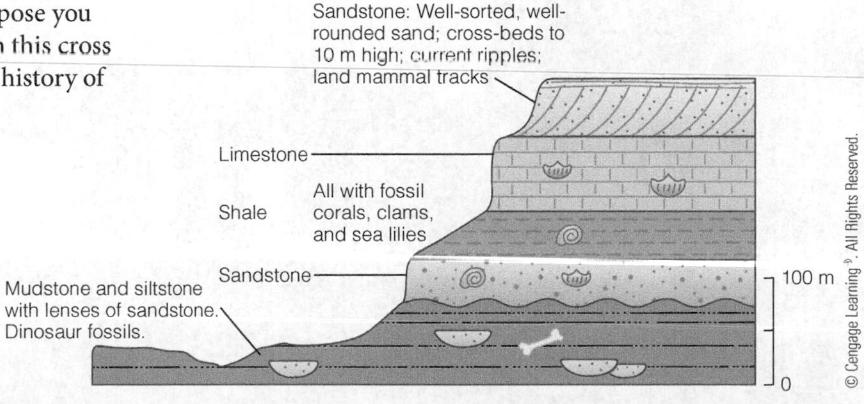

Figure 1 Cross section showing a succession of sedimentary rocks and their features.

Global GeoScience Watch On the GREENR database homepage, click on "Industrial Disasters" under the "Pollution" heading. Under the "Magazines" heading, click on the article titled "Drowning in Oil." Summarize what happened during the oil spill and what health problems it may have caused.

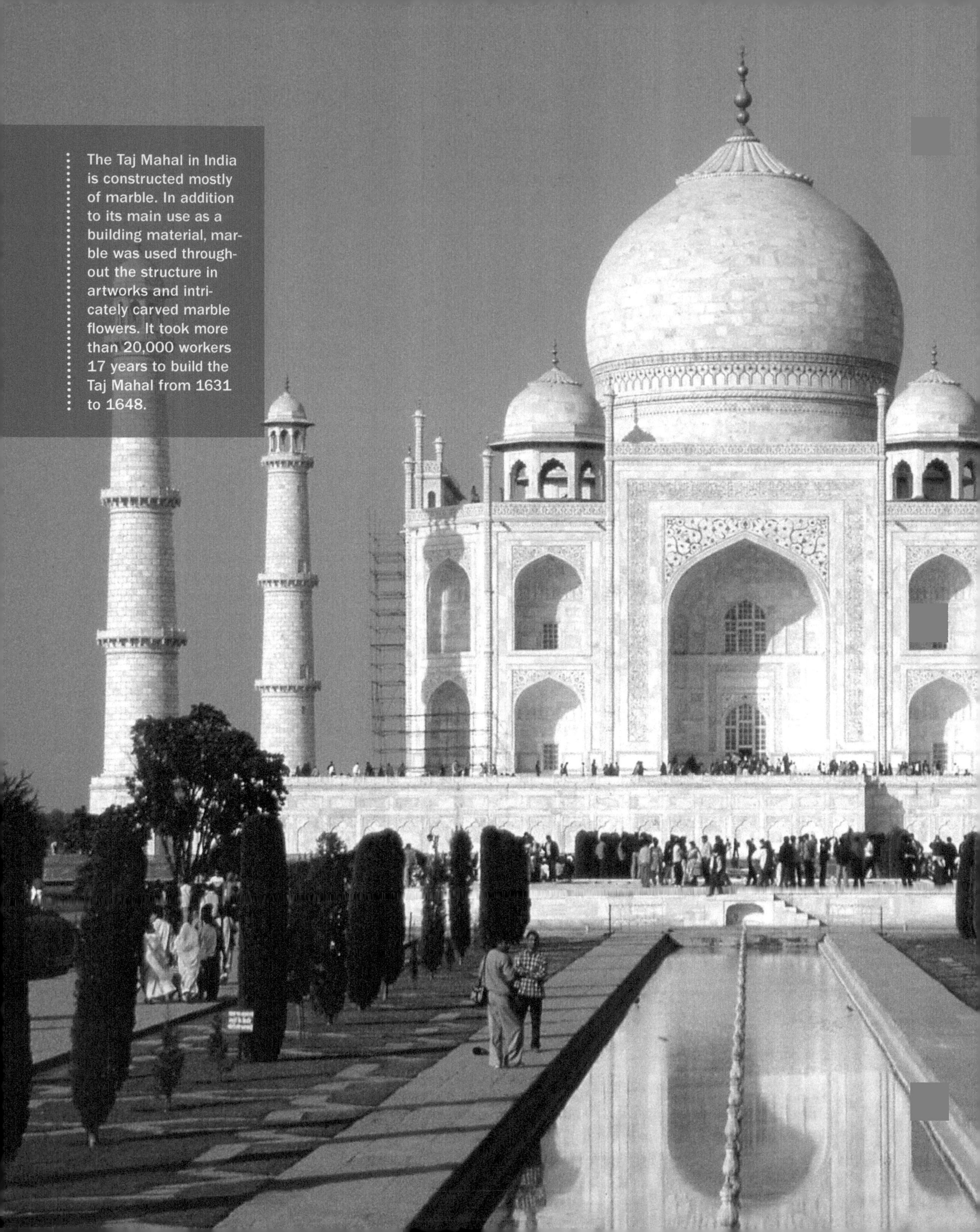

The Taj Mahal in India is constructed mostly of marble. In addition to its main use as a building material, marble was used throughout the structure in artworks and intricately carved marble flowers. It took more than 20,000 workers 17 years to build the Taj Mahal from 1631 to 1648.

Copyright and Photograph by Dr. Parvinder S. Sethi

Metamorphism and Metamorphic Rocks

OUTLINE

HAVE YOU EVER WONDERED?

- What causes metamorphism?
- How metamorphic rocks form?
- Where metamorphic rocks are most common and why?
- How metamorphic rocks are classified?
- Why marble is such a common rock in sculptures and structures?
- Why asbestos is an important industrial mineral for fireproofing and insulation in buildings and building materials, yet is also considered very dangerous to human health?

8.1 Introduction

Marble is a remarkable rock that has a variety of uses. Formed from limestone or dolostone by the metamorphic processes of heat and pressure, marble comes in a variety of colors and textures. It has been used by sculptors and architects for many centuries in statuary and monuments. For example, the marble statue *Aphrodite of Melos*, also known as *Venus de Milo*, is one of the most recognizable works of art in the world (▮ Figure 8.1). And the Peace Monument at Pennsylvania Avenue on the west side of the Capitol in Washington, D.C., is constructed from white marble from Carrara, Italy, a locality famous for its marble. Marble is also used as a facing and main stone in many buildings and various other structures. The Taj Mahal in India is constructed mostly of Makrana marble quarried from hills just southwest of Jaipur in Rajasthan (see the chapter opening photo). Moreover, marble is used in floor tiling and other ornamental and structural uses. In addition, ground marble is used in toothpaste and it is a source of lime in agricultural fertilizers.

▮ **Figure 8.1** ***Aphrodite of Melos*** *Aphrodite of Melos*, also known as *Venus de Milo*, is one of the most famous and recognizable statues in the world. Dated around 150 BC, *Venus de Milo* was created by an unknown artist during the Greek Hellenistic period and carved from the world-famous Parian marble from Papros in the Cyclades. Today, *Venus de Milo* attracts thousands of visitors a year to the Louvre Museum in Paris, where she can be viewed and appreciated.

Metamorphic rocks (from the Greek *meta*, "change," and *morpho*, "shape") like marble are the third major group of rocks we will be examining. They result from the transformation of other rocks by processes that typically occur beneath Earth's surface (see Figure 1.15). During **metamorphism**, rocks are subjected to sufficient heat, pressure, and fluid activity to change their mineral composition, texture, or both, thus forming new rocks that usually do not look anything like the original rock before it was metamorphosed. These transformations take place below the melting temperature of the rock; otherwise, an igneous rock would result.

A useful analogy for metamorphism is baking a cake. Just like a metamorphic rock, the resulting cake depends on the ingredients, their proportions, how they are mixed together, how much water or milk is added, and the temperature and length of time used for baking the cake.

Except for marble and slate, most people are not familiar with metamorphic rocks. Students frequently ask us why is it important to study metamorphic rocks and processes? The answer is always, "Just look around you."

A large portion of Earth's continental crust is composed of metamorphic and igneous rocks. Together, they form the crystalline basement rocks underlying the sedimentary rocks of a continent's surface. These basement rocks are widely exposed in regions of the continents known as *shields*, which have been very stable during the past 600 million years (see Figure 19.5). Some of the oldest known rocks, dated at about 4 billion years (▮ Figure 8.2) are metamorphic, which means that they formed from even older rocks.

ConnectionLink

To learn more about shields, see Chapter 19.

▮ **Figure 8.2** **Acasta Gneiss** This metamorphic rock, found in Canada, is estimated to be about 4 billion years old, making it one of the oldest known rocks on Earth.

Critical Thinking Question If the Acasta Gneiss (a metamorphic rock) is one of the oldest known rocks on Earth, why does Earth have to be older than 4 billion years?

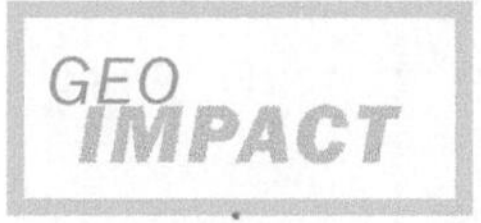

Asbestos and Public Policy

The problem of removing asbestos from public buildings is a continuing important national health and political issue. In 1986, Congress passed the Asbestos Hazard Emergency Response Act that required all schools be inspected for asbestos and take appropriate abatement action. This law, and the current policy of the United States Environmental Protection Agency (EPA), mandates that all forms of asbestos are treated as identical hazards. However, numerous studies indicate that only one form of asbestos is a known health hazard. Because the cost of asbestos removal has been estimated to be as high as $100 billion, many people are questioning whether it is cost effective to remove asbestos from all public buildings where it has been installed, especially in light of the fact that removal oftentimes results in higher levels of airborne asbestos particles. Furthermore, much of the asbestos is the less dangerous chrysotile variety, and if already stabilized such that it is not free to blow around, poses little or no threat to public health.

How would you address this issue in terms of formulating a policy that balances the risks and benefits of removing asbestos from public buildings? What role should geologists play in formulating this policy?

Metamorphic rocks such as the aforementioned marble, as well as slate, are used as building materials. In addition, many metamorphic minerals, such as garnets (used as gemstones or abrasives) and talc (used in cosmetics, in manufacturing paint, and as a lubricant), are economically important and useful. Asbestos, another valuable metamorphic mineral, is used for insulation and fireproofing and has been the subject of much debate over the danger it poses to the public's health (Geo-Focus 8.1).

8.2 The Agents of Metamorphism

The three principal agents of metamorphism are *heat*, *pressure*, and *fluid activity*. Time is also important to the metamorphic process, because chemical reactions proceed at different rates and thus require different amounts of time to complete. Reactions involving silicate compounds are particularly slow, and because most metamorphic rocks are composed of silicate minerals, it is thought that metamorphism is a very slow geologic process.

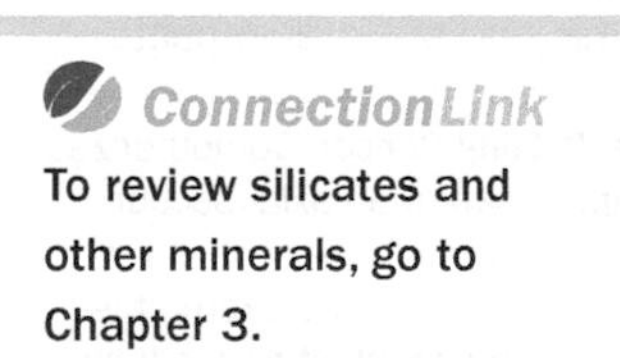

ConnectionLink

To review silicates and other minerals, go to Chapter 3.

During metamorphism, the original rock, which was in equilibrium with its environment—meaning that it was chemically and physically stable under those conditions—undergoes changes to achieve equilibrium with its new environment. These changes may result in the formation of new minerals, a change in the texture of the rock, or both. In some instances, the change is minor, and features of the original rock can still be recognized. In other cases, the rock changes so much that the identity of the original rock can be determined only with great difficulty, if at all.

Heat

Heat is an important agent of metamorphism because it increases the rate of chemical reactions that may produce minerals different from those in the original rock. Heat may come from lava, magma, or as a result of deep burial in the crust due to subduction along a convergent plate boundary.

When rocks are intruded by bodies of magma, they are subjected to intense heat that affects the surrounding rock. The most intense heating usually occurs adjacent to the magma body and gradually decreases with distance from the intrusion. The zone of metamorphosed rocks that forms in the country rock adjacent to an intrusive igneous body is usually distinct and easy to recognize.

It is known that temperature increases with depth. Some rocks that form at the surface may be transported to great depths by subduction along a convergent plate boundary and thus subjected to increasing temperature and pressure. During subduction, some minerals may be transformed into other minerals that are more stable under the higher temperature and pressure conditions.

Pressure

During burial, rocks are subjected to increasingly greater pressure, just as you feel greater pressure the deeper you dive into a body of water. Whereas the pressure you feel is known as *hydrostatic pressure*, because it comes from the water surrounding you, rocks undergo **lithostatic pressure**,

ConnectionLink

To learn more about stress and strain, and how they relate to the deformation of rocks, see Chapter 10.

GEO FOCUS 8.1 Asbestos: Good or Bad?

Asbestos (from the Latin, meaning "unquenchable") is a general term applied to any silicate mineral that easily separates into flexible fibers. The combination of such features as fire resistance and flexibility makes asbestos an important industrial material of considerable value. In fact, asbestos has more than 3,000 known uses, including brake linings, fireproof fabrics, and heat insulators.

Asbestos is divided into two broad groups: *serpentine asbestos* and *amphibole asbestos*. *Chrysotile* is the fibrous form of serpentine asbestos (▌Figure 1); it is the most valuable type and constitutes the bulk of all commercial asbestos. Its strong, silky fibers are easily spun and can withstand temperatures as high as 2,750°C.

The vast majority of chrysotile asbestos is in serpentine, a type of rock formed by the alteration of ultramafic igneous rocks, such as peridotite under low- and intermediate-grade metamorphic conditions. Other chrysotile results when the metamorphism of magnesium limestone or dolostone produces discontinuous serpentine bands within the carbonate beds.

Among the varieties of amphibole asbestos, *crocidolite* is the most common. Also known as blue asbestos, crocidolite is a long, coarse, spinning fiber that is stronger but more brittle than chrysotile and less resistant to heat. Crocidolite is found in such metamorphic rocks as slates and schists and forms by the solid-state alteration of other minerals as a result of deep burial.

Despite the widespread use of asbestos, the United States Environmental Protection Agency (EPA) instituted a gradual ban on all new asbestos products. The ban was imposed because some forms of asbestos can cause lung cancer and scarring of the lungs if fibers are inhaled.

Because the EPA apparently paid little attention to the issue of risks versus benefits when it enacted this rule, the United States Fifth Circuit Court of Appeals overturned the EPA ban on asbestos in 1991.

The threat of lung cancer has also resulted in legislation mandating the removal of asbestos already in place in all public buildings, including all public and private schools. However, important questions have been raised concerning the threat posed by asbestos and the additional potential hazards that may arise from its improper removal.

Current EPA policy mandates that all forms of asbestos are to be treated as identical hazards. Yet studies indicate that only the amphibole forms constitute a known health hazard. Chrysotile, whose fibers tend to be curly, does not become lodged in the lungs. Furthermore, its fibers are generally soluble and disappear in tissue. In contrast, crocidolite has long, straight, thin fibers that penetrate the lungs and stay there. These fibers irritate the lung tissue and over a long period of time can lead to lung cancer (▌Figure 2). Thus, crocidolite, and not chrysotile, is overwhelmingly responsible for asbestos-related lung cancer. Because approximately 95% of the asbestos in place in the United States is chrysotile, many people question whether the dangers from asbestos are exaggerated.

Copyright and Photograph by Dr. Parvinder S. Sethi

▌**Figure 1 Specimen of Chrysotile from Arizona** Chrysotile is the fibrous form of serpentine asbestos and the most commonly used in buildings and other structures.

Removing asbestos from buildings where it has been installed could cost as much as $100 billion. Unless the material containing the asbestos is disturbed, asbestos does not shed fibers and thus does not contribute to airborne asbestos that can be inhaled. Furthermore, improper removal of asbestos can lead to contamination. In most cases of improper removal, the concentration of airborne asbestos fibers is far higher than if the asbestos had been left in place.

The problem of asbestos contamination is a good example of how geology affects our lives and why we should have a basic knowledge of science before making decisions that could have broad economic and societal impacts.

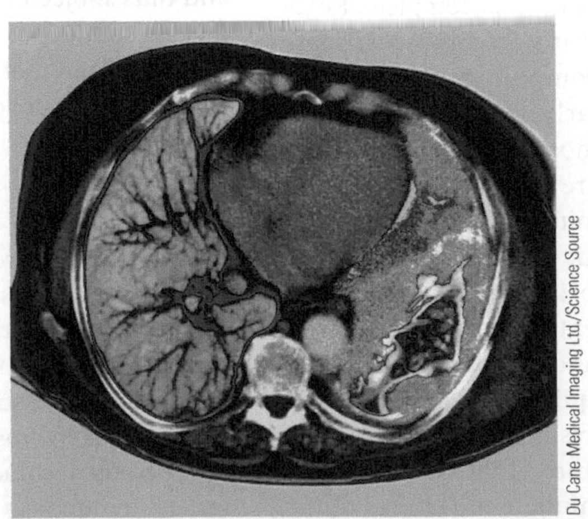

Du Cane Medical Imaging Ltd./Science Source

▌**Figure 2 Lung Cancer** Computerized tomography scan of an axial section through the chest of a patient with a mesothelioma cancer (light red). It is surrounding and constricting the lung at right (pink). The other lung (dark blue) has a healthy pleura (dark red). The spine (lower center, light blue), the descending aorta (green), and the heart (dark green between lungs) are also seen. Mesothelioma is a malignant cancer of the pleura, the membrane lining the chest cavity and lungs. It is usually caused by asbestos exposure. It often reaches a large size, as here, before diagnosis, and the prognosis is then poor.

Figure 8.3 Lithostatic Pressure

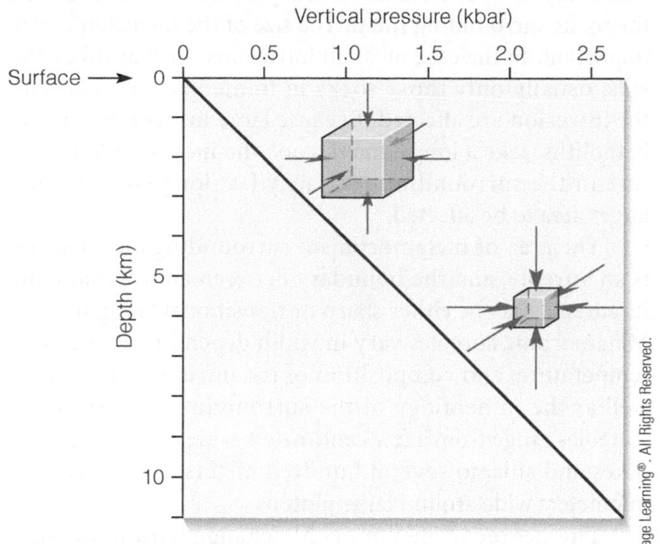

1 kilobar (kbar) = 1,000 bars
Atmospheric pressure at sea level = 1 bar

a Lithostatic pressure is applied equally in all directions in Earth's crust due to the weight of overlying rocks. Thus, pressure increases with depth, as indicated by the sloping black line.

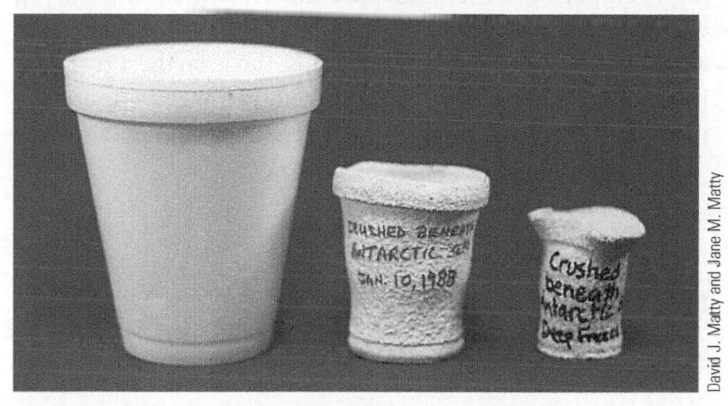

David J. Matty and Jane M. Matty

b A similar situation occurs when 200-ml cups composed of Styrofoam™ are lowered to ocean depths of approximately 750 m and 1,500 m. Increased water pressure is exerted equally in all directions on the cups, and they consequently decrease in volume while maintaining their general shape.

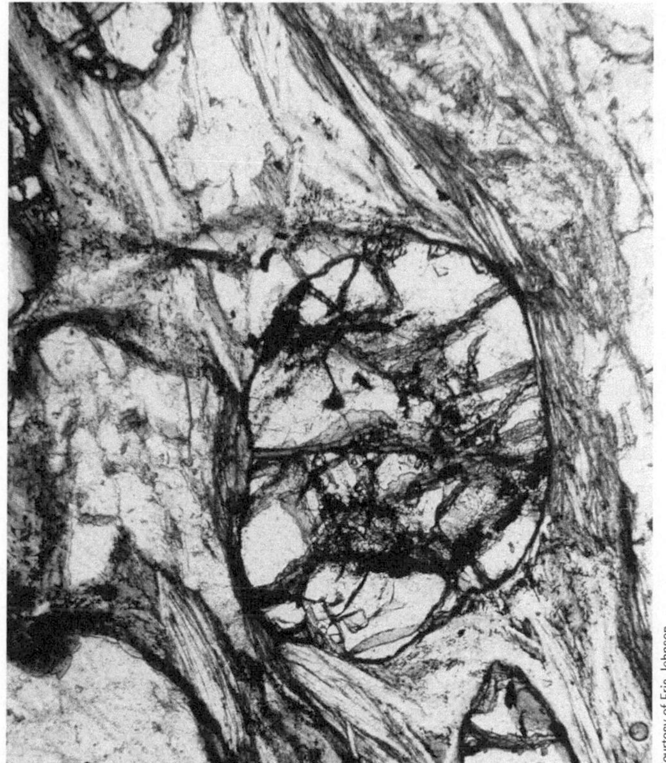

Courtesy of Eric Johnson

Figure 8.4 Differential Pressure Differential pressure results from stress that is unequally applied to an object. Rotated garnets are a good example of the effects of differential pressure applied to a rock during metamorphism. In this example from a schist in northeast Sardinia, stress was applied in opposite directions on the left and right sides of the garnet (center), causing it to rotate.

which means that the *stress* (force per unit area) on a rock in Earth's crust is the same in all directions (Figure 8.3a). A similar situation occurs when an object is immersed in water. For example, the deeper a cup composed of Styrofoam™ is submerged in the ocean, the smaller it gets because pressure increases with depth and is exerted on the cup equally in all directions, thereby compressing the cup (Figure 8.3b).

Along with lithostatic pressure resulting from burial, rocks may experience **differential pressure** (Figure 8.4). In this case, the stresses are not equal in all directions, but they are stronger from some directions than from others. Differential pressures typically occur when two plates collide, thus producing distinctive metamorphic textures and features.

Fluid Activity

In almost every region of metamorphism, water and carbon dioxide (CO_2) are present in varying amounts along mineral grain boundaries or in the pore spaces of rocks. These fluids, which may contain ions in solution, enhance metamorphism by increasing the rate of chemical reactions. Under dry conditions, most minerals react very slowly, but when even small amounts of fluid are introduced, reaction rates increase. This is mainly because ions can move readily through the fluid and thus enhance chemical reactions and the formation of new minerals.

The following reaction provides a good example of how new minerals can be formed by **fluid activity**. Seawater moving through hot basaltic rock in the oceanic crust transforms olivine into the metamorphic mineral serpentine:

$$\underset{\text{olivine}}{2Mg_2SiO_4} + \underset{\text{water}}{2H_2O} \rightarrow \underset{\text{serpentine}}{Mg_3Si_2O_5(OH)_4} + \underset{\text{carried away in solution}}{MgO}$$

The chemically active fluids important in the metamorphic process come primarily from three sources: (1) water trapped in the pore spaces of sedimentary rocks as they form, (2) the volatile fluids within magma, and (3) the dehydration of water-bearing minerals such as gypsum ($CaSO_4{\cdot}2H_2O$) and some clays.

8.3 The Three Types of Metamorphism

Geologists recognize three types of metamorphism: (1) *contact (thermal) metamorphism*, in which magmatic heat and fluids act to produce change; (2) *dynamic metamorphism*, which is principally the result of high differential pressures associated with intense deformation; and (3) *regional metamorphism*, which occurs within a large area and is associated with major mountain-building episodes. Even though we will discuss each type of metamorphism separately, the boundary between them is not always distinct and depends largely on which of the three metamorphic agents was dominant (▮ Figure 8.5).

Contact Metamorphism

Contact (thermal) metamorphism takes place when a body of magma alters the surrounding country rock. At shallow depths, intruding magma raises the temperature of the surrounding rock, causing thermal alteration. Furthermore, the release of hot fluids into the country rock by the cooling intrusion can aid in the formation of new minerals.

Important factors in contact metamorphism include the initial temperature, the size of the intrusion, the fluid content of the magma, and the nature of the country rock. The initial temperature of an intrusion depends, in part, on its composition; mafic magmas are hotter than felsic magmas (see Chapter 4) and hence have a greater thermal effect on the rocks surrounding them. The size of the intrusion is also important. In the case of small intrusions, such as dikes and sills, usually only those rocks in immediate contact with the intrusion are affected. Because large intrusions, such as batholiths, take a long time to cool, the increased temperature in the surrounding rock may last long enough for a larger area to be affected.

The area of metamorphism surrounding an intrusion is an **aureole**, and the boundary between an intrusion and its aureole may be either sharp or transitional (▮ Figure 8.6). Metamorphic aureoles vary in width depending on the size, temperature, and composition of the intruding magma, as well as the mineralogy of the surrounding country rock. Aureoles range from a few centimeters wide, bordering small dikes and sills, to several hundred meters or even several kilometers wide around large plutons.

The degree of metamorphic change within an aureole decreases with distance from the intrusion, reflecting the decrease in temperature from the original heat source. The region, or zone, closest to the intrusion, and hence subject to the highest temperatures, commonly contains high-temperature metamorphic minerals (i.e., minerals in equilibrium with the higher-temperature environment) such as sillimanite. The outer zones, that is, those farthest from the intrusion, are typically characterized by lower-temperature metamorphic minerals such as chlorite, talc, and epidote.

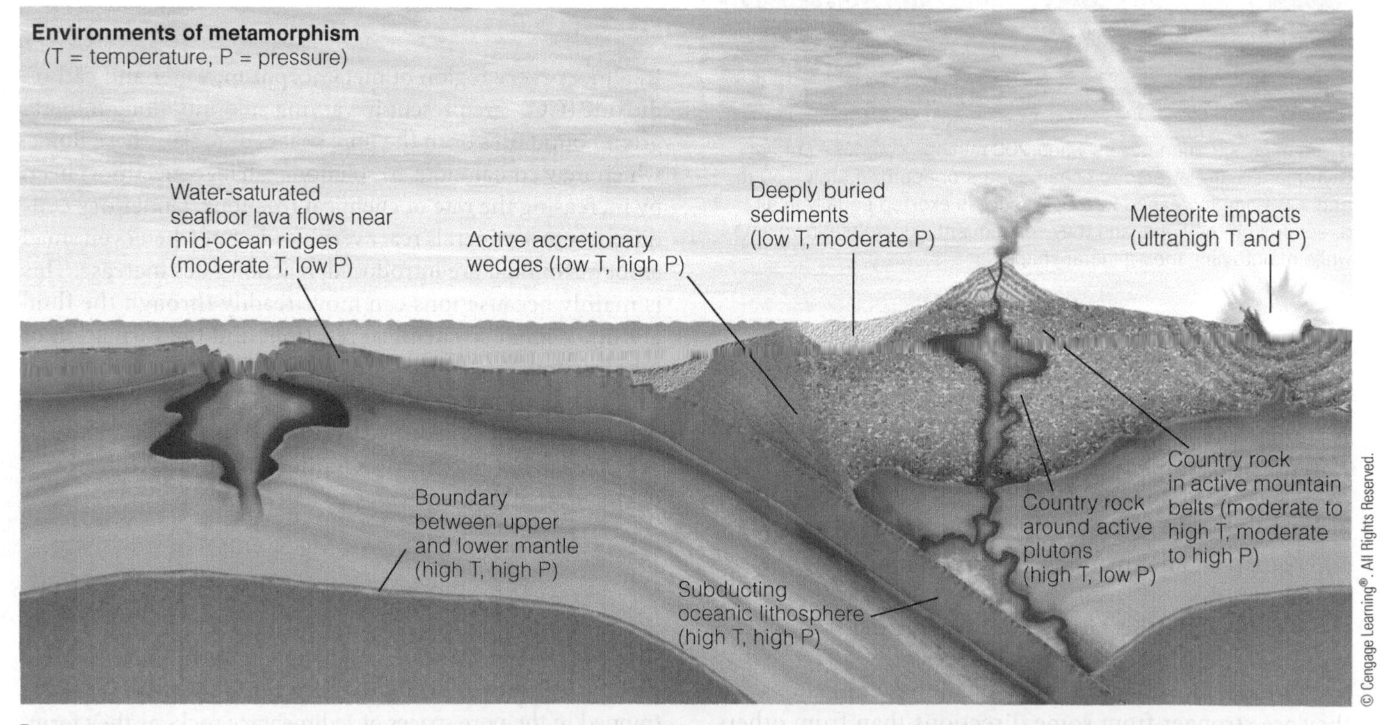

▮ **Figure 8.5 Environments of Metamorphism** The type of metamorphism that results depends largely on which of the three metamorphic agents was dominant. Illustrated here are some of the common metamorphic environments associated with plate movement, and whether the temperature and pressure in this environment is low, moderate, or high. The third agent of metamorphism, fluid activity, although important in metamorphism, is not shown here but is discussed in the text.

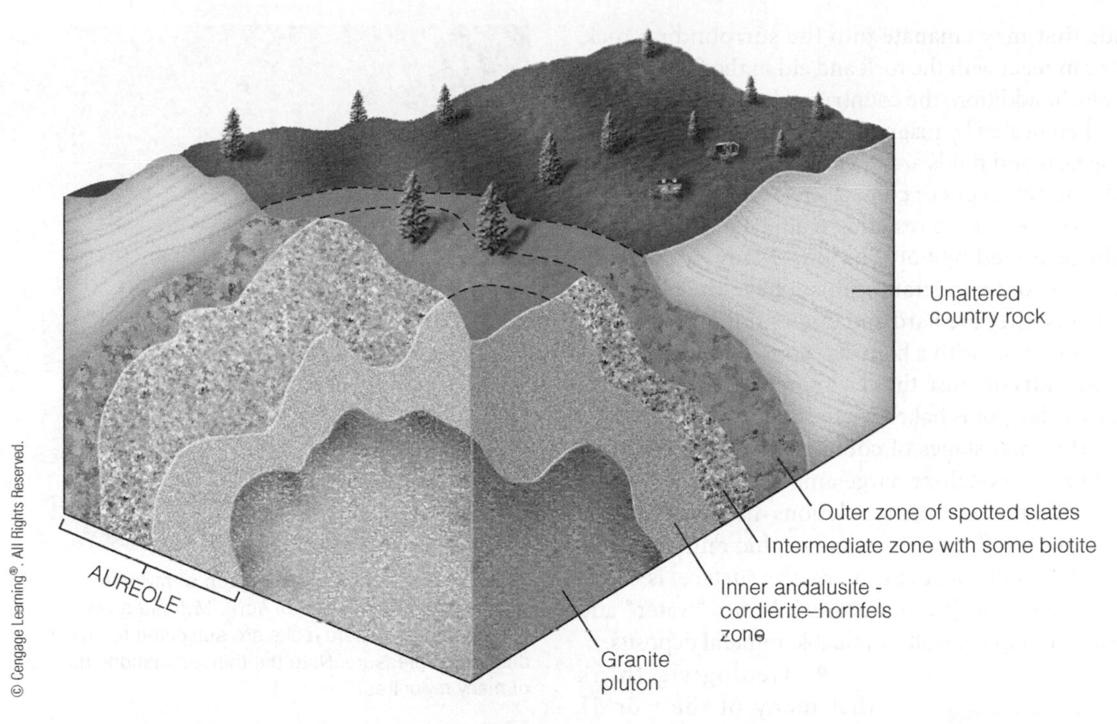

Figure 8.6 Metamorphic Aureole A metamorphic aureole, the area surrounding an intrusion, consists of zones that reflect the degree of metamorphism. The metamorphic aureole associated with this idealized granite pluton contains three zones of mineral assemblages reflecting the decreases in temperature with distance from the intrusion. An inner andalusite–cordierite hornfels zone forms adjacent to the pluton and indicates the high temperatures near the intrusion. This zone is followed by an intermediate zone of extensive recrystallization in which some biotite develops, and farthest from the intrusion is an outer zone characterized by spotted slates.

ConnectionLink

To learn more about how to differentiate between an intrusive and extrusive igneous body in an outcrop, and about its importance in working out the correct sequence of events, go to Figure 17.5 in Chapter 17.

Contact metamorphism can result not only from igneous intrusions but also from lava flows, either along mid-oceanic ridges (Figure 8.5) or from lava flowing over land and thermally altering the underlying rocks (Figure 8.7). Whereas recognizing a recent lava flow and the resulting contact metamorphism of the rocks below is easy, it is less obvious whether an igneous body is intrusive or extrusive in a rock outcrop where sedimentary rocks occur above and below the igneous body. Recognizing which sedimentary rock units have been metamorphosed enables geologists to determine whether the igneous body is intrusive (such as a sill or dike) or extrusive (lava flow). Such a determination is critical in reconstructing the geologic history of an area and may have important economic implications as well.

Fluids also play an important role in contact metamorphism. Magma is usually wet and contains hot, chemically

Figure 8.7 Contact Metamorphism from a Lava Flow A highly weathered basaltic lava flow near Susanville, California, has altered an underlying rhyolitic volcanic ash by contact metamorphism. The red zone below the lava flow has been baked by the heat of the lava when it flowed over the ash layer. The lava flow displays spheroidal weathering, a type of weathering common in fractured rocks (see Chapter 6).

Critical Thinking Question What criteria would you use to determine that what you see is indeed spheroidal weathering and not weathered pillow lavas?

active fluids that may emanate into the surrounding rock. These fluids can react with the rock and aid in the formation of new minerals. In addition, the country rock may contain pore fluids that, when heated by magma, increase reaction rates.

Because heat and fluids are the primary agents of contact metamorphism, two types of contact metamorphic rocks are generally recognized: those resulting from baking of country rock and those altered by hot solutions. Many of the rocks that result from contact metamorphism have the texture of porcelain; that is, they are hard and fine-grained. This is particularly true for rocks with a high clay content, such as shale. Such texture results because the clay minerals in the rock are baked, just as a clay pot is baked when fired in a kiln.

During the final stages of cooling, when an intruding magma begins to crystallize, large amounts of hot, watery solutions are often released. These solutions may react with the country rock and produce new metamorphic minerals. This process, which usually occurs near Earth's surface, is called *hydrothermal alteration* (from the Greek *hydro*, "water," and *therme*, "heat") and may result in valuable mineral deposits. Geologists think that many of the world's ore deposits result from the migration of metallic ions in hydrothermal solutions. Examples are copper, gold, iron ores, tin, and zinc in various localities, including Australia, Canada, China, Cyprus, Finland, the Russian Federation, and the western United States.

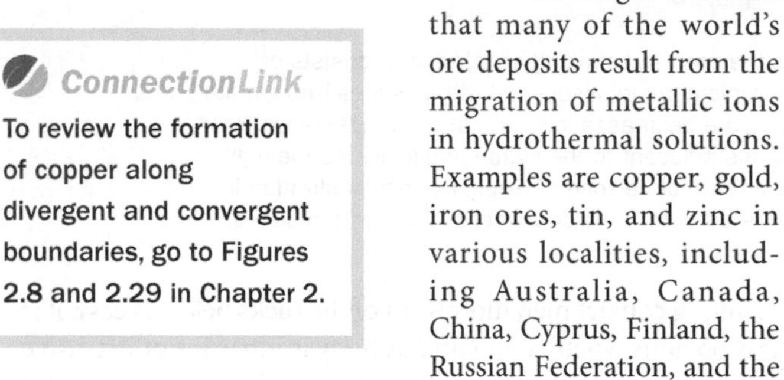

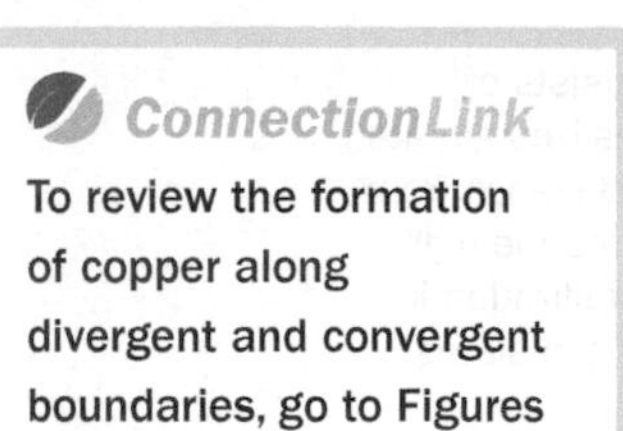

ConnectionLink

To review the formation of copper along divergent and convergent boundaries, go to Figures 2.8 and 2.29 in Chapter 2.

Dynamic Metamorphism

Most **dynamic metamorphism** is associated with *faults* (fractures along which movement has occurred) or fault zones, where rocks are subjected to high levels of differential pressure. The metamorphic rocks that result from pure dynamic metamorphism are called *mylonites*, and typically they are restricted to narrow zones adjacent to faults. Mylonites are hard, dense, fine-grained rocks, many of which are characterized by thin laminations (Figure 8.8). Tectonic settings where mylonites occur include the Moine Thrust Zone in northwest Scotland, the Adirondack Highlands in New York, and portions of the San Andreas Fault in California.

ConnectionLink

To learn more about the San Andreas Fault, see Figure 2.23 in Chapter 2.

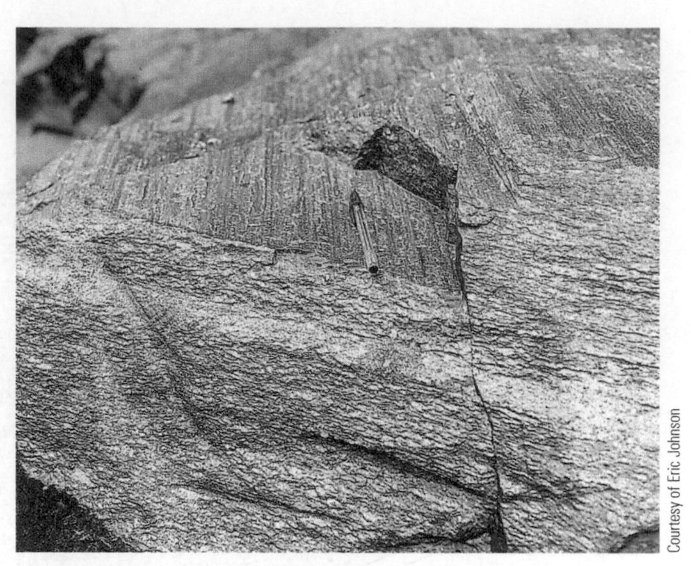

Figure 8.8 Mylonite An outcrop of mylonite from the Adirondack Highlands, New York. Mylonites result from dynamic metamorphism, where rocks are subjected to high levels of differential pressure. Note the thin laminations that are characteristic of many mylonites.

Courtesy of Eric Johnson

Regional Metamorphism

Most metamorphic rocks result from **regional metamorphism**, which occurs over a large area and is usually caused by tremendous temperatures, pressures, and deformation all occurring together within the deeper portions of the crust. Regional metamorphism is most obvious along convergent plate boundaries where rocks are intensely deformed and recrystallized during convergence and subduction (Figure 8.5). Within these metamorphic rocks, there is usually a gradation of metamorphic intensity from areas that were subjected to the most intense pressures, highest temperatures, or both intense pressure and high temperature, to areas of lower pressures and temperatures. Such a gradation in metamorphism can be recognized by the metamorphic minerals that are present.

Regional metamorphism is not, however, confined only to convergent margins. It can also occur in areas where plates diverge, although usually at much shallower depths because of the high geothermal gradient associated with these areas (Figure 8.5).

Index Minerals and Metamorphic Grade

From field studies and laboratory experiments, certain minerals are known to form only within specific temperature and pressure ranges. Such minerals are known as **index minerals** because their presence allows geologists to recognize low-, intermediate-, and high-grade metamorphism (Figure 8.9).

Metamorphic grade is a term that generally characterizes the degree to which a rock has undergone metamorphic change (Figure 8.9). Although the boundaries between the different metamorphic grades are not sharp, the distinction is nonetheless useful for communicating in a general way the degree to which rocks have been metamorphosed. The

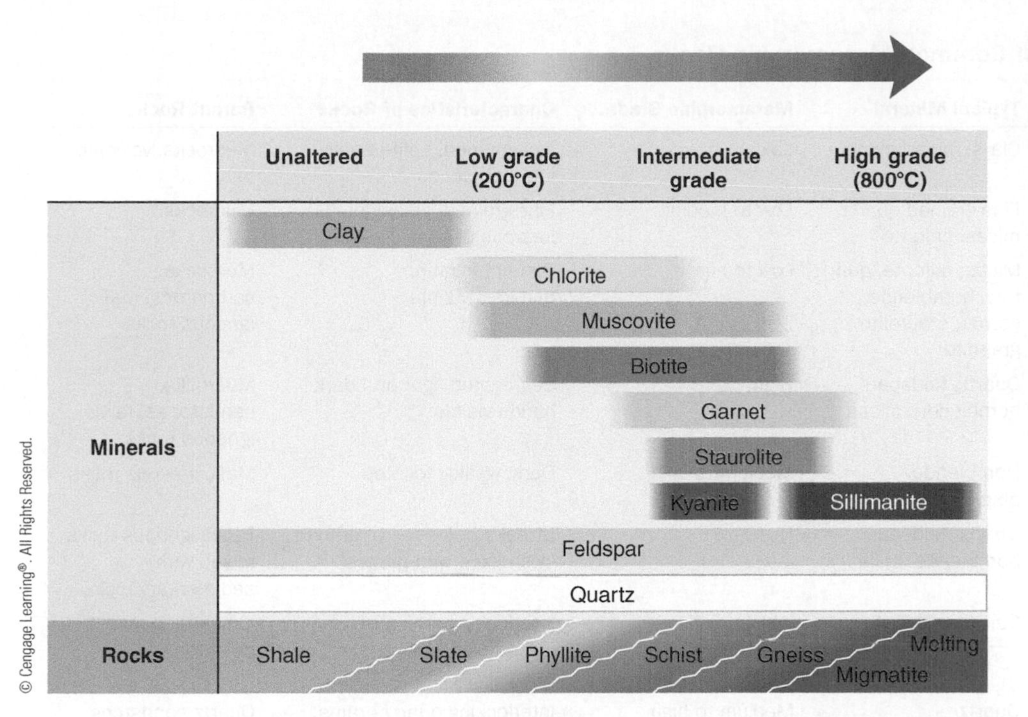

Figure 8.9 **Metamorphic Grade** Change in mineral assemblage and rock type with increasing metamorphism of shale. When a clay-rich rock such as shale is subjected to increasing metamorphism, new minerals form, as shown by the colored bars. The progressive appearance of certain minerals, known as index minerals, allows geologists to recognize low-, intermediate-, and high-grade metamorphism.

Critical Thinking Question Why aren't quartz or calcite index minerals that can be used to determine the metamorphic grade of a rock?

presence of index minerals thus helps determine metamorphic grade. For example, when a clay-rich rock such as shale undergoes regional metamorphism, the mineral chlorite first begins to crystallize under relatively low temperatures of about 200°C. Its presence in these rocks thus indicates low-grade metamorphism. If temperatures and pressures continue to increase, new minerals form to replace chlorite, because they are more stable under those new conditions. Thus, there is a progression in the appearance of new minerals from chlorite—whose presence indicates low-grade metamorphism—to biotite and garnet, which are good index minerals for intermediate-grade metamorphism, and then to sillimanite, whose presence indicates high-grade metamorphism and temperatures exceeding 500°C (Figure 8.9).

Different rock compositions develop different sets of index minerals. For example, clay-rich rocks such as shale will develop the index minerals shown in Figure 8.9, whereas a sandy dolomite will produce a different set of index minerals as metamorphism progresses, because it has a different mineral composition than shale. Thus, a particular set of index minerals will form based on the original composition of the parent rock undergoing metamorphism.

Although such common minerals as mica, quartz, and feldspar can occur in both igneous and metamorphic rocks, other minerals such as andalusite, sillimanite, and kyanite generally occur only in metamorphic rocks derived from clay-rich rocks such as shale. Whereas these three minerals all have the same chemical formula (Al_2SiO_5), they differ in crystal structure and other physical properties because each forms under a different range of pressures and temperatures and is thus reflective of a different metamorphic grade. For this reason, they are useful index minerals for metamorphic rocks that formed from clay-rich rocks.

8.4 How Are Metamorphic Rocks Classified?

For purposes of classification, metamorphic rocks are commonly divided into two groups: those exhibiting a *foliated texture* (from the Latin *folium*, "leaf") and those with a *nonfoliated texture* (Table 8.1).

Foliated Metamorphic Rocks

Rocks subjected to heat and differential pressure during metamorphism typically have minerals arranged in a parallel fashion, giving them a **foliated texture** (Figure 8.10). The size and shape of the mineral grains determine whether the foliation is fine or coarse. Low-grade metamorphic rocks, such as slate, have a finely foliated texture in which the mineral grains are so small that they cannot be distinguished without magnification. High-grade foliated rocks, such as gneiss, are coarse-grained, such that the individual grains can easily be seen with the unaided eye. Foliated metamorphic rocks can be arranged in order of increasingly coarse grain size and perfection of foliation.

Slate is a very fine-grained foliated metamorphic rock that commonly exhibits *slaty cleavage* (Figure 8.11). It results from regional metamorphism of shale or, more rarely, volcanic ash. Because it can easily be split along cleavage planes into flat pieces, slate is an excellent rock for roofing and floor tiles, billiard and pool table tops, and blackboards. The different colors of slate are caused by minute amounts of graphite (black), iron oxide (red and purple), and chlorite (green).

TABLE 8.1 Classification of Common Metamorphic Rocks

Texture	Metamorphic Rock	Typical Mineral	Metamorphic Grade	Characteristics of Rocks	Parent Rock
Foliated	Slate	Clays, micas, chlorite	Low	Fine-grained, splits easily into flat pieces	Mudrocks, volcanic ash
	Phyllite	Fine-grained quartz, micas, chlorite	Low to medium	Fine-grained, glossy or lustrous sheen	Mudrocks
	Schist	Micas, chlorite, quartz, talc, hornblende, garnet, staurolite, graphite	Low to high	Distinct foliation, minerals visible	Mudrocks, carbonates, mafic igneous rocks
	Gneiss	Quartz, feldspars, hornblende, micas	High	Segregated light and dark bands visible	Mudrocks, sandstones, felsic igneous rocks
	Amphibolite	Hornblende, plagioclase	Medium to high	Dark, weakly foliated	Mafic igneous rocks
	Migmatite	Quartz, feldspar, hornblende, micas	High	Streaks or lenses of granite intermixed with gneiss	Felsic igneous rocks mixed with sedimentary rocks
Nonfoliated	Marble	Calcite, dolomite	Low to high	Interlocking grains of calcite or dolomite, reacts with HCl	Limestone or dolostone
	Quartzite	Quartz	Medium to high	Interlocking quartz grains, hard, dense	Quartz sandstone
	Greenstone	Chlorite, epidote, hornblende	Low to high	Fine-grained, green	Mafic igneous rocks
	Hornfels	Micas, garnets, andalusite, cordierite, quartz	Low to medium	Fine-grained, equidimensional grains, hard, dense	Mudrocks
	Anthracite	Carbon	High	Black, lustrous, subconcoidal fracture	Coal

Figure 8.10 Foliated Texture

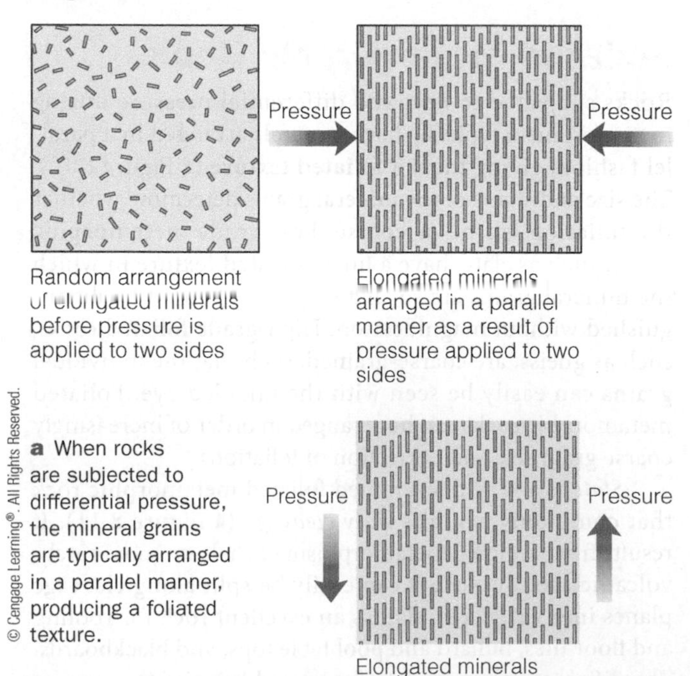

a When rocks are subjected to differential pressure, the mineral grains are typically arranged in a parallel manner, producing a foliated texture.

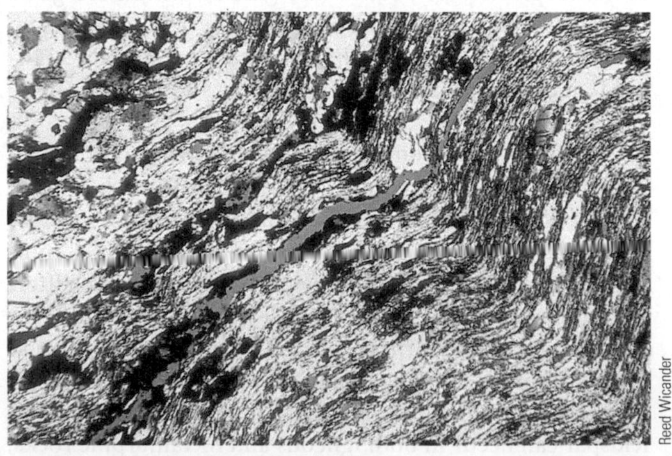

b Photomicrograph of a metamorphic rock with a foliated texture showing the parallel arrangement of mineral grains.

Figure 8.11 Slate

Copyright and Photograph by Dr. Parvinder S. Sethi

a Hand specimen of red slate.

James S. Monroe

b Slate roof of Chalet Enzian, Switzerland.

Phyllite is similar in composition to slate but is coarser grained. The minerals, however, are still too small to be identified without magnification. Phyllite can be distinguished from slate by its glossy or lustrous sheen (Figure 8.12) and represents an intermediate grain size between slate and schist.

Schist is most commonly produced by regional metamorphism. The type of schist formed depends on the intensity of metamorphism and the character of the original rock (Figure 8.13). Metamorphism of many rock types can yield schist, but most schist appears to have formed from clay-rich sedimentary rocks (Table 8.1).

All schists contain more than 50% platy and elongated minerals, all of which are large enough to be clearly visible. Their mineral composition imparts a *schistosity,* or *schistose foliation,* to the rock that usually produces a wavy type of parting when split. Schistosity is common in low- to high-grade metamorphic environments, and each type of schist is known by its most conspicuous mineral or minerals, such as mica schist, chlorite schist, or garnet-mica schist (Figure 8.13).

Copyright and Photograph by Dr. Parvinder S. Sethi

Figure 8.12 Phyllite Hand specimen of phyllite. Note the lustrous sheen as well as the bedding (upper left to lower right) at an angle to the cleavage of the specimen.

Critical Thinking Question How do you distinguish between bedding and cleavage in a metamorphic rock?

Figure 8.13 Schist

Copyright and Photograph by Dr. Parvinder S. Sethi

a Almandine garnet crystals in a mica schist.

Sue Monroe

b Hornblende–mica schist.

Figure 8.14 Gneiss Gneiss is characterized by segregated bands of light and dark minerals. This folded gneiss is exposed at Wawa, Ontario, Canada.

Gneiss is a high-grade metamorphic rock that is streaked or has segregated bands of light and dark minerals. Gneisses are composed of granular minerals, such as quartz, feldspar, or both, with lesser percentages of platy or elongated minerals, such as micas or amphiboles (Figure 8.14). Quartz and feldspar are the principal minerals of the light-colored mineral bands, and biotite and hornblende make up the dark-colored mineral bands. Gneiss typically breaks in an irregular manner, much like coarsely crystalline nonfoliated rocks.

Most gneiss probably results from recrystallization of clay-rich sedimentary rocks during regional metamorphism (Table 8.1). Gneiss also can form from igneous rocks such as granite or older metamorphic rocks.

Another fairly common foliated metamorphic rock is *amphibolite*. A dark rock, it is composed mainly of hornblende and plagioclase. The alignment of the hornblende crystals produces a slightly foliated texture. Many amphibolites result from intermediate- to high-grade metamorphism of basalt and ferromagnesian-rich mafic rocks.

In some areas of regional metamorphism, exposures of "mixed rocks" called *migmatites*, having both igneous and high-grade metamorphic characteristics, are present (Figure 8.15). Migmatites are thought to result from the extremely high temperatures produced during metamorphism. Part of the problem in determining the origin of migmatites, however, is explaining how the granitic component formed. According to one model, the granitic magma formed in place by the partial melting of rock during intense metamorphism. Such an origin is possible provided that the host rocks contained quartz and feldspars and that water was present.

Others argue that the characteristic layering or wavy appearance of migmatites arises by the redistribution of minerals during recrystallization in the solid state—that is, through purely metamorphic processes.

Nonfoliated Metamorphic Rocks

In some metamorphic rocks, the mineral grains do not show a discernible preferred orientation. Instead, these rocks consist of a mosaic of roughly equidimensional minerals and are characterized as having a **nonfoliated texture** (Figure 8.16). Most nonfoliated metamorphic rocks result from contact or regional metamorphism of rocks with no platy or elongate minerals. Frequently, the only indication

Figure 8.15 Migmatite A migmatite boulder in Rocky Mountain National Park, near Estes Park, Colorado. Migmatites consist of high-grade metamorphic rock intermixed with streaks or lenses of granite.

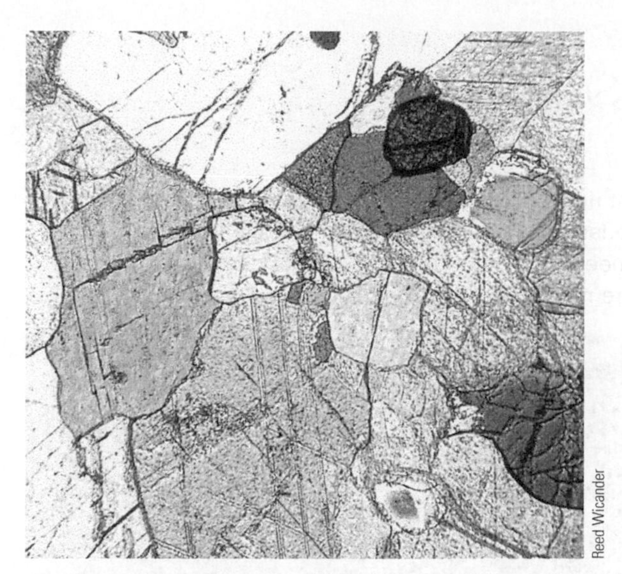

❚ Figure 8.16 **Nonfoliated Texture** Nonfoliated textures are characterized by a mosaic of roughly equidimensional minerals, as in this photomicrograph of marble.

that a granular rock has been metamorphosed is the large grain size resulting from recrystallization.

Nonfoliated metamorphic rocks are generally of two types: those composed of mainly one mineral—for example, marble or quartzite—and those in which the different mineral grains are too small to be seen without magnification, such as greenstone and hornfels.

Marble is a well-known metamorphic rock composed predominantly of calcite or dolomite; its grain size ranges from fine to coarsely granular. Marble results from either contact or regional metamorphism of limestones or dolostones (❚ Figure 8.17 and Table 8.1). Pure marble is snowy white or bluish; however, many color varieties exist because of the presence of mineral impurities in the original sedimentary rock. The softness of marble, its uniform texture, and its varying colors have made it the favorite rock of builders and sculptors throughout history.

Quartzite is a hard, compact rock typically formed from quartz sandstone under intermediate- to high-grade metamorphic conditions during contact or regional metamorphism (❚ Figure 8.18). Because recrystallization is so complete, metamorphic quartzite is of uniform strength and therefore usually breaks across the component quartz grains rather than around them when it is struck. Pure quartzite is white; however, iron and other impurities commonly impart a pinkish-red or other color to it. Quartzite is commonly used as foundation material for road and railway beds.

The name *greenstone* is applied to any compact, dark-green, altered, mafic igneous rock that formed under low- to high-grade metamorphic conditions. The green color results from the presence of chlorite, epidote, and hornblende.

Hornfels is a common, fine-grained, nonfoliated metamorphic rock resulting from contact metamorphism and composed of various equidimensional mineral grains. The composition of hornfels depends directly on the composition of the original rock, and many compositional varieties are known. Most hornfels, however, are apparently derived from contact metamorphism of clay-rich sedimentary rocks or impure dolostones.

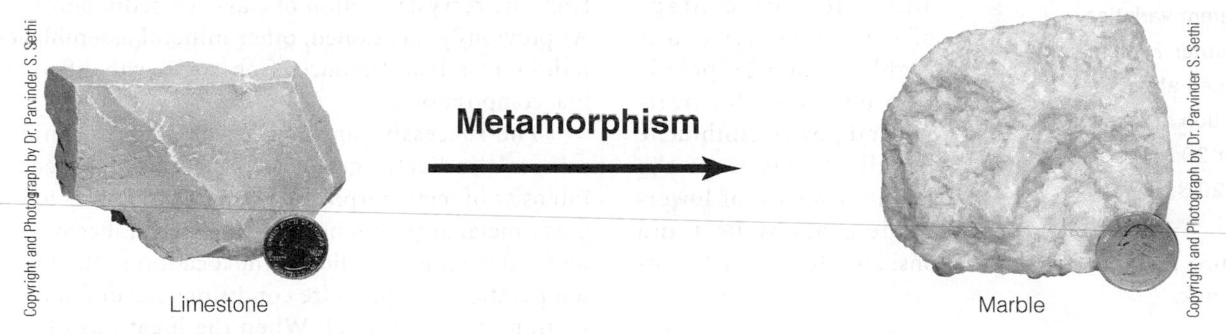

❚ Figure 8.17 **Marble** Metamorphism of the sedimentary rock limestone or dolostone yields marble.

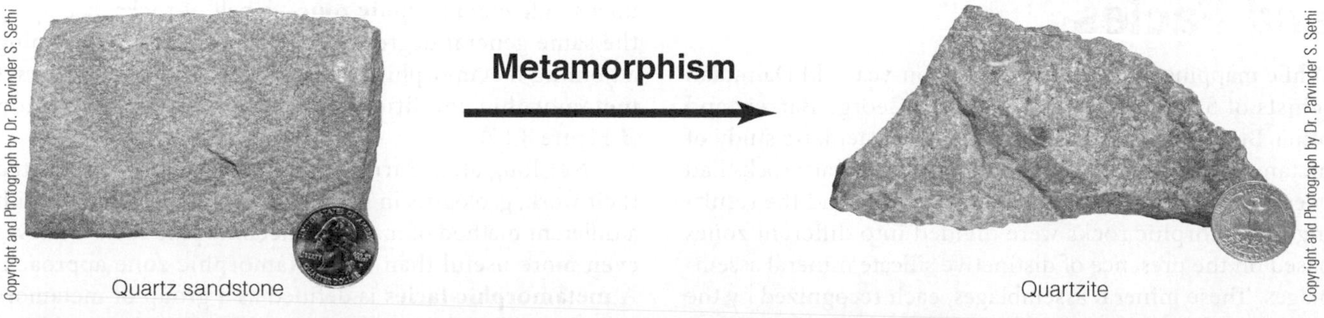

❚ Figure 8.18 **Quartzite** Metamorphism of the sedimentary rock quartz sandstone yields quartzite.

Marble as a Building Stone

Marble has been used for centuries as a facing and main stone in buildings and structures. A major utility in your state wants to build a dam and use the impounded water to run a hydroelectric plant to generate needed electricity for the region. The building site for the dam is in an area of metamorphic rocks, and specifically the dam will be built where the valley walls and subsurface rock are marble.

Because marble is a dense, nonfoliated metamorphic rock, and there has been no evidence of earthquake activity in the region, it seems that this would make an ideal site for the dam. However, there are many reasons why marble is not necessarily a good choice to use as the foundational material for a dam. What are some of the factors the engineers charged with designing this dam should consider before making the decision to go ahead with construction?

Migel/Shutterstock.com

The Vaiont Dam, located in Belluno Province, Italy, is the site of a major landslide that killed nearly 3,000 people on October 9, 1963. It was built in a glacial valley underlain by thick layers of folded and faulted limestone and clay layers as seen in this photo of the dam. Active solution of the limestones by slightly acidic groundwater helped weaken the rocks against which the dam was built. Marble results from the metamorphism of limestone or dolostone, and thus is equally susceptible to solution by acidic groundwater.

ConnectionLink

To learn more about the formation of coal and its different varieties, see Chapter 7. More can be learned about the distribution of coal in Chapter 20 and Figure 20.31.

Anthracite is a black, lustrous, hard coal that contains a high percentage of fixed carbon and a low percentage of volatile matter and is highly valued by people who burn coal for heating and power. Anthracite usually forms from the metamorphism of lower-grade coals by heat and pressure, and many geologists consider it to be a metamorphic rock.

8.5 Metamorphic Zones and Facies

While mapping the 440- to 400-million-year-old Dalradian schists of Scotland in the late 1800s, George Barrow and other British geologists made the first systematic study of metamorphic zones. Here, clay-rich sedimentary rocks had been subjected to regional metamorphism, and the resulting metamorphic rocks were divided into different zones based on the presence of distinctive silicate mineral assemblages. These mineral assemblages, each recognized by the presence of one or more index minerals, indicate different degrees of metamorphism. The index minerals that Barrow and his associates chose to represent increasing metamorphic intensity were chlorite, biotite, garnet, staurolite, kyanite, and sillimanite (Figure 8.9), which we now know all result from the recrystallization of clay-rich sedimentary rocks. As previously mentioned, other mineral assemblages and index minerals are produced from rocks with different original compositions.

The successive appearance of metamorphic index minerals indicates gradually increasing or decreasing intensity of metamorphism. Going from lower- to higher-grade metamorphic zones, the first appearance of a particular index mineral indicates the location of the minimum temperature and pressure conditions needed for the formation of that mineral. When the locations of the first appearances of that index mineral are connected on a map, the result is a line of equal metamorphic intensity or an *isograd*. The region between two adjacent isograds makes up a single **metamorphic zone**—a belt of rocks displaying the same general degree of metamorphism. By mapping adjoining metamorphic zones, geologists can reconstruct metamorphic conditions throughout an entire region (▮ Figure 8.19).

Not long after Barrow and his coworkers completed their work, geologists in Norway and Finland came up with a different method of mapping metamorphic rocks that was even more useful than the metamorphic zone approach. A **metamorphic facies** is defined as a group of metamorphic rocks characterized by particular mineral assemblages

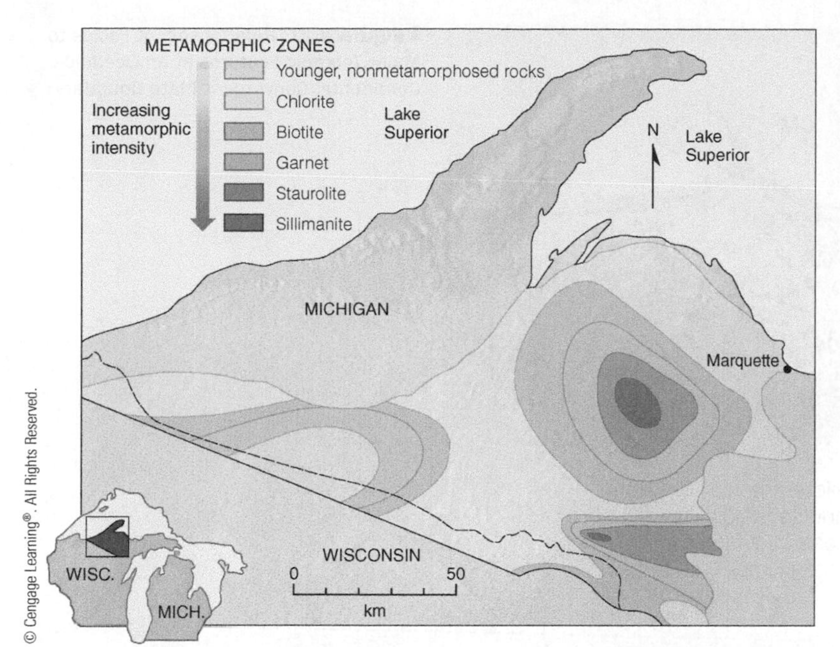

Figure 8.19 Metamorphic Zones in the Upper Peninsula of Michigan The zones in this region are based on the presence of distinctive silicate mineral assemblages resulting from the metamorphism of sedimentary rocks during an interval of mountain building and minor granitic intrusion during the Proterozoic Eon, about 1.5 billion years ago. The lines separating the different metamorphic zones are isograds.

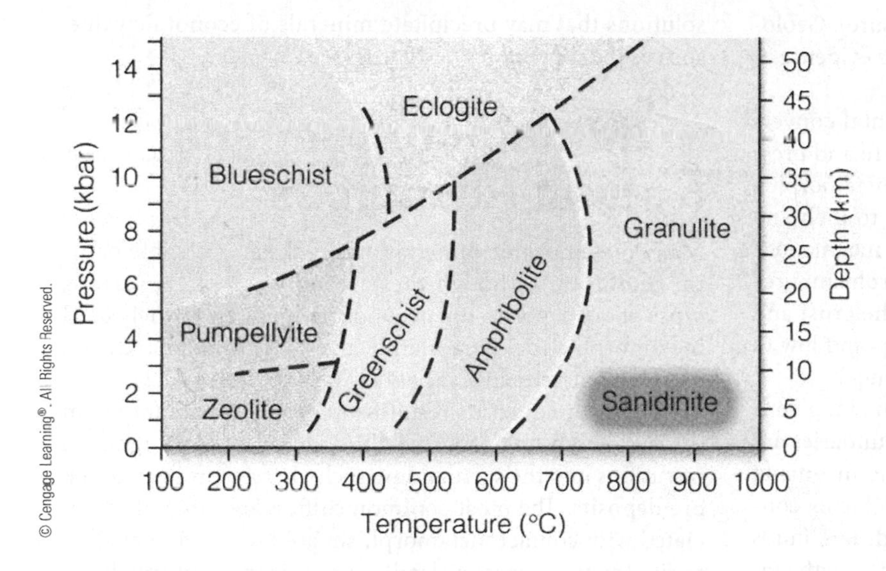

Figure 8.20 Metamorphic Facies and Their Associated Temperature–Pressure Conditions A temperature–pressure diagram showing under what conditions various metamorphic facies occur. A metamorphic facies is characterized by a particular mineral assemblage that formed under the same broad temperature pressure conditions. Each facies is named after its most characteristic rock or mineral.

Critical Thinking Question Go to a point that is represented by 200°C and 2 kbar of pressure. What metamorphic facies is represented by those conditions? If the pressure is raised to 12 kbar, what facies is represented by the new conditions? What change in depth of burial is required to effect the pressure change from 2 to 12 kbar?

formed under broadly similar temperature and pressure conditions (Figure 8.20). Each facies is named after its most characteristic rock or mineral. For example, the green metamorphic mineral chlorite, which forms under relatively low temperatures and pressures, yields rocks belonging to the *greenschist facies*. Under increasingly higher temperatures and pressures, mineral assemblages indicative of the *amphibolite* and *granulite facies* develop.

Although usually applied to areas where the original rocks were clay-rich, the concept of metamorphic facies can also be used with modification in other situations. It cannot, however, be used in areas where the original rocks were pure quartz sandstone or pure limestone or dolostone. Such rocks, regardless of the imposed temperature and pressure conditions, will yield only quartzite and marble, respectively. In such cases, all one can say is that "metamorphism has happened."

8.6 Plate Tectonics and Metamorphism

Although metamorphism is associated with all three types of plate boundaries (see Figure 1.14), it is most common along convergent plate margins. Metamorphic rocks form at convergent plate boundaries because temperature and pressure increase as a result of plate collisions.

Figure 8.21 illustrates the various metamorphic facies produced along a typical oceanic–continental convergent plate boundary. When an oceanic plate collides with a continental plate, tremendous pressure is generated as the oceanic plate is subducted. Because rock is a poor heat conductor, the cold, descending oceanic plate heats slowly, and metamorphism is caused mostly by increasing pressure with depth. Metamorphism in such an environment produces rocks typical of

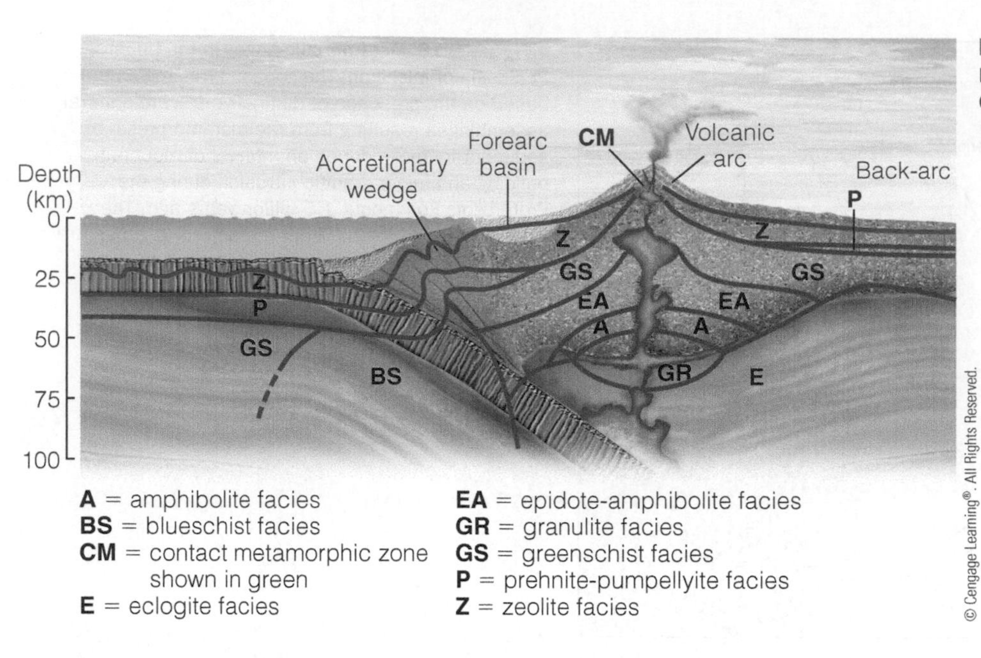

Figure 8.21 Relationship of Facies to Major Tectonic Features at an Oceanic–Continental Convergent Plate Boundary

the *blueschist facies* (low temperature, high pressure). Geologists use the presence of blueschist facies rocks as evidence of ancient subduction zones.

As subduction along the oceanic–continental convergent plate boundary continues, both temperature and pressure increase with depth and yield high-grade metamorphic rocks. Eventually, the descending plate begins to melt and generate magma that moves upward. This rising magma may alter the surrounding rock by contact metamorphism, producing migmatites in the deeper portions of the crust and hornfels at shallower depths. High temperatures and low to medium pressures characterize such an environment.

Although metamorphism is most common along convergent plate margins, many divergent plate boundaries are characterized by contact metamorphism. Rising magma at mid-oceanic ridges heats the adjacent rocks, producing contact metamorphic minerals and textures. In addition, fluids emanating from the rising magma—and its reaction with seawater—very commonly produce metal-bearing hydrothermal solutions that may precipitate minerals of economic value, such as the copper ores of Cyprus (see Chapter 2).

8.7 Metamorphism and Natural Resources

Many metamorphic minerals and rocks are valuable natural resources. Although these resources include various types of ore deposits, the two most familiar and widely used metamorphic rocks are marble and slate, which have been used for centuries in a variety of ways (Figure 8.22).

Many ore deposits result from contact metamorphism during which hot, ion-rich fluids migrate from igneous intrusions into the surrounding rock, thereby producing rich ore deposits. The most common sulfide ore minerals associated with contact metamorphism are bornite and chalcopyrite (copper), galena (lead), pyrite (iron), and sphalerite (zinc); two common iron oxide ore minerals are hematite

Figure 8.22 Slate Quarry in Wales Slate, which has a variety of uses, is the result of low-grade regional metamorphism of shale. These high-quality slates were formed by a mountain-building episode that took place approximately 400–440 million years ago in the present-day countries of Ireland, Scotland, Wales, and Norway.

TABLE 8.2 The Main Ore Deposits Resulting from Contact Matamorphism

Ore Deposit	Major Mineral	Formula	Use
Copper	Bornite, Chalcopyrite	Cu_5FeS_4, $CuFeS_2$	Important sources of copper, which is used in manufacturing, transportation, communications, and construction
Iron	Hematite, Magnetite	Fe_2O_3, Fe_3O_4	Major sources of iron for manufacture of steel, which is used in nearly every form of construction, manufacturing, transportation, and communications
Lead	Galena	PbS	Chief source of lead, which is used in batteries, pipes, solder, and elsewhere where resistance to corrosion is required
Tin	Cassiterite	SnO_2	Principal source of tin, which is used for tin plating, solder, alloys, and chemicals
Tungsten	Scheelite, Wolframite	$CaWO_4$, $(FeMn)WO_4$	Chief sources of tungsten, which is used in hardening metals and manufacturing carbides
Zinc	Sphalerite	(Zn,Fe)S	Major source of zinc, which is used in batteries and in galvanizing iron and making brass

and magnetite. Tin and tungsten are also important ores associated with contact metamorphism (Table 8.2).

Other economically important metamorphic minerals include asbestos, used for insulation and fire-proofing in buildings and building materials (Geo-Focus 8.1), talc for talcum powder, graphite for pencils and dry lubricants, and garnets and corundum, which are used as abrasives or gemstones, depending on their quality. In addition, andalusite, kyanite, and sillimanite, which all have the same chemical composition but differ in crystal structure, are used in manufacturing high-temperature porcelains and temperature-resistant materials for such products as sparkplugs and furnace linings.

Key Concepts Review

- Metamorphic rocks result from the transformation of other rocks, usually beneath Earth's surface, as a consequence of one, or a combination, of three agents: heat, pressure, and fluid activity.
- Heat for metamorphism comes from intrusive magmas, extrusive lava flows, or deep burial. Pressure is either lithostatic (uniformly applied stress) or differential (stress unequally applied from different directions). Fluids trapped in sedimentary rocks or emanating from intruding magmas can enhance chemical changes and the formation of new minerals.
- The three major types of metamorphism are contact, dynamic, and regional.
- Contact metamorphism results when magma or lava alters the surrounding country rock.
- Dynamic metamorphism is associated with fault zones where rocks are subjected to high differential pressure.
- Most metamorphic rocks result from regional metamorphism, which occurs over a large area and is usually caused by tremendous temperatures, pressures, and deformation within the deeper portions of the crust.
- Metamorphic grade generally characterizes the degree to which a rock has undergone metamorphic change.
- Index minerals—minerals that form only within specific temperature and pressure ranges—allow geologists to recognize low-, intermediate-, and high-grade metamorphism.
- Metamorphic rocks are primarily classified according to their texture. In a foliated texture, platy and elongate minerals have a preferred orientation. A nonfoliated texture does not exhibit any discernible preferred orientation of the mineral grains.
- Foliated metamorphic rocks can be arranged in order of increasing grain size, perfection of their foliation, or both. Slate is fine-grained, followed by (in increasingly larger grain size) phyllite and schist; gneiss displays segregated bands of minerals. Amphibolite is another fairly common foliated metamorphic rock. Migmatites have both igneous and high-grade metamorphic characteristics.
- Marble, quartzite, greenstone, hornfels, and anthracite are common nonfoliated metamorphic rocks.
- Metamorphic zones are based on index minerals and are areas of rock that all have similar grades of metamorphism; that is, they have all experienced the same intensity of metamorphism.
- A metamorphic facies is a group of metamorphic rocks whose minerals all formed under a particular range of temperatures and pressures. Each facies is named after its most characteristic rock or mineral.
- Metamorphism occurs along all three types of plate boundaries, but it is most common at convergent plate margins.
- Many metamorphic rocks and minerals, such as marble, slate, graphite, talc, and asbestos, are valuable natural resources. In addition, many ore deposits are the result of metamorphism and include copper, tin, tungsten, lead, iron, and zinc.

Important Terms

aureole (p. 186)
contact (thermal) metamorphism (p. 186)
differential pressure (p. 185)
dynamic metamorphism (p. 188)
fluid activity (p. 185)
foliated texture (p. 189)
heat (p. 183)
index mineral (p. 188)
lithostatic pressure (p. 183)
metamorphic facies (p. 194)
metamorphic grade (p. 188)
metamorphic rock (p. 182)
metamorphic zone (p. 194)
metamorphism (p. 182)
nonfoliated texture (p. 192)
regional metamorphism (p. 188)

Review Questions

1. From which of the following rock groups can metamorphic rocks form?
   a. ___ volcanic.
   b. ___ sedimentary.
   c. ___ plutonic.
   d. ___ metamorphic.
   e. ___ all of these.
2. Which is the correct metamorphic sequence of increasingly coarser grain size?
   a. ___ gneiss → schist → phyllite → slate.
   b. ___ phyllite → slate → schist → gneiss.
   c. ___ schist → slate → gneiss → phyllite.
   d. ___ slate → phyllite → schist → gneiss.
   e. ___ slate → schist → phyllite → gneiss.
3. Metamorphism is most common along what type of plate boundary?
   a. ___ divergent.
   b. ___ transform.
   c. ___ lithospheric.
   d. ___ aseismic.
   e. ___ convergent.
4. Which of the following are the three agents of metamorphism?
   a. ___ gravity, fluid activity, pressure.
   b. ___ heat, pressure, gravity.
   c. ___ gravity, heat, fluid activity.
   d. ___ heat, pressure, fluid activity.
   e. ___ none of these.
5. Concentric zones surrounding an igneous intrusion and characterized by distinctive mineral assemblages are
   a. ___ aureoles.
   b. ___ hydrothermal regions.
   c. ___ regional facies.
   d. ___ thermodynamic rings.
   e. ___ metamorphic zones.
6. How do metamorphic rocks record the influence of differential pressure in their structures and mineral textures?
7. Why is it important for people to know something about metamorphism, metamorphic rocks, and how they form?
8. If plate tectonic movement did not exist, could there be metamorphism?
9. Describe the two types of metamorphic texture, and discuss how they are produced.
10. **Creative Thinking Visual Question:** Foliated metamorphic rocks are characterized by having their minerals arranged in a parallel fashion, such that they can be split along foliation planes (Figure 8.10). In this photo (▌Figure 1) of 500-million- to 1-billion-year-old schists and gneisses of the Blue Ridge Belt in southwestern Virginia, the foliation planes are dipping in the same direction as the slope of the hillside. What problem does this present in terms of potential landslides along the roadway? Hint: Note the several large rocks along the base of the hillside.

▌**Figure 1** Schists and gneisses in the Blue Ridge Belt, Virginia.

Global GeoScience Watch Asbestos is a valuable metamorphic mineral that is used in building materials such as roof shingles and floor tiles, for fireproofing, and in friction products such as brake linings. Although some countries continue to use it, many countries have banned the use of asbestos because of its adverse health effects. Search for "asbestos" in the GREENR database, then look under the "Images" section and click on "The Threat from Deadly Asbestos" image. Who are the top producers, consumers, and exporters of asbestos? How many countries have banned or restricted the use of asbestos because it is a known carcinogen? Why do the top importers of asbestos still use it? Write a short report answering these questions.

The 7.0-magnitude earthquake that struck the island nation of Haiti on January 12, 2010, destroyed its capital city, Port-au-Prince, and devastated the surrounding areas, killing more than 222,500 people.

U.S. Navy

CHAPTER

9 Earthquakes and Earth's Interior

OUTLINE

HAVE YOU EVER WONDERED?

- What causes earthquakes?
- Why earthquakes occur where they do?
- How an earthquake's epicenter is located?
- How the strength of an earthquake is measured?
- Whether earthquakes can be predicted or even controlled?
- What you can do to protect yourself during an earthquake?
- How geologists know what Earth's internal structure is like?

9.1 Introduction

In the afternoon of January 12, 2010, a magnitude 7.0 earthquake struck the island nation of Haiti. According to official estimates, 222,570 people died, at least 300,000 were injured, and more than 285,000 residences and businesses were destroyed or severely damaged. Widespread devastation occurred in the capital of Port-au-Prince and elsewhere throughout the region, exacerbated by an almost total collapse of the vital infrastructure needed to respond to such a disaster, including medical, transportation, and communication systems. This was a disaster of truly epic proportions, yet it was not the first, nor will it be the last major devastating earthquake in this region or many other parts of the world.

A little more than a year later, on March 11, 2011, a 9.0-magnitude earthquake and tsunami struck Japan, causing more than 20,000 deaths and also resulting in tremendous destruction, including severe damage to a nuclear power plant in the northeastern part of the island. The cause of the Japanese earthquake resulted from movement along an oceanic–oceanic convergent plate boundary off the eastern coast of Japan, whereas the Haiti earthquake was movement along a strike-slip fault that is part of the boundary separating the Caribbean plate from the North American plate.

ConnectionLink

To review more about the causes and damage resulting from the 2010 and 2011 Haiti and Japanese earthquakes, see Geo-Insight 2.1.

Earthquakes, along with volcanic eruptions, are manifestations of Earth's dynamic and active makeup. As one of nature's most frightening and destructive phenomena, earthquakes have always aroused feelings of fear and have been the subject of myths and legends. What makes an earthquake so frightening is that when it begins, there is no way to tell how long it will last or how violent it will be. Approximately 13 million people have died in earthquakes during the past 4,000 years, with about 3 million of these deaths occurring during the last century (Table 9.1).

Geologists define an **earthquake** as the shaking or trembling of the ground caused by the sudden release of energy, usually as a result of faulting, which involves the displacement of rocks along fractures. After an earthquake, continuing adjustments along a fault may generate a series of earthquakes known as *aftershocks*. Most aftershocks are smaller than the main shock, but they can still cause considerable damage to already weakened structures.

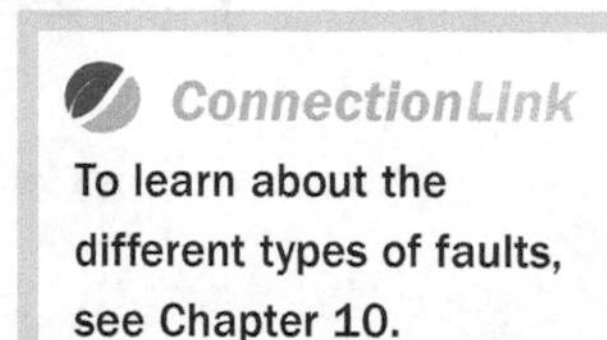

To learn about the different types of faults, see Chapter 10.

Although the geologic definition of an earthquake is accurate, it is not nearly as imaginative or colorful as the

TABLE 9.1 Some Significant Earthquakes

Year	Location	Magnitude (estimated before 1935)	Deaths (estimated)
1556	China (Shanxi Province)	8.0	1,000,000
1755	Portugal (Lisbon)	8.6	70,000
1906	U.S.A. (San Francisco, California)	8.3	3,000
1923	Japan (Tokyo)	8.3	143,000
1960	Chile	9.5	5,700
1964	U.S.A. (Anchorage, Alaska)	8.6	131
1976	China (Tangshan)	8.0	242,000
1985	Mexico (Mexico City)	8.1	9,500
1988	Armenia	6.9	25,000
1990	Iran	7.3	50,000
1993	India	6.4	30,000
1995	Japan (Kobe)	7.2	>6,000
1999	Turkey	7.4	17,000
2001	India	7.9	>14,000
2003	Iran	6.6	43,000
2004	Indonesia	9.0	>230,000
2005	Pakistan	7.6	>86,000
2006	Indonesia	6.3	>6,200
2008	China (Sichuan Province)	7.9	>69,000
2010	Haiti	7.0	>222,500
2011	New Zealand	6.3	181
2011	Japan	9.0	>20,000
2012	Iran	6.4	>300

explanations many people had in the past. Some cultures attributed the cause of earthquakes to movements of some kind of animal on which Earth rested. A legend from Mexico holds that earthquakes occur when the devil, El Diablo, rips open the crust so that he and his friends can reach the surface.

Even though we know that most earthquakes result from energy released along plate boundaries, you might still be asking, "Why should you study earthquakes, particularly if you do not live anywhere near a plate boundary or an area prone to earthquakes?" The obvious reason is that they are destructive and cause many deaths and injuries to the people living in earthquake-prone areas. Earthquakes also affect the economies of many countries in terms of cleanup costs, lost jobs, and lost business revenues. From a purely personal standpoint, you someday may be caught in an earthquake. Even if you do not live in an area subject to earthquakes, you may travel to places where the threat of earthquakes exists, and you should know what to do if you experience one. Such knowledge may help you avoid serious injury or even death.

9.2 Elastic Rebound Theory

Based on studies conducted after the 1906 San Francisco earthquake, H. F. Reid of The Johns Hopkins University proposed the **elastic rebound theory** to explain how energy is released during earthquakes. Reid studied three sets of measurements taken across a portion of the San Andreas Fault that had broken during the 1906 earthquake. The measurements revealed that points on opposite sides of the fault had moved 3.2 m during the 50-year period before breakage in 1906, with the west side moving northward (▌Figure 9.1).

According to Reid, rocks on opposite sides of the San Andreas Fault had been storing energy and bending slightly for at least 50 years before the 1906 earthquake. Any straight line, such as a fence or road that crossed the San Andreas Fault, was gradually bent, because rocks on one side of the fault moved relative to rocks on the other side (Figure 9.1).

▌**Figure 9.1** The Elastic Rebound Theory

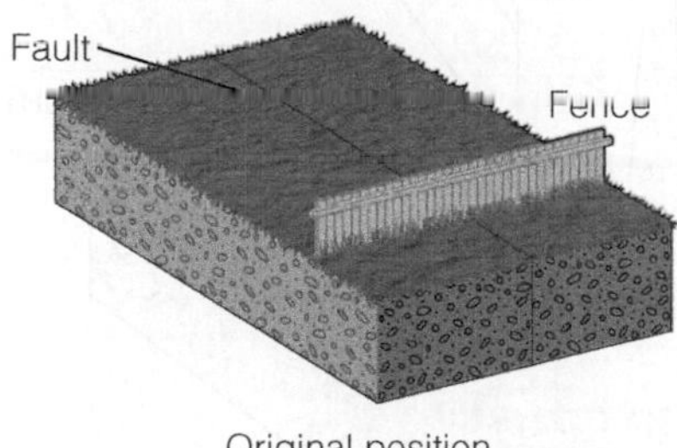

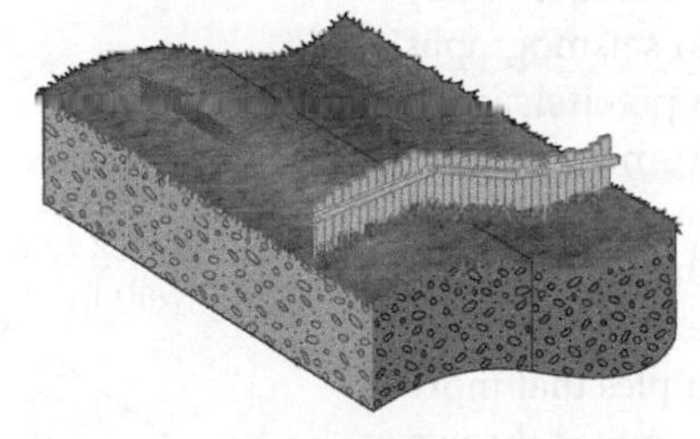

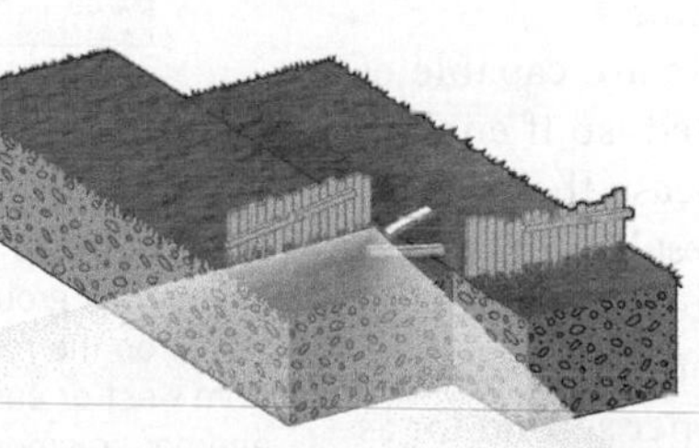

a According to the elastic rebound theory, rocks experiencing deformation store energy and bend. When the internal strength of the rocks is exceeded, they rupture, releasing their accumulated energy, and "snap back" or rebound to their former undeformed shape. This sudden release of energy is what causes an earthquake.

b During the 1906 San Francisco earthquake, this fence in Marin County was displaced by almost 5 m. Whereas many people would see a broken fence, a geologist sees that the fence has moved or been displaced and would look for evidence of a fault. A geologist would also notice that the ground has been displaced toward the right side, relative to his or her view. Regardless of what side of the fault you stand on, you must look to the right to see the other part of the fence. Try it!

Eventually, the strength of the rocks was exceeded, and they then fractured. After fracturing, the rocks on both sides of the fault rebounded or "snapped back" to their former undeformed shape, and the energy stored was released as earthquake waves radiating outward from the break (Figure 9.1). Additional field and laboratory studies conducted by Reid and others have confirmed that elastic rebound is the mechanism by which energy is released during earthquakes.

A useful analogy is that of bending a long, straight stick over your knee. As the stick bends, it deforms and eventually reaches the point at which it breaks. When this happens, the two pieces of the original stick snap back into their original straight position. Likewise, rocks subjected to intense forces bend until they break, and they then return to their original position, releasing energy in the process.

9.3 Seismology

Seismology, the study of earthquakes, emerged as a true science during the 1880s with the development of **seismographs**, instruments that detect, record, and measure the vibrations produced by an earthquake (▌Figure 9.2). The record made by a seismograph is called a *seismogram*. Modern seismographs have electronic sensors and record movements precisely using computers rather than simply relying on the drum strip charts commonly used on older seismographs.

When an earthquake occurs, energy in the form of *seismic waves* radiates out from the point of release (▌Figure 9.3a). These waves are somewhat analogous to the ripples that move out concentrically from the point at which a stone is thrown into a pond. Unlike waves on a pond, however, seismic waves move outward in all directions from their source.

Earthquakes take place because rocks are capable of storing energy, but their strength is limited, so if enough force is present, they rupture and thus release their stored energy. In other words, most earthquakes result when movement occurs along faults, most of which are related, at least indirectly, to plate movements. Once rupturing begins, it moves along the fault at several kilometers per second for as long as conditions for failure exist. The longer the fault along which movement occurs, the more time it takes for the stored energy to be released, and therefore the longer the ground will shake. During some very large earthquakes, the ground might shake for three minutes, a seemingly brief time, but interminable if you are experiencing the earthquake firsthand.

The Focus and Epicenter of an Earthquake

The location within Earth's lithosphere where rupturing begins—that is, the point at which energy is first released—is an earthquake's **focus**, or *hypocenter*. What we usually hear in news reports, however, is the location of the **epicenter**, the point on Earth's surface directly above the focus (Figure 9.3a).

▌**Figure 9.2** Seismographs

a Seismographs record ground motion during an earthquake. The record produced is a seismogram. This seismograph records earthquakes on a strip of paper attached to a rotating drum.

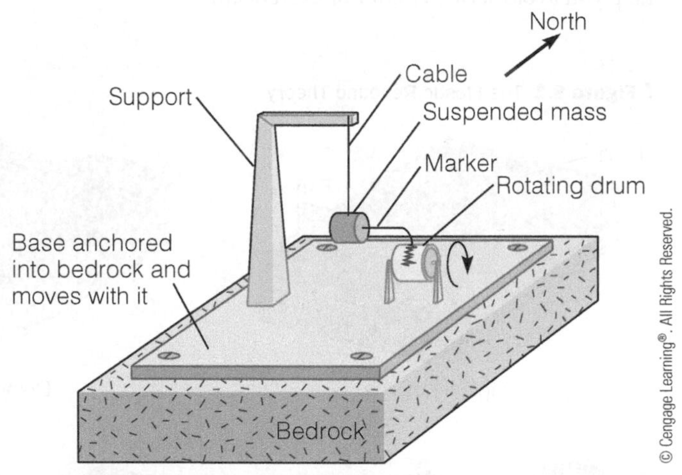

b A horizontal-motion seismograph. Because of its inertia, the suspended mass that contains the marker remains stationary, while the rest of the structure moves along with the ground during an earthquake. As long as the length of the arm is not parallel to the direction of ground movement, the marker will record the earthquake waves on the rotating drum. This seismograph would record waves from west or east, but to record waves from the north or south, another seismograph, at right angles to this one, is needed.

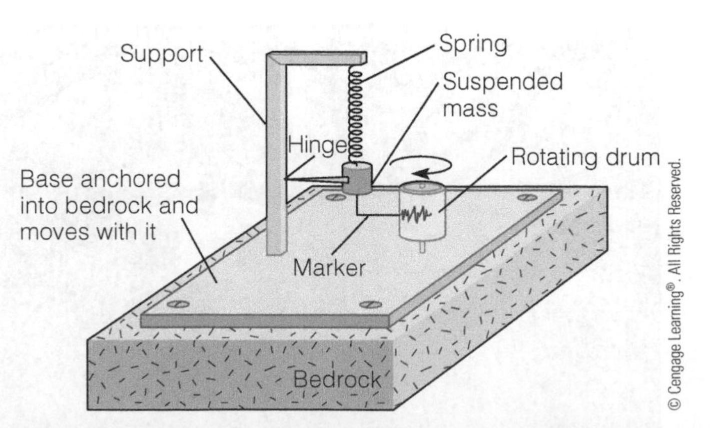

c A vertical-motion seismograph. This seismograph operates on the same principle as the horizontal-motion instrument and records vertical ground movement.

Figure 9.3 The Focus and Epicenter of an Earthquake

Critical Thinking Question Why isn't the epicenter located where the fault emerges at Earth's surface?

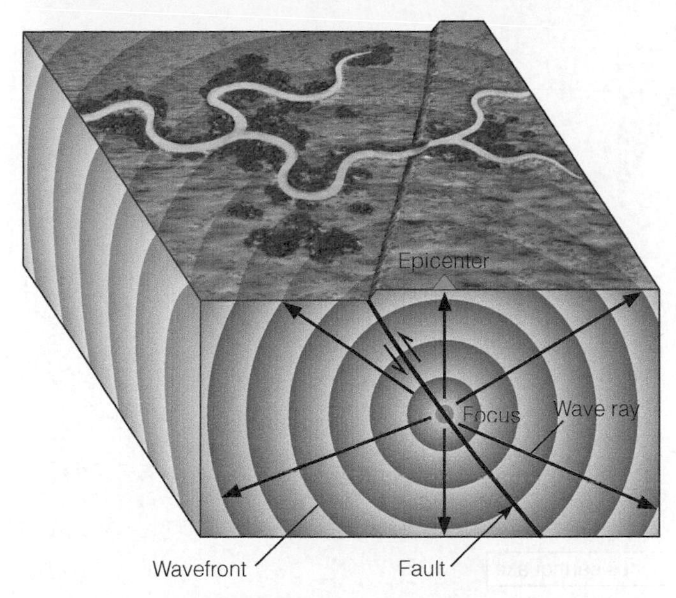

a The focus of an earthquake is the location where the rupture begins and energy is released. The place on the surface vertically above the focus is the epicenter. Seismic wavefronts move out in all directions from their source, the focus of an earthquake.

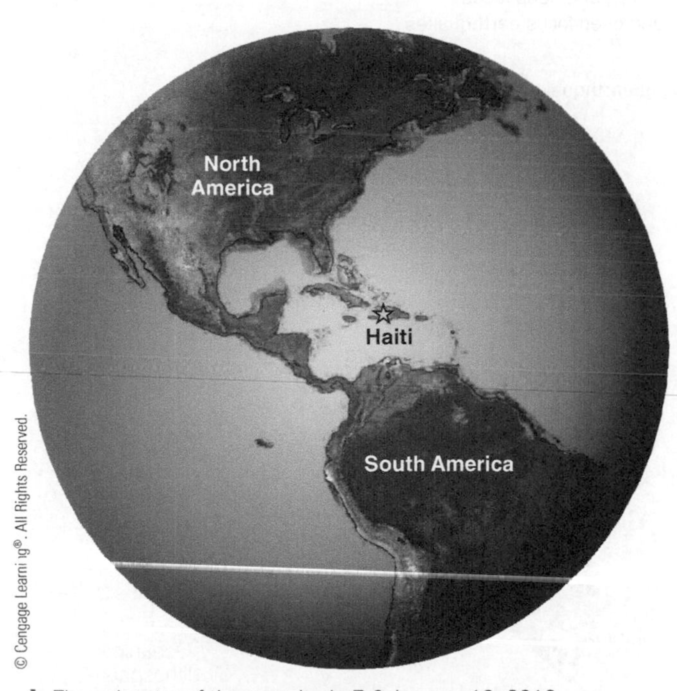

b The epicenter of the magnitude 7.0 January 12, 2010, earthquake that devastated Haiti was located approximately 25 km west-southwest of Port-au-Prince, and had a focal depth of 13 km.

Seismologists recognize three categories of earthquakes based on focal depth. *Shallow-focus* earthquakes have focal depths of less than 70 km from the surface, whereas those with foci between 70 and 300 km are *intermediate-focus earthquakes*, and the foci of those characterized as *deep-focus earthquakes* are more than 300 km deep. However, earthquakes are not evenly distributed among these three categories. Approximately 90% of all earthquake foci are at depths of less than 100 km, whereas only about 3% of all earthquakes are deep-focus. Shallow-focus earthquakes are, with few exceptions, the most destructive, because the energy they release has little time to dissipate before reaching the surface.

A definite relationship exists between earthquake foci and plate boundaries. Earthquakes generated along divergent or transform plate boundaries are invariably shallow-focus, whereas many shallow-focus earthquakes and nearly all intermediate- and deep-focus earthquakes occur along convergent margins (Figure 9.4). Furthermore, a pattern emerges when the focal depths of earthquakes near island arcs and their adjacent ocean trenches are plotted. Notice in Figure 9.5 that the focal depth increases beneath the Tonga Trench in a narrow, well-defined zone that dips approximately 45 degrees. Dipping seismic zones, called *Benioff* or *Benioff–Wadati zones*, are common to convergent plate boundaries where one plate is subducted beneath another. Such dipping seismic zones indicate the angle of plate descent along a convergent plate boundary.

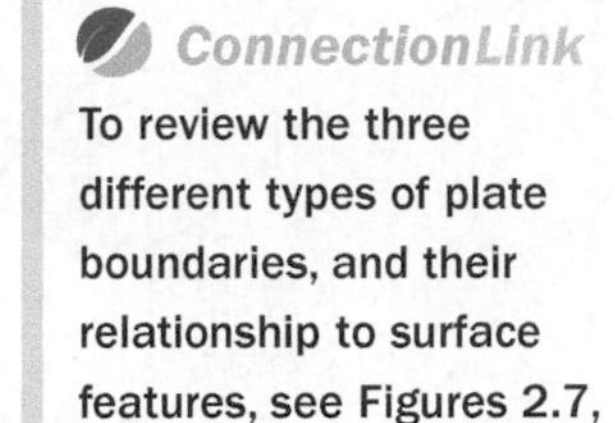

ConnectionLink

To review the three different types of plate boundaries, and their relationship to surface features, see Figures 2.7, 2.18, 2.20, and 2.22 in Chapter 2.

9.4 Where Do Earthquakes Occur, and How Often?

No place on Earth is immune to earthquakes, but almost 95% take place in seismic belts corresponding to plate boundaries where plates converge, diverge, and slide past each other. The relationship between plate margins and the distribution of earthquakes is readily apparent when the locations of earthquake epicenters are superimposed on a map showing the boundaries of Earth's plates (Figure 9.4).

The majority of all earthquakes (approximately 80%) occur in the *circum-Pacific belt*, a zone of seismic activity nearly encircling the Pacific Ocean basin. Most of these earthquakes result from convergence along plate margins, as in the case of the 1995 and 2011 Japanese earthquakes (Figure 9.6a). The earthquakes along the North American Pacific Coast, especially in California, are also in this belt, but these plates slide past one another rather than converge. The October 17, 1989, Loma Prieta earthquake in the San Francisco area and the January 17, 1994, Northridge earthquake occurred along this plate boundary (Figure 9.6b).

The second major seismic belt, accounting for 15% of all earthquakes, is the *Mediterranean–Asiatic belt*. This belt extends westward from Indonesia through the Himalayas,

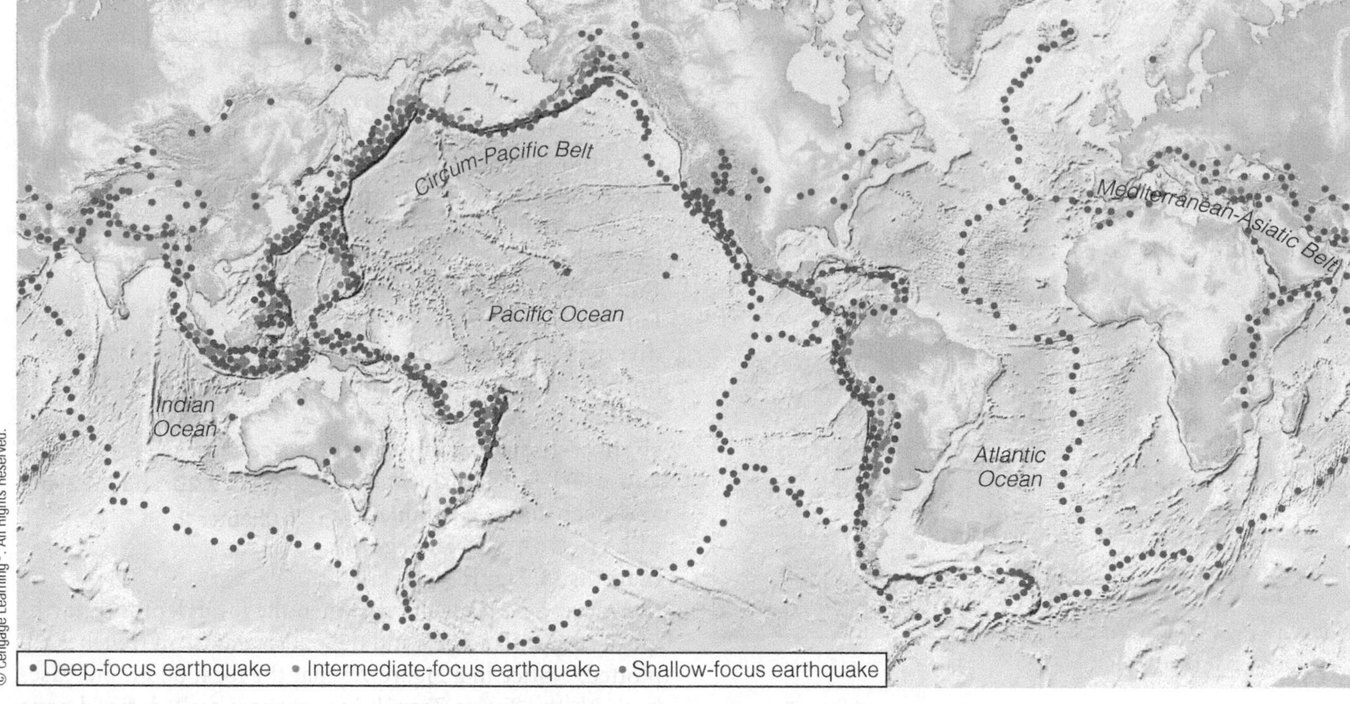

Figure 9.4 Earthquake Epicenters and Plate Boundaries This map of earthquake epicenters shows that most earthquakes occur within seismic zones that correspond closely to plate boundaries. Approximately 80% of earthquakes occur within the circum-Pacific belt, 15% within the Mediterranean–Asiatic belt, and the remaining 5% within plate interiors and along oceanic spreading ridges. The dots represent earthquake epicenters and are divided into shallow-, intermediate-, and deep-focus earthquakes. Along with shallow-focus earthquakes, nearly all intermediate- and deep-focus earthquakes occur along convergent plate boundaries.

Critical Thinking Question Why are nearly all intermediate- and deep-focus earthquakes associated with convergent plate boundaries?

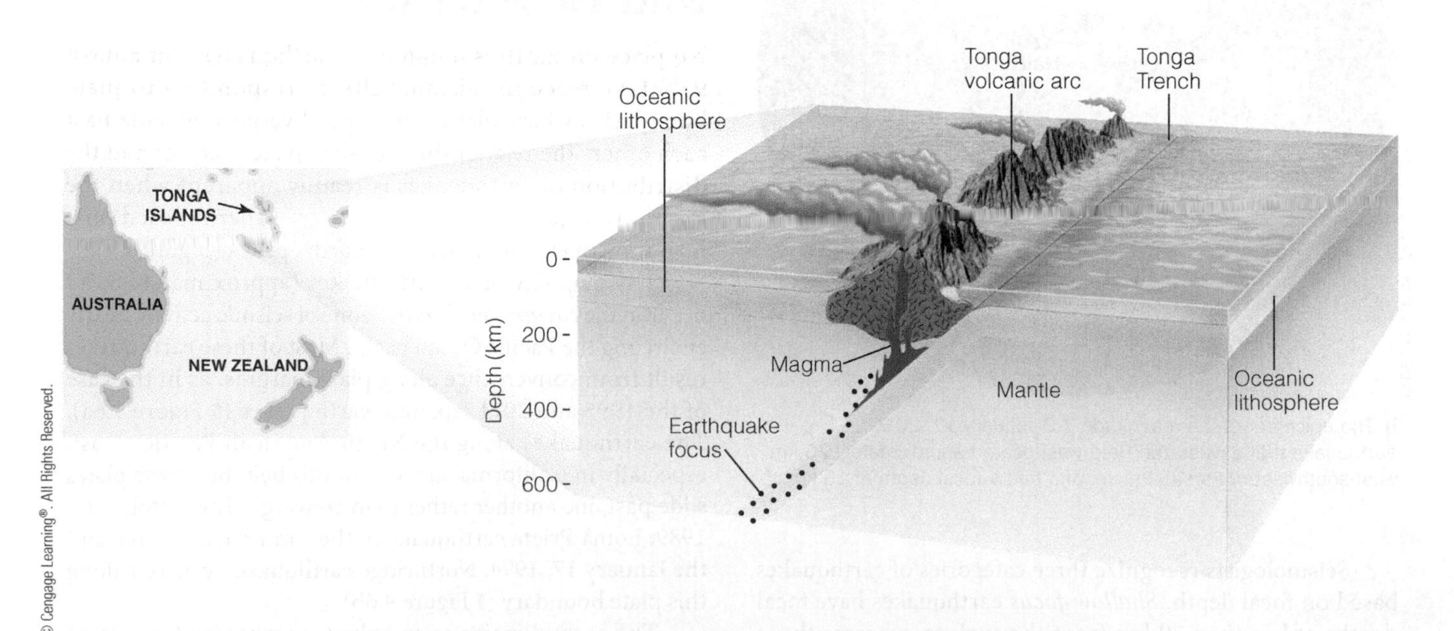

Figure 9.5 Benioff Zones Focal depth increases in a well-defined zone that dips approximately 45 degrees beneath the Tonga volcanic arc in the South Pacific. Dipping seismic zones are called *Benioff* or *Benioff-Wadati* zones.

Figure 9.6 Earthquake Damage in the Circum-Pacific Belt

Koichi Kamoshida/ZUMAPRESS/Newscom

a A collapsed road in Narahara, Fukushima Prefecture, caused by the 9.0-magnitude earthquake that struck Japan in 2011.

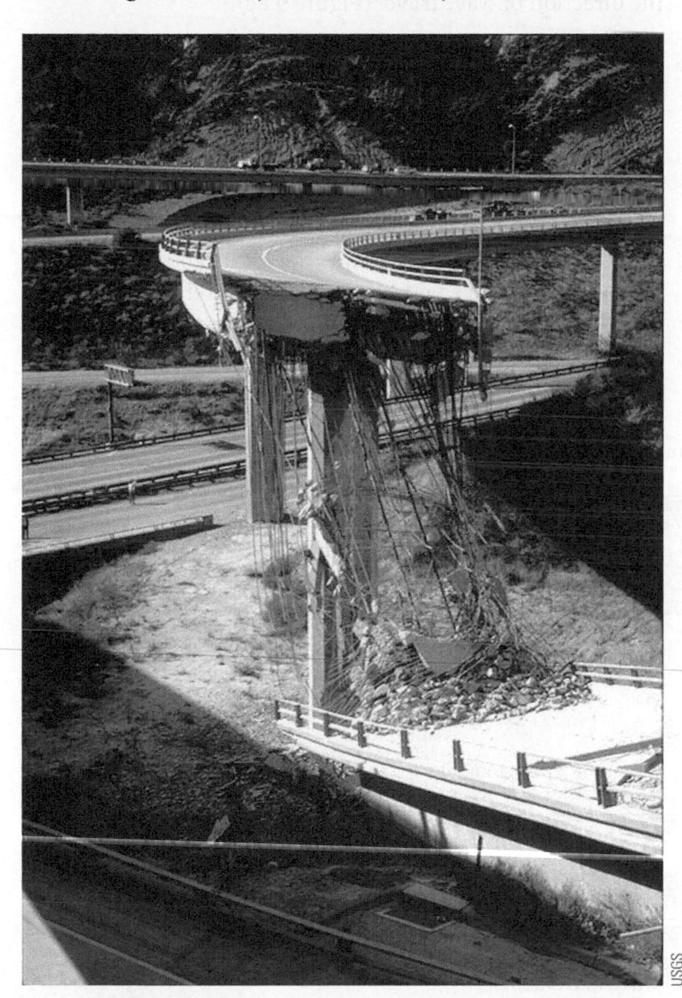

USGS

b One of several elevated freeway spans that collapsed during the January 1994 Northridge, California, earthquake in which 61 people were killed.

across Iran and Turkey, and westward through the Mediterranean region of Europe. The 2005 earthquake in Pakistan that killed more than 86,000 people is a recent example of the destructive earthquakes that strike this region (Table 9.1).

The remaining 5% of earthquakes occur mostly in the interiors of plates and along oceanic spreading-ridge systems. Most of these earthquakes are not strong, although there have been several major intraplate earthquakes of note. For example, the 1811 and 1812 earthquakes near New Madrid, Missouri, killed approximately 20 people and nearly destroyed the town. So strong were these earthquakes that they were felt from the Rocky Mountains to the Atlantic Ocean and from the Canadian border to the Gulf of Mexico. Within the immediate area, numerous buildings were destroyed and forests were flattened. The land sank several meters in some areas, causing flooding, and reportedly, the Mississippi River reversed its flow during the shaking and changed its course slightly.

Most recently, a magnitude 5.8 earthquake occurred 61 km northwest of Richmond, Virginia, on August 23, 2011, and was felt in a wide area along the East Coast. Although no casualties were reported, there was some minor damage to structures, including the Washington National Cathedral and the Smithsonian Castle in Washington, D.C. Albeit rare, earthquakes do occur on the East Coast of North America, and this earthquake was the largest to shake the eastern United States since 1897.

The cause of intraplate earthquakes is not well understood, but geologists think that they arise from localized stresses caused by the compression that most plates experience along their margins. A useful analogy is moving a house. Regardless of how careful the movers are, moving something so large without its internal parts shifting slightly is impossible. Similarly, plates are not likely to move without some internal stresses that occasionally cause earthquakes.

More than 900,000 earthquakes are recorded annually by the worldwide network of seismograph stations. Many of these are too small to be felt, but they are nonetheless recorded. These small earthquakes result from the energy released as continual adjustments take place between the various plates. On average, however, more than 31,000 earthquakes per year are strong enough to be felt and can cause various amounts of damage, depending on how strong they are and where they occur.

9.5 Seismic Waves

Many people have experienced an earthquake, but most are probably unaware that the shaking they feel and the damage to structures are caused by the arrival of *seismic waves*, a general term encompassing all waves generated by an earthquake. When movement on a fault takes place, energy is released in the form of two kinds of seismic waves that radiate outward in all directions from an earthquake's focus. *Body waves*, so called because they travel through the solid body of Earth, are somewhat like sound waves, and *surface waves*, which travel along the ground surface, are analogous to undulations or waves on water surfaces.

Body Waves

An earthquake generates two types of body waves: P-waves and S-waves (▮ Figure 9.7). **P-waves**, or *primary waves,* are the fastest seismic waves and can travel through solids, liquids, and gases. P-waves are compressional, or push-pull, waves and are similar to sound waves in that they move material forward and backward along a line in the same direction that the waves themselves are moving (Figure 9.7b). Thus, the material through which a P-wave travels is expanded and compressed as the wave moves through it and returns to its original size and shape after the wave passes by.

S-waves, or *secondary waves,* are somewhat slower than P-waves and can travel only through solids. S-waves are *shear waves* because they move the material perpendicular to the direction of travel, thereby producing shear stresses in the material they move through (Figure 9.7c). Because liquids (as well as gases) are not rigid, they have no shear strength, and S-waves cannot be transmitted through them.

The velocities of P- and S-waves are determined by the density and elasticity of the materials through which they travel. For example, seismic waves travel more slowly through rocks of greater density, but more rapidly through rocks with greater elasticity. *Elasticity* is a property of solids, such as rocks, and means that once they have been deformed by an applied force, they return to their original shape when the force is no longer present. Because P-wave velocity is greater than S-wave velocity in all materials, P-waves always arrive at seismic stations first.

Surface Waves

Surface waves travel along the surface of the ground, or just below it, and are slower than body waves (▮ Figure 9.8). Unlike the sharp jolting and shaking that body waves cause, surface waves generally produce a rolling or swaying motion, much like the experience of being on a boat.

Seismologists recognize several types of surface waves. The two most important are Rayleigh waves and Love waves, named after the British scientists who discovered them, Lord Rayleigh and A. E. H. Love. **Rayleigh waves (R-waves)** are generally the slower of the two and behave like water waves in that they move forward while the individual particles of material move in an elliptical path within a vertical plane oriented in the direction of wave movement (Figure 9.8b).

The motion of a **Love wave (L-wave)** is similar to that of an S-wave, but the individual particles of the material move only back and forth in a horizontal plane perpendicular to the direction of wave travel (Figure 9.8c).

9.6 Locating an Earthquake

We mentioned that news articles commonly report an earthquake's epicenter, but just how is the location of an epicenter determined? Once again, geologists rely on the study of seismic waves. We know that P-waves travel faster than S-waves,

▮ **Figure 9.7 Primary and Secondary Seismic Body Waves** Body waves travel through Earth.

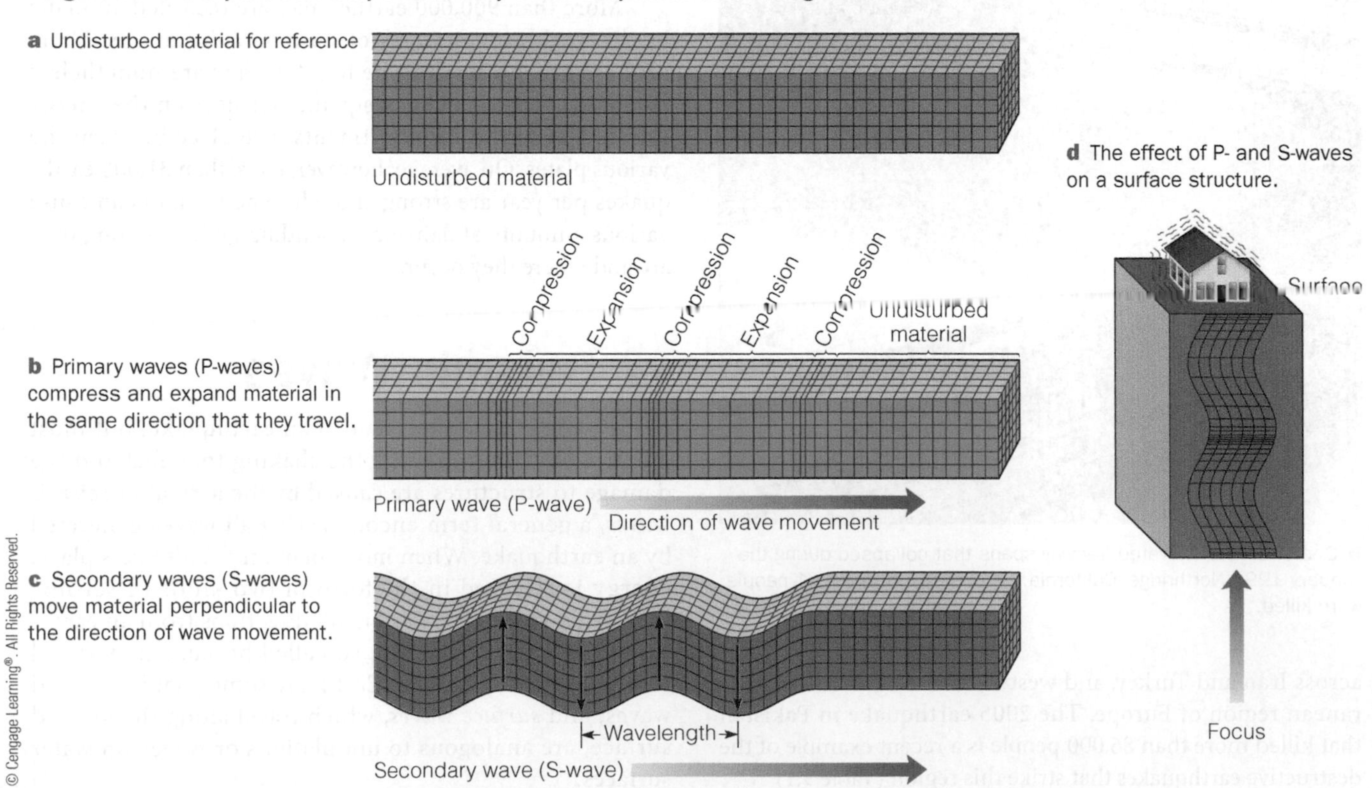

Figure 9.8 Rayleigh and Love Seismic Surface Waves Surface waves travel along Earth's surface or just below it.

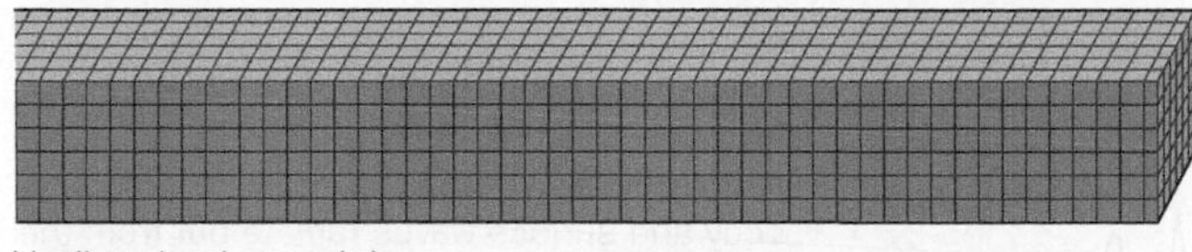

a Undisturbed material for reference.

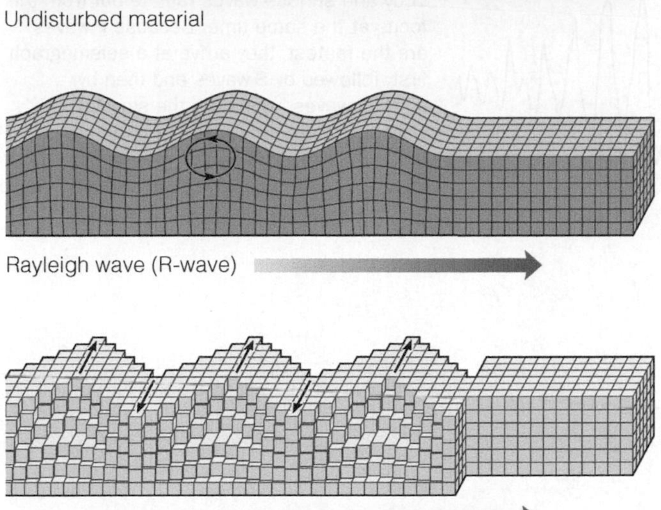

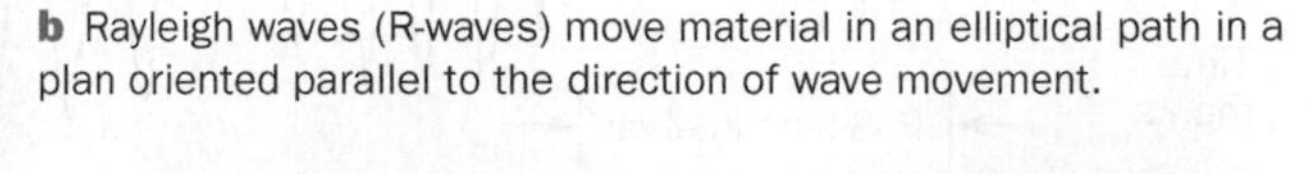

b Rayleigh waves (R-waves) move material in an elliptical path in a plan oriented parallel to the direction of wave movement.

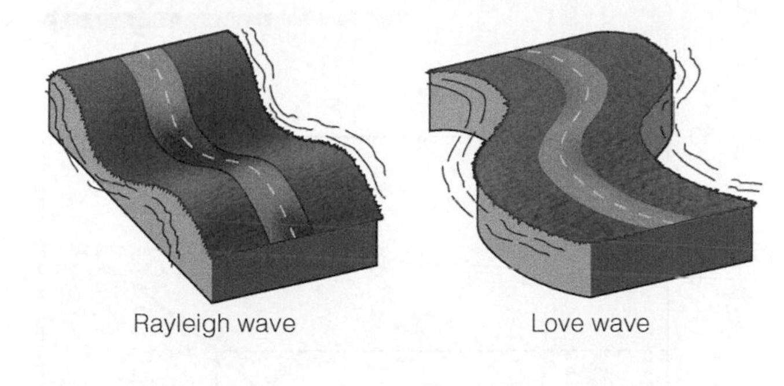

c Love waves (L-waves) move material back and forth in a horizontal plane perpendicular to the direction of wave movement.

d The arrival of R- and L-waves causes the surface to undulate and shake from side to side.

nearly twice as fast in all substances, so P-waves arrive at a seismograph station first, followed some time later by S-waves. Both P- and S-waves travel directly from the focus to the seismograph station through Earth's interior, but L- and R-waves arrive last because they are the slowest, and they also travel the longest route along the surface (❚ Figure 9.9a). Only the P- and S-waves need concern us, however, because they are the ones important in finding an epicenter.

Seismologists, who are geologists that study seismology, have accumulated tremendous amounts of data over the years and now know the average speeds of P- and S-waves for any specific distance from their source. These P- and S-wave travel times are published in *time–distance graphs* that illustrate the difference between the arrival times of the two waves as a function of the distance between a seismograph and an earthquake's focus (❚ Figure 9.9b). That is, the farther the waves travel, the greater the *P–S time interval*, which is simply the time difference between the arrivals of P- and S-waves (Figure 9.9a, b).

If the P–S time intervals are known from at least three seismograph stations, then the epicenter of any earthquake can be determined (❚ Figure 9.10). Here is how it works: subtracting the arrival time of the first P-wave from the arrival time of the first S-wave gives the P–S time interval for each seismic station. Each of these time intervals is plotted on a time–distance graph, and a line is drawn straight down to the distance axis of the graph, thus giving the distance from the focus to each seismic station (Figure 9.9b). Next, a circle whose radius equals the distance shown on the time–distance graph from each of the seismic stations is drawn on a map (Figure 9.10). The intersection of the three circles is the location of the earthquake's epicenter. It should be obvious from Figure 9.10 that P–S time intervals from at least three seismic stations are needed. If only one were used, the epicenter could be at any location on the circle drawn around that station, and using two stations would give two possible locations for the epicenter.

Determining the focal depth of an earthquake is much more difficult and considerably less precise than finding its epicenter. The focal depth is usually found by making computations based on several assumptions, comparing the results with those obtained at other seismic stations, and then recalculating and approximating the depth as closely as possible. Although the results are not highly accurate, they do tell us that most earthquakes, probably 75%, have foci no deeper than 10–15 km and that a few are as deep as 680 km.

9.7 Measuring the Strength of an Earthquake

Following any earthquake that causes extensive damage, fatalities, and injuries, graphic reports of the quake's violence and human suffering are common. Headlines tell us that thousands died, many more were injured or left homeless, and property damage was in the millions and possibly billions of dollars. Although such descriptions of fatalities and damage give some indication of the size of an earthquake, geologists are interested in more reliable methods for determining an earthquake's size.

Figure 9.9 Determining the Distance from an Earthquake

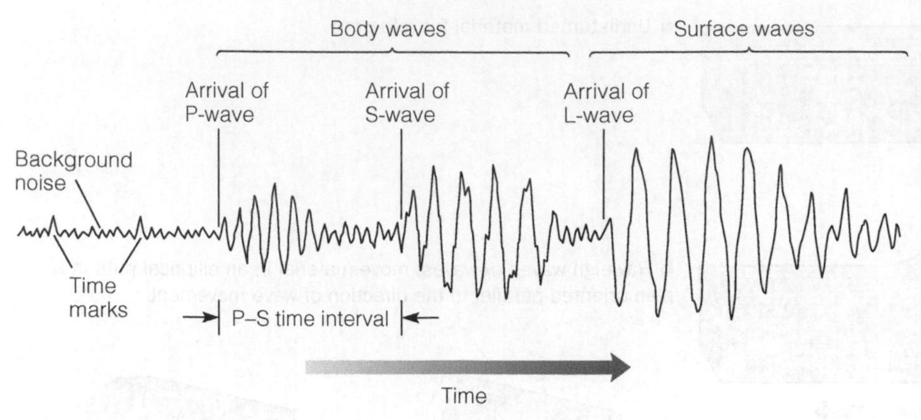

a A schematic seismogram showing the arrival order and pattern produced by P-, S-, and L-waves. When an earthquake occurs, body and surface waves radiate out from the focus at the same time. Because P-waves are the fastest, they arrive at a seismograph first, followed by S-waves and then by surface waves, which are the slowest waves. The difference between the arrival times of the P- and S-waves is the P–S time interval; it is a function of the distance the seismograph station is from the focus.

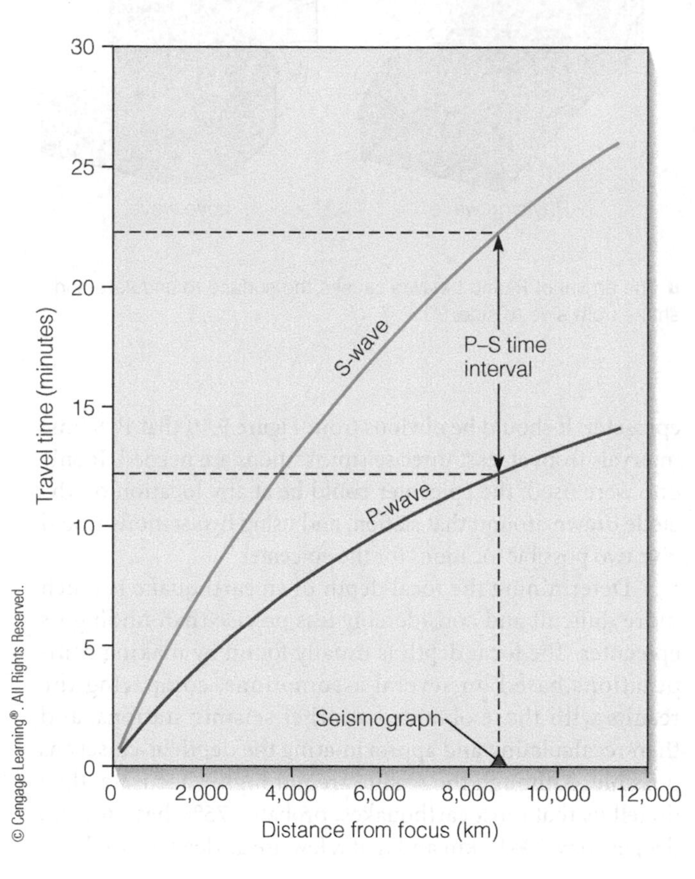

b A time–distance graph showing the average travel times for P- and S-waves. The farther away a seismograph station is from the focus of an earthquake, the longer the interval between the arrival of the P- and S-waves, and hence the greater the distance between the P- and S-wave curves on the time–distance graph as indicated by the P–S time interval. For example, let's assume the difference in arrival times between the P- and S-waves is 10 minutes (P–S time interval). Using the Travel time (minutes) scale, measure how long 10 minutes is (P–S time interval), and move that distance between the S-wave curve and the P-wave curve until the line touches both curves as shown. Then draw a line straight down to the Distance from focus (km) scale. That number is the distance the seismograph is from the Earth's focus. In this example, the distance is almost 9,000 km.

Two measures of an earthquake's strength are commonly used. One is *intensity*, a qualitative assessment of the kinds of damage done by an earthquake. The other, *magnitude*, is a quantitative measure of the amount of energy released by an earthquake. Each method provides important information that can be used to prepare for future earthquakes.

Intensity

Intensity is a subjective or qualitative measure of the kind of damage done by an earthquake as well as people's reaction to it. Since the mid-19th century, geologists have used intensity as a rough approximation of the size and strength of an earthquake. The most common intensity scale used in the United States is the **Modified Mercalli Intensity Scale**, which has values ranging from I to XII (Table 9.2).

Intensity maps can be constructed for regions hit by earthquakes by dividing the affected region into various intensity zones. The intensity value given for each zone is the maximum intensity that the earthquake produced for that zone. Even though intensity maps are not precise because of the subjective nature of the data, they do provide geologists with a rough approximation of the location of the earthquake, the kind and extent of the damage done, and the effects of local geology on different types of building construction (❚ Figure 9.11). Because intensity is a measure of

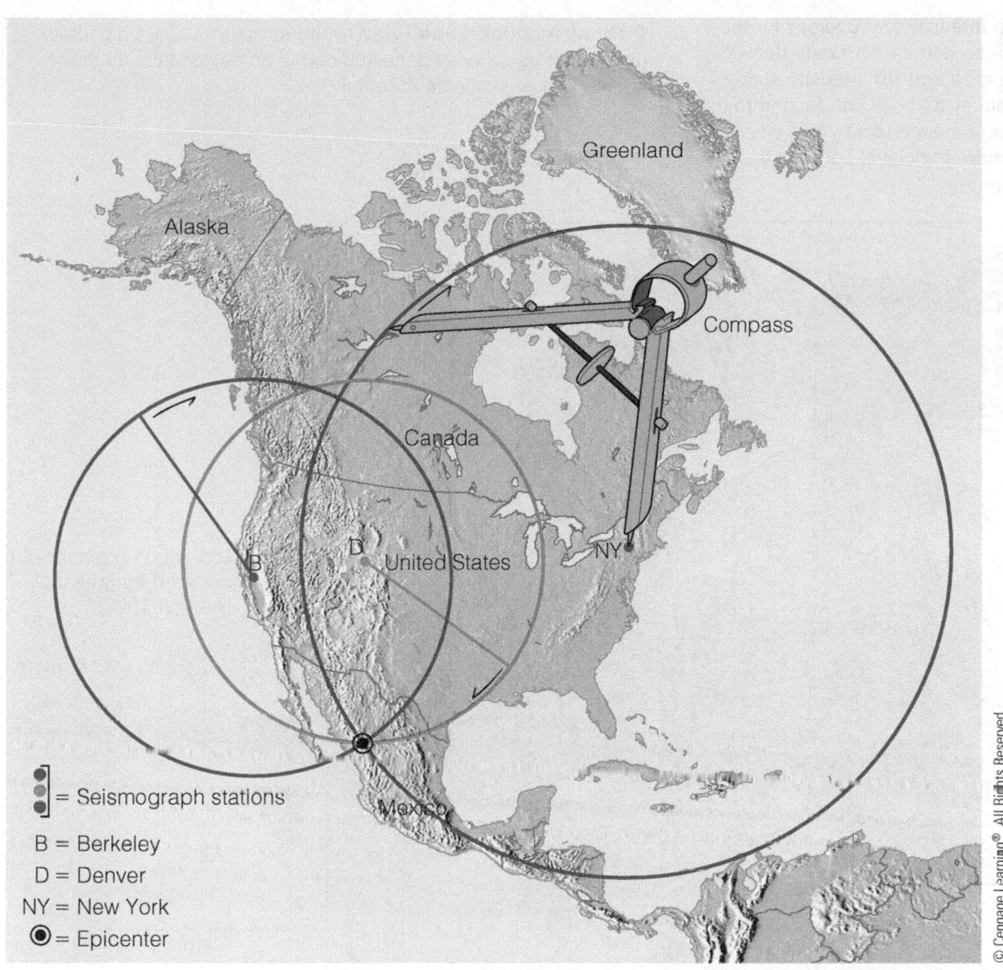

Figure 9.10 Determining the Epicenter of an Earthquake Three seismograph stations are needed to locate the epicenter of an earthquake. The P–S time interval is plotted on a time–distance graph for each seismograph station to determine the distance that station is from the epicenter. A circle with that radius is drawn from each station, and the intersection of the three circles is the epicenter of the earthquake.

TABLE 9.2 Modified Mercalli Intensity Scale

I Not felt except by very few under especially favorable circumstances.

II Felt by only a few people at rest, especially on upper floors of buildings.

III Felt quite noticeably indoors, especially on upper floors of buildings, but many people do not recognize it as an earthquake. Standing automobiles may rock slightly.

IV During the day felt indoors by many, outdoors by few. At night some awakened. Sensation like heavy truck striking building, standing automobiles rocked noticeably.

V Felt by nearly everyone, many awakened. Some dishes, windows, etc. broken, a few instances of cracked plaster. Disturbance of trees, poles, and other tall objects sometimes noticed.

VI Felt by all, many frightened and run outdoors. Some heavy furniture moved. A few instances of fallen plaster or damaged chimneys. Damage slight.

VII Everybody runs outdoors. Damage negligible in buildings of good design and construction; slight to moderate in well-built ordinary structures; considerable in poorly built or badly designed structures; some chimneys broken. Noticed by people driving automobiles.

VIII Damage slight in specially designed structures; considerable in normally constructed buildings with possible partial collapse; great in poorly built structures. Fall of chimneys, monuments, walls. Heavy furniture overturned. Sand and mud ejected in small amounts.

IX Damage considerable in specially designed structures. Buildings shifted off foundations. Ground noticeably cracked. Underground pipes broken.

X Some well-built wooden structures destroyed; most masonry and frame structures with foundations destroyed; ground badly cracked. Rails bent. Landslides considerable from river banks and steep slopes. Water splashed over river banks.

XI Few, if any (masonry) structures remain standing. Bridges destroyed. Broad fissures in ground. Underground pipelines completely out of service.

XII Damage total. Waves seen on ground surface. Objects thrown upward into the air.

Source: United States Geological Survey.

Figure 9.11 Relationship Between Intensity and Geology for the 1906 San Francisco Earthquake A close correlation exists between the **(a)** geology of the San Francisco area and **(b)** intensity during the 1906 earthquake. Areas underlain by bedrock correspond to the lowest intensity values, followed by areas underlain by thin alluvium (sediment) and thick alluvium. Bay mud, artificial fill, or both lie beneath the areas shaken most violently.

Critical Thinking Question Why are structures built on bedrock usually not as severely damaged during an earthquake as those sited on unconsolidated material?

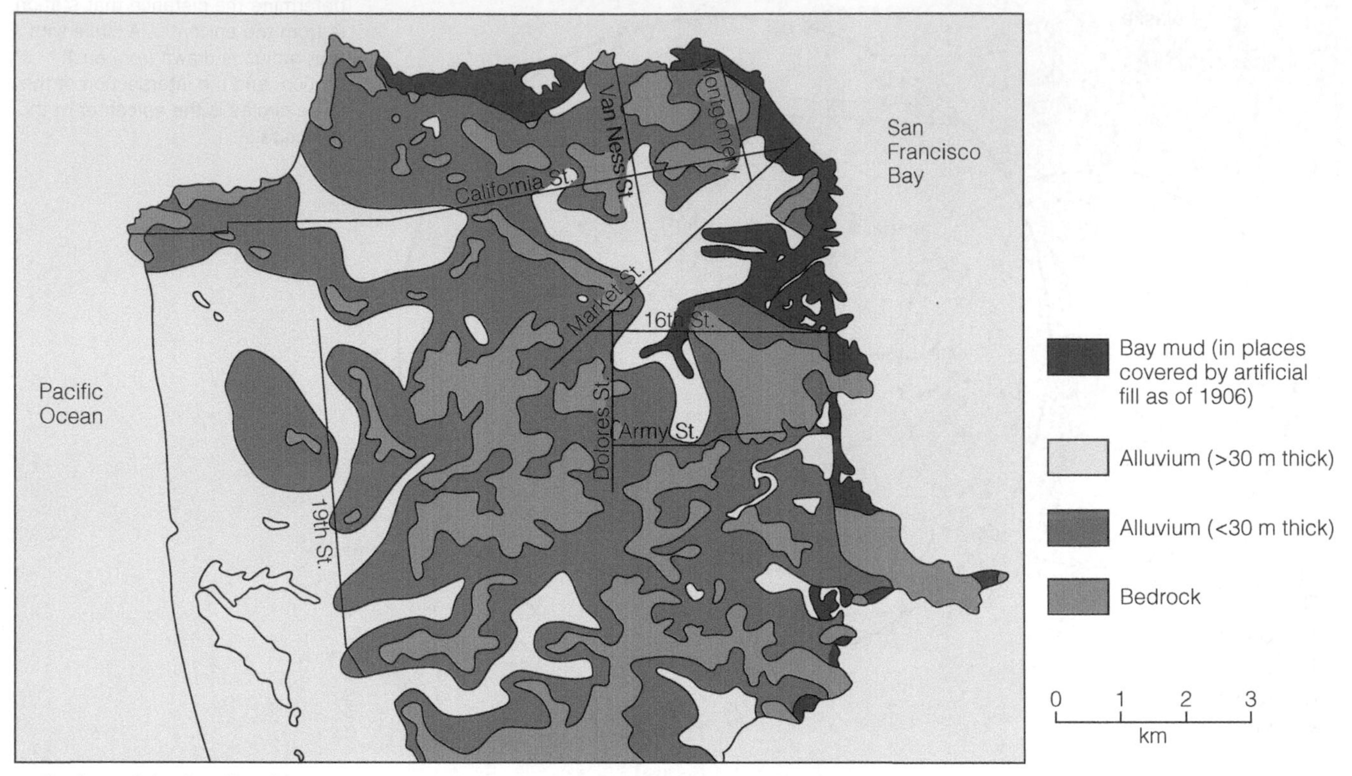

a Geology of the San Francisco area.

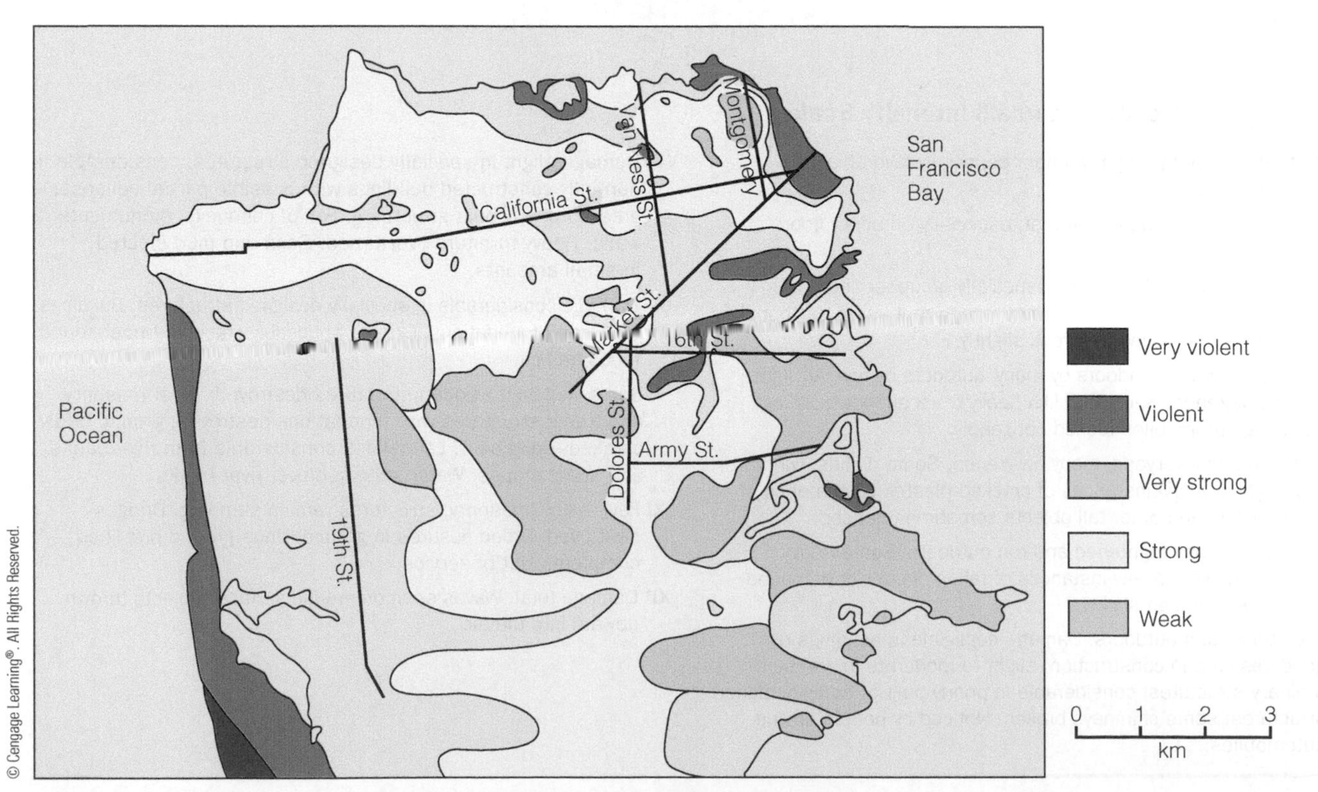

b Intensity of ground shaking in the San Francisco area during the 1906 earthquake.

the kind of damage done by an earthquake, insurance companies still classify earthquakes on the basis of intensity.

Generally, a large earthquake will produce higher intensity values than a small earthquake, but many other factors besides the amount of energy released by an earthquake also affect its intensity. These factors include distance from the epicenter, focal depth of the earthquake, population density, geology of the area, type of building construction employed, and duration of shaking.

A comparison of the intensity map for the 1906 San Francisco earthquake and a geologic map of the area shows a strong correlation between the amount of damage done and the underlying rock and soil conditions (Figure 9.11). Damage was greatest in those areas underlain by poorly consolidated material or artificial fill because the effects of shaking are amplified in these materials, whereas damage was less in areas of solid bedrock. The correlation between the geology and the amount of damage done by an earthquake was further reinforced by the 1989 Loma Prieta earthquake, when many of the same areas that were extensively damaged in the 1906 earthquake were once again heavily damaged.

Magnitude

If earthquakes are to be compared quantitatively, we must use a scale that measures the amount of energy released and is independent of intensity. Charles F. Richter, a seismologist at the California Institute of Technology, developed such a scale in 1935. The **Richter Magnitude Scale** measures earthquake **magnitude**, which is the total amount of energy released by an earthquake at its source. It is an open-ended scale with values beginning at zero. The largest magnitude recorded was a magnitude 9.5 earthquake in Chile on May 22, 1960 (Table 9.1).

Scientists determine the magnitude of an earthquake by measuring the amplitude of the largest seismic wave as recorded on a seismogram (❚ Figure 9.12). To avoid large numbers, Richter used a conventional base-10 logarithmic scale to convert the amplitude of the largest recorded seismic wave to a numerical magnitude value. Therefore, each whole-number increase in magnitude represents a 10-fold increase in wave amplitude. For example, the amplitude of the largest seismic wave for a magnitude 6 earthquake is 10 times that produced by a magnitude 5 earthquake, 100 times as large as a magnitude 4 earthquake, and 1,000 times that of a magnitude 3 earthquake (10 × 10 × 10 = 1,000).

A common misconception about the size of earthquakes is that an increase of one unit on the Richter Magnitude Scale—a 7 versus a 6, for instance—means a 10-fold increase in size. It is true that each whole-number increase in magnitude represents a 10-fold increase in the wave amplitude, but each magnitude increase of one unit corresponds to a roughly 30-fold increase in the amount of energy released (actually, it is 31.5, but 30 is close enough for our purposes). This means, for example, that the 2011 Japanese earthquake

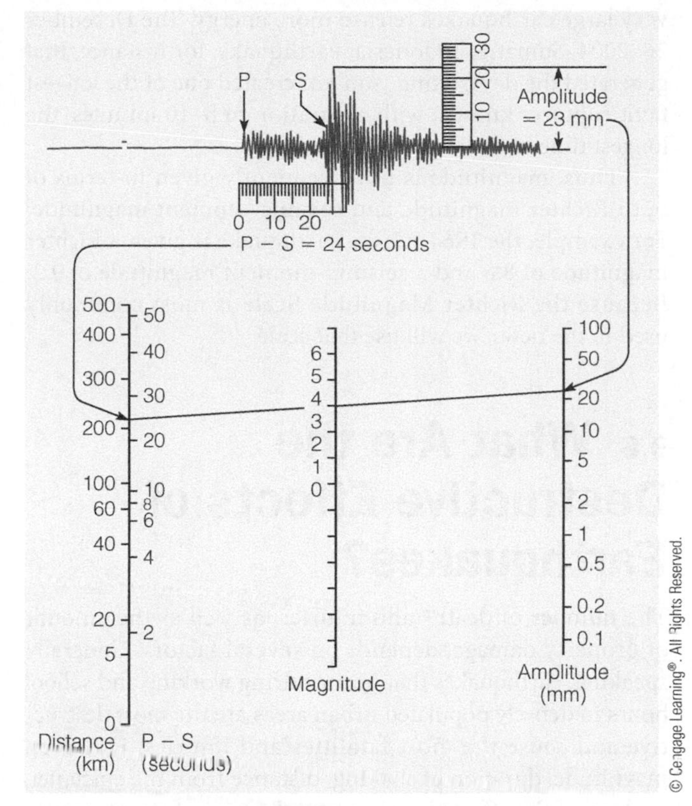

❚ **Figure 9.12** Richter Magnitude Scale The Richter Magnitude Scale measures the total amount of energy released by an earthquake at its source. The magnitude is determined by measuring the maximum amplitude of the largest seismic wave and marking it on the right-hand scale. The difference between the arrival times of the P- and S-waves (recorded in seconds) is marked on the left-hand scale. When a line is drawn between the two points, the magnitude of the earthquake is the point at which the line crosses the center scale.

with a magnitude of 9.0 released nearly 900 times more energy than the 2010 Haiti earthquake with a magnitude 7.0 (30 × 30 = 900).

The Richter Magnitude Scale was devised to measure earthquake waves on a particular seismograph and at a specific distance from an earthquake. One of its limitations is that it underestimates the energy of very large earthquakes because it measures the highest peak on a seismogram, which represents only an instant during an earthquake. For large earthquakes, though, the energy might be released over several minutes and along hundreds of kilometers of a fault. For example, during the 1857 Fort Tejon, California, earthquake, the ground shook for longer than two minutes, and energy was released for 360 km along the fault.

Seismologists now commonly use a somewhat-different scale to measure magnitude. Known as the *seismic-moment magnitude scale*, this scale takes into account the strength of the rocks, the area of a fault along which rupture occurs, and the amount of movement of rocks adjacent to the fault. Because larger earthquakes rupture more rocks than smaller earthquakes and rupture usually occurs along a longer segment of a fault and therefore for a longer duration, these very large earthquakes release more energy. The December 26, 2004, Sumatra, Indonesia, earthquake, for instance, that generated the devastating tsunami created one of the longest fault ruptures known, with a duration of 8–10 minutes, the longest time of faulting ever recorded.

Thus, magnitude is now frequently given in terms of both Richter magnitude and seismic-moment magnitude. For example, the 1964 Alaska earthquake is given a Richter magnitude of 8.6 and a seismic-moment magnitude of 9.2. Because the Richter Magnitude Scale is most commonly used in the news, we will use that scale.

9.8 What Are the Destructive Effects of Earthquakes?

The number of deaths and injuries, as well as the amount of property damage, depends on several factors. Generally speaking, earthquakes that occur during working and school hours in densely populated urban areas are the most destructive and cause the most fatalities and injuries. However, magnitude, duration of shaking, distance from the epicenter, geology of the affected region, and the type of structures are also important considerations. Given these variables, it should not be surprising that a comparatively small earthquake can have disastrous effects, whereas a much larger one might go unnoticed, except probably by seismologists.

The destructive effects of earthquakes include ground shaking, fire, seismic sea waves, and landslides, as well as panic, disruption of vital services, and psychological shock. In some cases, rescue attempts are hampered by inadequate resources or planning, conditions of civil unrest, or simply the enormity of the disaster.

Ground Shaking

Ground shaking, the most obvious and immediate effect of an earthquake, varies depending on the earthquake's magnitude, distance from the epicenter, and type of underlying materials in the area—unconsolidated sediment or fill versus bedrock, for instance. Certainly, ground shaking is terrifying, and it may be violent enough for fissures to open in the ground. Nevertheless, contrary to popular myth, fissures do not swallow up people and buildings and then close in on them. And although California will no doubt have big earthquakes in the future, rocks cannot store enough energy to displace a landmass as large as California into the Pacific Ocean, as some alarmists claim.

The effects of ground shaking, such as collapsing buildings, falling building facades and window glass, and toppling monuments and statues, cause more damage and result in more loss of life and injuries than any other earthquake hazard. Structures built on solid bedrock generally suffer less damage than those built on poorly consolidated

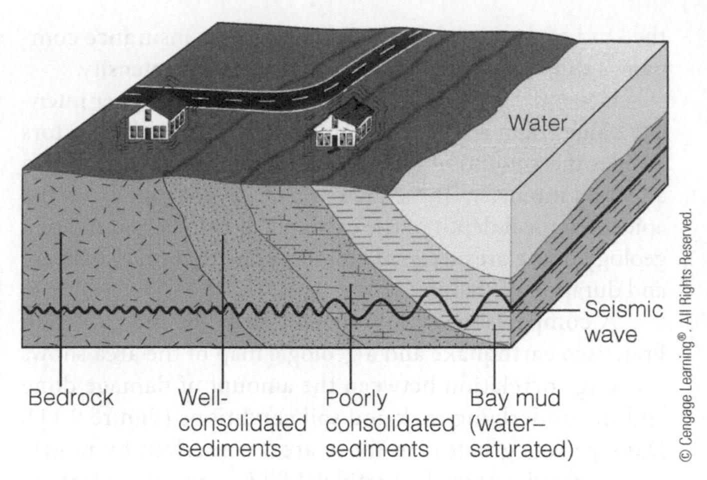

Figure 9.13 Relationship Between Seismic Wave Amplitude and Underlying Geology The amplitude and duration of seismic waves generally increase as the waves pass from bedrock into poorly consolidated or water-saturated material. Thus, structures built on weaker material typically suffer greater damage than similar structures built on bedrock because the shaking lasts longer.

material, such as water-saturated sediments or artificial fill (Geo-Insight 9.1).

Structures built on poorly consolidated or water-saturated material are subjected to ground shaking of longer duration and greater S-wave amplitude than structures built on bedrock (Figure 9.13). In addition, fill and water-saturated sediments tend to liquefy, or behave as a fluid, a process known as *liquefaction*. When shaken, the individual grains lose cohesion, and the ground flows. Two dramatic examples of damage resulting from liquefaction are Niigata, Japan, and Turnagain Heights, Alaska. In Niigata, Japan, large apartment buildings were tipped to their sides after the water-saturated soil of the hillside collapsed (Figure 9.14).

National Geophysical Data Center/NOAA

Figure 9.14 Liquefaction The effects of ground shaking on water-saturated soil are dramatically illustrated by the collapse of these buildings in Niigata, Japan, during a 1964 earthquake. The buildings were designed to be earthquake resistant and fell over on their sides intact when the ground below them underwent liquefaction.

Yann Arhus-Betrend/Peter Arnold/PhotoLibrary

Figure 9.15 Ground Shaking Most of the buildings collapsed or were severely damaged as a result of ground shaking during the August 17, 1999, Turkey earthquake, which killed more than 17,000 people.

In Turnagain Heights, Alaska, many homes were destroyed when the Bootlegger Cove Clay lost all its strength as it was shaken by the 1964 earthquake (see Figure 11.18).

Besides the magnitude of an earthquake and the underlying geology, the material used and the type of construction affect the amount of damage done. Adobe and mud-walled structures are the weakest and almost always collapse during an earthquake. Unreinforced brick structures and poorly built concrete structures are also particularly susceptible to collapse, as was the case in the 1999 Turkey earthquake in which an estimated 17,000 people died (Figure 9.15). The 1976 earthquake in Tangshan, China, completely leveled the city because hardly any structures were built to resist seismic forces. In fact, most had unreinforced brick walls, which have no flexibility, and consequently they collapsed during the shaking.

Fire

In many earthquakes, particularly in urban areas, fire is a major hazard. Nearly 90% of the damage done in the 1906 San Francisco earthquake was caused by fire. The shaking severed many of the electrical and gas lines, which touched off flames and started fires throughout the city (Figure 9.16a). Because the earthquake ruptured water mains, there was no effective way to fight the fires that raged out of control for three days, destroying much of the city.

Eighty-three years later, during the 1989 Loma Prieta earthquake, a fire broke out in the Marina district of San Francisco (Figure 9.16b). This time, however, the fire was contained within a small area because San Francisco had a system of valves throughout its water and gas pipeline system so that lines could be isolated from breaks.

During the September 1, 1923, earthquake in Japan, fires destroyed 71% of the houses in Tokyo and practically all of the houses in Yokohama. In all, 576,262 houses were destroyed by fire, and 143,000 people died, many as a result of fire. A horrible example occurred in Tokyo where thousands of people gathered along the banks of the Sumida River to escape the raging fires. Suddenly, a firestorm swept over the area, killing more than 38,000 people. The fires from this earthquake were particularly devastating because most of the buildings were constructed of wood and were fanned by 20-km/hr winds.

Tsunami: Killer Waves

On December 26, 2004, a magnitude 9.0 earthquake struck 160 km off the west coast of northern Sumatra, Indonesia, generating the deadliest tsunami in history. Within hours, walls of water as high as 10.5 m pounded the coasts of Indonesia, Sri Lanka, India, Thailand, Somalia, Myanmar, Malaysia, and the Maldives, killing more than 230,000 people and causing billions of dollars in damage.

Following the magnitude 9.0 earthquake that struck Japan on March 11, 2011, a massive tsunami was generated that resulted in tremendous property damage and loss of life along the coastline of Japan (Figure 9.17). Within minutes after the earthquake, walls of water, some as high as 37 m, inundated low-lying areas along the Japanese coast and extended as far as 10 km inland. Boats, vehicles, and structures were destroyed or swept aside as if they were toys, leaving a

GEO INSIGHT 9.1 Designing and Building Earthquake-Resistant Structures

One way to reduce property damage, injuries, and loss of life is to design and build structures that are as earthquake resistant as possible. Many things can be done to improve the safety of current structures and of new buildings.

To design earthquake-resistant structures, engineers must understand the dynamics and mechanics of earthquakes, including the type and duration of the ground motion and how rapidly the ground accelerates during an earthquake. An understanding of the area's geology is also important because certain ground materials such as water-saturated sediments or landfill can lose their strength and cohesiveness during an earthquake. Finally, engineers must be aware of how different structures behave under different earthquake conditions.

With the level of technology currently available, a well-designed, properly constructed building should be able to withstand small, short-duration earthquakes of less than magnitude 5.5 with little or no damage. In moderate earthquakes (magnitude 5.5–7.0), the damage suffered should not be serious and should be repairable. In a major earthquake of greater than magnitude 7.0, the building should not collapse, although it may later have to be demolished.

Many factors enter into the design of an earthquake-resistant structure, but the most important is that the building be tied together; that is, the foundation, walls, floors, and roof should all be joined together to create a structure that can withstand both horizontal and vertical shaking (▌Figure 1). Almost all

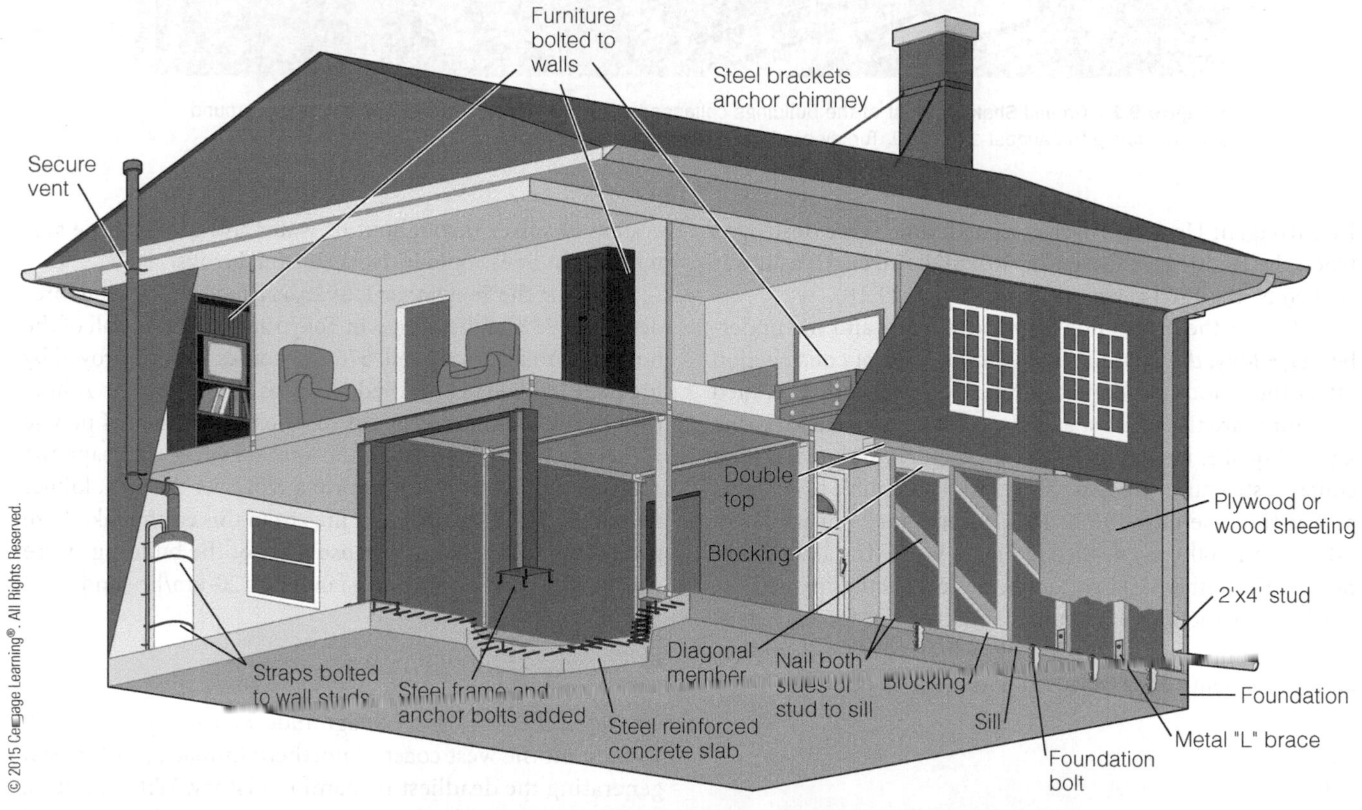

▌**Figure 1** Some of the useful things a homeowner can do to reduce damage to a building because of ground shaking during an earthquake. Notice that the structure must be solidly attached to its foundation and bracing the walls helps prevent damage from horizontal motion.

wake of destruction as the waters finally receded. Of concern also, was the fact that three nuclear power plants suffered damage from the tsunami, and at least one leaked radiation into the surrounding area as well as into the Pacific Ocean.

Both of these earthquakes generated what is popularly called a "tidal wave," but is more correctly termed a *seismic sea wave* or **tsunami**, a Japanese term meaning "harbor wave." The term *tidal wave* nevertheless persists in popular literature and some news accounts, but these waves are not caused by or related to tides. Indeed, tsunami are destructive sea waves generated when the seafloor undergoes sudden, vertical movements. Many result from submarine earthquakes, but volcanoes at sea or submarine landslides can also cause them. For example, the 1883 eruption of Krakatau

the structural failures resulting from earthquake ground movement occur at weak connections, where the various parts of a structure are not securely tied together. Buildings with open or unsupported first stories are particularly susceptible to damage. Some reinforcement must be done, or collapse is a distinct possibility (Figure 1).

Tall buildings, such as skyscrapers, must be designed so that a certain amount of swaying or flexing can occur, but not so much that they touch neighboring buildings during swaying. Besides designed flexibility, engineers must also ensure that a building does not vibrate at the same frequency as the ground, which depending on the material, can vibrate at different frequencies. When the vibration frequency of the ground and the building are the same, then the force applied by the seismic waves at ground level is multiplied several times by the time they reach the top of the building.

Damage to high-rise structures can be minimized or prevented by using diagonal steel beams to help prevent swaying. In addition, tall buildings in earthquake-prone areas are now commonly placed on layered steel and rubber structures and devices similar to shock absorbers that help decrease the amount of sway.

What about structures built many years ago? Just as in new buildings, the most important thing that can be done to increase the stability and safety of older structures is to tie together the different components of each building. Although such modifications are expensive, they are usually cheaper than having to replace a building that was destroyed by an earthquake.

Another problem related to both new and older buildings is that even with strict building codes, damage will still result if there is not adherence to construction standards for structures in earthquake-prone areas. Such was the case in the Izmit, Turkey, earthquake in 1999 (Figure 9.15), and the Sichuan, China, earthquake in 2008 (▮ Figure 2).

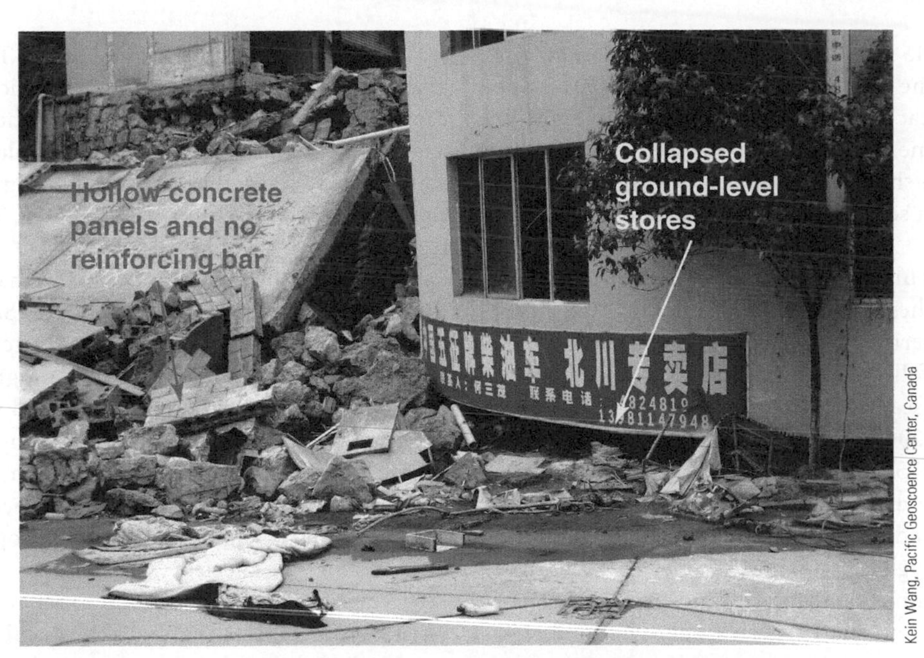

▮ **Figure 2** During the 2008 Sichuan, China earthquake, structures of masonry construction, coupled with little or no reinforcing collapsed, leading to tremendous damage and loss of life.

between Java and Sumatra generated a large sea wave that killed 36,000 on nearby islands.

Once a tsunami is generated, it can travel across an entire ocean and cause devastation far from its source. In the open sea, tsunami travel at several hundred kilometers per hour and commonly go unnoticed as they pass beneath ships because they are usually less than 1 m high and the distance between wave crests is typically hundreds of kilometers. When they enter shallow water, however, the waves slow down, and water piles up to heights anywhere from a meter or two to many

Connection Link

To learn more about how waves form and their terminology, see Chapter 16.

Figure 9.16 Fire In many urban areas, fire can be a major hazard following an earthquake.

University of California at Berkeley

a San Francisco following the 1906 earthquake. This view along Sacramento Street shows damaged buildings and the approaching fire. It is estimated that about 3,000 people died, and approximately 28,000 buildings were destroyed, many of them by the three-day fire that raged out of control.

The Marin Independent Journal

b San Francisco Marina district fire caused by broken gas lines during the 1989 Loma Prieta earthquake. Because of the system of valves throughout San Francisco's gas and water pipeline network, this fire was quickly contained before it could spread and do potentially greater damage.

meters high. The 1946 tsunami that struck Hilo, Hawaii, was 16.5 m high, and the one that struck Japan in 2011 was reported to have reached more than twice that height in some areas. The tremendous energy possessed by a tsunami is concentrated on a shoreline when it hits, either as a large breaking wave or, in some cases, what appears to be a very rapidly rising tide.

A common popular belief is that a tsunami is a single large wave that crashes onto a shoreline. But, in fact, a tsunami consists of a series of waves that pour onshore for as long as 30 minutes followed by an equal time during which water rushes back to sea. Furthermore, after the first wave hits, more waves follow at 20- to 60-minute intervals. Approximately 80 minutes after the 1755 Lisbon, Portugal, earthquake, the first of three tsunami, the largest, more than 12 m high, destroyed the waterfront area and killed thousands of people. Following the arrival of a 2-m-high tsunami in Crescent City, California, in 1964, curious people went to the waterfront to inspect the damage. Unfortunately, 10 were killed by a following 4-m-high wave.

One of nature's warning signs of an approaching tsunami is the sudden withdrawal of the sea from a coastal region. In reality, the sea might withdraw so far that it cannot be seen and the seafloor is laid bare over a huge area. On more than one occasion, people have rushed out to inspect exposed reefs or to collect fish and shells only to be swept away when the tsunami arrived.

During the December 2004 Indian Ocean tsunami, a 10-year-old British girl saved numerous lives because she recognized the warning signs that she had learned in a school lesson on tsunami only two weeks before. While vacationing with her mother on the island of Phuket, Thailand, a popular resort area, the girl noticed the water quickly receding from the beach. She immediately told her mother that she thought a tsunami was coming, and her mother, along with the resort staff, quickly warned everyone standing around watching the water recede to clear the beach area. Their quick action resulted in many lives being saved.

Following the tragic 1946 tsunami that hit Hilo, Hawaii, the United States Coast and Geodetic Survey established a Pacific Tsunami Early Warning System in Ewa Beach, Hawaii. This system combines seismographs and instruments that detect earthquake-generated sea waves. Whenever a strong earthquake takes place anywhere within the Pacific basin, its location is determined, and instruments are checked to see whether a tsunami has been generated. If it has, a warning is sent out to evacuate people from low-lying areas that may be affected.

Nevertheless, tsunami remain a threat to people in coastal areas, especially around the Pacific Ocean. Unfortunately, no such warning system existed in the Indian Ocean on December 26, 2004. If one had been in place, it is possible that the death toll from the tsunami on that date would have been significantly lower.

Ground Failure

Earthquake-triggered landslides are particularly dangerous in mountainous regions and have been responsible for tremendous amounts of damage and many deaths. The 1959 Hebgen Lake earthquake in Madison Canyon, Montana, for

Figure 9.17 2011 Japanese Tsunami

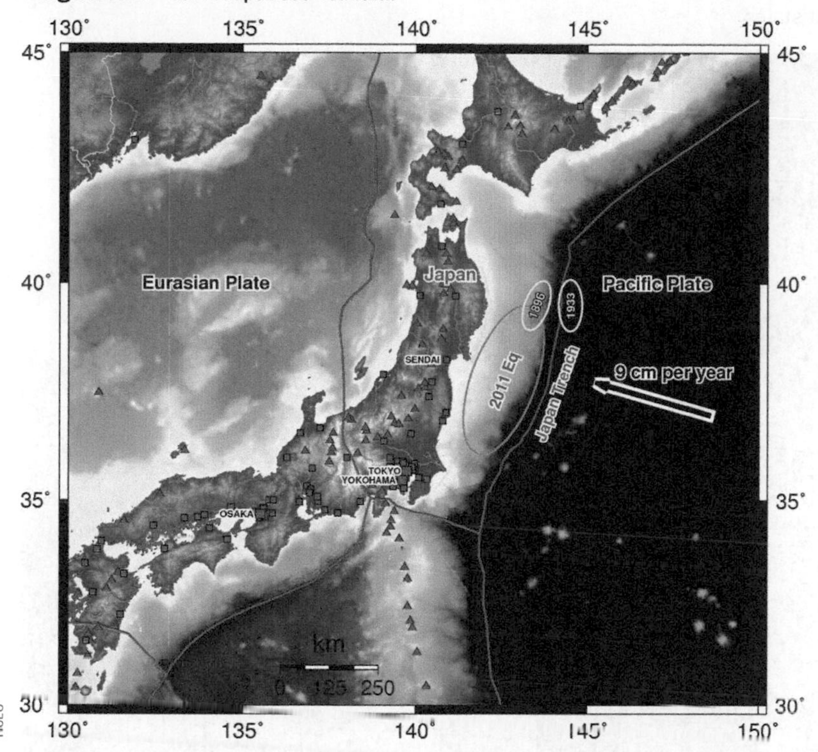

a Map showing the location of the epicenter (pink circled area, 2011 Eq) of the magnitude 9.0 earthquake that struck Japan on March 11, 2011, causing tremendous destruction and loss of life from both the earthquake and subsequent tsunami. The red lines are plate boundaries. Volcanoes are shown as red triangles. Cities with populations of greater than 250,000 are shown as red squares. The two yellow circled areas show previous large earthquakes (1896 and 1933) in the epicentral area.

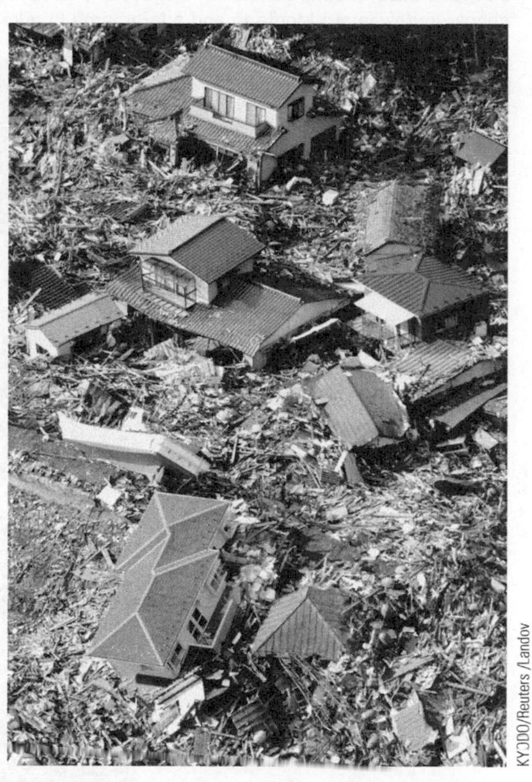

KYODO/Reuters /Landov

b Aerial view of Minamisanriku, Miyagi Prefecture in northeastern Japan, showing houses clogged with debris and a large boat that had been swept inland as a result of the tsunami.

KIM KYUNG-HOON/Reuters/Landov

c Vehicles, tossed like toys, mix with the rubble in floodwaters near the coastal town of Sendai, as a result of the tsunami that ravaged the northeastern portion of Japan, following the devastating magnitude 9.0 earthquake on March 11, 2011.

example, caused a huge rock slide (Figure 9.18), and the 1970 Peru earthquake caused an avalanche that destroyed the town of Yungay and killed an estimated 66,000 people (see Figure 11.23). Most of the 100,000 deaths from the 1920 earthquake in Gansu, China, resulted when cliffs composed of loess (wind-deposited silt) collapsed. More than 2,000 people were killed when two-thirds of the town of Port Royal, Jamaica, slid into the sea following an earthquake on June 7, 1692.

Figure 9.18 Ground Failure On August 17, 1959, an earthquake with a Richter magnitude of 7.3 shook southwestern Montana and a large area in adjacent states.

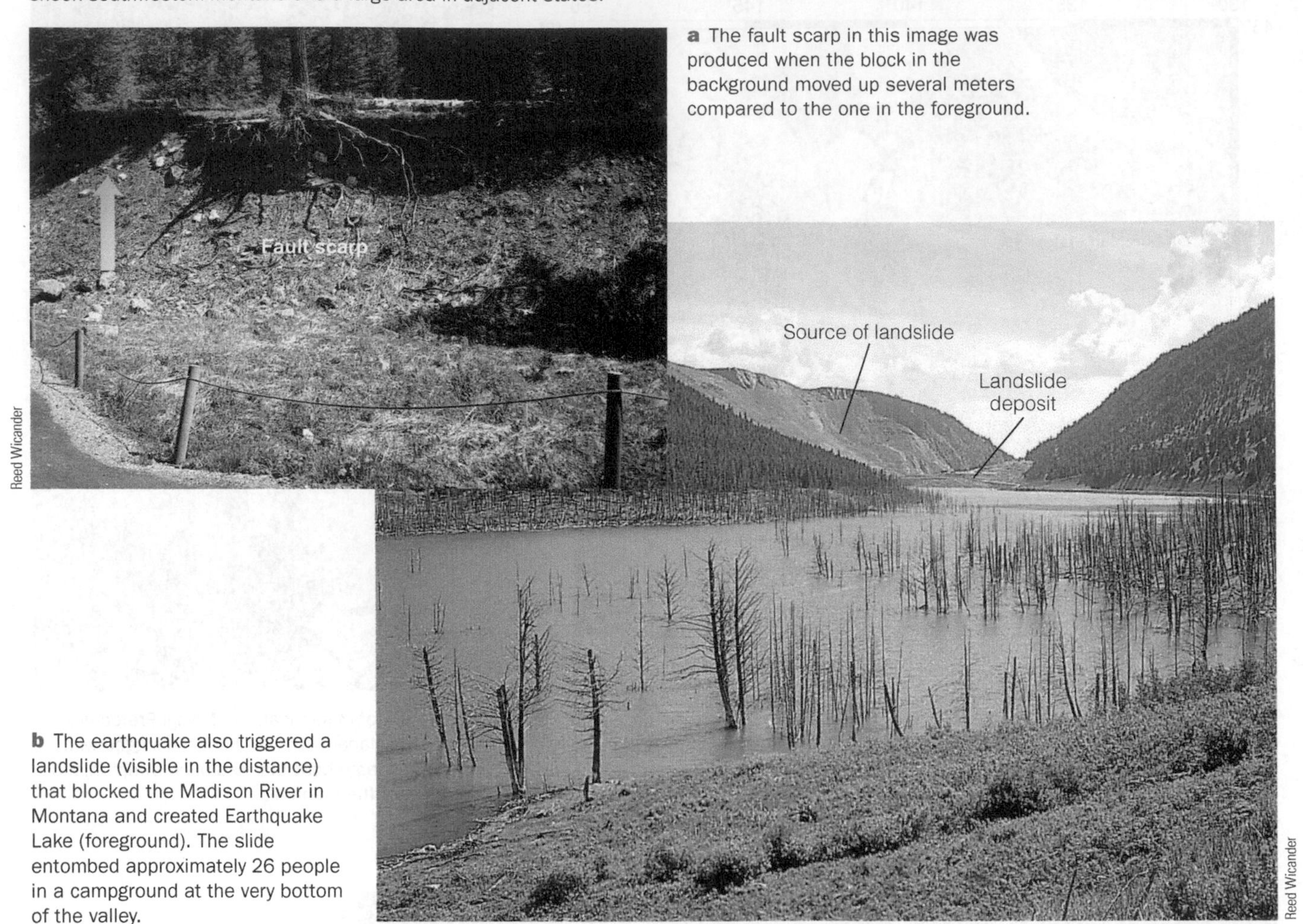

a The fault scarp in this image was produced when the block in the background moved up several meters compared to the one in the foreground.

b The earthquake also triggered a landslide (visible in the distance) that blocked the Madison River in Montana and created Earthquake Lake (foreground). The slide entombed approximately 26 people in a campground at the very bottom of the valley.

9.9 Earthquake Prediction

A successful prediction must include a time frame for the occurrence of an earthquake, its location, and its strength. Despite the tremendous amount of information geologists have gathered about the cause of earthquakes, successful predictions are still rare. Nevertheless, if reliable predictions can be made, they can greatly reduce the number of deaths and injuries.

From an analysis of historic records and the distribution of known faults, geologists construct *seismic risk maps* that indicate the likelihood and potential severity of future earthquakes based on the intensity of past earthquakes. An international effort by scientists from several countries resulted in the publication of the first Global Seismic Hazard Assessment Map in December 1999 (Figure 9.19). Although such maps cannot be used to predict when an earthquake will take place in any particular area, they are useful in anticipating future earthquakes and helping people plan and prepare for them (Geo-Focus 9.1).

Earthquake Precursors

Studies conducted during the past several decades indicate that most earthquakes are preceded by both short-term and long-term changes within Earth. Such changes are called *precursors* and may be useful in earthquake prediction.

One long-range prediction technique used in seismically active areas involves plotting the location of major earthquakes and their aftershocks to detect areas that have had major earthquakes in the past but are currently inactive. Such regions are said to be locked and not releasing energy. Nevertheless, pressure is continuing to accumulate in these regions because of plate motions, making these *seismic gaps* prime locations for future earthquakes. Several seismic gaps along the San Andreas Fault have the potential for future major earthquakes (Figure 9.20). A major earthquake that damaged Mexico City in 1985 occurred along a seismic gap in the convergence zone along the west coast of Mexico.

Earthquake precursors that may be useful in making short-term predictions include slight changes in elevation

Figure 9.19 Global Seismic Hazard Assessment Map The Global Seismic Hazard Assessment Program published this seismic hazard map showing peak ground accelerations. The values are based on a 90% probability that the indicated horizontal ground acceleration during an earthquake is not likely to be exceeded in 50 years. The higher the number, the greater the hazard. As expected, the greatest seismic risks are in the circum-Pacific belt and the Mediterranean–Asiatic belt.

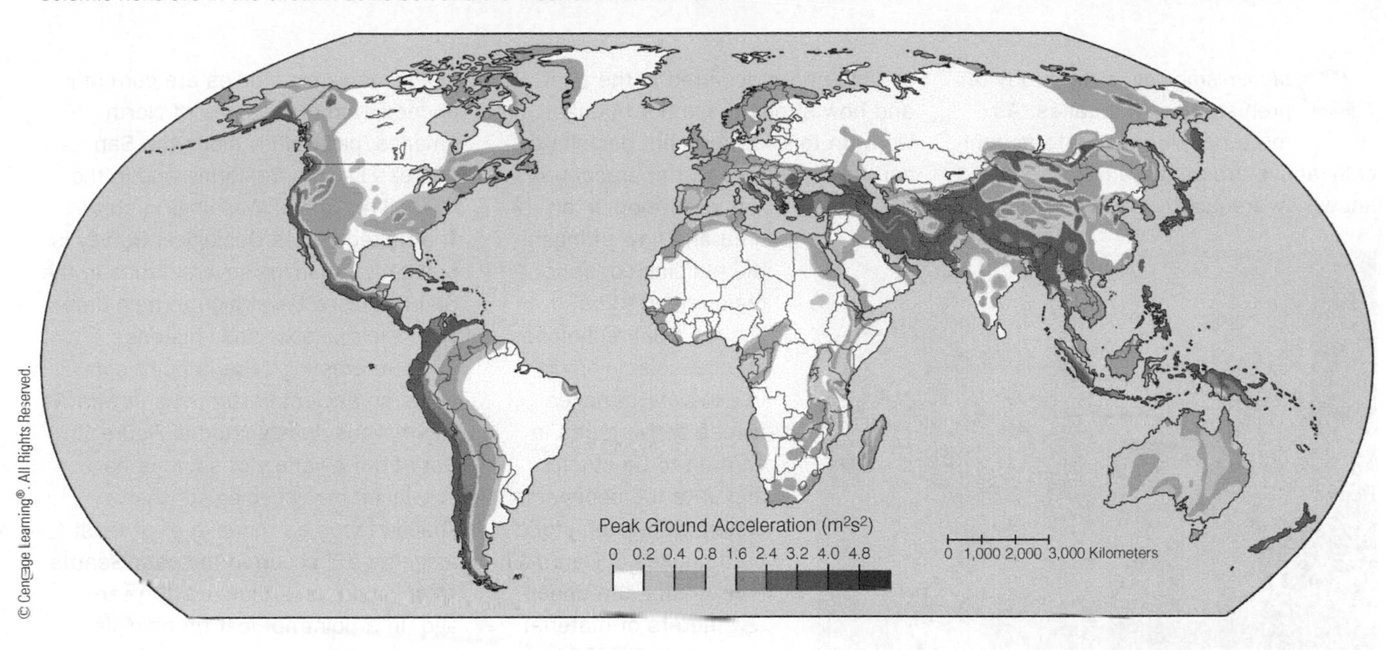

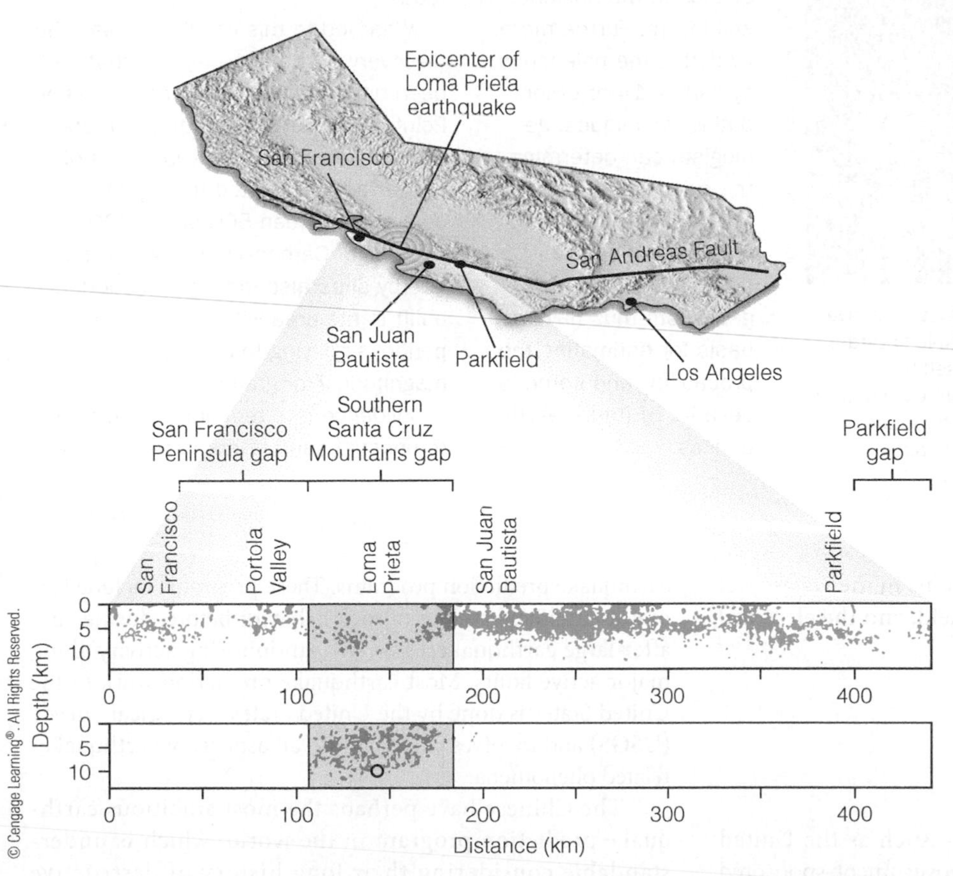

Figure 9.20 Earthquake Precursors Seismic gaps are one type of earthquake precursor that can indicate a potential earthquake in the future. Seismic gaps are regions along a fault that are locked; that is, they are not moving and releasing energy. Three seismic gaps are evident in this cross section along the San Andreas Fault from north of San Francisco to south of Parkfield. The first is between San Francisco and Portola Valley, the second is near Loma Prieta Mountain, and the third is southeast of Parkfield. The top section shows the epicenters of earthquakes between January 1969 and July 1989. The bottom section shows the southern Santa Cruz Mountains gap after it was filled by the October 17, 1989, Loma Prieta earthquake (open circle) and its aftershocks.

GEO FOCUS 9.1 Paleoseismology

Paleoseismology is the study of prehistoric earthquakes. As more people move into seismically active areas, it is important to know how frequently earthquakes in the area have occurred in the past, and how strong those earthquakes were. In this way, prudent decisions can be made about what precautions need to be taken in developing an area and how stringent the building codes for a region need to be.

A typical technique in paleoseismology is to excavate trenches across active faults in an area to be studied and date the sediments disturbed by prehistoric earthquakes (Figure 1). By exposing the upper few meters of material along an active fault, geologists can find evidence of previous earthquakes in the ancient soil layers. Furthermore, by dating the paleosoils by carbon-14 or other dating techniques, geologists can determine the frequency of past earthquakes and when the last earthquake occurred. In this way, geologists thus have a basis for estimating the probability, and potential severity, of future earthquakes.

© John Karachewski, Geoscapes Photography

Figure 1 Geologists examine a trench across an active fault in California to determine possible seismic hazards. Excavating trenches is a common method used by geologists to gather information about ancient earthquakes in a region and to help assess the potential for future earthquakes and the damage that they might cause.

Paleoseismic studies are currently under way in many areas of North America, particularly along the San Andreas Fault in California and in the coastal regions of Washington state. The United States Geological Survey is presently exhuming several faults in the San Francisco Bay region to help determine their paleoseismic histories.

An interesting case in point concerns an ancient earthquake in what is now Seattle, Washington (Figure 2). Data from a variety of sources have convinced many geologists that a shallow-focus earthquake of at least magnitude 7 occurred beneath Seattle, Washington, less than 1,100 years ago. In a point not lost on officials, they noted the catastrophic effects that a similar-sized earthquake would have if it occurred in the same area today.

What led to this conclusion was the discovery of a marine terrace that had been uplifted some 7 m at Restoration Point, 5 km west of Seattle. Carbon-14 analysis of peat within sediments of the terrace suggested that uplift occurred between 500 and 1,700 years ago. Carbon-14 dating of other nearby sites also indicates a sudden uplift in the area within the same time period, signifying to geologists a magnitude 7 or greater earthquake.

Evidence of a tsunami, in the form of unusual sand layers in nearby tidal

and tilting of the land surface, fluctuations in the water level in wells, changes in Earth's magnetic field, and the electrical resistance of the ground.

Earthquake Prediction Programs

Currently, only a handful of nations—such as the United States, Japan, Russia, and China—have government-sponsored earthquake prediction programs. These programs include laboratory and field studies of rock behavior before, during, and after large earthquakes, as well as monitoring activity along major active faults. Most earthquake prediction work in the United States is done by the United States Geological Survey (USGS) and involves research into all aspects of earthquake-related phenomena.

The Chinese have perhaps the most ambitious earthquake prediction program in the world, which is understandable considering their long history of destructive marsh deposits, was also discovered. Carbon-14 dating of organic matter associated with the sands yielded an age of 850–1,250 years ago, well within the time period during which the terrace was uplifted.

Geologists also found evidence of rock avalanches in the Olympic Mountains that dammed streams, thereby forming lakes. Drowned trees in the lakes were dated as having died between 1,000 and 1,300 years ago, again fitting in nicely with the date of the ancient earthquake.

Lastly, deposits found on the bottom of Lake Washington further supported an earthquake at this time. Ground shaking from the earthquake apparently caused the bottom sediment of the lake to be resuspended and to move downslope as a turbidity flow. Dating of these sediments indicates that they were deposited between 940 and 1,280 years ago, which is consistent with their being caused by an earthquake.

All evidence thus points to a large (magnitude-7 or greater) shallow-focus earthquake occurring in the Seattle area approximately 1,000 years ago. If history and the events of the geologic past are any guide, it is very likely that another large earthquake will hit the Seattle area in the foreseeable future. When this will occur cannot yet be predicted, but it would be wise to plan for such an eventuality. After all, metropolitan Seattle has a population of approximately 3.5 million people, and its entire port area is built on fill that would probably be hard hit by an earthquake. What the seismic future holds for Seattle has yet to be determined. However, geologists have provided a window on what has happened seismically in the past, and it is up to today's government, and its various agencies, to decide how they want to use this information.

Figure 2 Should another earthquake, similar to the one that occurred less than 1,100 years ago, strike the Seattle region, the results would be devastating.

earthquakes. Their earthquake prediction program was initiated soon after two large earthquakes occurred at Xingtai (300 km southwest of Beijing) in 1966. This program includes extensive study and monitoring of all possible earthquake precursors. Chinese seismologists successfully predicted the 1975 Haicheng earthquake, but unfortunately, they failed to predict the devastating 1976 Tangshan earthquake that killed at least 242,000 people, and the 2008 Sichuan earthquake that killed more than 69,000 people.

Progress is being made toward dependable, accurate earthquake predictions, and studies are under way to assess public reactions to long-, medium-, and short-term earthquake warnings. Unless short-term warnings are actually followed by an earthquake, most people will probably ignore the warnings as they frequently do now for hurricanes, tornadoes, and tsunami. Perhaps the best we can hope for is that people in seismically active areas will take measures to minimize their risk from the next major earthquake (Table 9.3).

TABLE 9.3 What You Can Do to Prepare for an Earthquake

Anyone who lives in an area that is subject to earthquakes or who will be visiting or moving to such an area can take certain precautions to reduce the risks and losses resulting from an earthquake.

Before an earthquake:

1. Become familiar with the geologic hazards of the area where you live and work.
2. Make sure your house is securely attached to the foundation by anchor bolts and that the walls, floors, and roof are all firmly connected together.
3. Heavy furniture such as bookcases should be bolted to the walls; semiflexible natural gas lines should be used so that they can give without breaking; water heaters and furnaces should be strapped and the straps bolted to wall studs to prevent gas-line rupture and fire. Brick chimneys should have a bracket or brace that can be anchored to the roof.
4. Maintain a several-day supply of fresh water and canned foods, and keep a fresh supply of flashlight and radio batteries, as well as a fire extinguisher.
5. Maintain a basic first-aid kit and have a working knowledge of first-aid procedures.
6. Learn how to turn off the various utilities at your house.
7. Above all, have a planned course of action for when an earthquake strikes.

During an earthquake:

1. Remain calm and avoid panic.
2. If you are indoors, get under a desk or table if possible, or stand in a room corner as that is the structurally strongest part of a room; avoid windows and falling debris.
3. In a tall building, do not rush for the stairwells or elevators.
4. In an unreinforced or other hazardous building, it may be better to get out of the building rather than to stay in it. Be on the alert for fallen power lines and the possibility of falling debris.
5. If you are outside, get to an open area away from buildings if possible.
6. If you are in an automobile, stay in the car, and avoid tall buildings, overpasses, and bridges if possible.

After an earthquake:

1. If you are uninjured, remain calm and assess the situation.
2. Help anyone who is injured.
3. Make sure there are no fires or fire hazards.
4. Check for damage to utilities and turn off gas valves if you smell gas.
5. Use your telephone only for emergencies.
6. Do not go sightseeing or move around the streets unnecessarily.
7. Avoid landslide and beach areas.
8. Be prepared for aftershocks.

9.10 Earthquake Control

Reliable earthquake prediction is still in the future, but can anything be done to control or at least partly control earthquakes? Because of the tremendous energy involved, it seems unlikely that humans will ever be able to prevent earthquakes. However, it may be possible to gradually release the energy stored in rocks, thus decreasing the probability of a large earthquake and extensive damage.

During the early- to mid-1960s, Denver, Colorado, experienced numerous small earthquakes, which was surprising because Denver had not been prone to earthquakes in the past. Geologist David M. Evans suggested that the earthquakes were directly related to the injection of contaminated wastewater into a 3,674-m-deep disposal well at the Rocky Mountain Arsenal, northeast of Denver (Figure 9.21a). A USGS study concluded that the pumping of waste fluids into the fractured rocks beneath the disposal well decreased the friction on opposite sides of fractures—in effect, lubricating them so that movement occurred, thus causing the earthquakes that Denver experienced.

Interestingly, a high degree of correlation was found when comparing the number of earthquakes in Denver against the average amount of contaminated fluids injected into the disposal well per month (Figure 9.21b). Furthermore, when no waste fluids were injected, earthquake activity decreased dramatically.

Experiments conducted in 1969 at an abandoned oil field near Rangely, Colorado, confirmed the hypothesis that fluids injected into the fractured bedrock at the Rocky Mountain Arsenal was causing Denver's earthquakes. When geologists pumped water into abandoned oil wells, seismographs recorded small earthquakes in the area. When fluids were pumped out, earthquake activity declined. Geologists were thus starting and stopping earthquakes at will and validating the relationship between pore-water pressures and earthquakes.

A similar situation took place in Basel, Switzerland, where more than 100 small earthquakes occurred in December 2006. An analysis of the injection of water into deep, hot rocks for geothermal power production revealed a correlation between injection of the water and the triggering of earthquakes.

More recently, the number of earthquakes in the Dallas–Ft. Worth area of Texas has dramatically increased, resulting in more earthquakes during 2008 and 2009 than in the previous 30 years combined. Beginning in 2004, thousands of gas wells have been drilled in the area, and almost all of them have undergone hydraulic fracturing, in

ConnectionLink

To learn more about hydraulic fracturing and its possible effects on the environment, see Geo-Focus 13.1 in Chapter 13.

Reducing the Threat of Earthquakes

Some geologists think that by pumping liquids into locked segments of active faults, they can generate small- to moderate-size earthquakes. These earthquakes would relieve the buildup of pressure along a fault and thus prevent very large earthquakes from taking place. What do you think of this proposal? What kind of social, political, and economic consequences would there be? Do you think such an effort will ever actually reduce the threat of earthquakes?

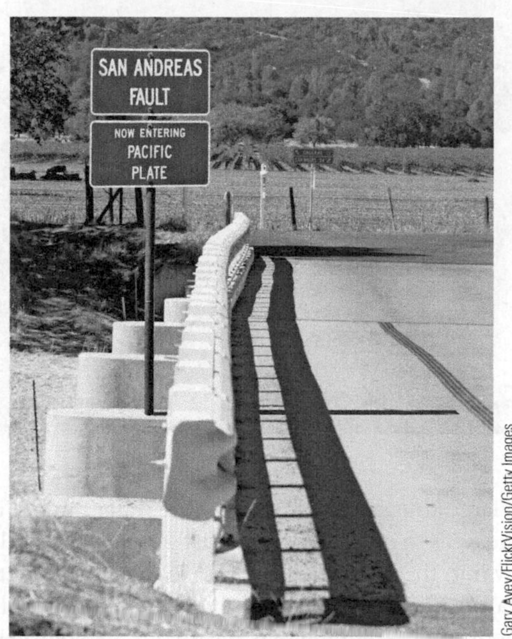

Movement along the San Andreas Fault, causing an earthquake on September 28, 2004, bent the Parkfield-Coalinga Road Bridge in Parkfield, California—a town that bills itself as the "earthquake capital of the world." The bridge was built to cross Little Cholame Creek, but it also straddles the San Andreas Fault, so when the fault moved, the bridge bent as is evident in this photo. Parkfield is the site of a long-term earthquake research project conducted by the United States Geological Survey designed to better understand the physics of earthquakes, which can lead to more accurate methods of earthquake prediction.

Figure 9.21 Controlling Earthquakes

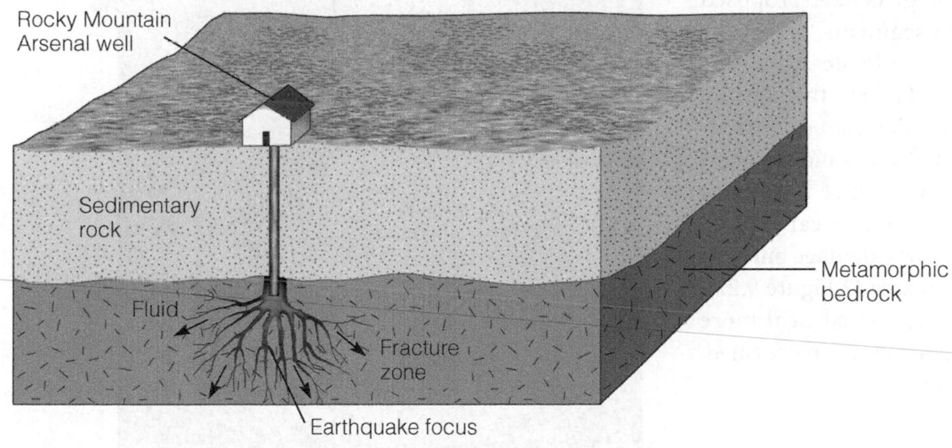

a A block diagram of the Rocky Mountain Arsenal well and the underlying geology.

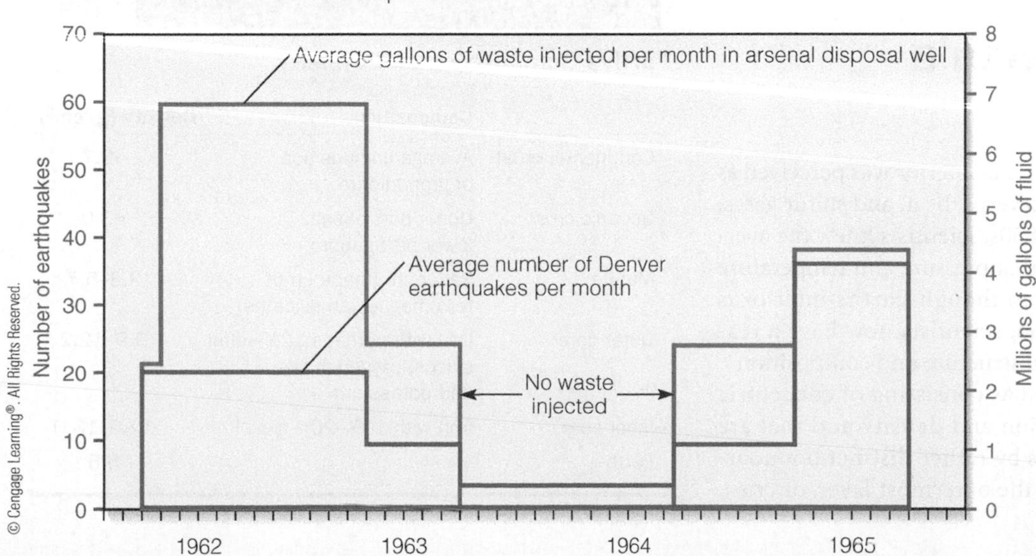

b A graph showing the relationship between the amount of wastewater injected into the well per month and the average number of Denver earthquakes per month. There have been no significant earthquakes in Denver since injection of wastewater into the disposal well ceased in 1965.

Copyright and Photograph by Dr. Parvinder S. Sethi

Figure 9.22 Population Density of San Francisco Downtown San Francisco sits near the active plate boundary between the Pacific plate and the North American plate. The high population density poses a risk to residents in the event of an earthquake. Although pumping fluids into a fault zone may relieve the pressure on a fault and prevent major earthquakes from occurring, people in populated areas are reluctant to risk earthquake control, lest a major earthquake be initiated by the process.

which large amounts of a high-pressure water mix are injected into the wells to open up the preexisting fractures to allow gas to flow into the wells. This increase in earthquake activity in the Dallas–Ft. Worth area, and in other areas where hydraulic fracturing is common, has resulted in geologists looking closely to see if there is a cause-and-effect relationship between hydraulic fracturing and the onset of small earthquakes.

Based on these results, some geologists have proposed that fluids be pumped into the locked segments, or seismic gaps, of active faults to cause small- to moderate-size earthquakes. They think that this action would relieve the pressure on the fault and prevent a major earthquake from occurring.

Although this plan is intriguing, it also has many potential problems. For instance, there is no guarantee that only a small earthquake might result. Instead, a major earthquake might occur, causing tremendous property damage and loss of life, especially in a densely populated area (Figure 9.22). Who would be responsible? Certainly, a great deal more research is needed before such an experiment is performed, even in an area of low population density.

Figure 9.23 Earth's Internal Structure The inset shows Earth's outer part in more detail. The asthenosphere is solid but behaves plastically and flows.

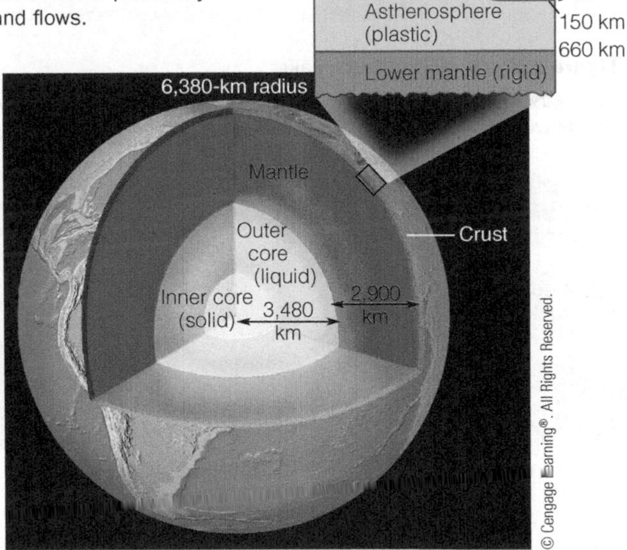

9.11 What Is Earth's Interior Like?

During most of historic time, Earth's interior was perceived as an underground world of vast caverns, heat, and sulfur gases, populated by demons. By the 1860s, scientists knew the average density of Earth and knew that pressure and temperature increased with depth. And even though Earth's interior is hidden from direct observation, scientists now have a reasonably good idea of its internal structure and composition.

Earth is generally depicted as consisting of concentric layers that differ in composition and density and that are separated from adjacent layers by rather distinct boundaries (Figure 9.23). Recall that the outermost layer, or *crust*,

Earth's Composition and Density

	Composition	Density (g/cm³)
Continental crust	Average composition of granodiorite	≈2.7
Oceanic crust	Upper part basalt, lower part gabbro	≈3.0
Mantle	Peridotite (made up of ferromagnesian silicates)	3.3–5.7
Outer core	Iron with perhaps 12% sulfur, silicon, oxygen, nickel, and potassium	9.9–12.2
Inner core	Iron with 10%–20% nickel	12.6–13.0
Earth		5.5

is Earth's thin skin. Below the crust and extending about halfway to Earth's center is the *mantle*, which accounts for more than 80% of Earth's volume. The central part of Earth consists of a *core*, which is divided into a solid inner portion and a liquid outer part (Figure 9.23).

The behavior and travel times of P- and S-waves provide geologists with information about Earth's internal structure. Seismic waves travel outward as wavefronts from their source areas, although it is most convenient to depict them as *wave rays*, which are lines showing the direction of movement of small parts of wavefronts (Figure 9.3a).

As noted earlier, P- and S-wave velocity is determined by the density and elasticity of the materials they travel through, both of which increase with depth. Wave velocity is slowed by increasing density, but it increases in materials with greater elasticity. Because elasticity increases with depth faster than density, a general increase in seismic wave velocity takes place as the waves penetrate to greater depths. P-waves travel faster than S-waves under all circumstances, but unlike P-waves, S-waves are not transmitted through liquids because liquids have no shear strength (rigidity); liquids simply flow in response to shear stress.

Because Earth is not a homogeneous body, seismic waves travel from one material into another of different density and elasticity, and thus their velocity and direction of travel change. That is, the waves are bent, a phenomenon known as *refraction*, in much the same way as light waves are refracted as they pass from air into more dense water. Because seismic waves pass through materials of differing density and elasticity, they are continually refracted so that their paths are curved. Wave rays travel in a straight line only when their direction of travel is perpendicular to a boundary (Figure 9.24).

In addition to refraction, seismic rays are *reflected*, much as light is reflected from a mirror. When seismic rays encounter a boundary separating materials of different density or elasticity, some of a wave's energy is reflected back to the surface (Figure 9.24). If we know the wave velocity and the time required for the wave to travel from its source to the boundary and back to the surface, we can calculate the depth of the reflecting boundary. Such information is useful in determining not only Earth's internal structure but also the depths of sedimentary rocks that may contain petroleum. Seismic reflection is a common tool used in petroleum exploration.

ConnectionLink

To review how petroleum forms and is trapped beneath Earth's surface, see Chapter 7.

Although changes in seismic wave velocity occur continuously with depth, P-wave velocity increases suddenly at the base of the crust and decreases abruptly at a depth of approximately 2,900 km (Figure 9.25). These marked changes in seismic wave velocity indicate a boundary called a **discontinuity** across which a significant change in Earth materials or their properties occurs. Discontinuities are the basis for subdividing Earth's interior into concentric layers.

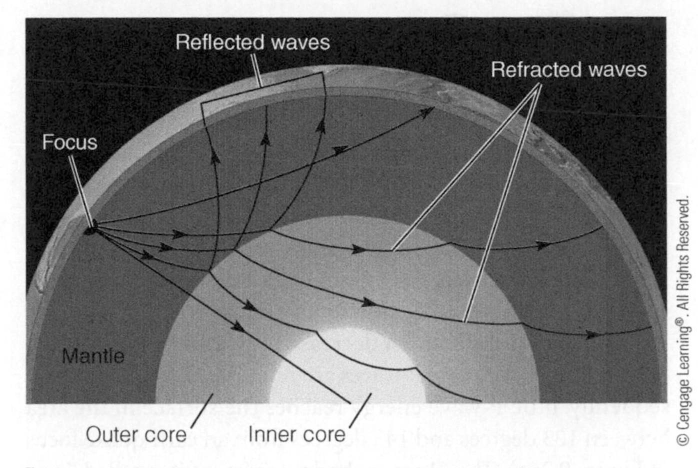

Figure 9.24 Refraction and Reflection of Seismic Waves Refraction and reflection of P-waves as they encounter boundaries separating materials of different density or elasticity. Notice that the only wave ray not refracted is the one perpendicular to boundaries.

Critical Thinking Question If Earth were a homogeneous body, how would seismic waves behave as they moved through Earth?

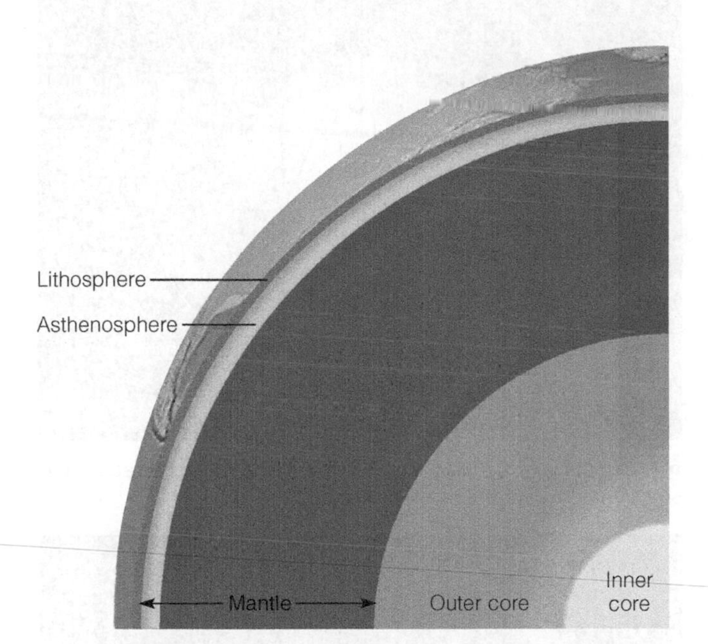

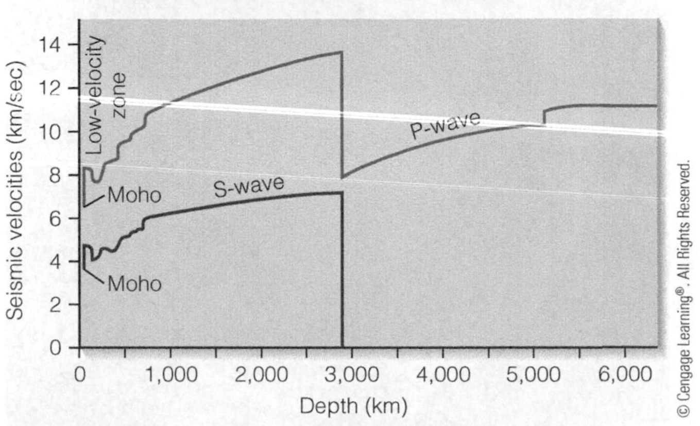

Figure 9.25 Seismic Wave Velocities Profiles showing seismic wave velocities versus depth. Several discontinuities are shown, across which seismic wave velocities change rapidly.

9.12 The Core

In 1906, R. D. Oldham of the Geological Survey of India realized that seismic waves arrived later than expected at seismic stations more than 130 degrees from an earthquake focus. He postulated that Earth has a core that transmits seismic waves more slowly than shallower Earth materials. We now know that P-wave velocity decreases markedly at a depth of 2,900 km, which indicates an important discontinuity recognized as the core–mantle boundary (Figure 9.25).

Because of the sudden decrease in P-wave velocity at the core–mantle boundary, P-waves are refracted in the core. Consequently, little P-wave energy reaches the surface in the area between 103 degrees and 143 degrees from an earthquake focus (▌Figure 9.26a). This **P-wave shadow zone**, as it is called, is an area in which little P-wave energy is recorded by seismographs.

▌Figure 9.26 P-Wave and S-Wave Shadow Zones

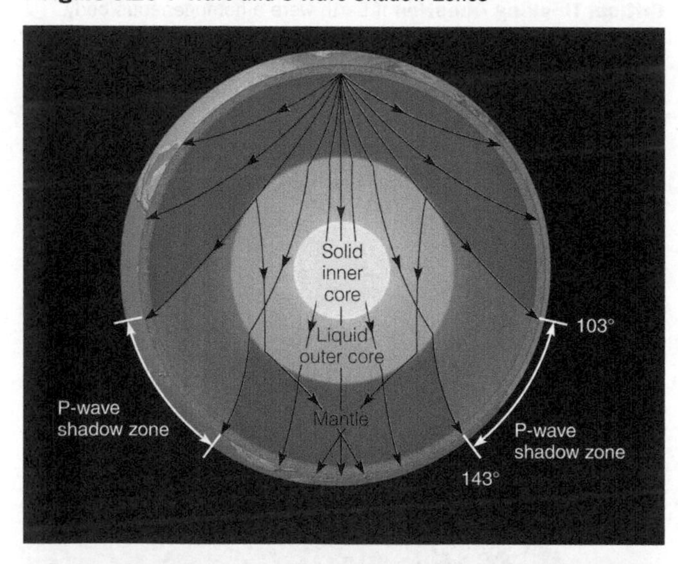

a P-waves are refracted so that no direct P-wave energy reaches the surface in the P-wave shadow zone.

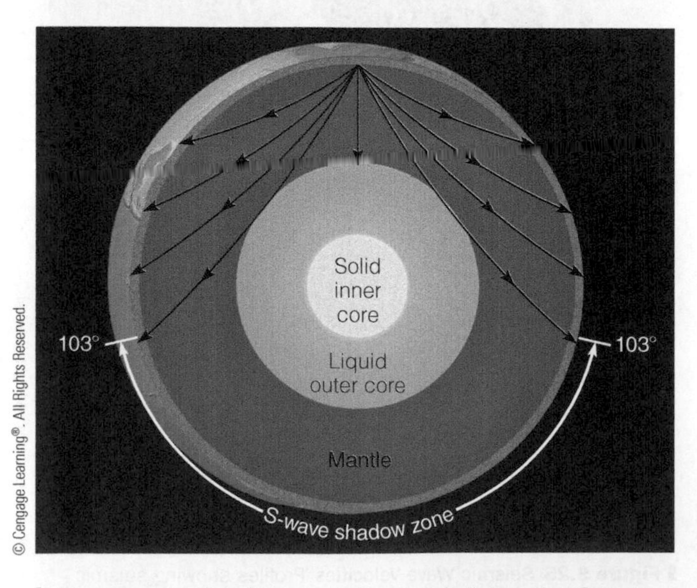

b The presence of an S-wave shadow zone indicates that S-waves are being blocked within Earth.

The P-wave shadow zone is not a perfect shadow zone because some weak P-wave energy is recorded within it. Scientists proposed several hypotheses to account for this observation, but all were rejected by the Danish seismologist Inge Lehmann, who in 1936 postulated that the core is not entirely liquid as previously thought. She proposed that seismic wave reflection from a solid inner core accounts for the arrival of weak P-wave energy in the P-wave shadow zone, a proposal that was quickly accepted by seismologists.

In 1926, the British physicist Harold Jeffreys realized that S-waves were not simply slowed by the core but were completely blocked by it. So, besides a P-wave shadow zone, a much larger and more complete **S-wave shadow zone** also exists (▌Figure 9.26b). At locations greater than 103 degrees from an earthquake focus, no S-waves are recorded, which indicates that S-waves cannot be transmitted through the core. S-waves will not pass through a liquid, so it seems that the outer core must be liquid or behave as a liquid. The inner core, however, is thought to be solid because P-wave velocity increases at the base of the outer core.

Density and Composition of the Core

The core constitutes 16.4% of Earth's volume and nearly one-third of its mass. Geologists can estimate the core's density and composition by using seismic evidence and laboratory experiments. Furthermore, meteorites, which are thought to represent remnants of the material from which the solar system formed, are used to make estimates of density and composition. From these studies, the density of the outer core has been determined to vary from 9.9 to 12.2 g/cm^3. At Earth's center, the pressure is equivalent to approximately 3.5 million times normal atmospheric pressure.

The core cannot be composed of minerals common at the surface because, even under the tremendous pressures at great depth, they still would not be dense enough to yield an average density of 5.5 g/cm^3 for Earth. Both the outer and inner cores are thought to be composed largely of iron, but pure iron is too dense to be the sole constituent of the outer core. It must be "diluted" with elements of lesser density. Laboratory experiments and comparisons with iron meteorites indicate that perhaps 12% of the outer core consists of sulfur and possibly some silicon, oxygen, nickel, and potassium (Figure 9.23).

In contrast, pure iron is not dense enough to account for the estimated density of the inner core, so perhaps 10%–20% of the inner core consists of nickel. These metals form an iron–nickel alloy thought to be sufficiently dense under the pressure at that depth to account for the density of the inner core.

 When the core formed during early Earth history, it was probably entirely molten and has since cooled so that its interior has crystallized. Indeed, the inner core

ConnectionLink
To review the early history of Earth and its place in our solar system, see Figure 1.10 in Chapter 1.

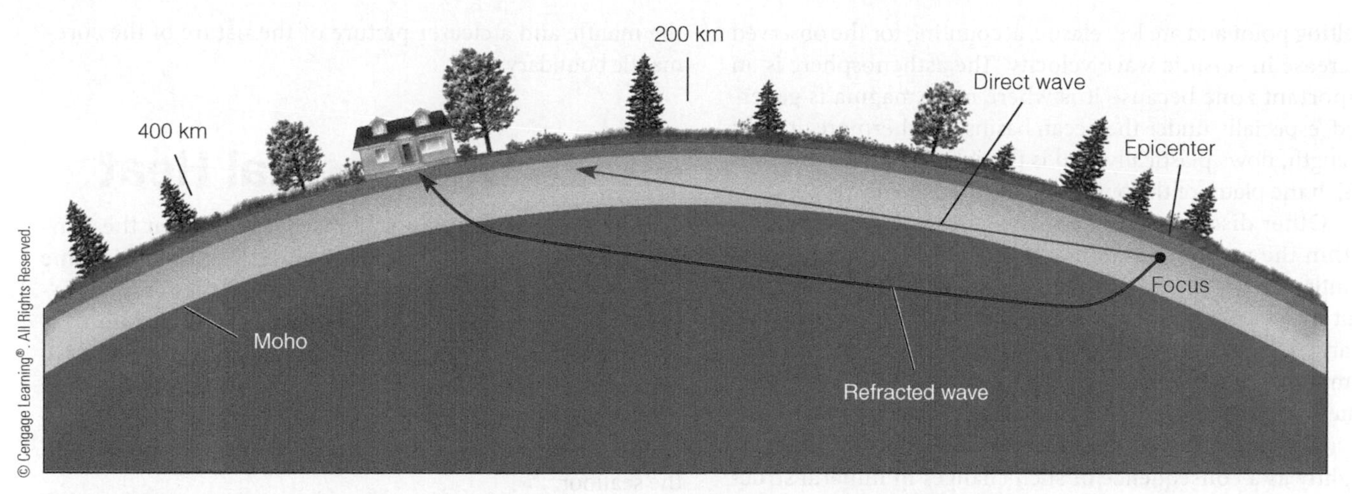

Figure 9.27 Seismic Discontinuity Andrija Mohorovičić studied seismic waves and detected a seismic discontinuity at a depth of about 30 km. The deeper, faster seismic waves arrive at seismic stations first, even though they travel farther. This discontinuity, now known as the Moho, is between the crust and mantle.

continues to grow as Earth slowly cools, and the liquid of the outer core crystallizes as iron. Recent evidence also indicates that at present, the inner core rotates faster than the outer core, moving approximately 20 km/yr relative to the outer core.

9.13 Earth's Mantle

Another significant discovery about Earth's interior was made in 1909 when the Yugoslavian seismologist Andrija Mohorovičić detected a seismic discontinuity at a depth of about 30 km. While studying the arrival times of seismic waves from Balkan (part of southeastern Europe) earthquakes, Mohorovičić noticed that seismic stations a few hundred kilometers from an earthquake's epicenter were recording two distinct sets of P- and S-waves.

From his observations, Mohorovičić concluded that a sharp boundary separates rocks with different properties at a depth of about 30 km. He postulated that P-waves below this boundary travel at 8 km/sec, whereas those above the boundary travel at 6.75 km/sec. When an earthquake occurs, some waves travel directly from the focus to a seismic station, whereas others travel through the deeper layer and some of their energy is refracted back to the surface (Figure 9.27). The waves traveling through the deeper layer (the mantle) travel farther to a seismic station, but they do so more rapidly and arrive before those that travel more slowly in the shallower layer.

The boundary identified by Mohorovičić separates the crust from the mantle and is now called the **Mohorovičić discontinuity**, or simply the **Moho**. It is present everywhere except beneath spreading ridges. However, its depth varies: beneath continents, it ranges from 20 to 90 km, with an average of 35 km; beneath the seafloor, it is 5 to 10 km deep.

The Mantle's Structure, Density, and Composition

Although seismic wave velocity in the mantle increases with depth, several discontinuities exist. Between depths of 100 and 250 km, both P- and S-wave velocities decrease markedly (Figure 9.28). This 100- to 250-km-deep layer is the *low-velocity zone*, which corresponds closely to the *asthenosphere* (Figure 9.23), a layer in which the rocks are close to their

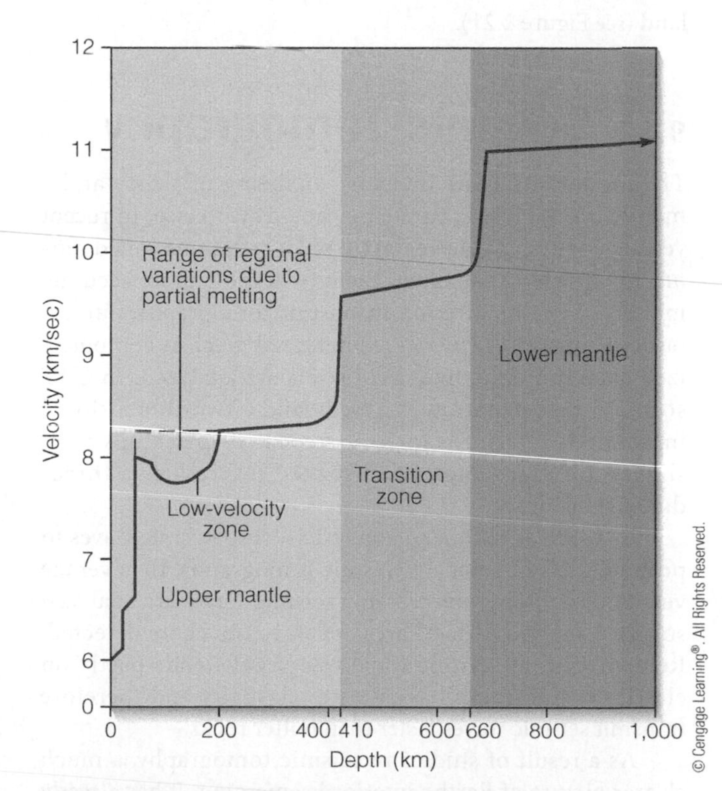

Figure 9.28 Variation in P-Wave Velocity Variations in P-wave velocity in the upper mantle and transition zone.

melting point and are less elastic, accounting for the observed decrease in seismic wave velocity. The asthenosphere is an important zone because it is where most magma is generated, especially under the ocean basins. Furthermore, it lacks strength, flows plastically, and is thought to be the layer over which the plates of the outer, rigid *lithosphere* move.

Other discontinuities are also present at deeper levels within the mantle. But unlike those between the crust and mantle or between the mantle and core, these probably represent structural changes in minerals rather than compositional changes. In other words, geologists think that the mantle is composed of the same material throughout, but the structural states of minerals such as olivine change with depth.

At a depth of 410 km, seismic wave velocity increases slightly as a consequence of such changes in mineral structure (Figure 9.28). Another velocity increase occurs at about 660 km, where the minerals break down into metal oxides, such as iron oxide (FeO) and magnesium oxide (MgO), and into silicon dioxide (SiO_2). These two discontinuities define the top and base of a *transition zone* separating the upper mantle from the lower mantle (Figure 9.28).

Although the mantle's density, which varies from 3.3 to 5.7 g/cm^3, can be inferred rather accurately from seismic waves, its composition is less certain. The igneous rock *peridotite*, containing mostly ferromagnesian silicates, is considered the most likely component of the upper mantle. Laboratory experiments indicate that it possesses physical properties that account for the mantle's density and observed rates of seismic wave transmissions. Peridotite also forms the lower parts of igneous rock sequences thought to be fragments of the oceanic crust and upper mantle emplaced on land (see Figure 2.21).

9.14 Seismic Tomography

The model of Earth's interior consisting of a core and a mantle is probably accurate but not very precise. In recent years, geophysicists have developed a technique called *seismic tomography* that allows them to develop more accurate models of Earth's interior. In seismic tomography, numerous crossing seismic waves are analyzed much as computerized axial tomography (CAT) scans are analyzed. In CAT scans, X-rays penetrate the body and a two-dimensional image of its interior is formed. Repeated CAT scans from slightly different angles are stacked to produce a three-dimensional image.

In a similar manner, geophysicists use seismic waves to probe Earth's interior. In seismic tomography, the average velocities of numerous crossing seismic waves are analyzed so that "slow" and "fast" areas of wave travel are detected. Remember that seismic wave velocity depends partly on elasticity; cold rocks have greater elasticity and therefore transmit seismic waves faster than hotter rocks.

As a result of studies in seismic tomography, a much clearer picture of Earth's interior is emerging. It has already given us a better understanding of complex convection within the mantle and a clearer picture of the nature of the core–mantle boundary.

9.15 Earth's Internal Heat

During the 19th century, scientists realized that the temperature in deep mines increases with depth, and this same trend has been observed in deep drill holes. This temperature increase with depth, or **geothermal gradient**, is approximately 25°C/km near the surface. In areas of active or recently active volcanism, the geothermal gradient is greater than in adjacent nonvolcanic areas, and temperature rises faster beneath spreading ridges than elsewhere beneath the seafloor.

Much of Earth's internal heat is generated by radioactive decay, especially the decay of isotopes of uranium and thorium and, to a lesser degree, potassium-40. When these isotopes decay, they emit energetic particles and gamma rays that heat surrounding rocks. Because rock is such a poor conductor of heat, it takes little radioactive decay to build up considerable heat, given enough time.

ConnectionLink
To learn more about radioactive decay and the different radioactive elements, see Table 17.1 in Chapter 17.

Unfortunately, the geothermal gradient is not useful for estimating temperatures at great depth. If we were simply to extrapolate from the surface downward, the temperature at 100 km would be so high that, despite the great pressure, all known rocks would melt. Yet except for pockets of magma, it appears that the mantle is solid rather than liquid because it transmits S-waves. Accordingly, the geothermal gradient must decrease markedly.

Current estimates of the temperature at the base of the crust are 800°–1,200°C. The latter figure seems to be an upper limit; if it were any higher, melting would be expected. Furthermore, fragments of mantle rock thought to have come from depths of 100–300 km appear to have reached equilibrium at these depths at a temperature of approximately 1,200°C. At the core–mantle boundary, the temperature is probably between 2,500° and 5,000°C; the wide range of values indicates the uncertainties of such estimates. If these figures are reasonably accurate, the geothermal gradient in the mantle is only about 1°C/km.

9.16 Earth's Crust

Our main concern in the latter part of this chapter is Earth's interior; however, to be complete, we must briefly discuss the crust, which along with the upper mantle constitutes the lithosphere.

Continental crust is complex, consisting of all rock types, but it is usually described as "granitic," meaning that its overall composition is similar to that of granitic rocks.

With the exception of metal-rich rocks such as iron ore deposits, most rocks of the continental crust have densities between 2.5 and 3.0 g/cm^3, with the average density of the crust being about 2.7 g/cm^3. P-wave velocity in continental crust is about 6.75 km/sec, but at the base of the crust, P-wave velocity abruptly increases to about 8 km/sec. Continental crust averages 35 km thick, but its thickness varies from 20 to 90 km. Beneath mountain ranges such as the Rocky Mountains, the Alps in Europe, and the Himalayas in Asia, continental crust is much thicker than it is in adjacent areas. In contrast, continental crust is much thinner than average beneath the Rift Valleys of East Africa and in a large area called the Basin and Range Province in the western United States and northern Mexico. The crust in these areas has been stretched and thinned in what appear to be the initial stages of rifting (see Chapter 2).

In contrast to continental crust, oceanic crust is simpler, consisting of gabbro in its lower part overlain by basalt. It is thinnest, about 5 km, at spreading ridges, and nowhere is it thicker than 10 km. Its average density of 3.0 g/cm^3 accounts for the fact that it transmits P-waves at about 7 km/sec. In fact, if oceanic crust is composed of basalt and gabbro, this P-wave velocity is exactly what one would expect.

Key Concepts Review

- Earthquakes are vibrations caused by the sudden release of energy, usually along a fault.
- The elastic rebound theory is an explanation for how energy is released during earthquakes. As rocks on opposite sides of a fault are subjected to force, they accumulate energy and slowly deform until their internal strength is exceeded. At that time, a sudden movement occurs along the fault, releasing the accumulated energy, and the rocks snap back to their original, undeformed shape.
- Seismology is the study of earthquakes. Earthquakes are recorded on seismographs, and the record of an earthquake is a seismogram.
- An earthquake's focus is the location where rupture within Earth's lithosphere occurs and energy is released. The epicenter is the point on Earth's surface directly above the focus.
- Approximately 80% of all earthquakes occur in the circum-Pacific belt, 15% within the Mediterranean–Asiatic belt, and the remaining 5% mostly in the interior of the plates and along oceanic spreading ridges.
- The two types of body waves are P-waves (primary waves), which are compressional (expanding and compressing) and the fastest seismic waves, traveling through all material, and S-waves (secondary waves), which are shear (moving material perpendicular to the direction of travel) and slower than P-waves and can travel only through solids.
- Rayleigh waves (R-waves) and Love waves (L-waves) move along or just below Earth's surface.
- An earthquake's epicenter is determined using a time–distance graph of the P- and S-waves to calculate how far away a seismic station is from an earthquake. The greater the difference in arrival times between the two waves, the farther away the seismic station is from the earthquake. Three seismic stations are needed to locate the epicenter.
- Intensity is a subjective, or qualitative, measure of the kind of damage done by an earthquake. It is expressed in values from I to XII in the Modified Mercalli Intensity Scale.
- The Richter Magnitude Scale measures an earthquake's magnitude, which is the total amount of energy released by an earthquake at its source. It is an open-ended scale with values beginning at 1. Each increase in magnitude number represents about a 30-fold increase in energy released.
- The seismic-moment magnitude scale more accurately measures the total energy released by very large earthquakes.
- The destructive effects of earthquakes include ground shaking, fire, tsunami, landslides, and disruption of vital services.
- Seismic risk maps help geologists determine the likelihood and potential severity of future earthquakes based on the intensity of past earthquakes.
- Earthquake precursors are changes preceding an earthquake and include seismic gaps, changes in surface elevations, and fluctuations of water levels in wells.
- A variety of earthquake research programs are under way in various countries. Studies indicate that most people would probably not heed a short-term earthquake warning.
- Although it is unlikely that earthquakes can ever be prevented, it might be possible to release small amounts of the energy stored in rocks and thus avoid a large, devastating earthquake.
- Various studies indicate that Earth has an outer layer of oceanic and continental crust below which lies a rocky mantle and an iron-rich core with a solid inner part and a liquid outer part.
- Density and elasticity of Earth materials determine the velocity of seismic waves. Seismic waves are refracted when their direction of travel changes. Wave reflection occurs at boundaries across which the properties of rocks change.
- Geologists use the behavior of P- and S-waves and the presence of the P- and S-wave shadow zones to estimate the density and composition of Earth's interior, as well as to estimate the size and depth of the core and mantle.

- Earth's inner core is probably made up of iron and nickel, whereas the outer core is mostly iron with 10%–20% other substances.
- Peridotite, an igneous rock composed mostly of ferromagnesian silicates, is the most likely rock making up Earth's mantle.
- Oceanic crust is composed of basalt and gabbro, whereas continental crust has an overall composition similar to that of granite. The Moho is the boundary between the crust and the mantle.
- The geothermal gradient of 25°C/km cannot continue to great depths; within the mantle and core, it is probably about 1°C/km. The temperature at Earth's center is estimated to be 6,500°C.

Important Terms

discontinuity (p. 227)
earthquake (p. 202)
elastic rebound theory (p. 203)
epicenter (p. 204)
focus (p. 204)
geothermal gradient (p. 230)
intensity (p. 210)
Love wave (L-wave) (p. 208)
magnitude (p. 213)
Modified Mercalli Intensity Scale (p. 210)
Mohorovičić discontinuity (Moho) (p. 229)
P-wave (p. 208)
P-wave shadow zone (p. 228)
Rayleigh wave (R-wave) (p. 208)
Richter Magnitude Scale (p. 213)
seismograph (p. 204)
seismology (p. 204)
S-wave (p. 208)
S-wave shadow zone (p. 228)
tsunami (p. 216)

Review Questions

1. Regions along a major fault that are locked and not releasing energy are prime sites for future earthquakes and are known as
   a. ______ epicenters.
   b. ______ foci.
   c. ______ hypocenters.
   d. ______ discontinuities.
   e. ______ seismic gaps.
2. The minimum number of seismographs needed to determine an earthquake's epicenter is
   a. ______ 1.
   b. ______ 2.
   c. ______ 3.
   d. ______ 4.
   e. ______ 5.
3. The majority of all earthquakes take place in the
   a. ______ spreading-ridge zone.
   b. ______ Mediterranean–Asiatic belt.
   c. ______ rifts in continental interiors.
   d. ______ circum-Pacific belt.
   e. ______ Appalachian fault zone.
4. How much more energy is released by a magnitude 5.0 earthquake than a magnitude 2.0 earthquake?
   a. ______ 4.
   b. ______ 810,000.
   c. ______ 27,000.
   d. ______ 90.
   e. ______ 2,500,000.
5. The seismic discontinuity at the base of the crust is known as the
   a. ______ transition zone.
   b. ______ magnetic reflection point.
   c. ______ low-velocity zone.
   d. ______ Moho.
   e. ______ high-velocity zone.
6. Refer to the graph in Figure 9.12. A seismograph in Berkeley, California, recorded the arrival time of an earthquake's P-waves as 6:59:54 p.m. and the S-waves as 7:00:02 p.m. The maximum amplitude of the S-waves as recorded on the seismogram was 75 mm. What was the magnitude of the earthquake, and how far away from Berkeley did it occur?
7. Why do scientists think that the inner core is solid and the outer core is liquid? What is the core composed of, and what is the evidence for this conclusion?
8. Why do insurance companies use the qualitative Modified Mercalli Intensity Scale instead of the quantitative Richter Magnitude Scale in classifying earthquakes?

9. From the arrival times of P- and S-waves shown in the accompanying chart, and the graph in Figure 9.9b, calculate how far away from each seismograph station the earthquake occurred. How would you determine the epicenter of this earthquake?

	Arrival Time of P-Wave	Arrival Time of S-Wave
Station A	2:59:03 p.m.	3:04:03 p.m.
Station B	2:51:16 p.m.	3:01:16 p.m.
Station C	2:48:25 p.m.	2:55:55 p.m.

10. **Creative Thinking Visual Question:** At 3:02 a.m., on August 17, 1999, violent shaking from a magnitude 7.4 earthquake awakened millions of people in Turkey. Unfortunately for many, their houses or apartment buildings collapsed, causing an estimated 17,000 deaths, at least 50,000 injuries, and leaving more than 150,000 buildings moderately to heavily damaged.

Like California, Turkey is situated in an earthquake-prone area. However, there are typically many more deaths and much greater destruction from earthquakes in Turkey than from earthquakes of similar size along the San Andreas Fault. Why is this so? What could some of the factors be that lead to so much more damage in Turkey than in California, even when the earthquakes are the same magnitudes? To answer this question, look at ▮ Figures 1a and b and think about how the types of construction, population density, building codes, the type of fault movement, and other factors generally lead to a greater number of deaths and more destruction in Turkey than in California.

▮ **Figure 1** The Izmit, Turkey, Earthquake of August 17, 1999

Mehmet Celebi, United States Geological Survey

a Some of the numerous severely damaged structures in Izmit, Turkey.

Charles Meuller, United States Geological Survey

b Damage ranged from minimal to complete in this area of Turkey following the magnitude 7.4 earthquake on August 17, 1999.

Global GeoScience Watch During the past century, approximately 3 million people have died as a result of earthquakes. Designing and building earthquake-resistant structures, as well as predicting where and when earthquakes will occur, could save thousands of lives and billions of dollars in property damage. Within the GREENR database, search for "earthquake prediction." In both the "Academic Journals" and "News" sections there are numerous articles related to earthquake prediction. Browse these articles and write a short report answering these questions: (a) Are there different types of earthquake prediction and do any of these types give citizens warnings of earthquakes that will occur within minutes, days, weeks, or months? (b) Which countries are most active in earthquake predictions? (c) What are some of the different methods scientists are using to predict earthquakes?

These strata at Mount Kidd in the Front Ranges of the Canadian Rocks have been contorted into folds that geologists call a syncline (the folded rocks on the right) and an anticline. Just imagine the tremendous forces that deformed these once horizontal or near horizontal layers of sedimentary rock.

Design Pics/Michael Interisano/Getty Images

Deformation, Mountain Building, and Earth's Crust

OUTLINE

HAVE YOU EVER WONDERED?

- What kinds of forces deform rocks?
- How geologists interpret rocks that have been crumpled and fractured?
- Whether deformed rocks are a problem in planning and building highways, dams, and other structures?
- How mountains differ from other parts of Earth's crust?
- Whether there is a relationship between plate tectonics and the origin of mountains?
- Why the continents stand higher than the ocean basins, and why mountains are higher than adjacent parts of continents?

10.1 Introduction

Recall from Chapter 3 that a solid is any rigid substance that resists changes in shape and volume, and yet we see many examples of solid rocks that have been crumpled into folds or fractured or both (❚ Figure 10.1 and see the chapter opening photo). It should be apparent from these figures that "solid as rock," which implies durability and permanence, is not true under all circumstances. You already know that all rocks disintegrate and decompose as they weather (Chapter 6), and they may change by metamorphism to different kinds of rocks (Chapter 8). Likewise, tremendous pressure and temperature cause **deformation,** a general term for all changes in the shape or volume of rocks. Keep in mind that when we mention rocks in this context, we are not referring to individual rocks or stones that you might pick up from a stream channel or a beach, but rather rock layers of considerable geographic extent and masses of rock within Earth's crust.

We mentioned earlier that Earth is a dynamic planet with internal heat that accounts for ongoing volcanism, plate movements, deformation, and mountain building. In contrast, the other terrestrial planets and Earth's Moon, with the possible exception of Venus, show no signs of active volcanism, deformation, and so on. In addition, on Earth mountain building is a continuing process, although it takes place at rates too slow for us to fully appreciate. Nevertheless, deformation and mountain building are occurring at measurable rates in the Andes in South America, the Himalayas of Asia, and elsewhere.

The vast mountain ranges on the continents form mostly at convergent plate boundaries where tremendous deformation takes place accompanied by volcanism, emplacement of plutons, and metamorphism. For instance, the Appalachian Mountains in North America formed when the eastern margin of the continent was the site of an oceanic–continental plate boundary that eventually became a continental–continental plate boundary as two plates collided. In a similar fashion, the Alps in Europe formed as the African plate moved northward and collided with the Eurasian plate, which is still taking place. In any case, in all episodes of mountain building at convergent plate boundaries, deformation is commonplace, so the rocks in these mountains are contorted, fractured, or both.

James S. Monroe

❚ **Figure 10.1** Deformed Rocks Deformation of rocks in Utah by fracturing. Notice that the original continuity of the rock layers has been disrupted by a fracture, or what geologists call a fault.

North America has not always had its present size and shape, nor have any of the other continents. In fact, continents grow by a process called *continental accretion* as new material is added along their margins at convergent plate boundaries. In the case of North America, it began evolving during the Archean Eon (4.0–2.5 billion years ago) and continued to do so as new material was added to the continent along deformation belts. Perhaps even older continental crust was present, but if so we have not identified it so far.

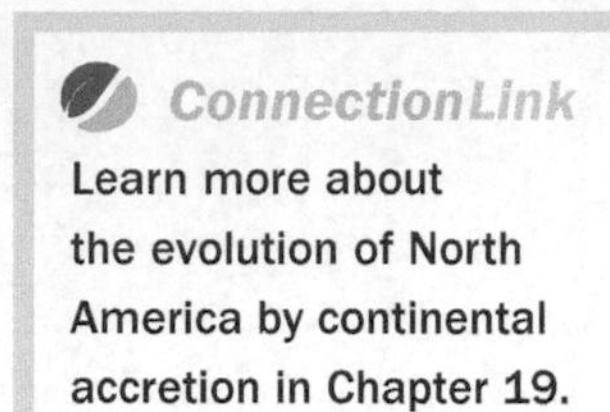

Much of this chapter is devoted to a review of geologic structures (folded and fractured rock layers) that result from deformation. We also emphasize the descriptive terminology for geologic structures, the forces responsible for them, and how to recognize them. The topic may seem rather esoteric, but there are practical reasons to study deformation and mountain building. For one thing, folded and fractured rock layers provide a record of the kinds and intensities of forces that shaped the continents. In other words, they are part of the geologic record, a record that properly interpreted is essential to many engineering projects, such as choosing sites for dams, highways, bridges, and nuclear power plants. In addition, many aspects of the search for and recovery of natural resources, such as iron ore, industrial minerals, coal, petroleum, and natural gas, depend on correctly identifying geologic structures.

10.2 Rock Deformation—How Does It Occur?

We defined the term *deformation* as any change in the volume or shape of rocks. The type of rock is irrelevant, although layered rocks show the effects of deformation most clearly. In any case, rock layers may be crumpled into folds or fractured as a result of **stress,** which results from force applied to a given area of rock. The rock's internal strength resists stress, but if the stress is great enough, the rock undergoes **strain,** which is simply deformation caused by stress. The terminology is a little confusing at first, but the following discussion and reference to ❚ Figure 10.2 will help you understand these terms.

Stress and Strain

Remember that stress is the force applied to a given area of rock, usually expressed in kilograms per square centimeter (kg/cm^2). For example, the stress, or force, exerted by a person

Figure 10.2 Stress and Strain Exerted on an Ice-Covered Pond

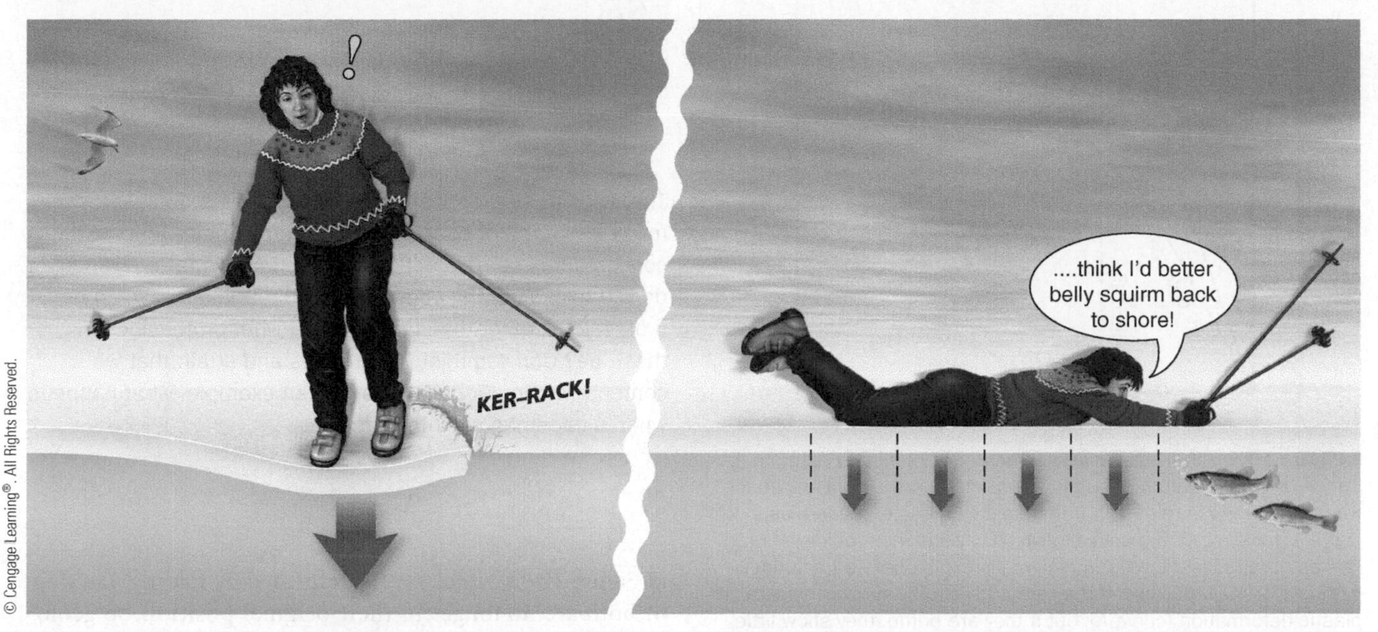

a The woman weighs 65 kg (6,500 g). Her weight is imparted to the ice through her feet, which have a contact area of 120 cm^2. The stress she exerts on the ice (6,500 g/120 cm^2) is 54 g/cm^2. This is now sufficient stress to cause the ice to crack

b To avoid plunging into the freezing water, the woman lays out flat, thereby decreasing the stress she exerts on the ice. Her weight remains the same, but her contact area with the ice is now 3,150 cm^2, so the stress is now only about 2 g/cm^2 (6,500 g/3,150 cm^2), which is well below the threshold needed to crack the ice.

walking on an ice-covered pond is a function of the person's weight and the area beneath her or his feet. The ice's internal strength resists the stress unless the stress is too great, in which case the ice may bend or crack as it is strained (deformed) (Figure 10.2). To avoid breaking through the ice, the person may lie down; this does not reduce the person's weight on the ice, but it does distribute it over a larger area, thus reducing the stress per unit area.

Although stress is force per unit area, it comes in three varieties: *compression, tension,* and *shear,* depending on the direction of the applied forces. In **compression,** rocks or any other object are squeezed or compressed by forces directed toward one another along the same line, as when you squeeze a rubber ball in your hand. Rock layers in compression tend to be shortened in the direction of stress by either folding or fracturing (❚ Figure 10.3a). **Tension** results from forces acting along the same line, but in opposite directions. Tension tends to lengthen rocks or pull them apart (❚ Figure 10.3b). Incidentally, rocks are much stronger in compression than they are in tension. In **shear stress,** forces act parallel to one another, but in opposite directions, resulting in deformation by displacement along closely spaced planes (❚ Figure 10.3c).

Types of Strain

Geologists characterize strain as **elastic strain** if deformed rocks return to their original shape when the deforming forces are relaxed. In Figure 10.2, the ice on the pond may bend under a person's weight but return to its original shape once the person leaves. As you might expect, rocks are not very elastic, but Earth's crust behaves elastically when loaded by glacial ice and depressed into the mantle.

As stress is applied, rocks respond first by elastic strain, but when strained beyond their elastic limit, they undergo **plastic strain** as when they yield by folding, or they behave like brittle solids and **fracture** (❚ Figure 10.4). In either folding or fracturing, the strain is permanent; that is, the rocks do not recover their original shape or volume even if the stress is removed.

Figure 10.3 Stress and Possible Types of Resulting Deformation

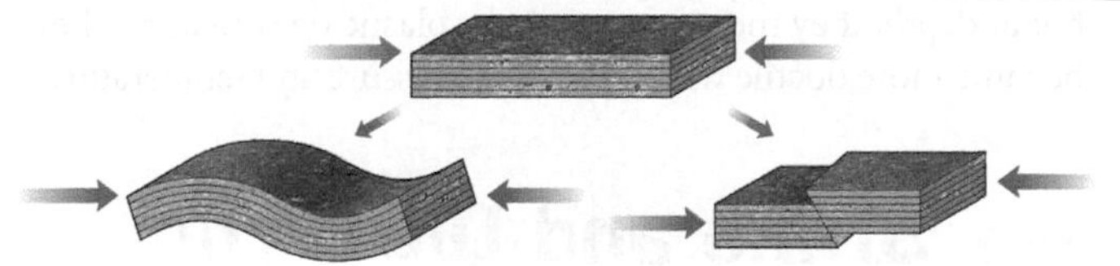

a Compression causes shortening of rock layers by folding or faulting.

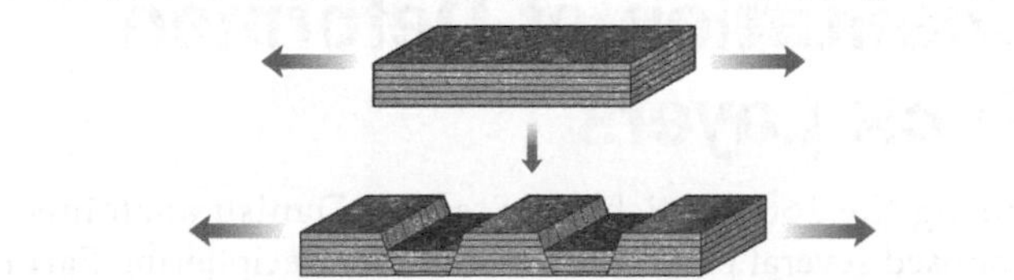

b Tension lengthens rock layers and causes faulting.

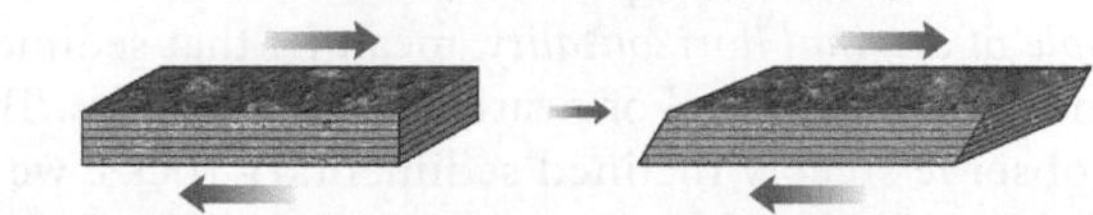

c Shear stress causes deformation by displacement along closely spaced planes.

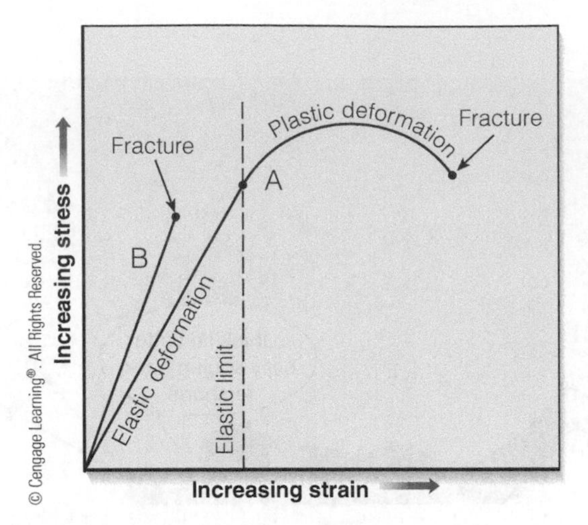

Figure 10.4 **Rock Responses to Stress** Rocks initially respond to stress by elastic deformation and then return to their original shape when the stress is released. If the elastic limit is exceeded, as in curve A, rocks deform plastically, which is permanent deformation. The amount of plastic deformation rocks exhibit before fracturing depends on their ductility. If they are ductile, they show considerable plastic deformation (curve A), but if they are brittle, they show little or no plastic deformation before failing by fracture (curve B).

Whether strain is elastic, plastic, or fracture depends on the kind of stress applied, pressure and temperature, rock type, and the length of time rocks are subjected to stress. A small stress applied over a long period, as on a mantelpiece supported only at its ends, will cause the rock to sag; that is, the rock deforms plastically (Figure 10.4). By contrast, a large stress applied rapidly to the same object, as when struck by a hammer, results in fracture. Rock type is important because not all rocks have the same internal strength and thus respond to stress differently. Some rocks are *ductile,* whereas others are *brittle,* depending on the amount of plastic strain they exhibit. Brittle rocks show little or no plastic strain before they fracture, but ductile rocks exhibit a great deal (Figure 10.4).

Many rocks show the effects of plastic deformation that must have taken place deep within the crust. At or near the surface, rocks commonly behave like brittle solids and fracture, but at depth, they more often yield by plastic deformation; they become more ductile with increasing pressure and temperature.

10.3 Strike and Dip—The Orientation of Deformed Rock Layers

During the 1660s, Nicholas Steno, a Danish anatomist, proposed several principles essential for deciphering Earth history from the record preserved in rocks. One is the *principle of original horizontality,* meaning that sediments accumulate in horizontal or nearly horizontal layers. Thus, if we observe steeply inclined sedimentary rocks, we are justified in inferring that they were deposited nearly horizontally, lithified, and then tilted into their present position

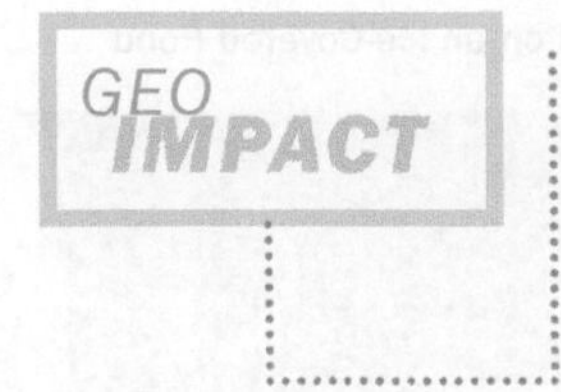

Stress, Strain, and Deformation

The types of stresses, as well as elastic versus plastic strain, might seem rather esoteric, but perhaps understanding these concepts has some practical applications. What relevance do you think knowing about stress and strain has to some professions, other than geology, and what professions might these be? Can you think of stresses and strain that we contend with in our daily lives? As an example, what happens when a car smashes into a tree?

(Figure 10.5a). Rock layers deformed by folding, faulting, or both are no longer in their original position, so geologists use *strike* and *dip* to describe their orientation with respect to a horizontal plane.

By definition, **strike** is the direction of a line formed by the intersection of a horizontal plane and an inclined plane. The surfaces of the rock layers in Figure 10.5b are good examples of inclined planes, whereas the ground surface is a horizontal plane. The direction of the line formed at the intersection of these planes is the strike of the rock layers. The strike line's orientation is determined by using a compass to measure its angle with respect to north. **Dip** is a measure of an inclined plane's deviation from horizontal, so it must be measured at right angles to strike direction (Figure 10.5b).

ConnectionLink

You can find more information about Nicholas Steno and his principle of original horizontality in Chapter 17.

Geologic maps showing the age, aerial distribution, and geologic structures in an area use a special symbol to indicate strike and dip. A long line oriented in the appropriate compass direction indicates strike, and a short line perpendicular to the strike line shows the direction of dip (Figure 10.5b). Adjacent to the strike and dip symbol is a number corresponding to the dip angle. The usefulness of strike and dip symbols will become apparent in the sections on folds and faults.

10.4 Deformation and Geologic Structures

Remember that deformation and its synonym strain refer to changes in the shape or volume of rocks. During deformation, rocks might be crumpled into folds, or

Figure 10.5 Strike and Dip of Deformed Rock Layers

Critical Thinking Question Suppose it is 1 km from the right side of the diagram in **(b)** to the point where the uppermost sandstone layer plunges beneath the surface. Can you predict the depth at which this sandstone would be encountered if you drilled a hole at the right side of the diagram in **(b)**? Explain.

a These layers of sedimentary rocks in Montana were deposited in nearly horizontal layers but have been deformed so that the layers are inclined from horizontal at about 70°. To define the present orientation of these rocks geologists use the concept of strike and dip.

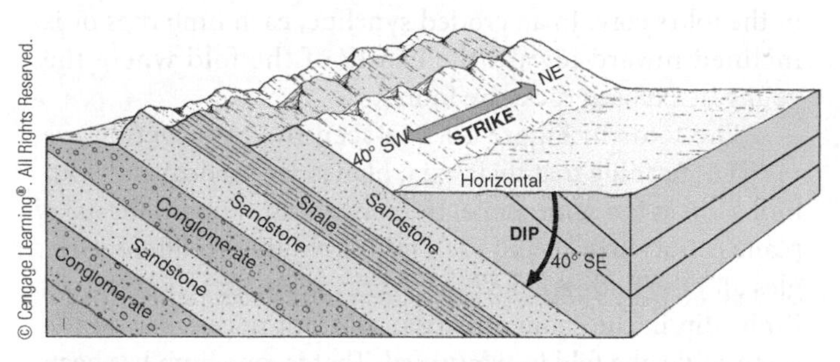

b Strike is the line formed by the intersection of an inclined plane by a horizontal plane. Dip is the maximum angular deviation of the inclined plane from horizontal. Notice the symbol that shows strike and dip.

they might be fractured, or perhaps folded and fractured (Figure 10.1 and see the chapter opening photo). Any of these features resulting from deformation is called a **geologic structure.** Geologic structures are present almost everywhere that rocks are exposed, and many are detected far below the surface by drilling and several geophysical techniques.

Folded Rock Layers

If you place your hands on a tablecloth and move them toward one another, the tablecloth is crumpled into a series of up- and down-arched folds. Rock layers behave similarly when in compression as they deform into **folds,** but in this case the folding is permanent. That is, plastic strain has taken place, so even if the stress is removed, the rock layers remain folded. Most folding probably occurs deep in the crust where pressure and temperature are high and rocks are more ductile than they are at or near the surface. The configuration of folds and the intensity of folding vary, but there are only three basic types of folds: *monoclines, anticlines,* and *synclines.*

Monoclines A simple bend or flexure in otherwise horizontal or uniformly dipping rock layers is a **monocline** (Figure 10.6a). The large monocline in Figure 10.6b formed when the Bighorn Mountains in Wyoming rose vertically along a fracture. The fracture did not penetrate to the surface, so as uplift of the mountains proceeded, the near-surface rocks were bent so that they now appear to be draped over the margin of the uplifted block. In a manner of speaking, a monocline is simply one-half of an anticline or syncline.

Figure 10.6 Monocline

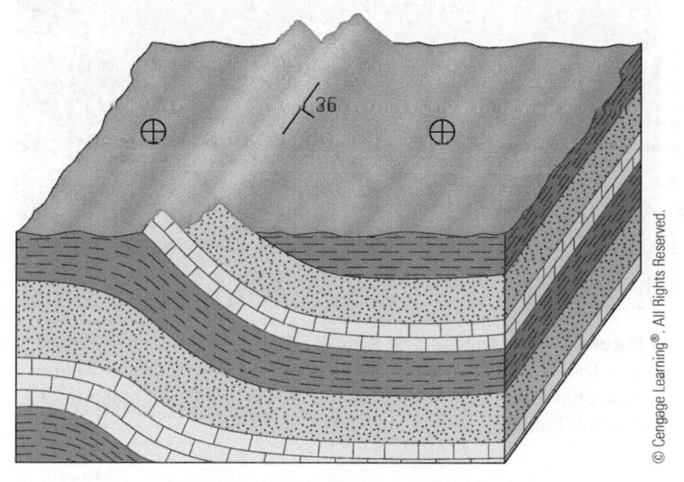

a A monocline. Notice the strike and dip symbol and the circled cross, which is the symbol for horizontal layers.

b A monocline in the Bighorn Mountains in Wyoming. Notice that the strata are horizontal along the crest of the Mountains, then increase in dip significantly, and then become nearly horizontal at the left side of the image.

Anticlines and Synclines Remember our analogy with a table cloth and compression-induced folds produced by moving your hands toward one another. Rock layers in compression respond similarly into up-arched (convex upward) folds called **anticlines** and down-arched (concave downward) folds known as **synclines** (▮ Figure 10.7). Anticlines and synclines also have an axial plane that connects the points of maximum curvature of each folded layer and divides the folds into halves, each of which is a *limb* (▮ Figure 10.8). Most often anticlines and synclines are found in a series with one alternating with the other, so two adjacent folds share a limb. The terms anticline and syncline refer only to folded rock layers, and they do not necessarily correspond to high and low areas on the surface (see the chapter opening photo).

In cross-sectional view, anticlines and synclines are easy to recognize, but most often they are exposed to our view from above only. For example, if ▮ Figure 10.9 showed only a surface view, how would we identify the folds? Actually, strike and dip and the relative ages of the rock layers are sufficient to make this distinction. Notice in Figure 10.9 that in the surface view of the anticline, each limb dips outward or away from the center of the fold, and the oldest exposed rocks are in the fold's core. In an eroded syncline, each limb dips or is inclined inward toward the center of the fold where the youngest exposed rocks are found.

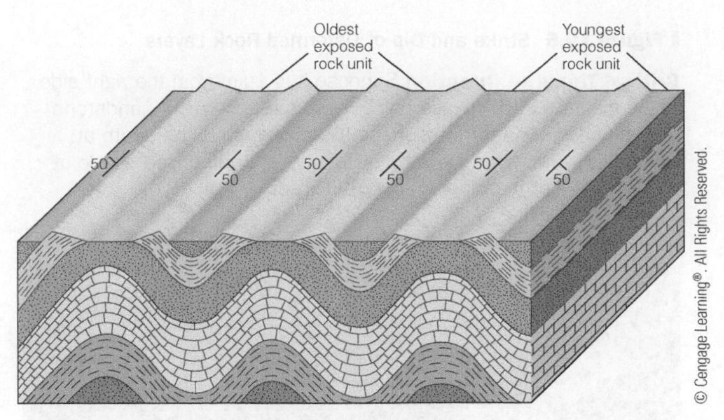

▮ **Figure 10.9** **Eroded Anticlines and Synclines** Geologists identify eroded anticlines and synclines by strike and dip and the relative ages of the folded rock layers.

Critical Thinking Question Show on this illustration the direction of the forces that caused deformation.

▮ **Figure 10.7** **Anticline and Syncline** This syncline and anticline are in the sedimentary rocks of the Old Red Sandstone in Wales in the United Kingdom. Compressive forces during an episode of mountain building caused the folding.

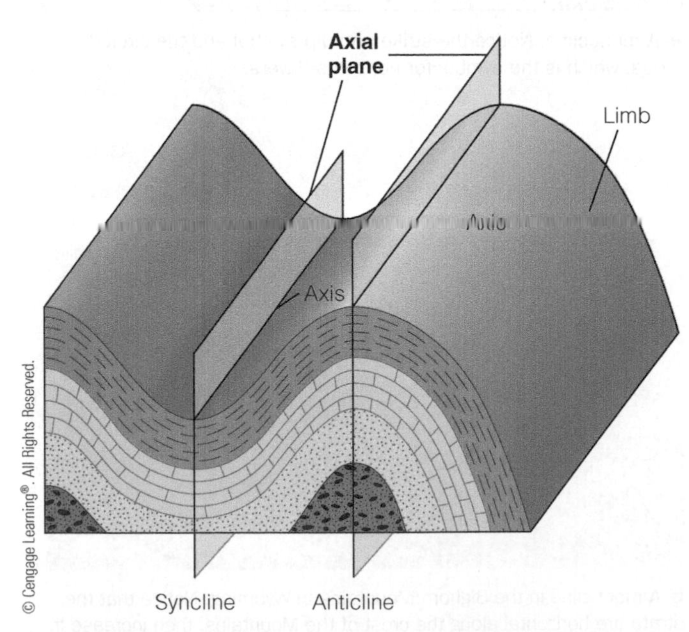

▮ **Figure 10.8** **Syncline and Anticline Axial Planes** Syncline and anticline showing the axial plane, axis, and fold limbs.

The folds in Figure 10.9 as seen in cross section are *upright,* meaning that their axial planes are vertical and both limbs dip at the same angle. In many folds, though, the axial plane is not vertical, and each fold limb dips at different angles and the folds are *inclined* (▮ Figure 10.10a). If both fold limbs dip in the same direction, but not necessarily at the same angle, the fold is *overturned.* That is, one limb has been rotated more than 90 degrees from its original orientation so that it is now upside down (▮ Figure 10.10b). If the axial planes of adjacent folds are horizontal, the folds are *recumbent* (▮ Figure 10.10c).

The distinction of anticlines from synclines is fairly straightforward for upright folds, but it is more difficult for those that are overturned or recumbent. Can you determine which of the folds are anticlines and synclines in 10.10c? Knowing the relative ages of the folded rock layers provides a solution. Remember that an anticline has the oldest rocks in its core, so the fold nearest the surface is an anticline and the lower fold is a syncline.

Plunging Anticlines and Synclines To complicate matters further, folds are characterized as *nonplunging* or *plunging.* All folds have a fold axis, a line formed by the intersection of the axial plane with the folded layers. If the fold axis is horizontal, the fold is nonplunging; more often, however, the fold axis is inclined so that it appears to plunge beneath adjacent rocks, and the fold is said to be plunging (▮ Figure 10.11a).

Figure 10.10 Inclined, Overturned, and Recumbent Folds

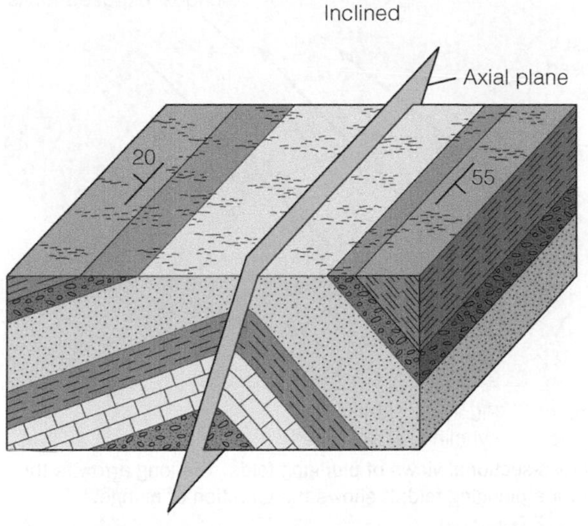

a An inclined fold. The axial plane is not vertical, and the fold limbs dip at different angles.

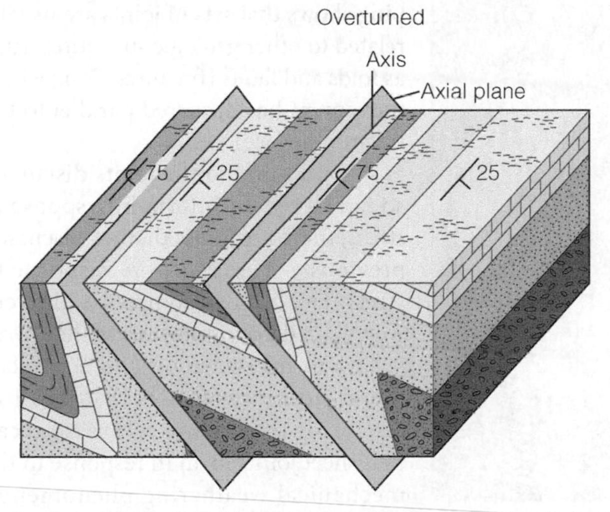

b Overturned folds. Both fold limbs dip in the same direction, but one limb is inverted. Notice the special strike-and-dip symbol to indicate overturned beds.

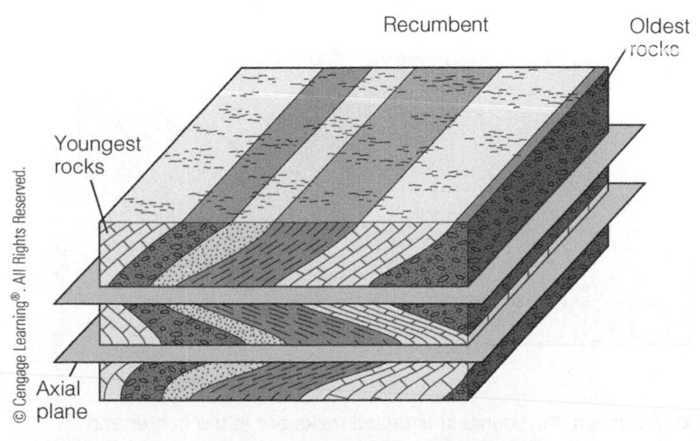

c The axial planes of recumbent folds are horizontal.

It might seem that this additional complication would make distinguishing plunging anticlines from plunging synclines much more difficult, but this is not the case. Geologists use exactly the same criteria that they used to identify nonplunging folds. Accordingly, a plunging anticline has all rock layers dipping outward or away from the center of the fold where the oldest exposed rock layers are found, and a plunging syncline has rock layers dipping inward toward the fold's center where the youngest rocks are exposed (Figure 10.11b, c).

Much of the worlds petroleum production comes from anticlines, although other types of structural traps as well as stratigraphic traps are important, too (see Figure 7.19). Accordingly, geologists are especially interested in correctly identifying geologic structures in areas of potential hydrocarbon production.

Domes and Basins Anticlines and synclines are elongate structures, meaning that their length greatly exceeds their width. In contrast, folds that are nearly equidimensional —that is, oval to circular—are *domes* and *basins.* In a **dome,** all of the folded strata dip outward from a central point (as opposed to outward from a line as in an anticline), and the oldest exposed rocks are at the center of the fold (Figure 10.12a), so we characterize a dome as the circular equivalent of an anticline. In contrast, a **basin,** the circular counterpart of a syncline, has all strata dipping inward toward a central point and the youngest exposed rocks are at the fold's center (Figure 10.12b). Many domes and basins are so large that they can be visualized only on geologic maps or aerial photographs.

Unfortunately, the terms *dome* and *basin* are also used to distinguish high and low areas of Earth's surface, but as with anticlines and synclines, domes and basins resulting from deformation do not necessarily correspond with mountains or valleys. In some of the following discussions, we will use these terms in other contexts, but we will be clear when we refer to surface elevations as opposed to geologic structures.

Joints

Under some circumstances, rock layers will fold as described in the previous sections, but if they are brittle, they may respond to stress by fracturing. **Joints** are fractures along which no movement has taken place parallel to the fracture surface (Figure 10.13). That is, a joint may open up (widen), but the rocks on opposite sides of the fracture do not move up, down, or laterally along the fracture. The term *joint* was coined long ago by coal miners who thought that cracks in rocks were the surfaces where adjacent rocks were "joined" together.

Almost all rocks exposed at the surface have joints that formed by compression, tension, or shearing. They vary in size from minute fractures to those that extend for many kilometers and are often arranged in two or three prominent sets. As we noted in Chapter 6, weathering takes place preferentially

Figure 10.11 Plunging Folds

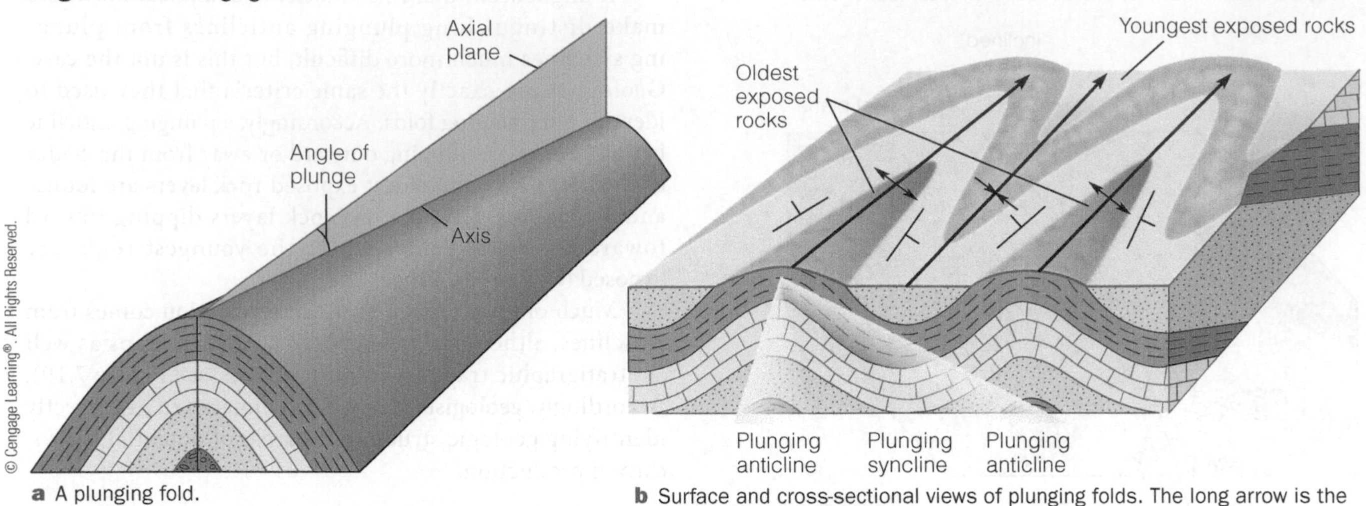

a A plunging fold.

b Surface and cross-sectional views of plunging folds. The long arrow is the geologic symbol for a plunging fold; it shows the direction of plunge.

c The Sheep Mountain Anticline in Wyoming You can tell that this is an anticline by the strike and dip symbols. The long line shows the axis of the anticline and its direction of plunge.

along joints because fluids can more easily penetrate fractures. Regional mapping shows that sets of joints are usually related to other geologic structures, such as folds and faults (fractures along which movement has occurred parallel to the fracture surface).

In addition to the joints discussed so far, all of which form in response to stress, there are joints that we discussed previously. These form in response to stress, too, but in a somewhat different way. Recall columnar joints that form as lava or magma cools and contracts, thereby causing stresses that result in fracturing (see Figure 5.7). Also recall that sheet joints form in response to the mechanical weathering phenomenon called pressure release (see Figure 6.4).

Figure 10.12 Domes and Basins

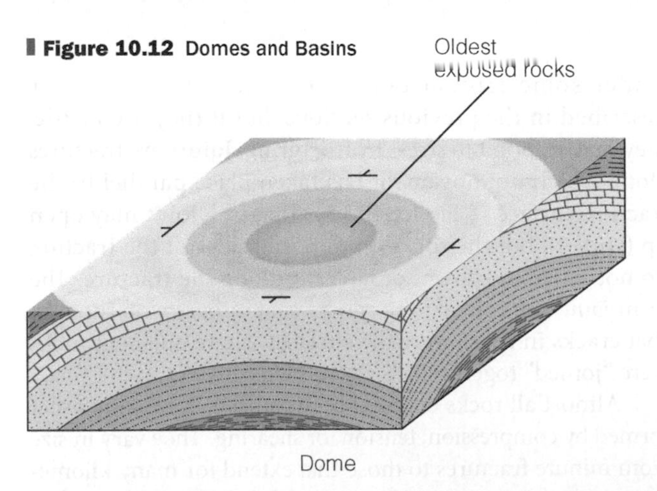

a Notice that in a dome, the oldest exposed rocks are in the center and all rocks dip outward from a central point.

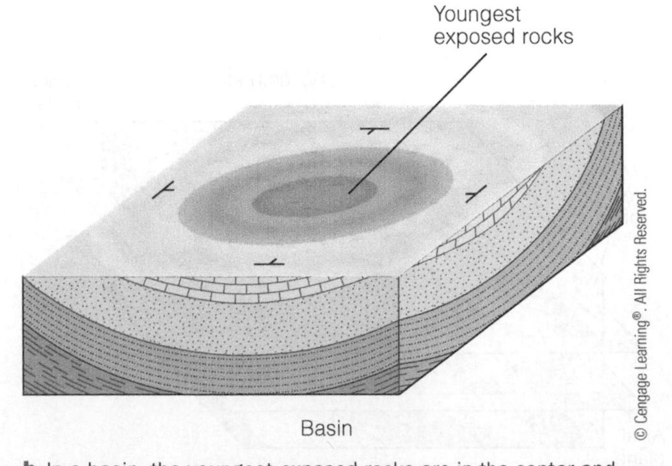

b In a basin, the youngest exposed rocks are in the center and all rocks dip inward toward a central point.

Figure 10.13 Joints Two sets of joints intersect at right angles to form a rectangular fracture pattern in these rocks in Tasmania, Australia. Notice that weathering has made the joints more prominent.

Faults

In Chapter 9, we noted that earthquakes occur when energy is suddenly released, usually as the result of displacement of rocks along fractures, or what geologists call **faults.** Remember that joints are also fractures, but on joints the only movement, if any, is perpendicular to the fracture surface, whereas on faults, blocks on opposite sides of the fracture move parallel with the fault surface, which is a **fault plane.** So, in the case of faults, blocks of rock move up, down, or along the fracture surface (Figure 10.14a).

Not all faults penetrate to the surface, but for those that do they may show a *fault scarp,* a bluff or cliff formed by vertical movement on the fault (Figure 10.14a). In some cases, the fault plane is scratched and polished, but in others, the movement of rocks on opposite sides of the fault grinds and pulverizes the rock into *fault breccia* (Figure 10.14b, c).

Figure 10.14a shows that the rocks overlying the fault plane is the **hanging wall block,** and those that lie beneath the fault plane make up the **footwall block**. The hanging wall and footwall blocks are easily identified on any fault except one that is vertical—that is, a fault that dips at 90 degrees. To identify some faults, you must recognize these two blocks

Figure 10.14 Faults Fractures along which movement has occurred parallel to the fracture surface are called faults.

Critical Thinking Question If the arrows were not shown in **(c)**, could you figure out the relative movement on this fault?

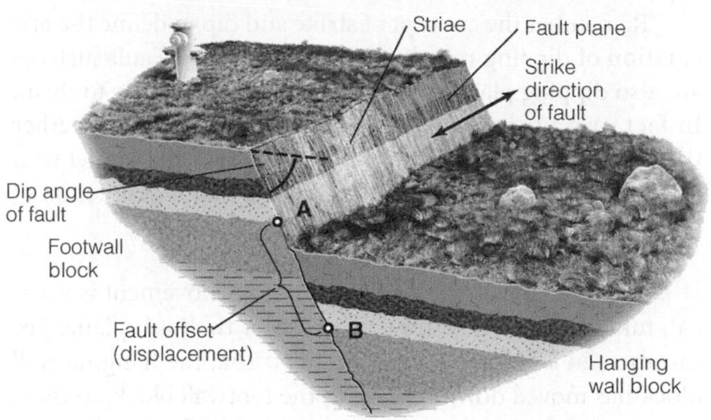

a Terms used to describe the orientation of a fault plane. Striae are scratch marks that form when one block slides past another. You can measure offset or displacement on a fault wherever the truncated end of one feature (point A) can be related to its equivalent across the fault (B).

b A polished fault plane and fault scarp in San Francisco, California. This polished, striated rock surface is called slickenside that resulted from friction as rocks on the fault plane moved past one another.

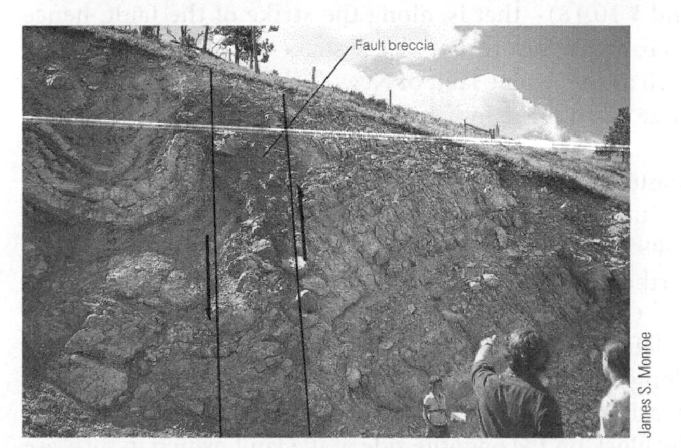

c Fault breccia, the zone of rubble along a fault in the Bighorn Mountains, Wyoming. The arrows show the direction of movement along the fault.

and determine the direction of *relative movement* of the blocks. In Figure 10.14a, the footwall block may have moved up, the hanging wall block may have moved down, or both blocks may have moved. Nevertheless, the hanging wall block appears to have moved down relative to the footwall block.

Remember the concept of strike and dip to define the orientation of dipping rock layers (Figure 10.5b). Fault surfaces are also dipping planes, so the same concept applies to them. In fact, two types of faults are identified based on whether the blocks on opposite sides of the fault moved parallel with the direction of dip (dip-slip faults) or along the direction of strike (strike-slip faults).

Dip-Slip Faults On **dip-slip faults,** all movement is vertical, taking place up or down the dip of the fault plane. For instance, in ▮ Figures 10.15a and ▮ 10.16a, the hanging wall block has moved down relative to the footwall block, so these are **normal faults**. In contrast, **reverse faults** are those on which the hanging wall block moves up relative to the footwall block (▮ Figures 10.15b and ▮ 10.16b). In ▮ Figure 10.15c, the hanging wall block has also moved up relative to the footwall block, but the fault dips at less than 45 degrees, so it is a special type of reverse fault called a **thrust fault**.

Normal faults are found along one of both sides of many mountain ranges in the Basin and Range Province of the western United States where the crust is being stretched and thinned (▮ Figure 10.17). The Sierra Nevada at the western margin of this province has risen more than 3,000 m above the lowlands to the east as the result of movement on normal faults. So, the normal faults in this entire region, or anywhere else for that matter, are caused by tensional stress. Just the opposite is true for reverse and thrust faults, which form in response to compression. Large-scale examples of reverse and thrust faults are found in mountain ranges that formed at convergent plate boundaries where one would expect compression (discussed later in this chapter).

Strike-Slip Faults Shear stress is responsible for movement on **strike-slip faults**. These faults differ from dip-slip faults in that all movement is horizontal (▮ Figures 10.15d and ▮ 10.18)—that is, along the strike of the fault, hence their name. Several very large strike-slip faults are known, such as the Alpine fault of New Zealand, which generated a large earthquake in February 2011. By far the best studied strike-slip fault is the San Andreas Fault that cuts through coastal California, where numerous earthquakes have occurred, including the tragic 1906 San Francisco earthquake, the 1989 Loma Prieta earthquake, and the 1994 Northridge earthquake (see Chapter 9).

Geologists characterize strike-slip faults as right lateral or left lateral, depending on the apparent direction of offset along the fault. Look at Figure 10.15d and imagine you are standing on the block on the left side of the fault. Look at the block on the opposite side of the fault, which in this case appears to have moved to the left, so it is a left-lateral strike-slip fault. Now try this same exercise from a vantage point

Geologic Maps and Land Use Planning

As a member of a planning commission, you are charged with developing zoning regulations and building codes for an area with known active faults, steep hills, and deep soils. A number of contractors, as well as developers and citizens in your community, are demanding action because they want to begin several badly needed housing developments. How might geologic maps and an appreciation of geologic structures influence you in this endeavor? Do you think, considering the probable economic gains for your community, that the regulations you draft should be rather lenient or very strict? If the latter, how would you explain why you favored regulations that would involve additional cost for houses?

on the right-hand block, and the block on the opposite side still appears to have moved to the left. Had this been a right-lateral strike-slip fault, the block on the opposite side of the fault from the observer would have appeared to have moved to his or her right. The San Andreas Fault is a right-lateral strike-slip fault, but the Alpine fault in New Zealand is a left-lateral strike-slip fault.

Oblique-Slip Faults Most faults that you will ever encounter are the ones just described, that is, dip-slip (normal and reverse) or strike-slip, but there are some that combine both dip-slip and strike-slip movements. These **oblique-slip faults** might be reverse and left lateral, normal and right lateral, or any other combination (▮ Figure 10.15e).

As we mentioned in the Introduction, the study of geologic structures is important in the exploration for minerals, oil, and natural gas. In addition, geologic engineers evaluate geologic structures in the planning stages for many projects, especially if they are in tectonically active areas (Geo-Focus 10.1).

10.5 Deformation and the Origin of Mountains

Mountains form in several ways, but the truly large mountains on continents result from compression-induced deformation at convergent plate boundaries. Before discussing mountain building, though, we should define what we mean by the term *mountain* and briefly discuss the types of mountains. *Mountain* is a designation for any area of land that stands significantly higher, at least 300 m, than the surrounding country and has a restricted summit area. Some mountains are single,

Figure 10.15 Types of Faults

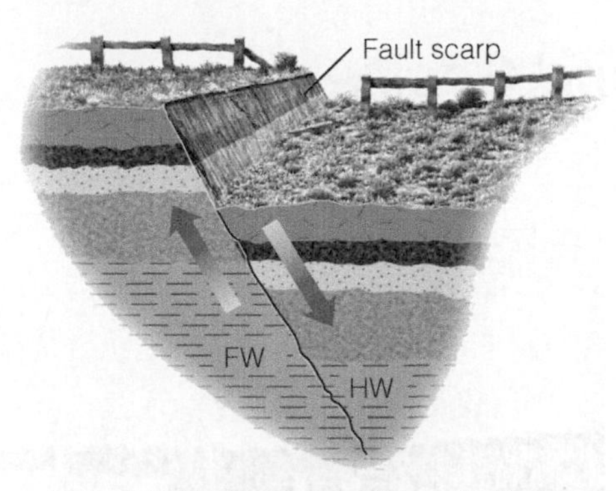

a Normal fault—hanging wall block (HW) moves down relative to the footwall block (FW)

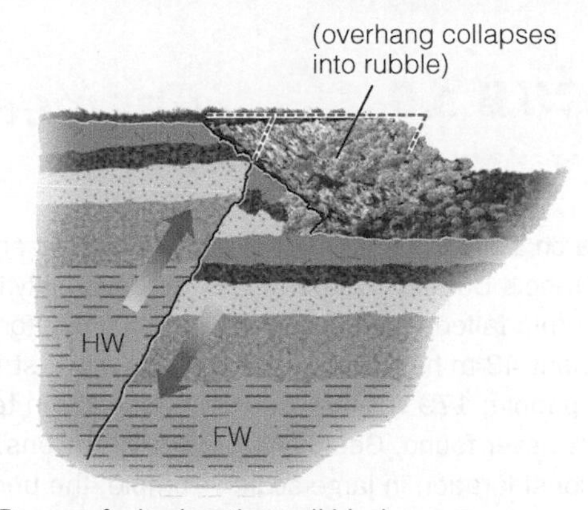

b Reverse fault—hanging wall block moves up relative to the footwall block.

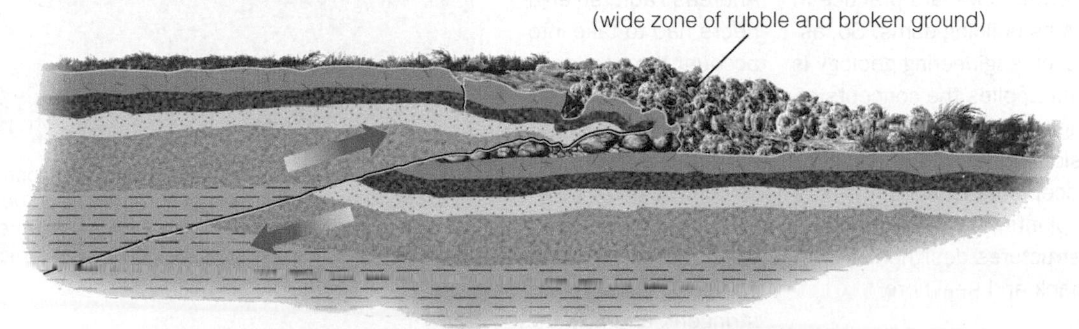

c A thrust is a type of reverse fault with a fault plane dipping at less than 45 degrees.

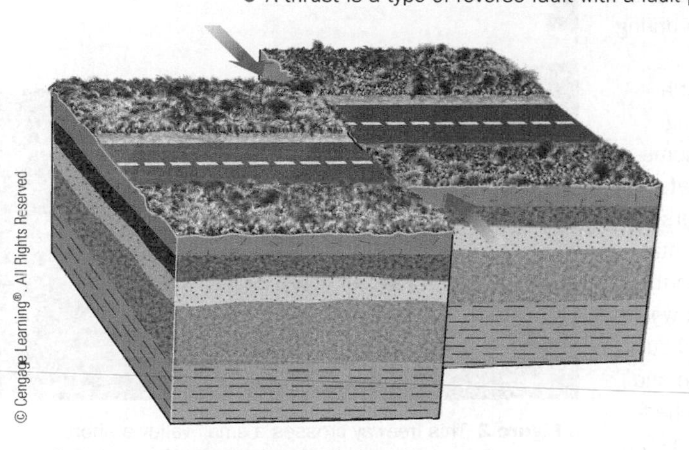

d All movement on strike-slip faults is horizontal.

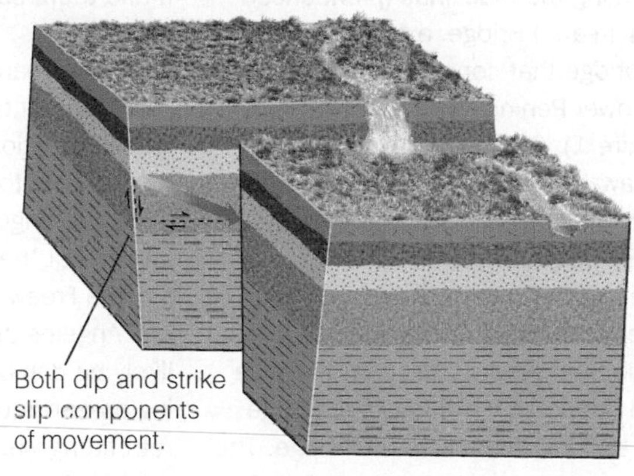

e An oblique-slip fault involves a combination of dip-slip and strike-slip movements.

isolated peaks, but more commonly they are parts of linear associations of peaks and ridges known as *mountain ranges* that are related in age and origin. A *mountain system* is a complex linear zone of deformation and crustal thickening that consists of several or many mountain ranges. The Bitterroot Mountains of Idaho and Montana, the Teton Range in Wyoming, the Wasatch Mountains of Utah, and the Sangre de Cristo Mountains of Colorado and New Mexico are a few of the many ranges in the Rocky Mountain system. The Appalachian Mountains of the eastern United States and Canada are a mountain system with such ranges as the Great Smoky Mountains of North Carolina and Tennessee, the Adirondack Mountains of New York, and the Green Mountains of Vermont.

Mountain Building

Mountains form in several ways, some involving little or no deformation. For example, differential weathering and erosion yield high areas with adjacent lowlands in the southwestern United States, but these erosional remnants are flat topped or pinnacle shaped and go by the names *mesa* and *butte*, and most are less than 300 m high. *Block-faulting* is another way that mountains form, but this is caused by deformation of the crust. It involves movement on normal faults in

ConnectionLink

You can find out more about mesas and buttes in Chapter 15.

GEO FOCUS 10.1

Engineering and Geology

On March 12, 1928, the St. Francis Dam in Southern California failed, resulting in a flood wave about 43 m high that killed at least 450 people; 179 victims of the flood were never found. Geology had been a consideration in large-scale projects for many years previously, but following this catastrophic event, geologic input became standard practice in ventures such as building dams. So, as you might expect, engineering geology is a specialty that applies the concepts of geology to engineering projects, which may involve slope stability studies, and studies for acceptable locations for dams, power plants, highways, tunnels, canals, and structures designed to protect riverbank and seashore communities.

A good example is the concern prior to building the Mackinac (pronounced "mack-in-aw") Bridge, a huge suspension bridge that connects the Upper and Lower Peninsulas of Michigan (▌Figure 1). Geologists and engineers were aware that some of the rock in the area, called Mackinac Breccia, was a collapse breccia, or rubble that formed when caverns collapsed. The concern was whether the breccia or any uncollapsed caverns beneath the area would support the weight of the huge piers and abutments for the bridge. The project was completed successfully in 1957, but detailed studies were done before construction began.

Geologic engineers are invariably involved in planning for any large-scale structure, especially in technically active regions. For example, the bridge in ▌Figure 2 is only a short distance from the San Andreas Fault, so engineers had to take into account the near certainty that it would be badly shaken during an earthquake. Many other structures on or near the San Andreas Fault were constructed when codes were much less stringent, and now they are being retrofitted to make them safer during earthquakes.

Being aware of a problem and taking remedial action sometimes come too late. For instance, engineers were aware that the Santa Monica Freeway in the Los Angeles area would likely be damaged during an earthquake and retrofitting was scheduled for February 1994. Unfortunately, the Northridge earthquake struck

Sue Monroe

▌**Figure 1** Before the Mackinac Bridge, that connects the Lower and Upper Peninsulas of Michigan, could be built, geologists and engineers had to determine whether the Mackinac Breccia could support such a large structure.

James S. Monroe

▌**Figure 2** This freeway crosses a small valley a short distance from the San Andreas Fault in California. Engineers had to take into account the near certainty that this structure would be badly shaken during an earthquake.

response to tension so that one or more blocks are elevated relative to adjacent blocks. A classic example is the Basin and Range Province, which is centered on Nevada, but extends into adjacent states and into Mexico. Differential movement on faults has produced uplifted blocks called *horsts* and down-dropped blocks called *grabens* (Figure 10.17). Erosion of the horsts has produced mountainous topography.

Volcanic outpourings form chains of volcanic mountains such as the Hawaiian Islands, where a plate moves over a hot spot (see Figure 2.24). Some mountains such as the Cascade Range of the Pacific Northwest are made up almost entirely of volcanic rocks (see Figure 5.18a), and the mid-ocean ridges are also mountains. However, most mountains on the continents are composed of all rock types and show clear evidence of deformation by compression.

on January 17, 1994, and part of the freeway collapsed. Of course, this and other similar events provide important information that can be incorporated into engineering practice to make freeways, buildings, and bridges safer during earthquakes.

Geologic hazards include flooding, landslides, volcanic eruptions, earthquakes, tsunami, land subsidence, soil creep, and radon gas. Geologic hazards account for thousands of fatalities and billions of dollars in property damage every year, and whereas the incidence of hazards has not increased, fatalities and damages have grown because more and more people live in disaster-prone areas.

We cannot eliminate geologic hazards, but we can better understand them, design structures to protect shoreline communities (▌Figure 3) and withstand shaking during earthquakes, enact zoning and land-use regulations, and, at the very least, decrease the amount of damage and human suffering.

Unfortunately, geologic information that is readily available is often ignored or overlooked. A case in point is the Turnagain Heights subdivision in Anchorage, Alaska, that was so heavily damaged when the soil liquefied during the 1964 earthquake (see Figure 11.18). Not only were reports on soil stability ignored or overlooked before homes were built there, but also since 1964, new homes were built on part of the same site.

Along rugged seacoasts and in mountainous areas, highways are notoriously unstable and slump or slide from hillsides (▌Figure 4). When this happens, engineering geologists are consulted for their recommendations for stabilizing slopes. They may suggest building retaining walls or drainage systems to keep the slopes dry, planting vegetation, or, in some cases, simply rerouting the highway if it is too costly to maintain in its present position.

As you read the following chapters on surface processes, keep in mind that engineering geologists are involved in many aspects of stabilizing slopes, designing and constructing dams and flood-control projects, and building structures to protect seaside communities.

Sue Monroe

▌**Figure 3** Following the devastation caused in 2005 by Hurricane Katrina, many of the floodwalls protecting New Orleans, Louisiana, had to be replaced.

Dr. Marli Miller/Visuals Unlimited/Getty Images

▌**Figure 4** The Devil's Slide area along the Pacific Coast of California. The thin line on the hillside is California Highway 1, which is periodically closed due to slides. Not only are the rocks here weak and susceptible to sliding and slumping, but the San Andreas Fault is nearby. The state built a tunnel, which was completed in March 2013, to bypass this area.

Plate Tectonics and Mountain Building

Geologists use the term **orogeny** for an episode of mountain building during which intense deformation takes place, generally accompanied by metamorphism, the emplacement of plutons, especially batholiths, and thickening of Earth's crust. The processes responsible for orogenies are still not fully understood, but it is known that mountain building is a consequence of plate movements.

Any theory that accounts for mountain building must adequately explain the characteristics of mountain ranges, such as their geometry and location; they tend to be long and narrow and at or near plate margins. Mountains also show intense deformation, especially compression-induced overturned and recumbent folds, as well as reverse and thrust faults.

Figure 10.16 Dip-Slip Faults

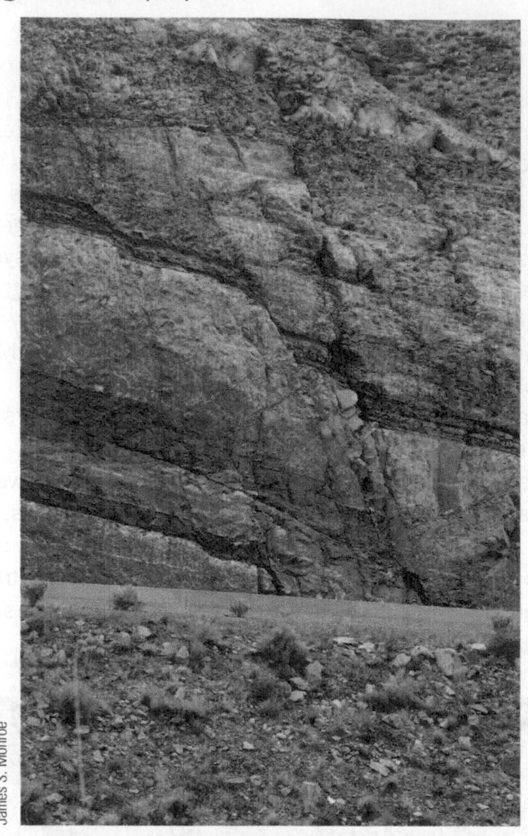

a Small, normal faults near Moab, Utah.

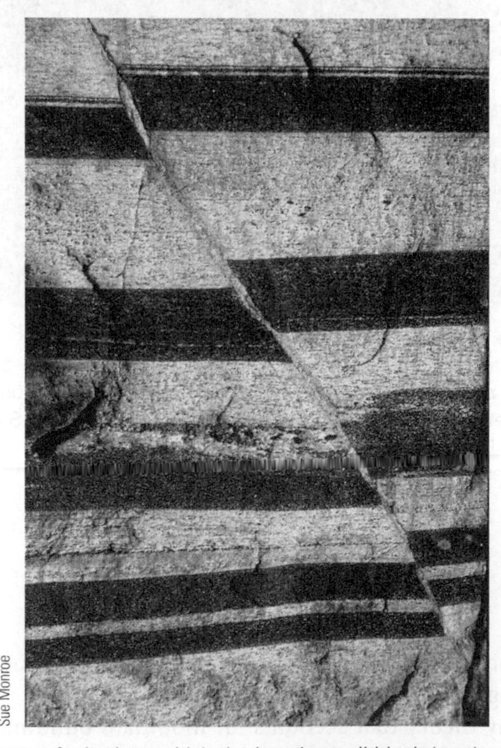

b A reverse fault along which the hanging wall block (on the right) has moved up relative to the footwall block.

Furthermore, granitic plutons and regional metamorphism characterize the interiors or cores of mountain ranges. Another feature is sedimentary rocks now far above sea level that were deposited in shallow and deep marine environments.

Figure 10.17 Origin of Horsts and Grabens

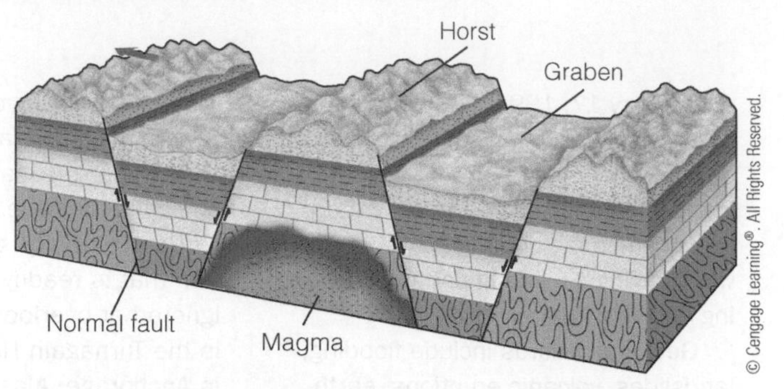

a Movement on parallel normal faults accounts for many of the mountain ranges in the Basin and Range Province of the western United States.

b The Egan Range in Nevada is one of many ranges in the Basin and Range Province bounded on one or both sides by normal faults.

Figure 10.18 Strike-slip fault This right-lateral strike-slip fault displaces strata in southern Nevada.

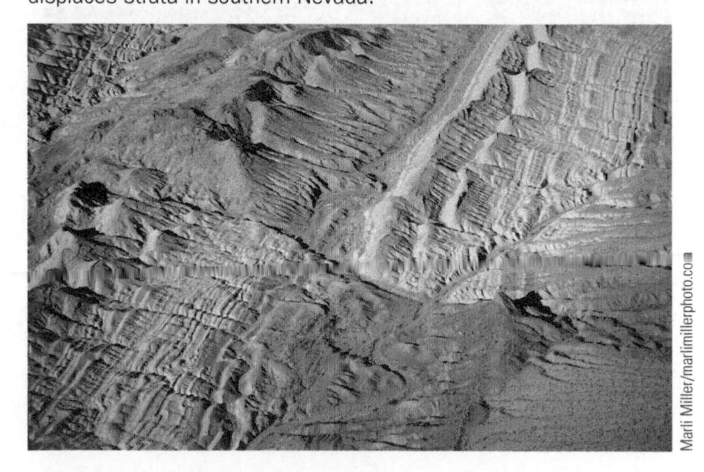

Deformation and associated activities at convergent plate boundaries are certainly important processes in mountain building. They account for a mountain system's location and geometry, as well as complex geologic structures, plutons, and metamorphism. Yet, the present-day topographic expression of mountains is also related to surface processes, such as mass wasting (gravity-driven processes including landslides), glaciers, and running water. In other words, erosion also plays an important role in the evolution of mountains.

Orogenies at Oceanic–Oceanic Plate Boundaries

Deformation, igneous activity, and the origin of a volcanic island arc characterize orogenies that take place where oceanic lithosphere is subducted beneath oceanic lithosphere. Sediments derived from the island arc are deposited in an adjacent oceanic trench and then are deformed and scraped off against the landward side of the trench (❚ Figure 10.19). These deformed sediments are part of a subduction complex, or an *accretionary wedge,* of intricately folded rocks cut by numerous thrust faults, resulting from compression. In addition, orogenies in this setting are characterized by low-temperature, high-pressure metamorphism of the blueschist facies (see Figure 8.20).

Deformation caused largely by the emplacement of plutons also takes place in the island arc system where many rocks show evidence of high-temperature, low-pressure metamorphism. The overall effect of an island arc orogeny is the origin of two more or less parallel orogenic belts consisting of a landward volcanic island arc underlain by batholiths and a seaward belt of deformed trench rocks (Figure 10.19). The Japanese Islands are a good example.

In the area between an island arc and its nearby continent, the back-arc basin, volcanic rocks, and sediments derived from the island arc and the adjacent continent are also deformed as the plates continue to converge. The sediments

❚ Figure 10.19 Orogeny and the Origin of a Volcanic Island Arc at an Oceanic–Oceanic Plate Boundary

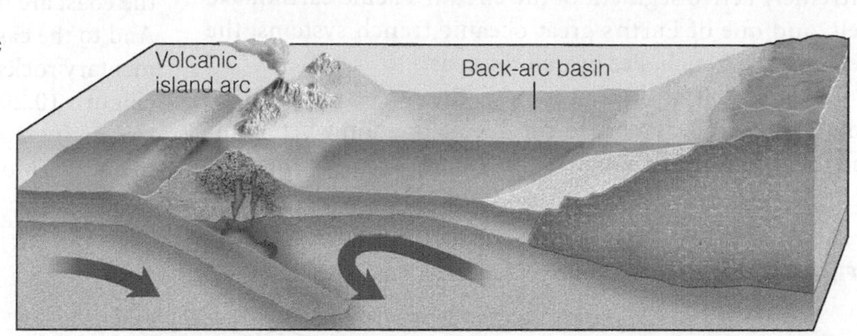

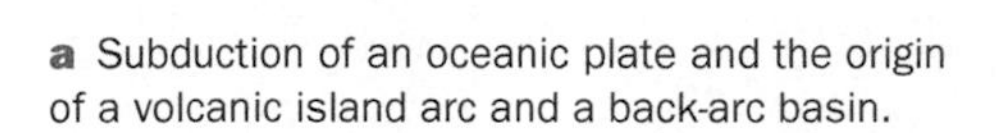

a Subduction of an oceanic plate and the origin of a volcanic island arc and a back-arc basin.

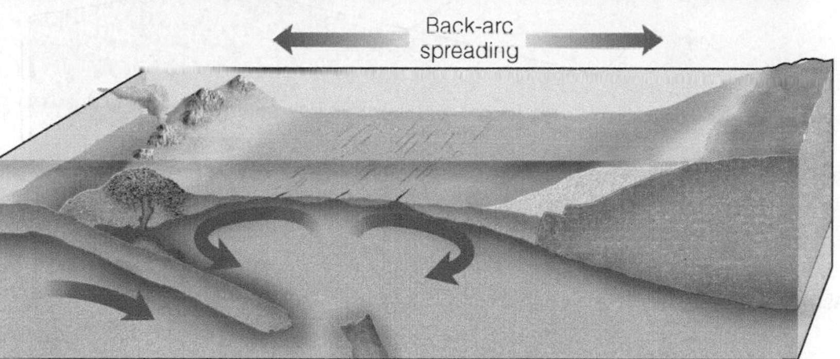

b Continued subduction and back-arc spreading.

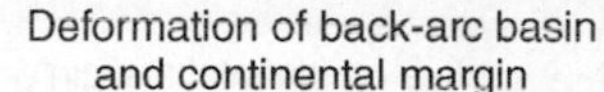

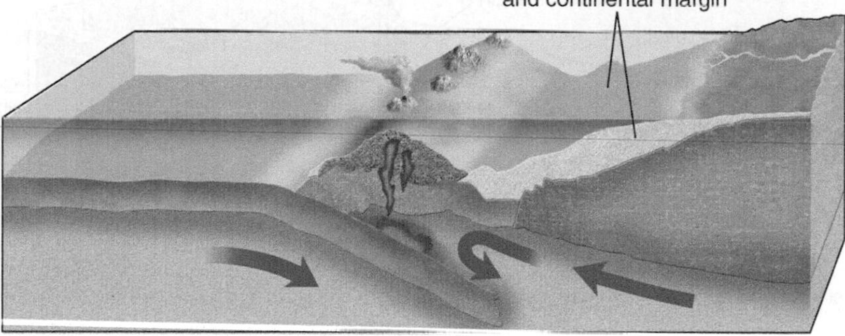

c Back-arc basin begins to close, resulting in deformation of back-arc basin and continental margin deposits.

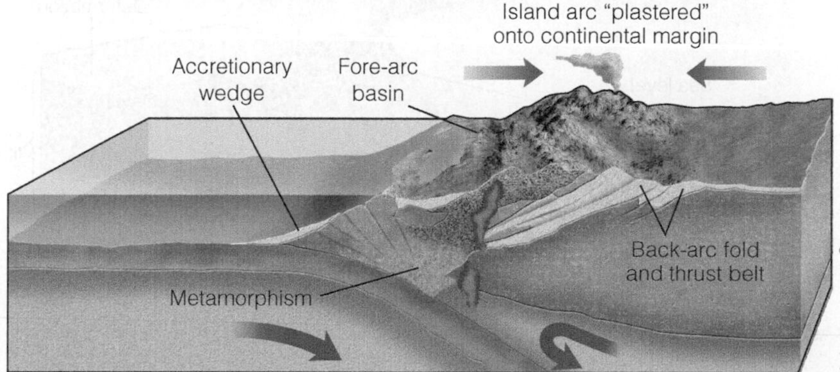

d Thrusting of back-arc sediments onto the adjacent continent and suturing of the island arc to the continent.

are intensely folded and displaced toward the continent along low-angle thrust faults. Eventually, the entire island arc complex is fused to the edge of the continent, and the back-arc basin sediments are thrust onto the continent, forming a thick stack of thrust sheets (Figure 10.19).

Orogenies at Oceanic–Continental Plate Boundaries The Andes of South America are the best example of continuing orogeny at an oceanic–continental plate boundary. Among the ranges of the Andes are the highest mountain peaks in the Americas and many active volcanoes. Furthermore, the west coast of South America is an extremely active segment of the circum-Pacific earthquake belt, and one of Earth's great oceanic trench systems, the Peru–Chile Trench, lies just off the coast.

Before 200 million years ago, the western margin of South America was a broad continental margin where sediments accumulated much as they do now along the East Coast of North America. However, when Pangaea split apart along what is now the Mid-Atlantic Ridge, the South American plate moved westward. As a consequence, the oceanic lithosphere west of South America began subducting beneath the continent (❚ Figure 10.20). Subduction resulted in partial melting of the descending plate, which produced the andesitic volcanic arc of composite volcanoes, and felsic magmas, mostly of granitic composition, were emplaced as large plutons beneath the arc (Figure 10.20).

As a result of the events just described, the Andes Mountains consist of a central core of granitic rocks capped by andesitic volcanoes. To the west of this central core along the coast are the deformed rocks of the accretionary wedge. And to the east of the central core are intensely folded sedimentary rocks that were thrust eastward onto the continent (Figure 10.20). Present-day subduction, volcanism, and seismicity along South America's west coast indicate that the Andes Mountains are still forming.

❚ Figure 10.20 The Andes Mountains in South America

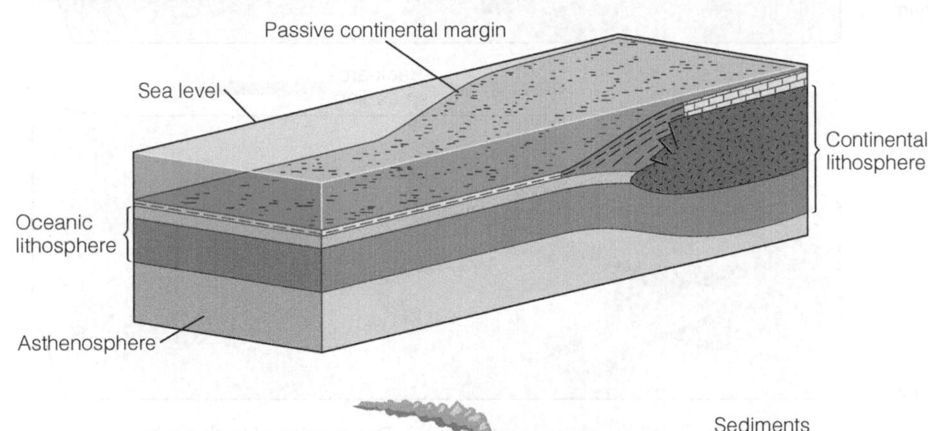

a Prior to 200 million years ago, the western margin of South America was a passive continental margin.

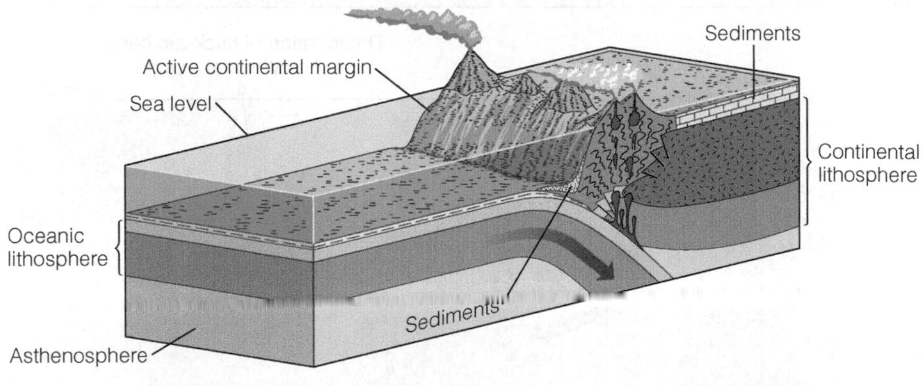

b Orogeny began when this area became an active continental margin as the South American plate moved to the west and collided with oceanic lithosphere.

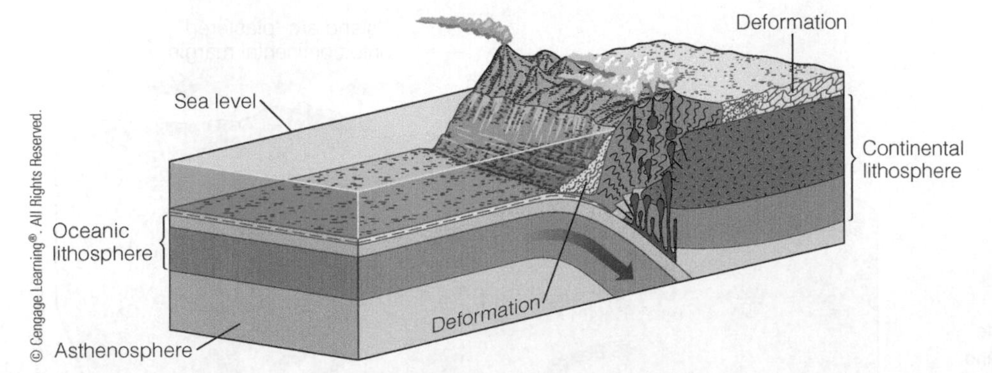

c Continued deformation, plutonism, and volcanism.

Orogenies at Continental–Continental Plate Boundaries The best example of an orogeny along a continental–continental plate boundary is the Himalayas of Asia. The Himalayas began forming when India collided with Asia about 40–50 million years ago. Before that time, India was far south of Asia and separated from it by an ocean basin (▌Figure 10.21a). As the Indian plate moved northward, a subduction zone formed along the southern margin of Asia where oceanic lithosphere was consumed. Partial melting generated magma, which rose to form a volcanic arc, and large granite plutons were emplaced into what is now Tibet. At this stage, the activity along Asia's southern margin was similar to what is now occurring along the west coast of South America.

The ocean separating India from Asia continued to close, and India eventually collided with Asia (Figure 10.21a). As a result, two continental plates were sutured, together. Thus, the Himalayas are now within a continent rather than along a continental margin. The exact time of India's collision with Asia is uncertain, but between 40 and 50 million years ago, India's rate of northward drift decreased abruptly from about 15–20 cm per year to about 5 cm per year. Because continental lithosphere is not dense enough to be subducted, this decrease seems to mark the time of collision and India's resistance to subduction. Consequently, the leading margin of India was thrust beneath Asia, causing crustal thickening, thrusting, and uplift. Sedimentary rocks that had been deposited in the sea south of Asia were thrust northward, and two major thrust faults carried rocks of Asian origin onto the Indian plate. Rocks deposited in the shallow seas along India's northern margin now form the higher parts of the Himalayas (▌Figure 10.21b). Since its collision with Asia, India has been thrust horizontally about 2,000 km beneath Asia and now moves north at several centimeters per year.

▌**Figure 10.21** Orogeny at a Continental-Continental Plate Boundary and the Origin of the Himalayas of Asia

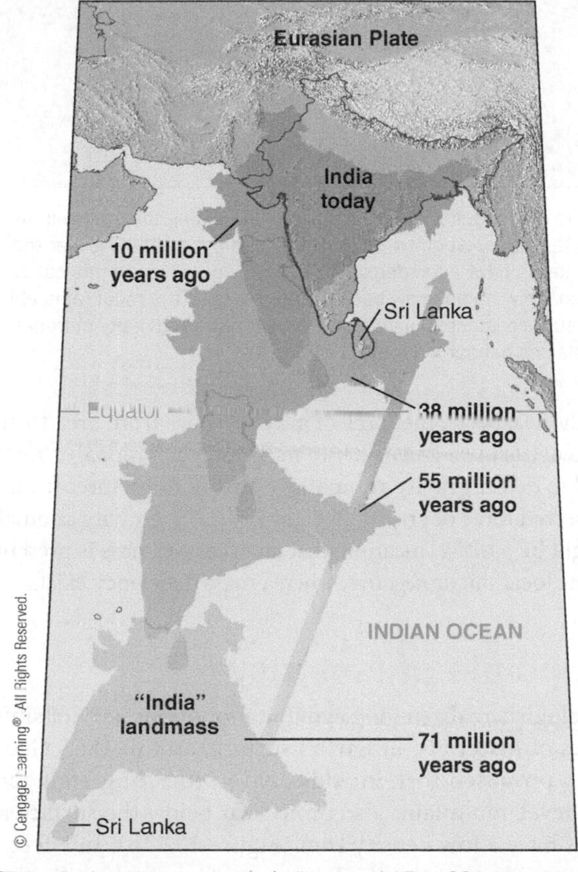

a During its long journey north, India moved 15 to 20 cm per year; however, beginning 40 to 50 million years ago, its rate of movement decreased markedly as it collided with the Eurasian plate.

b The Karakoram Range seen here from between Gilgit and Karimabad, Pakistan, is within the Himalayan orogen.

 Other mountain systems also formed as a result of collisions between two continental plates. The Urals in Russia and the Appalachians of North America formed by such collisions. In addition, the Arabian plate is now colliding with Asia along the Zagros Mountains of Iran.

Connection Link
You can learn more about the origin and evolution of the Appalachian Mountains in Chapters 20 and 22.

Terranes and the Origin of Mountains

In the preceding section, we discussed orogenies along convergent plate boundaries that result in adding material to a continent, a process termed **continental accretion.** Much of the material added to continental margins is eroded older continental crust, but some plutonic and volcanic rocks as well as oceanic sediments are new additions. During the 1970s and 1980s, however, geologists discovered that parts of many mountain systems are also made up of small, accreted lithospheric blocks that clearly originated elsewhere. These **terranes,*** as they are called, are fragments of seamounts, island arcs, and small pieces of continents that were carried on oceanic plates that collided with continental plates, thus adding them to the continental margins (see Chapter 22).

*Some geologists prefer the terms *suspect terrane, exotic terrane,* or *displaced terrane.* Notice also the spelling of terrane as opposed to the more familiar terrain, the latter a geographic term indicating a particular area of land.

Figure 10.22 Gravity Anomalies

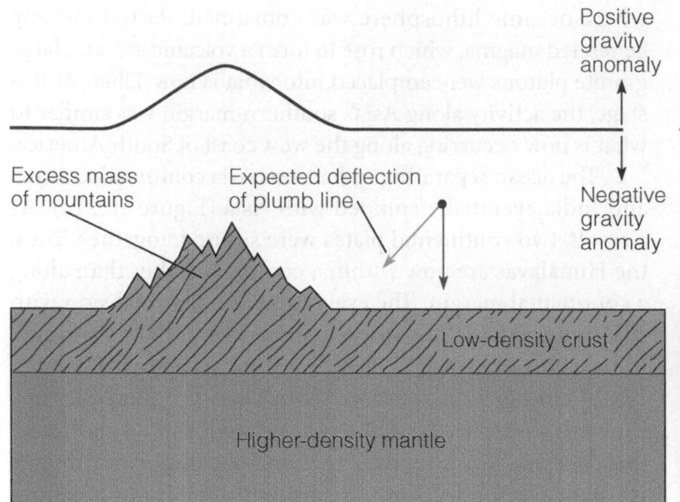

a A plumb line (a cord with a suspended weight) is normally vertical, pointing to Earth's center of gravity. Near a mountain range, the plumb line should be deflected as shown if the mountains are simply thicker, low-density material resting on denser material, and a gravity survey across the mountains would indicate a positive gravity anomaly.

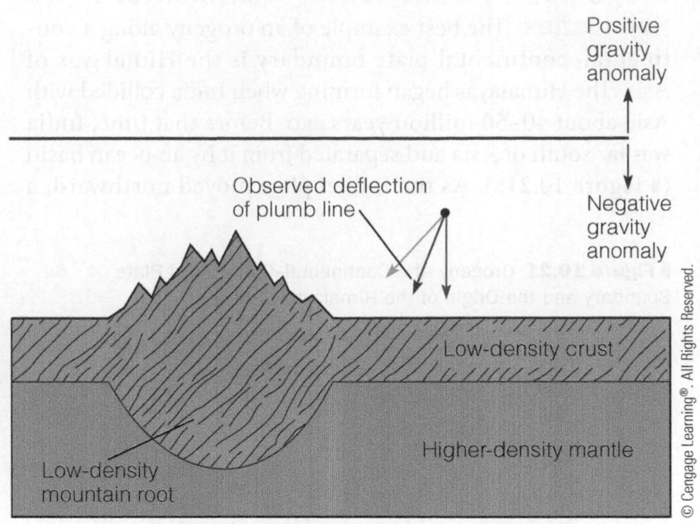

b The actual deflection of the plumb line during the survey in India was less than expected. It was explained by postulating that the Himalayas have a low-density root. A gravity survey, in this case, would show no anomaly because the mass of the mountains above the surface is compensated for at depth by low-density material displacing denser material.

10.6 Earth's Crust

Continental crust stands higher than oceanic crust, but why should this be so? Also, why do mountains stand higher than surrounding areas? To answer these questions, we must examine Earth's crust in more detail. You already know that continental crust is granitic with an overall density of 2.7 g/cm^3, whereas oceanic crust is made up of basalt and gabbro and its density is 3.0 g/cm^3 (see Chapter 9). In most places, continental crust is about 35 km thick except beneath mountain systems, where it is much thicker. Oceanic crust, in contrast, varies from only 5 to 10 km thick. So, these differences, as well as variations in crustal thickness, account for why mountains stand high and why continents stand higher than ocean basins (see the following section).

How is it possible for a solid (continental crust) to float in another solid (the mantle)? Floating brings to mind a ship at sea or a block of wood in water; however, continents do not behave in this manner. Or do they? Actually, they do float, in a manner of speaking, but a complete answer requires more discussion on the concept of gravity and on the principle of isostasy.

Isaac Newton formulated the law of universal gravitation in which the force of gravity (F) between two masses (m_1 and m_2) is directly proportional to the products of their masses and inversely proportional to the square of the distance between their centers of mass. This means that an attractive force exists between any two objects, and the magnitude of that force varies depending on the masses of the objects and the distance between their centers. We generally refer to the gravitational force between an object and Earth as its *weight.*

Gravitational attraction would be the same everywhere on the surface if Earth were perfectly spherical, homogeneous throughout, and not rotating. But because Earth varies in all of these aspects, the force of gravity varies from area to area. Geologists use a *gravimeter* to measure gravitational attraction and to detect **gravity anomalies**—that is, departures from the expected force of gravity (Figure 10.22b). Gravity anomalies might be *positive,* meaning that an excess of mass is present at some location, or *negative,* when a mass deficiency exists.

Principle of Isostasy

Geologists realized long ago that mountains are not simply piles of materials on Earth's surface, and in 1865 George Airy proposed that, in addition to projecting high above sea level, mountains also project far below the surface and thus have a low density root (Figure 10.22b). In effect, he was saying that the thicker crust of mountains floats on denser rock at depth, with their excess mass above sea level compensated for by low-density material at depth. Another explanation was proposed by J. H. Pratt, who thought that mountains were high because they were composed of rocks of lower density than those in adjacent regions.

Actually, both Airy and Pratt were correct, because there are places where density or thickness accounts for differences in the level of the crust. For example, Pratt's hyphothesis was confirmed because (1) continental crust is thicker and less dense than oceanic crust and thus stands high, and (2) the mid-oceanic ridges stand high because the crust there is hot and less dense than cooler crust elsewhere. Airy, on the other hand, was correct in his claim that the crust, continental or oceanic, "floats" on the mantle, which has a density of 3.3 g/cm^3 in its upper part. However, we have not yet explained what we mean by one solid floating in another solid.

This phenomenon of Earth's crust floating in the denser mantle is now known as the **principle of isostasy,**

Figure 10.23 The Principle of Isostasy An iceberg sinks to an equilibrium level with 10% of its mass above water level. The larger iceberg sinks farther below and rises higher above the water surface than the smaller one. If some of the ice above water level melts, the icebergs will rise to maintain the same proportions of ice above and below water level.

which is easy to understand by an analogy to an iceberg (Figure 10.23). Ice is slightly less dense than water, so it floats. According to Archimedes' principle of buoyancy, an iceberg sinks in water until it displaces a volume of water whose weight is equal to that of the ice. When the iceberg has sunk to an equilibrium position, only about 10% of its volume is above water level. If some of the ice above water level should melt, the iceberg rises to maintain equilibrium with the same proportion of ice above and below the water.

Earth's crust is similar to the iceberg in that it sinks into the mantle to its equilibrium level. Where the crust is thickest, as beneath mountains, it sinks farther down into the mantle and it also rises higher above the surface. And because continental crust is thicker and less dense than oceanic crust, it stands higher than the ocean basins. Remember, the mantle is hot, yet solid, and under tremendous pressure, so it behaves in a fluid-like manner.

Some of you might realize that crust floating on the mantle raises an apparent contradiction. In Chapter 9, we said that the mantle is a solid because it transmits S-waves, which do not move through fluids. But according to the principle of isostasy, the mantle behaves as a fluid. When considered in terms of the brief time required for S-waves to pass through it, the mantle is indeed solid. But when subjected to stress over long periods, it yields by flowage; thus, at this time scale, it is regarded as a viscous fluid.

Isostatic Rebound

What happens when a ship is loaded with cargo and then later unloaded? Of course, it first sinks lower in the water and then rises, but it always finds its equilibrium position. Earth's crust responds similarly to loading and unloading, but it does so much more slowly. For example, if the crust is loaded, as when widespread glaciers accumulate, the crust sinks farther into the mantle to maintain equilibrium. The crust behaves similarly in areas where huge quantities of sediment accumulate.

If loading by glacial ice or sediment depresses Earth's crust farther into the mantle, it follows that when vast glaciers melt or where deep erosion takes place, the crust should rise back up to its equilibrium level. And in fact it does. This phenomenon, known as **isostatic rebound,** is taking place in Scandinavia, which was covered by a thick ice sheet until about 10,000 years ago; it is now rebounding at about 1 m per century. In fact, coastal cities in Scandinavia have rebounded rapidly enough that docks constructed several centuries ago are now far from shore. Isostatic rebound has also occurred in eastern Canada where the crust has risen as much as 100 m in the past 6,000 years.

Figure 10.24 shows the response of Earth's continental crust to loading and unloading as mountains form and evolve.

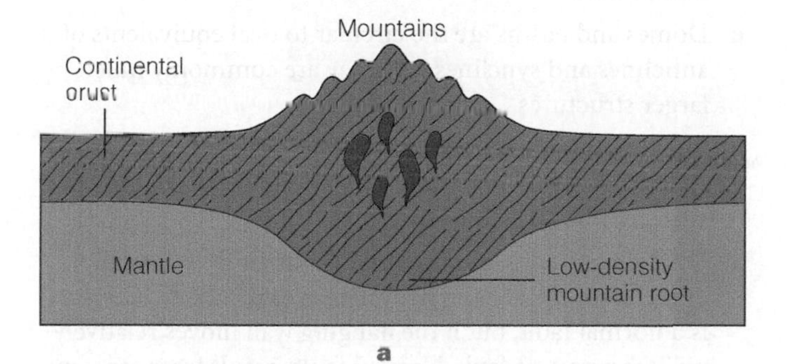

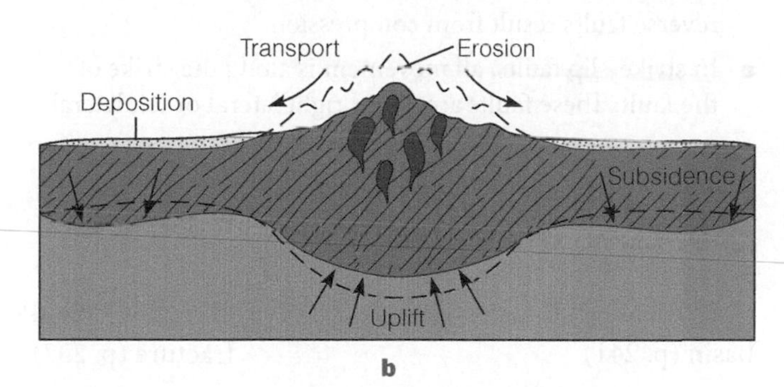

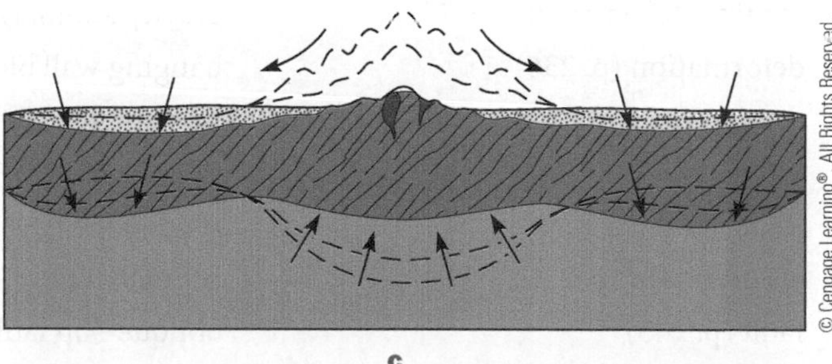

Figure 10.24 Isostatic Rebound A diagrammatic representation showing the isostatic response of the crust to erosion (unloading) and widespread deposition (loading).

Recall that during an orogeny, emplacement of plutons, metamorphism, and general thickening of the crust accompany deformation. However, as the mountains erode, isostatic rebound takes place and the mountains rise, whereas adjacent areas of sedimentation subside (Figure 10.24). If continued long enough, the mountains will disappear and then can be detected only by the plutons and metamorphic rocks that show their former existence.

Key Concepts Review

- Folded and fractured rocks have been deformed or strained by applied stresses.
- Stress is compression, tension, or shear. Elastic strain is not permanent, but plastic strain and fracture are, meaning that rocks do not return to their original shape or volume when the deforming forces are removed.
- Strike and dip are used to define the orientation of deformed rock layers. This same concept applies to other planar features, such as fault planes.
- Anticlines and synclines are up- and down-arched folds, respectively. They are identified by strike and dip of the folded rocks and the relative ages of rocks in these folds.
- Domes and basins are the circular to oval equivalents of anticlines and synclines, but they are commonly much larger structures.
- The two structures that result from fracture are joints and faults. Joints show no movement parallel with the fracture surface, whereas faults do.
- On dip-slip faults, all movement is up or down the dip of the fault. If the hanging wall moves relatively down, it is a normal fault, but if the hanging wall moves relatively up, it is a reverse fault. Normal faults result from tension; reverse faults result from compression.
- In strike-slip faults, all movement is along the strike of the fault. These faults are either right lateral or left lateral, depending on the apparent direction of offset of one block relative to the other.
- Oblique-slip faults show components of both dip-slip and strike-slip movement.
- A variety of processes account for the origin of mountains. Some involve little or no deformation, but the large mountain systems on the continents are the result of deformation at convergent plate boundaries.
- Subduction of an oceanic plate beneath another oceanic plate or beneath a continental plate causes an orogeny. At an oceanic-oceanic boundary, a volcanic island arc intruded by plutons forms, whereas at an oceanic-continental boundary, a volcanic arc forms on the continental plate. In both cases, deformation and metamorphism occur.
- Some mountain systems are within continents far from a present-day plate boundary. These mountains formed when two continental plates collided and became sutured.
- Geologists now realize that orogenies also involve collisions of terranes with continents.
- Continental crust is characterized as granitic, and it is much thicker and less dense than oceanic crust, which is composed of basalt and gabbro.
- According to the principle of isostasy, Earth's crust floats in equilibrium in the denser mantle below. Continental crust stands higher than oceanic crust because it is thicker and less dense.

Important Terms

anticline (p. 240)
basin (p. 241)
compression (p. 237)
continental accretion (p. 251)
deformation (p. 236)
dip (p. 238)
dip-slip fault (p. 244)
dome (p. 241)
elastic strain (p. 237)
fault (p. 243)
fault plane (p. 243)
fold (p. 239)
footwall block (p. 243)
fracture (p. 237)
geologic structure (p. 239)
gravity anomaly (p. 252)
hanging wall block (p. 243)
isostatic rebound (p. 253)
joint (p. 241)
monocline (p. 239)
normal fault (p. 244)
oblique-slip fault (p. 244)
orogeny (p. 247)
plastic strain (p. 237)
principle of isostasy (p. 252)
reverse fault (p. 244)
shear stress (p. 237)
strain (p. 236)
stress (p. 236)
strike (p. 238)
strike-slip fault (p. 244)
syncline (p. 240)
tension (p. 237)
terrane (p. 251)
thrust fault (p. 244)

Review Questions

1. If Earth's crust is first loaded by vast glaciers and then unloaded when the glaciers melt, the crust will rise, a phenomenon called
   a. ____ strike-slip faulting.
   b. ____ recumbent folding.
   c. ____ compressional release.
   d. ____ plastic stress.
   e. ____ isostatic rebound.
2. A fault on which the hanging wall block has moved down relative to the footwall block is a(n) ________ fault.
   a. ____ normal.
   b. ____ oblique.
   c. ____ strike-slip.
   d. ____ reverse.
   e. ____ thrust.
3. An ongoing orogeny at an oceanic–continental plate boundary is responsible for the
   a. ____ Himalayas.
   b. ____ Andes Mountains.
   c. ____ Basin and Range Province.
   d. ____ San Andreas Fault.
   e. ____ Michigan Basin.
4. Which one of the following statements is incorrect?
   a. ____ Most orogenies occur at convergent plate boundaries.
   b. ____ Shear stress involves forces operating along the same line, but in opposite directions.
   c. ____ Joints are fractures along which no movement has taken place parallel with the fracture surface.
   d. ____ Folds, joints, and faults are collectively known as geologic structures.
   e. ____ A strike-slip fault is one on which all movement is horizontal.
5. Rocks characterized as ductile
   a. ____ show a great amount of plastic strain.
   b. ____ are found along the crests of synclines.
   c. ____ make up the hanging wall blocks adjacent to faults.
   d. ____ fracture easily.
   e. ____ are stronger in tension than they are in compression.
6. Suppose that rocks along a strike-slip fault were displaced 200 km in 5 million years. What was the average rate of movement per year? Is this average likely to represent the actual rate of displacement on the fault? Explain.
7. How would you explain stress and strain to someone unfamiliar with the concept?
8. What sequence of events takes place during an orogeny at an oceanic–continental plate boundary?
9. Discuss how time, rock type, temperature, and pressure influence rock deformation.

10. **Creative Thinking Visual Question:** This geologic cross section (Figure 1) shows several rock layers and a fault. Your task is to decipher the history of the area. What event took place first, second, and so on? What kind of fold is present? What kind of fault? To fully answer these questions, recall the discussions of sedimentary structures in Chapter 7.

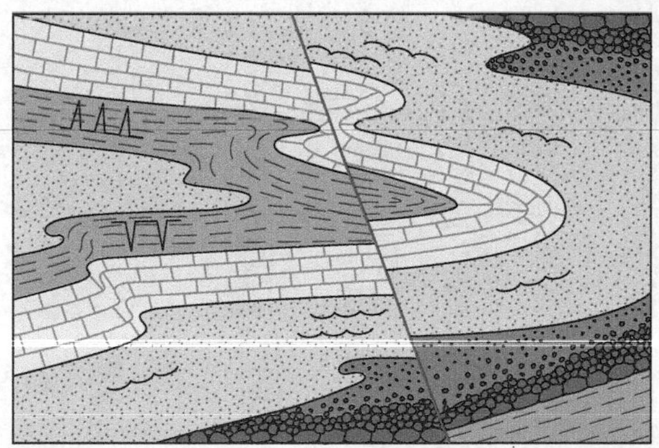

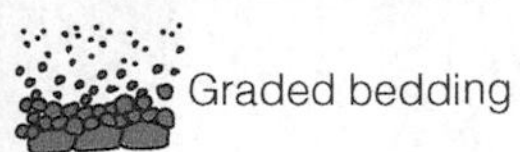

Figure 1 Cross section showing deformed strata and geologic structures in an area.

Global GeoScience Watch Within the GREENR database, search for the term "isostasy." Under the "Reference" section, click on the document titled "Isostasy." What is isostasy and how is Greenland "an example of isostasy in action?"

Floods and landslides in January 2011 devastated numerous towns in the mountainous areas around Rio de Janeiro, Brazil, killing more than 800 people and leaving at least 14,000 people homeless. As a result of several days of heavy rains, portions of this hillside in Nova Friburgo gave way, causing a landslide that sent mud, vegetation, and homes sliding down its slope.

Mass Wasting

OUTLINE

HAVE YOU EVER WONDERED?

- What causes landslides (known as mass movements or mass wasting)?
- Why landslides occur where they do?
- What, if anything, can be done to prevent some of the different types of mass wasting?
- How to minimize the destructive effects of certain types of mass wasting?
- What things you should look for when buying property or building in hilly or coastal areas to avoid, or at least minimize, potential mass-wasting effects to your property?

11.1 Introduction

In January 2011, southeast Brazil was struck by devastating floods and landslides triggered by unusually heavy torrential rains that dumped a month's worth of rain in just a few days. The hardest hit area was the mountainous Serrana region north of Rio de Janeiro, where relentless flooding and landslides destroyed hundreds of homes, killing more than 800 people and leaving at least 14,000 people homeless. The death toll easily surpassed that of the 1967 mudslides in Caraguatatuba, making this Brazil's worst natural disaster in more than four decades. Collapsed roads and bridges further exacerbated the situation, hampering rescue operations and emergency aid, and lack of power and telephone service contributed further to the misery of the survivors.

Although the floods and landslides swept away homes of both rich and poor alike, the poorer rural areas suffered the most because many structures were built in unstable areas. It is easy to cite the numbers of dead and homeless, but the human side of the disaster was vividly brought home in Terescopolis, where numerous bodies were lined up outside a police station awaiting identification by relatives because there was no more room at the morgue.

This terrible tragedy illustrates how geology affects all of our lives. The underlying causes of the mudslides in Brazil can be found anywhere in the world. In fact, worldwide, *landslides* (a general term for mass movements of Earth materials) cause an average of 7,500 deaths and approximately $25 billion in damages per year. In the United States, landslides result in 25 to 50 deaths per year and damages exceeding $2 billion annually. By being able to recognize and understand how landslides occur and what the results may be, we can find ways to reduce hazards and minimize damage in terms of both human life and property damage.

Mass wasting (also called *mass movement*) is defined as the downslope movement of material under the direct influence of gravity. Most types of mass wasting are aided by weathering and usually involve surficial material. The material moves at rates ranging from almost imperceptible, as in the case of creep, to extremely fast, as in a rockfall or slide. Although water can play an important role, the relentless pull of gravity is the major force behind mass wasting.

11.2 Factors That Influence Mass Wasting

Mass wasting is an important geologic process that can occur at any time and almost any place. Although all major landslides have natural causes, many smaller ones are the result of human activity and could have been prevented or their damage minimized.

When the gravitational force acting on a slope exceeds its resisting force, slope failure (mass wasting) occurs. The resisting forces that help maintain slope stability include the slope material's strength and cohesion, the amount of internal friction between grains, and any external support of the slope (❚ Figure 11.1). These factors collectively define a slope's **shear strength**.

❚ **Figure 11.1** Slope Shear Strength

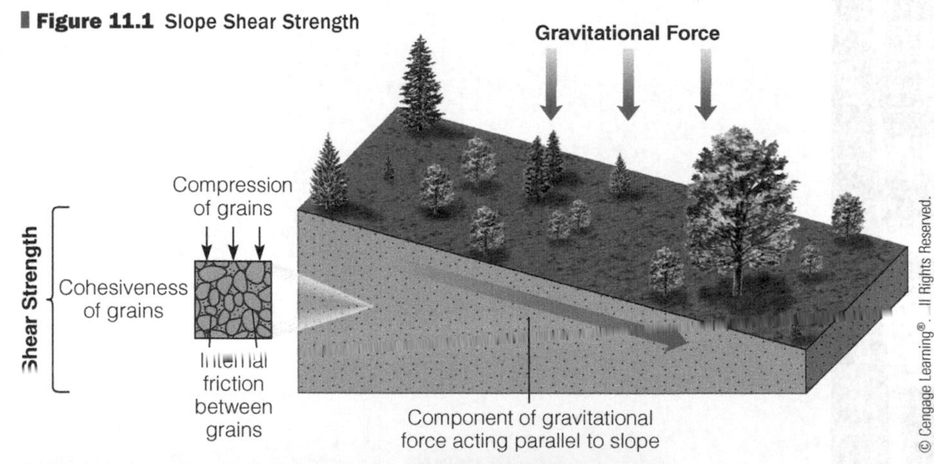

a A slope's shear strength depends on the slope material's strength and cohesion, the amount of internal friction between grains, and any external support of the slope. These factors promote slope stability. The force of gravity operates vertically, but it also has a component of force acting parallel to the slope. When this force, which promotes instability, exceeds a slope's shear strength, slope failure occurs.

b The angle of repose is a function of shear strength. Dry sand has an angle of repose of about 30 degrees. With wet sand, the shear strength is increased, and much steeper angles of repose are possible, such as this pile that is nearly vertical.

Opposing a slope's shear strength is the force of gravity. Gravity operates vertically, but it also has a component of force acting parallel to the slope, thereby causing instability (▌Figure 11.1a). The steeper a slope's angle is, the greater the component of force acting parallel to the slope and the greater the chance for mass wasting. The steepest angle that a slope can maintain without collapsing is its *angle of repose* (▌Figure 11.1b). At this angle, the shear strength of the slope's material exactly counterbalances the force of gravity. For unconsolidated material, such as, for example, dry sand grains, the angle of repose is usually no steeper than approximately 30 degrees. Wet sand, on the other hand, can stand almost vertical. However, slopes steeper than 30 degrees usually consist of unweathered solid rock.

All slopes are in a state of *dynamic equilibrium*, which means that they are constantly adjusting to new conditions. Although we tend to view mass wasting as a disruptive and usually destructive event, it is one of the ways that a slope adjusts to new conditions. Whenever a building or road is constructed on a hillside, the equilibrium of that slope is affected. The slope must then adjust, perhaps by mass wasting, to this new set of conditions.

Many factors can cause mass wasting: a change in slope angle, weakening of material by weathering, increased water content, changes in the vegetation cover, and overloading. Although most of these processes are interrelated, we will examine them separately for ease of discussion, but we also will show how they individually and collectively affect a slope's equilibrium.

Slope Angle

Slope angle is probably the major cause of mass wasting. Generally speaking, the steeper the slope, the less stable it is. Therefore, steep slopes are more likely to experience mass wasting than gentle ones.

A number of processes can oversteepen a slope. One of the most common is undercutting by stream or wave action (▌Figure 11.2). This process removes the slope's base, increases the slope angle, and thereby increases the gravitational force acting parallel to the slope. Wave action, especially during storms, often results in mass movements along the shores of oceans or large lakes (▌Figure 11.3).

Excavations for road cuts and hillside building sites are another major cause of slope failure (▌Figure 11.4). Grading the slope too steeply or cutting into its side increases the stress in the rock or soil until it is no longer strong enough to remain at the steeper angle, and mass movement ensues. Such action is analogous to undercutting by streams or waves and has the same result, thus explaining why so many mountain roads are plagued by frequent mass movements.

Weathering and Climate

Mass wasting is more likely to occur in loose or poorly consolidated slope material than in bedrock. As soon as rock is exposed at Earth's surface, weathering begins to disintegrate and decompose it, reducing its shear strength and increasing its susceptibility to mass wasting. The deeper the weathering zone extends, the greater the likelihood of some type of mass movement.

Recall that some rocks are more susceptible to weathering than others and that climate plays an important role in the rate and type of weathering (see Chapter 6). In the tropics,

▌Figure 11.2 Undercutting a Slope's Base by Stream Erosion

a Undercutting by stream erosion removes a slope's base,

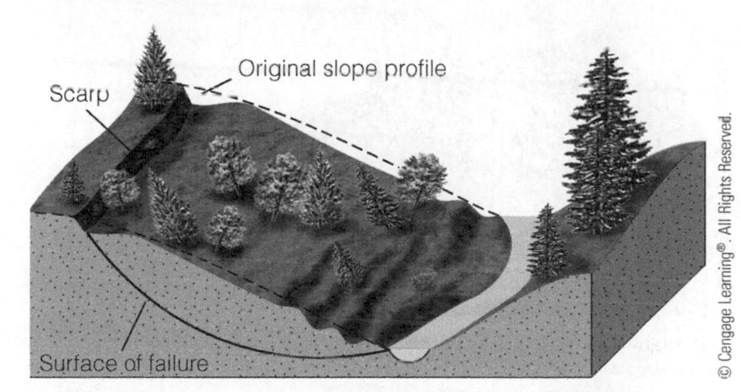

b which increases the slope angle, and can lead to slope failure.

c Undercutting by stream erosion caused slumping along this stream near Weidman, Michigan. Notice the scarp, which is the exposed surface of the underlying material following slumping.

Figure 11.3 Undercutting a Slope's Base by Wave Action This sea cliff north of Bodega Bay, California, was undercut by waves during the winter of 1997–1998. As a result, part of the land slid into the ocean, damaging several houses.

Figure 11.4 Highway Excavation

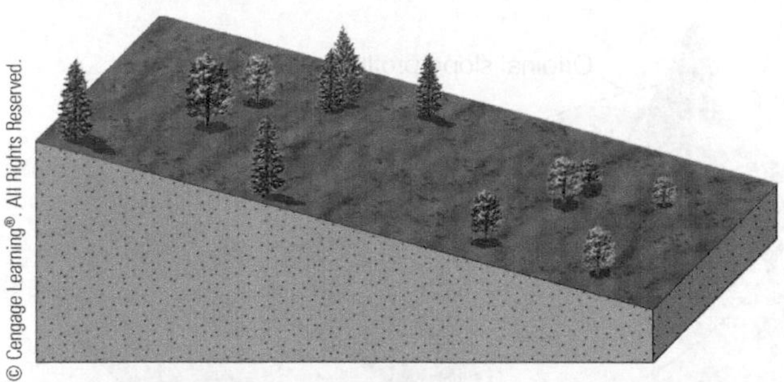

a Highway excavations disturb the equilibrium of a slope by

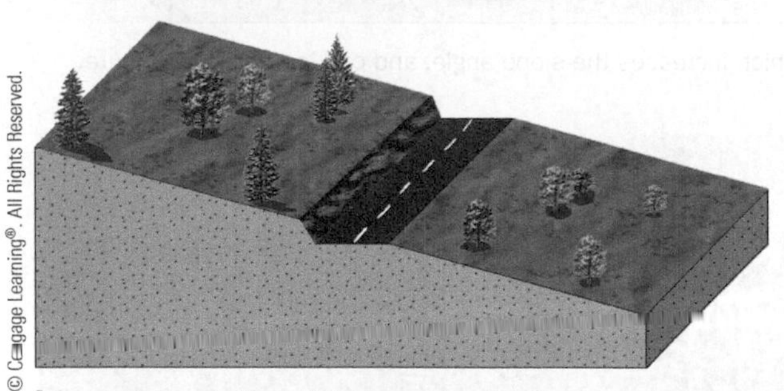

b removing a portion of its support, as well as oversteepening it at the point of excavation, which can result in

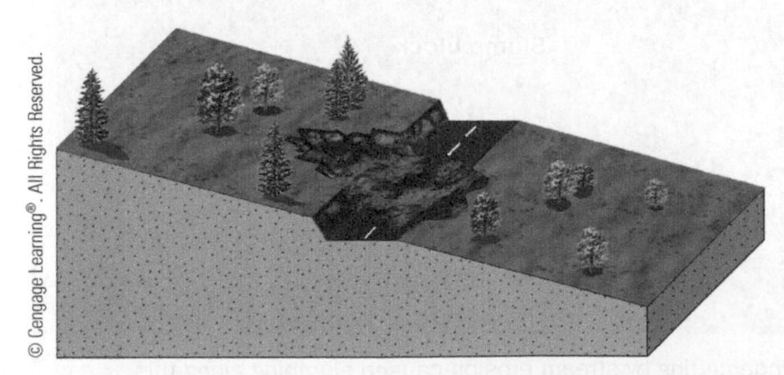

c landslides along the highway.

d Cutting into the hillside to construct this portion of the Pan-American Highway in Mexico resulted in a rockfall that completely blocked the road.

where temperatures are high and considerable rain falls, the effects of weathering extend to depths of several tens of meters, and mass movements most commonly occur in the deep weathering zone. In arid and semiarid regions, the weathering zone is usually considerably shallower. Nevertheless, intense, localized cloudbursts can drop large quantities of water on an area in a short time. With little vegetation to absorb this water, runoff is rapid and frequently results in mudflows.

Water Content

The amount of water in rock or soil influences slope stability. Large quantities of water from melting snow or heavy rainfall greatly increase the likelihood of slope failure. The additional weight that water adds to a slope can be enough to cause mass movement. Furthermore, water percolating through a slope's material helps to decrease friction between grains, contributing to a loss of cohesion. For example, slopes composed of dry clay are usually quite stable, but when wetted, they quickly lose cohesiveness and internal friction and become an unstable slurry. This occurs because clay, which can hold large quantities of water, consists of platy particles that easily slide over each other when wet. For this reason, clay beds are frequently the slippery layer along which overlying rock units slide downslope.

To learn more about how water moves through various Earth materials, see Chapter 13.

An excellent case in point is Point Fermin in Southern California. Dubbed the "sunken city" by residents of the area, Point Fermin is famous for its numerous examples of mass wasting. The area is underlain by fine-grained sedimentary rocks that are interbedded with diatomite layers (a rock composed of the shells of diatoms, which are siliceous algae) and volcanic ash. When these layers get wet from rains or percolating groundwater, they become slippery and tend to easily slide (▮ Figure 11.5). Furthermore, the rocks also dip slightly toward the ocean and form steep coastal bluffs that are being undercut by constant wave action at their base.

To learn more about how wave action can result in mass wasting, see Chapter 16.

Vegetation

Vegetation affects slope stability in several ways. By absorbing the water from a rainstorm, vegetation decreases water saturation of a slope's material that would otherwise lead to a loss of shear strength. Vegetation's root system also helps stabilize a slope by binding soil particles together and holding the soil to bedrock.

The removal of vegetation by either natural or human activity is a major cause of many mass movements. Summer brush and forest fires in southern California frequently leave the hillsides bare of vegetation. Fall rainstorms saturate the ground, causing mudslides that do tremendous damage and cost millions of dollars to clean up.

Overloading

Overloading is almost always the result of human activity and typically results from the dumping, filling, or piling up of material. Under natural conditions, a material's load is carried by its grain-to-grain contacts, with the friction between the grains maintaining a slope. The additional

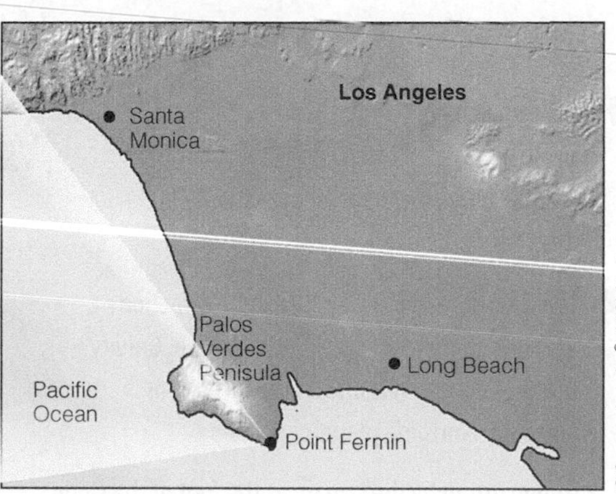

▮ **Figure 11.5 Landslides Caused by Sliding Along Slippery Rock Layers, Point Fermin, California** Homes dangerously close to an oversteepened, eroding coastal bluff. Water from rains and groundwater have resulted in slippery layers along which sliding has occurred. Two white drainage pipes that helped drain water out of the ground can be seen dangling from the top of the cliff. It is just a matter of time before these homes are destroyed by the effects of erosion and slumping of the unstable bluff.

weight created by overloading increases the water pressure within the material, which in turn decreases its shear strength, thereby weakening the slope material. If enough material is added, the slope will eventually fail, sometimes with tragic consequences.

Geology and Slope Stability

The relationship between the topography and the geology of an area is important in determining slope stability (❚ Figure 11.6). If the rocks underlying a slope dip in the same direction as the slope, mass wasting is more likely to occur than if the rocks are horizontal or dip in the opposite direction. When the rocks dip in the same direction as the slope, water can percolate along the various bedding planes and decrease the cohesiveness and friction between adjacent rock units (Figure 11.5). This is particularly true when clay layers are present because clay becomes slippery when wet.

Even if the rocks are horizontal or dip in a direction opposite to that of the slope, joints may dip in the same direction as the slope. Water migrating through them weathers the rock and expands these openings until the weight of the overlying rock causes it to fall.

Triggering Mechanisms

The factors discussed thus far all contribute to slope instability. Most, though not all, rapid mass movements are triggered by a force that temporarily disturbs slope equilibrium. The most common triggering mechanisms are strong vibrations from earthquakes and excessive amounts of water from a winter snowmelt or a heavy rainstorm.

Volcanic eruptions, explosions, and even loud claps of thunder may be enough to trigger a landslide if the slope is sufficiently unstable. Many *avalanches*, which are rapid movements of snow and ice down steep mountain slopes, are triggered by a loud gunshot or, in rare cases, even a person's shout.

❚ Figure 11.6 Geology, Slope Stability, and Mass Wasting Rocks dipping in the same direction as a hill's slope are particularly susceptible to mass wasting.

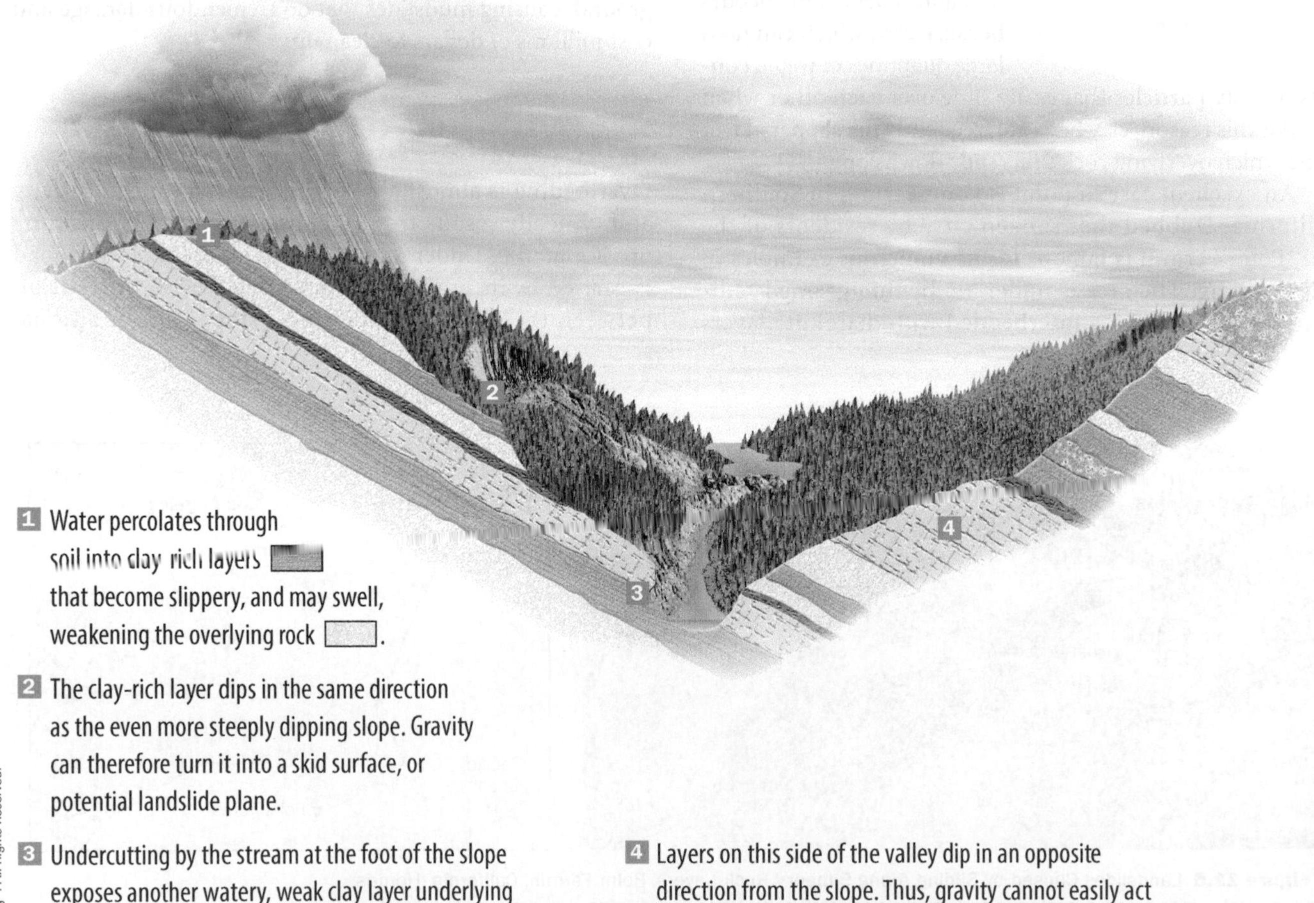

11.3 Types of Mass Wasting

Mass movements are generally classified on the basis of three major criteria (Table 11.1): (1) rate of movement (rapid or slow), (2) type of movement (primarily falling, sliding, or flowing), and (3) type of material involved (rock, soil, or debris). Even though many slope failures are combinations of different materials and movements, the resulting mass movements are typically classified according to their dominant behavior.

Rapid mass movements involve a visible movement of material. Such movements usually occur quite suddenly, and the material moves quickly downslope. Rapid mass movements are potentially dangerous and frequently result in loss of life and property damage. Most rapid mass movements occur on relatively steep slopes and can involve rock, soil, or debris.

Slow mass movements advance at an imperceptible rate and are usually detectable only by the effects of their movement, such as tilted trees and power poles or cracked foundations. Although rapid mass movements are more dramatic, slow mass movements are responsible for the downslope transport of a much greater volume of weathered material.

Falls

Rockfalls are a common type of extremely rapid mass movement in which rocks of any size fall through the air (▌Figure 11.7a). Rockfalls occur along steep canyons, cliffs, and road cuts and build up accumulations of loose rocks and rock fragments at their base called *talus* (see Figure 6.3b).

Rockfalls result from failure along joints or bedding planes in the bedrock and are commonly triggered by natural or human undercutting of slopes or by earthquakes. Many rockfalls in cold climates are the result of frost wedging (see Figure 6.3). Chemical weathering caused by water percolating through the fissures in carbonate rocks (limestone, dolostone, and marble) is also responsible for many rockfalls.

Rockfalls range in size from small rocks falling from a cliff to massive falls involving millions of cubic meters of debris that destroy buildings, bury towns, and block highways. Rockfalls are a particularly common hazard in mountainous areas where roads have been built by blasting and grading through steep hillsides of bedrock (▌Figure 11.7b). Anyone who has ever driven through the Appalachians, the Rocky Mountains, or the Sierra Nevada is familiar with the "Caution Falling Rocks" signs posted to warn drivers of the danger (▌Figure 11.7c). Slopes that are particularly susceptible to rockfalls are sometimes covered with wire mesh in an effort to prevent dislodged rocks from falling to the road below (▌Figure 11.8a). Another tactic is to put up wire mesh fences along the base of the slope to catch or slow down bouncing or rolling rocks (▌Figure 11.8b).

▌**Figure 11.7** Rockfalls

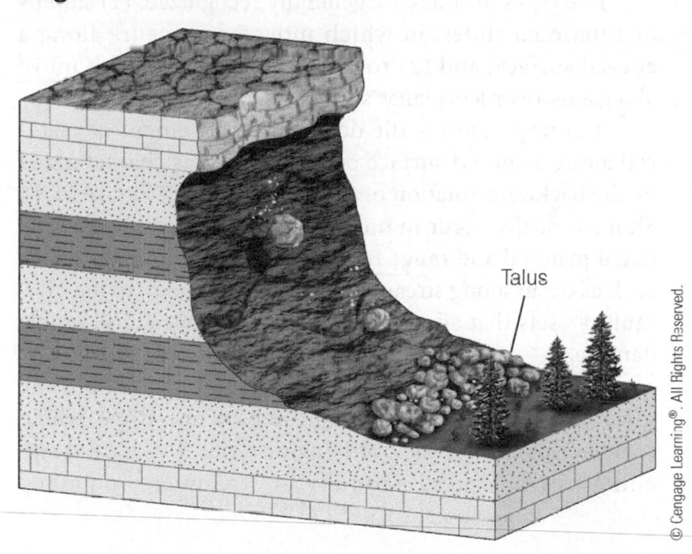

a Rockfalls result from failure along cracks, fractures, or bedding planes in the bedrock and are common features in the areas of steep cliffs.

b A recent rockfall of granite in Yosemite National Park, California.

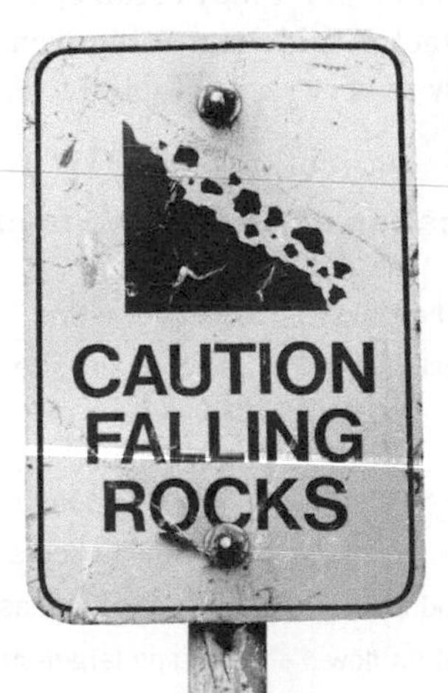

c Warning signs for falling rocks are commonly seen in mountainous areas.

Figure 11.8 Minimizing Damage from Rockfalls

Copyright and Photograph by Dr. Parvinder S. Sethi

a Wire mesh has been used to cover this steep slope near Narvik in northern Norway. This is a common practice in mountainous areas to prevent rocks from falling on the road.

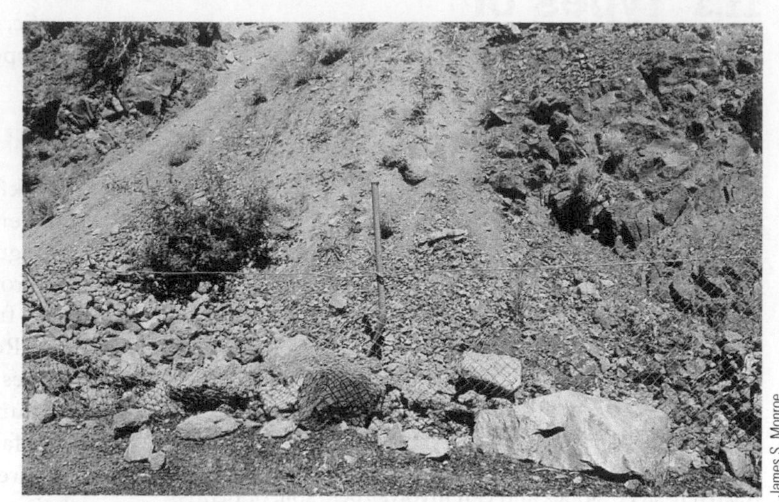

James S. Monroe

b A Wire mesh fence along the base of this hillside of Highway 44 in California has caught many boulders and prevented them from rolling onto the highway.

Slides

A **slide** involves movement of material along one or more surfaces of failure. The type of material may be soil, rock, or a combination of the two, and it may break apart during movement or remain intact. A slide's rate of movement can vary from extremely slow to very rapid (Table 11.1).

Two types of slides are generally recognized: (1) slumps or rotational slides, in which movement occurs along a curved surface; and (2) rock or block slides, which move along a more or less planar surface.

A **slump** involves the downward movement of material along a curved surface of rupture and is characterized by the backward rotation of the slump block (Figure 11.9). Slumps usually occur in unconsolidated or weakly consolidated material and range in size from small individual sets, such as occur along stream banks (Figure 11.2c), to massive, multiple sets that affect large areas and cause considerable damage.

Slumps can be caused by a variety of factors, but the most common is erosion along the base of a slope, which removes support for the overlying material. This local steepening may be caused naturally by stream erosion along

TABLE 11.1 Classification of Mass Movements and Their Characteristics

Type of Movement	Subdivision	Characteristic	Rate of Movement
Falls	Rockfall	Rocks of any size fall through the air from steep cliffs, canyons, and road cuts	Extremely rapid
Slides	Slump	Movement occurs along a curved surface of rupture; most commonly involves unconsolidated or weakly consolidated material	Extremely slow to moderate
Flows	Rock slide	Movement occurs along a generally planar surface	Rapid to very rapid
	Mudflow	Consists of at least 50% silt- and clay-sized particles and up to 30% water	Very rapid
	Debris flow	Contains larger-sized particles and less water than mudflows	Rapid to very rapid
	Earthflow	Thick, viscous, tongue-shaped mass of wet regolith	Slow to moderate
	Quick clays	Composed of fine silt and clay particles saturated with water; when disturbed by a sudden shock, lose their cohesiveness and flow like a liquid	Rapid to very rapid
	Solifluction	Water-saturated surface sediment	Slow
	Creep	Downslope movement of soil and rock	Extremely slow
Complex movements		Combination of different movement types	Slow to extremely rapid

Figure 11.9 Slumping In a slump, material moves downward along the curved surface of a rupture, causing the slump block to rotate backward. Most slumps involve unconsolidated or weakly consolidated material and are typically caused by erosion along the slope's base.

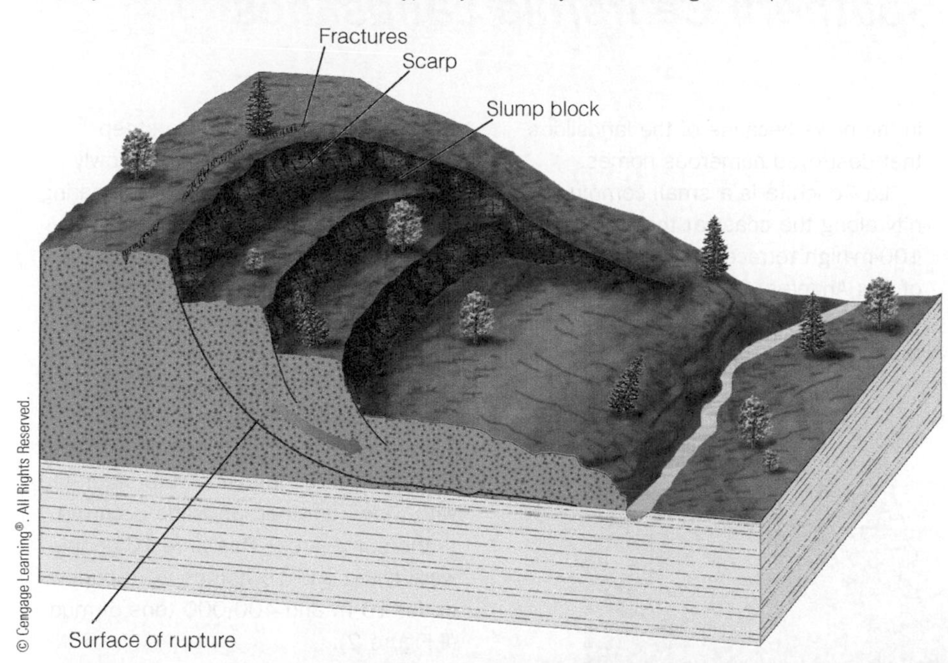

its banks (Figure 11.2c) or by wave action at the base of a coastal cliff (Figure 11.10).

Slope oversteepening can also be caused by human activity, such as the construction of highways and housing developments. Slumps are particularly prevalent along highway cuts, where they are generally the most frequent type of slope failure observed.

Although many slumps are merely a nuisance, large-scale slumps in populated areas and along highways can cause extensive damage. Such is the case in coastal southern California where slumping and sliding have been a constant problem, resulting in the destruction of many homes and the closing and relocation of numerous roads and highways (Geo-Focus 11.1).

Figure 11.10 Slumping in the Pacific Palisades, Southern California Undercutting of steep sea cliffs by wave action resulted in massive slumping in the Pacific Palisades area of southern California on March 31 and April 3, 1958. Highway 1 was completely blocked. Note the heavy earth-moving equipment for scale.

GEO FOCUS 11.1

Southern California Landslides

Southern California is no stranger to landslides. La Conchita, Point Fermin, Pacific Palisades, and Laguna Beach are all locations in southern California that have suffered damaging mass movements during the past 50 years. Two regions in particular, La Conchita and Laguna Beach, have been in the news because of the landslides that destroyed numerous homes.

La Conchita is a small community along the coast at the base of a 100-m-high terrace, 120 km northwest of Los Angeles, California. On March 4, 1995, following a period of heavy rains, some residents of this beach community noticed that the steep slope above their homes was slowly moving and that cracks were appearing in the walls of their houses, indicating that the homes were also moving. Shortly thereafter, a 200,000 m^3 slide destroyed or damaged nine homes in its path (▮ Figure 1).

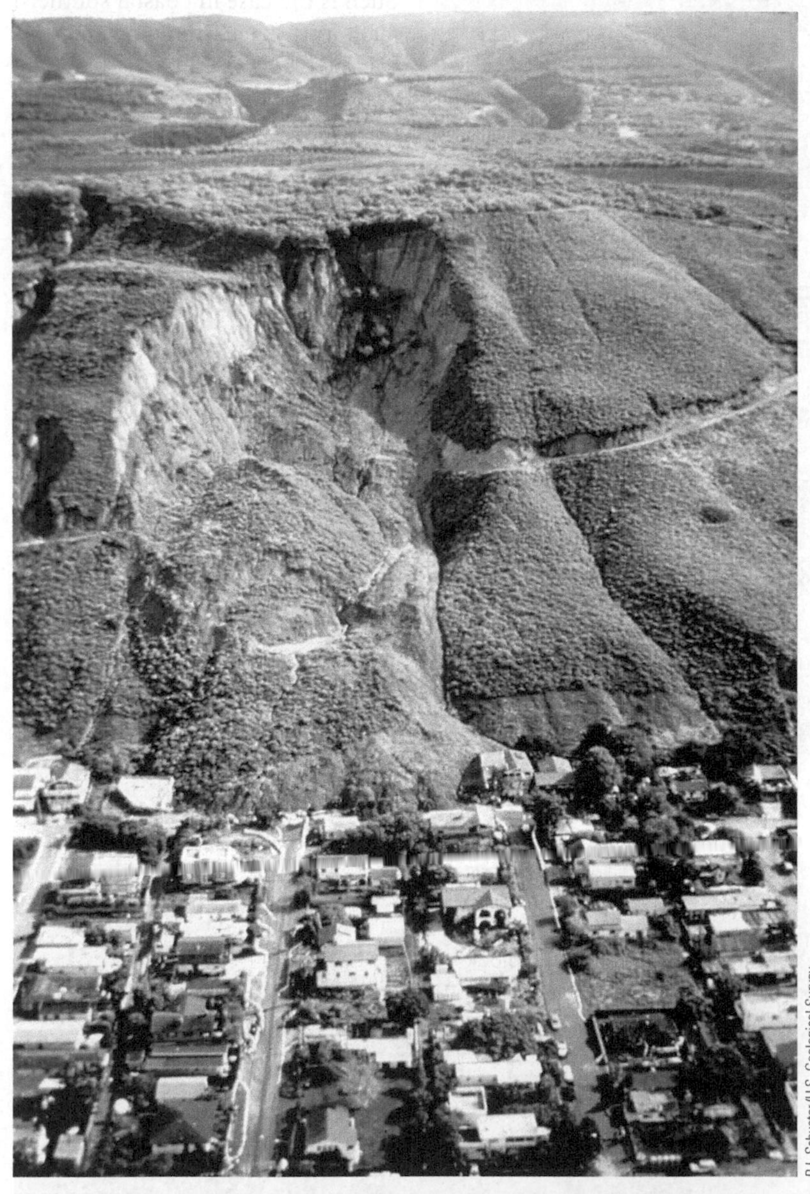

▮ **Figure 1** La Conchita, California, is located at the base of a steep-sloped terrace. Heavy rains and irrigation of an avocado orchard (visible at the top of the terrace) contributed to the landslide that destroyed nine homes in 1995.

Almost 10 years later, following a week of heavy rainfall in southern California, in which the hillside and previous landslide deposits were saturated with water, another landslide occurred in the same area. This time 10 people were killed and 15 homes were buried under 10 m and 400,000 tons of mud (▮ Figure 2).

What went wrong and why was this situation repeated? The rocks that make up the steep-sloped terrace behind La Conchita consist of soft, weak, and porous sediments that are not well lithified, and thus are easily weathered and susceptible to mass wasting. In addition, an irrigated avocado orchard sits on top of the hill, contributing to the water percolating through the porous sediments and rocks and contributing to the instability of the hillside. Add in heavy rainfall over an extended period, and you have all the ingredients for a landslide in the making. An ancient landslide area to begin with, a steep slope that has been undercut at its base by a road, well-saturated sediments decreasing the cohesion of the sediments that hold the hillside together, and continuing rains all contribute to the making of a landslide. And the potential is still there for another landslide, with no guarantee it will not happen again in the next 10 years.

Farther south in Laguna Beach, another landslide, in this case a rock slide, destroyed 18 expensive hillside homes and severely damaged approximately 20 others on June 1, 2005 (▮ Figure 3). Similar to what happened in 1978, the main triggering mechanism was probably unusually heavy winter rains, in this case the second-rainiest season on record. In this area of southern California, the rocks dip in the same direction as the slope and contain numerous clay beds interbedded with porous sandstones. Such conditions, when combined with heavy rainfall, are ideal for rock slides. It should come as no surprise that the area where both the 1978 and 2005 rock slides occurred is also part of an ancient slide complex.

Can anything be done to prevent future landslides? The short answer is probably no. Decreasing the slope, benching the hillside, and ensuring that there is sufficient drainage and a good cover of vegetation are all steps that can minimize future mass wasting. But the sad fact is that the geologic conditions are such that future landslides are inevitable as the landscape seeks equilibrium conditions by adjusting its slope. Add in the fact that the coastal terraces of Laguna Beach offer some of the most breathtaking views of the Pacific and people are willing to pay a premium to live here, and you have the formula for future landslides, loss of life, and property damage.

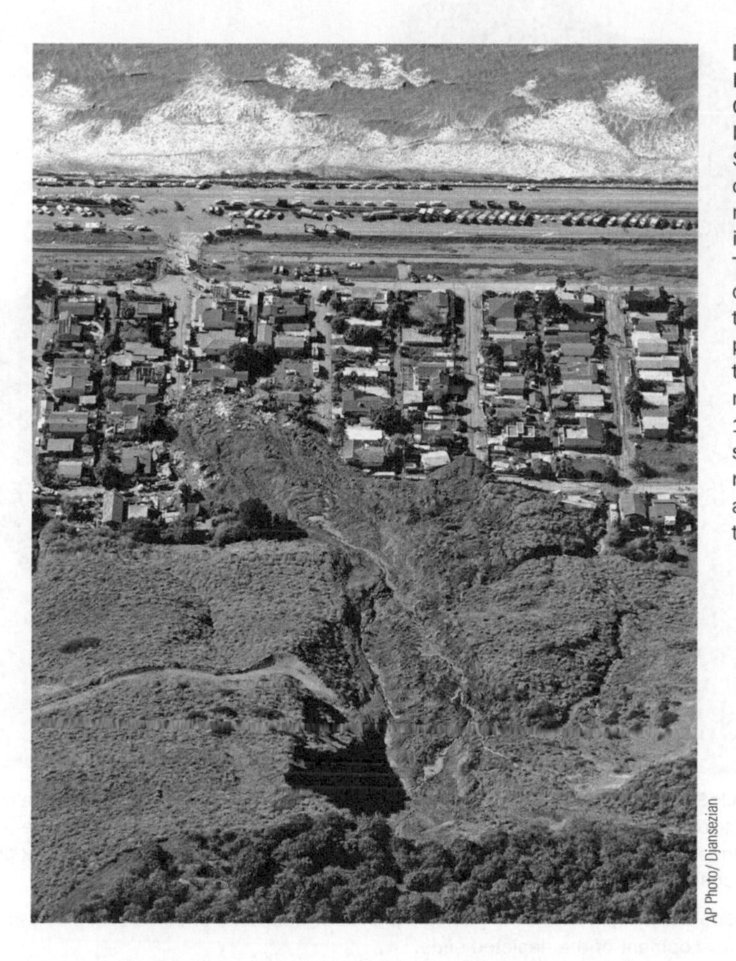

AP Photo/ Djansezian

Figure 2 La Conchita, California, 10 years later (2005). Similar factors caused another massive landslide in the same area. The landslide can clearly be seen in the center of this photograph, and the scarp and the remains of the 1995 landslide are still visible on the right side, as is the avocado orchard in the foreground.

AP Photo/Nick Ut

Figure 3 A rock slide on June 1, 2005, destroyed 18 expensive homes and damaged at least 20 others in Laguna Beach, California. Heavy rains, combined with unstable underlying geology, contributed to this most recent landslide in this area.

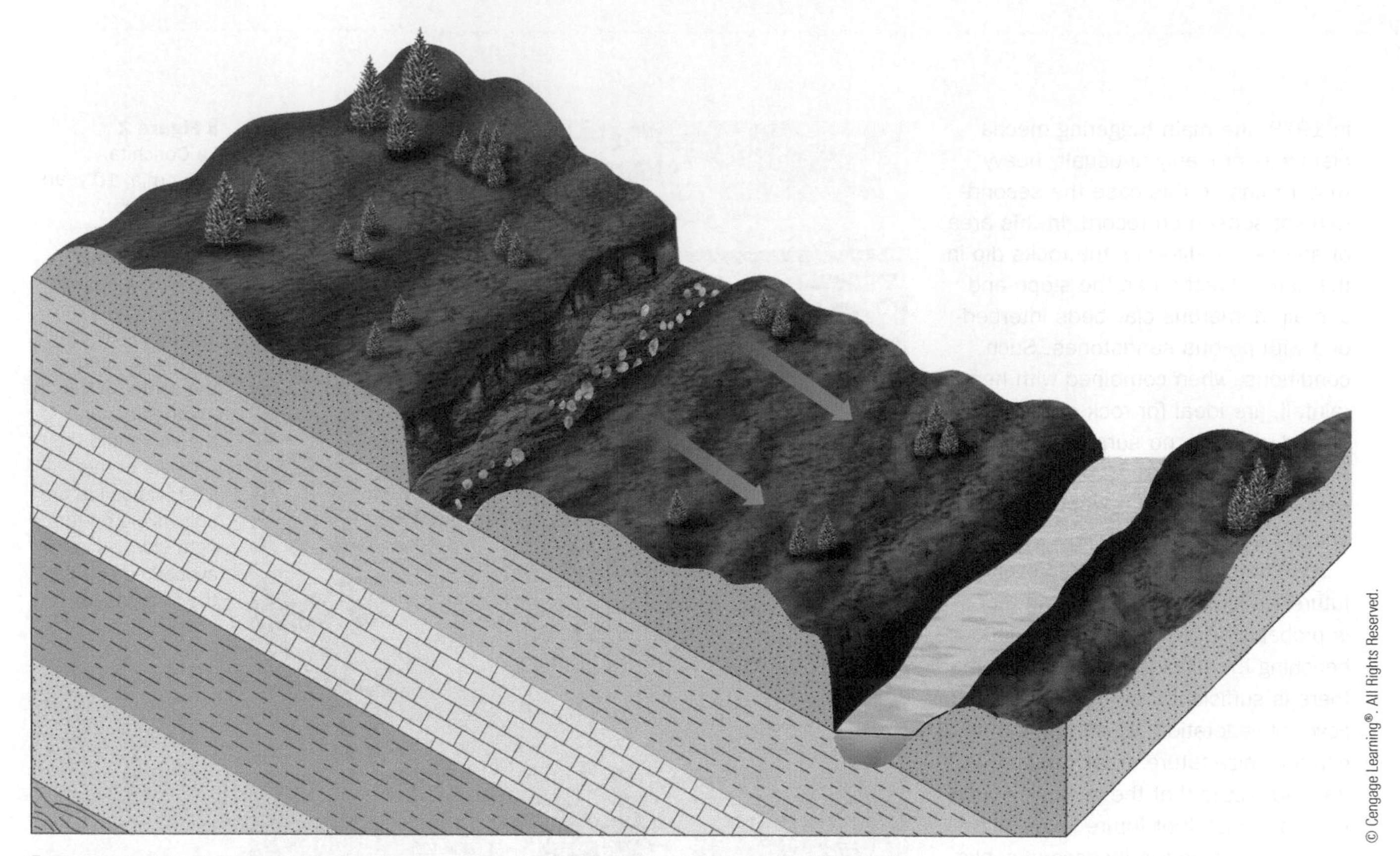

▌Figure 11.11 **Rock Slides** Rock slides occur when material moves downslope along a generally planar surface. Most rock slides result when the underlying rocks dip in the same general angle as the slope of the land. Undercutting along the base of the slope and clay layers beneath porous rock or soil layers increase the chance of rock slides.

Critical Thinking Question Explain how the geologic planes of weakness in this slope, plus water from rainfall, influenced development of the depicted slide.

A **rock slide** or *block slide*, occurs when rocks move downslope along a more or less planar surface. Most rock slides take place because the local slopes and rock layers dip in the same direction (▌Figure 11.11), although they can also occur along fractures parallel to a slope. Rock slides are also common occurrences along the southern California coast, such as at Point Fermin (Figure 11.5).

Farther south from Point Fermin is the town of Laguna Beach, where residents have been hit by rock slides and mudslides in 1978, 1998, and as recently as 2005 (▌Figure 11.12). Just as at Point Fermin, the rocks at Laguna Beach dip about 25 degrees in the same direction as the slope of the canyon walls and contain clay beds that "lubricate" the overlying rock layers, causing the rocks and the houses built on them to slide. Percolating water from heavy rains wets subsurface clayey siltstone, thus reducing its shear strength and helping to activate the slide. In addition, these slides are part of a larger ancient slide complex, which further worsens the situation.

Not all rock slides are the result of rocks dipping in the same direction as a hill's slope. The rock slide at Frank, Alberta, Canada, on April 29, 1903, illustrates how nature and human activity can combine to create a situation with tragic results (▌Figure 11.13).

It would appear at first glance that the coal-mining town of Frank, lying at the base of Turtle Mountain, was in no danger from a landslide (Figure 11.13). After all, many of the rocks dipped away from the mining town, unlike the situations at Point Fermin and Laguna Beach. The joints in the massive limestone composing Turtle Mountain, however, dip steeply toward the valley and are essentially parallel with the slope of the mountain itself. Furthermore, Turtle Mountain is supported by weak siltstones, shales, and coal layers that underwent slow plastic deformation from the weight of the overlying massive limestone. Coal mining along the base of the valley also contributed to the stress on the rocks by removing some of the underlying support. All of these factors, as well as the frost action and chemical weathering that widened the joints, finally resulted in a massive rock slide. Approximately 40 million m^3 of rock slid down Turtle Mountain along joint planes, killing 70 people and partially burying the town of Frank.

Building Along a Hilly Coastline

You are a member of a planning board for your seaside community. A developer wants to rezone some coastal property to build some expensive homes. This project would be a boon to the local economy because it would provide jobs and increase the tax base. However, because the area is somewhat hilly and fronts the ocean, you are concerned about how safe the homes would be. What types of studies would need to be done before any rezoning could take place? Is it possible to build safe structures along a hilly coastline? What specifically would you ask the environmental consulting firm that the planning board has hired to look for in terms of actual or potential geologic hazards if the approval is granted to build these homes?

Donald Hyndman

Homes in Laguna Beach, California, slid down water-saturated hillsides caused by El Niño winter storms, during a 2005 landslide.

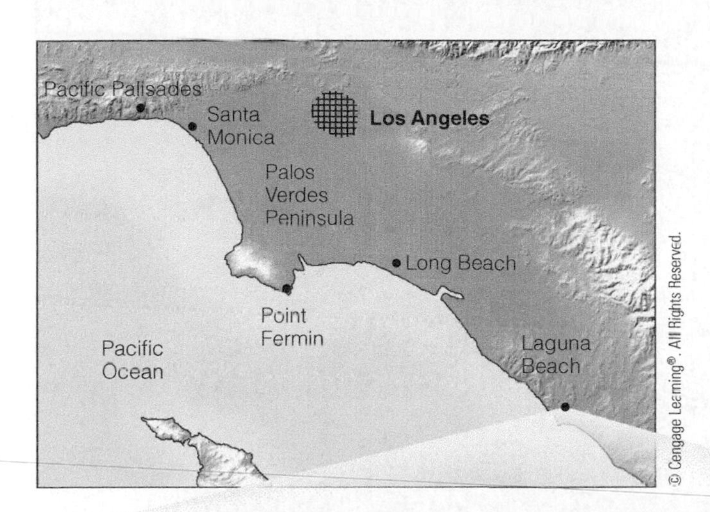

Steven R. Lower/GeoPhoto Publishing

Figure 11.12 Rock Slide, Laguna Beach, California A combination of interbedded clay beds that become slippery when wet, rocks dipping in the same direction as the slope of the sea cliffs, and undercutting of the sea cliffs by wave action activated a rock slide at Laguna Beach, California, that destroyed numerous homes and cars on October 2, 1978. This same area was hit by another rock slide in 2005.

Critical Thinking Question Is there anything that could have been done after the first rock slide that might have prevented the second one 27 years later? If not, should people be allowed to build on known active slide areas?

Figure 11.13 Rock Slide, Turtle Mountain, Canada

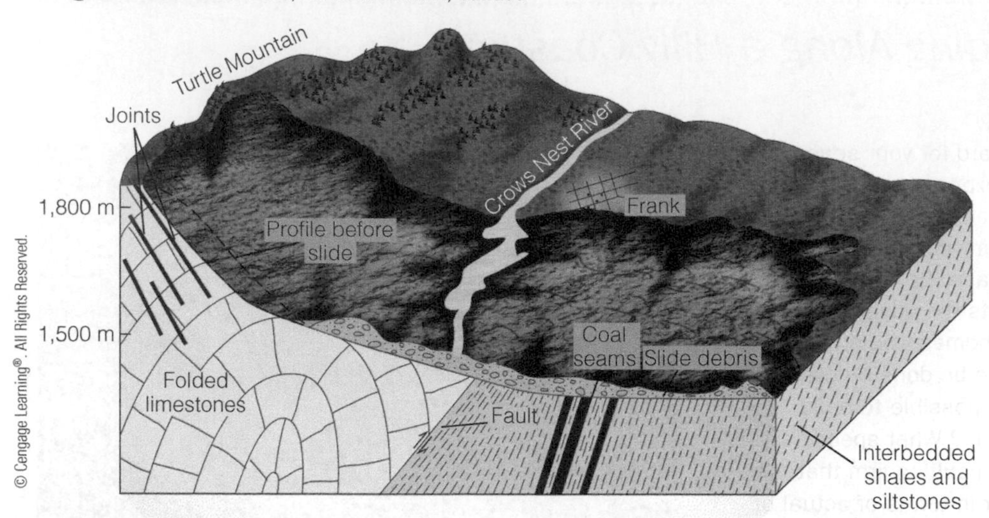

a The tragic Turtle Mountain rock slide that killed 70 people and partially buried the town of Frank, Alberta, Canada, on April 29, 1903, was caused by a combination of factors. These included joints that dipped in the same direction as the slope of Turtle Mountain, a fault partway down the mountain, weak shale and siltstone beds underlying the base of the mountain, and mined-out coal seams.

b Results of the 1903 rock slide at Frank, Canada.

Flows

Mass movements in which material moves as a viscous fluid or displays plastic movement are termed *flows*. Their rate of movement ranges from extremely slow to extremely rapid (Table 11.1). In many cases, mass movements begin as falls, slumps, or slides and change into flows farther downslope.

Of the major mass movement types, **mudflows** are the most fluid and move most rapidly (at speeds up to 80 km/hr). They consist of at least 50% silt- and clay-size material

Figure 11.14 Mudflow, Estes Park, Colorado Mudflows move swiftly downslope, engulfing everything in their path. Note how this mudflow in Rocky Mountain National Park has fanned out at the base of the hill. Also note the small lake adjacent to the mudflow that was formed after this mudflow created a dam across the stream.

combined with a significant amount of water (up to 30%). Mudflows are common in arid and semiarid environments where they are triggered by heavy rainstorms that quickly saturate the regolith, turning it into a raging flow of mud that engulfs everything in its path.

Mudflows can also occur in mountain regions (Figure 11.14) and in areas covered by volcanic ash, where they can be particularly destructive (see Chapter 5). Because mudflows are so fluid, they generally follow preexisting channels until the slope decreases or the channel widens, at which point they fan outward.

To review the destructive effects of volcanic mudflows (termed lahars), see Chapter 5.

Debris flows are composed of larger particles than mudflows and do not contain as much water. Consequently, they are usually more viscous than mudflows, typically do not move as rapidly, and rarely are confined to preexisting channels. Debris flows can be just as damaging, though, because they can transport large objects (Figure 11.15).

Earthflows move more slowly than either mudflows or debris flows. An earthflow slumps from the upper part of a hillside, leaving a scarp, and flows slowly downslope as a thick, viscous, tongue-shaped mass of wet regolith (Figure 11.16). Like mudflows and debris flows, earthflows can be of any size and are frequently destructive. They occur most commonly in humid climates on grassy, soil-covered slopes following heavy rains.

Some clays spontaneously liquefy and flow like water when they are disturbed. Such **quick clays** have caused serious damage and loss of lives in Sweden, Norway, eastern Canada (Figure 11.17), Alaska, United States, and elsewhere. Quick clays are composed of fine silt and clay particles made by the grinding action of glaciers. Geologists think that these fine sediments were originally deposited in a marine environment where their pore space was filled with saltwater. The ions in saltwater helped establish strong bonds between the clay particles, thus stabilizing and strengthening the clay. When the clays were subsequently uplifted above sea level, the saltwater was flushed out by fresh groundwater, reducing the effectiveness of the ionic bonds between the clay particles and thereby reducing the overall strength and cohesiveness of the clay. Consequently, when the clay is disturbed by a sudden shock or shaking, it essentially turns to a liquid and flows.

Figure 11.15 **Debris Flow, Ophir Creek, Nevada** A debris flow and damaged house in lower Ophir Creek, western Nevada. Note the many large boulders that are part of the debris flow. Debris flows do not contain as much water as mudflows and typically are composed of larger particles.

Critical Thinking Question Why are debris flows typically composed of larger particles than mudflows?

a Earthflows form tongue-shaped masses of wet regolith that move slowly downslope. They occur most commonly in humid climates on grassy, soil-covered slopes.

b An earthflow near Baraga, Michigan.

Figure 11.16 Earthflow

Figure 11.17 Quick-Clay Slide, Nicolet, Quebec, Canada The house on the slide (to the right of the bridge and circled in red) traveled several hundred meters with relatively little damage.

An excellent example of the damage that is done by quick clays occurred in the Turnagain Heights area of Anchorage, Alaska, in 1964 (Figure 11.18a). Underlying most of the Anchorage area is the Bootlegger Cove Clay, a massive clay unit of poor permeability. Because the Bootlegger Cove Clay forms a barrier that prevents groundwater from flowing through the adjacent glacial deposits to the sea, considerable hydraulic pressure builds up on the landward side of the clay. Some of this water has flushed out the saltwater in the clay and has saturated the lenses of sand and silt associated with the clay beds. When the 8.6-magnitude Good Friday earthquake struck on March 27, 1964, the shaking turned parts of the Bootlegger Cove Clay into a quick clay and precipitated a series of massive slides in the coastal bluffs that destroyed most of the homes in the Turnagain Heights subdivision (Figure 11.18b).

ConnectionLink

To review the destructive effects of ground shaking by earthquakes, see Chapter 9.

Solifluction is the slow downslope movement of water-saturated surface sediment. Solifluction can occur in any climate where the ground becomes saturated with water, but it is most common in areas of permafrost.

Permafrost, ground that remains permanently frozen, covers nearly 20% of the world's land surface (Figure 11.19a). During the warmer season when the upper portion of the permafrost thaws, water and surface sediment form a soggy mass that flows by solifluction and produces a characteristic lobate topography (Figure 11.19b).

As might be expected, many problems are associated with construction in a permafrost environment. For example, when an uninsulated building is constructed directly on permafrost, heat escapes through the floor, thaws the ground below, and turns it into a soggy, unstable mush. Because the ground is no longer solid, the building settles unevenly into the ground and numerous structural problems result (Figure 11.20).

Creep, the slowest type of flow, is the most widespread and significant mass-wasting process in terms of the total amount of material moved downslope and the monetary damage it does annually. Creep involves extremely slow downhill movement of soil or rock. Although it can occur anywhere and in any climate, it is most effective and significant as a geologic agent in humid regions.

Figure 11.18 Quick-Clay Slide, Anchorage, Alaska

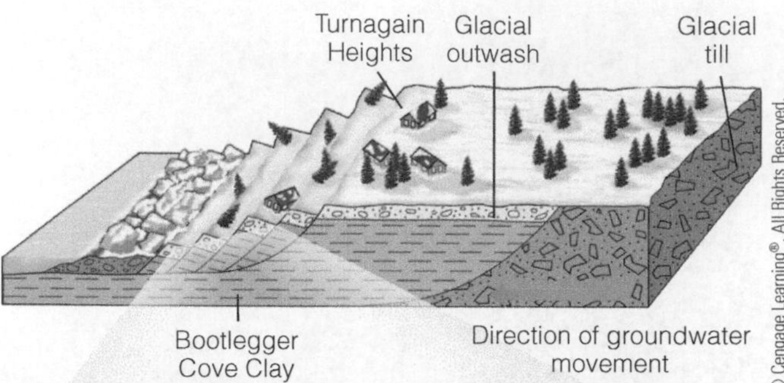

a Ground shaking by the 1964 Alaska earthquake turned parts of the Bootlegger Cove Clay into a quick clay, causing numerous slides.

b Low-altitude photograph of the Turnagain Heights subdivision of Anchorage shows some of the numerous landslide fissures that developed, as well as the extensive damage to buildings in the area. The remains of the Four Seasons apartment building can be seen in the background.

Figure 11.19 Permafrost and Solifluction

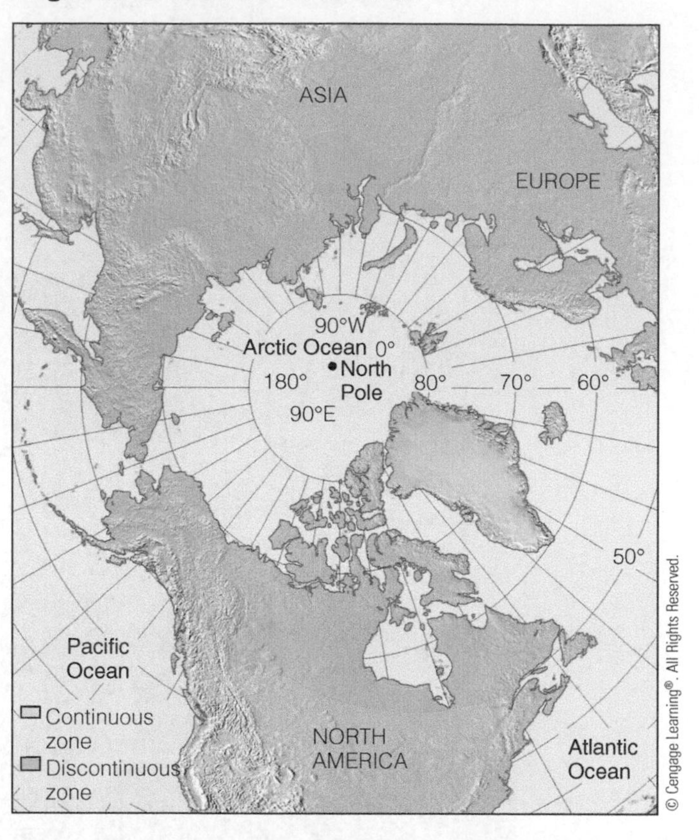

a Distribution of permafrost areas in the Northern Hemisphere.

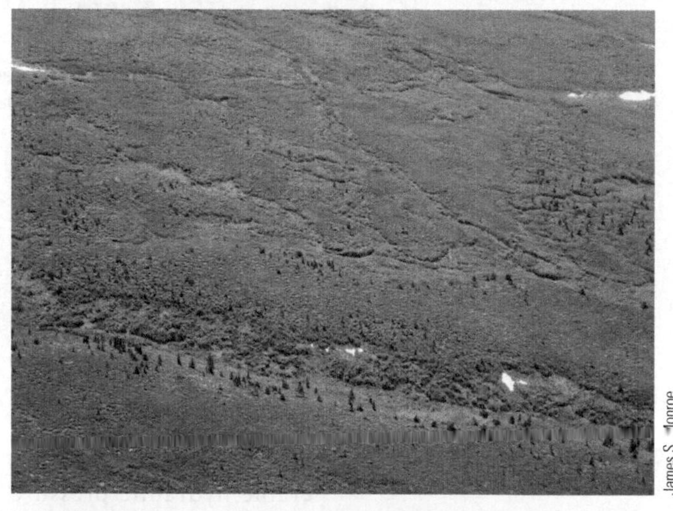

b Solifluction flows in Kluane National Park, Yukon Territory, Canada, show the typical lobate topography that is characteristic of solifluction conditions.

Because the rate of movement is essentially imperceptible, we are frequently unaware of creep's existence until we notice its effects: tilted trees and power poles, broken streets and sidewalks, or cracked retaining walls or foundations (Figure 11.21). Creep usually involves the whole hillside and probably occurs, to some extent, on any weathered or soil-covered, sloping surface.

Creep is not only difficult to recognize but also difficult to control. Although engineers can sometimes slow or stabilize creep, many times the only course of action is to simply avoid the area if at all possible or, if the zone of creep is relatively thin, to design structures that can be anchored into the underlying bedrock.

Complex Movements

Recall that many mass movements are combinations of different movement types. When one type is dominant, the movement can be classified as one of those described thus far. If several types are more or less equally involved, however, it is called a **complex movement**.

The most common type of complex movement is the *slide-flow*, in which there is sliding at the head and then some type of flowage farther along its course. Most slide-flow landslides involve well-defined slumping at the head, followed by a debris flow or earthflow (Figure 11.22). Any combination of different mass movement types is a complex movement.

Figure 11.20 **Permafrost Damage** This house, south of Fairbanks, Alaska, has settled unevenly because the underlying permafrost in fine-grained silts and sands has thawed.

Figure 11.21 Creep

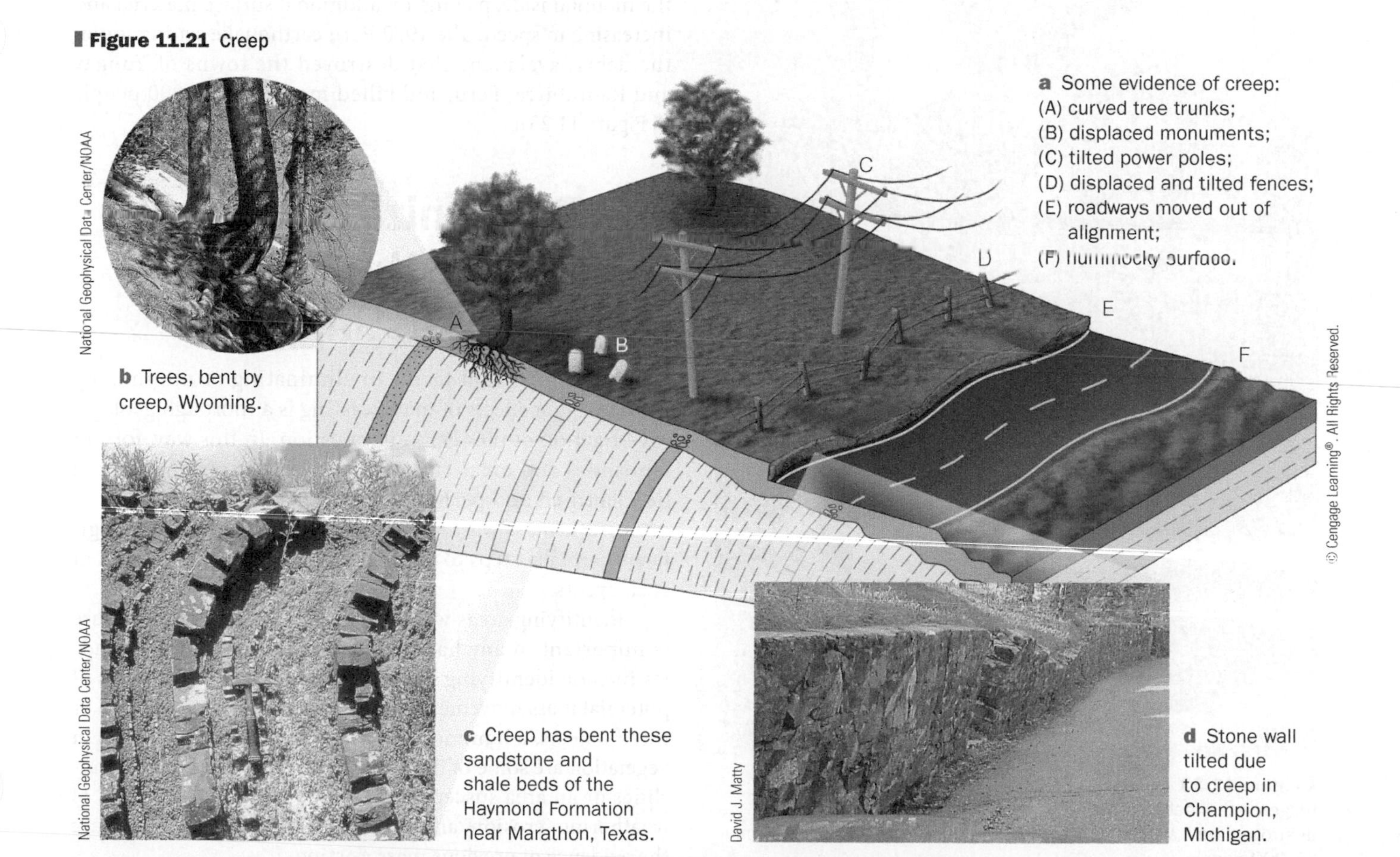

a Some evidence of creep: (A) curved tree trunks; (B) displaced monuments; (C) tilted power poles; (D) displaced and tilted fences; (E) roadways moved out of alignment; (F) hummocky surface.

b Trees, bent by creep, Wyoming.

c Creep has bent these sandstone and shale beds of the Haymond Formation near Marathon, Texas.

d Stone wall tilted due to creep in Champion, Michigan.

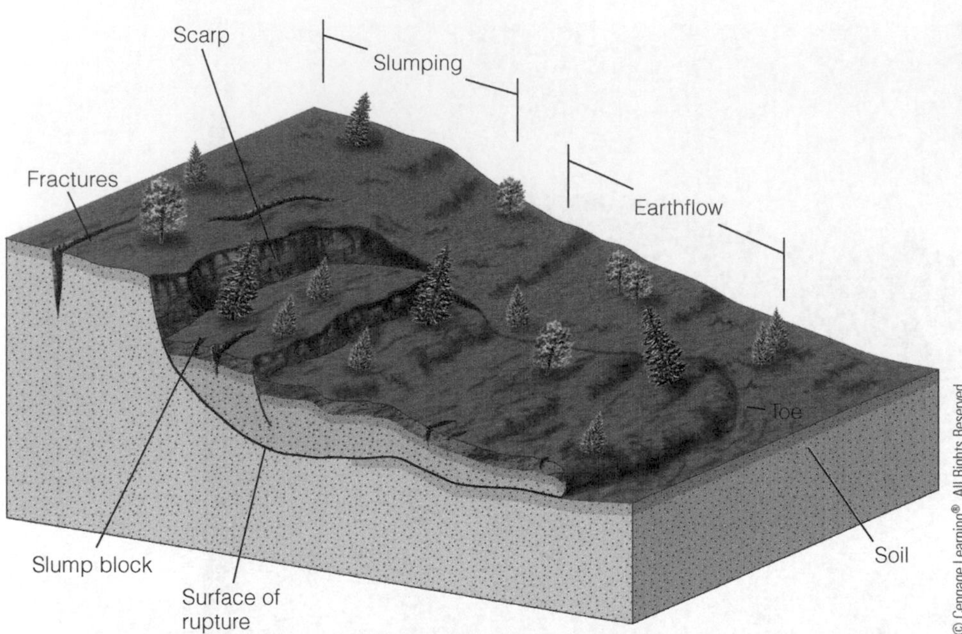

Figure 11.22 Complex Movement A complex movement is one in which several types of mass wasting are involved. In this example, slumping occurs at the head, followed by an earthflow.

Figure 11.23 Debris Avalanche An earthquake 65 km away triggered this debris avalanche on Nevado Huascarán, Peru, that destroyed the towns of Yungay and Ranrahirca and killed more than 25,000 people.

A *debris avalanche* is a complex movement that often occurs in very steep mountain ranges. Debris avalanches typically start out as rockfalls when large quantities of rock, ice, and snow are dislodged from a mountainside, frequently as a result of an earthquake. The material then slides or flows down the mountainside, picking up additional surface material and increasing in speed. The 1970 Peru earthquake set in motion the debris avalanche that destroyed the towns of Yungay and Ranrahirca, Peru, and killed more than 25,000 people (Figure 11.23).

11.4 Recognizing and Minimizing the Effects of Mass Wasting

The most important factor in eliminating or minimizing the damaging effects of mass wasting is a thorough geologic investigation of the region in question. In this way, former landslides and areas susceptible to mass movements can be identified and perhaps avoided. By assessing the risks of possible mass wasting before construction begins, engineers can take steps to eliminate or minimize the effects of such events.

Identifying areas with a high potential for slope failure is important in any hazard assessment study; these studies include identifying former landslides, as well as sites of potential mass movement. Scarps, open fissures, displaced or tilted objects, a hummocky surface, and sudden changes in vegetation are some of the features that indicate former landslides or an area susceptible to slope failure. The effects of weathering, erosion, and vegetation may, however, obscure the evidence of previous mass wasting.

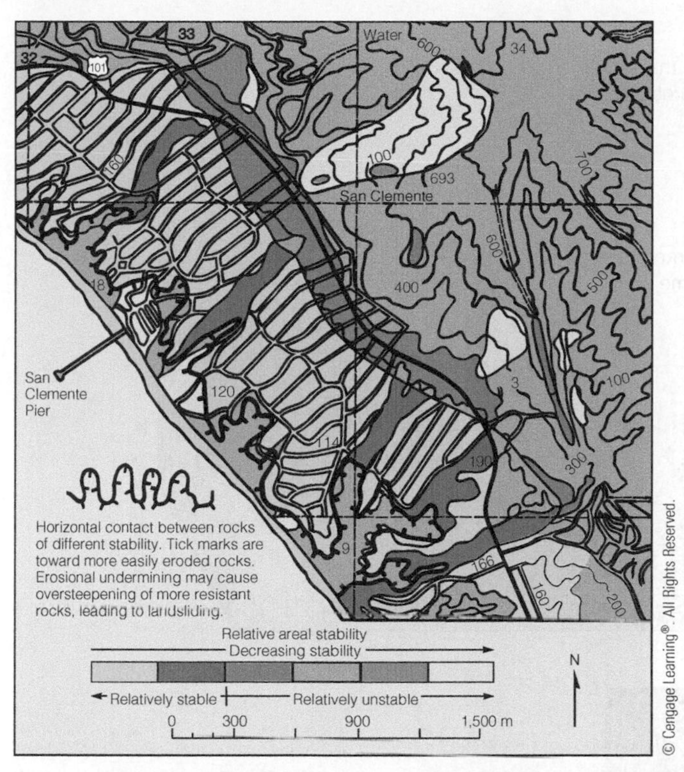

Figure 11.24 Slope-Stability Map This slope-stability map of part of San Clemente, California, shows areas delineated according to relative stability. Such maps help planners and developers make decisions about where to site roads, utility lines, buildings, and other structures.

Critical Thinking Question Locate the line that shows horizontal contact between rocks of different stability. What is the potential for mass wasting along this line, and why?

The information derived from a hazard assessment study can be used to produce *slope-stability maps* of the area (Figure 11.24). These maps allow planners and developers to make decisions about where to site roads, utility lines, and housing or industrial developments based on the relative stability or instability of a particular location. The maps also indicate the extent of an area's landslide problem and the type of mass movement that may occur.

Building codes, which spell out what types of site investigations need to be made and the manner in which structures must be built, also help determine how land will be developed and utilized. Of particular interest is the *International Building Code* that, in conjunction with local or state building codes, deals with all manner of construction, land use, and site development and has led to more stringent building codes, better land use, and decreased damage or losses to structures.

Although most large mass movements usually cannot be prevented, geologists and engineers can use various methods to minimize the danger and damage resulting from them. Because water plays such an important role in many landslides, one of the most effective and inexpensive ways to reduce the potential for slope failure or to increase existing slope stability is surface and subsurface drainage of a hillside. Drainage serves two purposes: (1) it reduces the weight of the material likely to slide, and (2) it increases the shear strength of the slope material by lowering pore pressure.

Surface waters can be drained and diverted by ditches, gutters, or culverts designed to direct water away from slopes. Drainpipes perforated along one surface and driven into a hillside can help remove subsurface water (Figure 11.25). Finally, planting vegetation on hillsides helps stabilize slopes by holding the soil together and reducing the amount of water in the soil.

Another way to help stabilize a hillside is to reduce its slope. Recall that overloading and oversteepening by grading are common causes of slope failure. Reducing the angle of a hillside decreases the potential for slope failure. Two methods are usually employed to reduce a slope's angle. In the *cut-and-fill* method, material is removed from the upper part of the slope and used as fill at the base, thus providing a flat surface for construction and reducing the slope (Figure 11.26).

The second method, which is called *benching*, involves cutting a series of benches or steps into a hillside (Figure 11.27). This process reduces the overall average slope, and the benches serve as collecting sites for small landslides or rockfalls that might occur. Benching is most commonly used on steep hillsides in conjunction with a system of surface drains to divert runoff.

In some situations, retaining walls are constructed to provide support for the base of the slope (Figure 11.28). The walls are usually anchored well into bedrock, backfilled with crushed rock, and provided with drain holes to prevent the buildup of water pressure in the hillside.

Rock bolts, similar to those employed in tunneling and mining, are sometimes used to fasten potentially unstable rock masses into the underlying stable bedrock (Figure 11.29). This technique has been used successfully on the hillsides of Rio de Janeiro, Brazil, and to help secure the slopes at the Glen Canyon Dam on the Colorado River.

Recognition, prevention, and control of landslide-prone areas are expensive, but not nearly as expensive as the damage can be when such warning signs are ignored or not recognized. Unfortunately, numerous examples of landfill and dam collapses serve as tragic reminders of the price paid in loss of lives and property damage when the warning signs of impending disaster are ignored.

Figure 11.25 Using Drainpipes to Remove Subsurface Water

Critical Thinking Question What two features in **(b)** can you identify that indicate mass wasting at this location is both a current and potential problem?

a Driving drainpipes that are perforated on one side into a hillside, with the perforated side up, can remove some subsurface water and help stabilize a hillside.

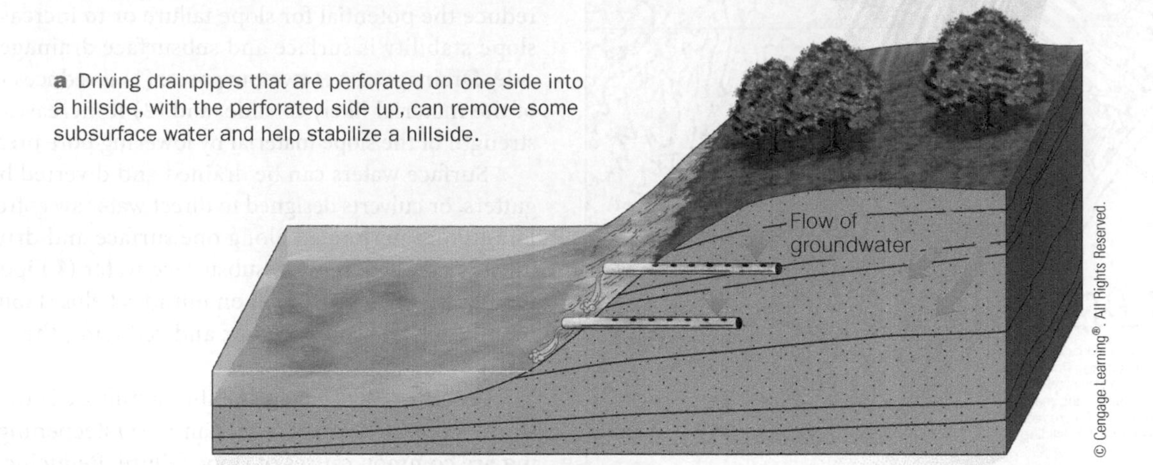

Reed Wicander

b A drainpipe driven into the hillside at Point Fermin, California, helps to reduce the amount of subsurface water in these porous beds.

Figure 11.26 Stabilizing a Hillside by the Cut-and-Fill Method One common method used to help stabilize a hillside and reduce its slope is the cut-and-fill method. Material from the steeper upper part of the hillside is removed, thereby decreasing the slope angle, and is used to fill in the base. This provides some additional support at the base of the slope.

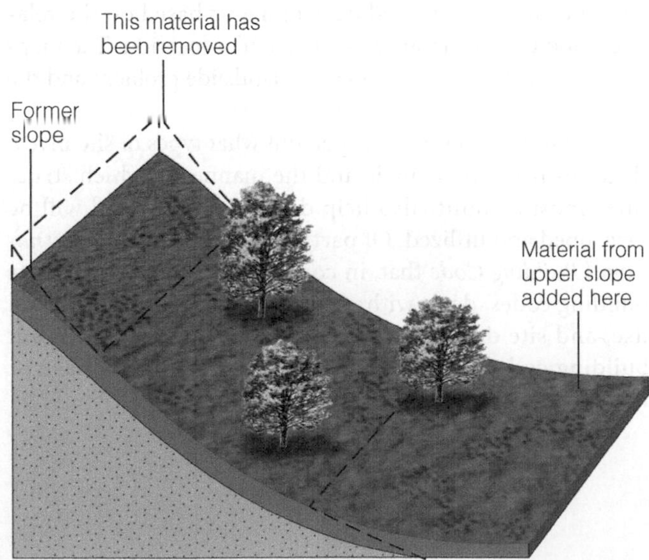

Figure 11.27 Stabilizing a Hillside by Benching

Critical Thinking Question Given the height of the road cut in **(b)**, how effective do you think benching will be in helping to stabilize this slope? What other measures can be taken to minimize the damage from potential mass wasting?

a Another common method used to stabilize a hillside and reduce its slope is benching. This process involves making several cuts along a hillside to reduce the overall slope. Furthermore, individual slope failures are now limited in size, and the material collects on the benches.

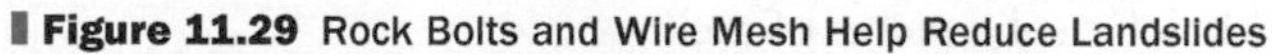

b Benching is used in many road cuts and can be clearly seen in this road cut along Interstate Highway 70 (I-70) west of Denver, in Golden, Colorado.

Figure 11.28 Retaining Walls Help Reduce Landslides

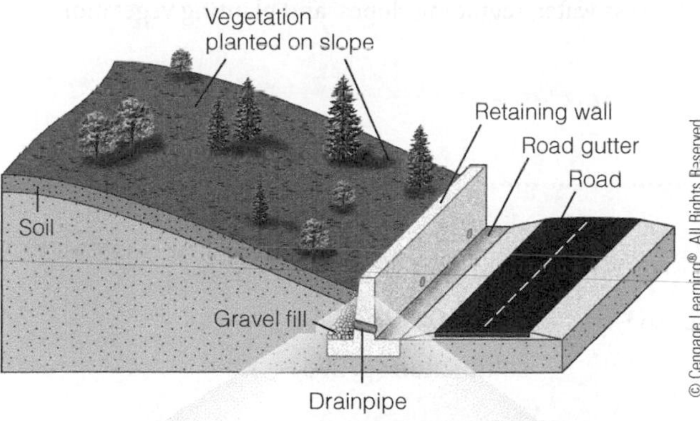

a Retaining walls anchored into bedrock, backfilled with gravel, and provided with drainpipes can support a slope's base and reduce landslides.

b A steel retaining wall built to stabilize the slope and keep falling and sliding rocks off the highway.

Figure 11.29 Rock Bolts and Wire Mesh Help Reduce Landslides

a Rock bolts secured in bedrock can help stabilize a slope and reduce landslides.

b Rock bolts and wire mesh are used to secure rock on a steep hillside in Brisbane, Australia.

Key Concepts Review

- Mass wasting is the downslope movement of material under the direct influence of gravity. It may result in loss of life, as well as millions to billions of dollars in damage annually.
- Mass wasting occurs when the gravitational force acting parallel to a slope exceeds the slope's shear strength (the resisting forces that help maintain slope stability).
- The major factors causing mass wasting include slope angle, weathering and climate, water content, overloading, and removal of vegetation. It is usually several of these factors in combination that results in slope failure.
- Mass movements are classified on the basis of their rate of movement (rapid versus slow), type of movement (falling, sliding, or flowing), and type of material (rock, soil, or debris).
- Rockfalls are a common mass movement in which rocks free-fall. They are common along steep canyons, cliffs, and road cuts.
- The two types of slides are slumps and rock slides. Slumps, or rotational slides, involve movement along a curved surface and are most common in poorly consolidated or unconsolidated material. Rock slides, also known as block slides, occur when movement takes place along a more or less planar surface, and they usually involve solid pieces of rock.
- Rate of movement (rapid versus slow), type of material (rock, sediment, or soil), and amount of water are the criteria used to recognize the several types of flows.
- Mudflows consist of mostly clay- and silt-size particles and contain up to 30% water. They are most common in semiarid and arid environments and generally follow preexisting channels.
- Debris flows are composed of larger particles and contain less water than mudflows.
- Earthflows move more slowly than either debris flows or mudflows and move downslope as thick, viscous, tongue-shaped masses of wet regolith.
- Quick clays are clays that spontaneously liquefy and flow like water when they are disturbed.
- Solifluction is the slow downslope movement of water-saturated surface material and is most common in areas of permafrost.
- Creep, the slowest type of flow, is the imperceptible downslope movement of soil or rock. It is the most widespread of all types of mass wasting.
- Complex movements are combinations of different types of mass movements in which no single type is dominant. Most complex movements involve sliding and flowing.
- The most important factor in reducing or eliminating the damaging effects of mass wasting is a thorough geologic investigation to outline areas susceptible to mass movements.
- Although mass movement cannot be eliminated, its effects can be minimized by building retaining walls, draining excess water, regrading slopes, and planting vegetation.

Important Terms

complex movement (p. 274)
creep (p. 273)
debris flow (p. 271)
earthflow (p. 271)
mass wasting (p. 258)
mudflow (p. 270)
permafrost (p. 273)
quick clay (p. 271)
rapid mass movement (p. 263)
rockfall (p. 263)
rock slide (p. 268)
shear strength (p. 258)
slide (p. 264)
slow mass movement (p. 263)
slump (p. 264)
solifluction (p. 273)

Review Questions

1. Mass wasting can occur
   a. ___ on gentle slopes.
   b. ___ on steep slopes.
   c. ___ in flat-lying areas.
   d. ___ all of these.
   e. ___ none of these.

2. Which of the following helps reduce the slope angle or provides support at the base of a hillside?
   a. ___ cut and fill.
   b. ___ retaining walls.
   c. ___ benching.
   d. ___ all of these.
   e. ___ none of these.

3. Which of the following factors influence mass wasting?
   a. ___ gravity.
   b. ___ weathering.
   c. ___ slope gradient.
   d. ___ water content.
   e. ___ all of these.
4. Which of the following factors can actually enhance slope stability?
   a. ___ increasing the slope angle.
   b. ___ vegetation.
   c. ___ overloading.
   d. ___ rocks dipping in the same direction as the slope.
   e. ___ none of these.
5. Former landslides and areas currently susceptible to slope failure can be identified by which of the following features?
   a. ___ tilted objects.
   b. ___ open fissures.
   c. ___ scarps.
   d. ___ hummocky surfaces.
   e. ___ all of these.
6. If an area has a documented history of mass wasting that has endangered or taken human life, how should people and governments prevent such events from happening again? Are most large mass-wasting events preventable or predictable?
7. What roles do climate and weathering play in mass wasting?
8. Discuss how the different factors that influence mass wasting are interconnected.
9. Why is creep so prevalent? Why does it do so much damage? What are some of the ways that creep might be controlled?
10. **Creative Thinking Visual Question:** What features of slope stabilization do you see in this photograph (▌Figure 1) of a housing development in Concord, California? You should be able to recognize at least three features.

John Karachewski/Geoscapes Photography

▌**Figure 1** Concord, California housing development.

Global GeoScience Watch A landslide is a general term used for mass movement of Earth materials, and it is one of many natural disasters that are responsible for many deaths and damage every year. Search "landslides" in the GREENR database, and then click on "VIEW ALL" next to the "Magazines" heading. What are the causes of landslides? Where are some of the landslide hotspots located? What are the things that can be done to manage and mitigate landslide risks? Write a short report to summarize your findings.

Toketee Falls on the North Umpqua River in Douglas County, Oregon, is a two-step falls that plunges a total of 34 m over a precipice of columnar basalt. Toketee is the Chinook word for graceful.

Sue Monroe

CHAPTER 12

Running Water—Streams and Rivers

OUTLINE

HAVE YOU EVER WONDERED?

- How water is continually recycled from the oceans to land and back to the oceans?
- How streams and rivers erode, transport, and deposit sediment?
- What kinds of streams and rivers there are and how they differ?
- Why floods occur and what might be done to predict and control them?
- How streams and rivers adjust to changes in climate?
- How valleys form and evolve?

12.1 Introduction

At 4:07 p.m. on May 21, 1889, residents of Johnstown, Pennsylvania, heard "a roar like thunder" and within 10 minutes the community was devastated by an 18-m-high wall of water that swept through the town at 60 km/hr sweeping up debris, houses, and entire families (▮ Figure 12.1). Heavy rainfall and the failure of a dam upstream from Johnstown caused the flood that killed at least 2,200 people, making it the most deadly river flood in United States history. A vivid account notes that "those caught up by the wave found themselves swept up in a torrent of oily, muddy water, surrounded by tons of grinding debris, which crushed some, provided rafts for others. Many became hopelessly entangled in miles of barbed wire from the destroyed wire works."*

The Johnstown flood was so devastating that many of the victims were never found and the town did not fully recover for five years. Within days of the flood, assistance began pouring in from concerned citizens elsewhere, volunteers, and the Red Cross. The failure of the dam that caused the flood was not the first nor the last dam failure in the United States. As tragic as this event was it does remind us of the power of running water, the single most important geologic agent in landscape modification in most areas on the continents, except some extremely arid regions and areas covered by glaciers.

Running water in stream and river channels is a tiny part of the hydrosphere that consists of all water on Earth, including water frozen in glaciers, in the atmosphere, and in groundwater, although most water (96.54%) is in the oceans (Table 12.1). Periodically, streams and rivers overtop their banks and flood, causing widespread destruction and, unfortunately, many fatalities. Nevertheless, we derive many benefits from running water, and even from some floods. Before the construction of the Aswan High Dam in 1970, Egyptian farmers depended on silt deposited on the Nile River floodplain to replenish their croplands. In fact, in ancient Egypt, taxes were levied based on the level of the Nile River.

TABLE 12.1 Water on Earth

Location	Volume km^3	Percentage of total
Oceans, seas, and bays	1,338,000,000	96.54
Ice caps, glaciers, and permanent snow	24,064,000	1.74
Groundwater (fresh and saline) and soil moisture	23,416,500	1.69
Ground ice and permafrost	300,000	0.022
Lakes (fresh and saline)	176,400	0.013
Atmosphere	12,900	0.001
Swamps	11,470	0.0008
Rivers	2,120	0.0002
Biological water	1,120	0.0001

Source: USGS (http://ga.water.usgs.gov/edu/earthhowmuch.html)

National Park Service (Nps.gov)

▮ **Figure 12.1** The Johnstown, Pennsylvania, Flood of 1889
On May 13, 1889, an 18-m-high wall of water destroyed Johnstown and killed at least 2,200 people.

All the terrestrial planets had a similar early history of accretion, differentiation, and volcanism, but their subsequent histories differ considerably. Mercury and Earth's Moon are too small and Venus is too hot to retain any surface water. Earth is the only one with abundant liquid water, although Mars had running water during its past and even now has some frozen water and a small amount of water vapor in its atmosphere. Perhaps data from the Mars Explorer *Curiosity* that landed on August 6, 2012, will provide additional information on the history of Mars.

On Earth, water is continually recycled from the oceans through stream and river channels and is one source of water for agriculture, domestic use, recreation, and industry. About 8% of the electricity used in North America is generated at hydroelectric plants. And when Europeans first explored North America, they followed such waterways, as the St. Lawrence, Mississippi, Missouri, and Ohio rivers. Indeed, the Lewis and Clark Expedition (1804–1806), the first to cross the continent, followed the Missouri River from Missouri to its headwaters in Montana where the explorers crossed the mountains by horse and then followed the Columbia River to the Pacific Ocean.

12.2 Water on Earth

Most of Earth's 1.338 billion km^3 of water is in the oceans, and nearly all of the rest is frozen in glaciers on land (Table 12.1). Only 1.72% of all water is in the atmosphere, groundwater, lakes, swamps, and bogs, and a tiny but important amount is in stream and river channels.

Much of our discussion of running water is descriptive, but always be aware that streams and rivers are dynamic systems that must continuously respond to change, natural or otherwise. For example, paving in urban areas increases

*National Park Service—United States Department of Interior; Johnstown Information Service Online.

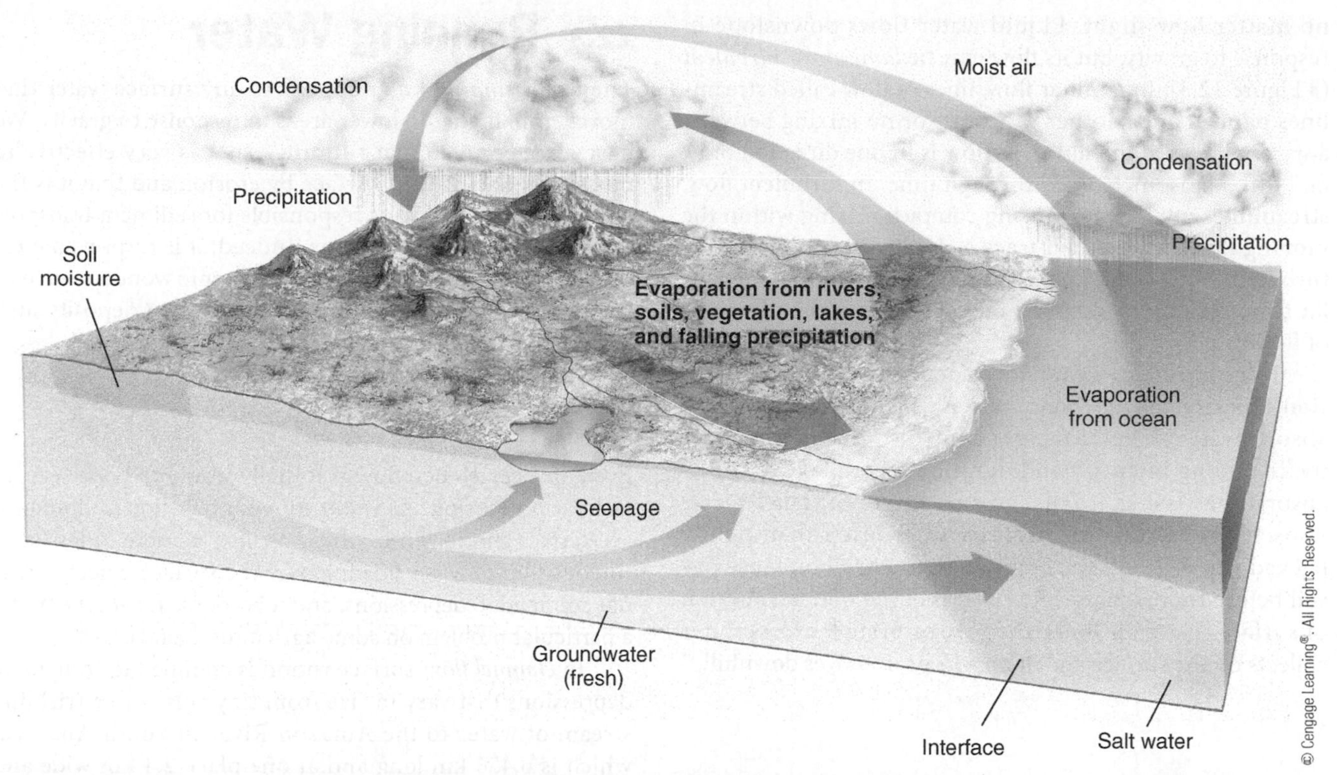

Figure 12.2 The Hydrologic Cycle During the hydrologic cycle, water evaporates from the oceans and rises as water vapor to form clouds that release their precipitation over the oceans or over land. Some of the precipitation falling on land enters the groundwater system, but much of it returns to the oceans by surface runoff, thus completing the cycle.

surface runoff to waterways, and other human activities such as building dams and impounding reservoirs also alter the dynamics of stream and river systems.

The Hydrologic Cycle

The connection between precipitation and clouds is obvious, but where does the moisture for rain and snow come from in the first place? Table 12.1 shows that 96.54% of all water on Earth is in the oceans, so you might immediately suspect that the oceans are the ultimate source of precipitation. In fact, water is continuously recycled from the oceans, through the atmosphere, to the continents, and back to the oceans. This **hydrologic cycle,** as it is called (Figure 12.2), is powered by solar radiation and is possible because water changes easily from liquid to gas (water vapor) under surface conditions. About 85% of all water entering the atmosphere comes from a layer about 1 m thick that evaporates from the oceans each year. The remaining 15% comes from water on land, but this water originally came from the oceans as well.

Regardless of its source, water vapor rises into the atmosphere where the complex processes of cloud formation and condensation take place. About 80% of all precipitation falls directly back into the oceans, in which case the hydrologic cycle is a three-step process of evaporation, condensation, and precipitation. For the 20% of precipitation that falls on land, the hydrologic cycle is more complex, involving evaporation, condensation, movement of water vapor from the oceans to land, precipitation, and runoff. Some precipitation evaporates as it falls and reenters the cycle, but about 36,000 km^3 of the precipitation that falls on land returns to the oceans by **runoff,** the surface flow in streams and rivers.

Not all precipitation returns directly to the oceans. Some is temporarily stored in lakes and swamps, or in snowfields and glaciers, or seeps below the surface where it enters the groundwater system. Water might remain in some of these reservoirs for thousands of years, but eventually glaciers melt, lakes and groundwater feed streams and rivers, and this water returns to the oceans. Even the water used by plants evaporates, a process known as *transpiration,* and returns to the atmosphere. In short, all water derived from the oceans eventually makes its way back to the oceans and can thus begin the hydrologic cycle again.

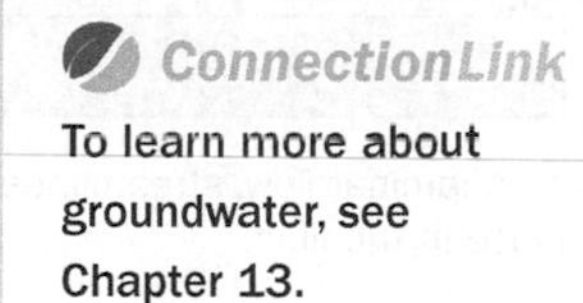

To learn more about groundwater, see Chapter 13.

Fluid Flow

Solids are rigid substances that retain their shapes unless deformed by a force, but fluids—that is, liquids and gases—have no strength, so they flow in response to any force, no matter how slight. Liquid water flows downslope in response to gravity, but its flow may be *laminar* or *turbulent* (▮ Figure 12.3). In laminar flow, lines of flow called streamlines parallel one another with little or no mixing between adjacent layers in the fluid. All flow is in one direction only, and it remains unchanged through time. In turbulent flow, streamlines intertwine, causing complex mixing within the moving fluid. If we could trace a single water molecule in turbulent flow, it may move in any direction at a particular time although its overall movement is in the direction of flow.

Runoff during a rainstorm depends on the **infiltration capacity,** the maximum rate at which surface materials absorb water. Several factors control infiltration capacity, including intensity and duration of rainfall. If rain is absorbed as fast as it falls, no surface runoff takes place. Loosely packed dry soil absorbs water faster than tightly packed wet soil, and thus more rain must fall on loose dry soil before runoff begins. Regardless of the initial condition of surface materials, once they are saturated, excess water collects on the surface and, if on a slope, it moves downhill.

▮ **Figure 12.3** Laminar and Turbulent Flow

a In laminar flow, streamlines are parallel and little mixing takes place in the fluid.

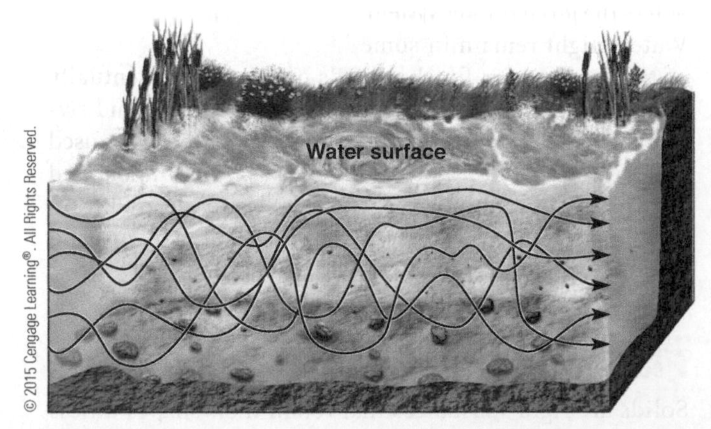

b In turbulent flow, streamlines intertwine, indicating mixing in the fluid.

12.3 Running Water

The term *running water* applies to any surface water that moves from higher to lower areas in response to gravity. We have already noted that running water is very effective in modifying Earth's land surface by erosion and that it is the primary geologic process responsible for sediment transport and deposition in many areas. Indeed, it is responsible for the tiniest rills in farmer's fields and scenic wonders, such as the Grand Canyon in Arizona, as well as vast deposits such as the Mississippi River delta.

Sheet Flow and Channel Flow

Even on steep slopes, flow is initially slow and hence causes little or no erosion. As water moves downslope, though, it accelerates and may move by *sheet flow,* a more or less continuous film of water flowing over the surface. Sheet flow is not confined to depressions, and it accounts for *sheet erosion,* a particular problem on some agricultural lands.

In *channel flow,* surface runoff is confined to troughlike depressions that vary in size from tiny rills with a trickling stream of water to the Amazon River in South America, which is 6,450 km long and at one place 2.4 km wide and 90 m deep. We describe flow in channels with terms such as *rill, brook, creek, stream,* and *river,* most of which are distinguished by size and volume. Here we use the terms *stream* and *river* more or less interchangeably, although the latter usually refers to a larger body of running water.

Streams and rivers receive water from several sources, including sheet flow and rain that falls directly into their channels. Far more important, though, is the water supplied by soil moisture and groundwater, both of which flow downslope and discharge into waterways. In areas where groundwater is plentiful, streams and rivers maintain a fairly stable flow year-round because their water supply is continuous. In contrast, the amount of water in streams and rivers of arid and semiarid regions fluctuates widely because they depend more on infrequent rainstorms and surface runoff for their water.

Gradient, Velocity, and Discharge

Water in any channel flows downhill over a slope known as its **gradient.** Suppose a river has its headwaters (source) 1,000 m above sea level and it flows 500 km to the sea, so it drops vertically 1,000 m over a horizontal distance of 500 km. Its gradient is found by dividing the vertical drop by the horizontal distance, which in this example is 1,000 m/500 km = 2 m/km (▮ Figure 12.4). On average, this river drops vertically 2 m for every kilometer along its course.

In this example, we calculated the average gradient for a hypothetical river, but gradients vary not only among channels but even along the course of a single channel. Rivers and streams are steeper in their upper reaches (near their headwaters) where they may have gradients of several

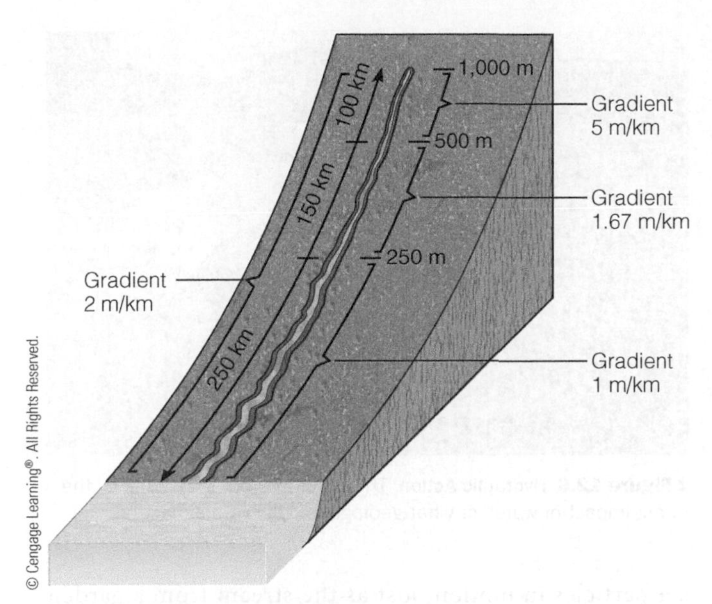

Figure 12.4 Calculating Gradient The average gradient for this stream is 2 m/km, but the gradient may be calculated for any segment of a stream system. Notice that the gradient is steepest in the headwaters area (where the stream originates) and decreases downstream. In the lower reaches of some large rivers, the gradient is as little as a few centimeters per kilometer.

tens of meters per kilometer, but they have gradients of only a few centimeters per kilometer where they discharge into the sea.

The **velocity** of running water is a measure of the downstream distance water travels in a given time. It is usually expressed in meters per second (m/sec) or feet per second (ft/sec), and it varies across a channel's width as well as along its length. Water moves more slowly and with greater turbulence near a channel's bed and banks because friction is greater there than it is some distance from these boundaries (Figure 12.5a). Channel shape and roughness also influence flow velocity. Broad, shallow channels and narrow, deep channels have proportionately more water in contact with their perimeters than channels with semicircular cross sections (Figure 12.5b). So, if other variables are the same, water flows faster in a semicircular channel because there is less frictional resistance. As one would expect, rough channels, such as those strewn with boulders, offer more frictional resistance to flow than do channels with a bed and banks composed of sand or mud.

Intuitively, you might think that gradient is the most important control on velocity—the steeper the gradient, the greater the velocity. In fact, a channel's average velocity actually increases downstream even though its gradient decreases! Keep in mind that we are talking about average

Figure 12.5 Velocity and Discharge Flow velocity in rivers and streams varies as a result of friction with their banks and beds.

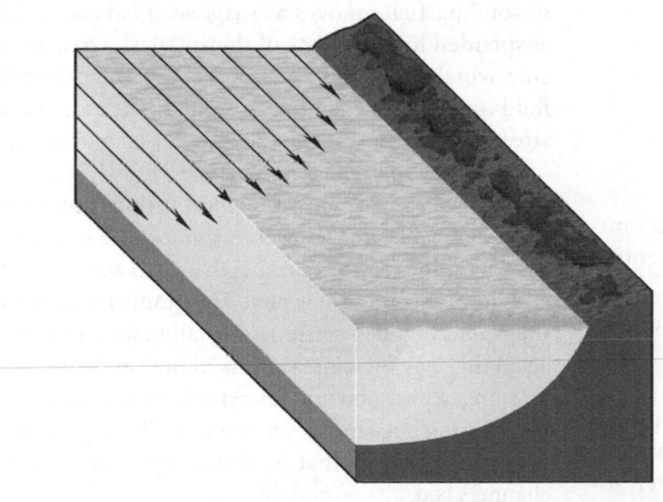

a The maximum flow velocity is near the center and top of a straight channel where the least friction takes place. The arrows are proportional to velocity.

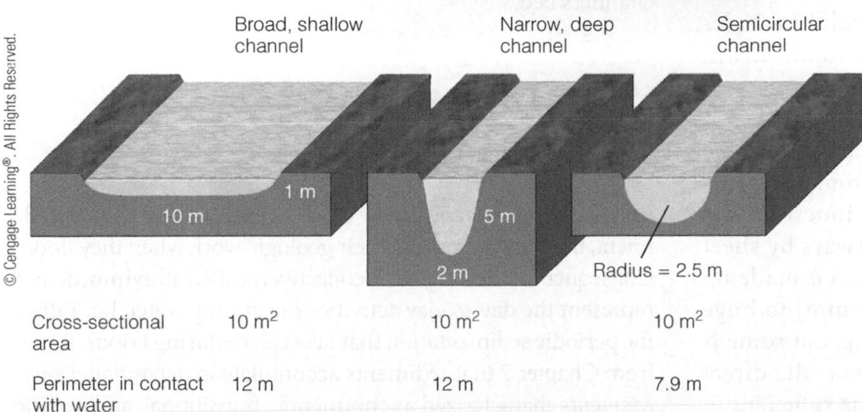

b These three differently shaped channels have the same cross-sectional area, but the semicircular one has less water in contact with its perimeter and thus less frictional resistance to flow.

velocity for a long segment of a channel, not velocity at a single point. Three factors account for this downstream increase in velocity. First, velocity increases even with decreasing gradient in response to the acceleration of gravity. Second, the upstream reaches of channels tend to be boulder-strewn, broad, and shallow, so frictional resistance to flow is high, whereas downstream segments of the same channels are more semicircular and have banks composed of finer materials. And finally, the number of smaller tributaries joining a larger channel increases downstream. Thus, the total volume of water (discharge) increases, and increasing discharge results in greater velocity.

Specifically, **discharge** is the volume of water that passes a particular point in a given period of time. Discharge is found from the dimensions of a water-filled channel—that is, its cross-sectional area (A) and flow velocity (V). Discharge (Q) is then calculated with the formula $Q = VA$ and is expressed in cubic meters per second (m^3/sec) or cubic feet per second (ft^3/sec). The Mississippi River has an average discharge of 18,000 m^3/sec, and the average discharge for the Amazon River in South America is 200,000 m^3/sec.

In most rivers and streams, discharge increases downstream as more and more water enters a channel, but there are a few exceptions. Because of high evaporation rates and infiltration, the flow in some desert waterways actually decreases downstream until the water disappears. And even in perennial rivers and streams, discharge is obviously highest during times of heavy rainfall and at a minimum during the dry season.

12.4 Running Water, Erosion, and Sediment Transport

Streams and rivers possess two kinds of energy to accomplish their tasks of erosion and sediment transport: potential and kinetic. *Potential energy* is the energy of position, such as the energy of water at high elevation. During stream flow, potential energy is converted to *kinetic energy,* which is the energy of motion. Much of this kinetic energy is used up in fluid turbulence, but some is available for erosion and transport. You already know that erosion involves the removal of weathered materials from their source area, so the materials transported by a stream include a load of solid particles (mud, sand, and gravel) and a dissolved load of ions in solution.

Because the **dissolved load** of a stream is invisible, it is commonly overlooked, but it is an important part of the total sediment load. Some of it is acquired from a stream's bed and banks where soluble rocks such as limestone are present, but much of it is carried into waterways by sheet flow and by groundwater. A stream's solid load is made up of particles ranging from clay size ($<$1/256 mm) to huge boulders, much of it supplied by mass wasting, but some is eroded directly from a stream's bed and banks. The direct impact of running water, **hydraulic action,** is sufficient to

Figure 12.6 Hydraulic Action This stream erodes its bank by the direct impact of water, or what geologists call hydraulic action.

set particles in motion, just as the stream from a garden hose gouges out a hole in soil (Figure 12.6).

Running water carrying sand and gravel erodes by **abrasion,** as exposed rock is worn and scraped by the impact of these particles. Circular to oval depressions called *potholes* in streambeds are one manifestation of abrasion (Figure 12.7), which form where swirling currents with sand and gravel erode the rock.

Once materials are eroded, they are transported for some distance from their source and eventually deposited. The dissolved load is transported in the water itself, but the load of solid particles moves as *suspended load* or *bed load.* The **suspended load** consists of the smallest particles of silt and clay, which are kept suspended above the channel's bed by fluid turbulence (Figure 12.8a). It is the suspended load of streams and rivers that gives the water its murky appearance.

The **bed load** of larger particles, mostly sand and gravel, cannot be kept suspended by fluid turbulence, so it is transported along the bed. However, some of the sand may be temporarily suspended by currents that swirl across the streambed and lift grains into the water. The grains move forward with the water, but also settle and finally come to rest and then again move by the same process of intermittent bouncing and skipping, a phenomenon known as *saltation* (Figure 12.8b). Particles too large to be even temporarily suspended are transported by traction; that is, they simply roll or slide along a channel's bed.

12.5 Deposition by Running Water

Rivers and streams constantly erode, transport, and deposit sediment, but they do most of their geologic work when they flood. Consequently, their deposits, collectively called **alluvium,** do not represent the day-to-day activities of running water, but rather the periodic sedimentation that takes place during floods. Recall from Chapter 7 that sediments accumulate in *depositional environments* characterized as continental, transitional, and marine

Figure 12.7 Abrasion by Running Water

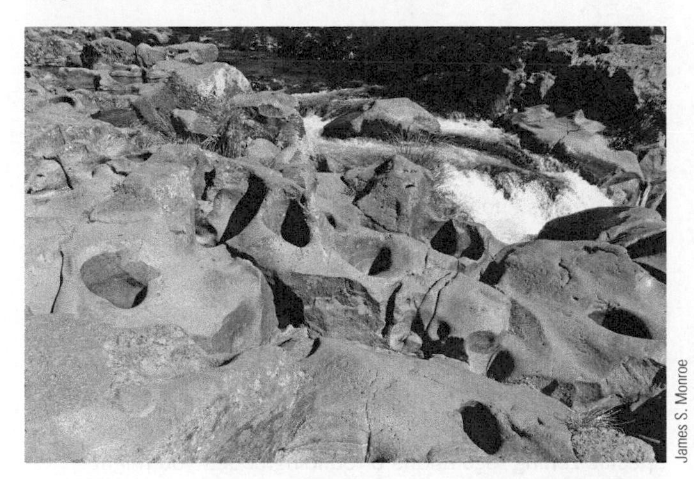

a Potholes in the bed of the McCloud River in California. Notice also that the rock surface has been smoothed and polished by abrasion.

b View into one of the potholes showing the sand and gravel that swirled around to erode the pothole.

Figure 12.8 Sediment Transport by Running Water

a You can tell by the water color that these two small streams in Glacier National Park in Montana carry sediment from two different sources.

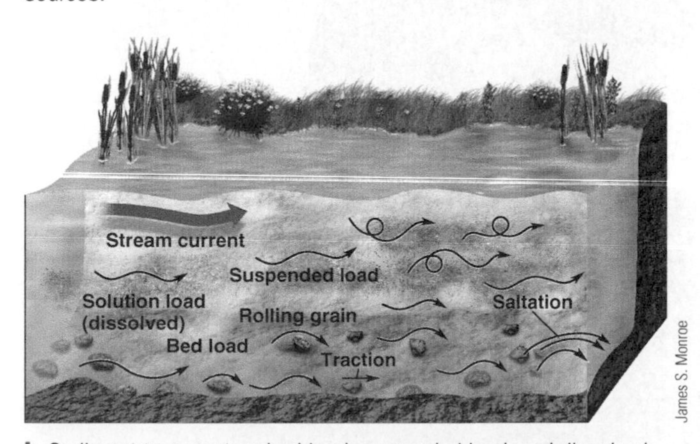

b Sediment transport as bed load, suspended load, and dissolved load. Flow velocity is highest near the surface, but gravel- and sand-size particles are too large to be lifted far from the streambed so they make up the bed load, whereas silt and clay are in the suspended load.

(see Figure 7.4). Deposits of rivers and streams are found in the first two of these settings; however, much of the detrital sediment on continental margins is derived from the land and transported to the oceans by running water.

Deposits of Braided and Meandering Rivers and Streams

A few rivers and streams have single, straight channels for some distance, but most waterways are either braided or meandering. **Braided streams** have intricate networks of dividing and rejoining broad, shallow channels separated from one another by sand and gravel bars (Figure 12.9). For instance, a braided stream may have a channel less than one meter deep but tens of meters across. Rivers and streams with this channel pattern develop when the supply of sediment exceeds the transport capacity of running water. During high-water stages, their sand and gravel bars are submerged, but when the water is low, these bars are exposed and divide the channel into multiple channels.

Braided streams are common in arid and semiarid regions with sparse vegetation and surface materials that are easily eroded. They are also common where sediment-laden water discharges from glaciers. In any case, braided channels are characterized as bed load transport streams because they transport and deposit mostly sand and gravel. Mud may accumulate in abandoned channels, but for the most part, braided stream deposits are sheets of sand and gravel with subordinate mud.

In marked contrast to braided streams, **meandering streams** have a single, sinuous channel with broadly looping curves called *meanders* (Figure 12.10). In cross section, meandering stream channels are semicircular along straight segments but are markedly asymmetric at meanders where they vary from shallow to deep across the channel. On the deeper side of the channel, called the *cut bank,*

Figure 12.9 Braided Stream This stream in Alaska is heavily laden with sediment it derives from the Exit Glacier in the background. As a result, it deposits sand and gravel bars, thus forming an intricate network of dividing and rejoining channels.

Zoning Regulations and River Erosion

Given what is known about the dynamics of running water in channels, it is remarkable that houses are still built on the cut banks of meandering rivers. No doubt, property owners think these locations provide good views because they sit high above the adjacent channel. Explain why you would or would not build a house in such a location. What recommendations would you make to a planning commission on land use for areas as described here? Are there any specific zoning regulations or building codes you might favor?

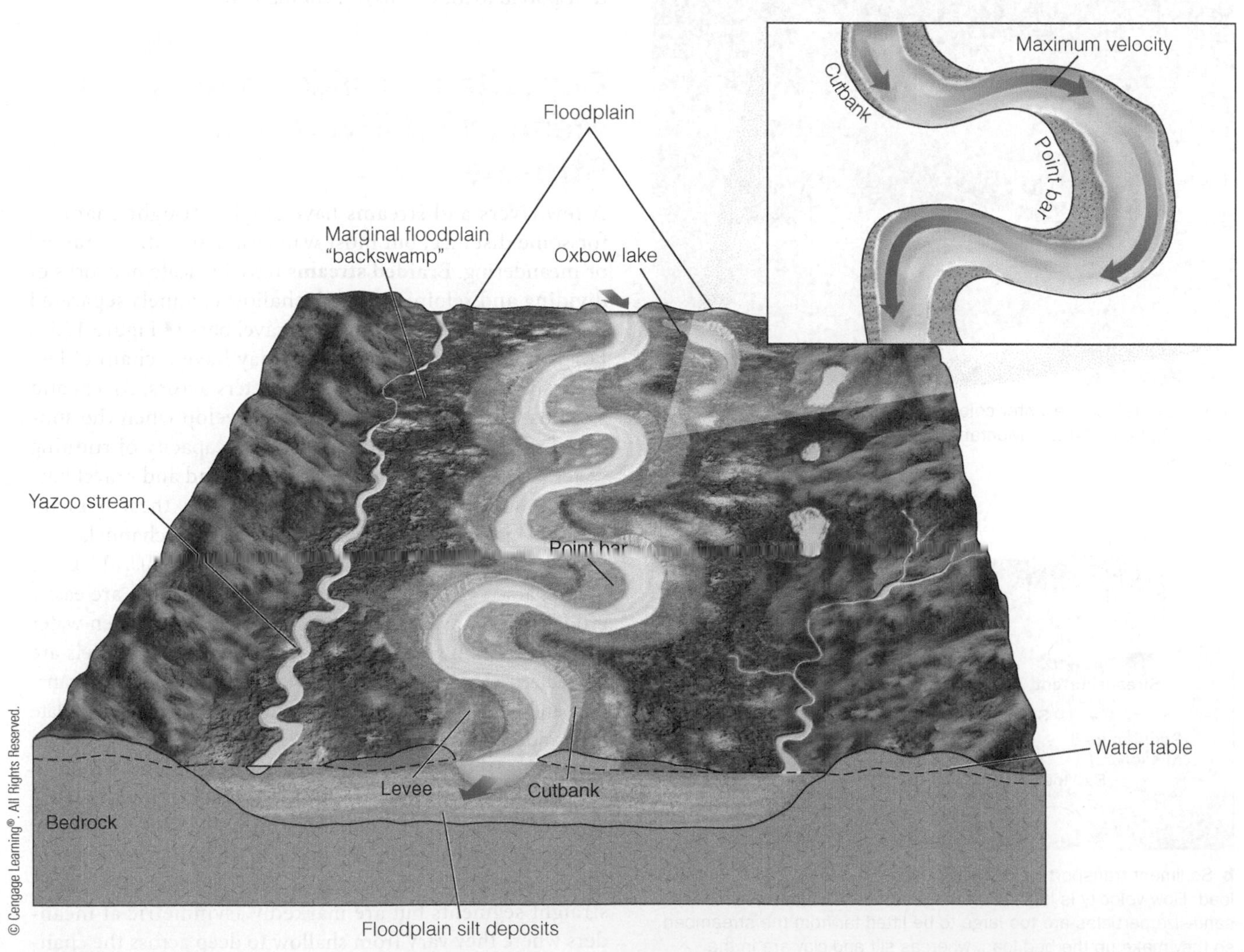

Figure 12.10 Diagrammatic View of a Meandering River

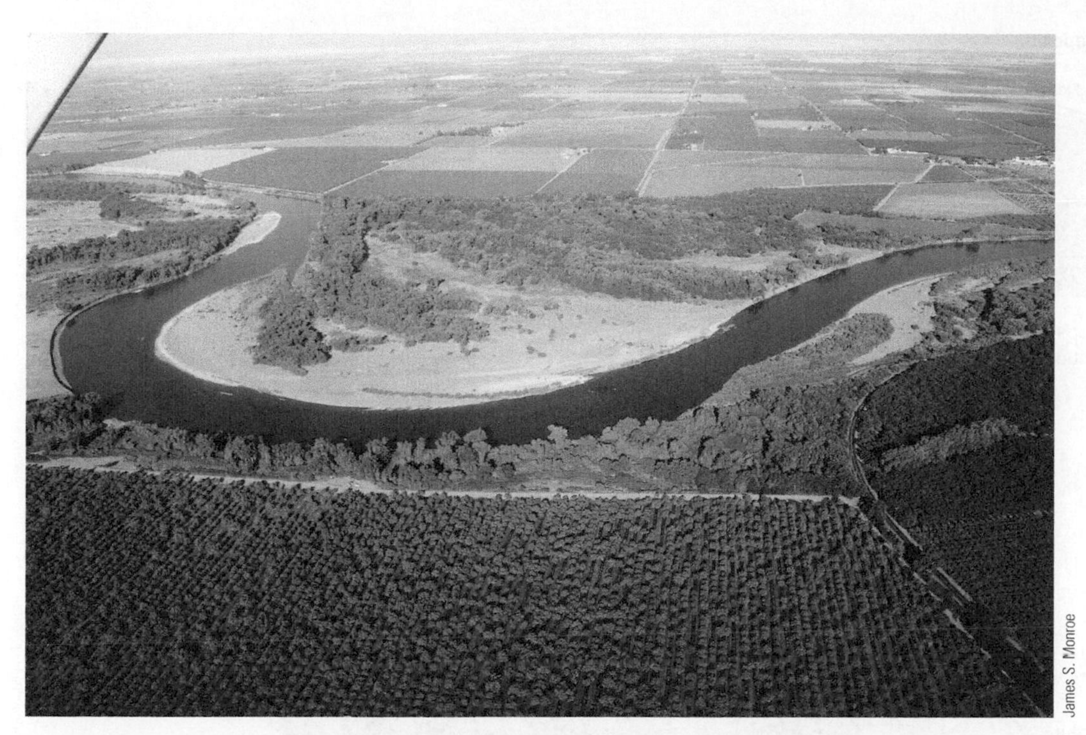

Figure 12.11 Point Bar This is a well-developed point bar on the Sacramento River near Hamilton City, California. The river here is about 150 m wide.

flow velocity and fluid turbulence are greatest and erosion occurs. On the opposite side of the channel, though, the water is shallow, flow velocity and fluid turbulence are less, and deposition of a sand body known as a **point bar** takes place (Figure 12.11). In the geologic record, point bar deposits are distinctive features that, taken with other rock properties, indicate deposition by meandering streams.

The broadly looping curves of meandering streams commonly become so sinuous that during a flood the thin neck of land separating adjacent meanders is cut off, thereby forming a crescent-shaped **oxbow lake** (Figures 12.10 and 12.12). Oxbow lakes may persist for many years, even decades, but eventually they fill in with organic matter and mud carried in during floods, and then they are recognized as *meander scars*. Meander cutoff and the origin of oxbow lakes are so common on large rivers that using these rivers as political boundaries is problematic.

Figure 12.12 Oxbow Lakes

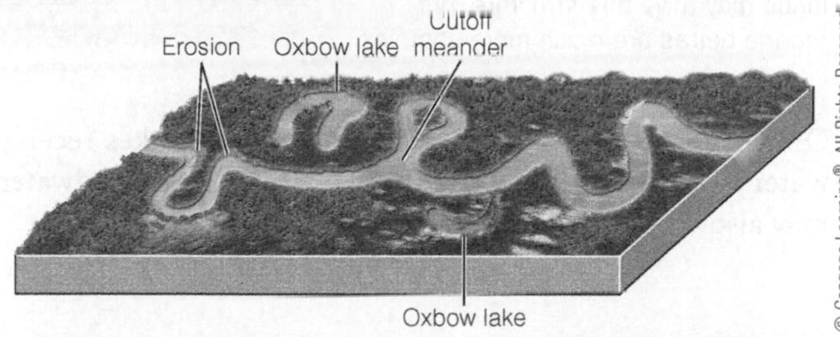

a A meandering stream showing various stages in the evolution of oxbow lakes.

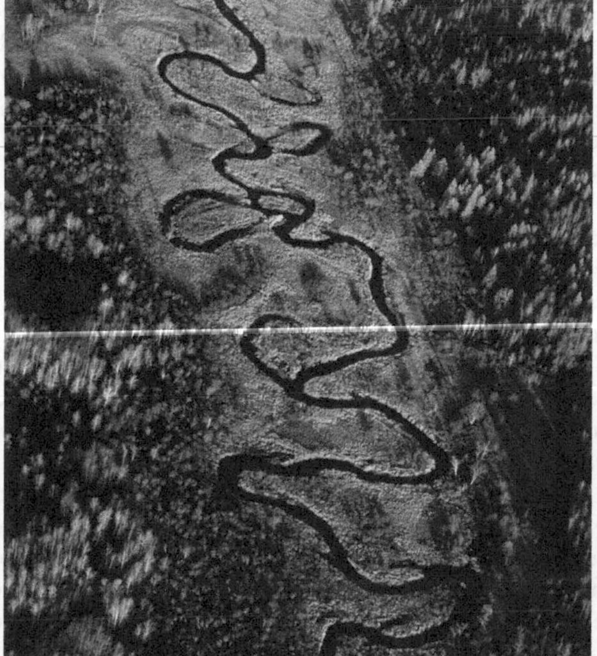

b Oxbow lakes along the Mississippi River in Minnesota.

Floodplain Deposits

Streams and rivers periodically receive more water than their channels can handle, so they overflow their banks and spread across adjacent low-lying, fairly flat **floodplains** (Figure 12.10). Sand and gravel might be deposited on floodplains where a stream or river bursts from its banks, but most of the deposits are silt and clay, or simply mud. During a flood, a stream or river overtops its banks, and water pours onto the floodplain, but as it does so, its depth and velocity rapidly decrease. As a result, ridges of sandy alluvium known as **natural levees** are deposited along the channel margins, and mud is carried beyond the natural levees into the floodplain where it settles from suspension (Figure 12.13).

Another feature found on floodplains is oxbow lakes; recall that oxbow lakes are cutoff meanders (Figure 12.12a).

Figure 12.13 Floodplain Deposits

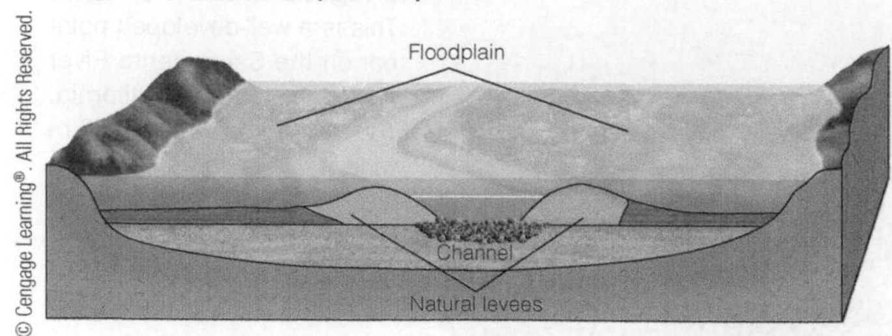

a The origin of floodplain deposits. During floods, streams deposit natural levees, and silt and mud settle from suspension on the floodplain.

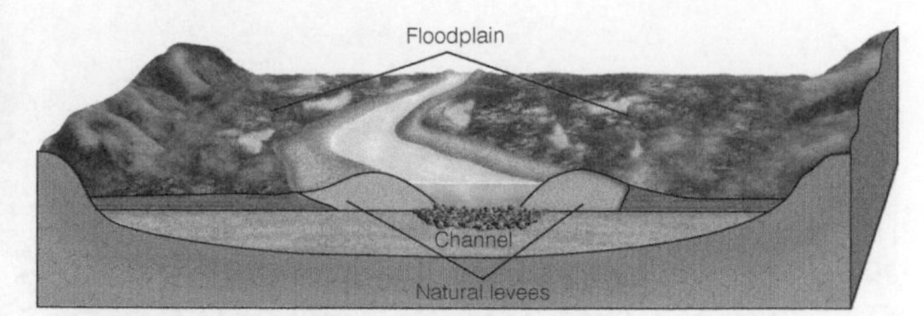

b After flooding.

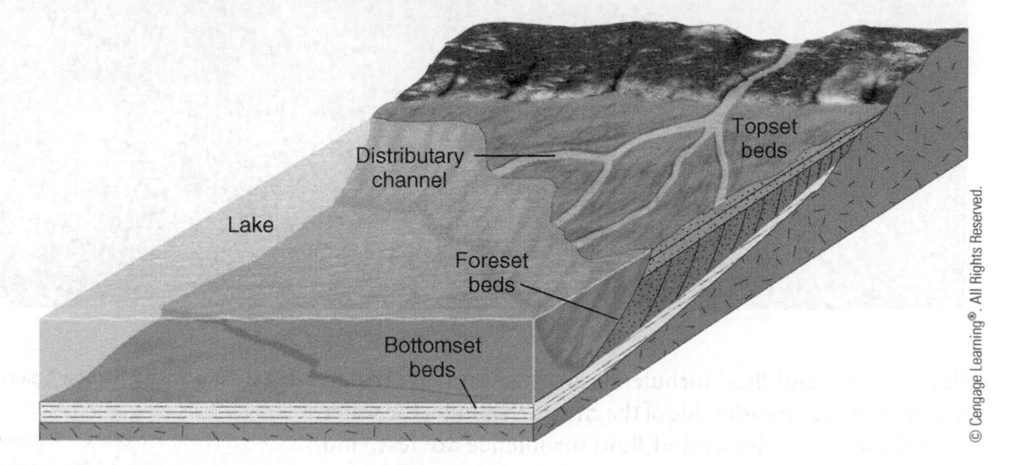

Figure 12.14 Prograding Delta
Internal structure of the simplest type of prograding delta. Small deltas in lakes may have this structure, but marine deltas are much more complex.

Once isolated from the main channel, oxbow lakes receive water mostly from periodic floods, although groundwater may also contribute.

Deltas

Where a river or stream flows into a standing body of water, such as a lake or the ocean, its flow velocity rapidly diminishes and any sediment in transport is deposited. Under some circumstances, this deposition creates a **delta,** an alluvial deposit that causes the shoreline to build outward into the lake or sea, a process called *progradation.* The simplest prograding deltas have a characteristic vertical sequence of *bottomset beds* overlain successively by *foreset beds* and *topset beds* (Figure 12.14). This vertical sequence develops when a river or stream enters another body of water where the finest sediment (silt and clay) is carried some distance out into the lake or sea where it settles to form bottomset beds. Nearer the shore, foreset beds are deposited as gently inclined layers, and topset beds, consisting of the coarsest sediments, are deposited in a network of *distributary channels* traversing the top of the delta (Figure 12.14).

Small deltas in lakes may have the three-part sequence described above, but deltas deposited along seacoasts are much larger, far more complex, and considerably more important as potential areas of natural resources. In fact, depending on the relative importance of running water, waves, and tides, geologists identify three main types of marine deltas. *Stream-dominated deltas* have long finger-like sand bodies, each deposited in a distributary channel that progrades far seaward. The Mississippi River delta is a good example (Geo-Focus 12.1 and Figure 12.15a). In contrast, the Nile River delta in Egypt is *wave dominated.* It also has distributary channels, but the seaward margin of the delta consists of islands reworked by waves, and the entire margin of the delta progrades (Figure 12.15b). *Tide-dominated deltas* are continuously modified into tidal sand bodies that parallel the direction of tidal flow (Figure 12.15c).

Alluvial Fans

Fan-shaped deposits of alluvium on land known as **alluvial fans** form best on lowlands with adjacent highlands in arid and semiarid regions where little vegetation exists to stabilize surface materials (Figure 12.16). During periodic rainstorms, surface materials are quickly saturated, and surface runoff is funneled into a mountain canyon leading to adjacent lowlands. In the mountain canyon, the runoff is confined so that it cannot spread laterally, but when it discharges onto the lowlands, it quickly spreads out, its velocity diminishes, and deposition ensues. Repeated episodes of sedimentation result in the accumulation of a fan-shaped body of alluvium.

ConnectionLink
To learn more about alluvial fans and other desert features, see Chapter 15.

Deposition by running water is responsible for many alluvial fans. In this case, they are composed mostly of sand and gravel, both of which contain a variety of sedimentary structures. In some cases, though, the water flowing through

Figure 12.15 Stream-, Wave-, and Tide-Dominated Deltas

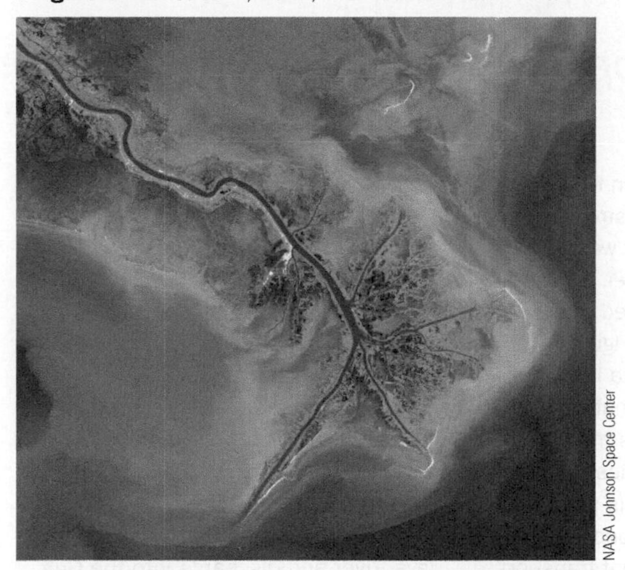

a The Mississippi River delta on the United States Gulf Coast is stream dominated.

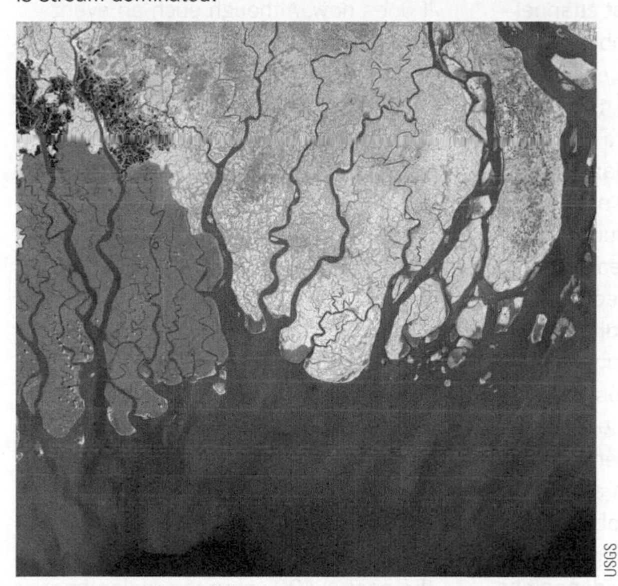

c The Ganges-Brahmaputra delta of Bangladesh is tide dominated.

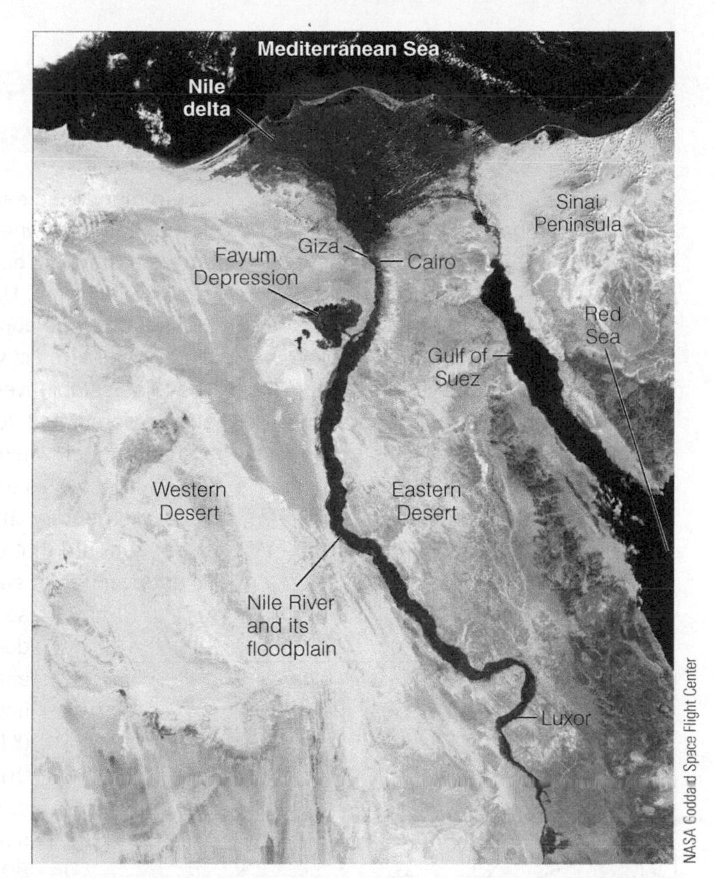

b The Nile River delta of Egypt is wave dominated.

Figure 12.16 Alluvial Fan This is the Badwater Fan in Death Valley National Park in California. It was deposited where streams and debris flows discharged from the mountain canyon visible in this image. Notice the road for scale.

a canyon picks up so much sediment that it becomes a viscous debris flow. Consequently, some alluvial fans consist mostly of debris-flow deposits that show little or no layering. Of course, the dominant type of deposition can change through time, so a particular fan might have both types of deposits.

12.6 Can Floods Be Predicted and Controlled?

When any waterway receives more water than its channel can handle, it floods, occupying part or all of its floodplain. Indeed, floods are so common that unless they cause extensive property damage or fatalities, they rate little more than a passing notice in the news. Dozens of floods occur in the United States every year, but the most extensive river flooding in recent history was the Great Flood of 1993 and the 2011 flooding on the lower Mississippi (Geo-Insight 12.1 and Figure 12.17).

GEO FOCUS 12.1 The Mississippi River Delta—Past and Present

In Chapter 23 we will note that deposition has taken place along the entire Gulf Coast since the beginning of the Late Jurassic. In fact, following the breakup of Pangaea, the area has had a complex history of transgressions and regressions as well as well as fluvial deposition, which continues on the Mississippi River delta. The lower Mississippi River basin is probably underlain by part of a failed rift that formed when the Gulf of Mexico formed (see Figure 22.2).

We mentioned earlier in this chapter that deltas form where a stream or river enters a standing body of water and deposits alluvium. However, a developing delta may be dominated by waves, tides, or stream activity (Figure 12.15). The Mississippi River delta is a classic stream-dominated delta that lies on a shallow shelf where it is mostly protected from waves and tides. On the delta, distributary channels prograde so far seaward that they become inefficient avenues of sediment transport, so the stream abandons that channel and establishes a new one elsewhere (▮ Figure 1). In fact, the entire coastline of Louisiana has built seaward as much as 80 km during the last 5,000 years as a result of sedimentation.

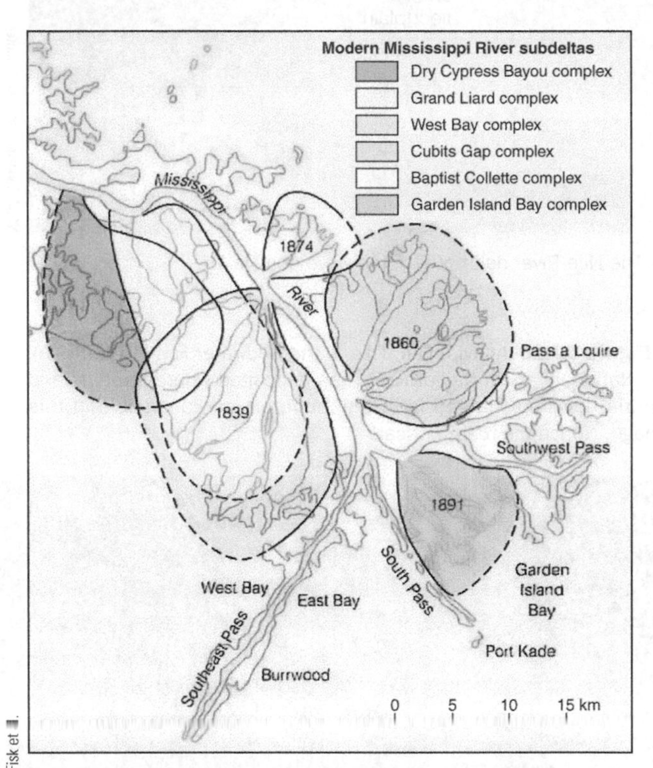

▮ **Figure 1** The present lobe of the Mississippi River delta consists of several subdeltas that formed when distributary channels were abandoned. The area shown here corresponds to the Balize lobe in Figure 2.

In addition to abandoning distributaries channels, stream-dominated deltas periodically abandon entire deltaic lobes. The Mississippi River is currently depositing sediment on a single lobe, but the entire delta is made up of several lobes that formed during the last few thousands of years (▮ Figure 2). As a matter of fact, if it were not for the efforts of the Army Corps of Engineers, the currently active lobe may have already been abandoned.

You may recall the flooding that took place on the lower Mississippi River during April and May 2011, when the Corp of Engineers diverted much of the Mississippi River into the Atchafalaya River at the Morganza Spillway. If the Mississippi were to abandon its present course, it would follow the Atchafalaya River and discharge into the Gulf of Mexico about 100 km west of where it does now. Although such an event would be economically catastrophic for New Orleans, the river would simply continue to build its delta, but in an area it occupied long ago (Figure 2).

Stream-dominated deltas grow seaward by progradation, but in the case of the Mississippi River, flood-control projects now trap much of the sediment so that it no longer reaches the delta. As a result, the delta is losing land along its margin due to wave erosion, and, more importantly, to compaction. Delta deposits, especially mud, compact under their own weight, and thus the surface subsides* and the sea invades the area where subsidence takes place. Geologists estimate that during the past 4,400 years the region has subsided about 12 cm per century.

Delta subsidence is not a problem of academic interest only; a large part of New Orleans lies below sea level because of subsidence. Much of the

*Pumping oil and natural gas from the delta sediments also contributes to subsidence.

More than 2,100 people died during river floods in 2007 in India and Bangladesh, including dozens killed by poisonous snakes that, like people, sought refuge on high ground.

In this section we concentrate on river floods as opposed to coastal flooding that takes place during hurricanes. River floods may take place rather slowly and persist for days or weeks, or they may occur rapidly and in this case they usually last for only a few days. Then there are channels that fill very rapidly and burst out in *flash floods* that last only minutes to hours. These are caused by local heavy

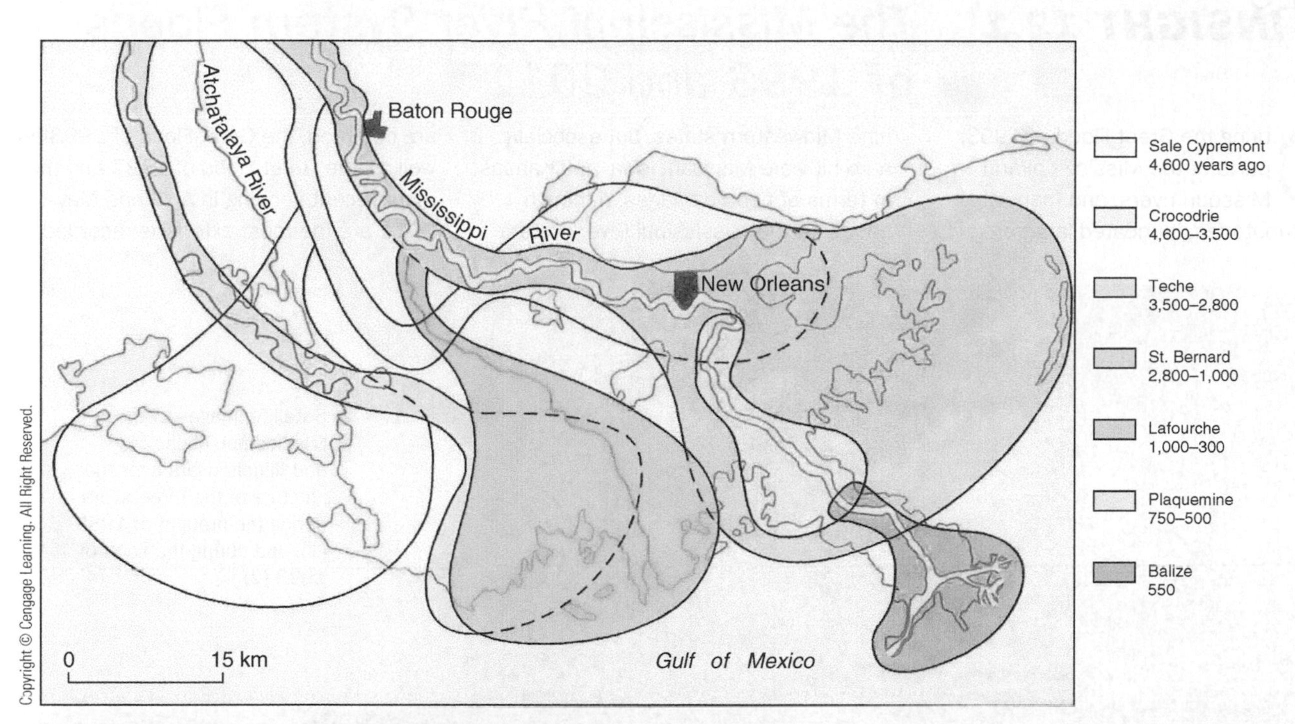

Figure 2 The Major Deltaic Lobes of the Mississippi River Deposition of the present-day Balize lobe began about 550 years ago.

Figure 3 The Deepwater Horizon drilling platform not long before it sank in the Gulf of Mexico on April 22, 2010.

city is located between Lake Pontchartrain and the Mississippi River, so when Hurricane Katrina roared ashore in 2005, nearly 80% of New Orleans was flooded. About 1,800 people died, and damages exceeded $100 billion, mostly in Louisiana and Mississippi.

Another rather recent event that had an impact on the Mississippi River delta is the *Deepwater Horizon* oil spill than began in April 2010. You no doubt recall that an oil rig exploded and sank about 65 km off the Louisiana coast, killing 11 workers and beginning the largest oil spill in the history of the petroleum industry (Figure 3). The spill caused considerable damage in the Gulf of Mexico and along the shoreline. Currently, about 4,000 wells in the Gulf supply 30% of all United States oil. Unfortunately, there are also some 27,000 abandoned wells in the Gulf, many of which are not properly sealed.

rainfall, dam failures (see the Introduction), and breaching of ice dams.

Flooding is a fact of life, but several practices can protect people and their property and minimize the impact of floods. Unfortunately, none of these measures are foolproof, and most are expensive. One solution is to simply move all or parts of a community to higher ground. Indeed, following the Great Flood of 1993, parts of some small communities in Missouri and Illinois were moved to higher ground. Martin, Kentucky, has been flooded 37 times since 1862, and parts

GEO INSIGHT 12.1 The Mississippi River System Floods of 1993 and 2011

During the Great Flood of 1993, parts of the Mississippi and Missouri rivers, and many of their tributaries, inundated large areas in nine Midwestern states, but especially hard hit were Missouri, Iowa, and Kansas in terms of flood damages. Although floods on the Mississippi River system are common, the Great Flood of 1993 as well as the Great Flood of 1927 and the more recent flooding in April and May of 2011 are the most extensive recorded.

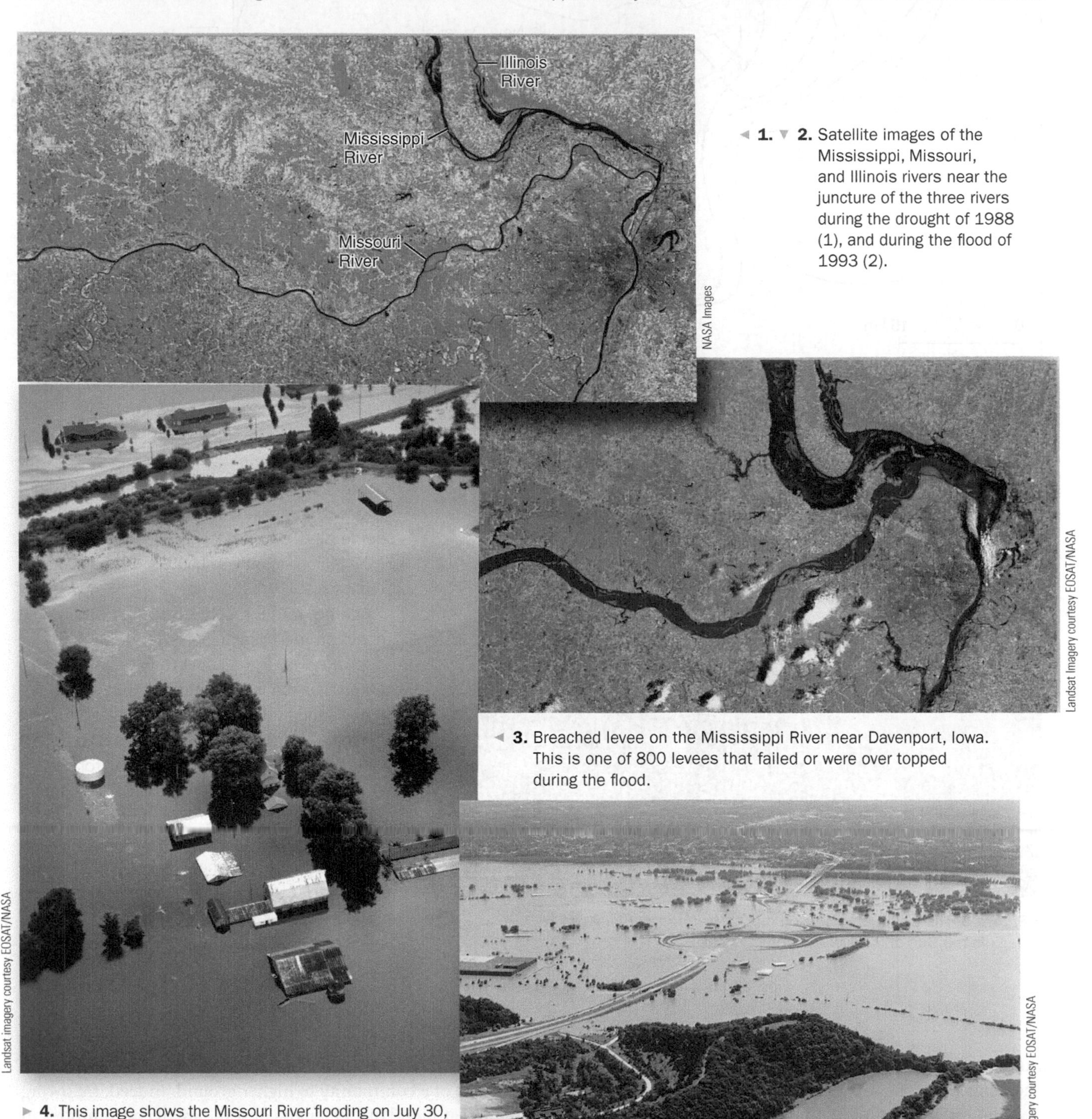

◄ **1.** ▼ **2.** Satellite images of the Mississippi, Missouri, and Illinois rivers near the juncture of the three rivers during the drought of 1988 (1), and during the flood of 1993 (2).

◄ **3.** Breached levee on the Mississippi River near Davenport, Iowa. This is one of 800 levees that failed or were over topped during the flood.

► **4.** This image shows the Missouri River flooding on July 30, 1993, near the Cedar City and Jefferson City Memorial Airport, which is just north of Jefferson City, Missouri.

AP Photo/The News-Star, Margaret Croft

▲ **5.** This levee in Louisiana protecting farmland failed on May 13, 2011, just as the governor flew over the area.

USGS/NASA

▲ **6.** Lansat images of the Mississippi River along the borders of Tennessee, Kentucky, Missouri, and Arkansas on May 12, 2006 (top) and on May 10, 2011 (bottom).

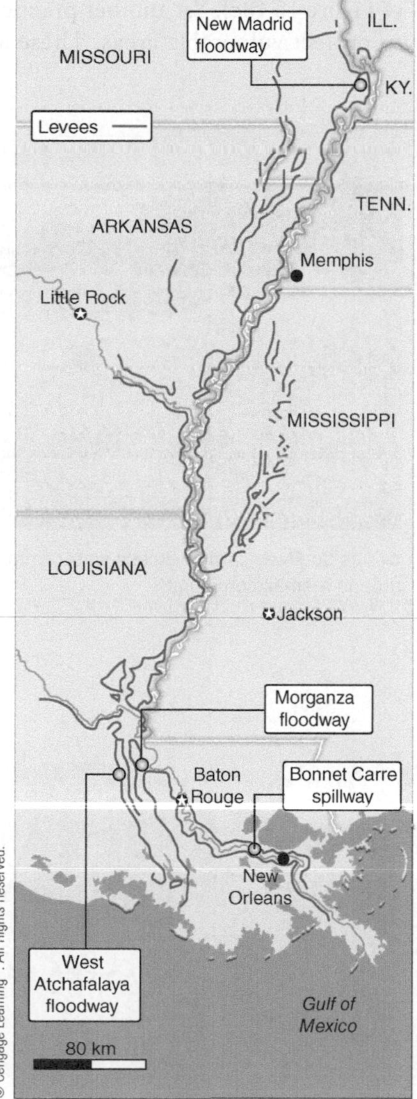

► **7.** Flood control structures on the Mississippi River in Louisiana. The Bonnet Carre Spillway diverts water into Lake Pontchartrain, whereas the Morganza Spillway diverts water into the Atchafalaya Floodway.

AP Photo/Patrick Semansky

▲ **8.** The Morganza Spillway was opened in May 2011 for the first time in 28 years. Water from the Mississippi River was diverted into the Atchafalya Floodway to protect New Orleans from flooding.

of the downtown area are being moved. Moving small communities is one option, but not for cities such as Des Moines, Iowa, which was extensively flooded in 1993.

One common flood control practice is to construct *dams* that impound excess water during floods (▌Figure 12.18a). Of course dams are expensive, they eventually fill in with sediment, unless dredged, and some fail. In 1976, the Teton Dam in Idaho collapsed, killing 11 people and 13,000 cattle. The total damage was about $2 billion. Another practice is to construct *levees* that raise the banks of streams and rivers, thereby increasing a channel's capacity (▌Figure 12.18b). Unfortunately, deposition within channels raises streambeds, making the levees less effective unless they are raised, too. Levees along the Huang He in China caused the streambed to rise more than 20 m above its surrounding floodplain in 4,000 years. In 1887 when the river breached its levees, more than 1 million people were killed.

In some areas, state or federal agencies build *floodways*, which are channels used to divert excess water around a community or an area of economic importance (▌Figure 12.18c). Yet another practice is to build *floodwalls* to protect vulnerable areas. These are vertical structures

Scott Olson/Getty Images

▌**Figure 12.17 Flooding on the Missississippi River System 2011** This house protected by a levee actually lies on the Yazoo River in Mississippi near its confluence with the Mississippi River. The 2011 flood compares with huge floods on this river system in 1927 and 1993.

▌**Figure 12.18 Flood Control** Dams and reservoirs, levees, floodways, and floodwalls are some of the structures used to control floods.

Critical Thinking Question What are some of the functions of a dam and its reservoir?

James S. Monroe

a Oroville Dam in California, at 235 m high, is the highest dam in the United States. It helps control floods, provides water for irrigation, and produces electricity at its power plant.

Sue Monroe

c This floodway carries excess water from a river (not visible) around a small community.

Sue Monroe

b This levee, an artificial embankment along a waterway, helps protect nearby areas from floods.

John Alexander

d This flood wall is along the banks of the Ohio River in Paducah, Kentucky. When the river rises, a gate is placed in the opening along the diagonal grooves.

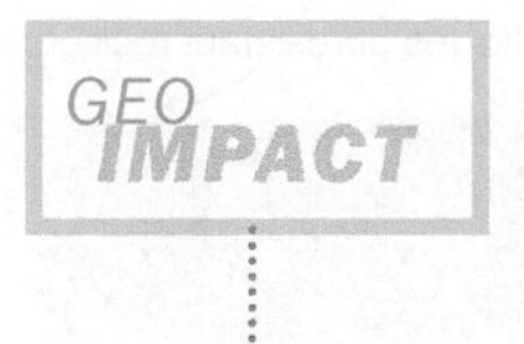

Sand and Gravel as Natural Resources

The largest part of many states' mineral revenues comes from sand and gravel, most of which is used in construction. You become aware of a sand and gravel deposit that you can acquire for a small investment. How would the proximity of the deposit to potential markets influence your decision? Assuming the deposit was stream deposited, would it be important to know whether deposition took place in braided or meandering channels? How could you tell one from the other?

placed along stream or river banks where levees are not practical, such as in cities (▮ Figure 12.18d). Reforestation of cleared land also reduces the potential for flooding because vegetated soil absorbs water and decreases runoff.

Flooding streams and rivers account for about 7.8 billion in property damage each year during the past 30 years in the United States, not including coastal flooding from hurricanes. And even though more and more flood control projects are completed, the damages from flooding are not decreasing. The combination of fertile soils, level surfaces, and proximity to water for agricultural, industrial, and domestic use makes floodplains popular places for development. Unfortunately, urbanization increases surface runoff because soils are compacted or covered by asphalt or concrete, reducing infiltration capacity. In addition, storm drains in urban areas quickly carry water to nearby channels, many of which flood more often than they did in the past.

As for predicting floods, the best that can be done is to monitor streams, evaluate their past behavior, and anticipate floods of a given size in a specified period. Most people have heard of 10-year floods, 20-year floods, and so on, but how are such determinations made? The United States Geological Survey as well as state agencies record and analyze stream behavior through time and anticipate floods of a specified size. So a 20-year flood, for example, is the period during which a flood of a given magnitude can be expected. It does not mean that the river in question will have a flood of that size every 20 years, only that over a long period of time, it will average 20 years. Or we can say that the chances of a 10-year flood taking place in any one year are 1 in 10 (1/10). In fact, it is possible that two 10-year floods could take place in successive years, but then not occur again for several decades.

12.7 Drainage Systems

Thousands of waterways are parts of larger systems consisting of a main channel with all its tributaries—that is, streams that contribute water to another stream. The Mississippi River and its tributaries such as the Ohio, Missouri, Arkansas, and Red rivers and thousands of smaller ones, or any other drainage system for that matter, carry runoff from an area known as a **drainage basin.** A topographically high area called a **divide** separates a drainage basin from adjoining ones (▮ Figure 12.19). The continental divide along the crest of the

▮ **Figure 12.19** Drainage Basins

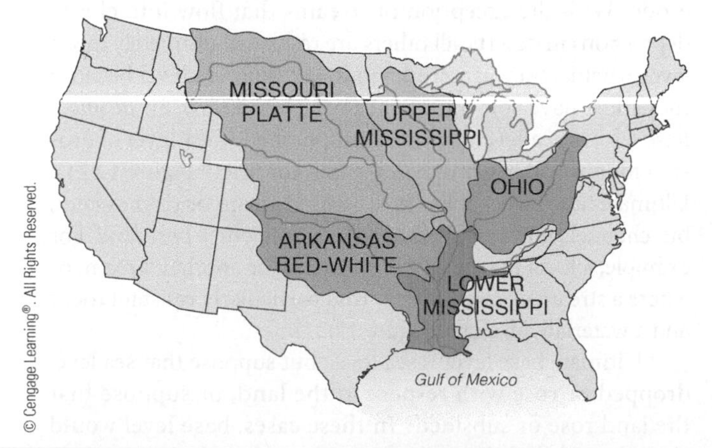

a The drainage basin of the Mississippi River and its main tributaries.

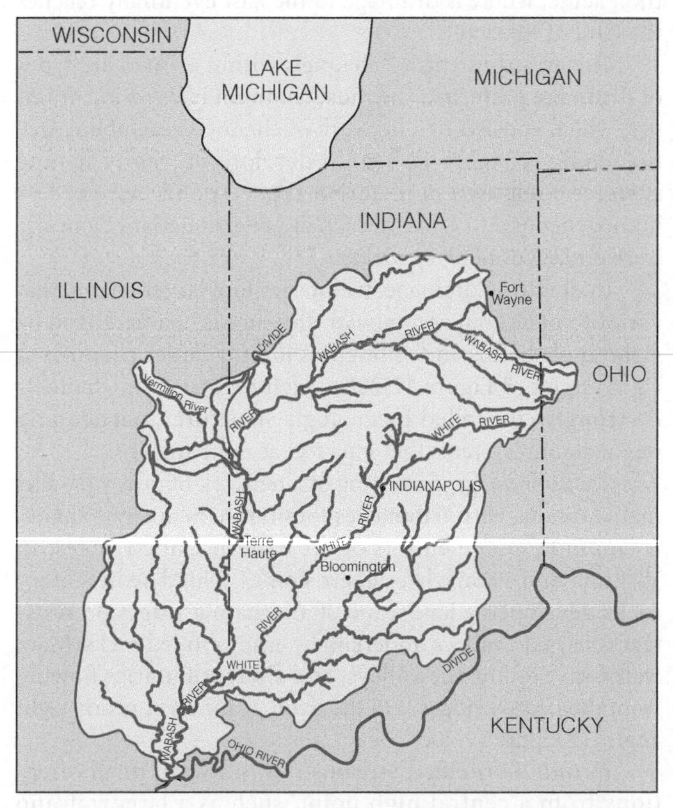

b A detailed view of the Wabash River's drainage basin, a tributary of the Ohio River. All tributary streams within the drainage basin, such as the Vermillion River, have their own smaller drainage basins. Divides are shown by red lines.

Figure 12.20 Drainage Patterns

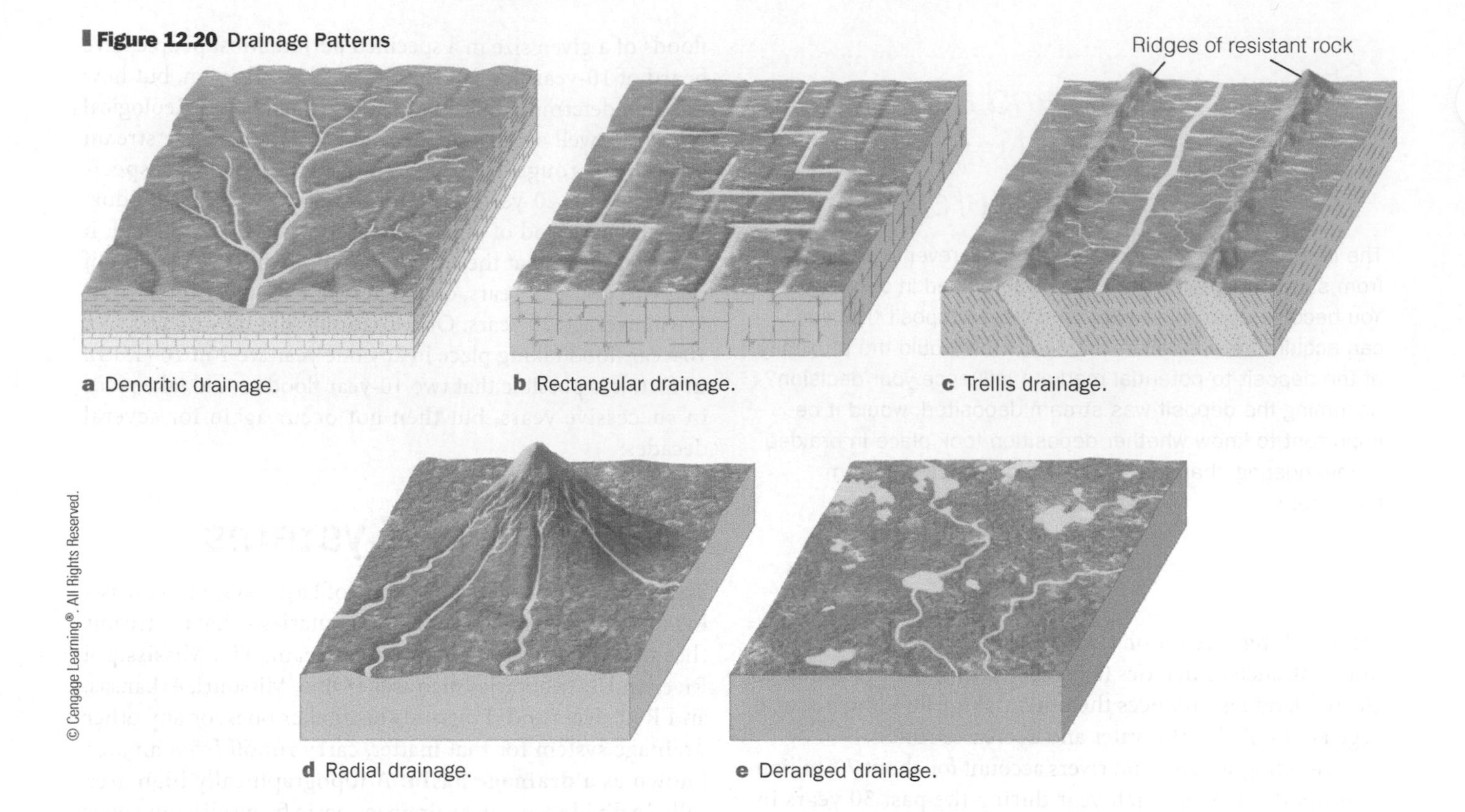

a Dendritic drainage. **b** Rectangular drainage. **c** Trellis drainage. **d** Radial drainage. **e** Deranged drainage.

Rocky Mountains in North America, for instance, separates drainage in opposite directions; drainage to the west goes to the Pacific, whereas drainage to the east eventually reaches the Gulf of Mexico.

The arrangements of channels within an area are types of **drainage patterns.** The most common is *dendritic drainage,* which consists of a network of channels resembling tree branching (Figure 12.20a). It develops on gently sloping surfaces composed of materials that respond more or less homogeneously to erosion, such as areas underlain by nearly horizontal sedimentary rocks.

In dendritic drainage, tributaries join larger channels at various angles, but *rectangular drainage* is characterized by right-angle bends and tributaries joining larger channels at right angles (Figure 12.20b). Such regularity in channels is strongly controlled by geologic structures, particularly regional joint systems that intersect at right angles.

Trellis drainage, consisting of a network of nearly parallel main streams with tributaries joining them at right angles, is common in some parts of the eastern United States. In Virginia and Pennsylvania, erosion of folded sedimentary rocks developed a landscape of alternating ridges on resistant rocks and valleys underlain by easily eroded rocks. Main waterways follow the valleys, and short tributaries flowing from the nearby ridges join the main channels at nearly right angles (Figure 12.20c).

In *radial drainage,* streams flow outward in all directions from a central high point, such as a large volcano (Figure 12.20d). Many of the volcanoes in the Cascade Range of western North America have radial drainage patterns.

In all of the types of drainage mentioned so far, some kind of pattern is easily recognized. *Deranged drainage,* in contrast, is characterized by irregularity, with streams flowing into and out of swamps and lakes, streams with only a few short tributaries, and vast swampy areas between channels (Figure 12.20e). This kind of drainage developed recently and has not yet formed a fully organized drainage system. In parts of Minnesota, Wisconsin, and Michigan, where glaciers obliterated the previous drainage, only 10,000 years have elapsed since the glaciers melted. As a result, drainage systems have not fully developed, and large areas remain undrained.

The Significance of Base Level

Base level is the lowest limit to which a stream or river can erode. With the exception of streams that flow into closed depressions in deserts, all others are restricted ultimately to sea level. That is, they can erode no lower than sea level because they must have some gradient to maintain flow. So *ultimate base level* is sea level, which is simply the lowest level of erosion for any waterway that flows into the sea (Figure 12.21). Ultimate base level applies to an entire stream or river system, but channels may also have a *local* or *temporary base level.* For example, a local base level may be a lake or another stream, or where a stream or river flows across particularly resistant rocks and a waterfall develops (Figure 12.21).

Ultimate base level is sea level, but suppose that sea level dropped or rose with respect to the land, or suppose that the land rose or subsided? In these cases, base level would change and bring about changes in stream and river systems. During the Pleistocene Epoch (Ice Age), sea level was about

Figure 12.21 Base Level and Graded Streams

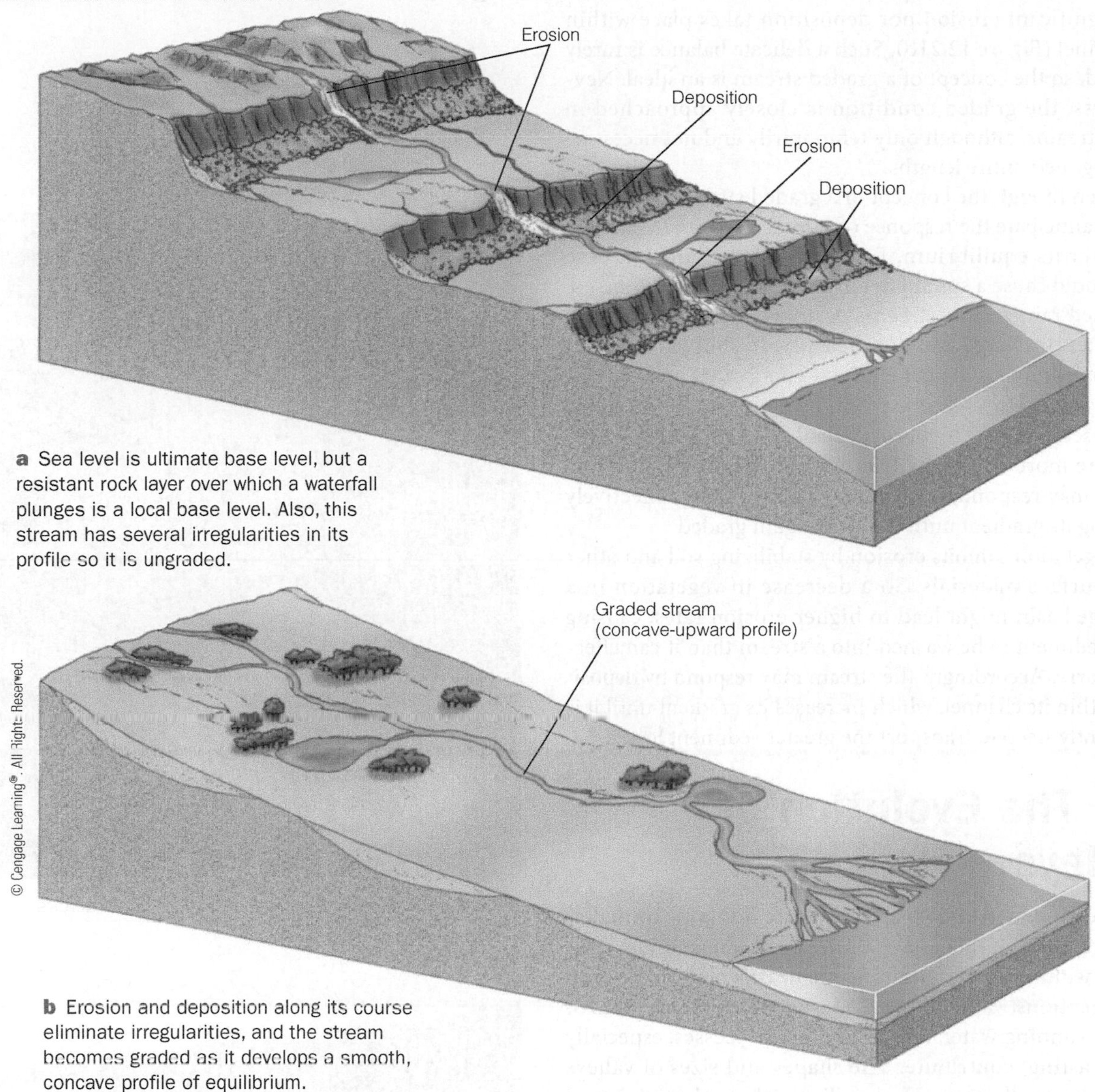

a Sea level is ultimate base level, but a resistant rock layer over which a waterfall plunges is a local base level. Also, this stream has several irregularities in its profile so it is ungraded.

b Erosion and deposition along its course eliminate irregularities, and the stream becomes graded as it develops a smooth, concave profile of equilibrium.

130 m lower than it is now, and streams adjusted by eroding deeper valleys and extending well out onto the continental shelves. Rising sea level at the end of the Ice Age accounted for a rising base level, decreased stream gradients, and deposition within channels.

Natural changes, such as fluctuations in sea level during the Pleistocene, alter the dynamics of rivers and streams, but so too does human intervention. Geologists and engineers are well aware that building a dam to impound a reservoir creates a local base level. A stream entering a reservoir deposits sediment, so unless dredged, reservoirs eventually fill with sediment. In addition, the water discharged at a dam is largely sediment free, but it still possesses energy to carry a sediment load. As a result, a stream may erode vigorously downstream from a dam to acquire a sediment load.

Draining a lake may seem like a small change and well worth the time and expense to expose dry land for agriculture or commercial development. But draining a lake eliminates a local base level, and a stream that originally flowed into the lake responds by rapidly eroding a deeper valley as it adjusts to a new base level.

What Is a Graded Stream?

The *longitudinal profile* of any waterway shows the elevations of a channel along its length as viewed in cross section (❚ Figure 12.21). For some rivers and streams, the longitudinal profile is smooth, but others show irregularities such as lakes and waterfalls, all of which are local base levels (Figure 12.21a). Over time, these irregularities tend to be eliminated because deposition takes place where the gradient is insufficient to maintain sediment transport, and erosion decreases the gradient where it is steep. So, given enough time, rivers and streams develop a smooth, concave longitudinal profile of equilibrium, meaning that all parts of the system dynamically adjust to one another.

A **graded stream** is one with an equilibrium profile in which a delicate balance exists among gradient, discharge, flow velocity, channel shape, and sediment load so that neither significant erosion nor deposition takes place within its channel (Figure 12.21b). Such a delicate balance is rarely attained, so the concept of a graded stream is an ideal. Nevertheless, the graded condition is closely approached in many streams, although only temporarily and not necessarily along their entire lengths.

Even though the concept of a graded stream is an ideal, we can anticipate the response of a graded stream to changes that alter its equilibrium. For instance, a change in base level would cause a stream to adjust as previously discussed. Increased rainfall in a stream's drainage basin would result in greater discharge and flow velocity. In short, the stream would now possess greater energy—energy that must be dissipated within the stream system by, for example, a change from a semicircular to a broad, shallow channel that would dissipate more energy by friction. On the other hand, the stream may respond by eroding a deeper valley, effectively reducing its gradient until it is once again graded.

Vegetation inhibits erosion by stabilizing soil and other loose surface materials. So a decrease in vegetation in a drainage basin might lead to higher erosion rates, causing more sediment to be washed into a stream than it can effectively carry. Accordingly, the stream may respond by deposition within its channel, which increases its gradient until it is sufficiently steep to transport the greater sediment load.

12.8 The Evolution of Valleys

Low areas on land known as **valleys** are bounded by higher land, and most of them have a river or stream running their length, with tributaries draining the nearby high areas. With few exceptions, valleys form and evolve in response to erosion by running water, although other processes, especially mass wasting, contribute. The shapes and sizes of valleys vary from small, steep-sided *gullies* to those that are broad with gently sloping valley walls (▮ Figure 12.22). Steep-walled, deep valleys of vast size are *canyons*, and particularly narrow and deep ones are *gorges*.

A valley might start to erode where runoff has sufficient energy to dislodge surface materials and excavate a small rill. Once formed, a rill collects more runoff and becomes deeper and wider and continues to do so until a full-fledged valley develops. Processes that contribute to valley formation include downcutting, lateral erosion, headward erosion, sheetwash, and mass wasting.

Downcutting takes place when a river or stream has more energy than it needs to transport sediment, so some of its excess energy is used to deepen its valley. If downcutting were the only process operating, valleys would be narrow and steep sided. In most cases, though, the valley walls are undercut, a process called *lateral erosion,* creating unstable slopes that may fail by *mass wasting.* Furthermore, erosion by sheetwash and by tributary streams carry materials from the valley walls into the main stream in the valley.

▮ **Figure 12.22** Gullies and Valleys

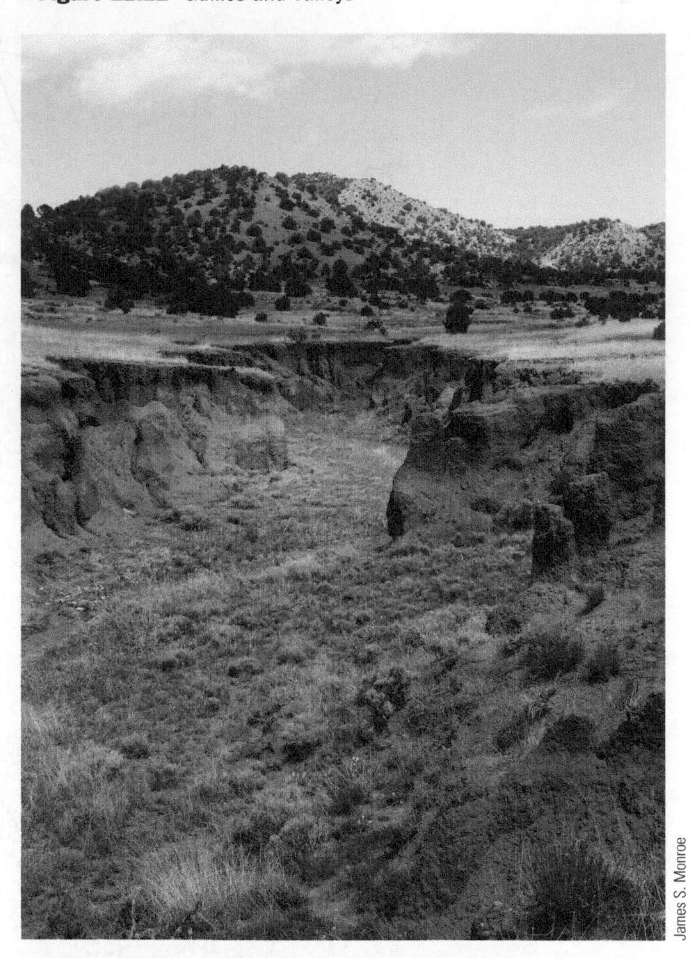

James S. Monroe

a A gully in New Mexico that measures about 15 m across.

James S. Monroe

b This valley has walls that descend to a narrow bottom.

Valleys not only become deeper and wider but also become longer by *headward erosion,* which involves erosion by entering runoff at the upstream end of a valley (▮ Figure 12.23a). Continued headward erosion may result in *stream piracy,* the breaching of a drainage divide and diversion of part of the drainage of another stream (▮ Figure 12.23b, c). Once stream piracy takes place, both drainage systems must adjust to these new conditions; one system now has greater discharge and the

Figure 12.23 The Evolution of a Valley by Headward Erosion, Lateral Erosion, and Stream Piracy

a This stream simultaneously extends its valley by headward erosion and widens its valley by lateral erosion.

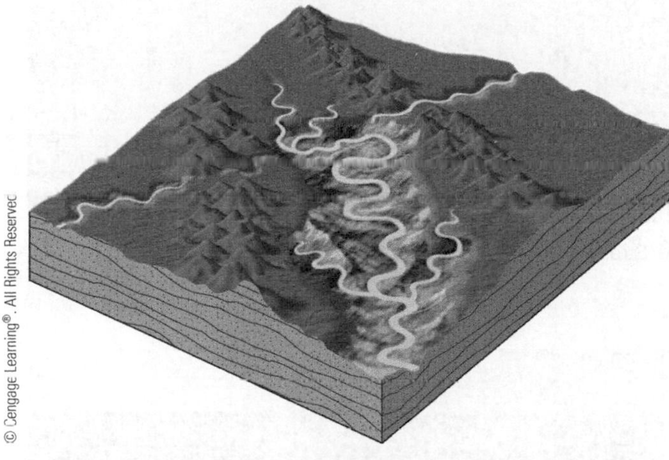

b As the larger stream continues to erode headward, stream piracy takes place when it captures some of the drainage of the smaller stream. Notice also that the valley is wider than it was in **(a)**.

Figure 12.24 Idealized Stages in the Development of a Stream and Its Associated Landforms

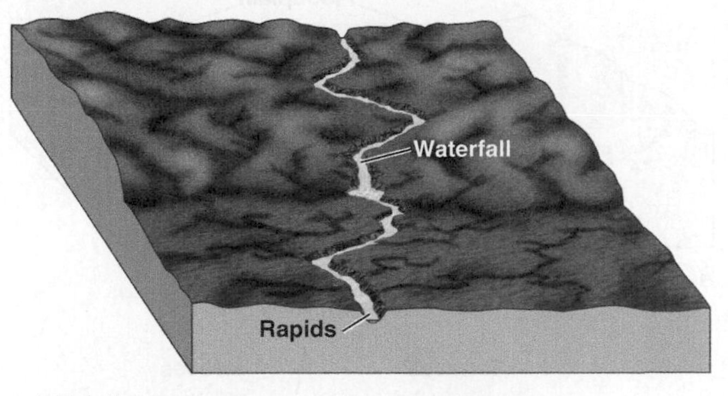

a Initial stage.

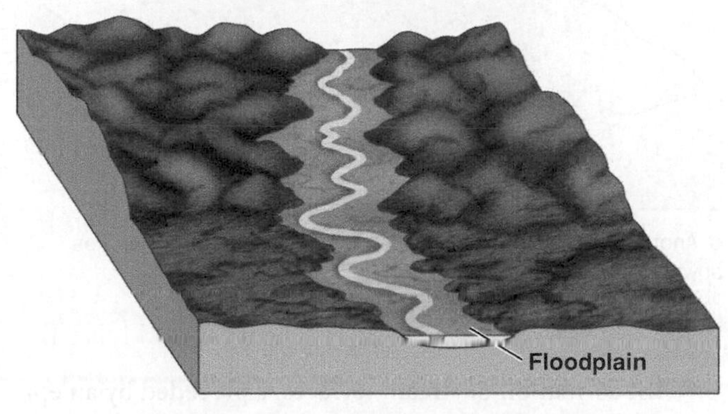

b Intermediate stage.

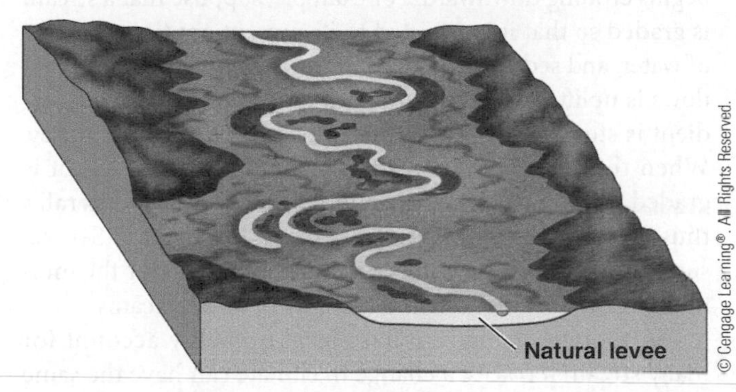

c Advanced stage.

potential to do more erosion and sediment transport, whereas the other is diminished in its ability to accomplish these tasks.

According to one concept, stream erosion of an area uplifted above sea level yields a distinctive series of landscapes. When erosion begins, streams erode downward; their valleys are deep, narrow, and V-shaped, and their profiles have a number of irregularities (▌Figure 12.24a). As streams cease eroding downward, they start eroding laterally, thereby establishing a meandering pattern and a broad floodplain (▌Figure 12.24b). Finally, with continued erosion, a vast, rather featureless plain develops (▌Figure 12.24c).

Many streams do indeed show the features typical of these stages. For instance, the Colorado River flows through the Grand Canyon and closely matches the features in the initial stage shown in Figure 12.24a. Streams in many areas approximate the second stage of development, and certainly the lower Mississippi closely resembles the last stage. Nevertheless, the idea of the sequential development of stream-eroded landscapes has been largely abandoned because there is no reason to think that streams necessarily follow this idealized progression. Indeed, a stream on a gently sloping surface near sea level could develop features of the last stage very early in its history. In addition, as long as the rate of uplift exceeds the rate of downcutting, a stream will continue to erode downward and be confined to a narrow canyon.

Stream Terraces Rather flat surfaces paralleling a stream or river, but at a higher level than the present-day floodplain, are **stream terraces.** These surfaces represent an older floodplain that was adjacent to the waterway when it flowed at a higher level but subsequently eroded down to a lower level (▌Figure 12.25). Some streams have several steplike surfaces above their present floodplains, indicating the terraces formed more than once.

▌Figure 12.25 Origin of Stream Terraces

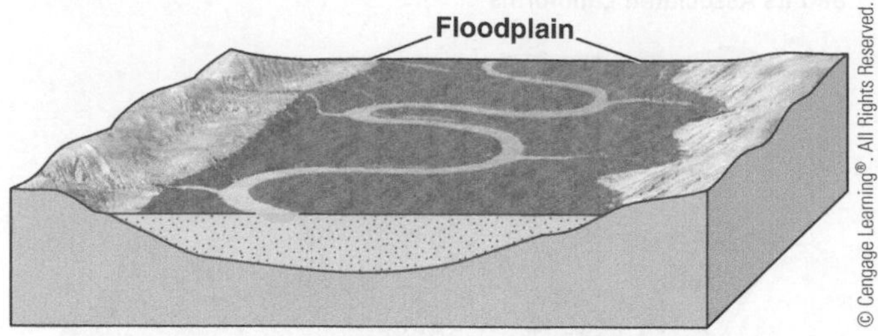

a A stream has a broad floodplain.

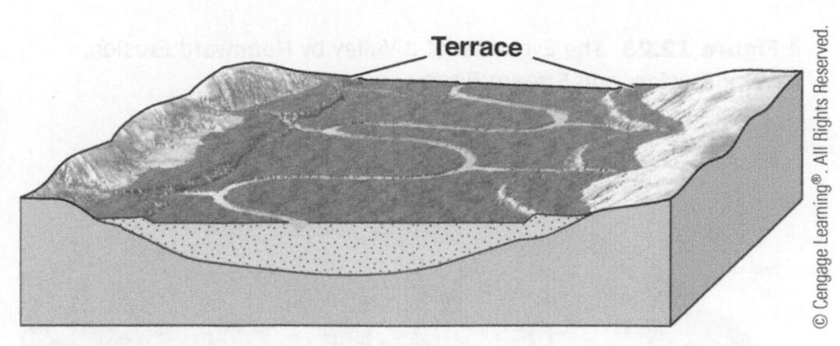

b The stream erodes downward and establishes a new floodplain at a lower level. Remnants of its old, higher floodplain are stream terraces.

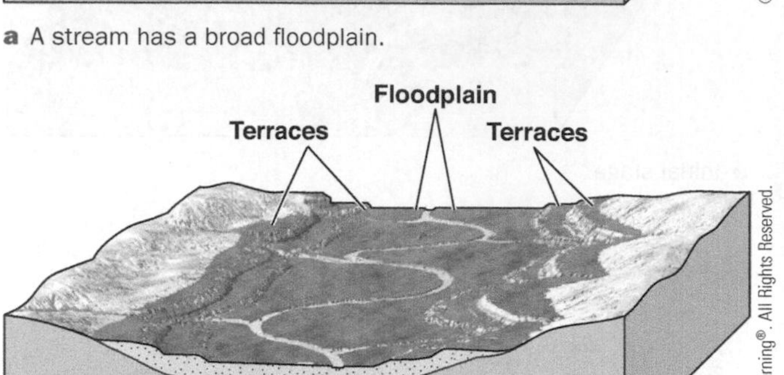

c Another level of stream terraces forms as the stream erodes downward again.

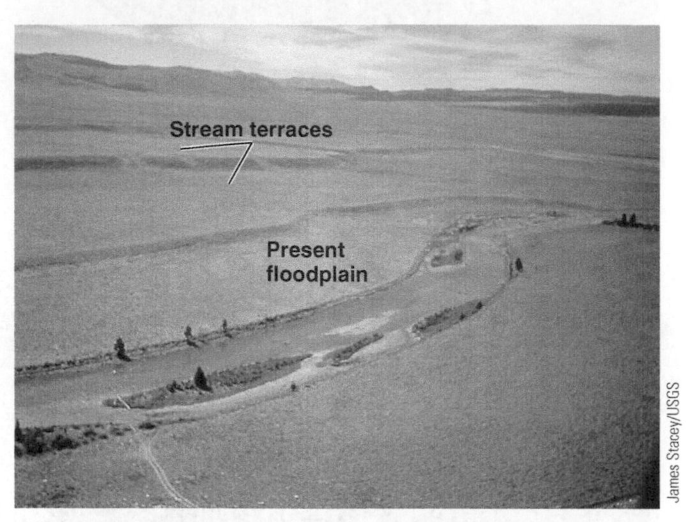

d Stream terraces along the Madison River in Montana.

James Stacey/USGS

The formation of stream terraces is preceded by an episode of deposition followed by erosion as the stream or river begins eroding downward. For example, suppose that a stream is graded so that it is adjusted to its present gradient, volume of water, and sediment load. If the land over which the stream flows is uplifted or sea level becomes lower, the stream's gradient is steeper, and it has more energy to deepen its valley. When this stream once again reaches a level at which it is graded, downcutting ceases, and it begins to erode laterally, thus establishing a new floodplain at a lower level. Several such episodes of deposition and erosion account for the multiple stream terraces seen in the valleys of some streams.

Although changes in base level probably account for many stream terraces, a change in climate can have the same result. If the amount of precipitation in a stream's drainage basin increases, the stream has a greater volume of water and the capacity to erode a deeper valley, perhaps leaving remnants of an older floodplain as stream terraces.

Incised Meanders

Some streams are restricted to deep, meandering canyons cut into bedrock, where they form features called **incised meanders.** For example, the San Juan River in Utah occupies a deep meandering canyon (▌Figure 12.26). Streams restricted by rock walls usually cannot erode laterally; thus, they lack a floodplain and occupy the entire width of the canyon floor.

It is not difficult to understand how a stream can cut downward into rock, but how a stream forms a meandering pattern in bedrock is another matter. Because lateral erosion is inhibited once downcutting begins, one must infer that the

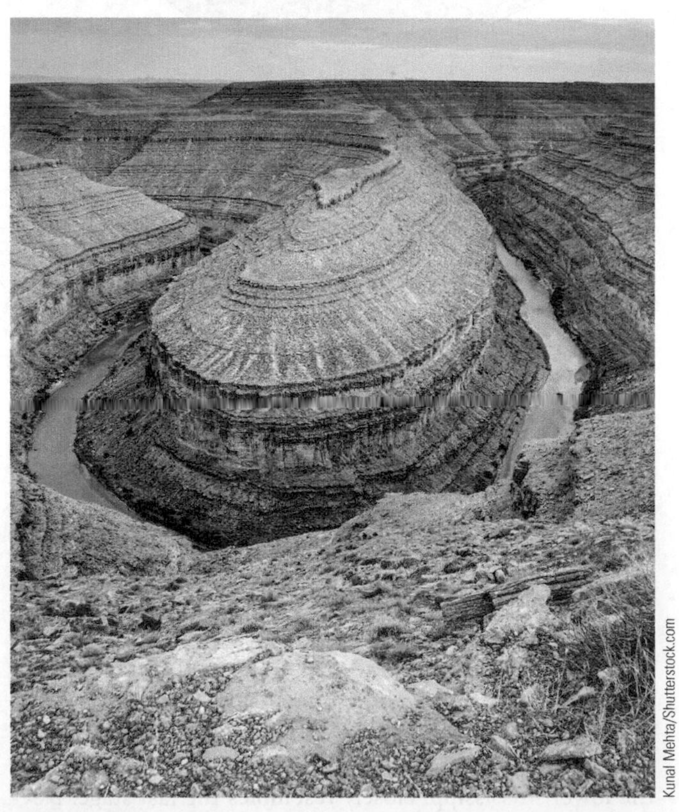

Kunal Mehta/Shutterstock.com

▌Figure 12.26 Incised Meanders The San Juan River in Utah has incised its meanders more than 300 m into sedimentary rocks.

meandering course was established when the stream flowed across an area covered by alluvium. For example, suppose that a stream near base level has established a meandering pattern. If the land that the stream flows over is uplifted, then erosion begins and the meanders become incised into the underlying bedrock.

Superposed Streams

Water flows downhill in response to gravity, so the direction of flow in streams and rivers is determined by topography. Yet a number of waterways seem, at first glance, to have defied this fundamental control. For instance, the Delaware, Potomac, and Susquehanna rivers in the eastern United States flow in valleys that cut directly through ridges that lie in their paths. These are examples of **superposed streams,** all of which once flowed on a surface at a higher level, but as they eroded downward, they eroded into resistant rocks and cut narrow canyons or what geologists call *water gaps* (❚ Figure 12.27).

A water gap has a stream flowing through it, but if the stream is diverted elsewhere, perhaps by stream piracy, the abandoned gap is then called a *wind gap.* The Cumberland Gap at the junction of Virginia, Tennessee, and Kentucky is a good example; it was the avenue through which settlers migrated from Virginia to Kentucky from 1790 until well into the 1800s. Furthermore, several water gaps and wind gaps played important strategic roles during the Civil War (1861–1865).

❚ Figure 12.27 Origin of a Superposed Stream

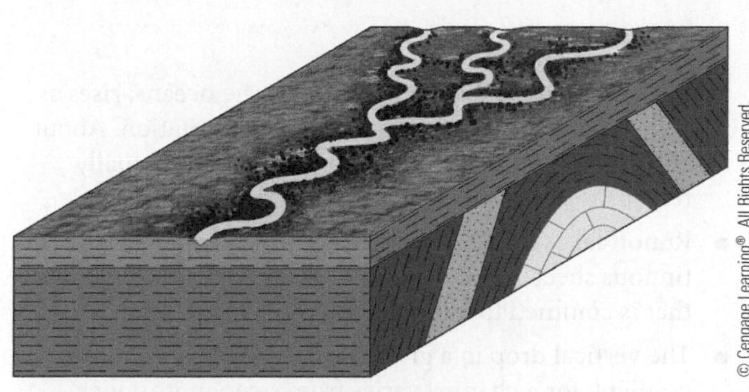

a As a stream erodes down and removes the surface layers of rock, it is lowered onto ridges that form when resistant rocks in the underlying structure are exposed.

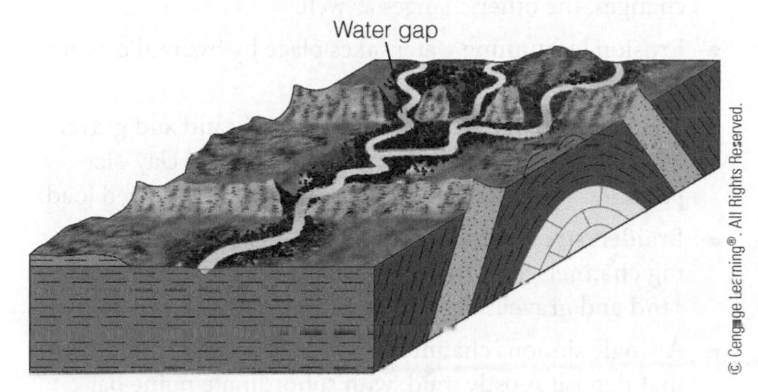

b The narrow valleys through the ridges are water gaps.

Sue Monroe

c The John Day River in Oregon cut this canyon through layers of basalt. It is a superposed stream that began eroding its canyon about 3 million years ago.

12.9 Water as a Natural Resource

The term *natural resource* most often brings to mind commodities such as gold, iron ore, and petroleum and natural gas, but certainly water is an essential resource for domestic use, industry, and agriculture. The amount of water on Earth is constant, although most of it is in the oceans and not usable, unless it is desalinated, which at present is expensive. About 20% of all water used in the United States comes from groundwater, with the remaining 80% from water taken directly from waterways, lakes, or reservoirs. Unfortunately, the amount of water available at any one time varies depending on precipitation as rain or snow, and how much is used. The Dust Bowl of the 1930s (see Figure 6.16) and the 2012 drought in the Midwestern United States remind us of how vulnerable we are to even small changes in weather patterns. And of course overuse also depletes the flow in many waterways, such as the lower Colorado River which is nearly dry because so much water is withdrawn by the United States and Mexico. Drought has the same effect, as during the summer of 2012, when freight barges could not navigate the lower Mississippi River because of low water levels.

According to the Energy Information Administration, 6% of all electricity used in the United States is generated at hydroelectric power plants, where water is used to spin a turbine connected to a generator where electricity is produced. It is true that most electricity is generated at coal-burning power plants, but even at these plants, water is heated to spin the turbines. And the same is true at nuclear power plants and those that use natural gas or petroleum, and in all of these examples, water is also needed as a coolant.

We cannot increase the amount of water on Earth, but we can use it more efficiently by reuse where possible, and conserving it as much as we can. We will have more to say about water as a resource as we investigate other reservoirs in the hydrological cycle, which includes groundwater (Chapter 13) and glaciers (Chapter 14).

Key Concepts Review

- Water continuously evaporates from the oceans, rises as water vapor, condenses, and falls as precipitation. About 20% of this precipitation falls on land and eventually returns to the oceans, mostly by surface runoff.
- Runoff takes place by sheet flow, a thin, more or less continuous sheet of water, and by channel flow, which is flow that is confined to long, trough-like depressions.
- The vertical drop in a given horizontal distance, or the gradient, for a channel varies from steep in its upper reaches to more gentle in its lower reaches.
- Flow velocity and discharge are related, so that if either changes, the other changes as well.
- Erosion by running water takes place by hydraulic action, abrasion, and solution.
- The bed load in channels is made up of sand and gravel, whereas suspended load consists of silt- and clay-size particles. Running water also transports a dissolved load.
- Braided streams have a complex of dividing and rejoining channels, and their deposits are mostly sheets of sand and gravel.
- A single sinuous channel is typical of meandering streams that deposit mostly mud, with subordinate point-bar deposits of sand or, more rarely, gravel.
- Broad, flat floodplains adjacent to channels are the sites of oxbow lakes, which are simply abandoned meanders.
- An alluvial deposit at a river's mouth is a delta. Some deltas conform to the three-part division of bottomset, foreset, and topset beds, but large marine deltas are much more complex and are characterized as stream-, wave-, or tide-dominated.
- Alluvial fans are fan-shaped deposits of sand and gravel on land that form best in semiarid regions. They form mostly by deposition from running water, but debris flows are also important.
- Rivers and streams carry runoff from their drainage basins, which are separated from one another by divides.
- Sea level is ultimate base level, the lowest level to which streams or rivers can erode. Local base levels may be lakes or where streams or rivers flow across resistant rocks.
- Water is an important natural resource, most of which is saline water in the oceans. Only a small percentage of all water is on land in lakes, swamps, groundwater, and streams and rivers.
- Graded streams tend to eliminate irregularities in their channels, so they develop a smooth, concave profile of equilibrium.
- A combination of processes, including downcutting, lateral erosion, sheetwash, mass wasting, and headward erosion, are responsible for the origin and evolution of valleys.
- Stream terraces and incised meanders usually form when a stream or river that was formerly in equilibrium, begins a new episode of downcutting.

Important Terms

abrasion (p. 288)
alluvial fan (p. 292)
alluvium (p. 288)
base level (p. 300)
bed load (p. 288)
braided stream (p. 289)
delta (p. 292)
discharge (p. 288)
dissolved load (p. 288)
divide (p. 299)
drainage basin (p. 299)
drainage pattern (p. 300)
floodplain (p. 291)
graded stream (p. 301)
gradient (p. 286)
hydraulic action (p. 288)
hydrologic cycle (p. 285)
incised meanders (p. 304)
infiltration capacity (p. 286)
meandering stream (p. 289)
natural levee (p. 291)
oxbow lake (p. 291)
point bar (p. 291)
runoff (p. 285)
stream terrace (p. 303)
superposed stream (p. 305)
suspended load (p. 288)
valley (p. 302)
velocity (p. 287)

Review Questions

1. Running water carrying sand and gravel effectively erodes by
   a. ____ solution.
   b. ____ sheetwash.
   c. ____ abrasion.
   d. ____ saltation.
   e. ____ alluviation.
2. The deposit that forms on the gently sloping side of a meander where current velocity is low is called a/an
   a. ____ point bar.
   b. ____ natural levee.
   c. ____ valley.
   d. ____ oxbow lake.
   e. ____ alluvial fan.
3. The recycling of water from the oceans to the land and back is known as the
   a. ____ hydraulic action.
   b. ____ profile of equilibrium.
   c. ____ drainage pattern.
   d. ____ hydrologic cycle.
   e. ____ suspended load.
4. A stream or river with multiple channels that divide and rejoin is characterized as
   a. ____ incised.
   b. ____ superposed.
   c. ____ braided.
   d. ____ sinuous.
   e. ____ meandering.
5. The suspended load of a stream or river is made up of
   a. ____ sand and gravel.
   b. ____ gravel and boulders.
   c. ____ sand and silt.
   d. ____ dissolved materials.
   e. ____ silt and clay.
6. What is a graded stream and how does it develop?
7. The discharge of most streams and rivers increases downstream, but in a few cases, it actually decreases and they eventually disappear. Explain why.
8. Calculate the daily discharge of a river 148 m wide and 2.6 m deep, with a flow velocity of 0.7 m/sec.
9. Where and how do point bars form?

10. **Creative Thinking Visual Question:** Explain how this stream in Canada cut this notch through bedrock (▮ Figure 1).

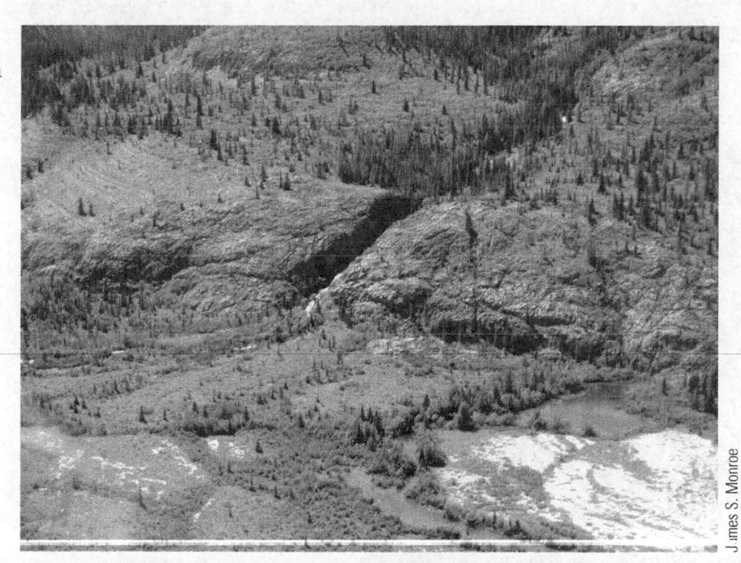

James S. Monroe

▮ **Figure 1** Erosion by a small stream in Canada.

Global GeoScience Watch In the GREENR database go to the "Floods" portal, and then within this portal search for "Deltas sink worldwide." Click on the article titled "Deltas Sink Worldwide, Increasing Flood Risk." How did scientists determine that deltas are sinking, what causes the sinking, and how has delta sinking influenced the frequency of flooding on deltas? Were any of the deltas mentioned in the article in the United States? If so, which one(s)?

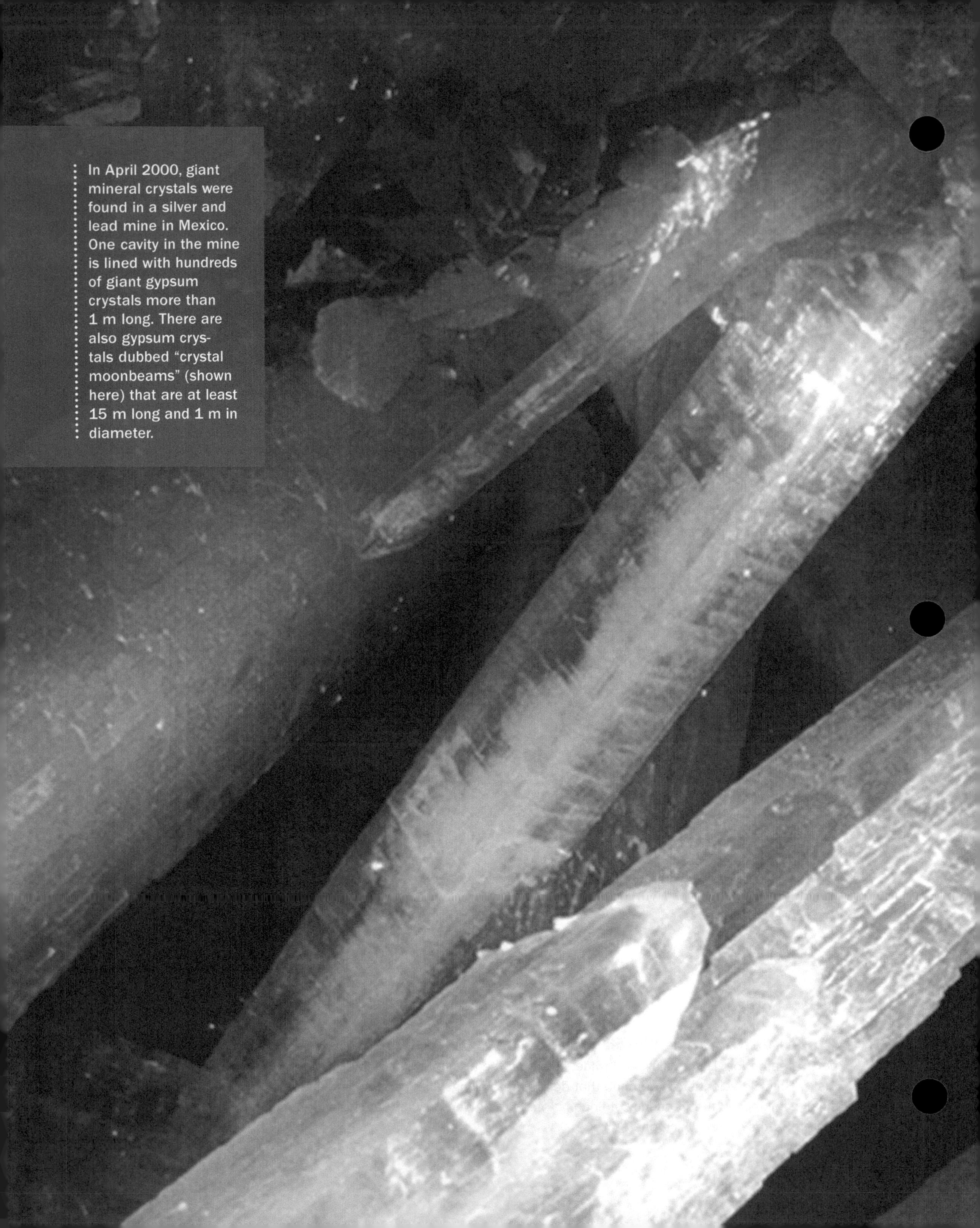

In April 2000, giant mineral crystals were found in a silver and lead mine in Mexico. One cavity in the mine is lined with hundreds of giant gypsum crystals more than 1 m long. There are also gypsum crystals dubbed “crystal moonbeams” (shown here) that are at least 15 m long and 1 m in diameter.

CHAPTER 13

Groundwater

OUTLINE

HAVE YOU EVER WONDERED?

- What percentage of the world's supply of freshwater comes from groundwater?
- What a water table is and where it is found?
- What an artesian system is and why some advertisers tout the superiority of artesian water in their product?
- How caves form?
- Why Old Faithful in Yellowstone National Park, Wyoming, erupts on a regular basis?
- What geothermal energy is and why it is important?

13.1 Introduction

Within the limestone region of western Kentucky lies the largest cave system in the world. In 1941, approximately 51,000 acres were set aside and designated as Mammoth Cave National Park; the park was designated a World Heritage Site in 1981. From ground level, the topography of the area is unimposing, with gently rolling hills. Beneath the surface, however, are more than 540 km of interconnected passageways whose spectacular geologic features have been enjoyed by millions of cave explorers and tourists.

During the War of 1812, approximately 180 metric tons of saltpeter (potassium nitrate—KNO_3), used in the manufacture of gunpowder, was mined from Mammoth Cave. At the end of the war, the saltpeter market collapsed, and Mammoth Cave was developed as a tourist attraction, easily overshadowing the other caves in the area. During the next 150 years, the discovery of new passageways and links to other caverns helped establish Mammoth Cave as the world's premier cave and the standard against which all others were measured.

The colorful cave deposits are the primary reason that millions of tourists have visited Mammoth Cave over the years. Hanging down from the ceiling and growing up from the floor are spectacular icicle-like structures, as well as columns and curtains in a variety of colors. Moreover, intricate passageways connect rooms of various sizes. The cave is also home to more than 200 species of insects and other animals, including about 45 blind species.

In addition to the beautiful caves, caverns, and cave deposits produced by groundwater movement, groundwater is also an important natural resource. Although groundwater constitutes only 1.69% of the world's water, it is, nonetheless, a significant source of freshwater for agricultural, industrial, and domestic use (see Table 12.1). More than 65% of the groundwater used in the United States each year goes for irrigation, with industrial use second, followed by domestic needs. These demands have severely depleted the groundwater supply in many areas and have led to such problems as ground subsidence and saltwater contamination. In other areas, pollution from landfills, toxic waste, and agriculture has rendered the groundwater supply unsafe.

As the world's population and industrial development expand, the demand for water, particularly groundwater, will increase. Not only must new groundwater sources be located, but once found, these sources must be protected from pollution and managed properly to ensure that users do not withdraw more water than can be replenished.

It is becoming vitally important that people become aware of what a valuable resource groundwater is, so they can ensure that future generations have a clean and adequate supply of water from this source.

13.2 Groundwater and the Hydrologic Cycle

Groundwater—water that fills open spaces in rocks, sediment, and soil beneath Earth's surface—is one reservoir in the hydrologic cycle. Like all other water in the hydrologic cycle, the ultimate source of groundwater is the oceans; however, its more immediate source is the precipitation that infiltrates the ground and seeps down through the voids in soil, sediment, and rocks. Groundwater may also come from water infiltrating from streams, lakes, swamps, artificial recharge ponds, and water-treatment systems.

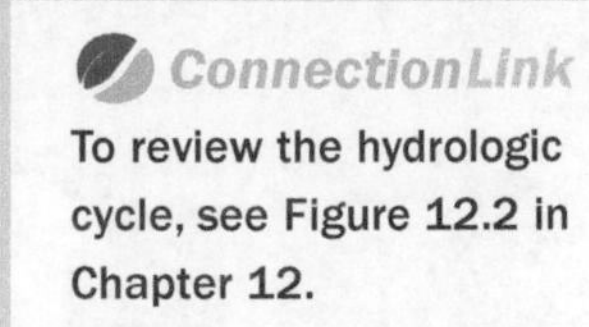

Regardless of its source, groundwater moving through the tiny openings between soil and sediment particles and the spaces in rocks filters out many impurities, such as disease-causing microorganisms and many pollutants. However, not all soils and rocks are good filters, and sometimes so much undesirable material may be present that it contaminates the groundwater. Groundwater movement and its recovery from wells depend on two critical aspects of the materials that it moves through: *porosity* and *permeability*.

13.3 Porosity and Permeability

Porosity and permeability are important physical properties of Earth materials and are largely responsible for the amount, availability, and movement of groundwater. Water soaks into the ground because soil, sediment, and rock have open spaces or pores. **Porosity** is the percentage of a material's total volume that is pore space. Porosity most often consists of the spaces between particles in soil, sediment, and sedimentary rocks, but other types of porosity include cracks, fractures, faults, and vesicles in volcanic rocks (Figure 13.1).

Porosity varies among different rock types and is dependent on the size, shape, and arrangement of the material composing the rock (Table 13.1). Most igneous and metamorphic rocks, as well as many limestones and dolostones, have very low porosity because they consist of tightly interlocking crystals. Their porosity can be increased, however, if they have been fractured or weathered by groundwater. This is particularly true for massive limestone and dolostone, whose fractures can be enlarged by acidic groundwater.

By contrast, detrital sedimentary rocks composed of well-sorted and well-rounded grains can have high porosity because any two grains touch at only a single point, leaving relatively large open spaces between the grains (Figure 13.1a). Poorly sorted sedimentary rocks, on the other hand, typically have low porosity because smaller grains fill in the spaces between the larger grains, further reducing porosity (Figure 13.1b). In addition, the amount of cement between grains can decrease porosity.

Porosity determines the amount of groundwater that Earth materials can hold, but it does not guarantee that the water can be easily extracted. So, in addition to being porous, Earth materials must have the capacity to transmit fluids, a property known as **permeability**. Thus, both porosity and permeability play important roles in groundwater movement and recovery.

Figure 13.1 **Porosity** A rock's porosity depends on the size, shape, and arrangement of the material composing the rock.

Critical Thinking Question How can some Earth materials be porous yet not permeable? Give an example.

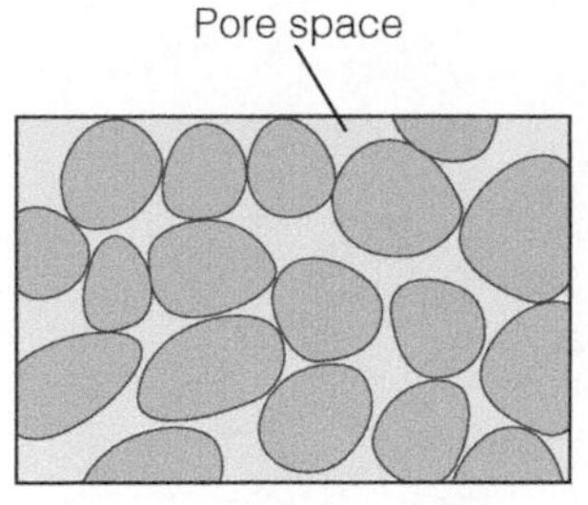

a A well-sorted sedimentary rock has high porosity, whereas

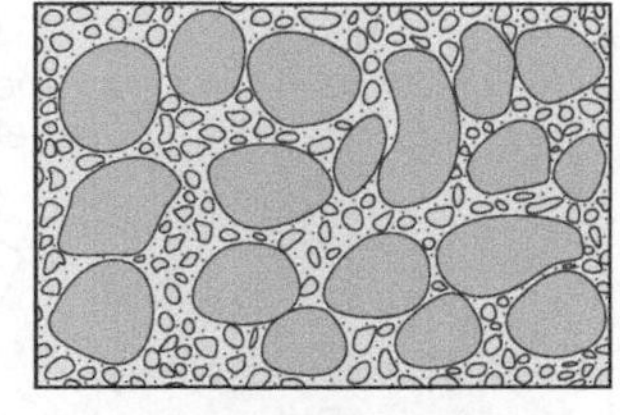

b a poorly sorted one has lower porosity.

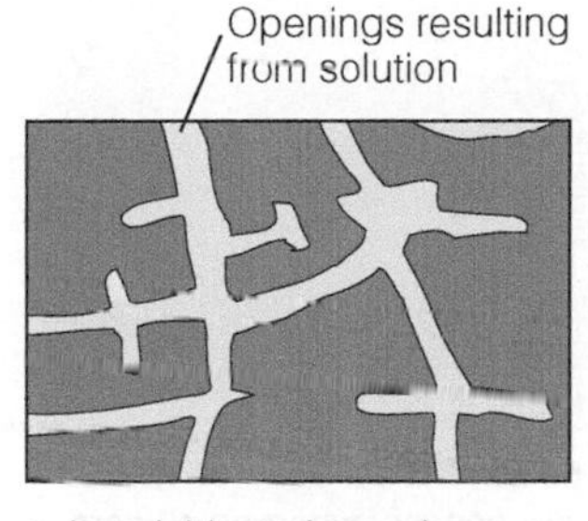

c In soluble rocks such as limestone, porosity can be increased by solution, whereas

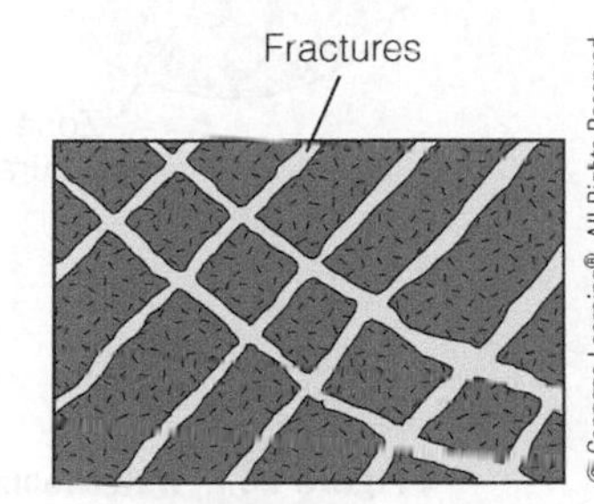

d crystalline metamorphic and igneous rocks can be rendered porous by fracturing.

TABLE 13.1 **Porosity Values for Different Materials**

Material	Percentage Porosity
Unconsolidated sediment	
Soil	55
Gravel	20–40
Sand	25–50
Silt	35–50
Clay	50–70
Rocks	
Sandstone	5–30
Shale	0–10
Solution activity in limestone, dolostone	10–30
Fractured basalt	5–40
Fractured granite	10

Source: United States Geological Survey, Water Supply Paper 2220 (1983) and others.

Permeability is dependent not only on porosity, but also on the size of the pores or fractures and their interconnections. For example, deposits of silt or clay are typically more porous than sand or gravel, but they have low permeability because the pores between the particles are very small and molecular attraction between the particles and water is great, thereby preventing movement of the water. In contrast, the pore spaces between grains in sandstone and conglomerate are much larger, and molecular attraction on the water is therefore low. Chemical and biochemical sedimentary rocks, such as limestone and dolostone, and many igneous and metamorphic rocks that are highly fractured, can also be very permeable provided that the fractures are interconnected.

The contrasting porosity and permeability of familiar substances are well demonstrated by sand versus clay. Pour some water on sand and it rapidly sinks in, whereas water poured on clay simply remains on the surface. Furthermore, wet sand dries quickly, but once clay absorbs water, it may take days to dry out because of its low permeability. Neither sand nor clay makes a good substance in which to grow crops, but a mixture of the two, plus some organic matter in the form of humus, makes an excellent soil for farming and gardening.

To review what makes a good soil, and the soil profile, see Figure 6.10 in Chapter 6.

A permeable layer transporting groundwater is an **aquifer,** from the Latin *aqua*, "water." The most effective aquifers are deposits of well-sorted and well-rounded sand and gravel. Limestones in which fractures and bedding planes have been enlarged by solution are also good aquifers. Shales and many igneous and metamorphic rocks make poor aquifers because they are typically impermeable, unless fractured. Rocks such as these and any other materials that prevent the movement of groundwater are *aquicludes*.

13.4 The Water Table

Some of the precipitation on land evaporates, and some enters streams and returns to the oceans by surface runoff; the remainder seeps into the ground. As this water moves down from the surface, a small amount adheres to the material it moves through and halts its downward progress. With the exception of this *suspended water*, however, the rest seeps further downward and collects until it fills all of the available pore spaces. Thus, two zones are defined by whether their pore spaces contain mostly air, the **zone of aeration**, or mostly water, the underlying **zone of saturation**. The surface that separates these two zones is the **water table** (Figure 13.2).

The base of the zone of saturation varies from place to place, but usually extends to a depth where an impermeable layer is encountered or to a depth where confining pressure closes all open space. Extending irregularly upward a few centimeters to several meters from the zone of saturation is the *capillary fringe*. Water moves upward in this region because of surface tension, much as water moves upward through a paper towel.

In general, the configuration of the water table is a subdued replica of the overlying land surface; that is, it rises beneath hills and has its lowest elevations beneath valleys.

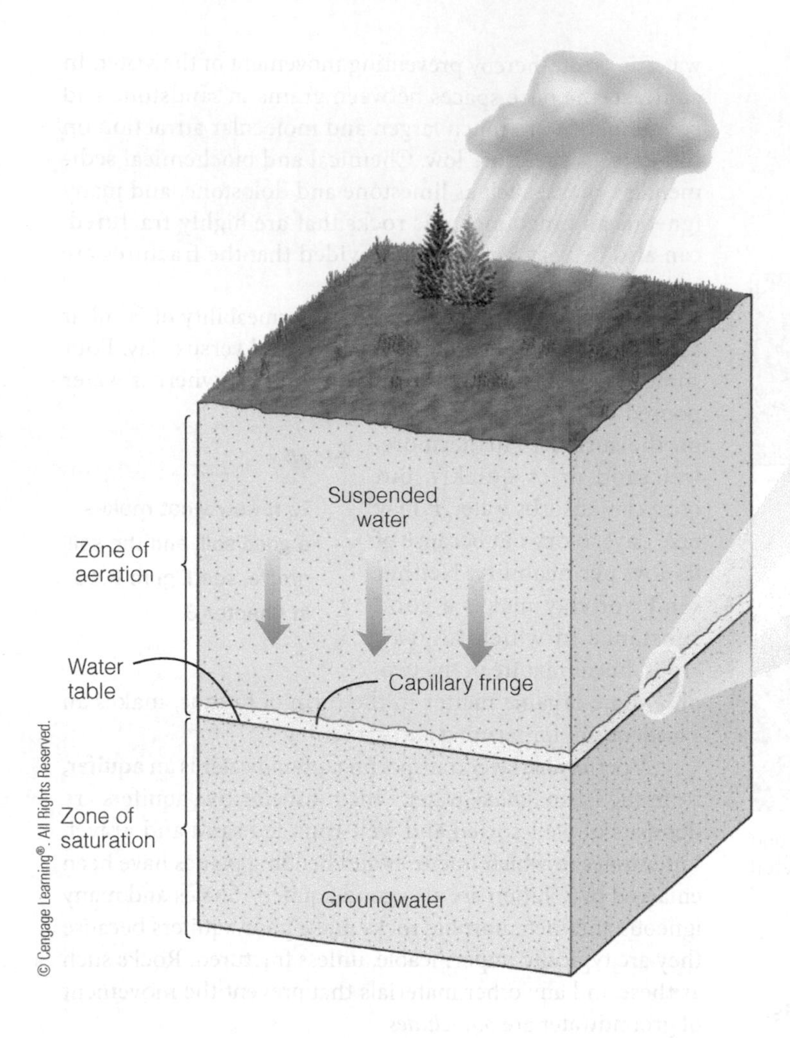

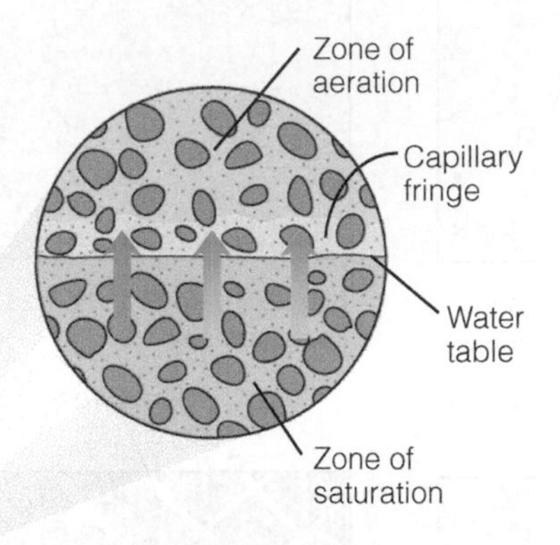

Figure 13.2 Water Table The zone of aeration contains both air and water within its pore spaces, whereas all pore spaces in the zone of saturation are filled with groundwater. The water table is the surface separating the zones of aeration and saturation. Within the capillary fringe, water rises by surface tension from the zone of saturation into the zone of aeration.

Several factors contribute to the surface configuration of a region's water table, including regional differences in the amount of rainfall, permeability, and rate of groundwater movement. During periods of high rainfall, groundwater tends to rise beneath hills because it cannot flow fast enough into adjacent valleys to maintain a level surface. During droughts, the water table falls and tends to flatten out because it is not being replenished. In arid and semiarid regions, the water table is usually quite flat regardless of the overlying land surface.

13.5 Groundwater Movement

Gravity provides the energy for the downward movement of groundwater. Water entering the ground moves through the zone of aeration to the zone of saturation (Figure 13.3). When water reaches the water table, it continues to move through the zone of saturation from areas where the water table is high toward areas where it is lower, such as streams, lakes, or swamps. Only some of the water follows the direct route along the slope of the water table. Most of it takes longer curving paths down and then enters a stream, lake, or swamp from below, because it moves from areas of high pressure toward areas of lower pressure within the saturated zone.

Groundwater velocity varies greatly and depends on many factors. Velocities range from 250 m per day in some extremely permeable material to less than a few centimeters per year in nearly impermeable material. In most ordinary aquifers, the average velocity of groundwater is a few centimeters per day.

13.6 Springs, Water Wells, and Artesian Systems

You can think of the water in the zone of saturation much like a reservoir whose surface rises or falls depending on additions as opposed to natural and artificial withdrawals. *Recharge*—that is, additions to the zone of saturation—comes from rainfall or melting snow, or water might be added artificially at wastewater-treatment plants or recharge ponds constructed for just this purpose. But if groundwater is discharged naturally or withdrawn at wells without sufficient recharge, the water table drops—just as a savings

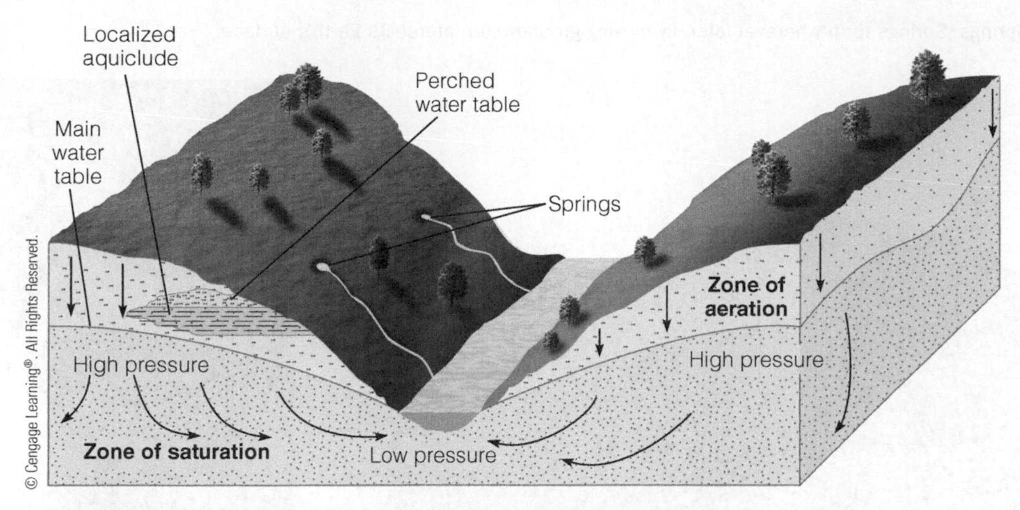

Figure 13.3 Groundwater Movement Groundwater moves down through the zone of aeration to the zone of saturation. Then some of it moves along the slope of the water table, and the rest moves through the zone of saturation from areas of high pressure toward areas of low pressure. Some water might collect over a local aquiclude, such as a shale layer, thus forming a perched water table.

Critical Thinking Question If you were drilling a water well on your property and struck water at a considerably shallower depth than your neighbors, should you consider drilling deeper, or just celebrate your good fortune at not having to pay for a deeper well?

account diminishes if withdrawals exceed deposits. Withdrawals from the groundwater system take place where groundwater flows laterally into streams, lakes, or swamps, where it discharges at the surface as *springs*, and where it is withdrawn from the system at water wells.

Springs

Places where groundwater flows or seeps out of the ground as **springs** have always fascinated people. The water flows out of the ground for no apparent reason and from no readily identifiable source. So it is not surprising that springs have long been regarded with superstition and revered for their supposed medicinal value and healing powers. Nevertheless, there is nothing mystical or mysterious about springs.

Although springs can occur under a wide variety of geologic conditions, they all form in basically the same way (Figure 13.4a). When percolating water reaches the water table or an impermeable layer, it flows laterally, and if this flow intersects the surface, the water discharges as a spring (Figure 13.4b). The Mammoth Cave area in Kentucky (see the Introduction) is underlain by fractured limestones whose fractures have been enlarged into caves by solution activity. In this geologic environment, springs occur where the fractures and caves intersect the ground surface, allowing groundwater to exit onto the surface.

Springs can also develop wherever a *perched water table*—a local aquiclude present within a larger aquifer, such as a lens of shale within sandstone—intersects the surface (Figure 13.3). As water migrates through the zone of aeration, it is stopped by the local aquiclude, and a localized zone of saturation "perched" above the main water table forms. Water moving laterally along the perched water table may intersect the surface to produce a spring.

Water Wells

Water wells are openings made by digging or drilling down into the zone of saturation. Once the zone of saturation has been penetrated, water percolates into the well, filling it to the level of the water table. A few wells are free flowing; for most, however, the water must be brought to the surface by pumping.

In some parts of the world, water is raised to the surface with nothing more than a bucket on a rope or a hand-operated pump. In many parts of the United States and Canada, one can see windmills from times past when wind power was used to pump water. Most of these are no longer in use, having been replaced by more efficient electric pumps.

ConnectionLink

To learn more about windmills and their role in pumping water and generating power, see Geo-Focus 15.1 in Chapter 15.

When groundwater is pumped from a well, the water table in the area around the well is lowered, forming a **cone of depression** (Figure 13.5). This happens when the rate of water withdrawal from the well exceeds the rate of water inflow to the well, thus lowering the water table around the well. A cone of depression's gradient—that is, whether it is steep or gentle—depends to a great extent on the permeability of the aquifer being pumped. A highly permeable aquifer produces a gentle gradient in the cone of depression, whereas a low-permeability aquifer results in a steep cone of depression because water cannot easily flow to the well to replace the water being withdrawn.

The formation of a cone of depression does not normally pose a problem for the average domestic well, provided that the well is drilled deep enough into the zone of saturation. However, the tremendous amounts of water used by industry

Figure 13.4 Springs Springs form wherever laterally moving groundwater intersects Earth's surface.

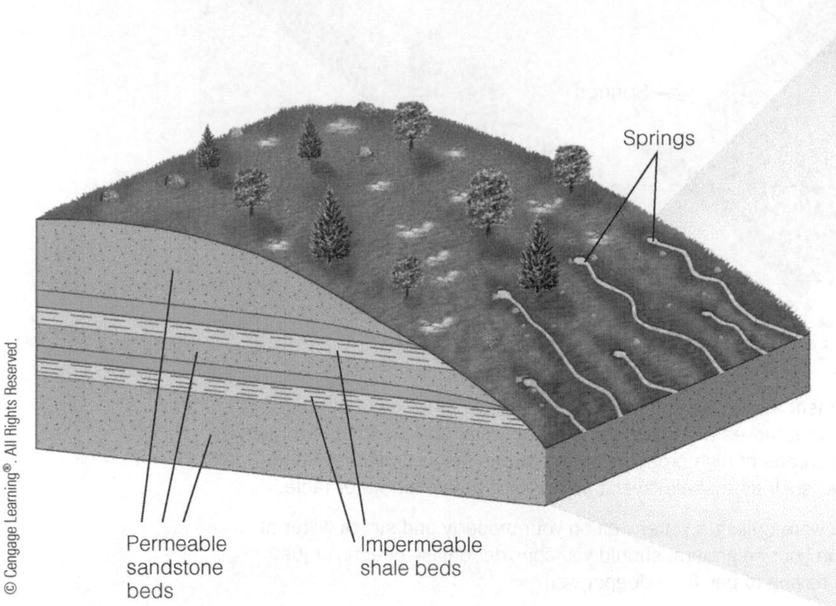

a Most commonly, springs form when percolating water reaches an impermeable layer and migrates laterally until it seeps out at the surface.

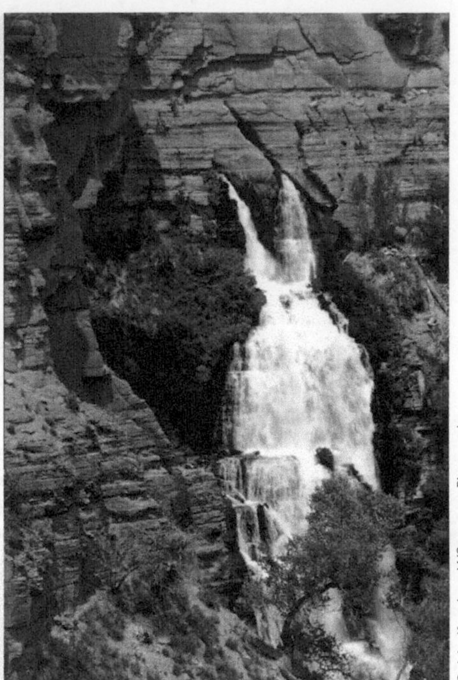

b Thunder River Spring in the Grand Canyon, Arizona, issues from rocks along a wall of the Grand Canyon. Water percolating downward through permeable rock is forced to move laterally when it encounters an impermeable zone, and thus gushes out along this cliff. Notice the vegetation parallel to and below the springs, indicating that enough water flows from springs along the cliff wall to support the vegetation.

and for irrigation may create a large cone of depression that lowers the water table sufficiently to cause shallow wells in the immediate area to go dry (Figure 13.5). This situation is not uncommon and frequently results in lawsuits by the owners of the shallow dry wells.

Lowering of the regional water table, because more groundwater is being withdrawn than is being replenished is becoming a serious problem in many areas, particularly in the southwestern United States, where rapid growth has placed tremendous demands on the groundwater system. As mentioned earlier, some of the largest cities in the United States depend entirely on groundwater for their municipal needs. Furthermore, some of the largest agricultural states are withdrawing groundwater from regional aquifers that are not being sufficiently replenished. Unrestricted withdrawal of groundwater cannot continue indefinitely, and the rising costs and decreasing supply of groundwater should soon limit the growth of some regions in the United States, such as the aforementioned Southwest.

Figure 13.5 Cone of Depression A cone of depression forms whenever water is withdrawn from a well. If water is withdrawn faster than it can be replenished, the cone of depression will grow in depth and circumference, lowering the water table in the area and causing nearby shallow wells to go dry.

Artesian Systems

The word *artesian* comes from the French town and province of Artois (called Artesium during Roman times) near Calais, where the first European artesian well was drilled in 1126 and is still flowing today. The term **artesian system** can be applied to any system in which groundwater is confined and builds up high hydrostatic (fluid) pressure (Figure 13.6). Water in such a system is able to rise above the level of the aquifer if a well is drilled through the confining layer, thereby reducing the pressure and forcing the water upward. An artesian system can develop when (1) an aquifer is confined above and below by aquicludes; (2) the rock sequence is (usually) tilted to build up hydrostatic pressure; and (3) the aquifer is exposed at the surface, thus enabling it to be recharged.

The elevation of the water table in the recharge area and the distance of the well from the recharge area determine the height to which artesian water rises in a well. The surface

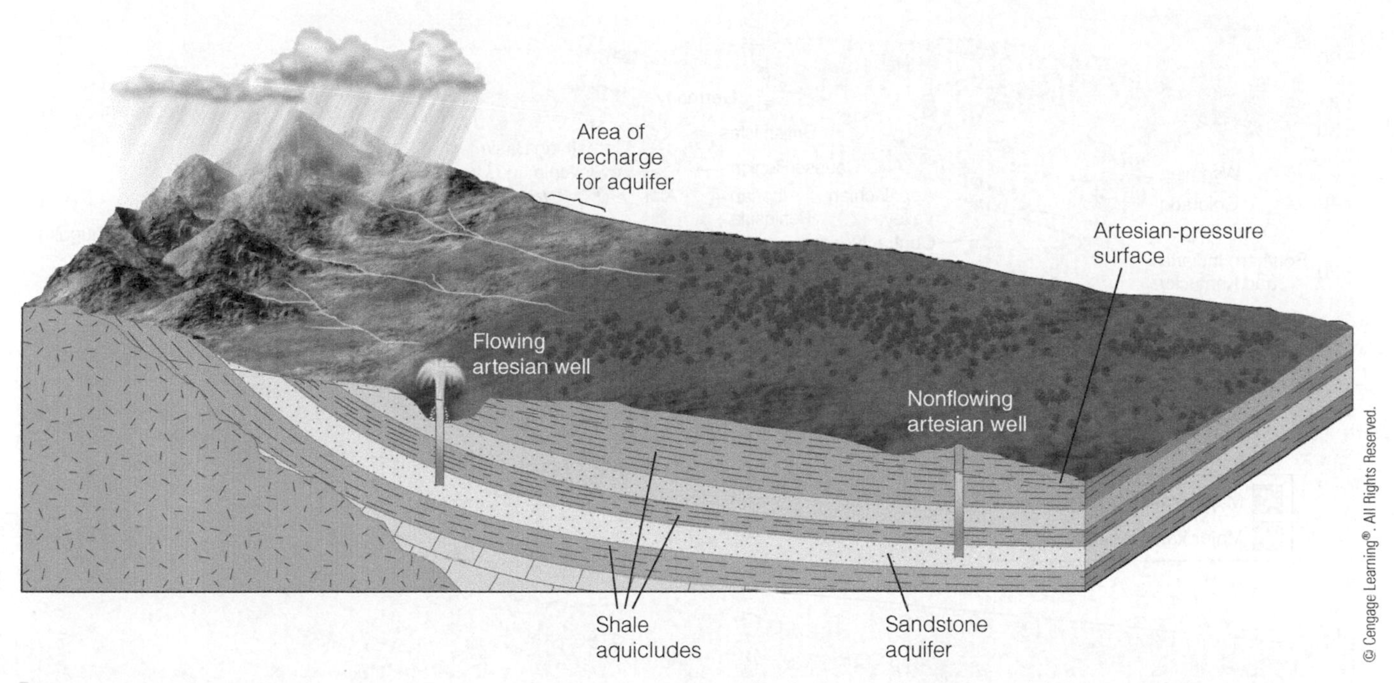

Figure 13.6 Artesian System An artesian system must have an aquifer confined above and below by aquicludes, the aquifer must be exposed at the surface, and there must be sufficient precipitation in the recharge area to keep the aquifer filled. The elevation of the water table in the recharge area, which is indicated by a sloping dashed line (the artesian-pressure surface), defines the highest level to which well water can rise.

Critical Thinking Question If the elevation of the wellhead is at or above that of the artesian-pressure surface, why then will the well be nonflowing?

defined by the water table in the recharge area, called the *artesian-pressure surface*, is indicated by the sloping dashed line in Figure 13.6. Because friction slightly reduces the pressure of the aquifer water, and consequently the level to which artesian water rises, the artesian-pressure surface therefore slopes.

An artesian well will flow freely at the ground surface only if the wellhead is at an elevation below the artesian-pressure surface. In this situation, the water flows out of the well because it rises toward the artesian-pressure surface, which is at a higher elevation than the wellhead. In a nonflowing artesian well, the wellhead is above the artesian-pressure surface, and the water will rise in the well only as high as the artesian-pressure surface.

One of the best-known artesian systems in the United States underlies South Dakota and extends southward to central Texas. The majority of the artesian water from this system is used for irrigation. The aquifer of this artesian system, the Dakota Sandstone, is recharged where it is exposed along the margins of the Black Hills of South Dakota. Originally, the hydrostatic pressure in this system was great enough to produce free-flowing wells and to operate waterwheels. However, because of the extensive use of this groundwater for irrigation over the years, the hydrostatic pressure in many of the wells is so low that they are no longer free flowing, and the water must be pumped.

Advertisers frequently tout the quality of artesian water as somehow being superior to other groundwater. Some artesian water might in fact be of excellent quality, but its quality is not dependent on the fact that water rises above the surface of an aquifer. Rather, its quality is a function of dissolved minerals and any introduced substances, so artesian water really is no different from any other groundwater. The myth of its superiority probably arises from the fact that people have always been fascinated by water that flows freely from the ground.

13.7 Groundwater Erosion and Deposition

When rainwater begins to seep into the ground, it immediately starts to react with the minerals it contacts, weathering them chemically. In an area underlain by soluble rock, groundwater is the principal agent of erosion and is responsible for the formation of many major features of the landscape.

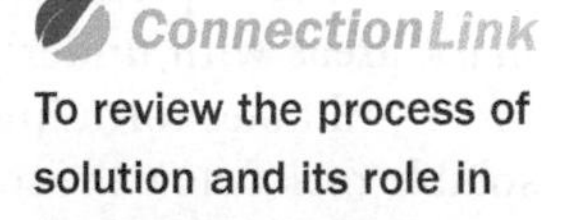

To review the process of solution and its role in chemical weathering, see Chapter 6.

Limestone, a common sedimentary rock composed primarily of the mineral calcite ($CaCO_3$), underlies large areas of Earth's surface (Figure 13.7). Although limestone

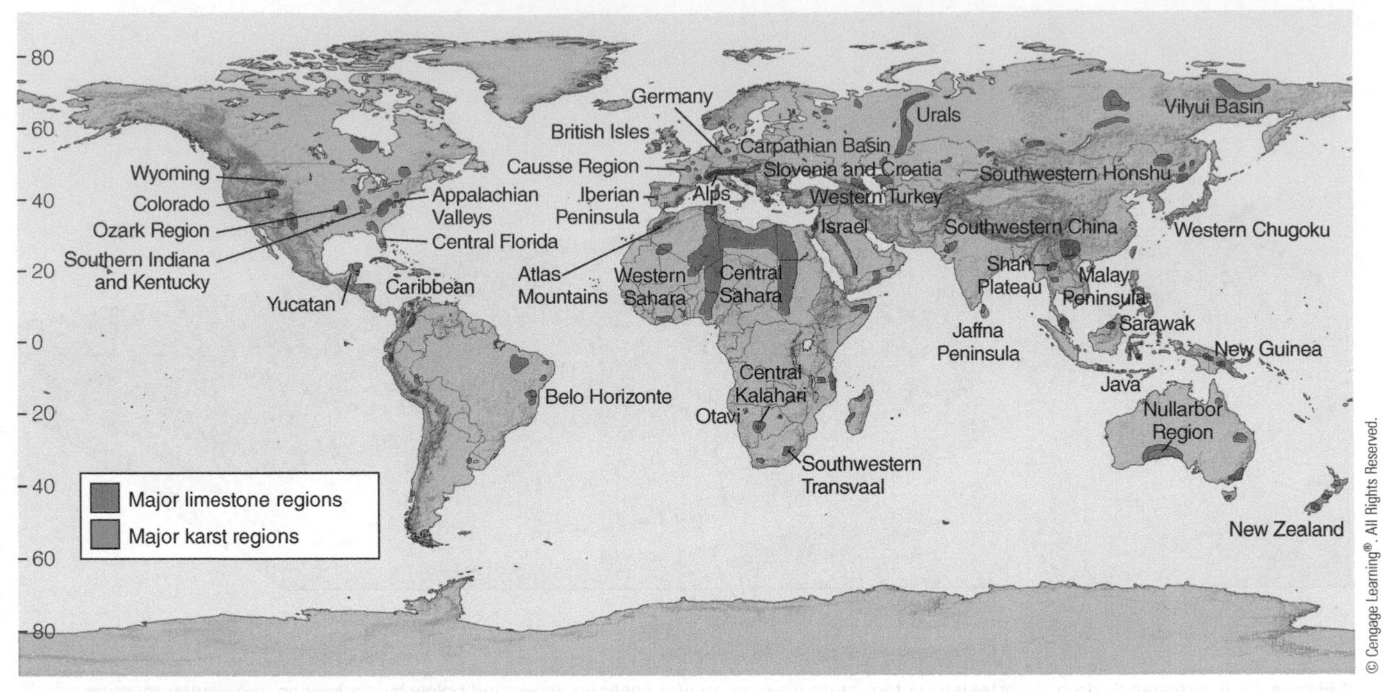

Figure 13.7 Distribution of the Major Limestone and Karst Areas of the World Karst topography develops largely by groundwater erosion in areas underlain by soluble rocks.

Critical Thinking Question Why is karst topography typically restricted to regions with humid and temperate climates?

is practically insoluble in pure water, it readily dissolves if a small amount of acid is present. Carbonic acid (H_2CO_3) is a weak acid that forms when carbon dioxide combines with water ($H_2O + CO_2 \rightarrow H_2CO_3$). Because the atmosphere contains a small amount of carbon dioxide (0.03%), and carbon dioxide is also produced in soil by the decay of organic matter, most groundwater is slightly acidic. When groundwater percolates through the various openings in limestone, the slightly acidic water readily reacts with the calcite to dissolve the rock by forming soluble calcium bicarbonate, which is carried away in solution (see Chapter 6).

Sinkholes and Karst Topography

In regions underlain by soluble rock, the ground surface may be pitted with numerous depressions that vary in size and shape. These depressions, called **sinkholes**, or merely *sinks*, mark areas with underlying soluble rock (Figure 13.8). Most sinkholes form in one of two ways. The first is when soluble rock below the soil is dissolved by seeping water, and openings in the rock are enlarged and filled in by the overlying soil. As the groundwater continues to dissolve the rock, the soil is eventually removed, leaving shallow depressions with gently sloping sides. When adjacent sinkholes merge, they form a network of larger, irregular, closed depressions called *solution valleys* (Figure 13.9).

Sinkholes also form when a cave's roof collapses, usually producing a steep-sided crater. Sinkholes formed in this way are a serious hazard, particularly in populated areas. In regions prone to sinkhole formation, extensive geologic and hydrogeologic investigation must be performed to determine the depth and extent of underlying cave systems before any site development to ensure that the underlying rocks are thick enough to support planned structures.

Karst topography, or simply *karst*, develops largely by groundwater erosion in many areas underlain by soluble rocks (Figure 13.9). The name karst is derived from the plateau region of the border area of Slovenia, Croatia, and northeastern Italy where this type of topography is well developed. In the United States, regions of karst topography include large areas of southwestern Illinois, southern Indiana, Kentucky, Tennessee, northern Missouri, Alabama, and central and northern Florida (Figure 13.7).

Karst topography is characterized by numerous caves, springs, sinkholes, solution valleys, and disappearing streams (Figure 13.9). *Disappearing streams* are so named because they typically flow only a short distance at the surface and then disappear into a sinkhole. The water continues flowing underground through fractures or caves until it surfaces again at a spring or other stream.

Karst topography varies from the spectacular high-relief landscapes of China to the subdued and pockmarked landforms of Kentucky (Figure 13.10). Common to all karst topography, though, is the presence of thick-bedded, readily soluble rock at the surface or just below the soil, and enough water for solution activity to occur. Karst topography is therefore typically restricted to humid and temperate climates.

Figure 13.8 Sinkholes

Critical Thinking Question Why are sink holes becoming more prevalent in Florida, especially in areas experiencing population growth?

AP Photo

a This sinkhole formed on May 8 and 9, 1981, in Winter Park, Florida. It formed in previously dissolved limestone following a drop in the water table. The 100-m-wide, 35-m-deep sinkhole destroyed a house, numerous cars, and a municipal swimming pool.

James S. Monroe

b A small sinkhole in Montana, now occupied by a lake. The water enters the lake from a hot spring so it remains warm year-round. In fact, tropical fish have been introduced into the lake and thus live in an otherwise inhospitable climate.

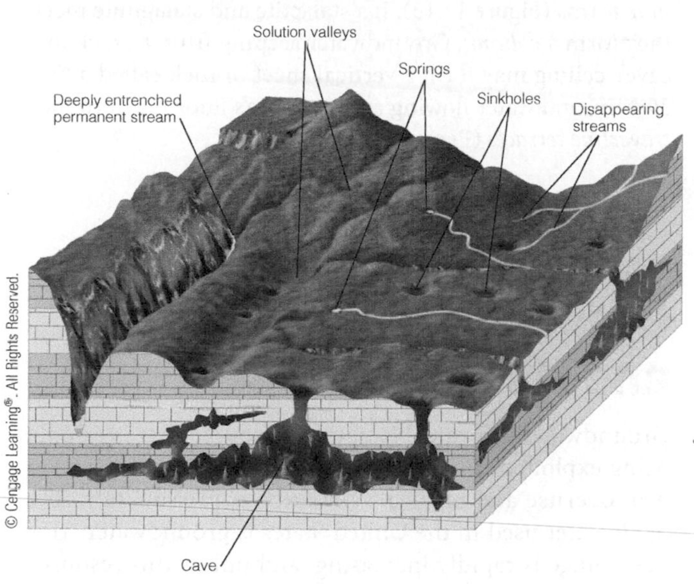

Figure 13.9 **Features of Karst Topography** Erosion of soluble rock by groundwater produces karst topography. Features commonly found include solution valleys, springs, sinkholes, and disappearing streams.

Figure 13.10 **Karst Landscape in Kunming, China, and Bowling Green, Kentucky**

Reed Wicander

a The Stone Forest, 125 km southeast of Kunming, China, is a high-relief karst landscape formed by the dissolution of carbonate rocks.

John S. Shelton/c/o University of Washington Librar

b Solution valleys, sinkholes, and sinkhole lakes dominate the subdued karst topography east of Bowling Green, Kentucky.

Caves and Cave Deposits

Caves are perhaps the most spectacular examples of the combined effects of weathering and erosion by groundwater. As groundwater percolates through carbonate rocks, it dissolves and enlarges fractures and openings to form a complex, interconnecting system of crevices, caves, caverns, and underground streams. A **cave** is usually defined as a naturally formed, subsurface opening that is generally connected to the surface and is large enough for a person to enter. A *cavern*, on the other hand, is a very large cave or a system of interconnected caves.

Bettmann/Corbis

Figure 13.11 Jesse James and His Outlaw Band Jesse James (left) and members of his outlaw band (probably two of the Younger brothers) pose in this ca. 1870s photo. Jesse James and his gang repeatedly used Meramec Caverns, Missouri, as a hideout in the 1870s.

More than 17,000 caves are known in the United States. Although most of them are small, a number are large and spectacular. Some of the more famous ones are Mammoth Cave, Kentucky; Carlsbad Caverns, New Mexico; Lewis and Clark Caverns, Montana; Lehman Cave, Nevada; and Meramec Caverns, Missouri, which Jesse James and his outlaw band often used as a hideout (Figure 13.11). Canada also has many famous caves, including the 536-m-deep Arctomys Cave in Mount Robson Provincial Park, British Columbia, the deepest known cave in North America. And Mexico is known for its Cueva de los Cristales, or Cave of Crystals, where some of the world's largest crystals of gypsum and selenite have been found (see the chapter opening photo).

Caves and caverns form as a result of the dissolution of carbonate rocks by weakly acidic groundwater (Figure 13.12). Groundwater percolating through the zone of aeration slowly dissolves the carbonate rock and enlarges its fractures and bedding planes. On reaching the water table, the groundwater migrates toward the region's surface streams. As the groundwater moves through the zone of saturation, it continues to dissolve the rock and gradually forms a system of horizontal passageways through which the dissolved rock is carried to the streams (Figure 13.12a).

As the surface streams erode deeper valleys, the water table drops in response to the lower elevation of the streams (Figure 13.12b). The water that flowed through the system of horizontal passageways now percolates to the lower water table where a new system of passageways begins to form. The abandoned channelways form an interconnecting system of caves and caverns. Caves eventually become unstable and collapse, littering the floor with fallen debris.

When most people think of caves, they think of the seemingly endless variety of colorful and bizarre-shaped deposits found in them. Although a great many different types of cave deposits exist, most form in essentially the same manner and are collectively known as *dripstone*. As water seeps into a cave, some of the dissolved carbon dioxide in the water escapes, and a small amount of calcite is precipitated. In this manner, the various dripstone deposits are formed (Figure 13.12c).

Stalactites are icicle-shaped structures hanging from cave ceilings that form as a result of precipitation from dripping water (Figure 13.13). With each drop of water, a thin layer of calcite is deposited over the previous layer, forming a cone-shaped projection that grows down from the ceiling. The water that drips from a cave's ceiling also precipitates a small amount of calcite when it hits the floor. As additional calcite is deposited, an upward-growing projection called a *stalagmite* forms (Figure 13.13). If a stalactite and stalagmite meet, they form a *column*. Groundwater seeping from a crack in a cave's ceiling may form a vertical sheet of rock called a *drip curtain*, and water flowing across a cave's floor may produce *travertine terraces* (Figure 13.12c).

13.8 Modification of the Groundwater System and Its Effects

Groundwater is a valuable natural resource that is rapidly being exploited with seemingly disregard to the effects on it of overuse and misuse. Currently, approximately 20% of all water used in the United States is groundwater. This percentage is rapidly increasing, and unless this resource is used more wisely, sufficient amounts of clean groundwater will not be available in the future. Modification of the groundwater system may have many consequences, including (1) lowering of the water table, causing wells to go dry; (2) saltwater incursion; (3) subsidence; and (4) contamination. In fact, a major ongoing debate today pits the benefits of hydraulic fracturing for the extraction of natural gas from organic-rich shales against the possible contamination of the groundwater system in the area where this exploration is occurring (Geo-Focus 13.1).

Lowering the Water Table

Withdrawing groundwater at a significantly greater rate than it is replaced by either natural or artificial recharge can have serious effects. For example, the High Plains aquifer is one of

▌Figure 13.12 Cave Formation

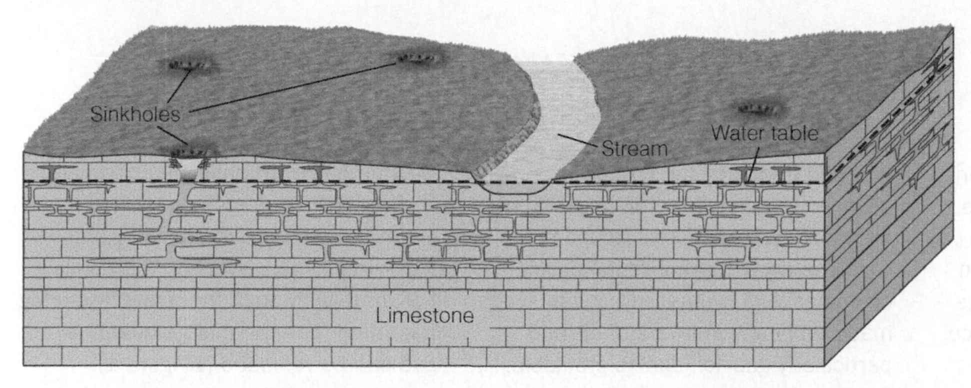

a As groundwater percolates through the zone of aeration and flows through the zone of saturation, it dissolves the carbonate rocks and gradually forms a system of passageways.

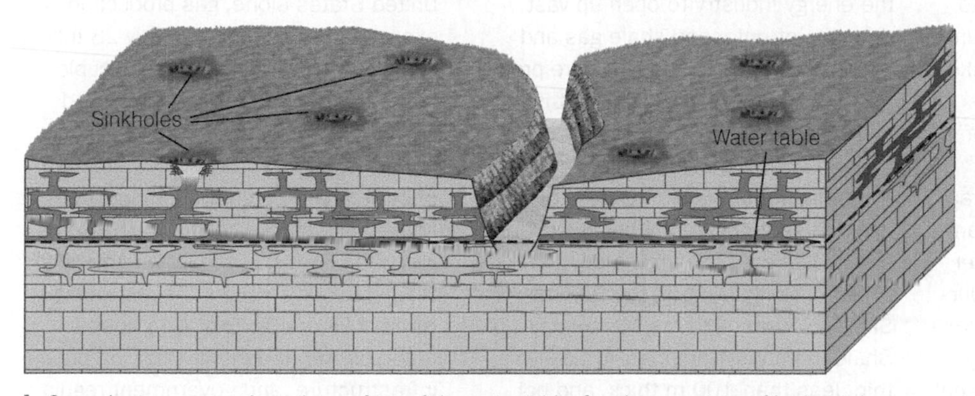

b Groundwater moves along the surface of the water table, forming a system of horizontal passageways through which dissolved rock is carried to the surface streams, thus enlarging the passageways.

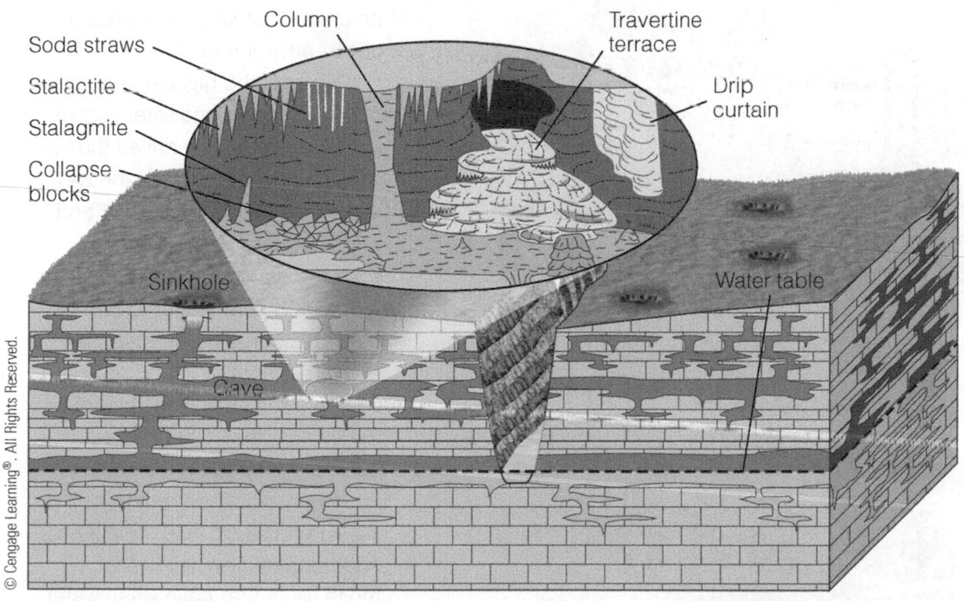

c As the surface streams erode deeper valleys, the water table drops, and the abandoned channelways form an interconnecting system of caves and caverns.

the most important aquifers in the United States. It underlies more than 450,000 km^2, including most of Nebraska, large parts of Colorado and Kansas, portions of South Dakota, Wyoming, and New Mexico, as well as the panhandle regions of Oklahoma and Texas, and accounts for approximately 30% of the groundwater used for irrigation in the United States (▌Figure 13.14).

Significant withdrawal of groundwater from the High Plains aquifer for irrigation began in the 1950s, and by 1980, the water table had dropped an average of 3 m. Irrigation from the High Plains aquifer is largely responsible for the region's agricultural productivity, including a significant percentage of the nation's corn, cotton, and wheat, and half of the beef cattle in the United States.

Although the High Plains aquifer has contributed to the high productivity of the region, it cannot continue to provide the quantities of water that it has in the past. In some parts of the High Plains, from 2 to 100 times more water is being pumped annually than is being recharged, causing a substantial drop in the water table in many areas (Figure 13.14).

It must be noted, however, that much of the aquifer's water infiltrated during wetter glacial climates more than 10,000 years ago. Consequently, most of the water being pumped is fossil water that is not being replenished at anywhere near the same rate as when it formed during the Pleistocene Epoch.

What will happen to this region's economy if long-term withdrawal of water from the High Plains aquifer continues to greatly exceed its recharge rate, and the aquifer can no longer supply the quantities of water necessary for irrigation? Solutions range from going back to farming without irrigation to diverting water from other regions such as the Great Lakes. Most users of the aquifer realize that they cannot continue to withdraw the quantities of groundwater that they have in the past, and thus they are turning to greater conservation, monitoring of the aquifer, and use of new technologies to try to better balance withdrawal with recharge rates.

GEO FOCUS 13.1

Hydraulic Fracturing: Pros and Cons

The debate over hydraulic fracturing, popularly known as "fracking," is certainly a contentious one with strong feelings on both sides. Those in favor of this new technology point to its benefits in reducing energy dependence, lowering energy costs, and creating new jobs in the energy sector. Those opposed to it cite environmental concerns, such as increased pollution to the environment, particularly to groundwater supplies, increased health risks, as well as a possible increase in global warming due to the leaking of methane, a potent greenhouse gas.

What is hydraulic fracturing? Hydraulic fracturing has been around for many years and involves forcing fluids under high pressure into an oil- or gas-producing rock formation to create fractures so that the oil or gas can more freely flow into the well. Drilling a conventional well and fracturing the producing zone only allows tapping the oil or gas from the small fractured area around the well. The advent of horizontal drilling, combined with hydraulic fracturing, soon made tapping thin layers of oil- and particularly gas-rich shales profitable. This combination of horizontal drilling and hydraulic fracturing has allowed the energy industry to open up vast deposits of untapped shale gas and oil in the United States that were previously unprofitable, and fuel what is now called the "shale revolution."

How does horizontal drilling and hydraulic fracturing work? Most of the gas-rich shale presently being explored and produced, such as the Bakken Shale in North Dakota, the Barnett Shale in Texas, and the Marcellus Shale in Pennsylvania, are typically thin, less than 100 m thick, and not worth the cost of drilling a conventional vertical well. These shales, however, extend over vast areas, and by drilling vertically to them, and then drilling horizontally through them and fracturing the shale formation, large quantities of gas can be released (❚ Figure 1).

The potential of these gas- and oil-producing shales is enormous. In the United States alone, gas production is expected to increase to nearly 28 trillion cubic feet, up from 21.6 trillion cubic feet in 2010 according to the United States Energy Administration's *Annual Energy Outlook 2012*. An examination of currently targeted and prospective shale plays shows huge areas of shale gas and oil reserves in North America (❚ Figure 2), as well as the rest of the world (❚ Figure 3). Exploiting these reserves will depend, in part, on price, infrastructure, and government regulations. There is no debate that there are vast oil and gas shales that can be produced by current technology, but at what potential cost to the environment?

Probably the biggest concern is polluting the groundwater system. Because the well is drilled through the groundwater aquifer, the well must be cased with cement and steel to guard against leakage from the pipe into the groundwater. Although this is standard procedure, accidents do happen, and over time, corrosion and pressure can lead to cracks or failure of the casing, that if not immediately sealed, can result in pollution of the groundwater system by drilling and hydraulic fluids (Figure 1). Another similar worry is that in those wells that have wastewater storage, the wastewater might leak into the ground, contaminating the soil, streams, and groundwater, as well as emitting noxious fumes into the surrounding area.

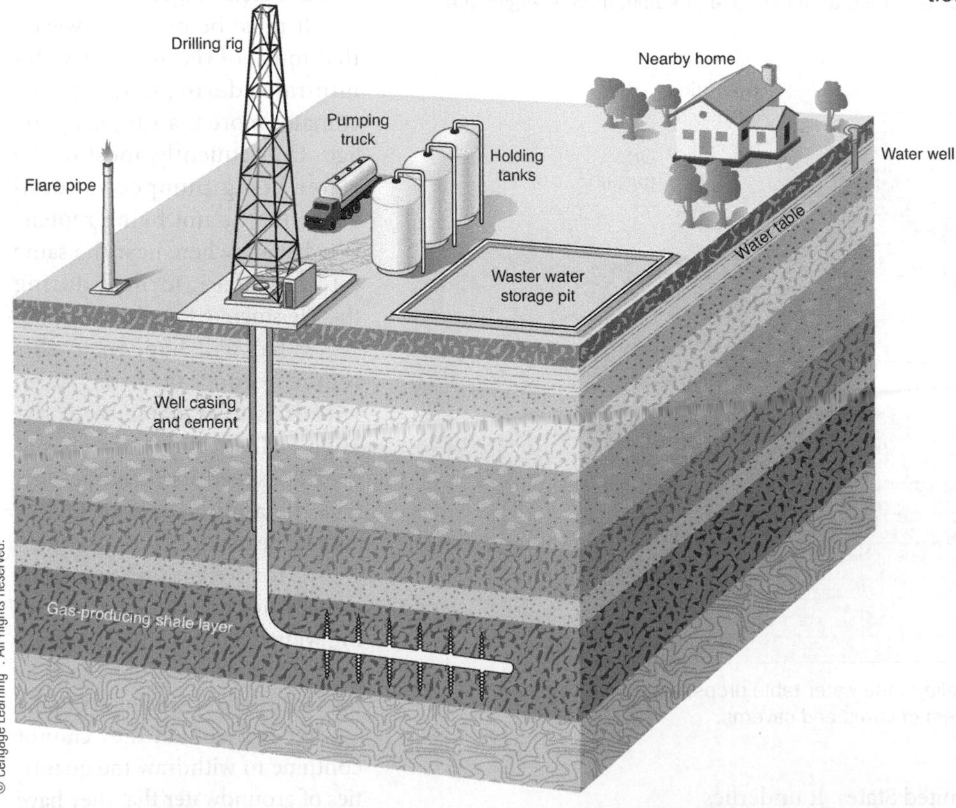

❚ Figure 1 A diagrammatic representation of how horizontal drilling and hydraulic fracturing are used to extract gas from a gas-rich shale layer.

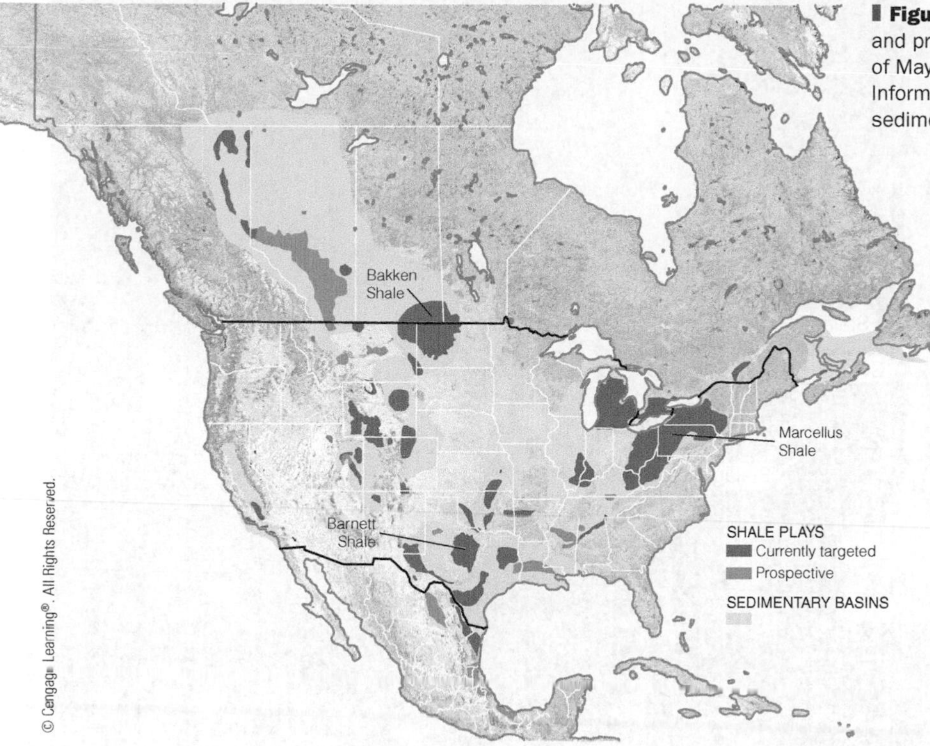

Figure 2 Map showing the current gas shale plays and prospective gas shale plays in North America as of May 2011, according to the United States Energy Information Administration. Also shown are the various sedimentary basins in which these shales are found.

The problems of methane leakage and pollution also have been raised. Methane is the primary gas in natural gas, and there are concerns about its leakage into the groundwater system, where it can build up to levels high enough that it can burn as water flows from the tap, or cause an explosion as once happened in a home. Likewise, when methane is released into the atmosphere, it acts as a potent greenhouse gas that can trap up to 20 times as much heat as carbon dioxide (see Figure 1.4 on the greenhouse effect and global warming).

Last, the question has been raised about whether hydraulic fracturing contributes to earthquakes in the region being drilled. As was discussed in the section on earthquake control in Chapter 9, it has been noted that the number of earthquakes in the Dallas–Ft. Worth area of Texas has dramatically increased recently, seemingly corresponding to the increased drilling activity associated with horizontal drilling and hydraulic fracturing of the Barnett Shale. Although still unresolved, a number of geologists are now looking at whether the increase in earthquake activity correlates to the fracked wells, or whether it is linked to wastewater injection wells situated nearby that are used to dispose of drilling wastewater.

What now needs to be asked is whether hydraulic fracturing is worth it. Like many technologies, it comes with both risks and benefits. The risks are damage to the environment in the form of pollution to the atmosphere, groundwater supplies, as well as the release of greenhouse gases, and health concerns resulting from the chemicals used in the fracking fluids, which are not always disclosed. The benefits are that oil and gas shales can go a long way toward achieving energy independence, provide a relatively inexpensive source of energy for the foreseeable future, and provide a boost to national economies. Whether the "shale revolution" is indeed the answer to our energy needs, and can be exploited with minimal damage to the environment, is yet to be determined.

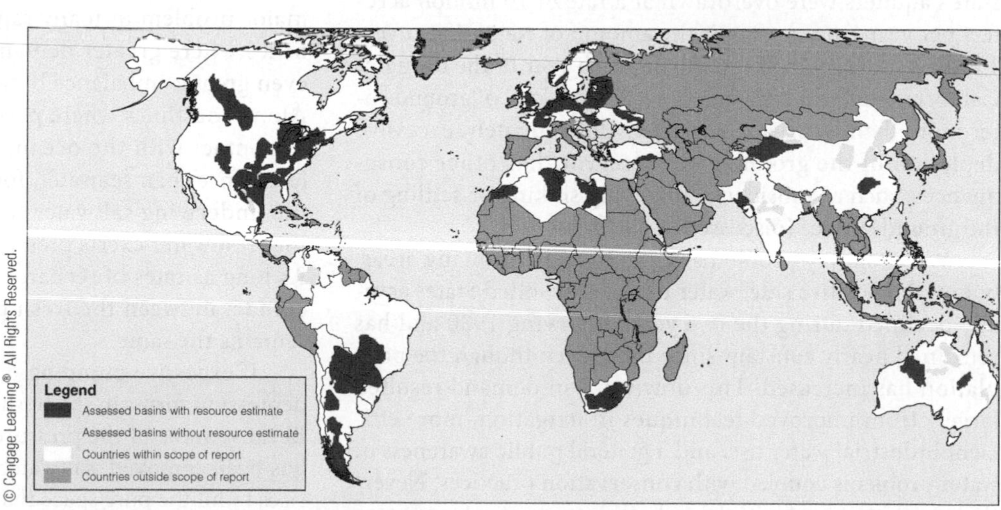

Figure 3 Map showing the analyzed major shale basins in 32 countries based on a 2011 report from the United States Energy Information Administration.

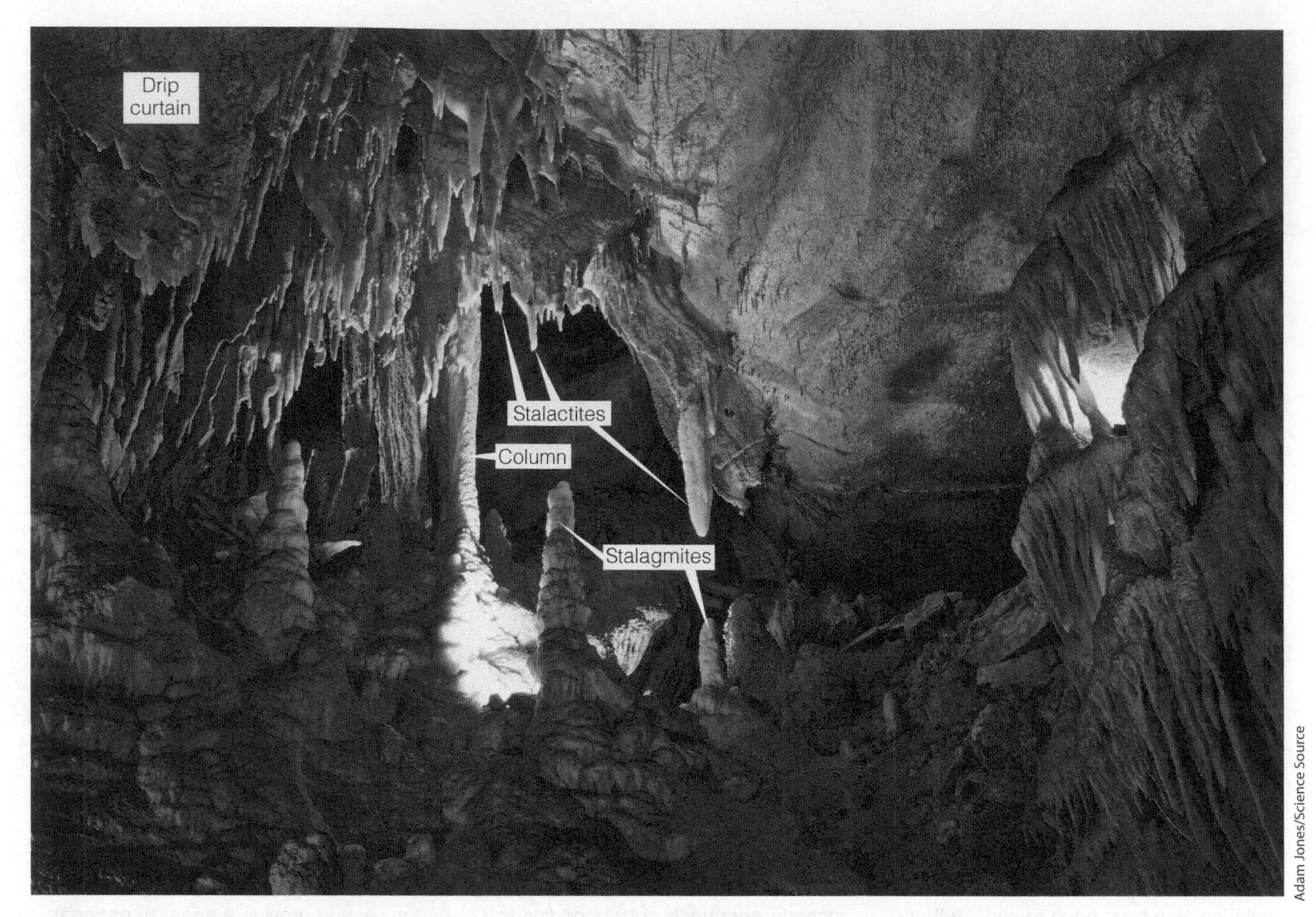

Adam Jones/Science Source

▌Figure 13.13 Cave Deposits Stalactites are the icicle-shaped structures hanging from the cave's ceiling, whereas the upward-pointing structures on the floor are stalagmites. Columns result when stalactites and stalagmites meet. A drip curtain is a vertical sheet of rock formed by groundwater seeping from a crack in the cave's ceiling. All four structures are present in Mammoth Cave, Kentucky.

Another excellent example of what we might call deficit spending with regard to groundwater took place in California during the drought of 1987–1992. During that time, the state's aquifers were overdrawn at a rate of 10 million acre-feet per year (an acre-foot is the amount of water that covers 1 acre, 1 ft deep). In short, during each year of the drought, California was withdrawing more than 12 km^3 of groundwater more than was being replaced. Unfortunately, excessive depletion of the groundwater reservoir has other consequences, such as subsidence involving sinking or settling of the ground surface (discussed in a later section).

Water supply problems certainly exist in many areas, but on the positive side, water use in the United States actually declined during the five years following 1980 and has remained nearly constant since then, even though the population has increased. This downturn in demand resulted largely from improved techniques in irrigation, more efficient industrial water use, and a general public awareness of water problems coupled with conservation practices. Nevertheless, the rates of withdrawal of groundwater from some aquifers still exceed their rates of recharge, and population growth in the arid to semiarid Southwest is continuing to put significant demands on an already limited water supply.

Saltwater Incursion

The excessive pumping of groundwater in coastal areas has resulted in *saltwater incursion*, which has become a major problem in many rapidly growing coastal communities where greater demand for groundwater creates an even greater imbalance between withdrawal and recharge. Along coastlines where permeable rocks or sediments are in contact with the ocean, the fresh groundwater, being less dense than seawater, forms a lens-shaped body above the underlying saltwater (▌Figure 13.15a). The weight of the freshwater exerts pressure on the underlying saltwater. As long as rates of recharge equal rates of withdrawal, the contact between the fresh groundwater and the seawater remains the same.

If excessive pumping occurs, however, a deep cone of depression forms in the fresh groundwater (▌Figure 13.15b). Because some of the pressure from the overlying freshwater has been removed, saltwater forms a *cone of ascension* as it rises to fill the pore space that formerly contained freshwater. When this occurs, wells become contaminated with saltwater and remain contaminated until recharge by freshwater restores the former level of the fresh groundwater water table.

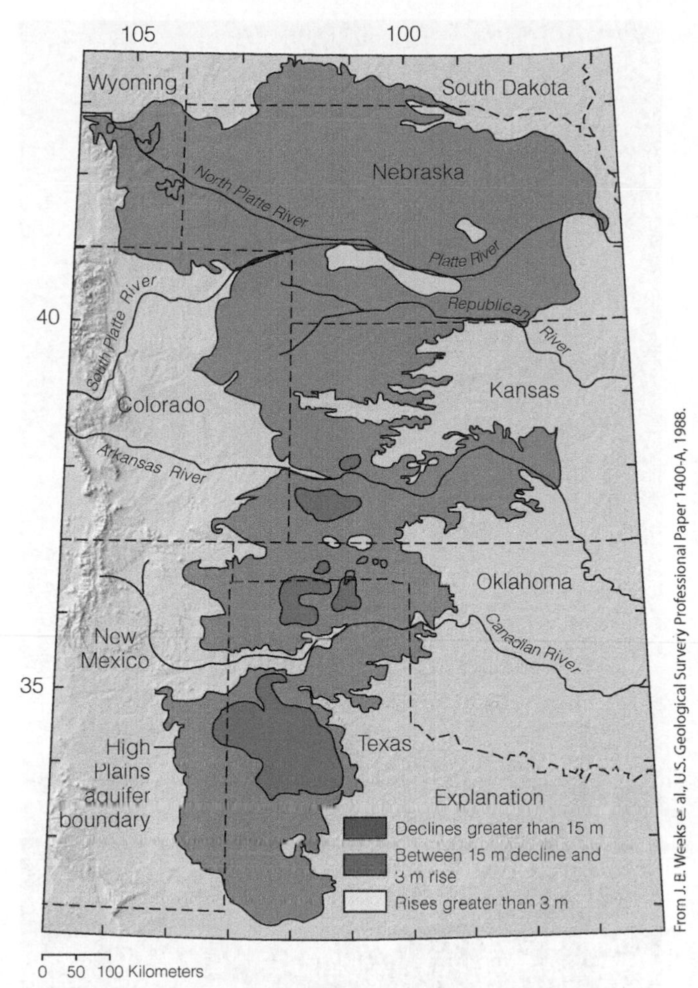

Figure 13.14 High Plains Aquifer The geographic extent of the High Plains aquifer and changes in water level from predevelopment through 1993. Irrigation from the High Plains aquifer is largely responsible for the region's agricultural productivity.

Figure 13.15 Saltwater Incursion

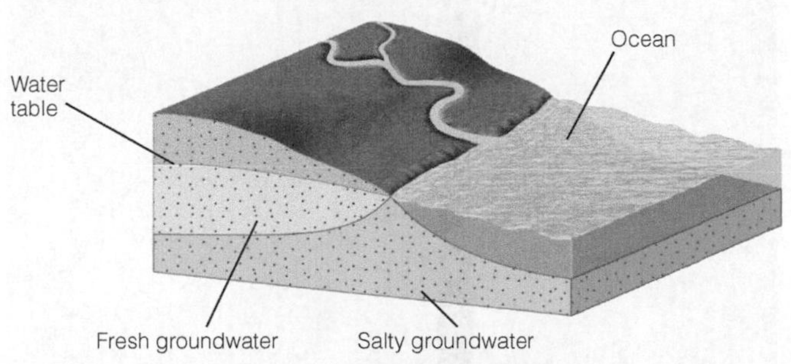

a Because freshwater is not as dense as saltwater, it forms a lens-shaped body above the underlying saltwater.

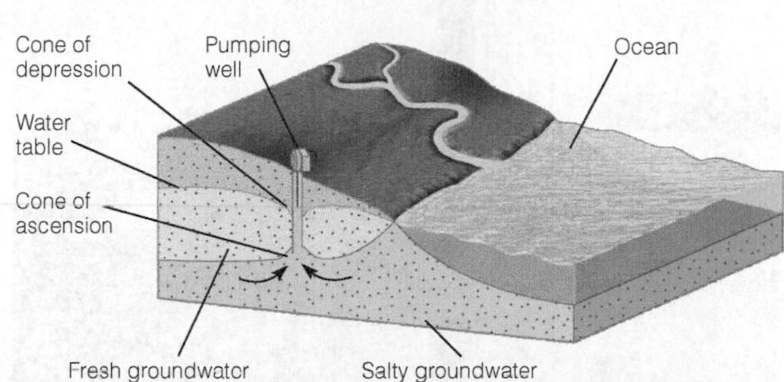

b If excessive pumping occurs, a cone of depression develops in the fresh groundwater, and a cone of ascension forms in the underlying salty groundwater, which may result in saltwater contamination of the well.

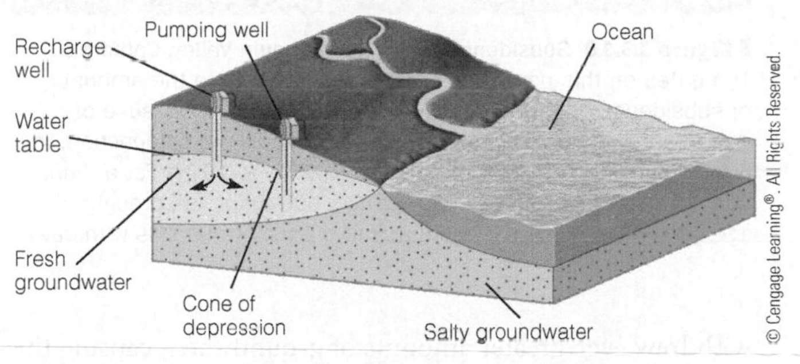

c Pumping water back into the groundwater system through recharge wells can help lower the interface between the fresh groundwater and the salty groundwater and reduce saltwater incursion.

To counteract the effects of saltwater incursion, recharge wells are often drilled to pump water back into the groundwater system (Figure 13.15c). Recharge ponds that allow large quantities of fresh surface water to infiltrate the groundwater supply may also be constructed.

Subsidence

As excessive amounts of groundwater are withdrawn from poorly consolidated sediments and sedimentary rocks, the water pressure between grains is reduced, and the weight of the overlying materials causes the grains to pack more closely together, resulting in *subsidence* of the ground. As greater amounts of groundwater are pumped to meet the increasing needs of agriculture, industry, and population growth, subsidence is becoming more prevalent, and an issue that is receiving increasing attention.

The San Joaquin Valley of California is a major agricultural region that relies largely on groundwater for irrigation. Between 1925 and 1977, groundwater withdrawals in parts of the valley caused subsidence of nearly 9 m (Figure 13.16). Other areas in the United States that have experienced subsidence due to groundwater withdrawal are New Orleans, Louisiana, and Houston, Texas, both of which have subsided more than 2 m, and Las Vegas, Nevada, where almost 9 m of subsidence has taken place.

Looking elsewhere in the world, the tilt of the Leaning Tower of Pisa in Italy is partly due to groundwater withdrawal (Figure 13.17). The tower started tilting soon after construction began in 1173 because of differential compaction of the foundation. During the 1960s, the city of Pisa

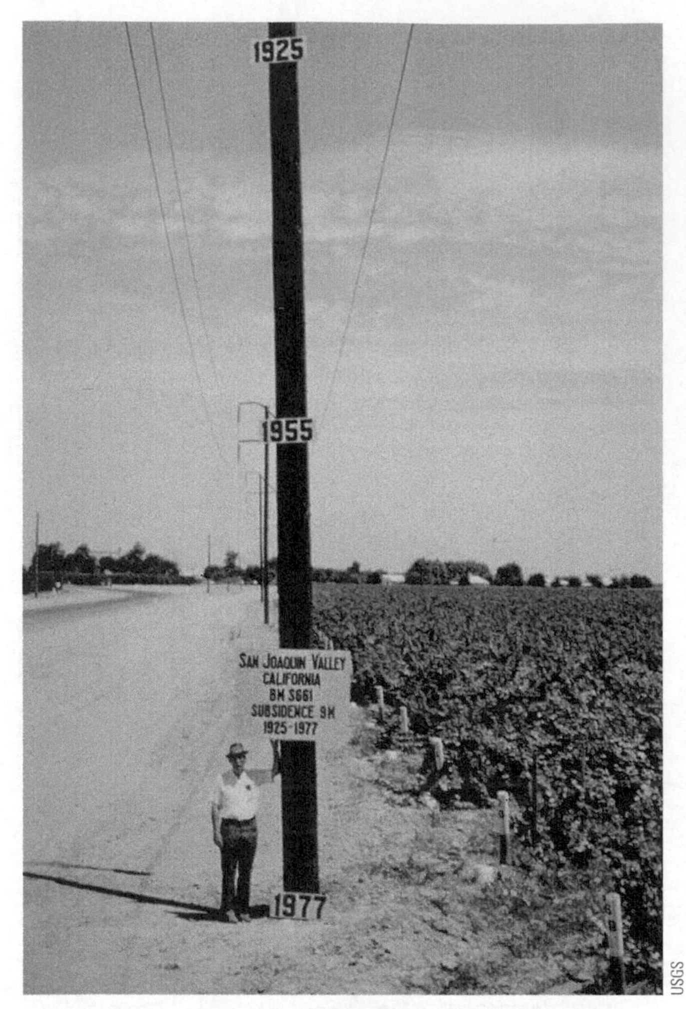

USGS

Figure 13.16 Subsidence in the San Joaquin Valley, California The dates on this power pole dramatically illustrate the amount of subsidence in the San Joaquin Valley, California. Because of groundwater withdrawals and subsequent sediment compaction, the ground subsided nearly 9 m between 1925 and 1977. For a time, surface water use reduced subsidence, but during the drought of 1987 to 1992, it started again as more groundwater was withdrawn.

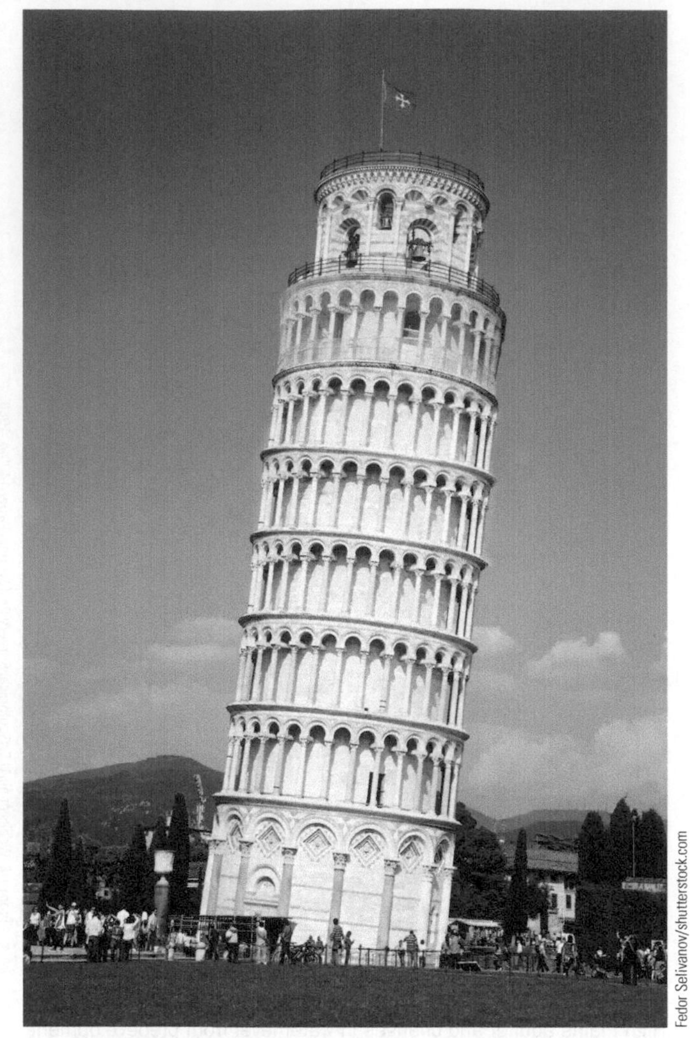

Fedor Selivanov/shutterstock.com

Figure 13.17 The Leaning Tower of Pisa, Italy The tilting is partly the result of subsidence due to the removal of groundwater. Strict control of groundwater withdrawal, stabilization of the foundation, and renovation of the structure itself have ensured that the Leaning Tower will continue leaning for many more centuries.

withdrew ever-greater amounts of groundwater, causing the ground to subside further; as a result, the tilt of the tower increased until it was in danger of falling over. Strict control of groundwater withdrawal, stabilization of the foundation, and renovations have reduced the amount of tilting to about 1 mm per year, thus ensuring that the tower should stand for several more centuries.

Mexico City, which is built on a former lakebed, has continuing subsidence problems. As groundwater is removed for the increasing needs of the city's 19.3 million people, the water table has been lowered up to 10 m. As a result, the fine-grained lake deposits are compacting, and Mexico City is slowly and unevenly subsiding (Figure 13.18). The fact that 72% of the city's water comes from the aquifer beneath the metropolitan area ensures that problems of subsidence will continue.

The extraction of oil can also cause subsidence. Long Beach, California, has subsided 9 m as a result of many decades of oil production. More than $100 million in damages was done to the pumping, transportation, and harbor facilities in this area because of subsidence and encroachment of the sea (Figure 13.19). Once water was pumped back into the oil reservoir, thus stabilizing it, subsidence virtually stopped.

Groundwater Contamination

A major problem facing our society is the safe disposal of the numerous pollutant by-products of an industrial economy. We are becoming increasingly aware that streams, lakes, and oceans are not unlimited reservoirs for waste and that we must find new, safe ways to dispose of pollutants.

The most common sources of groundwater contamination are sewage, landfills, toxic waste disposal sites, and agriculture. Once pollutants get into the groundwater system, they spread wherever groundwater travels, which can make their containment difficult (Geo-Insight 13.1). Furthermore,

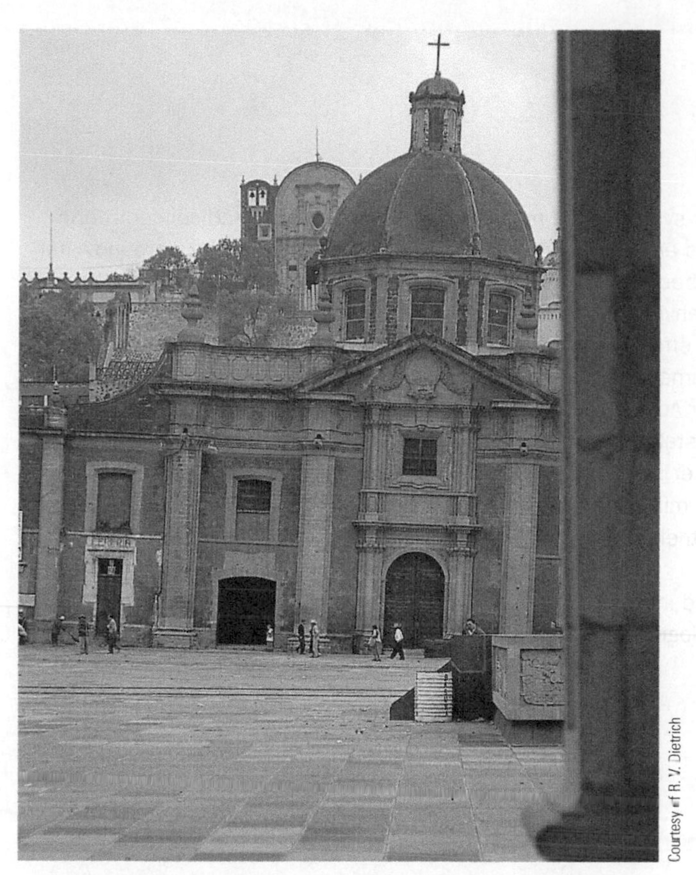

Courtesy of R. V. Dietrich

▮ Figure 13.18 Subsidence in Mexico City Excessive withdrawal of groundwater from beneath Mexico City has resulted in subsidence and uneven settling of buildings. The right side of this church (Our Lady of Guadalupe) has settled slightly more than 1 m.

City of Long Beach Department of Oil Properties

▮ Figure 13.19 Oil Field Subsidence, Long Beach, California The withdrawal of petroleum from the Long Beach, California, oil field resulted in up to 9 m of ground subsidence in some areas because of sediment compaction. In this photograph, note that the ground has settled around the well stems (the white "posts"), leaving the wellheads up above the ground. The levee on the left edge of the photo was built to keep seawater in the adjacent marina from flooding the oil field. It was not until water was pumped back into the reservoir to replace the petroleum that was being extracted that ground subsidence finally ceased.

because groundwater moves so slowly, it takes a long time to cleanse a groundwater reservoir once it has become contaminated.

In many areas, septic tanks are the most common way of disposing of sewage. A septic tank slowly releases sewage into the ground, where it is decomposed by oxidation and microorganisms and filtered by the sediment as it percolates through the zone of aeration. In most situations, by the time the water from the sewage reaches the zone of saturation, it has been cleansed of any impurities and is safe to use (▮ Figure 13.20a). If the water table is close to the surface, or if the rocks are very permeable, however, water entering the zone of saturation may still be contaminated and unfit to use.

Landfills are also potential sources of groundwater contamination (▮ Figure 13.20b). Not only does liquid waste seep into the ground, but also rainwater carries dissolved chemicals and other pollutants down into the groundwater reservoir. Unless the landfill is carefully designed and lined with an impermeable layer such as clay, many toxic compounds such as paints, solvents, cleansers, pesticides, and battery acid will find their way into the groundwater system.

Toxic waste sites where dangerous chemicals are either buried or pumped underground are an increasing source of groundwater contamination. The United States alone must dispose of several thousand metric tons of hazardous chemical waste per year. Unfortunately, much of this waste has been, and still is, being improperly dumped and is contaminating the surface water, soil, and groundwater.

Examples of indiscriminate dumping of dangerous and toxic chemicals can be found in every state. Perhaps the most famous is Love Canal, near Niagara Falls, New York. During the 1940s, the Hooker Chemical Company dumped approximately 19,000 tons of chemical waste into Love Canal. In 1953, Hooker covered one of the dump sites with dirt and sold it for $1.00 to the Niagara Falls Board of Education, which built an elementary school and playground on the site. Heavy rains and snow during the winter of 1976–1977 raised the water table and turned the area into a muddy swamp in the spring of 1977. Mixed with the mud were thousands of toxic, noxious chemicals that formed puddles in the playground, oozed into people's basements, and covered gardens and lawns. Trees, lawns, and gardens began to die, and many of the residents of the area suffered from serious illnesses. The cost of cleaning up the Love Canal site and relocating its residents exceeded $100 million, and the site and neighborhood are now vacant.

Groundwater Quality

Finding groundwater is rather easy because it is present beneath the land surface nearly everywhere, although the depth to the water table varies considerably. But just finding water is not enough. Sufficient amounts in porous and permeable materials must be located if groundwater is to be

GEO INSIGHT 13.1 Arsenic and Old Lace

Many people probably learned that arsenic is a poison from either reading or seeing the play *Arsenic and Old Lace*, written by Joseph Kesselring. In the play, the elderly Brewster sisters poison lonely old men by adding a small amount of arsenic to their homemade elderberry wine.

Arsenic is a naturally occurring toxic semimetal element found in the environment, and several types of cancers, such as bladder, kidney, liver, and lung, have been linked to arsenic in water. Arsenic also harms the central and peripheral nervous systems and may cause birth defects and reproductive problems. In fact, because of arsenic's prevalence in the environment and its adverse health effects, Congress included it in the amendments to the Safe Drinking Water Act in 1996. Arsenic is odorless and tasteless, and gets into the groundwater system mainly as arsenic-bearing minerals dissolve in the natural weathering process of rocks and soils.

A map published in 2001 by the United States Geological Survey (USGS) shows the extent and concentration of arsenic in the nation's groundwater supply (▌Figure 1). The highest concentrations of groundwater arsenic were found throughout the West and in parts of the Midwest and Northeast. Although the map is not intended to provide specific information for individual wells or even a locality within a county, it helps researchers and policy makers identify areas of high concentration so that they can make informed decisions about water use. We should point out, however, that a high degree of local

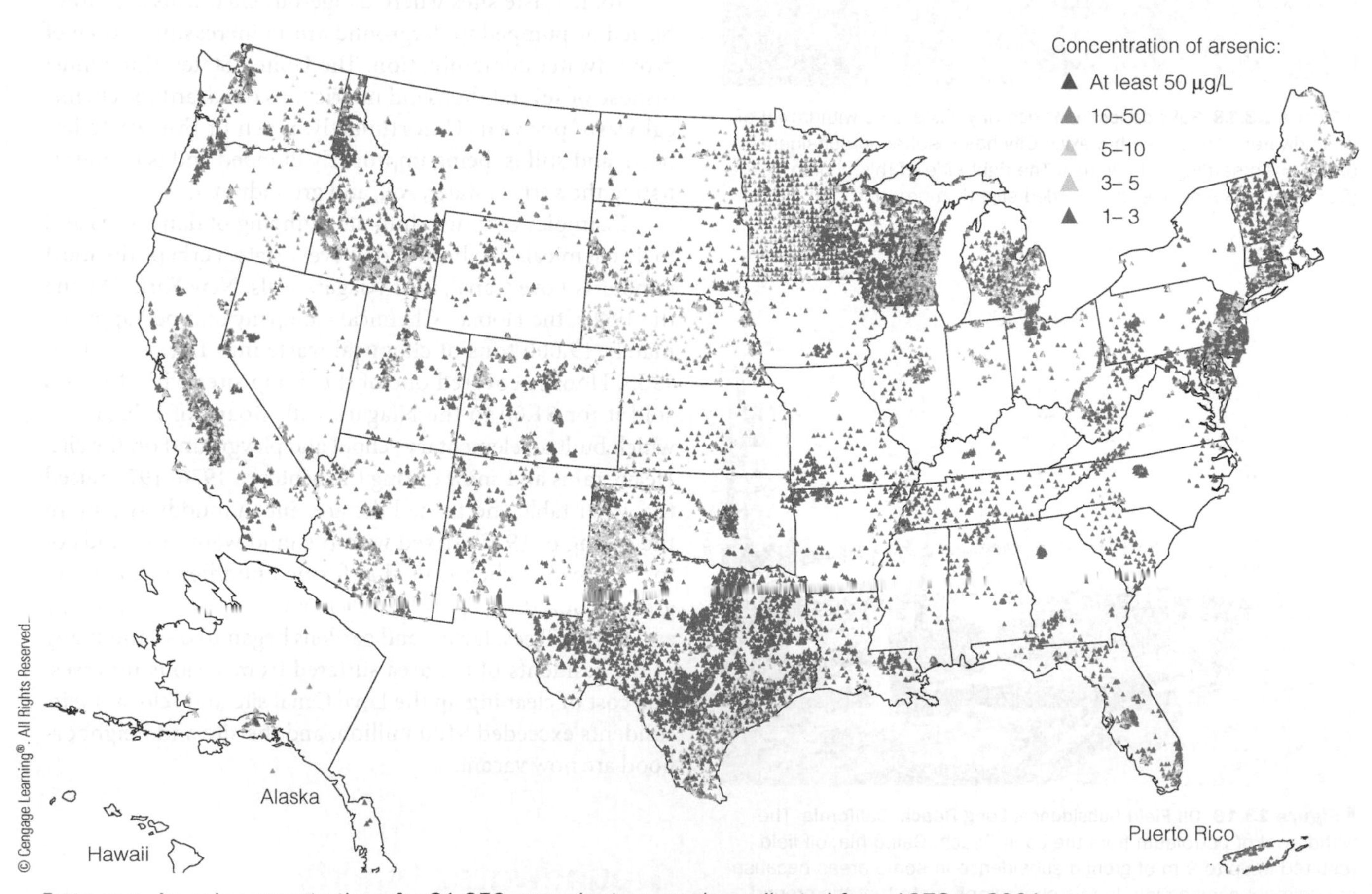

▌**Figure 1** Arsenic concentrations for 31,350 groundwater samples collected from 1973 to 2001.

variability in the amount of arsenic in the groundwater can be caused by local geology, type of aquifer, depth of well, and other factors. The only way to learn the arsenic concentration in any well is to have it tested.

What then is considered a safe level of arsenic in drinking water? In 2001, the United States Environmental Protection Agency (EPA) lowered the maximum level of arsenic permitted in drinking water to 0.010 ppm (parts per million) of arsenic, with a mandate for all United States drinking water systems to meet that level by 2006. Public water supply systems that exceeded the 2001 EPA arsenic standard were required to either treat the water to remove the arsenic or find an alternative supply by 2006. It is possible, however, for qualified public water systems to receive an exemption for an additional three years by their state, and for those qualified public water systems serving 3,300 people or less, to be granted up to an three additional two-year extensions.

On the basis of the data in Figure 1, additional maps of arsenic concentrations were created. ▮ Figure 2 shows which counties exceed various arsenic concentrations in at least 25% of the groundwater samples collected per county. Based on these data, approximately 10% of the samples in the USGS study exceed the EPA standard of 0.010 ppm of arsenic (= 10 μg [10 micrograms] or 10 μg/L [10 microgram per liter]) in drinking water.

Although reducing the acceptable level of arsenic in drinking water will surely increase the cost of water to consumers, it will also decrease their exposure to arsenic and the possible adverse health effects associated with this toxic element.

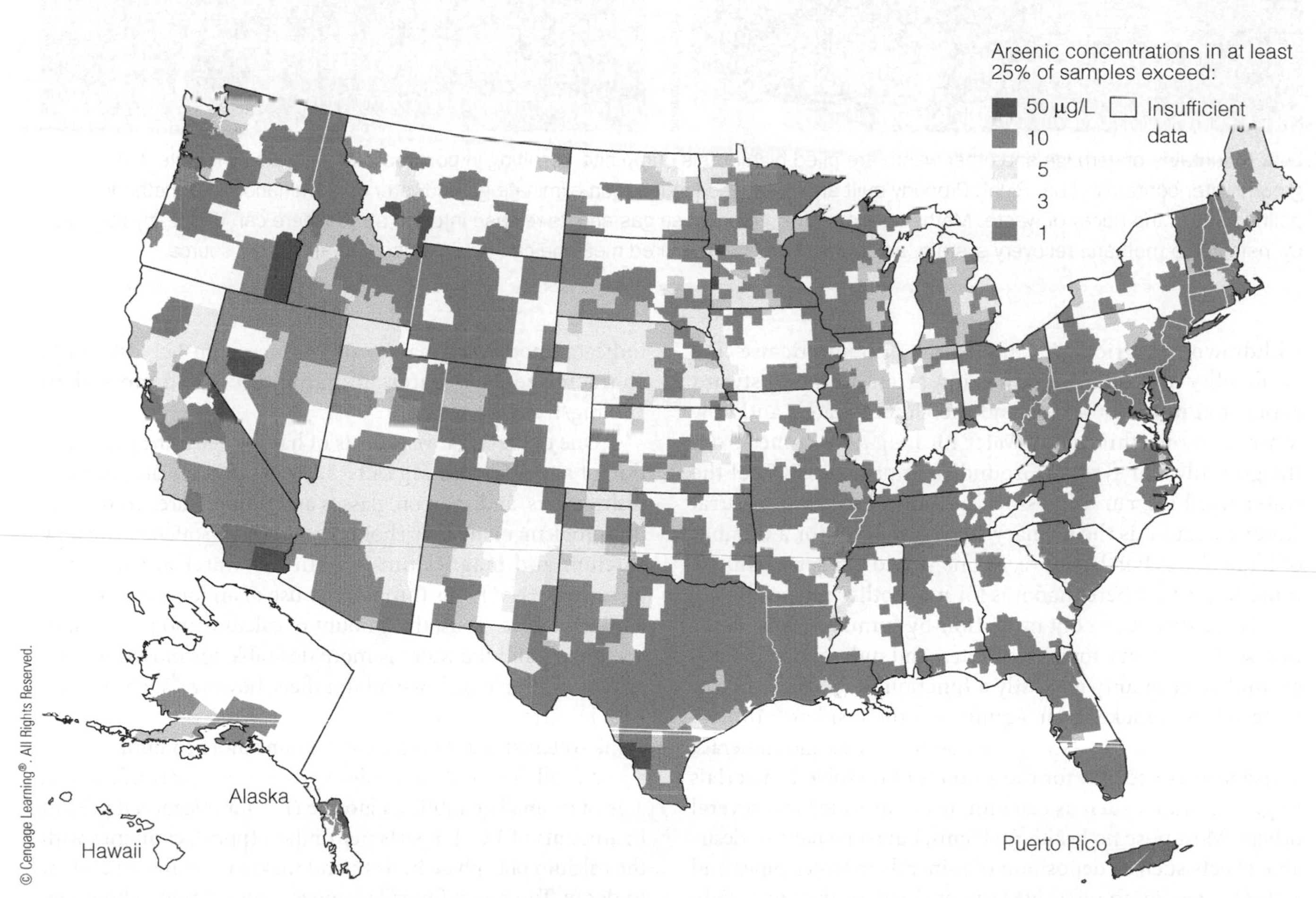

▮ **Figure 2** Arsenic concentrations found in at least 25% of groundwater samples per county. The map is based on 31,350 groundwater samples collected between 1973 and 2001.

Building an Environmentally Friendly Landfill

Americans generate tremendous amounts of waste. Some of this waste, such as battery acid, paint, cleaning agents, insecticides, and pesticides, can easily contaminate the groundwater system. Your community is planning to construct a landfill to contain waste products, but it simply wants to dig a hole, dump the waste in, and then bury it. What are the potential problems associated with this approach? How would you design a landfill to not only contain the waste products but also serve as a potential source of methane gas that can be piped to the local gas-fired electricity-generating plant in your county?

puwanai/Shutterstock.com

Jim West/Alamy

Left: Mountains of garbage and other waste are piled high at this dump site, resulting in polluted surface water and potential groundwater contamination. Right: Properly built and managed landfills can eliminate groundwater contamination and methane pollution from the decay of waste. Methane is a known greenhouse gas and its release into the atmosphere can be greatly reduced by installing a methane recovery system, as shown here. The captured methane can thus be used as an energy source.

withdrawn for agricultural, industrial, or domestic use. The availability of groundwater was important in the westward expansion in both Canada and the United States, and now more than one-third of all water for irrigation comes from the groundwater system. Groundwater provides 80% of the water used for rural livestock and domestic use in rural America, and it is the primary source of water for a number of large cities. Furthermore, as one would expect, quality is more important here than it is for most other purposes.

If we discount contamination by humans from landfills, septic systems, toxic waste sites, and industrial effluents, groundwater quality is mostly a function of (1) the kinds of materials that make up an aquifer, (2) the residence time of water in an aquifer, and (3) the solubility of rocks and minerals. These factors account for the amount of dissolved materials in groundwater, such as calcium, iron, fluoride, and several others. Most pose no health problems, but some have undesirable effects such as deposition of minerals in water pipes and water heaters or an offensive taste or smell, or they may stain clothing and fixtures or inhibit the effectiveness of detergents.

A good example that everyone is familiar with is *hard water*, which is a problem in many areas, especially those underlain by limestone and dolostone. Hard water is caused by dissolved calcium (Ca^{+2}) and magnesium (Mg^{+2}) ions. Water containing less than 60 milligrams of Ca^{+2} and Mg^{+2} per liter (mg/L) is considered soft, whereas 61–120 mg/L indicates moderately hard water, values from 121–180 mg/L characterize hard water, and any water with more than 180 mg/L is very hard.

One of the negative aspects of hard water is the precipitation of scale (Ca and Mg salts) in water pipes, water heaters, dishwashers, and even on glasses and dinnerware. To remedy this problem, many households have a water softener, whereby calcium and magnesium ions in the water are replaced by sodium (Na^{+}) ions through the use of an ion exchanger or a mineral sieve. Thus, the amount of calcium and magnesium is reduced, and the water is more desirable for most domestic purposes. People on low-sodium diets, however, such as those with hypertension (high blood pressure), are cautioned not to drink softened water because it contains more sodium.

Not all dissolved materials in groundwater are undesirable, at least in small quantities. Fluoride (F^{-}), for instance, if present in amounts of 1.0–1.5 parts per million (ppm), combines with the calcium phosphate in teeth and makes them more resistant to decay. Too much fluoride—more than 4.0 ppm—however, gives children's teeth a dark, blotchy appearance.

Fluoride in natural waters is rare, so few communities benefit from its presence. Thus, many cities and towns add fluorine to their drinking water, and many dentists routinely use fluoride treatments, so that the population as a whole has seen a reduction in cavities and tooth decay. Despite the beneficial effect of fluoridation in reducing tooth decay, there

Figure 13.20 Groundwater Contamination

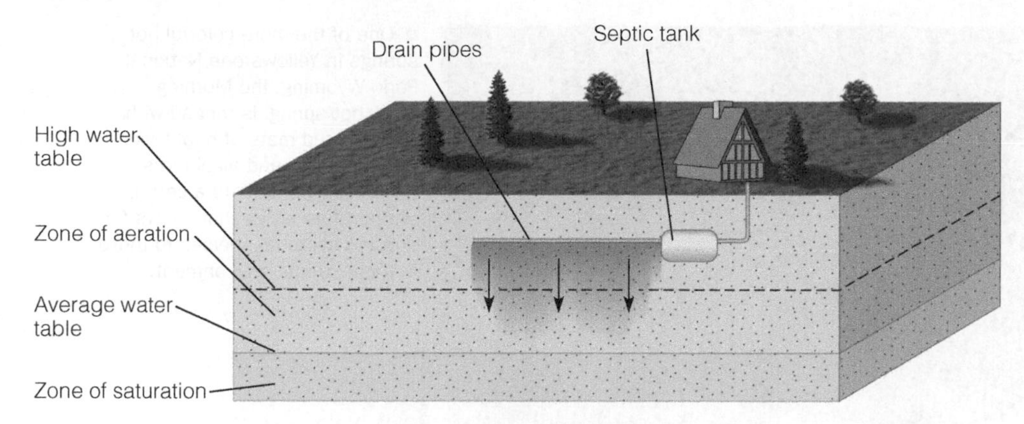

a A septic system slowly releases sewage into the zone of aeration. Oxidation, bacterial degradation, and filtering usually remove impurities before they reach the water table. However, if the rocks are very permeable or the water table is too close to the septic system, contamination of the groundwater can result.

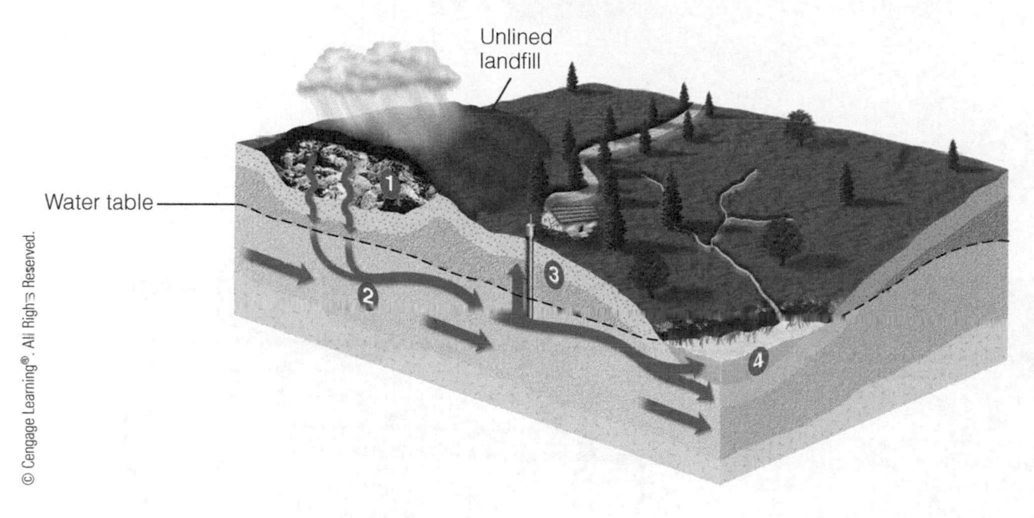

b Unless there is an impermeable barrier between a landfill and the water table, pollutants can be carried into the zone of saturation and contaminate the groundwater supply: (1) Infiltrating water leaches contaminants from the landfill; (2) the polluted water enters the water table and moves away from the landfill; (3) wells may tap the polluted water and thus contaminate drinking water supplies; and (4) the polluted water may emerge into streams and other water bodies downslope from the landfill.

has been an increase in opposition to fluoridating municipal water supplies because of perceived possible health risks.

13.9 Hydrothermal Activity

Hydrothermal is a term referring to hot water. Some geologists restrict the meaning to include only water heated by magma, but here we use it to refer to any hot subsurface water and the surface activity that results from its discharge. One manifestation of hydrothermal activity in areas of active or recently active volcanism is the discharge of gases, such as steam, at vents known as *fumeroles* (see Figure 5.3a). Of more immediate concern here, however, is the groundwater that rises to the surface as *hot springs* or *geysers*. It may be heated by its proximity to magma or by Earth's geothermal gradient because it circulates deeply.

Hot Springs

A **hot spring** (also called a *thermal spring* or *warm spring*) is any spring in which the water temperature is higher than 37°C, the temperature of the human body (▌Figure 13.21a). Some hot springs are much hotter, with temperatures up to the boiling point in many instances (▌Figure 13.21b). Another type of hot spring, called a *mud pot*, results when chemically altered rocks yield clays that bubble as hot water and steam rise through them (▌Figure 13.21c). Of the approximately 1,100 known hot springs in the United States, more than 1,000 are in the far West, with the others in the Black Hills of South Dakota, Georgia, the Ouachita region of Arkansas, and the Appalachian region.

Hot springs are also common in other parts of the world. One of the most famous is in Bath, England, where shortly after the Roman conquest of Britain in AD 43, numerous bathhouses and a temple were built around the hot springs (▌Figure 13.22).

The heat for most hot springs comes from magma or cooling igneous rocks. The geologically recent igneous activity in the western United States accounts for the large number of hot springs in that region. The water in some hot springs, however, circulates deep into Earth, where it is warmed by the normal increase in temperature, the geothermal gradient.

ConnectionLink

To review what the geothermal gradient is, and estimates of Earth's internal temperature at various depths, see Chapter 9.

Figure 13.21 Hot Springs

a One of the more colorful hot springs in Yellowstone National Park, Wyoming, the Morning Glory hot spring, is fringed with multicolored mats of heat-loving cyanobacteria and algal mats. Each color represents a certain temperature range that allows for specific bacterial species to thrive in this extreme environment.

b The water in this hot spring at Bumpass Hell in Lassen Volcanic National Park, California, is boiling.

c Mud pot at the Sulfur Works, also in Lassen Volcanic National Park.

d The United States National Park Service warns of the dangers in hydrothermal areas, but some people ignore the warnings and are injured or killed.

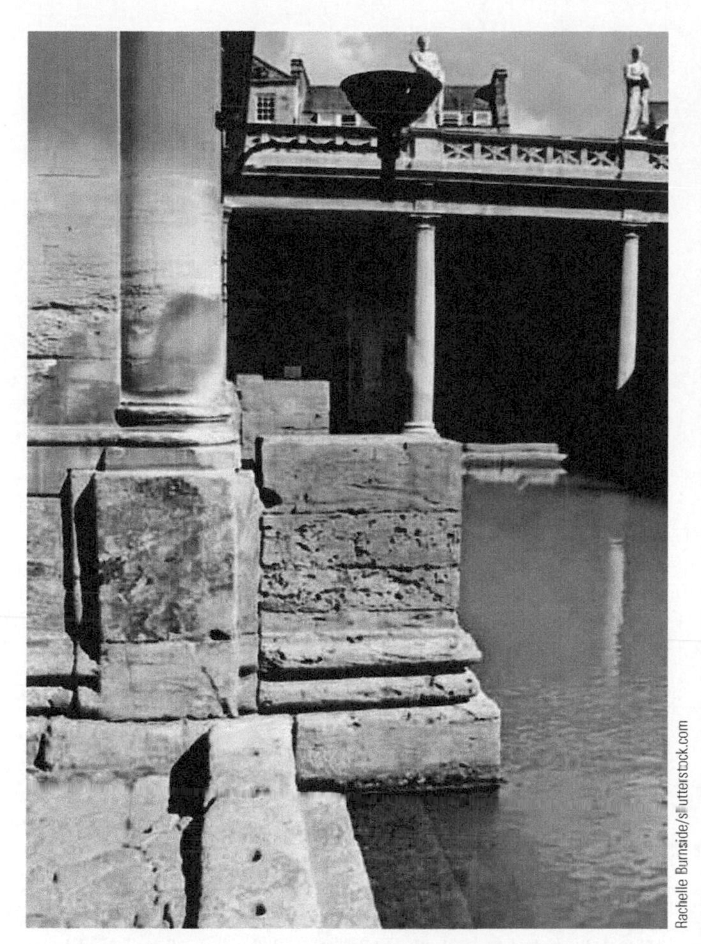

Rachelle Burnside/shutterstock.com

Figure 13.22 Bath, England One of the many bathhouses in Bath, England, that were built around hot springs shortly after the Roman conquest in AD 43.

James S. Monroe

Figure 13.23 Old Faithful Geyser Old Faithful Geyser in Yellowstone National Park, Wyoming, is one of the world's most famous geysers, erupting faithfully every 30–90 minutes and spewing water 32–56 m high.

For example, the spring water of Warm Springs, Georgia, is heated in this manner. This hot spring was a health and bathing resort long before the Civil War (1861–1865); later, with the establishment of the Georgia Warm Springs Foundation, it was used to help treat polio victims.

Geysers

Hot springs that intermittently eject hot water and steam with tremendous force are known as **geysers**. The word comes from the Icelandic geysir, "to gush" or "to rush forth." One of the most famous geysers in the world is Old Faithful in Yellowstone National Park, Wyoming (Figure 13.23). With a thunderous roar, it erupts a column of hot water and steam every 30–90 minutes. Other well-known geyser areas are found in Iceland and New Zealand.

Geysers are the surface expression of an extensive underground system of interconnected fractures within hot igneous rocks (Figure 13.24). Groundwater percolating down into the network of fractures is heated as it comes into contact with the hot rocks. Because the water near the bottom of the fracture system is under higher pressure than the water near the top, it must be heated to a higher temperature before it will boil. Thus, when the deeper water is heated to near the boiling point, a slight rise in temperature or a drop in pressure, such as from escaping gas, will instantly change it to steam. The expanding steam quickly pushes the water above it out of the ground and into the air, producing a geyser eruption. After the eruption, relatively cool groundwater starts to seep back into the fracture system where it heats to near its boiling temperature and the eruption cycle begins again. Such a process explains how geysers can erupt with some regularity.

Hot spring and geyser water typically contains large quantities of dissolved minerals, because most minerals dissolve more rapidly in warm water than in cold water. Because of this high mineral content, some people believe that the waters of many hot springs have medicinal properties. Numerous spas and bathhouses have been built at hot springs throughout the world to take advantage of these supposed healing properties.

When the highly mineralized water of hot springs or geysers cools at the surface, some of the material in solution is precipitated, forming various types of deposits. The amount

Figure 13.24 Anatomy of a Geyser

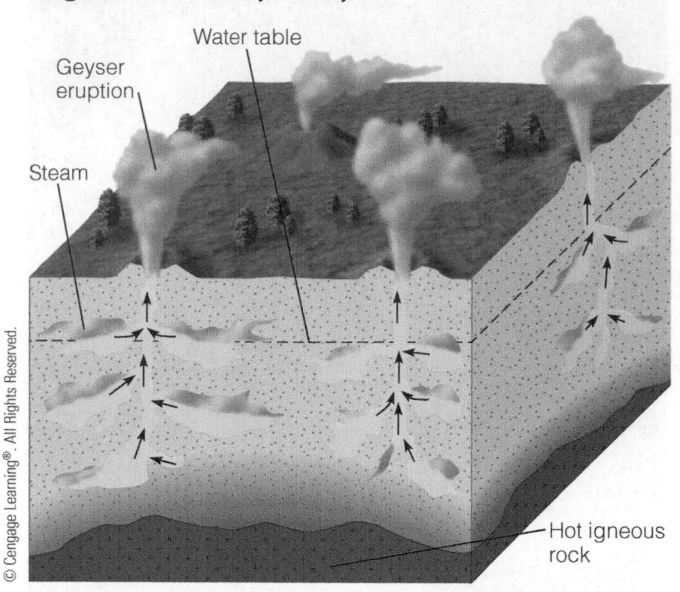

a The eruption of a geyser starts when groundwater percolates down into a network of interconnected openings and is heated by the hot igneous rocks. The water near the bottom of the fracture system is under higher pressure than the water near the top and consequently must be heated to a higher temperature before it will boil.

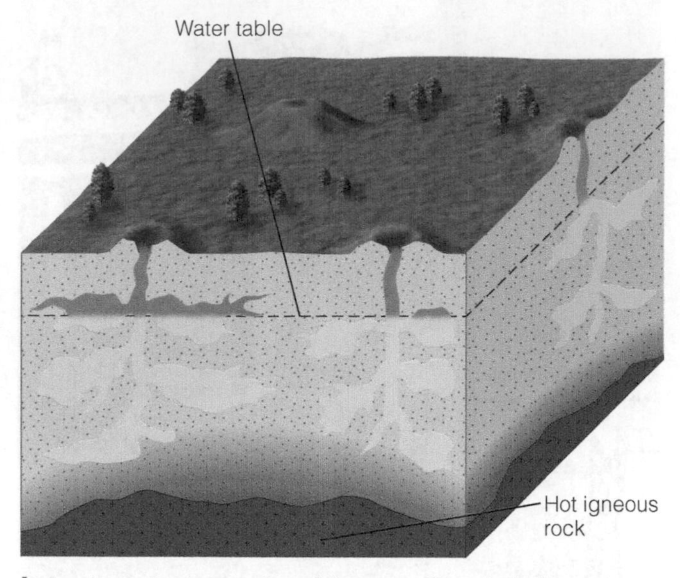

b Any rise in the temperature of the water above its boiling point or a drop in pressure will cause the water to change to steam, which quickly pushes the water above it up and out of the ground, producing a geyser eruption.

and type of precipitated minerals depend on the solubility and composition of the material that the groundwater flows through. If the groundwater contains dissolved calcium carbonate ($CaCO_3$), then *travertine* or *calcareous tufa* (both of which are varieties of limestone) are precipitated. Spectacular examples of hot spring travertine deposits are found at Pamukhale in Turkey and at Mammoth Hot Springs in Yellowstone National Park (Figure 13.25a). Groundwater containing dissolved silica will, upon reaching the surface, precipitate a soft, white, hydrated mineral called *siliceous sinter* or *geyserite*, which can accumulate around a geyser's opening (Figure 13.25b).

Figure 13.25 Hot Spring Deposits in Yellowstone National Park, Wyoming

a Minerva Terrace, formed when calcium-carbonate-rich hot-spring water cooled, precipitating travertine.

Geothermal Energy

Geothermal energy is any energy produced from Earth's internal heat. In fact, the term *geothermal* comes from *geo*, "Earth," and *thermal*, "heat." Several forms of internal heat

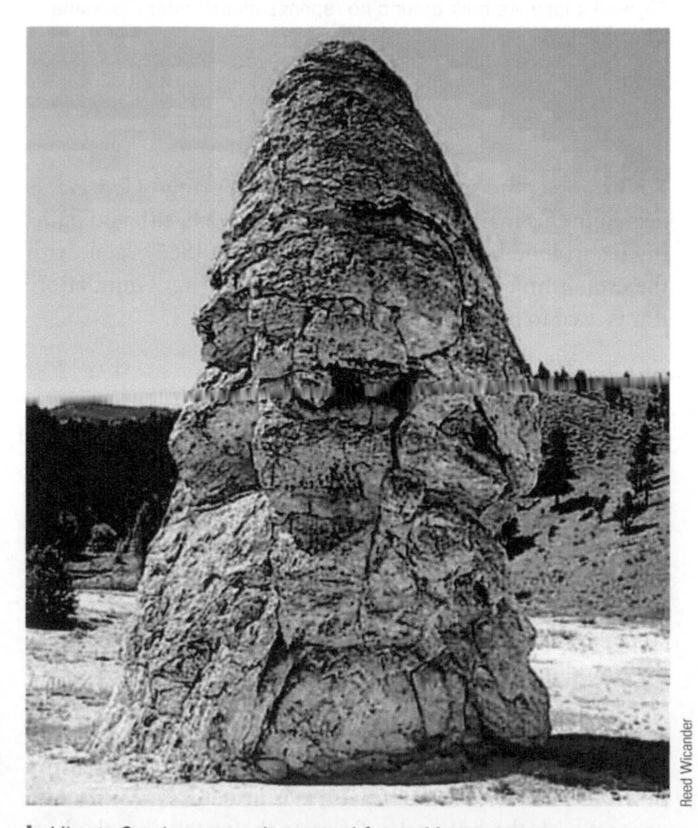

b Liberty Cap is a geyserite mound formed by numerous geyser eruptions of silicon-dioxide-rich hot spring water.

are known, such as hot dry rocks and magma, but so far, only hot water and steam are used.

As oil reserves decline, geothermal energy is becoming an attractive alternative, particularly where it is plentiful, such as in the western United States. Some of the countries currently using geothermal energy in one form or another are the United States, which led the world in geothermal electricity production in 2012, Iceland, Mexico, Italy, New Zealand, Japan, the Philippines, and Indonesia.

In the United States, the first commercial geothermal electricity-generating plant was built in 1960, and is still in operation, at The Geysers, about 120 km north of San Francisco, California. Here, wells were drilled into the numerous near-vertical fractures underlying the region. As pressure on the rising groundwater decreases, the water changes to steam, which is piped directly to electricity-generating turbines and generators (▌Figure 13.26).

© John Karachewski/Geoscapes Photography

▌**Figure 13.26** The Geysers, Sonoma County, California Steam rising from one of the geothermal power plants at The Geysers in Sonoma County, California. Steam from wells drilled into this geothermal region, about 120 km north of San Francisco, is piped directly to electricity-generating turbines to produce electricity that is distributed throughout the area.

Critical Thinking Question Although geothermal energy is a relatively nonpolluting form of energy used as a source of heat and to generate electricity, why is it typically a more expensive form of energy? (Hint: Which contains more dissolved minerals, hot or cold water?)

Key Concepts Review

- Groundwater is part of the hydrologic cycle and an important natural resource. It consists of all subsurface water trapped in the pores and other open spaces in rocks, sediment, and soil.
- Porosity is the percentage of a material's total volume that is pore space. Permeability is the capacity to transmit fluids, and is dependent on porosity, but also on the size of the pores or fractures and their interconnections.
- The water table is the surface separating the zone of aeration (in which the pores are filled with air and water) from the underlying zone of saturation (in which the pores are filled with water). The water table is a subdued replica of the overlying land surface in most places.
- Groundwater moves slowly downward under the influence of gravity through the zone of aeration to the zone of saturation. Some of it then moves along the surface of the water table, and the rest moves from areas of high pressure to areas of low pressure.
- Groundwater velocity varies greatly and depends on a number of factors. Generally, the average velocity of groundwater is a few centimeters per day.
- Springs are found wherever the water table intersects the ground surface. Some springs are the result of a perched water table—that is, a localized aquiclude within an aquifer and above the regional water table.
- Water wells are openings made by digging or drilling down into the zone of saturation. When water is pumped from a well, the water table in the area around the well is lowered, forming a cone of depression.
- In an artesian system, confined groundwater builds up high hydrostatic pressure. For an artesian system to develop, an aquifer must be confined above and below by aquicludes; the aquifer is usually tilted so that it can build up hydrostatic pressure; and the aquifer must be exposed at the surface so that it can be recharged.

- Karst topography develops by groundwater erosion in many areas underlain by soluble rocks. It is characterized by sinkholes, caves, solution valleys, and disappearing streams.
- Caves form when groundwater in the zone of saturation weathers and erodes soluble rock such as limestone. Common cave deposits include stalactites, stalagmites, columns, drip curtains, and travertine terraces.
- Modification of the groundwater system can cause serious problems such as lowering of the water table, saltwater incursion, subsidence, and contamination.
- Groundwater quality is mostly a function of the kinds of materials that make up an aquifer, the residence time of water in an aquifer, and the solubility of rocks and minerals.
- Contamination by humans is becoming a serious problem and can result from landfills, septic systems, toxic waste sites, and industrial effluents, all of which affect the quality of the groundwater.
- Hydrothermal refers to hot water, typically heated by magma, but also resulting from Earth's geothermal gradient as it circulates deeply beneath the surface. Manifestations of hydrothermal activity include fumaroles, hot springs, and geysers.
- Geothermal energy is energy produced from Earth's internal heat and comes from the steam and hot water trapped within Earth's crust. It is a relatively nonpolluting form of energy that is used as a source of heat and to generate electricity.

Important Terms

aquifer (p. 311)
artesian system (p. 314)
cave (p. 317)
cone of depression (p. 313)
geothermal energy (p. 332)
geyser (p. 331)
groundwater (p. 310)
hot spring (p. 329)
hydrothermal (p. 329)
karst topography (p. 316)
permeability (p. 310)
porosity (p. 310)
sinkhole (p. 316)
spring (p. 313)
water table (p. 311)
water well (p. 313)
zone of aeration (p. 311)
zone of saturation (p. 311)

Review Questions

1. Which of the following is not an example of groundwater erosion?
   a. _____ karst topography.
   b. _____ stalactites.
   c. _____ sinkholes.
   d. _____ caves.
   e. _____ disappearing streams.
2. The water table is a surface separating the
   a. _____ zone of porosity from the underlying zone of permeability.
   b. _____ capillary fringe from the underlying zone of aeration.
   c. _____ zone of aeration from the underlying zone of saturation.
   d. _____ capillary fringe from the underlying zone of aeration.
   e. _____ zone of saturation from the underlying zone of aeration.
3. The porosity of Earth materials is defined as
   a. _____ their ability to transmit fluids.
   b. _____ the depth of the zone of saturation.
   c. _____ the percentage of void space.
   d. _____ their solubility in the presence of weak acids.
   e. _____ the temperature of groundwater.
4. A cone of depression forms when
   a. _____ a stream flows into a sinkhole.
   b. _____ water in the zone of aeration is replaced by water from the zone of saturation.
   c. _____ a spring forms where a perched water table intersects the surface.
   d. _____ water is withdrawn from a well faster than it can be replaced.
   e. _____ the ceiling of a cave collapses, forming a steep-sided crater.

5. Which of the following conditions must exist for an artesian system to form?
   a. _____ Groundwater must circulate near magma.
   b. _____ The water table must be at or very near the surface.
   c. _____ The rocks below the surface must be especially resistant to solution.
   d. _____ Water must rise very high in the capillary fringe.
   e. _____ An aquifer must be confined above and below by aquicludes.
6. Explain how groundwater weathers and erodes Earth materials.
7. Why should we be concerned about how fast the groundwater supply is being depleted in some areas?
8. What is the difference between an artesian system and a water well? Is there any difference between the water obtained from an artesian system and the water from a water well?
9. One concern geologists have about burying nuclear waste in present-day arid regions such as Nevada is that the climate may change during the next several thousand years and become more humid, thus allowing more water to percolate through the zone of aeration. Why is this a concern? What would the average rate of groundwater movement have to be during the next 5,000 years to reach canisters containing radioactive waste buried at a depth of 400 m?
10. **Creative Thinking Visual Question:** Withdrawal of large quantities of groundwater in the Las Vegas, Nevada, area has resulted in differential subsidence and damage to roads and buildings. What evidence of subsidence is visible in this photograph (▌Figure 1)?

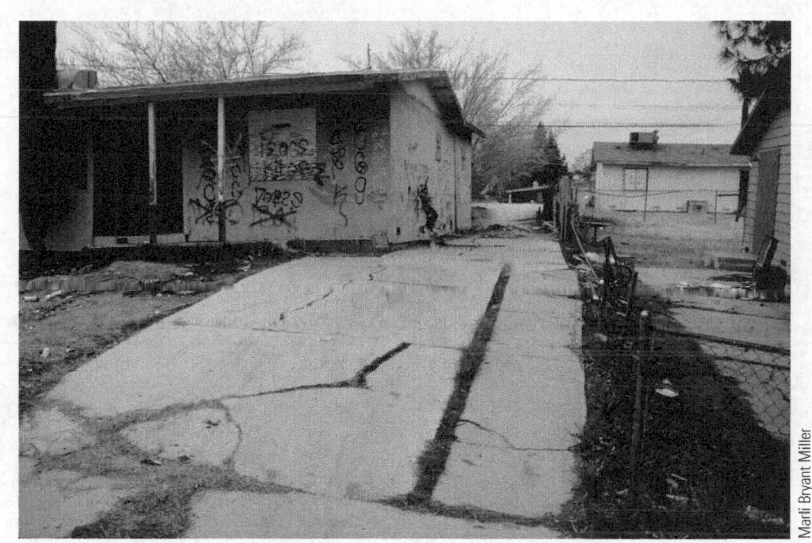

Marli Bryant Miller

▌**Figure 1** Differential subsidence in Las Vegas, Nevada.

Global GeoScience Watch Search for "hydraulic fracturing" within the GREENR database, and under the list of subjects in the left-hand column click on "Hydraulic Fracturing." Then click on "VIEW ALL" next to the "News" heading and browse some of the recent articles discussing the use of hydraulic fracturing outside of the United States, particularly in Europe. The use of hydraulic fracturing is currently one of the most contentious environmental issues in the United States. In Europe, despite reserves estimated at 10 percent of the world's gas shales, they have not begun using hydraulic fracturing to exploit these deposits. What are some of the reasons for this? What is the situation in regards to the use of hydraulic fracturing in other parts of the world?

The Bering Glacier in Alaska is the longest glacier in North America. At its greatest extent in 1996, it measured 190 km long, but since 2005 its terminus has retreated about 4.8 km and the glacier has thinned by more than 60 m. Only about 2% of all water on Earth is in glaciers, and most of that is in Antarctica and Greenland. Nevertheless, smaller glaciers like this one are important reservoirs in the hydrologic cycle.

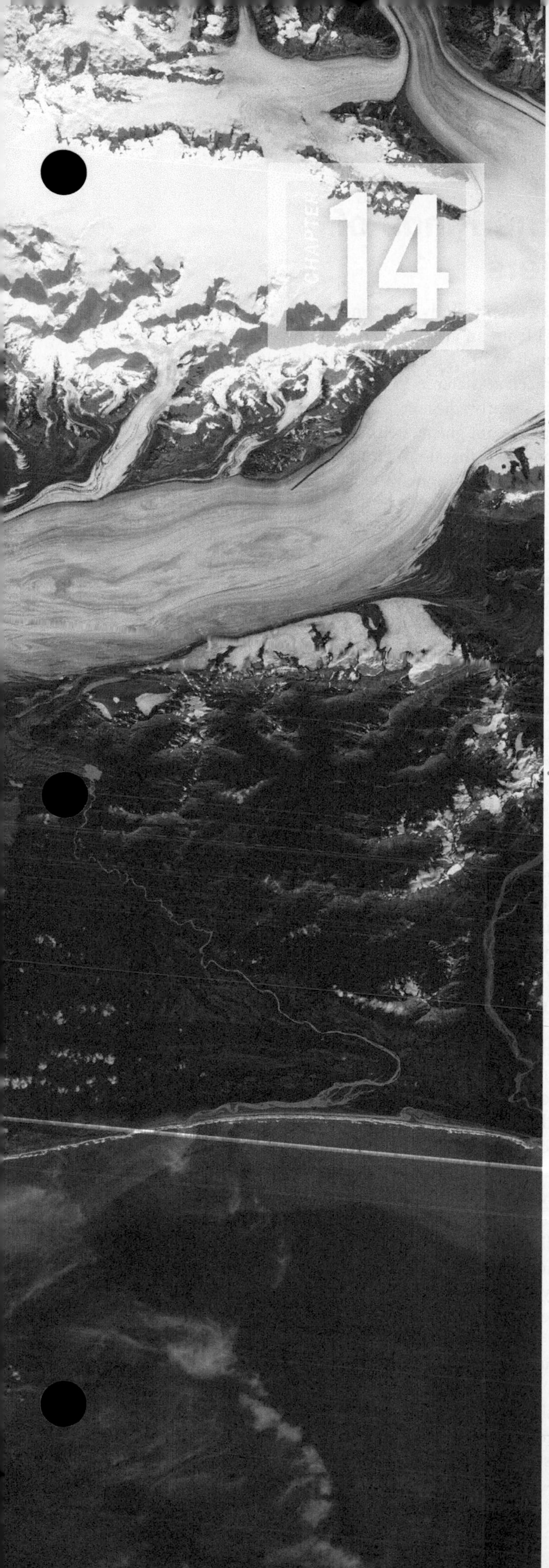

NASA Earth Observatory

Glaciers and Glaciation

OUTLINE

HAVE YOU EVER WONDERED?

- What glaciers are and where they are found?
- Why glaciers come in different sizes and shapes?
- How glaciers move?
- What controls the length and thickness of glaciers?
- How glaciers transport and deposit sediment?
- What kinds of landscapes result from glacial erosion?
- What the Ice Age was and what caused it?

14.1 Introduction

During the Ice Age, or what geologists call the Pleistocene Epoch, vast glaciers formed and expanded and small glaciers in mountain valleys were much more numerous and larger than they are now (see the chapter opening photo). The Ice Age was brief in the context of geologic time, having lasted from 2.6 million to 11,700 years ago, although the glaciers waxed and waned several times during this interval. Since then, Earth has gone through several climatic changes. During the Holocene Maximum (5,000 to 7,000 years ago), average temperatures were slightly higher than they are now, and some of today's arid regions were much more humid. Today, the only arable land in Egypt is along the Nile River, but during the Holocene Maximum, much of North Africa was covered by grasslands, swamps, and lakes.

Following the Holocene Maximum was a time of cooler temperatures, but from about AD 1000 to 1300, Europe went through the Medieval Warm Period when wine grapes grew 480 km farther north than they do now. Then a cooling trend began in about AD 1300 that led to the **Little Ice Age,** which lasted from 1500 to the middle or late 1800s. During this time, glaciers expanded to their greatest historic extent (Geo-Insight 14.1); summers were cooler and wetter; winters were colder; and sea ice persisted longer around Greenland, Iceland, and the Canadian Arctic islands. The cooler, wetter summers were a problem because the growing seasons were shorter, resulting in several famines.

During the coldest part of the Little Ice Age (1680–1730), the growing season in England was five weeks shorter than during the 1900s. Occasionally, Eskimos following the southern edge of the sea ice paddled their kayaks as far south as Scotland, and the canals in Holland froze over in some winters. In 1607, the first Frost Fair was held in London, England, on the Thames River, which began to freeze over nearly every winter. In the late 1700s, New York Harbor froze over, and 1816 is known as the "year without a summer" when unusually cold temperatures persisted into June and July in New England and northern Europe (the eruption of Tambora in 1815 contributed to the cold spring and summer of 1816).

Many of you have probably heard of the Ice Age and have some idea of what a glacier is, but you may not know about the dynamics of glaciers, how they form, and what may cause ice ages. In any case, *glaciers* are moving bodies of ice on land that are particularly effective at erosion, sediment transport, and deposition. They deeply scour the surfaces they move over, producing many easily recognizable landforms, and they deposit huge amounts of sediment.

Why study glaciers? If for no other reason, glaciers have produced some of the most stunning scenery on Earth, especially in mountains that are or were eroded by glaciers. And of course, glaciers are part of the hydrologic cycle, illustrating again the complex interactions among Earth's systems. In addition, a large part of the water in some countries such as Nepal, Tibet, and Pakistan comes from melting glaciers. Even in the United States and Canada, some areas depend partly on water stored in glaciers. And finally, glaciers are very sensitive to climatic changes, so scientists are interested in fluctuations in glaciers as possible indications of global warming.

14.2 The Kinds of Glaciers

A **glacier** is a moving body of ice on land that flows downslope or outward from an area of accumulation. We will discuss how glaciers flow in a later section. Our definition of a glacier excludes frozen seawater as in the North Polar region and sea ice that forms yearly adjacent to Greenland and Iceland. Drifting icebergs are not glaciers either, although they may have come from glaciers that flowed into lakes or the sea. The critical points in the definition are *moving* and *on land.* Accordingly, permanent snowfields in high mountains, though on land, are not glaciers because they do not move. All glaciers share several characteristics, but they differ enough in size and location for us to define two specific types—valley glaciers and continental glaciers—and several subvarieties.

Valley Glaciers

We use the term **valley glacier** for any glacier confined to a mountain valley where it flows from higher to lower elevations (▌Figure 14.1, chapter opening photo). Some geologists prefer the synonyms *alpine glacier* or *mountain glacier,* but whatever we call them, valley glaciers commonly have tributaries, just as streams and rivers do, thereby forming an interconnected system of glaciers in mountain valleys. In fact, some of the most spectacular scenery in mountains results from erosion by valley glaciers, a topic we cover more fully later in this chapter.

Valley glaciers are common in many of the mountain ranges of the world, such as in western North America, especially in Alaska and Canada, as well as the Andes of South America, the Alps in Europe, and the Himalaya of Asia. Some of the higher peaks in Africa, though near the equator, are high enough to support small valley glaciers. Australia is the only continent with no glaciers of any kind.

A valley glacier's shape is obviously controlled by the valley it occupies, so they tend to be long and narrow. However, where a valley glacier flows from a valley onto a plain and spreads to form a more extensive ice cover, it goes by the name of *piedmont glacier.* So-called *tidewater glaciers* are simply valley glaciers with their terminus in the ocean rather than on land (▌Figure 14.2).

Valley glaciers are rather small when compared with the much larger continental glacier, but even so, they may be several kilometers across, tens of kilometers long, and hundreds of meters thick. Bering Glacier in Alaska at 190 km long is the longest in North America (chapter opening photo). Erosion and deposition by valley glaciers were factors in yielding some of the scenery in several United States and Canadian national parks, most notably Yosemite in California, Glacier in Montana, and Banff-Jasper in Canada.

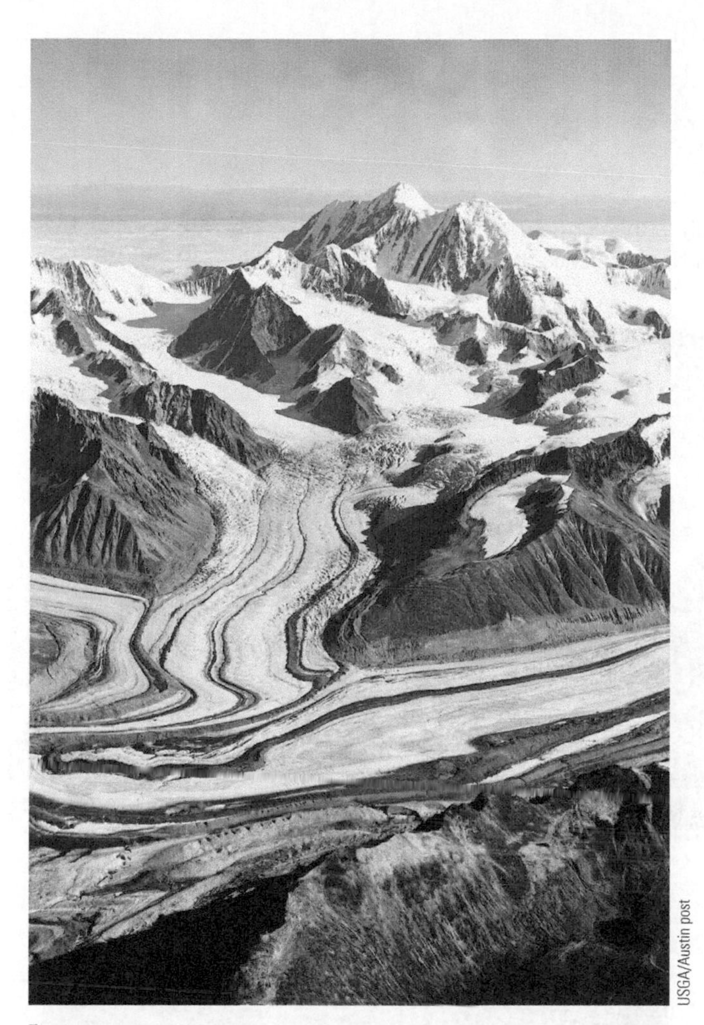

Figure 14.1 **Valley Glaciers in Alaska** Smaller glaciers join to form the Sustina Glacier in Alaska.

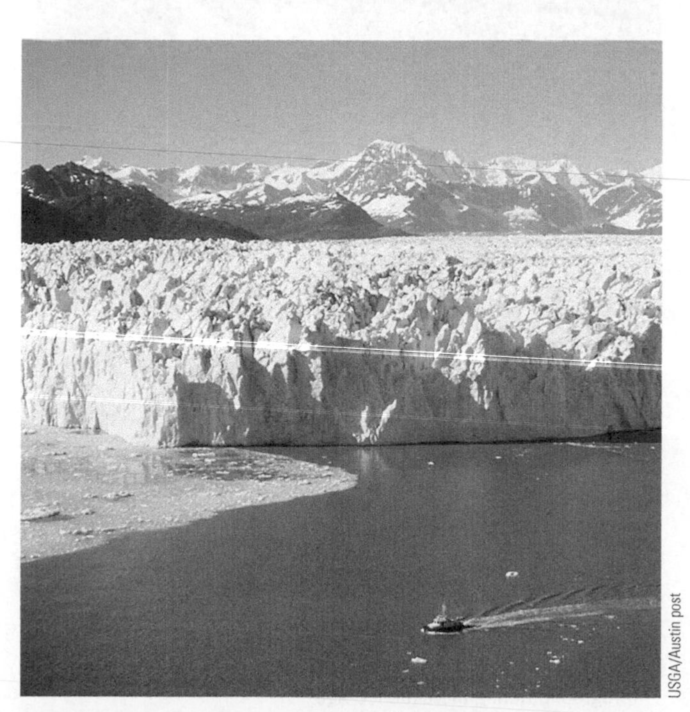

Figure 14.2 **A Tidewater Glacier** Columbia Glacier in Alaska terminates in the sea so it is a tidewater glacier.

Continental Glaciers

In contrast to valley glaciers, **continental glaciers**, also known as *ice sheets,* are vast, covering at least 50,000 km^2, and they are unconfined by topography. That is, their shape and movement are not controlled by the underlying landscape as is the case with valley glaciers, which tend to be long, narrow tongues of ice with the existing slope determining their direction of flow. Continental glaciers flow outward in all directions from a central area or areas of accumulation in response to variations in ice thickness.

Although some valley glaciers are as much as 1,500 m thick, present-day continental glaciers in Greenland and Antarctica are more than 3,000 m thick and cover all but the highest mountains (Figure 14.3a). The continental glacier in Greenland covers about 1,8090,00 km^2, but in Antarctica the East and West Antarctic glaciers merge to form an ice sheet covering more than 12,650,000 km^2. In short, about 90% of all glacial ice on Earth is in Antarctica, with most of the rest in Greenland. Antarctica's glaciers flow into the sea, where the buoyant effect of water causes the ice to float in vast *ice shelves* (Figure 14.3a).

Valley glaciers and continental glaciers are easily differentiated from one another, but there is also an intermediate type of glacier called an **ice cap**. These dome-shaped glaciers are similar to continental glaciers, covering less than 50,000 km^2. The Penny Ice Cap on Baffin Island, Canada, covers about 6,000 km^2, whereas the Vatnajökull Ice Cap in Iceland has an area of 8,100 km^2 (Figure 14.3b). Some ice caps form where valley glaciers grow and overtop the divides and passes between adjacent valleys and coalesce to form a larger ice cover, but some form on rather flat terrain as on some of the islands in the Canadian Arctic.

14.3 Glaciers—Moving Bodies of Ice on Land

The term **glaciation** includes all glacial activity, including the origin, expansion, and retreat of glaciers, as well as their impact on Earth's surface. Presently, glaciers cover nearly 15 million km^2, or about 10% of Earth's land surface. As a matter of fact, if all of Earth's glacial ice were in the United States and Canada, it would form a continuous ice cover about 1.5 km thick.

At first glance, glaciers appear static. Even briefly visiting a glacier may not dispel this impression because, although glaciers move, they usually do so slowly. Nevertheless, they do move, and just like other geologic agents such as running water, glaciers are dynamic systems that continuously adjust to changes. For example, a glacier may flow slower or more rapidly depending on decreased or increased amounts of snow or the absence or presence of water at its base. And glaciers may expand or contract depending on climatic changes.

GEO INSIGHT 14.1 The Little Ice Age

The Little Ice Age was a time of cooler temperatures, shorter growing seasons, and expansion of glaciers, especially in the northern parts of Europe and North America. It began in about 1300 and ended some time during the middle- to late-1800s.

1. 2. The Mer de Glace (Sea of Ice) at present is 7 km long, making it the longest glacier in the French Alps. The painting on the left by Samuel Birmann (1793–1847), a Swiss painter, shows the extent of the glacier not long after the Little Ice Age maximum. Notice the arrows that show the positions of the glacier's margin during the Little Ice Age and in 2000 (right).

Gugelmann Collection, Swiss National Library, Bern

James Tye/Dorling Kindersley/Getty Images

The Trustees of the British Museum/Art Resource, NY

3. The River Thames froze over regularly during the Little Ice Age. The first recorded Frost Fair was held on the Thames in 1608; the last one was in 1814 as shown in this image. Although the river froze over many times between these dates, the most extensive was during the winter of 1683–1684 when it was frozen for two months with ice 28 cm thick.

The Mer de Glace, Chamonix, 1881 (pencil & w/c heightened with white), Severn, Sarah Inger Louise (1812–1891) / Private Collection/ Photo © The Maas Gallery, London/The Bridgeman Art Library

4. This image depicts the Mer de Glace (Sea of Ice), which, at 12 km long, is the longest glacier in the French Alps. During the Little Ice Age it extended about 2,300 m farther than it does now.

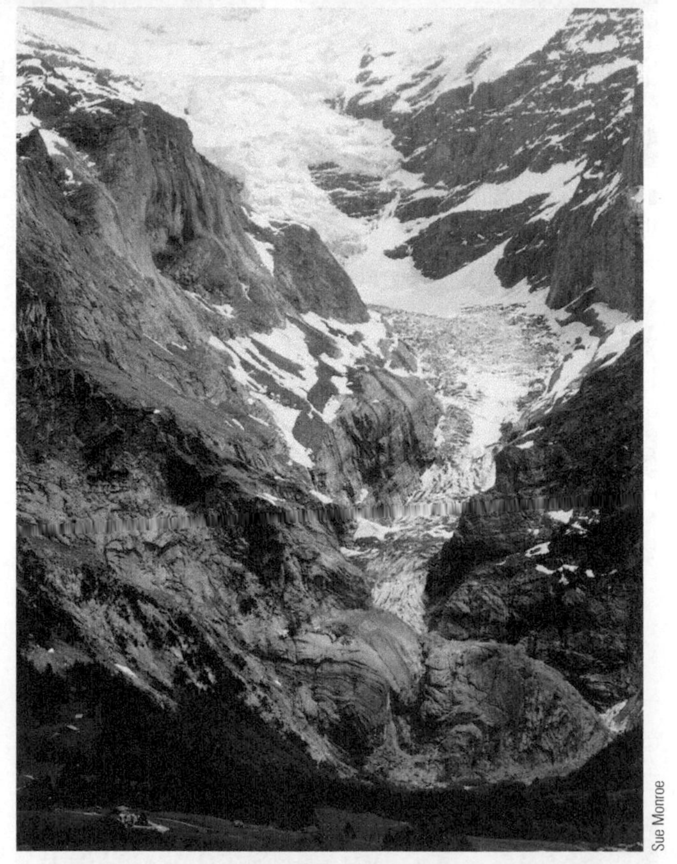

Sue Monroe

5. The terminus of the Unterer Grindelwald in Switzerland is hidden behind a large rock outcrop at the lower end of its valley. This glacier also extended much father into the valley in the foreground during the Little Ice Age.

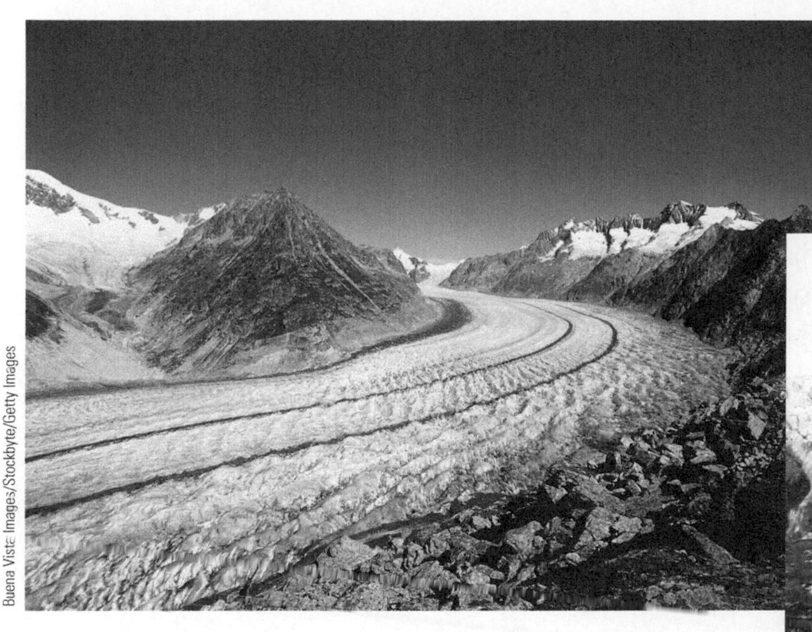

Buena Vista Images/Stockbyte/Getty Images

◄ **6.** The 23-km-long Aletsch Glacier in Switzerland is the country's longest. The glacier was much thicker during the Little Ice Age. Also note the valley directly across the glacier; this once contained a tributary glacier that flowed into the larger glacier.

Photo: Ó. Ingólfsson 2004. http://www3.hi.is/~oi/

► **7.** Gigjökull (Crater Glacier) in Iceland was more extensive during the Little Ice Age, but since 1900 its terminus has retreated. The people in the foreground are on the deposit that marks the maximum extent of the glacier, or what geologists call a terminal moraine. This glacier actually is an outlet from an ice cap that covers the active volcano Eyjafjallajökull, which erupted in April 2010 and disrupted air travel over the North Atlantic. During the eruption, Gigjökull's surface was covered by volcanic ash.

Washington Crossing the Delaware River, 25th December 1776, 1851 (oil on canvas) (copy of an original painted in 1848), Leutze, Emanuel Gottlieb (1816-68)/Metropolitan Museum of Art, New York, USA/ The Bridgeman Art Library

◄ **8.** Most of our information on the Little Ice Age comes from Europe, but the North American colonies were affected by this event, too. In this iconic painting of General George Washington crossing the Delaware River on December 25, 1777, large blocks of ice are shown in the river. On many occasions, the rivers in eastern North America froze over, and sea ice was present in many areas along the coastline.

Figure 14.3 Continental Glaciers and Ice Caps

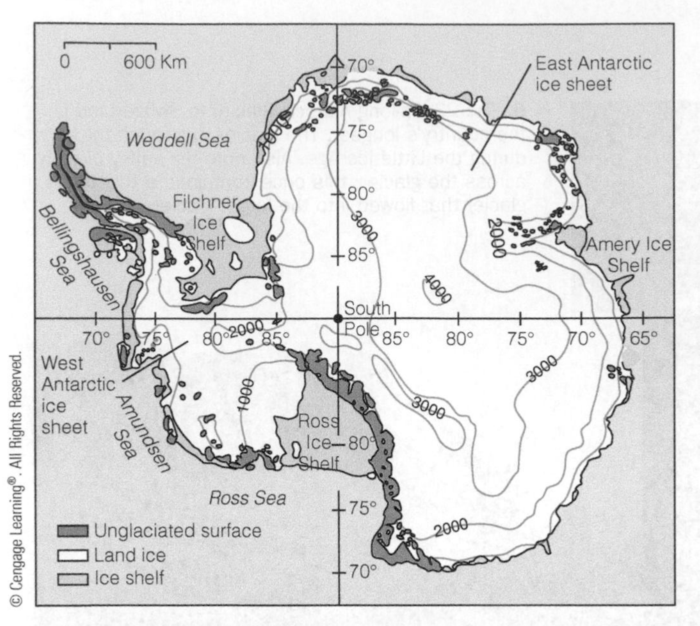

a The West and East Antarctic ice sheets merge to form a nearly continuous ice cover that averages 2,160 m thick. The blue lines are lines of equal thickness.

b The caldera of Öraefajökull Volcano in Iceland is occupied by a glacier that is part of the 8,100 km^2 Vatnajökull Ice Cap. The volcano last erupted in 1727–1728.

Glaciers—Part of the Hydrologic Cycle

In Chapter 12, you learned that the hydrosphere consists of all surface water, including water frozen in glaciers. So glaciers make up one reservoir in the hydrologic cycle where it is stored for long periods, but even this water eventually returns to its original source, the oceans (see Figure 12.2). Many glaciers at high latitudes, as in Alaska, northern Canada, and Scandinavia, flow directly into the oceans where they melt, or icebergs break off (a process known as calving) and drift out to sea where they eventually melt. At low latitudes or areas remote from the oceans, glaciers flow to lower elevations where they melt and the liquid water enters the groundwater system (another reservoir in the hydrologic cycle) or it returns to the seas by surface runoff.

In addition to melting, glaciers lose water by *sublimation,* when ice changes to water vapor without an intermediate liquid phase. Sublimation is not an exotic process; it occurs in the freezer compartment of a refrigerator. Because of sublimation, the older ice cubes at the bottom of the container are much smaller than the more recently formed ones. In any case, the water vapor derived by sublimation enters the atmosphere where it may condense and fall as rain or snow, but in the long run, this water also returns to the oceans.

How Do Glaciers Originate and Move?

Glaciers form in any area where more snow falls than melts during the warmer season and a net accumulation takes place. Freshly fallen snow has about 80% air-filled pore space and 20% solids, but it compacts as it accumulates, partially thaws, and refreezes, converting to a granular type of snow known as **firn.** As more snow accumulates, the firn is buried and further compacted and recrystallized until it is transformed into **glacial ice,** consisting of about 90% solids and 10% air (Figure 14.4).

Now you know how glacial ice forms, but we still have not addressed how glaciers move. At this time, it is useful to recall some of the information from previous chapters. In Chapter 3 we noted that ice is a crystalline solid with distinctive physical properties and a specific chemical composition and thus is a mineral. Accordingly, glacial ice is a type of metamorphic rock, but one that is easily deformed. In Chapter 10, we noted that *stress* is force per unit area and *strain* (or *deformation*) is a change in the shape or volume of solids. When accumulating ice and snow reach a critical thickness of about 40 m, the stress on the ice at depth is enough to induce **plastic flow,** a type of permanent deformation that involves no fracturing. Plastic flow is the most important process whereby glaciers move, but **basal slip** is also significant as glaciers slide over their underlying surfaces (Figure 14.5). Liquid water facilitates basal slip because it reduces the friction between the base of a glacier and the surface it moves over.

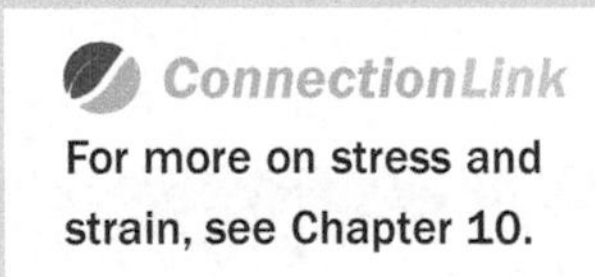

The total movement of a glacier in a given time is a consequence of plastic flow and basal slip, although the former occurs continuously, whereas the latter varies depending on the season, latitude, and elevation. Indeed, if a glacier is solidly frozen to the surface below, as in the case of many polar environments, it moves only by plastic flow. Basal slip is far

Figure 14.4 Glacial Ice

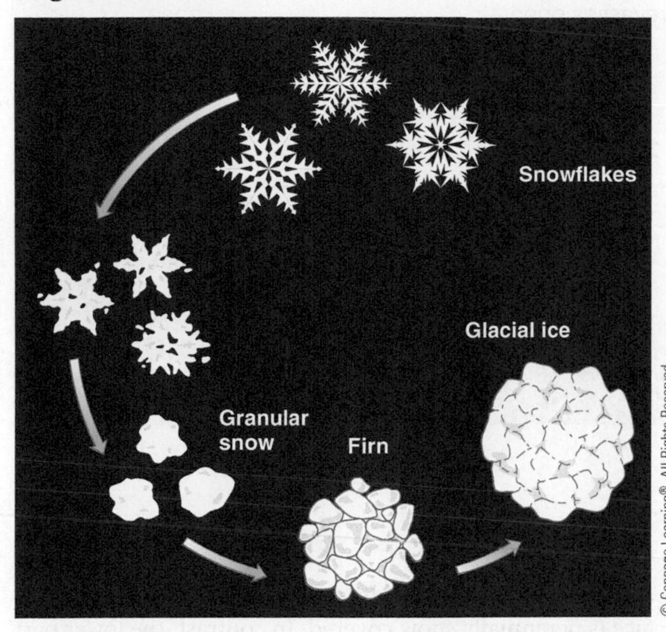

a The conversion of freshly fallen snow to firn and then to glacial ice.

b The Harriman Glacier in Alaska shows the blue color of glacial ice. The longer wavelengths of light are absorbed by the ice, but the shorter wavelength blue is scattered, accounting for the blue color. Also, since this glacier's terminus is in the sea, it is a tidewater glacier.

more important in valley glaciers as they flow from higher to lower elevations, whereas continental glaciers need no slope for flow.

Although glaciers move by plastic flow, the upper 40 m or so of ice behaves as a brittle solid and fractures if subjected to stress. Large crevasses commonly develop in glaciers where they flow over an increase in slope of the underlying surface or where they flow around a corner (Figure 14.6). In either case, the ice is stretched (subjected to tension) and crevasses open, which extend down to the zone of plastic flow. In some cases, a glacier descends over such a steep precipice that crevasses break up the ice into a jumble of blocks and spires, and an icefall develops.

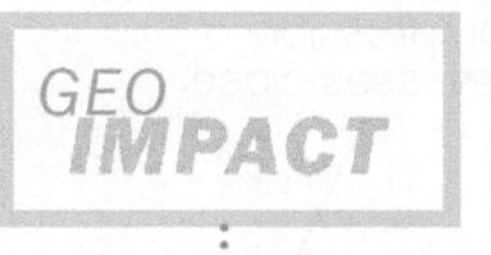

Ice: A Solid That Flows

Suppose you are a high school Earth science teacher trying to explain to students that ice is a mineral and rock, and how a solid like ice can flow like a fluid. Furthermore, you explain that the upper 40 m or so of a glacier is brittle and fractures, whereas ice below that depth simply flows when subjected to stress. Now that your students are thoroughly confused, how will you explain and demonstrate that ice can behave like a solid yet show properties of fluid flow?

Distribution of Glaciers

As you might suspect, the amount of snowfall and temperature are important factors in determining where glaciers form. Parts of northern Canada are cold enough to support glaciers but receive too little snowfall, whereas some mountain areas in California receive huge amounts of snow but are too warm for glaciers. Of course, temperature varies with elevation and latitude, so we would expect to find glaciers in high mountains and at high latitudes, if these areas receive enough snow.

Many small glaciers are present in the Sierra Nevada of California, but only at elevations exceeding 3,900 m. In fact, the high mountains in California, Oregon, and Washington all

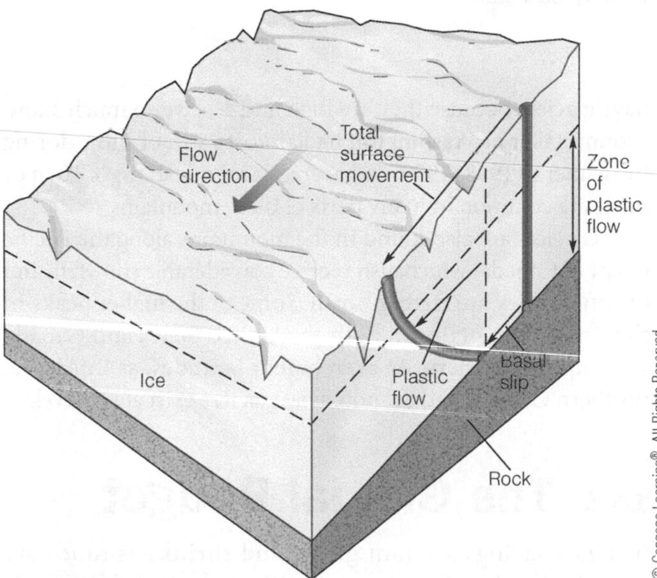

Figure 14.5 Part of a Glacier Showing Movement by a Combination of Plastic Flow and Basal Slip Plastic flow involves internal deformation within the ice, whereas basal slip is sliding over the underlying surface. If a glacier is solidly frozen to its bed, it moves only by plastic flow. Notice that the top of the glacier moves farther in a given time than the bottom does.

Figure 14.6 Crevasses Crevasses are common in the upper parts of glaciers when the ice is subjected to tension.

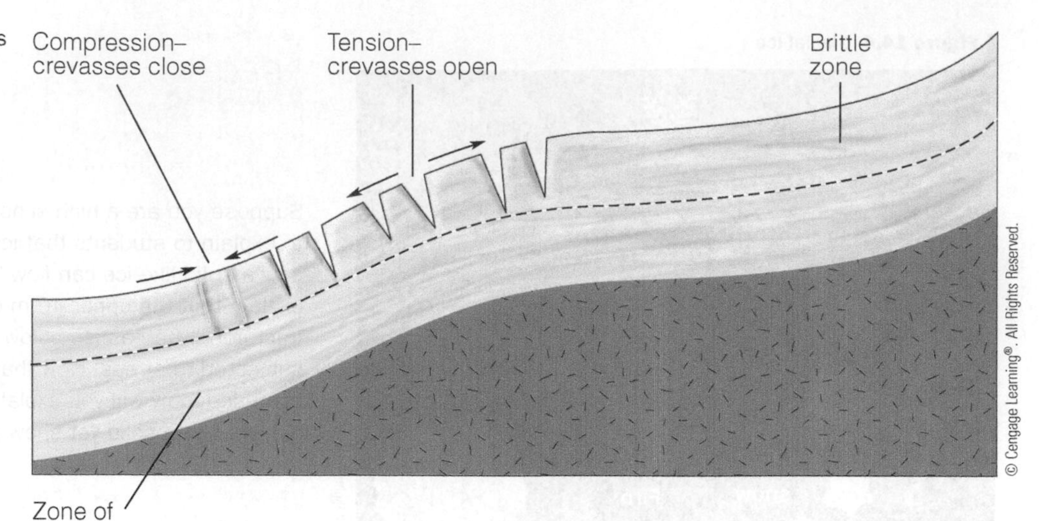

a Crevasses open where the brittle part of a glacier is stretched as it moves over a steeper slope in its valley.

Sue Monroe

b Crevasses on a small glacier in Kluane National Park, Yukon Territory, Canada.

have glaciers because they are high and receive so much snow. Mount Baker in Washington had almost 29 m of snow during the winter of 1998–1999, and average accumulations of 10 m or more are common in many parts of these mountains.

Glaciers are also found in the mountains along the Pacific Coast of Canada, which also receive considerable snowfall, and of course, they are farther north. Some of the higher peaks in the Rocky Mountains in both the United States and Canada also support glaciers. At even higher latitudes, as in Alaska, northern Canada, and Scandinavia, glaciers exist at sea level.

14.4 The Glacial Budget

Just as a savings account grows and shrinks as funds are deposited and withdrawn, a glacier expands and contracts in response to accumulation and wastage. We describe a glacier's behavior in terms of a **glacial budget,** which is essentially a balance sheet of accumulation and wastage. For instance, the upper part of a valley glacier is a **zone of accumulation,** where additions exceed losses and the surface is perennially snow covered. In contrast, the lower part of the same glacier is a **zone of wastage,** where losses from melting, sublimation, and calving of icebergs exceed the rate of accumulation (Figure 14.7).

At the end of winter, a valley glacier's surface is completely covered with the accumulated seasonal snowfall. During the spring and summer, the snow begins to melt, first at lower elevations and then progressively higher up the glacier. The elevation to which snow recedes during a wastage season is the *firn limit* (Figure 14.7). You can easily identify the zones of accumulation and wastage by noting the location of the firn limit.

The firn limit on a glacier may change yearly, but if it does not change or shows only minor fluctuations, the glacier has a balanced budget. That is, additions in the zone of accumulation are exactly balanced by losses in the zone of wastage, and the distal end, or terminus, of the glacier remains stationary (Figure 14.7). If the firn limit moves up the glacier, indicating a negative budget, the glacier's terminus retreats. If the firn limit moves down the glacier, however, the glacier has a positive budget, additions exceed losses, and its terminus advances.

Even though a glacier may have a negative budget and a retreating terminus, the glacial ice continues to move toward the terminus by plastic flow and basal slip. If a negative budget persists long enough, though, the glacier continues to recede, and it thins until it is no longer thick enough to maintain flow. It then ceases moving and becomes a *stagnant glacier;* if wastage continues, the glacier eventually disappears.

We used a valley glacier as an example, but the same budget considerations control the flow of ice caps and continental glaciers as well. The entire Antarctic ice sheet is in the zone of accumulation, but it flows into the ocean where wastage occurs.

We mentioned in the Introduction that glaciers are very sensitive to climate changes and that they may tell us about global warming (Geo-Focus 14.1). One alarming trend in glaciers is that most of those studied are retreating. In fact,

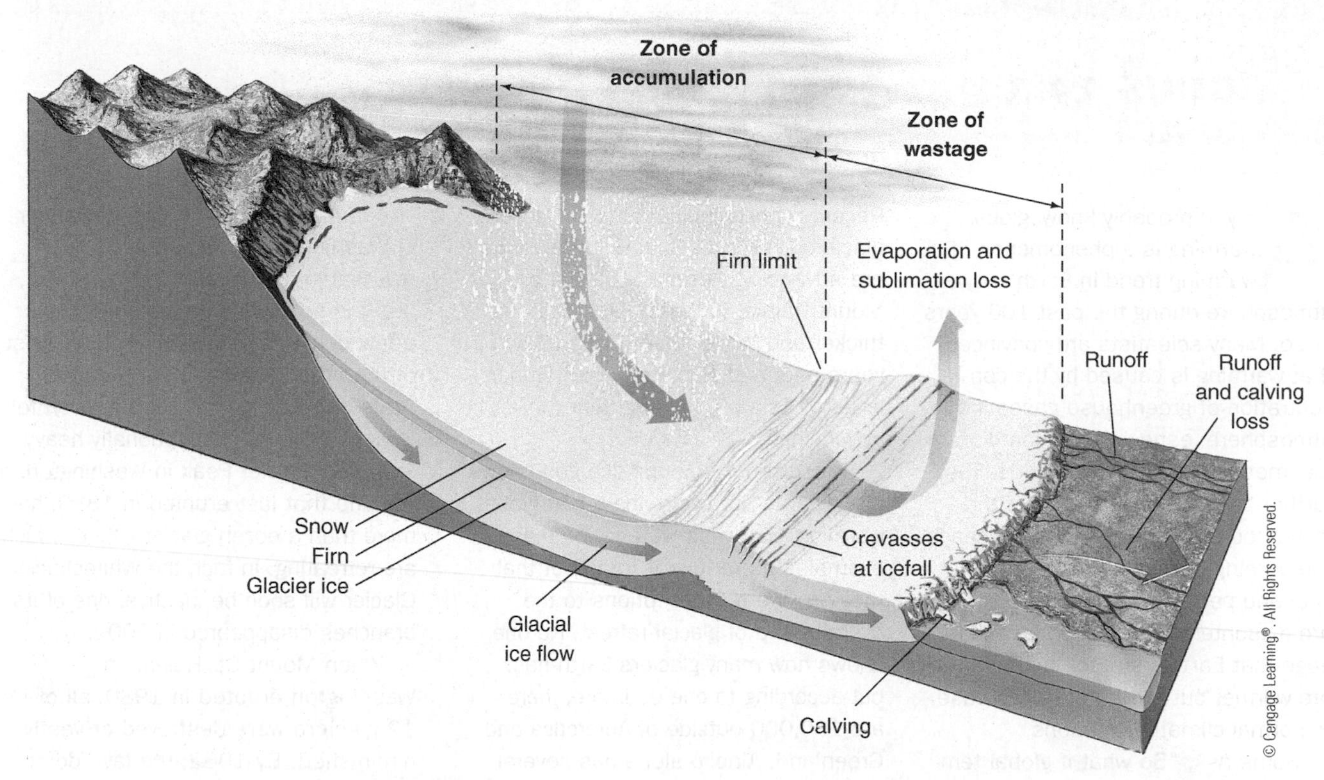

Figure 14.7 The Glacial Budget All glaciers have a zone of accumulation where additions exceed losses and the glacier's surface is perennially covered by snow. They also have a zone of wastage where losses from melting, calving of icebergs, evaporation, and sublimation exceed gains. If a glacier's budget is balanced, its terminus remains in the same position. However, should the budget be positive the terminus advances, and if the budget is negative the terminus retreats.

glaciers will probably be gone from Glacier National Park in Montana in only a few decades, and similar trends are taking place elsewhere.

How Fast Do Glaciers Move?

In general, valley glaciers move more rapidly than continental glaciers, but the rates for both vary from centimeters to tens of meters per day. Valley glaciers moving down steep slopes flow more rapidly than glaciers of comparable size on gentle slopes, assuming that all other variables are the same. The main glacier in a valley glacier system contains a greater volume of ice and thus has a greater discharge and flow velocity than its tributaries (Figure 14.1). Temperature exerts a seasonal control on valley glaciers because, although plastic flow remains rather constant year-round, basal slip is more important during warmer months when meltwater is abundant.

Flow rates also vary within the ice itself. For example, flow velocity increases downslope in the zone of accumulation until the firn limit is reached; from that point, the velocity becomes progressively slower toward the glacier's terminus. Valley glaciers are similar to streams, in that the valley walls and floor cause frictional resistance to flow, so the ice in contact with the walls and floor moves more slowly than the ice some distance away (Figure 14.8a).

Notice in Figure 14.8a that flow velocity in the interior of a glacier increases upward until the top few tens of meters of ice are reached, but little or no additional increase occurs after that point. This upper ice layer constitutes the rigid part of the glacier that is moving as a result of basal slip and plastic flow below.

Continental glaciers ordinarily flow at a rate of centimeters to meters per day. One reason continental glaciers move comparatively slowly is that they exist at higher latitudes and are frozen to the underlying surface most of the time, which limits the amount of basal slip. But some basal slip does occur even beneath the Antarctic ice sheet, although most of its movement is by plastic flow. Nevertheless, some parts of continental glaciers manage to achieve extremely high flow rates. Near the margins of the Greenland ice sheet, the ice is forced between mountains in what are called *outlet glaciers*. In some of these outlets, flow velocities exceed 100 m per day.

In parts of the continental glacier covering West Antarctica, scientists have identified ice streams in which flow rates are considerably higher than in adjacent glacial ice. Drilling has revealed a 5-m-thick layer of water-saturated sediment beneath these ice streams, which acts to facilitate movement of the ice above. Some geologists think that geothermal heat from subglacial volcanism melts the underside of the ice, thus accounting for the layer of water-saturated sediment.

GEO FOCUS 14.1 Glaciers and Global Warming

As you probably know, global warming is a phenomenon of a warming trend in Earth's lower atmosphere during the past 100 years or so. Many scientists are convinced that warming is caused by the concentration of greenhouse gases in the atmosphere, especially carbon dioxide, methane, and nitrous oxide. They further think that the source of the increased quantities of these gases is the burning of fossil fuels, especially coal and petroleum. Of course, there are dissenters, most of whom acknowledge that Earth's surface temperatures are warmer but attribute the increase to normal climatic variations.

Some ask, "So what if global temperatures increase by a few degrees?" After all, residents of Fairbanks, Alaska, might welcome slightly warmer winter temperatures, and it is unlikely that the same trend will make the summers in Phoenix, Arizona, or Death Valley, California, much more uncomfortable than they are now. However, temperature increase has far-reaching effects. Remember that during the Little Ice Age, temperatures were only a little cooler than they are now, and yet the cooler weather, especially during the summers, had tragic consequences.

Whatever the cause of global warming, no one doubts that glaciers are good indicators of short-term climatic changes. Two factors account for the health of a glacier—the amount of snowfall and temperature. If there is considerable snowfall and cold temperatures, glaciers thicken and advance (they have a positive budget), whereas if the opposite conditions prevail, they lose mass and their termini retreat. Recent reports indicate that all seven glaciers on Mount Shasta in California are advancing, Nisqually Glacier on Mount Rainier in Washington State is thicker and longer than it was a few years ago, and glaciers in the Himalayas of Asia are holding their own or advancing.

These reports sound like good news and we can stop worrying about global warming, but when we analyze these examples more fully, it turns out that they are the few exceptions to the overall trend of glacial retreat. No one knows how many glaciers Earth has, but according to one estimate, there are 160,000 outside of Antarctica and Greenland. Alaska alone has several tens of thousands, and several other western states have many more but most of these are not very large. Washington State has the most followed by California, Wyoming, Montana, and Oregon; only in these states are the glaciers large enough to evaluate from Landsat images from space, although Colorado has some glaciers and Nevada and Utah have one each. Of course Canada, too, has numerous glaciers, but no one knows how many.

So, what do glaciers tell us about climate? We already noted that snowfall and temperature are controls on a glacier's budget (Figure 14.7). It is true that not many of the 160,000 or so glaciers on Earth have been studied, but those that have show an alarming trend: many are thinning, their termini are retreating, and in many cases, they have nearly or completely disappeared. It is true that the seven small glaciers on Mount Shasta in California are advancing, but all other California glaciers are shrinking.

Nisqually Glacier on Mount Rainier in Washington State has shown a reversal in its overall trend of retreat, but it is not nearly as large as it was a few decades ago. Furthermore, most of the other glaciers in Washington State show an overall retreat, even following years with exceptionally heavy snowfall. Glacier Peak in Washington, a volcano that last erupted in 1880, has more than a dozen glaciers, all of which are retreating. In fact, the Whitechuck Glacier will soon be inactive; one of its branches disappeared in 2002.

When Mount St. Helens in Washington erupted in 1980, all of its 12 glaciers were destroyed or vastly diminished. By 1982, the lava dome in the volcano had cooled sufficiently for snow to accumulate yearly, which transformed into glacial ice that is now about 190 m thick. In any case, the ice is thick enough to flow, so a glacier formed where none was present in a mere 28 years. Keep in mind, however, that Mount St. Helens already had conditions favorable for glaciers before it erupted. The factors that contributed to the origin of this glacier are very heavy snowfall, protection from the Sun by the crater walls, and rockfalls from the crater walls that help insulate this newly formed glacier.

Of the 150 or so glaciers that were in Glacier National Park in Montana about a century ago, perhaps 25 remain, and many of those are such shrunken remnants that it is difficult to determine if they are active glaciers or not (Figure 1). Indeed, the consensus is that they will all be gone in a decade or two. The warming in Glacier National Park is slight, but fewer days per year

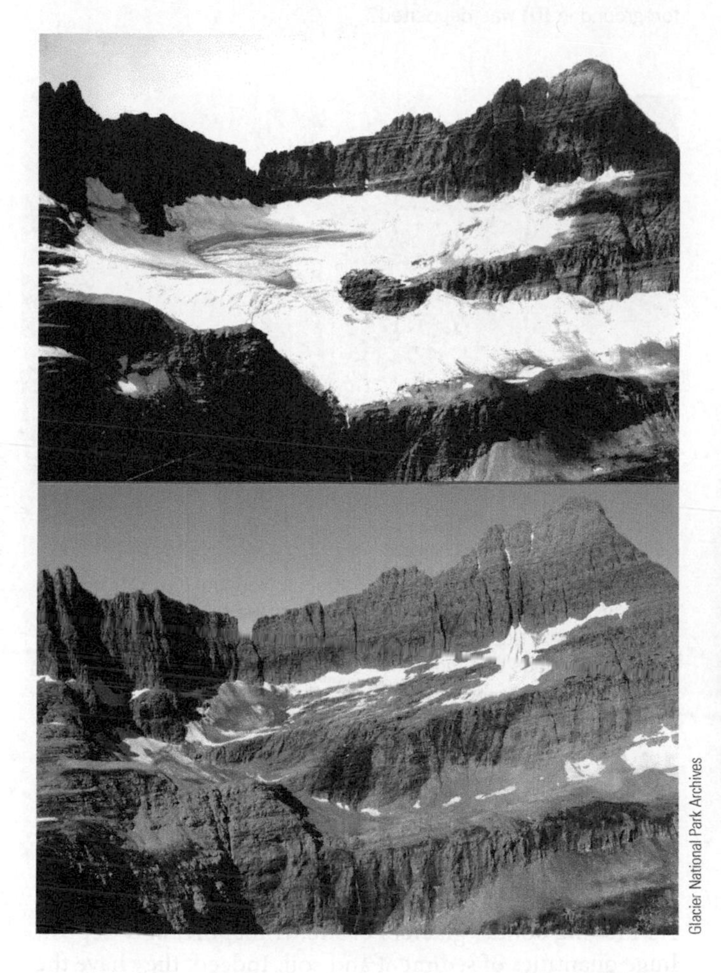

Glacier National Park Archives

Figure 1 Views of Shepherd Glacier in Glacier National Park, Montana taken in 1913 (top) and 2005 (bottom). By 2010, Shepherd Glacier was too small to qualify as a glacier, as was another glacier in the park.

James S. Monroe

Figure 2 Exit Glacier as it descends from the Harding Icefield in the background. The terminus of the glacier has receded 488 m in recent years and it has thinned by more than 90 m.

have below-freezing temperatures, and there are more warmer days.

In Africa, the area covered by glaciers on Mount Kilimanjaro has decreased by 85%, and a similar trend is seen on Mount Kenya. In the Himalayas of Asia, it is true that some glaciers are advancing or holding their own, but mostly in the Karakoram Range, whereas those in other parts of this mountain system are retreating. In Alaska, the glaciers at higher elevations are not retreating, but those near sea level certainly are. For example the Exit Glacier (Figure 2), which descends from the Harding Icefield, is not nearly as large as it was a few decades ago. Indeed, according to a pamphlet published by the Alaska Natural History Association:

> Fifty years ago, [Exit] glacier's terminus was 488 m farther down valley than it is today. It is also thinning. The lower reaches have lost 91 m of thickness in that same period. The entire Harding Icefield has lost an average of 21 m of ice thickness, releasing more than 33.5 cubic kilometers of water into the sea since 1950.

From 2000 to the present, Earth has experienced many of the hottest years on record. And 2012 is no exception; it will turn out to be the hottest or nearly hottest ever. One consequence of higher temperatures is, of course, retreating glaciers as we have discussed. But remember that glaciers are one reservoir in the hydrologic cycle, so as they waste away, their water returns to the oceans and sea level rises. We know that during the Pleistocene Epoch (Ice Age), sea level was as much as 130 m lower than it is today, and we know that it has risen on average nearly 2 mm per year for the past 100 years; the current rate is estimated at about 3 mm per year. This increase may not sound like much, but even a slight rise in sea level will eventually be a real problem for many of Earth's coastal regions.

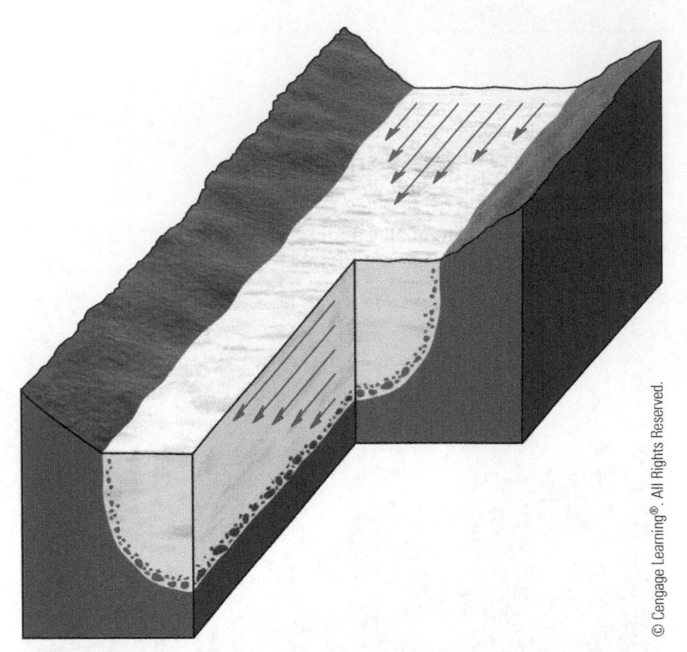

a Flow velocity in a valley glacier varies horizontally and vertically. Velocity is greatest at the top center because friction with the walls and floor of the trough slows the flow adjacent to these boundaries. The lengths of the arrows are proportional to velocity.

▮ Figure 14.8 Flow in Valley Glaciers

Critical Thinking Question How do you think the debris in the foreground in **(b)** was deposited?

b Terminus of the Lowell Glacier in Kluane National Park, Yukon Territory, Canada. The glacier was surging when this image was taken on July 2, 2010. Its surge probably began in October 2009 when its terminus was more than 1.5 km farther up the valley.

Glacial Surges A **glacial surge** is a short-lived episode of accelerated flow during which the glacier's surface breaks into a maze of crevasses and its terminus advances noticeably. These brief episodes are best known in valley glaciers, but they also take place in ice caps and even in continental glaciers. In 1995, a huge ice shelf in Antarctica broke apart, and several ice streams from the Antarctic ice sheet surged toward the ocean.

During a surge, a glacier's terminus may advance several tens of meters per day for weeks or months and then return to its previous flow rate. Not many glaciers surge, and none does so in the United States outside Alaska. Even in Canada, they are found only in the Yukon Territory and the Queen Elizabeth Islands. Lowell Glacier in Kluane National Park in the Yukon Territory began surging probably in late 2009, and by May 2010, its terminus had advanced 1.5 km (▮ Figure 14.8b). The fastest surge ever recorded was in 1953 in the Kutiah Glacier in Pakistan; the glacier's terminus advanced 12 km in three months, or 130 m per day on average.

The onset of a glacial surge is heralded by a thickened bulge in the upper part of a glacier that begins to move toward the terminus at several times the glacier's normal velocity. When the bulge reaches the terminus, it causes rapid movement and displacement of the terminus by as much as 20 km. Surges are also probably related to accelerated rates of basal slip rather than more rapid plastic flow. One theory holds that thickening in the zone of accumulation with concurrent thinning in the zone of wastage increases the glacier's slope and accounts for accelerated flow. According to another theory, pressure on soft sediment beneath a glacier squeezes fluids through the sediment, thereby allowing the overlying glacier to slide more effectively.

14.5 Erosion and Sediment Transport by Glaciers

As moving solids, glaciers erode, transport, and deposit huge quantities of sediment and soil. Indeed, they have the capacity to transport any size of sediment, up to boulders the size of houses. Important processes of erosion include bulldozing, plucking, and abrasion.

Although *bulldozing* is not a formal geologic term, it is fairly self-explanatory; glaciers shove or push unconsolidated materials in their paths. This effective process was aptly described in 1744 during the Little Ice Age by an observer in Norway:

> When at times [the glacier] pushes forward a great sound is heard, like that of an organ and it pushes in front of it unmeasurable masses of soil, grit and rocks bigger than any house could be, which it then crushes small like sand.*

Plucking, also called *quarrying,* results when glacial ice freezes in the cracks and crevices of a bedrock projection and eventually pulls it loose. One manifestation of plucking is a landform called a *roche moutonnée,* a French term for "rock sheep." As shown in ▮ Figure 14.9, a glacier smoothes the "upstream" side of an obstacle, such as a small hill, and

---

*Quoted in C. Officer and J. Page, *Tales of the Earth* (New York: Oxford University Press, 1993), p. 99.

Figure 14.9 Roche Moutonnée A French term meaning "rock sheep," a roche moutonnée is a bedrock projection that was shaped by glacial abrasion and plucking.

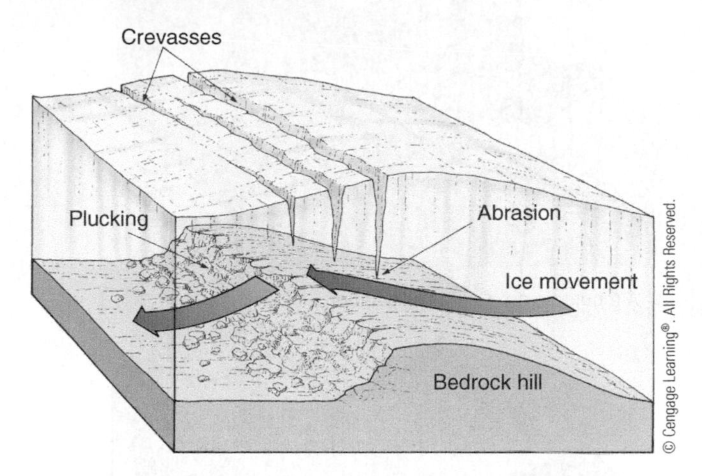

a Origin of a roche moutonnée. A glacier abrades and polishes the "upstream" side of a bedrock projection and shapes its "downstream" side by plucking.

b Lambert Dome in Yosemite National Park in California is a large roche moutonnée. It stands about 240 m above the surrounding meadow.

plucks pieces of rock from the "downstream" side by repeatedly freezing to and pulling away from the obstacle.

Bedrock over which sediment-laden glacial ice moves is effectively eroded by **abrasion** and develops a **glacial polish,** a smooth surface that glistens in reflected light (Figure 14.10a). Abrasion also yields **glacial striations,** rather straight scratches rarely more than a few millimeters deep on rock surfaces. Abrasion thoroughly pulverizes rocks, yielding an aggregate of clay- and silt-size particles that have the consistency of flour—hence, the name *rock flour.* Rock flour is so common in streams discharging from glaciers that the water has a milky appearance (Figure 14.10b).

Continental glaciers derive sediment from mountains projecting into or through them, and wind-blown dust settles on their surfaces, but most of their sediment comes from the surface they move over. As a result, most sediment is transported in the lower part of the ice sheet. In contrast, valley

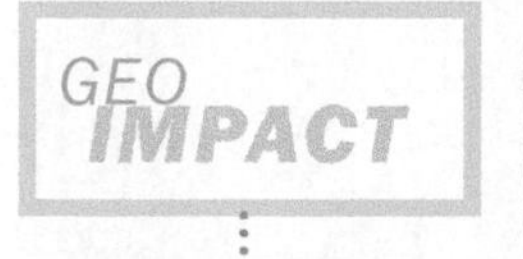

Glaciers and Global Warming

After carefully observing the same valley glacier for several years, you conclude (1) that its terminus has retreated at least 1 km and (2) that debris on its surface has obviously moved several hundreds of meters toward the glacier's terminus. Can you think of an explanation for these observations? Also, why do you think studying glaciers might have some implications for the debate on global warming?

Figure 14.10 Glacial Striations, Polish, and Rock Flour

Critical Thinking Question Why is the basalt in image **(a)** broken into 5- and 6-sided polygons?

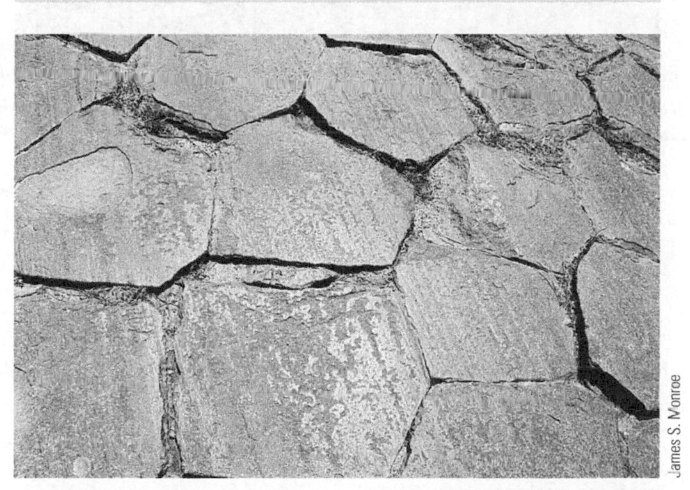

a Glacial polish and striations, the straight scratches, on basalt at Devils Postpile National Monument in California.

b The water in this stream in Switzerland is discolored by rock flour, small particles produced by glacial abrasion.

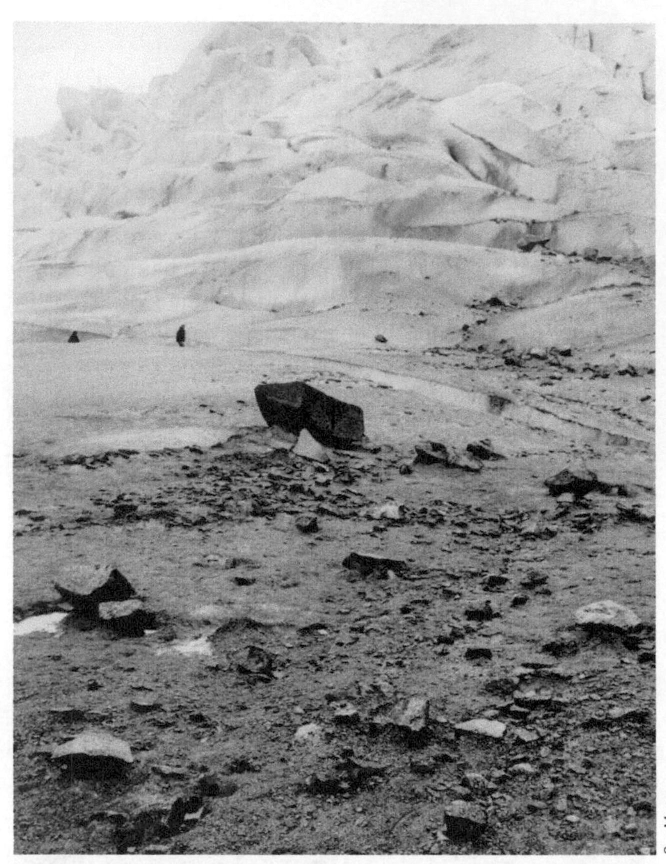

Figure 14.11 Sediment Transport by Valley Glaciers Debris on the surface of the Mendenhall Glacier in Alaska. The largest boulder is about 2 m across. Notice the icefall in the background. The person left of center provides scale.

glaciers carry sediment in all parts of the ice, but it is concentrated at the base and along the margins (Figure 14.11). Some of the marginal sediment is derived by abrasion and plucking, but much of it is supplied by mass-wasting processes, as when soil, sediment, or rock falls or slides onto the glacier's surface.

Figure 14.12 Erosional Landforms Produced by Valley Glaciers

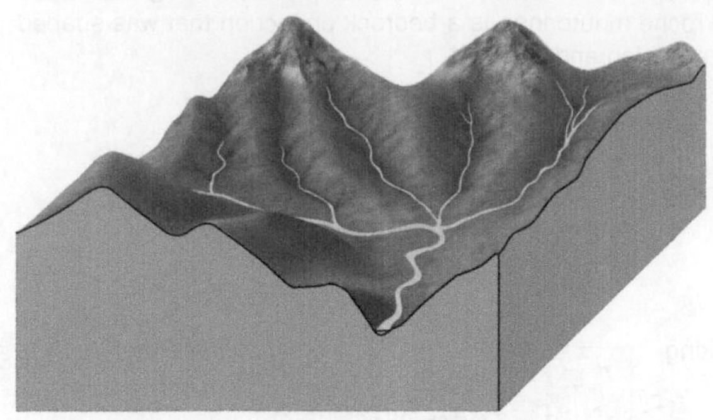

a A mountain area before glaciation.

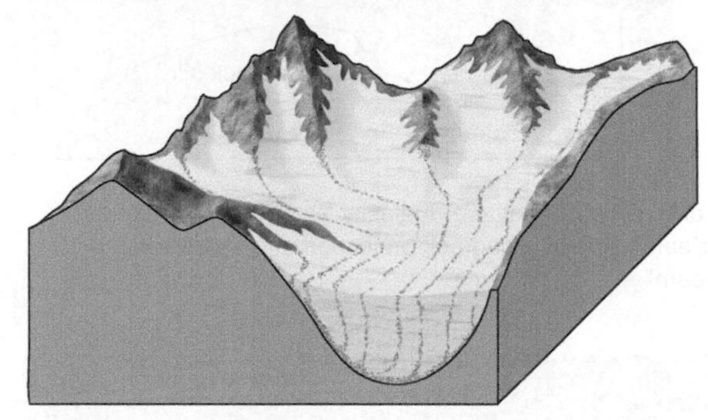

b The same area during the maximum extent of valley glaciers.

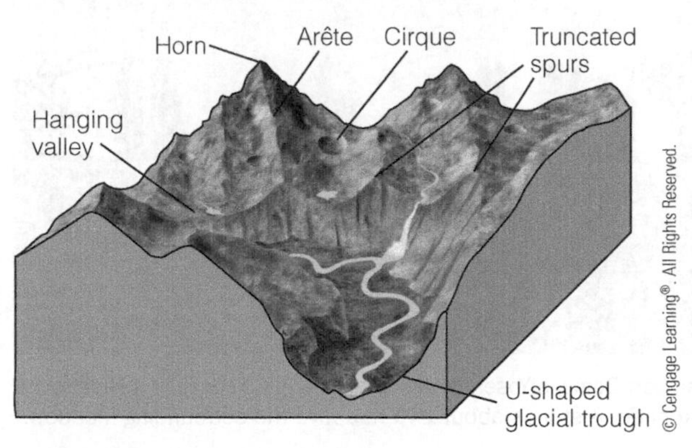

c After glaciation.

Erosion by Valley Glaciers

When mountain ranges are eroded by valley glaciers, they take on a unique appearance of angular ridges and peaks in the midst of broad, smooth valleys with near-vertical walls. The erosional landforms produced by valley glaciers include U-shaped glacial troughs, hanging valleys, cirques, arêtes, and horns, most of which are easily recognized.

U-Shaped Glacial Troughs A **U-shaped glacial trough** is one of the most distinctive features of valley glaciation. Mountain valleys eroded by running water are typically V-shaped in cross section; that is, they have valley walls that descend to a narrow valley bottom (Figure 14.12a). In contrast, valleys scoured by glaciers are deepened, widened, and straightened so that they have very steep or vertical walls, but broad, rather flat valley floors; thus they exhibit a U-shaped profile (Figures 14.12c, 14.13a).

Many glacial troughs contain triangular-shaped *truncated spurs*, which are cutoff or truncated ridges that extend into the preglacial valley (Figure 14.12c). Another common feature is a series of steps or rock basins in the valley floor where the glacier eroded rocks of varying resistance; many of the basins now contain small lakes.

During the Pleistocene, when glaciers were more extensive, sea level was as much as 130 m lower than at present, so glaciers that flowed into the sea eroded their valleys below present sea level. When the glaciers melted at the end of the Pleistocene, sea level rose, and the ocean filled the lower ends of the glacial troughs so that now they are long, steep-walled embayments called **fiords** (Figure 14.13b).

Fiords are restricted to high latitudes where glaciers exist at low elevations, such as Alaska, western Canada, Scandinavia, Greenland, southern New Zealand, and southern Chile. Lower sea level during the Pleistocene was not entirely responsible for

Figure 14.13 U-Shaped Glacial Trough and Fiord

Critical Thinking Question How does a U-shaped glacial trough differ from a mountain valley eroded by running water?

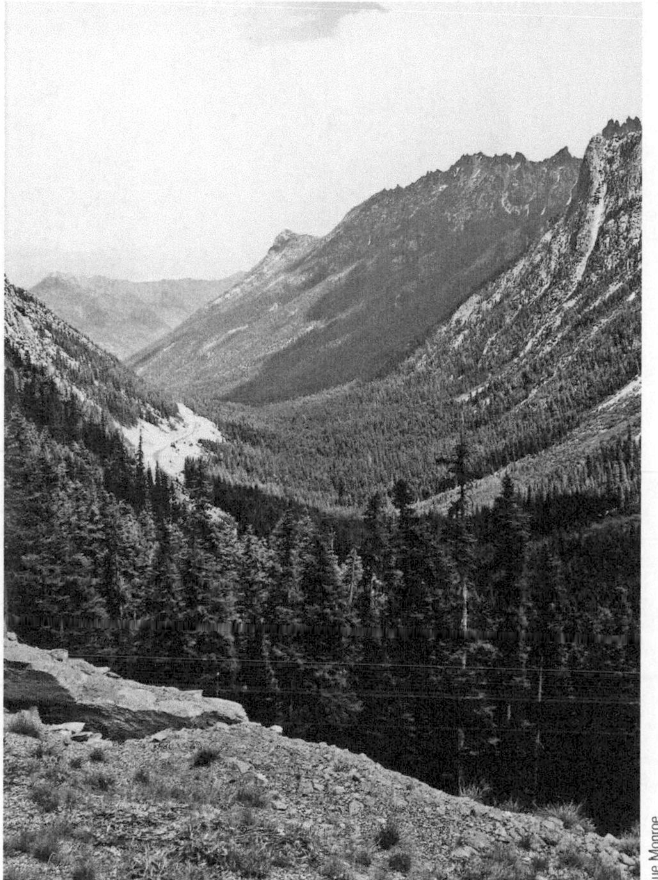

a This U-shaped glacial trough is in the Cascade Range in Washington State.

Sue Monroe

b This steep-walled U-shaped glacial trough in Norway extends below sea level, so it is a fiord.

James S. Monroe

the formation of all fiords. Unlike running water, glaciers can erode a considerable distance below sea level. In fact, a glacier 500 m thick can stay in contact with the seafloor and effectively erode it to a depth of about 450 m before the buoyant effects of water cause the glacial ice to float. The depth of some fiords is impressive; some in Norway and southern Chile are about 1,300 m deep.

Hanging Valleys In a landscape shaped by valley glacier erosion, a main valley may have one or more smaller valleys entering part way up its side. These so-called **hanging valleys** form when the main valley is deepened more rapidly than its tributary-valley because of its greater volume of ice (Figure 14.12c, Figure 14.14). In Figure 14.14a, for example, a huge glacier eroded the valley in the foreground, whereas a much smaller one eroded the hanging valley, which is perched high above the floor of the main valley. Not all hanging valleys have waterfalls plunging from them, but those that do have some of the highest and most spectacular ones. A good example is Bridalveil Falls in Yosemite National Park in California (Figure 14.14b).

Cirques, Arêtes, and Horns Perhaps the most spectacular erosional landforms in areas of valley glaciation are at the upper ends of glacial troughs and along the divides that separate adjacent glacial troughs. Valley glaciers form and move out from steep-walled, bowl-shaped depressions called **cirques** at the upper end of their troughs (Figure 14.15). Cirques are typically steep-walled on three sides, but one side is open and leads into the glacial trough. Some cirques slope continuously into the glacial trough, but many have a lip or threshold at their lower end.

The details of cirque origin are not fully understood, but they probably form by erosion of a preexisting depression on a mountainside. As snow and ice accumulate in the depression, frost wedging and plucking, combined with glacial erosion, enlarge and transform the head of a steep mountain valley into a typical amphitheater-shaped cirque. Tension in the upper part of the glacier may reduce the erosive power of the ice on the immediate, downslope side of a cirque, leaving a lip or threshold in the valley floor after the ice melts away. Small lakes of meltwater, called *tarns,* often form on the floors of cirques behind such thresholds.

Cirques become wider and are cut deeper into mountainsides by headward erosion as a result of abrasion, plucking, and several mass-wasting processes. For example, part of a steep cirque headwall may collapse while frost wedging continues to pry loose rocks that tumble downslope, so a combination of processes erode a small mountainside depression into a large cirque.

Arêtes—narrow, serrated ridges—form in two ways. In many cases, cirques form on opposite sides of a ridge, and headward erosion reduces the ridge until only a thin partition of rock remains (Figure 14.16a). The same effect occurs when erosion in two parallel glacial troughs reduces the intervening ridge to a thin spine of rock.

The most majestic of all mountain peaks are **horns,** steep-walled, pyramidal peaks formed by headward erosion of cirques. For a horn to form, a mountain peak must have at least three cirques on its flanks, all of which erode headward (Figure 14.12). Excellent examples of horns are Mount Assiniboine in the Canadian Rockies, the Grand Teton in

Figure 14.14 Hanging Valley

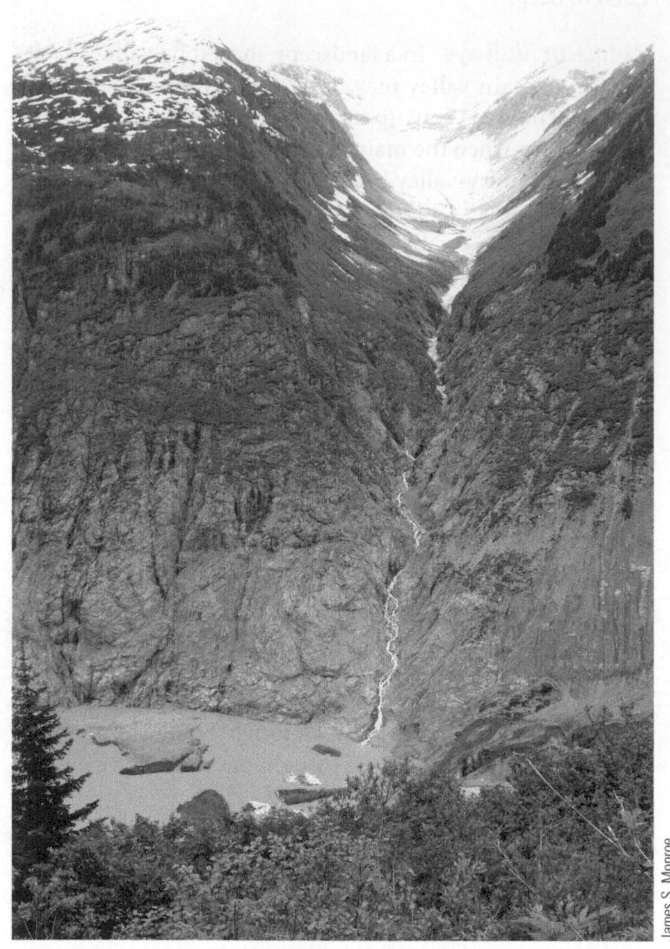

James S. Monroe

a Notice the small U-shaped glacial trough high up on the side of the larger glacial trough in the foreground. The larger trough has an active glacier that you can just see at the lower right part of the image, but its surface is obscured by sediment.

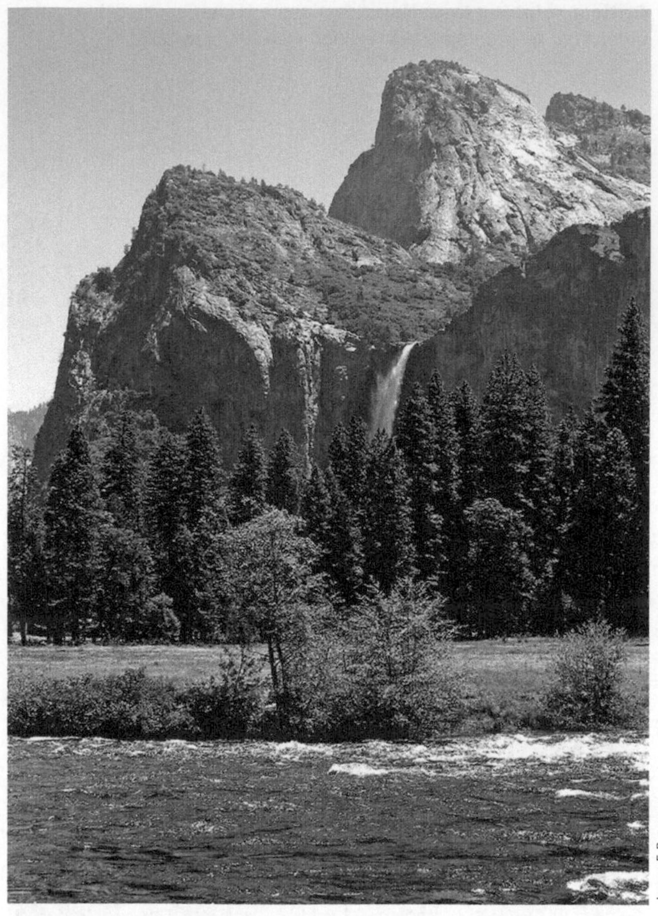

James F. Petersen

b Bridalveil Falls plunge 188 m from a hanging valley in Yosemite National Park in California. The valley in the foreground is a large U-shaped glacial trough.

Figure 14.15 Cirques in Canada and Alaska

James S. Monroe

a A cirque in the St. Elias Mountains of the Yukon Territory of Canada.

Sue Monroe

b A small glacier in a cirque in the Chugach Mountains near Girdwood, Alaska.

Figure 14.16 Arête and Horn Peak

Sue Monroe

a This knifelike ridge in Canada is an arête. It is the ridge between two U-shaped glacial troughs.

Swiss National Tourist Office

b The Matterhorn in Switzerland Is a Well-Known Horn Horns are pyramid-shaped peaks eroded by three or more glaciers on a single mountain.

Wyoming, and the most famous of all, the Matterhorn in Switzerland (Figure 14.16b).

Continental Glaciers and Erosional Landforms

Areas eroded by continental glaciers tend to be smooth and rounded because these glaciers bevel and abrade high areas that project into the ice. Rather than yielding the sharp, angular landforms typical of valley glaciation, they produce a landscape of subdued topography interrupted by rounded hills because they bury landscapes during their development.

In a large part of Canada, particularly the vast Canadian Shield region, continental glaciers have stripped off the soil and unconsolidated surface sediment, revealing extensive exposures of striated and polished bedrock. In addition to exposed bedrock, these areas, called **ice-scoured plains,** have deranged drainage (Figure 14.17), numerous lakes and swamps, low relief, and little or no soil. Similar though smaller bedrock exposures are also found in the northern United States from Maine through Minnesota.

Allan Kellelhelm/Mary Pat Ziter, JLM Visuals

Figure 14.17 Ice-Scoured Plain This ice-scoured plain in the Northwest Territories of Canada shows little relief, numerous lakes, little or no soil, and extensive bedrock exposures, which is typical of areas eroded by continental glaciers.

14.6 Glacial Deposits

Given that glaciers erode and transport gravel, sand, and mud, it follows that they must also deposit sediment. All glacial deposits go by the general term **glacial drift,** but geologists recognize two types of drift—till and stratified drift. **Till** is made up of any sediment deposited directly by glacial ice, as, for example, at the terminus of a glacier. Till is not sorted by particle size, and it shows no stratification. The till of both valley and continental glaciers is similar, but that of continental glaciers is much more extensive and usually has been transported much farther. As opposed to till, **stratified drift,** as the name implies, is layered or stratified, and invariably it shows some degree of sorting by particle size. As a matter of fact, most stratified drift is layers of gravel and sand or mixtures thereof deposited in braided stream channels that discharge from melting glaciers.

The appearance of till and stratified drift may not be as inspiring as some landforms resulting from glacial erosion, but they are important groundwater reservoirs, and they are exploited for their sand and gravel. In fact, sand and gravel, mostly for construction, are the most valuable mineral commodities in many areas. It is true that all glaciers deposit sediment, but as you would expect, the deposits of continental glaciers are far more extensive and more varied (Figure 14.18). In contrast, the deposits of a valley glacier tend to be restricted to the lower parts of the valley occupied by the glacier.

One conspicuous aspect of glacial drift is rock fragments of various sizes that were obviously not derived from the underlying bedrock. These fragments, called **glacial erratics,** came from some distant source and were transported and then deposited in their present location (Figure 14.19). Some glacial erratics are gigantic—the 15,000-metric-ton Big Rock, or Okotoks erratic, in Alberta, Canada, is the largest one known.

Figure 14.18 Development of Features Associated with Past Continental Glaciation

Critical Thinking Question Was running water important in the origin of any of the features in this illustration?

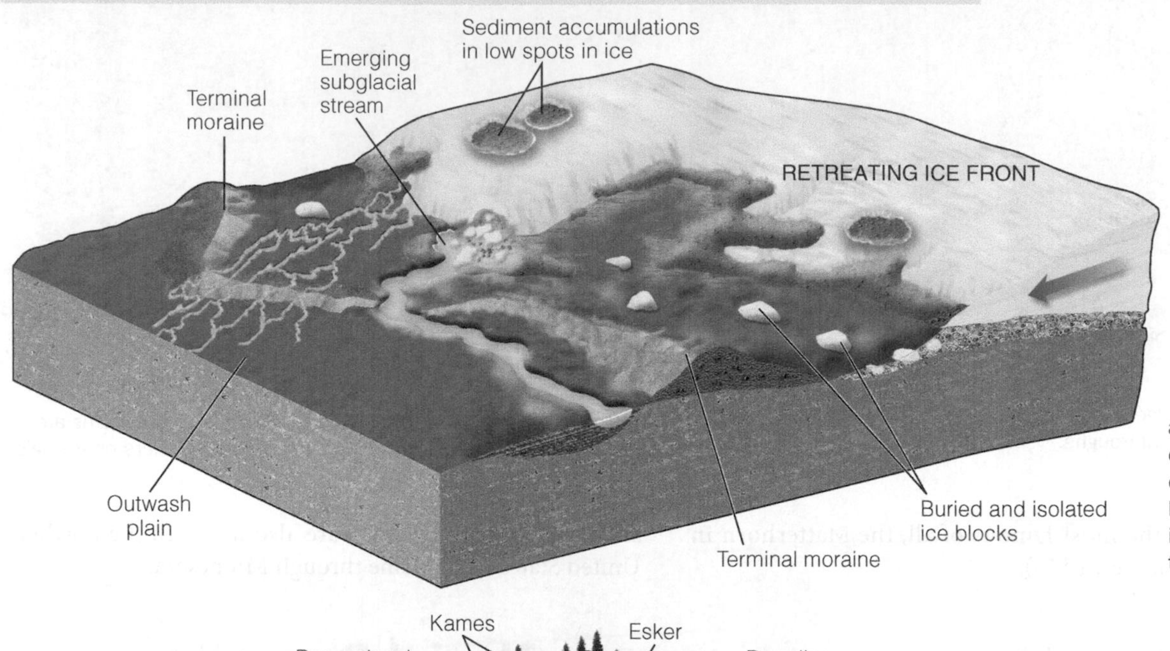

a This retreating continental glacier once covered a larger area as indicated by the terminal moraines.

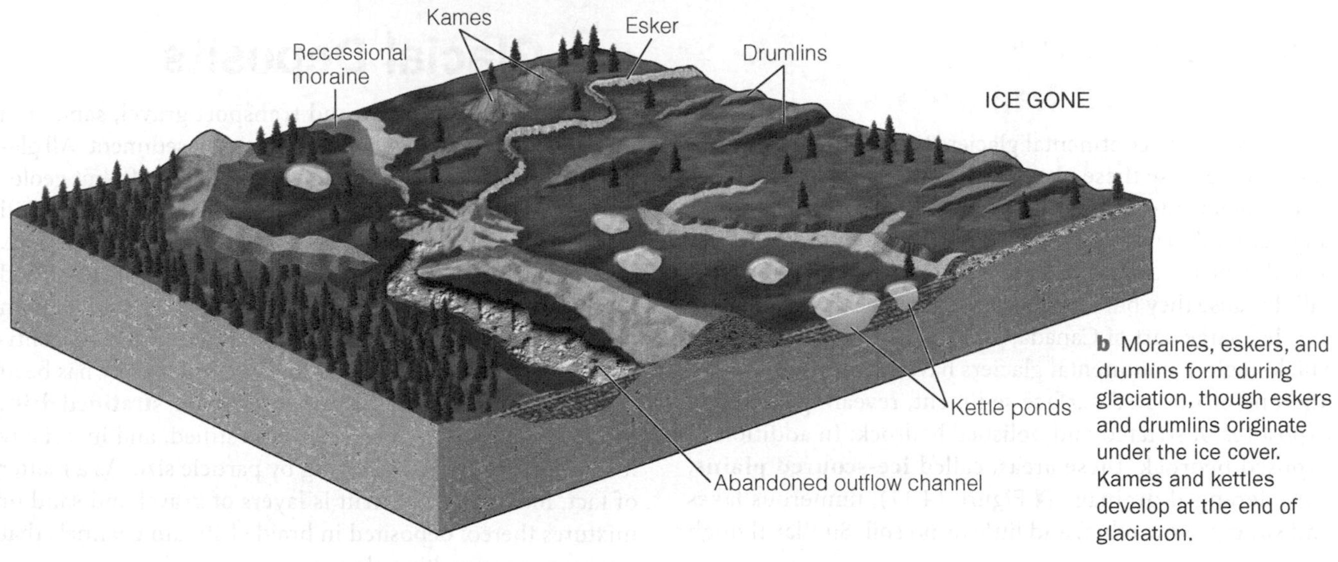

b Moraines, eskers, and drumlins form during glaciation, though eskers and drumlins originate under the ice cover. Kames and kettles develop at the end of glaciation.

Figure 14.19 Glacial Erratics

Critical Thinking Question How can you determine how far the erratic in **(b)** was transported?

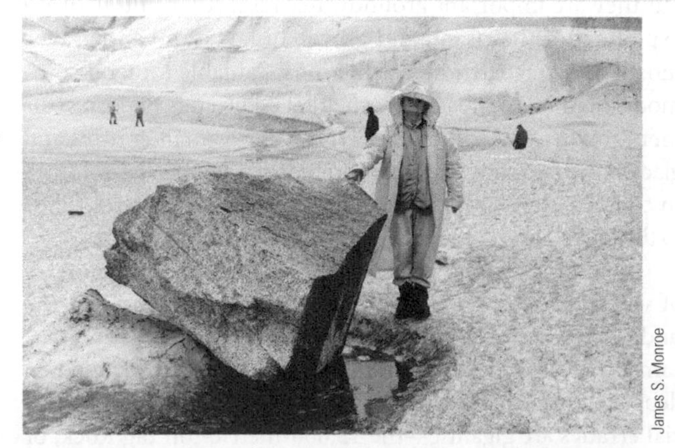

James S. Monroe

a A glacial erratic in the making. This boulder on the Mendenhall Glacier in Alaska will be deposited far from its source.

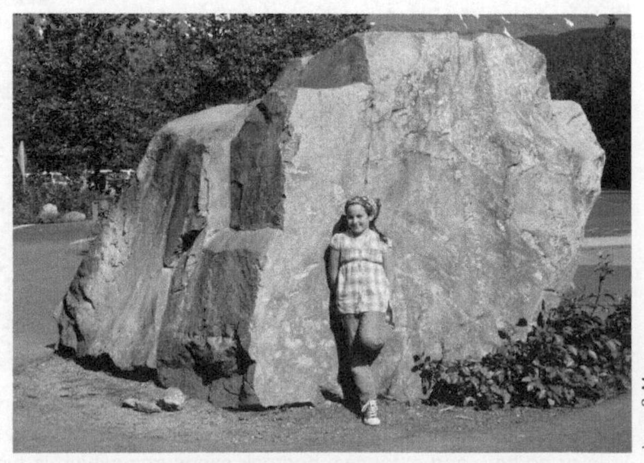

James S. Monroe

b Glacial erratic that was deposited by the Exit Glacier in Kenai Fjords National Park near Seward, Alaska.

Landforms Composed of Till

Remember, till is deposited directly by glacial ice, and it shows no sorting or stratification. Till is found in several types of deposits collectively called moraines and in elongated hills known as drumlins.

End Moraines The terminus of any glacier may become stabilized in one position for some period of time, perhaps a few years or even decades. Stabilization of the ice front does not mean that the glacier has ceased to flow, only that it has a balanced budget. When an ice front is stationary, flow within the glacier continues, and any sediment transported within or upon the ice is dumped as a pile of rubble at the glacier's terminus (▌Figure 14.20). These deposits are **end moraines,** which continue to grow as long as the ice front remains stationary. End moraines of valley glaciers are crescent-shaped ridges of till spanning the valley occupied by the glacier. Those of continental glaciers similarly parallel the ice front but are much more extensive.

Following a period of stabilization, a glacier may advance or retreat, depending on changes in its budget. If it advances, the ice front overrides and modifies its former moraine. If it has a negative budget, though, the ice front retreats toward the zone of accumulation. As the ice front recedes, till is deposited as it is liberated from the melting ice and forms a layer of **ground moraine.** Ground moraine has an irregular, rolling topography, whereas end moraine consists of long, ridgelike accumulations of sediment.

After a glacier has retreated for some time, its terminus may once again stabilize, and it deposits another end moraine. Because the ice front has receded, such moraines are called **recessional moraines** (Figure 14.18b). During the Pleistocene Epoch, continental glaciers in the mid-continent region extended as far south as southern Ohio, Indiana, and Illinois. Their outermost end moraines, marking the greatest extent of the glaciers, go by the special name **terminal moraine** (valley glaciers also deposit terminal moraines). As the glaciers retreated from the positions where their terminal moraines were deposited, they temporarily stopped retreating numerous times and deposited dozens of recessional moraines.

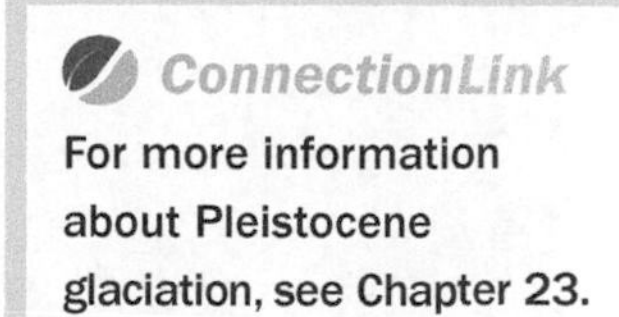

ConnectionLink
For more information about Pleistocene glaciation, see Chapter 23.

James S. Monroe

▌**Figure 14.20 End Moraine** This end moraine, consisting of unsorted, nonstratified sediment, is at the end of the Salmon Glacier in the Coast Mountains of British Columbia, Canada. Notice the dark material on the glacier's surface, which is sediment that will become part of the end moraine.

Lateral and Medial Moraines Valley glaciers transport considerable sediment along their margins, much of it abraded and plucked from the valley walls, but a significant amount falls or slides onto the glacier's surface by mass-wasting processes. In any case, this sediment is transported and deposited as long ridges of till called **lateral moraines** along the margin of the glacier (▌Figure 14.21).

Where two lateral moraines merge, as when a tributary glacier flows into a larger glacier, a **medial moraine** forms (Figure 14.21). A large glacier will often have several dark stripes of sediment on its surface, each of which is a medial moraine. One can determine how many tributaries a valley glacier has by the number of its medial moraines.

Drumlins In many areas where continental glaciers deposited till, the till has been reshaped into elongated hills known as **drumlins** (Figure 14.18b). Some drumlins are as much as 50 m high and 1 km long, but most are much smaller. From the side, a drumlin looks like an inverted spoon, with the steep end on the side from which the glacial ice advanced and the gently sloping end pointing in the direction of ice movement. Drumlins are rarely found as single, isolated hills; instead, they occur in *drumlin fields* that contain hundreds or thousands of drumlins. Drumlin fields are found in several states and Ontario, Canada, but perhaps the finest example is near Palmyra, New York.

According to one hypothesis, drumlins form when till beneath a glacier is reshaped into streamlined hills as the ice moves over it by plastic flow. Another hypothesis holds that huge floods of glacial meltwater modify till into drumlins.

Landforms Composed of Stratified Drift

Stratified drift exhibits sorting and layering, both indications that it was deposited by running water. Stratified drift is deposited by streams discharging from both valley and continental glaciers, but as you would expect, it is more extensive in areas of continental glaciation.

Outwash Plains and Valley Trains Glaciers discharge meltwater laden with sediment most of the time, except perhaps during the coldest months. This meltwater forms a series of braided streams that radiate out from the front of continental glaciers over a wide region.

ConnectionLink
For more on braided streams and their deposits, see Chapter 12.

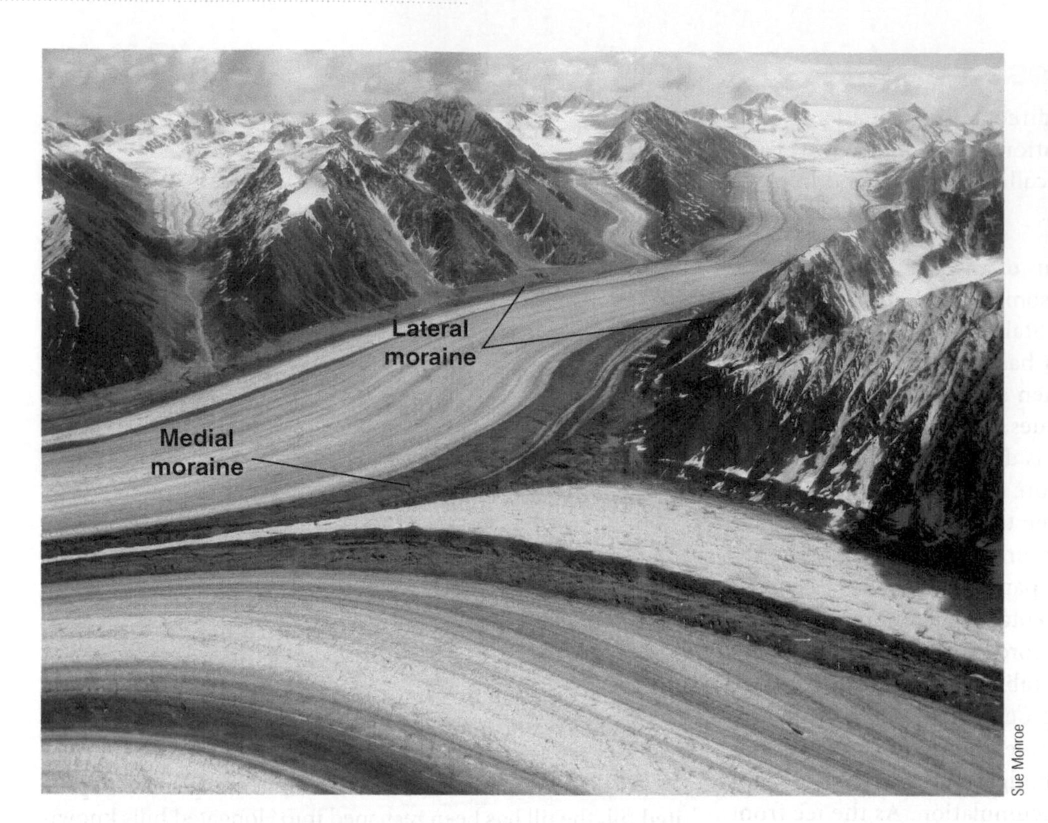

Figure 14.21 Types of Moraines A moraine is a mound or ridge of unstratified till. Moraines are called end, lateral, or medial depending on their position. End moraines are deposited at a glacier's terminus (Figure 14.20), but a lateral moraine is found along the margin of a glacier, and a medial moraine, which forms where two lateral moraines merge, is on the more central part of a glacier.

Critical Thinking Question In addition to moraines, what other features of glacial erosion can you identify?

So much sediment is supplied to these streams that much of it is deposited within their channels as sand and gravel bars. The vast blanket of sediment so formed is an **outwash plain** (Figure 14.18).

Valley glaciers also discharge large amounts of meltwater and, like continental glaciers, have braided streams extending from them. However, these streams are confined to the lower parts of glacial troughs, and their long, narrow deposits of stratified drift are known as **valley trains** (Figure 14.22).

Outwash plains, valley trains, and some moraines commonly contain numerous circular to oval depressions, many of which contain small lakes. These depressions are *kettles* that form when a retreating glacier leaves a block of ice that is subsequently partly or wholly buried (Figure 14.22). When the ice block eventually melts, it leaves a depression; if the depression extends below the water table, it becomes the site of a small lake. Some outwash plains have so many kettles that they are called *pitted outwash plains*.

Kames and Eskers **Kames** are conical hills of stratified drift up to 50 m high (Figure 14.18 and Figure 14.23a). Many form when a stream deposits sediment in a depression on a glacier's surface; as the ice melts, the deposit is lowered to the land surface. Kames also form in cavities within or beneath stagnant ice.

Long sinuous ridges of stratified drift, many of which meander and have tributaries, are **eskers** (Figures 14.18 and 14.23b). Most eskers have sharp crests and sides that slope at about 30 degrees. Some are as high as 100 m and can be traced for more than 500 km. The sorting and stratification of the sediments in eskers clearly indicate deposition by running water. The features of ancient eskers and observations of present-day glaciers show that they form in tunnels beneath stagnant ice. Excellent examples of eskers can be seen at Kettle Moraine State Park in Wisconsin and in several other states, but the most extensive eskers in the world are in northern Canada.

Glacial Lake Deposits

Glacial lakes can form in several ways, but in all cases, they received some or all of their water from melting glaciers. For example, a small lake may be present in a kettle, where water accumulates behind a moraine, or where glacial erosion scours out a depression. In fact, some glacial lakes known as *proglacial lakes* are actually in contact with the ice margin. Regardless of how they formed, glacial lakes, just as all lakes, are areas of sedimentation. Sediment may be carried in by a stream and deposited as a small delta similar to the one shown in Figure 12.14, but here we are interested mostly in glacial lake deposits composed of mud.

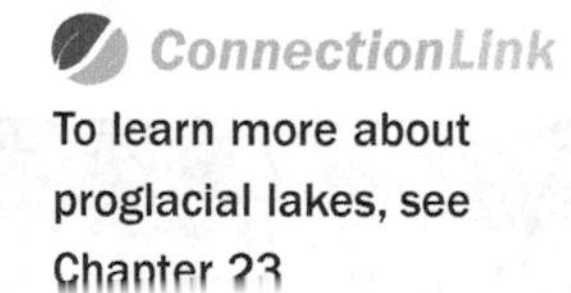

ConnectionLink
To learn more about proglacial lakes, see Chapter 23

Mud deposits in glacial lakes are commonly finely laminated (having layers less than 1 cm thick) and consist of alternating light and dark layers known as *varves* (Figure 14.24a), which represents an annual episode of deposition. The light layer formed during the spring and summer and consists of silt and clay; the dark layer formed during the winter when the smallest particles of clay and organic matter settled from

James S. Monroe

Figure 14.22 **Valley Train and Kettle** This circular depression is a kettle that formed when a block of glacial ice was buried in sand and gravel of a valley train. The stream in the upper part of the image discharges from the nearby Salmon Glacier and is now depositing more sand and gravel in its valley train.

Critical Thinking Question Why do you think that the water in the stream has a milky appearance?

suspension as the lake froze over. The number of varves indicates how many years a glacial lake has existed.

Another distinctive feature of glacial lakes with varves is *dropstones* (Figure 14.24b). These are pieces of gravel, some of boulder size, in otherwise very fine-grained deposits. The presence of varves indicates that currents and turbulence in these lakes were minimal; otherwise, clay and organic matter would not have settled from suspension. Most of them were probably carried into the lakes by icebergs that eventually melted and released sediment contained in the ice.

Figure 14.23 Kames and Eskers

Courtesy of B.M.C. Pape

a This small hill in Wisconsin is a kame.

Christian Zdanovich/William W. Shilts/Illinois State Geological Survey

b This esker was exposed in 1992 from a melting glacier on Bylot Island, Nunavut, Canada.

a Each of these pairs of dark and light layers makes up a varve, an annual deposit in a glacial lake.

b These varves have a dropstone that was probably liberated from floating ice.

Figure 14.24 Varves and a Dropstone in Glacial Deposits

14.7 What Causes Ice Ages?

We discussed the conditions necessary for a glacier to form earlier in this chapter: More snow falls than melts during the warm season, thus accounting for a net accumulation of snow and ice over the years. But this really does not address the broader question of what causes ice ages—that is, times of much more extensive glaciation. Actually, we need to address not only what causes ice ages but also why there have been so few episodes of widespread glaciation in all of Earth history. Only during the Late Proterozoic Eon, the Late Ordovician, Carboniferous, and Permian periods, and the Pleistocene Epoch has Earth had glaciers on a grand scale. Additionally, widespread glaciation occurred several times during the Pleistocene, with each glacial episode separated by a long *interglacial stage* during which glaciers were restricted in their distribution.

ConnectionLink

To learn more about Proterozoic and Paleozoic glaciation, see Chapter 19 and Chapter 20.

For more than a century, scientists have attempted to develop a comprehensive theory explaining all aspects of ice ages, but they have not yet been completely successful. One reason for their lack of success is that the climatic changes responsible for glaciation, the cyclic occurrence of glacial–interglacial episodes, and short-term events such as the Little Ice Age operate on vastly different time scales.

Only a few periods of glaciation are recognized in the geologic record, each separated from the others by long intervals of mild climate. Such long-term climatic changes probably result from slow geographic changes related to plate tectonic activity. Moving plates carry continents to high latitudes where glaciers exist, provided they receive enough precipitation as snow. Plate collisions, the subsequent uplift of vast areas far above sea level, and the changing atmospheric and oceanic circulation patterns caused by the changing shapes and positions of plates also contribute to long-term climate change.

The Milankovitch Theory

Changes in Earth's orbit as a cause of intermediate-term climatic events were first proposed during the mid-1800s, but the idea was made popular during the 1920s by the Serbian astronomer Milutin Milankovitch. He proposed that minor irregularities in Earth's rotation and orbit are sufficient to alter the amount of solar radiation received at any given latitude and hence bring about climate changes. Now called the **Milankovitch theory,** the proposal was initially ignored, but it has received renewed interest since the 1970s and is now widely accepted.

Milankovitch attributed the onset of glacial episodes to variations in three aspects of Earth's orbit. The first is *orbital eccentricity,* which is the degree to which Earth's orbit around the Sun changes over time (Figure 14.25a). When the orbit is nearly circular, both the Northern and Southern Hemispheres have similar contrasts between the seasons. However, if the orbit is more elliptic, hot summers and cold winters will occur in one hemisphere, whereas warm summers and cool winters will take place in the other hemisphere. Calculations indicate a roughly 100,000-year cycle between times of maximum eccentricity, which corresponds closely to the 20 warm–cold climatic cycles that took place during the Pleistocene.

Milankovitch also pointed out that the angle between Earth's axis and a line perpendicular to the plane of Earth's orbit shifts about 1.5 degrees from its current value of 23.5 degrees during a 41,000-year cycle (Figure 14.25b).

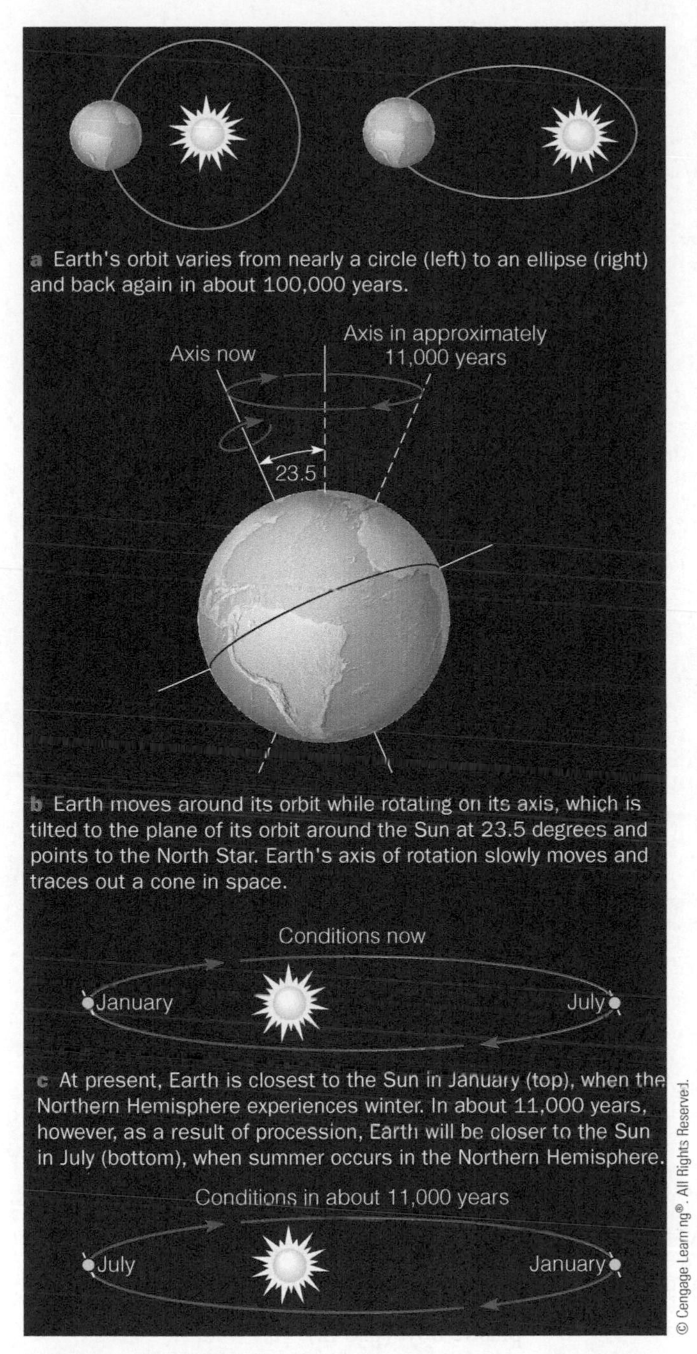

a Earth's orbit varies from nearly a circle (left) to an ellipse (right) and back again in about 100,000 years.

b Earth moves around its orbit while rotating on its axis, which is tilted to the plane of its orbit around the Sun at 23.5 degrees and points to the North Star. Earth's axis of rotation slowly moves and traces out a cone in space.

c At present, Earth is closest to the Sun in January (top), when the Northern Hemisphere experiences winter. In about 11,000 years, however, as a result of procession, Earth will be closer to the Sun in July (bottom), when summer occurs in the Northern Hemisphere.

▌Figure 14.25 The Milankovitch Theory According to the Milankovitch theory, minor irregularities in Earth's rotation and orbit may affect climatic changes.

Although changes in *axial tilt* have little effect on equatorial latitudes, they strongly affect the amount of solar radiation received at high latitudes and the duration of the dark period at and near Earth's poles. Coupled with the third aspect of Earth's orbit, precession of the equinoxes, high latitudes might receive as much as 15% less solar radiation, certainly enough to affect glacial growth and melting.

Precession of the equinoxes, the last aspect of Earth's orbit that Milankovitch cited, refers to a change in the time of the equinoxes. At present, the equinoxes take place on or about March 21 and September 21 when the Sun is directly over the equator. But as Earth rotates on its axis, it also wobbles as its axial tilt varies 1.5 degrees from its current value, thus changing the time of the equinoxes. Taken alone, the time of the equinoxes has little climatic effect, but changes in Earth's axial tilt also change the times of *aphelion* and *perihelion,* which are, respectively, when Earth is farthest from and closest to the Sun during its orbit (▌Figure 14.25c). Earth is now at perihelion, closest to the Sun, during Northern Hemisphere winters, but in about 11,000 years perihelion will be in July. Accordingly, Earth will be at aphelion, farthest from the Sun, in January and have colder winters.

Continuous variations in Earth's orbit and axial tilt cause the amount of solar heat received at any latitude to vary slightly through time. The total heat received by the planet changes little, but according to Milankovitch, and now many scientists agree, these changes cause complex climatic variations and provided the triggering mechanism for the glacial–interglacial episodes of the Pleistocene.

Short-Term Climatic Events

Climatic events with durations of several centuries, such as the Little Ice Age, are too short to be accounted for by plate tectonics or Milankovitch cycles. Several hypotheses have been proposed, including variations in solar energy and volcanism.

Variations in solar energy could result from changes within the Sun itself or from anything that would reduce the amount of energy Earth receives from the Sun. The latter could result from the solar system passing through clouds of interstellar dust and gas or from substances in the atmosphere reflecting solar radiation back into space. Records kept over the past 90 years indicate that during this time the amount of solar radiation has varied only slightly. Although variations in solar energy may influence short-term climatic events, such a correlation has not been demonstrated.

During large volcanic eruptions, tremendous amounts of ash and gases are spewed into the atmosphere, where they reflect incoming solar radiation and thus reduce atmospheric temperatures. Small droplets of sulfur gases remain in the atmosphere for years and can have a significant effect on climate. Several large-scale volcanic events have occurred, such as the 1815 eruption of Tambora, and are known to have had climatic effects. However, no relationship between periods of volcanic activity and periods of glaciation has yet been established.

Key Concepts Review

- Glaciers currently cover about 10% of the land surface and contain about 1.74% of all water on Earth.
- A glacier forms when winter snowfall exceeds summer melt and accumulates year after year. Snow is compacted and converted to glacial ice, and when the ice is about 40 m thick, pressure causes it to flow.
- Glaciers move by plastic flow and basal slip.
- Valley glaciers are confined to mountain valleys and flow from higher to lower elevations, whereas continental glaciers cover vast areas and flow outward in all directions from a zone of accumulation.
- The behavior of a glacier depends on its budget, which is the relationship between accumulation and wastage. If a glacier has a balanced budget, its terminus remains stationary; a positive or negative budget results in the advance or retreat of the terminus, respectively.
- Glaciers move at varying rates depending on slope, discharge, and season. Valley glaciers tend to flow more rapidly than continental glaciers.
- Glaciers effectively erode and transport because they are solids in motion. They are particularly effective at eroding soil and unconsolidated sediment, and they can transport any size sediment supplied to them.
- Continental glaciers transport most of their sediment in the lower part of the ice, whereas valley glaciers may carry sediment in all parts of the ice.
- Erosion of mountains by valley glaciers yields several sharp, angular landforms, including cirques, arêtes, and horns. U-shaped glacial troughs, fiords, and hanging valleys are also products of valley glaciation.
- Continental glaciers abrade and bevel high areas, producing a smooth, rounded landscape known as an ice-scoured plain.
- Depositional landforms include moraines, which are ridgelike accumulations of till. The several types of moraines are terminal, recessional, lateral, and medial.
- Drumlins are composed of till that was apparently reshaped into streamlined hills by continental glaciers or floods.
- Stratified drift in outwash plains and valley trains consists of sand and gravel deposited by meltwater streams issuing from glaciers. Ridges known as eskers, and conical hills called kames are also composed of stratified drift.
- Major glacial intervals separated by tens or hundreds of millions of years probably occur as a result of the changing positions of tectonic plates, which in turn cause changes in oceanic and atmospheric circulation patterns.
- Currently, the Milankovitch theory is widely accepted as the explanation for glacial–interglacial intervals.
- The reasons for short-term climatic changes, such as the Little Ice Age, are not understood. Two proposed causes are changes in the amount of solar energy received by Earth and volcanism.

Important Terms

abrasion (p. 349)
arête (p. 351)
basal slip (p. 342)
cirque (p. 351)
continental glacier (p. 339)
drumlin (p. 355)
end moraine (p. 355)
esker (p. 356)
fiord (p. 350)
firn (p. 342)
glacial budget (p. 344)
glacial drift (p. 353)
glacial erratic (p. 353)
glacial ice (p. 342)
glacial polish (p. 349)
glacial striations (p. 349)
glacial surge (p. 348)
glaciation (p. 339)
glacier (p. 338)
ground moraine (p. 355)
hanging valley (p. 351)
horn (p. 351)
ice cap (p. 339)
ice-scoured plain (p. 353)
kame (p. 356)
lateral moraine (p. 355)
Little Ice Age (p. 338)
medial moraine (p. 355)
Milankovitch theory (p. 358)
outwash plain (p. 356)
plastic flow (p. 342)
recessional moraine (p. 355)
stratified drift (p. 353)
terminal moraine (p. 355)
till (p. 353)
U-shaped glacial trough (p. 350)
valley glacier (p. 338)
valley train (p. 356)
zone of accumulation (p. 344)
zone of wastage (p. 344)

Review Questions

1. A medial moraine forms where
   a. ___ snow is converted to glacial ice.
   b. ___ erosion by a valley glacier forms a cirque.
   c. ___ two lateral moraines merge.
   d ___ isolated blocks of ice are buried in outwash.
   e. ___ a continental glacier abrades its underlying surface.
2. The line on a glacier that separates the zone of accumulation from the zone of wastage is the
   a. ___ drift area.
   b. ___ firm limit.
   c. ___ lateral termination.
   d. ___ ground moraine.
   e. ___ drumlin.
3. Glaciers move mostly by
   a. ___ plastic flow.
   b. ___ abrasion.
   c. ___ plucking.
   d. ___ basal surge.
   e. ___ elastic rebound.
4. A distinctive erosional feature of mountains that have been eroded by valley glaciers is a/an
   a. ___ valley train.
   b. ___ kame.
   c. ___ ice-scoured plain.
   d. ___ cirque.
   e. ___ stratified drift.
5. When freshly fallen snow compacts and partly melts and refreezes, it form granular ice known as
   a. ___ till.
   b. ___ outwash.
   c. ___ glacial frost.
   d. ___ firn.
   e. ___ glacial drift.
6. How does the Milankovitch theory account for the onset of glacial episodes?
7. What kinds of evidence would indicate that an ice-free mountain area was once glaciated?
8. A valley glacier has a cross-sectional area of 400,000 m^2 and a flow velocity of 2 m per day. How long will it take for 1 km^3 of ice to move past a given point?
9. Explain in terms of the glacial budget how a once-active glacier becomes stagnant.

10. **Creative Thinking Visual Question:** This image (▌Figure 1) shows the Chugach Mountains in Alaska. Identify the land forms that resulted from valley glaciation.

Photograph by Bruce F. Molina

▌**Figure 1** Glaciers in the Chugach Mountains of Alaska.

Global GeoScience Watch Search for "Antarctic" within the GREENR database, and under "Magazines" click on "Antarctic Fjord Formation Dated." How did minerals in a sediment core from the Antarctic help scientists determine when that continent became glacier-covered? When did this event take place?

Picture a house surrounded by palm trees, nestled in an oasis, and surrounded by large sand dunes in the Sahara Desert of Morocco. Contrast this view with the fact that the Sahara was a fertile savannah more than 6,000 years ago, thus illustrating the importance of studying how global climate change, in conjunction with human activities, plays a significant role in changes to the ecosystem of a region.

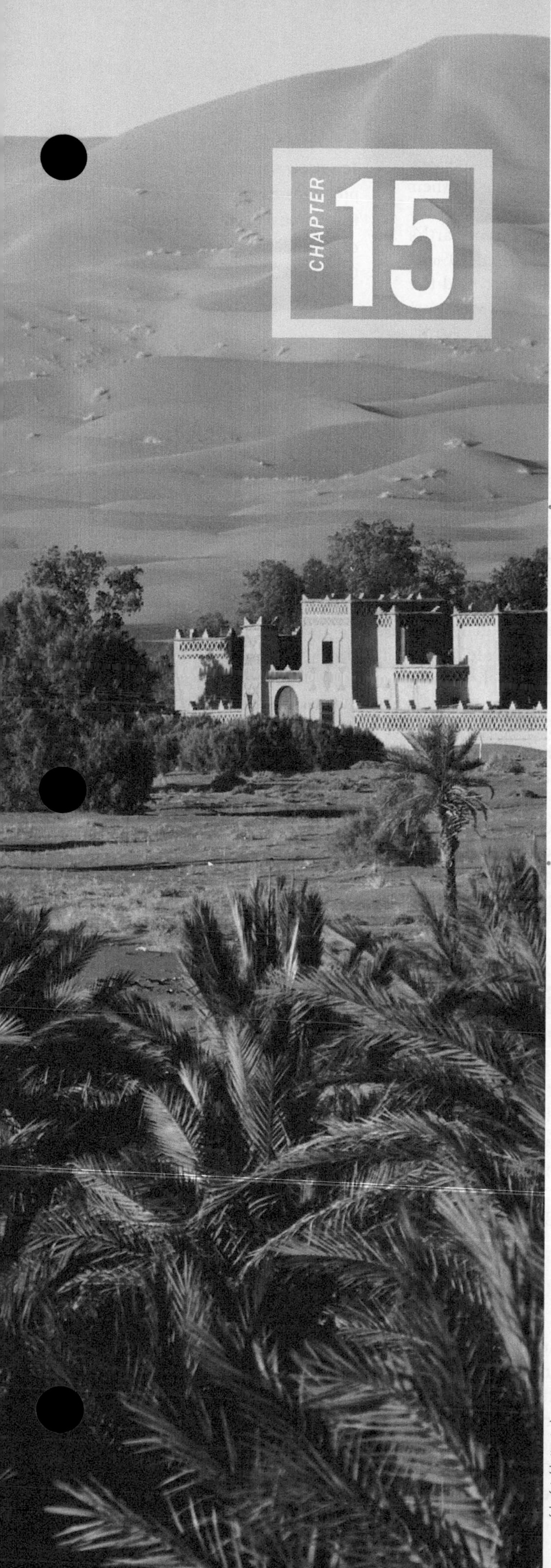

CHAPTER 15

The Work of Wind and Deserts

OUTLINE

HAVE YOU EVER WONDERED?

- How wind transports and erodes sediment, thus modifying the landscape?
- How dunes form, and how they are classified?
- What loess is, and why it forms some of the world's most fertile soils?
- Why deserts are so hot and dry, and why they are distributed where they are in the world today and in the past?
- If running water is important in the formation of desert landforms?
- Why buttes and mesas rise pillar-like in deserts?
- What petroglyphs are, and why they are an important cultural resource?

15.1 Introduction

During the past several decades, deserts have been advancing across millions of acres of productive land, destroying rangelands, croplands, and even villages. Such expansion, estimated at a rate of 48 km per year southward for the Sahara Desert alone, has exacted a terrible toll in human suffering. Because of the relentless advance of deserts, hundreds of thousands of people have died of starvation or been forced to migrate as "environmental refugees" from their homelands to camps, where the majority are severely malnourished.

This expansion of deserts into formerly productive lands is called **desertification**, and it affects about 70% of Earth's dry lands, which is approximately 30% of Earth's land surface. Desertification not only is a major problem in many countries but also is becoming a global issue due to the confluence of climate change and such human activities as destructive agricultural practices and overgrazing by livestock.

Most regions undergoing desertification lie along the margins of existing deserts where a delicately balanced ecosystem serves as a buffer between the desert on one side and a more humid environment on the other. These regions have limited potential to adjust to increasing environmental pressures from natural causes as well as human activity. Ordinarily, desert regions expand and contract gradually in response to natural processes such as climatic change, but much of recent desertification has been greatly accelerated by human activities.

In many areas, the natural vegetation has been cleared as crop cultivation has expanded into increasingly drier fringes to support growing populations. Because grasses are the dominant natural vegetation in most desert fringe areas, raising livestock is a common economic activity. Increasing numbers of livestock in many areas, however, have greatly exceeded the land's ability to support them. Consequently, the vegetation cover that protects the soil has diminished, causing the soil to crumble and to be stripped away by wind and water, which results in increased desertification and the many problems associated with it.

One particularly hard-hit area of desertification is the Sahel of Africa (a belt 300–1,100 km wide, lying south of the Sahara). Because drought is common in the Sahel, the region can support only a limited population of livestock and humans. Unfortunately, expanding human and animal populations and more intensive agriculture have increased the demands on the land. Plagued with periodic droughts, this region has suffered tremendously as crops have failed and livestock has overgrazed the natural vegetation, resulting in thousands of deaths, displaced people, and the encroachment of the Sahara.

The tragedy of the Sahel and prolonged droughts in other desert fringe areas remind us of the delicate equilibrium of ecosystems in such regions. Once the fragile soil cover has been removed by erosion, it takes centuries for new soil to form (see Chapter 6).

There are many important reasons to study deserts and the processes that are responsible for their formation. First, deserts cover large regions of Earth's surface. More than 40% of Australia is desert, and the Sahara occupies a vast part of northern Africa. Additionally, many desert regions are experiencing increases in population growth, such as the high desert area of southern California and parts of Nevada, as well as various locations in Arizona. Many of these places already have problems associated with increasing population and the strains it places on the environment, particularly the need for great amounts of groundwater.

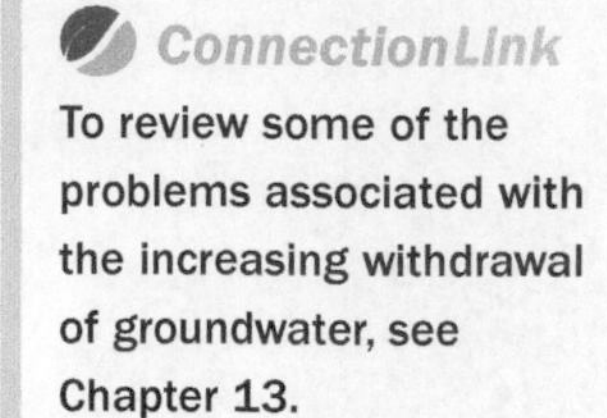
ConnectionLink
To review some of the problems associated with the increasing withdrawal of groundwater, see Chapter 13.

Furthermore, with the current debate about global warming, it is important to understand how desert processes operate and how global climate changes affect the various Earth systems and subsystems. Learning about the underlying causes of climate change by examining ancient desert regions may provide insight into the possible duration and severity of future climatic changes.

As an example, more than 6,000 years ago, the Sahara was a fertile savannah supporting a diverse fauna and flora, including humans. Then the climate changed, and the area became a desert. How did this happen? Will this region change back again in the future? These are some of the questions geoscientists hope to answer by studying deserts. By understanding the underlying causes of desertification, it might be possible to implement steps to reduce the destruction done by desertification, particularly in terms of human suffering.

15.2 Sediment Transport by Wind

Wind is a turbulent fluid and therefore transports sediment in much the same way as running water. Although wind typically flows at a greater velocity than water, it has a lower density and thus can carry only clay- and silt-size particles as *suspended load*. Sand and larger particles are moved along the ground as *bed load*.

Bed Load

Sediments that are too large or heavy to be carried in suspension by water or wind are moved as bed load either by *saltation* or by rolling and sliding. Saltation on land occurs when wind starts sand grains rolling and lifts and carries some grains short distances before they fall back to the surface. As the descending sand grains hit the surface, they strike other grains, causing them to bounce along (▮ Figure 15.1). Wind-tunnel experiments show that once sand grains begin moving,

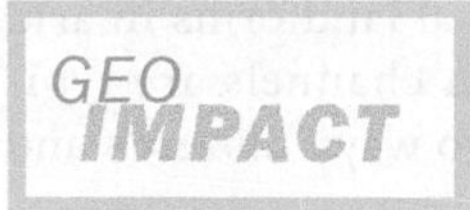

What Can Be Done with Nuclear Waste?

One problem of the nuclear age is finding safe storage sites for radioactive waste from nuclear power plants, the manufacture of nuclear weapons, and the radioactive by-products of nuclear medicine. Desert regions, such as Nevada's Yucca Mountain, have been suggested as possible repositories of nuclear waste because of their dry conditions, very deep water table, and remoteness from population centers. Even though the proposed Yucca Mountain Nuclear Waste Repository, after more than 23 years of study, was terminated in 2011 because of political reasons, nuclear waste must be stored somewhere. Discuss what the possible advantages are to constructing such storage sites in desert areas, such as Nevada's Yucca Mountain. What are the possible disadvantages?

Note: At the time of this writing (August 2013), the United States District Court of Appeals in Washington, D.C., has ordered the reopening of the Department of Energy's license application for a nuclear waste repository at Yucca Mountain, Nevada.

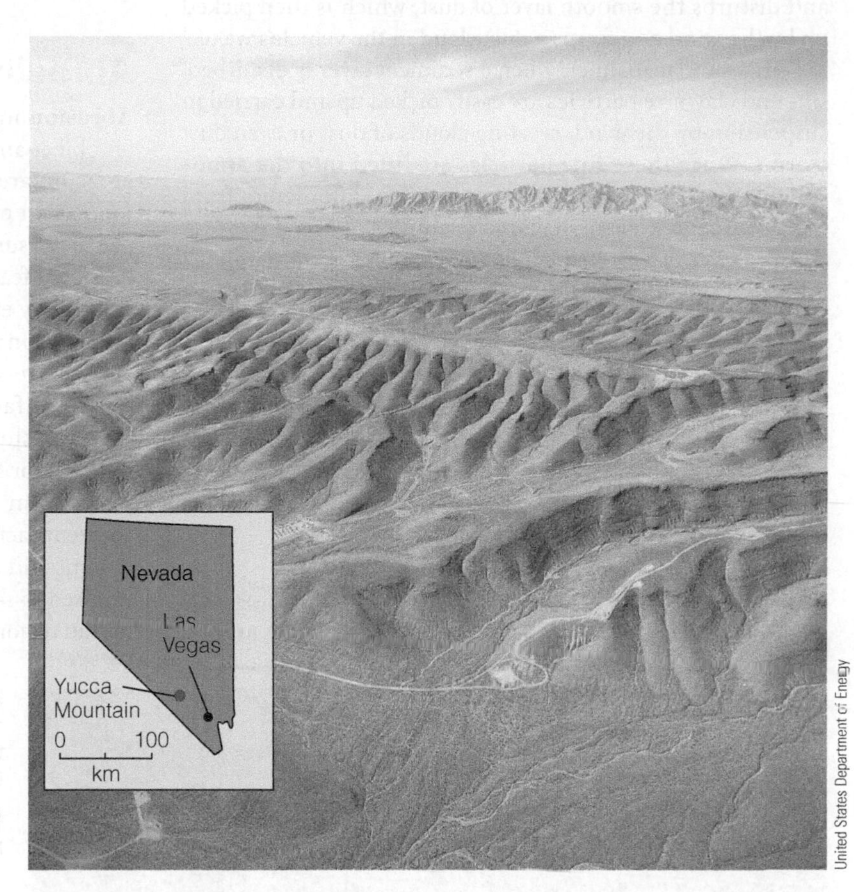

Location and aerial view of Nevada's Yucca Mountain, a once-proposed high-level radioactive waste site.

they continue to move, even if the wind drops below the speed necessary to start them moving. This happens because once saltation begins, it sets off a chain reaction of collisions between sand grains that keeps the grains in constant motion.

Saltating sand usually moves near the surface, and even when winds are strong, grains are rarely lifted higher than about a meter. If the winds are very strong, these wind-whipped grains can cause extensive abrasion. It is not uncommon during a severe sandstorm to have a car's paint removed by sandblasting in a short time, and its windshield become completely frosted and translucent from pitting.

Suspended Load

Silt- and clay-size particles constitute most of a wind's suspended load. Even though these particles are much smaller and lighter than sand-size particles, wind usually starts the latter moving first. The reason for this phenomenon is that a very thin layer of motionless air lies next to the ground where the small silt and clay particles remain undisturbed. The larger sand grains, however, stick up into the turbulent air zone where they can be moved. Unless the stationary air layer is disrupted, the silt and clay particles remain on the ground, providing a smooth surface.

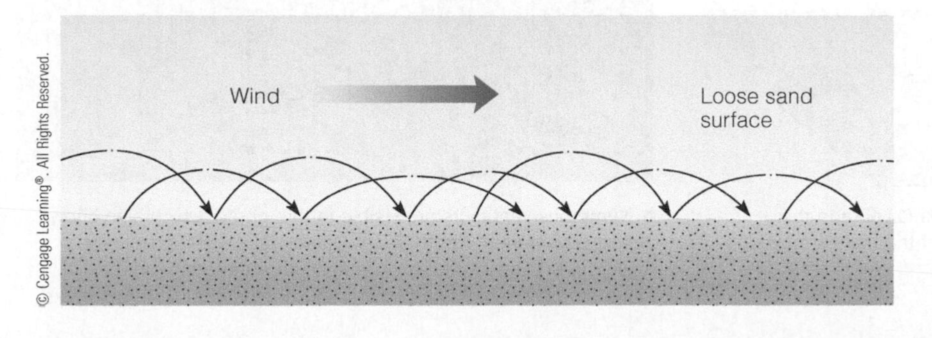

Figure 15.1 Saltation Most sand is moved near the ground surface by saltation. Sand grains are picked up by the wind and carried a short distance before falling back to the ground, where they usually hit other grains, causing them to bounce and move in the direction of the wind.

This phenomenon can be observed on a dirt road on a windy day. Unless a vehicle travels over the road, little dust is raised even though it is windy. When a vehicle moves over the road, however, it breaks the calm boundary layer of air and disturbs the smooth layer of dust, which is then picked up by the wind and forms a dust cloud in the vehicle's wake.

In a similar manner, when a sediment layer is disturbed, silt- and clay-size particles are easily picked up and carried in suspension by the wind, creating clouds of dust or even dust storms. Once these fine particles are lifted into the atmosphere, they may be carried thousands of kilometers from their source. For example, large quantities of fine dust from the southwestern United States were blown eastward and fell on New England during the dust storms of the 1930s (see Figure 6.16). Most recently, and reminiscent of the 1930s dust storm, a massive dust cloud, also known as a "haboob," up to 3,100 m high, swept through the Phoenix, Arizona, area in July 2011, depositing sand and dust, delaying flights, and causing numerous power outages.

15.3 Wind Erosion

Although wind action produces many distinctive erosional features and is an extremely efficient sorting agent, running water is responsible for most erosional landforms in arid regions, despite the fact that stream channels are typically dry. Wind erodes material in two ways: *abrasion* and *deflation*.

Abrasion

Abrasion involves the impact of saltating sand grains on an object and is analogous to sandblasting. The effects of abrasion are usually minor because sand, the most common agent of abrasion, is rarely carried more than a meter above the surface (▌Figure 15.2). Rather than creating major erosional features, wind abrasion typically modifies existing features by etching, pitting, smoothing, or polishing. Nonetheless, wind abrasion can produce many strange-looking and bizarre-shaped features.

Ventifacts are a common product of wind abrasion; these are stones whose surfaces have been polished, pitted, grooved, or faceted by the wind (▌Figure 15.3). If the wind blows from different directions, or if the stone is moved, the ventifact will have multiple facets. Ventifacts are most common in deserts, yet they can form wherever stones are exposed to saltating sand grains—for example, on beaches in humid regions and on some outwash plains.

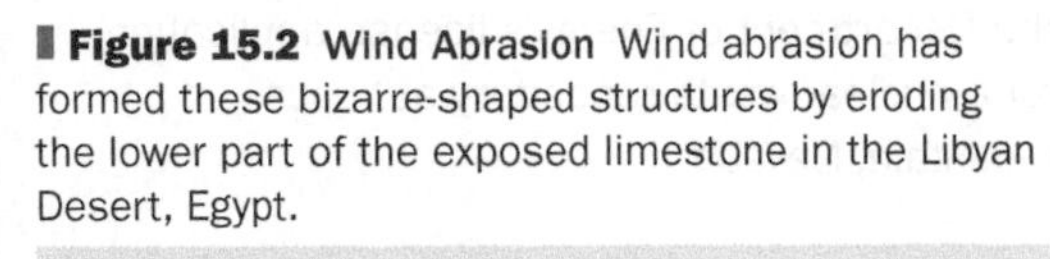

▌**Figure 15.2 Wind Abrasion** Wind abrasion has formed these bizarre-shaped structures by eroding the lower part of the exposed limestone in the Libyan Desert, Egypt.

Critical Thinking Question Why is only the lower part of these outcrops eroded by wind abrasion?

▌**Figure 15.3 Ventifacts**

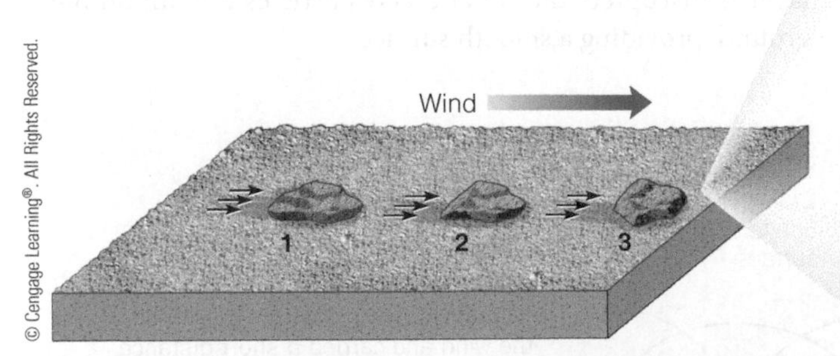

a A ventifact forms when windborne particles (1) abrade the surface of a rock, (2) forming a flat surface. If the rock is moved, (3) additional flat surfaces are formed.

b Numerous ventifacts are visible in this photo, which also shows desert pavement in Death Valley, California. Desert pavement prevents further erosion and transport of a desert's surface materials by forming a protective layer of close-fitting, larger rocks.

Figure 15.4 **Yardang** A profile view of a streamlined yardang in the Roman playa deposits of the Kharga Depression, Egypt. Yardangs form by wind erosion.

Yardangs are larger features than ventifacts and also result from wind erosion (Figure 15.4). They are elongated, streamlined ridges that look like an overturned ship's hull. Yardangs are typically found in clusters aligned parallel to the prevailing winds. They probably form by differential erosion in which depressions, parallel to the direction of wind, are carved out of a rock body, leaving sharp, elongated ridges. These ridges may then be further modified by wind abrasion into their characteristic shape.

Deflation

Another important mechanism of wind erosion is **deflation**, which is the removal of loose surface sediment by the wind. Among the characteristic features of deflation in many arid and semiarid regions are *deflation hollows*, or *blowouts* (Figure 15.5). These shallow depressions of variable dimensions result from differential erosion of surface materials. Ranging in size from several kilometers in diameter and tens of meters deep to small depressions only a few meters wide and less than a meter deep, deflation hollows are common in the southern Great Plains region of the United States.

In many dry regions, the removal of sand-size and smaller particles by wind leaves a surface of pebbles, cobbles, and boulders. As the wind removes the fine-grained material from the surface, the effects of gravity and occasional heavy rain, and even the swelling of clay grains, rearrange the remaining coarse particles into a mosaic of close-fitting rocks called **desert pavement** (Figure 15.3b and Figure 15.6). Once desert pavement forms, it protects the underlying material from further deflation.

Figure 15.5 **Deflation Hollow** A deflation hollow, shown here as the low area between two sand dunes in Death Valley, California, results when loose surface sediment is differentially removed by wind.

15.4 Wind Deposits

Although wind is of minor importance as an erosional agent, it is responsible for impressive deposits, which are primarily of two types. The first, *dunes*, occur in several distinctive types, all of which consist of sand-size particles that are usually deposited near their source. The second, *loess*, consists of layers of wind-blown silt and clay deposited over large areas downwind and commonly far from their source.

Figure 15.6 Desert Pavement

Critical Thinking Question Why is desert pavement important in desert environments?

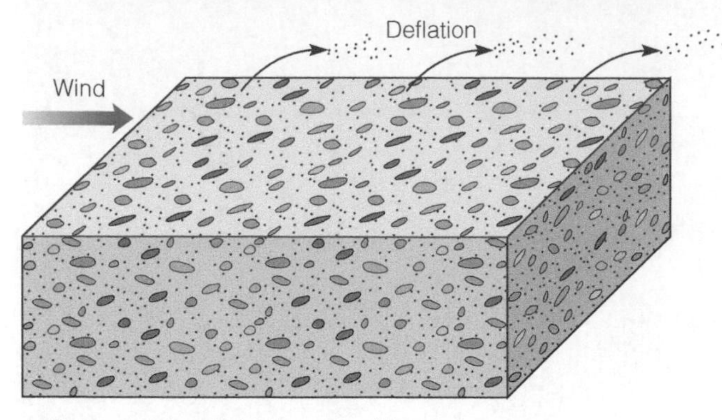

a Fine-grained material is removed by wind.

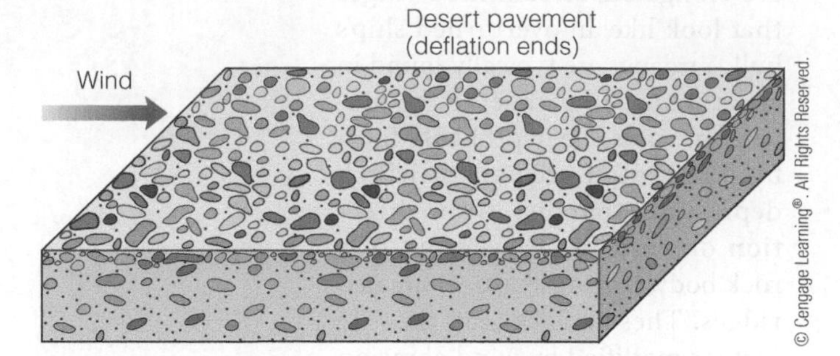

b leaving a concentration of larger particles that form desert pavement.

The Formation and Migration of Dunes

The most characteristic features in sand-covered regions are **dunes**, which are mounds or ridges of wind-deposited sand (Figure 15.7). Dunes form when wind flows over and around an obstruction, resulting in the deposition of sand grains, which accumulate and build up a deposit of sand. As they grow, these sand deposits become self-generating in that they form ever-larger wind barriers that further reduce the wind's velocity, resulting in more sand deposition and growth of the dune.

Most dunes have an asymmetrical profile, with a gentle windward slope and a steeper downwind or leeward slope that is inclined in the direction of the prevailing wind (Figure 15.8a). Sand grains move up the gentle windward slope by saltation and accumulate on the leeward side, forming an angle of 30 to 34 degrees from the horizontal,

Figure 15.7 Sand Dunes Large sand dunes in Death Valley, California. The prevailing wind direction is from left to right as indicated by the sand dunes in which the gentle windward side is on the left and the steeper leeward slope is on the right.

Figure 15.8 Dune Migration

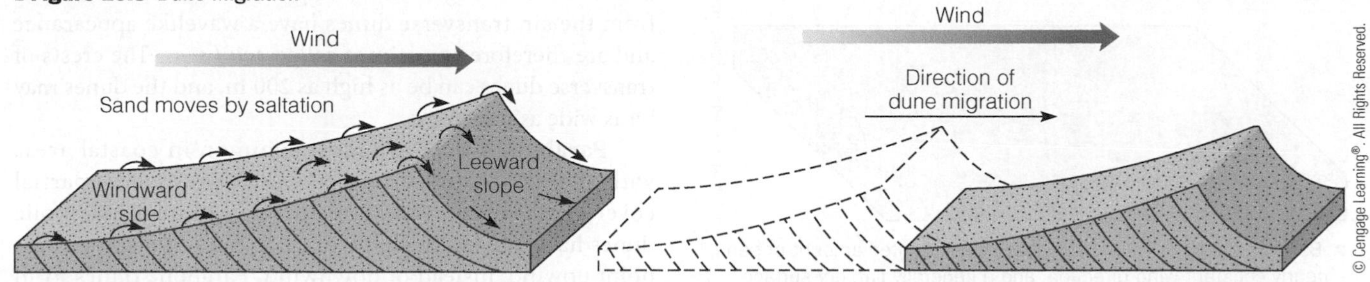

a Profile of a sand dune.

b Dunes migrate when sand moves up the windward side and slides down the leeward slope. Such movement of the sand grains produces a series of cross-beds that slope in the direction of wind movement.

which is the angle of repose of dry sand. When this angle is exceeded by accumulating sand, the slope collapses, and the sand slides down the leeward slope, coming to rest at its base. As sand moves from a dune's windward side and periodically slides down its leeward slope, the dune slowly migrates in the direction of the prevailing wind (Figure 15.8b). When preserved in the geologic record, dunes help geologists determine the prevailing direction of ancient winds (Figure 15.9).

To review the sedimentary structure cross-bedding and what it can tell geologists about ancient depositional environments, see Chapter 7.

Dune Types

Geologists recognize four major dune types (barchan, longitudinal, transverse, and parabolic), although intermediate forms also exist. The size, shape, and arrangement of dunes result from the interaction of such factors as sand supply, the direction and velocity of the prevailing wind, and the amount of vegetation. Although dunes are usually found in deserts, they can also develop wherever sand is abundant, such as along the upper parts of many beaches.

Barchan dunes are crescent-shaped dunes whose tips point downwind (Figure 15.10). They form in areas that have a generally flat, dry surface with little vegetation, a limited supply of sand, and a nearly constant wind direction. Most barchans are small, with the largest reaching about 30 m high. Barchans are the most mobile of the major dune types, moving at rates that can exceed 10 m per year.

Longitudinal dunes (also called *seif dunes*) are long, parallel ridges of sand aligned generally parallel to the direction of the prevailing winds; they form where the sand supply is somewhat limited (Figure 15.11). Longitudinal dunes result when winds converge from slightly different directions to produce the prevailing wind. They range in height from about 3 m to more than 100 m, and some stretch for more than 100 km. Longitudinal dunes are especially well developed in central Australia, where they cover nearly one-fourth of the continent. They also cover extensive areas in Saudi Arabia, Egypt, and Iran.

Transverse dunes form long ridges perpendicular to the prevailing wind direction in areas that have abundant sand

Figure 15.9 Cross-Bedding Ancient cross-bedding in sandstone beds in Zion National Park, Utah, helps geologists determine the prevailing direction of the wind that formed these ancient sand dunes.

Critical Thinking Question Did the prevailing direction of wind change during the time that the sand forming these sandstone beds was deposited? Explain.

Figure 15.10 Barchan Dunes

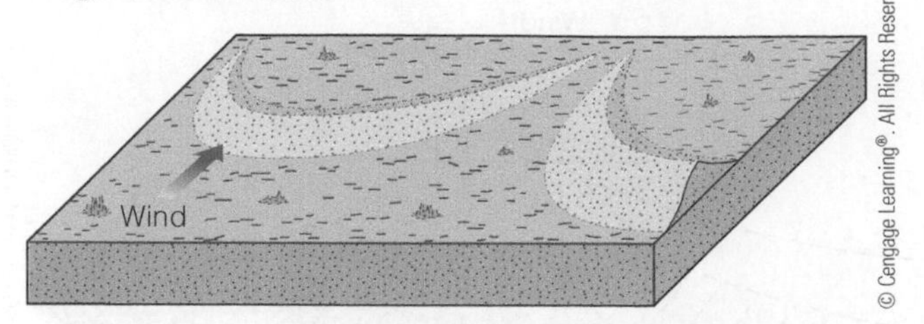

a Barchan dunes form in areas that have a limited amount of sand, a nearly constant wind direction, and a generally flat, dry surface with little vegetation. The tips of barchan dunes point downward.

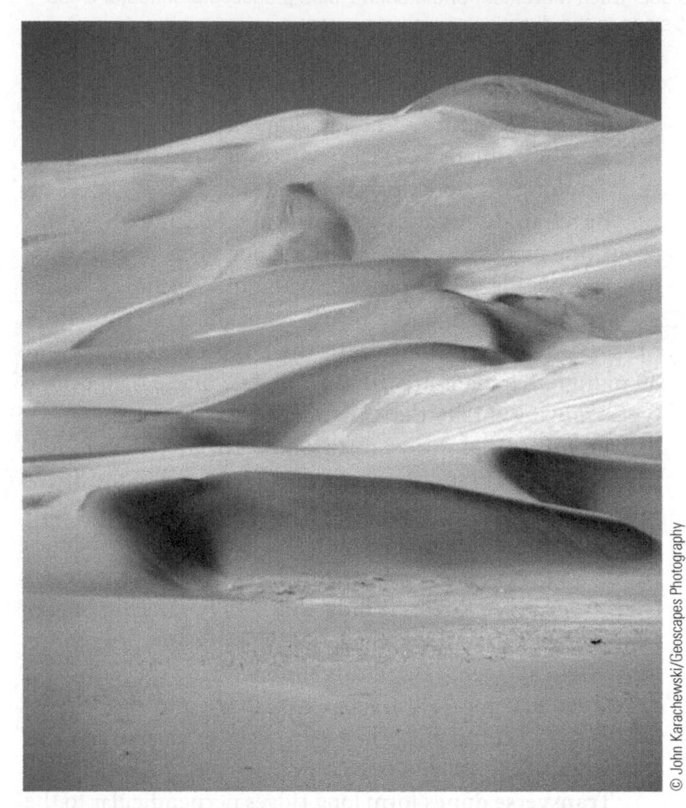

b A ground-level view of several barchan dunes.

Figure 15.11 Longitudinal Dunes

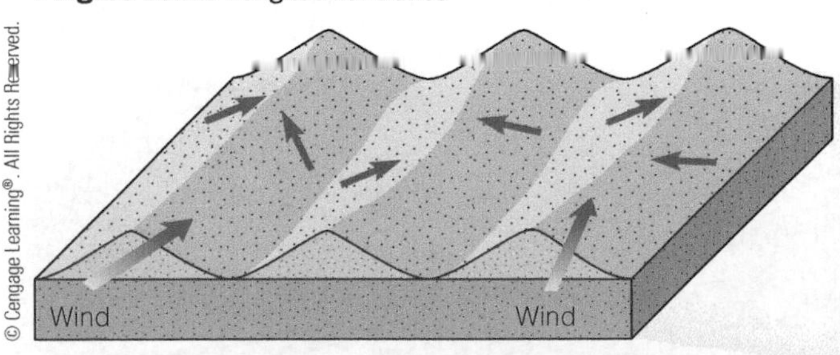

a Longitudinal dunes form long, parallel ridges of sand aligned roughly parallel to the prevailing wind direction. They typically form where sand supplies are limited.

b Longitudinal dunes, 15 m high, in the Gibson Desert, west central Australia. The bright blue areas between the dunes are shallow pools of rainwater, and the darkest patches are areas where the Aborigines have set fires to encourage the growth of spring grasses.

and little or no vegetation (Figure 15.12). When viewed from the air, transverse dunes have a wavelike appearance and are therefore sometimes called *sand seas*. The crests of transverse dunes can be as high as 200 m, and the dunes may be as wide as 3 km.

Parabolic dunes are most common in coastal areas with abundant sand, strong onshore winds, and a partial cover of vegetation (Figure 15.13). Although parabolic dunes have a crescent shape like barchan dunes, their tips point upwind, instead of downwind. Parabolic dunes form when the vegetation cover is broken and deflation produces a deflation hollow or blowout. As the wind transports the sand out of the depression, it builds up on the convex downwind dune crest. The central part of the dune is excavated by the wind, while vegetation holds the ends and sides fairly well in place.

Another type of dune commonly found in the deserts of North Africa and Saudi Arabia is the *star dune*, so named because of its resemblance to a multipointed star (Figure 15.14). Star dunes are among the taller dunes in the world, rising, in some cases, more than 100 m above the surrounding desert plain. They consist of pyramidal hills of sand, from which radiate several ridges of sand, and they develop where the wind direction is variable. Star dunes can remain stationary for centuries and have served as desert landmarks for nomadic peoples.

Loess

Wind-blown silt and clay deposits composed of angular quartz grains, feldspar, micas, and calcite are known as **loess**. The distribution of loess shows that it is derived from three main sources: deserts, Pleistocene glacial outwash deposits, and the floodplains of rivers in semiarid regions. Loess must be stabilized by moisture and vegetation in order to accumulate. Consequently, loess is not found in deserts, even though deserts provide much of its material. Because of its unconsolidated nature, loess is easily eroded, and as a result, eroded loess areas are characterized by steep cliffs and rapid lateral and headward stream erosion (Figure 15.15).

Figure 15.12 Transverse Dunes

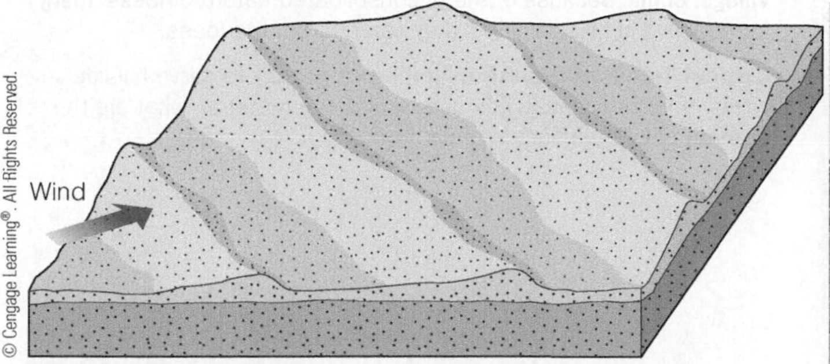

a Transverse dunes form long ridges of sand that are perpendicular to the prevailing wind direction in areas of little or no vegetation and abundant sand.

b Transverse dunes, Great Sand Dunes National Park, Colorado.

Figure 15.13 Parabolic Dunes

Critical Thinking Question Why do the tips of a parabolic dune point upwind and those of a barchan dune point downwind?

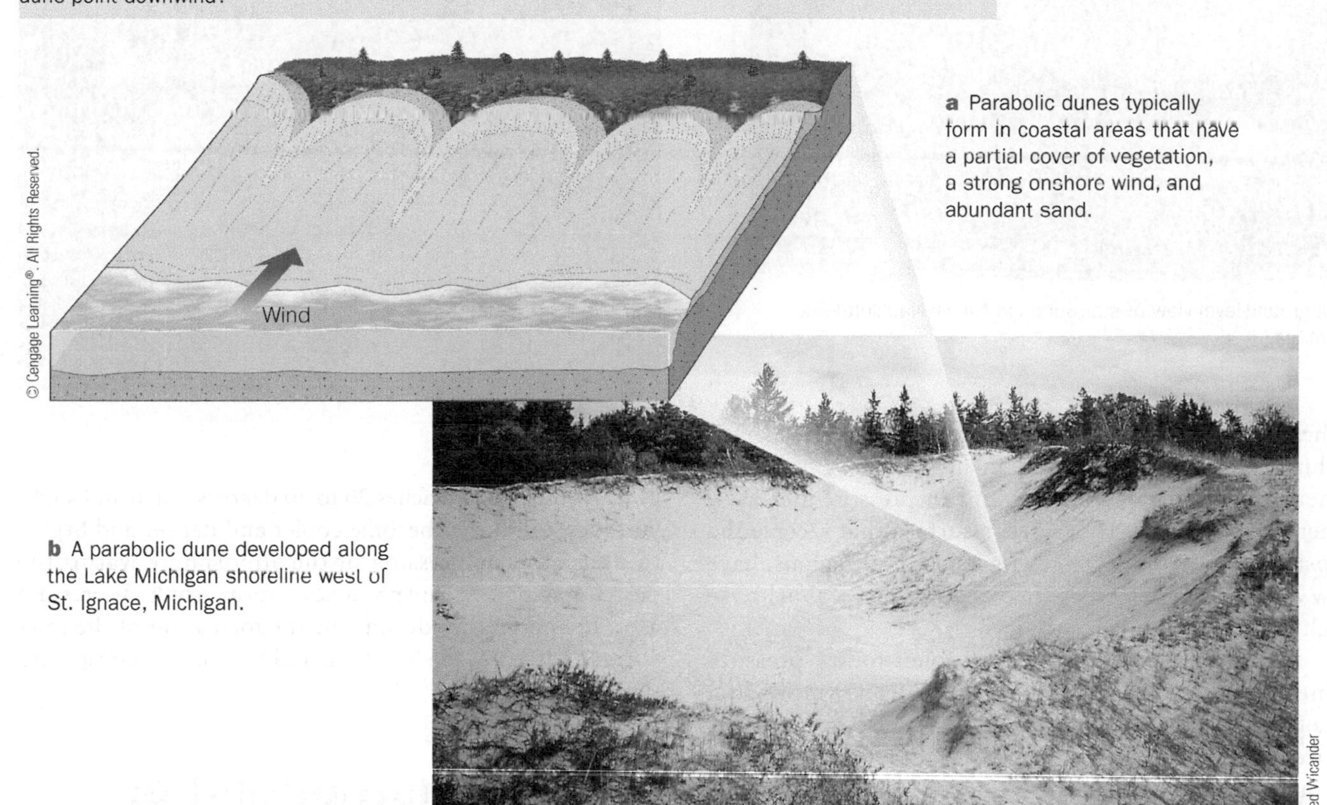

a Parabolic dunes typically form in coastal areas that have a partial cover of vegetation, a strong onshore wind, and abundant sand.

b A parabolic dune developed along the Lake Michigan shoreline west of St. Ignace, Michigan.

Presently, loess deposits cover approximately 10% of Earth's land surface and 30% of the United States. The most extensive and thickest loess deposits are found in northeast China, where accumulations greater than 30 m thick are common. Loess-derived soils are some of the world's most fertile (Figure 15.15). It is therefore not surprising that the world's major grain-producing regions correspond to large loess deposits, such as the North European Plain, Ukraine, and the Great Plains of North America.

15.5 Air-Pressure Belts and Global Wind Patterns

To understand the work of wind and the distribution of deserts, we need to consider the global pattern of air-pressure belts and winds, which are responsible for Earth's atmospheric circulation patterns. Air pressure is the density of air exerted on its surroundings (i.e., its weight). When air

Figure 15.14 Star Dunes

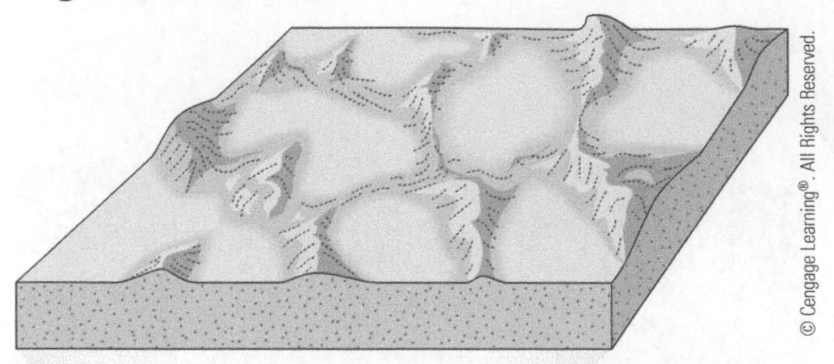

a Star dunes are pyramidal hills of sand that develop where the wind direction is variable.

b A ground-level view of star dunes in Namib-Naukluft Park, Namibia.

Figure 15.15 **Terraced Wheat Fields in the Loess Soil at Tangwa Village, China** Because of the unconsolidated nature of loess, many farmers live in hillside caves they carved from the loess.

Critical Thinking Question Although it is easy to carve hillside caves in loess because it is unconsolidated material, what are the hazards of living in such a structure?

is heated, it expands and rises, reducing its mass for a given volume and causing a decrease in air pressure. Conversely, when air is cooled, it contracts and air pressure increases. Therefore, those areas of Earth's surface that receive the most solar radiation, such as the equatorial regions, have low air pressure, whereas the colder, polar regions have high air pressure.

Air flows from high-pressure zones to low-pressure zones. If Earth did not rotate, winds would move in a straight line from one zone to another. Because Earth rotates, however, winds are deflected to the right of their direction of motion (clockwise) in the Northern Hemisphere and to the left of their direction of motion (counterclockwise) in the Southern Hemisphere. This deflection of air between latitudinal zones resulting from Earth's rotation is known as the **Coriolis effect**. The combination of latitudinal pressure differences and the Coriolis effect produces a worldwide pattern of east–west–oriented wind belts (Figure 15.16).

Earth's equatorial zone receives the most solar energy, which heats the surface air and causes it to rise. As the air rises, it cools and releases moisture that falls as rain in the equatorial region (Figure 15.16). The rising air is now much drier as it moves northward and southward toward each pole. By the time it reaches 20 to 30 degrees north and south latitudes, the air has become cooler and denser and begins to descend. Compression of the atmosphere warms the descending air mass and produces a warm, dry, high-pressure area, the perfect conditions for the formation of the low-latitude deserts of the Northern and Southern Hemispheres (Figure 15.17).

15.6 The Distribution of Deserts

Dry climates occur in the low and middle latitudes where the potential loss of water by evaporation may exceed the yearly precipitation (Figure 15.17). Dry climates cover about 30% of Earth's land surface and are subdivided into semiarid and arid regions. *Semiarid regions* receive more precipitation than arid regions, yet are moderately dry. Their soils are usually well developed and fertile, and support a natural grass cover. *Arid regions*, generally described as **deserts**, are dry;

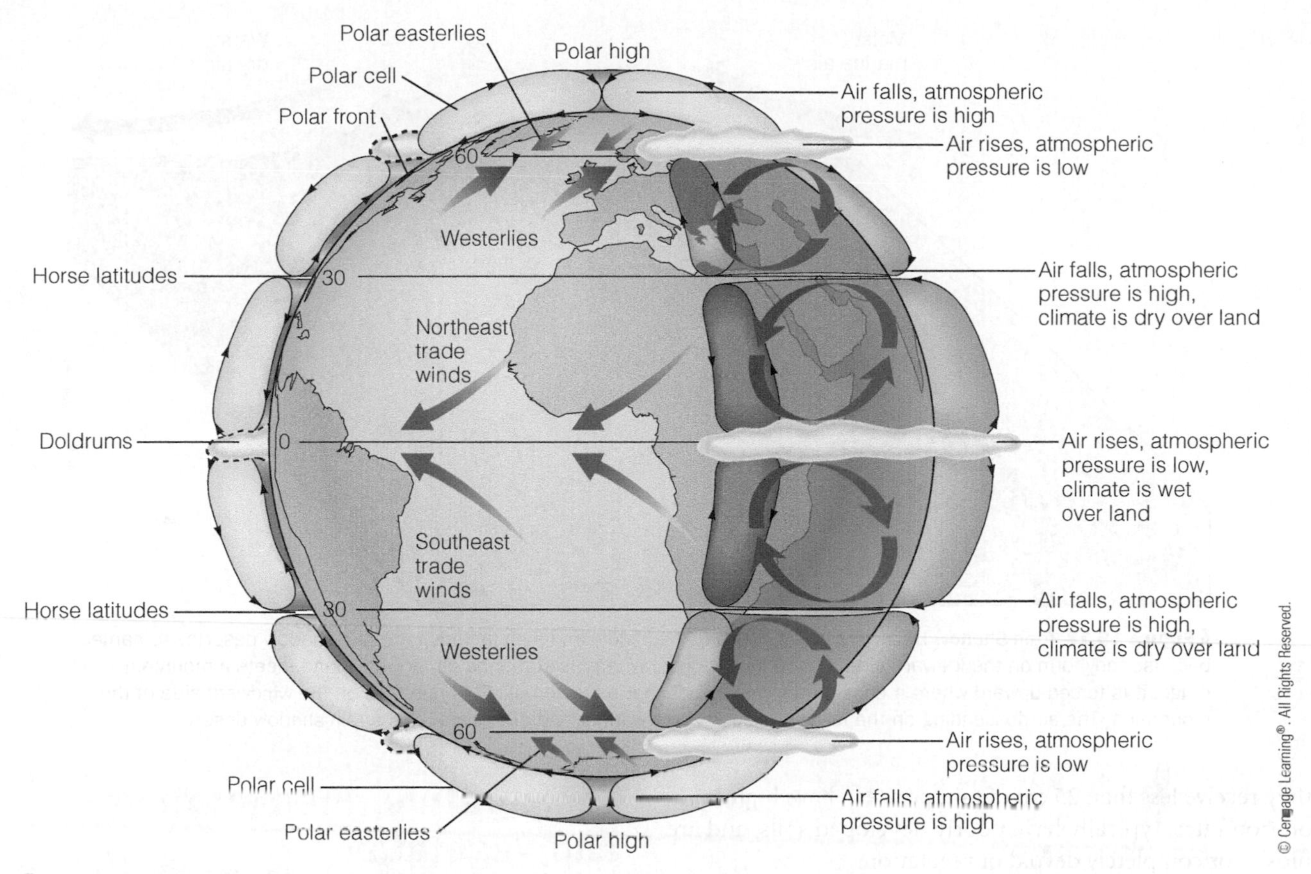

Figure 15.16 The General Circulation Pattern of Earth's Atmosphere Air flows from high-pressure zones to low-pressure zones, and the resulting winds are deflected to the right of their direction of movement (clockwise) in the Northern Hemisphere, and to the left of their direction of movement (counterclockwise) in the Southern Hemisphere. This deflection of air between latitudinal zones resulting from Earth's rotation is known as the Coriolis effect.

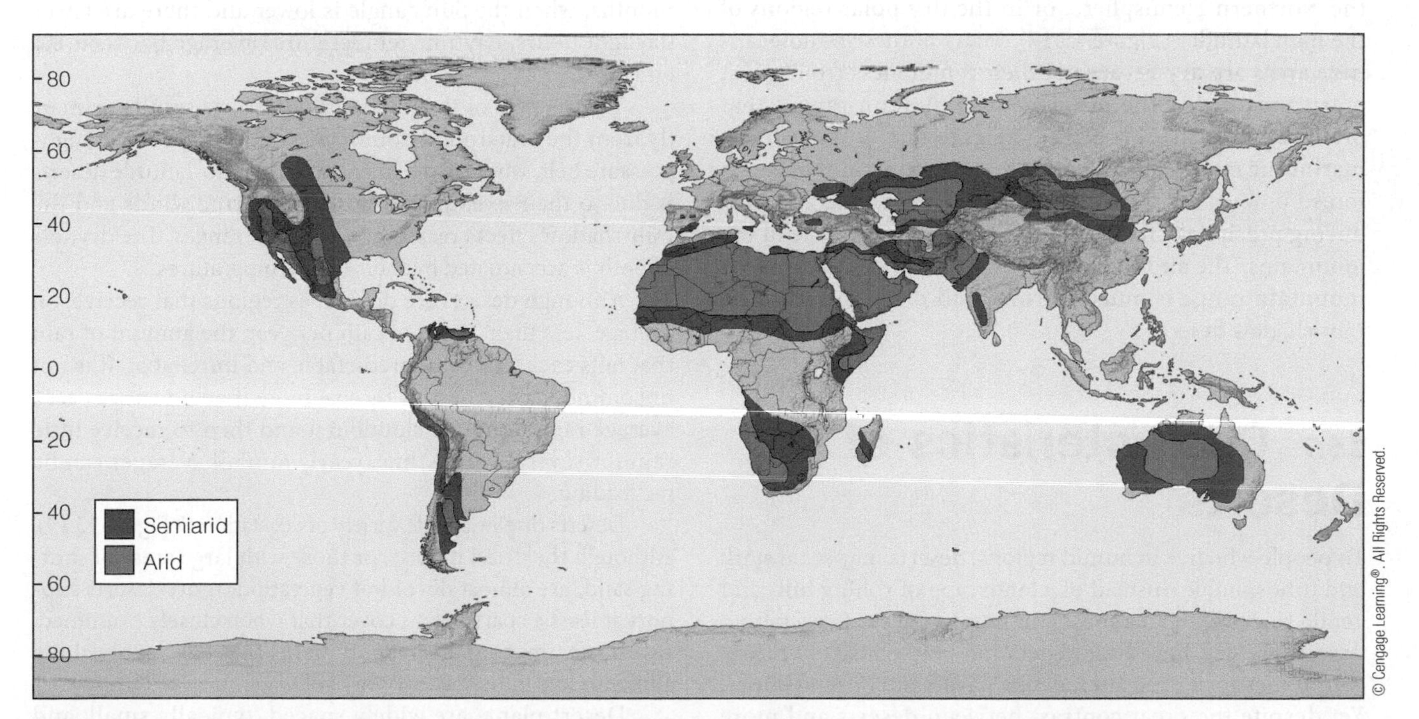

Figure 15.17 The Distribution of Earth's Arid and Semiarid Regions Semiarid regions receive more precipitation than arid regions, yet they are still moderately dry. Arid regions, generally described as deserts, are dry and receive less than 25 cm of rain per year. The majority of the world's deserts are located in the dry climates of the low and middle latitudes.

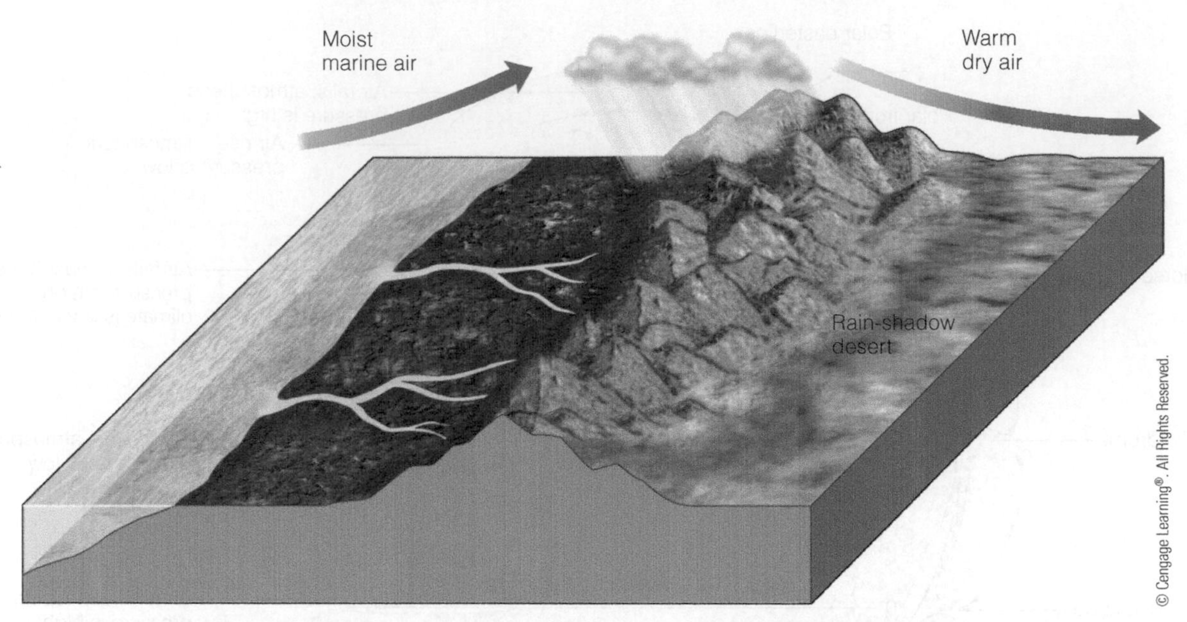

Figure 15.18 Rain-Shadow Deserts Many deserts in the middle and high latitudes are rain-shadow deserts, so named because they form on the leeward side of mountain ranges. When moist marine air moving inland meets a mountain range, it is forced upward where it cools and forms clouds that produce rain. This rain falls on the windward side of the mountains. The air descending on the leeward side is much warmer and drier, producing a rain-shadow desert.

they receive less than 25 cm of rain per year, have high evaporation rates, typically have poorly developed soils, and are mostly or completely devoid of vegetation.

The majority of the world's deserts are in the low latitude, dry-climate zone between 20 and 30 degrees north and south latitudes (Figure 15.17). The remaining deserts of the world are found in the middle or high latitudes, mostly within the continental interiors in the middle latitudes of the Northern Hemisphere, or in the dry polar regions of the high latitudes (Figure 15.17). Many of these middle latitude areas are dry because of their remoteness from moist maritime air and the presence of mountain ranges that produce a **rain-shadow desert** (Figure 15.18). When moist marine air moves inland and meets a mountain range, it is forced upward. As it rises, it cools, forming clouds and producing precipitation that falls on the windward side of the mountains. The air that descends on the leeward side of the mountain range is much warmer and drier, producing a rain-shadow desert.

15.7 Characteristics of Deserts

To people who live in humid regions, deserts may seem stark and inhospitable. Instead of a landscape of rolling hills and gentle slopes with an almost continuous cover of vegetation, deserts are dry, have little vegetation, and consist of nearly continuous rock exposures, desert pavement, or sand dunes. Yet despite the great contrast between deserts and more humid areas, the same geologic processes are at work, only operating under different climatic conditions.

Temperature, Precipitation, and Vegetation

The heat and dryness of deserts are well known. Many of the deserts of the low latitudes have average summer temperatures that range between 32° and 38°C. It is not uncommon for some low-elevation inland deserts to record daytime highs of 46°–50°C for weeks at a time. During the winter months, when the Sun's angle is lower and there are fewer daylight hours, daytime temperatures average between 10° and 18°C.

The dryness of the low-latitude deserts results primarily from the year-round dominance of the subtropical high-pressure belt, whereas the dryness of the mid-latitude deserts is due to their isolation from moist marine winds and the rain-shadow effect created by mountain ranges. The dryness of both is accentuated by their high temperatures.

Although deserts are defined as regions that receive, on average, less than 25 cm of rain per year, the amount of rain that falls each year is unpredictable and unreliable. It is not uncommon for an area to receive more than an entire year's average rainfall in one cloudburst and then to receive little rain for several years. Thus, yearly rainfall averages can be misleading.

Deserts display a wide variety of vegetation (Figure 15.19). Although the driest deserts, or those with large areas of shifting sand, are almost devoid of vegetation, most deserts support at least a sparse plant cover that when closely examined, reveals an amazing diversity of plants that have evolved the ability to live in the near absence of water.

Desert plants are widely spaced, typically small, and grow slowly. Their stems and leaves are usually hard and waxy to minimize water loss by evaporation and to protect

Copyright and Photograph by Dr. Parvinder S. Sethi

Figure 15.19 Desert Vegetation Desert vegetation is typically sparse, widely spaced, and characterized by slow growth rates. The vegetation shown here is in Death Valley, California.

the plant from sand erosion. Most plants have a widespread shallow root system to absorb the dew that forms each morning in all but the driest deserts and to help anchor the plant in what little soil there may be. In extreme cases, many plants lie dormant during particularly dry years and spring to life after the first rain shower with a beautiful profusion of flowers.

Weathering and Soils

Mechanical weathering is dominant in desert regions. Daily temperature fluctuations and frost wedging are the primary forms of mechanical weathering. The breakdown of rocks by roots and from salt crystal growth is of minor importance. Some chemical weathering does occur, but its rate is greatly reduced by aridity and the scarcity of organic acids produced by the sparse vegetation. Most chemical weathering takes place during the winter months when there is more precipitation, particularly in the mid-latitude deserts.

ConnectionLink

To review the processes of mechanical and chemical weathering, see Chapter 6.

An interesting feature seen in many deserts is a thin, red, brown, or black, shiny coating on the surface of many rocks. This coating, called *rock varnish*, is composed of iron and manganese oxides. Because many of the varnished rocks contain little or no iron and manganese oxides, the varnish is thought to result either from windblown iron and manganese dust that settles on the ground or from the precipitated waste of microorganisms.

In many desert regions, such as the Southwest and the Great Basin area of North America, a common form of rock art, known as *petroglyphs*, is especially abundant. Petroglyphs (from the Greek *petro*, meaning "rock," and *glyph*, meaning "carving or engraving") are the abraded, pecked, incised, or scratched marks made by humans on boulders, cliffs, and cave walls. Many petroglyphs are the result of removing the thin, brown- or black-colored rock varnish, revealing the underlying lighter-colored natural rock surface, thus providing an excellent contrast for petroglyphs (Figure 15.20). Petroglyphs, which in North America extend back to about 2000 BC, are a fragile and nonrenewable cultural resource that cannot be replaced if they are damaged or destroyed.

Desert soils, if developed at all, are usually thin and patchy because the limited rainfall and the resultant scarcity of vegetation reduce the efficiency of chemical weathering and hence soil formation. Furthermore, the sparseness of the vegetative cover enhances wind and water erosion of what little soil actually forms.

Reed Wicander

Figure 15.20 Rock Varnish A human-like petroglyph exposed at an outcrop along Cub Creek Road in Dinosaur National Monument, Utah. Note the contrast between the fresh exposure of the rock where the upper part of the petroglyph's head has been removed, the weathered brown surface of the rest of the petroglyph, and the black rock varnish coating the rock surface.

Mass Wasting, Streams, and Groundwater

When traveling through a desert, most people are impressed by such wind-formed features as moving sand, sand dunes, and sand and dust storms. They may also notice the dry washes and dry streambeds. Because of the lack of running water, most people would conclude that wind is the most important erosional agent in deserts. They would be wrong! Running water, even though it occurs infrequently, causes most of the erosion in deserts. The dry conditions and sparse vegetation characteristic of deserts enhance water erosion.

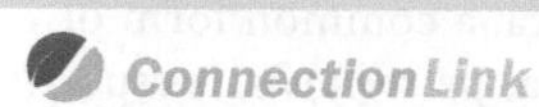

To review the various ways running water erodes and shapes a landscape, see Chapter 12.

As we have previously mentioned, most of a desert's average annual rainfall of 25 cm or less comes in brief, heavy, localized cloudbursts. During these times, considerable erosion takes place, because the ground cannot absorb all of the rainwater. With so little vegetation to hinder the flow of water, runoff is rapid, especially on moderately to steeply sloping surfaces, resulting in flash floods and sheet flows. Dry stream channels quickly fill with raging torrents of muddy water and mudflows, which carve out steep-sided gullies and overflow their banks. During these times, a tremendous amount of sediment is rapidly transported and deposited far downstream.

Although water is the major erosive agent in deserts today, it was even more important during the Pleistocene Epoch when these regions were more humid. During that time, many of the major topographic features of deserts were forming. Today, that topography is modified by wind and infrequently flowing streams.

Most desert streams are poorly integrated and flow only intermittently. Many of them never reach the sea because the water table is usually far deeper than the channels of most streams, so they cannot draw upon groundwater to replace water lost to evaporation and absorption into the ground. This type of drainage in which a stream's load is deposited within the desert is called *internal drainage* and is common in most arid regions.

Although most deserts have internal drainage, some deserts have permanent through-flowing streams, such as the Nile and Niger rivers in Africa, the Rio Grande and Colorado rivers in the southwestern United States, and the Indus River in Asia. These streams can flow through desert regions because (1) their headwaters are well outside the desert and (2) water is plentiful enough to offset losses resulting from evaporation and infiltration.

Wind

Although running water does most of the erosional work in deserts, wind can also be an effective geologic agent capable of producing a variety of distinctive erosional (Figures 15.2 and 15.4) and depositional features (Figures 15.10 through 15.15). Wind is effective in transporting and depositing unconsolidated sand-, silt-, and clay- (dust) size particles. Contrary to popular belief, most deserts are not sand-covered wastelands, but rather vast areas of rock exposures and desert pavement. Sand-covered regions, or sandy deserts, constitute less than 25% of the world's deserts. The sand in these areas has accumulated primarily by the action of wind.

Wind is not only an effective erosional and depositional agent in deserts, it is also becoming an important resource in generating electricity in many parts of the world (Geo-Focus 15.1). The same wind that erodes, transports, and deposits materials is increasingly being harnessed to produce electricity for an ever more energy-hungry world.

15.8 Desert Landforms

Because of differences in temperature, precipitation, and wind, as well as the underlying rocks and recent tectonic events, landforms in arid regions vary considerably. Running water, although infrequent in deserts, is responsible for producing and modifying many distinctive landforms found there.

After an infrequent and particularly intense rainstorm, excess water not absorbed by the ground may accumulate in low areas and form *playa lakes* (Figure 15.21a). These lakes are temporary, lasting from a few hours to several months. Most of them are shallow and have rapidly shifting boundaries as water flows in or leaves by evaporation and seepage into the ground, and the water in the playa lake is often very saline.

When a playa lake evaporates, the dry lake bed is called a **playa** or *salt pan* and is characterized by mud cracks and precipitated salt crystals (Figure 15.21b). Salts in some playas are thick enough to be mined commercially. For example, borates have been mined in Death Valley, California, for more than 100 years.

Other common features of deserts, particularly in the Basin and Range Province of the western United States, are alluvial fans and bajadas. **Alluvial fans** form when sediment-laden streams flowing out from the generally straight, steep mountain fronts deposit their load on the relatively flat desert floor. Once beyond the mountain front, where no valley walls confine streams, the sediment spreads out laterally, forming a gently sloping and poorly sorted fan-shaped sedimentary deposit (Figure 15.22).

Although alluvial fans are similar in origin and shape to deltas (see Chapter 12), they are formed entirely on land. Alluvial fans may coalesce to form a *bajada*, a broad alluvial apron that typically has an undulating surface resulting from the overlap of adjacent fans.

Large alluvial fans and bajadas are frequently important sources of groundwater for domestic and agricultural use (see Chapter 13). Their outer portions are typically

Figure 15.21 Playas and Playa Lakes

a A playa lake formed after a rainstorm near Badwater, Death Valley National Park, California. Playa lakes are ephemeral features, lasting from a few hours to several months.

b Salt deposits and salt ridges cover the floor of this playa in the Mojave Desert, California. Salt crystals and mud cracks are characteristic features of playas.

Figure 15.22 Alluvial Fan A ground view of an alluvial fan, Death Valley, California. Alluvial fans form when sediment-laden streams flowing out from a mountain deposit their load on the relatively flat desert floor, forming a gently sloping, fan-shaped, sedimentary deposit.

Critical Thinking Question
Why are alluvial fans that form on land the same general shape as deltas that form in water?

GEO FOCUS 15.1 Windmills and Wind Power

Whoosh, whoosh, whoosh. Ah, the gentle sound of a windmill's blades turning in the wind. The image most people associate with windmills is one of a pastoral landscape in Holland dominated by a classic Dutch windmill crafted of wood (▮ Figure 1), or perhaps Don Quixote tilting at windmills in the famous novel *Don Quixote de la Mancha* by Miguel Cervantes. Today, instead of a whoosh, whoosh, whoosh, the sound of modern electricity-generating windmills in a wind farm is more like a woomph, woomph, woomph (▮ Figure 2).

As early as 5000 BC, people began to harness the power of wind to propel boats along the Nile River. The Chinese used windmills to pump water for irrigating crops as long ago as 2000 BC. Wind power was used in the Middle Ages in Europe, particularly in Holland, where windmills have played an important role in society. Windmills were first used to grind corn, which is where the term *windmill* originally came from. Later, windmills were used to drain lakes and marshes from the low-lying districts and to saw timber. Settlers in the United States in the late 19th and early 20th centuries used this technology to pump water and generate electricity in the Great Plains.

With the application of steam power and industrialization in Europe, and later the United States, the use of windmills rapidly declined. However, industrialization led to the development of larger and more efficient windmills exclusively designed to generate electricity. Denmark began using such windmills as early as 1890, and other countries soon followed suit. Interest in electricity-generating windmills has always mirrored the price of fossil fuels. When the price of petroleum, natural gas, and coal is

© Larry Lee Photography/CORBIS

▮ Figure 1 Five traditional Dutch windmills lined up along a canal at Kinderdiik, the Netherlands.

Copyright and Photograph by Dr. Parvinder S. Sethi

▮ Figure 2 A wind farm in the Coachella Valley, California. California has more windmills than any other state in the United States.

composed of fine-grained sediments suitable for cultivation, and their gentle slopes allow good drainage of water. Many alluvial fans and bajadas are also the sites of large towns and cities, such as San Bernardino, California; Salt Lake City, Utah; and Teheran, Iran.

Most mountains in desert regions, including those of the Basin and Range Province, rise abruptly from gently sloping surfaces called **pediments**. Pediments are erosional bedrock surfaces of low relief that slope gently away from mountain bases, and are typically covered by a thin layer of debris, alluvial fans, or bajadas (▮ Figure 15.23).

The origin of pediments has been the subject of much controversy. Most geologists agree that they are erosional features developed on bedrock in association with the erosion and retreat of a mountain front (Figure 15.23a). The disagreement concerns how the erosion has occurred. Although not all geologists would agree, it appears that pediments are produced by the combined activities of lateral erosion by streams, sheet flooding, and various weathering processes along the retreating mountain front. Thus, pediments grow at the expense of the mountains, and they continue to expand as the mountains are eroded away or partially buried.

low, it is cheaper to use these fuels to generate electricity. When the price of fossil fuels goes up, interest in wind power also increases. Today, the use of wind power to generate electricity is increasing, both in the United States and elsewhere in the world, particularly in Europe. As wind turbine technology has increased the efficiency of wind-generated electricity, the cost of producing electricity has decreased greatly, such that wind farms are no longer a novelty. However, it should be pointed out that without federal subsidies, wind-generated electric power is still too expensive in most areas to compete with traditional fossil fuel–burning plants, particularly those using natural gas.

How do windmills produce electricity? Simply stated, wind turbines (the term commonly used to describe electricity-producing windmills) convert the kinetic energy (the energy of an object due to motion) of the wind into mechanical power, in this case, the generation of electricity. This electricity is then sent to the local power grid where it is distributed throughout the area.

To be effective, numerous wind turbines are clustered together in wind farms that are located in areas with relatively strong, steady winds. The number of turbines on wind farms can range from several to thousands, as in California (Figure 2). In addition to wind farms on land, wind farms offshore are becoming more common because they are out of sight and people cannot hear the blades turning. In fact, more than 25% of Denmark's electricity came from wind power in 2011, thus ranking it as the world's leader in wind-generated electrical production. In the United States, however, wind power production contributed just 2.9% of all electrical generation in 2011.

What are the advantages and disadvantages of wind power? First of all, the wind is a free, renewable energy source, so it cannot be used up. It is also a clean source of energy that does not pollute the water or atmosphere, or contribute to greenhouse gases. Thus, it reduces the consumption of fossil fuels. The land on which windmills are sited can still be used for farming and ranching, thereby increasing the productivity of the land and providing an additional source of income to the landowner who leases the land to utilities. In addition, wind farms can benefit the local economy of rural and remote areas by supplying wind energy for local consumption.

There are some disadvantages to wind-generated electricity. The major disadvantage is that wind does not always blow with sufficient strength to be totally reliable, thereby necessitating backup generation. Furthermore, good wind sites are frequently located in remote areas, far from the areas where large quantities of electricity are needed, or in coastal areas, where land is expensive and local residents do not want large wind turbines as neighbors. The initial start-up cost of a wind farm is usually higher than the cost of building a conventional power plant. However, as the cost of wind power has decreased because of better technology, wind-generated electricity is beginning to compete favorably with traditional power plants in many areas.

The "not-in-my-backyard" opposition to wind farms can make siting a wind farm difficult. The major objection to wind farms is the noise generated by the turbines, although as the windmills are built taller and the turbines are more efficient at noise reduction, that is not the major concern it once was. Of course, aesthetics still play a role in many people's objection to wind farms, and the fact that some birds are killed by the rotating blades is another issue. Studies have indicated, however, that the impact of wind turbines on bird mortality and injury is less than that of many other structures, such as buildings, power lines, and communication towers.

As the price of fossil fuel continues to rise, the use of a centuries-old staple, the windmill, albeit modernized, will continue to gain in popularity.

Rising conspicuously above the flat plains of many deserts are isolated, steep-sided erosional remnants called *inselbergs*, a German word meaning "island mountain." Inselbergs have survived for a longer period of time than other mountains because of their greater resistance to weathering. Uluru (formerly known as Ayers Rock, Australia) is an excellent example of an inselberg (▌Figure 15.24).

Other easily recognized erosional remnants common to arid and semiarid regions are mesas and buttes (▌Figure 15.25). A **mesa** is a broad, flat-topped erosional remnant bounded on all sides by steep slopes. Continued weathering and stream erosion form isolated, pillar-like structures known as **buttes**. Buttes and mesas consist of relatively easily weathered sedimentary rocks capped by nearly horizontal, resistant rocks such as sandstone, limestone, or basalt. They form when the resistant rock layer is breached, which allows rapid erosion of the less-resistant underlying sediment. One of the best-known areas of mesas and buttes in the United States is Monument Valley on the Arizona–Utah border (▌Figure 15.25).

Figure 15.23 Pediment

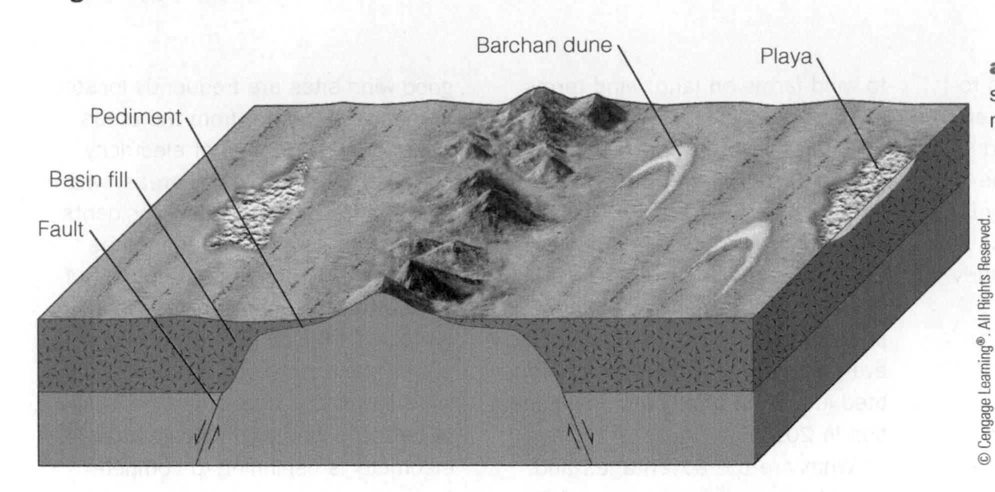

a Pediments are erosional bedrock surfaces formed by erosion along a mountain front.

b A pediment north of Mesquite, Nevada.

Figure 15.24 Uluru at Sunset Contrary to popular belief, Uluru, Australia, is not a giant boulder. Rather, it is the exposed portion of the nearly vertically tilted Uluru Arkose, which can clearly be seen in this image. Differential weathering of the sedimentary layers has produced the distinct parallel ridges and other features characteristic of Uluru.

Figure 15.25 Buttes and Mesas

Joseph Sohm/Spirit/Corbis

a Several mesas and buttes can be seen in this aerial view of Monument Valley Navajo Tribal Park, Navajo Indian Reservation, Arizona. The mesas are the broad, flat-topped structures, one of which is prominent in the foreground, whereas the buttes are more pillar-like structures.

Copyright and Photograph by Dr. Parvinder S. Sethi

b Left Mitten Butte and Right Mitten Butte in Monument Valley on the border of Arizona and Utah.

Key Concepts Review

- Desertification is the expansion of deserts into formerly productive lands. It destroys croplands and rangelands, causing massive starvation and forcing hundreds of thousands of people to migrate from their homelands.
- Wind transports sediment in suspension or as bed load. Suspended load is the material that is carried in suspension by water or wind. Silt- and clay-size particles constitute most of a wind's suspended load. Bed load is the material that is too large or heavy to be carried in suspension and is thus moved along the surface by saltation, and by rolling or sliding.
- Wind erodes material by either abrasion or deflation. Abrasion is the impact of saltating sand grains on an object. Ventifacts are common products of wind abrasion.
- Deflation is the removal of loose surface material by wind. Deflation hollows resulting from differential erosion of surface material are common features of many deserts, as is desert pavement, which effectively protects the underlying surface from additional deflation.
- Dunes are mounds or ridges of wind-deposited sand that form when wind flows over and around an obstruction, resulting in the deposition of sand grains, which accumulate and build up a deposit of sand.
- Barchan, longitudinal, transverse, and parabolic are the four major dune types. The amount of sand available, the prevailing wind direction and velocity, and the amount of vegetation determine which type of dune will form.
- Loess consists of wind-blown deposits of silt and clay that is derived from deserts, Pleistocene glacial outwash deposits, or river floodplains in semiarid regions. It covers approximately 10% of Earth's land surface and weathers to a rich, productive soil.
- The winds of the major air-pressure belts, oriented east–west, result from the rising and cooling of air. The winds are deflected clockwise in the Northern Hemisphere and counterclockwise in the Southern Hemisphere by the Coriolis effect to produce Earth's global wind patterns.
- Dry climates, located in the low and middle latitudes where the potential loss of water by evaporation exceeds the yearly precipitation, cover 30% of Earth's land surface and are subdivided into semiarid and arid regions. Semiarid regions receive more precipitation than arid regions, yet are moderately dry. Arid regions, generally described as deserts, are dry and receive less than 25 cm of rain per year.
- The majority of the world's deserts are in the low-latitude, dry-climate zone between 20 and 30 degrees north and south latitudes. Their dry climate results from a high-pressure belt of descending dry air. The remaining deserts are in the middle latitudes, where their distribution is related to the rain-shadow effect, and also in the dry polar regions.
- Deserts are characterized by high temperatures, little precipitation, and sparse plant cover. Rainfall is unpredictable and, when it does occur, tends to be intense and of short duration.
- Mechanical weathering is the dominant form of weathering in deserts and, coupled with slow rates of chemical weathering, results in poorly developed soils.
- Running water is the major agent of erosion in deserts and was even more important during the Pleistocene, when wetter climates resulted in humid conditions.
- Wind is also an erosional agent in deserts and is very effective in transporting and depositing unconsolidated, fine-grained sediments.
- Desert landforms include playas, which are dry lakebeds, but when temporarily filled with water, form playa lakes. Alluvial fans are fan-shaped sedimentary deposits that may coalesce to form bajadas. Pediments are erosional bedrock surfaces of low relief that slope gently away from mountain bases, and which are covered by alluvial fans or bajadas.
- Inselbergs are isolated, steep-sided erosional remnants that rise above the surrounding desert plains. Buttes and mesas are, respectively, pinnacle-like and flat-topped erosional remnants with steep sides.

Important Terms

abrasion (p. 366)
alluvial fan (p. 376)
barchan dune (p. 369)
butte (p. 379)
Coriolis effect (p. 372)
deflation (p. 367)
desert (p. 372)
desertification (p. 364)
desert pavement (p. 367)
dune (p. 368)
loess (p. 370)
longitudinal dune (p. 369)
mesa (p. 379)
parabolic dune (p. 370)
pediment (p. 378)
playa (p. 376)
rain-shadow desert (p. 374)
transverse dune (p. 369)
ventifact (p. 366)

Review Questions

1. What type of dune is crescent-shaped, and its tips point downwind?
   a. ___ parabolic.
   b. ___ star.
   c. ___ longitudinal.
   d. ___ barchan.
   e. ___ transverse.
2. The majority of the world's deserts are located between what latitudes?
   a. ___ 10 and 20 degrees.
   b. ___ 20 and 30 degrees.
   c. ___ 30 and 40 degrees.
   d. ___ 40 and 60 degrees.
   e. ___ 60 and 80 degrees.
3. Which of the following is a feature produced by wind deposition?
   a. ___ ventifact.
   b. ___ loess.
   c. ___ butte.
   d. ___ yardang.
   e. ___ inselberg.
4. The Coriolis effect causes wind to be deflected
   a. ___ to the right in the Northern Hemisphere and to the left in the Southern Hemisphere.
   b. ___ to the left in the Northern Hemisphere and to the right in the Southern Hemisphere.
   c. ___ only to the left in both hemispheres.
   d. ___ only to the right in both hemispheres.
   e. ___ not at all.
5. The major agent of erosion in deserts today is
   a. ___ glaciers.
   b. ___ wind.
   c. ___ abrasion.
   d. ___ running water.
   e. ___ none of the previous answers.
6. Desertification is caused by climate change and exacerbated by human activities. Although we cannot effectively affect short-term climate change on a human time scale, what steps could be taken to reduce the desertification resulting primarily from human activities?
7. As more people move into arid and semiarid areas such as Las Vegas, Nevada, Phoenix, Arizona, and southern California, an increasing strain is placed on the environment of these areas. What are some of the issues that government entities must face in dealing with these population increases? What are some of the problems that must be dealt with from a geologic perspective?
8. Using what you now know about deserts, their location, how they form, and the various landforms found in them, what evidence in the rock record might allow you to determine where deserts existed in the past?
9. As more images of Mars surface are sent back to Earth from the various landers, a number of features suggest that running water, as well as wind deposition and erosion, have played a major role in shaping the Martian landscape. What type of evidence would you look for in these images that show wind has been an important agent in the formation of the Martian surface?
10. **Creative Thinking Visual Question:** What desert landforms or features of a desert can you identify in this image (Figure 1) of the Mesquite Flat sand dunes in Death Valley, California? What types of sand dunes are present?

Figure 1 Mesquite Flat sand dunes, Death Valley, California.

Global GeoScience Watch Within the GREENR database, go to the "Renewable Energy" portal. Under the related portals on the left-hand side, click on "Wind Energy." Next to the "Magazines" heading, click on "VIEW ALL." What is the outlook in the United States and elsewhere for wind energy in the foreseeable future? What are the advantages of wind energy? What are some of the problems associated with wind energy? How has the "shale revolution" (i.e., oil- and gas-shale production) affected the outlook for renewable energy?

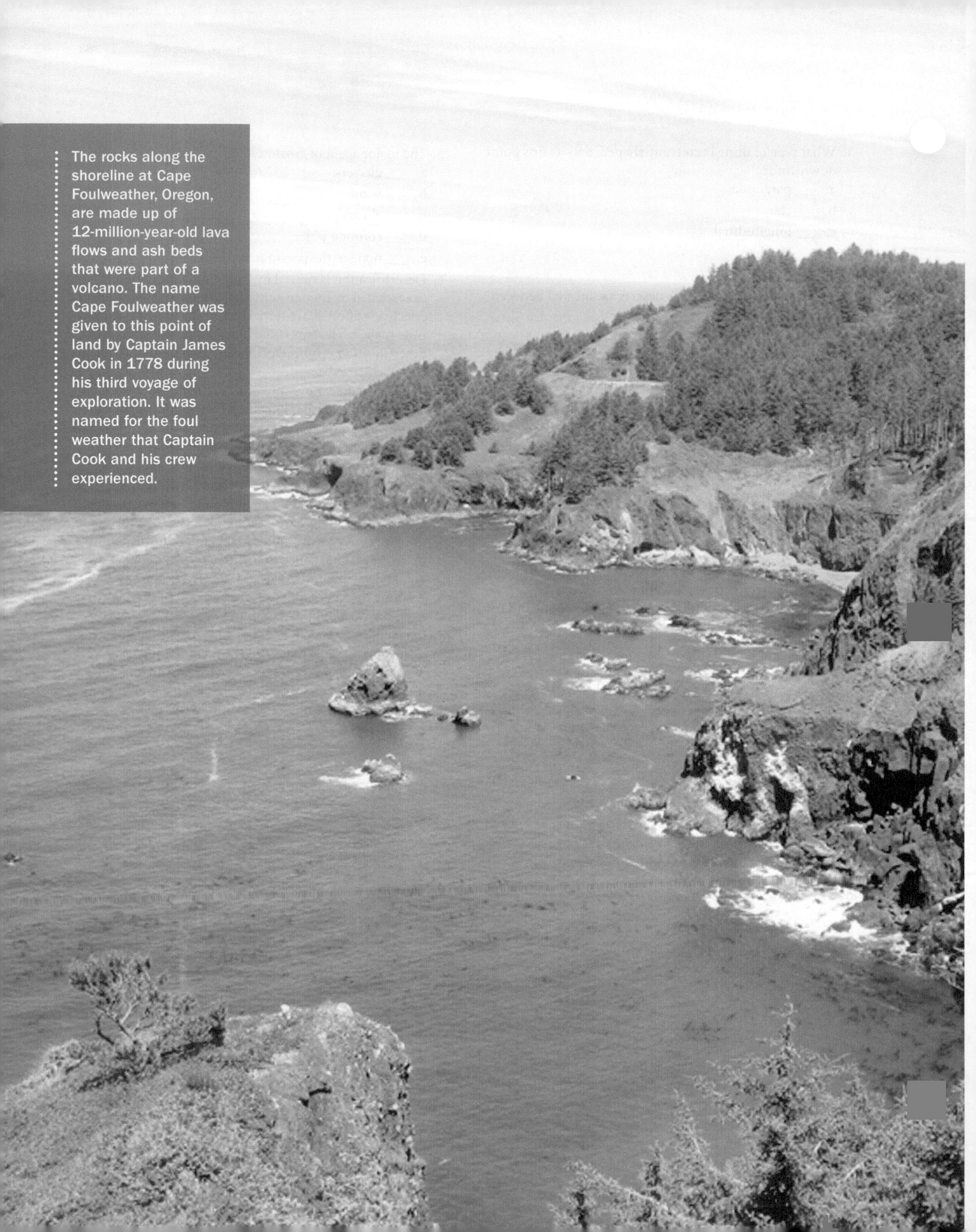

The rocks along the shoreline at Cape Foulweather, Oregon, are made up of 12-million-year-old lava flows and ash beds that were part of a volcano. The name Cape Foulweather was given to this point of land by Captain James Cook in 1778 during his third voyage of exploration. It was named for the foul weather that Captain Cook and his crew experienced.

James S. Monroe

Oceans, Shorelines, and Shoreline Processes

OUTLINE

HAVE YOU EVER WONDERED?

- How scientists study the oceans?
- Why seawater is salty, and whether it is getting saltier with time?
- What causes ocean currents, waves, and tides?
- How waves and currents erode shorelines?
- Why some shorelines have long, sandy beaches and others do not?
- What causes beaches to vary with the seasons?
- How shoreline communities deal with storm waves and coastal flooding?
- What kinds of resources come from the oceans?

16.1 Introduction

In Chapter 2 we discussed several aspects of the seafloor, including oceanic ridges, deep-sea hydrothermal vents (black smokers), seamounts, and guyots. We also discussed the nature of the continental margins, which include the continental shelf, slope, and rise. Our interest here is in the oceanic waters themselves and their impact on shorelines, although we will also discuss the sediment found on the deep seafloor as well as resources from the oceans.

You already know that the hydrosphere consists of all water on Earth, most of which is in the oceans (see Table 12.1). This vast, interconnected body of saltwater covers 71% of Earth's surface, but parts of it are distinct enough for us to recognize the Pacific, Atlantic, Indian, and Arctic oceans (❚ Figure 16.1). Seas, on the other hand, are marginal parts of oceans, as in the Sea of Japan, the Mediterranean Sea, and the Black Sea. The oceans and seas are underlain by oceanic crust, but the same is not true of the Dead Sea, the Caspian Sea, or the Salton Sea; these are actually saline lakes on the continents.

What lies beneath the ocean's surface is hidden from view, which accounts for why so many sensational stories about it have persisted for centuries. According to two dialogues written about 350 BC by the Greek Philosopher Plato, a continent called *Atlantis* existed in the Atlantic Ocean west of what we now call the Strait of Gibraltar (❚ Figure 16.2). Plato claimed that following the conquest of Atlantis by Athens, this vast continent sank, and now only "mud shallows" mark its former location. In fact, no geologic evidence intdicates that Atlantis ever existed, not even the "mud shallows" that Plato mentioned, so why has this story persisted for so long? One reason is that stories of lost civilizations are popular, but another is that until fairly recently little was known about what lay beneath the ocean's surface.

In this enormous body of water we call oceans, wave energy is transferred through the water to shorelines where it has a tremendous impact. Accordingly, understanding shoreline processes is important to many people including oceanographers, geologists, and coastal engineers, as well as elected officials and city planners of coastal communities. Indeed, tourism is an important part of the economies of many coastal communities such as Myrtle Beach, South Carolina; Fort Lauderdale, Florida; and Padre Island, Texas.

Another important aspect of shorelines is rising sea level, because buildings or even entire communities that were once far inland are now in peril, must be protected or moved, or have already been destroyed. Furthermore, hurricanes expend much of their energy on shorelines resulting in extensive

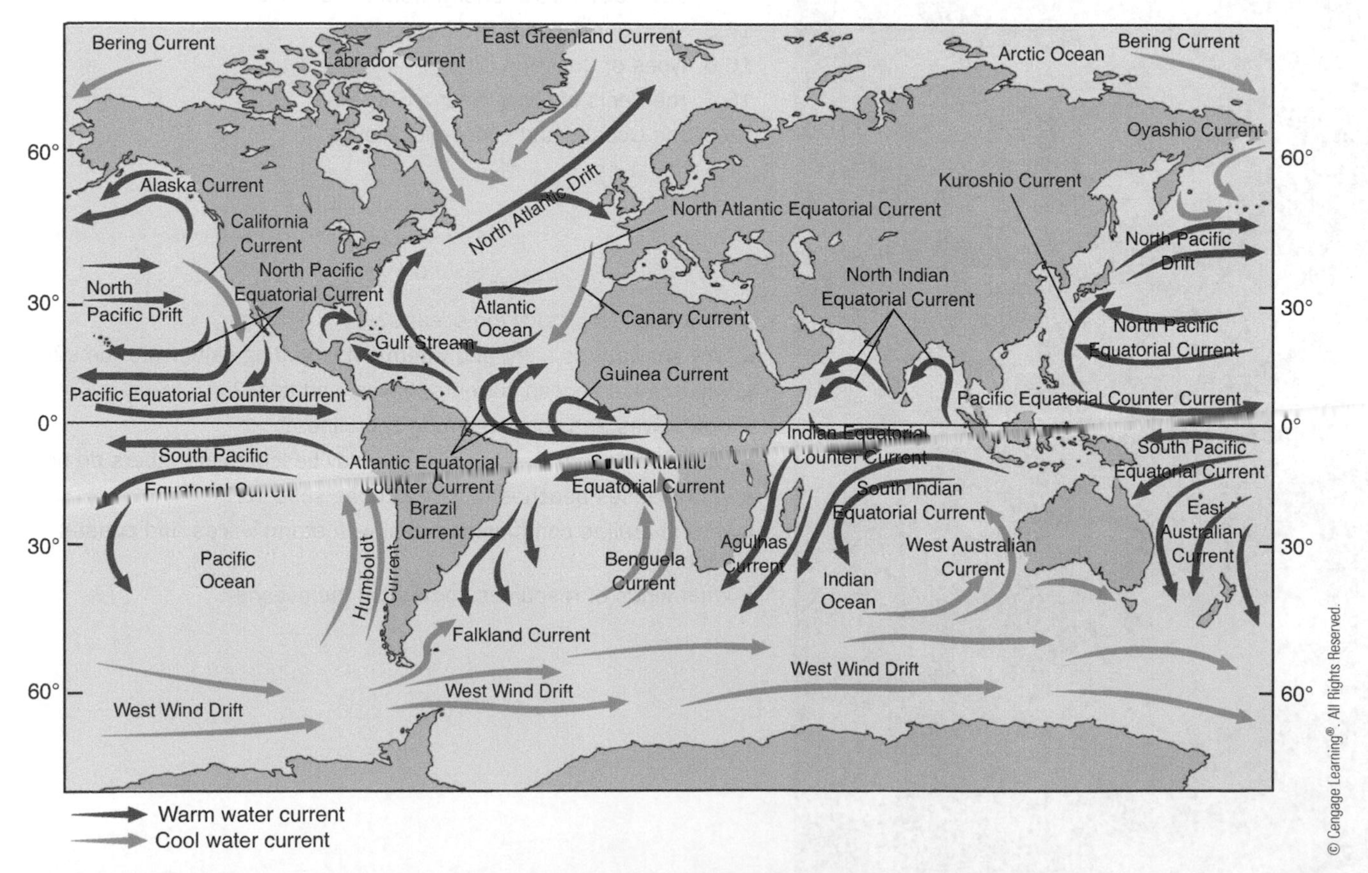

❚ Figure 16.1 The Oceans Map showing the Atlantic, Pacific, Indian, and Arctic oceans, and the ocean currents.

Critical Thinking Question How do ocean currents modify Earth's climate?

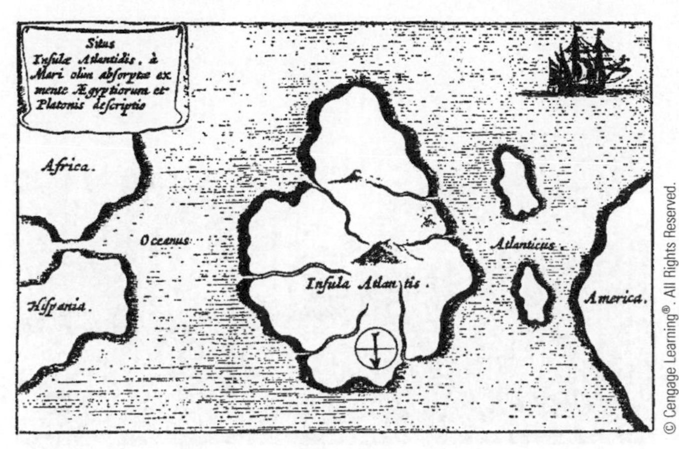

Figure 16.2 **Atlantis** According to Plato, Atlantis was a continent west of the Pillars of Hercules, now called the Strait of Gibraltar. In this map from Anthanasium Kircher's *Mundus Subterraneus* (1664), north is toward the bottom of the map. The Strait of Gibraltar is the narrow area between Hispania (Spain) and Africa.

coastal flooding, numerous fatalities, and widespread property damage. Unfortunately, we were once again reminded of just how vulnerable coastlines are when Hurricane Sandy caused extensive coastal flooding in New Jersey and New York during October 2012.

The study of oceans and shorelines provides another example of systems interactions—in this case, between the hydrosphere and the solid Earth. The atmosphere is also involved because energy is transferred from wind to water, thereby causing waves, which in turn generate nearshore currents. And, of course, the gravitational attraction of the Moon and Sun on the ocean waters is responsible for the rhythmic rise and fall of tides. As dynamic systems, shorelines continually adjust to all changes, such as increased wave energy or an increase or a decrease in sediment supply from the land.

16.2 Exploring the Oceans

About 2,200 years ago the Greeks had determined Earth's size and shape rather accurately, but Europeans were unaware of the vastness of the oceans until the 1400s and 1500s, when explorers sought trade routes to the Indies. In 1492 when Christopher Columbus set sail to find a route to the Indies, he greatly underestimated the width of the Atlantic Ocean. He was not, as it is popularly portrayed, attempting to demonstrate that Earth is spherical; its shape was well accepted by then. The controversy was over Earth's circumference and the shortest route to the Indies; on this point Columbus's critics were correct.

Columbus's voyages and similar ones added considerably to our knowledge of the oceans, but truly scientific investigations did not begin until well into the 1700s. During the 1760s and 1770s the British financed oceanographic expeditions led by Captain James Cook. Charles Darwin was aboard the research vessel HMS *Beagle* as an unpaid naturalist during its around-the-world voyage from 1831 to 1836. As a result of his observations during this journey, he proposed a theory on the evolution of coral reefs. During these 18th- and 19th-century expeditions, researchers determined oceanic depths, sampled seawater, classified marine organisms, sampled the seafloor, and visited many islands previously unknown to Europeans.

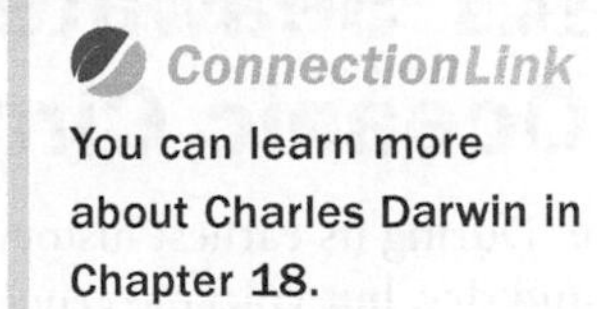
ConnectionLink
You can learn more about Charles Darwin in Chapter 18.

Measuring the length of a weighted line lowered to the seafloor was the first way to determine oceanic depths. Scientists now use an *echo sounder,* which detects sound waves that travel from a ship to the seafloor and back. They calculate depth by knowing the velocity of sound in water and the time for the waves to reach the seafloor and return to the ship. *Seismic profiling* is even more useful. Strong waves from an energy source reflect from the seafloor or from layers beneath the seafloor (Figure 16.3).

International programs for oceanographic research have been carried out by a succession of ships equipped to drill into the seafloor and retrieve samples of seafloor sediments and oceanic crust. The most recent of these, the R/V *Chikyu* ("Earth"), is a Japanese ship that can drill as deep as 11 km into the seafloor. In addition to surface vessels, submersibles are used in oceanographic research, two of which, the *Trieste* (1960) and the *Deepsea Challenger* (2012), descended to the greatest known oceanic depth (10,924 m). One of the most famous of these submersibles is *Alvin,* a vessel that has carried scientists to the seafloor in many places to make observations and collect samples. Other submersibles are remotely controlled or towed by surface ships. In 1985, the *Argo,* equipped with sonar and radar systems, provided the first images of the British ocean liner HMS *Titanic* since it sank in 1912 after hitting an iceberg.

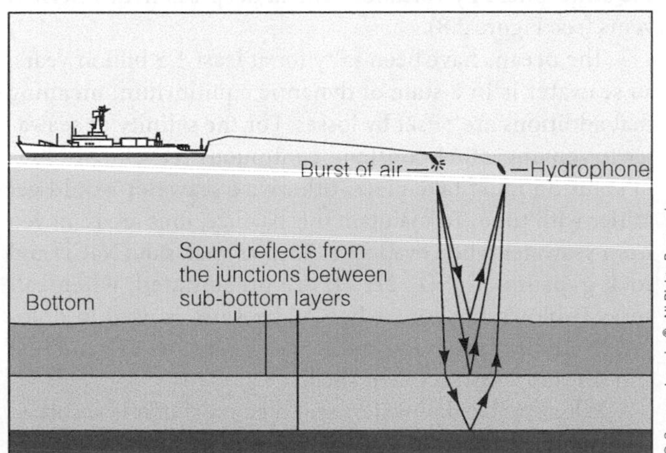

Figure 16.3 **Seismic Profiling** In seismic profiling, the energy generated at a source is reflected from various layers back to the surface where it is detected by hydrophones.

16.3 Seawater and Oceanic Circulation

During its earliest history, Earth was probably hot, airless, and dry, but volcanic eruptions were ubiquitous. Volcanoes emit several gases, but the most abundant one is water vapor, which accumulated in the atmosphere. As Earth cooled, the water vapor condensed, fell as rain, and began collecting on the surface. Geologic evidence indicates that oceans were present by at least 3.5 billion years ago, although their volumes and extent are unknown.

ConnectionLink

For more information about the origin and evolution of the oceans, see Chapter 19.

James S. Monroe

Figure 16.4 The Color of Seawater Notice in this view along the United States Pacific Coast that the water near the shore is greenish whereas the water offshore is blue. The terms shoreline and coast are used more or less interchangeably, but the former refers to the restricted area where the sea meets the land, whereas the latter is more inclusive, including the sea cliffs and the land some distance onshore.

Seawater—Its Composition and Color

Seawater contains more than 70 chemical elements in solution, but the most common ones are chloride ions and sodium ions, which make up 85.6% of all dissolved substances and give seawater its most distinctive feature—its saltiness. Seawater's saltiness, or its **salinity**, is a measure of the total quantity of dissolved solids. On average, 1 kg (1,000 g) of seawater contains 35 g of dissolved solids, or 35 parts per thousand, which is symbolized as 35‰. In the open ocean the salinity varies from 32‰ to 37‰, but in some marginal seas, especially in dry, hot areas nearly isolated from the open ocean, values may exceed 40‰.

One source of chemical elements in seawater is the continents. During chemical weathering, minerals yield ions in solution, which are transported as the dissolved load of streams and rivers. Runoff from the continents adds about 4 billion tons of dissolved solids to the oceans each year. Another important source of elements is *outgassing,* in which gases from within Earth are released into the oceans and the atmosphere by volcanoes and at deep-sea hydrothermal vents (see Figure 2.8).

The oceans have been salty for at least 1.5 billion years, so seawater is in a state of dynamic equilibrium, meaning that additions are offset by losses. For the salinity of seawater to remain rather constant, continuous recycling of ions in solution must take place, otherwise seawater would get saltier with time. To maintain this balance, ions are removed from seawater when evaporites such as rock salt (NaCl) and rock gypsum ($CaSO_4 \cdot 2H_2O$) are precipitated, when salt spray is blown onshore, when magnesium is used in dolomite and clay minerals, and when organisms use calcium and silica to construct their shells.

When visible light enters seawater, most of it is absorbed in the upper 1 m and elevates the water temperature. However, seawater absorbs the longer wavelengths of visible light, the reds and yellows, more readily than the shorter wavelengths, the greens and blues. The blue color of water in the open ocean is related to the selective absorption of light, but in shallow nearshore waters, the water is green or yellow because suspended fine-grained sediment and organic matter reflect the green and yellow wavelengths (Figure 16.4).

Based on the decreasing intensity of light with depth, scientists define two layers in seawater. The upper layer, called the **photic zone,** is usually 100 m or less thick and receives enough light for plants to photosynthesize. Below is the **aphotic zone** where too little light is available for photosynthesis. In this region, most organisms depend directly or indirectly on organic substances that "rain" down from the photic zone. The most notable exceptions are the organisms living adjacent to deep-sea hydrothermal vents such as black smokers, and those living near methane seeps.

You can find more about black smokers in Figure 2.8 in Chapter 2.

Oceanic Circulation

The ocean's surface is nearly always in motion as a result of surface currents, waves, and tides. Even the deep ocean waters are constantly in motion, but unlike wind-driven surface currents and waves, deep-ocean circulation is driven mostly by density differences between adjacent water masses. Much of the circulation in both surface currents and deep-ocean currents is horizontal, but some circulation is vertical when water is transferred from depth to the surface or from the surface to depth. Circulation from depth to the surface, called *upwelling,* has important biological and economic consequences.

As wind blows over a water surface, some of its energy is transferred to the water, which generates surface currents and waves, which we discuss in section 16.4. Figure 16.1 shows the global surface-water current patterns averaged over a long time. Notice in Figure 16.1 that surface currents in the Northern Hemisphere rotate clockwise and counterclockwise in the Southern Hemisphere. This deflection is the **Coriolis effect**, which results from Earth's rotation. The combination of wind and the Coriolis effect produces large-scale water circulation systems known as **gyres** between the 60° parallels in the Atlantic, Pacific, and Indian oceans (Figure 16.1). One of the best-known ocean currents is the *Gulf Stream,* which is actually part of the much larger North Atlantic gyre.

These wind-driven gyres are important as a world temperature control because seawater near the equator absorbs huge amounts of heat and transports it to high latitudes in warm currents. Cold currents originate at high latitudes and flow toward the equator. Indeed, the reason that some rather northerly countries, such as Scotland, have mild climates is related to these currents. On the other hand, the current along the northern and central coasts of California is cold, which keeps that area much cooler than the inland parts of the state.

Horizontal circulation of water takes place in the deep ocean basins because of differences in temperature and density of adjacent water masses. Circulation resulting from density differences operates on a simple principle—a water mass of greater density (colder or saltier) will displace and flow beneath a water mass of lesser density. Deep-ocean circulation affects about 90% of all ocean water, but because studying it is expensive and time-consuming, scientists know less about it than about other circulation patterns.

Vertical circulation takes place in the oceans when **upwelling** slowly transfers cold water from depth to the surface and when **downwelling** carries warm water from the surface to depth. Most upwelling takes place along the west coasts of continents where water moves offshore and colder water rises from depth. Upwelling is by far the most important, because it not only transfers water upward but also carries nutrients, particularly nitrates and phosphate, into the photic zone. Here, high concentrations of plankton are sustained that in turn support other organisms. In fact, areas of upwelling are very productive biologically; less than 1% of the ocean surface is in regions of upwelling, yet it supports more than 50% by weight of all fishes. In addition, most of Earth's sedimentary rocks containing phosphate were deposited along continental margins where upwelling takes place (see section 16.8).

Sediments and Sedimentary Rocks on the Seafloor

Much of the sediment eroded from continents is deposited on the continental margins—that is, the continental shelf, slope, and rise—and forms layers of sandstone, shale, and limestone. However, some sediment, mostly silt- and clay-size particles, is carried into the deep ocean basins where it is deposited. Most of this sediment is *pelagic,* meaning that it settled from suspension far from land. **Pelagic clay** is brown or red and composed mostly of clay-size particles, whereas **ooze** is made up of the tiny shells of marine organisms. If dominated by calcium carbonate ($CaCO_3$) skeletons, it is *calcarious ooze,* or, if mostly silica (SiO_2) skeletons, it is *siliceous ooze* (▮ Figure 16.5).

The term **reef** has many meanings, such as shallowly submerged rocks that pose a hazard to navigation. Here, however, we are concerned with reefs defined as moundlike, wave-resistant structures composed of the shells of marine organisms (▮ Figure 16.6). Although commonly called coral reefs, they actually have a solid framework of skeletons of corals and clams and encrusting organisms such as sponges and algae. Most reefs are found in tropical seas where the water temperature does not fall below about 20°C, and rarely at depths of more than 50 m because many corals rely on symbiotic algae that must have enough sunlight for photosynthesis.

Most reefs are one of three basic types: fringing, barrier, and atoll. *Fringing reefs* are up to 1 km wide, are solidly attached to a landmass, have a rough, table-like surface, and on their seaward side slope steeply down to the seafloor. *Barrier reefs* are similar except that a lagoon separates them from the mainland. The 2,000-km-long Great Barrier Reef of Australia is a good example. Circular to oval reefs surrounding a lagoon are *atolls.* They form around volcanic islands that subside below sea level as the plate they rest on moves into progressively deeper water. As subsidence occurs, the reef-building organisms grow upward so that the living part of the reef remains in shallow water (Figure 16.6). Atolls are particularly common in the western Pacific Ocean basin, although many of them began as fringing reefs, evolved first into barrier reefs, and finally to atolls.

16.4 Shorelines and Shoreline Processes

A **shoreline** is the area of land in contact with the ocean or a lake, but we can expand this definition by noting that ocean shorelines include the land between low tide and the highest level on land affected by storm waves. How does a shoreline differ from a coast? Actually, the terms are commonly used interchangeably, but *coast* is more inclusive and includes the shoreline as well as an area of indefinite width seaward and landward of the shoreline. Our main concern in this section is with ocean shorelines, or seashores, but waves and nearshore currents are also effective in large lakes. However, waves and nearshore currents are much more vigorous along seashores, and even the largest lakes have insignificant tides.

Tides, Waves, and Nearshore Currents

In the marine realm, several biological, chemical, and physical processes are operating continuously. Organisms change the local chemistry of seawater and contribute their skeletons to nearshore sediments, and temperature and salinity changes and internal waves occur in the oceans. However, the processes most important for modifying shorelines are purely physical ones, especially waves, tides, and nearshore currents.

Figure 16.5 Sediments on the Deep Seafloor Most of the sediments on the seafloor are pelagic clay and calcarious ooze and siliceous ooze, both of which are composed of the tiny skeletons of single-celled, floating animals and plants.

Critical Thinking Question Why are there no coarse-grained (gravel and sand) sediments on the deep seafloor far from land?

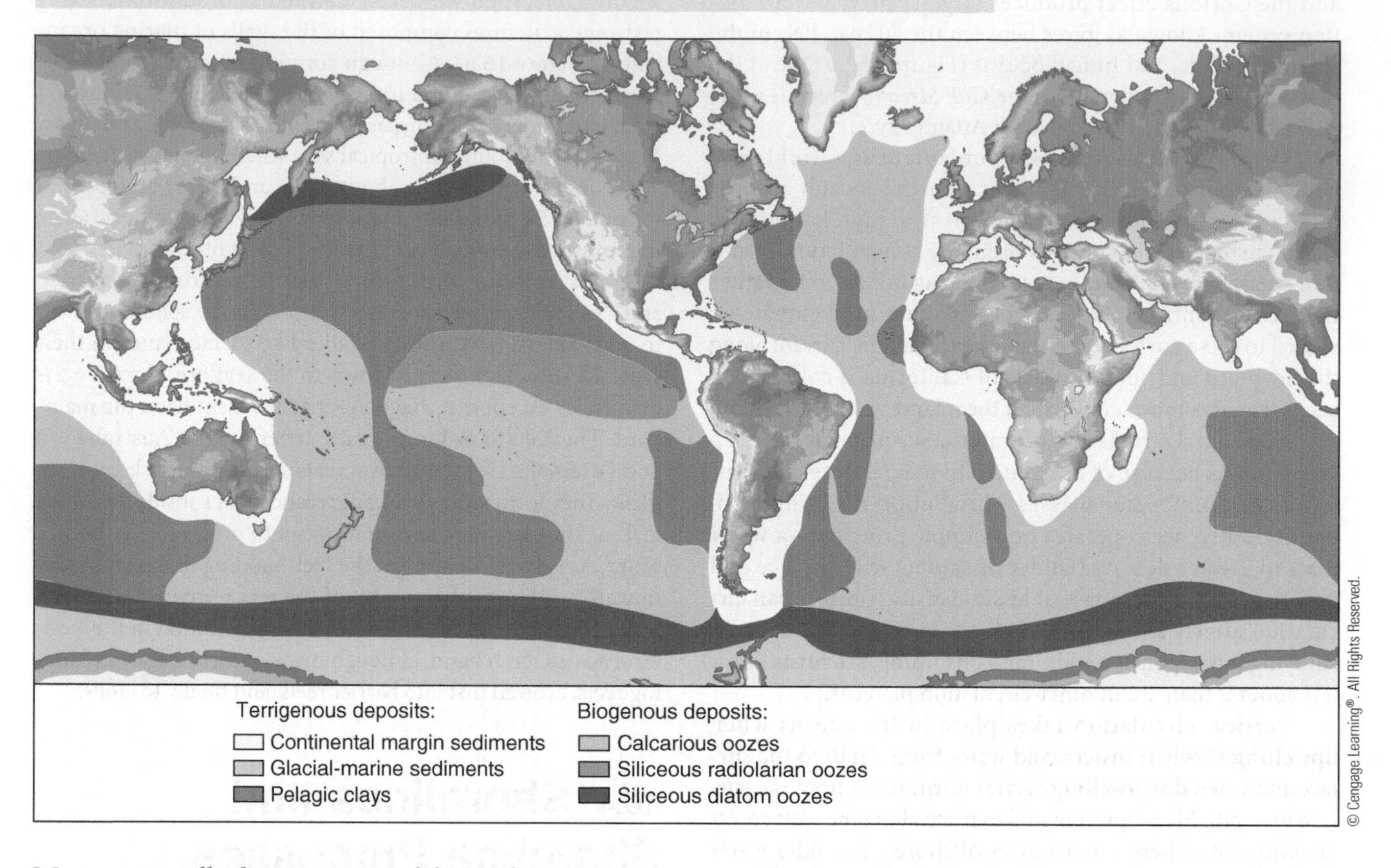

We cannot totally discount some of these other processes, though; offshore reefs composed of the skeletons of organisms, for instance, may protect a shoreline from most of the energy of waves.

Tides The surface of the oceans rises and falls twice daily in response to the gravitational attraction of the Moon and Sun. These regular fluctuations in the ocean's surface, or **tides,** result in most seashores having two daily high tides and two low tides as sea level rises and falls anywhere from a few **centimeters** to more than 15 m (Figure 16.7). A complete tidal cycle includes a *flood tide* that progressively covers more and more of a nearshore area until high tide is reached, followed by *ebb tide,* during which the nearshore area is once again exposed (Figure 16.7). These regular fluctuations in sea level constitute one largely untapped source of energy as do waves, ocean currents, and temperature differences in seawater (Geo-Focus 16.1).

Both the Moon and the Sun have sufficient gravitational attraction to exert tide-generating forces strong enough to deform the solid body of Earth, but they have a much greater influence on the oceans. The Sun is 27 million times more massive than the Moon, but it is 390 times as far from Earth, and its tide-generating force is only 46% as strong as that of the Moon. Accordingly, the tides are dominated by the Moon, but the Sun plays an important role as well.

If we consider only the Moon acting on a spherical, water-covered Earth, its tide-generating forces produce two bulges on the ocean surface (Figure 16.8a). One bulge points toward the Moon because it is on the side of Earth where the Moon's gravitational attraction is greatest. The other bulge is on the opposite side of Earth; it points away from the Moon because of centrifugal force due to Earth's rotation, and the Moon's gravitational attraction is less. These two bulges always point toward and away from the Moon (Figure 16.8a), so as Earth rotates and the Moon's position changes, an observer at a particular shoreline location experiences the rhythmic rise and fall of tides twice daily, but the heights of two successive high tides may vary depending on the Moon's inclination with respect to the equator.

The Moon revolves around Earth every 28 days, so its position with respect to any latitude changes slightly each day. That is, as the Moon moves in its orbit and Earth rotates on its axis, it takes the Moon 50 minutes longer each day to return to the same position it was in the previous day. Thus, an observer would experience a high tide at 1:00 p.m. on one day, for example, and at 1:50 p.m. on the following day.

Even though the Sun's tide-generating force is weaker than the Moon's, when the two are aligned every two weeks, their forces added together generate *spring tides* about 20% higher than average tides (Figure 16.8b). When the Moon and Sun are at right angles to each other, also at two-week

Figure 16.6 The Evolution of Coral Reefs The origin of a fringing reef, barrier reef, and atoll as the plate upon which the fringing reef sits is carried into deeper water.

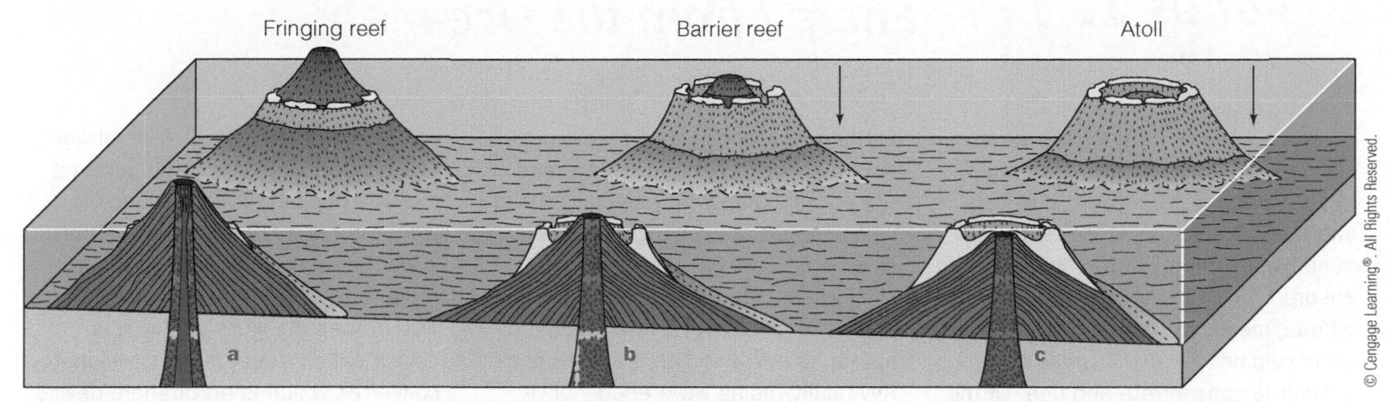

a A fringing reef forms around an island in the tropics.

b The island sinks as the oceanic plate on which it rides moves away from a spreading ridge. In this case, the island does not sink at a rate faster than coral organisms can build upward.

c The island eventually disappears beneath the surface, but the coral remains at the surface as an atoll.

d An underwater view of a coral reef in St. Croix, United States, Virgin Islands.

intervals, the Sun's tide-generating force cancels some of the Moon's, and *neap tides* about 20% lower than average occur (Figure 16.8c).

Tidal ranges are also affected by shoreline configuration. Broad, gently sloping continental shelves as in the Gulf of Mexico have low tidal ranges, whereas steep, irregular shorelines experience much greater rise and fall of tides. Tidal ranges are greatest in some narrow, funnel-shaped bays and inlets. The Bay of Fundy in Nova Scotia has a tidal range of 16.5 m and ranges greater than 10 m occur in several other areas.

Tides have an important impact on shorelines because the area of wave attack constantly shifts onshore and offshore as the tides rise and fall. Tidal currents themselves, however, have little modifying effect on shorelines, except in narrow passages where tidal current velocity is great enough to erode and transport sediment.

Figure 16.7 Low and High Tides Low tide **(a)** and high tide **(b)** in Turnagain Arm, part of Cook Inlet in Alaska. The tidal range here is about 10 m. Turnagain Arm is a huge fiord now being filled with sediment carried in by rivers. Notice the mudflats in **(a)**.

a Low tide.

b High tide.

GEO FOCUS 16.1 Energy from the Oceans

If we could harness the energy of waves, ocean currents, temperature differences in oceanic waters, and tides, an almost limitless, largely nonpolluting energy supply would be ensured. Unfortunately, ocean energy is diffuse, meaning that the energy for a given volume of water is small and thus difficult to concentrate and use. Of the several sources of ocean energy, only tides show much promise for the near future.

Ocean thermal energy conversion (OTEC) exploits the temperature difference between surface waters and those at depth to run turbines and generate electricity. The amount of energy from this source is enormous, but a number of practical problems must be solved. For one thing, large quantities of water must be circulated through a power plant, which requires large surface areas devoted to this purpose. Despite several decades of research, only a few facilities have been tested in Hawaii and Japan.

Ocean currents also possess energy that might be tapped to generate electricity. Unfortunately, these currents flow at only a few kilometers per hour at most, whereas hydroelectric power plants on land rely on water moving rapidly from higher to lower elevations. Furthermore, ocean currents cannot be dammed, their energy is diffuse, and any power plant would have to contend with unpredictable changes in flow direction.

Harnessing wave energy to generate electricity is not a new idea, and, in fact, is now used on a limited scale. Any facility using wave energy obviously must be designed to withstand the impact of waves, especially storm waves; be resistant to saltwater corrosion; and be situated where wave energy is vigorous enough for the task. No large-scale wave-energy plants are now operating, but the Japanese have developed devices to power buoys and lighthouses. The world's first commercial wave power plant called Limpet began operating in 2000 on the Scottish island of Islay. This is a shoreline device that uses the energy of oscillating waves to generate electricity to power about 300 homes (▮ Figure 1). Another device called the Pelamis wave-energy converter, which is an offshore device, became operational in 2004 in the United Kingdom.

Tidal power has been used for centuries in some coastal areas to run mills, but its use at present for electrical generation is limited. One limitation is that the tidal range must be at last 5 m, and there must also be a coastal region where water can be stored following

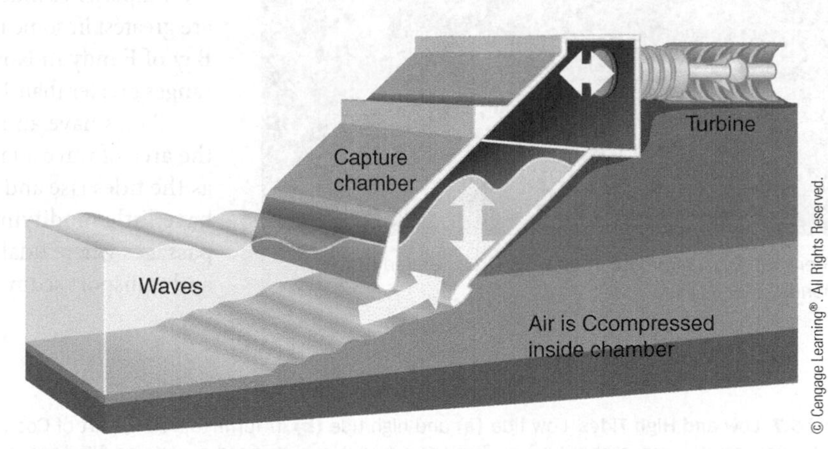

▮ **Figure 1** Diagram showing the Limpet (Land Installed Marine Powered Energy Transformer). The Limpit, which is set into a rock face on a shoreline, has a chamber with a turbine that operates when waves alternately compress and decompress air.

high tide. Suitable sites for tidal power plants are limited not only by tidal range but also by location.

Many areas along the United States Gulf Coast would certainly benefit from a tidal power plant, but the tidal range of generally less than 1 m precludes the possibility of development. However, an area with an appropriate tidal range in some remote area such as southern Chile or the Arctic islands of Canada offers little potential because of the great distance from population centers. Accordingly, in North America, only a few areas show much potential for developing tidal energy.

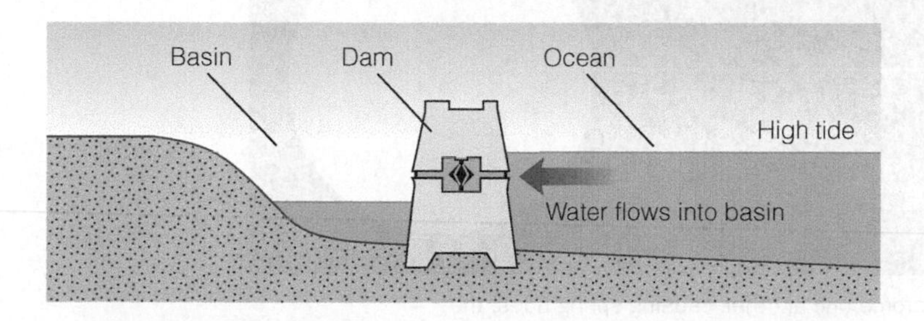

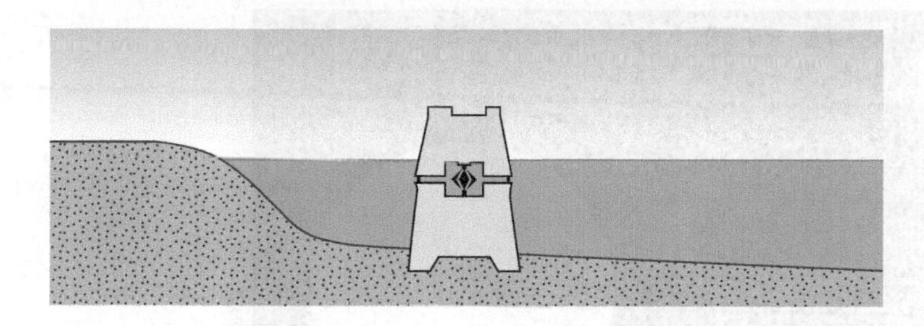

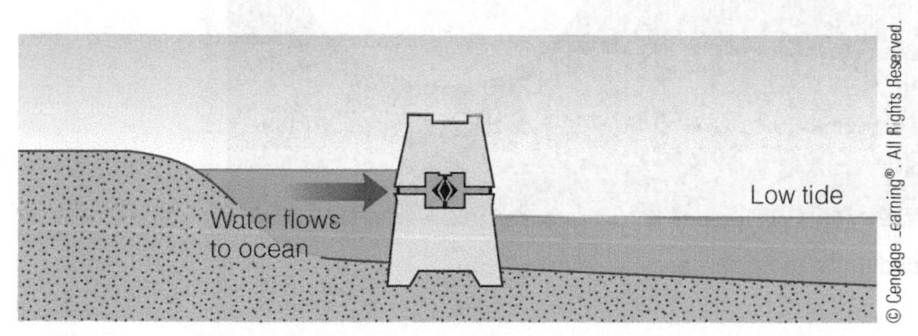

Figure 2 Rising and falling tides produce electricity by spinning turbines connected to generators, just as at hydroelectric plants. This view in the foreground is a cross section showing how water flows into and out of the basin, but the basin would actually be closed off here by land.

The idea behind tidal power is simple, although putting it into practice is not easy. First, a dam with sluice gates to regulate water flow must be built across the entrance to a bay or estuary. When the water level has risen sufficiently high during flood tide, the sluice gates are closed. Water held on the landward side of the dam is then released and electricity is generated just as it is at a hydroelectric dam. Actually, a tidal power plant can operate during both flood and ebb tides (Figure 2).

The first tidal power-generating facility was constructed in 1966 at the La Ranee River estuary in France. In North America, a much smaller tidal power plant has been operating in the Bay of Fundy, Nova Scotia, where the tidal range, the greatest in the world, exceeds 16 m.

Although tidal power shows some promise, it will not solve our energy needs even if developed to its fullest potential. Most analysts think that only 100 to 150 sites worldwide have sufficiently high tidal ranges and the appropriate coastal configuration to exploit this energy resource. This coupled with the facts that construction costs are high and tidal energy *systems can* have disastrous effects on the ecology (biosphere) of estuaries makes it unlikely that tidal energy will ever contribute more than a small percentage of all energy production.

a Tidal bulges if only the Moon caused them.

Figure 16.8 Tidal Bulges The gravitational attraction of the Moon and Sun causes tides. The sizes of the tidal bulges are greatly exaggerated.

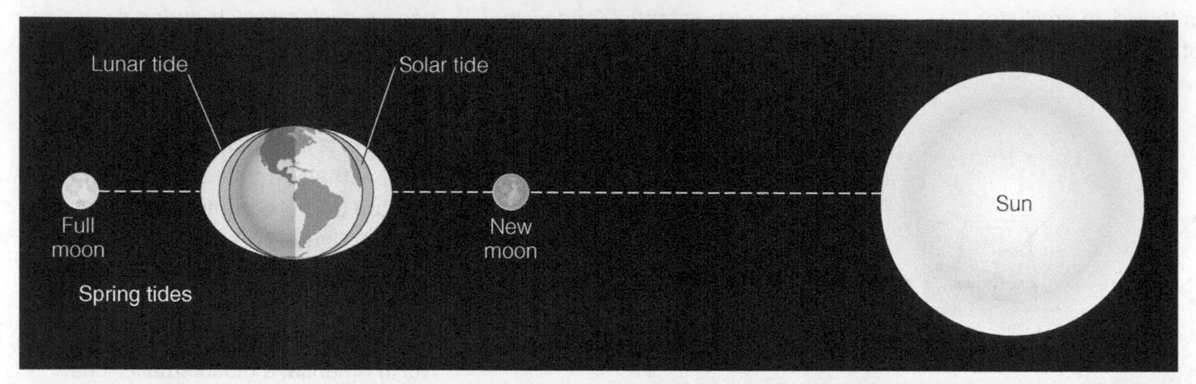

b When the Moon is new or full, the solar and lunar tides reinforce one another, causing spring tides, the highest high tides and lowest low tides.

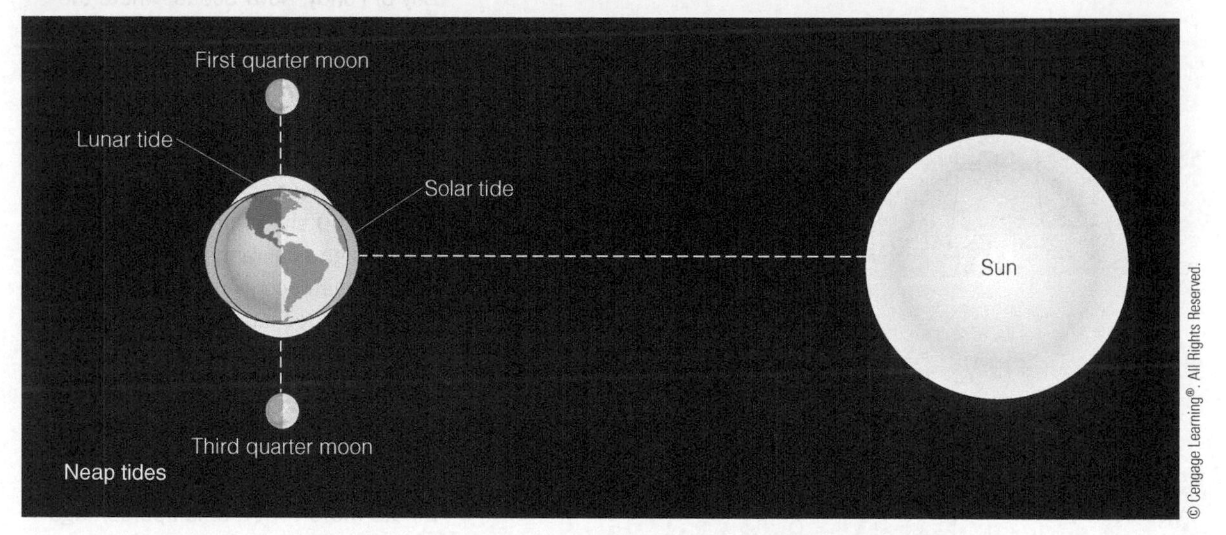

c During the Moon's first and third quarters, the Moon, Sun, and Earth form right angles, causing neap tides, the lowest high tides and highest low tides.

Waves You can see **waves,** or oscillations of a water surface, on all bodies of water, but they are best developed in the oceans. In fact, waves are directly or indirectly responsible for most erosion, sediment transport, and deposition in coastal areas. Wave terminology is illustrated with a typical series of waves in ▌Figure 16.9a. A *crest,* as you would expect, is the highest part of a wave, whereas the low area between crests is a *trough.* The distance from crest to crest (or trough to trough) is the *wavelength,* and the vertical distance from trough to crest is *wave height.* You can calculate the speed at which a wave advances, called celerity (C), by the formula

$$C = L/T$$

where L is wavelength and T is wave period—that is, the time it takes for two successive wave crests, or troughs, to pass a given point.

The speed of wave advance (C) is actually a measure of the velocity of the wave form rather than the speed of the molecules of water in a wave. When waves move across a water surface, the water moves in circular orbits but shows little or no net forward movement (Figure 16.9a). Only the wave form moves forward, and as it does so it transfers energy in the direction of wave movement.

The diameters of the orbits that water follows in waves diminish rapidly with depth, and at a depth of about one-half wavelength ($L/2$), called **wave base,** they are essentially zero. Thus, at a depth exceeding wave base, the water and seafloor

Figure 16.9 Waves and Wave Terminology

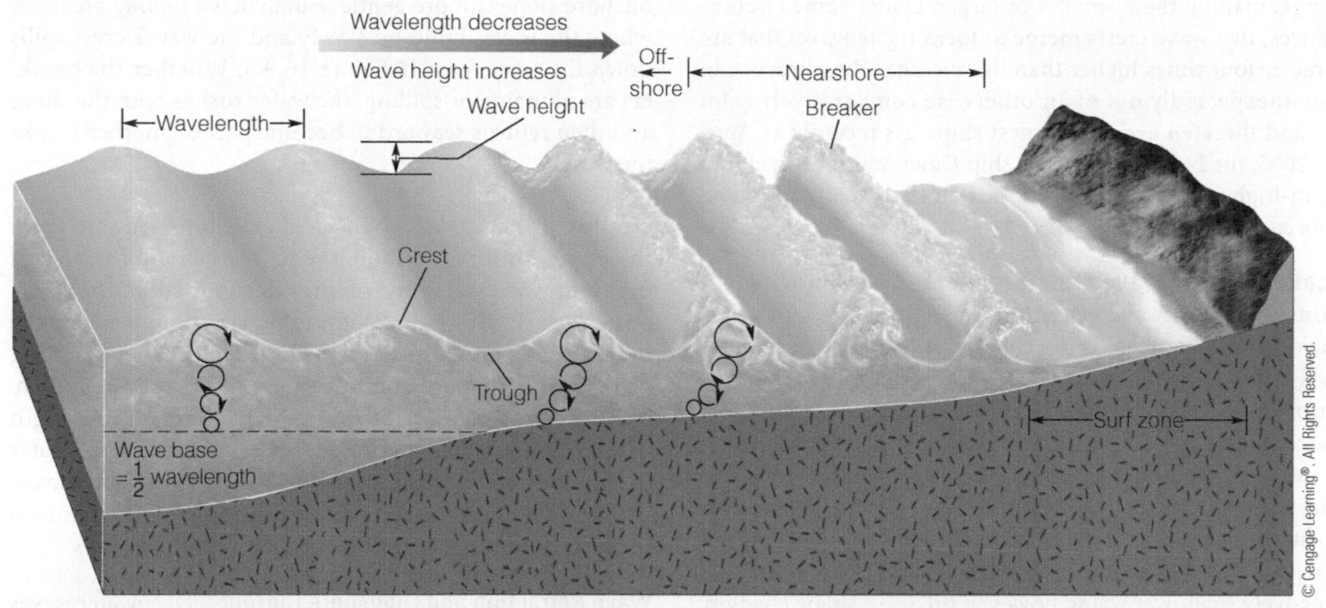

a Waves and the terminology applied to them. Note that waves are disrupted when they encounter water shallower than wave base, and they eventually form breakers.

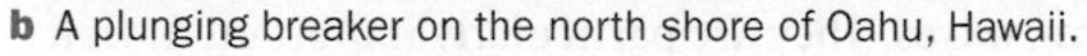

b A plunging breaker on the north shore of Oahu, Hawaii.

c A spilling breaker.

or lake floor are unaffected by surface waves (Figure 16.9a). We will explore wave base more fully in later sections of this chapter.

Wave Generation Several processes such as landslides and earthquakes cause waves, but most geologic work on shorelines is accomplished by wind-generated waves, especially storm waves. When wind blows over water—that is, one fluid (air) moves over another fluid (water)—friction between the two transfers energy to the water, causing the water surface to oscillate.

In areas where waves are produced, such as beneath a storm center at sea, sharp-crested, irregular waves called *seas* develop. Seas have various heights and lengths, and one wave cannot be easily distinguished from another. But as seas move out from their area of generation, they are sorted into broad *swells* with rounded, long crests and all are about the same size.

The harder and longer the wind blows, the larger are the waves, but these are not the only factors that control wave size. High-velocity wind blowing over a small pond will never generate large waves regardless of how long it blows. In fact, waves on ponds and most lakes appear only while the wind is blowing; once the wind stops, the water quickly smooths out. In contrast, the surface of the ocean is always in motion, and waves with heights of 34 m have been recorded during storms in the open sea.

The reason for the disparity between wave sizes on ponds and lakes and on the oceans is the **fetch,** which is the distance the wind blows over a continuous water surface. So on ponds and lakes the fetch corresponds to their length or width, depending on wind direction. To produce waves of greater length and height, more energy must be transferred from wind to water; hence large waves form beneath large storms at sea.

Waves with different lengths, heights, and periods may merge, making them smaller or larger. Under some circumstances, two wave crests merge to form *rogue waves* that are three or four times higher than the average. These waves can rise unexpectedly out of an otherwise comparatively calm sea and threaten even the largest ships. As recently as April 16, 2005, the Norwegian cruise ship *Dawn* was damaged by a 21-m-high rogue wave that flooded more than 60 cabins and injured four passengers.

Shallow-Water Waves and Breakers Swells moving out from an area of wave generation lose little energy as they travel long distances across the ocean. In these deepwater swells, the water surface oscillates and water moves in circular orbits, but little net displacement of water takes place in the direction of wave travel (Figure 16.9a). Of course, wind blows some water from wave crests, thus forming whitecaps with foamy white crests, and surface currents transport water great distances; but deepwater waves themselves accomplish little actual water movement. When these waves enter progressively shallower water, however, the wave shape changes, and water is displaced in the direction of wave advance.

Broad, undulating deepwater waves are transformed into sharp-crested waves as they enter shallow water. This transformation begins at a water depth corresponding to wave base—that is, one-half wavelength. At this point, the waves "feel" the seafloor, and the orbital motion of water within the waves is disrupted (Figure 16.9a). As waves continue moving shoreward, the speed of wave advance and wavelength decrease, but wave height increases. Thus, as they enter shallow water, waves become oversteepened as the wave crest advances faster than the wave form, and eventually the crest plunges forward as a **breaker** (Figure 16.9). Breaking waves might be several times higher than their deepwater counterparts, and when they break, they expend their kinetic energy on the shoreline.

The waves just described are the classic *plunging breakers* that crash onto shorelines with steep offshore slopes, such as those on the north shore of Oahu in the Hawaiian Islands (▌Figure 16.9b). In contrast, shorelines where the offshore slope is more gentle usually have *spilling breakers*, where the waves build up slowly and the wave's crest spills down the wave front (▌Figure 16.9c). Whether the breakers are plunging or spilling, the water rushes onto the shore and then returns seaward to become part of another breaking wave.

Nearshore Currents The area extending seaward from the upper limit of the shoreline to just beyond the area of breaking waves is conveniently designated as the *nearshore zone* (Figure 16.9). Within the nearshore zone are the breaker zone and a surf zone, where water from breaking waves rushes forward and then flows seaward as backwash. The nearshore zone's width varies depending on the length of approaching waves because long waves break at a greater depth, and thus farther offshore, than do short waves. Incoming waves are responsible for two types of currents in the nearshore zone: *longshore currents* and *rip currents.*

Wave Refraction and Longshore Currents Deepwater waves have long, continuous crests, but rarely are their crests parallel with the shoreline (▌Figure 16.10). In other words, they seldom approach a shoreline head-on, but rather at some angle. Thus, one part of a wave enters shallow water where it encounters wave base and begins breaking before other parts of the same wave. As a wave begins to break, its velocity diminishes, but the part of the wave still in deep water races ahead until it too encounters wave base. The net effect of this oblique approach is that waves bend so that they more nearly parallel the shoreline, a phenomenon known as **wave refraction** (Figure 16.10).

Even though waves are refracted, they still usually strike the shoreline at some angle, causing the water between the breaker zone and the beach to flow parallel to the shoreline. These **longshore currents,** as they are called, are long and narrow and flow in the same general direction as the approaching waves. They are particularly important in transporting and depositing sediment in the nearshore zone.

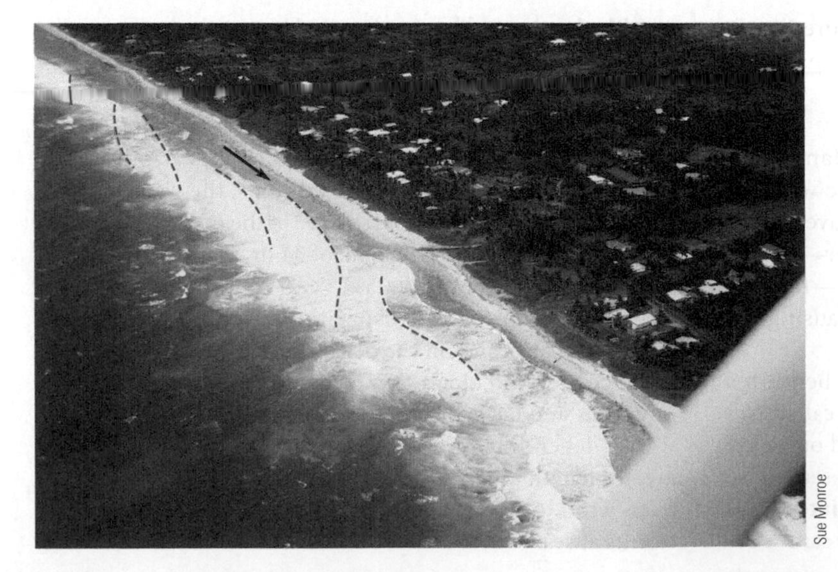

▌**Figure 16.10 Wave Refraction** Wave refraction (wave crests are indicated by dashed lines). These waves are refracted as they enter shallow water and more nearly parallel the shoreline. The waves generate a longshore current that flows in the direction of wave approach, from upper left to lower right (arrow) in this example.

Rip Currents Waves carry water into the nearshore zone, so there must be a mechanism for mass transfer of water back out to sea. One way in which water moves seaward from the nearshore zone is in **rip currents,** narrow surface currents that flow out to sea through the breaker zone (▮ Figure 16.11). Surfers commonly take advantage of rip currents for an easy ride out beyond the breaker zone, but these currents pose a danger to inexperienced swimmers. Some rip currents flow at several kilometers per hour, so if a swimmer is caught in one, it is useless to try to swim directly back to shore. Instead, because rip currents are narrow and usually nearly perpendicular to the shore, one can swim parallel to the shoreline for a short distance and then turn shoreward with little difficulty.

Rip currents are circulating cells fed by longshore currents that increase in velocity from midway between each rip current. When waves approach a shoreline, the amount of water builds up until the excess moves out to sea through the breaker zone.

Rip currents commonly develop where wave heights are lower than in adjacent areas, and differences in wave height are controlled by variations in water depth. For instance, if waves move over a depression, the height of the waves over the depression tends to be less than in adjacent areas, forming the ideal environment for rip currents.

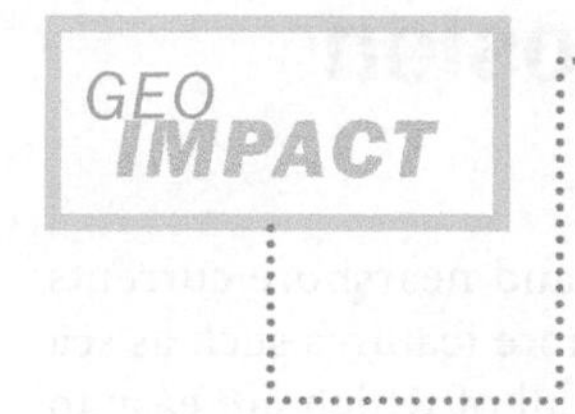

Nearshore Currents and Swimming Safety

While swimming at your favorite beach, you suddenly notice that you are far down the beach from the point where you entered the water. Furthermore, the size of the waves you were swimming in has diminished considerably. You decide to swim back to shore and then walk back to your original starting place, but no matter how hard you swim, you are carried farther and farther away from the shore. Assuming that you survive this incident, explain what happened and what you did to remedy the situation.

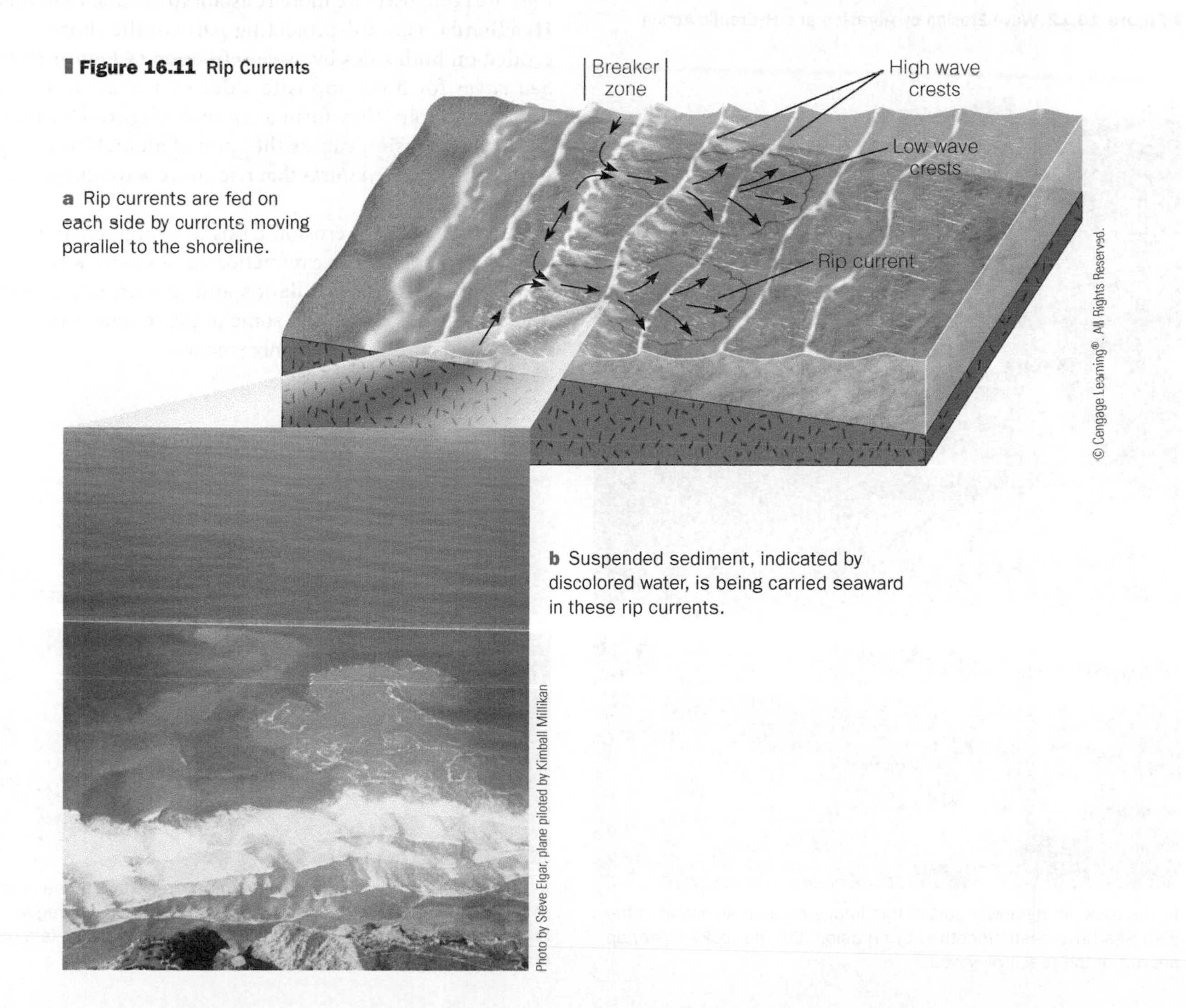

▮ **Figure 16.11** Rip Currents

a Rip currents are fed on each side by currents moving parallel to the shoreline.

b Suspended sediment, indicated by discolored water, is being carried seaward in these rip currents.

16.5 Shoreline Erosion and Deposition

Erosion and deposition by waves and nearshore currents account for many interesting seashore features such as sea stacks, arches, beaches, and spits, all of which are easy to recognize. In fact, many of these features are present along the Pacific, Gulf, and Atlantic coasts as well as along the shorelines of large lakes.

Erosion

On many seashores, erosion creates steep or vertical slopes known as *sea cliffs* that during storms are pounded by waves (hydraulic action), worn by the impact of sand and gravel (abrasion) (▮ Figure 16.12), and eroded by dissolution involving the chemical breakdown of rocks by the solvent action of seawater. Tremendous energy from waves is concentrated on the lower parts of sea cliffs and is most effective on those composed of sediments or highly fractured rocks. In any case, the net effect is erosion of the sea cliff and retreat of the cliff face landward.

▮ Figure 16.12 Wave Erosion by Abrasion and Hydraulic Action

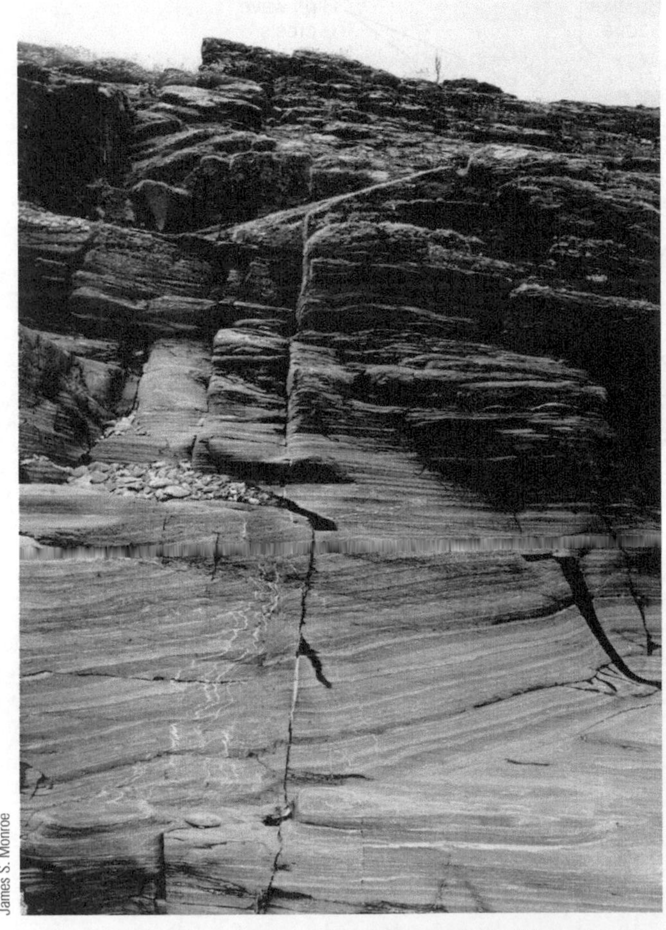

James S. Monroe

a The rocks in the lower part of this image on a small island in the Irish Sea have been smoothed by abrasion, but the rocks higher up are out of the reach of waves.

Sue Monroe

b During storms, hydraulic action and abrasion erode these layers of sandstone at Shore Acres State Park near Coos Bay, Oregon. Storm waves impact the entire rock face, the top of which is more than 25 m above the sea.

Wave-Cut Platforms Wave intensity and the resistance of shoreline materials to erosion determine the rate at which a sea cliff retreats landward. A sea cliff of glacial drift on Cape Cod, Massachusetts, erodes as much as 30 m per century. By comparison, sea cliffs of dense igneous or metamorphic rocks erode and retreat much more slowly.

Sea cliffs erode mostly as a result of hydraulic action and abrasion at their bases. As a sea cliff is undercut by erosion, the upper part is left unsupported and susceptible to mass-wasting processes. Thus, sea cliffs retreat little by little, and as they do, they leave a beveled surface called a **wave-cut platform** that slopes gently seaward (▮ Figure 16.13). The water over broad wave-cut platforms is invariably shallow because the abrasive planing action of waves is effective to a depth of only about 10 m. The sediment eroded from sea cliffs is transported seaward until it reaches deeper water at the edge of the wave-cut platform. There it is deposited and forms a *wave-built platform,* which is a seaward extension of the wave-cut platform (Figure 16.13). Wave-cut platforms now above sea level are known as **marine terraces.**

Sea Caves, Arches, and Stacks Sea cliffs do not retreat uniformly because some of the materials of which they are composed are more resistant to erosion than others. **Headlands,** seaward-projecting parts of the shoreline, are eroded on both sides by wave refraction (▮ Figure 16.14a). *Sea caves* form on opposite sides of a headland, and if these caves join, they form a *sea arch* (Figure 16.14a, ▮ b). Continued erosion causes the span of an arch to collapse, creating isolated *sea stacks* that rise above wave-cut platforms (Figure 16.14a, ▮ c).

In the long run, erosion tends to straighten an initially irregular shoreline. Wave refraction causes more wave energy to be expended on headlands and less on embayments. Thus, headlands erode, and some of the sediment yielded by erosion is deposited in the embayments.

Figure 16.13 Origin of a Wave-Cut Platform

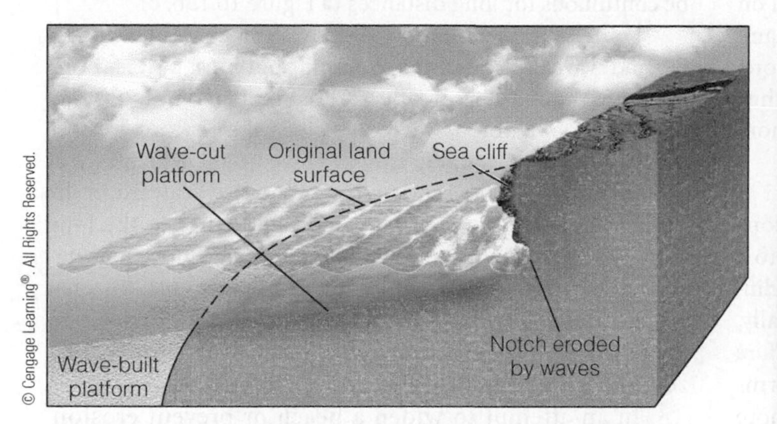

a Erosion causes a sea cliff to migrate landward, leaving a gently sloping surface, called a wave-cut platform. A wave-built platform forms by deposition at the seaward margin of the wave-cut platform.

b A wave-cut platform at Cape Foulweather, Oregon.

Deposition

In a previous section, we mentioned that longshore currents are effective at transporting sediment. In fact, we can think of the area from the breaker zone to the upper limit of wave swash as a "river" that flows along the shoreline in the direction of wave approach. Unlike rivers on land, though, its direction of flow changes if waves approach from a different direction. Nevertheless, the analogy is apt, and just like rivers on land, a longshore current's capacity for transport varies with flow velocity and water depth.

Figure 16.14 Erosion of a Headland

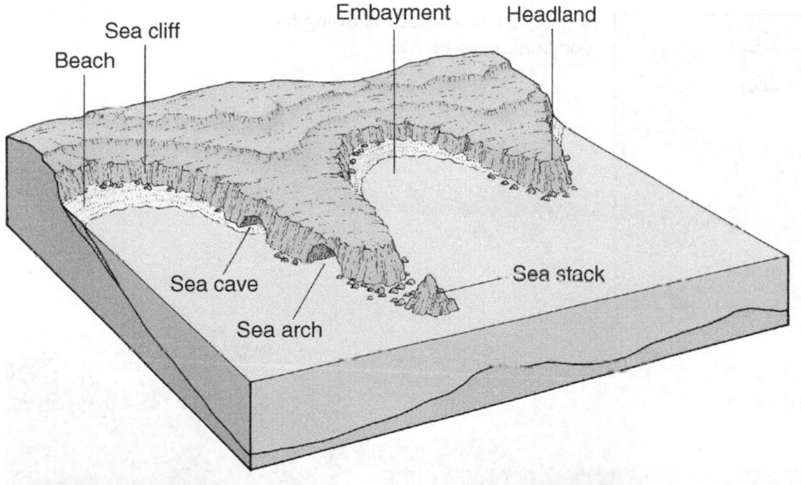

a Erosion of a headland and the origin of a sea cave, a sea arch, and sea stacks.

b A sea arch at Samuel H. Boardman State Park, Oregon.

c Sea stacks at Shell Beach along the California coast.

Wave refraction and the resulting longshore currents are the primary agents of sediment transport and deposition on shorelines, but tides also play a role because, as they rise and fall, the position of wave attack shifts onshore and offshore. Rip currents play no role in shoreline deposition, but they do transport fine-grained sediment (silt and clay) offshore through the breaker zone.

Beaches By definition, a **beach** is a deposit of unconsolidated sediment extending landward from low tide to a change in topography, such as a line of sand dunes, a sea cliff, or the point where permanent vegetation begins. Typically, a beach has several component parts, including a *backshore* that is usually dry, being covered by water only during storms or exceptionally high tides (❚ Figure 16.15a). The backshore consists of one or more *berms,* platforms composed of sediment deposited by waves; the berms are nearly horizontal or slope gently landward. The sloping area below a berm exposed to wave swash is the *beach face.* The beach face is part of the *foreshore,* an area covered by water during high tide but exposed during low tide.

Depending on shoreline materials and wave intensity, beaches may be discontinuous, existing as only *pocket beaches* in protected areas such as embayments, or they may be continuous for long distances (❚ Figure 16.15b, c).

Some of the sediment on beaches comes from weathering and wave erosion of the shoreline, but most of it is transported to the coast by streams and redistributed by longshore currents. As we noted, waves usually strike beaches at some angle, causing the sand grains to move up the beach face at a similar angle; as the sand grains are carried seaward in the backwash, however, they move perpendicular to the long axis of the beach. Thus, individual sand grains move in a zigzag pattern in the direction of longshore currents, by what is called *longshore drift.* This movement is not restricted to the beach; it extends seaward to the outer edge of the breaker zone (❚ Figure 16.16a).

In an attempt to widen a beach or prevent erosion, shoreline residents often build *groins,* structures that project seaward at right angles from the shoreline. A groin interrupts the flow of longshore currents, causing sand deposition on the upcurrent side and widening of the beach at that location. However, erosion inevitably occurs on the downcurrent side of a groin (❚ Figure 16.16b).

Quartz is the most common mineral in most beach sands, but there are some notable exceptions. For example,

❚ **Figure 16.15** Beaches

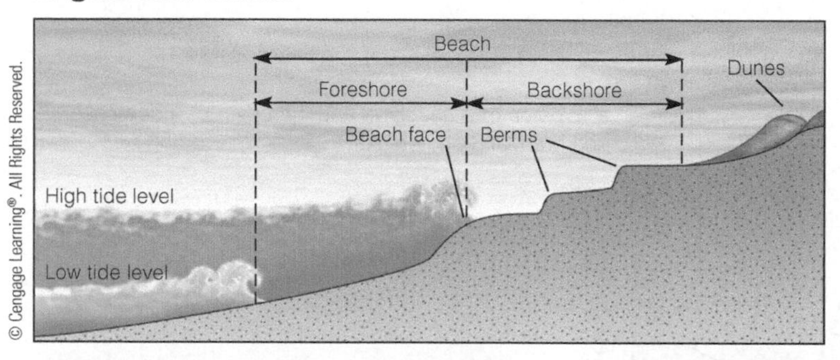

a Diagram of a beach showing its component parts.

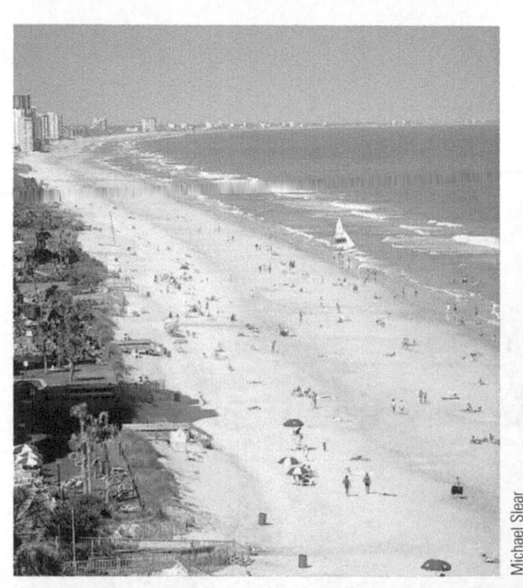
Michael Slear

b The Grand Strand of South Carolina, shown here at Myrtle Beach, is 100 km of nearly continuous beach.

James S. Monroe

c A pocket beach near Seal Rock, Oregon.

▌Figure 16.16 Longshore Currents and Longshore Drift

Critical Thinking Question What is the direction of the longshore currents in **(b)**?

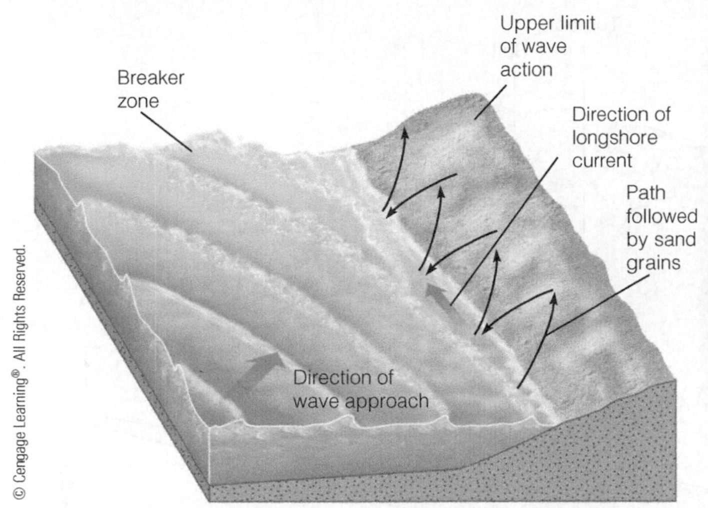

a Longshore currents transport sediment along the shoreline between the breaker zone and the upper limit of wave action.

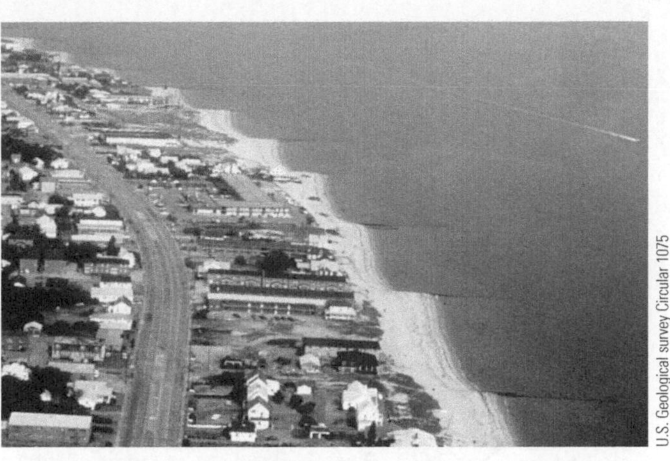

b These groins along the shoreline at Norfolk, Virginia, trap sand to maintain the beach in this area.

the black sand beaches of Hawaii are composed of sand- and gravel-size basalt rock fragments or small grains of volcanic glass, and some Florida beaches are composed of the fragmented calcium carbonate shells of marine organisms. In short, beaches are composed of whatever material is available.

Seasonal Changes in Beaches The loose grains on beaches are constantly moved by waves, but the overall configuration of a beach remains unchanged as long as equilibrium conditions persist. We can think of the beach profile consisting of a berm or berms and a beach face as a profile of equilibrium (Figure 16.15a); that is, all parts of the beach are adjusted to the prevailing conditions of wave intensity, nearshore currents, sediment supply, and materials composing the beach.

Tides and longshore currents affect the configuration of beaches to some degree, but storm waves are by far the most important agent modifying their equilibrium profile. In many areas, beach profiles change with the seasons; so we recognize *summer beaches* and *winter beaches,* each of which is adjusted to the conditions prevailing at those times. Summer beaches are sand covered and have a wide berm, a gently sloping beach face, and a smooth offshore profile. Winter beaches, in contrast, tend to be coarser grained and steeper; they have a small berm or none at all, and their offshore profiles reveal sandbars paralleling the shoreline.

Seasonal changes in beach profiles are related to changing wave intensity. During the winter, energetic storm waves erode the sand from beaches and transport it offshore where it is stored in sandbars. The same sand that was eroded from a beach during the winter returns the next summer when it is driven onshore by more gentle swells. The volume of sand in the system remains more or less constant; it simply moves farther offshore or onshore depending on wave energy.

Spits, Baymouth Bars, and Tombolos Beaches are the most familiar depositional features of coasts, but spits, baymouth bars, and tombolos are common, too. In fact, these features are simply continuations of a beach. A **spit,** for instance, is a finger-like projection of a beach into a body of water such as a bay, and a **baymouth bar** is a spit that has grown until it completely closes off a bay from the open sea (▌Figures 16.17, ▌16.18a, b). Both are composed of sand, more rarely gravel, that was transported and deposited by longshore currents where they weakened as they entered the deeper water of a bay's opening. Some spits are modified by waves so that their free ends are curved; they go by the name *hook* or *recurved spit* (Figure 16.17).

A **tombolo** is a type of spit that extends out from the shoreline to an offshore feature such as an island or sea stack (Figures 16.17, ▌16.18c). A tombolo forms on the shoreward side of an island as wave refraction around the island creates converging currents that turn seaward and deposit a sandbar. So, the long axes of tombolos are nearly at right angles to the shoreline.

Spits, baymouth bars, and tombolos are most common along irregular seashores, but they can also be found in large lakes. Regardless of their setting, spits and baymouth bars constitute a continuing problem where bays must be kept open for pleasure boating, commercial shipping, or both. Obviously, a bay closed off by a sandbar is of little use for either endeavor, so a bay must be regularly dredged or protected from deposition by longshore currents. In some areas, *jetties* are constructed that extend seaward (or lakeward) to interrupt the flow of longshore currents and thus protect the opening to a bay.

Barrier Islands Long, narrow islands of sand lying a short distance offshore from the mainland are **barrier islands** (▌Figure 16.19). On their seaward sides, they are smoothed

Figure 16.17 Spits, Baymouth Bars, and Tombolos Spits form where longshore currents deposit sand in deeper water, as at the entrance to a bay. A baymouth bar is a spit that has grown until it closes off the mouth of a bay. Wave refraction around an island causes converging currents to deposit a sand bar nearly perpendicular to the shoreline, forming a tombolo.

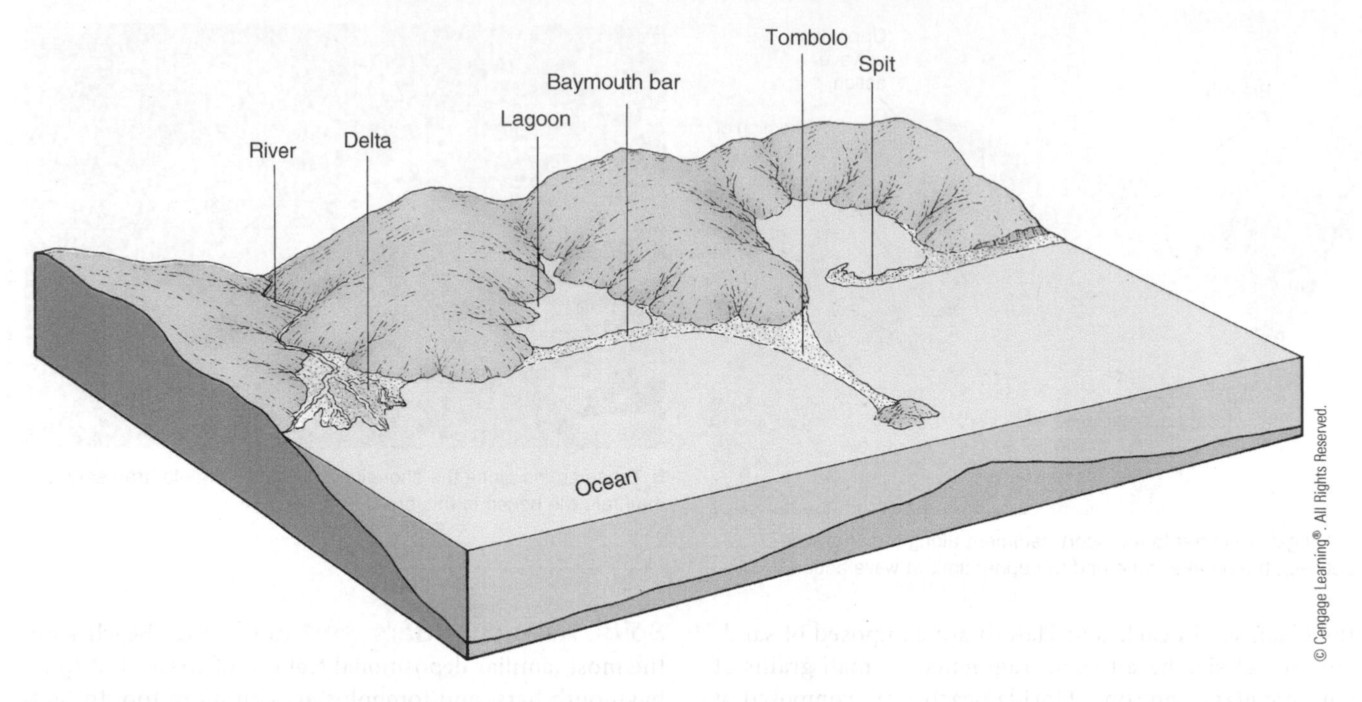

Figure 16.18 Spits, Baymouth Bars, and Tombolos

a A spit at the mouth of the Russian River near Jenner, California.

b Rodeo Beach north of San Francisco, California, is a baymouth bar.

c A tombolo along the Oregon Coast.

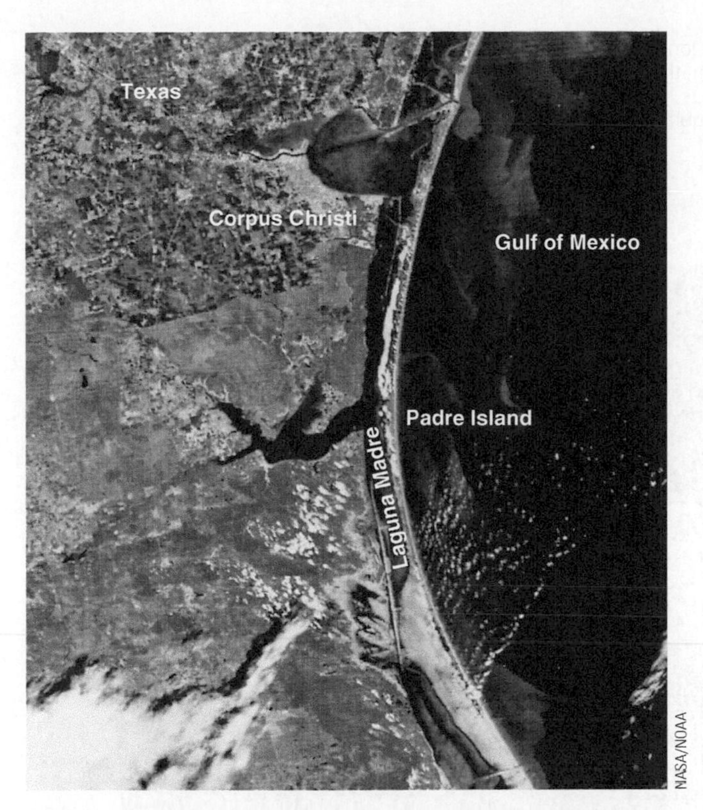

Figure 16.19 Barrier Islands View from space of the barrier islands along the Gulf Coast of Texas. Notice that a lagoon up to 20 km wide separates the long, narrow barrier islands from the mainland.

by waves, but their landward margins are irregular because storm waves carry sediment over the island and deposit it in a lagoon where it is little modified by further wave activity. The component parts of a barrier island include a beach, windblown sand dunes, and a marshy area on their landward sides.

Everyone agrees that barrier islands form on gently sloping continental shelves where abundant sand is available and where both wave energy and the tidal range are low. In fact, these are the reasons that so many are along the United States' Atlantic and Gulf coasts. But even though it is well known where barrier islands form, the details of their origin are still unresolved. According to one model, they formed as spits that became detached from land, whereas another model holds that they formed as beach ridges that subsequently subsided.

Most barrier islands are migrating landward as a result of erosion on their seaward sides and deposition on their landward sides. This is a natural part of barrier island evolution, and it takes place rather slowly, but fast enough to cause many problems for island residents and communities.

The Nearshore Sediment Budget

We can think of the gains and losses of sediment in the nearshore zone in terms of a **nearshore sediment budget** (Figure 16.20). If a nearshore system has a balanced budget, sediment is supplied to it as fast as it is removed, and the volume of sediment remains more or less constant, although sand may shift offshore and onshore with the changing seasons. A positive budget means that gains exceed losses, whereas a negative budget means that losses exceed gains. If a negative budget prevails long enough, a nearshore system is depleted and beaches may disappear.

Erosion of sea cliffs provides some sediment to beaches, but in most areas probably no more than 5%–10% of the total sediment comes from this source. There are exceptions, though; almost all of the sediment on the beaches of Maine is derived from the erosion of shoreline rocks. Most sediment on typical beaches is transported to the shoreline by streams and then redistributed along the shoreline by longshore currents. Thus, longshore currents also play a role in the nearshore sediment budget because they continuously move sediment into and away from beach systems (Figure 16.20).

The primary ways that a nearshore system loses sediment are offshore transport, wind, and deposition in submarine canyons. Offshore transport mostly involves fine-grained sediment carried seaward where it eventually settles in deeper water. Wind is an important process because it removes sand from beaches and blows it inland where it piles up as sand dunes.

If the heads of submarine canyons are nearshore, huge quantities of sand are funneled into them and deposited in deeper water. La Jolla and Scripps submarine canyons off the coast of southern California funnel off an estimated 2 million m^3 of sand each year. In most areas, however, submarine canyons are too far offshore to interrupt the flow of sand in the nearshore zone.

It should be apparent that if a nearshore system is in equilibrium, its incoming supply of sediment exactly offsets its losses. Such a delicate balance tends to continue unless the system is somehow disrupted. One change that affects this balance is the construction of dams across the streams that supply sand. Once dams have been built, all sediment from the upper reaches of the drainage systems is trapped in reservoirs and thus cannot reach the shoreline.

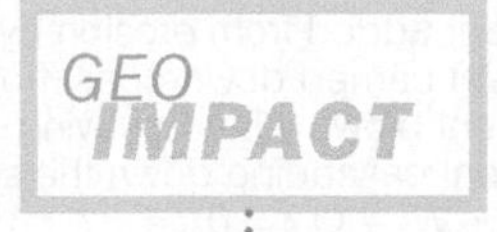

Barrier Island Migration

While visiting a barrier island, you notice some fine-grained, organic-rich deposits along the beach. In fact, there is so much organic matter that the sediments are black and have a foul odor. Their presence on the beach does not seem to make sense because you know that sediments like these were almost certainly deposited in a marsh on the landward (opposite) side of the island. How can you explain your observations?

Figure 16.20 The Nearshore Sediment Budget The long-term sediment budget can be assessed by considering inputs versus outputs. If inputs and outputs are equal, the system is in a steady state, or equilibrium. If outputs exceed inputs, however, the beach has a negative budget and erosion occurs. Accretion takes place when the beach has a positive budget with inputs exceeding outputs.

Critical Thinking Question What will happen to the beaches in this area if a dam is built across the river?

16.6 Types of Coasts

Coasts are difficult to classify because of variations in the factors that control their development and variations in their composition and configuration. Rather than attempt to categorize all coasts, we shall simply note that two types of coasts have already been discussed: those dominated by deposition and those dominated by erosion.

In addition, we will examine coasts in terms of their changing relationship to sea level. But note that although some coasts, such as those in southern California, are described as emergent (uplifted), these same coasts may be erosional as well. In other words, coasts commonly have features that allow them to be classified in more than one way.

Depositional and Erosional Coasts

Depositional coasts, such as the United States Gulf Coast, are characterized by an abundance of detrital sediment and such depositional landforms as wide sandy beaches, deltas, and barrier islands. In contrast, erosional coasts are steep and irregular and typically lack well-developed beaches, except in protected areas (Figure 16.15c). They are further characterized by sea cliffs, wave-cut platforms, and sea stacks. Many of the coasts along the West Coast of North America fall into this category.

Submergent and Emergent Coasts

If sea level rises with respect to the land or the land subsides, coastal regions are flooded and said to be **submergent** or *drowned* **coasts** (Figure 16.21a). Much of the East Coast of North America from Maine southward through South Carolina was flooded during the rise in sea level following the Pleistocene Epoch, so it is extremely irregular. Recall that during the expansion of glaciers during the Pleistocene, sea level was as much as 130 m lower than at present, and that streams eroded their valleys more deeply and extended across continental shelves. When sea level rose, the lower ends of these valleys were drowned, forming *estuaries* such as Delaware and Chesapeake bays (Figure 16.21a). Estuaries are simply the seaward ends of river valleys where seawater and freshwater mix. The divides between adjacent drainage systems on submergent coasts project seaward as broad headlands or a line of islands.

Connection Link

You can review Pleistocene glaciation and its causes in Chapter 14.

Emergent coasts are found where the land has risen with respect to sea level (Figure 16.21b). Emergence takes place when water is withdrawn from the oceans, as occurred during the Pleistocene expansion of glaciers. Presently, coasts are emerging as a result of isostasy or tectonism. In northeastern Canada and the Scandinavian countries, for instance, the coasts are irregular because isostatic rebound is elevating formerly glaciated terrain from beneath the sea.

Coasts that form in response to tectonism, on the other hand, tend to be straighter because the seafloor topography being exposed by uplift is smooth. The west coasts of North and South America are rising as a consequence of plate tectonics. Distinctive features of these coasts are marine terraces, which are wave-cut platforms now elevated above sea level (Figure 16.21b). Uplift in these areas appears to be episodic rather than continuous, as indicated by the multiple levels of terraces in some places. In southern California, several terrace levels are present, each of which probably represents a period of tectonic stability followed by uplift. The highest terrace is now approximately 425 m above sea level.

Figure 16.21 Submergent and Emergent Coasts

Critical Thinking Question Do you think a wave-cut platform is forming in **(b)**? If so, where?

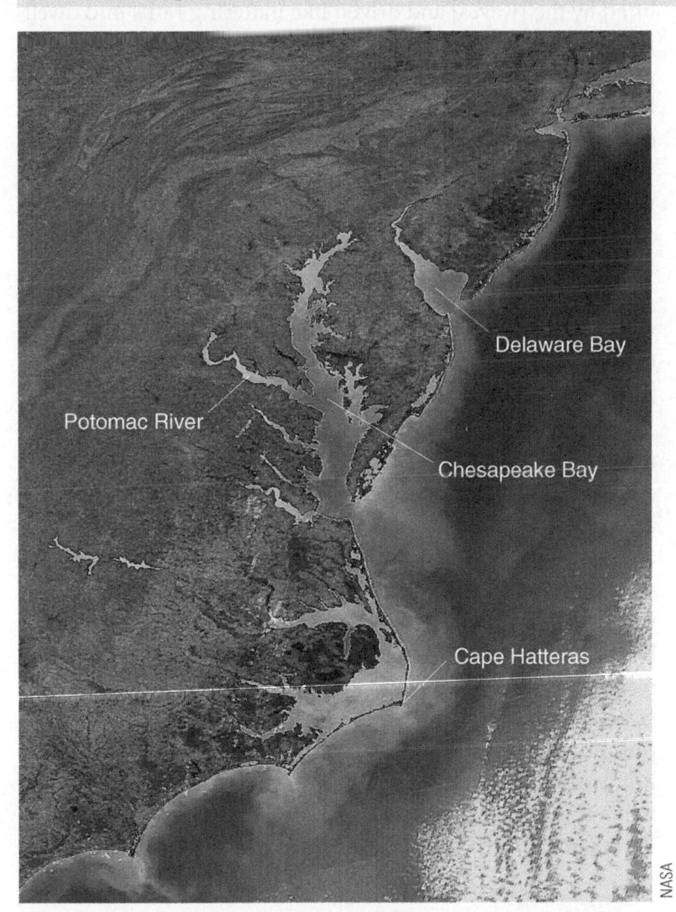

a Submergent coasts tend to be extremely irregular, with estuaries such as Chesapeake and Delaware bays. They formed when the East Coast of the United States was flooded as sea level rose following the Pleistocene Epoch.

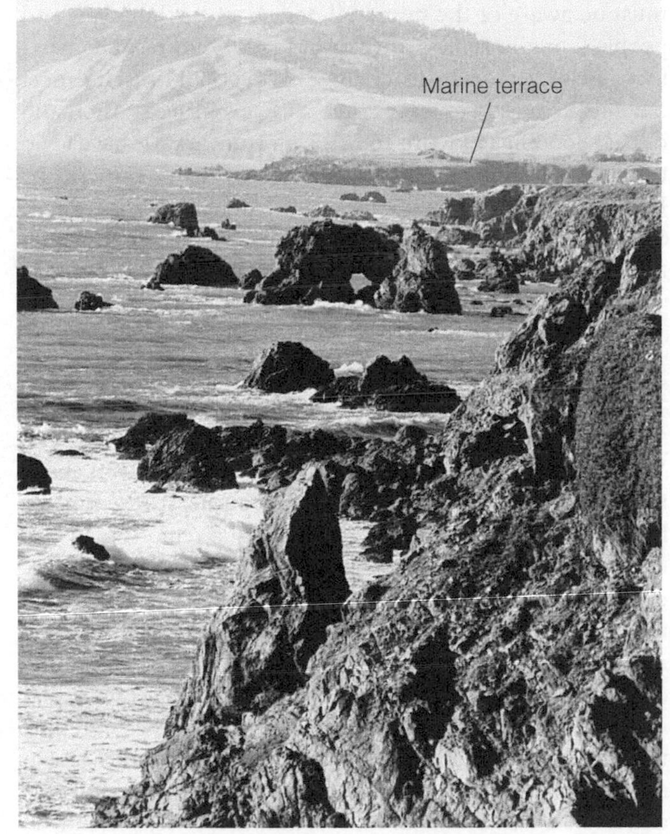

b Emergent coasts tend to be steep and straighter than submergent coasts. Notice the several sea stacks and the sea arch. Also, a marine terrace is visible in the distance.

Figure 16.22 Storm Waves and Coastal Flooding

Sue Monroe

a Storm waves were particularly effective in eroding these sea cliffs near Bodega Bay, California, during February 1998. Attempts to stabilize the shoreline failed and these homes are gone, and others are now threatened.

MC Images/Alamy

b Coastal flooding caused by Hurricane Sandy on November 1, 2012, along the New Jersey shoreline. The coasts of states from Florida to Maine were affected by the storm, but the greatest damages were seen in New Jersey and New York.

16.7 The Perils of Living Near a Shoreline

No matter where we live there are risks. Great Plains residents are concerned about tornadoes, those living near active faults must be aware of the potential danger of earthquakes, and people living in forests are cognizant of wildfires. Likewise, living on or very near a shoreline poses certain risks, the most obvious being storm waves that cause extensive erosion and coastal flooding (Figure 16.22). Unfortunately, we have vivid reminders of this risk from large storms such as Hurricane Isaac in August 2012 and Hurricane Sandy in October 2012. Strong winds during hurricanes also pose a risk, but much of the damage and most fatalities results from coastal flooding.

Storm Waves and Coastal Flooding

We mentioned in Chapter 12 that if a river or stream channel receives more water than it can handle, it overtops its banks and floods. Communities in coastal regions may also experience river or stream floods for exactly the same reason, especially during hurricanes when as much as 60 cm of rain might fall in as little as 24 hours. In addition, as a hurricane moves over the ocean, low atmospheric pressure beneath the eye of the storm causes the ocean's surface to bulge upward as much as 0.5 m. When the storm's eye reaches the shoreline, the bulge coupled with wind-driven waves, piles up in a **storm surge** that may rise several meters above normal high tide and flood areas far inland. Indeed, during Hurricane Sandy in October 2012, the storm surge took place shortly before high tide, which added to the magnitude of flooding. Hurricane Sandy resulted in nearly 100 deaths in the United States, and the damages may total $50 billion, making it the most expensive hurricane since Hurricane Katrina of 2005.

In 1900, hurricane-driven waves surged inland, eventually covering the entire island that Galveston, Texas, occupied. As the waves swept into the city, buildings near the shoreline were battered to pieces and "great beams and railway ties were lifted by the [waves] and driven like battering rams into dwellings and business houses."* Between 6,000 and 8,000 people perished. In an effort to protect the city from future storms, a huge seawall was constructed, and the entire city was elevated to the level of the top of the seawall (Figure 16.23).

The seawall has largely protected Galveston from more recent storm surges, but the same is not true for several other areas. Large-scale coastal flooding took place when Hurricane Isabel hit the Outer Banks of North Carolina in 2003, and in 2004, Florida was hit by four hurricanes causing widespread wind damage and coastal flooding. The United States Gulf Coast was hardest hit in 2005, first in August by Hurricane Katrina and then again in September by Hurricane Rita. When Hurricane Katrina roared ashore on August 29, high winds, a huge storm surge, and coastal flooding destroyed nearly everything in an area covering 230,000 km^2. Parts of Gulfport and Biloxi, Mississippi, were leveled, but most of the public's attention focused on New Orleans, Louisiana (Figure 16.24).

When Hurricane Katrina came ashore, the levees initially held, but on the next day some of the floodwalls were breached and about 80% of New Orleans was flooded. And because New Orleans is mostly below sea level, the floodwaters could not drain out naturally. In fact, the city has 22 pumping stations to remove water from normal rainstorms, but as the city flooded, the pumps were overwhelmed, and when the electricity failed, the pumps were useless. All in all, Hurricane Katrina was the most expensive natural disaster

---

*L. W. Bates, Jr., "Galveston—A City Built upon Sand," *Scientific American*, 95 (1906), p. 64.

Figure 16.23 Seawall Construction Construction on this seawall to protect Galveston, Texas, from storm waves began in 1902. Notice that the wall is curved to deflect waves upward.

Courtesy of the Rosenberg Library, Galveston, Texas

in the history of the United States; the property damages exceeded $100 billion and more than 1,800 people died, mostly in Louisiana. Then, in September, Hurricane Rita hit the Gulf Coast of Texas, and caused about 120 more fatalities and nearly $12 billion in damages.

Scientists, engineers, and some politicians had warned of a Katrina-type disaster for New Orleans for many years. They were aware that during a large hurricane the levees or floodwalls would likely fail. Most of you reading this book will not become geologists, but perhaps you will be engineers, city planners, members of planning commissions, politicians, or simply concerned citizens of coastal communities and will have to deal with these or similar problems.

MENAHEM KAHANA/AFP/Getty Images

Figure 16.24 Hurricane Katrina, 2005 Although many areas were hit hard by Katrina, New Orleans was extensively flooded when the floodwalls that were built to protect the city from Lake Pontchartrain and the Mississippi River failed.

Coastal Management as Sea Level Rises

During the past century, sea level rose about 12 cm worldwide, and all indications are that it will continue to rise. The absolute rate of sea-level rise in a shoreline region depends on two factors. The first is the volume of water in the ocean basins, which is increasing as a result of glacial ice melting and the thermal expansion of near-surface seawater. Many scientists think that sea level will continue to rise because of global warming caused by increasing concentrations of greenhouse gases in the atmosphere.

The second factor that controls sea level is the rate of uplift or subsidence of a coastal area. In some areas, uplift is occurring fast enough that sea level is actually falling with respect to the land. In other areas, sea level is rising while the coastal region is simultaneously subsiding, resulting in a net change in sea level of as much as 30 cm per century. Perhaps such a "slow" rate of sea-level change seems insignificant; after all, it amounts to only a few millimeters per year. But in gently sloping coastal areas, as in the eastern United States from New Jersey southward, even a slight rise in sea level will have widespread effects.

Many of the nearly 300 barrier islands along the East and Gulf coasts of the United States are migrating landward as sea level rises (Figure 16.25). Landward migration of barrier islands would pose few problems if it were not for the numerous communities, resorts, and vacation homes located on them. Moreover, barrier islands are not the only threatened areas. For example, Louisiana's coastal wetlands, an important wildlife habitat and seafood-producing area, are currently being lost at a rate of about 90 km^2 per year. Much of this loss results from sediment compaction, but rising sea level exacerbates the problem (Geo-Focus 12.1).

Figure 16.25 Barrier Island Migration

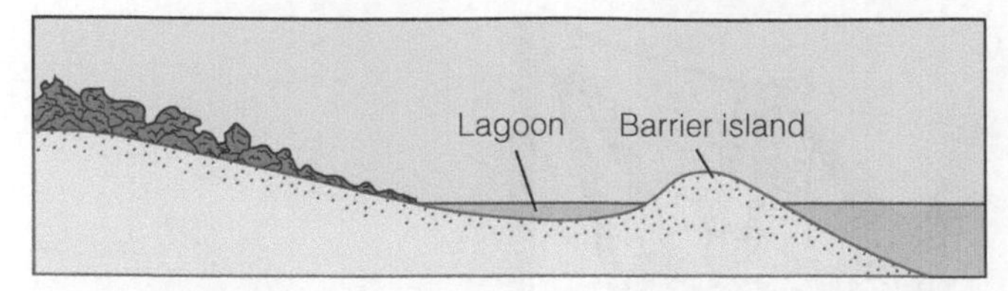

a A barrier island.

b A barrier island migrates landward as sea level rises and storm waves carry sand from its seaward side into its lagoon.

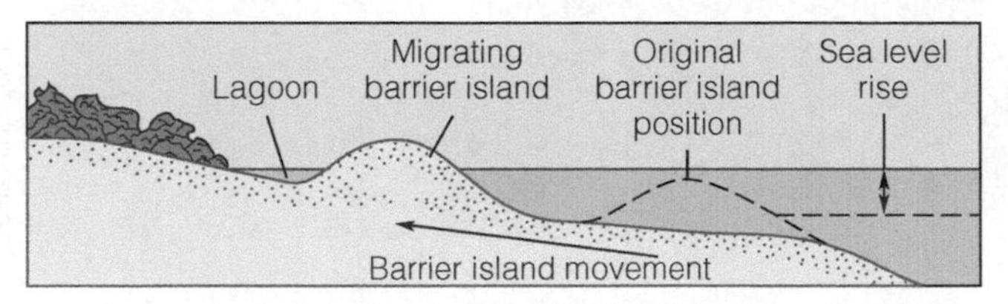

c Over time, the entire island shifts toward the land.

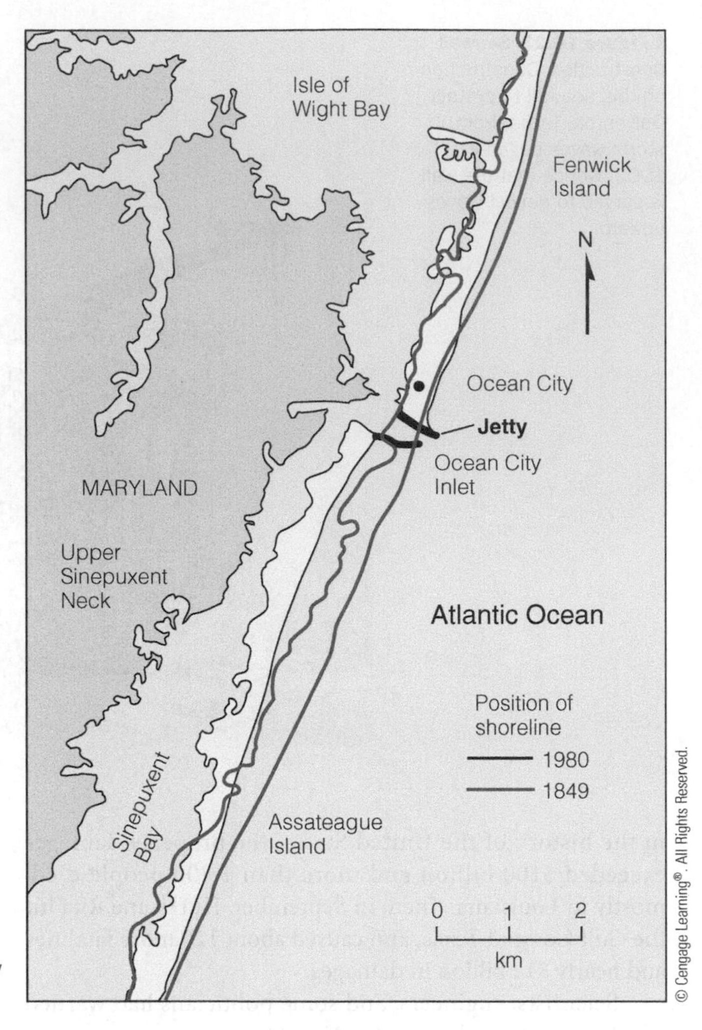

d Jetties were constructed during the 1930s to protect the inlet at Ocean City, Maryland, but they disrupted the net southerly longshore drift and Assateague Island, starved of sediment, has migrated 500 m landward. Beginning in the fall of 2002, beach sand was artificially replenished in an effort to stabilize the island.

Rising sea level also directly threatens many beaches upon which communities depend for revenue. The beach at Miami Beach, Florida, for instance, was disappearing at an alarming rate until the Army Corps of Engineers began replacing the eroded beach sand. The problem is even more serious in other countries. A rise in sea level of only 2 m would inundate large areas of the United States' East and Gulf coasts, but would cover 20% of the entire country of Bangladesh.

Armoring shorelines with *seawalls* (embankments of reinforced concrete or stone) (Figure 16.23) and using *riprap* (piles of stones) (▌Figure 16.26a) protect beachfront structures, but both are initially expensive and during large storms are commonly damaged or destroyed. Seawalls do afford some protection and are seen in many coastal areas along the oceans and large lakes, but some states, including North and South Carolina, Rhode Island, Oregon, and Maine, no longer allow their construction. The futility of artificially maintaining beaches is aptly shown by efforts to protect homes on a South Carolina barrier island. After each spring tide, heavy equipment is used to build a sand berm to protect homes from the next spring tide (▌Figure 16.26b), only to have the berm disappear and then be rebuilt in a never-ending cycle of erosion and expensive artificial replacement.

Because sea level is rising, engineers, scientists, planners, and political leaders must examine what they can do to prevent or minimize the effects of shoreline erosion. At present, only a few viable options exist. One is to put strict controls on coastal development. North Carolina, for example, permits large structures no closer to the shoreline than 60 times the annual erosion rate. Although a growing awareness of shoreline processes has resulted in similar legislation elsewhere, some states have virtually no restrictions on coastal development.

Regulating coastal development is commendable, but it has no impact on existing structures and coastal communities. A general retreat from the shoreline may be possible, but expensive, for individual dwellings and small communities, but it is impractical for large population centers. Such communities as Atlantic City, New Jersey; Miami Beach, Florida; and Galveston, Texas, have adopted one of two strategies to combat coastal erosion. One is to build protective barriers such as seawalls. Seawalls, such as the one at Galveston, Texas, are effective, but they are tremendously expensive to construct and maintain. Furthermore, barriers retard erosion only in the area directly behind them; Galveston Island west of its seawall has been eroded back about 45 m.

Figure 16.26 Riprap and Sand Berm

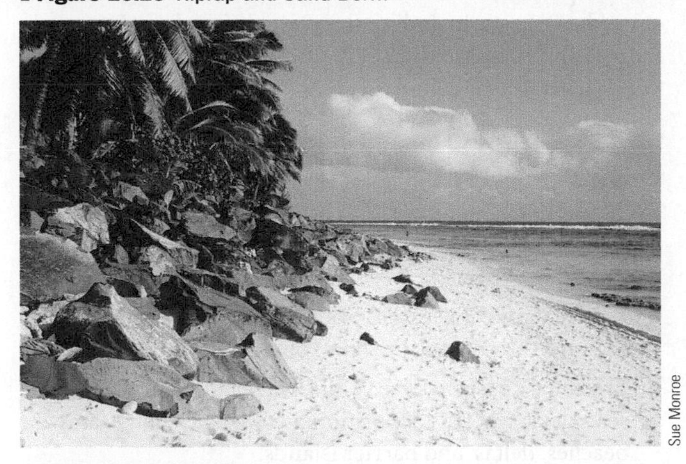

a Riprap made up of large pieces of basalt was piled on this beach to protect a luxury hotel just to the left of this image.

b Heavy equipment builds a berm, an embankment of sand, on the seaward side of beach homes on the Isle of Palms, South Carolina, to protect them from waves. The berm must be rebuilt every two weeks after each spring tide.

Another option, adopted by both Atlantic City, New Jersey, and Miami Beach, Florida, is to pump sand onto the beaches to replace that lost to erosion. This, too, is expensive as the sand must be replenished periodically because erosion is a continuing process.

16.8 The Oceans and Economic Geology

Seawater contains many elements in solution, the most common being sodium (Na) and chlorine (Cl), both of which are extracted for salt by evaporation or mining salt deposits. Much of Earth's supply of magnesium (Mg) comes from seawater, and potassium (K) is derived from evaporite deposits. In fact, the mineral sylvite (KCl) is used in salt substitutes for those on a low-salt diet for health reasons. Also, recall from previous chapters that most limestone (used in cement) and gypsum (used for wallboard) come mostly from deposits that formed from ocean waters. And of course oceanic water itself is the source of precipitation (recall the hydrologic cycle), it has a modifying effect on climate, and the seas are avenues of commerce.

Seafloor deposits are becoming important, and as a result, most nations bordering the oceans claim those resources within their adjacent waters. The United States, by presidential proclamation issued in 1983, claims sovereign rights over an area designed the **Exclusive Economic Zone (EEZ),** which extends 200 nautical miles (370 km) seaward. In addition to the waters adjacent to the continental United States and Alaska, the EEZ also includes those waters adjoining United States territories such as Guam, Puerto Rico, and American Samoa. Other ocean-bordering nations make similar claims.

Resources within the United States EEZ include sand and gravel for construction, and about one-third of all oil production in the United States comes from wells on the continental margins, especially in the Gulf of Mexico, but also adjacent to the coastlines of Alaska and California. Although this is an important source of oil and natural gas, offshore drilling is not without risk. Recall the *Deepwater Horizon* oil rig in the Gulf of Mexico that exploded and sank in August 2010, killing 11 workers and beginning the largest oil spill in the history of the petroleum industry (see Geo-Focus 12.1, Figure 3).

Many areas of the seafloors are littered with spherical objects known as *manganese nodules* which are made up of manganese, iron oxides, as well as cobalt, copper, and nickel. The United States imports most of the manganese and cobalt it needs, so these nodules have attracted some attention as a potential resource. Also, recall the submarine hydrothermal vents (black smokers) we discussed in Chapter 2 (see Figure 2.8). The hot waters at these vents, which are at or very near spreading ridges, precipitate iron, copper, and zinc sulfides, and other minerals. These deposits may also be a future source of some metals.

We noted in the section titled Oceanic Circulation earlier in this chapter that upwelling is responsible for much of the phosphate deposits called *phosphorite* that we use for chemical fertilizers, animal feed supplements, matches, metallurgy, preserved foods, and ceramics. In addition, even though areas of upwelling constitute only a tiny part of the oceans they support 50% by weight of all fishes.

A potential resource within the EEZ is methane hydrate, consisting of single methane molecules bound up in networks formed by frozen water. It is also found in permafrost soils. Methane hydrate is stable at water depths of more than 500 m and near-freezing temperatures. According to one estimate, the carbon in these deposits is double that in all coal, oil, and natural gas. However, no one knows if methane hydrate can be effectively recovered and used as an energy source. Another factor to consider is that the volume of methane hydrate is about 3,000 times greater than the volume of methane in the atmosphere—and methane is 10 times more effective than carbon dioxide as a greenhouse gas.

In Geo-Focus 16.1 in this chapter, we discussed the potential for generation of electricity by waves, tides, and thermal differences in ocean water masses. The first two of these alternatives are now in use, although on a limited scale, but the potential is enormous. In addition to these sources, several countries, including the United Kingdom, Denmark, China, and several others, are currently using or planning wind-generating facilities located at sea.

Key Concepts Review

- Present-day research vessels investigate the seafloor by sampling, drilling, echo sounding, and seismic profiling. Scientists also use submersibles in their studies.
- The upper 100 m or so of the oceans is the photic zone where sunlight is sufficient for photosynthesizing organisms. The aphotic zone lies below.
- Oceanic circulation is mostly horizontal in surface currents and deep-sea currents, but vertical circulation also takes place.
- The sediments on the seafloor are mostly pelagic clay and ooze consisting of the skeletons of tiny organisms.
- Moundlike, wave-resistant structures consisting of animal skeletons are reefs. Most reefs are fringing reefs, barrier reefs, or atolls.
- Tides are caused by the combined effects of the Moon and the Sun on the oceans.
- As wind-generated waves enter shallow water, they become oversteepened and plunge forward as breakers or spill onto the shoreline, thus expending their kinetic energy.
- Longshore currents resulting from waves approaching a shoreline at an angle erode, transport, and deposit sediment.
- Rip currents carry excess water from the nearshore zone seaward through the breaker zone.
- Erosional coasts have sea cliffs, wave-cut platforms, and sea stacks, whereas depositional coasts have long sandy beaches, deltas, and barrier islands.
- Beaches are continuously modified by waves and nearshore currents, and their profiles usually show seasonal changes.
- Spits, baymouth bars, and tombolos all form and grow as a result of longshore transport and deposition.
- The sediment budget of a nearshore system remains rather constant unless the system is disrupted, as when dams are built across streams supplying sand to the system.
- Submergent and emergent coasts are defined on the basis of their relationship to changes in sea level.
- Coastal flooding during storms by waves and storm surges is an ongoing problem in many areas.
- The United States claims rights to all resources within 200 nautical miles (370 km) of its shorelines, or what is called its Exclusive Economic Zone (EEZ).

Important Terms

aphotic zone (p. 388)
barrier island (p. 401)
baymouth bar (p. 401)
beach (p. 400)
breaker (p. 396)
Coriolis effect (p. 389)
downwelling (p. 389)
emergent coast (p. 405)
Exclusive Economic Zone (EEZ) (p. 409)
fetch (p. 395)
gyre (p. 389)
headland (p. 398)
longshore current (p. 396)
marine terrace (p. 398)
nearshore sediment budget (p. 403)
ooze (p. 389)
pelagic clay (p. 389)
photic zone (p. 388)
reef (p. 389)
rip current (p. 397)
salinity (p. 388)
shoreline (p. 389)
spit (p. 401)
storm surge (p. 406)
submergent coast (p. 405)
tide (p. 390)
tombolo (p. 401)
upwelling (p. 389)
wave (p. 394)
wave base (p. 394)
wave-cut platform (p. 398)
wave refraction (p. 396)

Review Questions

1. The depth in the oceans below about 100 m where there is too little sunlight for photosynthesis is called the
   a. _____ bathosphere.
   b. _____ aphotic zone.
   c. _____ marine terrace.
   d. _____ submergent area.
   e. _____ solution depth.
2. Because most waves approach a shoreline at an angle, they generate
   a. _____ an undertow.
   b. _____ gyres.
   c. _____ longshore currents.
   d. _____ offshore canyons.
   e. _____ the Coriolis effect.
3. Waves approaching a shoreline have a wavelength of 15 m, so the depth of the wave base is about
   a. _____ 30 m.
   b. _____ 5 m.
   c. _____ 45 m.
   d. _____ 7.5 m.
   e. _____ 150 m.
4. Erosion by waves is caused by the impact of water on shorelines and the wearing action of water carrying sand and gravel. These processes are
   a. _____ hydraulic action/abrasion.
   b. _____ chemical weathering/dissolution.
   c. _____ fetch/surge.
   d. _____ downwelling/gyre.
   e. _____ longshore drift/salinity.
5. The distance that wind blows over a continuous water surface is called the
   a. _____ berm.
   b. _____ celerity.
   c. _____ fetch.
   d. _____ wavelength.
   e. _____ photic zone.
6. While driving along North America's West Coast, you notice a broad surface above sea level that slopes gently toward the ocean and has several masses of rock rising above it. How would you explain the origin of this landform to your children?
7. What is upwelling and why is it important?
8. What is wave base and how does it affect waves as they enter shallow water?
9. As the member of a planning commission for your coastal community where shoreline erosion is a continuing problem, what recommendations would you make to remedy or at least mitigate the problem?

10. **Creative Thinking Visual Question:** In this image (▮ Figure 1), identify a headland and a pocket beach. Explain where you would expect a sea cave, sea arch, and sea stack to form.

James S. Monroe

▮ **Figure 1** Aerial view of the Pacific coast of the United States.

Global GeoScience Watch Within the GREENR database, search for "methane hydrate." Next to the "News" heading, click on "VIEW ALL," and then scroll to find the article titled "Japan Pioneers Gas Extraction from Seafloor Methane Hydrate." How much methane hydrate is available to the Japanese, and how might their exploitation of this resource have an impact on the economy of Russia? Does the United States Geological Survey think that methane hydrate is a viable source to replace fossil fuels?

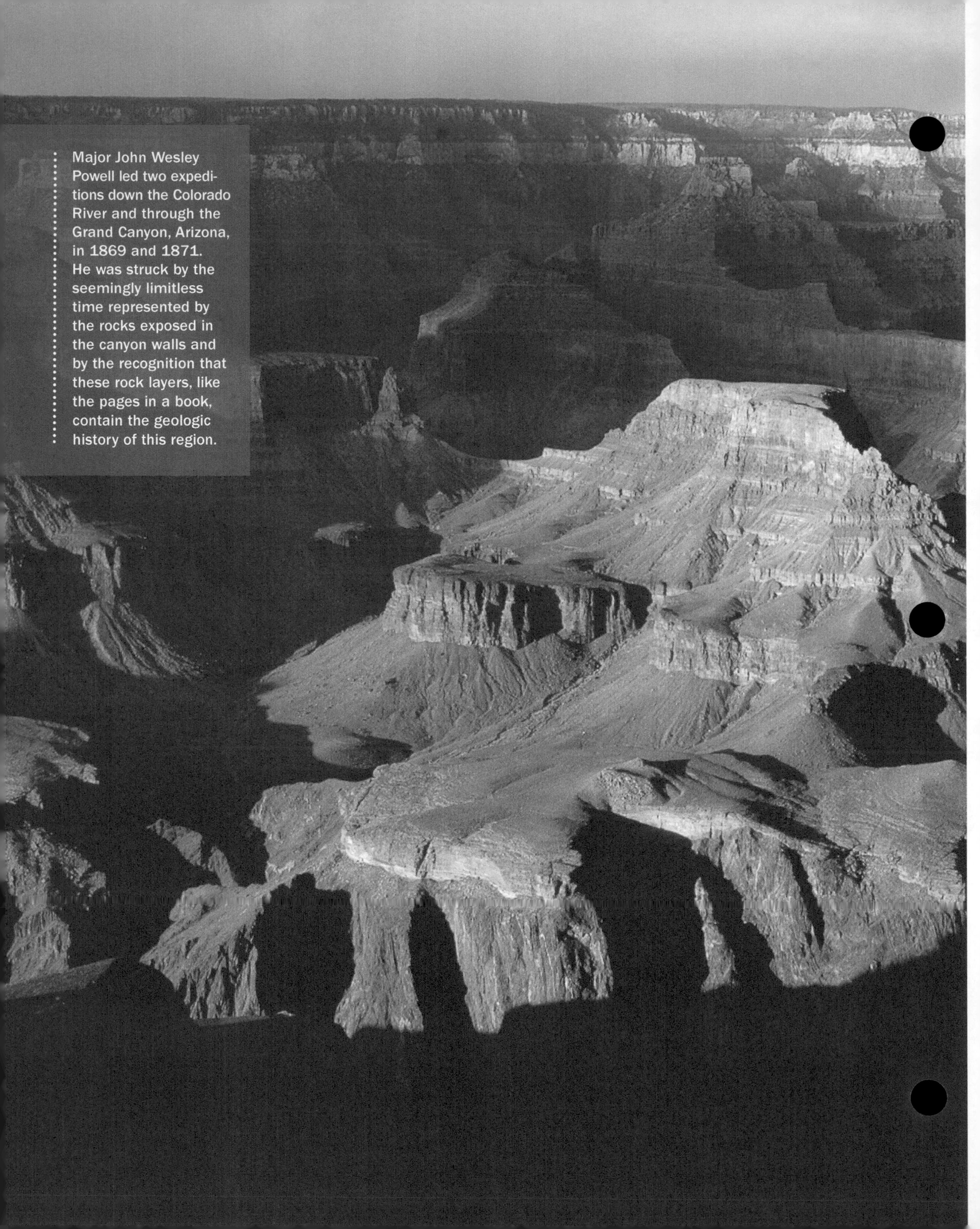

Major John Wesley Powell led two expeditions down the Colorado River and through the Grand Canyon, Arizona, in 1869 and 1871. He was struck by the seemingly limitless time represented by the rocks exposed in the canyon walls and by the recognition that these rock layers, like the pages in a book, contain the geologic history of this region.

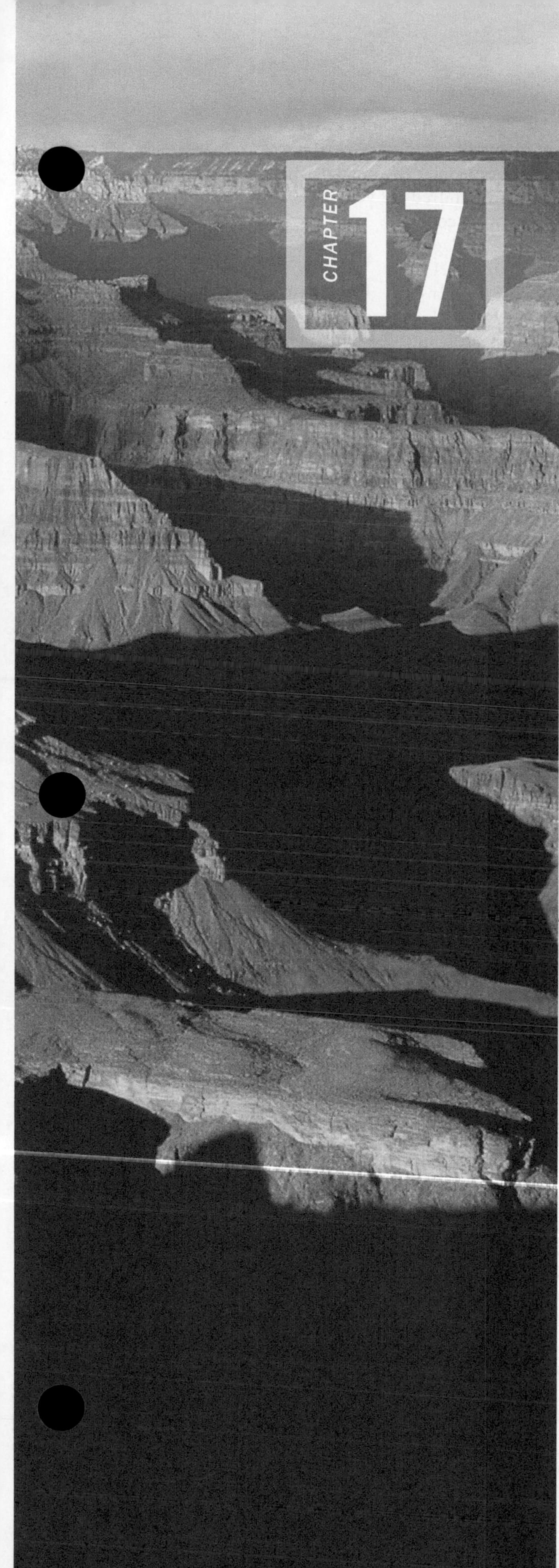

Copyright and Photograph by Dr. Parvinder S. Sethi

Geologic Time: Concepts and Principles

OUTLINE

HAVE YOU EVER WONDERED?

- How geologic time is measured?
- What the age of Earth is?
- Why it is important to have an appreciation and understanding of geologic time?
- How geologists determine that geologic events in different parts of the world took place at the same time?
- What radioactivity is, and how it can be used to determine the age of different rocks?
- How the geologic time scale came into being?
- How geologists reconstruct past climatic regimes and changes?

17.1 Introduction

In 1869, Major John Wesley Powell, a Civil War veteran who lost his right arm in the battle of Shiloh, led a group of hardy explorers down the uncharted Colorado River through the Grand Canyon. With no maps or other information, Powell and his group ran the many rapids of the Colorado River in fragile wooden boats, hastily recording what they saw. Powell wrote in his diary that "all about me are interesting geologic records. The book is open and I read as I run."

Probably no one has contributed as much to the understanding of the Grand Canyon as Major Powell. In recognition of his contributions, the Powell Memorial was erected on the South Rim of the Grand Canyon in 1969 to commemorate the 100th anniversary of this history-making first expedition.

Most tourists today, like Powell and his fellow explorers in 1869, are astonished by the seemingly limitless time represented by the rocks exposed in the walls of the Grand Canyon. For most visitors, viewing a 1.5-km-deep cut into Earth's crust is the only encounter they'll ever have with the magnitude of geologic time. When standing on the rim and looking down into the Grand Canyon, we are really looking far back in time—all the way back to the early history of our planet. In fact, more than 1 billion years of history are preserved in the rocks of the Grand Canyon, and reading what is preserved in those rocks is, just as Powell noted in his diary more than 100 years ago, like reading the pages in a history book.

Vast periods of time set geology apart from most of the other sciences, and an appreciation of the immensity of geologic time is fundamental to understanding the physical and biological history of our planet. In fact, understanding and accepting the magnitude of geologic time is one of the major contributions geology has made to the sciences.

Besides providing an appreciation for the immensity of geologic time, why is the study of geologic time important? One reason is that Earth has undergone periods of warmer and colder conditions in the past, and an understanding of what caused these climate oscillations might be helpful in the current debate over global climate change. Second, one of the most valuable lessons you will learn in this chapter is how to reason and apply some of geology's fundamental principles to interpret past geologic events. The logic used in applying these principles to interpret the geologic history of an area involves basic reasoning skills that can be transferred and used in almost any profession or discipline.

17.2 How Geologic Time Is Measured

In some respects, time is defined by the methods used to measure it. Geologists use two different frames of reference when discussing geologic time. **Relative dating** is placing geologic events in a sequential order as determined from their position in the geologic record. Relative dating does not tell us how long ago a particular event took place, only that one event preceded another.

The various principles used to determine relative dating were discovered hundreds of years ago, and since then, they have been used to construct the *relative geologic time scale* (Figure 17.1). Furthermore, these principles are still widely used by geologists today, especially in reconstructing the geologic history of the terrestrial planets and their moons.

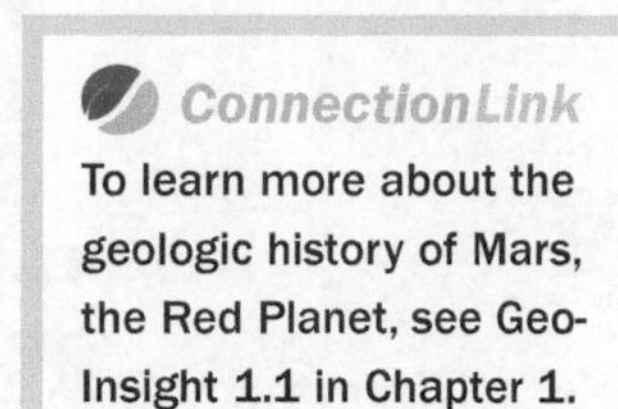

Absolute dating (also called *numerical dating*) provides specific dates for rock units or events expressed in years before the present. *Radiometric dating* is the most common method of obtaining absolute ages. Dates are calculated from the natural decay rates of various radioactive elements present in trace amounts in some rocks. It was not until the discovery of radioactivity near the end of the 19th century that absolute ages could be accurately applied to the relative geologic time scale. Today, the geologic time scale is really a dual scale—a relative scale based on rock sequences with radiometric dates expressed as years before the present (Figure 17.1).

Advances and refinements in absolute dating techniques during the 20th century have changed the way we view Earth in terms of when events occurred in the past and the rates of geologic change through time. The ability to accurately determine past climatic changes and their causes has important implication for the current debate on global warming and its effects on humans. In addition, not only does global warming affect humans, but humans are also affecting the planet in a myriad of ways, such that we might be responsible for a new geologic epoch (Geo-Focus 17.1).

17.3 Early Concepts of Geologic Time and Earth's Age

The concept of geologic time and its measurement have changed throughout human history. Some early Christian scholars and clerics tried to establish the date of creation by analyzing historical records and the genealogies found in Scripture. One of the most influential of these scholars was James Ussher (1581–1656), Archbishop of Armagh, Ireland, who, based on Old Testament genealogy, asserted that God created Earth on Sunday, October 23, 4004 BC. In 1701, an authorized version of the Bible made this date accepted Church doctrine. For nearly a century thereafter, it was considered heresy to assume that Earth and all of its features were more than about 6,000 years old. Thus, the idea of a very young Earth provided the basis for most Western chronologies of Earth history before the 18th century.

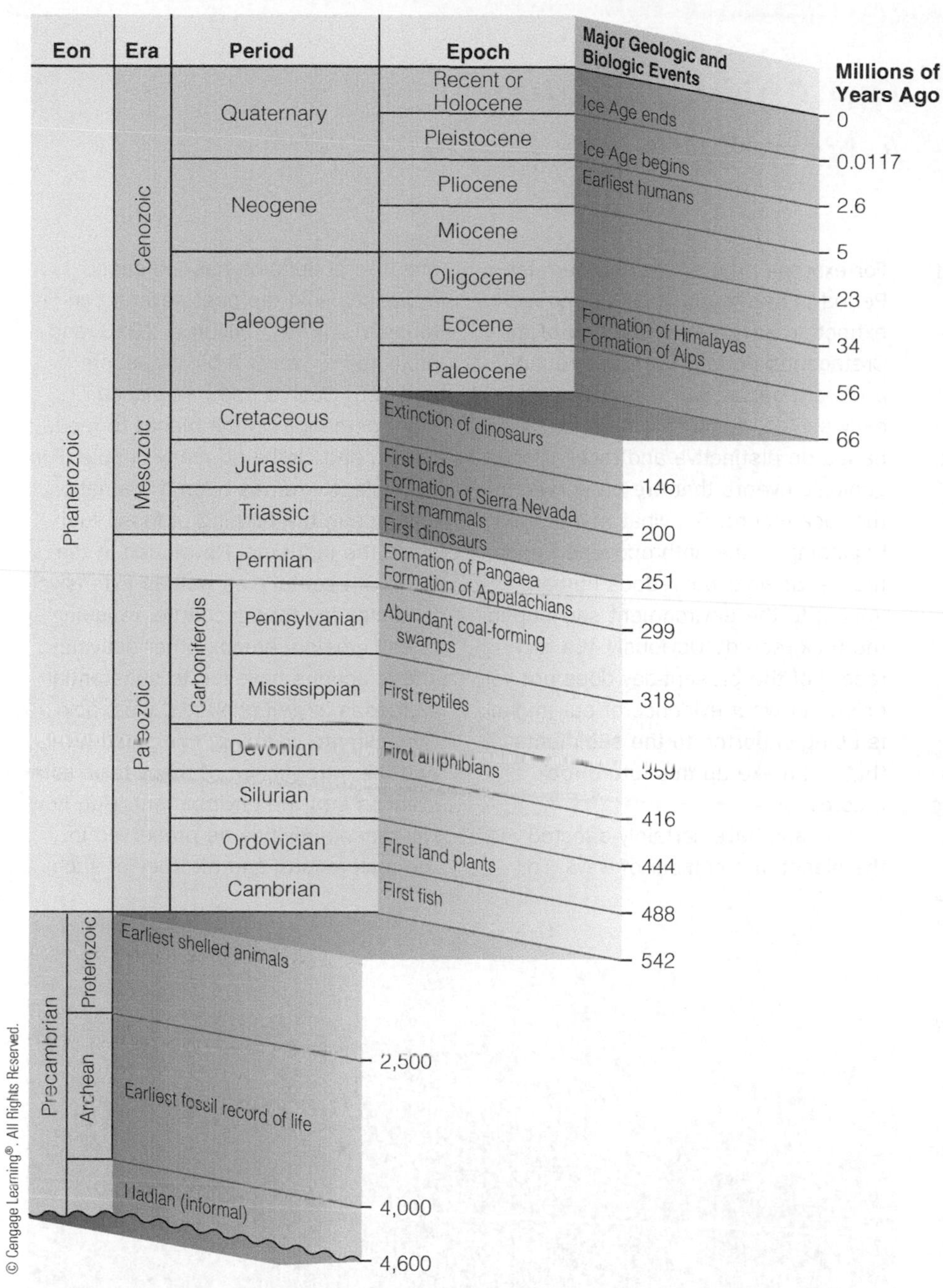

Figure 17.1 The Geologic Time Scale Some of the major geologic and biologic events are indicated for the various eras, periods, and epochs. Dates are from the 2009 International Stratigraphic Chart, © 2009 by the International Commission on Stratigraphy.

During the 18th and 19th centuries, several attempts were made to determine Earth's age on the basis of scientific evidence rather than revelation. The French zoologist Georges Louis de Buffon (1707–1788) assumed that Earth has gradually cooled to its present condition from a molten beginning. To simulate this history, he melted iron balls of various diameters and allowed them to cool to the surrounding temperature. By extrapolating their cooling rate to a ball the size of Earth, he determined that Earth was at least 75,000 years old. Although this age was much older than that derived from Scripture, it was still vastly younger than we now know our planet to be.

Other scholars were equally ingenious in attempting to calculate Earth's age. For example, if deposition rates could be determined for various sediments, geologists reasoned that they could calculate how long it would take to deposit any rock layer. They could then extrapolate how old Earth was from the total thickness of sedimentary rock in its crust. Rates of deposition vary, however, even for the same type of rock. Furthermore, it is impossible to estimate how much of a rock has been removed by erosion, or how much a rock sequence has been reduced by compaction. As a result of these variables, estimates of Earth's age ranged from younger than 1 million years to older than 2 billion years.

Another method involved ocean salinity. John Joly, a 19th-century Irish geologist, measured the amount of salt currently in the world's streams. He then calculated that it would have taken at least 90 million years for the oceans to reach their present salinity level, assuming Earth's oceans were originally freshwater. This was still much younger than the now-accepted age of 4.6 billion years, in part, because Joly had no way to calculate how much salt has been recycled or the amount of salt that is stored in continental salt deposits and seafloor clay deposits.

Besides trying to determine Earth's age, the naturalists of the 18th and 19th centuries were also formulating some of the fundamental geologic principles that are used in deciphering Earth history. From the evidence preserved in the geologic record, it was clear to them that Earth is very old and that geologic processes have operated over long periods of time.

17.4 James Hutton and the Recognition of Geologic Time

Many consider the Scottish geologist James Hutton (1726–1797) to be the founder of modern geology (Figure 17.2). His detailed studies and observations of rock exposures

GEO FOCUS 17.1

The Anthropocene: A New Geologic Epoch?

Since being coined on the spur of the moment at a scientific conference in 2000 by Nobel Prize–winning Dutch atmospheric chemist Paul Crutzen, the term "anthropocene" has caught on in both the scientific community, as well as spreading to the much wider general public. The idea of formalizing a new geologic epoch, the Anthropocene Epoch, is being widely discussed in geologic circles and is based on the increasing impact humans are having on the planet.

Interestingly, as far back as 1873, Antonio Stoppani, an Italian geologist, proposed that humans had introduced a new geologic era that he called the "anthropozoic," and over the years, other terms followed. Why then has Crutzen's "anthropocene," which comes from the Latin *antropos*, meaning man or human, and the geologic epoch suffix *cene*, as in Paleocene or Holocene, gained so much favor? Perhaps it is because of an increasing realization that humans are altering the environment, and at an alarmingly rapid rate.

Just because a term is widely used in the scientific literature does not mean that it will be formally accepted as the name for a new geologic epoch. For this to happen, many criteria and formal steps must be taken and met. Because the geologic time scale is fundamental to the work geologists do and is the framework on which Earth history is based, adding another epoch is not to be taken lightly.

Subdivisions of the geologic time scale are based on the recognition of distinctive events that are preserved in the rock record, such as changes in fossil assemblages, which represent changes in Earth's biota. For example, the end of the Permian Period is marked by a global mass extinction event, as is the end of the Cretaceous Period. Even the boundaries of epochs, which are, geologically speaking, short in duration, are based on distinctive and recognizable geologic events that are preserved in the rock record. So, what marks the beginning of the Anthropocene Epoch, that is, at what point does human impact to the environment show up in the rock record? Obviously, the rock record of the present-day does not yet exist, but what evidence of our impact is being imparted to the sediments that will make up this future rock record?

Humans have certainly affected the planet in a number of ways. For one, the population has increased dramatically in the past several centuries, rising to 7 billion in 2013, and projected to reach 8 billion people by 2025. Such a rise has placed great demands on the planet to feed, house, and clothe so many people. For example, there has been a dramatic increase in the burning of fossil fuel since the Industrial Revolution in the late 18th century, as well as extensive land clearing for agriculture, resulting in soil erosion, among other activities. These actions have led to substantial increases in greenhouse gases, acid rain, climate warming, rising sea levels, and the introduction of many toxic substances into the environment. But, how will these activities be preserved in the rock record, and are they of such

Figure 1 Although cities have changed the landscape by expropriating the land surface on which they are built, using large amounts of energy and natural resources, as well as contributing to pollution, they will leave little, if any, direct evidence of their existence in the future geologic record. Dubai, in the United Arab Emirates, houses some 2 million people and has the world's tallest building, which depends on air conditioning and desalinated seawater, both of which involve high energy consumption of fossil fuels, as well as contributing to atmospheric pollution.

Warwick Sweeney/Photographer's Choice

Figure 2 Large coal-fired power plants, like this one in England, have contributed to the sizable surge in carbon dioxide emissions into the atmosphere, which has exacerbated climate warming. Although cities, factories, and power plants will probably not leave any direct evidence of their existence in the geologic record, the emissions from the burning of coal, oil, and natural gas are likely to be discernible in atmospheric gases trapped in glacial ice and will probably be indirectly noticeable by changes in the fossil record.

consequence that they will stand out above the background "noise" of the geologic record?

Some of the most plainly visible evidence of human activities, such as cities (Figure 1), factories, power plants (Figure 2), and farmland, will leave barely, if any, traces in the rock record because they occur on land, that is, for the most part, the site of erosion, rather than deposition. However, changes in the pollen record, beginning with the rise of farming, will be visible by the sharp increase in the pollen of corn, wheat, soy, and rice, as opposed to the more diverse pollen record of rain forests, prairies, and other plant biomes.

Probably the most significant change, from a geologic perspective, will be the change in the composition of the atmosphere, especially in carbon dioxide levels, which can be detected in the air trapped in polar ice. Changes in the climate can also be deduced by the latitudinal shifts of some plants and animals, as they extend their geographic range closer to the poles. Furthermore, increasing carbon dioxide levels help acidify the world's ocean, which can cause environmental stress such that corals no longer construct reefs (Figure 3), which would also show up in the fossil record.

If we have, indeed, entered a new epoch, when did it begin? On this point, there is still disagreement. Some scientists favor the Anthropocene beginning approximately 8,000 years ago with the rise of farming and the subsequent clearing of land, soil erosion, as well as the global spread and increase in the human population. Others, including Paul Crutzen, consider the beginning of the Industrial Revolution, with its increasing levels of carbon dioxide and methane, resulting from the exploitation of coal, oil, and natural gas to fuel a burgeoning population and industrialized society, as marking the start of the Anthropocene Epoch.

Although much work remains to be done before formalizing the Anthropocene Epoch, for now and the foreseeable future, humans and their global impact will remain firmly part of the Holocene Epoch.

Tim Laman/National Geographic Stock

Figure 3 This bleached, white coral (*Acropora* sp.) is evidence of increasing water temperatures caused by global warming. High water temperatures kill the individual coral animals, leaving behind dead, white-colored reefs, such as seen here.

Figure 17.2 James Hutton James Hutton, a Scottish geologist, originated the principle of uniformitarianism and, through his writings, profoundly influenced the course of geologic thinking.

and present-day geologic processes formed the foundation for the *principle of uniformitarianism* (see Chapter 1), the concept that the same processes seen today have operated throughout geologic time. Because Hutton relied on known processes to account for Earth history, he concluded that Earth must be very old and wrote that "we find no vestige of a beginning, and no prospect of an end."

In 1830, Charles Lyell published a landmark book, *Principles of Geology*, in which he championed Hutton's concept of uniformitarianism. Instead of relying on catastrophic events to explain various Earth features, Lyell recognized that imperceptible changes brought about by present-day processes could, over long periods of time, have tremendous cumulative effects. Through his writings, Lyell firmly established uniformitarianism as the guiding principle of geology. Furthermore, the recognition of vastly long periods of time was also necessary for, and instrumental in, the acceptance of Darwin's 1859 theory of organic evolution.

ConnectionLink

To learn more about Darwin's theory of organic evolution and its implications, see Chapter 18.

After establishing that present-day processes have operated over vast periods of time, geologists were nevertheless nearly forced to accept a very young age for Earth. Lord Kelvin (1824–1907), a highly respected English physicist, claimed, in a paper written in 1866, to have destroyed the uniformitarian foundation on which Huttonian–Lyellian geology was based. Starting with the generally accepted belief that Earth was originally molten, Kelvin assumed that it has gradually been losing heat and that, by measuring this heat loss, he could determine its age.

Kelvin knew from deep mines in Europe that Earth's temperature increases with depth, and he reasoned that Earth is losing heat from its interior. By knowing the size of Earth, the melting temperatures of rocks, and the rate of heat loss, Kelvin calculated the age at which Earth was entirely molten. From these calculations, he concluded that Earth could not be older than 400 million years or younger than 20 million years. This wide discrepancy in age reflected uncertainties in average temperature increases with depth and the various melting points of Earth's constituent materials.

After establishing that Earth was very old and that present-day processes operating over long periods of time account for geologic features, geologists were in a quandary. If they accepted Kelvin's dates, they would have to abandon the concept of seemingly limitless time that was the underpinning of uniformitarian geology and one of the foundations of Darwinian organic evolution, and squeeze events into a shorter time frame.

Kelvin's reasoning and calculations were sound, but his basic premises were false, thereby invalidating his conclusions. Kelvin was unaware that Earth has an internal heat source—radioactivity—that has allowed it to maintain a fairly constant temperature through time.* His 40-year campaign for a young Earth ended with the discovery of radioactivity near the end of the 19th century and the insight in 1905 that natural radioactive decay can be used in many cases to date how long ago a rock formed. His "unassailable calculations" were no longer valid, and his proof for a geologically young Earth collapsed.

Although the discovery of radioactivity destroyed Kelvin's arguments, it provided geologists with a clock that could measure Earth's age and validate what geologists had long thought—namely, that Earth was indeed very old.

*Actually, Earth's temperature has decreased through time because the original amount of radioactive materials has been decreasing and thus is not supplying as much heat. However, the temperature is decreasing at a rate considerably slower than would be required to lend any credence to Kelvin's calculations.

17.5 Relative Dating Methods

Before the development of radiometric dating techniques, geologists had no reliable means of absolute dating and therefore depended solely on relative dating methods. Recall that relative dating places events in sequential order, but it does not tell us how long ago an event took place. Although the principles of relative dating may now seem self-evident, their discovery was an important scientific achievement because they provided geologists with a means to interpret geologic history and develop a *relative geologic time scale*.

Fundamental Principles of Relative Dating

The 17th century was an important time in the development of geology as a science because of the widely circulated writings of the Danish anatomist Nicolas Steno (1638–1686). Steno observed that when streams flood, they spread out across their floodplains and deposit layers of sediment that bury organisms dwelling on the floodplain. Subsequent floods produce new layers of sediments that are deposited, or superposed, over previous deposits. When lithified, these layers of sediment become sedimentary rock.

Thus, in an undisturbed succession of sedimentary rock layers, the oldest layer is at the bottom and the youngest layer is at the top. This **principle of superposition** is the basis for relative-age determinations of strata and their contained fossils (❚ Figure 17.3a).

Steno also observed that, because sedimentary particles settle from water under the influence of gravity, sediment is deposited in essentially horizontal layers, thus illustrating the **principle of original horizontality** (Figure 17.3a and chapter opening photograph). Therefore, a sequence of sedimentary rock layers that is steeply inclined from the horizontal must have been tilted after deposition and lithification (❚ Figure 17.3b).

Steno's third principle, the **principle of lateral continuity**, states that sediment extends laterally in all directions until it thins and pinches out or terminates against the edge of the depositional basin (Figure 17.3a).

James Hutton is credited with formulating the **principle of cross-cutting relationships**. On the basis of his detailed studies and observations of rock exposures in Scotland, Hutton recognized that an igneous intrusion or a fault must be younger than the rocks that it intrudes or displaces (❚ Figure 17.4).

Although this principle illustrates that an intrusive igneous structure is younger than the rocks it intrudes, the association of sedimentary and igneous rocks may cause problems in relative dating. Buried lava flows and sills look very similar in a sequence of strata (❚ Figure 17.5). A buried lava flow, however, is older than the rocks above it (principle

❚ Figure 17.3 The Principles of Original Horizontality, Superposition, and Lateral Continuity

a The sedimentary rocks of Bryce Canyon National Park, Utah, illustrate three of the six fundamental principles of relative dating. These rocks were originally deposited horizontally in a variety of continental environments (principle of original horizontality). The oldest rocks are at the bottom of this highly dissected landscape, and the youngest rocks are at the top, forming the rims (principle of superposition). The exposed rock layers extend laterally in all directions for some distance (principle of lateral continuity).

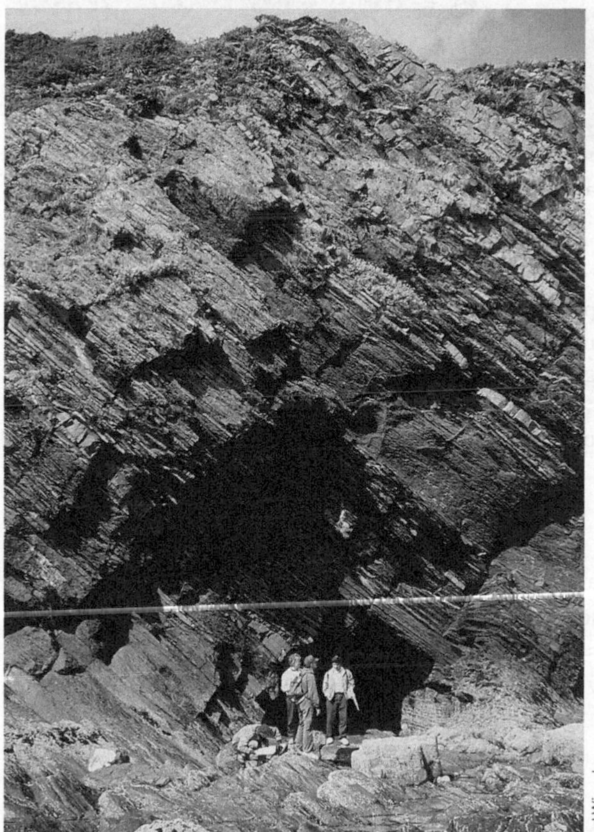

b These shales and limestones of the Postolonnec Formation, at Postolonnec Beach, Crozon Peninsula, France, were originally deposited horizontally, but have been significantly tilted since their formation.

Figure 17.4 The Principle of Cross-Cutting Relationships

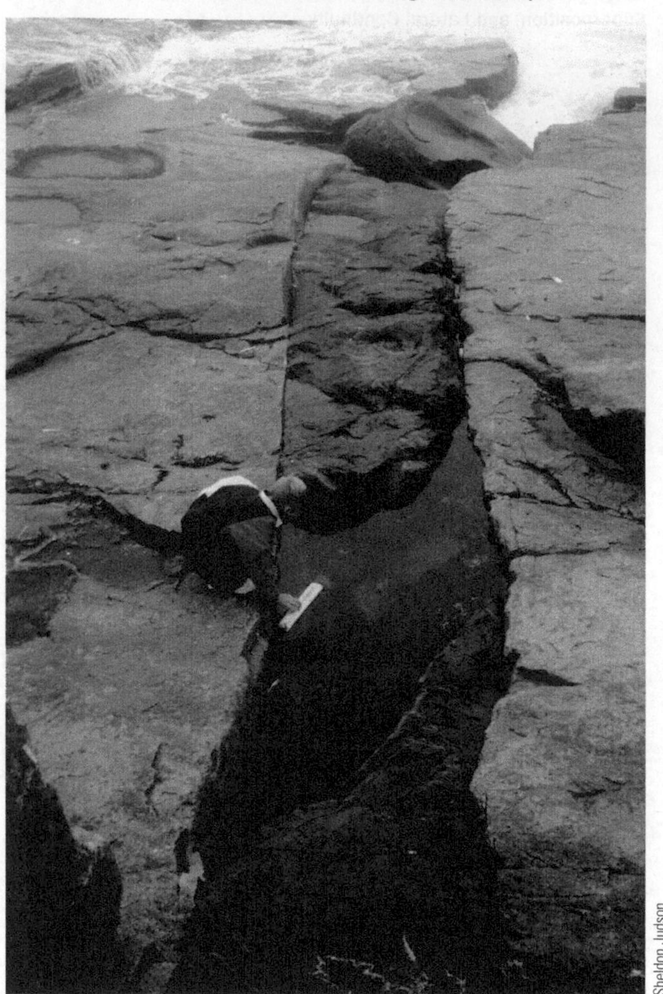

Sheldon Judson

a A dark gabbro dike cuts across granite in Acadia National Park, Maine. The dike is younger than the granite that it intrudes.

Reed Wicander

b A small fault (arrows show direction of movement) cuts across, and thus displaces, tilted sedimentary beds along the Templin Highway in Castaic, California. The fault is therefore younger than the youngest beds that are displaced.

of superposition), whereas a sill, resulting from later igneous intrusion, is younger than all of the beds below it and younger than the immediately overlying bed as well.

To resolve such relative-age problems as these, geologists observe whether the sedimentary rocks in contact with the igneous rocks show signs of baking or alteration by heat (contact metamorphism). A sedimentary rock that shows such effects must be older than the igneous rock with which it is in contact. In Figure 17.5, for example, a sill produces a zone of baking immediately above and below it because it intruded into previously existing sedimentary rocks. A lava flow, in contrast, bakes only those rocks below it.

ConnectionLink

To review contact metamorphism and how it alters the surrounding country rocks, see Figure 8.7 in Chapter 8.

The **principle of inclusions** is yet another way to determine relative ages, because inclusions, or fragments of one rock within a layer of another, are older than the rock itself. The batholith shown in Figure 17.6a contains sandstone inclusions, and the sandstone unit shows the effects of baking. Accordingly, we conclude that the sandstone is older than the batholith. In Figure 17.6b, however, the sandstone contains granite rock fragments, indicating that the batholith was the source rock for the inclusions and is therefore older than the sandstone.

Fossils have been known for centuries, yet their utility in relative dating and geologic mapping was not fully appreciated until the early 19th century. William Smith (1769–1839), an English civil engineer involved in surveying and building canals in southern England, independently recognized the principle of superposition by reasoning that the fossils at the bottom of a sequence of strata are older than those at the top of the sequence. This recognition served as the basis for the **principle of fossil succession**, or the *principle of faunal and floral succession*, as it is sometimes called (Figure 17.7).

ConnectionLink

To learn more about fossils and their importance in evolution, see Chapter 18.

According to this principle, fossil assemblages succeed one another through time in a regular and predictable order. The validity and successful use of this principle depend on three points: (1) life has varied through time, (2) fossil assemblages are recognizably different from one another, and (3) the relative ages of the fossil assemblages can be determined. Observations of fossils in older versus younger strata clearly demonstrate that life-forms have changed.

Figure 17.5 Differentiating Between a Buried Lava Flow and a Sill In **(a)** and **(b)** below, the ages of sedimentary strata are shown by numbering from 1–6, with 1 being the oldest sedimentary rock layer in each frame and 6 the youngest. Remember that the lava flow took place during deposition of an ordinary sedimentary sequence, whereas the sill intruded the sedimentary strata after they had accumulated. The notes in both frames highlight the physical features one would look for to distinguish between a lava flow and a sill, which can look quite similar in an outcrop.

Critical Thinking Question In **(b)**, it is stated that the sill is younger than beds 2 and 3, but its age relative to beds 4–6 cannot be determined. Why is this?

a A buried lava flow has baked underlying bed 2 when it flowed over it. Clasts of the lava were deposited along with other sediments during deposition of bed 3. The lava flow is younger than bed 2 and older than beds 3, 4, and 5.

b The rock units above and below the sill have been baked, indicating that the sill is younger than beds 2 and 3, but its age relative to beds 4–6 cannot be determined.

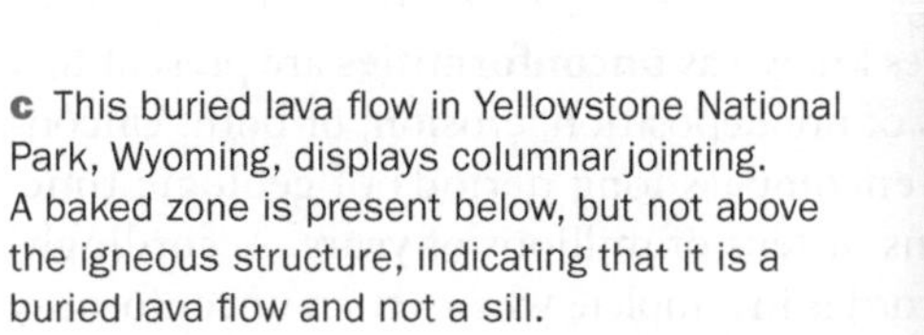

c This buried lava flow in Yellowstone National Park, Wyoming, displays columnar jointing. A baked zone is present below, but not above the igneous structure, indicating that it is a buried lava flow and not a sill.

Figure 17.6 The Principle of Inclusions

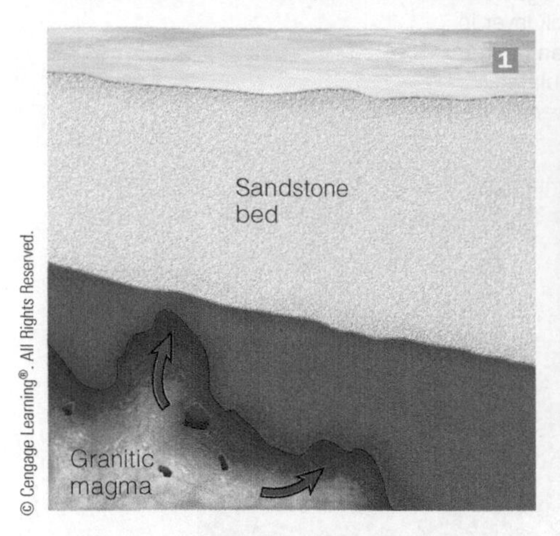

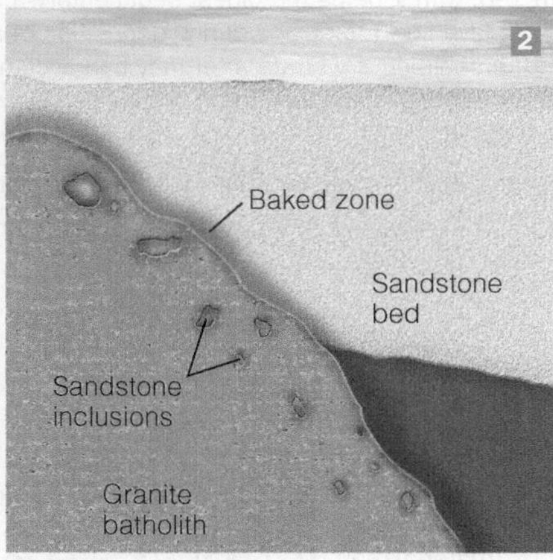

a The sandstone is older than the granite batholith because there are inclusions of sandstone inside the granite. The sandstone also shows evidence of having been baked along its contact with the granite batholith when the granitic magma intruded the overlying sedimentary beds.

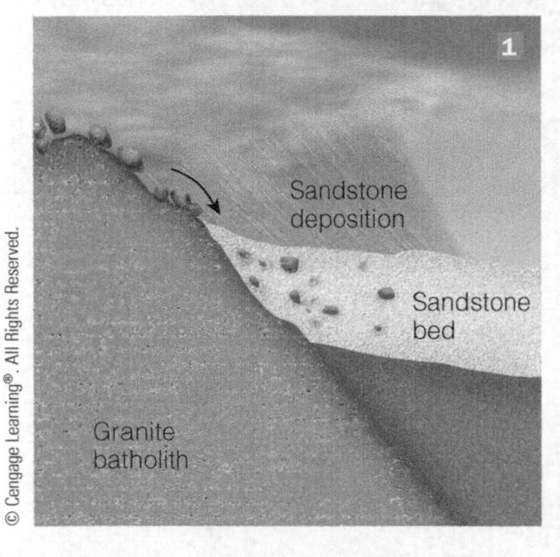

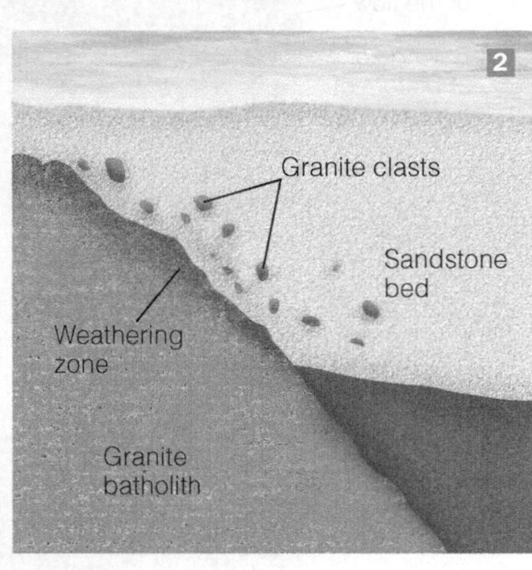

b The sandstone is younger than the granite batholith because it contains pieces (clasts) of granite. The granite is also weathered along the contact with the sandstone, indicating that it was the source of the granite clasts, and must therefore be older than the sandstone.

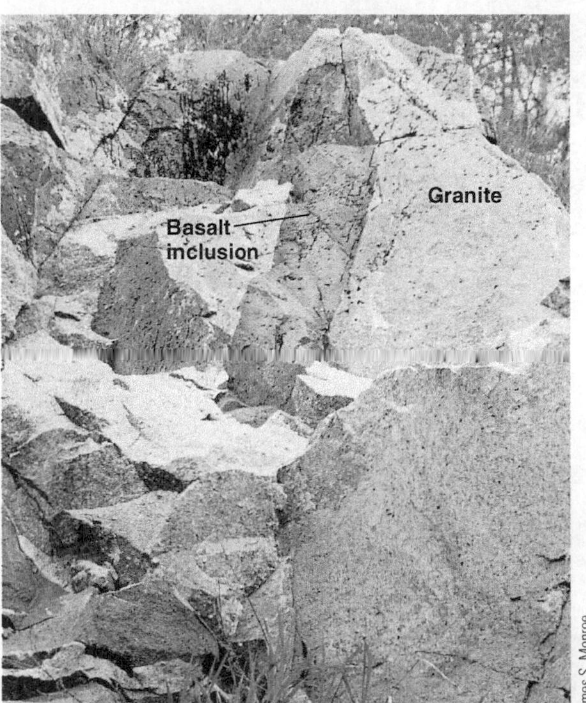

c Outcrop in northern Wisconsin showing basalt inclusions (dark gray) in granite (white). Accordingly, the basalt inclusions are older than the granite.

Because this is true, fossil assemblages (point 2) are recognizably different. Furthermore, superposition can be used to demonstrate the relative ages of the fossil assemblages.

Unconformities

Our discussion so far has been concerned with conformable strata—that is, sequences of rocks in which deposition was more or less continuous. A bedding plane between strata may represent a depositional break of anywhere from minutes to tens or hundreds of years, but it is inconsequential in the context of geologic time. However, in many sequences of strata, surfaces known as **unconformities** are present that represent times of nondeposition, erosion, or both. Unconformities thus encompass long periods of geologic time, perhaps millions or tens of millions of years. Accordingly, the geologic record is incomplete wherever an unconformity is present, just as a book with missing pages is incomplete, and the interval of geologic time not represented by strata is called a *hiatus* (▌Figure 17.8).

Figure 17.7 The Principle of Fossil Succession This generalized diagram shows how geologists use the principle of fossil succession to determine the relative ages of rocks in widely separated areas. The rocks encompassed by the dashed lines contain similar fossils and are therefore the same age. Note that the youngest rocks in this region are in section B, whereas the oldest rocks are in section C.

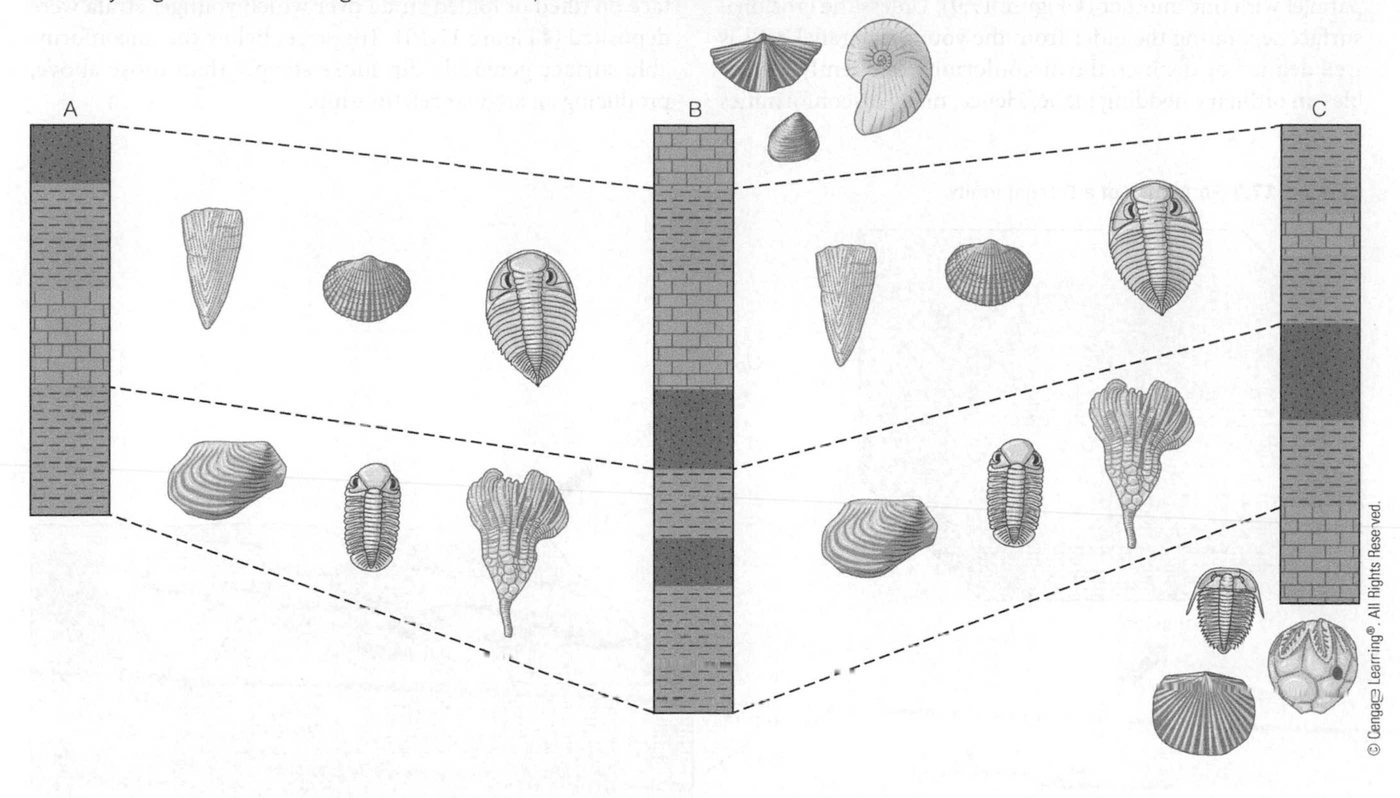

Figure 17.8 The Development of a Hiatus and an Unconformity

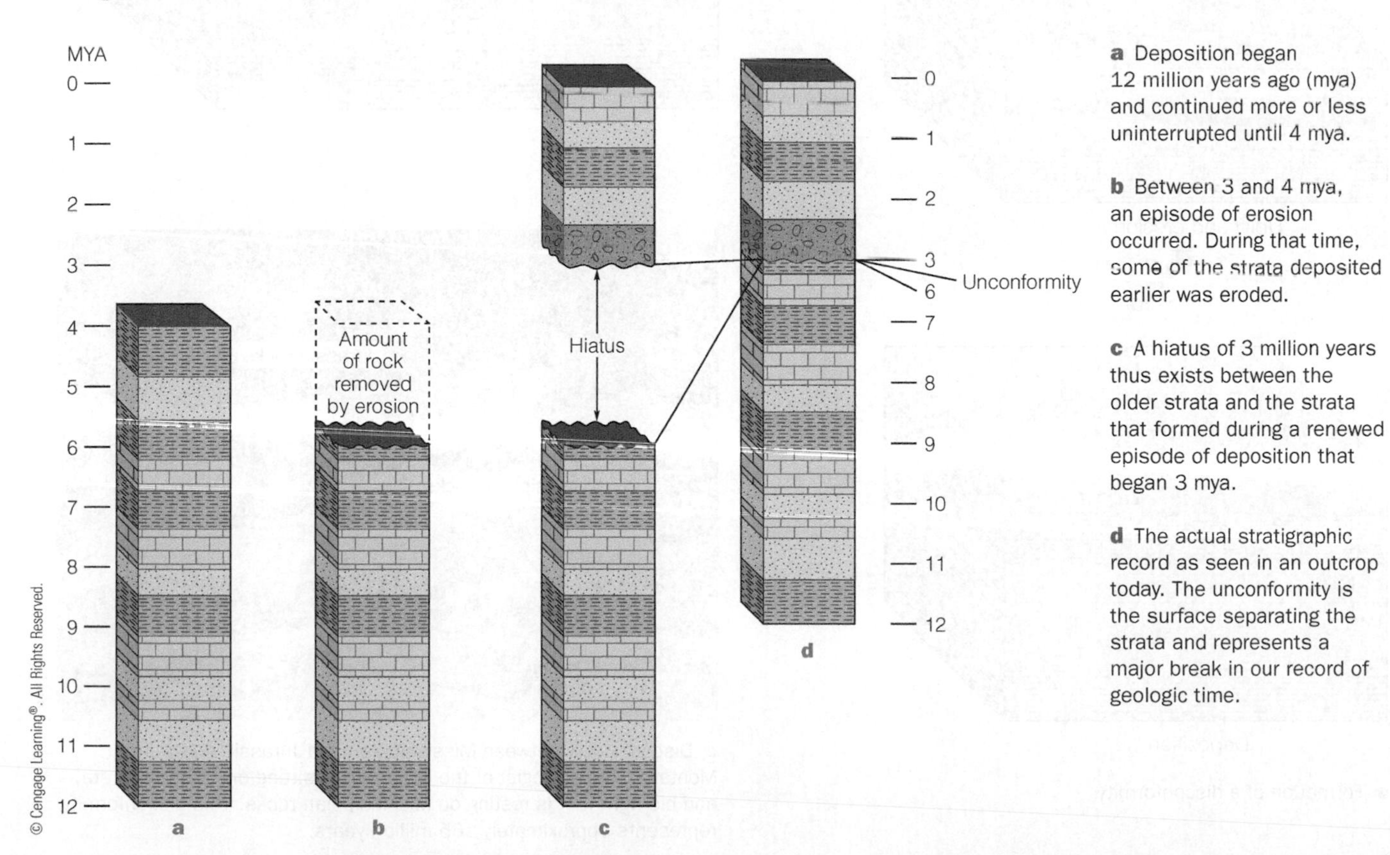

a Deposition began 12 million years ago (mya) and continued more or less uninterrupted until 4 mya.

b Between 3 and 4 mya, an episode of erosion occurred. During that time, some of the strata deposited earlier was eroded.

c A hiatus of 3 million years thus exists between the older strata and the strata that formed during a renewed episode of deposition that began 3 mya.

d The actual stratigraphic record as seen in an outcrop today. The unconformity is the surface separating the strata and represents a major break in our record of geologic time.

Geologists recognize three types of unconformities. First, a **disconformity** is a surface of erosion or nondeposition separating younger from older rocks, both of which are parallel with one another (▌Figure 17.9). Unless the erosional surface separating the older from the younger parallel beds is well defined or distinct, the disconformity frequently resembles an ordinary bedding plane. Hence, many disconformities are difficult to recognize and must be identified on the basis of fossil assemblages.

Second, an **angular unconformity** is an erosional surface on tilted or folded strata over which younger strata were deposited (▌Figure 17.10). The strata below the unconformable surface generally dip more steeply than those above, producing an angular relationship.

▌Figure 17.9 Formation of a Disconformity

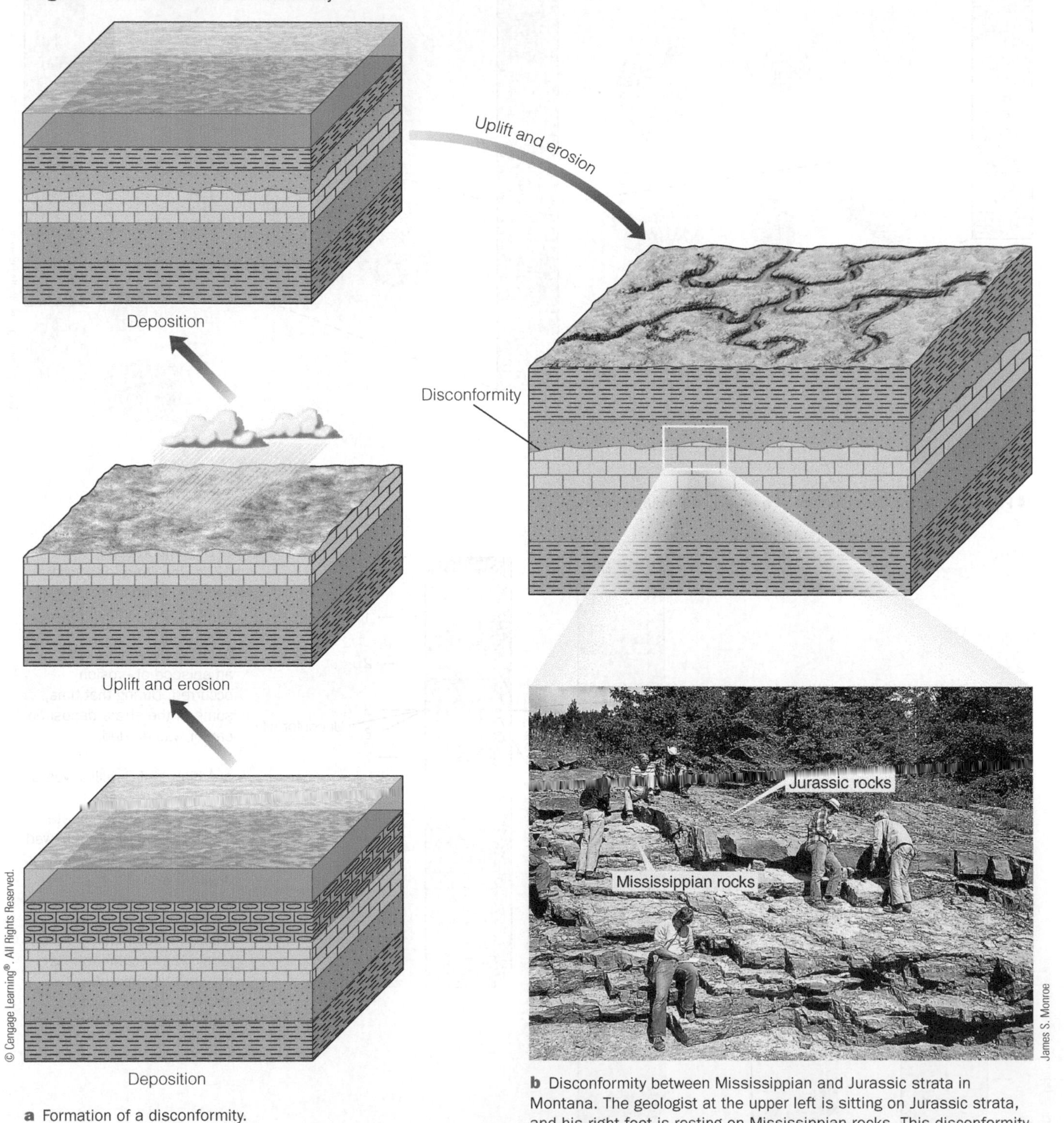

a Formation of a disconformity.

b Disconformity between Mississippian and Jurassic strata in Montana. The geologist at the upper left is sitting on Jurassic strata, and his right foot is resting on Mississippian rocks. This disconformity represents approximately 165 million years.

Figure 17.10 Formation of an Angular Unconformity

Critical Thinking Question From what you can see in the photo in **(b)**, which is the same scene Hutton viewed in 1788, what evidence is there that this outcrop represents an ancient mountain range? Were the forces compressional or tensional?

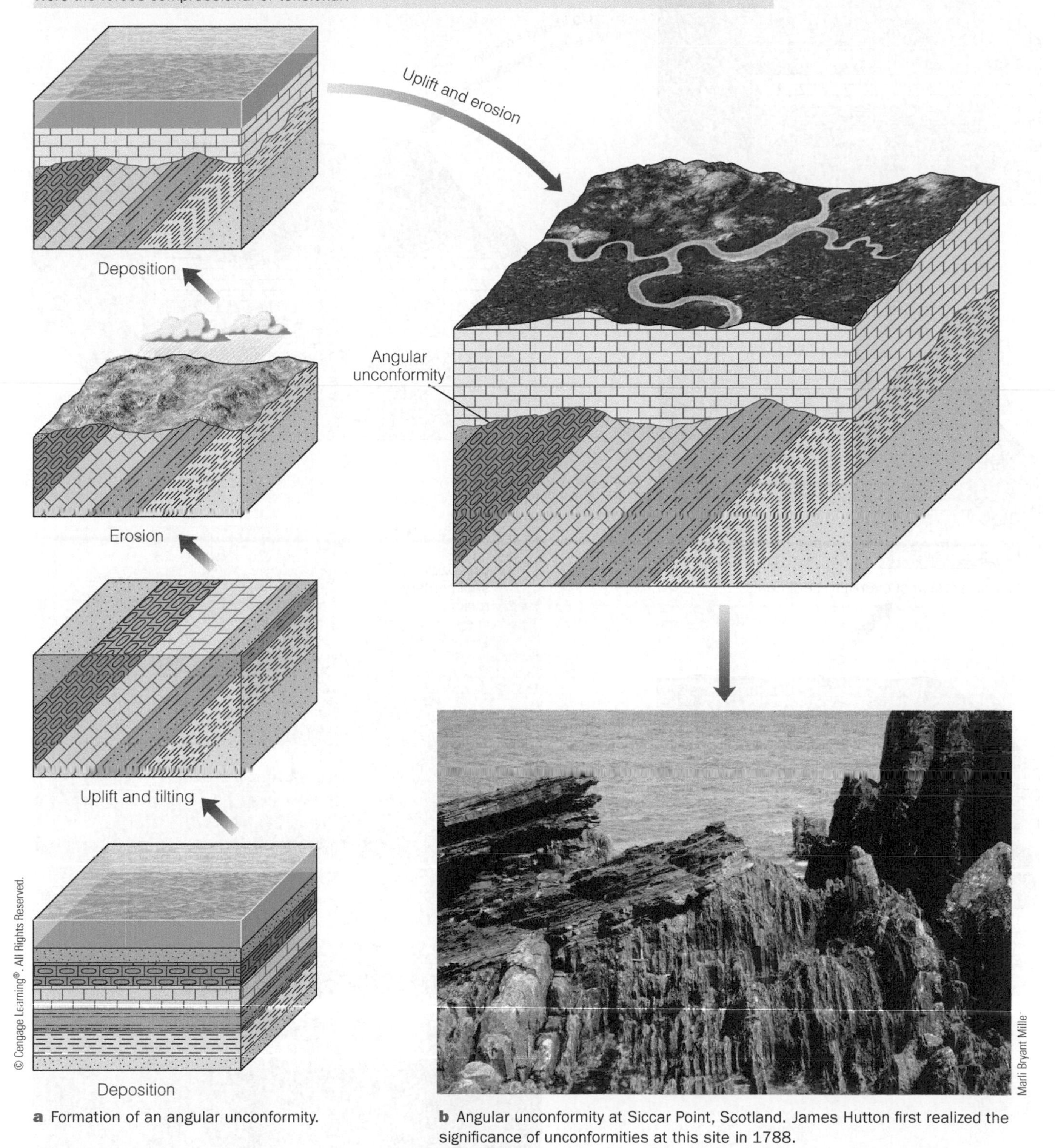

a Formation of an angular unconformity.

b Angular unconformity at Siccar Point, Scotland. James Hutton first realized the significance of unconformities at this site in 1788.

The angular unconformity illustrated in Figure 17.10b is probably the most famous in the world. It was here at Siccar Point, Scotland, that James Hutton realized that severe upheavals had tilted the lower rocks and formed mountains that were then worn away and covered by younger, flat-lying rocks. The erosional surface between the older tilted rocks and the younger flat-lying strata meant that a significant gap existed in the geologic record. Although Hutton did not use the term

Figure 17.11 Formation of a Nonconformity

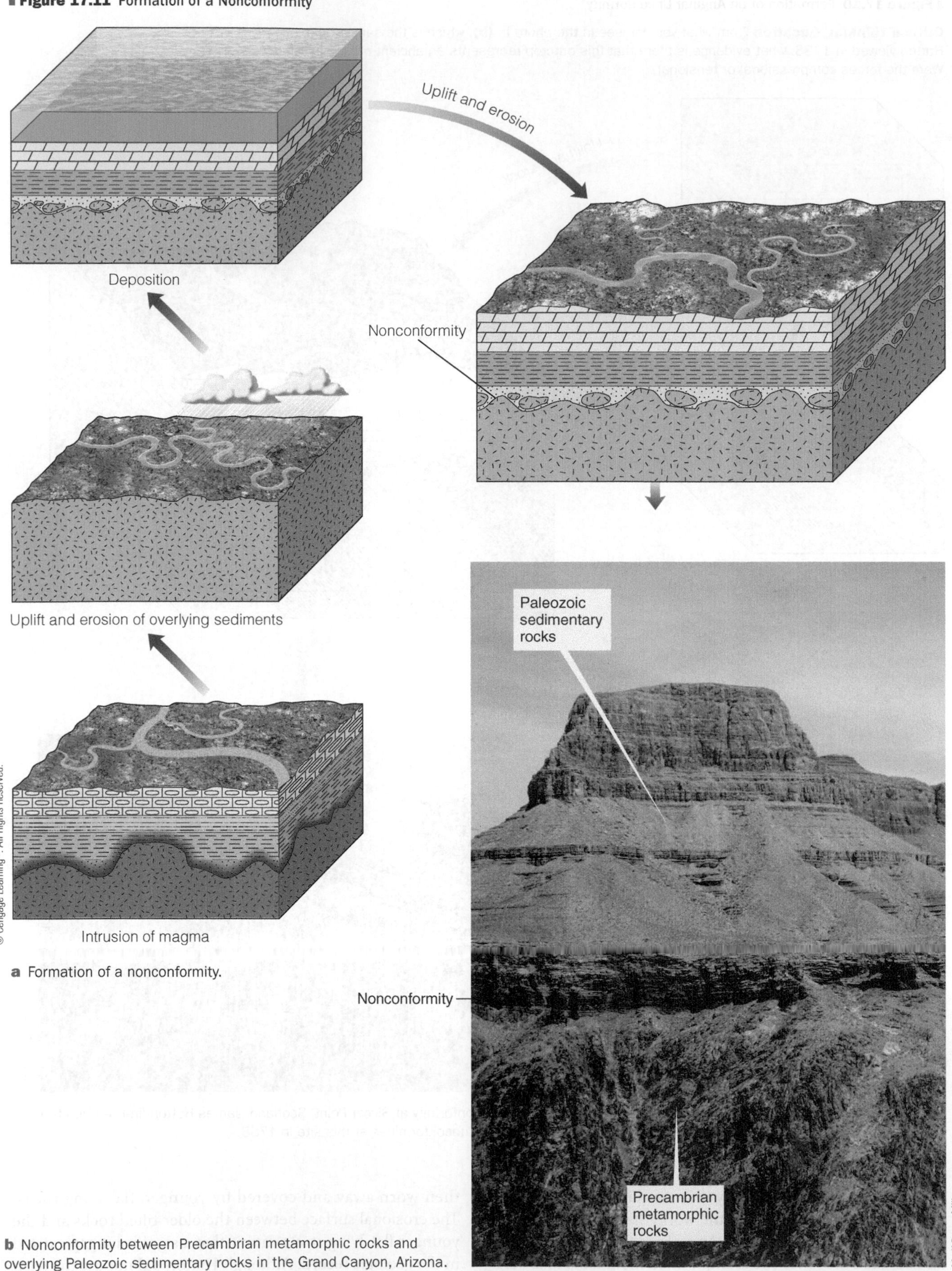

a Formation of a nonconformity.

b Nonconformity between Precambrian metamorphic rocks and overlying Paleozoic sedimentary rocks in the Grand Canyon, Arizona.

unconformity, he was the first to understand and explain the significance of such discontinuities in the geologic record.

A **nonconformity** is the third type of unconformity. Here, an erosional surface cut into metamorphic or igneous rocks is covered by sedimentary rocks (❚ Figure 17.11). This type of unconformity closely resembles an intrusive igneous contact with sedimentary rocks. The principle of inclusions (Figure 17.6) is helpful in determining whether the relationship between the underlying igneous rocks and the overlying sedimentary rocks is the result of an intrusion or erosion. A nonconformity is also marked in many places by an ancient zone of weathering, or even a reddened, brick-like soil horizon, or paleosol. In the case of an intrusion, the igneous rocks are younger, whereas in the case of erosion, the sedimentary rocks are younger. Being able to distinguish between a nonconformity and an intrusive contact is important because they represent different sequences of events.

Applying the Principles of Relative Dating

We can decipher the geologic history of the area represented by the block diagram in ❚ Figure 17.12 by applying the various relative-dating principles just discussed. The methods and logic used in this example are the same as those applied by 19th-century geologists in constructing the geologic time scale.

According to the principles of superposition and original horizontality, beds A–G were deposited horizontally; then either they were tilted, faulted (H), and eroded, or after deposition, they were faulted (H), tilted, and then eroded (❚ Figure 17.13a–c). Because the fault cuts beds A–G, it must be younger than the beds according to the principle of cross-cutting relationships.

Beds J–L were then deposited horizontally over this erosional surface, producing an angular unconformity (I) (❚ Figure 17.13d). Following deposition of these three beds, the entire sequence was intruded by a dike (M), which, according to the principle of cross-cutting relationships, must be younger than all of the rocks that it intrudes (❚ Figure 17.13e).

The entire area was then uplifted and eroded; next, beds P and Q were deposited, producing a disconformity (N) between beds L and P and a nonconformity (O) between the igneous intrusion M and the sedimentary bed P (❚ Figure 17.13f, g). We know that the relationship between igneous intrusion M and the overlying sedimentary bed P is a nonconformity because of the inclusions of M in P (principle of inclusions).

At this point, there are several possibilities for reconstructing the geologic history of this area. According to the principle of cross-cutting relationships, dike R must be

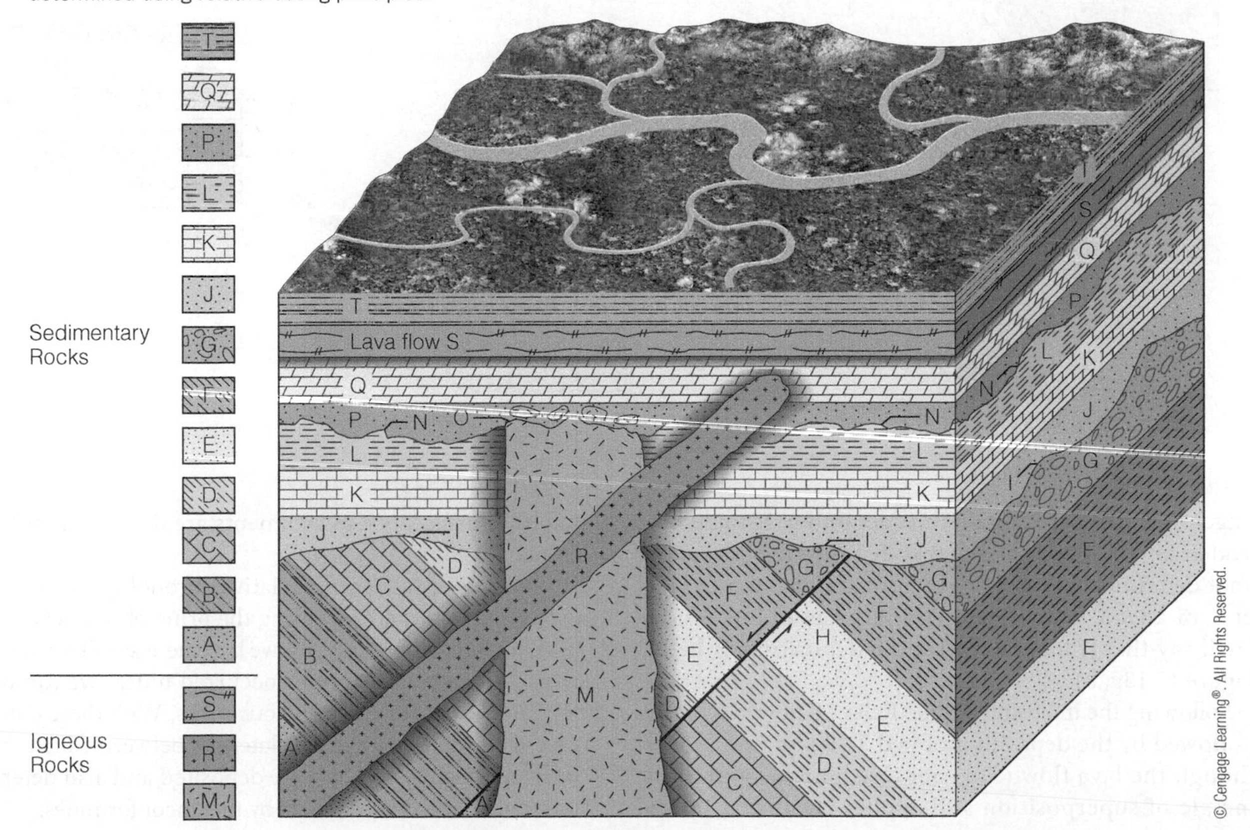

❚ **Figure 17.12 Block Diagram of a Hypothetical Area** A block diagram of a hypothetical area in which the various relative dating principles can be applied to determine its geologic history. See Figure 17.13 to learn how the geologic history was determined using relative dating principles.

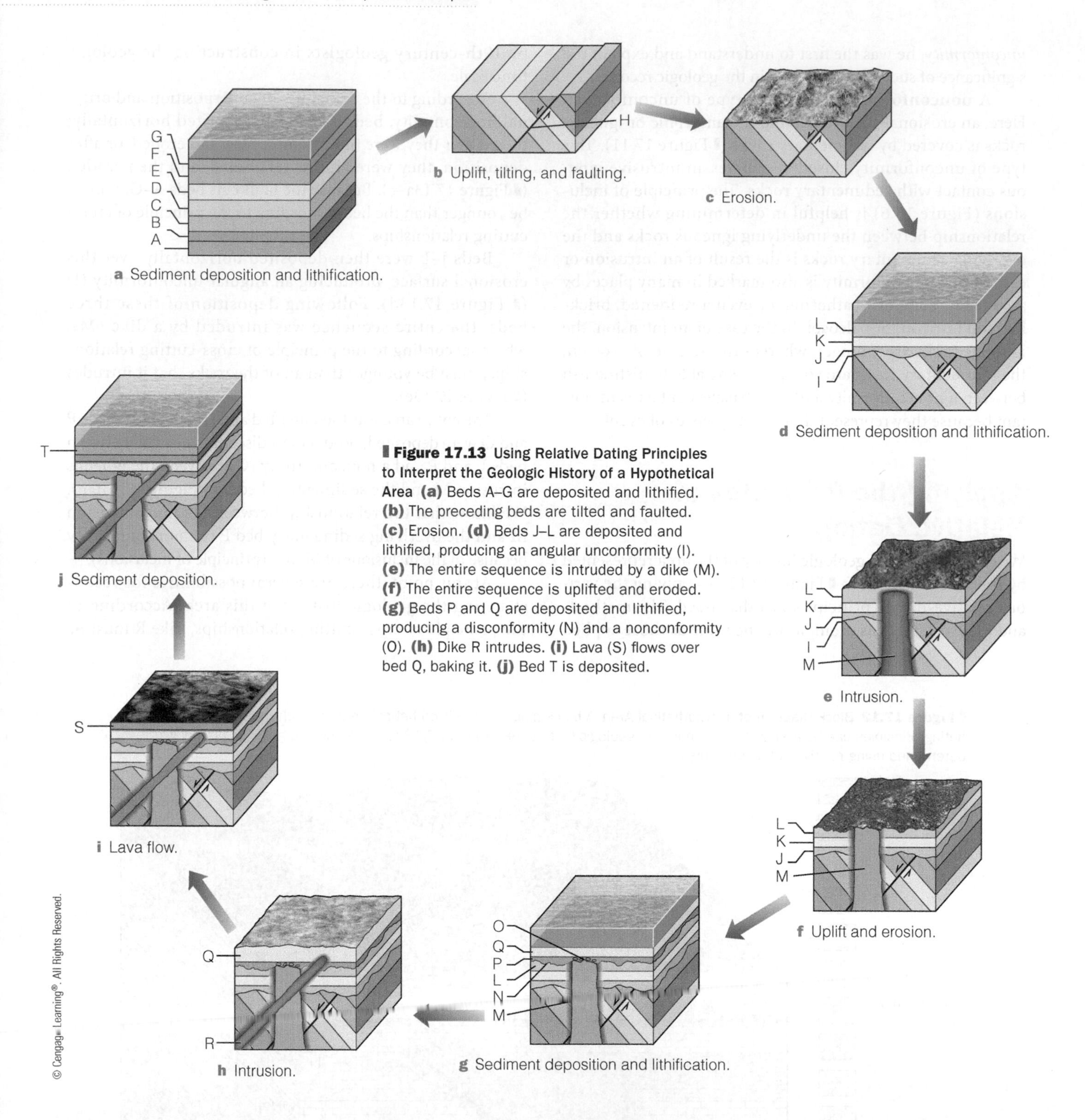

▌Figure 17.13 Using Relative Dating Principles to Interpret the Geologic History of a Hypothetical Area (a) Beds A–G are deposited and lithified. **(b)** The preceding beds are tilted and faulted. **(c)** Erosion. **(d)** Beds J–L are deposited and lithified, producing an angular unconformity (I). **(e)** The entire sequence is intruded by a dike (M). **(f)** The entire sequence is uplifted and eroded. **(g)** Beds P and Q are deposited and lithified, producing a disconformity (N) and a nonconformity (O). **(h)** Dike R intrudes. **(i)** Lava (S) flows over bed Q, baking it. **(j)** Bed T is deposited.

younger than bed Q because it intrudes into it. It could have intruded anytime *after* bed Q was deposited; however, we cannot determine whether R was formed right after Q, right after S, or after T was formed. For purposes of this history, we will say that it intruded after the deposition of bed Q (▌Figure 17.13g, h).

Following the intrusion of dike R, lava S flowed over bed Q, followed by the deposition of bed T (▌Figure 17.13i, j). Although the lava flow (S) is not a sedimentary unit, the principle of superposition still applies because it flowed onto the surface, just as sediments are deposited on Earth's surface.

We have established a relative chronology for the rocks and events of this area by using the principles of relative dating. Remember, however, that we have no way of knowing how many years ago these events occurred unless we can obtain radiometric dates for the igneous rocks. With these dates, we can establish the range of absolute ages between which the different sedimentary units were deposited and also determine how much time is represented by the unconformities.

17.6 Correlating Rock Units

To decipher Earth history, geologists must demonstrate the time equivalency of rock units in different areas. This process is known as **correlation**.

If surface exposures are adequate, units may simply be traced laterally (principle of lateral continuity), even if occasional gaps exist (▌Figure 17.14). Other criteria used to correlate units are similarity of rock type, position in a sequence, and key beds. *Key beds* are units, such as coal beds or volcanic ash layers, that are sufficiently distinctive to allow identification of the same unit in different areas (Figure 17.14).

Generally, no single location in a region has a geologic record of all events that occurred during its history; therefore, geologists must correlate from one area to another to determine the complete geologic history of the region. An excellent example is the history of the Colorado Plateau (▌Figure 17.15). This region provides a record of events occurring over approximately 2 billion years. Because of the forces of erosion, the entire record is not preserved at any single location. Within the walls of the Grand Canyon are rocks of the Precambrian and Paleozoic eras, whereas Paleozoic and Mesozoic Era rocks are found in Zion National Park, and Mesozoic and Cenozoic Era rocks are exposed in Bryce Canyon (Figure 17.15). By correlating the uppermost rocks at one location with the lowermost equivalent rocks of another area, geologists can decipher the history of the entire region.

Although geologists match up rocks on the basis of similar rock type and superposition, correlation of this type can be done only in a limited area where beds can be traced from one site to another. To correlate rock units over a large area or to correlate age-equivalent units of different composition, fossils and the principle of fossil succession must be used.

Fossils are useful as relative time indicators because they are the remains of organisms that lived for a certain length of time during the geologic past. Fossils that are easily identified, are geographically widespread, and existed for a rather short interval of geologic time are particularly useful. Such fossils are **guide fossils**, or *index fossils* (▌Figure 17.16). The trilobite *Paradoxides* and the brachiopod *Atrypa* meet these criteria and are therefore good guide fossils. In contrast, the brachiopod *Lingula* is easily identified and widespread, but its long geologic range of Ordovician to Recent makes it of little use in correlation.

Because most fossils have fairly long geologic ranges, geologists construct *concurrent range zones* to determine the age of the sedimentary rocks containing the fossils. Concurrent range zones are established by plotting the overlapping geologic ranges of two or more fossils that have different geologic ranges (▌Figure 17.17). The first and last occurrences of fossils are used to determine zone boundaries. Correlating concurrent range zones is probably the most accurate method of determining time equivalence.

▌Figure 17.14 Correlating Rock Units In areas of adequate exposures, rock units can be traced laterally, even if occasional gaps exist, and correlated on the basis of similarity in rock type and position in a sequence. Rocks can also be correlated by a key bed—in this case, volcanic ash.

Critical Thinking Question Can you give at least one explanation for why the lower tongue of sandstone present in the left and center columns does not reach all the way to the column on the right?

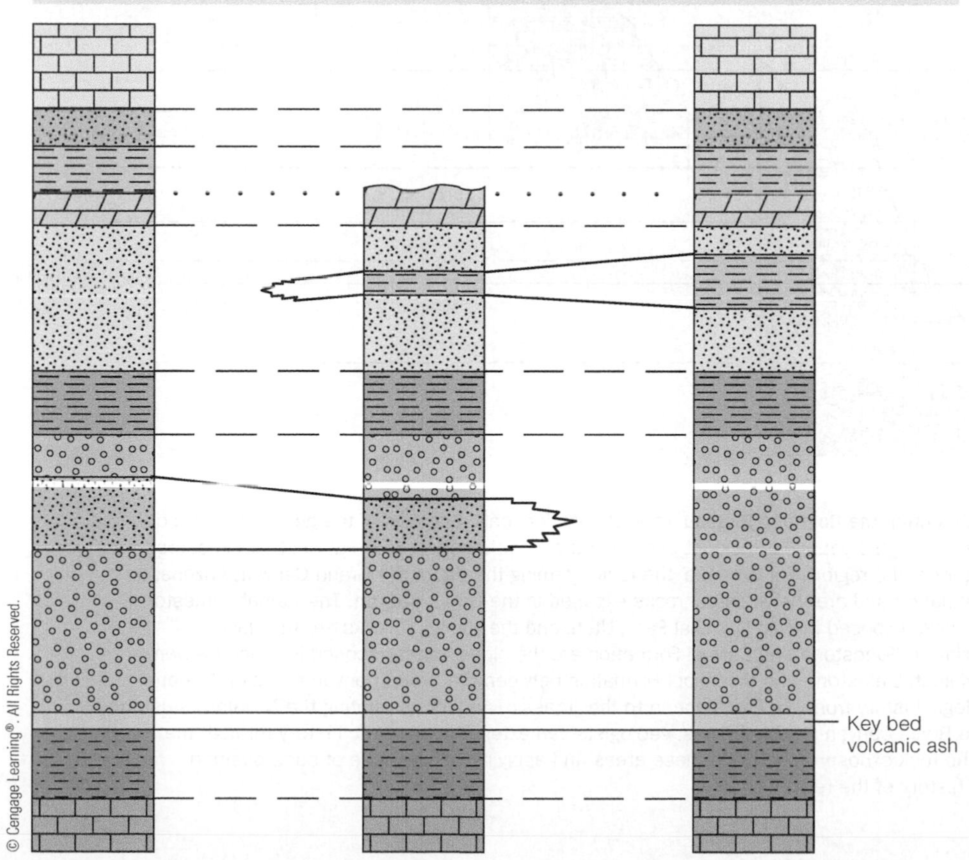

Subsurface Correlation

In addition to surface geology, geologists are interested in subsurface geology because it provides additional information about geologic features beneath Earth's surface. A variety of techniques and methods are used to acquire and interpret data about the subsurface geology of an area.

When drilling is done for oil or natural gas, cores or rock chips called *well cuttings* are commonly recovered from the drill hole.

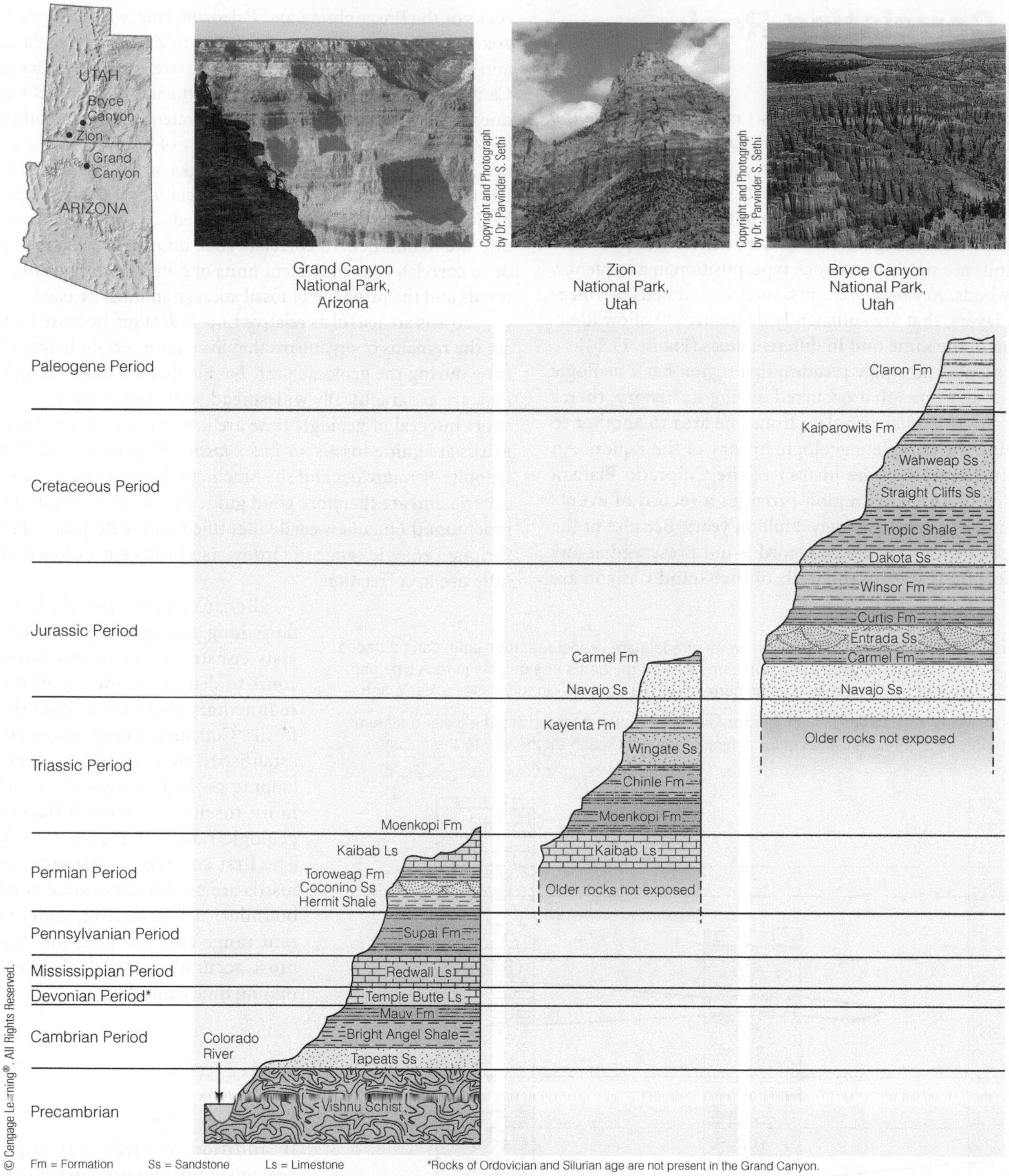

Figure 17.15 Correlation of Rock Units Within the Colorado Plateau At each location, only a portion of the geologic record of the Colorado Plateau is exposed. By correlating the youngest rocks at one exposure with the oldest rocks at another exposure, geologists can determine the entire history of the region. For example, the rocks forming the rim of the Grand Canyon, Arizona, are the Kaibab Limestone and Moenkopi Formation and are the youngest rocks exposed in the Grand Canyon. The Kaibab Limestone and Moenkopi Formation are the oldest rocks exposed in Zion National Park, Utah, and the youngest rocks are the Navajo Sandstone and Carmel Formation. The Navajo Sandstone and Carmel Formation are the oldest rocks exposed in Bryce Canyon National Park, Utah. By correlating the Kaibab Limestone and Moenkopi Formation between the Grand Canyon and Zion National Park, geologists have extended the geologic history from the Precambrian to the Jurassic. And, by correlating the Navajo Sandstone and Carmel Formation between Zion and Bryce Canyon National Parks, geologists can extend the geologic history through the Paleogene Period. Thus, by correlating the rock exposures between these areas and applying the principle of superposition, geologists can reconstruct the geologic history of the region.

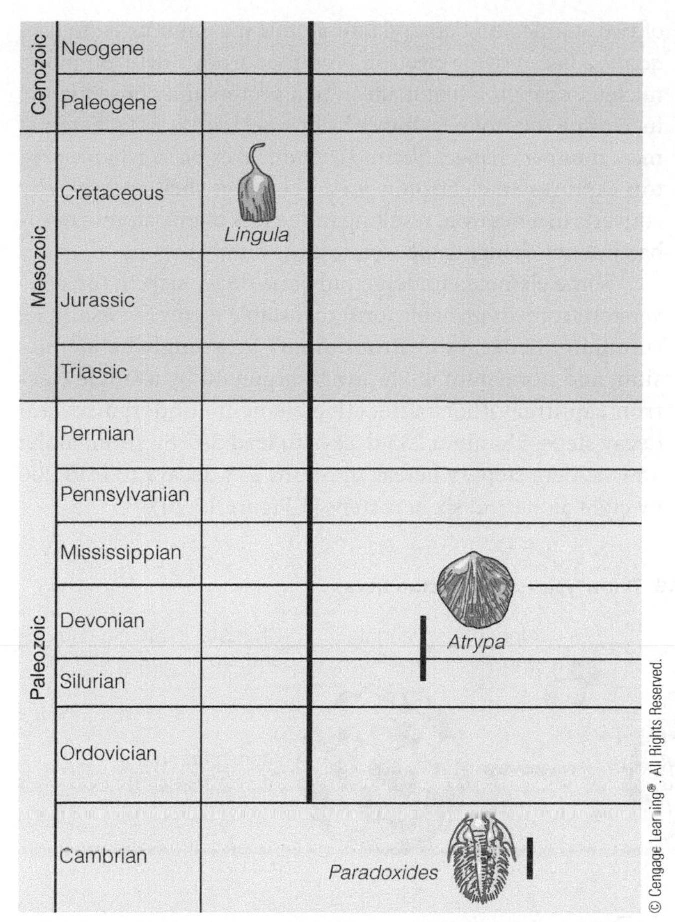

Figure 17.16 Guide Fossils Comparison of the geologic ranges (heavy vertical lines) of three marine invertebrate animals. *Lingula* is of little use in correlation because it has such a long geologic range. However, *Atrypa* and *Paradoxides* are good guide fossils because both are widespread, are easily identified, and have short geologic ranges. Thus, both can be used to correlate rock units that are widely separated and to establish the relative age of a rock that contains them.

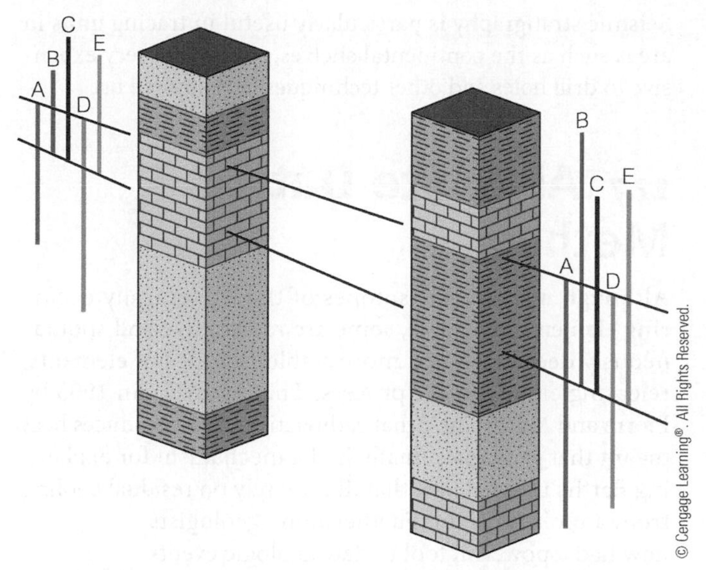

Figure 17.17 Correlation of Two Sections Using Concurrent Range Zones This concurrent range zone was established by the overlapping geologic ranges of fossils symbolized here by the letters A through E. The concurrent range zone is of shorter duration than any of the individual fossil geologic ranges. Correlating by concurrent range zones is probably the most accurate method of determining time equivalence.

ConnectionLink

To review what porosity and permeability are and their effect on the movement of groundwater in different rocks, see Chapter 13.

These samples are studied under the microscope and reveal such important information as rock type, porosity (the amount of pore space), permeability (the ability to transmit fluids), and the presence of oil stains. In addition, the samples can be processed for a variety of microfossils that aid in determining the geologic age of the rock and the environment of deposition.

Geophysical instruments may be lowered down the drill hole to record such rock properties as electrical resistivity and radioactivity, thus providing a record or *well log* of the rocks penetrated. Cores, well cuttings, and well logs are all extremely useful in making subsurface correlations (Figure 17.18).

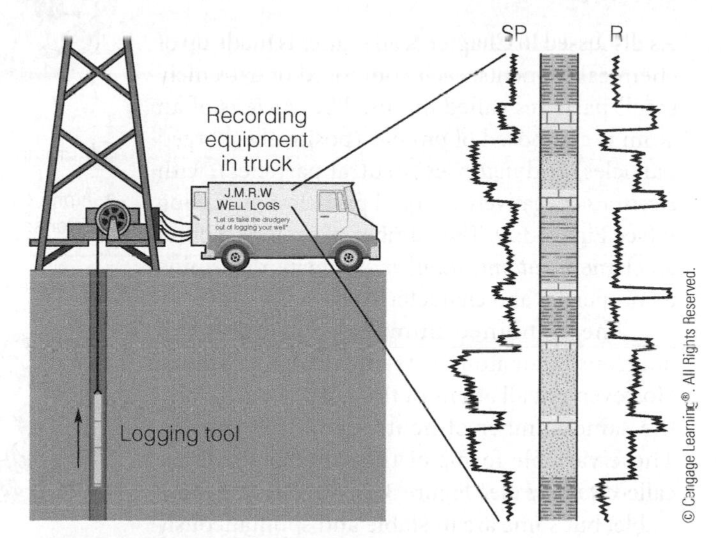

Figure 17.18 Well Logs A schematic diagram showing how well logs are made. As the logging tool is withdrawn from the drill hole, data are transmitted to the surface, where they are recorded and printed as a well log. The curve labeled SP in this diagrammatic electric log is a plot of self-potential (electrical potential caused by different conductors in a solution that conducts electricity) with depth. The curve labeled R is a plot of electrical resistivity with depth. Electric logs yield information about the rock type and fluid content of subsurface formations. Electric logs are also used to correlate from well to well.

Subsurface rock units may also be detected and traced by the study of seismic profiles. Energy pulses, such as those from explosions, travel through rocks at a velocity determined by rock density, and some of this energy is reflected from various horizons (contacts between contrasting layers) back to the surface, where it is recorded (see Figure 16.3).

Seismic stratigraphy is particularly useful in tracing units in areas such as the continental shelves, where it is very expensive to drill holes and other techniques have limited use.

17.7 Absolute Dating Methods

Although most of the isotopes of the 92 naturally occurring elements are stable, some are radioactive and spontaneously decay to other, more stable isotopes of elements, releasing energy in the process. The discovery in 1903 by Pierre and Marie Curie that radioactive decay produces heat meant that geologists finally had a mechanism for explaining Earth's internal heat that did not rely on residual cooling from a molten origin. Furthermore, geologists now had a powerful tool to date geologic events accurately and to verify the long time periods postulated by Hutton and Lyell.

Atoms, Elements, and Isotopes

As discussed in Chapter 3, all matter is made up of chemical elements, each composed of extremely small particles called *atoms*. The *nucleus* of an atom is composed of *protons* (positively charged particles) and *neutrons* (neutral particles), with *electrons* (negatively charged particles) encircling it (see Figure 3.2). The number of protons defines an element's *atomic number* and helps determine its properties and characteristics.

The combined number of protons and neutrons in an atom is its *atomic mass number*. However, not all atoms of the same element have the same number of neutrons in their nuclei. These variable forms of the same element are called *isotopes* (see Figure 3.4). Most isotopes are stable, but some are unstable and spontaneously decay to a more stable form. It is the decay rate of unstable isotopes that geologists measure to determine the absolute ages of rocks.

Radioactive Decay and Half-Lives

Radioactive decay is the process whereby an unstable atomic nucleus is spontaneously transformed into an atomic nucleus of a different element. Three types of radioactive decay are recognized, all of which result in a change of atomic structure (▌Figure 17.19).

In *alpha decay*, two protons and two neutrons are emitted from the nucleus, resulting in the loss of two atomic numbers and four atomic mass numbers. In *beta decay*, a fast-moving electron is emitted from a neutron in the nucleus, changing that neutron to a proton and consequently increasing the atomic number by one, with no resultant atomic mass number change. *Electron capture* takes place when a proton captures an electron from an electron shell and thereby converts to a neutron, resulting in the loss of one atomic number, but not changing the atomic mass number.

Some elements undergo only one decay step in the conversion from an unstable form to a stable form. For example, rubidium 87 decays to strontium 87 by a single beta emission, and potassium 40 decays to argon 40 by a single electron capture. Other radioactive elements undergo several decay steps. Uranium 235 decays to lead 207 by seven alpha and six beta steps, whereas uranium 238 decays to lead 206 by eight alpha and six beta steps (▌Figure 17.20).

▌Figure 17.19 Three Types of Radioactive Decay

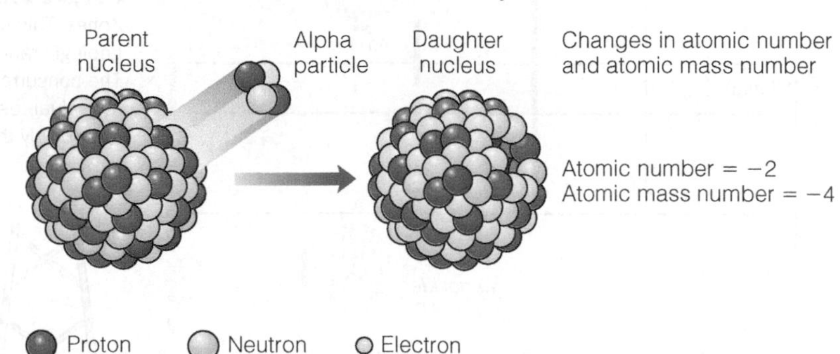

a Alpha decay, in which an unstable parent nucleus emits two protons and two neutrons.

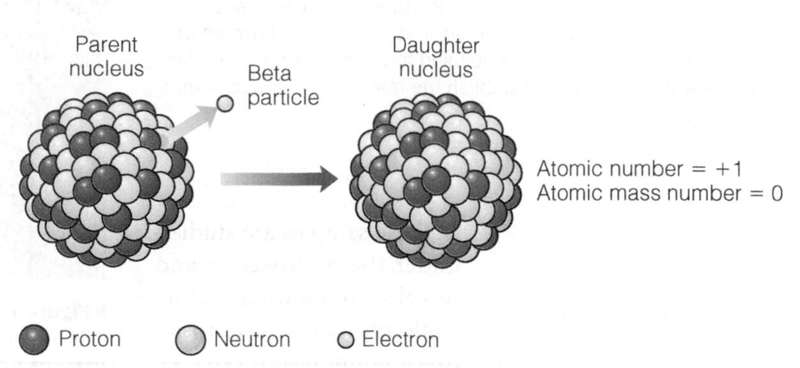

b Beta decay, in which an electron is emitted from the nucleus.

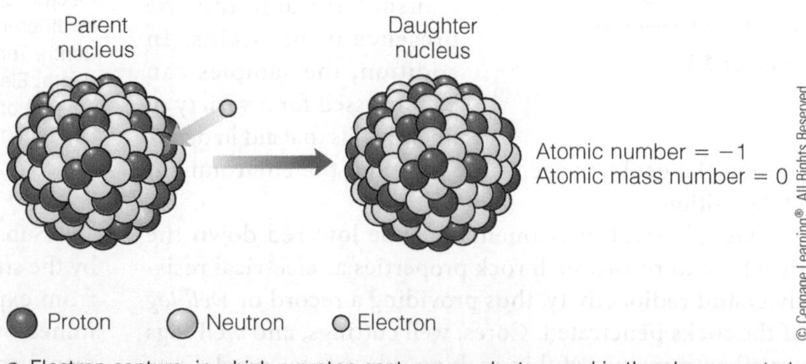

c Electron capture, in which a proton captures an electron and is thereby converted to a neutron.

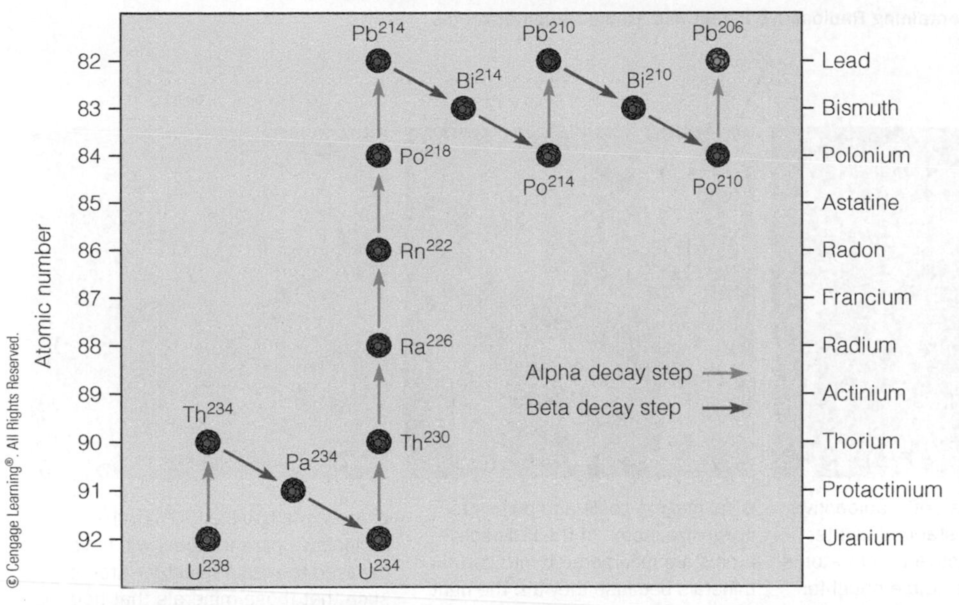

Figure 17.20 Radioactive Decay Series for Uranium 238 to Lead 206 Radioactive uranium 238 decays to its stable daughter product, lead 206, by eight alpha and six beta decay steps. A number of different isotopes are produced as intermediate steps in the decay series.

When discussing decay rates, it is convenient to refer to them in terms of half-lives. The **half-life** of a radioactive element is the time it takes for one-half of the atoms of the original unstable *parent element* to decay to atoms of a new, more stable *daughter element*. The half-life of a given radioactive element is constant and can be precisely measured. Half-lives of various radioactive elements range from less than a billionth of a second to 49 billion years.

Radioactive decay occurs at a geometric rate rather than a linear rate. Therefore, a graph of the decay rate produces a curve rather than a straight line (Figure 17.21). For example, an element with *1,000,000* parent atoms will have *500,000* parent atoms and *500,000* daughter atoms after one half-life. After two half-lives, it will have *250,000* parent atoms (one-half of the previous parent atoms, which is equivalent to one-fourth of the original parent atoms) and *750,000* daughter atoms. After three half-lives, it will have *125,000* parent atoms (one-half of the previous parent atoms, or one-eighth of the original parent atoms) and *875,000* daughter atoms, and so on, until the number of parent atoms remaining is so few that they cannot be accurately measured by present-day instruments.

By measuring the parent–daughter ratio and knowing the half-life of the parent (which has been determined in the laboratory), geologists can calculate the age of a sample that contains the radioactive element. The parent–daughter ratio is usually determined by a *mass spectrometer*, an instrument that measures the proportions of atoms of different masses.

Sources of Uncertainty

The most accurate radiometric dates are obtained from igneous rocks. As magma cools and begins to crystallize, radioactive parent atoms are separated from previously

Figure 17.21 Uniform, Linear Change Compared with Geometric Radioactive Decay

a Uniform, linear change is characteristic of many familiar processes. In this example, water is being added to a glass at a constant rate.

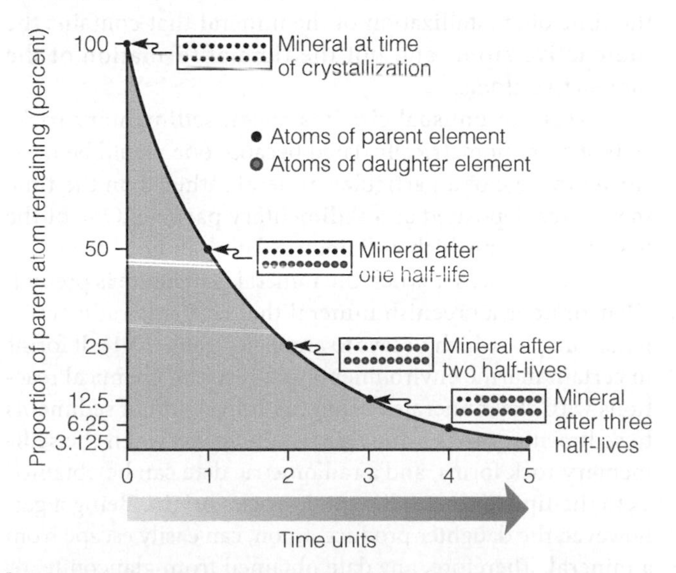

b A geometric radioactive decay curve, in which each time unit represents one half-life, and each half-life is the time it takes for half of the parent element to decay to the daughter element.

▌Figure 17.22 **Crystallization of Magma Containing Radioactive Parent and Stable Daughter Atoms**

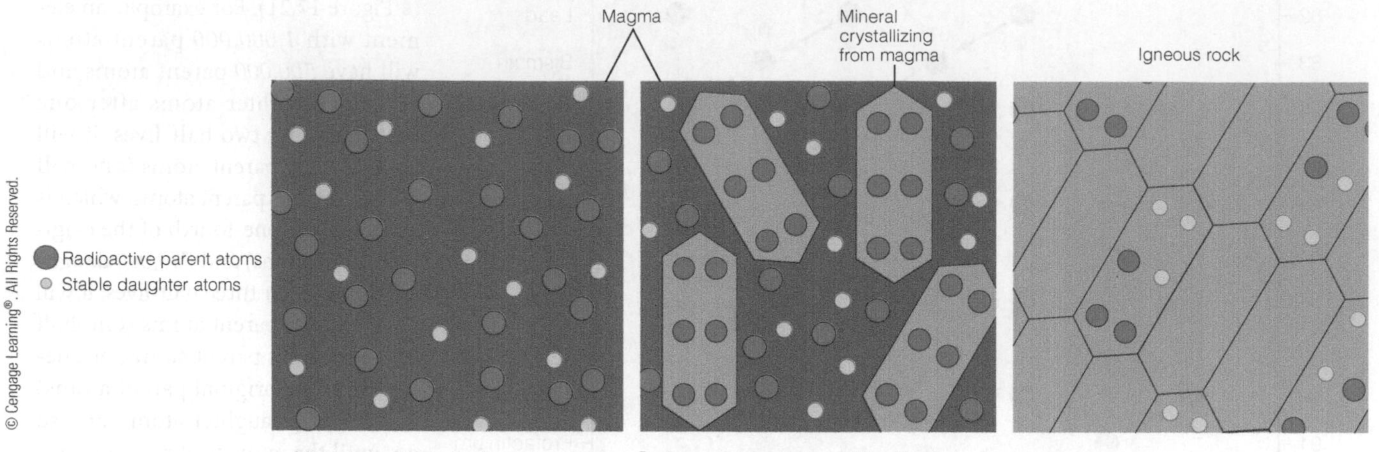

a Magma contains both radioactive parent atoms and stable daughter atoms. The radioactive parent atoms are larger than the stable daughter atoms.

b As magma cools and begins to crystallize, some of the radioactive atoms are incorporated into certain minerals because they are the right size and can fit into the crystal structure. In this example, only the larger radioactive parent atoms fit into the crystal structure. Therefore, at the time of crystallization, minerals in which the radioactive parent atom can fit into the crystal structure will contain 100% radioactive parent atoms and 0% stable daughter atoms.

c After one half-life, 50% of the radioactive parent atoms will have decayed to stable daughter atoms, such that those minerals that had radioactive parent atoms in their crystal structure will now have 50% radioactive parent atoms and 50% stable daughter atoms.

formed daughter atoms. Because they are the right size, some radioactive parent atoms are incorporated into the crystal structure of certain minerals. The stable daughter atoms, though, are a different size from the radioactive parent atoms and consequently cannot fit into the crystal structure of the same mineral as the parent atoms. Therefore, a mineral crystallizing in cooling magma will contain radioactive parent atoms but no stable daughter atoms (▌Figure 17.22). Thus, the time that is being measured is the time of crystallization of the mineral that contains the radioactive atoms and not the time of formation of the radioactive atoms.

Except in unusual circumstances, sedimentary rocks cannot be radiometrically dated because one would be measuring the age of a particular mineral rather than the time that it was deposited as a sedimentary particle. One of the few instances in which radiometric dates can be obtained on sedimentary rocks is when the mineral glauconite is present. Glauconite is a greenish mineral that contains radioactive potassium 40, which decays to argon 40 (Table 17.1). It forms in certain marine environments as a result of chemical reactions with clay minerals during the conversion of sediments to sedimentary rock. Thus, glauconite forms when the sedimentary rock forms, and a radiometric date can be obtained as to the time of the sedimentary rock's origin. Being a gas, however, the daughter product, argon, can easily escape from a mineral. Therefore, any date obtained from glauconite, or any other mineral containing the potassium 40 and argon 40 pair, must be considered a minimum age.

To obtain accurate radiometric dates, geologists must be sure that they are dealing with a *closed system*, meaning that neither parent nor daughter atoms have been added or removed from the system since crystallization and that the ratio between them results only from radioactive decay. Otherwise, an inaccurate date will result. If daughter atoms have leaked out of the mineral being analyzed, then the calculated age will be too young; if parent atoms have been removed, then the calculated age will be too old.

Leakage may take place if the rock is heated or subjected to intense pressure as can sometimes occur during metamorphism. If this happens, some of the parent or daughter atoms may be driven from the mineral being analyzed, resulting in an inaccurate age determination. If the daughter product were completely removed, then one would be measuring the time since metamorphism (a useful measurement itself) and not the time since crystallization of the mineral (▌Figure 17.23).

Because heat and pressure affect the parent–daughter ratio, metamorphic rocks are difficult to date accurately. Remember that although the resulting parent–daughter ratio of the sample being analyzed may have been affected by heat, the decay rate of the parent element remains constant, regardless of any physical or chemical changes.

To obtain an accurate radiometric date, geologists must make sure that the sample is fresh and unweathered and that it has not been subjected to high temperature or intense pressures after crystallization. Furthermore, it is sometimes possible to cross-check the radiometric date obtained by

TABLE 17.1 Five of the Principal Long-Lived Radioactive Isotope Pairs Used in Radiometric Dating

ISOTOPES				
Parent	Daughter	Half-Life of Parent (years)	Effective Dating Range (years)	Minerals and Rocks That Can Be Dated
Uranium 238	Lead 206	4.5 billion	10 million to 4.6 billion	Zircon Uraninite
Uranium 235	Lead 207	704 million		
Thorium 232	Lead 208	14 billion		
Rubidium 87	Strontium 87	48.8 billion	10 million to 4.6 billion	Muscovite Biotite Potassium feldspar Whole metamorphic or igneous rock
Potassium 40	Argon 40	1.3 billion	100,000 to 4.6 billion	Glauconite Muscovite Biotite Hornblende Whole volcanic rock

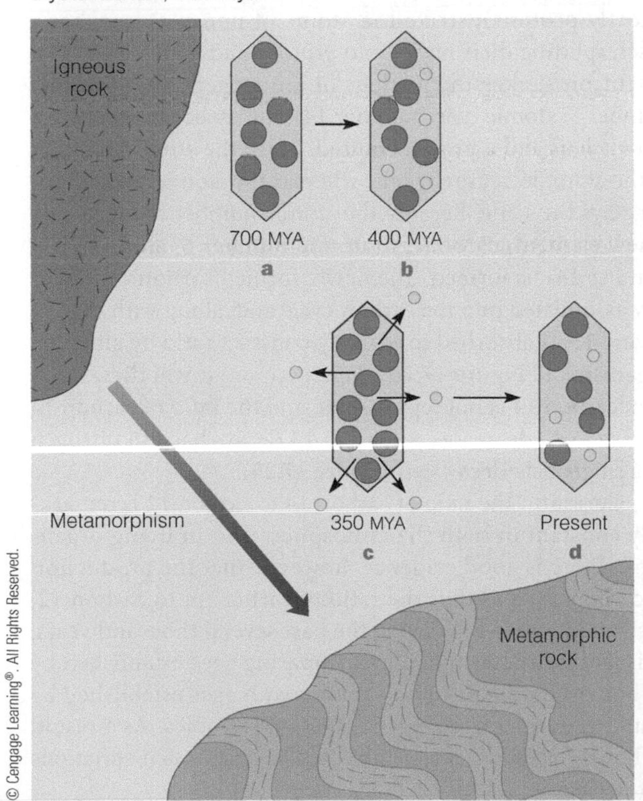

Figure 17.23 Effects of Metamorphism on Radiometric Dating The effect of metamorphism in driving out daughter atoms from a mineral that crystallized 700 million years ago (mya). The mineral is shown **(a)** immediately after crystallization, **(b)** then at 400 million years, when some of the parent atoms had decayed to daughter atoms. **(c)** Metamorphism at 350 mya drives the daughter atoms out of the mineral into the surrounding rock. **(d)** If the rock has remained a closed chemical system throughout its history, dating the mineral today yields the time of metamorphism, whereas dating the whole rock provides the time of its crystallization, 700 mya.

measuring the parent–daughter ratio of two different radioactive elements in the same mineral.

For example, naturally occurring uranium consists of both uranium 235 and uranium 238 isotopes. Through various decay steps, uranium 235 decays to lead 207, whereas uranium 238 decays to lead 206 (Figure 17.20). If the minerals that contain both uranium isotopes have remained closed systems, the ages obtained from each parent–daughter ratio should agree closely. If they do, they are said to be *concordant*, thus reflecting the time of crystallization of the magma. If the ages do not closely agree, then they are said to be *discordant*, and other samples must be taken and ratios measured to see which, if either, date is correct.

Recent advances and the development of new techniques and instruments for measuring various isotope ratios have enabled geologists not only to analyze increasingly smaller samples, but also to analyze samples with a greater precision than ever before. Presently, the measurement error for many radiometric dates is typically less than 0.5% of the age, and in some cases it is even better than 0.1%. Thus, for a rock 540 million years old (near the beginning of the Cambrian Period), the possible error could range from nearly 2.7 million years to as low as less than 540,000 years.

Long-Lived Radioactive Isotope Pairs

Table 17.1 shows the five common, long-lived parent–daughter isotope pairs used in radiometric dating. Long-lived pairs have half-lives of millions or billions of years. All of these were present when Earth formed and are still present in measurable quantities. Other, shorter-lived radioactive isotope pairs have decayed to the point that only small quantities near the limit of detection remain.

To learn more about Earth's oldest rocks, see Chapter 19.

The most commonly used isotope pairs are the uranium–lead and thorium–lead series, which are used principally to date ancient igneous intrusives, lunar samples, and some meteorites. The rubidium–strontium pair is also used for very old samples and has been effective in dating the oldest rocks on Earth, as well as meteorites.

The potassium–argon method is typically used for dating fine-grained volcanic rocks from which individual crystals cannot be separated; hence, the whole rock is analyzed. Because argon is a gas, great care must be taken to ensure that the sample has not been subjected to heat, which would allow argon to escape; such a sample would yield an age that is too young. Other long-lived radioactive isotope pairs exist, but they are rather rare and are used only in special geologic dating situations.

Fission-Track Dating

The emission of atomic particles resulting from the spontaneous decay of uranium within a mineral damages its crystal structure. The damage appears as microscopic linear tracks that are visible only after the mineral has been etched with hydrofluoric acid, an acid so powerful that without careful handling, its vapors can destroy one's sense of smell. The age of the sample is determined from the number of fission tracks present and the amount of uranium that the sample contains: the older the sample, the greater the number of tracks (Figure 17.24).

Fission-track dating is of particular interest to archaeologists and geologists because the technique can be used to date samples ranging from only a few hundred to hundreds of millions of years old. It is most useful for dating samples between approximately 40,000 and 1.5 million years ago, a period for which other dating techniques are not always particularly suitable. One of the problems in fission-track dating occurs when the rocks have later been subjected to high temperatures. If this happens, the damaged crystal structures are repaired by annealing, and consequently the tracks disappear. In such instances, the calculated age will be younger than the actual age.

Figure 17.24 Fission-Track Dating Each fission track (about 16 microns [= 16/1,000 mm] long) in this apatite crystal is the result of the radioactive decay of a uranium atom. The apatite crystal, which has been etched with hydrofluoric acid to make the fission tracks visible, comes from one of the dikes at Shiprock, New Mexico, and has a calculated age of 27 million years.

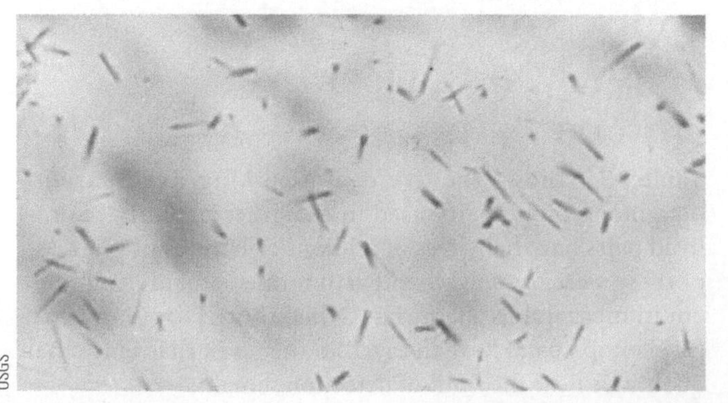

USGS

Radiocarbon and Tree-Ring Dating Methods

Carbon is an important element in nature and is one of the basic elements found in all forms of life. It has three isotopes; two of these, carbon 12 and 13, are stable, whereas carbon 14 is radioactive (see Figure 3.4). Carbon 14 has a half-life of 5,730 years plus or minus 30 years. The **carbon-14 dating technique** is based on the ratio of carbon 14 to carbon 12 and is generally used to date formerly living material.

The short half-life of carbon 14 makes this dating technique practical only for specimens younger than about 70,000 years. Consequently, the carbon-14 dating method is especially useful in archaeology and has greatly helped unravel the events of the latter portion of the Pleistocene Epoch such as where the first center for maize domestication in Mesoamerica arose, or when humans began populating North America.

To learn more about the events of the Pleistocene Epoch, see Chapter 23.

Carbon 14 is constantly formed in the upper atmosphere when cosmic rays, which are high-energy particles (mostly protons), strike the atoms of upper-atmospheric gases, splitting their nuclei into protons and neutrons. When a neutron strikes the nucleus of a nitrogen atom (atomic number 7, atomic mass number 14), it may be absorbed into the nucleus and a proton emitted. Thus, the atomic number of the atom decreases by one, whereas the atomic mass number stays the same. Because the atomic number has changed, a new element, carbon 14 (atomic number 6, atomic mass number 14), is formed. The newly formed carbon 14 is rapidly assimilated into the carbon cycle and, along with carbon 12 and 13, is absorbed in a nearly constant ratio by all living organisms (Figure 17.25). When an organism dies, however, carbon 14 is not replenished, and the ratio of carbon 14 to carbon 12 decreases as carbon 14 decays back to nitrogen by a single beta decay step (Figure 17.25).

Currently, the ratio of carbon 14 to carbon 12 is remarkably constant in both the atmosphere and in living organisms. There is good evidence, however, that the production of carbon 14, and thus the ratio of carbon 14 to carbon 12, has varied somewhat during the past several thousand years. This fact was determined by comparing ages established by carbon-14 dating of wood samples with ages established by counting annual tree rings in the same samples. As a result, carbon-14 ages have been corrected to reflect such variations in the past.

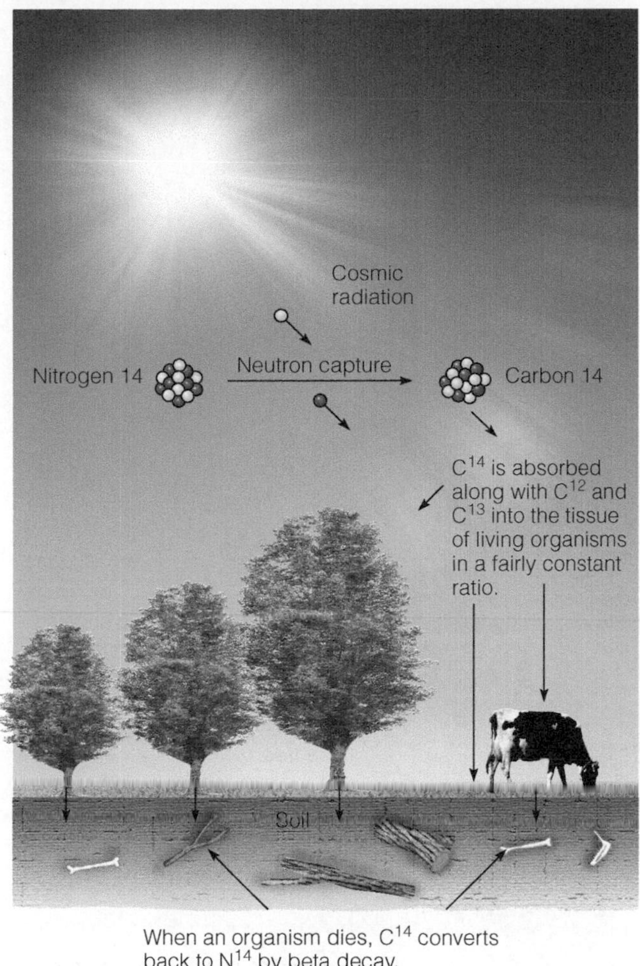

Figure 17.25 Carbon-14 Dating Method The carbon cycle involves the formation of carbon 14 in the upper atmosphere, its dispersal and incorporation into the tissue of all living organisms, and its decay back to nitrogen 14 by beta decay.

Critical Thinking Question Why do you think the ratio of carbon 14 to carbon 12 is constant in living organisms?

Tree-ring dating is another useful method for dating geologically recent events. The age of a tree can be determined by counting the growth rings in the lower part of the trunk. Each ring represents one year's growth, and the pattern of wide and narrow rings can be compared among trees to establish the exact year in which the rings were formed. The procedure of matching ring patterns from numerous trees and wood fragments in a given area is called *cross-dating*.

By correlating distinctive tree-ring sequences from living and nearby dead trees, scientists can construct a composite ring scale, and thus a time scale, that extends back approximately 14,000 years (Figure 17.26). Wood samples whose ages are not known, can be accurately dated by comparing their ring patterns to a composite ring scale that has been developed for that area.

The applicability of tree-ring dating is somewhat limited because it can be used only where continuous tree records are found. It is therefore most useful in arid regions, particularly the southwestern United States, where trees live a very long time.

17.8 Development of the Geologic Time Scale

The geologic time scale is a hierarchical scale in which the 4.6-billion-year history of Earth is divided into time units of varying duration (Figure 17.1). It did not result from the work of any one individual, but rather evolved, primarily during the 19th century, through the efforts of many people.

By applying relative-dating methods to rock outcrops, geologists in England and western Europe defined the major geologic time units without the benefit of radiometric dating techniques. Using the principles of superposition and fossil succession, they correlated various rock exposures and pieced together a composite geologic section. This composite section is, in effect, a relative time scale because the rocks are arranged in their correct sequential order.

By the beginning of the 20th century, geologists had developed a relative geologic time scale, but they did not yet have any absolute dates for the various time-unit boundaries. Following the discovery of radioactivity near the end of the 19th century, radiometric dates were added to the relative geologic time scale (Figure 17.1).

Because sedimentary rocks, with rare exceptions, cannot be radiometrically dated, geologists have had to rely on interbedded volcanic rocks and igneous intrusions to apply absolute dates to the boundaries of the various subdivisions of the geologic time scale (Figure 17.27).

An ash fall or lava flow provides an excellent marker bed that is a time-equivalent surface, supplying a minimum age for the sedimentary rocks below and a maximum age for the rocks above. Ash falls are particularly useful because they may take place over both marine and nonmarine sedimentary environments and can provide a connection between these different environments.

Thousands of absolute ages are now known for sedimentary rocks of known relative ages, and these absolute dates have been added to the relative time scale. In this way, geologists have been able to determine both the absolute ages of the various geologic periods and their durations (Figure 17.1). In fact, the dates for the era, period, and epoch boundaries of the geologic time scale are still being refined as more accurate dating methods are developed and new exposures are dated. The ages shown in Figures 1.18 and 17.1 are the most recently published (2009).

Figure 17.26 Tree-Ring Dating Method In the cross-dating method, tree-ring patterns from different woods are compared to establish a ring-width chronology backward in time.

a Living tree.

b Cross section of tree showing annular tree rings. Each tree ring represents one year's growth. The ring widths vary depending on the growing conditions during the life of the tree.

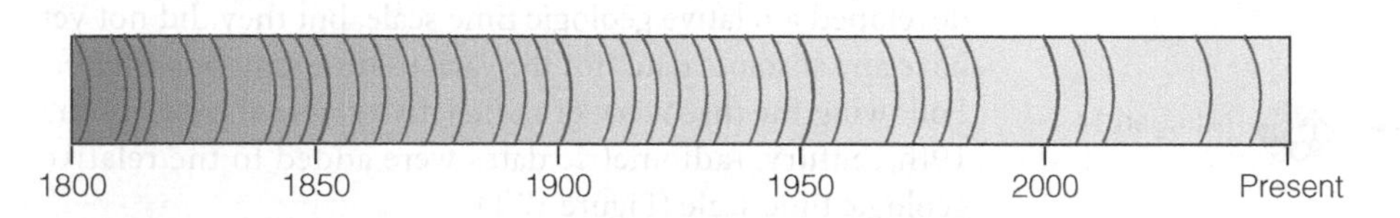

c By matching up tree-ring patterns from numerous trees in a region, a master tree-ring chronology can be constructed. Note that these tree rings are not to scale, but are diagrammatic to show how a master tree-ring chronology would work.

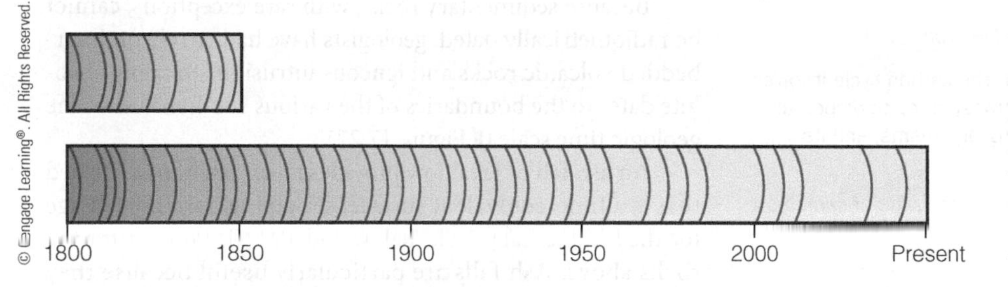

d By matching the tree-ring pattern from a wood beam found at an ancient site to the same pattern of the master tree-ring chronology for that region, the age of the site can be determined, at least to the time the tree died or was cut down.

17.9 Stratigraphy and Stratigraphic Terminology

The recognition of a relative geologic time scale brought some order to *stratigraphy* (the study of the composition, origin, areal distribution, and age relationships of layered rocks); however, problems remained because many sedimentary rock units are time transgressive. This means that they were deposited during one geologic period in a particular area and during another period elsewhere (see Figure 7.11). Therefore, modern stratigraphic terminology includes two fundamentally different kinds of units to deal with both rocks and time: those defined by their content and those related to geologic time (Table 17.2).

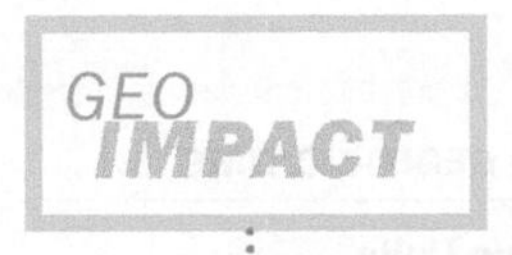

Building a Dam over a Fault?

You are a member of a regional planning commission that is considering a proposal to construct a dam across a river to impound water behind it, thus creating a recreational lake that could increase tourism, and hence improve the region's economy. At the first public meeting to discuss this proposal and receive input from the citizenry, opponents presented a geologic report and map showing that a fault underlies the area of the proposed dam. This faction is worried that the fault may be active and someday move, causing the dam to burst and sending a wall of water downstream. They also provide a number of examples in which the underlying geology of the dam site was not taken into consideration, consequently resulting in dam failure.

To be fair to both groups and ensure that the proposed location of the dam is safe, you seek the advice of a local geologist who has worked in the potential dam area. She tells you that she found a lava flow covering the fault, less than a kilometer from the proposed dam site. Why would a radiometric date on the lava flow, along with the fact the lava flow overlies the fault, be important in deciding if this buried fault poses a risk? Depending on the results of the radiometric date, and a careful analysis of the geology and surface features of the area, how can this information be used to make a reasonable decision on the advisability of building a dam at this location?

View of the remains of the St. Francis Dam, California, following its collapse on March 12, 1928.

Figure 17.27 Determining Absolute Dates for Sedimentary Rocks The absolute ages of sedimentary rocks can be determined by dating associated igneous rocks. In **(a)** and **(b)**, sedimentary rocks are bracketed by rock bodies for which absolute ages have been determined.

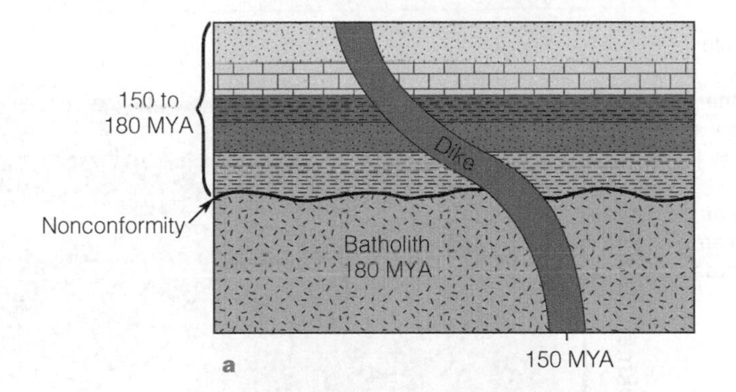

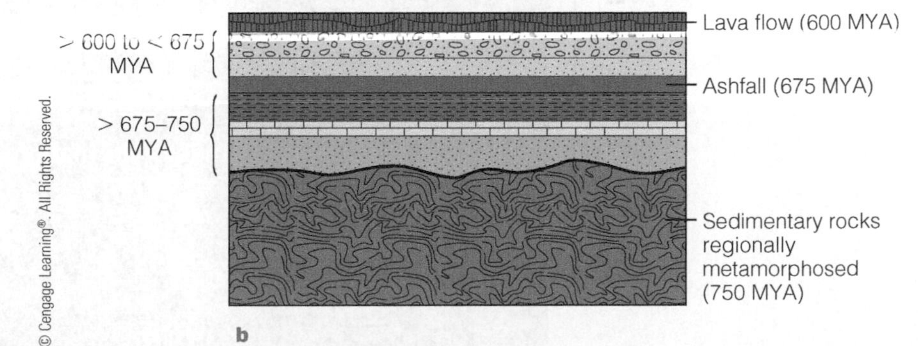

TABLE 17.2 Classification of Stratigraphic Units

UNITS DEFINED BY CONTENT		UNITS EXPRESSING OR RELATED TO GEOLOGIC TIME	
Lithostratigraphic Units	**Biostratigraphic Units**	**Time-Stratigraphic Units**	**Time Units**
Supergroup	Biozones	Eonothem	Eon
Group		Erathem	Era
Formation		System	Period
Member		Series	Epoch
Bed		Stage	Age

Units defined by their content include lithostratigraphic and biostratigraphic units. **Lithostratigraphic units** (lith- and litho- are prefixes meaning "stone" or "stonelike") are defined by the physical attributes of the rocks, such as rock type (e.g., sandstone or limestone), with no consideration of time of origin. The basic lithostratigraphic unit is the *formation*, which is a mapable body of rock with distinctive upper and lower boundaries (▌Figure 17.28). Formations may consist of a single rock type, for example, the Redwall Limestone (Figure 17.15), or a variety of related rock types such as the Morrison Formation (see Figure 22.16). Formations are commonly subdivided into smaller units known as *members* and *beds*, and they may be parts of larger units known as *groups* and *supergroups* (Table 17.2).

A body of strata recognized only on the basis of its fossil content is a **biostratigraphic unit**, the boundaries of which do not necessarily correspond to those of lithostratigraphic boundaries (▌Figure 17.29). The fundamental biostratigraphic unit is the *biozone*. Several types of biozones are recognized, one of which, the *concurrent range zone*, was discussed in the section on correlation (Figure 17.17).

The category of units expressing or related to geologic time includes (1) time-stratigraphic units (also known as *chronostratigraphic units*) and (2) time units (Table 17.2). **Time-stratigraphic units** consist of rocks deposited during a particular interval of geologic time. The *system*, the basic time-stratigraphic unit, is based on a *stratotype*, which consists of rocks in an area where the system was first described. Systems are recognized beyond their stratotype area by their fossil content.

Time units are simply designations for certain parts of geologic time. The basic time unit is the *period*; however, two or more periods may be designated as an *era*, and two or more eras constitute an *eon*. Periods also consist of

▌**Figure 17.28** **Lithostratigraphic Units** Triassic Peak in Palo Duro Canyon State Park, Texas, is composed of the Quartermaster Formation (Permian), followed by the Tecovas Formation and Trujillo Sandstone, both of Triassic age. The Dockman Group includes the Tecovas Formation and Trujillo Sandstone.

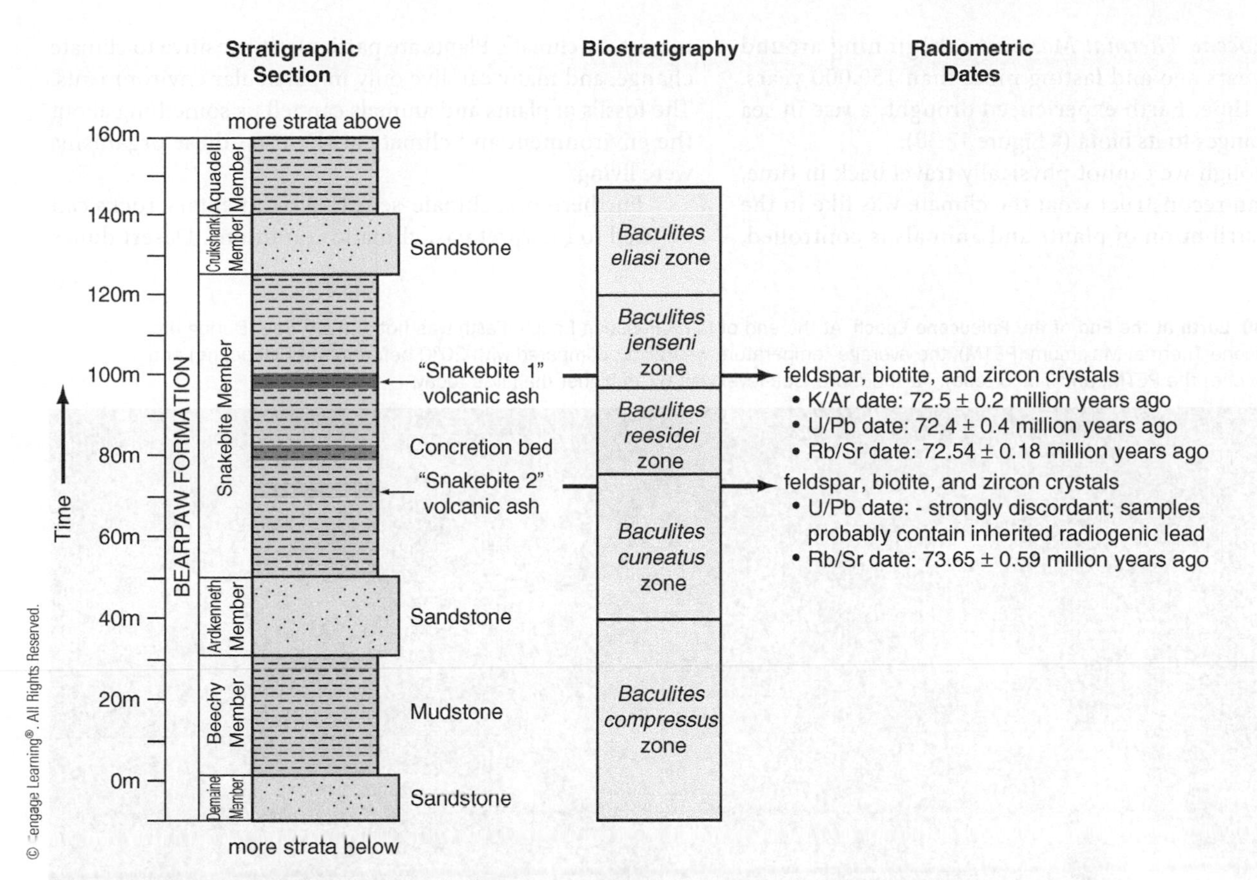

Figure 17.29 Rocks and Fossils of the Bearpaw Formation in Saskatchewan, Canada The column on the left shows formation and members that are lithostratigraphic units. Notice that the biozone boundaries do not correspond with lithostratigraphic boundaries. The absolute ages for the two volcanic ash layers indicate that the *Baculites reesidei* zone is about 72 to 73 million years old.

Critical Thinking Question Why don't the biozone boundaries correspond with the lithostratigraphic boundaries?

shorter designations such as *epoch* and *age*. The time units known as period, epoch, and age correspond to the time-stratigraphic units known as system, series, and stage, respectively (Table 17.2).

These two types of units referring to time and their relationship are particularly confusing to beginning students. Remember though, that time-stratigraphic units are material bodies of rock that occupy a position in a sequence of strata (e.g., the Devonian System), whereas time units refer only to time (the Devonian Period). Thus, a system is a body of rock with lower, medial, and upper parts. In contrast, a time unit, such as the Devonian Period, is the time during which rocks of the Devonian System were deposited, and we divide time units into early, middle, and late subdivisions.

17.10 Geologic Time and Climate Change

Given the debate concerning global warming and its possible implications, it is extremely important to be able to reconstruct past climatic regimes as accurately as possible. To model how Earth's climate system has responded to changes in the past and to use that information for simulations of future climate scenarios, geologists must have a geologic calendar that is as precise and accurate as possible.

New dating techniques with greater precision are providing geologists with more accurate dates for when and how long ago past climate changes occurred. The ability to accurately determine when past climate changes occurred helps geologists correlate these changes with regional and global geologic events to see whether there are any possible connections.

In addition to short-term climatic changes, it is also important to step back a bit and look at climate change from a geologic perspective. We know that Earth has experienced periods of glaciation in the past—for instance, during the Proterozoic, at the end of the Ordovician Period, and most recently during the Pleistocene Epoch (see Figure 2 in Review Question #10, Chapter 1). Earth has also undergone large-scale periods of soaring global temperature, such as during what is known as the

For a review of the causes of Earth's ice ages, see Chapter 14.

Paleocene-Eocene Thermal Maximum, beginning around 56 million years ago and lasting more than 150,000 years. During this time, Earth experienced drought, a rise in sea level, and changes to its biota (▌Figure 17.30).

Even though we cannot physically travel back in time, geologists can reconstruct what the climate was like in the past. The distribution of plants and animals is controlled, in part, by climate. Plants are particularly sensitive to climate change, and many can live only in particular environments. The fossils of plants and animals can tell us something about the environment and climate at the time these organisms were living.

Furthermore, climate-sensitive sedimentary rocks can be used to interpret past climatic conditions. Desert dunes

▌**Figure 17.30 Earth at the End of the Paleocene Epoch** At the end of the Paleocene Epoch, Earth was hot and ice free. During the Paleocene–Eocene Thermal Maximum (PETM), the average temperature was 25°C, compared with 20°C before (Late Paleocene) and after (Early Eocene) the PETM, and 7.2°C today. Furthermore, sea level was 67 m higher than it is today.

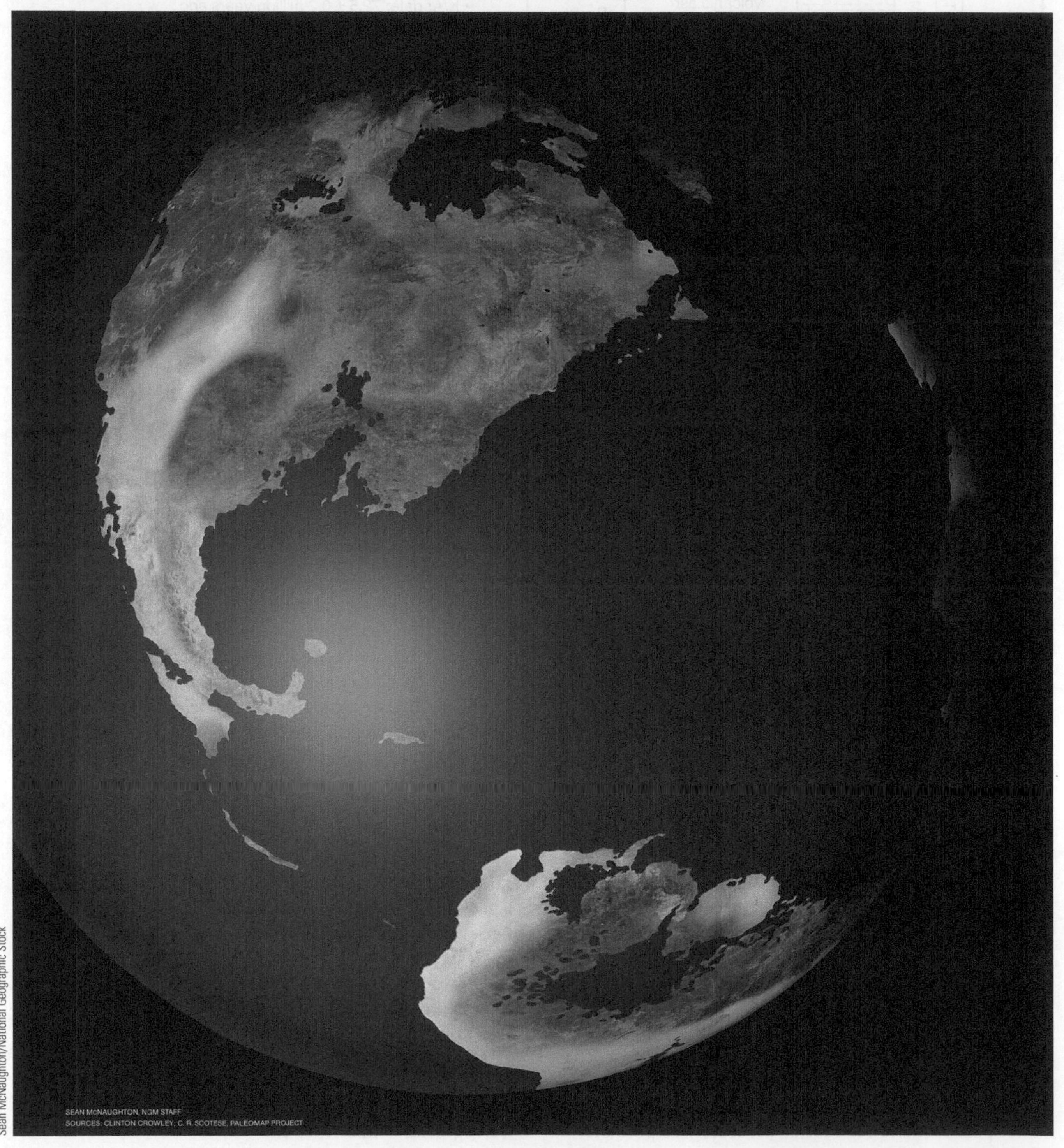

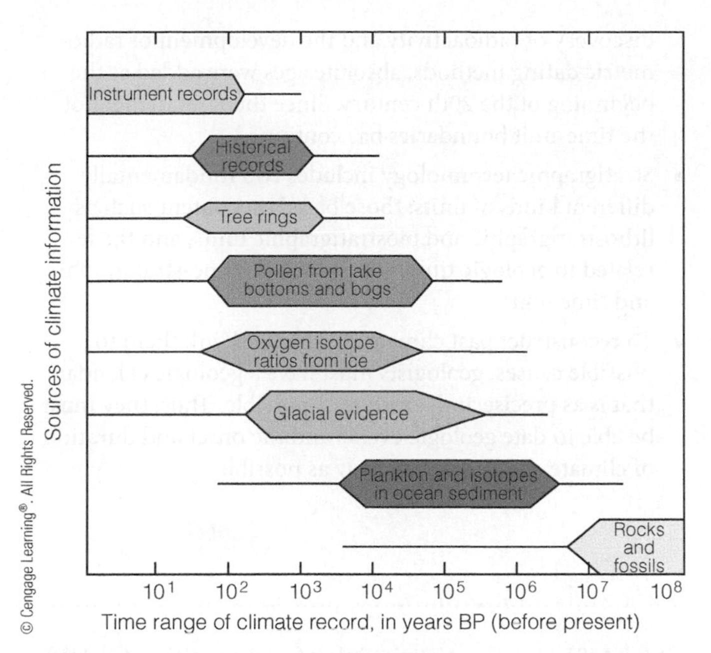

Figure 17.31 Sources of Climate Data Some of the methods used by scientists to determine historical and ancient climates. Each method has its own useful time range and accuracy within that time range.

are typically well sorted and exhibit large-scale cross-bedding (see Figures 7.18 and 15.9). Coals form in fresh-water swamps where climatic conditions promote abundant plant growth (see Figure 20.19). Evaporites such as rock salt result when evaporation exceeds precipitation, such as in desert regions or along hot, dry shorelines. Tillites (glacial sediments) result from glacial activity and indicate cold, wet environments.

By combining all relevant geologic and paleontologic information (Figure 17.31), geologists can reconstruct what the climate was like in the past, and how it has changed over time in a particular area. Furthermore, geologists hope that by analysis of this data, they can, at sometime in the near future, predict and even possibly modify regional climate changes.

Key Concepts Review

- Time is defined by the methods used to measure it. Relative dating places geologic events in sequential order as determined from their position in the geologic record. Absolute dating provides specific dates for geologic rock units or events that are expressed in years before the present.
- During the 18th and 19th centuries, attempts were made to determine Earth's age based on scientific evidence rather than revelation. Although some attempts were ingenious, they yielded a variety of ages that now are known to be much too young.
- James Hutton, considered by many to be the founder of modern geology, thought that present-day processes operating over long periods of time could explain all of the geologic features of Earth. His observations were instrumental in establishing the principle of uniformitarianism and the fact that Earth was much older than earlier scientists thought.
- Uniformitarianism, as articulated by Charles Lyell, soon became the guiding principle of geology. It holds that the laws of nature have been constant through time and that the same processes operating today have also operated in the past, although not necessarily at the same rates.
- Besides uniformitarianism, the principles of superposition, original horizontality, lateral continuity, cross-cutting relationships, inclusions, and fossil succession are basic for determining relative geologic ages and for interpreting Earth history.
- An unconformity is a surface of erosion, nondeposition, or both, separating younger strata from older strata. These surfaces encompass long periods of geologic time for which there is no geologic record at that location.
- Three types of unconformities are recognized. A disconformity separates younger from older sedimentary strata that are parallel to each other. An angular unconformity is an erosional surface on tilted or folded rocks, over which younger sedimentary rocks were deposited. A nonconformity is an erosional surface cut into igneous or metamorphic rocks and overlain by younger sedimentary rocks.
- Correlation is the demonstration of time equivalency of rock units in different areas. Similarity of rock type, position within a rock sequence, key beds, and fossil assemblages can all be used to correlate rock units.
- Radioactivity was discovered during the late 19th century, and soon thereafter, radiometric dating techniques enabled geologists to determine absolute ages for rock units and geologic events.
- Absolute (numerical) dates for rocks are usually obtained by determining how many half-lives of a radioactive parent element have elapsed since the sample originally crystallized. A half-life is the time it takes for one-half of the original unstable radioactive parent element to decay into a new, more stable daughter element.

- The most accurate radiometric dates are obtained from long-lived radioactive isotope pairs in igneous rocks. The most reliable dates are those obtained by using at least two different radioactive decay series in the same rock.
- Carbon-14 dating can be used only on organic matter such as wood, bones, and shells and is effective back to approximately 70,000 years ago. Unlike the long-lived radioactive isotopic pairs, the carbon-14 dating technique determines age by the ratio of radioactive carbon-14 to stable carbon-12.
- The geologic time scale was developed primarily during the 19th century through the efforts of many people. It was originally a relative geologic time scale, but with the discovery of radioactivity and the development of radiometric dating methods, absolute ages were added at the beginning of the 20th century. Since then, refinement of the time-unit boundaries has continued.
- Stratigraphic terminology includes two fundamentally different kinds of units: those based on content such as lithostratigraphic and biostratigraphic units, and those related to geologic time, which include time-stratigraphic and time units.
- To reconstruct past climate changes and link them to possible causes, geologists must have a geologic calendar that is as precise and accurate as possible. Thus, they must be able to date geologic events and the onset and duration of climate changes as precisely as possible.

Important Terms

absolute dating (p. 414)
angular unconformity (p. 424)
biostratigraphic unit (p. 440)
carbon-14 dating technique (p. 436)
correlation (p. 429)
disconformity (p. 424)
fission-track dating (p. 436)
guide fossil (p. 429)
half-life (p. 433)
lithostratigraphic unit (p. 440)
nonconformity (p. 427)
principle of cross-cutting relationships (p. 419)
principle of fossil succession (p. 420)
principle of inclusions (p. 420)
principle of lateral continuity (p. 419)
principle of original horizontality (p. 419)
principle of superposition (p. 419)
radioactive decay (p. 432)
relative dating (p. 414)
time-stratigraphic unit (p. 440)
time unit (p. 440)
tree-ring dating (p. 437)
unconformity (p. 422)

Review Questions

1. The basic lithostratigraphic unit is the
   a. _____ group.
   b. _____ formation.
   c. _____ bed.
   d. _____ member.
   e. _____ supergroup.
2. If a rock is heated during metamorphism and the daughter atoms migrate out of a mineral that is subsequently radiometrically dated, an inaccurate date will be obtained. This date therefore
   a. _____ will be the same as the actual date.
   b. _____ will be younger than the actual date.
   c. _____ will be older than the actual date.
   d. _____ cannot be determined.
   e. _____ none of these.
3. Placing geologic events in sequential order as determined by their position in the geologic record is
   a. _____ absolute dating.
   b. _____ correlation.
   c. _____ historical dating.
   d. _____ relative dating.
   e. _____ uniformitarianism.
4. If a radioactive element has a half-life of 32 million years, what fraction of the original amount of parent material will remain after 96 million years?
   a. _____ one-half.
   b. _____ one-eighth.
   c. _____ one-sixteenth.
   d. _____ one-fourth.
   e. _____ one-thirty-second.
5. If a flake of biotite within a sedimentary rock (such as a sandstone) is radiometrically dated, the date obtained indicates when
   a. _____ the biotite crystal formed.
   b. _____ the sedimentary rock formed.
   c. _____ the parent radioactive isotope formed.
   d. _____ the daughter radioactive isotope(s) formed.
   e. _____ none of these.

6. When geologists reconstruct the geologic history of an area, why is it important for them to differentiate between a sill and a lava flow? How could you tell the difference between a sill and a lava flow at an outcrop if both structures consisted of basalt? What features would you look for in an outcrop to positively identify the structure as either a sill or a lava flow?
7. Why were Lord Kelvin's arguments and calculations so compelling, and what was the basic flaw in his assumption? What do you think the course of geology would have been if radioactivity had not been discovered?
8. Given the current debate over global warming and the many short-term consequences for humans, can you visualize how the world might look in 10,000 years or even 1 million years from now? Use what you have learned about plate tectonics and the direction and rate of movement of plates, as well as how plate movement and global warming will affect ocean currents, weather patterns, weathering rates, and other factors, to make your prediction. Do you think such short-term changes can be extrapolated to long-term trends in trying to predict what Earth will be like using a geologic time perspective?
9. A volcanic ash fall was radiometrically dated using the potassium 40 to argon 40 and rubidium 87 to strontium 87 isotope pairs. The isotope pairs yielded distinctly different ages. What possible explanation could be offered as to why these two isotope pairs yielded different ages? What would you do to rectify the discrepancy in ages?

10. **Creative Thinking Visual Question:** This photograph (Figure 1) was taken on Fish Creek Trail in the Sierra Nevada, California, and shows two black metavolcanic erratics on granodiorite bedrock that has been intruded by a white dike. Using your knowledge of igneous rocks and processes, glacial features, and relative dating principles, provide a geologic history of this area based on what is shown in this image.

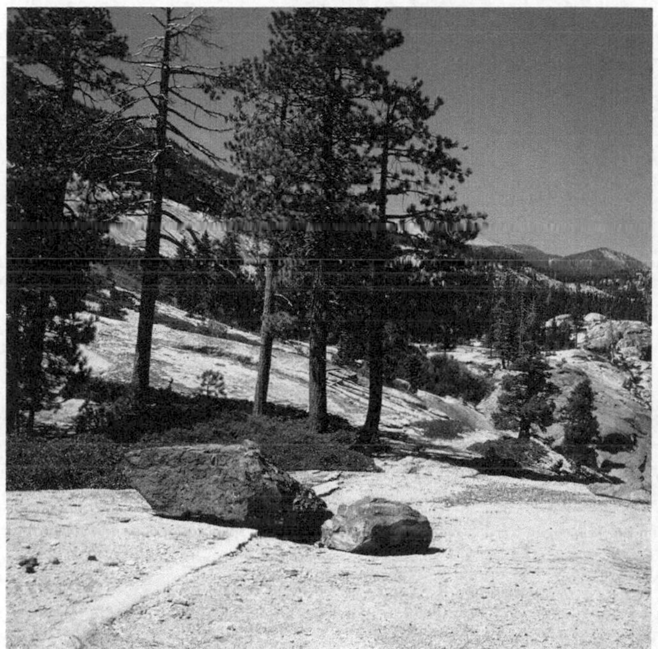

Reed Wicander

Figure 1 Sierra Nevada, California.

Global GeoScience Watch The Anthropocene is a term coined by Paul Crutzen for a new geologic epoch that is based on the human impact on the planet. Within the GREENR database, search for "Anthropocene." Under the "Magazines" section there are a number of articles relating to the Anthropocene. What is the Anthropocene and why is it being proposed as a new geologic epoch? How will it be defined? Why, in some people's opinion, is it the wrong word to use?

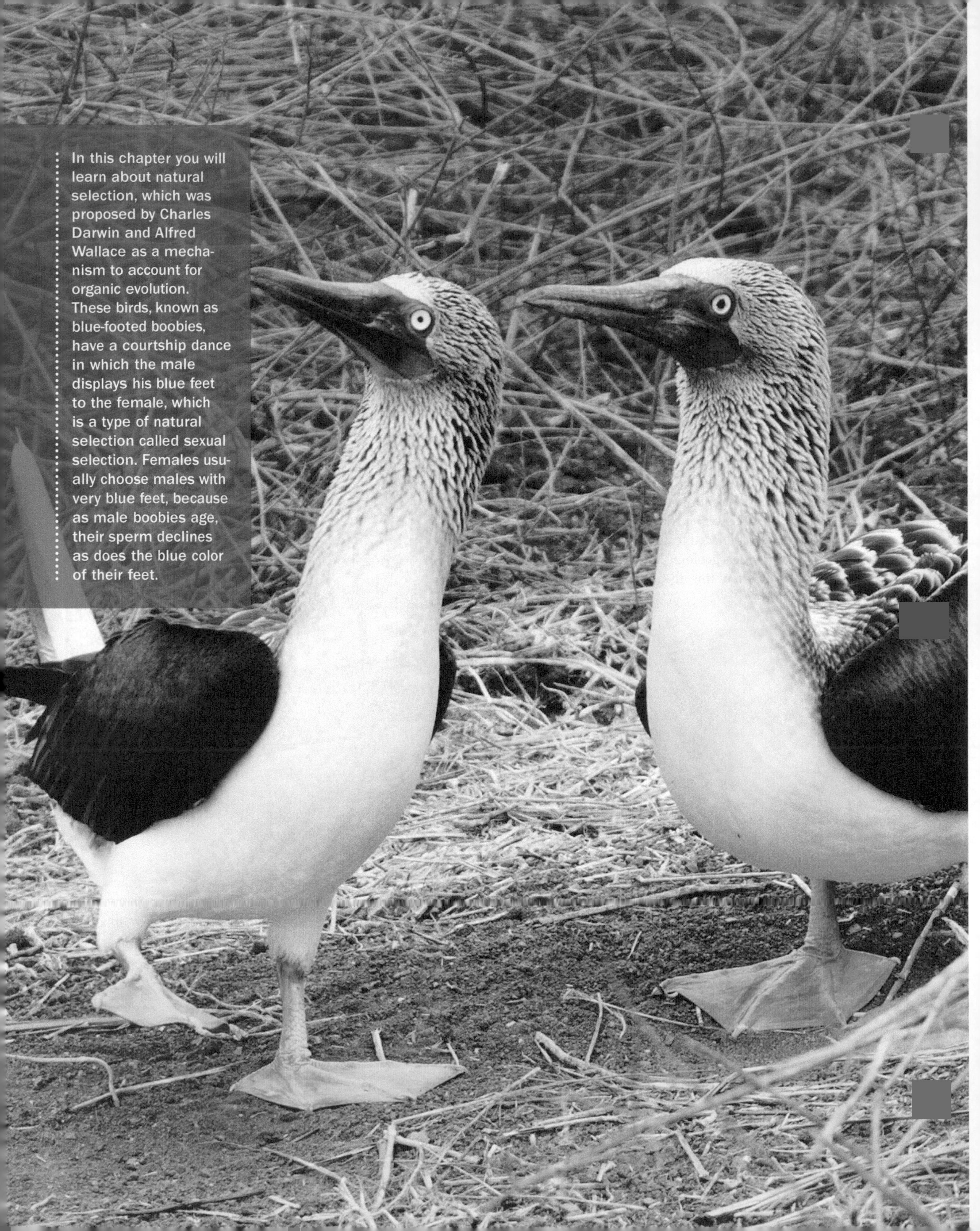

In this chapter you will learn about natural selection, which was proposed by Charles Darwin and Alfred Wallace as a mechanism to account for organic evolution. These birds, known as blue-footed boobies, have a courtship dance in which the male displays his blue feet to the female, which is a type of natural selection called sexual selection. Females usually choose males with very blue feet, because as male boobies age, their sperm declines as does the blue color of their feet.

tamara bizjak/Shutterstock.com

Organic Evolution—The Theory and Its Supporting Evidence

OUTLINE

HAVE YOU EVER WONDERED?

- What the central claim of evolutionary theory is?
- How the theory of organic evolution developed, and who contributed to it?
- Why inheritance and mutations are important in evolutionary thinking?
- What the modern view of organic evolution is?
- How fossils and studies in biology and biochemistry support evolutionary theory?
- Whether predictive statements can be made from evolutionary theory so that its validity may be assessed?

18.1 Introduction

A rugged group of 13 large islands, 8 smaller ones, and 40 islets, all belonging to Ecuador, lies in the Pacific Ocean about 1,000 km west of South America. Called the Archipelago de Colon after Christopher Columbus, the group is better known as the Galápagos Islands. During Charles Robert Darwin's five-year voyage (1831–1836) as an unpaid naturalist aboard the research vessel HMS *Beagle,* he visited the Galápagos Islands, where he made important observations that changed his ideas about the then widely held concept called *fixity of species.* According to this idea, all present-day species had been created in their present form and had changed little or not at all.

Darwin began his voyage not long after graduating from Christ's College of Cambridge University with a degree in theology, and although he was rather indifferent to religion, he fully accepted the biblical account of creation. During the voyage, though, his ideas began to change. For example, some of the fossil mammals he collected in South America were similar to present-day llamas, sloths, and armadillos yet also differed from them. He realized that these animals had descended with modification from ancestral species, and so he began to question the idea of fixity of species.

Darwin postulated that the 13 species of finches living on the Galápagos Islands had evolved from a common ancestor species that somehow reached the islands as an accidental immigrant from South America. Indeed, their ancestor was very likely a single species resembling the blue-back grassquit finch now living along South America's Pacific Coast. The islands' scarcity of food accounts for the ancestral species evolving different physical characteristics, especially beak shape, to survive (❚ Figure 18.1).

Charles Darwin became convinced that organisms descended with modification from their ancestors, which is the central claim of the **theory of evolution**. Many things evolve, or simply change with time, but biologic or organic evolution is not concerned with changes in nonorganic systems such as mountains, rivers, and continents, nor does it include changes in the life cycle of organisms (growth, aging, and metamorphosis).

One reason to study the theory of evolution involving inheritable variations is that it is fundamental to biology and **paleontology**, the study of life history as revealed by fossils. In addition, like plate tectonic theory, evolution is a unifying theory that explains an otherwise encyclopedic collection of facts. And finally, it serves as the framework for the discussions of life history in the following chapters.

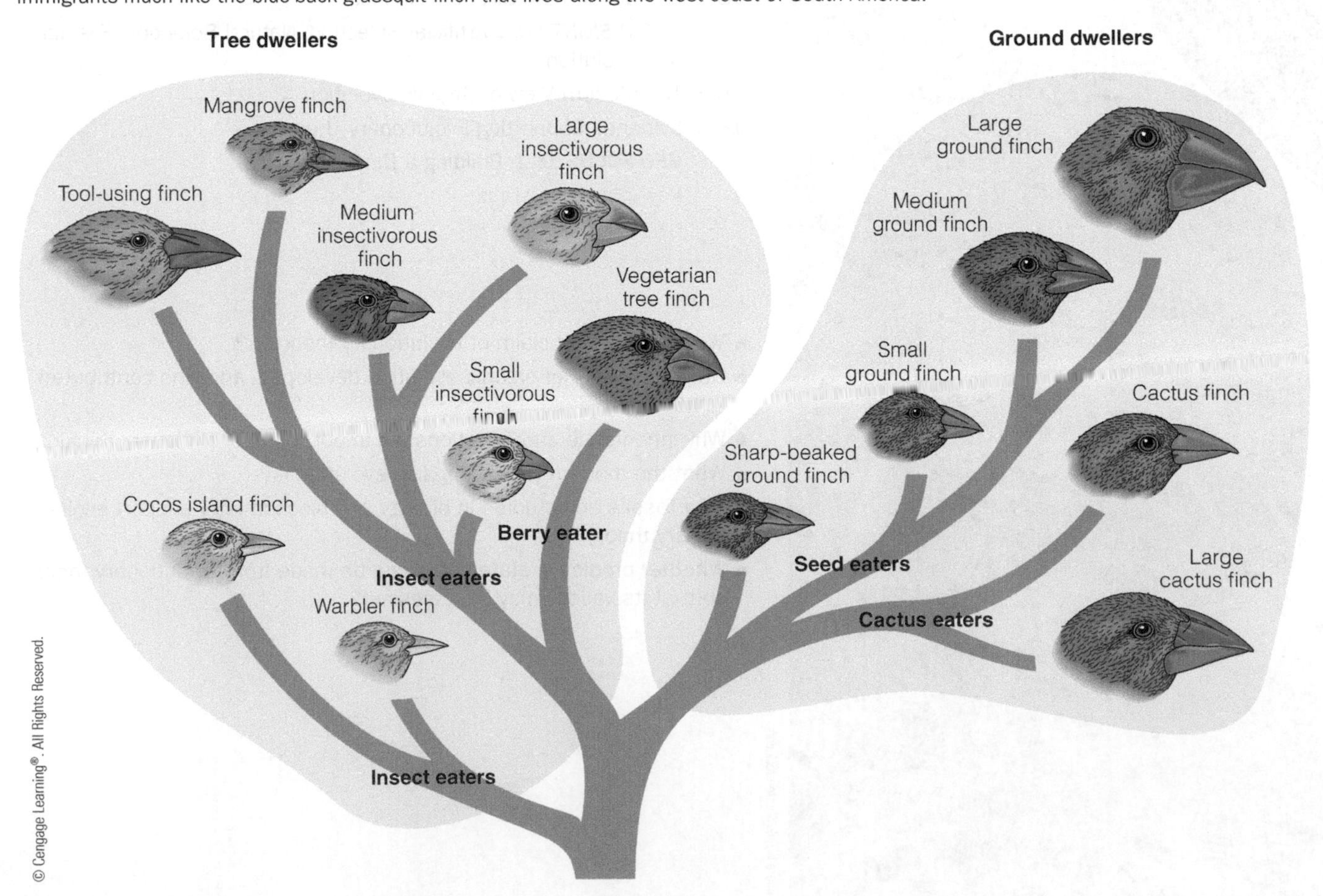

❚ **Figure 18.1** **Darwin's Finches of the Galápagos Islands** These birds probably evolved from accidental immigrants much like the blue-back grassquit finch that lives along the west coast of South America.

Even though everyone has heard of the theory of evolution, few understand the concept very well. For example, many think that the theory attempts to explain how life originated. It does not. There are theories about life's origin, but they must be evaluated on their own merits (see Chapter 19). Another misconception is that all evolutionary change takes place strictly by chance. This is not true either. One sometimes hears the claim that evolution is not true because we do not see monkeys changing into humans. This too is an incorrect assessment of the theory. In fact, no scientist has ever claimed that monkeys were the ancestors of humans. We will address many of these misconceptions in this and some of the following chapters.

18.2 Organic Evolution: What Does It Mean?

The idea of evolution is usually attributed solely to Charles Darwin, but it was seriously considered long before he was born, even by some ancient Greeks and by philosophers and theologians during the Middle Ages. Nevertheless, among Europeans, the prevailing belief well into the 1700s was that the works of the Greek philosopher Aristotle (384–322 BC) and the first two chapters of the book of Genesis contained all important knowledge. Literally interpreted, Genesis was taken as the final word on the origin and diversity of life, as well as much of Earth history. To question any aspect of this interpretation was heresy, which was usually dealt with harshly.

The social and intellectual climate changed in 18th-century Europe, when the absolute authority of the church in all matters was questioned. Ironically, the very naturalists who were trying to find evidence supporting Genesis found more and more evidence that could not be reconciled with a literal reading of Scripture. In this changing intellectual atmosphere, scientists gradually accepted the principle of uniformitarianism and Earth's great age, and the French zoologist Georges Cuvier demonstrated that many plants and animals had become extinct. In view of the accumulating fossil evidence, as well as studies of living organisms, scientists became convinced that organic evolution had occurred, but they lacked a theoretical framework to explain evolution.

Jean-Baptiste de Lamarck and His Ideas on Evolution

Jean-Baptiste de Lamarck (1744–1829) was not the first to propose a mechanism to account for evolution, but in 1809, he was the first to be taken seriously. Lamarck contributed greatly to our understanding of the natural world, but unfortunately he is best remembered for his theory of **inheritance of acquired characteristics.** According to this idea, new traits arise in organisms because of their needs and somehow these characteristics are passed on to their descendants. In an ancestral population of short-necked giraffes, for instance, neck stretching to browse in trees gave them the capacity to have offspring with longer necks. In short, Lamarck thought that characteristics acquired during an individual's lifetime were inheritable.

Given the information available at the time, Lamarck's theory seemed logical and was widely accepted as a viable mechanism for evolution. Indeed, it was not totally refuted until decades later when scientists discovered that the units of heredity known as *genes* cannot be altered by any effort by an organism during its lifetime.

Despite many attempts to demonstrate inheritance of acquired characteristics, none have been successful. One effort took place in the former Soviet Union when Trofim Denisovich Lysenko (1898–1976) exposed plants and animals to dry conditions or to cold, reasoning that they would acquire resistance to drought or cold weather, and these traits would be inherited by future generations. The results were widespread crop failures and famine and the dismantling of Soviet genetic research in agriculture. Lysenko's ideas on inheritance were endorsed by the Central Committee of the Soviet Union because they seemed comparable with Marxist–Leninist philosophy, not because of their scientific merit. In short, it is absurd to base scientific theories on philosophical or political beliefs. A government mandate does not validate a scientific theory, nor does a popular vote or a decree by some ecclesiastic body.

The Contributions of Charles Robert Darwin and Alfred Russel Wallace

In 1859, Charles Robert Darwin (1809–1882) (▮ Figure 18.2) published *On the Origin of Species,* in which he detailed his ideas on organic evolution and proposed a mechanism whereby evolution could take place. Darwin had concluded during his 1831–1836 voyage aboard the *Beagle* that species are not immutable and fixed (see the Introduction), but he had no idea what might bring about change in organisms through time. However, his observations of selection practiced by plant and animal breeders and a chance reading of Thomas Malthus's essay on population gave him the ideas necessary to formulate his theory.

Plant and animal breeders practice **artificial selection** by selecting those traits they deem desirable and then breeding plants and animals with those traits, thereby bringing about a great amount of change (▮ Figure 18.3: Geo-Insight 18.1). The fantastic variety of plants and animals so produced made Darwin wonder whether a process selecting among variant types in nature could also bring about change. He came to fully appreciate the power of selection when he read in Malthus's essay that far more animals are born than reach maturity, yet the adult populations remain rather constant. Malthus reasoned that competition

DEA PICTURE LIBRARY/Getty Images

Figure 18.2 Charles Robert Darwin in 1840 Darwin was convinced that organisms evolved by natural selection, but he did not publish his ideas until 1859.

for resources resulted in a high infant mortality rate, thus limiting population size.

In 1858, Darwin received a letter from Alfred Russel Wallace (1823–1913), a naturalist working in southern Asia, who had also read Malthus's essay and came to exactly the same conclusion that a natural process was selecting only a few individuals for survival. Darwin and Wallace presented their idea, called *natural selection,* simultaneously in 1859 to the Linnaean Society in London.

Natural Selection—What Is Its Significance?

We can summarize the salient points of **natural selection,** a mechanism that accounts for evolution, as follows:

1. Organisms in all populations possess heritable variations—size, speed, agility, visual acuity, digestive enzymes, color, and so forth.

AgStock Images/Flirt/Corbis

George D. Lepp/Documentary Value/Corbis

Media Images/SuperStock

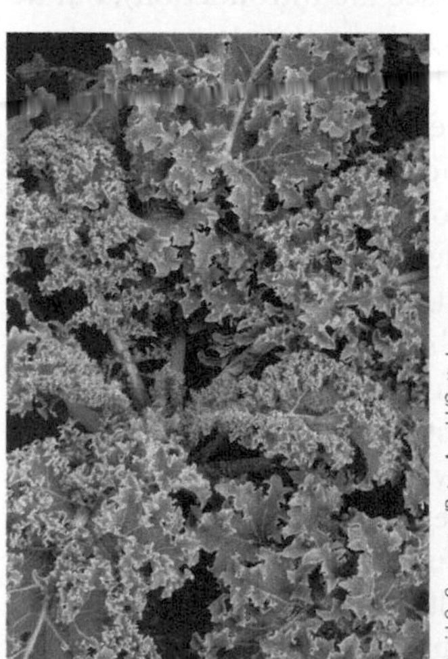

David Q. Cavagnaro/Peter Arnold/Getty Images

Michael P. Gadomski/Science Source

Figure 18.3 Artificial Selection Humans have practiced artificial selection for thousands of years, thereby giving rise to dozens of varieties of domestic dogs, pigeons, sheep, cereal crops, and vegetables. Wild mustard was selectively bred to yield broccoli, cauliflower, kale, and cabbage, and two other vegetables, Brussels sprouts and kohlrabi, not shown here.

Critical Thinking Question Can you think of other examples of artificial selection in plants and animals?

2. Some variations are more favorable than others; that is, some variant types have a competitive edge in acquiring resources, attracting mates, and/or avoiding predators.
3. Those with favorable variations are *more likely* to survive to reproductive maturity and pass on their favorable variations.

We should note that *sexual selection* is a special type of natural selection in which males compete for mates or females select mates based on such traits as showy tail features, bright colors, or courtship rituals (see the chapter opening photo).

In colloquial usage, natural selection is sometimes expressed as "survival of the fittest," which is misleading because it reduces to "the fittest are those that survive and are thus the fittest." But it actually involves differential rates of survival and reproduction. Therefore, it is largely a matter of reproductive success. Having favorable variations does not guarantee that an individual will live long enough to reproduce, but in a population of perhaps thousands, those with favorable variations are more likely to survive and reproduce (Geo-Insight 18.1).

Natural selection works on the existing variation in a population, thus simply giving a competitive edge to some individuals. So evolution by natural selection and Lamarck's evolution by inheritance of acquired characteristics are both testable theories, but evidence supports the former, whereas attempts to verify the latter have failed.

Darwin was not unaware of potential problems for his theory of natural selection. In fact, in *On the Origin of Species,* he wrote:

> If it could be demonstrated that any complex organ existed which could not possibly have been formed by numerous, successive, slight modifications, my theory would absolutely break down. (p. 171)

One common criticism of evolution by natural selection by some outside the sciences is to pose the question, "Of what use is half an eye or half a wing?" The implication is that anything less than eyes or wings as they exist now would be useless, and thus they could not have evolved. First, there is no such thing as half an eye or anything else; eyes and wings of any organism are fully evolved and functional at any one time. Accordingly, all eyes even if they are no more than light-sensitive spots, crude image makers, or image makers as in birds and mammals are fully formed and useful. And second, the precursors of wings need not have served the same function as they do now. The earliest birds may have used wings for gliding or capturing prey. Furthermore, wings that cannot be used for flying are quite useful. Young partridges cannot fly, but they flap their tiny wings vigorously so they can run up trees to avoid predators. Juvenile partridges with their wings taped down cannot climb trees.

One persistent misconception about natural selection, stemming from its popularization as "survival of the fittest," is that among animals only the biggest, strongest, and fastest are likely to survive. In some circumstances these attributes may indeed be favorable, but under different circumstances natural selection may favor the smallest, the most easily concealed, or those having the ability to detoxify some natural or human-made substance.

18.3 Mendel and the Birth of Genetics

Once a formal theory of evolution was proposed it was accepted rather quickly, but the Darwin–Wallace theory of natural selection faced some resistance, mainly because so little was known about inheritance. Critics were quick to point out that Darwin and Wallace could not account for the origin of variations and how variations were maintained in populations. The critics reasoned that should a variant trait arise, it would blend with other traits and be lost. Actually, the answers to these critics existed even then, but would remain in obscurity until 1900.

Mendel's Experiments

During the 1860s, Gregor Mendel, an Austrian monk, performed a series of controlled experiments with true breeding strains of garden peas (strains that when self-fertilized always display the same trait, such as flower color). He concluded from these experiments that traits are controlled by a pair of factors, or what we now call **genes,** and that these factors (genes) controlling the same trait occur in alternative forms, or **alleles.** He further realized that one allele may be dominant over another and that offspring receive one allele of each pair from each parent. For example, in ▮ Figure 18.4, A represents the allele for red flower color, and a represents white, so any offspring may inherit the combinations of alleles symbolized as AA, Aa, or aa. And since A is dominant over a, only those offspring with the aa combination will have white flowers.

We can summarize the most important aspects of Mendel's work as follows: The factors (genes) that control traits do not blend during inheritance, and even though traits may not be expressed in each generation, they are not lost. Therefore, alternative expressions of genes account for some variation in populations.

Our discussion has focused on a single gene controlling a trait, but most traits are controlled by many genes, and some genes show incomplete dominance. Mendel was unaware of mutations (changes in the genetic material), chromosomes, and the fact that some genes control the expression of other genes. Nor had he heard of HOX genes that regulate development of major body segments. Nevertheless, Mendel's work provided the answers Darwin and Wallace needed, but his research was published in an obscure journal and went unnoticed until three independent researchers rediscovered it.

GEO INSIGHT 18.1 Artificial Selection, Natural Selection, Fossils, and Evolution

Darwin and Wallace used artificial selection as a key element in formulating the theory of natural selection. Fossils provide some of the evidence that organisms have descended with modification (evolved) from ancestors that lived during the past.

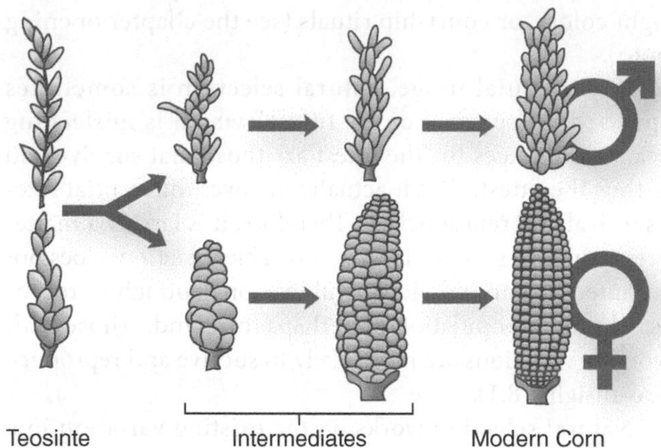

▸ **1.** The evolution of corn from teosinte. Wild teosinte has ears 5–7 cm long with 5–12 kernels per ear. Between 6,000 and 10,000 years ago, Native Americans in Central America selectively bred teosinte, and now a 30-cm ear of corn has about 500 kernels.

▾ **2.** Artificial selection of captive populations of rock pigeons has yielded hundreds of varieties of domestic pigeons.

▾ **3.** Natural selection at work. In preindustrial England, most peppered moths were gray and blended well with the lichens on trees. Black peppered moths were known but rare. With industrialization and pollution, the trees became soot-covered and dark, so dark-colored moths became more common. In rural areas unaffected by pollution, the light–dark frequency did not change.

◂ **4.** Natural selection and the coat color of pocket mice. These mice live in the Sonoran Desert, which has areas of light-brown granite and dark patches of basalt (left). As you might expect, mice living on the light-brown granite have light-brown coats, whereas those living on the basalt have dark-gray coats. The images of the two mice show them on the background that contrasts with their coat color.

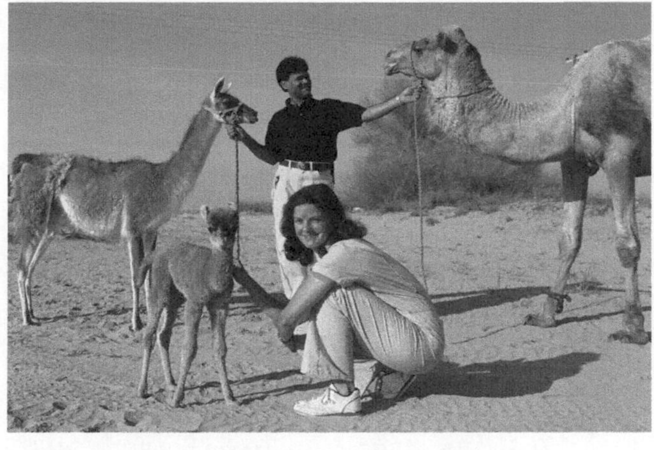

AP Photo/Kamran Jabreili

▸ **5.** Evidence for evolutionary relationships. You would not likely confuse a llama with a dromedary camel, but in fact, both are camels. They evolved from a common ancestor and even now share several characteristics, although camels live in Asia and llamas live in South America. The animal in the center is a cama, a dromedary camel–llama hybrid. This animal resulted from the artificial insemination of a female llama with sperm from a male dromedary camel.

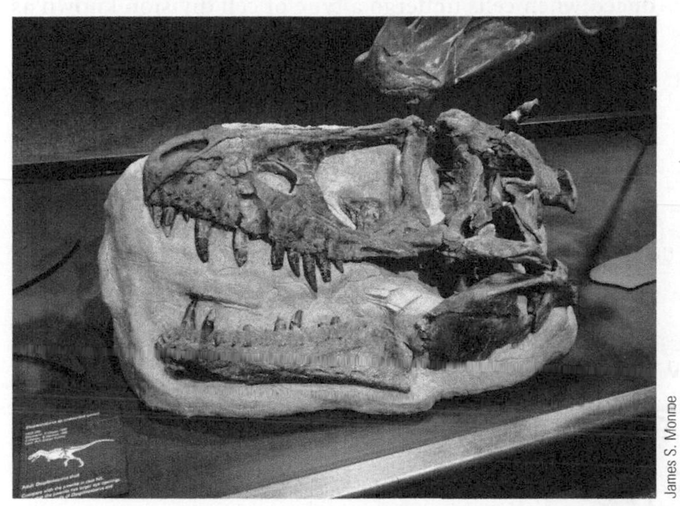

James S. Monroe

◂ **6.** Fossils provide some of the evidence for biological evolution. This dinosaur skull of *Daspletosaurus* is on display in the Museum of the Rockies in Bozeman, Montana. This carnivorous dinosaur lived in western North America during the Late Cretaceous. It measured 8–9 m long and probably weighed 2.5 metric tons.

▾ **7.** These fossil trilobites are on display in the Mesalands Community College's Dinosaur Museum in Tucumcari, New Mexico. Although distantly related to today's insects, spiders, crabs, and lobsters, trilobites went extinct at the end of the Paleozoic Era, 251 million years ago.

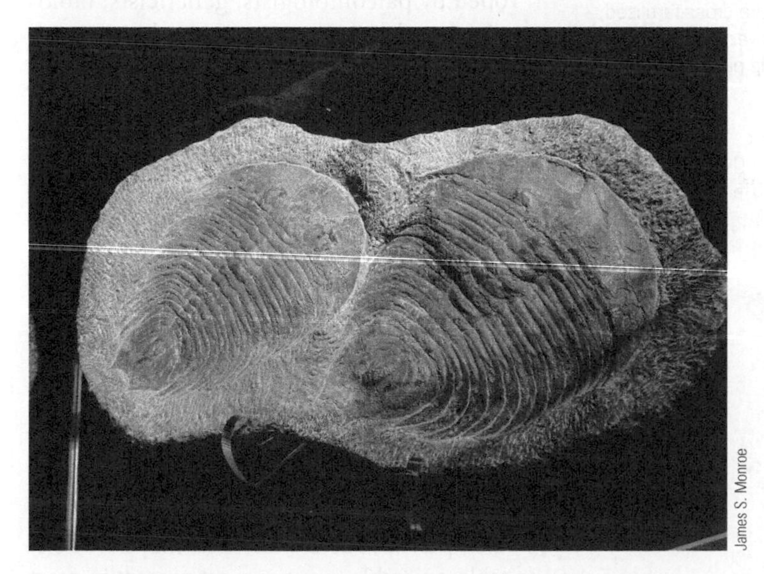

James S. Monroe

James S. Monroe

▴ **8.** About 3 million years ago, a blast from an erupting volcano knocked down numerous redwood trees now exposed at The Petrified Forest near Calistoga, California. The trees were buried in silica-rich volcanic ash and petrified. The largest part of a tree preserved is about 20 m long.

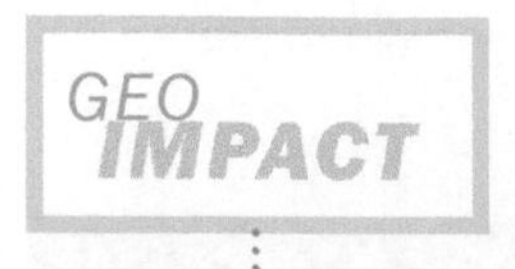

Legislation and Science

Suppose that a powerful group in Congress mandated that all future genetic research had to conform to strict guidelines—specifically, that plants and animals should be exposed to particular environments so that they would acquire characteristics that would allow them to live in otherwise-inhospitable areas. Furthermore, this group enacted legislation that prohibited any other type of genetic research. Why would it be unwise to implement this research program?

Figure 18.4 Mendel's Experiment with Flower Color in Garden Peas

Critical Thinking Question Using the third generation, what would be the possible fourth generation offspring if Aa and aa are cross-fertilized?

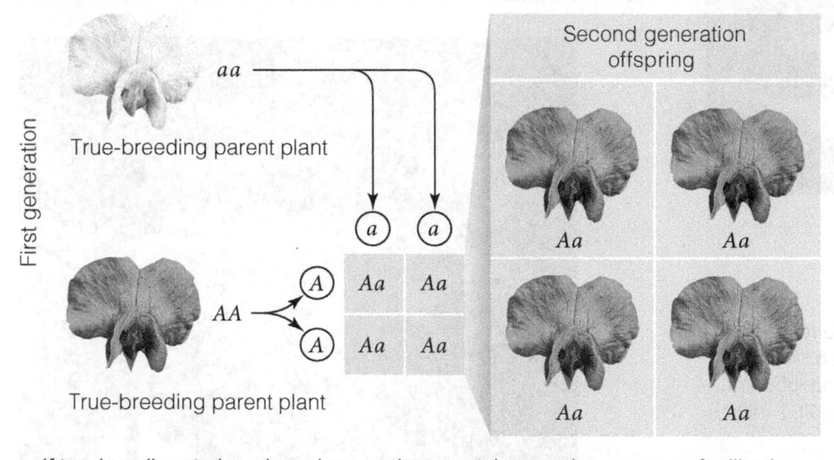

a If true-breeding strains, shown here as the parental generation, are cross-fertilized, they yield offspring with the Aa combination alleles. A is dominant over a, so all offspring in the second generation have red flowers as does the true-breeding parental generation.

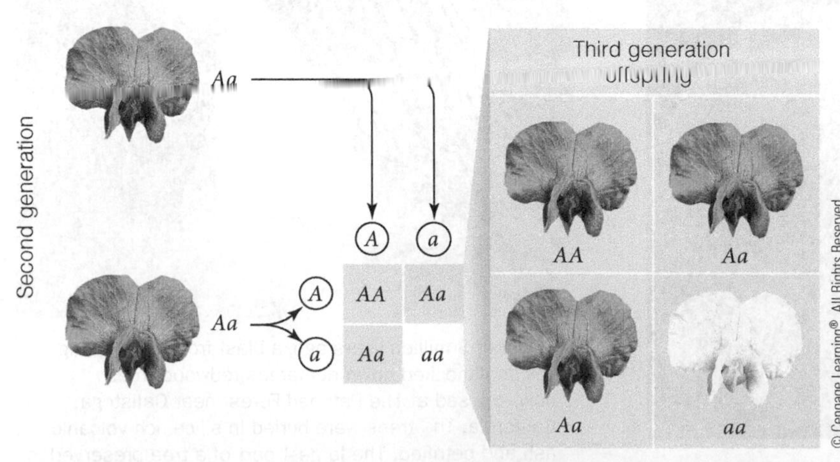

b If the plants from the second generation are cross-fertilized, the third generation has the allele combinations AA, Aa, and aa. Accordingly, this generation has a ratio of three plants with red flowers to one plant with white flowers.

Genes and Chromosomes

Complex, double-stranded, helical molecules of **deoxyribonucleic acid (DNA)** called **chromosomes** are found in the cells of all organisms. Specific segments or regions of the DNA molecule are the hereditary units, the genes. The number of chromosomes is specific for a single species, but varies among species. For instance, fruit flies have 8 chromosomes (4 pairs), humans have 46, and domestic horses have 64; chromosomes are always found in pairs carrying genes that control the same traits.

In sexually reproducing organisms, *sex cells* (pollen and ovules in plants and sperm and eggs in animals) are produced when cells undergo a type of cell division known as **meiosis.** This process yields cells with only one chromosome of each pair, so all sex cells have only half the chromosome number of the parent cell (▌Figure 18.5a). During reproduction, a sperm fertilizes an egg (or pollen fertilizes an ovule), yielding an egg (or ovule) with the full set of chromosomes typical for that species (▌Figure 18.5b).

As Mendel deduced from his experiments, half of the genetic makeup of a fertilized egg comes from each parent. The fertilized egg, however, develops and grows by a cell-division process called **mitosis** during which cells are simply duplicated; that is, there is no reduction in the chromosome number as in meiosis (▌Figure 18.5c).

18.4 The Modern View of Organic Evolution

During the 1930s and 1940s, the ideas developed by paleontologists, geneticists, biologists, and others were merged to form a **modern synthesis** or neo-Darwinian view of evolution. The chromosome theory of inheritance was incorporated into evolutionary thinking, as were changes in genes (mutations) as a source of variation in populations. Lamarck's idea of inheritance of acquired characteristics was completely rejected, and the importance of natural selection was reaffirmed, although the authors of the modern synthesis knew that other processes were also involved.

It is important to realize that in modern evolutionary theory, populations rather than individuals evolve. Individuals develop according to the genes they inherited from their parents, but an individual may have heritable, favorable variations not present in most members of its population. In short, individuals with favorable variations in genetically

Figure 18.5 Meiosis and Mitosis

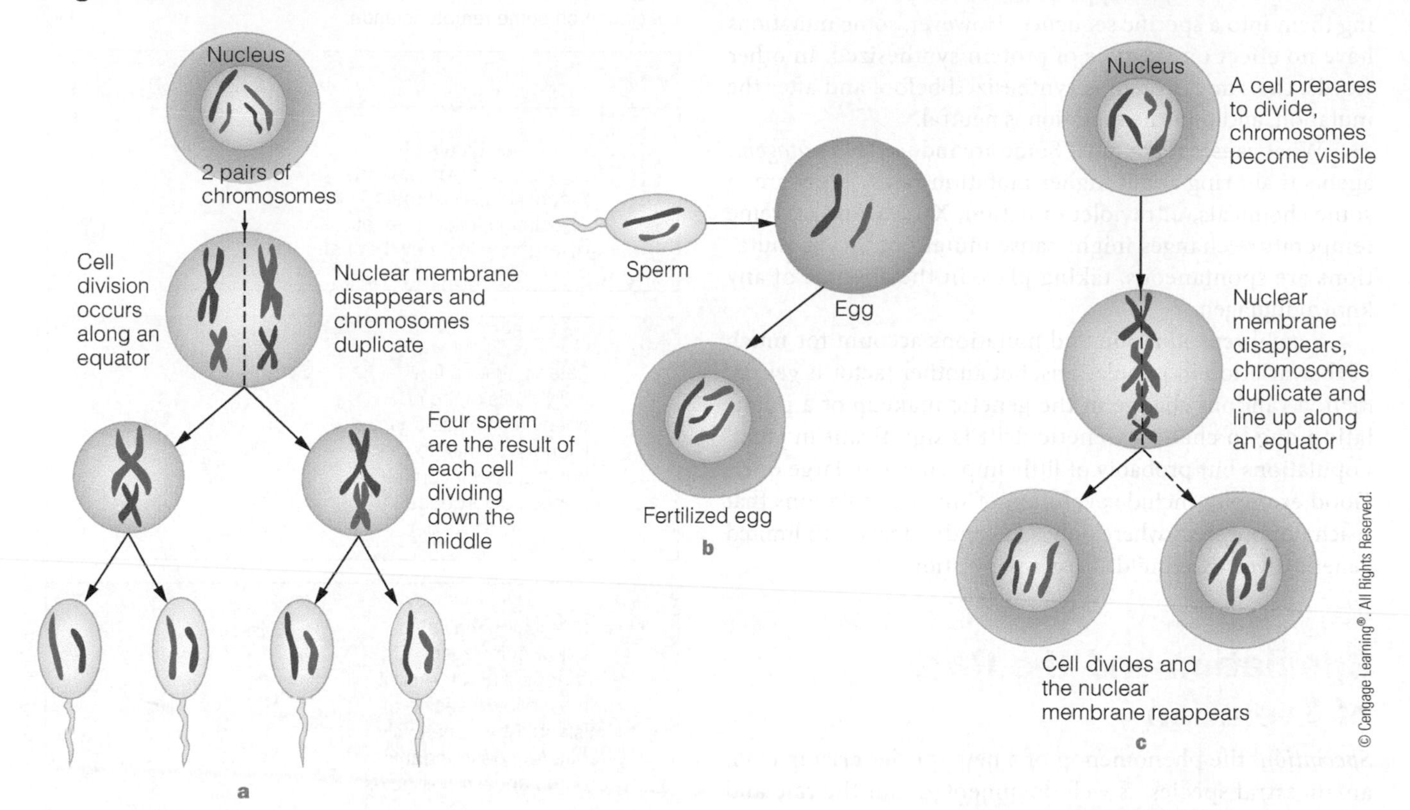

a During meiosis, sex cells form that contain one member of each chromosome pair. The formation of sperm is shown here; eggs form the same way, but only one of the four final eggs is functional.

b The full number of chromosomes is restored when a sperm fertilizes an egg.

c Mitosis results in the complete duplication of a cell. In this example, a cell with four chromosomes (two pairs) produces two cells, each with four chromosomes. Mitosis takes place in all cells except sex cells. Once an egg has been fertilized, the developing embryo grows by mitosis.

varying populations are more likely to leave more descendants than others in their populations, and thus descendent populations possess these variations in greater frequency. So, an individual does not evolve during its lifetime, but evolution does take place in a succession of populations through time. Accordingly, there never was a first dinosaur or a first horse, only successions of populations of organisms developing features we now associate with these groups.

The theory of evolution is not static. New concepts now incorporated into the theory include genetic drift (random changes in populations due to chance) and new ideas on how speciation occurs (see the next sections). Until fairly recently, biologists and paleontologists thought that all genetic transfer was vertical—that is, from one generation to the next. They now know about *lateral gene transfer* in single-celled organisms such as bacteria, in which one bacterium might incorporate genes from another bacterium but not be the offspring of that bacterium.

What Brings About Variation?

Natural selection works on variations in populations, most of which result from the reshuffling of alleles from generation to generation during sexual reproduction. Given that each of thousands of genes might have several alleles, and that offspring receive half of their genes from each parent, the potential for variation is enormous. Any new variations, though, arise by **mutations** involving some kind of change in the chromosomes or genes—that is, a change in the hereditary information. Whether a *chromosomal mutation* (affecting a large segment of a chromosome) or a *point mutation* (a change in a particular gene), as long as it takes place in a sex cell, it is inheritable.

To fully understand mutations, we must explore them further. For one thing, they are random with respect to fitness, meaning they may be beneficial, neutral, or harmful. If a species is well adapted to its environment, most mutations would not be particularly useful and might be harmful. But what was a harmful mutation can become a useful one if the environment changes. For instance, some plants have developed a resistance to contaminated soils around mines. Plants of the same species from the normal environment do poorly or die in contaminated soils, whereas contaminant-resistant plants do very poorly in the normal environment. Mutations for contaminant resistance probably occurred repeatedly in the population, but they were not beneficial until contaminated soils were present.

How can a mutation be neutral? In cells, information carried on chromosomes directs the formation of proteins in cells by selecting the appropriate amino acids and arranging them into a specific sequence. However, some mutations have no effect on the type of protein synthesized. In other words, the same protein is synthesized before and after the mutation, and thus the mutation is neutral.

What causes mutations? Some are induced by *mutagens,* agents that bring about higher mutation rates. Exposure to some chemicals, ultraviolet radiation, X-rays, and extreme temperature changes might cause mutations. Some mutations are spontaneous, taking place in the absence of any known mutagen.

Sexual reproduction and mutations account for much of the variation in populations, but another factor is *genetic drift,* a random change in the genetic makeup of a population due to chance. Genetic drift is significant in small populations but probably of little importance in large ones. Good examples include evolution of small populations that reach remote areas where only a few individuals with limited genetic diversity rebuild a larger population.

Speciation and the Rate of Evolution

Speciation, the phenomenon of a new species arising from an ancestral species, is well documented, but the rate and ways in which it takes place vary. First, though, let us be clear on what we mean by **species,** a biological term for a population of similar individuals that in nature interbreed and produce fertile offspring. Thus, a species is reproductively isolated from other species. This definition does not apply to organisms such as bacteria that reproduce asexually, but it is nevertheless useful for our discussion of plants, animals, fungi, and single-cell organisms called protistans.

Goats and sheep are distinguished by physical characteristics, and they do not interbreed in nature; thus, they are separate species. Yet in captivity they can produce fertile offspring. Lions and tigers can also interbreed in captivity, although they do not interbreed in nature and their offspring are sterile, so they too are separate species. Domestic horses that have gone wild can interbreed with zebras to yield a *zebroid,* which is sterile; thus, horses and zebras are separate species. It should be obvious from these examples that reproductive barriers are not complete in some species, indicating varying degrees of change from a common ancestral species.

The process of speciation involves a change in the genetic makeup of a population, which also may bring about changes in form and structure. Scientists recognize several ways in which new species may arise, but the most common is **allopatric speciation,** which occurs when a new species evolves from a small population that became isolated from its parent population (❚ Figure 18.6). Once isolated, a population no longer exchanges genes with the parent population. Given these conditions and the fact that different selective pressures are likely, a reproductively isolated species may evolve.

❚ Figure 18.6 Allopatric Speciation An example of allopatric speciation on some remote islands.

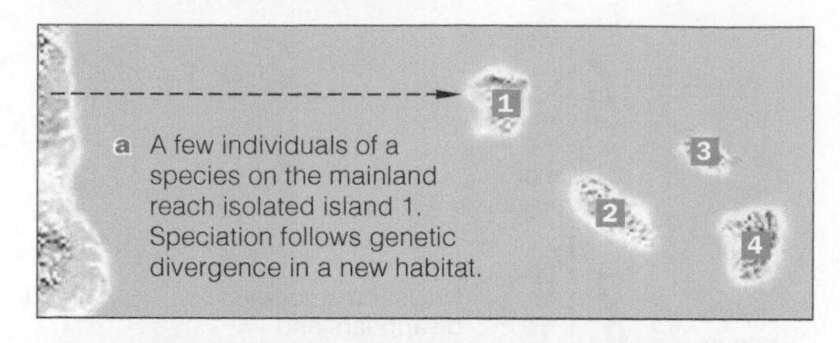

a A few individuals of a species on the mainland reach isolated island 1. Speciation follows genetic divergence in a new habitat.

b Later in time, a few individuals of the new species colonize nearby island 2. In this new habitat, speciation follows genetic divergence.

c Speciation may also follow colonization of islands 3 and 4. And it may follow invasion of island 1 by genetically different descendants of the ancestral species.

Akepa (*Loxous coccineus*)

Palila (*Loxioides bailleui*)

Akohekohe (*Palmeria doli*)

Iiwi (*Vestiaria coccinea*)

d More than 20 species of Hawaiian honeycreepers have evolved from a common ancestor, four of which are shown here, as they adapted to diverse resources on the islands.

According to Darwin and reaffirmed by the modern synthesis, the gradual accumulation of minor changes eventually brings about the origin of a new species, a phenomenon called **phyletic gradualism.** Another view, known as **punctuated equilibrium,** holds that little or no change takes place in a species during most of its existence, and then evolution occurs rapidly, giving rise to a new species in perhaps as little as a few thousands of years.

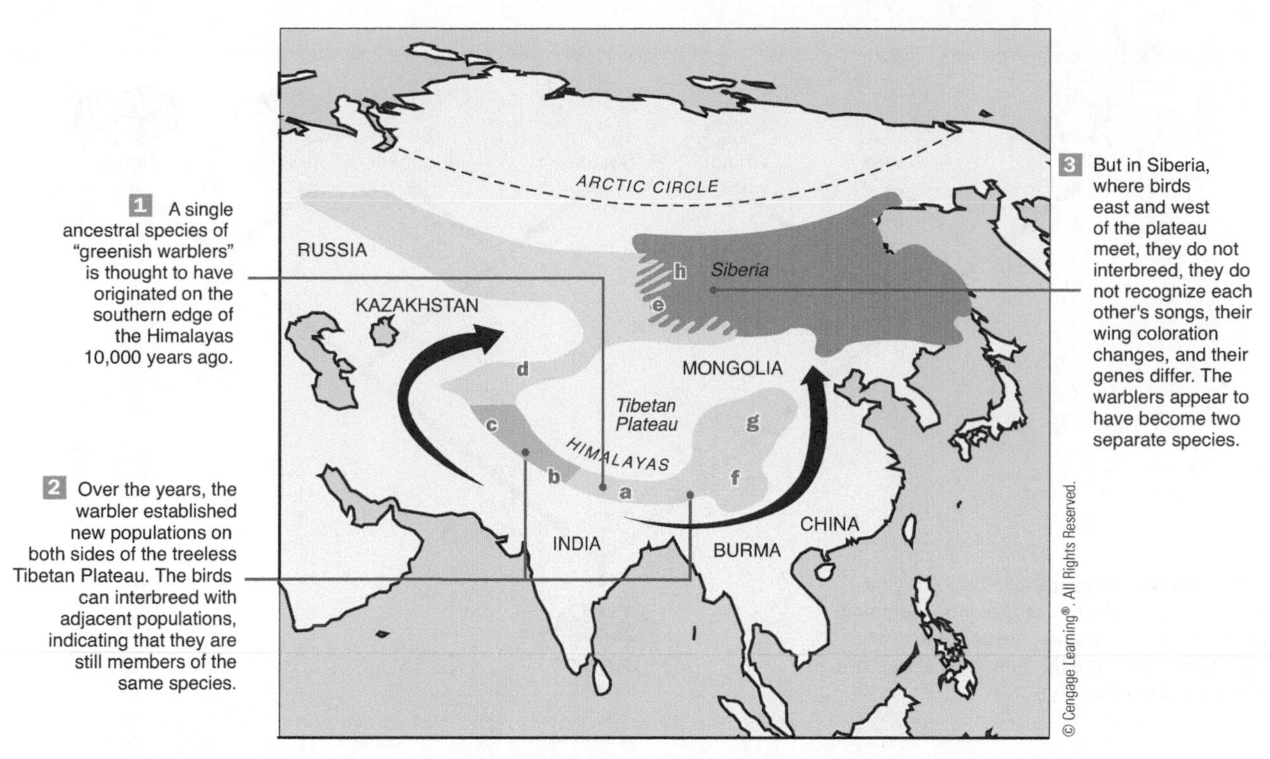

Figure 18.7 Speciation in Songbirds Two species of Eurasian songbirds appear to have evolved from an ancestral species. Adjacent populations of these birds, a and b, or f and g, for instance, can interbreed even though they differ slightly. However, where populations e and h overlap in Siberia, they cannot interbreed.

Proponents of punctuated equilibrium argue that few examples of gradual transitions from one species to another are found in the fossil record. Critics, however, point out that neither Darwin nor those who formulated the modern synthesis insisted that all evolutionary change was gradual and continuous, a view shared by many present-day biologists and paleontologists. Furthermore, deposition of sediments in most environments is not continuous; thus, the lack of gradual transitions in many cases is simply an artifact of the fossil record. And finally, despite the incomplete nature of the fossil record, a number of examples of gradual transitions from ancestral to descendant species are well known.

If speciation occurs as we have described, evidence of it taking place now should be available, and in fact, it is. Speciation, or at least the first steps in speciation, has taken place and continues in populations of mosquitoes, bees, mice, salamanders, fish, and birds that are isolated or partly isolated from one another (Figure 18.7).

Divergent, Convergent, and Parallel Evolution

Divergent evolution occurs when an ancestral species gives rise to diverse descendant species. An impressive example involves the mammals whose diversification from a common ancestor during the Late Mesozoic resulted in such varied animals as platypuses, armadillos, rodents, bats, primates, whales, and rhinoceroses (Figure 18.8).

Divergent evolution leads to descendants that differ markedly from their ancestors. *Convergent evolution* and *parallel evolution* are processes whereby similar adaptations arise in different groups. Unfortunately, they differ in degree and are not always easy to distinguish. **Convergent evolution** involves the development of similar characteristics in distantly related organisms, whereas similar characteristics arising in closely related organisms is **parallel evolution.** In both cases, similar characteristics develop independently because the organisms in question adapt to comparable environments (Figure 18.9).

During much of the Cenozoic, South America was an island continent with a unique fauna of mammals that evolved in isolation. Nevertheless, several mammals in South and North America adapted in similar ways so that they superficially resembled one another (Figure 18.9a)—a good example of convergent evolution. Parallel evolution, in contrast, involves closely related organisms, such as jerboas and kangaroo rats, that independently evolved comparable features (Figure 18.9b).

Microevolution and Macroevolution

Micro and *macro* are prefixes that mean "small" and "large," so **microevolution** is any change in the genetic makeup of a species. For example, house sparrows were introduced into North America in 1852 and have evolved so that members

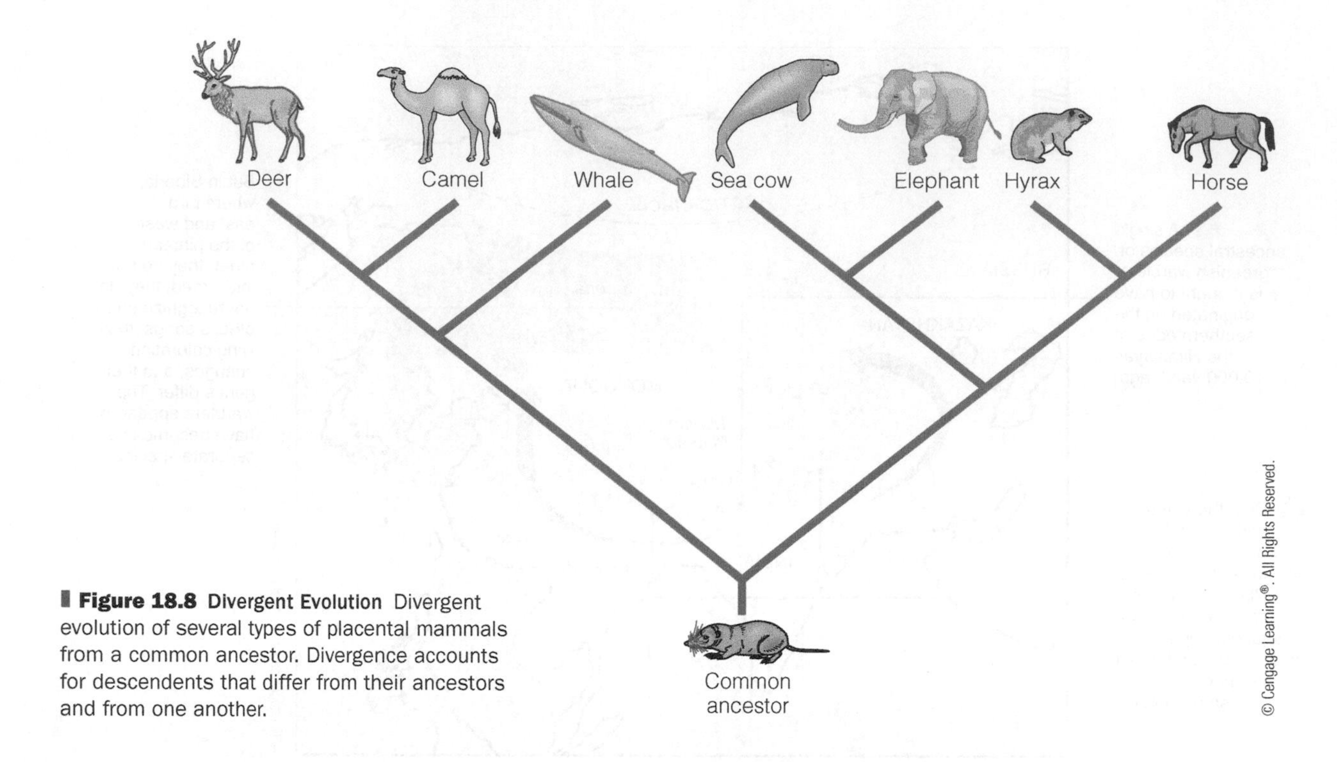

Figure 18.8 Divergent Evolution Divergent evolution of several types of placental mammals from a common ancestor. Divergence accounts for descendents that differ from their ancestors and from one another.

of northern populations are larger than those of southern populations, probably a response to climate. Likewise, organisms that develop resistance to insecticides and pesticides, as well as plants that adapt to contaminated soils, are examples of microevolution.

In contrast, **macroevolution** entails changes such as the origin of a new species or changes at even higher levels in the classification of organisms, such as the origin of new genera, families, orders, and classes. The fossil record provides many good examples of macroevolution—the origin of birds from reptiles, the evolution of whales from land-dwelling ancestors, and many others. Although macroevolution encompasses greater changes than microevolution, the cumulative effects of microevolution are responsible for macroevolution. They differ only in the degree of change.

Mosaic Evolution and Evolutionary Trends

Evolutionary changes do not involve all aspects of an organism simultaneously because selection pressure may be greater on some features than on others. As a result, a key feature we associate with a descendant group may appear before other features typical of that group. For example, the oldest known bird had feathers and the typical fused clavicles of birds, but most of its other features were more like

ConnectionLink

You can find more information on the evolution of birds in Chapter 22.

Figure 18.9 Convergent and Parallel Evolution

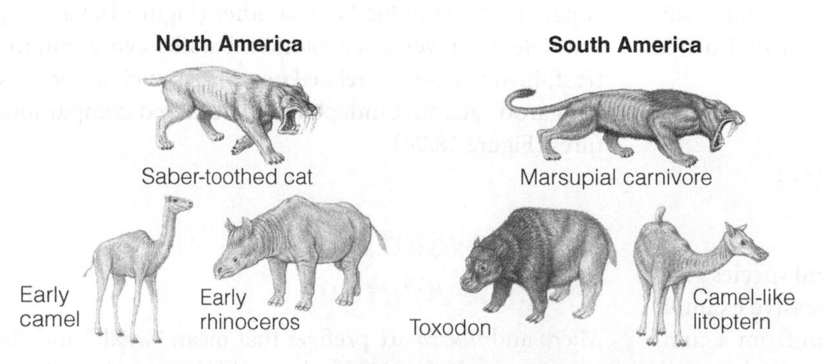

a Convergent evolution takes place when distantly related organisms give rise to species that resemble one another because they adapt in comparable ways.

b Parallel evolution involves the independent origin of similar features in closely related organisms.

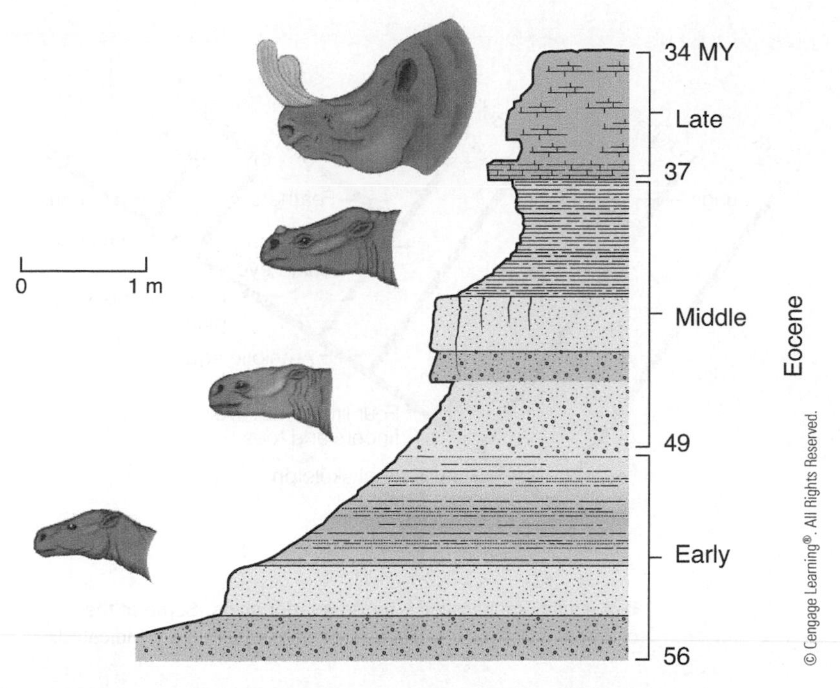

Figure 18.10 Evolutionary Trends in Titanotheres These relatives of horses and rhinoceroses existed for about 20 million years during the Eocene Epoch. During that time, they evolved from small ancestors to giants standing 2.4 m at the shoulder. In addition, they developed large horns, and the shape of the skull changed. Only 4 of the 16 known genera of titanotheres are shown here.

Critical Thinking Question Is this sequence of fossil titanotheres consistent with any of the predictions in Table 18.1? If so, which one(s)?

those of small carnivorous dinosaurs. Accordingly, it represents the concept of **mosaic evolution,** meaning that it had recently evolved characteristics, as well as some features of its ancestral group.

Paleontologists determine in some detail the *phylogeny,* or evolutionary history, and *evolutionary trends* for groups of organisms if sufficient fossil material is available. Size increase is one of the most common evolutionary trends; but trends are extremely complex, they may be reversed, and several trends taking place may not all proceed at the same rate. Abundant fossils show that the Eocene mammals called *titanotheres* not only increased in size but also developed large nasal horns, and the shape of their skull changed (Figure 18.10). The evolution of horses is well documented by fossils that show a general increase in size to living horses, although some extinct species show a size decrease.

Isn't evolution by natural selection a random process? If so, how is it possible for a trend to continue long enough to account, just by chance, for such complex structures as eyes, wings, and hands? Actually, evolution by natural selection is a two-step process, and only the first step involves chance. First, variation must be present or arise in a population. Whether or not variations arising by mutations are favorable is indeed a matter of chance, but the second step involving natural selection is not, because only individuals with favorable variations are most likely to survive and reproduce.

Cladistics and Cladograms

Cladistics is a type of biological analysis in which organisms are grouped together based on derived, as opposed to primitive, characteristics. For instance, all land-dwelling vertebrates have bone and paired limbs, so these characteristics are primitive and of no use in establishing relationships among them. Hair and three middle ear bones, on the other hand, are derived characteristics, sometimes called *evolutionary novelties,* because they served to differentiate mammals from other vertebrate animals. If we consider only mammals, hair and three middle ear bones are of no further use because all mammals have them, so in this context they are primitive. However, live birth, as opposed to laying eggs, is a derived characteristic that serves to distinguish most mammals from the egg-laying mammals.

Traditionally, scientists have depicted evolutionary relationships with *phylogenetic trees,* in which the horizontal axis shows anatomic differences and the vertical axis denotes time (Figure 18.11a). The patterns of ancestor–descendant relationships are based on shared features, but the ones used are rarely specified. In contrast, a **cladogram** resulting from cladistic analysis shows the relationships among members of a *clade,* a group of organisms including its most recent common ancestor (Figure 18.11b). Thus, a cladogram includes organisms with shared derived characteristics.

Any number of organisms can be depicted in a cladogram, but the more shown, the more complex and difficult it is to construct. Let's use an example of three animals—bats, dogs, and birds. Figure 18.12 shows three cladograms, each a different interpretation of their relationship. Bats and birds fly, so we might conclude they are more closely related to each other than to dogs (Figure 18.12a). On the other hand, perhaps birds and dogs are more closely related to each other than either is to bats (Figure 18.12b). But if we concentrate on evolutionary novelties, such as hair and giving birth to live young, we conclude that bats and dogs are more closely related than either is to birds (Figure 18.12c).

Cladistics and cladograms work well for living organisms, but when applied to fossils, care must be taken to determine what are primitive versus derived characteristics, especially in groups with poor fossil records. Furthermore, cladistic analysis depends solely on characteristics inherited from a common ancestor, so paleontologists must be especially careful of characteristics that result from convergent evolution. Nevertheless, cladistics is a powerful tool that has more clearly elucidated the relationships among many fossil lineages and is now used extensively by paleontologists.

Figure 18.11 Phylogenetic Tree and Cladogram

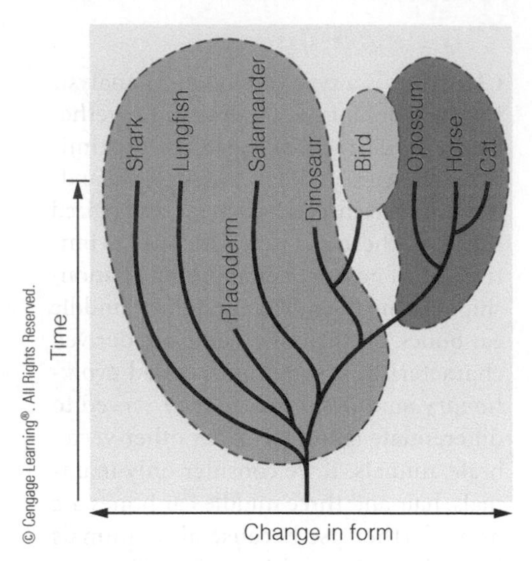

a A phylogenetic tree showing the relationships among various vertebrate animals.

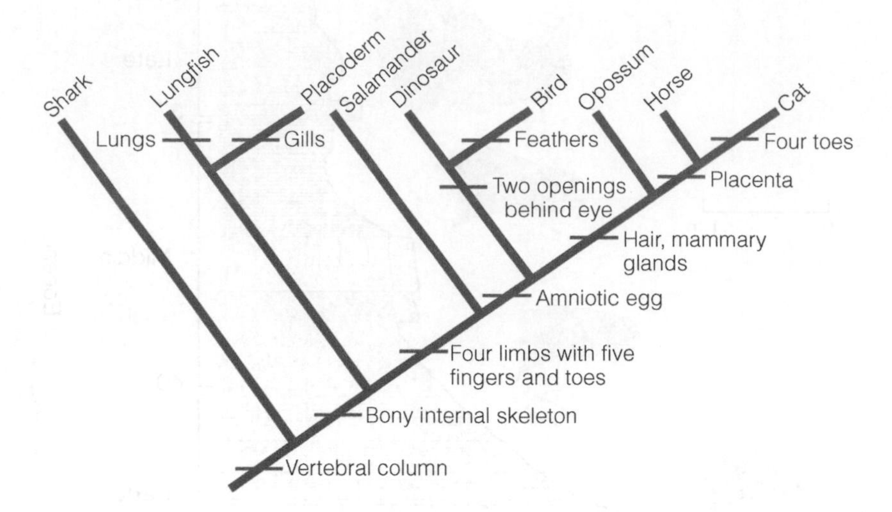

b A cladogram showing inferred relationships. Some of the characteristics used to construct this cladogram are indicated.

Extinctions

Judging from the fossil record, most organisms that ever existed are now extinct—perhaps as many as 99% of all species. Now, if species actually evolve as natural selection favors certain traits, shouldn't organisms be evolving toward some kind of higher order of perfection or greater complexity? Certainly, vertebrates are more complex, at least in overall organization, than are bacteria, but complexity does not necessarily mean that they are superior in some survival sense—after all, bacteria have persisted for at least 3.5 billion years. Actually, natural selection does not yield some kind of perfect organism, but rather those adapted to a specific set of circumstances at a particular time. Thus, a clam or lizard existing now is not somehow superior to those that lived millions of years ago.

The continual extinction of species is referred to as *background extinction,* to clearly differentiate it from a **mass extinction,** during which accelerated extinction rates sharply reduce Earth's biotic diversity. Extinction is a continuous occurrence, but so too is the evolution of new species that usually, but not always, quickly exploit the opportunities created by another species' extinction.

Everyone is familiar with the mass extinction of dinosaurs and other animals at the end of the Mesozoic Era. The greatest extinction, though, was at the end of the Paleozoic Era.

ConnectionLink

You can find more information on dinosaurs and their extinction in Chapter 22.

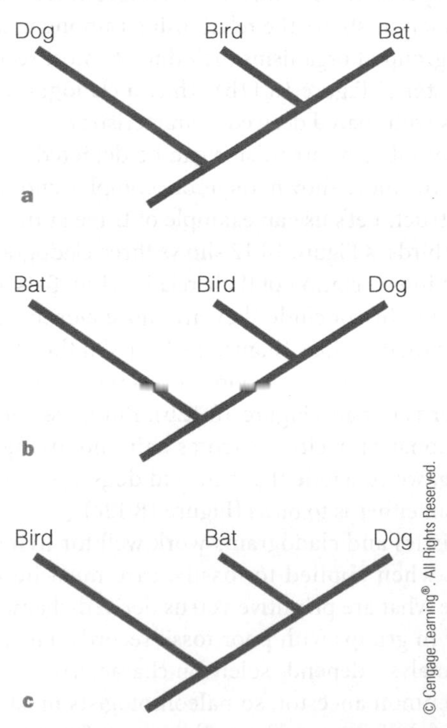

Figure 18.12 Cladograms for Dogs, Birds, and Bats Cladograms showing three hypotheses for the relationships among dogs, birds, and bats. Derived characteristics such as hair and giving birth to live young indicate that dogs and bats are most closely related, as shown in **(c)**.

18.5 Evidence Supporting Evolutionary Theory

When Charles Darwin proposed his theory of evolution, he cited supporting evidence such as classification, embryology, comparative anatomy, geographic distribution, and, to a limited extent, the fossil record. He had little knowledge of the

TABLE 18.1 Some Predictions from the Theory of Evolution

1. If evolution has taken place, the oldest fossil-bearing rocks should have remains of organisms very different from those existing now, and more recent rocks should have fossils more similar to today's organisms.
2. If evolution by natural selection actually occurred, Earth must be very old, perhaps many millions or billions of years old.
3. If today's organisms descended with modification from ones in the past, there should be fossils showing characteristics that connect orders, classes, and so on.
4. If evolution is true, closely related species should be similar not only in details of their anatomy but also in their biochemistry, genetics, and embryonic development, whereas distantly related species should show fewer similarities.
5. If the theory of evolution is correct—that is, living organisms descended from a common ancestor—classification of organisms should show a nested pattern of similarities.
6. If evolution actually took place, we would expect cave-dwelling plants and animals to most closely resemble those immediately outside their respective caves rather than being most similar to those in caves elsewhere.
7. If evolution actually took place, we would expect land-dwelling organisms on oceanic islands to most closely resemble those of nearby continents rather than those on other distant islands.
8. If evolution has taken place, a mechanism should exist that accounts for the evolution of one species to another.
9. If evolution occurred, we would expect mammals to appear in the fossil record long after the appearance of the first fish. Likewise, we would expect reptiles to appear before the first mammals or birds.
10. If we examine the fossil record of presumably related organisms such as horses and rhinoceroses, we should find that they were quite similar when they diverged from a common ancestor but became increasingly different as their divergence continued.

mechanism of inheritance, and both biochemistry and molecular biology were unknown during his time. Studies in these areas, coupled with a more complete and much better understood fossil record, have convinced scientists that the theory is as well supported by evidence as any other major theory. Of course, scientists disagree on many details, but the central claims of the theory are well established and widely accepted.

But is the theory of evolution truly scientific? That is, can testable predictive statements be made from it? First, we must be clear on what a theory is and what we mean by "predictive." Scientists propose hypotheses to explain natural phenomena, test them, and, in some cases, raise them to the status of a theory—an explanation of some natural phenomenon well supported by evidence from experiments, observations, or both.

Almost everything in the sciences has some kind of theoretical underpinning—optics, the nature of matter, inheritance, the present distribution of continents, diversity in the organic world, and so on. Of course, no theory is ever proven in some final sense, although it might be supported by substantial evidence; all are always open to question, to revision, and occasionally to replacement by a more comprehensive theory. In his book *Why Darwin Matters*, Michael Shermer noted that

> a "theory" is not just someone's opinion or a wild guess made by a scientist. A theory is a well-supported and well-tested generalization that explains a set of observations. Science without theory is useless. (pp. 1–2)

Prediction is commonly taken to mean "to foresee an event that has not yet occurred," as in predicting the next solar eclipse. However, not all predictions are about future events. For example, one prediction of seafloor spreading theory is that oceanic crust should be younger at spreading ridges and become progressively older with increasing distance from ridges, which, in fact, it does. Likewise, the theory of evolution allows us to make many predictions about what we should see in the present-day world and in the fossil record if the theory is correct (Table 18.1).

If the theory of evolution is correct, closely related species such as wolves and coyotes should be similar not only in their overall anatomy but also in terms of their biochemistry, genetics, and embryonic development (point 4 in Table 18.1). Suppose they differed in their biochemical mechanisms as well as embryology. Obviously, our prediction would fail, and we would at least have to modify the theory. If other predictions also failed—for example, mammals appeared in the fossil record before the first fishes—we would have to abandon the theory and find a better explanation for our observations. Accordingly, the theory of evolution is truly scientific because it can, at least in principle, "be falsified"—that is, proven wrong.

Classification—A Nested Pattern of Similarities

Carolus Linnaeus (1707–1778) proposed a classification system in which organisms are given a two-part genus and species name; the coyote, for instance, is *Canis latrans*. Table 18.2 shows Linnaeus's classification, which is a hierarchy of categories that becomes more inclusive as one proceeds up the list. The coyote (*Canis latrans*) and the wolf (*Canis lupus*) share numerous characteristics, so they are members of the same genus, whereas both share some, but fewer, characteristics with the red fox (*Vulpes fulva*), and all three are members of the family Canidae. All canids share some characteristics with cats, bears, and weasels and are grouped together in the order Carnivora, which is 1 of 18 living orders of the class Mammalia, all of whom are warm-blooded, possess fur or hair, and have mammary glands.

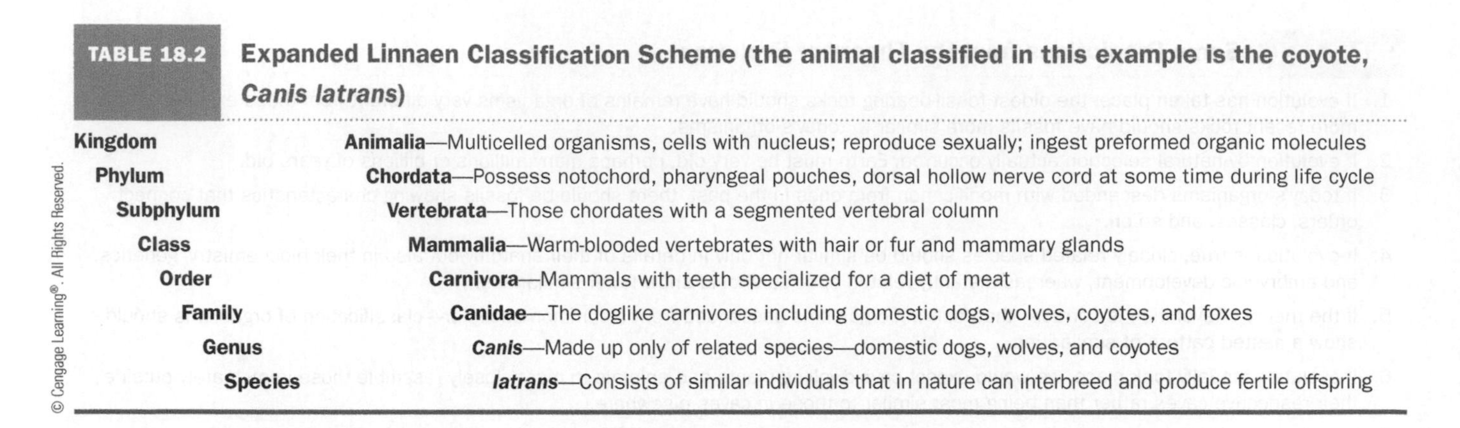

TABLE 18.2 Expanded Linnaen Classification Scheme (the animal classified in this example is the coyote, *Canis latrans*)

Kingdom	**Animalia**—Multicelled organisms, cells with nucleus; reproduce sexually; ingest preformed organic molecules
Phylum	**Chordata**—Possess notochord, pharyngeal pouches, dorsal hollow nerve cord at some time during life cycle
Subphylum	**Vertebrata**—Those chordates with a segmented vertebral column
Class	**Mammalia**—Warm-blooded vertebrates with hair or fur and mammary glands
Order	**Carnivora**—Mammals with teeth specialized for a diet of meat
Family	**Canidae**—The doglike carnivores including domestic dogs, wolves, coyotes, and foxes
Genus	***Canis***—Made up only of related species—domestic dogs, wolves, and coyotes
Species	***latrans***—Consists of similar individuals that in nature can interbreed and produce fertile offspring

Linnaeus certainly recognized shared characteristics among organisms, but his intent was simply to categorize species that he thought were specially created and immutable. Following the publication of Darwin's *On the Origin of Species* in 1859, however, scientists quickly realized that shared characteristics constituted a strong argument for evolution.

Linnaeus's classification communicates information about the biological world, but it is arbitrary when assigning organisms to categories above the species level. Also, it does not in all cases reflect evolutionary relationships. As a consequence, scientists now more commonly use evolutionary relationships, or phylogeny, to group organisms. Recall from our discussion of cladistics that a clade is a group of organisms including its most recent common ancestor (Figure 18.11b). Accordingly, a well-constructed cladogram shows a nested pattern of similarities (❚ Figure 18.13).

❚ **Figure 18.13 Cladograms and Shared Characteristics** This example shows the nested pattern of similarities among vertebrate animals—that is, those animals with a segmented vertebral column.

Biological Evidence Supporting Evolution

Life is incredibly diverse—scientists have described about 1.7 million species and estimate that there may be millions not yet recognized. And yet, if existing organisms evolved from common ancestors, there should be similarities among all life-forms. As a matter of fact, all living things—be they bacteria, redwood trees, or whales—are composed mostly of carbon, nitrogen, hydrogen, and oxygen. Furthermore, their chromosomes consist of DNA, and all cells synthesize proteins in essentially the same way.

Evidence from biochemistry includes blood proteins that are similar among all mammals, but also indicates that humans are most closely related to great apes, followed in order by Old World monkeys, New World monkeys, and other primates such as lemurs. Biochemical tests support the idea that birds descended from reptiles, a conclusion supported by comparative anatomy and evidence in the fossil record.

The forelimbs of humans, whales, dogs, and birds are superficially dissimilar. Yet all are made up of the same bones; all have basically the same arrangement of muscles, nerves, and blood vessels; all are similarly arranged with respect to other structures; and all have a similar pattern of embryonic development (❚ Figure 18.14). These **homologous structures,** as they are called, are similar even though they may serve different functions—flying versus running, for example—but all were derived from a common ancestor and therefore indicate evolutionary relationships. However, there are some similarities that do not indicate descent from a common ancestor. **Analogous structures** may be superficially similar and serve the same function, as in the wings of insects and birds, but they are dissimilar in structure and development (❚ Figure 18.15).

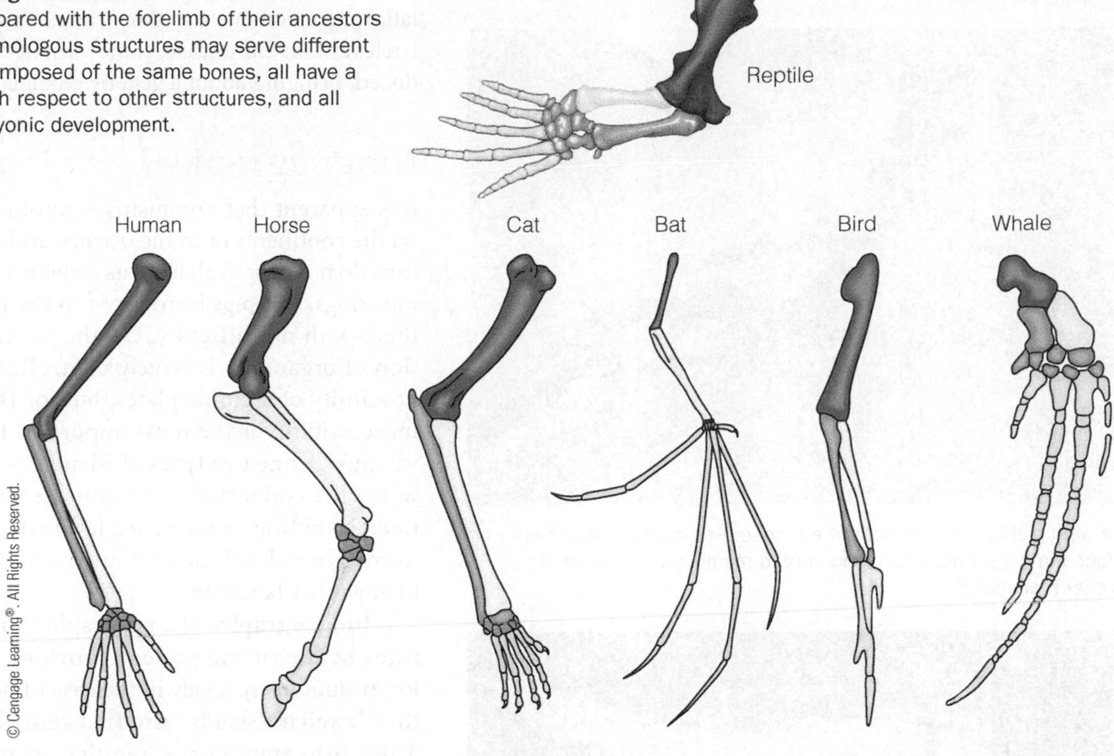

Figure 18.14 **Homologous Structures** The forelimbs of several mammals, compared with the forelimb of their ancestors among the reptiles. Homologous structures may serve different functions, but all are composed of the same bones, all have a similar arrangement with respect to other structures, and all undergo a similar embryonic development.

Why do dogs have tiny, functionless toes called dewclaws? Dewclaws in dogs, and many other mammals, are examples of **vestigial structures**, features that were fully functional in their ancestors but now only partly functional, nonfunctional, or serve a different function (Figure 18.16). The ancestors of dogs had five toes on each foot, all of which contacted the ground, but as they evolved, they became toe walkers. As a result, only four digits contacted the ground, and the big toes and thumbs were lost or reduced to their present state. Snakes and whales have the remnants of a pelvis but no rear limbs, and the wisdom teeth of humans cause problems for most people that have them (some people never develop wisdom teeth) because the jaw is too short to accommodate them. The incus and malleus of the mammalian middle ear were derived from the articular and quadrate bones that formed the jaw–skull joint in mammal-like reptiles, so they are functional but perform a different function.

As opposed to vestigial structures, an **atavism** is an anomaly that looks like the reappearance of an ancestral trait. Rather than being common, as dewclaws in dogs or wisdom teeth in humans, they are found only rarely. Good examples include the rare whale with hind limbs much

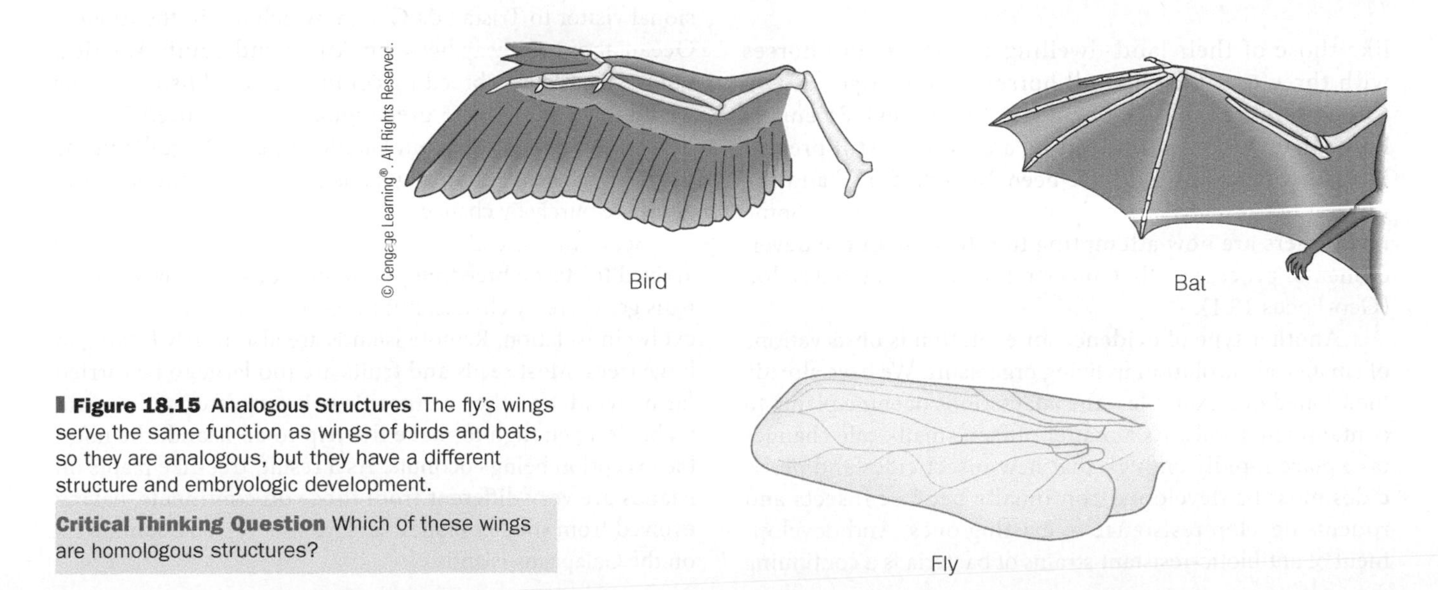

Figure 18.15 **Analogous Structures** The fly's wings serve the same function as wings of birds and bats, so they are analogous, but they have a different structure and embryologic development.

Critical Thinking Question Which of these wings are homologous structures?

Figure 18.16 Vestigial Structures

a Vestigial toes on the foot of a moose. Remnants of toes are found in dogs, cats, even-toed hoofed mammals, and several other mammals.

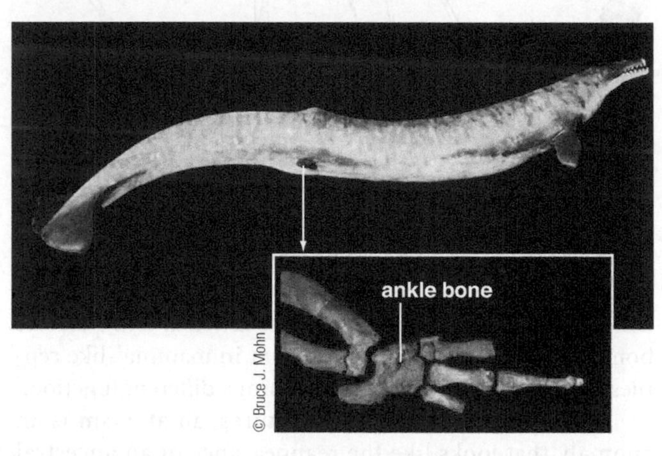

b The Eocene-age whale *Basilosaurus* had tiny vestigial back limbs, but it did not use limbs to support its body weight. Even today, whales have a vestige of a pelvis, and on a few occasions, whales with rear limbs have been caught.

like those of their land-dwelling ancestors, and horses with three toes (actually all horses have vestiges of two toes but only rarely are born with three toes). Scientists know that the genes controlling atavism are still present in many organisms but have been "switched off" and the features they control are only infrequently expressed. Some researchers are now attempting to intervene in the developmental process so that ancestral features may redevelop (Geo-Focus 18.1).

Another type of evidence for evolution is observations of small-scale evolution in living organisms. We have already mentioned one example—the adaptations of some plants to contaminated soils. As a matter of fact, small-scale changes take place rapidly enough that new insecticides and pesticides must be developed continually because insects and rodents develop resistance to existing ones. And development of antibiotic-resistant strains of bacteria is a continuing problem in medicine. Whether the variation in these populations previously existed or was established by mutations is irrelevant. In either case, some variant types lived and reproduced, bringing about a genetic change.

Biogeography

It is apparent that organisms are not randomly distributed on the continents or in the oceans, and it is a fact that organisms do not exist in all habitats suitable for them. For example, rats, dogs, and pigs introduced to the Hawaiian Islands live there with no difficulty. On the continents, the distribution of organisms is largely controlled by climate and the proximity of tectonic plates, but for islands in the oceans, inaccessibility is the most important factor. First, though, we must define two types of islands: *continental islands* that were once connected to a continent but have become separated by rifting or rising sea level; and *oceanic islands* that formed by volcanic activity and have never been connected to any other landmass.

Biogeography, the geographic distribution of organisms, both past and present, provides compelling evidence for evolution, especially in the case of oceanic islands. When the Hawaiian Islands were first settled, they had no freshwater fish, amphibians, reptiles, or mammals other than bats and monk seals. Likewise, the Galápagos Islands had no freshwater fish, amphibians, or mammals. In fact, oceanic islands everywhere have peculiar biotas that in almost all details differ from, but most closely resemble, those of the nearest continent. In addition, the more remote an oceanic island, the less diverse its biota, and in a chain of islands, the flora and fauna of each island differs slightly from nearby islands.

It should come as no surprise that most organisms that colonize these remote islands are the ones that are most mobile and capable of doing so. Present-day land-dwelling birds are sometimes seen far out over the oceans. Many probably perish, but it takes only a few to reach an island and start a breeding colony. The purple gallinule was an occasional visitor to Tristan da Chuna, which lies in the Atlantic Ocean about midway between Africa and South America, but in the 1950s, a breeding colony was established on the island. And in 1995, 15 green iguanas were rafted 280 km on floating vegetation from Guadeloupe to Anguilla in the Caribbean Sea. Of course iguanas are not very mobile; they got there purely by chance.

So, in the case of oceanic islands, a few hardy migrants arrived to start a breeding population or even fewer organisms get there by chance, but in either case, once there they evolve in isolation. Remote islands are also nearly lacking in large trees. Most seeds and fruits are too large to be carried far by wind, and they cannot float in seawater and remain viable long enough to make the trip to an island, the notable exception being coconuts. As a result, tree-like plants on islands are very different from those on continents, having evolved from smaller plants, such as the tree-size sunflowers on the Galápagos Islands.

GEO FOCUS 18.1 Building a Dinochicken

In the text we defined a *vestigial structure* as some attribute of an organism that is no longer functional, only partially functional, or performs a different function than it did in the ancestors of the organism in question. The popular misconception is that to be vestigial a structure must be functionless, but even Darwin recognized that this is not necessarily true. In contrast, an *atavism* is the rare reappearance of some ancestral feature that has been lost by most members of a population. For instance, horses may be born with and develop into adults with three toes similar to those of their distance ancestors, but that are not present in their parents.

So what does this all have to do with building a dinochicken, or, as some call it, chicasaurus? First, let us be clear that no one is claiming that any amount of developmental modification of a chicken will produce a dinosaur, only that it is possible to modify the developmental pathways in chickens so that they redevelop ancestral features such as teeth, a three-clawed hand, and a long tail. In short, chickens, and all other birds, descended from dinosaurs, and as descendants they retain genes that if reactivated may produce these features. The process, called *atavism reactivation*, is being investigated by Jack Horner at The Museum of the Rockies in Bozeman, Montana, and his colleagues.

All of the features in question—teeth, three-clawed hand, and long tail—are present in the developing embryos of chickens, but the genes controlling these features have been switched off so that they do not develop further. In fact, chicken embryos have been induced to develop teeth. Horner and his colleagues are attempting to reactivate these genes; in short, they are not trying to change the basic heredity (the DNA) of chickens, but rather to intervene in the developmental process. If successful, the dinochicken will still be a chicken, although certainly a modified one, and if it were to breed with other chickens, the offspring would be normal chickens, with no teeth, a bird's wing, and a short tail.

The fact that developing embryos provide evidence for descent with modification has been known for more than a century. Common descent accounts for the remarkable similarities in embryos of fish, amphibians, reptiles, birds, and mammals during their earliest development, but as development progresses, they differ more and more from one another. If you compare the embryos of chickens and the adult skeletons of pigeons and small theropod (meat-eating) dinosaurs, the similarities in the arm–wing are obvious as are the long tails in each of these animals.

Should the attempt to produce a dinochicken be successful it would not mean that dinosaurs have been reintroduced into the world. In the words of Horner, "We would have brought back some of the characteristics of the dinosaurs. We would have used the signature left by evolution in the chicken's DNA to rewind evolution. But we would have re-created ancestral traits, not the ancestor itself."*

*Horner, Jack and James Gorman, *How to Build a Dinosaur* (New York, NY: A Plume Book, 2009), p. 193.

Continental islands became separated by moving plates (Madagascar) or by rising sea level (the British Isles). When these islands became separated, they already had a biota much like that of the continent from which they became separated. However, the longer a continental island has been separated, the more its biota differs from that of its nearby landmass. Madagascar separated from Africa 160 million years ago and consequently has many species that do not exist in Africa. The British Isles, however, were isolated from mainland Europe by rising sea level only about 14,000 years ago and retains a biota much like that of mainland Europe.

Fossils: What Do We Learn from Them?

In Chapter 7 we defined **fossils** as the remains or traces of organisms preserved in rocks, and we also noted that they are useful for determining environments of deposition and for relative age determinations (see Geo-Insight 7.1). Geologists use the term **body fossil** for remains, mostly shells, bones, and teeth, and rarely mummies and frozen animals, whereas a **trace fossil** is any indication of organic activity such as burrows and footprints (see Geo-Insight 18.1). The fossil record is an important repository of life history; in fact, if it were not for fossils, we would have no idea that trilobites and dinosaurs ever existed. In addition, fossils provide some of the evidence for organic evolution.

Some fossils are quite common, especially those of marine-dwelling organisms that have hard skeletons such as corals and clams. After all, so many billions of clams have existed through hundreds of millions of years that if only a tiny fraction of them were preserved as fossils the total number is enormous. On the other hand, the fossil records for worms, jellyfish, and bats are not nearly as good because either the organisms in question did not live where burial in sediment was likely, or they had very delicate skeletons or no skeleton at all.

Fossil marine invertebrates found far from the sea, and even high in mountains, led early naturalists to conclude that the fossil-bearing rocks were deposited during a worldwide flood. In 1508, Leonardo da Vinci realized that the fossil distribution was not what one would expect from a rising flood, but the flood explanation persisted, and John Woodward (1665–1728) proposed a testable hypothesis. According to him, the size, shape and density of organic remains determined the order in which they settled from floodwaters. Woodward's hypothesis was quickly rejected because observations did not support it; fossils of various sizes, shapes, and densities are found throughout the fossil record.

The fossil record does show a sequence of different organisms, but not one based on density, size, shape, mobility, or habitat. Rather, the sequence consists of first appearances of groups of organisms through time. One-celled organisms appeared before multicelled ones, plants before animals, and invertebrates before vertebrates. Among vertebrates, fish appeared first, followed in succession by amphibians, reptiles, mammals, and birds.

If fossils provide evidence for evolution, we might ask, where are the fossils, the so-called missing links, that connect descendants with ancestors? Paleontologists call these "missing links" *transitional fossils* to emphasize the fact that they show characteristics of both ancestors and descendants, and there are, in fact, many of them. In the chapters on Earth and life history, we will point out several such as those between amphibians and reptiles, reptiles and birds, and reptiles and mammals. Excellent examples are the Jurassic-age fossils from Germany that have anatomic features much like those of small, carnivorous dinosaurs, and yet they have feathers and the fused clavicles (wishbone) so typical of birds.

To learn more about amphibian, reptile, mammal, and bird evolution, see Chapters 22 and 23.

Evolution, Theories, and Fossils

Suppose someone told you that evolution is "only a theory that has never been proven" and that "the fossil record shows a sequence of organisms in older to younger rocks that was determined by their density and habitat." Why is the first statement irrelevant to theories in general? What kinds of evidence could you cite to refute the second statement?

Horses, rhinoceroses, and tapirs may seem an odd assortment of mammals, but fossils and studies of living animals clearly show they evolved from a common ancestor (Figure 18.17). If this assessment is correct, we can predict that as we trace these animals back in the fossil record, telling one from the other should become increasingly difficult. And, in fact, the earliest members of each group are remarkably similar, differing mostly in size and minor details of their teeth. As diversification proceeded, though, the differences became more and more apparent.

Of course we will never have enough fossils to document the evolutionary histories of all living organisms,

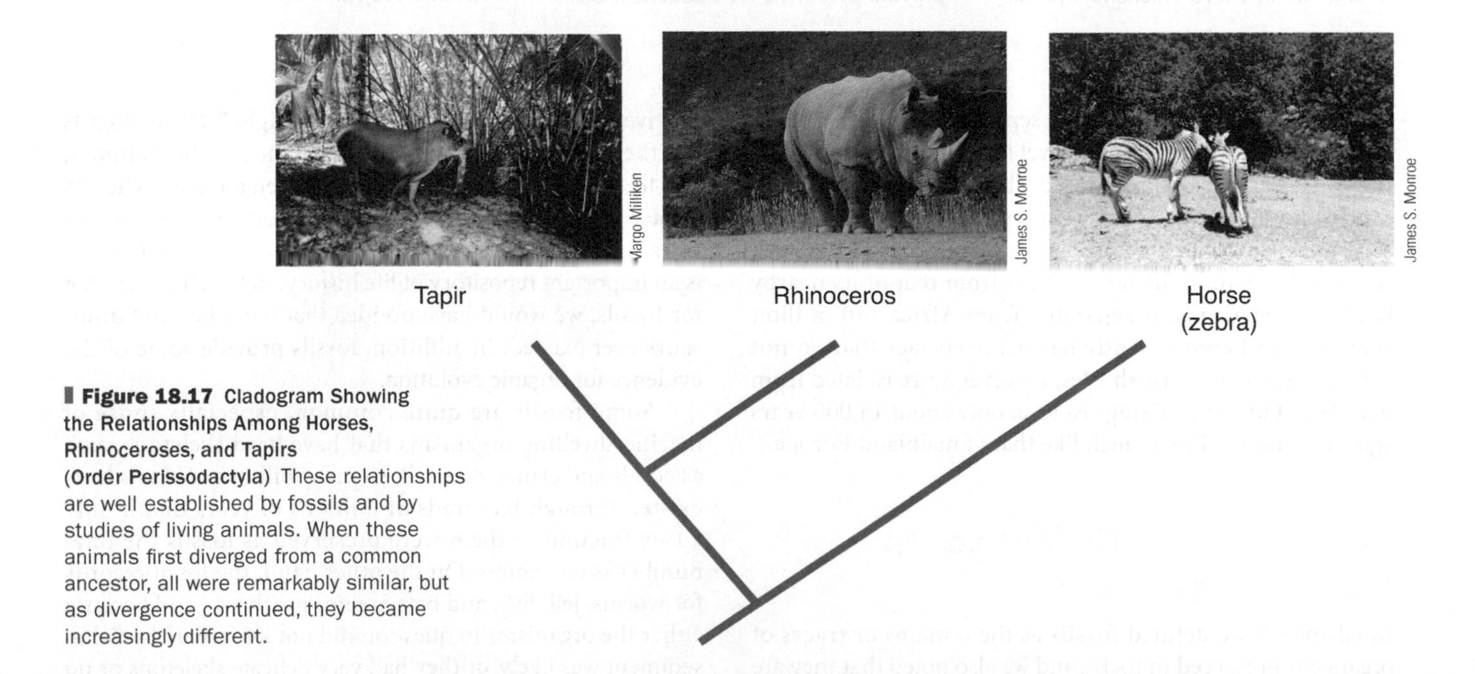

Figure 18.17 Cladogram Showing the Relationships Among Horses, Rhinoceroses, and Tapirs (Order Perissodactyla) These relationships are well established by fossils and by studies of living animals. When these animals first diverged from a common ancestor, all were remarkably similar, but as divergence continued, they became increasingly different.

because the remains of some are more likely to be preserved than others, and the accumulation of sediment varies in time and space. Nevertheless, the histories of many groups are well known, and more and more fossils are being found that fill in some of the gaps in our knowledge.

The Evidence—A Summary

Scientists are convinced that evolutionary theory is as well supported by evidence as any major theory. Indeed, studies as diverse as molecular biology and paleontology all confirm evolutionary theory—that is, organisms living today descended with modification from ancestors that lived during the past. Certainly, scientists disagree on specific issues such as rates of evolution, the significance of some fossils, and precise relationships among evolutionary lineages, but overall they are overwhelmingly united in their support for the theory. And despite stories holding that scientists are unyieldingly committed to this idea and unwilling to investigate alternative explanations for the same data, just the opposite is true. If another scientific (testable) hypothesis offered a more fruitful approach, scientists would be scrambling to investigate it. There is no better way in the sciences to gain respect and lasting recognition than to modify or replace an existing widely accepted theory.

Key Concepts Review

- Jean-Baptiste de Lamarck's proposal of inheritance of acquired characteristics was the first formal explanation of evolution to be taken seriously.
- In 1859, Charles Robert Darwin and Alfred Russel Wallace published their views on evolution and proposed natural selection as the mechanism for evolutionary change.
- Darwin's observations of variation in natural populations and artificial selection, as well as his reading of Thomas Malthus's essay on population, helped him formulate the idea that natural processes select favorable variants for survival.
- Gregor Mendel's breeding experiments with garden peas provided some of the answers regarding how variation is maintained and passed on.
- Genes are the hereditary determinants in all organisms. This genetic information is carried in the chromosomes of cells, but only the genes in the chromosomes of sex cells are inheritable.
- Sexual reproduction and mutations account for most variation in populations.
- Evolution by natural selection is a two-step process. First, variation must be produced and maintained in interbreeding populations, and second, favorable variants must be selected for survival.
- An important way in which new species evolve is by allopatric speciation. When a group is isolated from its parent population, gene flow is restricted or eliminated, and the isolated group is subjected to different selection pressures.
- Divergent evolution involves an ancestral stock giving rise to diverse species. The development of similar adaptive types in different groups of organisms results from parallel and convergent evolution.
- Microevolution involves changes within a species, whereas macroevolution encompasses all changes above the species level. Macroevolution is simply the outcome of microevolution over time.
- Scientists are increasingly using cladistic analyses to determine relationships among organisms.
- Extinctions take place continually, and times of past mass extinctions that resulted in marked decreases in Earth's biologic diversity have occurred several times.
- The theory of evolution is truly scientific because we can think of observations and experiments that could support it or falsify it.
- Much of the evidence supporting the theory of evolution comes from classification, embryology, genetics, biochemistry, molecular biology, and present-day small-scale evolution.
- The fossil record also provides evidence for evolution in that it shows a sequence of different groups appearing through time, and some fossils show features that we would expect in the ancestors of birds or mammals, and so on.

Important Terms

allele (p. 451)
allopatric speciation (p. 456)
analogous structure (p. 462)
artificial selection (p. 449)
atavism (p. 463)
biogeography (p. 464)
body fossil (p. 465)
chromosome (p. 454)
cladistics (p. 459)
cladogram (p. 459)
convergent evolution (p. 457)
deoxyribonucleic acid (DNA) (p. 454)
divergent evolution (p. 457)
fossil (p. 465)
gene (p. 451)
homologous structure (p. 462)
inheritance of acquired characteristics (p. 449)
macroevolution (p. 458)
mass extinction (p. 460)
meiosis (p. 454)
microevolution (p. 457)
mitosis (p. 454)
modern synthesis (p. 454)
mosaic evolution (p. 459)
mutation (p. 455)
natural selection (p. 450)
paleontology (p. 448)
parallel evolution (p. 457)
phyletic gradualism (p. 456)
punctuated equilibrium (p. 456)
species (p. 456)
theory of evolution (p. 448)
trace fossil (p. 465)
vestigial structure (p. 463)

Review Questions

1. Many mammals possess ________ that are no long . functional or only partly functional.
   a. _____ analogous features.
   b. _____ cladistic anomalies.
   c. _____ phylogenetic remnants.
   d. _____ fortuitous organs.
   e. _____ vestigial structures.
2. The type of cell division that yields sperm and eggs is
   a. _____ allopatric.
   b. _____ conjunction.
   c. _____ mosaic evolution.
   d. _____ meiosis.
   e. _____ microevolution.
3. Structures in dogs, bats, and deer that are made up of the same bones, have the same relationship to other structures, and develop in a similar embryonic manner are called
   a. _____ homologous structures.
   b. _____ ambiguous structures.
   c. _____ convergent structures.
   d. _____ cladistic structures.
   e. _____ embryonic anomalies.
4. One prediction of evolutionary theory is that
   a. _____ humans evolved from monkeys.
   b. _____ features acquired during an animal's lifetime are inheritable.
   c. _____ the size, shape, and density of fossils determines where they will be found in the fossil record.
   d. _____ species are fixed and immutable.
   e. _____ mammals should appear in the fossil record long after the first fish.
5. The idea that a species evolves continuously by small changes is
   a. _____ punctuated equilibrium.
   b. _____ phyletic gradualism.
   c. _____ parallel evolution.
   d. _____ chromosome accumulation.
   e. _____ artificial selection.
6. What is/are the difference(s) between inheritance of acquired characteristics and natural selection?
7. Explain what is meant by macroevolution and give two examples from the fossil record.
8. Give three examples of predictions from evolutionary theory.
9. Does natural selection really mean only the biggest, strongest, and fastest will survive and reproduce?

10. **Creative Thinking Visual Question:** This cladogram (▮ Figure 1) shows that among carnivorous mammals, cats, hyenas, and mongooses are more closely related to one another than to other carnivores. What kinds of evidence from living animals and the fossil record would lend credence to the conclusion that these three animals constitute a clade?

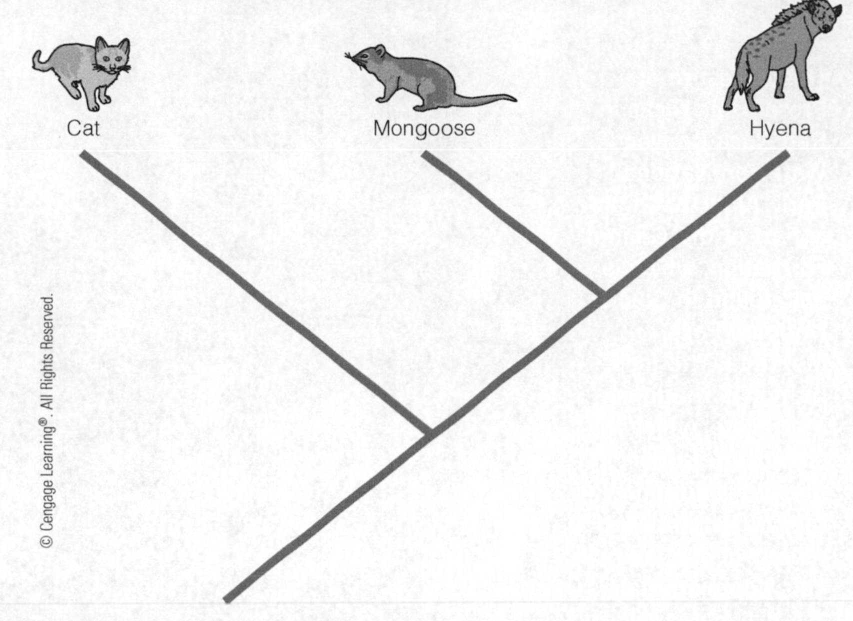

▮ **Figure 1** A cladogram showing the inferred relationships among three mammals.

Global GeoScience Watch Search for "whales" in the GREENR database, and then within this portal search for "evolution." Next to the "Magazines" heading, click on "VIEW ALL," then scroll to find the article titled "Evolution of whales." Whales are mammals, but to which mammals are they most closely related? Where have fossils been found that tell the story of whale evolution? What kinds of evidence, other than fossils, provide information on the relationship of whales to other mammals?

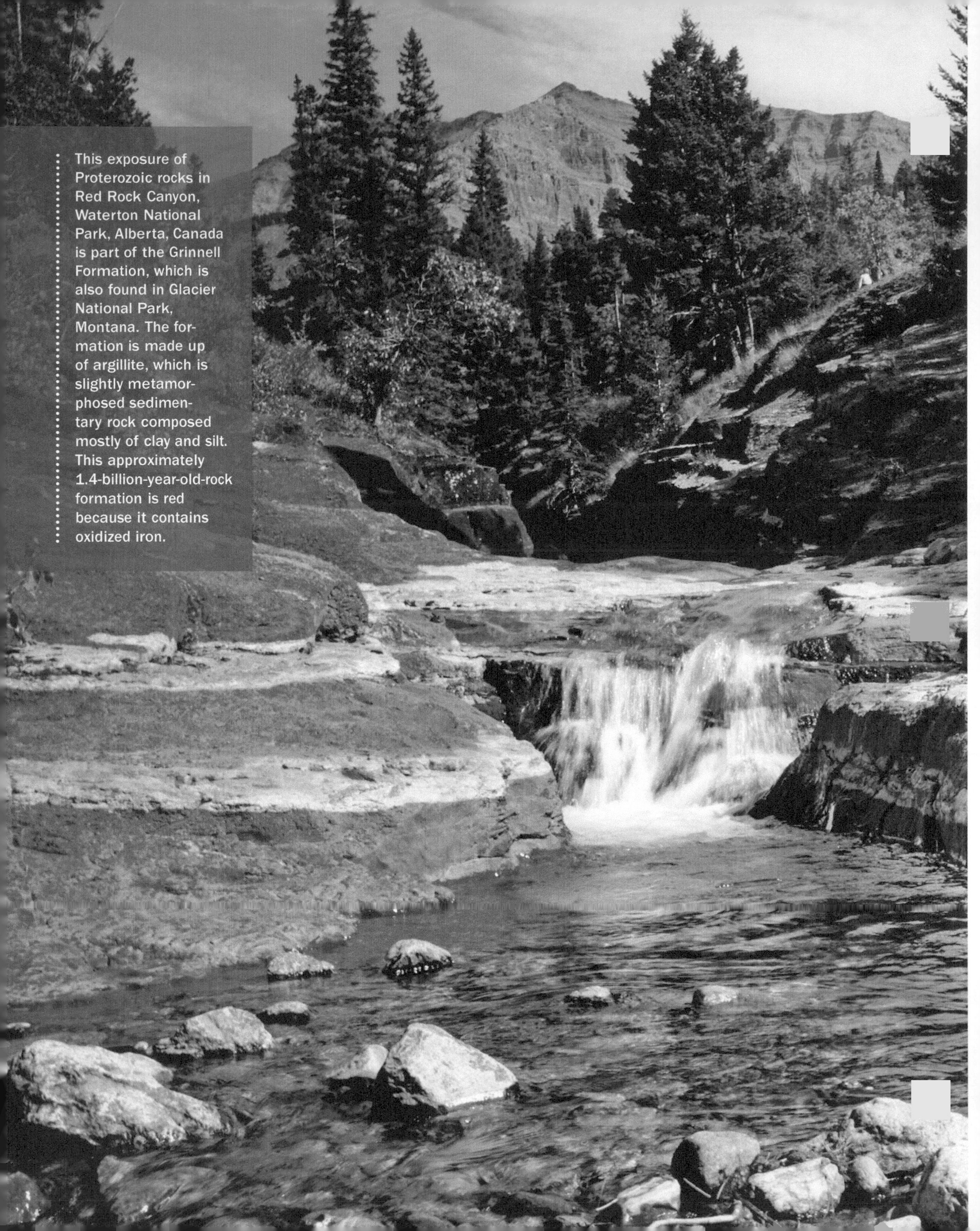

This exposure of Proterozoic rocks in Red Rock Canyon, Waterton National Park, Alberta, Canada is part of the Grinnell Formation, which is also found in Glacier National Park, Montana. The formation is made up of argillite, which is slightly metamorphosed sedimentary rock composed mostly of clay and silt. This approximately 1.4-billion-year-old-rock formation is red because it contains oxidized iron.

BGSmith/Shutterstock.com

CHAPTER 19

Precambrian Earth and Life History

OUTLINE

HAVE YOU EVER WONDERED?

- When and how Earth formed?
- What Earth was like during its earliest history?
- How old the oldest known rocks are?
- How the continents formed and evolved?
- Whether Earth's atmosphere has always been like it is now?
- Where the oceans came from and whether they are getting saltier with time?
- What kinds of organisms existed on early Earth?
- Whether Earth's most ancient rocks contain any important resources?

19.1 Introduction

The oldest known written record dates back to about 3,500 years ago, a mere moment in the time that has transpired since Earth's origin 4.6 billion years ago. We all understand the concept of time from the human perspective of days, months, and years, and we are aware that we use time to specify the duration of events and to place events in a chronological order. But geologic time, or *deep time*, as some call it, is beyond our comprehension. Perhaps you can more fully appreciate the magnitude of geologic time if you suppose that 1 second equals 1 year and you count out Earth's history. Should you take on this task, you and your descendants will be counting for nearly 146 years.

In this chapter we are concerned only with that part of geologic time designated *Precambrian,* 4.6 billion to 542 million years ago. If all geologic time were represented by a 24-hour clock, the Precambrian alone would be more than 21 hours long and constitute 88% of all geologic time (▌Figure 19.1). And yet we discuss this incredibly long interval in a single chapter. Perhaps this treatment seems disproportionate, but we know much more about the Phanerozoic geologic record, which is the more recent 542 million years of geologic time; therefore, we discuss it in more depth.

Precambrian is a widely used term that refers to both time and rocks. As a time term it encompasses all geologic time from Earth's origin until the onset of the Phanerozoic Eon (▌Figure 19.2). The term also refers to all rocks lying below Cambrian-age rocks, hence the name *Precambrian.* Unfortunately, no rocks are known for the first 600 million years of geologic time, so our geologic record begins

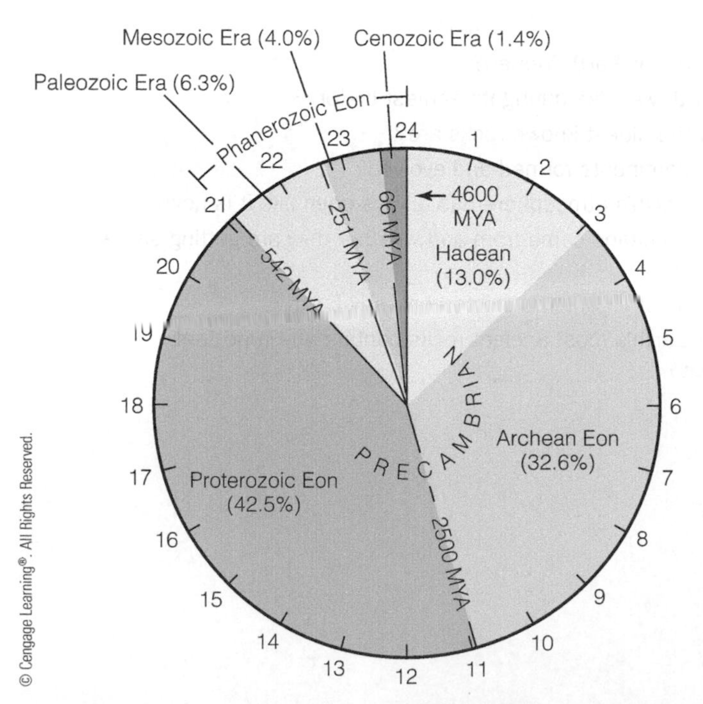

▌**Figure 19.1 Geologic Time Represented by a 24-Hour Clock** If 24 hours represented all geologic time, the Precambrian would be more than 21 hours long, thus more than 88% of the total.

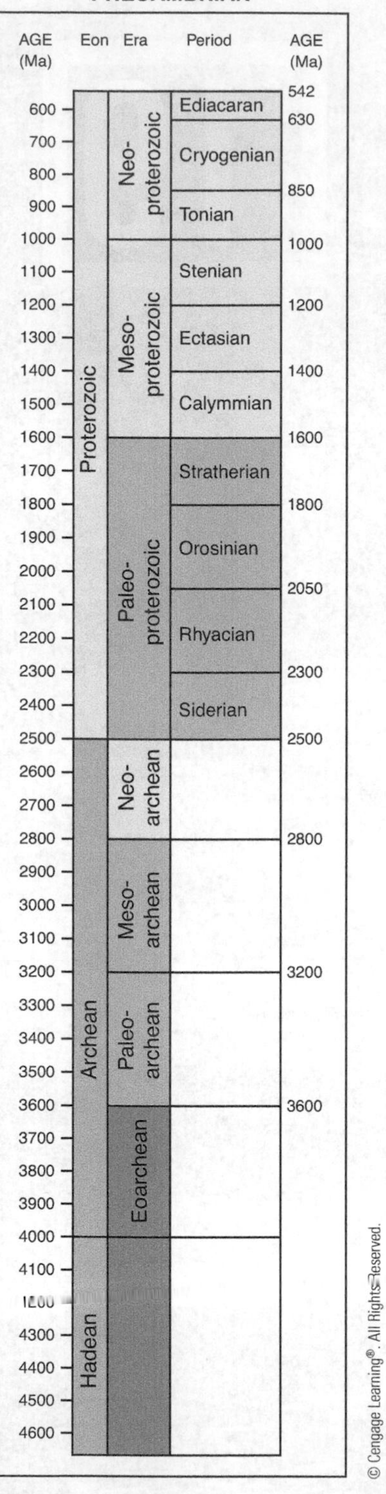

▌**Figure 19.2 The Precambrian Geologic Time Scale** This most recent revision of the geologic time scale was published by the International Commission on Stratigraphy (ICS) in 2009. See Figure 1.18 for the complete time scale. Notice the use of the prefixes *eo* (early or dawn), *paleo* (old or ancient), *meso* (middle), and *neo* (new or recent). The age columns on the left and right sides of the time scale are in hundreds and thousands of millions of years (1,800 million years = 1.8 billion years, for example).

about 4.0 billion years ago with the oldest known rocks on Earth except meteorites. The geologic record we do have for the Precambrian, especially for the earliest part designated Archean (Figure 19.2), is difficult to decipher because (1) many of the rocks have been metamorphosed and complexly deformed; (2) in many places they are deeply buried; and (3) they contain few fossils.

Establishing formal subdivisions of the Precambrian has proved difficult and has varied through the years. However, in 2009, the International Commission on Stratigraphy recommended the use of *Archean Eon* and *Proterozoic Eon* for most of Precambrian time, but it also suggested the informal term *Hadean* for the earliest part of this interval for which there are no known rocks (Figure 19.2). The subdivisions in Figure 19.2 are based on absolute dates rather than time-stratigraphic units, which is a departure from standard practice.

Many of the events that took place during the Precambrian set the stage for further evolution of the planet and life. It was during this time that Earth's systems (see Figure 1.1) became operative, although not all at the same time or in their present form. Earth did not differentiate into a core and mantle until millions of years after it formed, but once it did, internal heat was responsible for plate movements and the evolution of the continents. This was also the time during which the atmosphere formed (although it was very different than it is now), surface waters began to accumulate, and the first organisms appeared at least 3.5 billion years ago. In short, Earth was very different when it first formed, but during the Precambrian it began to evolve and became increasingly more like it is now.

19.2 What Happened During the Hadean?

Hadean is an informal term that encompasses all geologic time from 4.6 to 4.0 billion years ago. Indeed, the oldest known rocks other than meteorites come from the Acasta Gneiss of Canada, which is 3.95 to 4.04 billion years old. So we have no geologic record for the Hadean, but we can make some reasonable inferences about the events that took place then. Of course, Earth accreted as planetesimals collided, it differentiated into a core and mantle (see Figures 1.9 and 1.10), and some continental crust formed, perhaps as long ago as 4.4 billion years.

As the accreting planet grew, it swept up the debris in its vicinity, and just like the other terrestrial planets, it was bombarded by meteorites and comets until about 4 billion years ago. Unlike Mercury, Mars, and Earth's Moon, however, the evidence of this period of impacts has been obliterated by weathering, erosion, volcanism, plate movement, and mountain building. In addition to bombardment by meteorites, Earth was probably hit by a Mars-size planetesimal 4.4 to 4.6 billion years ago, causing the ejection of a huge mass of hot material that coalesced to form the Moon.

After it first formed, Earth retained considerable heat from its origin, and much more heat was generated by radioactive decay; as a result, volcanism was ubiquitous (▮ Figure 19.3). Gases emitted by volcanoes formed an atmosphere, but it was very unlike the oxygen-rich one present now, and when the planet cooled sufficiently, surface waters began to accumulate. If we could somehow go back and visit early Earth, we would see a rapidly rotating, hot, barren, waterless planet bombarded by meteorites and comets. There were no continents, cosmic radiation would have been intense, and, of course, you would see no organisms.

The age of the oldest continental crust is uncertain, but we can be sure that at least some was present by 3.8 billion years ago, and detrital sedimentary rocks in Australia have zircons ($ZrSiO_4$) 4.4 billion years old, indicating that source rocks that old must have existed. In fact, about 4-billion-year-old rocks are known from several areas, some of which are metamorphic so their parent rock must be even older.

The friction caused by the Moon on the oceans as well as the continents causes the rate of Earth's rotation to slow very slightly every year. When Earth first formed, it probably

▮ **Figure 19.3 Earth as It May Have Appeared Shortly After Forming** No rocks are known from this earliest time in Earth's history, but geologists can make some reasonable inferences about the nature of the newly formed planet.

Figure 19.4 Origin or Granitic Continental Crust

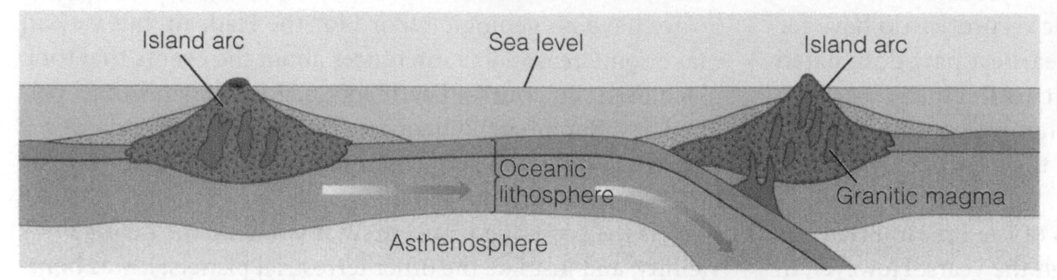

a An andesitic island arc forms by subduction of oceanic lithosphere and partial melting of basaltic oceanic crust. Partial melting of andesite yields granitic magma.

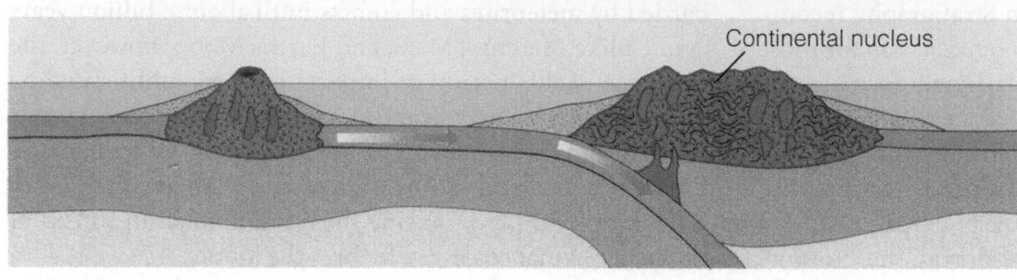

b The island arc in **(a)** collides with a previously formed island arc, thereby forming a continental core.

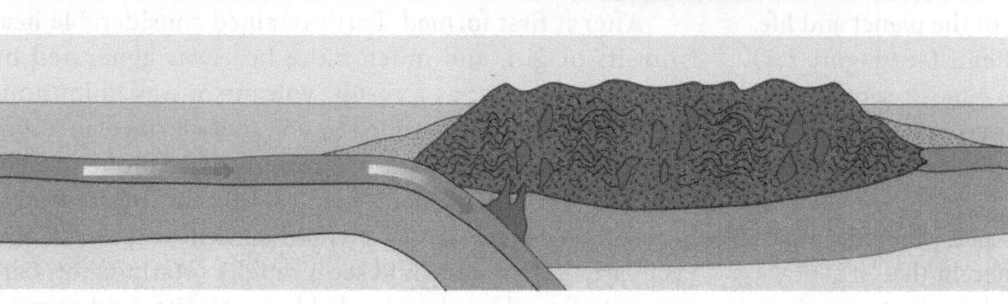

c The process occurs again when the island arc in **(b)** collides with the evolving continent, thereby forming a craton, the nucleus of a continent.

rotated in as little as 10 hours, so there were many more days in a year; there is no evidence indicating that Earth's orbital period around the Sun has decreased. Another effect of the Earth–Moon tidal interaction is the recession of the Moon from Earth at a few centimeters per year. Thus, during the Hadean, the view of the Moon would have been spectacular.

Geologists agree that when Earth formed, it was exceedingly hot, at least hot enough that it partially melted and differentiated into a core and mantle. Rather than Earth being a fiery orb for more than a half billion years, as was formerly accepted, some geologists now think that it had cooled enough for surface water to accumulate by 4.4 billion years ago (Geo-Focus 19.1). They base this conclusion on oxygen 18 to oxygen 16 ratios in tiny inclusions in zircon crystals that indicate reactions with surface waters.

When Earth differentiated during the Hadean, the core and mantle formed, but we have made little mention so far of the crust. Remember that we have defined two types of crust, oceanic and continental, which differ in composition, density, and thickness. The first crust probably formed as upwelling mantle currents of mafic magma disrupted the surface, subduction zones formed, and the first island arcs developed (Figure 19.4a). Weathering of these island arcs yielded sediments richer in silica, and partial melting of mafic rocks yielded magma richer in silica. Collisions between island arcs formed a few continental nuclei as silica-rich materials were metamorphosed and intruded by magma (Figure 19.4b). As these larger island arcs collided, the first protocontinents formed and continued to grow by accretion along their margins (Figure 19.4c).

19.3 The Foundations of Continents—Shields, Platforms, and Cratons

Continents are more than simply land above sea level. Indeed, they consist of rocks with an overall composition similar to granite, and continental crust is thicker and less dense than oceanic crust, which is made up of basalt and gabbro. Furthermore, a **shield** consisting of a vast area or areas of exposed ancient rocks is found on all continents. Continuing outward from the shields are broad **platforms** of buried Precambrian rocks that underlie much of each continent. Collectively, a shield and platform make up a **craton,** which we can think of as a continent's ancient nucleus.

The cratons are the foundations of continents, and along their margins more continental crust was added, a process called **continental accretion,** as they evolved to their present sizes and shapes. Both Archean and Proterozoic rocks are present in cratons, many of which indicate several episodes of deformation accompanied by metamorphism, igneous activity, and mountain building. However, the cratons have experienced remarkably little deformation since the Precambrian.

In North America, the exposed part of the craton is the **Canadian shield,** which occupies most of northeastern Canada, a large part of Greenland, the Adirondack Mountains of New York, and parts of the Lake Superior region in

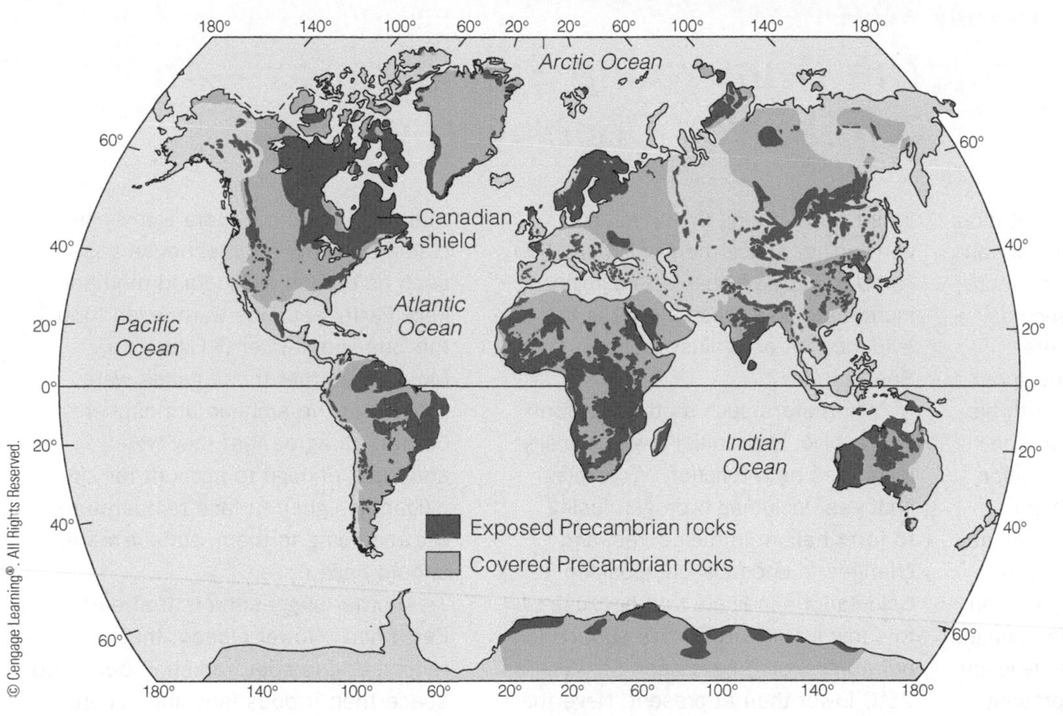

Figure 19.5 The Distribution of Precambrian Rocks Areas of exposed Precambrian rocks constitute the shields, whereas the platforms consist of buried Precambrian rocks. A shield and its adjoining platform make up a craton.

Minnesota, Wisconsin, and Michigan (Figure 19.5) Overall, the Canadian shield is an area of subdued topography, numerous lakes, and exposed Archean and Proterozoic rocks thinly covered in places by Pleistocene glacial deposits. The rocks themselves are plutonic, volcanic, and sedimentary, and metamorphic equivalents of all of these (Figure 19.6).

Actually, the Canadian shield, as well as the adjacent platform, are made up of numerous units or smaller cratons that amalgamated along deformation belts during the Paleoproterozoic. Absolute ages and structural trends differentiate these smaller cratons from one another

Drilling and geophysical evidence indicate that Precambrian rocks underlie much of North America, but beyond the Canadian shield, they are exposed only in areas of deep erosion and uplift. For instance, Precambrian rocks are present in the deeper parts of the Grand Canyon of Arizona, as well as in the Appalachian Mountains and the Rocky Mountains.

Figure 19.6 Archean Rocks in North America Archean rocks in North America are exposed mostly in the Canadian shield and elsewhere in areas of uplift and deep erosion.

Critical Thinking Question What kind of unconformity is present between the Precambrian and Cambrian rocks in **(b)**, and what is the duration of the hiatus?

Stephen J. Mojzsis/University of Colorado

a Outcrop of the Acasta Gneiss in the Northwest Territories of Canada. At about 4.0 billion years old, these are the oldest known rocks on Earth except for meteorites.

James S. Monroe

b Archean granite (2.9 billion years old) overlain by the Cambrian-age Flathead Sandstone in the Bighorn Mountains of Wyoming.

GEO FOCUS 19.1 The Faint Young Sun Paradox—An Unresolved Controversy

It should not be surprising that older events in written human history are more poorly understood than more recent ones, because older records, even if accurate, are incomplete, fragmentary, and difficult to decipher. A similar situation holds for Earth history, especially for the Archean Eon for reasons already noted; alteration of these rocks by metamorphism, deep burial in many areas, as well as few fossils. Nevertheless, on some aspects of early Earth history we can be reasonably certain: Archean oceans existed; the atmosphere was deficient in free oxygen; rocks that represent continental crust were present; and organisms appeared at least 3.0 billion years ago, and perhaps as early as 3.5 billion years ago.

However, according to the standard model for the evolution of stars, Earth's early Sun was only about 70% to 75% as luminous as it is now. If this is correct, Earth's surface should have been cold enough for all water to have frozen but the evidence for liquid water during the Archean is convincing, hence the *Faint Young Sun Paradox*. First, why has solar luminosity increased, and second, what is the evidence for or against the Faint Young Sun Paradox?

When stars such as the Sun form and evolve, their initial low luminosity increases as a function of complex changes, including hydrogen fusing to form helium in their cores, and changes in chemical composition. Calculations indicate that because of this low luminosity Earth's surface temperature should have been as much as 25°C lower than at present. Nevertheless, evidence from the geologic record indicates that liquid water was present and surface temperatures were not notably different than they are now.

The Faint Young Sun Paradox was proposed in 1972 by astronomer Carl Sagan and his colleague George Mullen. Since then, scientists have proposed several solutions to the paradox, but so far none are without critics. One of the most obvious solutions is that an early atmosphere with appreciable amounts of greenhouse gases such as carbon dioxide and methane kept Earth's surface warm even though the Sun was fainter (Figure 1a). Most agree that these gases were present in the Archean atmosphere, but not all agree that they were abundant enough to account for significantly higher surface temperatures, so, according to them, surface waters should have frozen.

Another suggestion is that early Earth had a lower albedo; that is, it reflected less solar radiation back into space than it does now and accordingly was warm enough for water to remain liquid. We know that there were vast oceans during the Archean, but we have little evidence for extensive continents that have a higher albedo than water. According to this hypothesis, oceanic waters absorbed more heat and kept Earth's surface temperature high enough to inhibit freezing (Figure 1b). Also, a proposed thinner cloud cover would have

19.4 Archean Earth History

Geologists place the beginning of the Archean Eon at 4.0 billion years ago (Figure 19.2), which is the age of the oldest known rocks, the Acasta Gneiss in Canada (Figure 19.6a). The end of the Archean Eon and the beginning of the Proterozoic Eon 2.5 billion years ago is arbitrary, but it does correspond to a time when the style of crustal evolution changed and when rock assemblages more like those of the present appeared in abundance.

The Archean Eon alone accounts for 32.3% of all geologic time, yet we review both its physical and biological history in just a few pages. One should not assume, however, that it was an unimportant time in Earth history. The geologic record is more complete for more recent intervals of geologic time, and because Earth is so active, ancient rocks are more likely to have been eroded or changed by metamorphism. In short, older intervals of geologic time are represented by smaller volumes of rock, especially sedimentary rock, and even if preserved, they are more difficult to find and interpret (Table 19.1).

Archean Rocks

Only 22% of Earth's exposed Precambrian crust is Archean, with the most extensive exposures in Africa and North America. Archean crust is made up of many types of rocks, but we characterize most of them as greenstone belts and granite–gneiss complexes, the latter being by far the most common. **Granite–gneiss complexes** are composed of allowed more solar energy to reach the surface.

Estimates of the rate at which continents grow by accretion along their margins vary widely. Nevertheless, given that more residual heat and radiogenic heat were available during the Archean, most geologists think that the continents grew more rapidly during that time. In fact, some estimate that the continents had reached about 75% of their present volume by the beginning of the Phanerozoic Eon. And, of course, at the same time, the Sun had increased its luminosity by about 30%. In the final analysis, geologists think that Earth's surface temperatures have remained fairly constant through most of geologic time.

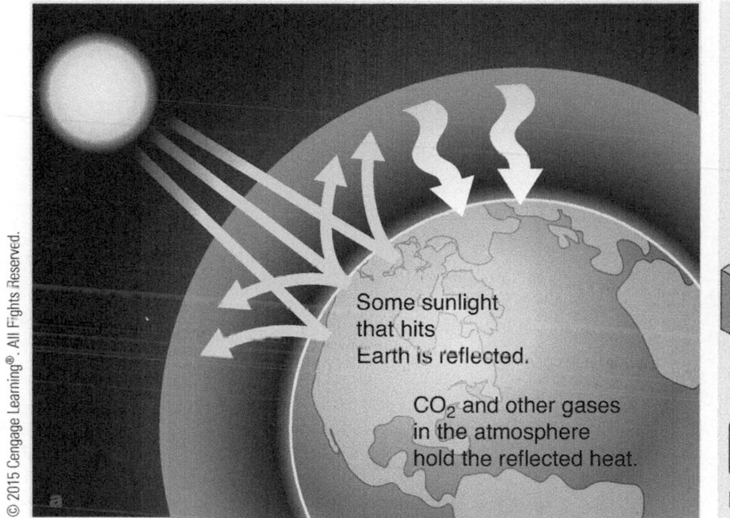

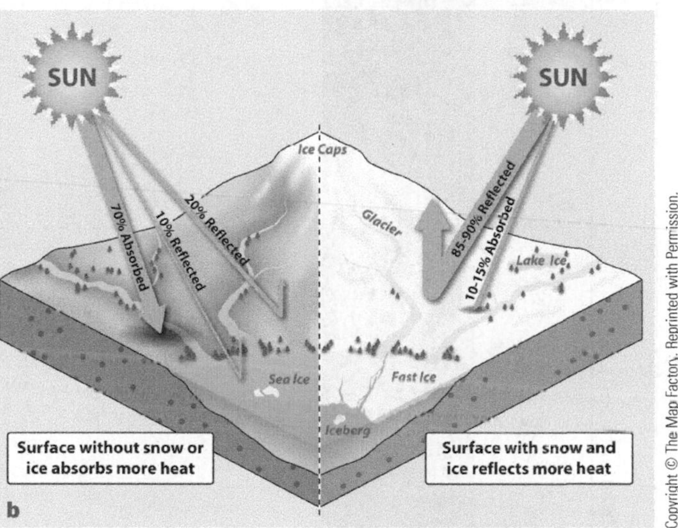

Figure 1 Some scientists think that a greenhouse effect on early Earth accounts for the Faint Young Sun Paradox. **(a)** Early Earth and a strong greenhouse effect. **(b)** Earth's albedo varies depending on the surface materials sunlight encounters. One proposal for the Faint Young Sun Paradox is that early Earth had vast oceans that had a low albedo and thus absorbed heat, thereby offsetting the weaker energy from the Sun.

rocks as varied as peridotite and sedimentary rocks; but granitic gneiss and granitic plutonic rocks are the most common, both of which were probably derived from plutons emplaced in volcanic island arcs. Greenstone belts are subordinate, accounting for only 10% of Archean rocks, and yet they are important in unraveling some of the complexities of Archean tectonic events.

A **greenstone belt** has three main components—its lower and middle parts are mostly volcanic rocks, whereas the upper rocks are mostly sedimentary (▮ Figure 19.7a). The origin of the minerals chlorite, actinolite, and epidote during low-grade metamorphism gives the rocks their greenish color, hence the name *greenstone*. Greenstone belts have a syncline-like structure and measure 40–250 km wide and 120–800 km long. In addition, they have been intruded by granitic magma and complexly folded and cut by thrust faults.

Much of the volcanism responsible for the lower and middle units in greenstone belts must have been subaqueous because pillow lavas are common (▮ Figure 19.7b), but some large volcanic centers built above sea level. Some of the most interesting volcanic rocks are ultramafic lava flows, or *komatiites*. We noted previously that eruptions producing these flows are rare in rocks younger than Archean because Earth's radiogenic heat production has decreased and near-surface temperatures are not high enough for ultramafic magma to reach the surface (see Chapter 4).

Sedimentary rocks are a minor component in the lower parts of greenstone belts, but they become increasingly abundant toward the top (▮ Figure 19.7c). Associations of graywacke, a sandstone with abundant clay, quartz, feldspars, and rock fragments, and argillite (slightly metamorphosed mudrock) are particularly common. Small-scale

TABLE 19.1 Summary of Archean Geologic and Biologic Events Discussed in the Text

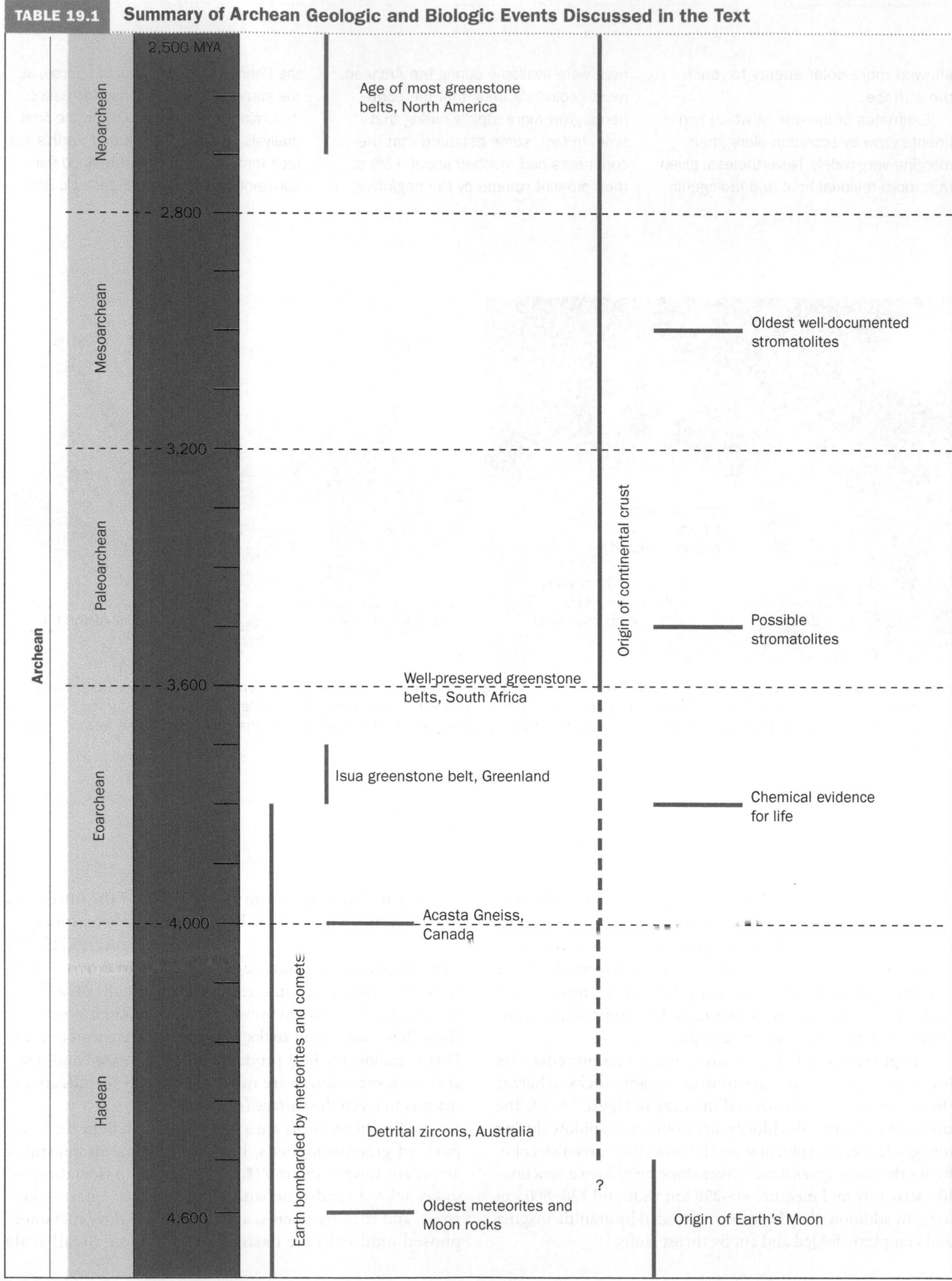

Figure 19.7 Greenstone Belts and Granite–Gneiss Complexes

Critical Thinking Question How did the pillow lava in **(b)** form?

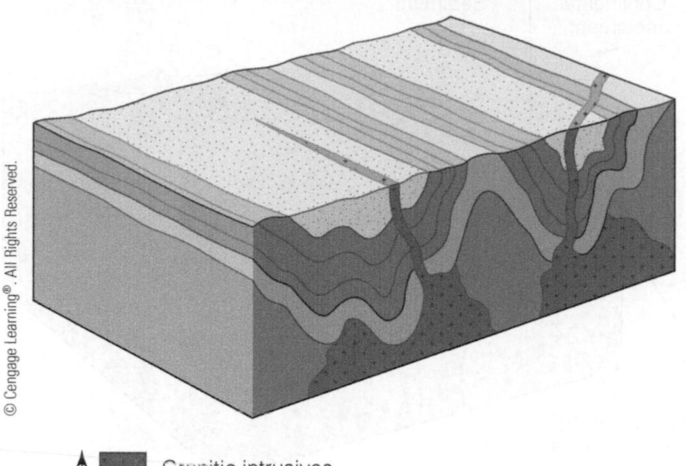

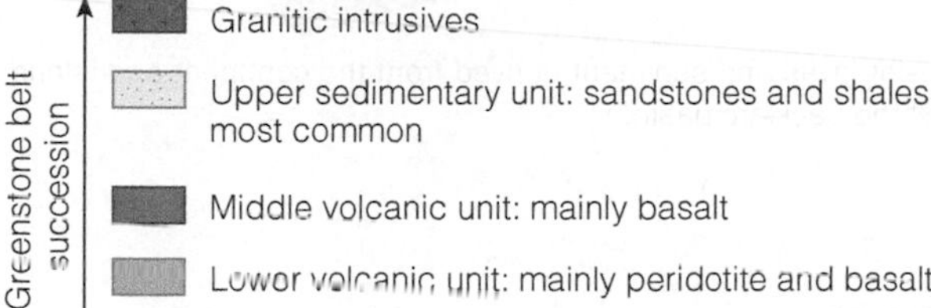

a Two adjacent greenstone belts. Older belts—those more than 2.8 billion years old—have an ultramafic unit overlain by a basaltic unit. In younger belts, the succession is from a basaltic lower unit to an andesite-rhyolite unit. In both cases, the upper unit is made up mostly of sedimentary rocks.

James S. Monroe

b Pillow lava of the Ispheming greenstone belt in Michigan.

R.V. Dietrich

c Gneiss from a granite–gneiss complex in Ontario, Canada.

graded bedding and cross-bedding indicate that deposition by turbidity currents accounts for the graywacke–argillite successions. Some sedimentary rocks were deposited in delta, tidal flat, barrier island, and shallow marine shelf environments.

Other sedimentary rocks present in Archean greenstone belts include conglomerate, carbonates, and chert. Banded iron formations (BIFs) are also found in Archean greenstone belts, but BIFs are much more common in areas of Proterozoic rocks.

Greenstone belts probably developed in several tectonic settings, but an appealing model for the origin of some relies on Archean plate tectonics and the development of a **back-arc basin** that subsequently closed (Figure 19.8). Recall that a back-arc basin, such as the Sea of Japan, is found between a continent and a volcanic island arc. There is an early stage in which the basin opens, accompanied by volcanism, the emplacement of plutons, and sedimentation, followed by an episode of compression when the basin closes. During this latter phase, the evolving greenstone belt rocks are deformed, metamorphosed, and intruded by granitic magma.

Archean Plate Tectonics and the Origin of Cratons

Certainly, the present plate tectonic regime of opening and closing ocean basins has been a primary agent in Earth's evolution since the Paleoproterozoic. Most geologists are convinced that some kind of plate tectonic activity took place during the Archean as well, but it differed in detail from what is going on now. With more residual heat from Earth's origin and more radiogenic heat, plates must have moved faster and magma was generated more rapidly. As a result, continents no doubt grew more rapidly along their margins as plates collided with island arcs and other plates.

There were, however, marked differences between the Archean world and the one that followed. We have little evidence of Archean rocks deposited on broad, passive continental margins, but they become widespread by the Proterozoic. Deformation belts between colliding cratons indicate that Archean plate tectonics was active, but the *ophiolites* so typical of younger convergent plate boundaries are rare, although Archean-age ophiolites are now known in several areas, including one in Greenland that is 3.8 billion years old.

Figure 19.8 Origin of a Greenstone Belt in a Back-Arc Basin

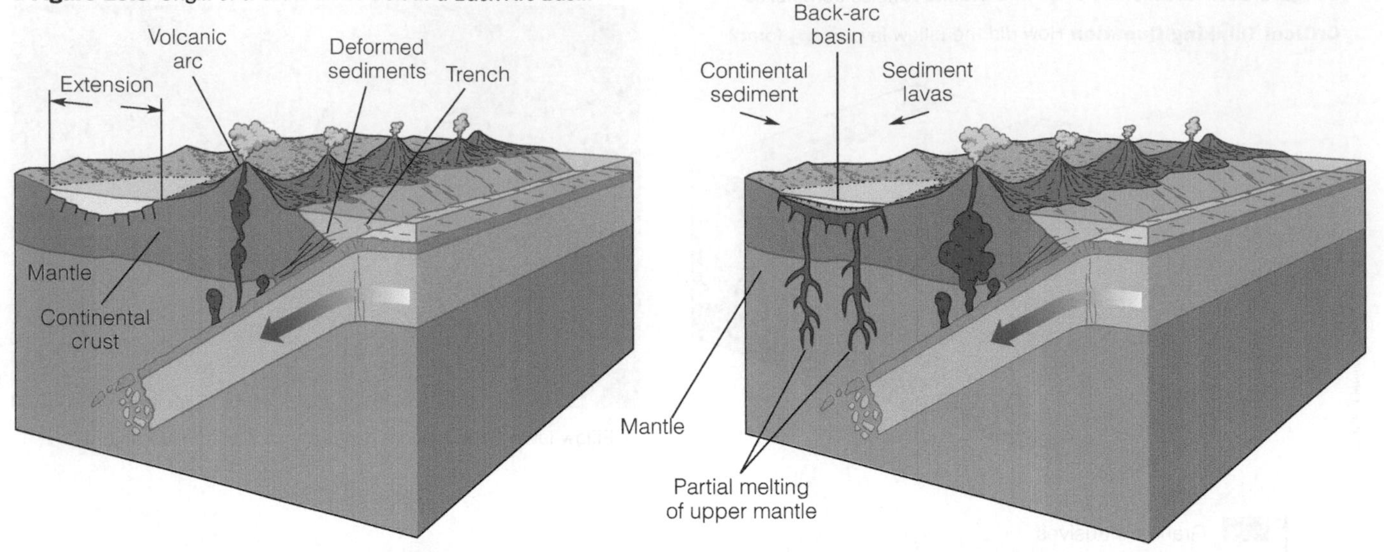

a Rifting on the continental side of the volcanic arc forms a back-arc basin. Partial melting of subducted oceanic crust supplies andesite and dioritic magmas to the island arc.

b Basalt lavas and sediment derived from the continent and island arc fill the back-arc basin.

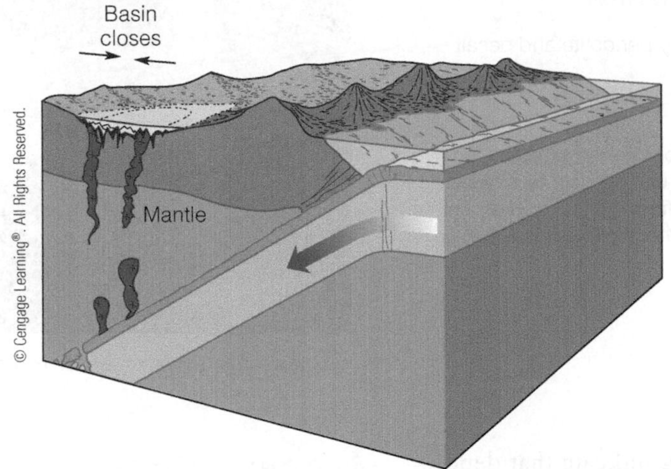

c Closure of the back-arc basin causes compression and deformation. The evolving greenstone belt is deformed into a syncline-like structure into which granitic magmas intrude.

Certainly, several small cratons existed during the Archean and grew by continental accretion periodically during the rest of that eon, and they amalgamated into a much larger unit during the Proterozoic. By the end of the Archean, perhaps 30%–40% of the present volume of continental crust existed. A plate tectonic model for the Archean crustal evolution of granite-genesis complexes and greenstone belts relies on a series of events, including greenstone belt evolution, plutonism, and deformation (▮ Figure 19.9). We can take this as a provisional model for Archean crustal evolution in general.

The events leading to the origin of the craton (Figure 19.9) are part of a more extensive episode of deformation that took place near the end of the Archean. Deformation at this time was responsible for the formation of the Superior and Slave cratons and the origin of some Archean rocks in other parts of the Canadian shield, as well as in Wyoming, Montana, and the Mississippi River Valley. By the time it was over, several sizable cratons existed that are now found in the older parts of the Canadian shield.

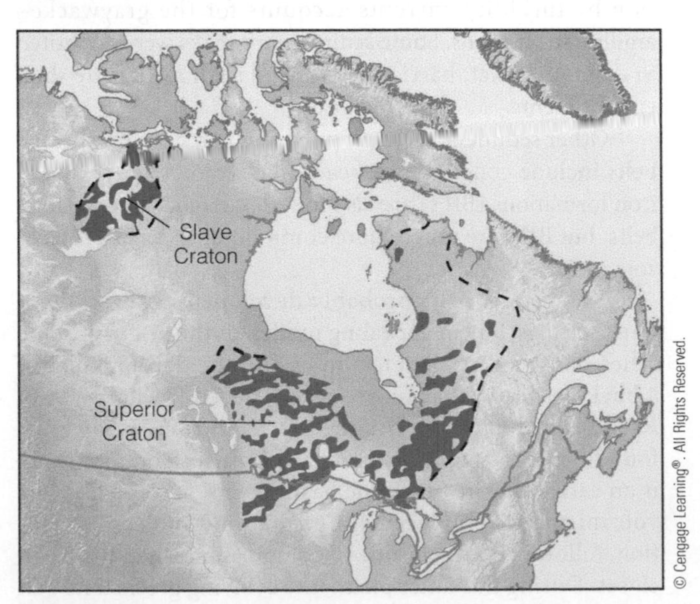

Figure 19.9 Greenstone Belts in North America Archean greenstone belts (shown in dark green) of the Canadian shield are mostly in the Superior and Slave cratons.

19.5 Proterozoic Earth History

The main difference between Archean and Proterozoic Earth history is the style of crustal evolution. Crust-forming processes that generated greenstone belts and granite–gneiss complexes during the Archean continued into the Proterozoic, but at a considerably reduced rate. Most Archean rocks were altered by metamorphism, but many Proterozoic rocks have been little altered. And finally, widespread assemblages of sedimentary rocks deposited on passive continental margins are common in the Proterozoic but rare in the Archean.

We noted that Archean cratons assembled through a series of island-arc and minicontinent collisions. These provided the nuclei around which Proterozoic continental crust accreted, thereby forming much larger cratons. One large landmass so formed, called **Laurentia,** consisted mostly of North America and Greenland, parts of northwestern Scotland, and perhaps parts of the Baltic shield of Scandinavia. We discuss the geologic evolution of Laurentia in the following sections (Table 19.2).

TABLE 19.2 Summary of the Proterozoic Geologic and Biologic Events Discussed in the Text

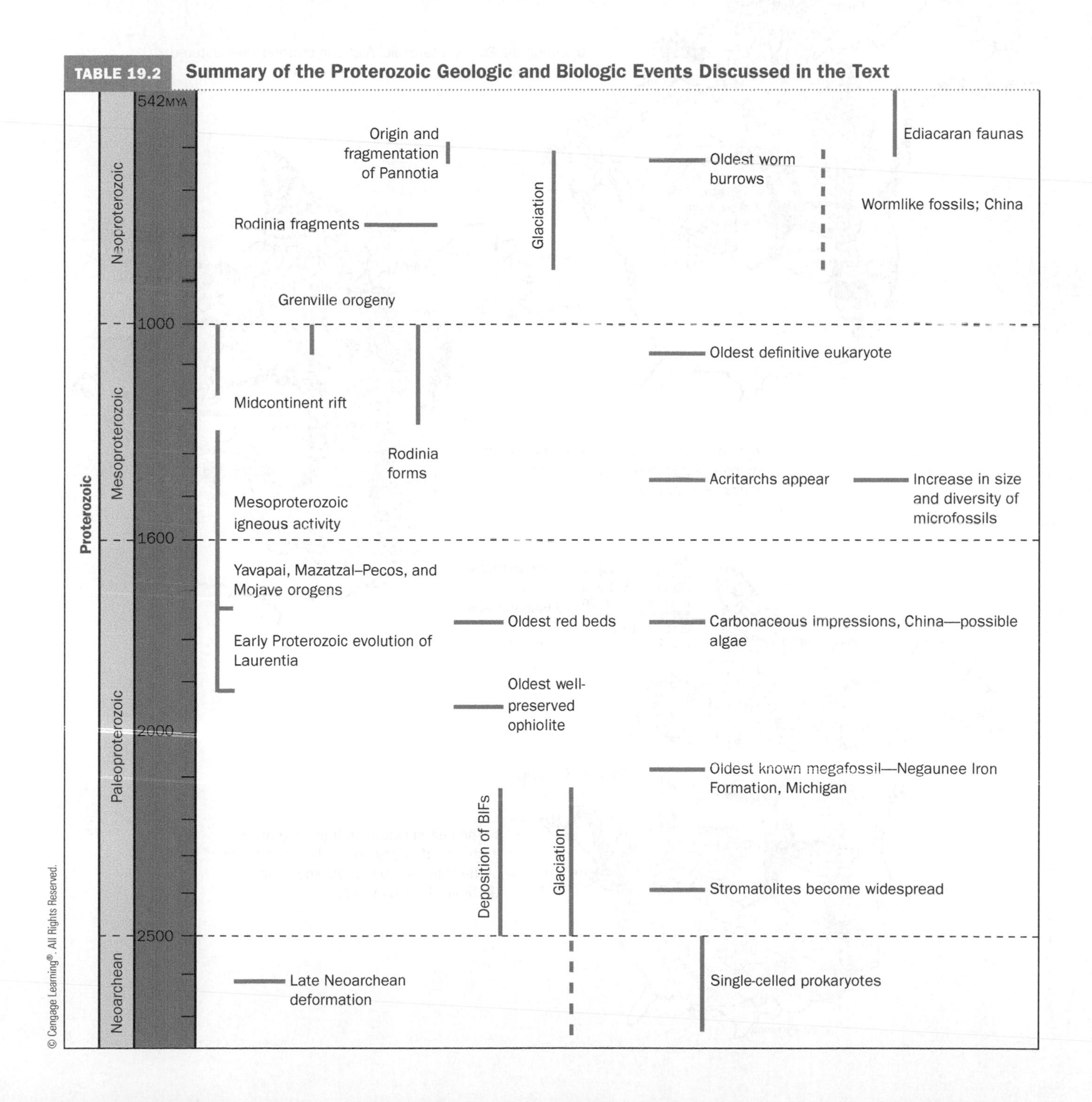

Paleoproterozoic History of Laurentia

During the time between 2.0 and 1.8 billion years ago, collisions among plates formed several **orogens,** which are linear or arcuate deformation belts in which many of the rocks have been metamorphosed and intruded by magma, thus forming plutons, especially batholiths. Accordingly, Archean cratons were sutured along these deformation belts, thereby forming a larger landmass (▌Figure 19.10a). By 1.8 billion years ago, much of what is now Greenland, central Canada, and the north-central United States existed.

Good examples of these craton-forming processes are recorded in rocks of the Thelon orogen in northwestern

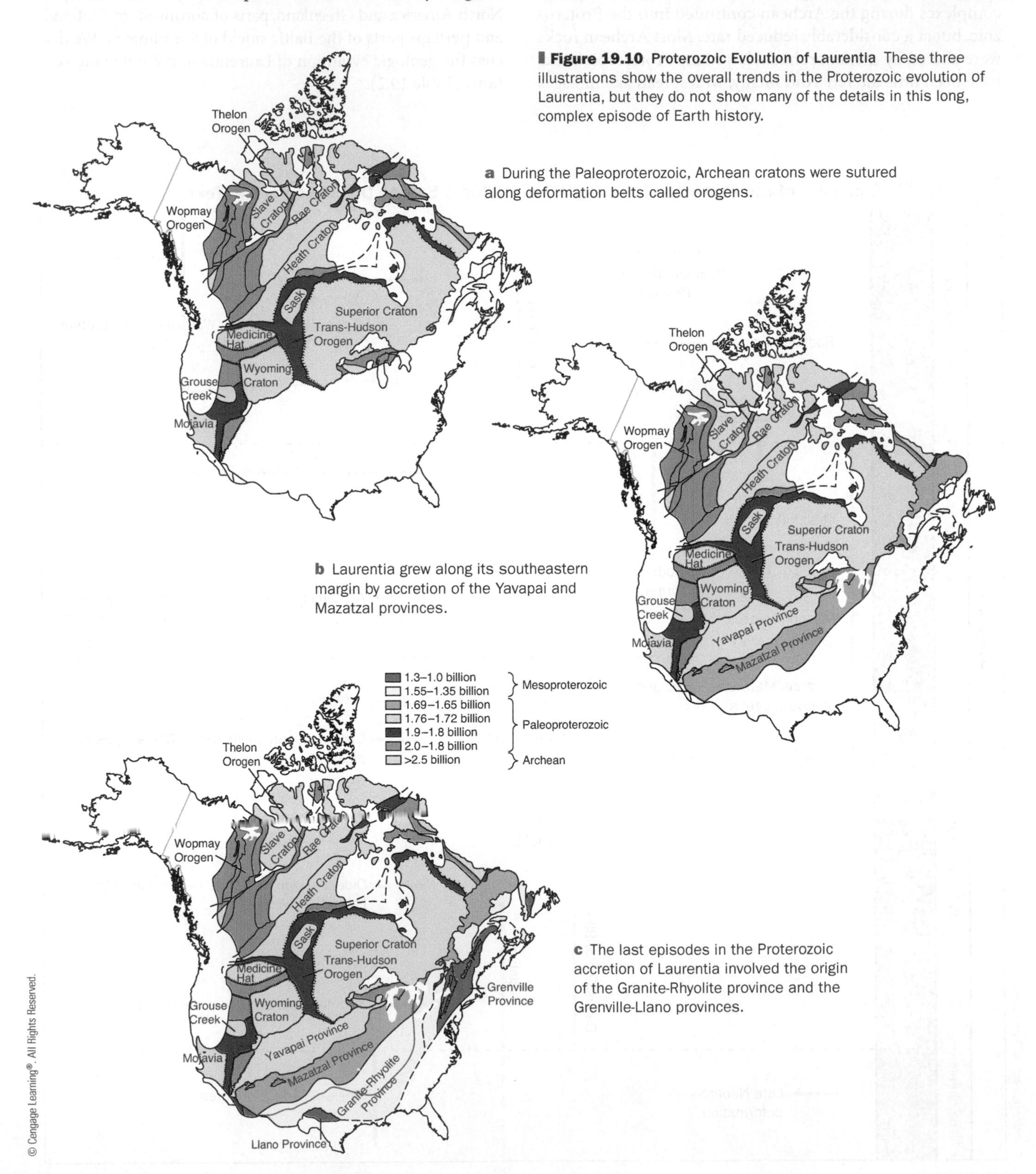

▌**Figure 19.10 Proterozoic Evolution of Laurentia** These three illustrations show the overall trends in the Proterozoic evolution of Laurentia, but they do not show many of the details in this long, complex episode of Earth history.

a During the Paleoproterozoic, Archean cratons were sutured along deformation belts called orogens.

b Laurentia grew along its southeastern margin by accretion of the Yavapai and Mazatzal provinces.

c The last episodes in the Proterozoic accretion of Laurentia involved the origin of the Granite-Rhyolite province and the Grenville-Llano provinces.

Canada where the Slave and Rae cratons collided, and the Trans-Hudson orogen, in Canada and the United States, where the Superior, Hearne, and Wyoming cratons were sutured (Figure 19.10a). Sedimentary rocks of the Wopmay orogen in northwestern Canada are important because they record the opening and closing of an ocean basin, or what is called a **Wilson cycle.** A complete Wilson cycle, named after the Canadian geologist J. Tuzo Wilson, involves rifting of a continent and the opening of an ocean basin with passive continental margins on both sides. As rifting proceeds, an expansive ocean basin forms, but it eventually begins to close, and subduction zones and volcanic island arcs develop on one or both sides of the ocean basin. The final stage involves a continent–continent collision, during which the rocks that formed along the previous passive continental margins and seafloor are deformed during an orogeny. Some of the rocks in the Wopmay orogen belong to a suite of sedimentary rocks called a **sandstone–carbonate–shale assemblage** that is deposited on passive continental margins. Paleoproterozoic sandstone–carbonate–shale assemblages are widespread in the Great Lakes region of the United States and Canada (▌Figure 19.11). The sandstones (now quartzites) have sedimentary structures such as ripple marks and cross-beds, whereas some of the carbonate rocks, now mostly dolostone, contain abundant stromatolites (Figure 19.11b).

Following the initial episode of amalgamation of Archean cratons, accretion took place along Laurentia's southern margin. From 1.8 to 1.6 billion years ago, continental accretion continued in what is now the southwestern and central United States as successively younger belts were sutured to Laurentia, forming the Yavapai and Mazatzal–Pecos orogens (▌Figure 19.10b).

▌**Figure 19.11** Proterozoic Rocks of North America

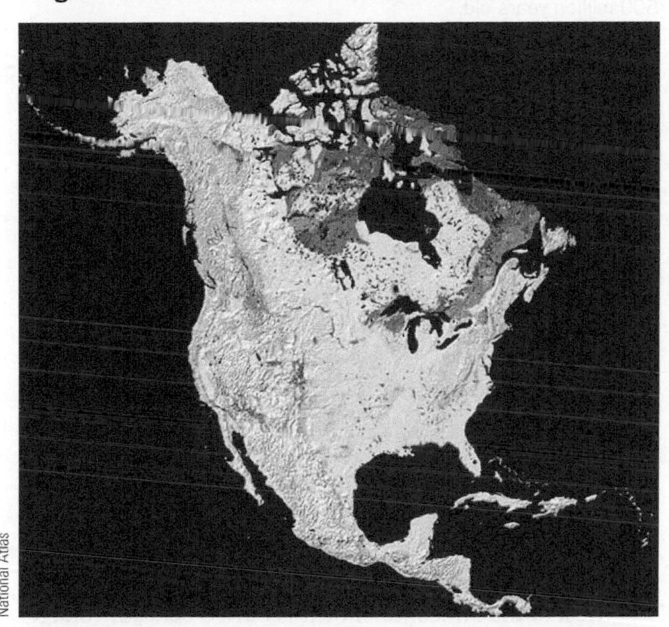

National Atlas

a The distribution of Proterozoic rocks in North America.

James S. Monroe

b Outcrop of the Paleoproterozoic-age Kona Dolomite in Michigan. The bulbous structures are stromatolites that resulted from the activities of cyanobacteria (blue-green algae).

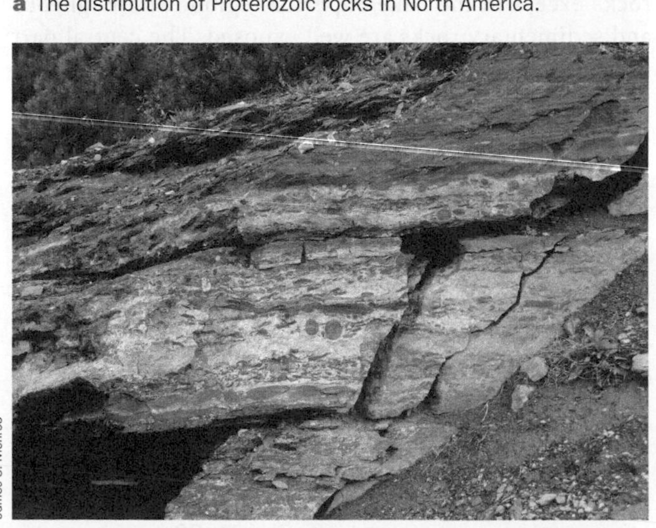

James S. Monroe

c This 1.0-billion-year-old sandstone and mudstone in Glacier National Park in Montana has been only slightly metamorphosed.

Copyright and Photograph by Dr. Parvinder S. Sethi

d Neoproterozoic sedimentary rocks in the Grand Canyon of Arizona. The rocks are reddish because of small amounts of iron oxides.

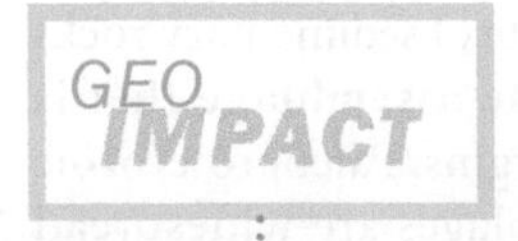

Planetary Geology

Suppose you visit a planet that, like Earth, has continents and vast oceans. What evidence would indicate that this hypothetical planet's continents formed and evolved like those on Earth?

The Paleoproterozoic was also the time during which most of Earth's *BIFs* were deposited. And the first *continental red beds*—sandstone and shale with oxidized iron—were deposited about 1.8 billion years ago. And finally, we should mention that some Paleoproterozoic-age rocks and associated features provide excellent evidence for widespread glaciation.

Paleo- and Mesoproterozoic Igneous Activity

From 1.8 to 1.1 billion years ago, extensive igneous activity took place that was unrelated to orogenic processes. Although quite widespread, much of the magma was either intruded into or erupted onto existing continental crust. These igneous rocks are exposed in eastern Canada, extend across Greenland, and are also found in the Baltic shield of Scandinavia. However, they are deeply buried in most areas.

The origins of these granitic and anorthosite* plutons, calderas and their fill, and vast sheets of rhyolite and ash flows are the subject of debate. According to one hypothesis, large-scale upwelling of magma beneath a Proterozoic supercontinent was responsible for these rocks. According to this hypothesis, the mantle temperature beneath a Proterozoic supercontinent would have been considerably higher than beneath later supercontinents because radiogenic heat production within Earth has decreased. Accordingly, nonorogenic igneous activity would have occurred following the amalgamation of the first supercontinent.

Mesoproterozoic Orogeny and Rifting

The only Mesoproterozoic orogenic event in Laurentia was the 1.3- to 1.0-billion-year-old **Grenville orogeny** (▮ Figure 19.10c). Grenville rocks are well exposed in the present-day northern Appalachian Mountains, as well as in eastern Canada, Greenland, and Scandinavia. Many geologists think that the Grenville orogen resulted from closure of an ocean basin, the final stage in a Wilson cycle. Others disagree, and think that intracontinental deformation or major shearing was responsible for deformation.

Cole Kingsbury

▮ **Figure 19.12 Rocks of the Grenville Orogen** Grenville gneiss overlain by the Paleozoic-age Nepean Sandstone in Quebec, Canada. The gneiss is about 1.0 billion years old, whereas the sandstone is 500 million years old.

Whatever the cause of the Grenville orogeny, it was the final stage in the Proterozoic continental accretion of Laurentia (▮ Figure 19.12). By then, about 75% of present-day North America existed. The remaining 25% accreted along its margins, particularly its eastern and western margins, during the Phanerozoic Eon.

Beginning about 1.1 billion years ago, tensional forces opened the **Midcontinent rift,** a long, narrow trough bounded by faults that outline two branches (▮ Figure 19.13). Although not all geologists agree, many think that the Midcontinent rift is a failed rift where Laurentia began splitting apart. Had rifting continued, Laurentia would have split into two separate landmasses, but the rifting ceased after about 20 million years.

Most of the Midcontinent rift is buried beneath younger rocks except in the Lake Superior region, where igneous and sedimentary rocks are well exposed. The central part of the rift contains numerous overlapping basalt lava flows, forming a volcanic pile several kilometers thick. Along the rift's margins, coarse-grained sediments were deposited in large alluvial fans that grade into sandstone and shale with increasing distance from the sediment source.

Meso- and Neoproterozoic Sedimentation

 The Grenville orogeny between 1.3 and 1.0 billion years ago (Figure 19.10c) was the final episode of continental accretion in Laurentia until the Ordovician Taconic orogeny.

ConnectionLink

You can find more about the Taconic orogeny in Chapter 20.

*Anorthosite is a plutonic rock composed almost entirely of plagioclase feldspars.

Figure 19.13 The Midcontinent Rift

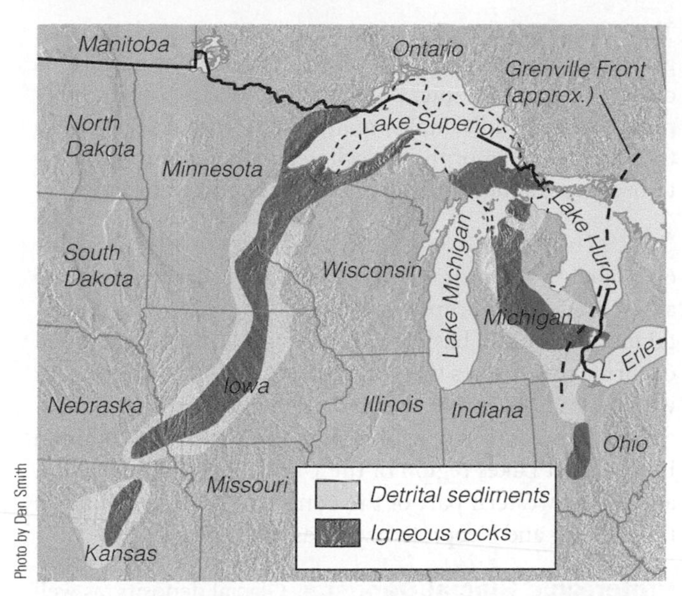

a Rocks filling the Midcontinent rift are well exposed around Lake Superior in the United States and Canada, but they are deeply buried elsewhere.

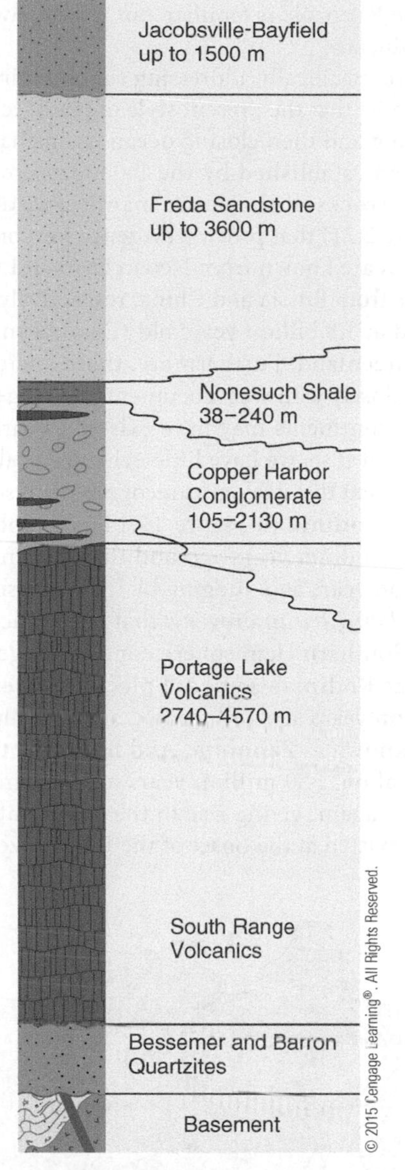

b Section showing the vertical relationships among the rocks in the rift in the Lake Superior region.

c The Nonesuch Shale along the shore of Lake Superior in Michigan.

Nevertheless, important geologic events were taking place. Meso- and Neoproterozoic-age sedimentary rocks are exceptionally well exposed in the northern Rocky Mountains of Montana, Idaho, and Alberta, Canada (Figure 19.11c). Like the rocks in the Great Lakes region, they are mostly sandstones, shales, and stromatolite-bearing carbonates.

Sedimentary rocks of Proterozoic age are also found in Utah and the Grand Canyon in Arizona (Figure 19.11d). The rocks, consisting mostly of sandstone, shale, and dolostone, were deposited in shallow-water marine and fluvial environments. The presence of stromatolites and carbonaceous impressions of algae in some of the rocks indicate probable marine deposition.

Proterozoic Supercontinents

A *continent* is one of Earth's major landmasses having an overall composition similar to granite. At present, the continents are mostly above sea level, but at times they have been shallowly submerged during marine transgressions (see Figure 7.11). A *supercontinent* consists of two or more continents sutured together, or, in some cases, most of the

For more information about the origin and eventual breakup of the supercontinent Pangaea, see the discussions in Chapters 20, 22, and 23.

continents. But other than its size, a supercontinent is the same as a continent. The supercontinent Pangaea that existed at the end of the Paleozoic is familiar, but few are aware of earlier supercontinents.

Before specifically addressing supercontinents, though, we must note that the present style of plate tectonics involving opening and then closing ocean basins had almost certainly been established by the Paleoproterozoic. In fact, *ophiolites* (rocks of the upper mantle and oceanic crust) (see Figure 2.21) that provide evidence for convergent plate boundaries are known from Neoarchean and Paleoproterozoic rocks from Russia and China, respectively, and recently one dated at 3.8 billion years old (Eoarchean) was discovered in Greenland. Furthermore, these ancient ophiolites compare closely with well-documented younger ones.

Supercontinents may have existed as early as the Late Archean, but if so, we have little evidence of them. The first supercontinent that geologists recognize with some certainty, known as **Rodinia** (❚ Figure 19.14), assembled between 1.3 and 1.0 billion years ago and then began fragmenting 750 million years ago. Judging by the large-scale deformation, the *Pan-African orogeny,* that took place in what are now the Southern Hemisphere continents, geologists conclude that Rodinia's separate pieces reassembled about 650 million years ago and formed another supercontinent, this one known as **Pannotia.** And finally, by the latest Proterozoic, about 550 million years ago, fragmentation was under way again, giving rise to the continental configuration that existed at the onset of the Phanerozoic Eon.

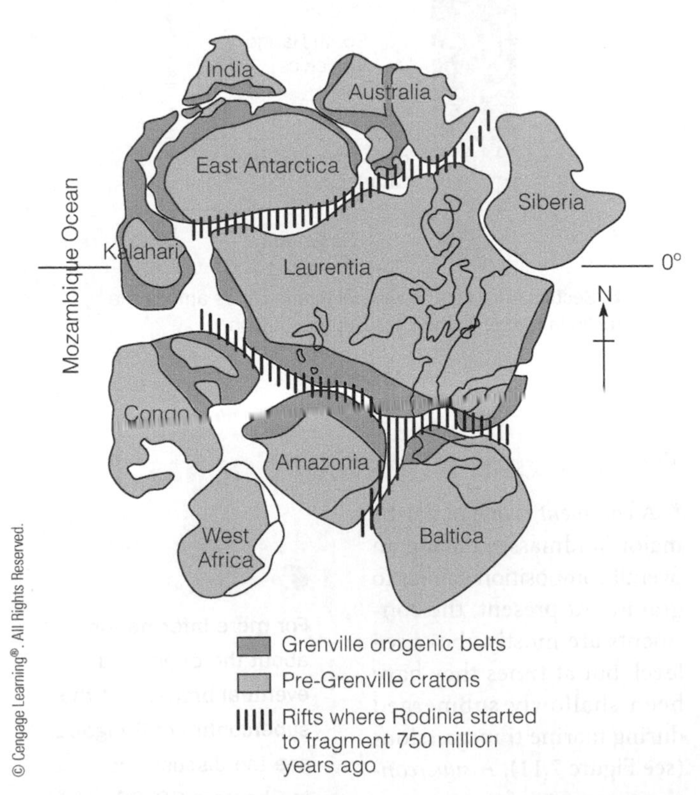

❚ **Figure 19.14 Rodinia** Possible configuration of the Neoproterozoic supercontinent Rodinia before it began fragmenting about 750 million years ago.

Proterozoic Rocks

The same crust-forming processes that generated Archean granite–gneiss complexes and greenstone belts continued during the Proterozoic, but they did so at a considerably reduced rate. Additionally, many Archean rocks have been thoroughly metamorphosed, whereas vast exposures of Proterozoic rocks show little or no effects of metamorphism.

Sandstone–Carbonate–Shale Assemblages Fully 60% of all Proterozoic rocks are sandstone–carbonate–shale assemblages that were deposited along rifted continental margins and in basins within cratons. Their widespread occurrence indicates that large, stable cratons were present with depositional environments much like those of the present. Paleoproterozoic assemblages of these rocks are common in the Great Lakes region of the United States and Canada, and in the western part of the continent they are found in three Meso- and Neoproterozoic basins (Figure 19.11).

Proterozoic Glacial Deposits Glacial deposits, as well as striated and polished bedrock, indicate that two major episodes of Proterozoic glaciation took place. North America probably had an extensive ice sheet centered southwest of Hudson Bay during the Paleoproterozoic. Similar deposits of about the same age are present in the United States, Australia, and South Africa, but their ages are not known precisely enough to determine whether there was a single widespread period of glaciation, or a number of glacial events in different areas at different times. Another time of widespread glaciation is indicated by Neoproterozoic-age deposits on all of the continents except Antarctica (Table 19.2; ❚ Figure 19.15). It seems that these glaciers existed even in near-equatorial areas.

❚ **Figure 19.15 Evidence for Proterozoic Glaciation in Death Valley, California** Geologists think this deposit in Death Valley is glacial till, that is, sediment deposited directly by glacial ice.

Based on the fact that Neoproterozoic glacial deposits are so widespread, some geologists think that glaciers covered all land and the seas were frozen—a *snowball Earth,* as it has come to be known. The snowball Earth hypothesis is controversial, but proponents claim that the onset of this glacial episode may have been triggered by the near-equatorial location of all continents, and as a result, accelerated weathering would absorb huge quantities of carbon dioxide from the atmosphere. With little CO_2 in the atmosphere, glaciers would form and reflect solar radiation back into space and more glacial ice would form.

So if there actually was a snowball Earth, why wouldn't it stay frozen? Of course, volcanoes would continue to erupt, spewing volcanic gases, which includes the greenhouse gases carbon dioxide and methane, which would warm the atmosphere and end the glacial episode. In fact, proponents of this hypothesis note that several such snowball Earths may have occurred until the continents moved into higher latitudes. One criticism of the hypothesis is that if all land was ice covered and the sea froze, how would life survive? Several suggestions have been made to account for this—life persisted at hydrothermal vents on the seafloor; even photosynthesis can take place beneath thin glacial ice; perhaps life persisted in subglacial lakes as it does now in Antarctica; and there may have been pools of liquid water near active volcanoes.

Two other Proterozoic-age sedimentary rocks were banded iron formations and continental red beds, both of which have important implications for the evolving atmosphere. Both are considered in some detail in the following section.

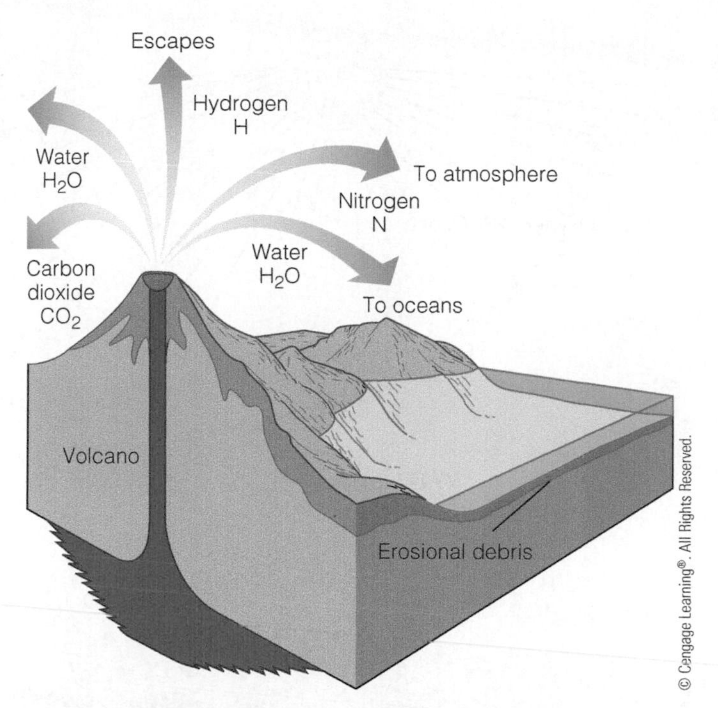

Figure 19.16 Outgassing and Earth's Early Atmosphere. Notice that the atmosphere contains several gases but no free oxygen.

19.6 Origin and Evolution of the Atmosphere and Hydrosphere

Early Earth was waterless, its atmosphere lacked free oxygen—that is, oxygen not combined with other elements as in water vapor (H_2O) and carbon dioxide (CO_2)—and no ozone layer was present, so ultraviolet radiation was intense. The atmosphere now contains nearly 21% free oxygen and other important gases in trace amounts, especially carbon dioxide (CO_2) and water vapor. An ozone layer in the upper atmosphere blocks out most ultraviolet radiation, and now 71% of Earth's surface is covered by water. Both the atmosphere and hydrosphere have played important roles in the development of the biosphere. The obvious question is, What brought about such remarkable changes?

The Atmosphere

Before Earth had a differentiated core, it lacked a magnetic field and a *magnetosphere,* the area around the planet in which the magnetic field is confined. Accordingly, a strong solar wind, an outflow of ions from the Sun, would have swept away any gases that otherwise might have formed an atmosphere. Once a magnetosphere was established, though, gases from Earth's interior were released during volcanic eruptions and began to accumulate—a phenomenon known as **outgassing** (Figure 19.16).

Archean volcanoes emitted the same gases as today's volcanoes do—mostly water vapor and lesser amounts of carbon dioxide, sulfur dioxide, carbon monoxide, nitrogen, hydrogen, and several others. These gases accumulated rapidly because Earth possessed more residual heat and more radiogenic heat, and as a result, volcanism was common. The atmosphere so formed was rich in carbon dioxide, but contained little or no free oxygen, and with no free oxygen, there was no ozone layer. Furthermore, as these volcanic gases reacted chemically, they very likely yielded atmospheric ammonia (NH_3) and methane (CH_4).

Archean sedimentary deposits with detrital minerals such as pyrite (FeS_2) and uraninite (UO_2) indicate an oxygen-deficient atmosphere because both minerals oxidize rapidly when free oxygen is present. However, oxidized iron is common in Paleoproterozoic-age rocks, so the atmosphere had at least some free oxygen by that time. Two processes account for the introduction of free oxygen into the atmosphere: photochemical dissociation and photosynthesis.

Photochemical dissociation is a process in which water vapor (H_2O) molecules are broken up by ultraviolet radiation in the upper atmosphere to yield hydrogen (H_2) and free oxygen (O_2) (Figure 19.17). However, photochemical dissociation is a self-limiting process and probably accounts for no more than 2% of the present-day level of free oxygen. With this much free oxygen, ozone (O_3) forms a barrier against incoming ultraviolet radiation, so another process must account for most of the atmosphere's free oxygen.

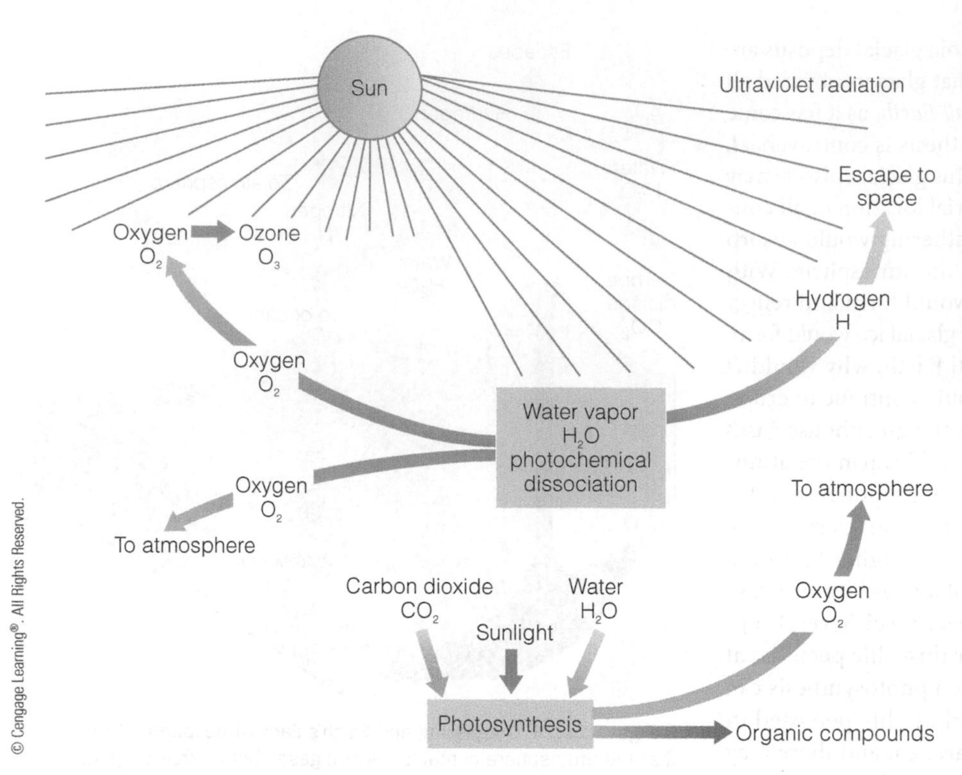

Figure 19.17 Evolution of the Atmosphere Photochemical dissociation occurs when ultraviolet radiation disrupts water molecules that release hydrogen (H_2) and free oxygen (O_2), some of which is converted to ozone (O_3) that blocks most of the ultraviolet radiation. In the presence of sunlight, photosynthesizing organisms use carbon dioxide (CO_2) and water (H_2O) to make organic molecules, and in the processes, release free oxygen as a waste product.

Critical Thinking Question Which of these processes, photochemical dissociation or photosynthesis, was the most important in contributing free oxygen to the Precambrian atmosphere?

Photosynthesis is much more important in releasing free oxygen into the atmosphere, but obviously it could not take place until organisms practicing this metabolic process had evolved. Organisms that carry out photosynthesis combine carbon dioxide and water into the organic molecules they need for survival, and they release free oxygen as a waste product (Figure 19.17). We know from the fossil record that organisms that almost certainly practiced photosynthesis were present more than 3.0 billion years ago. Even so, at the end of the Archean Eon, the atmosphere may have had no more than 1% of its present free oxygen level.

During the Proterozoic Eon, the free oxygen level increased from 1% to perhaps 10% of its present level, but probably not until well into the Paleozoic Era, about 400 million years ago, did it reach its current concentration of 21% of the atmosphere. The cessation of deposition of BIFs and the presence of continental red beds (Figure 19.11c) provide compelling evidence that Earth's Proterozoic atmosphere was an oxidizing one.

The Great Oxygenation Event

As a result of photochemical dissociation and, more importantly, photosynthesis, Earth's atmosphere became richer in free oxygen. This so-called **Great Oxygenation Event** took place about 2.4 billion years ago and had a profound impact on the biosphere. But what evidence indicates that free oxygen was accumulating in the atmosphere at this time? Much of the evidence comes from two types of Proterozoic sedimentary rocks: banded iron formations and continental red beds.

In Chapter 7 we noted that **banded iron formations (BIFs)** are made up of alternating layers of chert and the iron oxide minerals hematite (Fe_2O_3) and magnetite (Fe_3O_4) (Figure 19.18). BIFs are known from the Archean, but those of Proterozoic age are much thicker, more extensive geographically, and were likely deposited in shallow marine environments. Fully 92% of all BIFs formed during the interval from 2.3 to 2.5 billion years ago.

Iron is a highly reactive element that in the presence of oxygen forms rustlike oxides that do not readily dissolve in water. In the absence of free oxygen, though, iron goes

Photo Courtesy of Megan Babich

Figure 19.18 Proterozoic-Age Banded Iron Formation (BIF) At this outcrop on Jasper Knob in Ishpeming, Michigan, the Negaunee Iron Formation rocks are made up of alternating layers of red chert and silvery iron minerals.

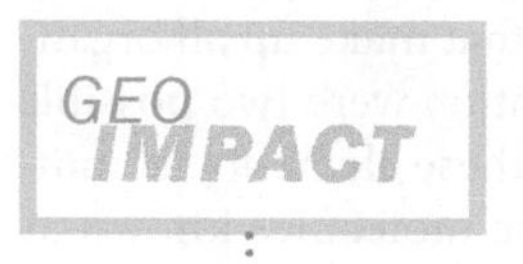

Ancient Glaciation

As a working geologist, you encounter a Proterozoic-age conglomerate that you are convinced is a glacial deposit. Others, however, think that it is simply stream-deposited gravel or perhaps an ancient landslide deposit. What kinds of evidence, if present, would verify your interpretation? That is, what attributes of the rock itself and its associated deposits might lead to your analysis?

into solution and accumulates in the oceans, as it must have done during the Archean when the atmosphere had little or no free oxygen. Then about 2.3 billion years ago, there was a notable increase in the free oxygen released by photosynthesizing bacteria into the atmosphere and oceans and the beginning of precipitation of iron to form BIFs. Precipitation of BIFs continued until the iron in seawater was largely depleted, and because the atmosphere had some free oxygen, iron was no longer taken into solution in large quantities.

Of course, the term **continental red beds** means red rocks on the continents, but more specifically it refers to red sandstone and mudrocks colored or stained by iron oxide cement, especially hematite (Fe_2O_3) (see the chapter opening photo). Continental red beds first appeared about 2.3 billion years ago, increased in abundance during the rest of the Proterozoic, and remain common in Phanerozoic-age rocks. Iron oxide cement forms under chemically oxidizing conditions, which means that Earth's atmosphere had some free oxygen at this time.

The Hydrosphere

We have outlined the history of the evolving atmosphere, and now we turn our attention to the hydrosphere, another of Earth's major systems. All water on Earth is part of the hydrosphere, but most of it—nearly 97%— is in the oceans. Where did it come from, how has it changed, and is it still accumulating?

Certainly, outgassing released water vapor from Earth's interior (Figure 19.16), and once the planet cooled sufficiently, water vapor condensed and surface waters began to accumulate. Another source of water vapor, and eventually liquid water, was meteorites and especially icy comets. It is not known which of these—outgassing or meteorites and comets—was most important, but we do know that oceans were present by the Eoarchean, although their volumes and geographic extent cannot be determined. Nevertheless, we can envision an early Earth with numerous erupting volcanoes and an early episode of intense meteorite and comet bombardment accounting for a rapid rate of surface water accumulation.

Following Earth's early episode of meteorite and comet bombardment, which ended about 3.8 billion years ago, these extraterrestrial bodies have added little to the accumulating surface waters. However, volcanoes continue to erupt and expel water vapor (much of it recycled surface water), so is the volume of the oceans increasing? Probably it is, but at a considerably reduced rate. The reason is that much of Earth's residual heat from when it formed has dissipated and the amount of radioactive decay to generate internal heat has diminished, so volcanism is not nearly as commonplace. Accordingly, the amount of water added to the oceans now is trivial compared with their volumes.

Recall from Chapter 17 that one early attempt to determine Earth's age was to calculate how long it took for the oceans to reach their current salinity level—assuming, of course, that the oceans formed soon after Earth did, that they were freshwater to begin with, and that their salinity increased at a uniform rate. None of these assumptions is correct, so the ages determined were vastly different. We now know that the very early oceans were salty, probably about as salty as they are now. That is, very early in their history, the oceans reached chemical equilibrium and have remained in near-equilibrium conditions ever since.

19.7 Life—Its Origin and Early History

Today's biosphere is made up of millions of species of organisms assigned to five kingdoms—animals, plants, protistans, fungi, and monera (bacteria and archaea).* Scientists have found probable fossils in rocks 3.5 billion years old, and chemical evidence in 3.8-billion-year-old rocks in Greenland convinces some that organisms were present then. Certainly by 3.0 billion years ago (Mesoarchean) organisms were present, but they represent only one of the kingdoms just listed—monera. In Chapter 18, we discussed the theory of evolution, which accounts for how organisms have changed through time, but this theory does not address how life originated in the first place. Here, we are concerned with **abiogenesis**, that is, how life formed from nonliving matter. Before proceeding, however, note that abiogenesis does not hold that a living organism such as a bacterium, or even a complex organic molecule, sprang fully developed from nonliving material. Rather than one huge step occurring, abiogenesis holds that several small steps took place, each leading to an increase in organization and complexity.

But what is living and nonliving? The distinction is clear in most cases: dogs and trees are alive, rocks and water are not. A biologist might use several criteria to make the

*Archaea include microscopic organisms that resemble bacteria, but differ from them genetically and biochemically.

distinction, including growth and reaction to stimuli, but minimally, a living thing must practice some kind of chemical activity (metabolism) to maintain itself, and it must be capable of reproduction to ensure the long-term survival of the group to which it belongs. This metabolism–reproduction criterion might seem sufficient to decide whether something is living or not, and yet the distinction is not always easy to make.

Bacteria are living, but under some circumstances, they can go for long periods during which they show no signs of living and then go on living again. Are viruses living? They behave like living organisms in the appropriate host cell, but when outside a host cell, they neither metabolize nor reproduce. Some biologists think that viruses represent another way of living, but others disagree. Comparatively simple organic molecules called *microspheres* form spontaneously and grow and divide in a somewhat-organism-like manner, but these processes are more like random chemical reactions, so they are not living.

So what do viruses and microspheres have to do with the origin of life? First, they show that the living versus nonliving distinction is not always clear. And second, if life originated by natural processes, it must have passed through prebiotic stages—that is, stages in which the entities would have shown signs of living organisms but were not truly living.

The Origin of Life

Investigators agree that for abiogenesis to occur, an energy source must have acted upon the appropriate chemical elements from which organic molecules could synthesize. The early atmosphere, composed of carbon dioxide (CO_2), water vapor (H_2O), nitrogen (N_2), and probably ammonia (NH_3) and methane (CH_4), provided carbon, oxygen, hydrogen, and nitrogen, the primary elements that make up all organisms. Lightning and ultraviolet radiation were two possible sources of the energy necessary for these elements to combine and form rather simple organic molecules known as **monomers,** such as amino acids.

Monomers are needed as basic building blocks of more complex organic molecules, but is it plausible that they formed naturally? Experimental evidence indicates that it is. During the 1950s, Stanley Miller circulated gases approximating Earth's early atmosphere through a closed glass vessel and subjected the mixture to an electric spark to simulate lightning. Within a few days, the mixture became turbid and an analysis showed that Miller had synthesized several amino acids common in organisms. More recent experiments using different gases have successfully synthesized many of the 20 amino acids in organisms.

Making monomers in a test tube is one thing, but organisms are composed of more complex molecules called **polymers,** such as nucleic acids (DNA and RNA) and proteins consisting of monomers linked together in a specific sequence. So how did this linking of monomers, or polymerization, take place? Researchers have successfully synthesized small molecules called *thermal proteins,* consisting of more than 200 linked amino acids (❚ Figure 19.19). In fact, when heated, dehydrated, and concentrated, amino acids spontaneously polymerize and form thermal proteins.

We call these artificially synthesized thermal proteins *protobionts,* meaning they have characteristics between those of inorganic chemical compounds and living organisms. Suppose that these protobionts came into existence as outlined previously. They would have been diluted and would have ceased to exist if they had not developed some kind of outer covering. In other words, they had to be self-contained chemical systems as today's cells are. In the experiments just mentioned, proteinoids have spontaneously aggregated into

❚ **Figure 19.19** Experimental Production of Thermal Proteins and Microspheres

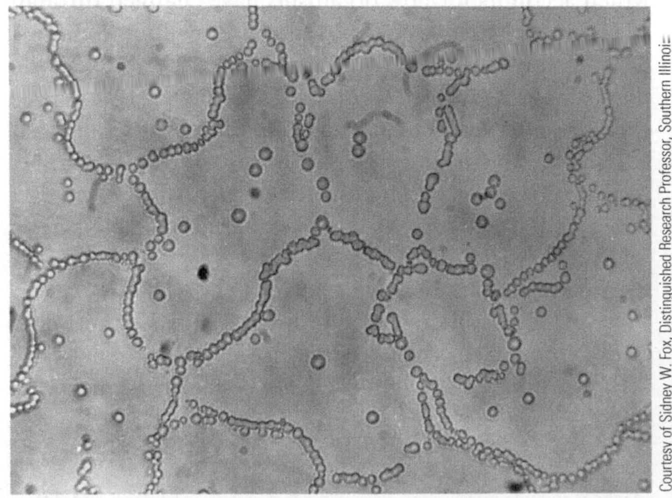

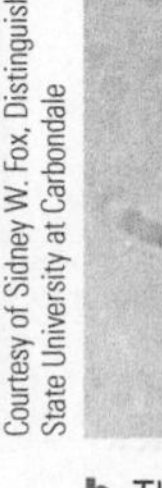

Courtesy of Sidney W. Fox, Distinguished Research Professor, Southern Illinois State University at Carbondale

a Bacteria-like thermal proteins.

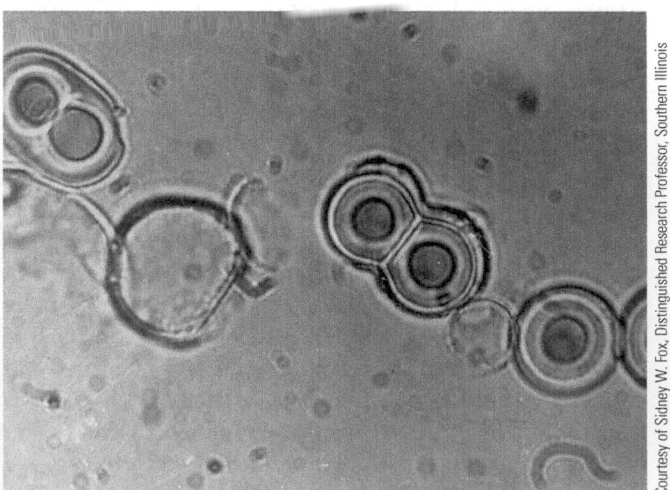

Courtesy of Sidney W. Fox, Distinguished Research Professor, Southern Illinois State University at Carbondale

b Thermal proteins aggregate to form microspheres.

microspheres that have a cell-like outer covering and grow and divide somewhat like bacteria do (Figure 19.19b).

Perhaps the first steps leading to life took place as monomers formed in great abundance and polymerized, but little is known about how a reproductive mechanism came about. Microspheres divide but do so in a nonbiologic manner. In fact, for some time, researchers were baffled because in present-day organisms either DNA or RNA is necessary for reproduction, but these nucleic acids cannot replicate without protein enzymes, yet protein enzymes cannot be made without nucleic acids. Or so it seemed until the 1980s, when researchers discovered that some small RNA molecules can in fact reproduce without the aid of protein enzymes. So, the first replicating system might have been an RNA molecule. Just how these molecules were synthesized under conditions that existed on early Earth has not been resolved.

A common theme is that life originated when organic molecules were synthesized from atmospheric gases, but this idea has been challenged by those who think that the same sequence of events took place near hydrothermal vents (black smokers) on the seafloor (see Figure 2.8b). The necessary elements (carbon, hydrogen, nitrogen, and oxygen) are present in seawater, and heat may have been the energy necessary for monomers to form; in fact, amino acids have been detected in some of these vent waters. Those endorsing this view hold that polymerization took place on the surfaces of clay minerals, and finally protocells were deposited on the seafloor.

Archean Organisms

Prior to the mid-1950s, scientists assumed that the fossils so abundant in Cambrian rocks had a long earlier history, but they had little direct knowledge of Precambrian life. During the early 1900s, Charles Walcott described layered moundlike structures from the Paleoproterozoic-age Gunflint Iron Formation of Canada that he proposed were constructed by algae. We now call these structures **stromatolites,** but not until 1954 did scientists demonstrate that they are actually the result of organic activity. In fact, stromatolites still form today in a few areas where they originate by trapping sediment on sticky mats of photosynthesizing cyanobacteria, more commonly known as blue-green algae (❚ Figure 19.20a). Although widespread in Proterozoic rocks, they are now restricted to aquatic environments with especially high salinity where snails cannot live and graze on them.

❚ Figure 19.20 Stromatolites and Archean Fossils

Critical Thinking Question Stromatolites were widespread during the Proterozoic Eon, but now they have a restricted distribution. Why?

Courtesy of Geoffrey Playford, The University of Queensland, Brisbane, Australia

a Present-day stromatolites, Shark Bay, Australia.

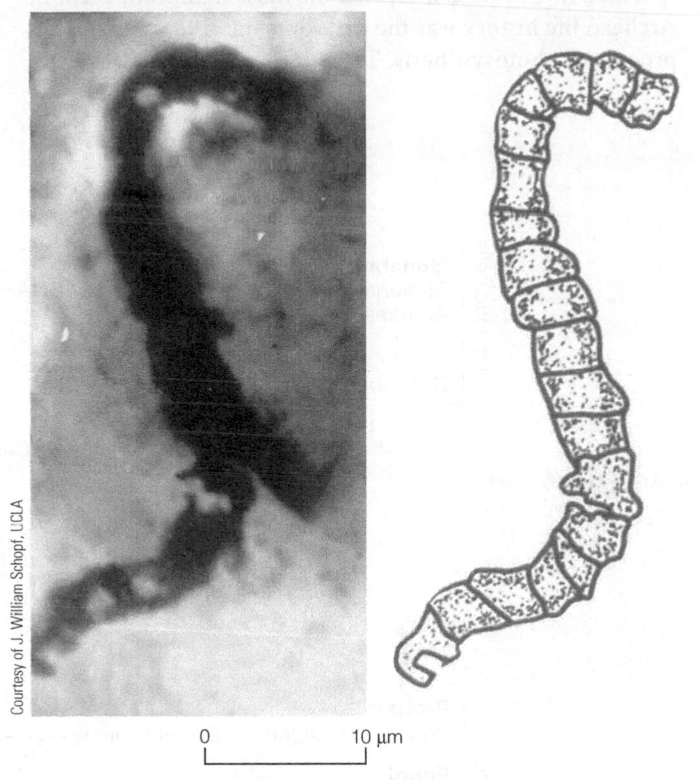

Courtesy of J. William Schopf, UCLA

b Photomicrograph and schematic restoration of fossil prokaryote from the 3.3- to 3.5-billion-year-old Warrawoona Group, Western Australia.

Currently, the oldest known stromatolites are from 3.0-billion-year-old rocks in South Africa, but other probable ones have been discovered in 3.3- to 3.5-billion-year-old rocks near North Pole, Australia. Even more ancient evidence comes from 3.8-billion-year-old rocks in Greenland that have tiny particles of carbon that indicate organic activity, perhaps photosynthesis; but the evidence is not conclusive.

Cyanobacteria, the oldest known fossils, practice photosynthesis, a complex metabolic process that must have been preceded by a simpler process. So it seems reasonable that cyanobacteria were preceded by nonphotosynthesizing organisms for which we so far have no fossils. They probably resembled tiny bacteria, and because the atmosphere lacked free oxygen, they were **anaerobic,** meaning they required no free oxygen. They also were probably **heterotrophic,** dependent on an external source of nutrients, rather than **autotrophic,** capable of manufacturing their own nutrients as in photosynthesis. And finally, they were **prokaryotic cells,**

cells lacking a cell nucleus and other internal structures typical of more advanced *eukaryotic cells* (discussed in a later section).

We can characterize these earliest organisms as single-celled, anaerobic, heterotrophic prokaryotes (▮ Figure 19.20b). Furthermore, they reproduced asexually. Their energy source was probably adenosine triphosphate (ATP), which can be synthesized from simple gases and phosphate, so it was no doubt available in the early Earth environment. These cells may have simply acquired their ATP directly from their surroundings, but this situation could not have persisted for long because as more and more cells competed for the same resources, the supply should have diminished. Thus, a more sophisticated metabolic process developed, probably *fermentation,* an anaerobic process during which molecules such as sugars are split and release carbon dioxide, alcohol, and energy. As a matter of fact, most living prokaryotes practice fermentation.

Of course, the nature of the earliest cells is informed speculation, but we can say that the most significant event in Archean life history was the development of the autotrophic process of photosynthesis. These more advanced cells were still anaerobic and prokaryotic, but as autotrophs they no longer depended on an external source of preformed organic molecules for their nutrients. The Archean fossil in Figure 19.20b belongs to the kingdom Monera, which is represented today by bacteria and archaea.

Life of the Proterozoic

The Paleoproterozoic, like the Archean, is characterized mostly by a biota of single-celled, prokaryotic archea and bacteria. No doubt thousands of varieties existed, but none of the more familiar organisms such as plants and animals were present. Before the appearance of cells capable of sexual reproduction, evolution was a comparatively slow process, accounting for the low organic diversity.

A New Type of Cell Appears The origin of **eukaryotic cells** marks one of the most important events in life history (Table 19.2). These cells are much larger than prokaryotic cells; they have a membrane-bounded nucleus containing the genetic material, and most reproduce sexually (▮ Figure 19.21). And many eukaryotes—that is, organisms

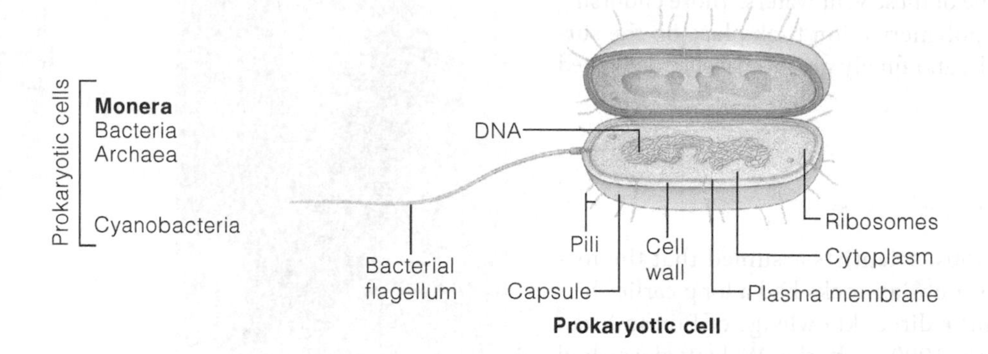

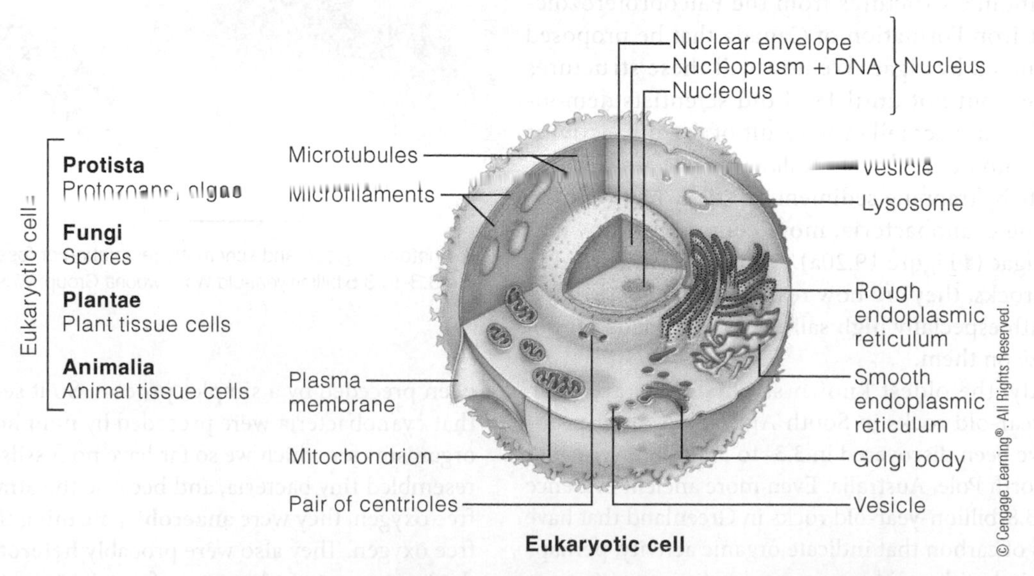

▮ **Figure 19.21 Prokaryotic and Eukaryotic Cells** Eukaryotic cells have a cell nucleus containing the genetic material and organelles such as mitochondria and plastids. In contrast, prokaryotic cells are smaller and not nearly as complex as eukaryotic cells.

Figure 19.22 The Oldest Known Eukaryote and Megafossil

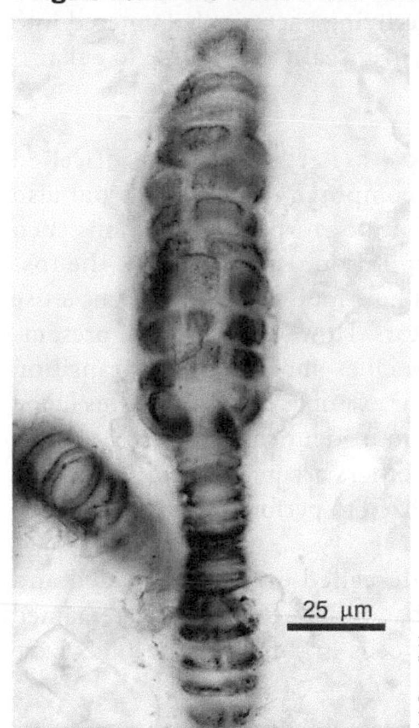

N.J. Butterfield

a At 1.2 billion years old, *Bangiomorpha* is the oldest known eukaryotic organism. It is only a few millimeters long.

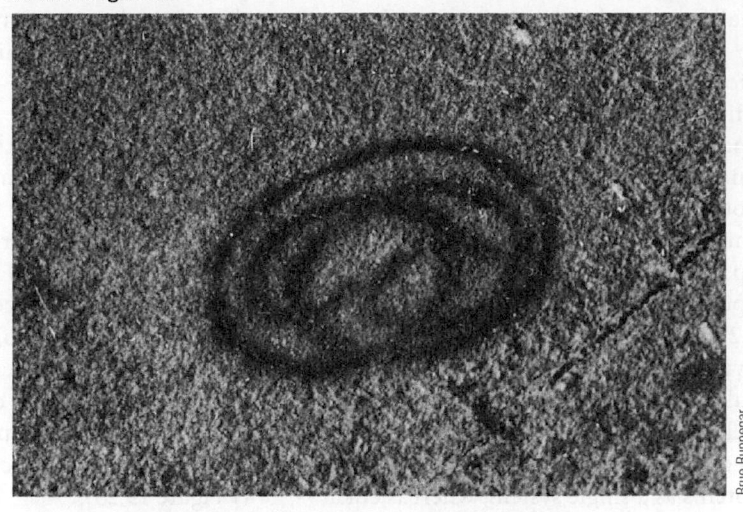

Brue Runnegar

b *Grypania,* at 2.1 billion years old, is the oldest known megafossil. It was probably a bacterium or some kind of algae. It is about 1 cm across.

made up of eukaryotic cells—are multicellular and aerobic, so they could not have existed until some free oxygen was present in the atmosphere.

No one doubts that eukaryotes were present by the Mesoproterozoic. Currently, the oldest known eukaryote is *Bangiomorph* from 1.2-billion-year-old rocks in Canada; it was multicelled and reproduced sexually (Figure 19.22a). Other Proterozoic rocks have yielded fossils of unknown affinities. For example, the 2.1-billion-year-old Negaunee Iron Formation of Michigan has fossils known as *Grypania,* which is the oldest known megafossil, but it was probably a single-celled bacterium or some kind of algae (Figure 19.22b).

Cells larger than 60 microns appear in abundance at least 1.4 billion years ago, and many show an increased degree of organizational complexity compared with that of prokaryotic cells. An internal membrane-bounded cell nucleus is present in some, for instance. Furthermore, microscopic, hollow fossils known as *acritarchs* that probably represent cysts of planktonic algae become common during the Meso- and Neoproterozoic (Figure 19.23a, b). And vase-shaped microfossils from rocks in the Grand Canyon have also been tentatively identified as cysts of some kind of algae (Figure 19.23c).

Endosymbiosis and the Origin of Eukaryotic Cells According to a widely accepted theory, eukaryotic cells formed from several prokaryotic cells that entered into a symbiotic relationship. Symbiosis, involving a prolonged association of two or more dissimilar organisms, is common today. In many cases, both symbionts benefit from

Figure 19.23 Proterozoic Fossils These are probably from eukaryotic organisms and measure only 40 to 50 microns across.

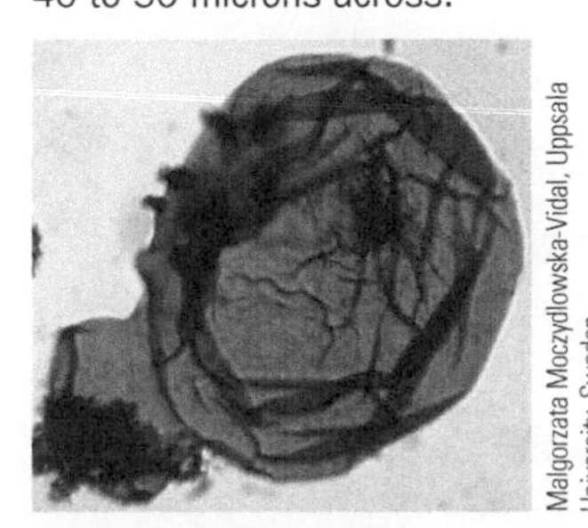

Malgorzata Moczydlowska-Vidal, Uppsala University, Sweden

a The acritarch *Tappania plana* is from Mesoproterozoic rocks in China.

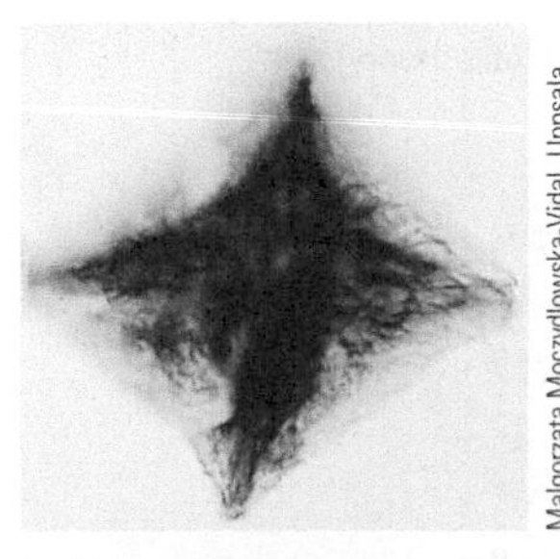

Malgorzata Moczydlowska-Vidal, Uppsala University, Sweden

b This acritarch, known as *Octoedryxium truncatum,* was found in Neoproterozoic rocks in Sweden.

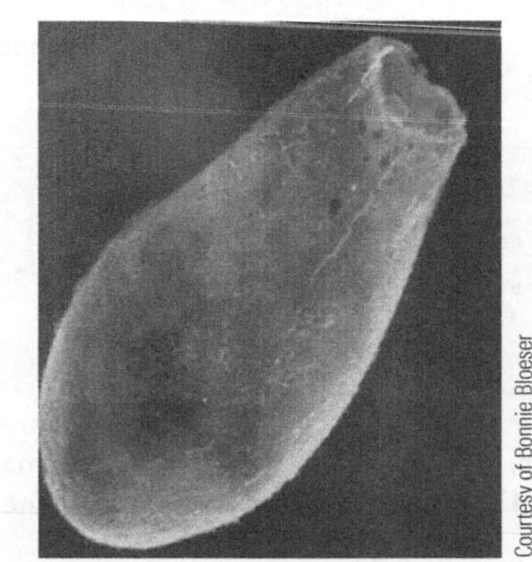

Courtesy of Bonnie Bloeser

c This vase-shaped microfossil from Neoproterozoic rocks in the Grand Canyon in Arizona is probably a cyst from some kind of algae.

the association, as in lichens, which were once thought to be plants but actually are symbiotic associations between fungi and algae.

In a symbiotic relationship, each symbiont must be capable of metabolism and reproduction, but the degree of dependence in some relationships is such that one symbiont cannot live independently. This may have been the case with Proterozoic symbiotic prokaryotes that became increasingly interdependent until the unit could exist only as a whole. In this relationship, though, one symbiont lived within the other, which is a special type of symbiosis called **endosymbiosis** (❚ Figure 19.24).

Supporting evidence for endosymbiosis comes from studies of living eukaryotic cells containing internal structures called organelles, such as mitochondria and plastids, that have their own genetic material. In addition, prokaryotic cells synthesize proteins as a single system, whereas eukaryotic cells are a combination of protein-synthesizing systems. That is, some of the organelles within eukaryotic cells are capable of protein synthesis. These organelles with their own genetic material and protein-synthesizing capabilities are thought to have been free-living bacteria that entered into a symbiotic relationship, eventually giving rise to eukaryotic cells.

The Dawn of Multicelled Organisms

Multicelled organisms not only are composed of many cells but also have cells specialized to perform specific functions such as reproduction and respiration. Unfortunately, the fossil record does not tell us how multicelled organisms arose from single-celled ancestors. However, studies of present-day organisms give some clues about how this transition might have taken place. For example, some organisms living now consist of as few as four identical cells, each capable of living on its own, and others are made up of many cells some of which are specialized to perform specific functions (❚ Figure 19.25).

Suppose that a single-celled organism divided and formed a group of cells that did not disperse, but remained together as a colony. The cells in some colonies may have

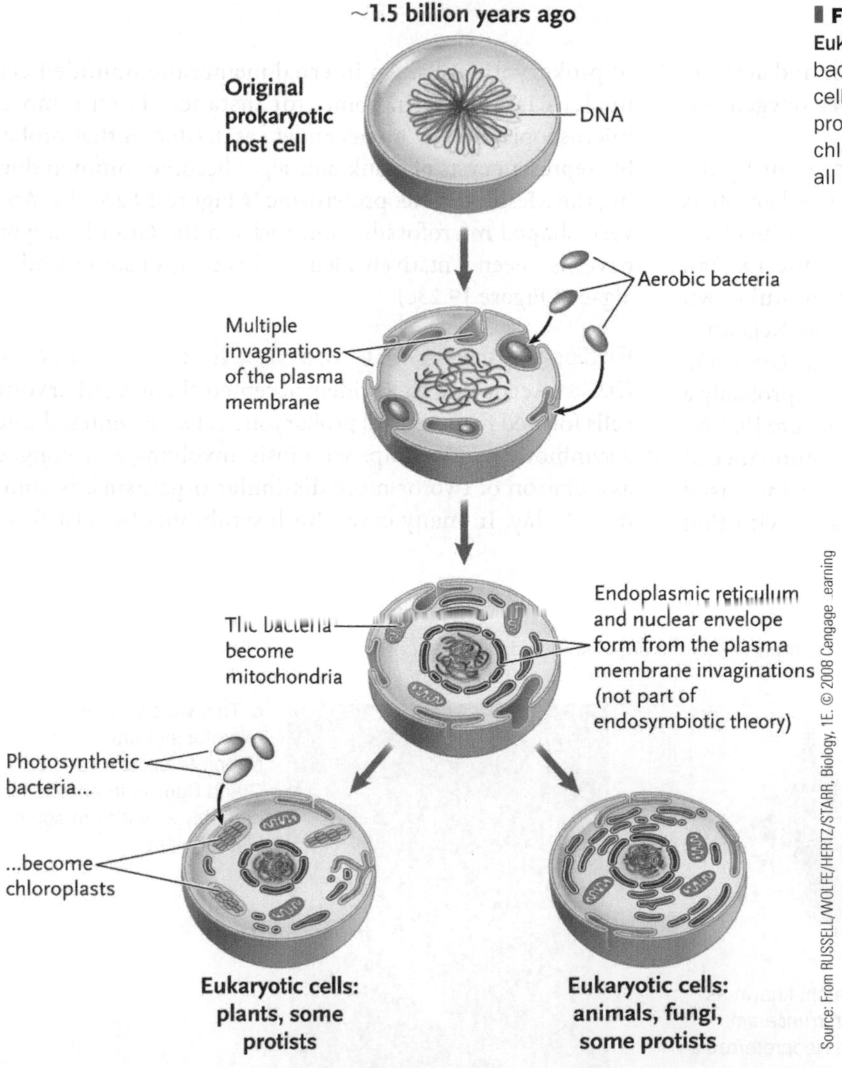

❚ Figure 19.24 Endosymbiosis and the Origin of Eukaryotic Cells An aerobic bacteria and a larger host bacterium unite and form a mitochondria-containing cell. This cell gives rise to the animal, plant, and protistan kingdoms, whereas cells that acquire chloroplasts give rise to plants and some protists, all of which are eukaryotes.

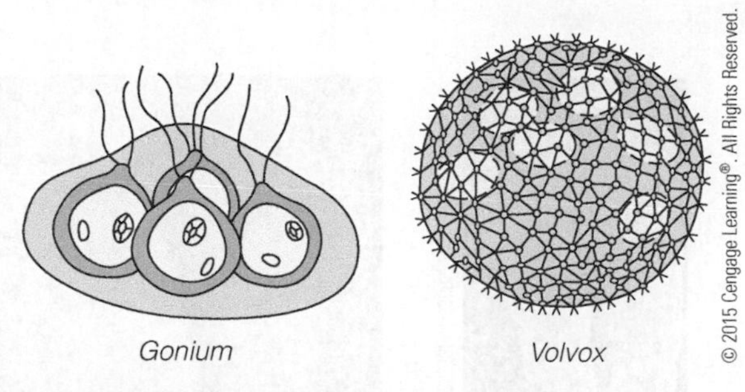

Figure 19.25 Single-Celled and Multicelled Organisms *Gonium* consists of as few as four cells. All cells are alike and can reproduce a new colony. *Volvox* has some cells specialized for specific functions, so it has crossed the threshold that separates single-celled from multicelled organisms.

become somewhat specialized, similar to the situation in some living *colonial organisms* (Figure 19.25b). Further specialization might have led to simple multicelled organisms such as sponges, consisting of cells that carry out functions such as reproduction, respiration, and food gathering. Carbonaceous impressions of Proterozoic multicelled algae are known from many areas.

The Ediacaran Fauna In 1947, the Australian geologist R. C. Sprigg discovered a unique assemblage of multicelled, soft-bodied animals preserved as molds and casts on the undersides of sandstone layers (Figure 19.26). Some investigators think that these fossils represent at least three present-day invertebrate phyla: jellyfish and sea pens (phylum Cnidaria), segmented worms (phylum Annelida), and primitive members of the phylum Arthropoda. One wormlike Ediacaran fossil, *Spriggina*, has been cited as a possible ancestor of trilobites, and another, *Tribachidium*, may be a primitive echinoderm (Figure 19.26a, b). On the other hand, some think Ediacaran fossils represent an early evolutionary radiation distinct from the ancestry of existing invertebrates. These **Ediacaran faunas** existed between 635 and 542 million years ago and are now known on all continents except South America and Antarctica. The animals were widespread, but their fossils are not very common because all of them lacked durable skeletons.

Other Proterozoic Animal Fossils Although scarce, a few animal fossils older than those of the Ediacaran fauna are known from Neoproterozoic rocks (Figure 19.27). A jellyfish-like impression is present in rocks 2,000 m below the Pound Quartzite, and in many areas, burrows, presumably made by worms, are found in rocks at least 700 million years old. Wormlike fossils, as well as fossil algae, come from 700- to 900-million-year-old rocks in China, but the identity and age of these "fossils" have been questioned.

All known Proterozoic animals were soft-bodied, but there is some evidence that the earliest stages in the origin of skeletons were already under way. Even some Ediacaran animals may have had a chitinous carapace, and others appear to have had areas of calcium carbonate. The odd creature known as *Kimberella*, from the latest Proterozoic of Russia, had a tough outer covering similar to that of some present-day marine invertebrates (Figure 19.27c). Exactly what *Kimberella* was remains uncertain; some think that it was a mollusk.

ConnectionLink

For more information about the emergence of a shelly fauna in the Late Proterozoic and the beginning of the Paleozoic, see Chapter 21.

Neoproterozoic fossils of minute scraps of shell-like material, as well as small toothlike denticles and spicules, presumably from sponges, indicate that several animals with skeletons or at least partial skeletons existed. But more durable skeletons of silica, calcium carbonate,

Figure 19.26 The Ediacaran Fauna of Australia

a The affinities of *Tribachidium* remain uncertain. It may be either a primitive echinoderm or a cnidarian.

b *Spriggina* was originally thought to be a segmented worm (annelid), but now it appears more closely related to arthropods, possibly even an ancestor of trilobites.

c Shield-shaped *Parvanconrina* is perhaps related to the arthropods.

Figure 19.27 Neoproterozoic Fossils

Dr. G. M. Narbonne

a These fossils from Newfoundland have not been given scientific names. The one at the upper center is called a *feather duster,* whereas the other fossils are called *spindles*.

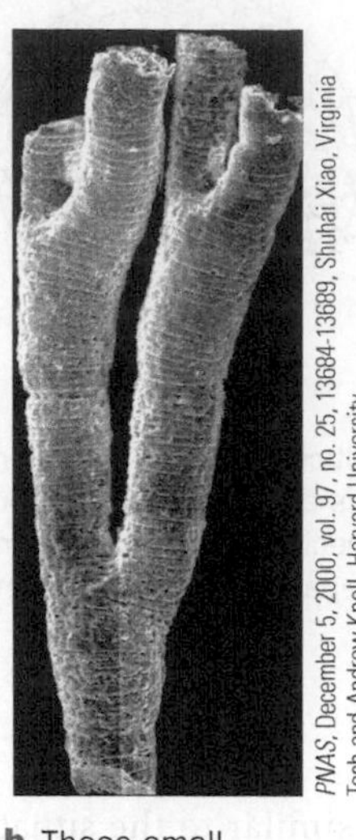

PNAS, December 5, 2000, vol. 97, no. 25, 13684-13689, Shuhai Xiao, Virginia Tech and Andrew Knoll, Harvard University

b These small branching tubes from China may have been made by early relatives of corals. The tubes measure 0.1 to 0.3 mm across.

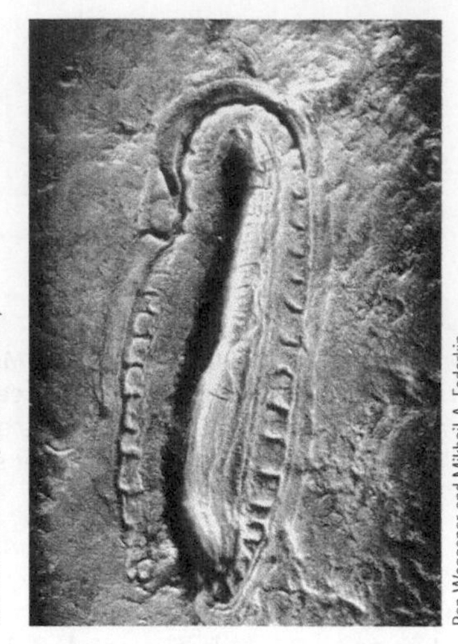

Ben Waggoner and Mikhail A. Fedarkin

c All agree that *Kimberella* from Russia was an animal; some think it was a mollusk. Specimens measure 3 to 10 cm long.

and chitin (a complex organic substance) did not appear in abundance until the beginning of the Phanerozoic Eon 542 million years ago. Nevertheless, by the end of the Proterozoic, Earth had a diverse biota that included archea, bacteria, sponges, jellyfish, worms, algae (including acritarchs), arthropods, possible corals, and several others of unknown affinities.

19.8 Resources in Precambrian Rocks

Many resources are found in Precambrian rocks, but the one most often associated with rocks of Archean age is gold, whereas iron ore is the one that comes to mind when we discuss Proterozoic resources.

Archean Resources

Archean and Proterozoic rocks near Johannesburg, South Africa, have yielded more than 50% of the world's gold since 1886; but Archean gold is or has also been mined in the Superior craton in Canada and the Homestake Formation in South Dakota. Several Archean-age massive sulfide deposits of zinc, copper, and nickel are known in Australia, Zimbabwe, and Canada, and similar deposits are currently forming adjacent to black smokers on the seafloor.

About one-fourth of the chrome reserves are in Archean rocks, especially in Zimbabwe; they probably formed along with crystals and settled in mafic and ultramafic plutons. The Stillwater Complex in Montana has low-grade chrome and platinum ores that have been mined and stockpiled during the past, but so far they have not been refined for their chrome or platinum.

Archean-age pegmatites are mostly granitic and of little economic importance. However, some in Manitoba, Canada, and the Rhodesian Province in Africa contain valuable minerals. In addition to gem-quality minerals, a few Archean pegmatites are mined for lithium, beryllium, and rubidium.

Proterozoic Resources

BIFs, the world's major source of iron ores, are present on all continents, and 92% of all BIFs were deposited during the Paleoproterozoic (**Figure 19.10**). Most of the iron ore mined in North America comes from mines in the Lake Superior region of Minnesota and Michigan and in eastern Canada.

The Sudbury mining district in Ontario, Canada, is an important area of nickel and platinum production. Nickel is essential for the manufacture of nickel alloys, such as stainless steel and Monel metal (nickel plus copper), which are valued for their strength and resistance to corrosion and heat. The United States imports 47% of all the nickel it uses, most from the Sudbury mining district in Canada. Some platinum for jewelry, surgical instruments, and chemical and electrical equipment is also exported from Canada to the United States, but the largest exporter is South Africa. The United States also depends on South Africa for much of its chromite, the ore of chromium.

The 2.0-billion-year-old Bushveld complex in South Africa is a huge layered igneous body with rich ores of chromium, platinum, and several other metals. It probably originated as multiple intrusions of magma slowly crystallized so that different minerals were concentrated in nearly horizontal layers.

Economically recoverable oil and natural gas were discovered in Proterozoic rocks in China and Siberia, arousing interest in the Midcontinent rift (Figure 19.13) as a potential source of hydrocarbons. Some rocks within the rift are known to contain oil, but so far no oil or gas wells are operating.

Key Concepts Review

- Geologists use a threefold division of the Precambrian—the Hadean, Archean, and Proterozoic.
- Each continent has an ancient, stable craton made up of a Precambrian shield and platform. The Canadian shield in North America is made up of several subunits.
- Archean rocks are mostly granite–gneiss complexes and subordinate greenstone belts. One model for the origin of greenstone belts holds that they formed in back-arc basins.
- The amalgamation of Archean cratons and continental accretion along their margins account for the origin of a large landmass known as Laurentia.
- Many geologists think that Archean plates moved faster than plates do now because Earth possessed more residual and radiogenic heat.
- Sandstone–carbonate–shale assemblages deposited on passive continental margins and in intracratonic basins are the most common Proterozoic-age rocks.
- Widespread glaciers were present during the Paleoproterozoic and Neoproterozoic.
- Earth's earliest atmosphere lacked free oxygen, but it was rich in carbon dioxide. It was derived by the release of gases during volcanism, a process called outgassing. Meteorite and comet impacts and outgassing yielded the hydrosphere.
- Deposition of widespread banded iron formations between 2.5 and 2.3 billion years ago and the first red beds about 2.3 billion years ago indicate that some free oxygen was present in the atmosphere.
- Energy such as lightning and ultraviolet radiation acting on chemical elements present on early Earth may have yielded the first living things. Some investigators think that RNA molecules were the first molecules capable of reproduction.
- All known Archean fossils represent prokaryotic bacteria. Stromatolites formed by photosynthesizing bacteria may date from 3.5 billion years ago.
- Endosymbiosis practiced by prokaryotic cells was probably responsible for the first eukaryotic cells.
- The oldest megafossils are carbonaceous impressions, probably of algae, in rocks more than 2 billion years old.
- The Neoproterozoic Ediacaran faunas include the oldest well-documented animal fossils. None had durable skeletons, so their fossils are not common.
- Gold and iron ore are the best known Precambrian natural resources.

Important Terms

abiogenesis (p. 489)
anaerobic (p. 491)
autotrophic (p. 491)
back-arc basin (p. 479)
banded iron formation (BIF) (p. 488)
Canadian shield (p. 474)
continental accretion (p. 474)
continental red bed (p. 489)
craton (p. 474)
Ediacaran faunas (p. 495)
endosymbiosis (p. 494)
eukaryotic cell (p. 492)
granite–gneiss complex (p. 476)
Great Oxygenation Event (p. 488)
greenstone belt (p. 477)
Grenville orogeny (p. 484)
heterotrophic (p. 491)
Laurentia (p. 481)
Midcontinent rift (p. 484)
monomer (p. 490)
multicelled organism (p. 494)
orogen (p. 482)
outgassing (p. 487)
Pannotia (p. 486)
photochemical dissociation (p. 487)
photosynthesis (p. 488)
platform (p. 474)
polymer (p. 490)
prokaryotic cell (p. 491)
Rodinia (p. 486)
sandstone–carbonate–shale assemblage (p. 483)
shield (p. 474)
stromatolite (p. 491)
Wilson cycle (p. 483)

Review Questions

1. The term abiogenesis is used for
   a. ____ the amalgamation of cratons.
   b. ____ a process whereby banded iron formations were deposited.
   c. ____ the origin of life from nonliving matter.
   d. ____ a metabolic process involving fermentation.
   e. ____ layered, moundlike structures produced by cyanobacteria.
2. Which one of the following sequences of geologic time designations is in the correct order from oldest to youngest?
   a. ____ Phanerozoic-Archean-Mesozoic.
   b. ____ Hadean-Archean-Proterozoic.
   c. ____ Archean-Hadean-Paleozoic.
   d. ____ Cenozoic-Proterozoic-Phanerozoic.
   e. ____ Proterozoic-Cenozoic-Precambrian.
3. The origin of greenstone belts has not been fully resolved, but many geologists think that they formed in
   a. ____ transverse shear zones.
   b. ____ carbonate-evaporite depositional systems.
   c. ____ continental shelf environments.
   d. ____ river floodplains.
   e. ____ back-arc basins.
4. A large landmass composed mostly of North America and Greenland that evolved during the Proterozoic Eon is called
   a. ____ Atlantis.
   b. ____ Mesoamerica.
   c. ____ Wilsonia.
   d. ____ Laurentia.
   e. ____ Grenvillia.
5. Photochemical dissociation is a process whereby
   a. ____ water molecules are disrupted to form hydrogen and oxygen.
   b. ____ plants synthesize organic molecules.
   c. ____ continents grow along their margins by accretion.
   d. ____ methane and chlorine are emitted during volcanic eruptions.
   e. ____ monomers link together in a specific sequence to form polymers.
6. Proterozoic rocks in the northern Rocky Mountains are 4,000 m thick and were deposited between 1.4 billion and 850 million years ago. What was the average rate of sediment accumulation in centimeters per year? Why is this figure unlikely to represent the actual rate of sedimentation?

7. Outline the events that led to the evolution of Laurentia during the Proterozoic.
8. What are the most common types of Archean rocks and how are they thought to have formed?
9. How do continental red beds and banded iron formations provide evidence about changes in the Proterozoic atmosphere?

10. **Creative Thinking Visual Question:** The illustration below (Figure 1) shows the 40- to 90-m-thick, 1.2-billion-year-old Purcell Sill in the Northern Rocky Mountains. What evidence convinces you that this is in fact a sill rather than a buried lava flow? Also, what can you say about the numeric ages of the rocks below and above the sill? The minerals in the central part of the sill are larger than those near its upper and lower boundaries. Why?

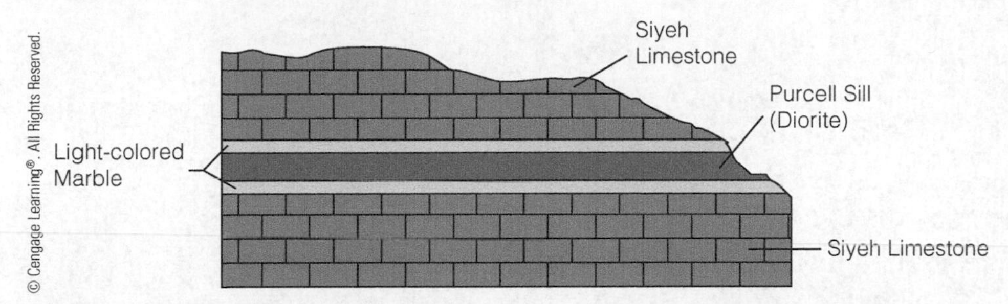

Figure 1 Diagram of the Purcell Sill and associated rocks in the Northern Rocky Mountains.

Global GeoScience Watch In the GREENR database search for "fossils," then within this portal search for "3.5 billion." Under "News" click on the article titled "Fossils May Date from 3.5 Billion Years Ago." Where and what fossil organisms were found in these ancient rocks? How do they differ from most other fossils, say dinosaur bones, and what implications do they have for the history of life?

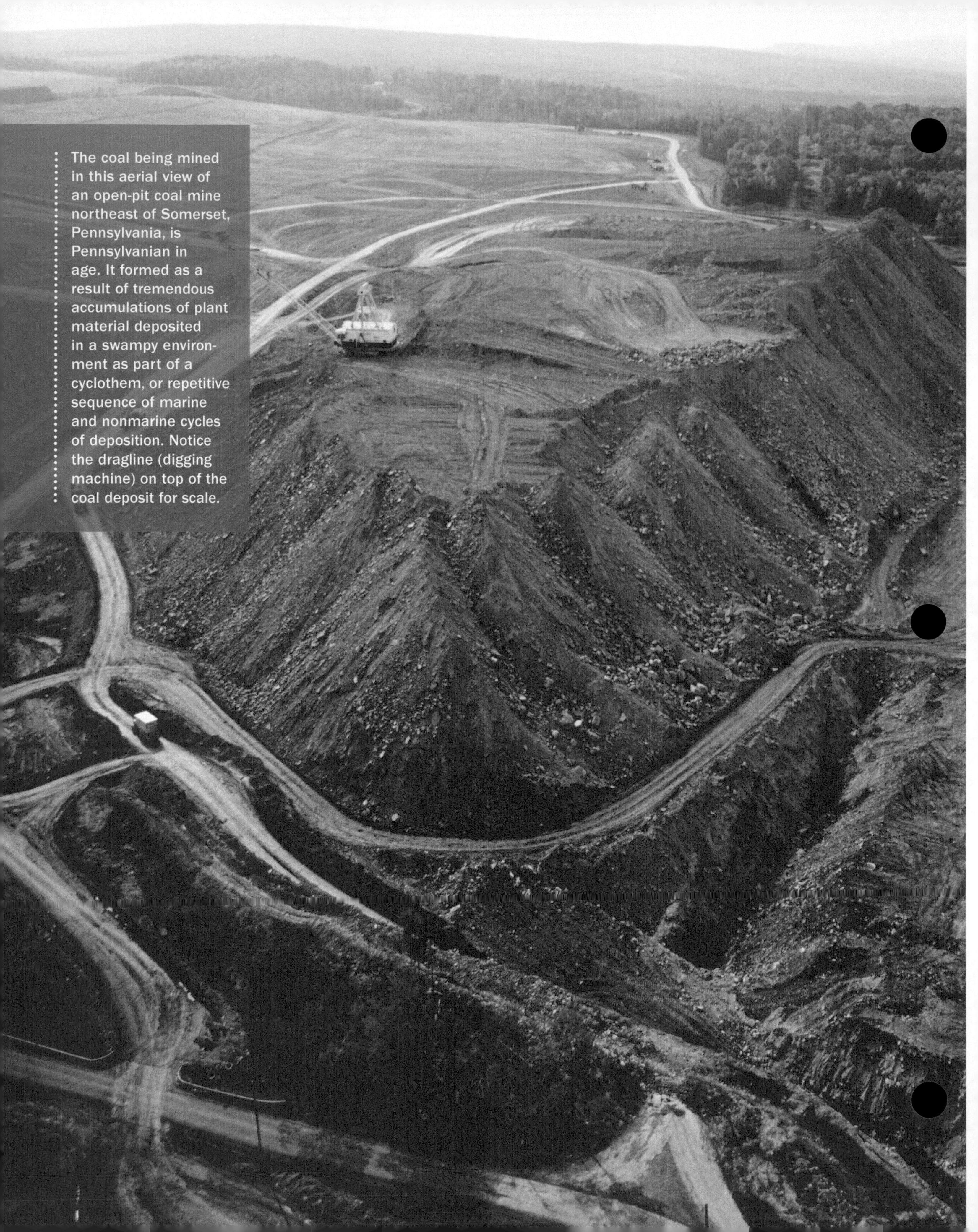

The coal being mined in this aerial view of an open-pit coal mine northeast of Somerset, Pennsylvania, is Pennsylvanian in age. It formed as a result of tremendous accumulations of plant material deposited in a swampy environment as part of a cyclothem, or repetitive sequence of marine and nonmarine cycles of deposition. Notice the dragline (digging machine) on top of the coal deposit for scale.

© H. Mark Weidman Photography/Alamy

Paleozoic Earth History

OUTLINE

HAVE YOU EVER WONDERED?

- What the geography of Earth looked like during the Paleozoic Era?
- What the climate of Earth was during the Paleozoic Era?
- What North America looked like during the Paleozoic Era, and how it changed from the beginning of the Cambrian Period to the end of the Permian Period?
- How Pangaea came into existence, and how its formation at the end of the Paleozoic Era affected the world's climate?
- How the Appalachian Mountains formed?
- Why the rocks of the Catskill Mountains look the same as many of the rocks that were used to build the castles of England and Scotland?
- About the origin of the vast coal deposits found in the Midwestern states and the Appalachian region of the United States?

20.1 Introduction

Coal. The name conjures up various images. For some, it is what helped fuel the Industrial Revolution in England during the 18th and 19th centuries. For others, it brings to mind the many hazards associated with underground coal mining, and the denudation of the landscape caused by open-pit surface mining. And, more recently, coal has been linked closely with global warming as a result of its use in industry and power plants.

Coal is a biochemical sedimentary rock composed of the partially altered, compressed remains of land plants (see Figure 7.10). It forms in anaerobic swamps and bogs, where decomposition of the plant matter is interrupted and the accumulating vegetation is altered into peat. If the peat is later covered and buried by sediments, it is then converted into coal.

Much of the world's coal formed during the Late Carboniferous (Pennsylvanian Period), when widespread swampy conditions provided the ideal conditions for coal formation. It is the coal mined from these Carboniferous deposits that has fueled the economies of many of the world's nations. But coal did not become an important commodity until it was needed in quantity to fuel the steam engines and furnaces of factories during the Industrial Revolution.

William Smith (1769–1839), an English civil engineer involved in surveying and building canals in southern England, and who also published the world's first geologic map, got his start mapping various coal mines. He soon turned his attention to finding the most efficient canal routes for bringing coal to market. In fact, coal production in England increased from 2.7 million tons in 1700 to 10 million tons by 1800 as a direct result of the Industrial Revolution.

ConnectionLink
To review William Smith's principle of fossil succession, as well as how rocks are correlated, see Chapter 17.

Today, coal is primarily used to generate electricity. In the United States, coal-fired power plants provided 37.4% of the electricity generated in 2012. Other uses of coal are for iron and steel production; liquefaction, in which it is converted into liquid fuels such as diesel or gasoline; and home heating, as well as any industrial application requiring large amounts of heat, such as in cement manufacturing.

Coal is generally mined in two ways, and the method used depends on the geology of the coal deposit. In underground mining, shafts are drilled into the *coal seam* (a layer or bed of coal), and the seam is then followed until as much of the coal that can be extracted is removed. Underground mining currently accounts for about 60% of the world's coal production.

When the coal is close to the surface, and of widespread extent, then open-pit, or surface mining is used. This method involves removing the surface rocks and exposing the coal seam, where it can be drilled or blasted and loaded onto trucks or conveyers (see chapter opening photo). One type of surface mining, called mountaintop removal mining, literally involves the removal of the top of a mountain, and filling in the adjacent valleys with excavated material. This method of mining is highly controversial because of the damage done to the surrounding environment as a result of the denudation of the surface.

As of 2011, world production of coal amounted to more than 6,637 million tons of hard coal, that is, bituminous grade or higher. The top producer of coal, and one of its largest consumers is China, which according to 2011 statistics, mined 3,162 million tons of hard coal, followed by the United States, at 932 million tons, India, Australia, and South Africa. As impressive as these numbers are, it is estimated that, based on current usage, the United States has an approximately 300-year supply of coal.

Despite its abundance and contribution to the world's energy supply, coal has disadvantages. The extraction and burning of coal releases many heavy metals into the environment that are harmful to human health. Furthermore, the burning of coal, especially in power plants, releases tremendous quantities of carbon dioxide, which cause climate change and global warming.

The Paleozoic history of most continents involves major mountain-building activity along their margins and numerous shallow-water marine transgressions and regressions over their interiors. These transgressions and regressions were caused by global changes in sea level that most probably were related to plate movement and glaciation.

This chapter first provides an overview of the geologic history of the world during the Paleozoic Era to place into context the geologic events taking place in North America during this time. It then focuses attention on the geologic history of North America—not in a period-by-period chronology, but rather in terms of the major transgressions and regressions taking place on the continents, as well as the mountain-building activity occurring during this time. Such an approach allows us to place North American geologic events within a global context.

20.2 Continental Architecture: Cratons and Mobile Belts

During the Precambrian, continental accretion and orogenic activity led to the formation of sizable continents. Movement of these continents during the Mesoproterozoic led to the first supercontinent known as Rodinia. Following fragmentation of Rodinia sometime around 750 million years ago (Neoproterozoic), a second Pangaea-like supercontinent, referred to as Pannotia, formed. This supercontinent began breaking apart during the

ConnectionLink
To review Proterozoic supercontinents and their histories, see Chapter 19.

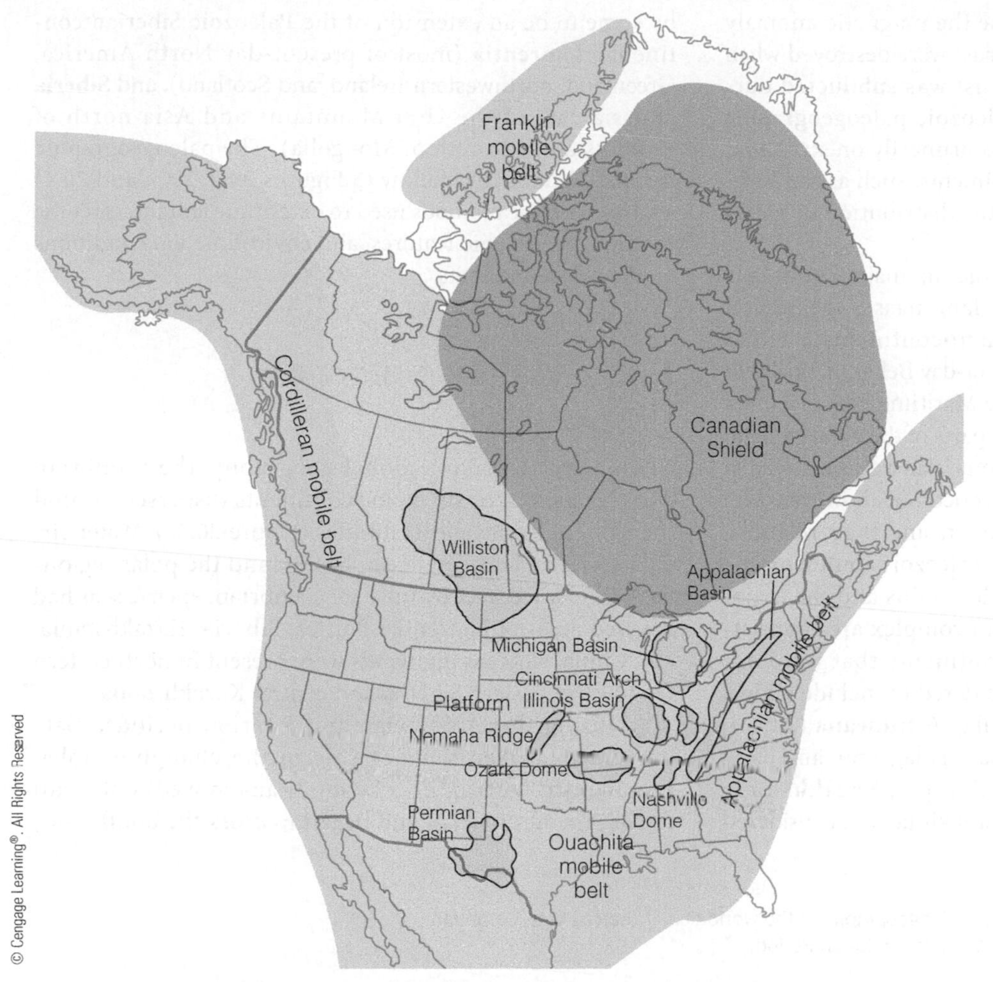

Figure 20.1 Major Cratonic Structures and Mobile Belts The major cratonic structures and mobile belts of North America that formed during the Paleozoic Era.

latest Neoproterozoic (550 million years ago), and by the beginning of the Paleozoic Era, six major continents were present. Each continent can be divided into two major components: a craton and one or more mobile belts.

Cratons are the relatively stable and immobile parts of continents and form the foundation on which Phanerozoic sediments were deposited (see Figure 19.5). Cratons typically consist of two parts: a shield and a platform. *Shields* are the exposed portions of the crystalline basement rocks of a continent and are composed of Precambrian metamorphic and igneous rocks that reveal a history of extensive orogenic activity during the Precambrian (see Figure 19.5). During the Phanerozoic, however, shields were extremely stable and formed the foundation of the continents.

Extending outward from the shields are buried Precambrian rocks that constitute a *platform*, another part of the craton (see Figure 19.5). Overlying the platform are flat-lying or gently dipping Phanerozoic detrital and chemical sedimentary rocks that were deposited in widespread shallow seas that transgressed and regressed over the craton. These seas, called **epeiric seas**, were a common feature of most Paleozoic cratonic histories. Changes in sea level, caused primarily by continental glaciation, as well as by plate movement, were responsible for the advance and retreat of these epeiric seas.

Whereas most Paleozoic platform rocks are still essentially flat-lying, in some places, they were gently folded into regional arches, domes, and basins (Figure 20.1). In many cases, some of these structures stood out as low islands during the Paleozoic Era and supplied sediments to the surrounding epeiric seas.

Mobile belts are elongated areas of mountain-building activity. They are located along the margins of continents where sediments are deposited in the relatively shallow waters of the continental shelf and the deeper waters at the base of the continental slope. During plate convergence along these margins, the sediments are deformed and intruded by magma, creating mountain ranges (see Figures 10.19, 10.20, and 10.21).

Four mobile belts formed around the margin of the North American craton during the Paleozoic: the Franklin, **Cordilleran**, **Ouachita**, and **Appalachian mobile belts** (Figure 20.1). Each was the site of mountain building in response to compressional forces along a convergent plate boundary and formed mountain ranges, such as the Appalachians and Ouachitas.

20.3 Paleozoic Paleogeography

One result of plate tectonics is that Earth's geography is constantly changing. The present-day configuration of the continents and ocean basins is merely a snapshot in time. One of the goals of historical geology is to provide paleogeographic reconstructions of the world during the geologic past. By synthesizing all of the pertinent paleoclimatic, paleomagnetic, paleontologic, sedimentologic, stratigraphic, and tectonic data available, geologists can construct paleogeographic maps. Such maps are simply interpretations of the geography of an area for a particular time in the geologic past and usually show the distribution of land and sea, possible climatic regimes, and such geographic features as mountain ranges, swamps, and glaciers.

The paleogeographic history of the Paleozoic Era, for example, is not as precisely known as that of the Mesozoic and Cenozoic eras, in part because the magnetic anomaly patterns preserved in the oceanic crust were destroyed when much of the Paleozoic oceanic crust was subducted during the formation of Pangaea. Paleozoic paleogeographic reconstructions are therefore based primarily on structural relationships; climate-sensitive sediments, such as red beds, evaporites, tillites, and coals; and the distribution of plants and animals.

At the beginning of the Paleozoic, six major continents were present. Besides these large landmasses, geologists have also identified numerous microcontinents, such as *Avalonia* (composed of parts of present-day Belgium, northern France, England, Wales, Ireland, the Maritime Provinces and Newfoundland of Canada, as well as parts of the New England area of the United States), and various island arcs associated with microplates. We are primarily concerned, however, with the history of the six major continents and their relationships to one another. The six major Paleozoic continents are **Baltica** (Russia west of the Ural Mountains and the major part of northern Europe), **China** (a complex area consisting of at least three Paleozoic continents that were not widely separated and are here considered to include China, Indochina, and the Malay Peninsula), **Gondwana** (Africa, Antarctica, Australia, Florida, India, Madagascar, and parts of the Middle East and southern Europe), **Kazakhstania** (a triangular continent centered on Kazakhstan but considered by some to be an extension of the Paleozoic Siberian continent), **Laurentia** (most of present-day North America, Greenland, northwestern Ireland, and Scotland), and **Siberia** (Russia east of the Ural Mountains and Asia north of Kazakhstan and south of Mongolia). The paleogeographic reconstructions that follow (❚ Figures 20.2, 20.3, and 20.4) are based on the methods used to determine and interpret the location, geographic features, and environmental conditions on the paleocontinents.

Early-Middle Paleozoic Global History

In contrast to today's global geography, the Cambrian world consisted of six major continents dispersed around the globe at low tropical latitudes (Figure 20.2a). Water circulated freely between ocean basins, and the polar regions were mostly ice free. By the Late Cambrian, epeiric seas had covered areas of Laurentia, Baltica, Siberia, Kazakhstania, and China, whereas highlands were present in northeastern Gondwana, eastern Siberia, and central Kazakhstania.

During the Ordovician and Silurian periods, plate movement played a major role in the changing global geography (Figure 20.2b, c). Gondwana moved southward during the Ordovician and began to cross the South Pole,

❚ **Figure 20.2 Paleozoic Paleogeography** Paleogeography of the world during the **(a)** Late Cambrian Period, **(b)** Late Ordovician Period, and **(c)** Middle Silurian Period.

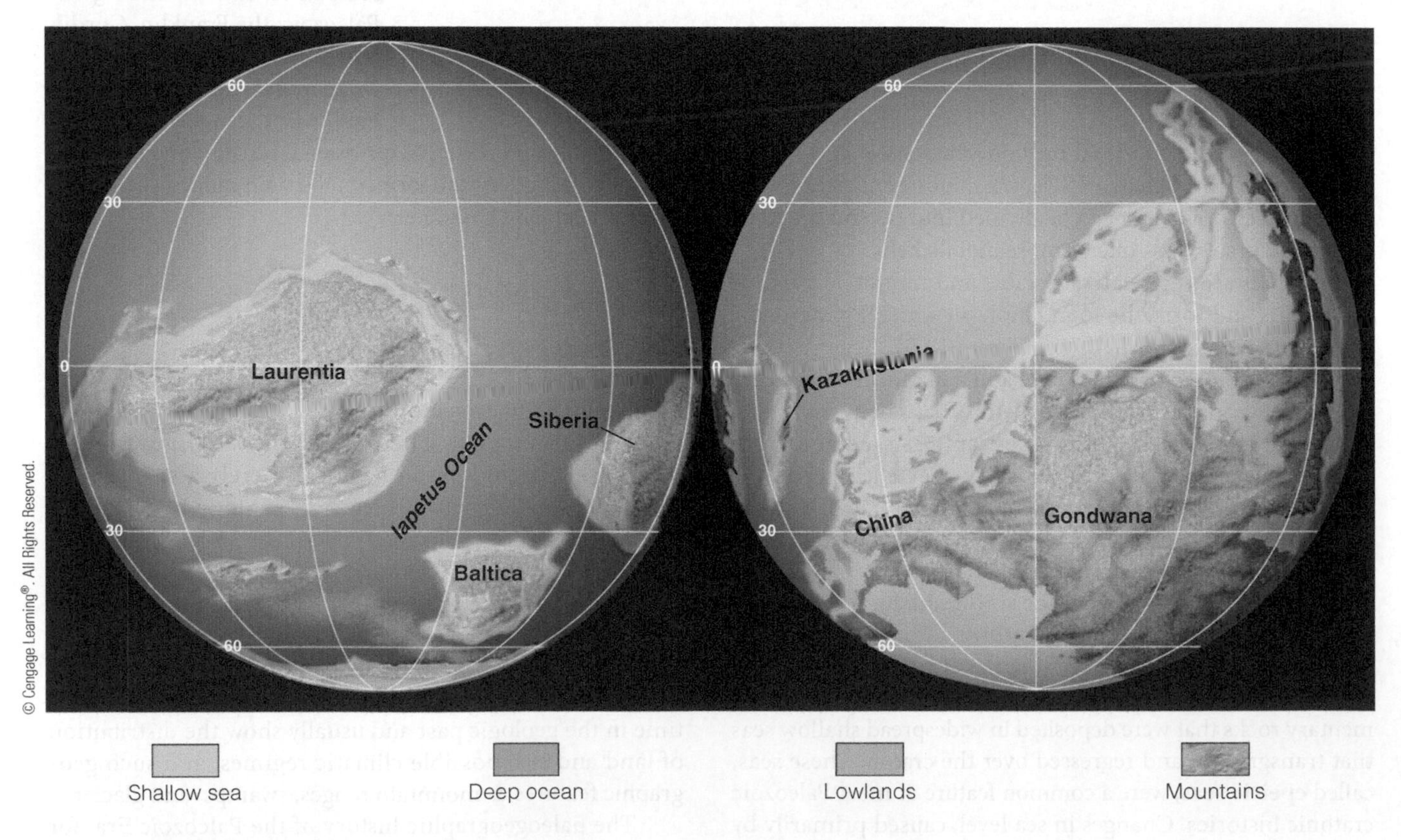

a Late Cambrian Period.

Figure 20.2 Paleozoic Paleogeography *(continued)*

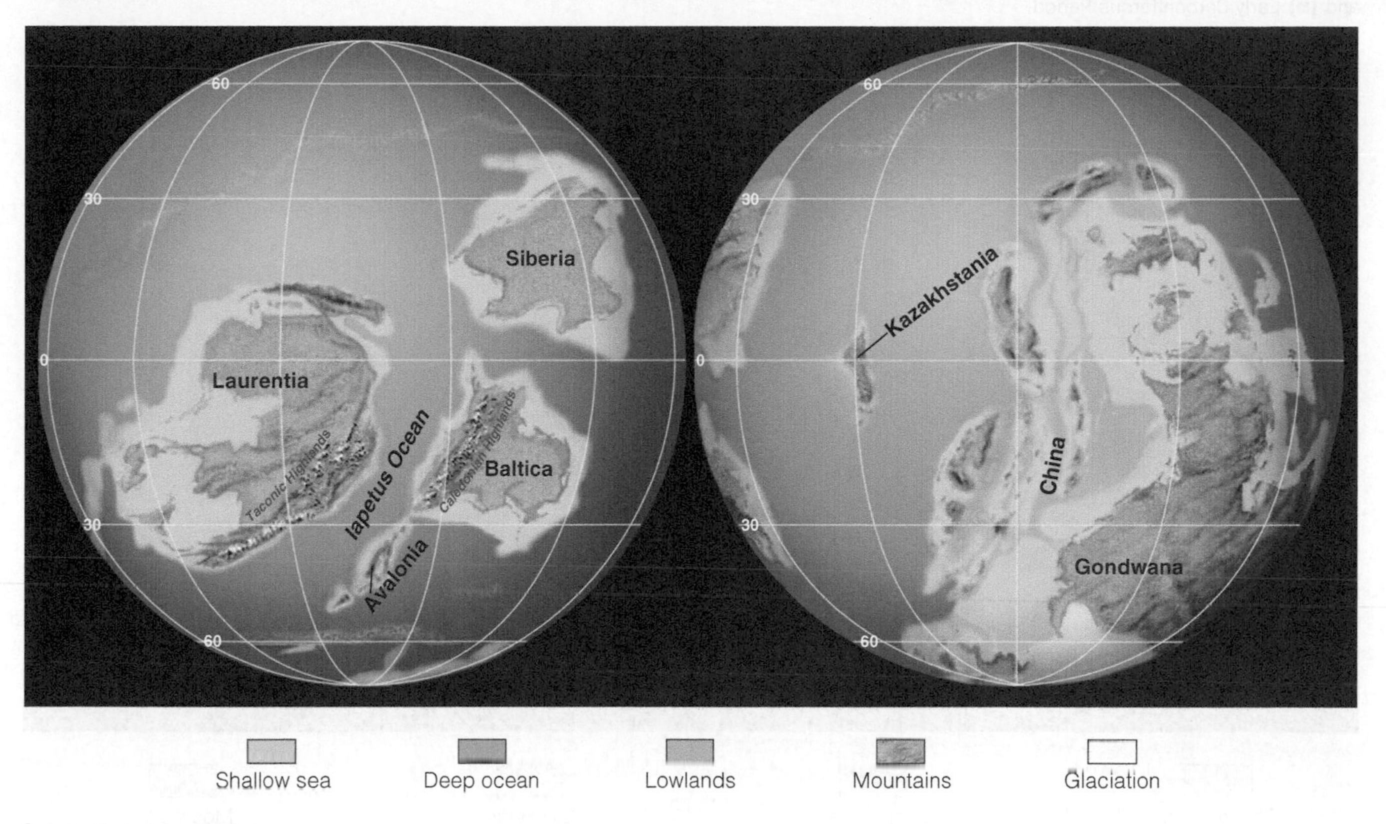

b Late Ordovician Period.

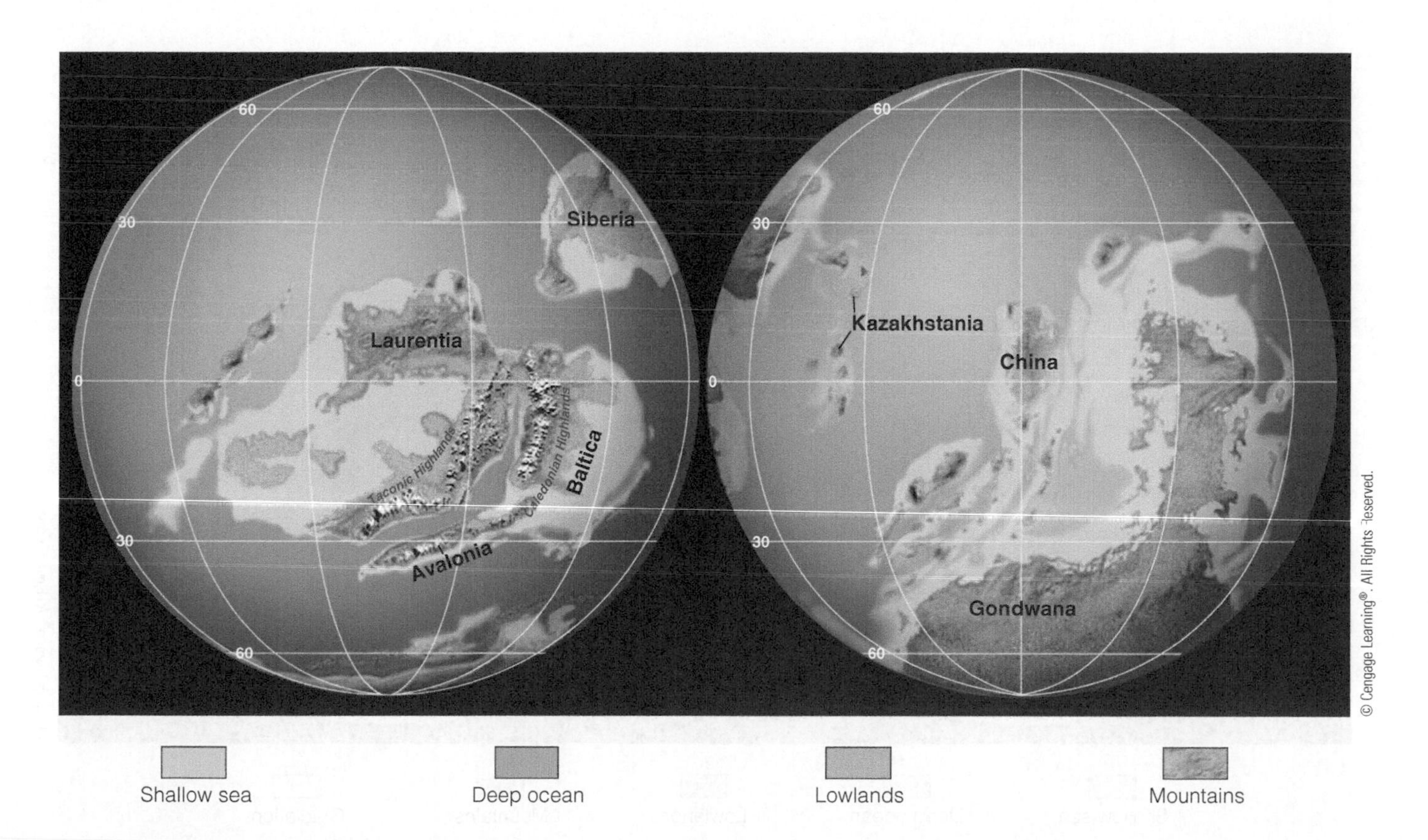

c Middle Silurian Period.

Figure 20.3 Paleozoic Paleogeography Paleogeography of the world for the **(a)** Late Devonian Period and **(b)** Early Carboniferous Period.

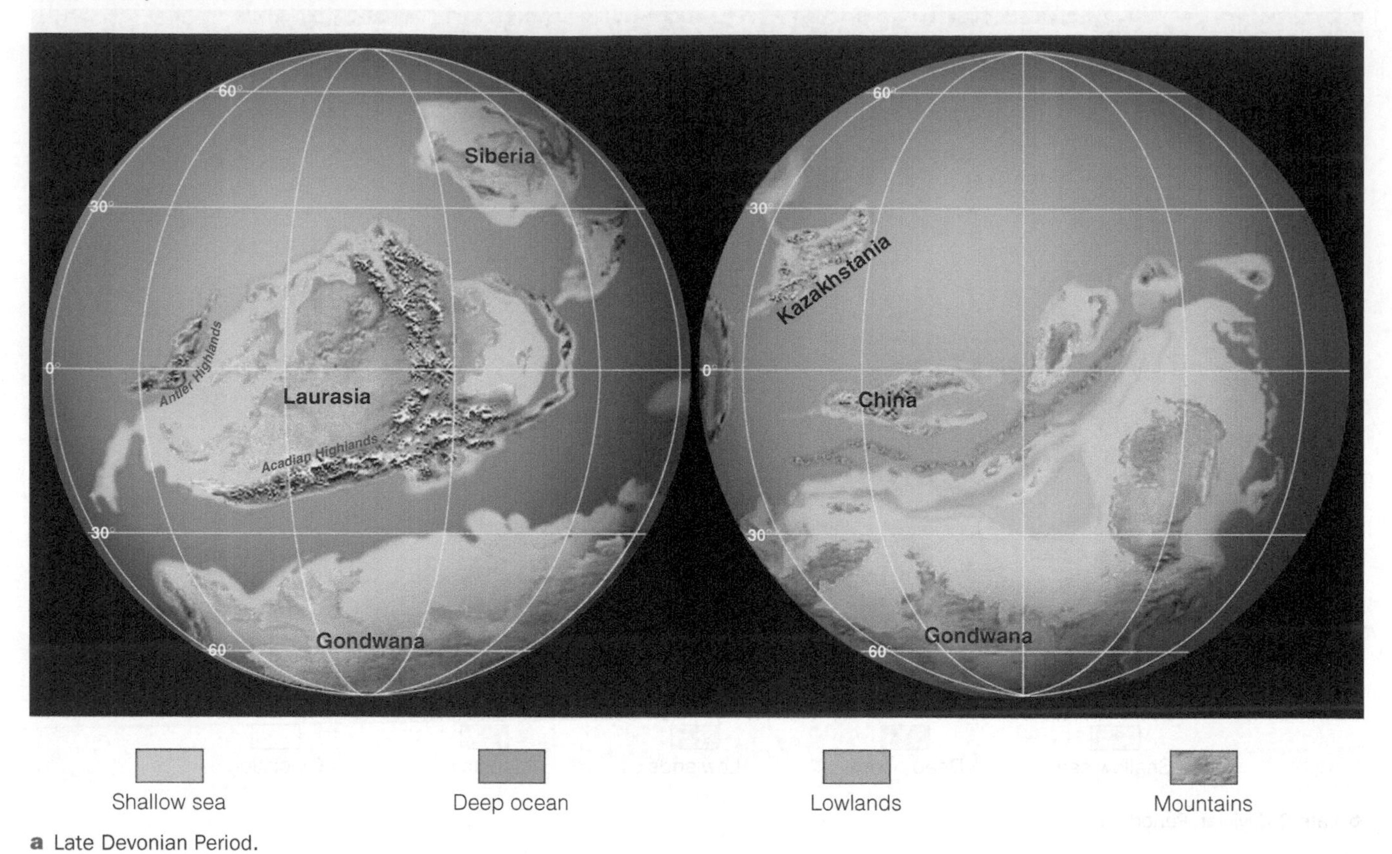

a Late Devonian Period.

Siberia
Laurasia
Kazakhstania
China
Gondwana
Gondwana

Shallow sea | Deep ocean | Lowlands | Mountains | Glaciation

b Early Carboniferous Period.

Figure 20.4 Paleozoic Paleogeography Paleogeography of the world for the **(a)** Late Carboniferous Period and **(b)** Late Permian Period.

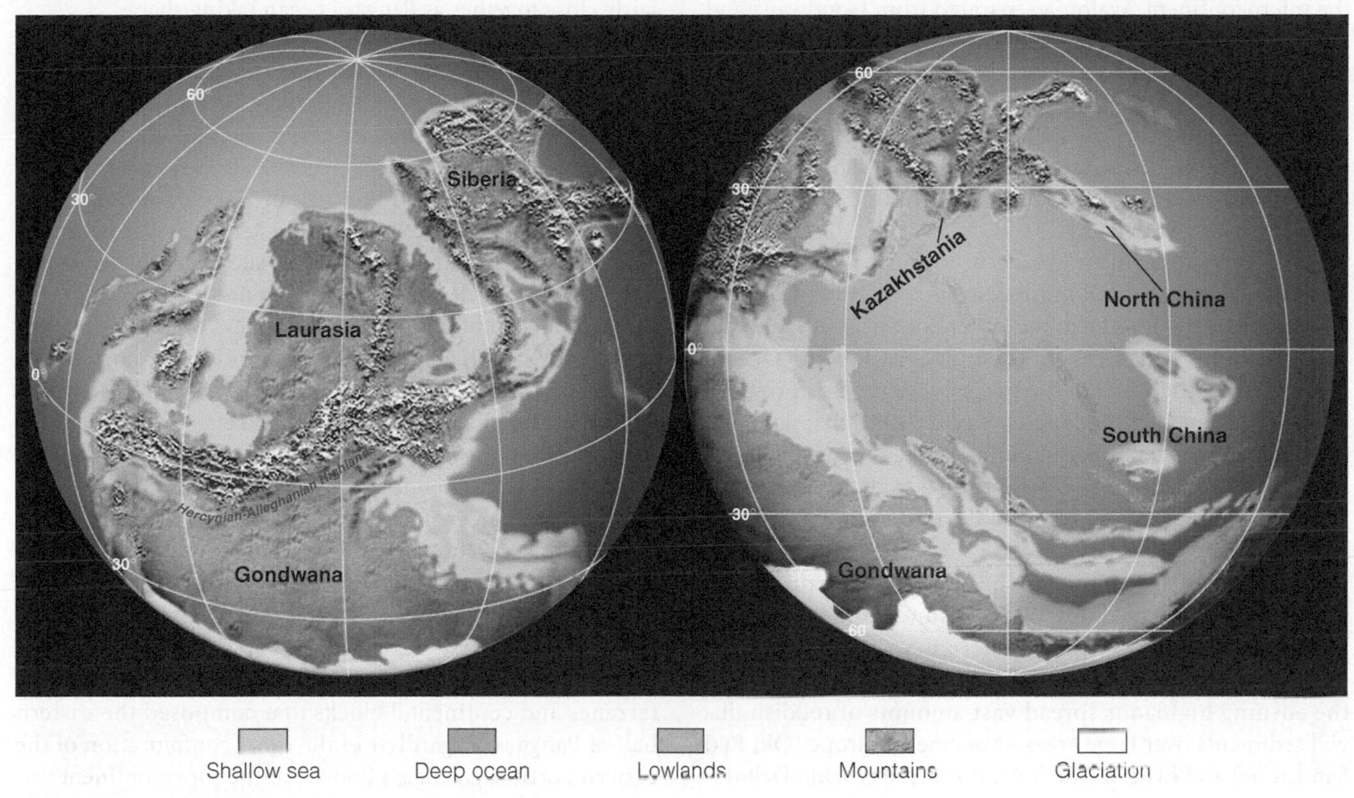

a Late Carboniferous Period.

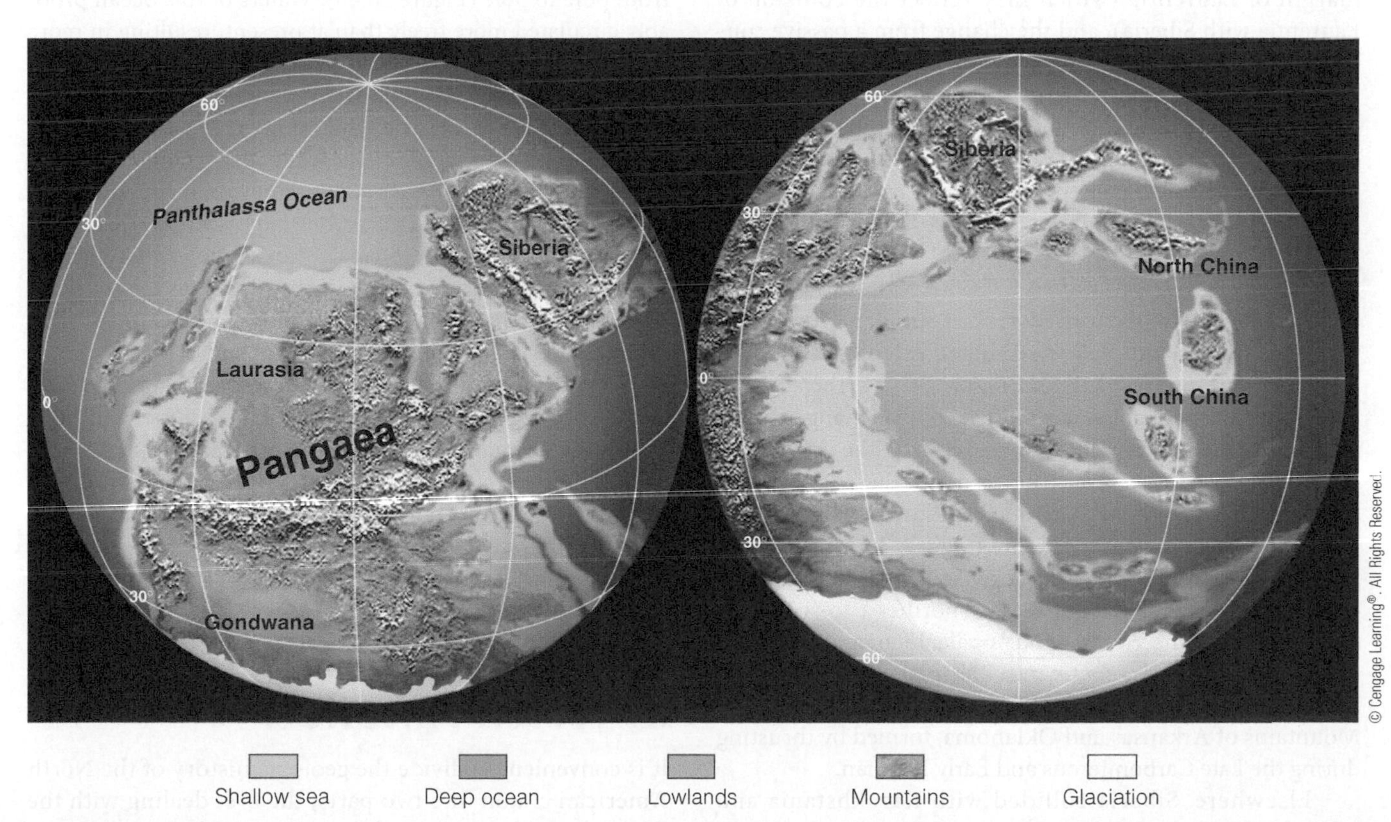

b Late Permian Period.

as indicated by Upper Ordovician glacial deposits found today in the Sahara Desert. During the Early Ordovician, the microcontinent Avalonia separated from Gondwana and began moving northeastward, where it would finally collide with Baltica during the Late Ordovician–Early Silurian. In contrast to the passive continental margin Laurentia exhibited during the Cambrian, an active convergent plate boundary formed along its eastern margin during the Ordovician, as indicated by the Late Ordovician *Taconic orogeny* that occurred in New England.

During the Silurian, Baltica, along with the newly attached Avalonia, moved northwestward relative to Laurentia and collided with it to form the larger continent of **Laurasia**. This collision, which closed the northern Iapetus Ocean, is marked by the *Caledonian orogeny*. After this orogeny, the southern part of the Iapetus Ocean still remained open between Laurentia and Avalonia–Baltica (Figure 20.2c). Siberia and Kazakhstania moved from a southern equatorial position during the Cambrian to north temperate latitudes by the end of the Silurian Period.

During the Devonian, as the southern Iapetus Ocean narrowed between Laurasia and Gondwana, mountain building continued along the eastern margin of Laurasia as a result of the *Acadian orogeny* (Figure 20.3a). Erosion of the ensuing highlands spread vast amounts of reddish fluvial sediments over large areas of northern Europe (Old Red Sandstone) and eastern North America (the Catskill Delta).

Other Devonian tectonic events, probably related to the collision of Laurentia and Baltica, include the Cordilleran *Antler orogeny,* the *Ellesmere orogeny* along the northern margin of Laurentia (which may reflect the collision of Laurentia with Siberia), and the change from a passive continental margin to an active convergent plate boundary in the Uralian mobile belt of eastern Baltica. The distribution of reefs, evaporites, and red beds, as well as the existence of similar floras throughout the world, suggests a rather uniform global climate during the Devonian Period.

Late Paleozoic Global History

During the Carboniferous Period, southern Gondwana moved over the South Pole, resulting in extensive continental glaciation (Figure 20.3b and Figure 20.4a). The advance and retreat of these glaciers produced global changes in sea level that affected sedimentation patterns on the cratons. As Gondwana moved northward, it began colliding with Laurasia during the Early Carboniferous and continued suturing with it during the rest of the Carboniferous (Figures 20.3b and 20.4a). Because Gondwana rotated clockwise relative to Laurasia, deformation generally progressed in a northeast-to-southwest direction along the Hercynian, Appalachian, and Ouachita mobile belts of the two continents. The final phase of collision between Gondwana and Laurasia is indicated by the Ouachita Mountains of Arkansas and Oklahoma, formed by thrusting during the Late Carboniferous and Early Permian.

Elsewhere, Siberia collided with Kazakhstania and moved toward the Uralian margin of Laurasia (Baltica), colliding with it during the Early Permian. By the end of the Carboniferous, the various continental landmasses were fairly close together as Pangaea began taking shape.

The Carboniferous coal basins of eastern North America, western Europe, and the Donets Basin of Ukraine all lay in the equatorial zone, where rainfall was high and temperatures were consistently warm. The absence of strong seasonal growth rings in fossil plants from these coal basins indicates such a climate. The fossil plants found in the coals of Siberia, however, show well-developed growth rings, signifying seasonal growth with abundant rainfall and distinct seasons such as in the temperate zones (latitude 40 degrees to 60 degrees north).

Glacial conditions and the movement of large continental ice sheets in the high southern latitudes are indicated by widespread glacial deposits and glacial striations in southern Gondwana (see Figure 2.5). These ice sheets spread toward the equator and, at their maximum growth, extended well into the middle temperate latitudes.

The assembly of Pangaea was essentially completed during the Permian as a result of the many continental collisions that began during the Carboniferous (Figure 20.4b). Although geologists generally agree on the configuration and location of the western half of the supercontinent, there is no consensus on the number or configuration of the various terranes and continental blocks that composed the eastern half of Pangaea. Regardless of the exact configuration of the eastern portion, geologists know that the supercontinent was surrounded by subduction zones and moved steadily northward during the Permian. Furthermore, an enormous single ocean, **Panthalassa**, surrounded Pangaea and spanned Earth from pole to pole (Figure 20.4b). Waters of this ocean probably circulated more freely than at present, resulting in more equable water temperatures.

The formation of a single large landmass had climatic consequences for the terrestrial environment as well. Terrestrial Permian sediments indicate that arid and semiarid conditions were widespread over Pangaea. The mountain ranges produced by the *Hercynian* (*Variscan*), *Alleghenian,* and *Ouachita orogenies* were high enough to create rain shadows that blocked the moist, subtropical, easterly winds—much as the southern Andes Mountains do in western South America today. This produced very dry conditions in North America and Europe, as evident from the extensive Permian red beds and evaporites found in western North America, central Europe, and parts of Russia. Permian coals, indicating abundant rainfall, were mostly limited to the northern temperate belts (latitude 40 degrees to 60 degrees north), whereas the last remnants of the Carboniferous ice sheets continued their recession.

20.4 Paleozoic Evolution of North America

It is convenient to divide the geologic history of the North American craton into two parts, the first dealing with the relatively stable continental interior over which epeiric seas

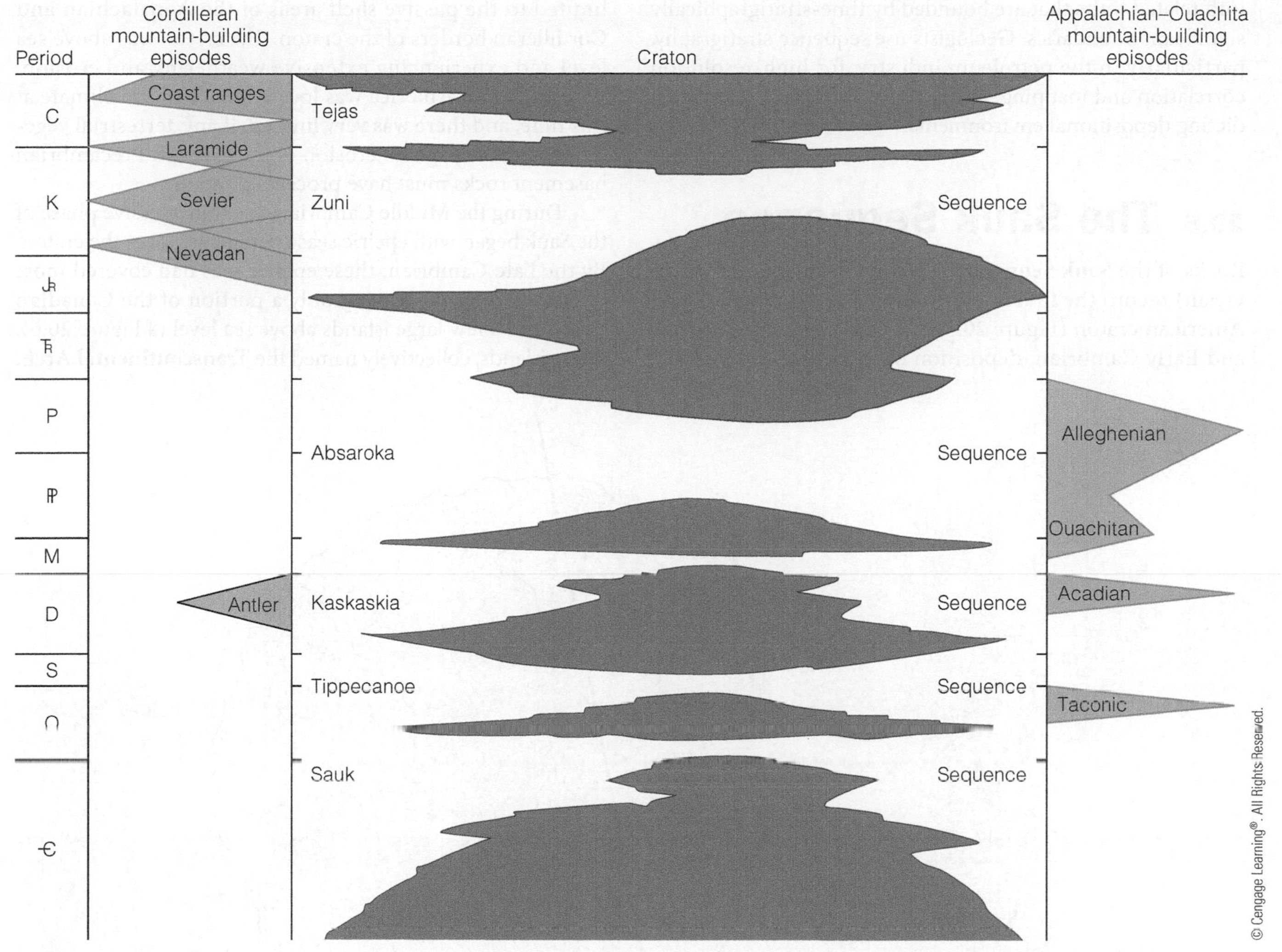

Figure 20.5 **Cratonic Sequences of North America** A cratonic sequence is a large-scale lithostratigraphic unit representing a major transgressive–regressive cycle and bounded by craton-wide unconformities. The white areas represent sequences of rocks separated by large-scale unconformities (brown areas). The major Cordilleran orogenies are shown on the left side, and the major Appalachian orogenies are on the right.

transgressed and regressed, and the other with the mobile belts where mountain building occurred.

The Phanerozoic sedimentary rock record of North America is divided into six cratonic sequences. A **cratonic sequence** is a large-scale (greater than supergroup) lithostratigraphic unit representing a major transgressive–regressive cycle bounded by craton-wide unconformities (Figure 20.5). The transgressive phase, which is usually covered by younger sediments, commonly is well preserved, whereas the regressive phase of each sequence is marked by an unconformity. Where rocks of the appropriate age are preserved, each of the six unconformities extend across the various sedimentary basins of the North American craton and into the mobile belts along the cratonic margin.

ConnectionLink

To review transgressions and regressions and how they relate to sedimentary facies, see Figure 7.11 in Chapter 7.

Geologists have also recognized major unconformity-bounded sequences in cratonic areas outside North America. Such global transgressive–regressive cycles of sea-level changes are thought to result from major tectonic and glacial events.

The realization that rock units can be divided into cratonic sequences, and that these sequences can be further subdivided into smaller units and correlated, provides the foundation for an important concept in geology that allows for high-resolution analysis of time and facies relationships within sedimentary rocks. **Sequence stratigraphy** is the study of rock relationships within a time-stratigraphic framework of related facies bounded by erosional or nondepositional surfaces. The basic unit of sequence stratigraphy is the *sequence*, which is a succession of rocks bounded by unconformities and their equivalent conformable strata. Sequence boundaries result from a relative drop in sea level. Sequence stratigraphy is an important tool in geology because it allows geologists to subdivide sedimentary rocks into related units that are bounded by time-stratigraphically significant boundaries. Geologists use sequence stratigraphy, particularly in the petroleum industry, for high-resolution correlation and mapping, as well as for interpreting and predicting depositional environments.

20.5 The Sauk Sequence

Rocks of the **Sauk Sequence** (Neoproterozoic–Early Ordovician) record the first major transgression onto the North American craton (Figure 20.5). During the Neoproterozoic and Early Cambrian, deposition of marine sediments was limited to the passive shelf areas of the Appalachian and Cordilleran borders of the craton. The craton was above sea level and experiencing extensive weathering and erosion. Because North America was located in a tropical climate at this time, and there was very limited, if any, terrestrial vegetation, weathering and erosion of the exposed Precambrian basement rocks must have proceeded rapidly.

During the Middle Cambrian, the transgressive phase of the Sauk began with epeiric seas encroaching over the craton. By the Late Cambrian, these epeiric seas had covered most of North America, leaving only a portion of the Canadian shield and a few large islands above sea level (▌Figure 20.6). These islands, collectively named the **Transcontinental Arch**,

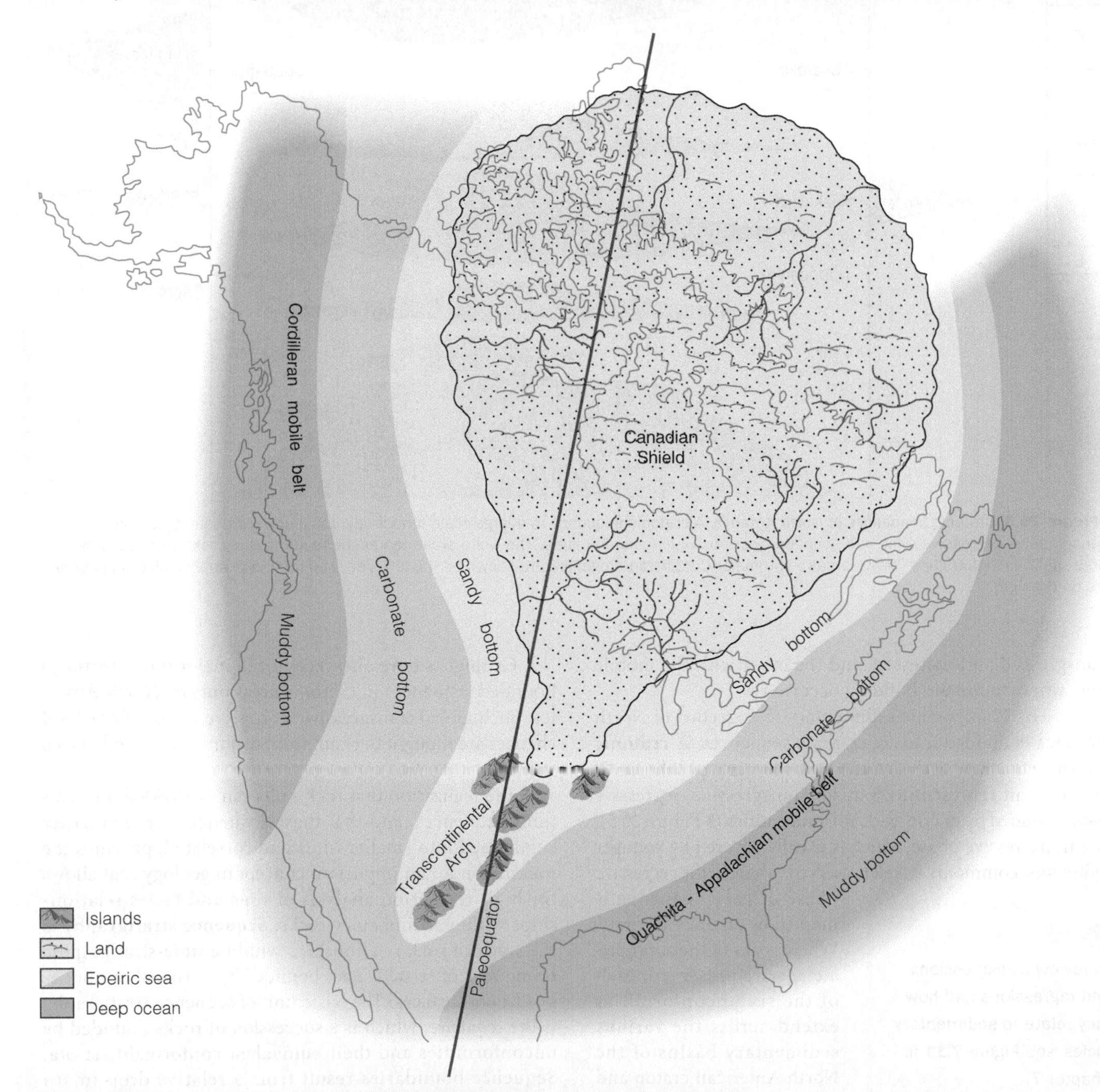

▌**Figure 20.6 Cambrian Paleogeography of North America** Note the position of the Cambrian paleoequator. During this time, North America straddled the equator as indicated in Figure 20.2a.

Figure 20.7 Cambrian Rocks of the Grand Canyon, Arizona

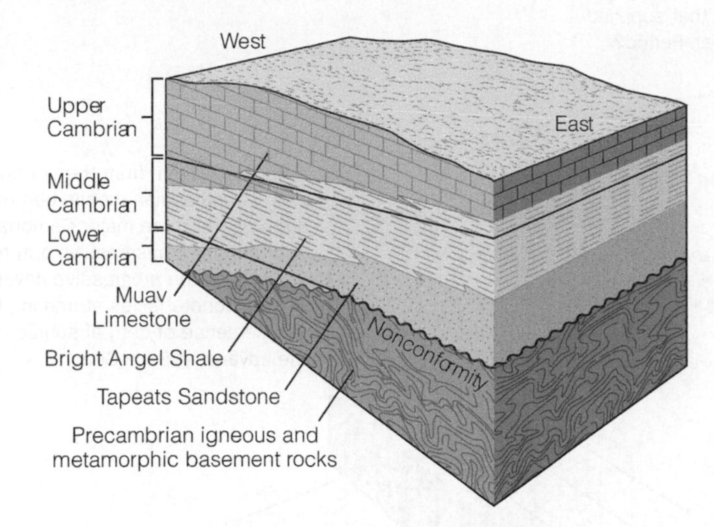

a Block diagram of Cambrian strata exposed in the Grand Canyon, Arizona, illustrating the transgressive nature of the three formations.

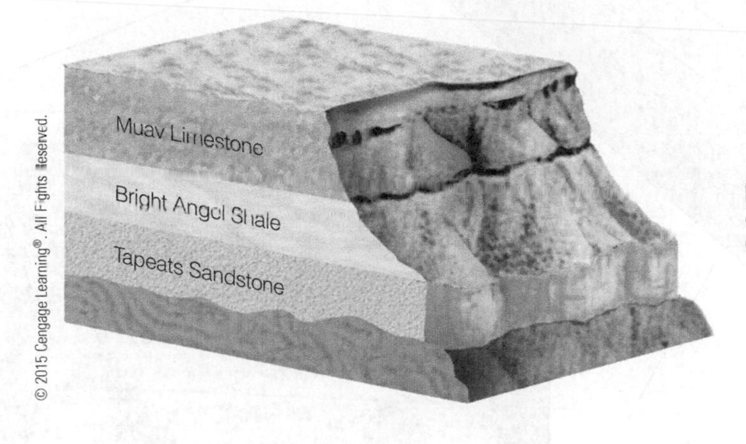

b Block diagram of Cambrian rocks exposed along the Bright Angel Trail, Grand Canyon, Arizona.

extended from New Mexico to Minnesota and the Lake Superior region.

The sediments deposited both on the craton and along the shelf area of the craton margin show abundant evidence of shallow-water deposition. The only difference between the shelf and craton deposits is that the shelf deposits are thicker. In both areas, the sands are generally clean and well sorted, and commonly contain ripple marks and small-scale cross-bedding. Many of the carbonates are bioclastic (composed of fragments of shells), contain stromatolites, or have oolitic (small, spherical calcium carbonate grains) textures. Such sedimentary structures and textures indicate shallow-water deposition.

Recall from Chapter 7 that sediments become increasingly finer the farther away from land they are. Therefore, in a stable environment where sea level remains the same, coarse detrital sediments are typically deposited in the nearshore environment, and finer-grained sediments are deposited in the offshore environment. Carbonates form farthest from land, in the area beyond the reach of detrital sediments. During a transgression, these facies (sediments that represent a particular environment) migrate in a landward direction (see Figure 7.11).

This time-transgressive facies relationship occurred not only in the Grand Canyon area of the craton during the Sauk Sequence (▌Figure 20.7), but elsewhere on the craton as well, as the seas encroached from the Appalachian and Ouachita mobile belts onto the craton interior (▌Figure 20.8). Carbonate deposition dominated on the craton as the Sauk transgression continued during the Early Ordovician, and the advancing Sauk Sea soon covered the islands of the Transcontinental Arch. By the end of Sauk time, much of the craton was submerged beneath a warm, equatorial epeiric sea (Figure 20.2a).

20.6 The Tippecanoe Sequence

As the Sauk Sea regressed from the craton during the Early Ordovician, a landscape of low relief emerged. The exposed rocks were predominantly limestones and dolostones deposited earlier as part of the Sauk transgression. Because North America was still located in a tropical environment when the seas regressed, these carbonates experienced extensive erosion at that time (▌Figure 20.9). The resulting

Figure 20.8 Time-Transgressive Cambrian Facies

Critical Thinking Question: What were the sources that supplied detrital sediments to the Sauk Sea during the Cambrian Period?

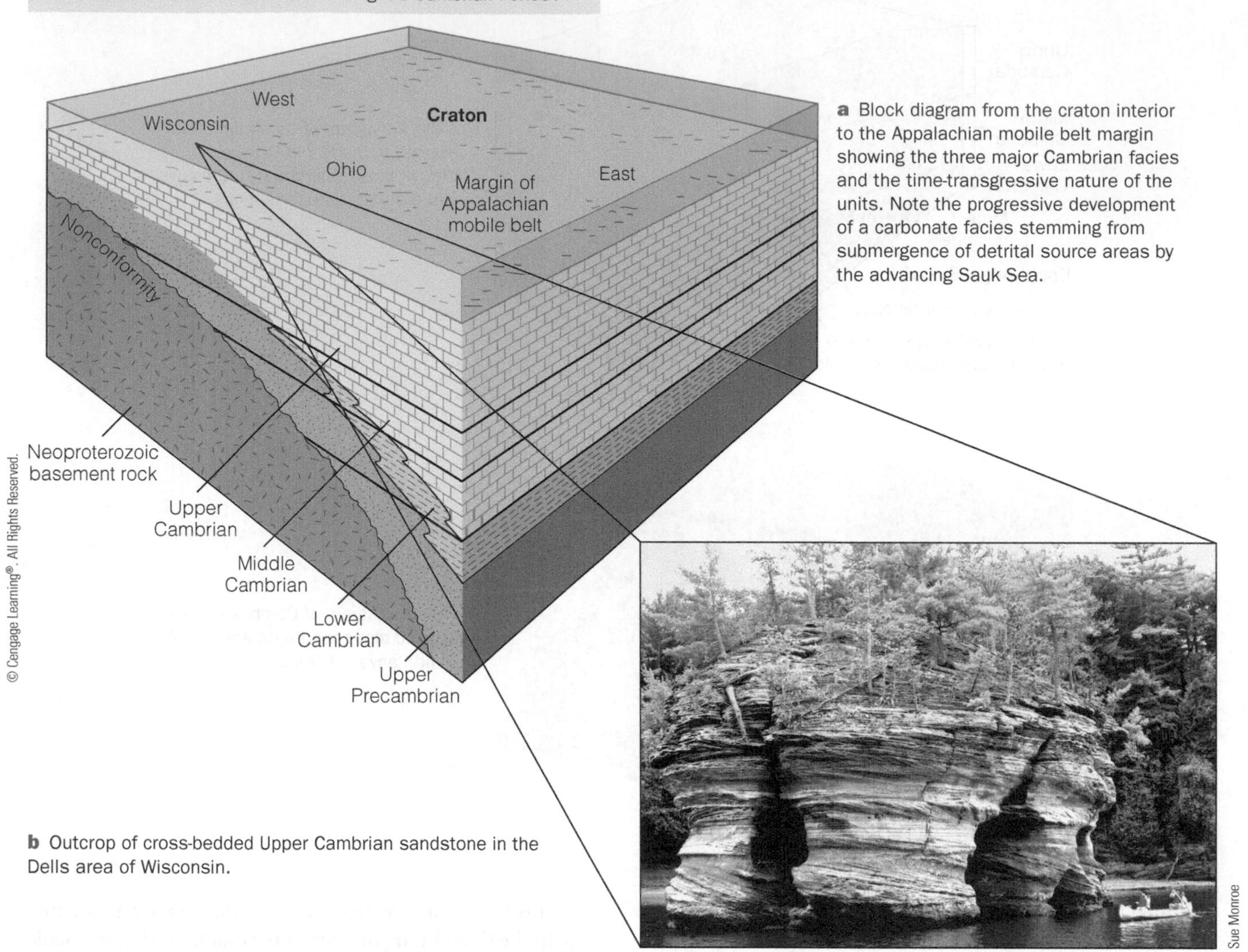

a Block diagram from the craton interior to the Appalachian mobile belt margin showing the three major Cambrian facies and the time-transgressive nature of the units. Note the progressive development of a carbonate facies stemming from submergence of detrital source areas by the advancing Sauk Sea.

b Outcrop of cross-bedded Upper Cambrian sandstone in the Dells area of Wisconsin.

craton-wide unconformity marks the boundary between the Sauk and Tippecanoe sequences.

Just as in the Sauk Sequence, deposition of the **Tippecanoe Sequence** (Middle Ordovician–Early Devonian) began with a major transgression onto the craton. This transgressing sea deposited clean and well-sorted quartz sands over most of the craton. The best known of the Tippecanoe basal sandstones is the St. Peter Sandstone, an almost-pure quartz sandstone used in manufacturing glass. It occurs throughout much of the mid-continent and resulted from numerous cycles of weathering and erosion of Proterozoic and Cambrian sandstones deposited during the Sauk transgression.

The Tippecanoe basal sandstones were followed by widespread carbonate deposition (Figure 20.9). The limestones were generally the result of deposition by calcium-carbonate–secreting organisms such as corals, brachiopods, stromatoporoids, and bryozoans. Besides the limestones, there were also many dolostones. Most of the dolostones formed as a result of magnesium replacing calcium in calcite, thus converting the limestones into dolostones.

In the eastern portion of the craton, the carbonates grade laterally into shales. These shales mark the farthest extent of detrital sediments derived from weathering and erosion of the Taconic Highlands, which resulted from a tectonic event taking place in the Appalachian mobile belt, and which we will discuss later.

Tippecanoe Reefs and Evaporites

Organic reefs are limestone structures constructed by living organisms, some of which contribute skeletal materials to the reef framework (Figure 20.10). Today, corals and calcareous algae are the most prominent reef builders, but in the geologic past other organisms played a major role.

ConnectionLink

To review the formation and distribution of reefs, see Chapter 16.

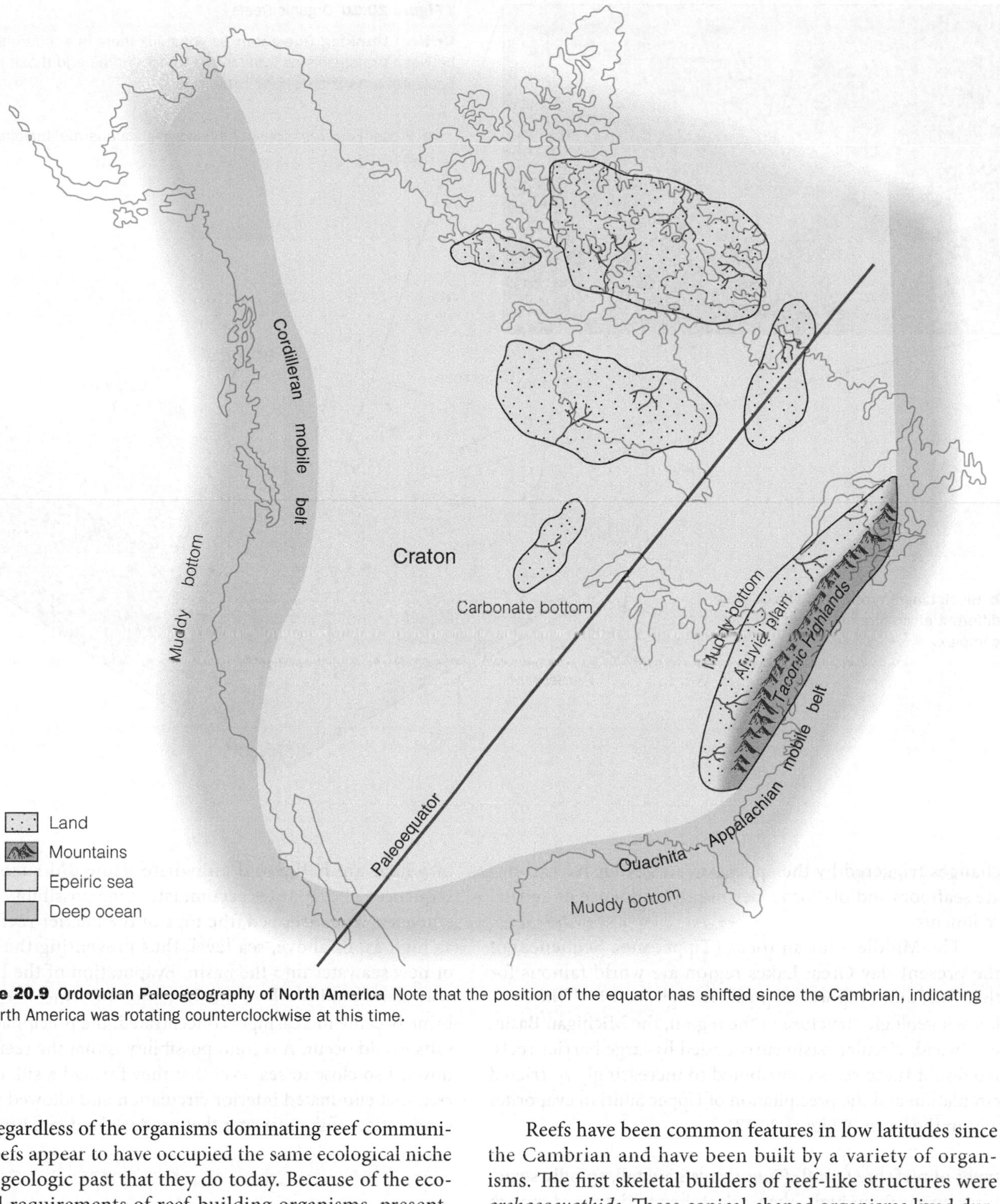

Figure 20.9 Ordovician Paleogeography of North America Note that the position of the equator has shifted since the Cambrian, indicating that North America was rotating counterclockwise at this time.

Regardless of the organisms dominating reef communities, reefs appear to have occupied the same ecological niche in the geologic past that they do today. Because of the ecological requirements of reef-building organisms, present-day reefs are confined to a narrow latitudinal belt between 30 degrees north and south of the equator. Corals, the major reef-building organisms today, require warm, clear, shallow water of normal salinity for optimal growth.

The size and shape of a reef are mostly the result of the interactions among the reef-building organisms, the bottom topography, wind and wave action, and subsidence of the seafloor. Reefs also alter the area around them by forming barriers to water circulation or wave action.

Reefs have been common features in low latitudes since the Cambrian and have been built by a variety of organisms. The first skeletal builders of reef-like structures were *archaeocyathids*. These conical-shaped organisms lived during the Cambrian and had double, perforated, calcareous shell walls. Archaeocyathids built small mounds that have been found on all continents except South America (see Figure 21.5).

Beginning in the Middle Ordovician, stromatoporoid–coral reefs became common in the low latitudes, and similar reefs remained so throughout the rest of the Phanerozoic Eon. The burst of reef building seen in the Late Ordovician through Devonian probably occurred in response to evolutionary

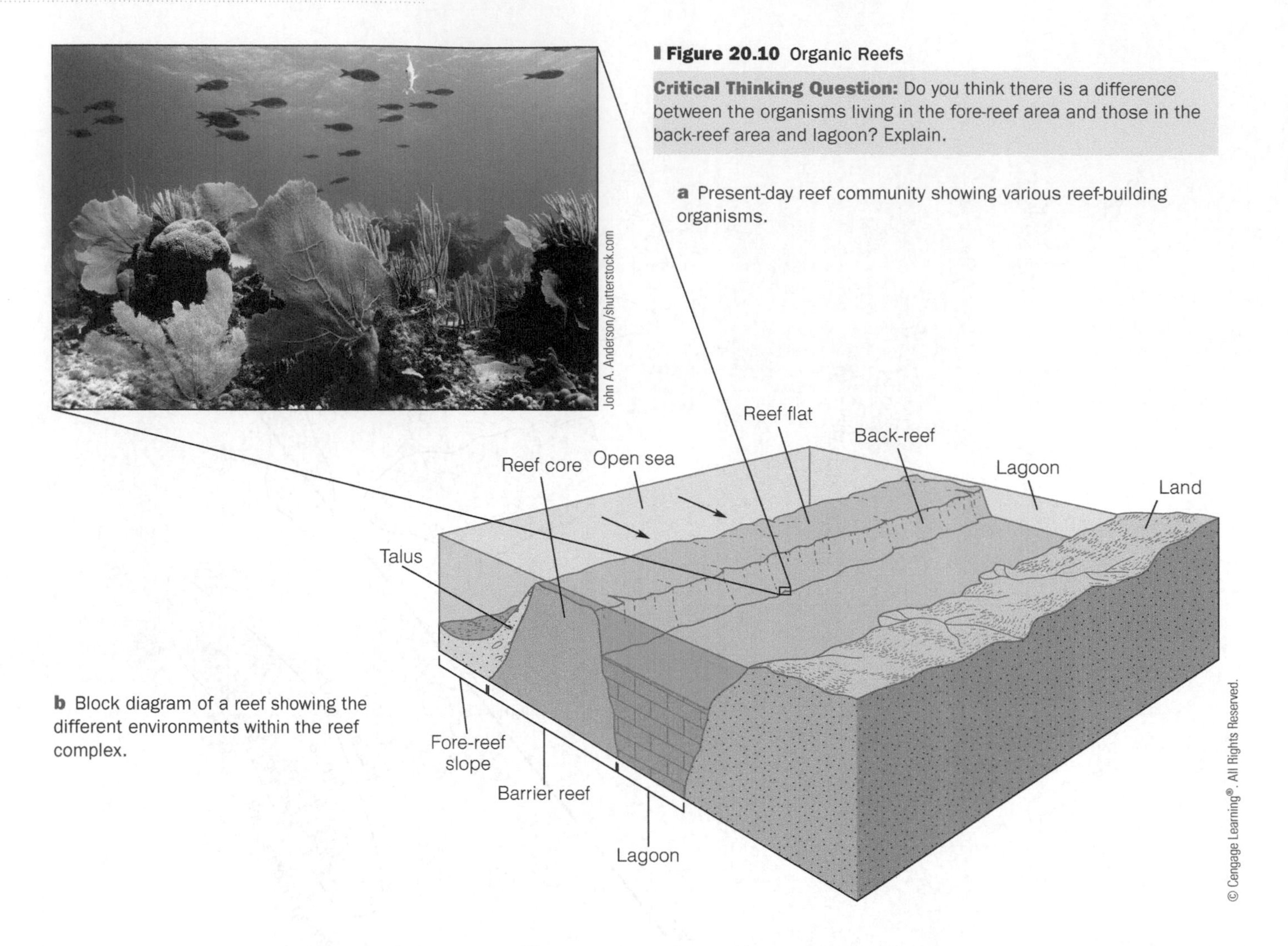

Figure 20.10 Organic Reefs

Critical Thinking Question: Do you think there is a difference between the organisms living in the fore-reef area and those in the back-reef area and lagoon? Explain.

a Present-day reef community showing various reef-building organisms.

b Block diagram of a reef showing the different environments within the reef complex.

changes triggered by the appearance of extensive carbonate seafloors and platforms beyond the influence of detrital sediments.

The Middle Silurian rocks (Tippecanoe Sequence) of the present-day Great Lakes region are world famous for their reef and evaporite deposits (❚ Figure 20.11). The best-known geologic structure in the region, the Michigan Basin, is a broad, circular basin surrounded by large barrier reefs. No doubt these reefs contributed to increasingly restricted circulation and the precipitation of Upper Silurian evaporites within the basin (❚ Figure 20.12).

Within the rapidly subsiding interior of the basin, other types of reefs are found. *Pinnacle reefs* are tall, spindly structures up to 100 m high. They reflect the rapid upward growth needed to maintain themselves near sea level during subsidence of the basin (Figure 20.12a). Besides the pinnacle reefs, bedded carbonates and thick sequences of salt and anhydrite are also found in the Michigan Basin (Figure 20.12a, b, c).

As the Tippecanoe Sea gradually regressed from the craton during the Late Silurian, precipitation of evaporite minerals occurred in the Appalachian, Ohio, and Michigan basins (Figure 20.11). In the Michigan Basin alone, approximately 1,500 m of sediments were deposited, nearly half of which are halite and anhydrite. How did such thick sequences of evaporites accumulate? One possibility is that when sea level dropped, the tops of the barrier reefs were as high as, or above, sea level, thus preventing the influx of new seawater into the basin. Evaporation of the basinal seawater would result in the formation of brine, and as the brine became increasingly concentrated, the precipitation of salts would occur. A second possibility is that the reefs grew upward so close to sea level that they formed a sill, or barrier, that eliminated interior circulation and allowed for the evaporation of the seawater that produced a dense brine that eventually resulted in evaporite deposits (❚ Figure 20.13).

The End of the Tippecanoe Sequence

By the Early Devonian, the regressing Tippecanoe Sea had retreated to the craton margin, exposing an extensive lowland topography. During this regression, marine deposition was initially restricted to a few interconnected cratonic basins and, finally, by the end of the Tippecanoe Sequence, to only the mobile belts surrounding the craton.

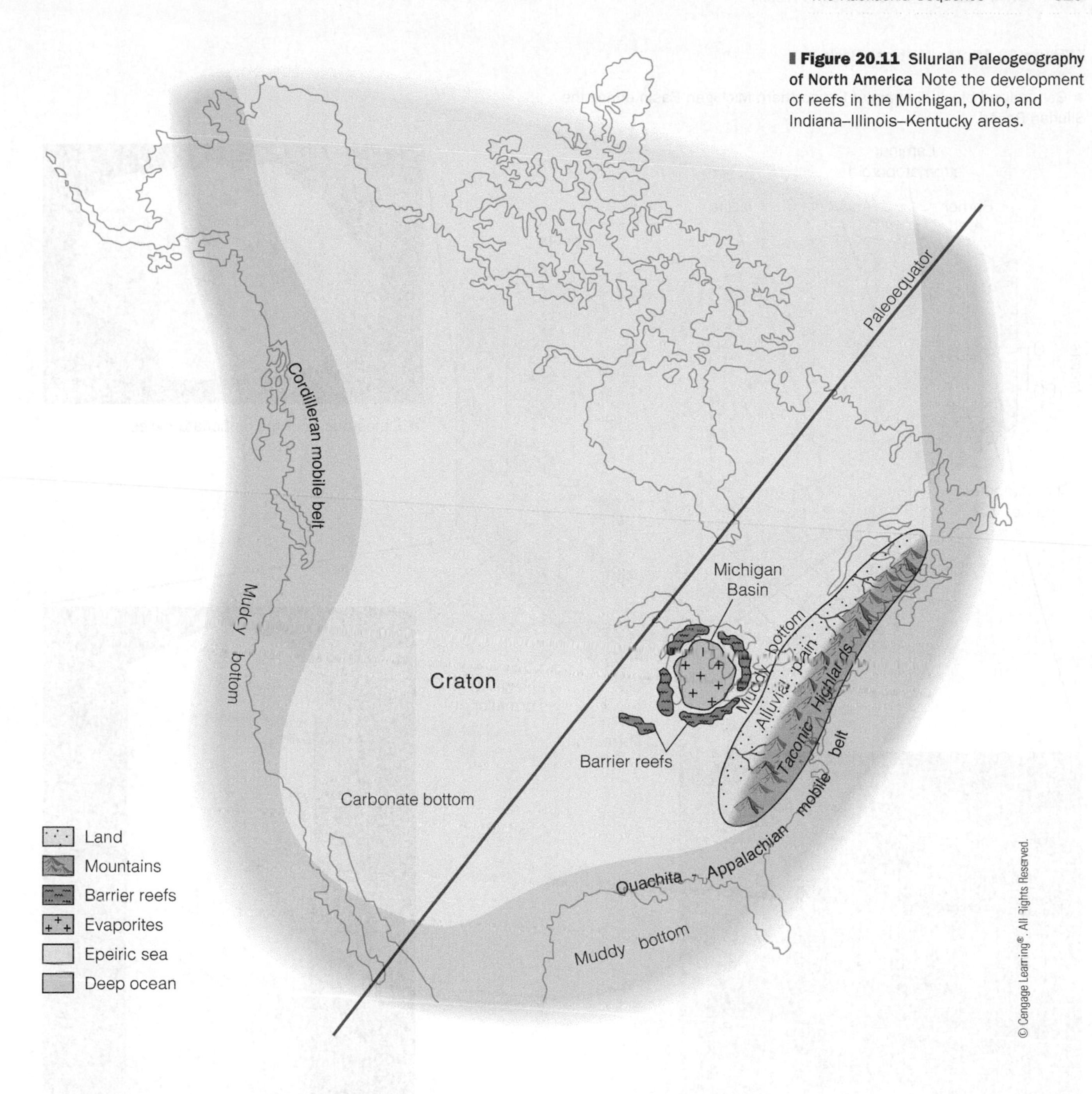

Figure 20.11 Silurian Paleogeography of North America Note the development of reefs in the Michigan, Ohio, and Indiana–Illinois–Kentucky areas.

As the Tippecanoe Sea regressed during the Early Devonian, the craton experienced mild deformation, forming many domes, arches, and basins (Figure 20.1). These structures were mostly eroded during the time that the craton was exposed, and deposits from the ensuing and encroaching Kaskaskia Sea eventually covered them.

20.7 The Kaskaskia Sequence

The boundary between the Tippecanoe Sequence and the overlying **Kaskaskia Sequence** (Middle Devonian–Late Mississippian) is marked by a major unconformity. As the Kaskaskia Sea transgressed over the low-relief landscape of the craton, most basal beds deposited consisted of clean, well-sorted quartz sandstones.

The source areas for the basal Kaskaskia sandstones were primarily the eroding highlands of the Appalachian mobile belt area (Acadian Highlands) (Figure 20.14), exhumed Cambrian and Ordovician sandstones cropping out along the flanks of the Ozark Dome, and exposures of the Canadian shield in the Wisconsin area. The lack of similar sands in the Silurian carbonate beds below the Tippecanoe–Kaskaskia unconformity indicates that the source areas of the basal Kaskaskia detrital rocks were submerged when the Tippecanoe Sequence was deposited. Stratigraphic studies indicate that these source areas were uplifted and the Tippecanoe carbonates removed by erosion before the Kaskaskia transgression.

Figure 20.12 The Michigan Basin

a Generalized block diagram of the northern Michigan Basin during the Silurian Period.

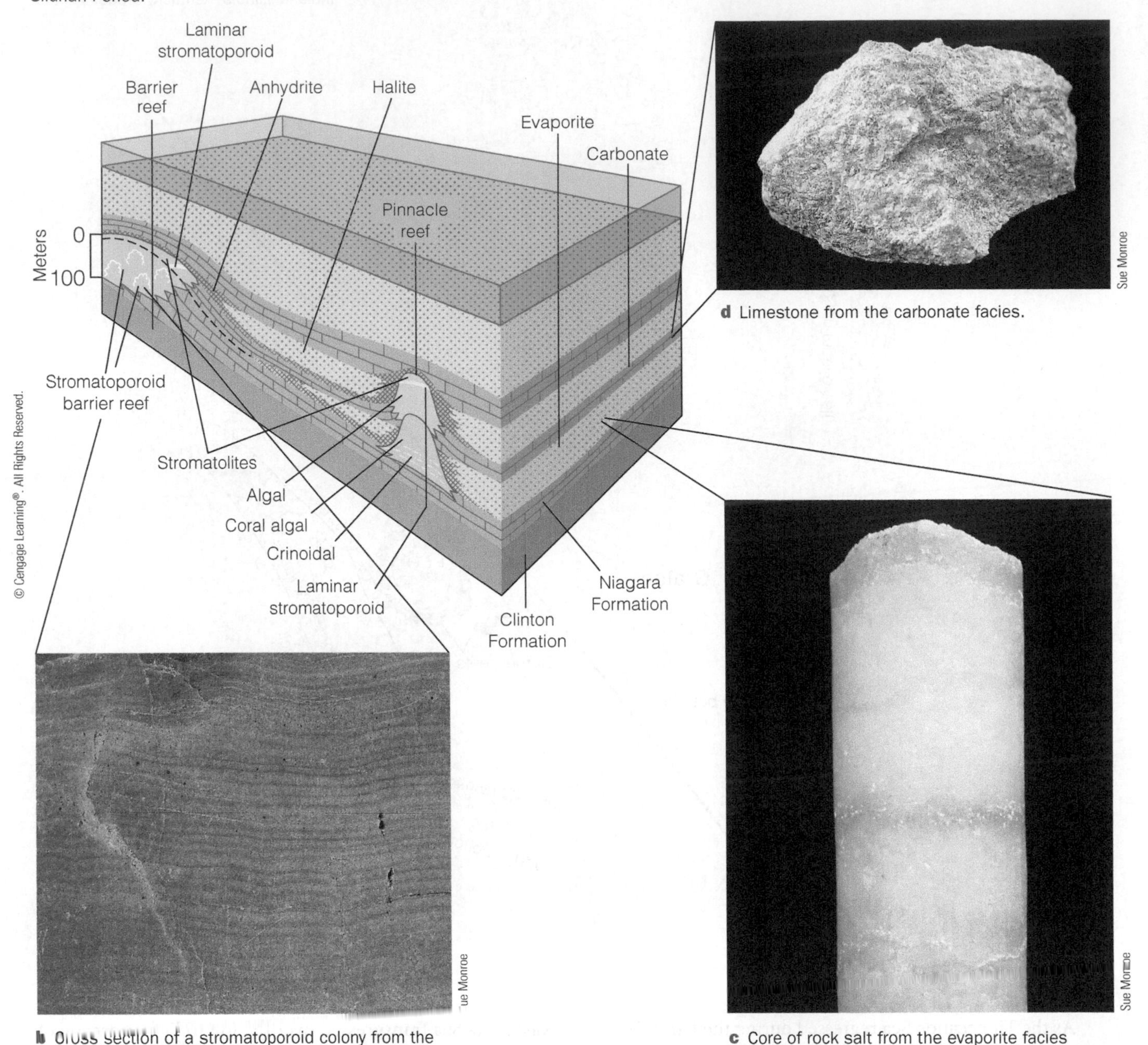

d Limestone from the carbonate facies.

b Cross section of a stromatoporoid colony from the stromatoporoid barrier reef facies.

c Core of rock salt from the evaporite facies

Kaskaskian basal rocks elsewhere on the craton consist of carbonates that are frequently difficult to differentiate from the underlying Tippecanoe carbonates unless they are fossiliferous.

Except for widespread Upper Devonian and Lower Mississippian black shales, the majority of Kaskaskian rocks are carbonates, including reefs, and associated evaporite deposits. In many other parts of the world, such as southern England, Belgium, Central Europe, Australia, and Russia, the Middle and early Late Devonian epochs were times of major reef building.

Reef Development in Western Canada

The Middle and Late Devonian reefs of western Canada contain large reserves of petroleum and have been widely studied from outcrops and in the subsurface (**▮** Figure 20.15). These reefs began forming as the Kaskaskia Sea transgressed southward into western Canada. By the end of the Middle Devonian, they had coalesced into a large barrier-reef system that restricted the flow of oceanic water into the

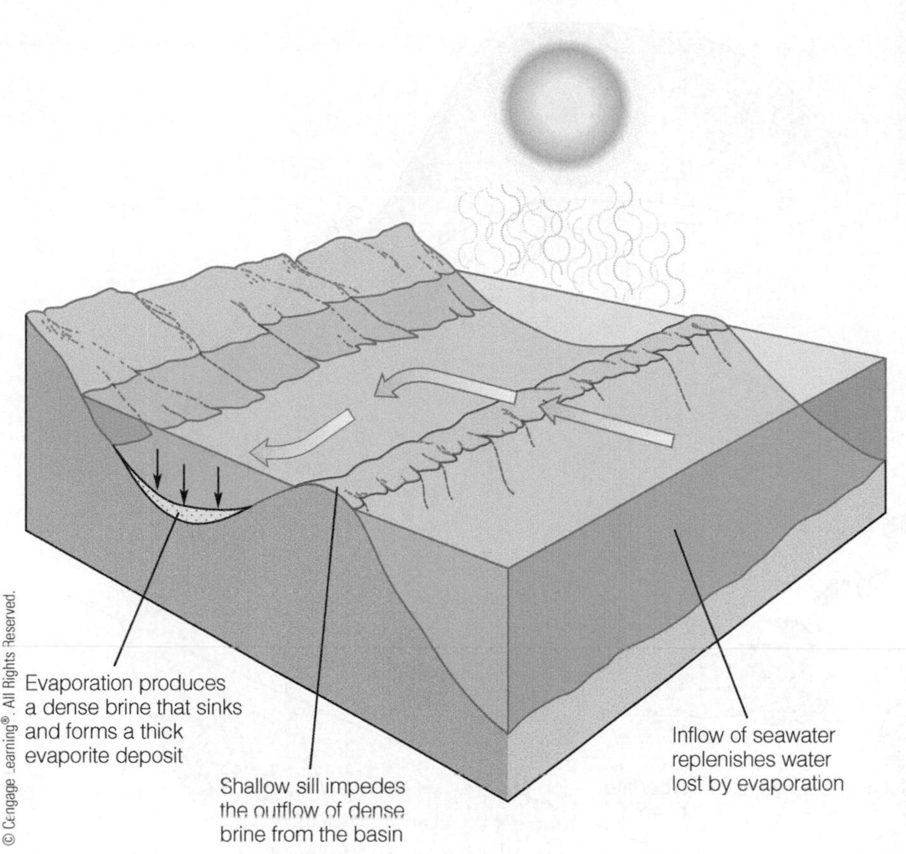

Figure 20.13 Evaporite Sedimentation Silled basin model for evaporite sedimentation by direct precipitation from seawater. Vertical scale is greatly exaggerated.

back-reef platform, creating conditions for evaporite precipitation (Figure 20.15). In the back-reef area, up to 300 m of evaporites precipitated in much the same way as in the Michigan Basin during the Silurian (Figure 20.12). More than half of the world's potash, which is used in fertilizers, comes from these Devonian evaporites. By the middle of the Late Devonian, reef growth had stopped in western Canada, although nonreef carbonate deposition continued.

Black Shales

In North America, many areas of carbonate–evaporite deposition gave way to a greater proportion of shales and coarser detrital rocks beginning in the Middle Devonian and continuing into the Late Devonian. This change to detrital deposition resulted from the formation of new source areas brought about by the mountain-building activity associated with the Acadian orogeny in North America (Figure 20.14).

As the Devonian Period ended, a conspicuous change in sedimentation took place over the North American craton with the appearance of widespread black shales. In the eastern United States, these black shales are commonly called the Chattanooga Shale, but they are known by a variety of local names elsewhere (e.g., New Albany Shale and Antrim Shale). Although these black shales are best developed from the cratonic margins along the Appalachian mobile belt to the Mississippi Valley, correlative units can also be found in many western states and in western Canada (Figure 20.16a).

The Upper Devonian–Lower Mississippian black shales of North America are typically noncalcareous, thinly bedded, and less than 10 m thick (Figure 20.16b). Fossils are usually rare, but some Upper Devonian black shales do contain rich conodont (microscopic animal) faunas. Because most black shales lack body fossils, they are difficult to date and correlate. However, in places where conodonts, acritarchs (microscopic algae), or plant spores are found, these fossils indicate that the lower beds are Late Devonian and the upper beds are Early Mississippian in age.

Although the origin of these extensive black shales is still being debated, the essential features required to produce them include undisturbed anaerobic bottom water, a reduced supply of coarser detrital sediment, and high organic productivity in the overlying oxygenated waters. High productivity in the surface waters leads to a shower of organic material, which decomposes on the undisturbed seafloor and depletes the dissolved oxygen at the sediment–water interface.

The wide extent of such apparently shallow-water black shales in North America remains puzzling. Nonetheless, these shales are rich in uranium and are an important source rock of oil and gas in the Appalachian region.

Connection Link
To review the exploration and production of organic-rich shales using horizontal drilling and hydraulic fracturing, see Chapter 13, Geo-Focus 13.1.

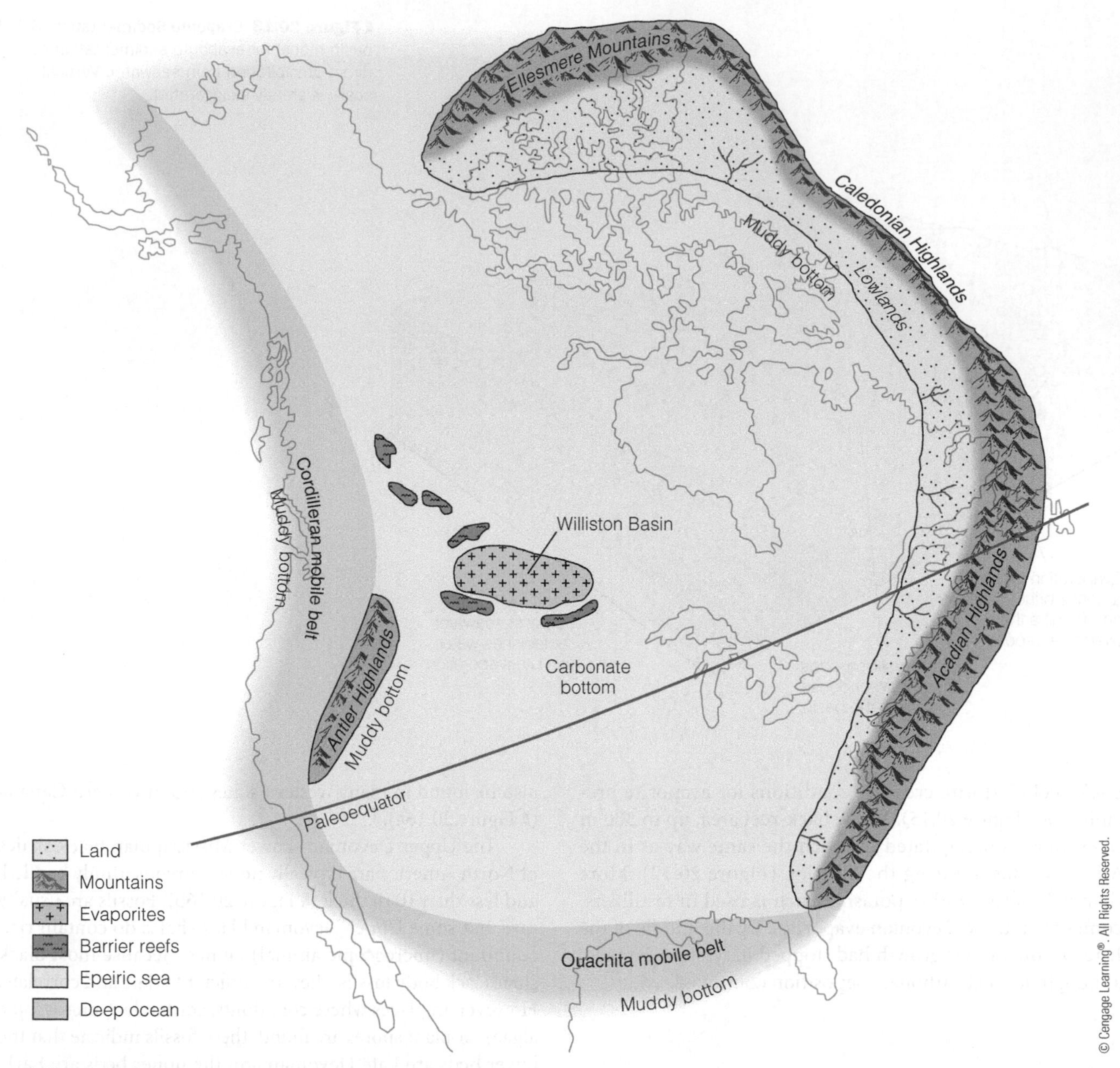

Figure 20.14 Devonian Paleogeography of North America

The Late Kaskaskia—A Return to Extensive Carbonate Deposition

Following deposition of the widespread Upper Devonian–Lower Mississippian black shales, carbonate sedimentation on the craton dominated the remainder of the Mississippian Period (Figure 20.17). During this time, a variety of carbonate sediments were deposited in the epeiric sea, as indicated by the extensive deposits of crinoidal limestones (rich in crinoid fragments), oolitic limestones, and various other limestones and dolostones. These Mississippian carbonates display cross-bedding, ripple marks, and well-sorted fossil fragments, all of which indicate a shallow-water environment. Analogous features can be observed on the present-day Bahama Banks. In addition, numerous small organic reefs occurred throughout the craton during the Mississippian. These were all much smaller, however, than the large barrier-reef complexes that dominated the earlier Paleozoic seas.

During the Late Mississippian regression of the Kaskaskia Sea from the craton, vast quantities of detrital sediments replaced carbonate deposition. The resulting sandstones, particularly in the Illinois Basin, have been studied in great detail because they are excellent petroleum reservoirs. Before the end of the Mississippian, the epeiric sea had retreated to the craton margin, once again exposing the craton to widespread weathering and erosion, resulting in a craton-wide unconformity at the end of the Kaskaskia Sequence.

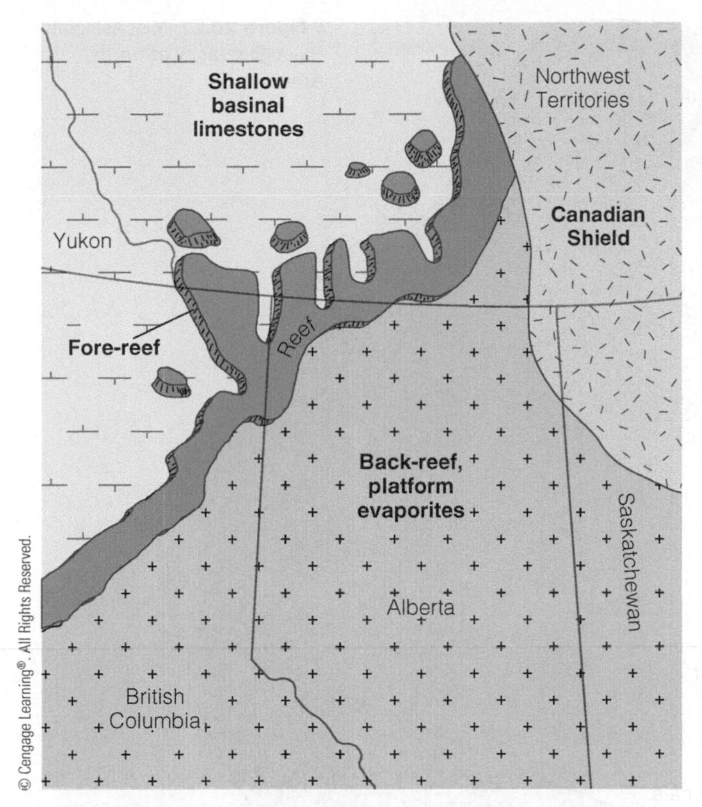

Figure 20.15 Devonian Reef Complex of Western Canada
Reconstruction of the extensive Devonian Reef complex of western Canada. These extensive reefs controlled the regional facies of the Devonian epeiric seas.

Critical Thinking Question: How do you think the reefs controlled the regional facies in this area?

20.8 The Absaroka Sequence

The **Absaroka Sequence** includes rocks deposited during the Pennsylvanian through Early Jurassic. In this chapter, however, we are concerned only with the Paleozoic rocks of the Absaroka Sequence. The extensive unconformity separating the Kaskaskia and Absaroka sequences essentially divides the strata into the North American Mississippian and Pennsylvanian systems. These two systems are closely equivalent to the European Lower and Upper Carboniferous systems, respectively. The rocks of the Absaroka Sequence not only differ from those of the Kaskaskia Sequence, but also result from different tectonic regimes.

The lowermost sediments of the Absaroka Sequence are confined to the margins of the craton. These deposits are generally thickest in the east and southeast, near the emerging highlands of the Appalachian and Ouachita mobile belts, and thin westward onto the craton. The rocks also reveal lateral changes from nonmarine detrital rocks and coals in the east, through transitional marine–nonmarine beds, to largely marine detrital rocks and limestones farther west (Figure 20.18).

What Are Cyclothems, and Why Are They Important?

One characteristic feature of Pennsylvanian rocks is their repetitive pattern of alternating marine and nonmarine strata known as **cyclothems**. They result from repeated alternations of marine and nonmarine environments, usually in

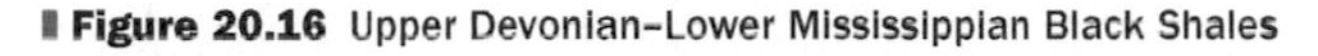
Figure 20.16 Upper Devonian–Lower Mississippian Black Shales

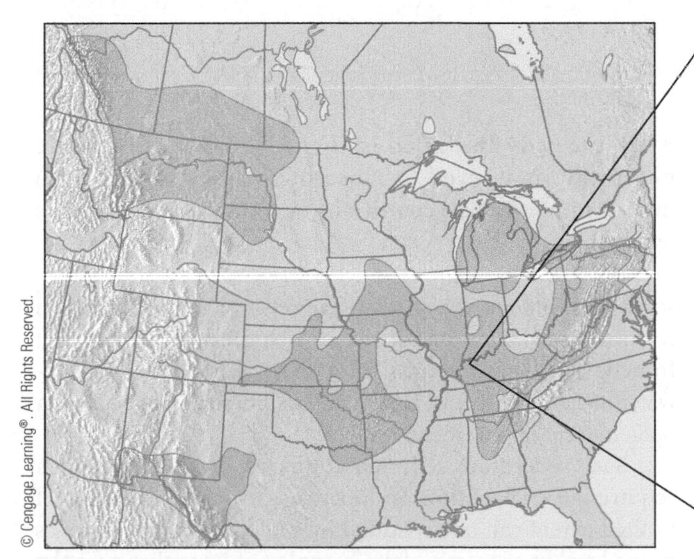

a The extent of the Upper Devonian to Lower Mississippian Chattanooga Shale and its equivalent units (such as the Antrim Shale and New Albany Shale) in North America.

b Upper Devonian New Albany Shale, Button Mold Knob Quarry, Kentucky.

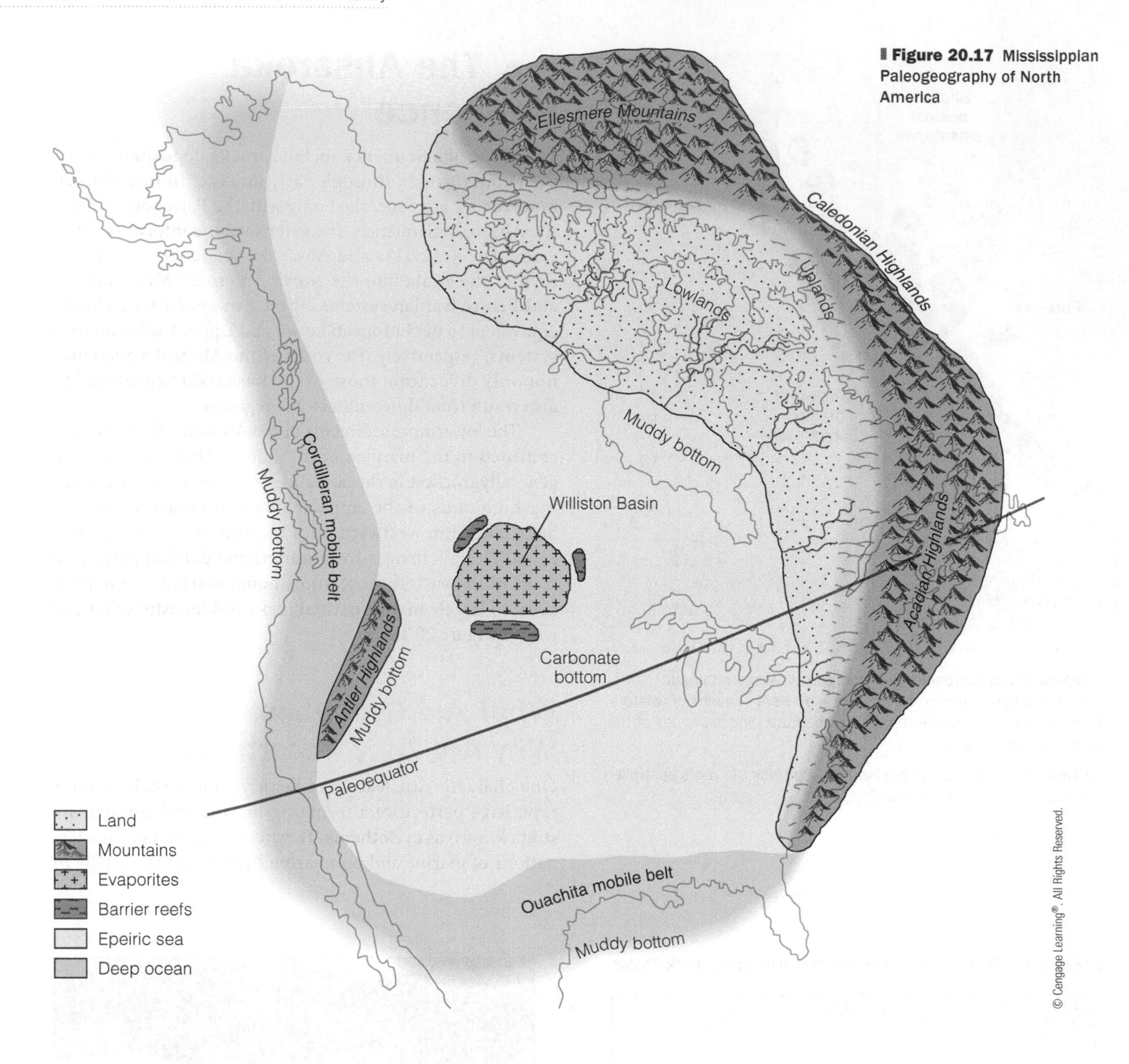

Figure 20.17 Mississippian Paleogeography of North America

areas of low relief. Although seemingly simple, cyclothems reflect a delicate interplay between nonmarine deltaic and shallow-marine interdeltaic and shelf environments.

To illustrate this point, look at a typical coal-bearing cyclothem from the Illinois Basin (Figure 20.19a). Such a cyclothem contains nonmarine units, capped by a coal unit and overlain by marine units. Figure 20.19a shows the depositional environments that produced the cyclothem. The initial units represent deltaic and fluvial deposits. Above them is an underclay that frequently contains root casts from the plants and trees that make up the overlying coal. The coal bed results from accumulations of plant material and is overlain by marine units of alternating limestones and shales, usually with an abundant marine invertebrate fauna. The marine cycle ends with an erosion surface. A new cyclothem begins with a nonmarine deltaic sandstone. All of the beds illustrated in the idealized cyclothem are not always preserved because of abrupt changes from marine to nonmarine conditions or removal of some units by erosion.

Cyclothems represent transgressive and regressive sequences with an erosional surface separating one cyclothem from another. Thus, an idealized cyclothem passes upward from fluvial–deltaic deposits, through coals, to detrital shallow-water marine sediments, and finally to limestones typical of an open marine environment.

Such repetitious sedimentation over a widespread area requires an explanation. In most cases, local cyclothems of limited extent can be explained by rapid, but slight, changes in sea level in a swamp–delta complex of low relief near the sea such as by progradation (see Figure 12.14) or by localized crustal movement.

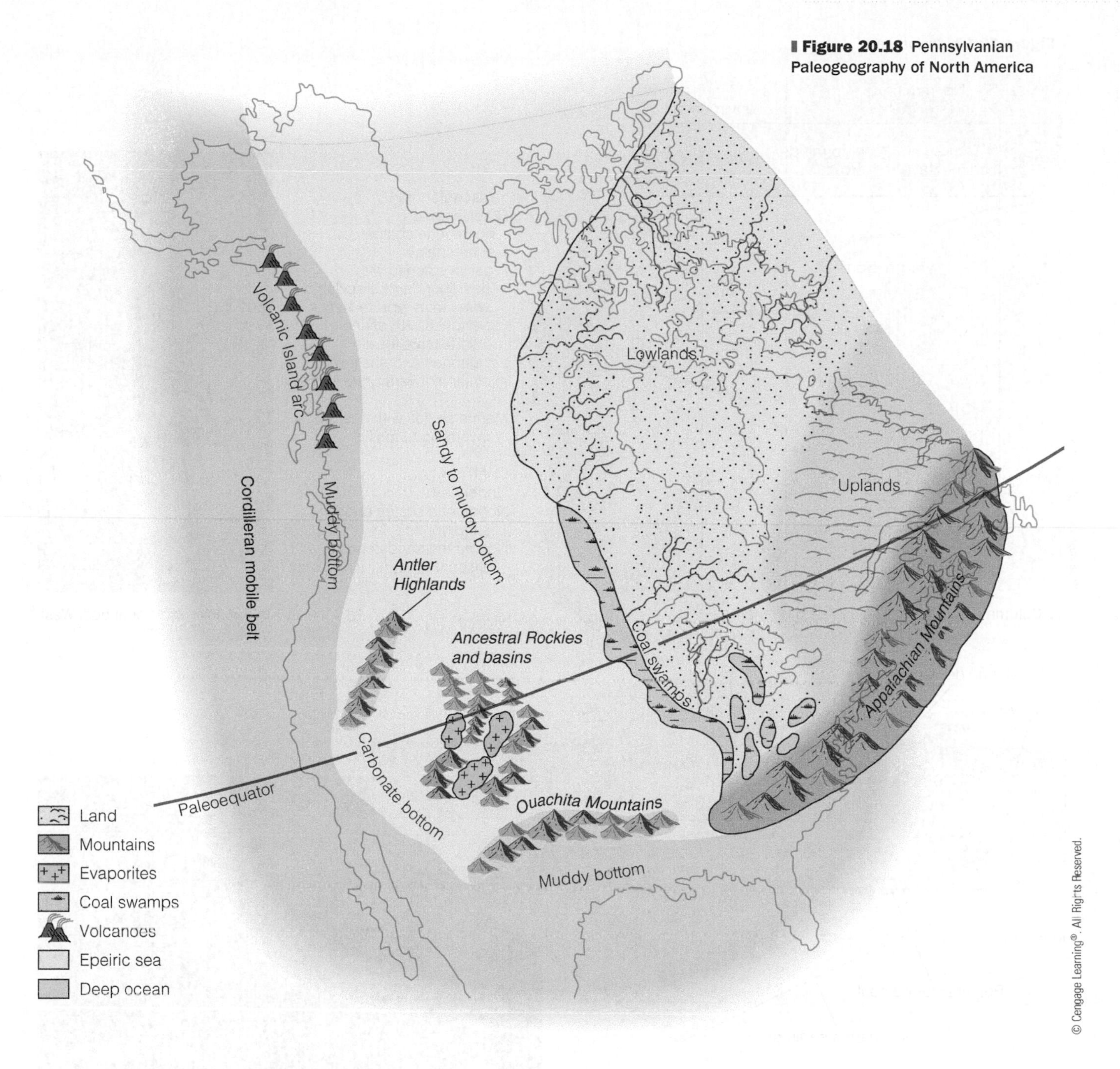

Figure 20.18 Pennsylvanian Paleogeography of North America

Explaining widespread cyclothems is more difficult. The hypothesis favored by many geologists is a rise and fall of sea level related to advances and retreats of Gondwanan continental glaciers. When the Gondwanan ice sheets advanced, sea level dropped; when they melted, sea level rose. Late Paleozoic cyclothem activity on all of the cratons closely corresponds to Gondwanan glacial–interglacial cycles.

Cratonic Uplift—The Ancestral Rockies

Recall that cratons are stable areas, and what deformation they do experience is usually mild. The Pennsylvanian Period, however, was a time of unusually severe cratonic deformation, resulting in uplifts of sufficient magnitude to expose Precambrian basement rocks. In addition to newly formed highlands and basins, many previously formed arches and domes, such as the Cincinnati Arch, Nashville Dome, and Ozark Dome, were also reactivated (Figure 20.1).

During the Late Absaroka (Pennsylvanian), the area of greatest deformation was in the southwestern part of the North American craton, where a series of fault-bounded uplifted blocks formed the **Ancestral Rockies** (Figure 20.20a). These mountain ranges had diverse geologic histories and were not all elevated at the same time. Uplift of these mountains, some of which were elevated more than 2 km along near-vertical faults, resulted in erosion of overlying Paleozoic sediments and exposure of the Precambrian igneous and metamorphic basement rocks (Figure 20.20b). As the mountains

Figure 20.19 Cyclothems

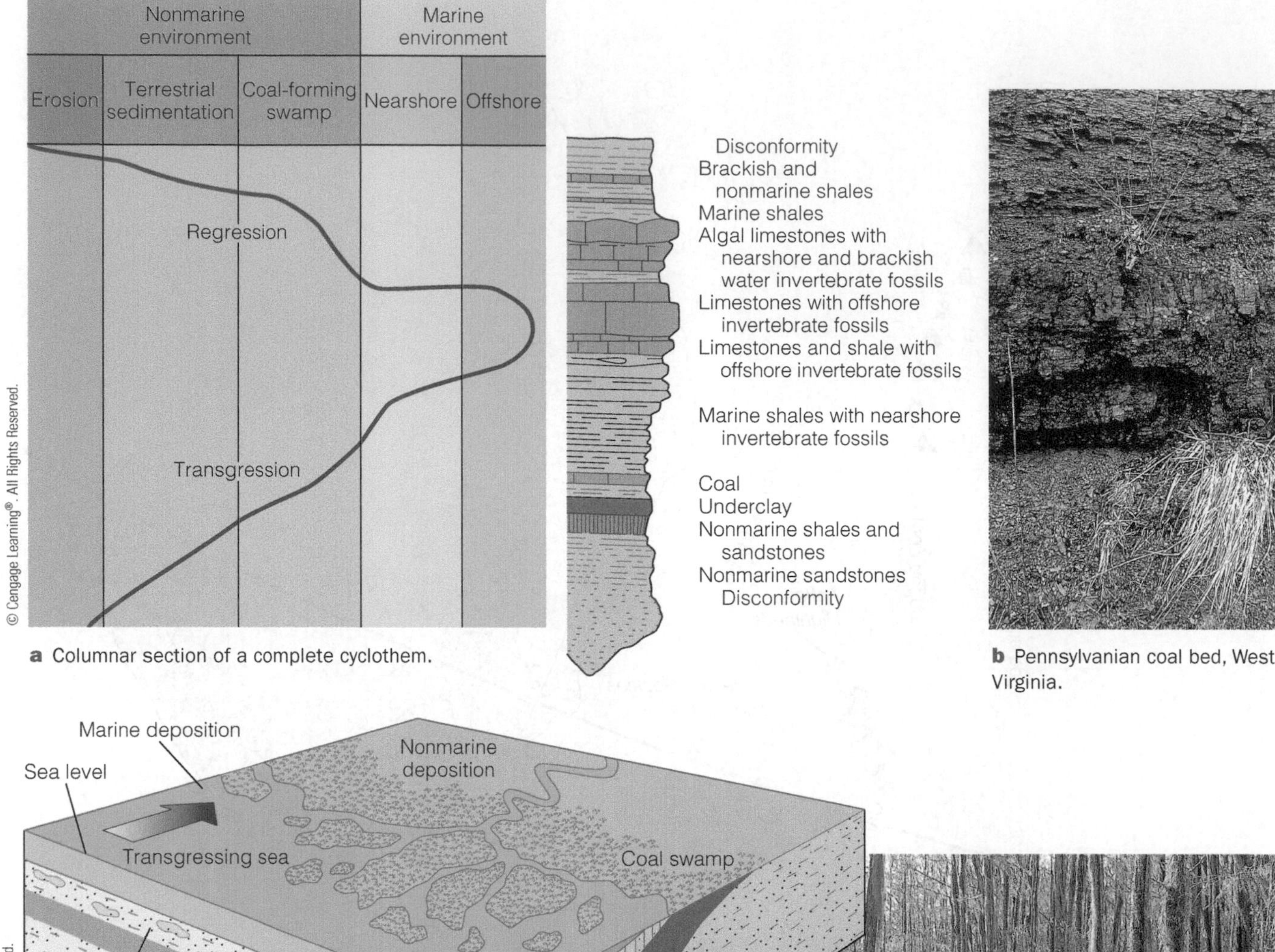

a Columnar section of a complete cyclothem.

b Pennsylvanian coal bed, West Virginia.

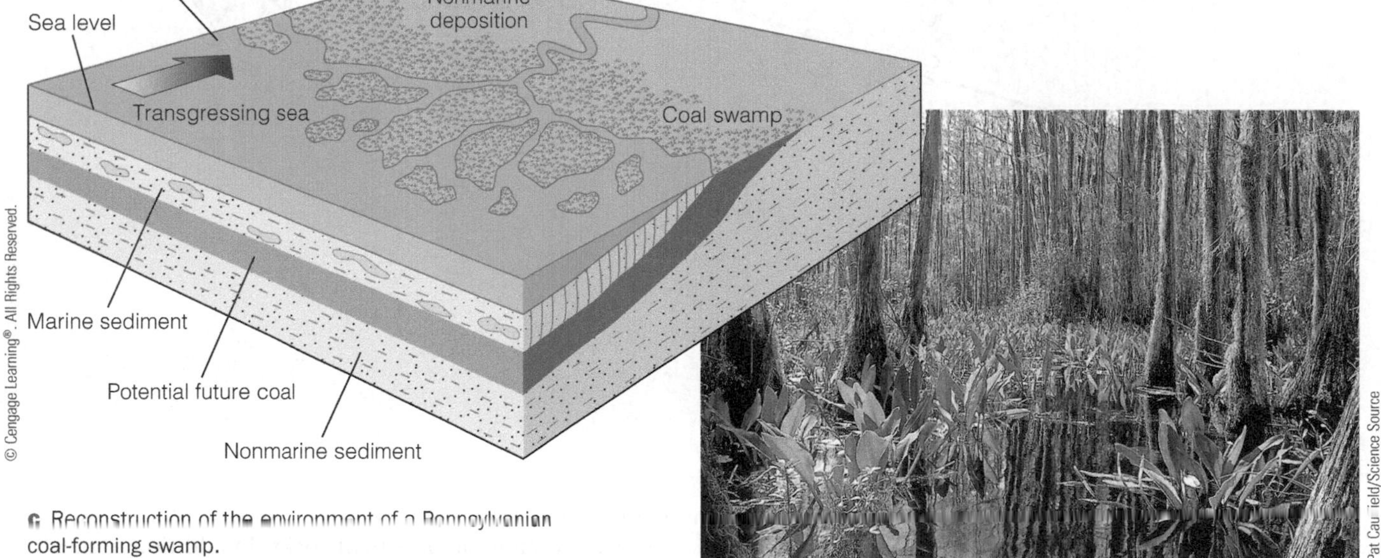

c Reconstruction of the environment of a Pennsylvanian coal-forming swamp.

d The Okefenokee Swamp, Georgia, is a modern example of a coal-forming environment, similar to those occurring during the Pennsylvanian Period.

eroded, tremendous quantities of coarse, red arkosic sand and conglomerate were deposited in the surrounding basins. These sediments are preserved in many areas, including the rocks of the Garden of the Gods near Colorado Springs (**❙** Figure 20.20c; see Figure 7.6a, b) and at the Red Rocks Amphitheatre near Morrison, Colorado.

Intracratonic mountain ranges are unusual, and their cause has long been debated. It is currently thought that the collision of Gondwana with Laurasia along the Ouachita mobile belt (Figure 20.4a) generated great stresses in the southwestern region of the North American craton. These crustal stresses were relieved by faulting. Movement along these faults

Figure 20.20 The Ancestral Rockies

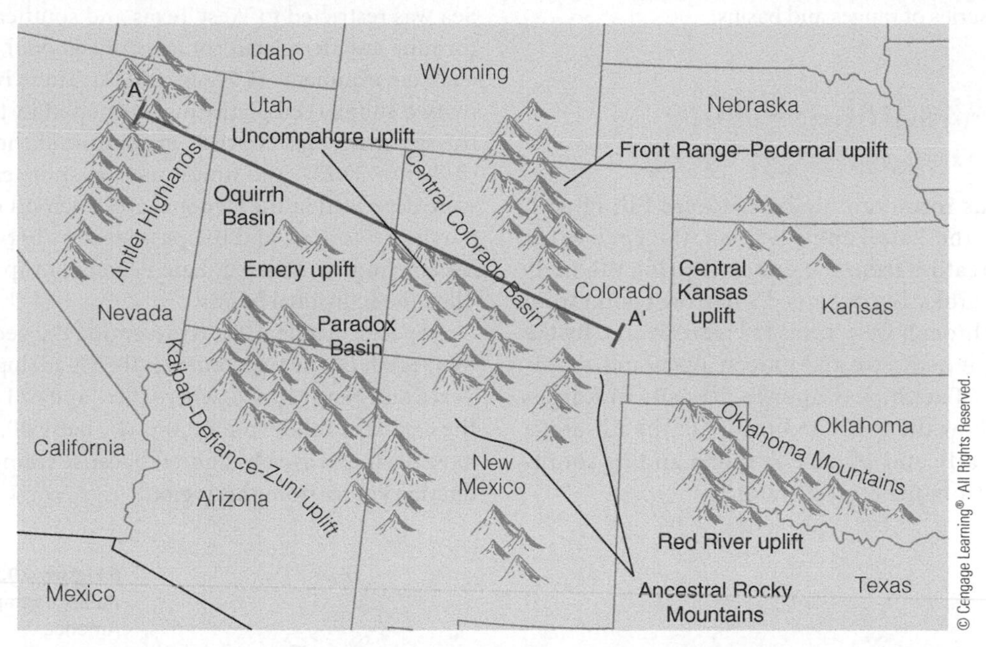

a Location of the principal Pennsylvanian highland areas and basins of the southwestern part of the craton.

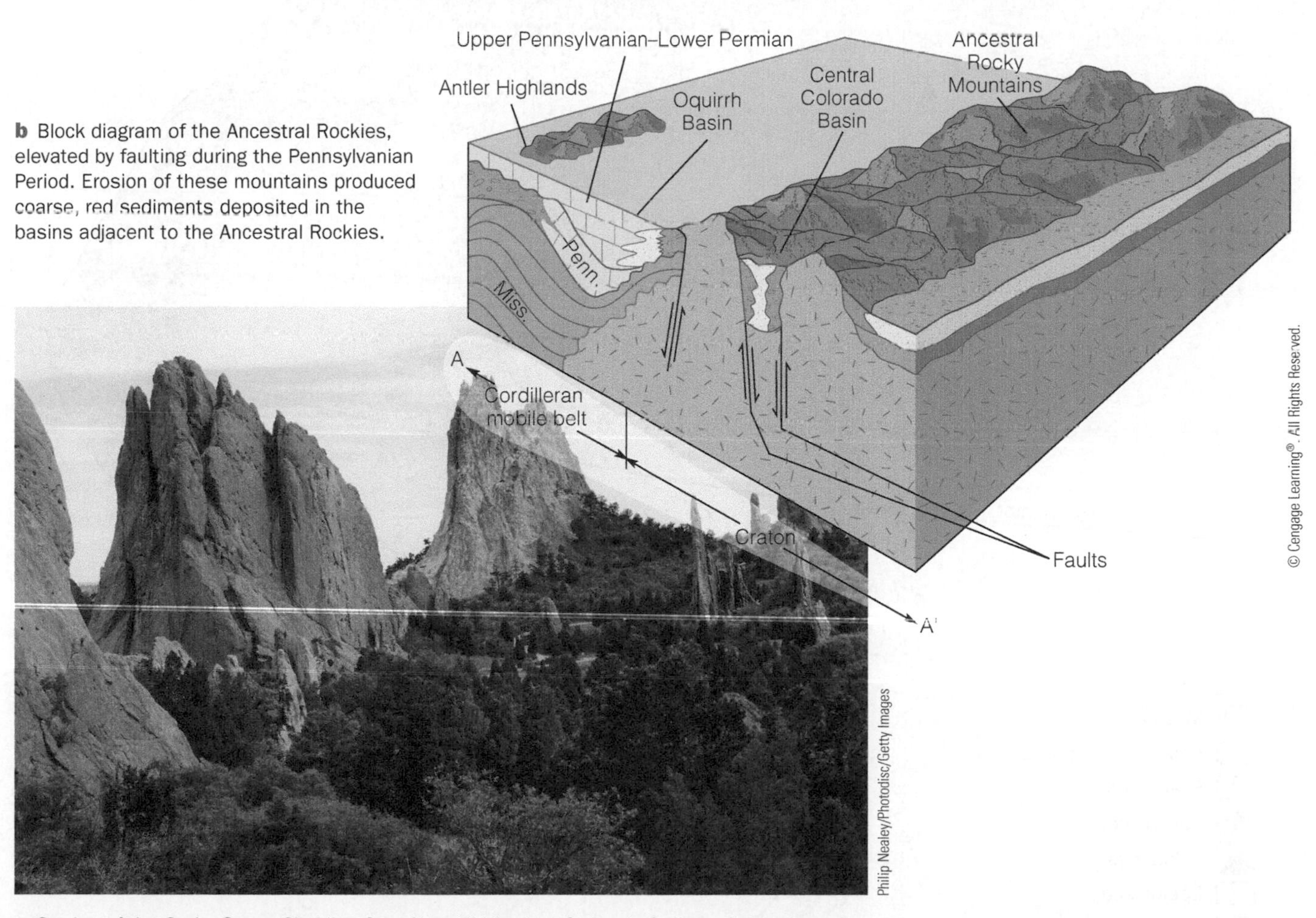

b Block diagram of the Ancestral Rockies, elevated by faulting during the Pennsylvanian Period. Erosion of these mountains produced coarse, red sediments deposited in the basins adjacent to the Ancestral Rockies.

c Garden of the Gods, Storm Sky View from Near Hidden Inn, Colorado Springs, Colorado.

produced uplifted cratonic blocks and downwarped adjacent basins, forming a series of ranges and basins.

The Late Absaroka—More Evaporite Deposits and Reefs

While the various intracratonic basins were filling with sediment during the Late Pennsylvanian, the epeiric sea slowly began retreating from the craton. During the Early Permian, the Absaroka Sea occupied a narrow region from Nebraska, south through west Texas (▮ Figure 20.21). By the Middle Permian, it had retreated to west Texas and southern New Mexico. The thick evaporite deposits in Kansas and Oklahoma show the restricted nature of the Absaroka Sea during the Early and Middle Permian and its southwestward retreat from the central craton.

During the Middle and Late Permian, the Absaroka Sea was restricted to West Texas and southern New Mexico, forming an interrelated complex of lagoonal, reef, and open-shelf environments (▮ Figure 20.22). Three basins separated by two submerged platforms developed in this area during the Permian. Massive reefs grew around the basin margins (▮ Figure 20.23), and limestones, evaporites, and red beds were deposited in the lagoonal areas behind the reefs. As the barrier reefs grew and the passageways between the basins became more restricted, Late Permian evaporites gradually filled the individual basins.

Spectacular deposits representing the geologic history of this region can be seen today in the Guadalupe Mountains of Texas and New Mexico, where the Capitan Limestone forms the caprock of these mountains (▮ Figure 20.24). These reefs have been extensively studied because tremendous oil production comes from this region.

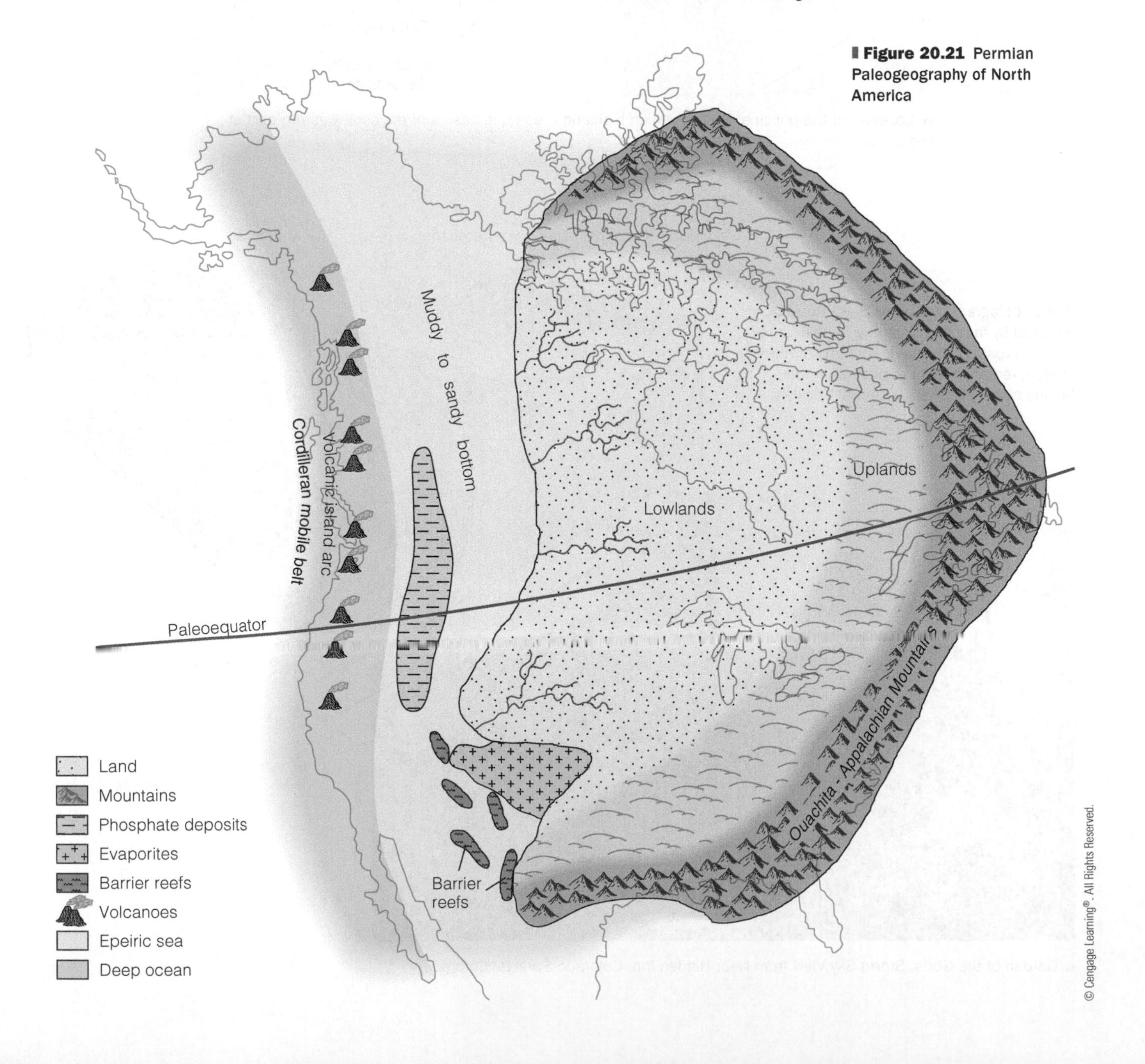

▮ **Figure 20.21** Permian Paleogeography of North America

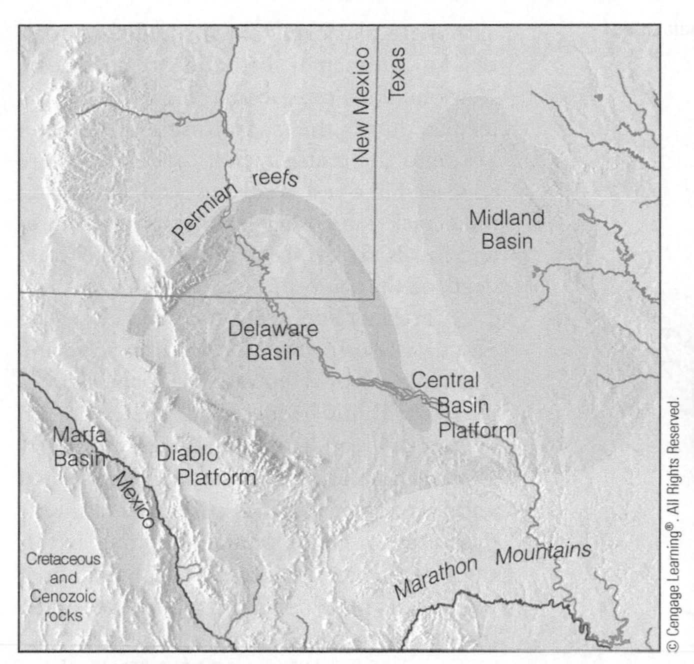

Figure 20.22 West Texas Permian Basins and Surrounding Reefs During the Middle and Late Permian, an interrelated complex of lagoonal, barrier-reef, and open-shelf environments formed in the West Texas and southern New Mexico area. Much of the tremendous oil production in this region comes from these reefs.

Critical Thinking Question: Why is so much oil found in reef structures?

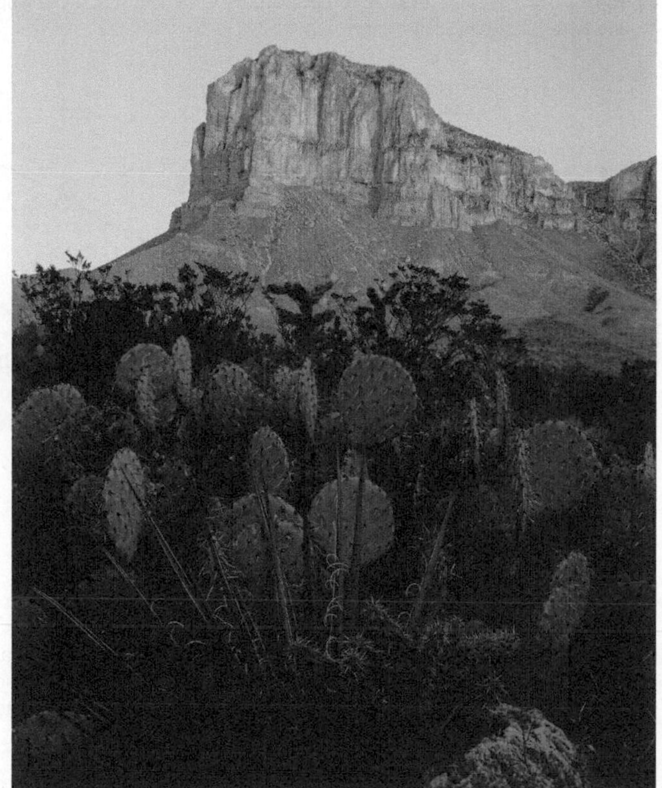

Figure 20.24 Guadalupe Mountains, Texas The prominent Capitan Limestone forms the caprock of the Guadalupe Mountains. The Capitan Limestone is rich in fossil corals and associated reef organisms.

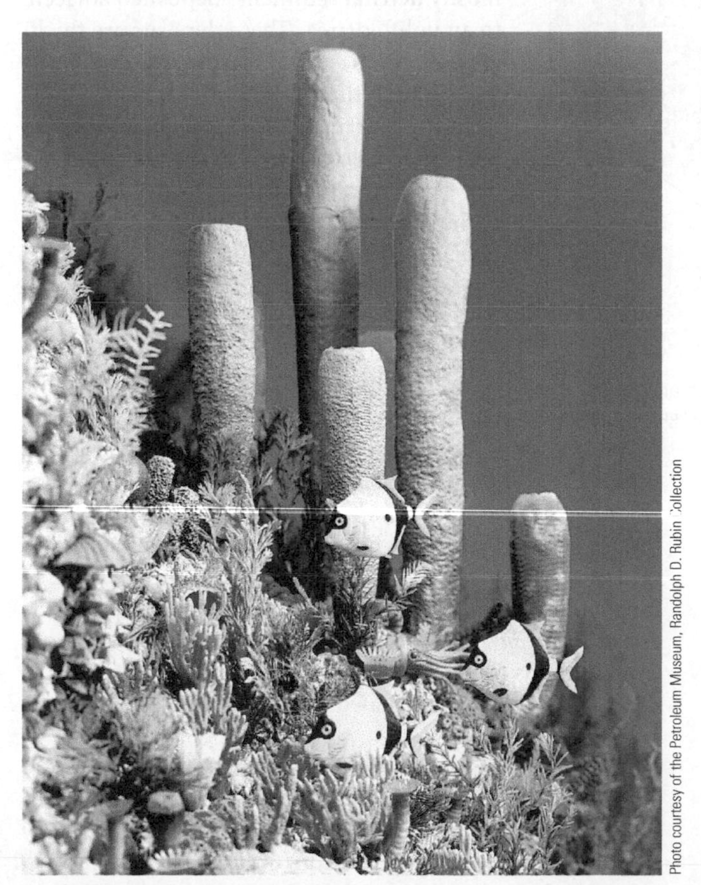

Figure 20.23 Middle Permian Capitan Limestone Reef Environment A reconstruction of the Middle Permian Capitan Limestone reef environment. Shown are brachiopods, corals, bryozoans, and large glass sponges.

By the end of the Permian Period, the Absaroka Sea had retreated from the craton, exposing continental red beds that had been deposited over most of the southwestern and eastern region (Figure 20.4b).

20.9 History of the Paleozoic Mobile Belts

Having examined the Paleozoic history of the craton, we now turn our attention to the orogenic activity in the mobile belts. The mountain building occurring during the Paleozoic Era had a profound influence on the climate and sedimentary history of the craton. In addition, it was part of the global tectonic regime that sutured the continents together, forming Pangaea by the end of the Paleozoic Era.

Appalachian Mobile Belt

Throughout Sauk time (Neoproterozoic–Early Ordovician), the Appalachian region was a broad, passive, continental margin. Sedimentation was closely balanced by

Figure 20.25 Neoproterozoic to Late Ordovician Evolution of the Appalachian Mobile Belt

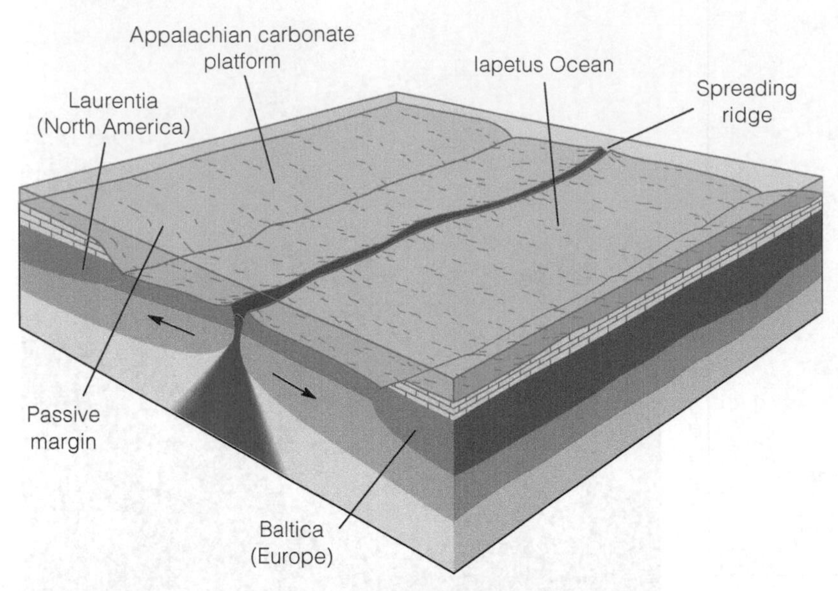

a During the Neoproterozoic to the Early Ordovician, the Iapetus Ocean was opening along a divergent plate boundary. Both the east coast of Laurentia and the west coast of Baltica were passive continental margins with large carbonate platforms.

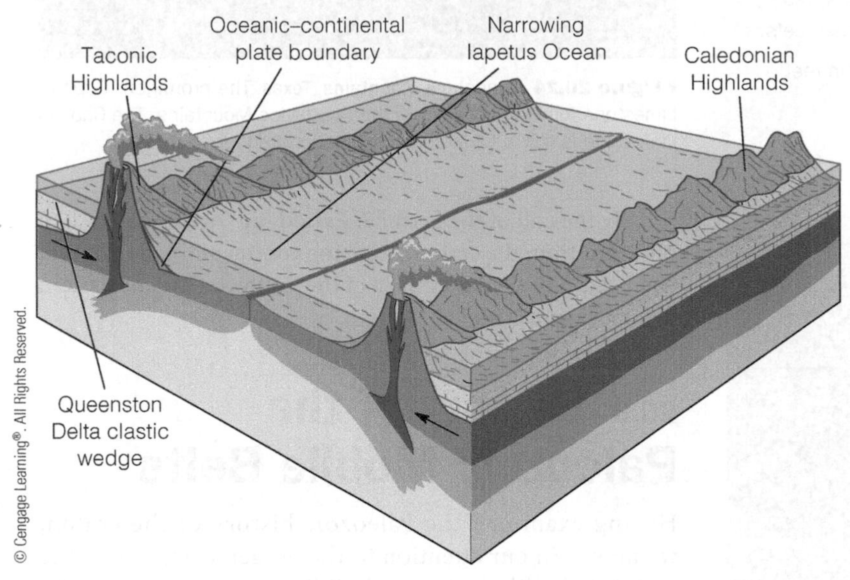

b Beginning in the Middle Ordovician, the passive margins of Laurentia and Baltica changed to active oceanic–continental plate boundaries, resulting in orogenic activity.

subsidence as extensive carbonate deposits succeeded thick, shallow marine sands. During this time, movement along a divergent plate boundary was widening the **Iapetus Ocean** (❚ Figure 20.25a).

Taconic Orogeny Beginning with the subduction of the Iapetus plate beneath Laurentia (an oceanic–continental convergent plate boundary), the Appalachian mobile belt was born (❚ Figure 20.25b). The resulting **Taconic orogeny**—named after the present-day Taconic Mountains of eastern New York, central Massachusetts, and Vermont—was the first of several orogenies to affect the Appalachian region.

The Appalachian mobile belt can be divided into two depositional environments. The first is the extensive, shallow-water carbonate platform that formed the broad eastern continental shelf and stretched from Newfoundland to Alabama (Figure 20.25a). It formed during the transgression of the Sauk Sea onto the craton when carbonates were deposited in a vast, shallow sea. Stromatolites, mud cracks, and other sedimentary structures and fossils are evidence of the shallow water depth on the platform.

Carbonate deposition ceased along the East Coast during the Middle Ordovician and was replaced by deepwater deposits characterized by thinly bedded black shales, graded beds, coarse sandstones, graywackes (poorly sorted, immature sandstones), and associated volcanic material. This suite of sediments marks the onset of mountain building, in this case, the Taconic orogeny. The subduction of the Iapetus plate beneath Laurentia resulted in volcanism and downwarping of the carbonate platform (Figure 20.25b). Throughout the Appalachian mobile belt, facies patterns, paleocurrents, and sedimentary structures all indicate that these deposits were derived from the east, where the Taconic Highlands and associated volcanoes were rising.

The final piece of evidence for the Taconic orogeny is the development of a large **clastic wedge**, an extensive accumulation of mostly detrital sediments deposited adjacent to an uplifted area. These deposits are thickest and coarsest nearest the highland area and become thinner and finer grained away from the source area, eventually grading into the carbonate cratonic facies (❚ Figure 20.26). The clastic wedge resulting from the erosion of the Taconic Highlands is referred to as the **Queenston Delta**. Careful mapping and correlation of these deposits indicate that more than 600,000 km^3 of rock were eroded from the Taconic Highlands. On the basis of this figure, geologists estimate that the Taconic Highlands were at least 4,000 m high.

The Taconic orogeny marked the first pulse of mountain building in the Appalachian mobile belt and was a response to the subduction taking place beneath the east coast of Laurentia. As the Iapetus Ocean narrowed and closed, another orogeny occurred in Europe during the Silurian. The Caledonian orogeny was essentially a mirror image of the Taconic and Acadian orogenies and was part of the global mountain-building episode that occurred during the Paleozoic Era.

Caledonian Orogeny The Caledonian mobile belt stretches along the western border of Baltica and includes the present-day countries of Scotland, Ireland, and Norway (Figure 20.2c). During the Middle Ordovician, subduction along the boundary between the Iapetus plate and Baltica

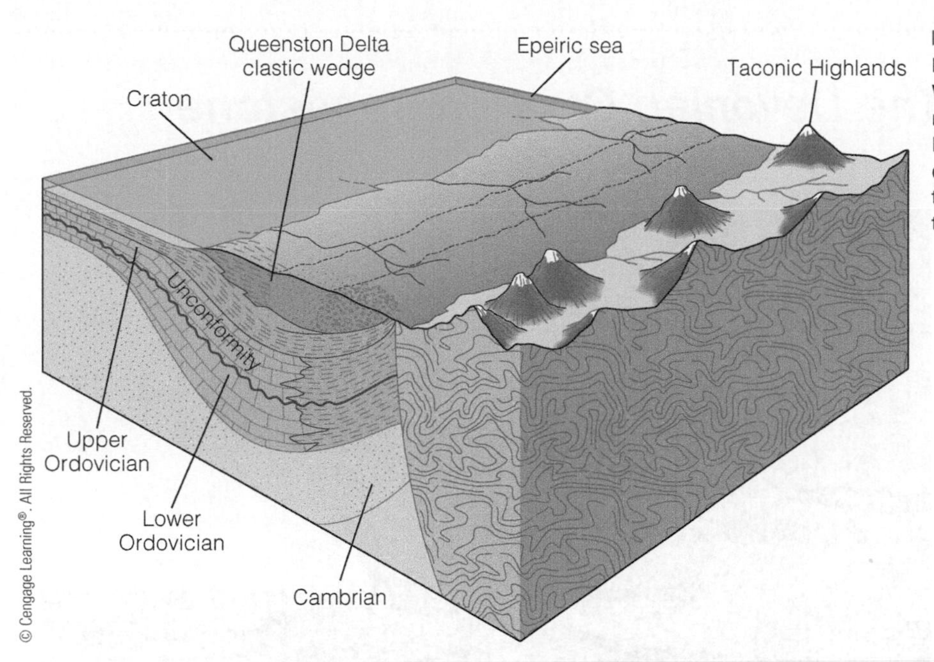

Figure 20.26 Reconstruction of the Taconic Highlands and Queenston Delta Clastic Wedge The Queenston Delta clastic wedge, resulting from the erosion of the Taconic Highlands, consists of thick, coarse-grained detrital sediments nearest the highlands and thins laterally into fine-grained sediments in the epeiric seas covering the craton.

(Europe) began, forming a mirror image of the convergent plate boundary off the east coast of Laurentia (North America).

The culmination of the **Caledonian orogeny** occurred during the Late Silurian and Early Devonian with the formation of a mountain range along the western margin of Baltica. Red-colored sediments deposited along the front of the Caledonian Highlands formed a large clastic wedge known as the *Old Red Sandstone* (Geo-Insight 20.1; see Figure 10.7).

Acadian Orogeny The third Paleozoic orogeny to affect Laurentia and Baltica began during the Late Silurian and concluded at the end of the Devonian Period. The **Acadian orogeny** affected the Appalachian mobile belt from Newfoundland to Pennsylvania as sedimentary rocks were folded and thrust against the craton.

As with the preceding Taconic and Caledonian orogenies, the Acadian orogeny occurred along an oceanic–continental convergent plate boundary. As the northern Iapetus Ocean continued to close during the Devonian, the plate carrying Baltica finally collided with Laurentia, forming a continental–continental convergent plate boundary along the zone of collision (Figure 20.3a).

Weathering and erosion of the Acadian Highlands produced the **Catskill Delta**, a thick clastic wedge named for the Catskill Mountains in upstate New York where it is well exposed. Composed of red, coarse conglomerates, sandstones, and shales, the Catskill Delta contains nearly three times as much sediment as the Queenston Delta (Figure 20.26).

The Devonian rocks of New York are among the best studied on the continent. A cross section of the Devonian strata clearly reflects an eastern source (Acadian Highlands) for the Catskill facies (Figure 20.27). These detrital rocks can be traced from eastern Pennsylvania, where the coarse-grained deposits are approximately 3 km thick, to Ohio, where the deltaic facies are only about 100 m thick and consist of cratonic shales and carbonates.

The red beds of the Catskill Delta derive their color from the hematite (Fe_2O_3) in the sediments. Plant fossils and oxidation of the hematite indicate that the beds were deposited in a continental environment. Toward the west, the red beds grade laterally into gray sandstones and shales containing fossil tree trunks, which indicate a swamp or marsh environment.

The Old Red Sandstone

The red beds of the Catskill Delta have a European counterpart in the Devonian Old Red Sandstone of the British Isles (Figure 20.27).  This formation was a Devonian clastic wedge that grew eastward from the Caledonian Highlands onto the Baltica craton, and, like its North American Catskill counterpart, contains numerous fossils of freshwater fish, early amphibians, and land plants.

ConnectionLink
To learn more about the fish, amphibians, reptiles, and land plants of the Paleozoic, see the respective sections in Chapter 21.

By the end of the Devonian Period, Baltica and Laurentia were sutured together, forming Laurasia (Figure 20.27). The red beds of the Catskill Delta can be traced north, through Canada and Greenland, to the Old Red Sandstone of the British Isles, and into northern Europe. These beds were all deposited in similar environments along the flanks of developing mountain chains formed at convergent plate boundaries.

The Taconic, Caledonian, and Acadian orogenies were all part of the same major orogenic event related to the closing of the Iapetus Ocean (Figures 20.25 and 20.27). This event began with paired oceanic–continental convergent plate boundaries during the Taconic and Caledonian orogenies and culminated along a continental–continental convergent plate boundary during the Acadian orogeny as Laurentia and Baltica became sutured. After this, the Hercynian (Variscan)–Alleghenian orogeny began, followed by orogenic activity in the Ouachita mobile belt.

GEO INSIGHT 20.1 The Devonian Old Red Sandstone

The Devonian System was named by Roderick Murchison and Adam Sedgwick in 1839 for rock exposures in Devon County, England (▌Figure 1). The Old Red Sandstone, which was deposited during the Devonian Period, and crops out extensively throughout the British Isles, is predominantly of terrestrial origin.

The Old Red Sandstone consists of upward of 11,000 m of clastic sediments deposited in various structural basins throughout Scandinavia, the British Isles, Greenland, and northeastern Canada. The sediments are poorly sorted and variable, and consist of conglomerates, coarse-grained sandstones, cross-bedded sandstones, siltstones, shales, and thin limestones, deposited in several continental environments. The predominant red color comes from the presence of oxidized iron minerals, but colors also range from gray and green, through orange and purple.

The collision of Baltica and Laurentia during the Caledonian orogeny (Late Silurian and Early Devonian), produced a mountain range along the western margin of Baltica (Figures 20.2c and 20.3a) that is referred to as the Caledonian Mountain chain, or Caledonian Highlands. The weathering and erosion of these highlands resulted in the formation of a large clastic wedge, whose sediments were named the Old Red Sandstone (Figure 20.27).

It was in 1788 at Siccar Point, Scotland, that James Hutton first recognized the significance of unconformities in interpreting Earth history. At this location he observed the Devonian Old Red Sandstone overlying the steeply dipping Silurian graywackes (see Figure 17.10b).

Not only is The Old Red Sandstone important in British and global stratigraphy, but it is also important for the many fossils of jawless armored fish (ostracoderms), jawed armored fish (placoderms), early bony fish, as well as arthropods, and primitive seedless vascular plants found throughout its extent (see Chapter 21).

Lastly, the Old Red Sandstone has been widely used as a building stone. For example, many of the castles, cathedrals, and abbeys in Scotland, Wales, and England were built using Old Red Sandstone. Some examples include Muchalls Castle, Aberdeenshire, Scotland, St. Magnus Cathedral, Orkney, Scotland, and Goodrich Castle, Herefordshire, England (▌Figure 2). Interestingly, the New York Life Insurance Building (also known as the Quebec Bank Building) in Montreal, Canada, was built in 1887–1889 using the Old Red Sandstone, imported from Dumfriesshire, Scotland.

kolo5/Shutterstock.com

▌**Figure 1** Red cliffs composed of the Old Red Sandstone crop out along the shoreline in Devon County, England.

David Hughes/shutterstock.com

▌**Figure 2** Remains of Goodrich Castle in Herefordshire, England, dated to the period between 1160 and 1270. Goodrich Castle is one of the many castles built from rocks quarried from the Old Red Sandstone, a Devonian-age formation. Identical red sandstone is found in the Catskill Mountains of New York. It too has been used as a building stone for many structures in the New York area.

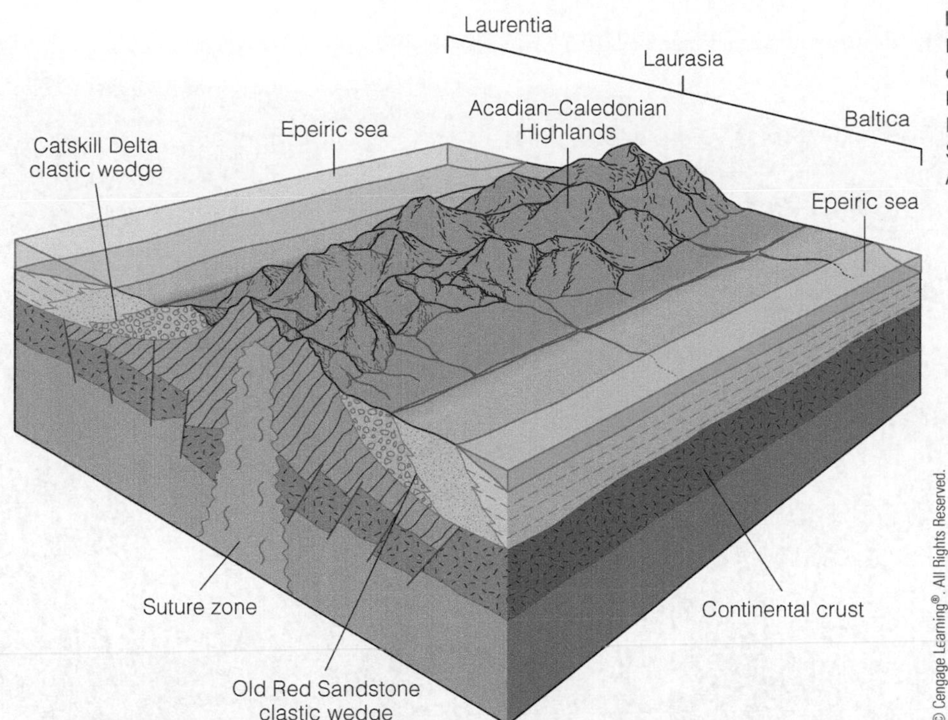

Figure 20.27 Formation of Laurasia Block diagram showing the area of collision between Laurentia and Baltica. Note the bilateral symmetry of the Catskill Delta clastic wedge and the Old Red Sandstone and their relationship to the Acadian and Caledonian Highlands.

Hercynian (Variscan)–Alleghenian Orogeny The Hercynian mobile belt of southern Europe and the Appalachian and Ouachita mobile belts of North America mark the zone along which Europe (part of Laurasia) collided with Gondwana (Figure 20.3). While Gondwana and southern Laurasia collided during the Pennsylvanian and Permian periods in the area of the Ouachita mobile belt, eastern Laurasia (Europe and southeastern North America) joined together with Gondwana (Africa) as part of the **Hercynian (Variscan)–Alleghenian orogeny** (Figure 20.4a). The terms *Hercynian* and *Variscan* are frequently used interchangeably, and originally referred to the direction of fold belts in Europe.

Initial contact between eastern Laurasia and Gondwana began during the Mississippian Period along the Hercynian mobile belt. The greatest deformation occurred during the Pennsylvanian and Permian periods and is referred to as the *Hercynian* or *Variscan orogeny* (Figure 20.28). Conversely, and contemporaneously, the central and southern parts of the Appalachian mobile belt (from New York to Alabama) were folded and thrust toward the craton as eastern Laurasia and Gondwana were sutured. This tectonic event in North America is referred to as the *Alleghenian orogeny*.

These three Late Paleozoic orogenies (Hercynian [Variscan], Alleghenian, and Ouachita) represent the final joining of Laurasia and Gondwana into the supercontinent Pangaea during the Permian.

Cordilleran Mobile Belt

During the Neoproterozoic and Early Paleozoic, the Cordilleran area was a passive continental margin along which extensive continental shelf sediments were deposited. Thick sections of marine sediment graded laterally into thin cratonic units as the Sauk Sea transgressed onto the craton. Beginning in the Middle Paleozoic, an island arc formed off the western margin of the craton. A collision between this eastward-moving island arc and the western border of the craton took place during the Late Devonian and Early Mississippian, producing a highland area.

This orogenic event, the **Antler orogeny**, was caused by subduction, resulting in deep-water deposits and oceanic crustal rocks being thrust eastward over shallow-water continental shelf sediments, thus closing the narrow ocean basin separating the island arc from the craton (Figure 20.29). Erosion of the resulting Antler Highlands produced large quantities of sediments that were deposited to the east in the epeiric sea covering the craton and to the west in the deep sea. The Antler orogeny was the first in a series of orogenic events to affect the Cordilleran mobile belt. During the Mesozoic and Cenozoic, this area was the site of major tectonic activity caused by oceanic–continental convergence and accretion of various terranes.

To learn more about Mesozoic and Cenozoic tectonic activity, see Chapters 22 and 23.

Ouachita Mobile Belt

The Ouachita mobile belt extends for approximately 2,100 km from the subsurface of Mississippi to the Marathon region of Texas. Approximately 80% of the former mobile belt is buried beneath a Mesozoic and Cenozoic sedimentary cover. The two major exposed areas in this region are the Ouachita Mountains of Oklahoma and Arkansas and the Marathon Mountains of Texas.

Figure 20.28 The Variscan Orogeny The Variscan unconformity exposed at Telheiro Beach in southwestern Portugal shows the highly folded rocks of the Pennsylvanian Brejeira Formation, unconformably overlain by the red-colored, terrestrial beds of the Upper Triassic "Grés de Silves." The folded strata of the Brejeira Formation were deformed during the Hercynian (Variscan)–Alleghenian orogeny.

Critical Thinking Question What type of unconformity is shown here? Why is this type of unconformity evidence of mountain building?

During the Neoproterozoic to Early Mississippian, shallow-water detrital and carbonate sediments were deposited on a broad continental shelf, and in the deeper-water portion of the adjoining mobile belt, bedded cherts and shales were accumulating (Figure 20.30a). Beginning in the Mississippian Period, the rate of sedimentation increased dramatically as the region changed from a passive continental margin to an active convergent plate boundary, marking the beginning of the **Ouachita orogeny** (Figure 20.30b).

Thrusting of sediments continued throughout the Pennsylvanian and Early Permian, driven by the compressive forces generated along the zone of subduction as Gondwana collided with Laurasia (Figure 20.30c). The collision of Gondwana and Laurasia is marked by the formation of a large mountain range, most of which eroded during the Mesozoic Era. Only the rejuvenated Ouachita and Marathon Mountains remain of this once lofty mountain range.

The Ouachita deformation was part of the general worldwide tectonic activity that occurred when Gondwana united with Laurasia. The Hercynian, Appalachian, and Ouachita mobile belts were continuous and

Volcanic arc
Antler Highlands
Continental shelf
Oceanic crust
Continental crust

Figure 20.29 Antler Orogeny Reconstruction of the Cordilleran mobile belt during the Early Mississippian in which deep-water continental slope deposits were thrust eastward over shallow-water continental shelf carbonates, forming the Antler Highlands.

Figure 20.30 Ouachita Mobile Belt Plate tectonic model for deformation of the Ouachita mobile belt.

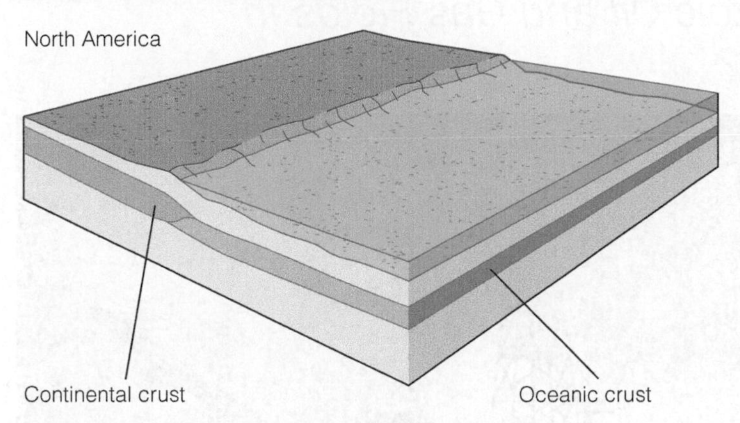

a Depositional environment prior to the beginning of the orogenic activity.

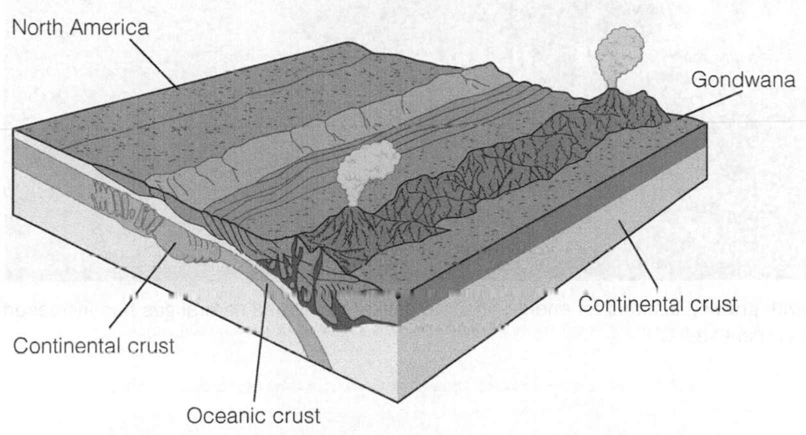

b Incipient continental collision between North America and Gondwana began during the Mississippian Period.

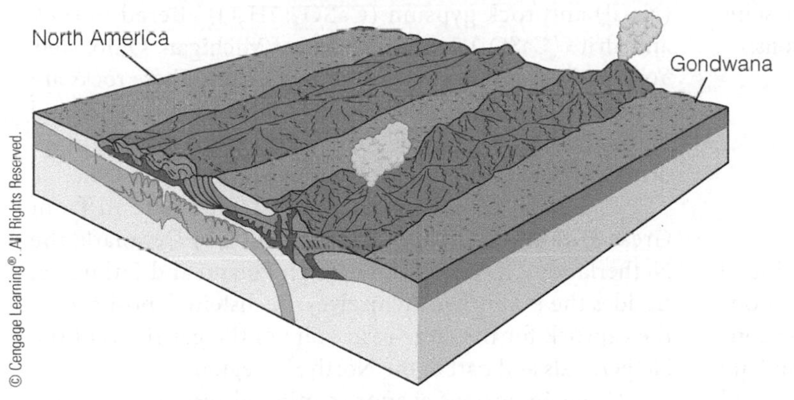

c Continental collision continued during the Pennsylvanian and Permian periods.

marked the southern boundary of Laurasia (Figure 20.4). The tectonic activity that uplifted the Ouachita mobile belt was very complex and involved not only the collision of Laurasia and Gondwana but also several microplates and terranes between the continents that eventually became part of Central America. The compressive forces impinging on the Ouachita mobile belt also affected the craton by broadly uplifting the southwestern part of North America.

20.10 What Role Did Microplates and Terranes Play in the Formation of Pangaea?

We have presented the geologic history of the mobile belts bordering the Paleozoic continents in terms of subduction along convergent plate boundaries. It is becoming increasingly clear, however, that accretion along the continental margins is more complicated than the somewhat simple, large-scale plate interactions that we have described. Geologists now recognize that numerous terranes or microplates existed during the Paleozoic and were involved in the orogenic events that occurred during that time.

In this chapter, we have been concerned with only the six major Paleozoic continents. Terranes and microplates of varying sizes, however, were present during the Paleozoic and participated in the formation of Pangaea. For example, the microcontinent of Avalonia consisted of some coastal parts of New England, southern New Brunswick, much of Nova Scotia, the Avalon Peninsula of eastern Newfoundland, southeastern Ireland, Wales, England, and parts of Belgium and northern France. This microcontinent separated from Gondwana in the Early Ordovician and existed as a small, separate continent until it collided with Baltica during the Late Ordovician–Early Silurian and then with Laurentia (as part of Baltica) during the Silurian (Figures 20.2b, c and 20.3a).

Other terranes and microplates include *Iberia–Armorica* (a portion of southern France, Sardinia, and most of the Iberian peninsula), *Perunica* (Bohemia), and numerous Alpine fragments (especially in Austria), as well as many other bits and pieces of island arcs and suture zones. Not only did these terranes and microplates separate and move away from the larger continental landmasses during the Paleozoic, but they usually developed their own unique faunal and floral assemblages.

 Thus, although the basic history of the formation of Pangaea during the Paleozoic remains essentially the same, geologists now realize

Connection Link

To learn more about the role that terranes played in the growth of western North America during the Mesozoic Era, see Figure 22.17 and section 22.4 in Chapter 22.

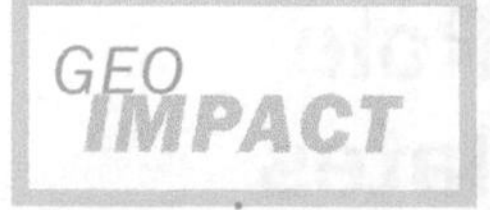

Offshore Paleozoic Oil and Gas Fields in Northern Brazil?

The multinational oil company you work for holds the exploration and drilling rights for vast offshore acreage in northern Brazil. Onshore, the Solimões Basin (450,000 km^2) is the only commercial Brazilian basin producing from Paleozoic rocks. Your job is to evaluate the potential of these offshore leases and make recommendations as to the feasibility of drilling exploratory wells in the region. What types of information do you need to help you evaluate these offshore prospects? Can the onshore geology of the Paleozoic Solimões Basin help in correlating not only the geology, but also the structural trends offshore? How do you think your knowledge of Paleozoic geology and global tectonics might be useful in a broad overall assessment of the area?

With growing demand for energy, offshore drilling for oil and natural gas has increased in recent years.

that microplates and terranes also played an important role in the formation of Pangaea. Furthermore, they help explain some previously anomalous geologic and paleontologic situations.

20.11 Paleozoic Mineral Resources

Paleozoic-age rocks contain a variety of important mineral resources, including energy resources, metallic and non-metallic mineral deposits, and sand and gravel for construction. Important sources of industrial or silica sand are the Upper Cambrian Jordan Sandstone of Minnesota and Wisconsin; the Middle Ordovician St. Peter Sandstone; the Lower Silurian Tuscarora Sandstone in Pennsylvania and Virginia; the Devonian Ridgeley Formation in West Virginia, Maryland, and Pennsylvania; and the Devonian Sylvania Sandstone in Michigan.

Silica sand has several uses, including the manufacture of glass, refractory bricks for blast furnaces, and molds for casting iron, aluminum, and copper alloys. Some silica sands, called *hydraulic fracturing sands*, are pumped into wells to fracture oil- or gas-bearing rocks and provide permeable passageways for the oil or gas to migrate to the well (see Geo-Focus 13.1).

Thick deposits of Silurian evaporites, mostly rock salt (NaCl) and rock gypsum ($CaSO_4 \cdot 2H_2O$) altered to rock anhydrite ($CaSO_4$), underlie parts of Michigan, Ohio, New York, and adjacent areas in Ontario, Canada. These rocks are important sources of various salts. In addition, barrier and pinnacle reefs in carbonate rocks associated with these evaporites are reservoirs for oil and gas in Michigan and Ohio.

The Zechstein evaporites of Europe extend from Great Britain across the North Sea and into Denmark, the Netherlands, Germany, and eastern Poland and Lithuania. Besides the evaporites themselves, Zechstein deposits form the caprock for the large reservoirs of the gas fields of the Netherlands and part of the North Sea region.

Other important evaporite mineral resources include those of the Permian Delaware Basin of West Texas and New Mexico, and Devonian evaporites in the Elk Point Basin of Canada. In Michigan, gypsum is mined and used in the production of sheetrock. Late Paleozoic-age limestones from many areas in North America are used in the manufacturing of cement. Limestone is also mined and used in blast furnaces for steel production.

Metallic mineral resources, including tin, copper, gold, and silver, are known from Late Paleozoic-age rocks, especially those deformed during mountain building. The host rocks for deposits of lead and zinc in southeast Missouri are Cambrian dolostones, although some Ordovician rocks contain these

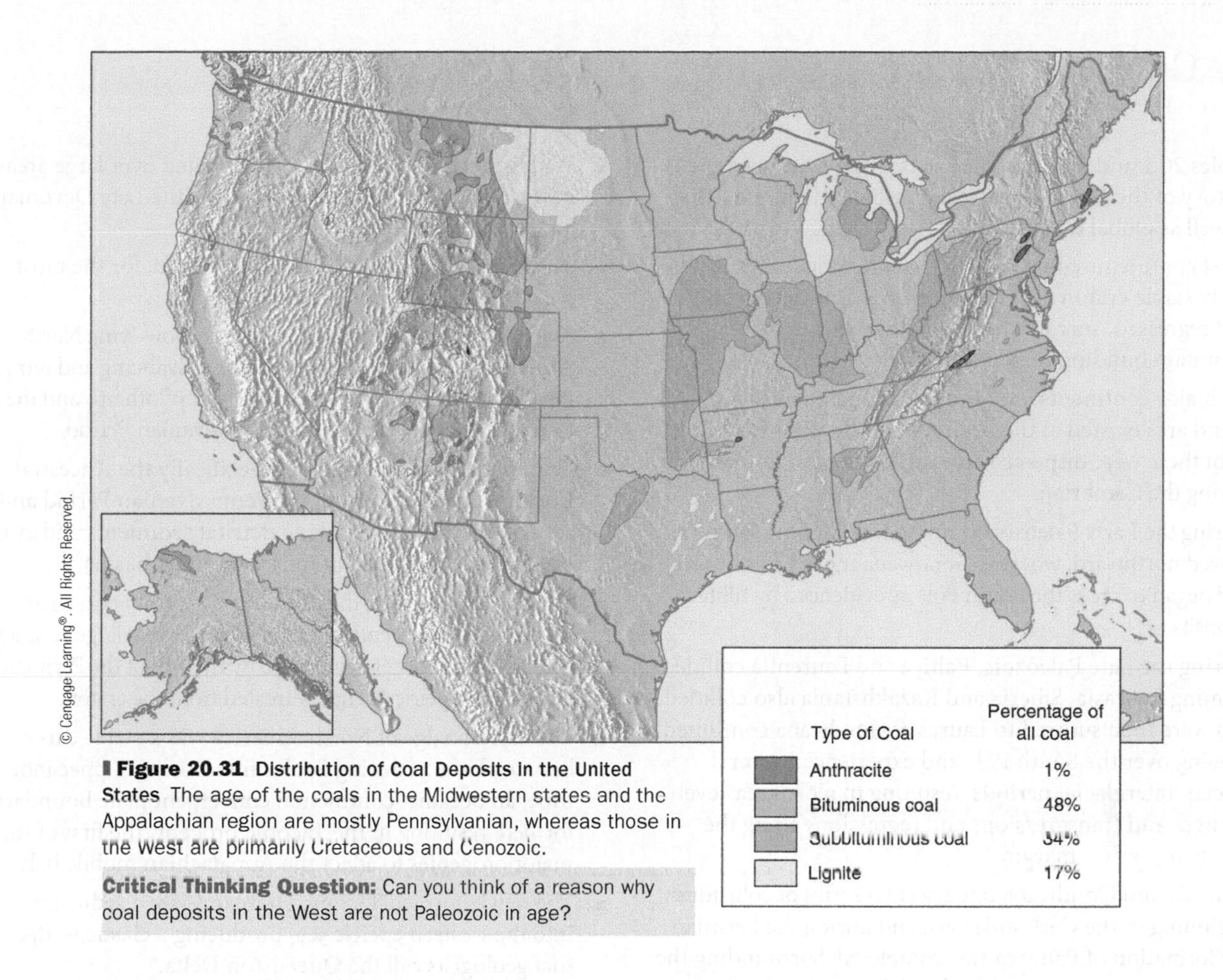

▌Figure 20.31 Distribution of Coal Deposits in the United States The age of the coals in the Midwestern states and the Appalachian region are mostly Pennsylvanian, whereas those in the West are primarily Cretaceous and Cenozoic.

Critical Thinking Question: Can you think of a reason why coal deposits in the West are not Paleozoic in age?

metals as well. These deposits have been mined since 1720, but they have been largely depleted. Now most lead and zinc mined in Missouri comes from Mississippian-age sedimentary rocks.

The Silurian Clinton Formation crops out from Alabama north to New York, and equivalent rocks are found in Newfoundland. This formation has been mined for iron in many places. In the United States, the richest ores and most extensive mining occurred near Birmingham, Alabama, but only a small amount of ore is currently produced in that area.

Petroleum and natural gas are recovered in commercial quantities from rocks ranging in age from the Devonian through Permian. For example, Devonian rocks in the Michigan Basin, Illinois Basin, and the Williston Basin of Montana, South Dakota, and adjacent parts of Alberta, Canada, have yielded considerable amounts of hydrocarbons. Middle Devonian black shales, such as the Marcellus Shale in the Appalachian Basin, are being actively explored because of their potential to yield enormous quantities of natural gas. Permian reefs and other strata in the western United States, particularly Texas, have also been important producers.

Although Permian coal beds are known from several areas, including Asia, Africa, and Australia, much of the coal in North America and Europe comes from Pennsylvanian (Upper Carboniferous) deposits. Large areas in the Appalachian region and the Midwestern United States are underlain by vast coal deposits (▌Figure 20.31). These coal deposits formed from the lush vegetation that flourished in Pennsylvanian coal-forming swamps (Figure 20.19).

Much of this coal is bituminous coal, which contains about 80% carbon. It is a dense, black coal that has been so thoroughly altered that plant remains can be seen only rarely (see Figure 7.10c). Bituminous coal is used to make *coke*, a hard, gray substance made up of the fused ash of bituminous coal. Coke is used to fire blast furnaces for steel production.

Some Pennsylvanian coal from North America is *anthracite*, a metamorphic type of coal containing up to 98% carbon. Most anthracite is in the Appalachian region (Figure 20.31). It is especially desirable because it burns with a smokeless flame and yields more heat per unit volume than other types of coal. Unfortunately, it is the least common type—much of the coal used in the United States is bituminous.

Key Concepts Review

- Tables 20.1 and 20.2 summarize the Paleozoic geologic history of the North American craton and mobile belts, as well as global events and sea-level changes.
- Most continents consist of two major components: a relatively stable craton over which epeiric seas transgressed and regressed, surrounded by mobile belts in which mountain building took place.
- Six major continents and numerous microcontinents and island arcs existed at the beginning of the Paleozoic Era; all of these were dispersed around the globe at low latitudes during the Cambrian.
- During the Early Paleozoic (Cambrian–Silurian), Laurentia moved northward, whereas Gondwana moved southward and began to cross the South Pole as evidenced by tillite deposits.
- During the Late Paleozoic, Baltica and Laurentia collided, forming Laurasia. Siberia and Kazakhstania also collided and were then sutured to Laurasia. Gondwana continued moving over the South Pole and experienced several glacial–interglacial periods, resulting in global sea-level changes and transgressions and regressions along the low-lying craton margins.
- Laurasia and Gondwana underwent a series of collisions beginning in the Carboniferous, and during the Permian the formation of Pangaea was completed. Surrounding the supercontinent Pangaea was the global ocean, Panthalassa.
- Geologists divide the geologic history of North America into cratonic sequences that formed as a result of craton-wide transgressions and regressions.
- The Sauk Sequence began with a major marine transgression onto the craton. At its maximum, the Sauk Sea covered the craton except for parts of the Canadian shield and the Transcontinental Arch, a series of large, northeast–southwest trending islands.
- The Tippecanoe Sequence began with deposition of an extensive sand unit over the exposed and eroded Sauk landscape and was followed by extensive carbonate deposition. In addition, large barrier reefs surrounded many cratonic basins, resulting in evaporite deposition within these basins.
- The basal beds of the Kaskaskia Sequence that were deposited on the exposed Tippecanoe surface consisted either of sandstones derived from the eroding Taconic Highlands or of carbonate rocks.
- Most of the Kaskaskia Sequence is dominated by carbonates and associated evaporites. The Devonian Period was a time of major reef building in western Canada, southern England, Belgium, Australia, and Russia.
- Widespread black shales were deposited over large areas of the North American craton during the Late Devonian and Early Mississippian.
- The Mississippian Period was dominated, for the most part, by carbonate deposition.
- Transgressions and regressions over the low-lying North American craton, probably caused by advancing and retreating Gondwanan ice sheets, resulted in cyclothems and the formation of coals during the Pennsylvanian Period.
- Cratonic mountain building, specifically the Ancestral Rockies, occurred during the Pennsylvanian Period and resulted in thick, nonmarine detrital sediments and evaporites being deposited in the intervening basins.
- By the Early Permian, the Absaroka Sea occupied a narrow zone of the south-central craton. Here, several large reefs and associated evaporites developed. By the end of the Permian Period, this epeiric sea had retreated from the craton.
- The eastern edge of North America was a stable carbonate platform during Sauk time. During Tippecanoe time, an oceanic–continental convergent plate boundary formed, resulting in the Taconic orogeny, the first of three major orogenies to affect the Appalachian mobile belt.
- The newly formed Taconic Highlands shed sediments into the western epeiric sea, producing a clastic wedge that geologists call the Queenston Delta.
- The Caledonian, Acadian, and Hercynian (Variscan)–Alleghenian orogenies were all part of the global tectonic activity that resulted in the assembly of Pangaea.
- The Cordilleran mobile belt was the site of the Antler orogeny, a minor Devonian orogeny, during which deep-water sediments were thrust eastward over shallow-water sediments.
- During the Pennsylvanian and Early Permian, mountain building occurred in the Ouachita mobile belt. In addition to the mountain range the Ouachita orogeny produced, this tectonic activity was partly responsible for the cratonic uplift in the southwest, which resulted in the Ancestral Rockies.
- During the Paleozoic Era, numerous microplates and terranes—such as Avalonia, Iberia–Armorica, and Perunica—existed and played an important role in the formation of Pangaea.
- Paleozoic-age rocks contain a variety of mineral resources, including petroleum, coal, evaporites, silica sand, lead, zinc, and other metallic deposits.

TABLE 20.1 Summary of Early Paleozoic Geologic and Evolutionary Events

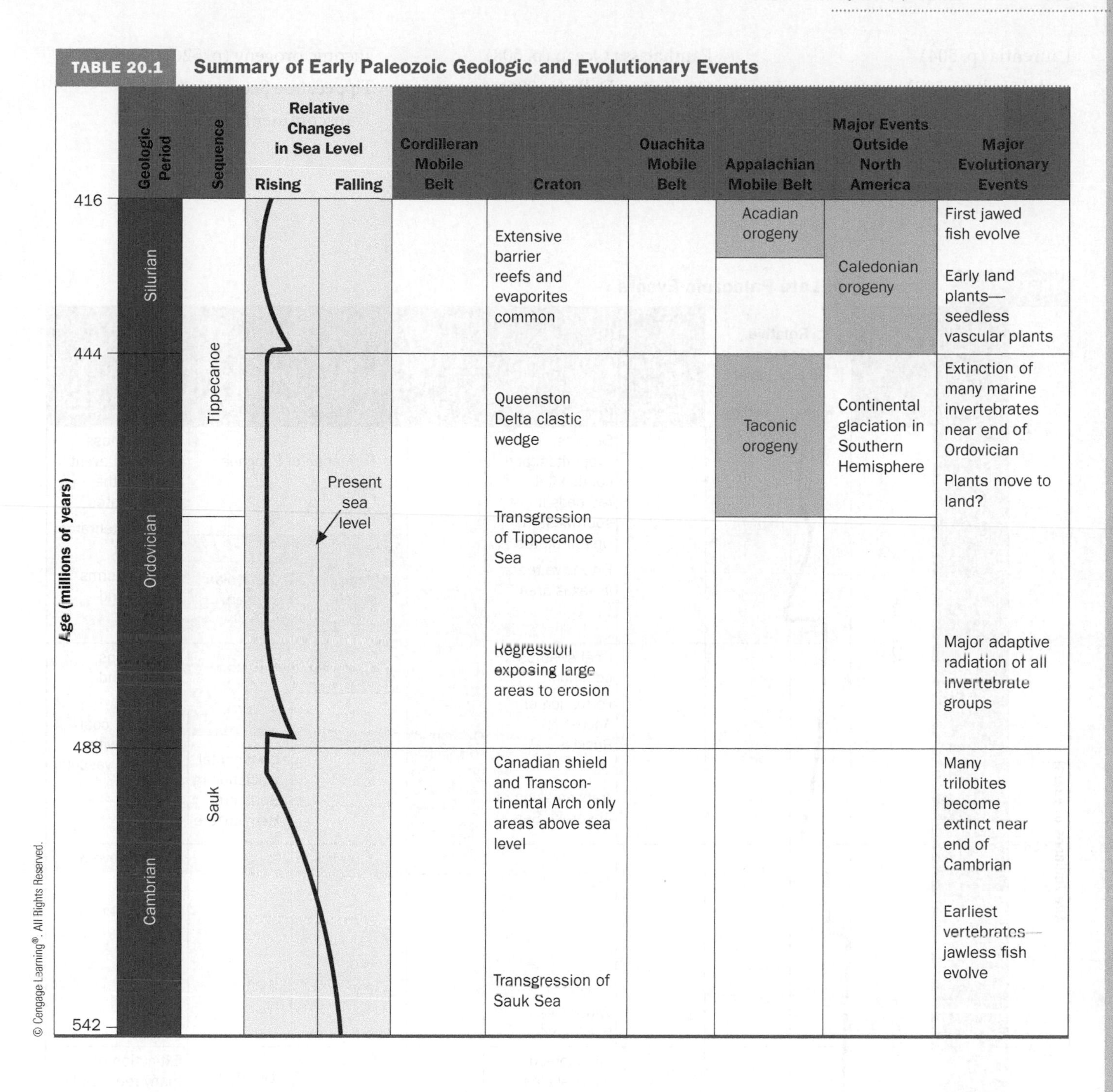

Age (millions of years)	Geologic Period	Sequence	Relative Changes in Sea Level (Rising / Falling)	Cordilleran Mobile Belt	Craton	Ouachita Mobile Belt	Appalachian Mobile Belt	Major Events Outside North America	Major Evolutionary Events
416–444	Silurian	Tippecanoe			Extensive barrier reefs and evaporites common		Acadian orogeny	Caledonian orogeny	First jawed fish evolve Early land plants—seedless vascular plants
444–488	Ordovician	Tippecanoe (upper); Sauk (lower)	Present sea level		Queenston Delta clastic wedge Transgression of Tippecanoe Sea Regression exposing large areas to erosion		Taconic orogeny	Continental glaciation in Southern Hemisphere	Extinction of many marine invertebrates near end of Ordovician Plants move to land? Major adaptive radiation of all invertebrate groups
488–542	Cambrian	Sauk			Canadian shield and Transcontinental Arch only areas above sea level Transgression of Sauk Sea				Many trilobites become extinct near end of Cambrian Earliest vertebrates—jawless fish evolve

Important Terms

Absaroka Sequence (p. 519)
Acadian orogeny (p. 527)
Ancestral Rockies (p. 521)
Antler orogeny (p. 529)
Appalachian mobile belt (p. 503)
Baltica (p. 504)
Caledonian orogeny (p. 527)
Catskill Delta (p. 527)
China (p. 504)
clastic wedge (p. 526)
Cordilleran mobile belt (p. 503)
cratonic sequence (p. 509)
cyclothem (p. 519)
epeiric sea (p. 503)
Gondwana (p. 504)
Hercynian (Variscan)–Alleghenian orogeny (p. 529)
Iapetus Ocean (p. 526)
Kaskaskia Sequence (p. 515)
Kazakhstania (p. 504)
Laurasia (p. 508)

Laurentia (p. 504)
mobile belt (p. 503)
organic reef (p. 512)
Ouachita mobile belt (p. 503)
Ouachita orogeny (p. 530)
Panthalassa Ocean (p. 508)
Queenston Delta (p. 526)
Sauk Sequence (p. 510)
sequence stratigraphy (p. 509)
Siberia (p. 504)
Taconic orogeny (p. 526)
Tippecanoe Sequence (p. 512)
Transcontinental Arch (p. 510)

TABLE 20.2 **Summary of Late Paleozoic Events**

Age (millions of years)	Geologic Period	Sequence	Relative Changes in Sea Level (Rising / Falling)	Cordilleran Mobile Belt	Craton	Ouachita Mobile Belt	Appalachian Mobile Belt	Major Events Outside North America	Major Evolutionary Events
251–299	Permian	Absaroka			Deserts, evaporites, and continental red beds in southwestern United States Extensive reefs in Texas area	Ouachita orogeny	Formation of Pangaea Allegheny orogeny	Formation of Pangaea Hercynian orogeny	Largest mass extinction event to affect the invertebrates Many vertebrates go extinct Gymnosperms diverse and abundant
299–318	Carboniferous: Pennsylvanian	Absaroka			Coal swamps common Formation of Ancestral Rockies Transgression of Absaroka Sea	Ouachita orogeny	Allegheny orogeny	Hercynian orogeny Continental glaciation in Southern Hemisphere	Amphibians diverse and abundant Abundant coal swamps with seedless vascular plants
318–359	Carboniferous: Mississippian	Kaskaskia			Widespread black shales	Ouachita orogeny			Reptiles evolve Gymnosperms evolve
359–416	Devonian	Kaskaskia Tippecanoe	Present sea level	Antler orogeny	Widespread black shales Catskill Delta clastic wedge Extensive barrier reef formation in Western Canada Transgression of Kaskaskia Sea		Acadian orogeny	Old Red Sandstone clastic wedge in British Isles Caledonian orogeny	Extinction of many reef-building invertebrates Amphibians evolve All major groups of fish present—Age of Fish Early land plants-seedless vascular plants
416–	Silurian	Tippecanoe					Acadian orogeny	Caledonian orogeny	

Review Questions

1. An elongated area marking the site of mountain building is a
   a. ___ cyclothem.
   b. ___ mobile belt.
   c. ___ platform.
   d. ___ shield.
   e. ___ craton.
2. A major transgressive–regressive cycle bounded by craton-wide unconformities is a(n)
   a. ___ cratonic sequence.
   b. ___ biostratigraphic unit.
   c. ___ orogeny.
   d. ___ shallow sea.
   e. ___ cyclothem.
3. Which orogeny was *not* involved in the closing of the Iapetus Ocean?
   a. ___ Acadian.
   b. ___ Alleghenian.
   c. ___ Antler.
   d. ___ Caledonian.
   e. ___ Taconic.
4. The European Old Red Sandstone is the equivalent of the North American
   a. ___ Queenston Delta.
   b. ___ Capitan Limestone.
   c. ___ Phosphoria Formation.
   d. ___ Oriskany Sandstone.
   e. ___ Catskill Delta.
5. The economically valuable deposit in a cyclothem is
   a. ___ gravel.
   b. ___ metallic ore.
   c. ___ coal.
   d. ___ carbonate.
   e. ___ evaporites.
6. According to estimates made from mapping and correlation, the Queenston Delta contains more than 600,000 km^3 of rock eroded from the Taconic Highlands. Based on this figure, geologists estimate that the Taconic Highlands were at least 4,000 m high. They also estimate that the Catskill Delta contains three times as much sediment as the Queenston Delta. From what you know about the geographic distribution of the Taconic Highlands and the Acadian Highlands, can you estimate how high the Acadian Highlands might have been?
7. Paleogeographic maps of what the world looked like during the Paleozoic Era can be found in almost every Earth history book and in numerous scientific journals. What criteria are used to determine the location of ancient continents and ocean basins, and why are there minor differences in the location and size of these paleocontinents among the various books and articles?
8. Based on the discussion of Milankovitch cycles and their role in causing glacial–interglacial cycles (see Chapter 14), could these cycles be partly responsible for the transgressive–regressive cycles that resulted in cyclothems during the Pennsylvanian Period?
9. Discuss how sequence stratigraphy can be used to make global correlations and why it is so useful in reconstructing past events.

10. **Creative Thinking Visual Question:** This close-up (▮ Figure 1) of a Devonian red rock from a building in Glasgow, Scotland, shows a distinctive sedimentary structure. Identify the sedimentary structure, indicate what type of environment you think it was deposited in, and give the name of the formation this rock comes from.

James S. Monroe

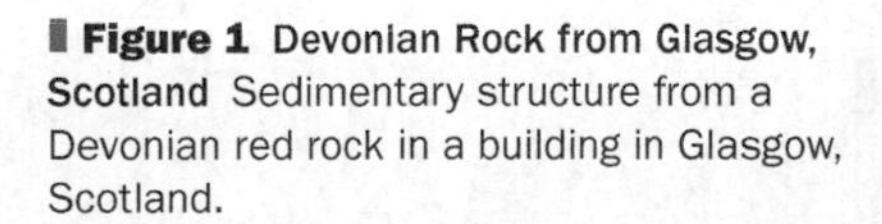

▮ **Figure 1 Devonian Rock from Glasgow, Scotland** Sedimentary structure from a Devonian red rock in a building in Glasgow, Scotland.

Global GeoScience Watch On the GREENR database home page, click on "Coal Energy" under the "Energy" heading on the left side of the page. In this portal, read the "Overview," then click on "Coal Resource Use" under the "Reference" heading. What is coal and what are its different grades? How does coal damage the environment? How does the burning of coal affect climate change? Why is coal losing some of its luster as an energy source in many parts of the world?

The biota of the Middle Cambrian Burgess Shale, British Columbia, Canada, was composed of a number of strange-looking and now extinct animals that inhabited the seas of North America at that time.

Publiphoto/Science Source/Photo researchers, Inc

CHAPTER 21 Paleozoic Life History

OUTLINE

HAVE YOU EVER WONDERED?

- Why animals with skeletons appeared abruptly at the beginning of the Paleozoic Era?
- What caused the mass extinction at the end of the Permian Period?
- What the first vertebrates were like and when they evolved?
- What the evolutionary history of fish was, and which group gave rise to the amphibians?
- What problems had to be solved by both animals and plants to make the transition from water onto land, and how they solved these problems?
- What allowed the reptiles to fully exploit the terrestrial environment, and which group of reptiles gave rise to the mammals?
- Why both the amphibians and seedless vascular plants were so abundant during the Pennsylvanian Period?
- What role the onset of arid conditions during the assembly of Pangaea played in gymnosperms (flowerless seed plants) becoming the dominant element of the world's flora during the Permian Period?

21.1 Introduction

On August 30 and 31, 1909, near the end of the summer field season, Charles D. Walcott, geologist and head of the Smithsonian Institution, was searching for fossils along a trail on Burgess Ridge between Mount Field and Mount Wapta, near Field, British Columbia, Canada. On the west slope of this ridge, he discovered the first soft-bodied fossils from the Burgess Shale, a discovery of immense importance in deciphering the early history of life. During the following week, Walcott and his collecting party split open numerous blocks of shale, many of which yielded carbonized impressions of a number of soft-bodied organisms beautifully preserved on bedding planes. Walcott returned to the site the following summer and located the shale stratum that was the source of his fossil-bearing rocks in the steep slope above the trail. He quarried the site and shipped back thousands of fossil specimens to the United States National Museum of Natural History, where he later cataloged and studied them.

The importance of Walcott's discovery is not that it was just another collection of well-preserved Cambrian fossils, but that it allowed geologists a rare glimpse into a world previously almost unknown: that of the soft-bodied animals that lived during the Middle Cambrian. The beautifully preserved fossils from the Burgess Shale present a much more complete picture of a Middle Cambrian community than do deposits containing only fossils of the hard parts of organisms. In fact, 60% of the total fossil assemblage is composed of soft-bodied animals, a percentage comparable to present-day marine communities.

In this chapter, we examine the history of Paleozoic life as a system in which its parts consist of a series of interconnected biologic and geologic events. The underlying processes of evolution and plate tectonics are the forces that drove this system. The opening and closing of ocean basins, transgressions and regressions of epeiric seas, the formation of mountain ranges, and the changing positions of the continents profoundly affected the evolution of the marine and terrestrial communities.

A time of tremendous biologic change began with the appearance of skeletonized animals near the Precambrian–Cambrian boundary. Following this event, marine invertebrates began a period of adaptive radiation and evolution, during which the Paleozoic marine invertebrate community greatly diversified.

The earliest fossil record of vertebrates is the fish, which evolved during the Cambrian Period and rapidly diversified during the Devonian Period, a time popularly called "the Age of Fish." The Devonian was also the time when amphibians became established on land, having evolved from an ancestral fish group. Reptiles evolved from the amphibians during the Mississippian Period, and soon established themselves as the dominant vertebrate animals on land.

Plants preceded animals onto land during the Middle to Late Ordovician, evolving a variety of innovations that allowed subsequent radiations and diversification within the terrestrial environment. One of the striking parallels between plants and animals is that they had to solve the same basic problems in making the transition from water to land. For both groups, the method of reproduction proved to be the major barrier to expansion into the various terrestrial environments. But, with the evolution of the seed in plants and the amniote egg in animals, that limitation was removed, allowing for expansion into all terrestrial habitats.

The end of the Paleozoic Era witnessed the greatest mass extinction in Earth's history. Some scientists estimate that more than 90% of marine invertebrate species, and upward of 70% of all terrestrial vertebrate species became extinct. Thus, by the end of the Permian, a near collapse of both the marine and terrestrial ecosystems had occurred.

21.2 What Was the Cambrian Explosion?

At the beginning of the Paleozoic Era, animals with skeletons appeared rather abruptly in the fossil record. In fact, their appearance is described as an explosive development of new types of animals and is referred to as the "Cambrian explosion" by most scientists. This sudden appearance of new animals in the fossil record is rapid, however, only in the context of geologic time, having taken place over millions of years during the Early Cambrian Period.

Moreover, it is not a recent discovery. Early geologists observed that the remains of skeletonized animals appeared rather abruptly in the fossil record. Charles Darwin addressed this problem in *On the Origin of Species by Means of Natural Selection* (1859) and observed that, without a convincing explanation, such an event was difficult to reconcile with his newly expounded evolutionary theory.

To review the contributions of Charles Darwin and the mechanism of natural selection, see Chapter 18.

The sudden appearance of shelled animals during the Early Cambrian contrasts sharply with the biota that lived during the preceding Proterozoic Eon. Up until the evolution of the Ediacaran fauna, Earth was populated primarily by single-celled organisms. Recall from Chapter 19 that the Ediacaran fauna, which is found on all continents except Antarctica, consists primarily of multicelled, soft-bodied organisms (see Figure 19.26). Microscopic calcareous tubes, presumably housing wormlike suspension-feeding organisms, have also been found at some localities. In addition, trails and burrows, which represent the activities of worms and other sluglike animals, are associated with Ediacaran faunas throughout the world. The trails and burrows are similar to those made by present-day soft-bodied organisms.

Until recently, it appeared that there was a fairly long period between the extinction of the Ediacaran fauna and the appearance of the first Cambrian fossils. That gap has been considerably narrowed in recent years with the discovery of new Proterozoic fossiliferous localities. Now, Proterozoic fossil assemblages continue right to the base of the Cambrian.

Nonetheless, the cause of the sudden appearance of so many different animal phyla during the Early Cambrian is still hotly debated. Newly developed molecular techniques that allow evolutionary biologists to compare molecular sequences of the same gene from different species are being applied to the phylogeny, or evolutionary history, of many organisms. In addition, new fossil sites and detailed stratigraphic studies are shedding light on the early history and ancestry of the various invertebrate phyla.

The Cambrian explosion probably had its roots firmly planted in the Proterozoic. The mechanism or mechanisms that triggered this event, however, are still being investigated. Although some would argue for a single causal event, it is more likely that the Cambrian explosion was a combination of factors, both biological and geological. For example, geologic evidence indicates that Earth was glaciated one or more times during the Proterozoic, followed by global warming during the Cambrian. These global environmental changes may have stimulated evolution and contributed to the Cambrian explosion.

ConnectionLink

To review Proterozoic glaciation, see Chapter 19.

Others would argue that a change in the chemistry of the oceans favored the evolution of a mineralized skeleton. In this scenario, an increase in the concentration of calcium from the Neoproterozoic through the Early Cambrian, in part due to volcanism along mid-oceanic ridges, allowed for the precipitation of calcium carbonate and calcium phosphate, compounds that make up the shells of most invertebrates.

Another hypothesis is that the rapid evolution of a skeletonized fauna was a response to the evolution of predators. A shell or mineralized covering would provide protection against predation by the various predators evolving during this time. The conflict between predators, whose goal is to consume prey, and prey, whose goal is to avoid becoming a meal for a predator, is one of the strongest factors in natural selection.

A further line of research related to the Cambrian explosion involves what are known as *homeobox* or, simply *Hox* genes. These are sequences of regulatory genes that control the development of individual regions of the body and thus the basic body plan in organisms. Studies indicate that the basic body plans for all animals were apparently established by the end of the Cambrian explosion, and only minor modifications have occurred since then. Whatever the ultimate cause of the Cambrian explosion, the appearance of a skeletonized fauna and the rapid diversification of that fauna during the Early Cambrian were major events in life history.

21.3 The Emergence of a Shelly Fauna

The earliest organisms with hard parts had small, calcareous tubes found associated with Ediacaran faunas (Neoproterozoic) from several locations throughout the world. These were followed by other microscopic skeletonized fossils found in Lower Cambrian rocks and are collectively referred to as the *small shelly fauna* or *small shelly fossils* (Figure 21.1). This small shelly fauna was in turn followed by the appearance of large skeletonized animals during the Cambrian explosion. Along with the question of why animals appeared so suddenly in the fossil record is the equally intriguing question of why they initially acquired skeletons and what selective advantage skeletons provided. Various explanations about why marine organisms evolved skeletons have been proposed, but none is completely satisfactory or universally accepted.

The formation of an exoskeleton, or shell, however, does confer many advantages on an organism, such as (1) providing protection against ultraviolet (UV) radiation, thus allowing animals to move into shallower waters; (2) helping to prevent drying out in an intertidal environment; (3) allowing animals to increase their size and providing attachment sites for muscles; and (4) providing protection against predators. Evidence of actual fossils of predators and specimens of damaged prey, as well as antipredatory adaptations in some animals, indicates that the impact of predation during the Cambrian was great, leading some scientists to hypothesize that the rapid evolution of a shelly invertebrate fauna was a response to the rise of predators. With predators playing an important role in the Cambrian marine ecosystem, any mechanism or feature

Figure 21.1 Lower Cambrian Shelly Fossils Two small (several millimeters in size) Lower Cambrian shelly fossils.

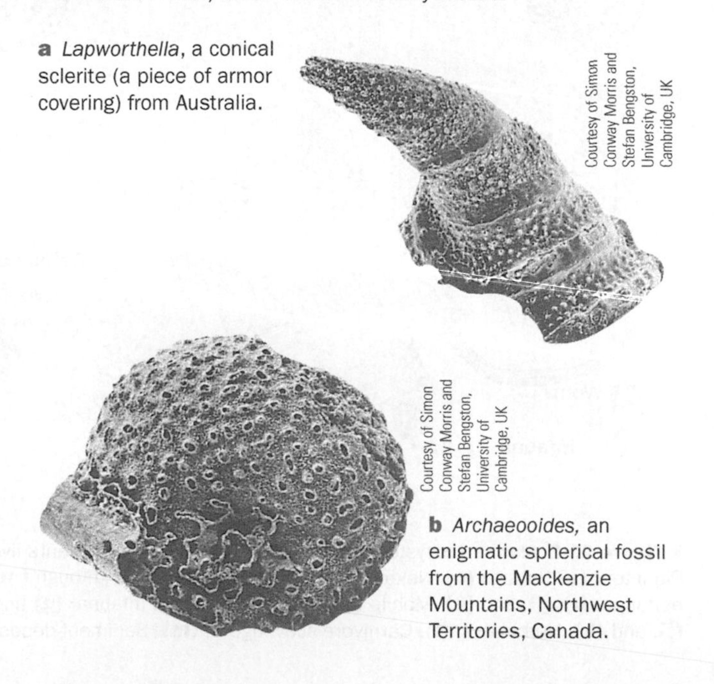

a *Lapworthella*, a conical sclerite (a piece of armor covering) from Australia.

b *Archaeooides*, an enigmatic spherical fossil from the Mackenzie Mountains, Northwest Territories, Canada.

that protected an animal would certainly be advantageous and confer an adaptive advantage to the organism.

Scientists currently have no clear answer as to why marine organisms evolved mineralized skeletons (small shelly fauna) during the Proterozoic and Early Cambrian, followed by a variety of large shelly invertebrates during the Cambrian explosion and shortly thereafter. They undoubtedly evolved because of a variety of biologic and environmental factors. Whatever the reason, the acquisition of a mineralized skeleton was a major evolutionary innovation that allowed invertebrates to successfully occupy a wide variety of marine habitats.

21.4 The Present Marine Ecosystem

Before we can begin to understand ancient ecosystems, we must first analyze the present-day marine ecosystem and look at where organisms live, how they get around, and how they feed (Figure 21.2). Organisms that live in the water column above the seafloor are called *pelagic*. They are divided into two main groups: the floaters, or **plankton**, and the swimmers, or **nekton**.

Plankton are mostly passive and go where currents carry them. Plant plankton, such as diatoms, dinoflagellates, and various algae, are called *phytoplankton* and are mostly microscopic. Animal plankton are called *zooplankton* and are also mostly microscopic. Examples of zooplankton include foraminifera, radiolarians, and jellyfish. The nekton are swimmers and are mainly vertebrates, such as fish; the invertebrate nekton include cephalopods.

Organisms that live on or in the seafloor make up the **benthos**. They are characterized as *epifauna* (animals) or *epiflora* (plants)—those that live on the seafloor—or as *infauna*—animals that live in and move through the sediments. The benthos are further divided into those organisms that stay in one place, called *sessile*, and those that move around on or in the seafloor, called *mobile*.

The feeding strategies of organisms are also important in terms of their relationships with other organisms in the marine ecosystem. There are basically four feeding groups: **suspension feeders** remove or consume microscopic plants

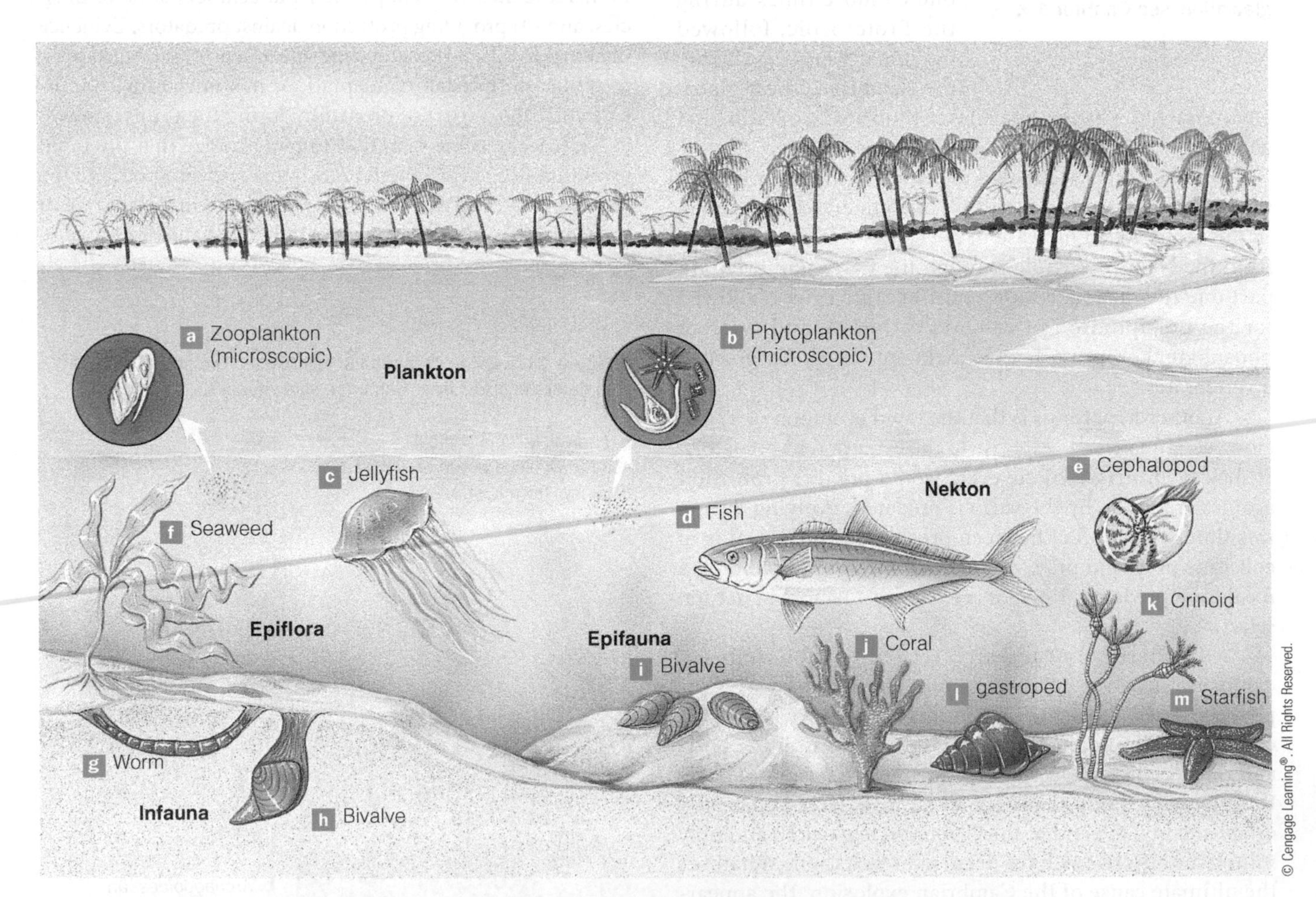

Figure 21.2 Marine Ecosystem Where and how animals and plants live in the marine ecosystem. Plankton: **(a)** through **(c)**. Nekton: **(d)** and **(e)**. Benthos: **(f)** through **(m)**. Sessile epiflora: **(f)**. Sessile epifauna: **(i)**, **(j)**, and **(k)**. Mobile epifauna: **(l)** and **(m)**. Infauna: **(g)** and **(h)**. Suspension feeders: **(i)**, **(j)**, and **(k)**. Herbivores: **(l)**. Carnivore-scavengers: **(m)**. Sediment-deposit feeders: **(g)**.

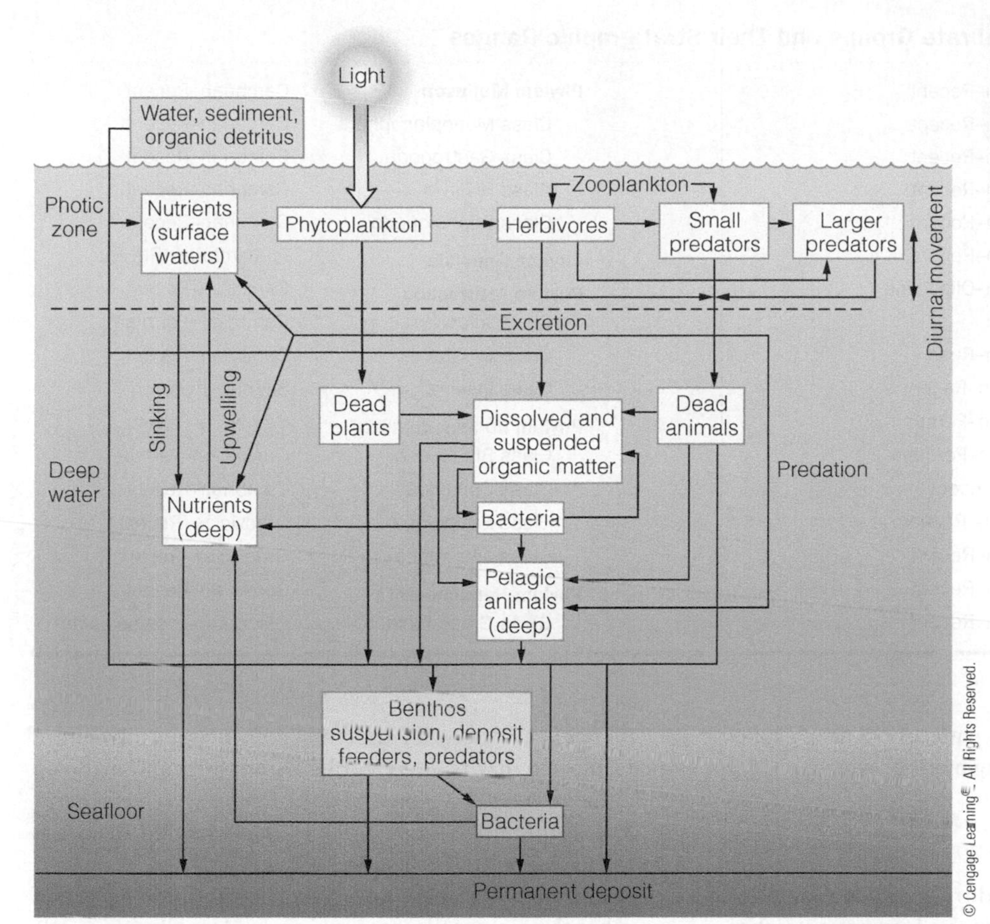

Figure 21.3 Marine Food Web Marine food web showing the relationship between the producers (phytoplankton), consumers (herbivores, small and large predators, pelagic animals, and benthos), and decomposers (bacteria).

and animals as well as dissolved nutrients from the water; **herbivores** are plant eaters; **carnivore-scavengers** are meat eaters; and **sediment-deposit feeders** ingest sediment and extract the nutrients from it.

We can define an organism's place in the marine ecosystem by where it lives and how it eats. For example, a brachiopod is a benthonic, epifaunal suspension feeder, whereas a cephalopod is a nektonic carnivore.

An ecosystem includes several *trophic levels*, which are tiers of food production and consumption within a feeding hierarchy. The feeding hierarchy, and hence energy flow, in an ecosystem make up a food web of complex interrelationships among the producers, consumers, and decomposers (Figure 21.3). The **primary producers**, or *autotrophs*, are those organisms that manufacture their own food. Virtually all marine primary producers are phytoplankton. Feeding on the primary producers are the *primary consumers*, which are mostly suspension feeders. *Secondary consumers* feed on the primary consumers and thus are predators, whereas *tertiary consumers*, which are also predators, feed on the secondary consumers. Besides the producers and consumers, there are also transformers and decomposers. These are bacteria that break down the dead organisms that have not been consumed into organic compounds, which are then recycled.

When we look at the marine realm today, we see a complex organization of organisms interrelated by trophic interactions and affected by changes in the physical environment. When one part of the system changes, the whole structure changes, sometimes almost insignificantly, other times catastrophically.

21.5 Paleozoic Invertebrate Marine Life

Having already considered the origin, differentiation, and evolution of the Proterozoic–Cambrian marine biota, and having examined the nature and structure of living marine communities for context, we will now explore the changes that occurred in the marine invertebrate community during the Paleozoic Era. Rather than focusing on the history of each invertebrate phylum (Table 21.1), we will survey the evolution of the Paleozoic marine invertebrate community through time, concentrating on the major features and changes that took place.

As we study the evolution of the Paleozoic marine ecosystem, keep in mind how geologic and evolutionary changes have had a significant impact on its composition and structure. For example, the major marine transgressions over the world's cratons opened up vast areas of shallow seas that could be inhabited and exploited. Furthermore, the movement of continents

TABLE 21.1 The Major Invertebrate Groups and Their Stratigraphic Ranges

Phylum Protozoa	Cambrian–Recent
Class Sarcodina	Cambrian–Recent
Order Foraminifera	Cambrian–Recent
Order Radiolaria	Cambrian–Recent
Phylum Porifera	Cambrian–Recent
Class Demospongea	Cambrian–Recent
Order Stromatoporoida	Cambrian–Oligocene
Phylum Archaeocyatha	Cambrian
Phylum Cnidaria	Cambrian–Recent
Class Anthozoa	Ordovician–Recent
Order Tabulata	Ordovician–Permian
Order Rugosa	Ordovician–Permian
Order Scleractinia	Triassic–Recent
Phylum Bryozoa	Ordovician–Recent
Phylum Brachiopoda	Cambrian–Recent
Class Inarticulata	Cambrian–Recent
Class Articulata	Cambrian–Recent
Phylum Mollusca	Cambrian–Recent
Class Monoplacophora	Cambrian–Recent
Class Gastropoda	Cambrian–Recent
Class Bivalvia	Cambrian–Recent
Class Cephalopoda	Cambrian–Recent
Phylum Annelida	Precambrian–Recent
Phylum Arthropoda	Cambrian–Recent
Class Trilobita	Cambrian–Permian
Class Crustacea	Cambrian–Recent
Class Insecta	Silurian–Recent
Phylum Enchinodermata	Cambrian–Recent
Class Blastoidea	Ordovician–Permian
Class Crinoidea	Cambrian–Recent
Class Echinoidea	Ordovician–Recent
Class Asteroidea	Ordovician–Recent
Phylum Hemichordata	Cambrian–Recent
Class Graptolithina	Cambrian–Mississippian

not only affected oceanic circulation patterns but also resulted in environmental changes and shifting biotic provinces.

Cambrian Marine Community

The Cambrian Period was a time during which many new body plans evolved and animals moved into new niches. Although almost all of the major invertebrate phyla evolved during the Cambrian (Table 21.1), many were represented by only a few species. Whereas trace fossils are common and echinoderms diverse, trilobites, brachiopods, and archaeocyathids made up the majority of Cambrian skeletonized life (▮ Figure 21.4).

Trilobites were by far the most conspicuous element of the Cambrian marine invertebrate community and made up about half of the total fauna. Most trilobites were benthonic, mobile, sediment-deposit feeders that crawled or swam along the seafloor (Geo-Insight 21.1).

Cambrian *brachiopods* were mostly primitive types and were benthonic, sessile suspension feeders. Brachiopods became an abundant member of the marine invertebrate community during the Ordovician Period.

The third major group of Cambrian organisms was the *archaeocyathids* (▮ Figure 21.5). These organisms were benthonic, sessile, presumably suspension feeders that constructed reeflike structures beginning in the Early Cambrian,

▮ **Figure 21.4 Cambrian Marine Community** Reconstruction of a Cambrian marine community showing floating jellyfish, swimming arthropods, benthonic sponges, and scavenging trilobites.

Museum Victoria. Photo by Benjamin Healey

▌Figure 21.5 Archaeocyathids Models of archaeocyathids in life position and lying on the seafloor after dying and falling over. Archaeocyathids were vase-shaped, benthonic, sessile, presumably suspension feeders that constructed reeflike structures during the Cambrian. Visible in these models is the porate nature of the walls, and the holdfasts that kept the organism anchored to the substrate.

were abundant during the Early and Middle Cambrian, and went extinct at the end of the Cambrian.

The remainder of the Cambrian fauna consisted of representatives of most of the other major phyla, including many organisms that were short-lived evolutionary experiments. As might be expected during times of adaptive radiation and evolutionary experimentation, many of the invertebrates that evolved during the Cambrian soon became extinct.

The Burgess Shale Biota

No discussion of Cambrian life would be complete without mentioning one of the best examples of a preserved soft-bodied fauna and flora: the Burgess Shale biota. As the Sauk Sea transgressed from the Cordilleran shelf onto the western edge of the craton, Early Cambrian-age sands were covered by Middle Cambrian-age black muds that preserved members of this diverse soft-bodied benthonic community as carbonaceous impressions (▌Figure 21.6).

▌Figure 21.6 Fossils from the Burgess Shale Some of the fossil animals preserved in the Burgess Shale.

De Agostini/Getty Images

a *Anomalocaris*, a predator from the Early and Middle Cambrian Period, is shown feeding on *Opabinia*, one of many extinct invertebrates found in the Burgess Shale. *Anomalocaris* was approximately 45 cm long and used its gripping appendages to bring its prey to its circular mouth structure.

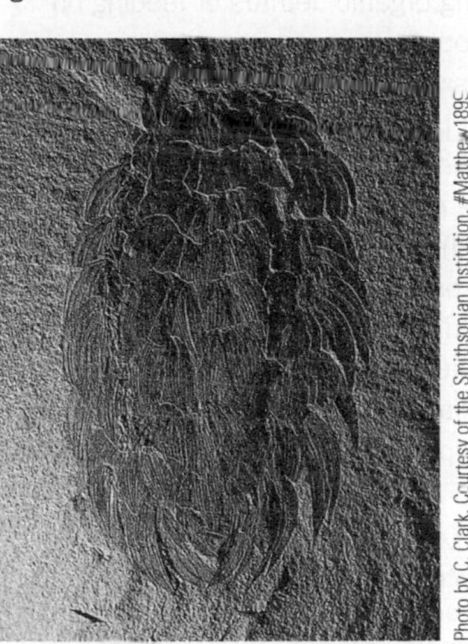

Photo by C. Clark. Courtesy of the Smithsonian Institution, #Matthew189C

b *Wiwaxia*, a scaly armored sluglike creature whose affinities remain controversial.

Photo by C. Clark. Courtesy of the Smithsonian Institution, #Walcott1911d

c *Hallucigenia*, a velvet worm.

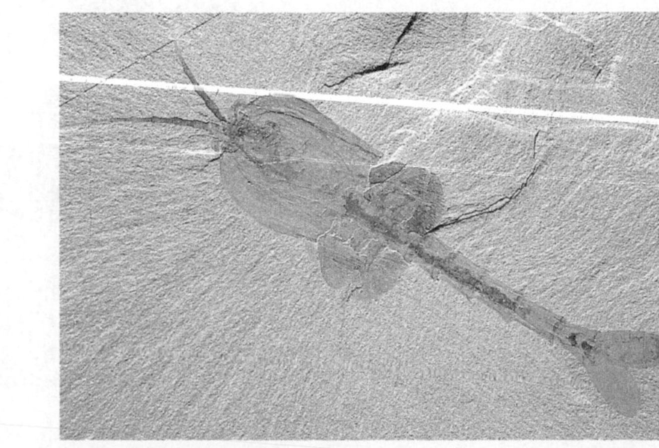

Photo by C. Clark. Courtesy of the Smithsonian Institut on, #Walcott1912a

d *Waptia*, an arthropod.

GEO INSIGHT 21.1 Trilobites—Paleozoic Arthropods

Trilobites, an extinct class of arthropods, are probably the favorite and most sought after of invertebrate fossils. They lived from the Early Cambrian until the end of the Permian and were most diverse during the Late Cambrian. More than 15,000 species of trilobites have been described, and they are currently grouped into nine orders.

Trilobites had a worldwide distribution throughout the Paleozoic, and they lived in all marine environments, from shallow, nearshore waters to deep oceanic settings. They occupied a wide variety of habitats. Most were bottom-dwellers, crawling around the seafloor and scavenging organic detritus or feeding on microorganisms or algae. Others were free-swimming predators or filter-feeders living throughout the water column.

The trilobite body is divided into three parts (▮ Figure 1): a cephalon (head), containing the eyes, mouth, and sensory organs; the thorax (body), composed of individual segments; and the pygidium (tail). Interestingly, the name *trilobite* does not refer to its three main body parts, but means *three-lobed*, which corresponds to the three longitudinal lobes of the thorax—the axial lobe and the two flanking pleural lobes.

The appendages of trilobites are rarely preserved. Paleontologists know, however, from specimens in which the soft-part anatomy is preserved as an impression, that beneath each thoracic segment was a two-part appendage consisting of a gill-bearing outer branch used for respiration and an inner branch or walking leg, composed of articulating limb segments (▮ Figure 2).

Trilobites range in size from a few millimeters in length, as found in many of the agnostid trilobites (▮ Figure 3), to more than 70 cm, as recorded by the world's largest trilobite (▮ Figure 4). Most trilobites, however, range between 3 and 10 cm long.

Trilobites were either blind, as in the agnostids (Figure 3), or possessed holochroal or schizochroal paired eyes. *Holochroal* eyes are characterized by generally hexagonal-shaped, biconvex lenses that are packed closely together, and covered by a single corneal layer

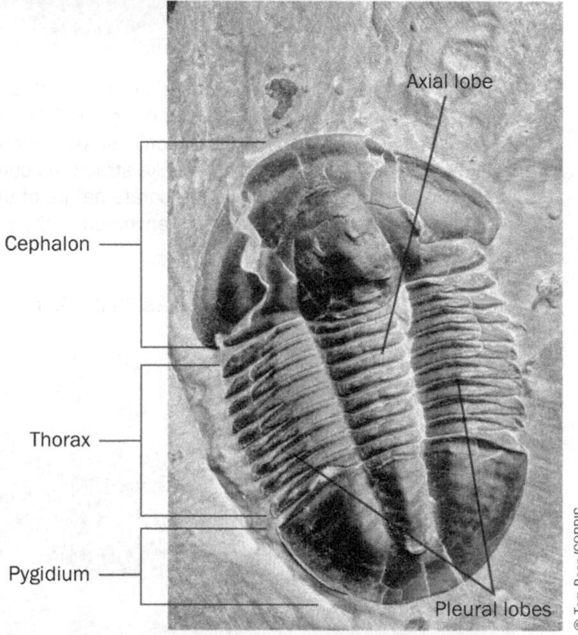

▲ **1.** *Elrathia kingii*, from the Middle Cambrian Wheeler Shale, Utah, illustrates the major body parts of a trilobite.

► **2.** Model of the dorsal **(a)** and ventral **(b)** anatomy of *Triarthus eatoni*, a Late Ordovician trilobite from the Frankfort Shale, New York.

a Dorsal view.

b Ventral view.

(▌Figure 5). This type of eye is similar to a modern compound eye and results in mosaic vision.

Schizochroal eyes, in contrast, are composed of relatively large, individual lenses, each of which is covered by its own cornea and separated from adjacent lenses by a thick dividing wall (▌Figure 6). Schizochroal lenses are arranged in rows and columns and most likely produced a visual field consisting of shadows.

Although trilobites had a hard dorsal exoskeleton, useful, in part, as protection against predators, some trilobites also had the ability to enroll, presumably to protect their antennae, limbs, and soft ventral appendages. By doing so, trilobites could view their surroundings, while still protecting their soft ventral anatomy (▌Figure 7).

Trilobites were a very successful group. They first appeared in the Early Cambrian, rapidly diversified, reached their maximum diversity in the Late Cambrian, and then suffered major reductions in diversity near the end of the Cambrian, and again at the end of the Ordovician, and once again near the end of the Devonian.

As yet, no consensus exists on what caused the trilobite extinctions, but a combination of factors was likely involved, possibly including a reduction of shelf space, increased competition, and a rise in predators. Some scientists have also suggested that a cooling of the seas may have played a role, particularly for the extinctions at the end of the Ordovician Period. The trilobites, like most of their Permian marine invertebrate contemporaries, finally fell victim to the Permian mass extinction event.

▼ **3.** Several specimens of *Peronopsis interstricta*, a small (1 cm long), Cambrian-age agnostid trilobite from Utah.

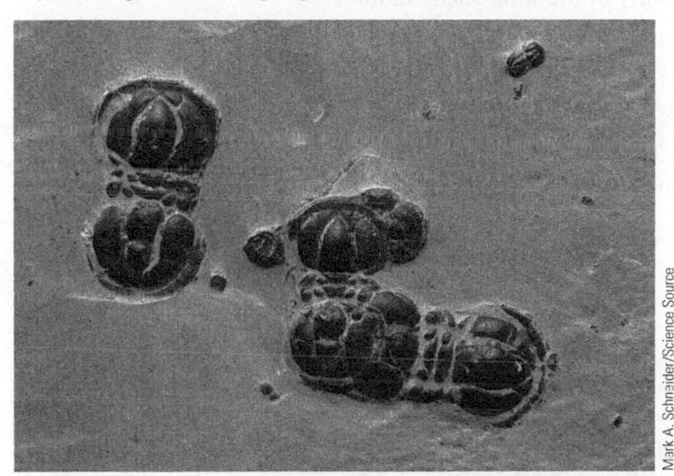

Mark A. Schneider/Science Source

► **4.** *Isotellus rex*, the world's largest trilobite (more than 70 cm long), from the Ordovician Churchill River Group, Manitoba, Canada.

© The Manitoba Museum/Dr. Graham Young

◄ **5.** Holochroal eyes from the trilobite *Scutellum campaniferum*.

Courtesy of E. N. K. Clarkson, used with permission, University of Edinburgh

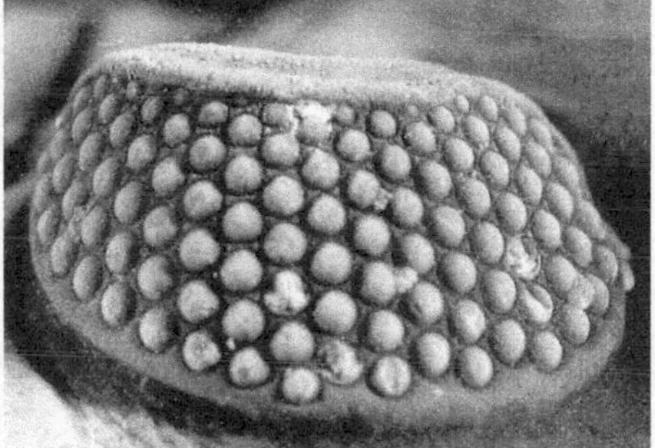

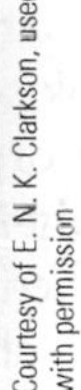

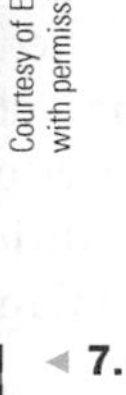

◄ **6.** Schizochroal eyes from the trilobite *Eophacops trapeziceps*.

Courtesy of E. N. K. Clarkson, used with permission

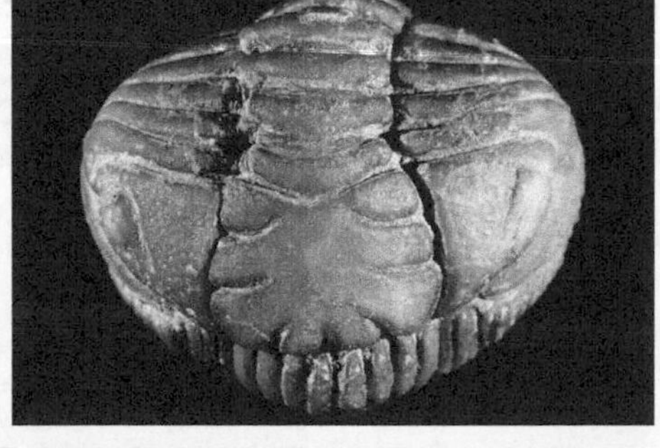

◄ **7.** An enrolled specimen of the Ordovician trilobite *Pilomera fisheri* from Putilowa, Poland.

© Sinclair Stammers/SPL/Photo Researchers, Inc.

In recent years, the reconstruction, classification, and interpretation of many of the Burgess Shale fossils have undergone a major change that has led to new theories and explanations of the Cambrian explosion of life. Recall that during the Neoproterozoic, multicelled organisms evolved, and shortly thereafter animals with hard parts made their first appearance. These were soon followed by a plethora of different invertebrates during the Cambrian explosion, most of which are now extinct. These Cambrian organisms represent the root-stock and basic body plans from which all present-day invertebrates evolved.

The question that paleontologists are still debating is how many phyla arose during the Cambrian, and at the center of that debate are the Burgess Shale fossils. For years, most paleontologists placed the bulk of the Burgess Shale organisms into existing phyla, with only a few assigned to phyla that are now extinct. Thus, the phyla of the Cambrian world were viewed as being essentially the same in number as the phyla of the present-day world, but with fewer species in each phylum. According to this view, the history of life has been simply a gradual increase in the diversity of species within each phylum through time, and thus, the number of basic body plans has remained more or less constant since the initial radiation of multicelled organisms.

This view, however, has been challenged by other paleontologists who think that the initial explosion of varied life-forms in the Cambrian was promptly followed by a short period of experimentation and then extinction of many phyla. According to this opposing view, the richness and diversity of modern life-forms are the result of repeated variations of the basic body plans that survived the Cambrian extinctions. In other words, life was much more diverse in terms of phyla during the Cambrian than it is today. The reason why members of the Burgess Shale biota look so strange to us is that no living organisms possess their basic body plans, and therefore many of them have been reassigned into new phyla.

Discoveries of new Cambrian fossils at localities such as Sirius Passet, Greenland, and Yunnan, China, have resulted in reassignment of some Burgess Shale specimens back into extant phyla. If these reassignments to known phyla prove to be correct, then no massive extinction followed the Cambrian explosion, and life has gradually increased in diversity through time. Currently, there is no clear resolution to this debate, and the outcome will probably be decided as more fossil discoveries are made.

Ordovician Marine Community

A major transgression that began during the Middle Ordovician (Tippecanoe Sequence) resulted in the most widespread inundation of the craton. This vast epeiric sea, which was uniformly warm during this time, opened many new marine habitats that were soon filled by a variety of organisms.

ConnectionLink

To review cratonic sequences, and the Tippecanoe Sequence in particular, see Chapter 20.

Not only did sedimentation patterns change dramatically from the Cambrian to the Ordovician, but the fauna underwent equally striking changes. Whereas trilobites, brachiopods, and archaeocyathids dominated the Cambrian invertebrate community, the Ordovician was characterized by the continued diversification of brachiopods and by the adaptive radiation of many other animal phyla (such as bryozoans and corals), with a consequent dramatic increase in the diversity of the total shelly fauna (▮ Figure 21.7). In fact, so spectacular was this increase in marine invertebrate diversity, it is rightly called "The Great Ordovician Biodiversification Event."

ConnectionLink

To review the formation and distribution of modern reefs, see Chapter 16, and for Paleozoic reefs, see Chapter 20.

During the Cambrian, archaeocyathids were the main builders of reeflike

▮ **Figure 21.7 Middle Ordovician Marine Community** Reconstruction of a Middle Ordovician seafloor fauna. Cephalopods, crinoids, colonial corals, bryozoans, trilobites, and brachiopods are shown.

Critical Thinking Question What role did the transgressing seas during the Tippecanoe Sequence play in terms of opening up new ecologic niches for the marine invertebrate community, and what was the contribution of the invertebrates to the sedimentary rocks of the Tippecanoe Sequence?

structures, but bryozoans, stromatoporoids, and tabulate and rugose corals assumed that role beginning in the Middle Ordovician. Many of these reefs were small patch reefs similar in size to those of the Cambrian but of a different composition, whereas others were quite large. As with present-day reefs, Ordovician reefs showed a high diversity of organisms and were dominated by suspension feeders.

The end of the Ordovician was a time of mass extinctions in the marine realm. More than 100 families of marine invertebrates became extinct, and in North America alone, about half of the brachiopods and bryozoans died out. What caused such an event? Many geologists think that these extinctions were the result of the extensive glaciation at the end of the Ordovician Period as Gondwana moved over the south pole region (see Chapter 20). This widespread glaciation resulted in decreased sea level, and a cooling of the surface waters, thus creating stressful conditions for the near-surface phytoplankton community and the shallow-water marine faunas, especially the suspension-feeding organisms.

Silurian and Devonian Marine Communities

The mass extinction at the end of the Ordovician was followed by rediversification and recovery of many of the decimated groups. Brachiopods, bryozoans, gastropods, bivalves, corals, crinoids, and graptolites were just some of the groups that rediversified beginning in the Silurian.

As discussed in Chapter 20, the Silurian and Devonian were times of major reef building. Whereas most of the Silurian radiations of invertebrates represented the repopulation of niches, organic reef builders diversified in new ways, building massive reefs larger than any produced during the Cambrian or Ordovician. This repopulation was probably caused, in part, by renewed transgressions over the cratons. Although a major drop in sea level occurred at the end of the Silurian, the Middle Paleozoic sea level was generally high (see Table 20.1).

The Silurian and Devonian reefs were dominated by tabulate and colonial rugose corals and stromatoporoids (▮ Figure 21.8). Whereas the fauna of these Silurian and Devonian reefs was somewhat different from that of earlier reefs and reeflike structures, the general composition and structure are the same as in present-day reefs (see Figure 20.10).

The Silurian and Devonian periods were also the time when *eurypterids* (arthropods with scorpion-like bodies and impressive pincers) were abundant. Unlike many other marine invertebrates, eurypterids expanded into brackish and freshwater habitats (▮ Figure 21.9). *Ammonoids*, a subclass of the cephalopods, evolved from nautiloids during the Early Devonian and rapidly diversified. With their distinctive suture patterns, short stratigraphic ranges, and widespread distribution, ammonoids are excellent guide fossils for the Devonian through Cretaceous periods (▮ Figure 21.10).

▮ **Figure 21.8 Middle Devonian Marine Reef Community** Reconstruction of a Middle Devonian reef from the Great Lakes area of North America. Shown are corals, bryozoans, cephalopods, trilobites, crinoids, and brachiopods.

Critical Thinking Question Why does a reef make a good oil reservoir?

Figure 21.9 Silurian Brackish Water Community Restoration of a Silurian brackish water scene near Buffalo, New York. Shown are algae, eurypterids, gastropods, worms, and shrimp.

Figure 21.10 Ammonoid Cephalopod A Late Devonian-age ammonoid cephalopod from Erfoud, Morocco. The distinctive suture pattern, short stratigraphic range, and wide geographic distribution make ammonoids excellent guide fossils.

Near the end of the Devonian, another mass extinction occurred that resulted in a worldwide near-total collapse of the massive reef communities. On land, however, the seedless vascular plants were seemingly unaffected. Thus, extinctions at this time were most extensive among marine life, particularly in the reef and pelagic communities.

The demise of the Middle Paleozoic reef communities highlights the geographic aspects of the Late Devonian mass extinction. The tropical groups were most severely affected; in contrast, the higher-latitude communities were seemingly minimally affected. Apparently, an episode of global cooling was largely responsible for the extinctions near the end of the Devonian. During such a cooling, the disappearance of tropical conditions would have had a severe effect on reef and other warm-water organisms. Cool-water species, in contrast, could have simply migrated toward the equator. Although cooling temperatures certainly played an important role in the Late Devonian extinctions, the closing of the Iapetus Ocean and the orogenic events of this time (see Figure 20.3a) undoubtedly also played a role by reducing the area of shallow shelf environments where many marine invertebrates lived.

Carboniferous and Permian Marine Communities

The Carboniferous invertebrate marine community responded to the Late Devonian extinctions in much the same way that the Silurian invertebrate marine community

Figure 21.11 Late Mississippian Marine Community Reconstruction of marine life during the Mississippian, based on an Upper Mississippian fossil site at Crawfordsville, Indiana. The majority of the invertebrate animals shown are crinoids and blastoids.

responded to the Late Ordovician extinctions—that is, by renewed adaptive radiation and rediversification. The brachiopods and ammonoids quickly recovered and again assumed important ecologic roles. Other groups, such as the lacy bryozoans and crinoids, reached their greatest diversity during the Carboniferous. With the decline of the stromatoporoids and the tabulate and rugose corals, large organic reefs such as those existing earlier in the Paleozoic virtually disappeared and were replaced by small patch reefs. These reefs were dominated by crinoids, blastoids, lacy bryozoans, brachiopods, and calcareous algae and flourished during the Late Paleozoic (Figure 21.11). In addition, bryozoans and crinoids contributed large amounts of skeletal debris to the formation of the vast bedded limestones that constitute the majority of Mississippian sedimentary rocks.

The Permian invertebrate marine faunas resembled those of the Carboniferous, but they were not as widely distributed because of the restricted size of the shallow seas on the cratons and the reduced shelf space along the continental margins (see Figure 20.21). The spiny and odd-shaped *productids* dominated the brachiopod assemblage and constituted an important part of reef complexes during the Permian (Figure 21.12). Bryozoans, sponges, and some types of calcareous algae also were common elements of the Permian invertebrate fauna.

The Permian Mass Extinction Event

The greatest recorded mass extinction to affect Earth occurred at the end of the Permian Period (Figure 21.13). By the time the Permian ended, roughly 50% of all marine invertebrate families and about 96% of all marine invertebrate species became extinct. Fusulinids (spindle-shaped foraminifera), rugose and tabulate corals, and several bryozoan and brachiopod orders, as well as trilobites and blastoids, did not survive the end of the Permian. All of these groups had been very successful during the Paleozoic Era. In addition, approximately 70% of all terrestrial vertebrate species, as well as nearly 33% of insects on land, also became extinct.

What caused such a crisis for both marine and land-dwelling organisms? Various hypotheses have been proposed, but no completely satisfactory answer has yet been found. Because the extinction event extended over many millions of years at the end of the Permian, a meteorite impact such as occurred at the end of the Cretaceous Period can be reasonably discounted.

ConnectionLink

To learn more about the Cretaceous mass extinction and whether a meteorite impact played a role, see Chapter 22.

Figure 21.12 Permian Patch-Reef Marine Community Reconstruction of a Permian patch-reef community from the Glass Mountains of West Texas. Shown are productid brachiopods, cephalopods, sponges, and corals.

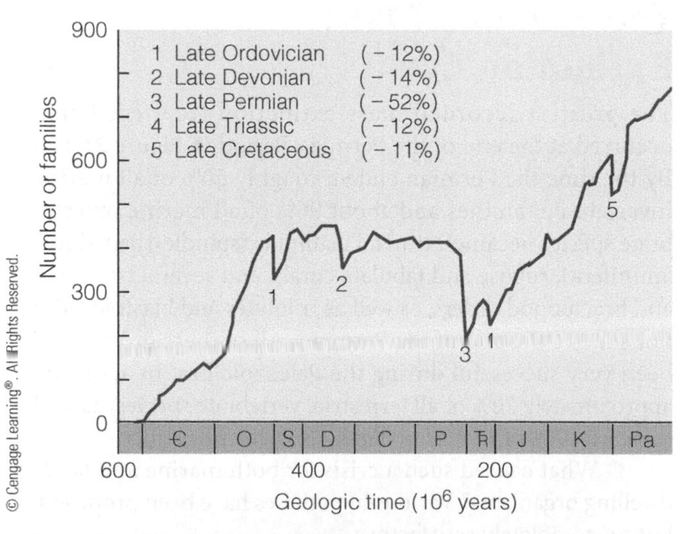

Figure 21.13 Phanerozoic Marine Diversity Phanerozoic diversity for marine invertebrate and vertebrate families. Note the three episodes of Paleozoic mass extinctions, with the greatest occurring at the end of the Permian Period.

Critical Thinking Question Note that the greatest extinction in Earth history occurred at the end of the Permian, yet the Late Cretaceous mass extinction is the one most people are familiar with. Why do you think most people have heard of the Late Cretaceous mass extinction but not the one at the end of the Permian?

A reduction in the habitable marine shelf area caused by the formation of Pangaea and a widespread marine regression resulting from glaciation can also be rejected as the primary cause of the Permian extinctions. By the end of the Permian, most collisions of the continents had already taken place, such that the reduction in the shelf area had already occurred before the mass extinctions began in earnest. Furthermore, the widespread glaciation that took place during the Carboniferous was now waning in the Permian.

Many scientists think that an episode of deep-sea anoxia and increased oceanic CO_2 levels resulted in a highly stratified ocean during the Late Permian. In other words, there was very little, if any, circulation of oxygen-rich surface waters into the deep ocean. Stagnant waters also covered the shelf regions during this time, thus affecting the shallow marine fauna.

There is also evidence of increased global warming during the Late Permian. This factor would contribute to a stratified global ocean because warming in the high latitudes would significantly reduce or even eliminate the downwelling of cold, dense, oxygenated waters from the polar areas into the deep oceans at lower latitudes that occurs today. Increased global warming would also result in stagnant, stratified oceans rather than a well-mixed, oxygenated oceanic system.

Global Warming and Mass Extinctions—A New Television Series

A major debate today is global warming and its effects, not only on humans, but also on Earth's total biota. In addition to contributing to global warming, increases in atmospheric carbon dioxide can lead to higher acidity levels in the oceans, which can have deleterious effects on the marine ecosystem as well. A major concern is that the cascading effects of global warming could lead to major extinctions of Earth's flora and fauna, similar to the mass extinctions that have occurred a number of times in the geologic past.

You are a highly respected paleontology professor who has published numerous academic papers on mass extinctions, as well as articles in magazines and newspapers explaining to a general readership what the potential effects of global warming might be on the world. You have been asked by a major television studio to be the scientific advisor for an educational series that will look at global warming in terms of what it is, what the potential impact on the world's biota might be should global warming continue, and most importantly, what, if anything, can the geologic past tell us about the future. The producers of the program are particularly interested in showing what has happened in previous episodes of global warming.

From your research on past mass extinctions, such as those that occurred at the end of the Permian, the end of the Cretaceous, as well as what the effects were during and after the Paleocene-Eocene Thermal Maximum (see Chapter 23), you are well aware of the possible causes of these extinctions. How can the producers of this series explain to a lay audience the perspective of geologic time, what has happened globally in the geologic past, and what the possible implications are in the future to Earth's biota, if global warming continues?

iStockphoto.com/Ryan Balderas; maukun/Shutterstock.com

During the Late Permian, widespread volcanic and continental fissure eruptions, such as the Siberian Traps, where lava flows covered more than 2 million km^2 were also taking place. These eruptions released not only large amounts of additional carbon dioxide into the atmosphere, but also high amounts of fluorine and chlorine, which could have damaged the ozone layer and helped contribute to increased climatic instability and ecologic collapse.

Several recent studies have examined the role that the Siberian Traps flood basalts may have played in causing widespread ozone depletion, thus allowing an increase in UV-B radiation at the end of the Permian. These studies have included using a two-dimensional atmospheric chemistry-transport model to assess the impact of the Siberian Traps eruption in altering the end-Permian stratospheric ozone layer, as well as an analysis of the relationship of morphologic abnormalities in gymnosperm (flowerless seed plant) pollen grains (see the section "Plant Evolution" later in this chapter) to deteriorating conditions around the Permian–Triassic boundary.

By the end of the Permian, a near collapse of both the marine and terrestrial ecosystems had occurred. Although the ultimate cause of such devastation is still being debated and investigated, it is safe to say that it was probably a combination of interconnected and related geologic and biologic events.

21.6 Vertebrate Evolution

A **chordate** (phylum Chordata) is an animal that has, at least during part of its life cycle, a notochord, a dorsal hollow nerve cord, and pharyngeal slits (▮ Figure 21.14). **Vertebrates**, which are animals with backbones, are simply a subphylum of chordates.

The ancestors and early members of the phylum Chordata were soft-bodied organisms that left few fossils (▮ Figure 21.15). Consequently, we know little about the early evolutionary history of the chordates, and likewise, little about the early history of the subphylum Vertebrata, but we have a fairly good record of its subsequent history. Surprisingly, a close relationship exists between echinoderms (Table 21.1) and chordates, and they may even have shared a common ancestor. This is because the development of the embryos of echinoderms and chordates are the same in both groups—that is, the cells divide by radial cleavage so that the cells are aligned directly above each other (▮ Figure 21.16a). In all other invertebrates, cells undergo spiral cleavage, which results in having cells nested between each other in successive rows (▮ Figure 21.16b). Furthermore, the biochemistry of muscle activity and blood proteins, and the larval stages, are similar in both echinoderms and chordates.

The evolutionary pathway to vertebrates thus appears to have taken place much earlier and more rapidly than many scientists have long thought. On the basis of fossil evidence and recent advances in molecular biology, one scenario suggests that vertebrates evolved shortly after an ancestral chordate acquired a second set of genes. According to this hypothesis, a random mutation produced a duplicate set of genes, letting the ancestral vertebrate animal evolve entirely new body structures that proved to be evolutionarily advantageous. Not all scientists accept this hypothesis, and the origin of vertebrates is still hotly debated.

▮ **Figure 21.14 Three Characteristics of a Chordate** The structure of the lancelet *Amphioxus* (approximately 5 cm long) illustrates the three characteristics of a chordate: a notochord, a dorsal hollow nerve cord, and pharyngeal slits.

▮ **Figure 21.15** ***Yunnanozoon lividum*** Found in 525-million-year-old rocks in Yunnan province, China, *Yunnanozoon lividum*, a 5-cm-long animal, is one of the oldest known chordates.

21.7 Fish

The most primitive vertebrates are fish. Although carbonaceous impressions and small vertebral elements that are thought to belong to members of the class Agnatha (jawless fish) have been reported from Lower Cambrian rocks in Yunnan province, China, some of the oldest, and undisputed fish remains, are found in the Upper Cambrian Deadwood Formation in northeastern Wyoming. Here, phosphatic scales and plates of *Anatolepis*, a primitive member of the agnathids, have been recovered from marine sediments.

Presently, all known Cambrian and Ordovician fossil fish have been found in shallow, nearshore marine deposits, whereas the earliest nonmarine (freshwater) fish remains have been found in Silurian strata. This does not prove that fish originated in the oceans, but it does lend strong support to the idea.

As a group, and excluding the Lower Cambrian potential fish fossils from China, fish range from the Late Cambrian to

▮ **Figure 21.16** Cell Cleavage

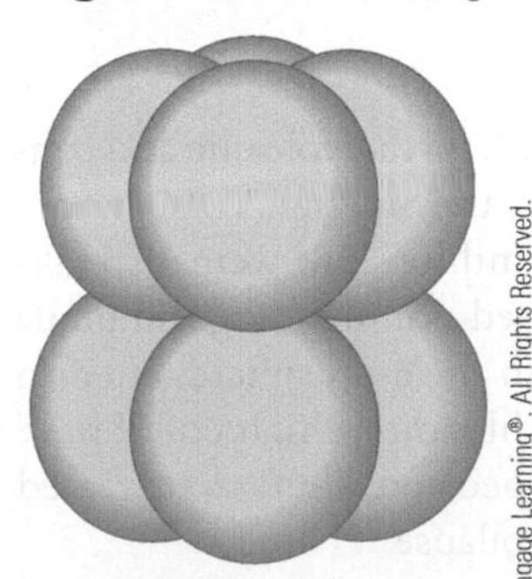

a Arrangement of cells resulting from radial cleavage is characteristic of chordates and echinoderms. In this configuration, cells are directly above each other.

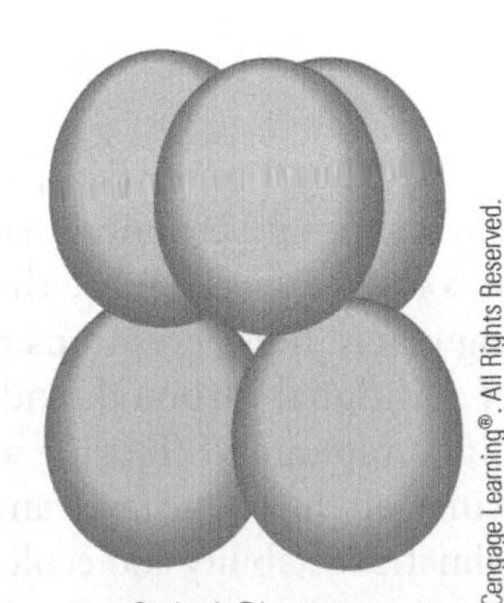

b Arrangement of cells resulting from spiral cleavage. In this arrangement, cells in successive rows are nested between each other. Spiral cleavage is characteristic of all invertebrates except echinoderms.

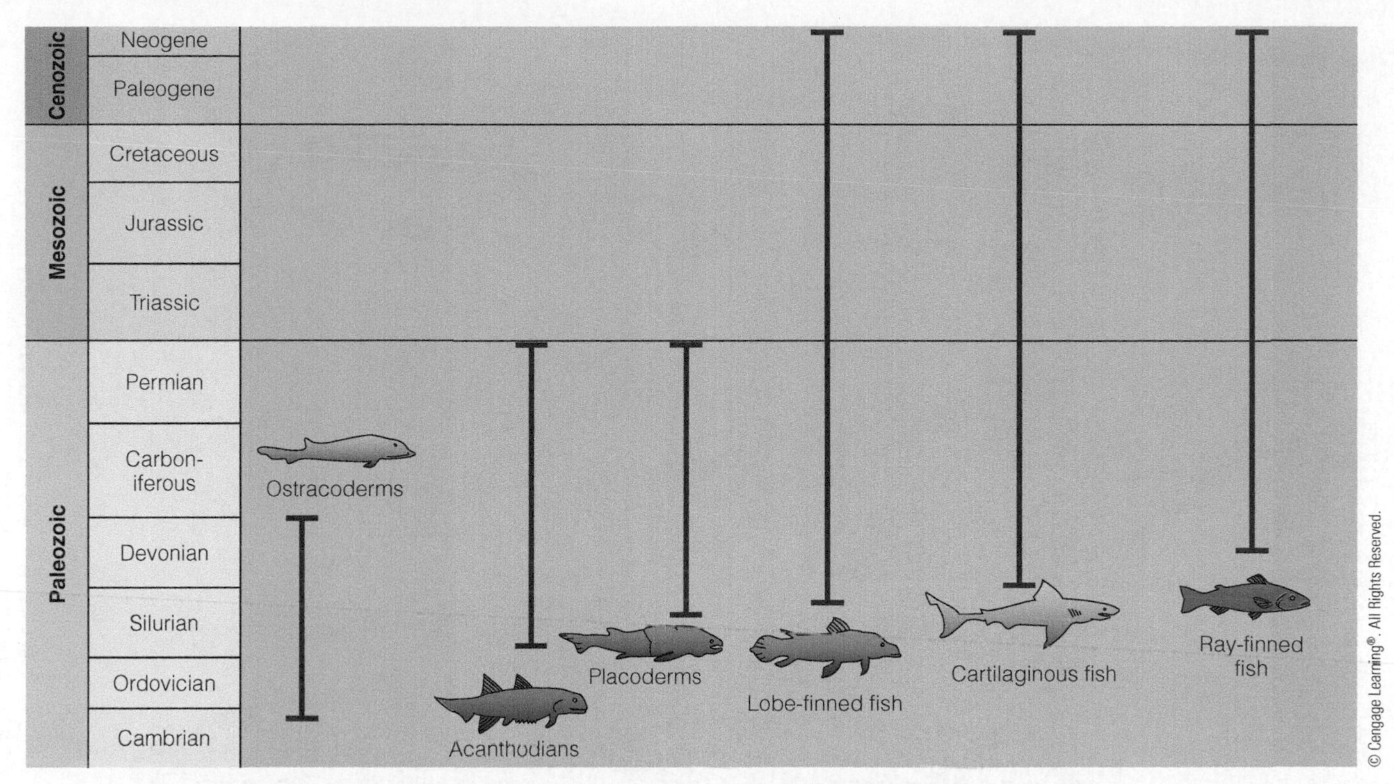

Figure 21.17 Geologic Ranges of the Major Fish Groups Ostracoderms are early members of the class Agnatha (jawless fish). Acanthodians (class Acanthodii) are the first fish with jaws. Placoderms (class Placodermii) are armored, jawed fish. Lobe-finned (subclass Sarcopterygii) and ray-finned (subclass Actinopterygii) fish are members of the class Osteichthyes (bony fish), whereas cartilaginous fish belong to the class Chondrichthyes.

the present (Figure 21.17). The oldest and most primitive of the class Agnatha are the **ostracoderms**, whose name means "bony skin" (Table 21.2). These are armored, jawless fish that first evolved during the Late Cambrian, reached their zenith during the Silurian and Devonian, and then became extinct.

The majority of ostracoderms lived on the seafloor. *Hemicyclaspis* is a good example of a bottom-dwelling ostracoderm (Figure 21.18a). Vertical scales allowed *Hemicyclaspis* to wiggle sideways, propelling itself along the seafloor, and the eyes on top of its head allowed it to see

TABLE 21.2 Brief Classification of Fish Groups Referred to in the Text

Classification	Geologic Range	Living Example
Class Agnatha (jawless fish)	Late Cambrian–Recent	Lampery, hagfish
Early members of the class are called ostracoderms	Late Cambrian–Devonian	No living ostracoderms
Class Acanthodii (the first fish with jaws)	Early Silurian–Permian	None
Class Placodermii (armored, jawed fish)	Late Silurian–Permian	None
Class Chondrichthyes (cartilaginous fish)	Devonian–Recent	Sharks, rays, skates
Class Osteichthyes (bony fish)	Devonian–Recent	Tuna, perch, bass, pike, catfish, trout, salmon, lungfish, *Latimeria*
Subclass Actinopterygii (ray-finned fish)	Devonian–Recent	Tuna, perch, bass, pike, catfish, trout, salmon
Subclass Sarcopterygii (lobe-finned fish)	Devonian–Recent	Lungfish, *Latimeria*
Order Coelacanthimorpha	Devonian–Recent	*Latimeria*
Order Dipnoi	Devonian–Recent	Lungfish
Order Crossopterygii	Devonian–Permian	None
Suborder Rhipidistia	Devonian–Permian	None

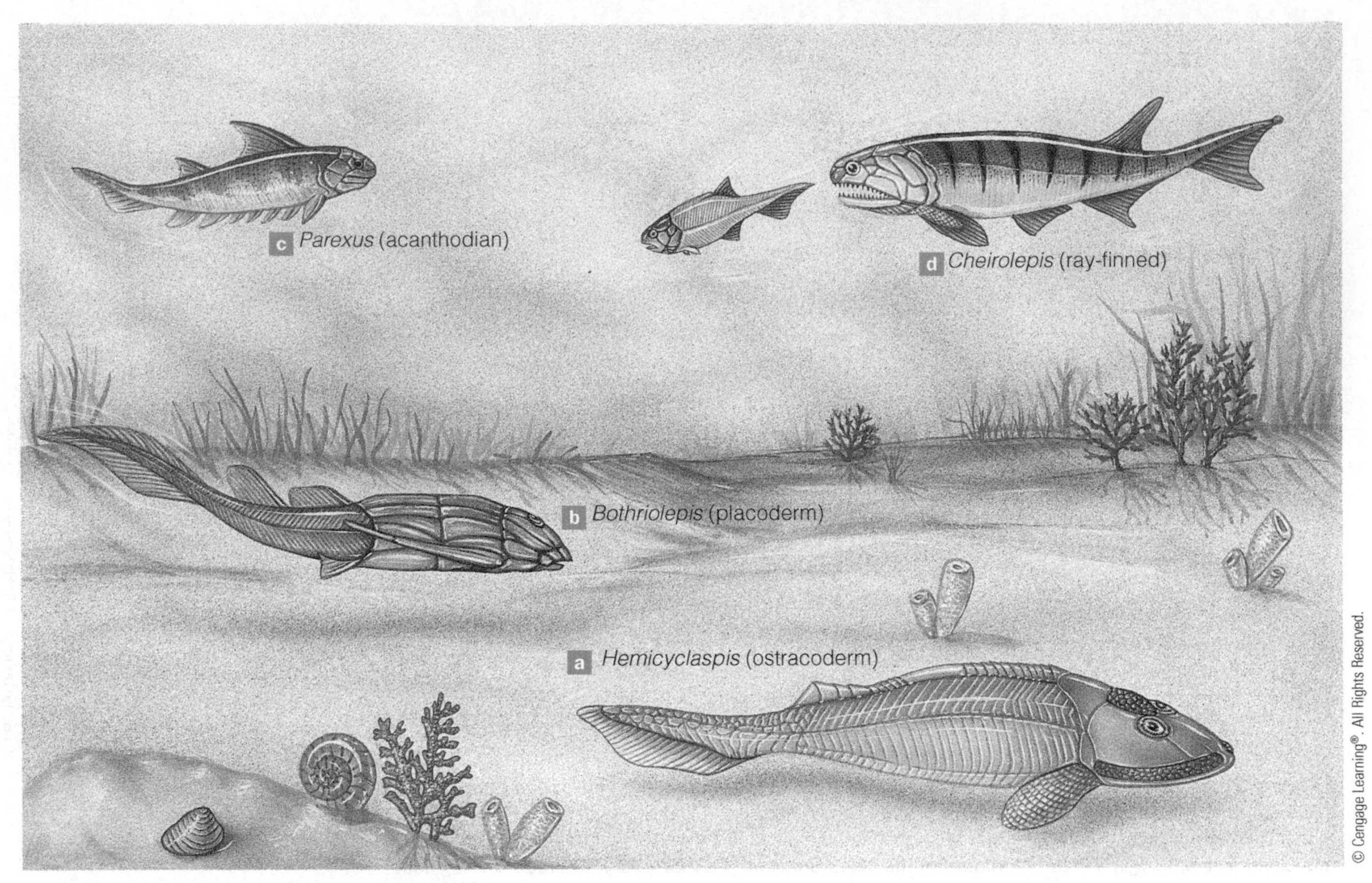

Figure 21.18 Devonian Seafloor Recreation of a Devonian seafloor showing **(a)** an ostracoderm, **(b)** a placoderm, **(c)** an acanthodian, and **(d)** a ray-finned fish.

such predators as cephalopods and jawed fish approaching from above. While moving along the sea bottom, it probably sucked up small bits of food and sediment through its jawless mouth.

The evolution of jaws was a major evolutionary advance among primitive vertebrates. Although their jawless ancestors could only feed on detritus, jawed fish could chew food and become active predators, thus opening many new ecological niches.

Indeed, the vertebrate jaw is an excellent example of evolutionary opportunism. Various studies suggest that the jaw originally evolved from the first two or three anterior gill arches of jawless fish. Because the gills are soft, they are supported by gill arches of bone or cartilage. The evolution of the jaw may thus have been related to respiration rather than to feeding (Figure 21.19). By evolving joints in the forward gill arches, jawless fish could open their mouths wider. Every time a fish opened and closed its mouth, it would pump more water past the gills, thereby increasing the oxygen intake. The modification from rigid to hinged forward gill arches let fish increase both their food consumption and oxygen intake, and the evolution of the jaw as a feeding structure rapidly followed.

The fossil remains of the first jawed fish are found in Lower Silurian rocks and belong to the *acanthodians*, a group of small, enigmatic fish characterized by large spines, paired fins, scales covering much of the body, jaws, teeth, and greatly reduced body armor (Figure 21.18c and Table 21.2). Although the relationship of acanthodians to other fish is not yet well established, many scientists think

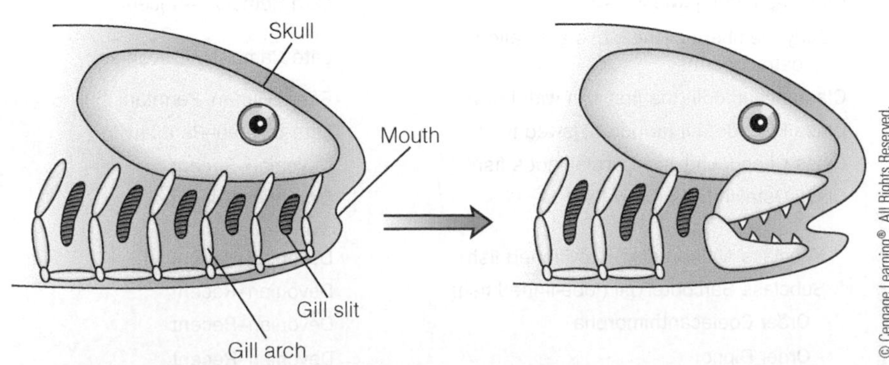

Figure 21.19 Evolution of the Vertebrate Jaw The evolution of the vertebrate jaw is thought to have begun from the modification of the first two or three anterior gill arches. This theory is based on the comparative anatomy of living vertebrates.

that the acanthodians included the ancestors of the present-day bony and cartilaginous fish groups. The acanthodians were most abundant during the Devonian, declined in importance through the Carboniferous, and became extinct in the Permian.

The other jawed fish, the **placoderms** (class Placodermii), whose name means "plate-skinned," evolved during the Silurian (Table 21.2). Placoderms were heavily armored, jawed fish that lived in both freshwater and the ocean. Like the acanthodians, placoderms reached their peak of abundance and diversity during the Devonian.

The placoderms showed considerable variety, including small bottom dwellers (▌Figure 21.18b) as well as large major predators such as *Dunkleosteus*, a Late Devonian fish that lived in the mid-continental North American epeiric seas (▌Figure 21.20a). *Dunkleosteus* was by far the largest fish of the time, reaching a length of more than 12 m. It had a heavily armored head and shoulder region, a huge jaw lined with razor-sharp bony teeth, and a flexible tail, all features consistent with its status as a ferocious predator.

Besides the abundant acanthodians, placoderms, and ostracoderms, other fish groups, such as the cartilaginous and bony fish, also evolved during the Devonian Period. Small wonder, then, that the Devonian is informally called the "Age of Fish," because all major fish groups were present during this time period.

The **cartilaginous fish** (class Chondrichthyes) (Table 21.2), represented today by sharks, rays, and skates, first evolved during the Early Devonian, and by the Late Devonian, primitive marine sharks such as *Cladoselache* were abundant (▌Figure 21.20b). Cartilaginous fish have never been as numerous or as diverse as their cousins, the bony fish, but they were, and still are, important members of the marine vertebrate fauna.

Along with the cartilaginous fish, the **bony fish** (class Osteichthyes) (Table 21.2) also first evolved during the Devonian. Because bony fish are the most varied and numerous of all the fishes, and because the amphibians evolved from them, their evolutionary history is particularly important. There are two groups of bony fish: the common *ray-finned fish* (subclass Actinopterygii) (▌Figures 21.18d and ▌21.20d) and the less familiar *lobe-finned fish* (subclass Sarcopterygii) (Table 21.2).

The term *ray-finned* refers to the way that the fins are supported by thin bones that spread away from the body (▌Figure 21.21a). From a modest freshwater beginning during the Devonian, ray-finned fish, which include most of the familiar fish such as trout, bass, perch, salmon, and tuna, rapidly diversified to dominate the Mesozoic and Cenozoic seas.

Present-day lobe-finned fish are characterized by muscular fins. The fins do not have radiating bones, but, rather, have articulating bones with the fin attached to the body by a fleshy shaft (▌Figure 21.21b). Such an arrangement allows for a powerful stroke of the fin, making the fish an effective swimmer. Three orders of lobe-finned fish are recognized: *coelacanths*, *lungfish*, and *crossopterygians* (Table 21.2).

Coelacanths (order Coelacanthimorpha) are marine lobe-finned fish that evolved during the Middle Devonian and were thought to have gone extinct at the end of the Cretaceous.

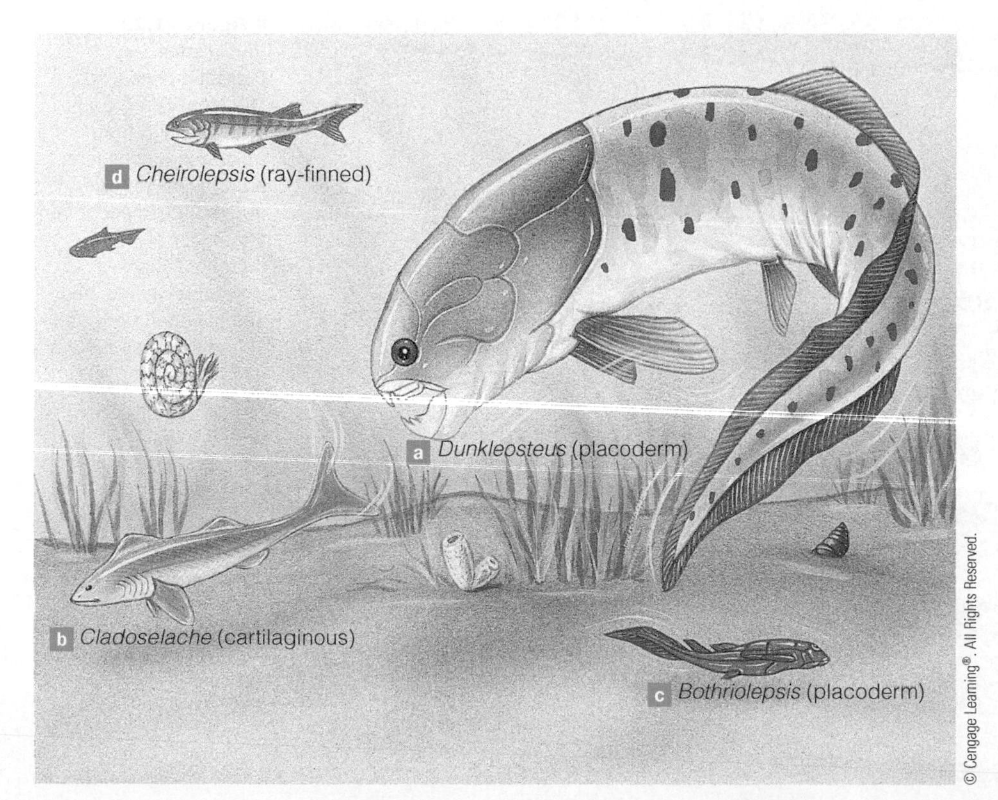

▌**Figure 21.20 Late Devonian Seascape** A Late Devonian marine scene from the mid-continent of North America. **(a)** The giant placoderm *Dunkleosteus* (length more than 12 m) is pursuing **(b)** the shark *Cladoselache* (length up to 1.2 m), a cartilaginous fish. Also shown are **(c)** the bottom-dwelling placoderm *Bothriolepsis* and **(d)** the swimming ray-finned fish *Cheirolepsis*, both of which reached a length of 40–50 cm.

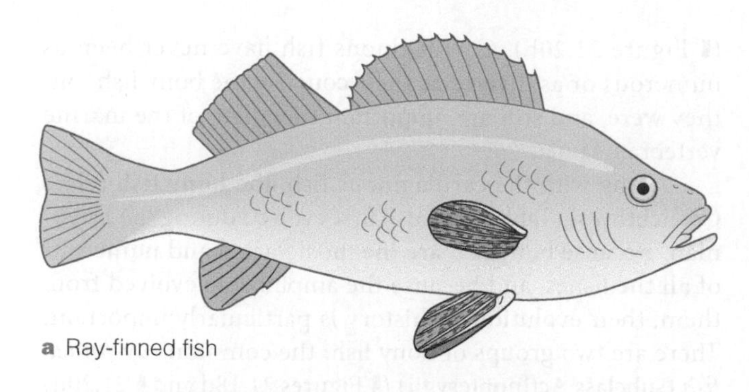

a Ray-finned fish

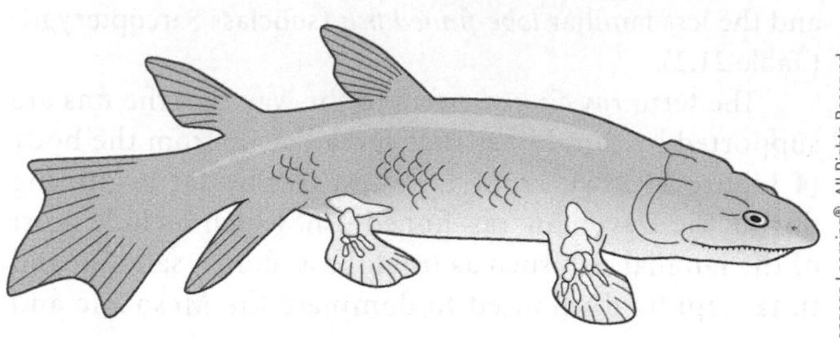

b Lobe-finned fish

Figure 21.21 Ray-finned and Lobe-finned Fish Arrangement of fin bones for **(a)** a typical ray-finned fish and **(b)** a lobe-finned fish. The muscles extend into the fin of the lobe-finned fish, allowing greater flexibility of movement than that of the ray-finned fish.

In 1938, however, a fisherman caught a coelacanth in the deep waters off Madagascar, and since then, several dozen more have been caught, both there and in Indonesia.

Lungfish (order Dipnoi) were fairly abundant during the Devonian, but today only three freshwater genera exist, one each in South America, Africa, and Australia.  Their present-day distribution presumably reflects the Mesozoic breakup of Gondwana.

ConnectionLink

To learn more about the breakup of Gondwana during the Mesozoic, and its effect on global climate and ocean circulation patterns, see Chapter 22.

The "lung" of a modern-day lungfish is actually a modified swim bladder that most fish use for buoyancy in swimming. In lungfish, this structure absorbs oxygen, allowing them to breathe air when the lakes or streams in which they live become stagnant and dry up. During such times, they burrow into the sediment to prevent dehydration and breathe through their swim bladder until the stream begins flowing or the lake that they were living in fills with water. When they are back in the water, lungfish then rely on gill respiration.

The **crossopterygians** (order Crossopterygii) are an important group of lobe-finned fish, because it is probably from them that amphibians evolved. However, the transition between crossopterygians and true amphibians is not as simple as it was once portrayed. Among the crossopterygians, the *rhipidistians* appear to be the ancestral group (Table 21.2). These fish, reaching lengths of more than 2 m, were the dominant freshwater predators during the Late Paleozoic. *Eusthenopteron*, a good example of a rhipidistian crossopterygian and a classic example of a transitional form between fish and amphibians, had an elongated body that helped it move swiftly in the water and had paired muscular fins that many scientists thought could be used for moving on land (Figure 21.22).

Figure 21.22 Rhipidistian Crossopterygian The crossopterygians are the group from which the amphibians are thought to have evolved. *Eusthenopteron*, a member of the rhipidistian crossopterygians, had an elongated body and paired fins that could be used for moving about on land.

The structural similarity between crossopterygian fish and the earliest amphibians is striking and is one of the most widely cited examples of a transition from one major group to another (▮ Figure 21.23). Recent discoveries of older lobe-finned fish and newly published findings of tetrapod footprints and tetrapod-like fish, however, are now filling in the gaps in the evolution from fish to tetrapods (animals with four limbs).

▮ Figure 21.23 Similarities Between Crossopterygians and Labyrinthodonts Similarities between the crossopterygian lobe-finned fish and the labyrinthodont amphibians.

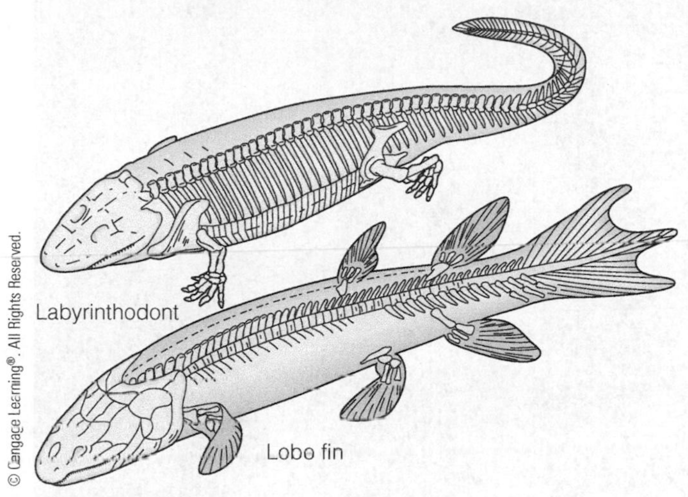

a Skeletal similarity.

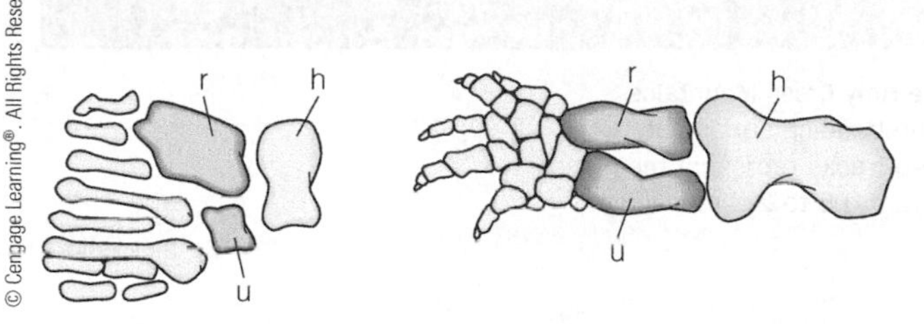

b Comparison of the limb bones of a crossopterygian (*left*) and amphibian (*right*); color identifies the bones (u = ulna, shown in blue, r = radius, mauve, h = humerus, gold) that the two groups have in common.

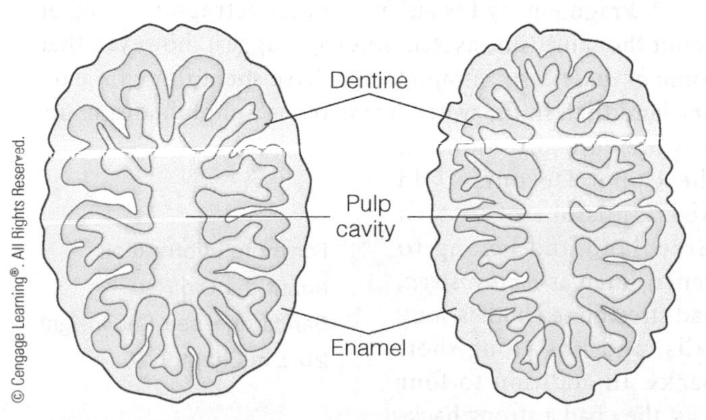

c Comparison of tooth cross sections shows the complex and distinctive structure found in both the crossopterygians (*left*) and labyrinthodont amphibians (*right*).

Before discussing this transition and the evolution of amphibians, it is useful to place the evolutionary history of Paleozoic fish in the larger context of Paleozoic evolutionary events. Certainly, the evolution and diversification of jawed fish, as well as eurypterids and ammonoid cephalopods, had a profound effect on the marine ecosystem. Previously defenseless organisms either evolved defensive mechanisms or suffered great losses, possibly even extinction.

For example, ostracoderms, although armored, would have been easy prey for the swifter jawed fish. Ostracoderms became extinct by the end of the Devonian, a time that coincides with the rapid evolution of jawed fish. Placoderms, like acanthodians, greatly decreased in abundance after the Devonian and became extinct by the end of the Paleozoic Era. In contrast, cartilaginous fish and ray-finned bony fish expanded during the Late Paleozoic, as did the ammonoids (Figure 21.10), the other major predators of the Late Paleozoic seas.

21.8 Amphibians—Vertebrates Invade the Land

Although amphibians were the first vertebrates to live on land, they were not the first land-living organisms. Land plants, which probably evolved from green algae, first evolved during the Ordovician. Furthermore, insects, millipedes, spiders, and even snails invaded the land before amphibians. Fossil evidence indicates that such land-dwelling arthropods as scorpions and flightless insects had also evolved by at least the Devonian.

The transition from water to land required animals to surmount several barriers. The most critical were desiccation, reproduction, the effects of gravity, and the extraction of oxygen from the atmosphere by lungs rather than from water by gills. Until the 1990s, the traditional evolutionary sequence had a rhipidistian crossopterygian, like *Eusthenopteron*, evolving into a primitive amphibian such as *Ichthyostega*. At that time, fossils of those two genera were about all that paleontologists had to work with, and although there were gaps in morphology, the link between crossopterygians and these earliest amphibians was easy to see (Figure 21.23). A recent discovery, however, of fossilized tetrapod footprints in rocks dated at 395 million years old (earliest Middle Devonian) from the Holy Cross Mountains of Poland, as well as additional tetrapod fossil specimens, has forced paleontologists to rethink when and where animals emerged onto land (▮ Figure 21.24).

An equally intriguing question is why limbs evolved in the first place. The answer to that question is that they probably were not for walking on land. The current thinking of many scientists is that aquatic limbs made it easier for animals to move around in streams, lakes, or swamps that were choked with water plants or other debris.

Figure 21.24 Tetrapod Trackway Exposed at the Zachelmie Quarry in the Holy Cross Mountains of Poland The tracks, which measure up to 26 cm wide, are exposed on the bedding surface of 395-million-year-old (earliest Middle Devonian) shallow, marine rocks. These tracks represent the oldest known evidence of four-legged animals, estimated to have ranged in size from 0.5 to 2.5 m in length.

The fossil evidence that began to emerge in the 1990s also seems to support this hypothesis. However, the discovery of fossilized tetrapod footprints from the mud to coral-reef lagoon deposits in Poland (Figure 21.24) is forcing paleontologists to rethink their long-held ideas of a terrestrial origin for the tetrapods.

Fossils of *Acanthostega*, a tetrapod found in the 360-million-year-old Old Red Sandstone from Greenland, reveal an animal that had limbs, but was clearly unable to walk on land. Paleontologist Jennifer Clack, who has recovered and analyzed hundreds of specimens of *Acanthostega*, has pointed out that *Acanthostega*'s limbs were not strong enough to support its weight on land, and its rib cage was too small for the necessary muscles needed to hold its body off the ground. In addition, *Acanthostega* had gills and lungs, meaning that it could survive on land, but it was more suited for the water. Clack has suggested that *Acanthostega* used its limbs to maneuver around in swampy, plant-filled waters, where swimming would be difficult and limbs were an advantage.

 Fragmentary fossils from other tetrapods living at about the same time as *Acanthostega* suggest, however, that some of these early tetrapods may have spent more time on dry land than in the water. These oldest amphibians, many of which are also found in the Upper Devonian Old Red Sandstone of eastern Greenland and belong to genera such as *Ichthyostega*, had streamlined bodies, long tails, and fins along their backs. In addition to four legs, they had a strong backbone, a rib cage, and pelvic

ConnectionLink

For more information about the Old Red Sandstone, see Geo-Insight 20.1 in Chapter 20.

Figure 21.25 Late Devonian Landscape Shown is *Ichthyostega*, both in the water and on land. *Ichthyostega* was an amphibian that grew to a length of about 1 m. The flora at this time was diverse and consisted of a variety of small and large seedless vascular plants.

and pectoral girdles, all of which were structural adaptations for walking on land (Figure 21.25).

These earliest amphibians thus appear to have inherited many characteristics from the crossopterygians with little modification (Figure 21.23). However, with the discovery of such fossils as *Acanthostega* and others like it, the transition between fish and amphibians involves a number of new genera that are intermediary between the two groups and fills in some of the gaps between the earlier postulated rhipidistian crossopterygian amphibian phylogeny.

Tiktaalik roseae (Figure 21.26a), a 1.2- to 2.8-m-long, 375-million-year-old (Late Devonian) "fishapod," discovered on Ellesmere Island, Canada, has been hailed as an intermediary between the lobe-finned fish and the earliest tetrapod *Acanthostega*. Featuring a mixture of both fish and tetrapod characteristics (Figure 21.26b), *Tiktaalik roseae*, is truly a "fishapod." On the one hand, it has gills and fish scales, but it also has a broad skull, eyes on top of its head, a flexible neck, and large rib cage that could support its body on land or in shallow water, and lungs, all of which are tetrapod features.

What really excites scientists, however, is that *Tiktaalik roseae* has the beginnings of a true tetrapod forelimb, complete with functional wrist bones and five digits, as well as a modified ear region. Sedimentological evidence also suggests that *Tiktaalik roseae* lived in a shallow-water habitat associated with the Late Devonian floodplains of Laurasia.

As previously mentioned, the oldest known amphibians, such as *Ichthyostega*, had skeletal features that allowed it to spend its life on land. Because amphibians did not evolve until the Late Devonian, they were a minor element of the Devonian terrestrial ecosystem. Like other groups that moved into new and previously unoccupied niches, amphibians underwent rapid adaptive radiation and became abundant during the Carboniferous and Early Permian.

The Late Paleozoic amphibians did not at all resemble the familiar frogs, toads, newts, and salamanders that make up the modern amphibian fauna. Rather, they displayed a broad spectrum of sizes, shapes, and modes of life (Figure 21.27). One group of amphibians was the **labyrinthodonts**, so named for the labyrinthine wrinkling and folding of the chewing surface of their teeth (Figure 21.23c). Most labyrinthodonts were large animals, as much as 2 m in length. These typically sluggish creatures lived in swamps and streams, eating fish, vegetation, insects, and other small amphibians (Figure 21.27).

Labyrinthodonts were abundant during the Carboniferous, when swampy conditions were widespread (see Chapter 20), but they soon declined in abundance during the Permian, perhaps in response to changing climatic conditions. Only a few species survived into the Triassic.

21.9 Evolution of the Reptiles—The Land Is Conquered

Amphibians were limited in colonizing the land because they had to return to water to lay their gelatinous eggs. The evolution of the **amniote egg** (Figure 21.28) freed reptiles from this constraint. In such an egg, the developing embryo is surrounded by a liquid-filled sac called the *amnion* and provided with both a *yolk*, or food sac, and an *allantois*, or waste sac. In this way, the emerging reptile is in essence, a miniature adult, bypassing the need for a larval stage in the water. The evolution of the amniote egg allowed vertebrates to colonize all parts of the land because they no longer had to return to the water as part of their reproductive cycle.

Figure 21.26 ***Tiktaalik roseae*** *Tiktaalik roseae*, a "fishapod," has been hailed as an intermediary between lobe-finned fish and tetrapods because it has characteristics of both fish and tetrapods.

a Skull and skeleton of *Tiktaalik roseae*.

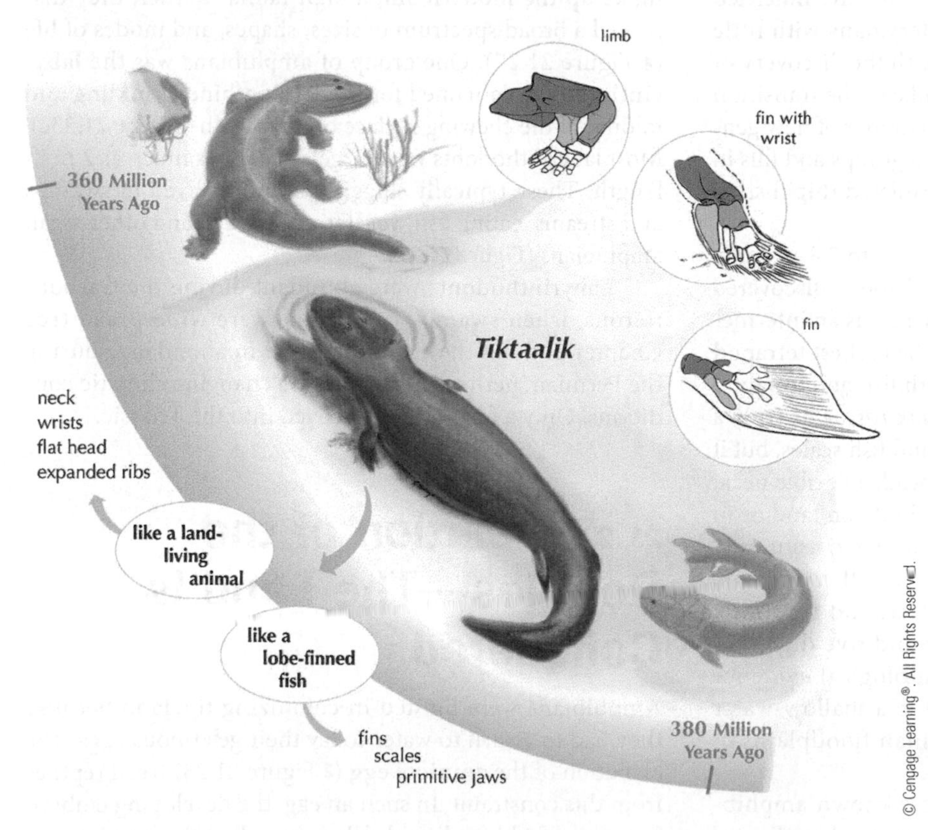

b Diagram illustrating how *Tiktaalik roseae* is a transitional species between lobe-finned fish and tetrapods.

Many of the differences between amphibians and reptiles are physiologic and are not preserved in the fossil record. Nevertheless, amphibians and reptiles differ sufficiently in skull structure, jawbones, ear location, and limb and vertebral construction to suggest that reptiles evolved from labyrinthodont ancestors by the Late Mississippian. This assessment is based on the discovery of a well-preserved fossil skeleton of the oldest known reptile, *Westlothiana*, and other fossil reptile skeletons from Late Mississippian-age rocks in Scotland.

Other early reptile fossils occur in the Lower Pennsylvania Joggins Formation in Nova Scotia, Canada. Here, remains of *Hylonomus* are found in the sediments filling in tree trunks. These earliest reptiles from Scotland and Canada were small and agile and fed largely on grubs and insects. They are loosely grouped together as **protorothyrids**, whose members include the earliest reptiles (▌Figure 21.29). During the Permian Period, reptiles diversified and began displacing many amphibians. The reptiles succeeded partly because of their advanced method of reproduction and their more advanced jaws and teeth, as well as their tough skin and scales to prevent desiccation, and their ability to move rapidly on land.

The **pelycosaurs**, or finback reptiles, evolved from the protorothyrids during the Pennsylvanian and were the dominant reptile group by the Early Permian. They evolved into a diverse assemblage of herbivores, exemplified by *Edaphosaurus*, and carnivores such as *Dimetrodon* (▌Figure 21.30). An interesting feature of the pelycosaurs is their sail. It was formed by vertebral spines that, in life, were covered with skin. The sail has been variously explained as a type of sexual display, a means of protection, and a display to look more ferocious. The current consensus seems to be that the sail served as some type of thermoregulatory device, raising the reptile's temperature by catching the sun's rays or cooling it by facing the wind. Because pelycosaurs are considered to be the group from which therapsids (mammal-like reptiles)

Figure 21.27 Carboniferous Coal Swamp The amphibian fauna during the Carboniferous was varied. Shown is the serpentlike *Dolichosoma* (foreground) and the large labyrinthodont amphibian *Eryops*.

Critical Thinking Question What were the conditions that made the Late Carboniferous (Pennsylvanian) Period so favorable for the proliferation of amphibians and seedless vascular plants?

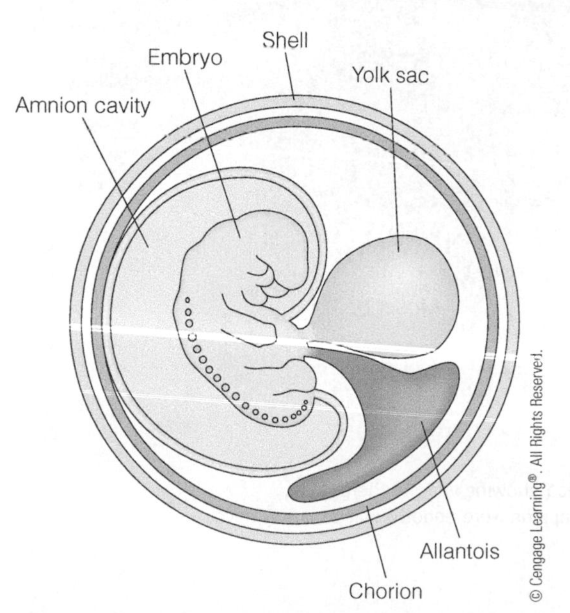

Figure 21.28 Amniote Egg In an amniote egg, the embryo is surrounded by a liquid sac (amnion cavity) and provided with a food source (yolk sac) and waste sac (allantois). The evolution of the amniote egg freed reptiles from having to return to the water for reproduction and let them inhabit all parts of the land.

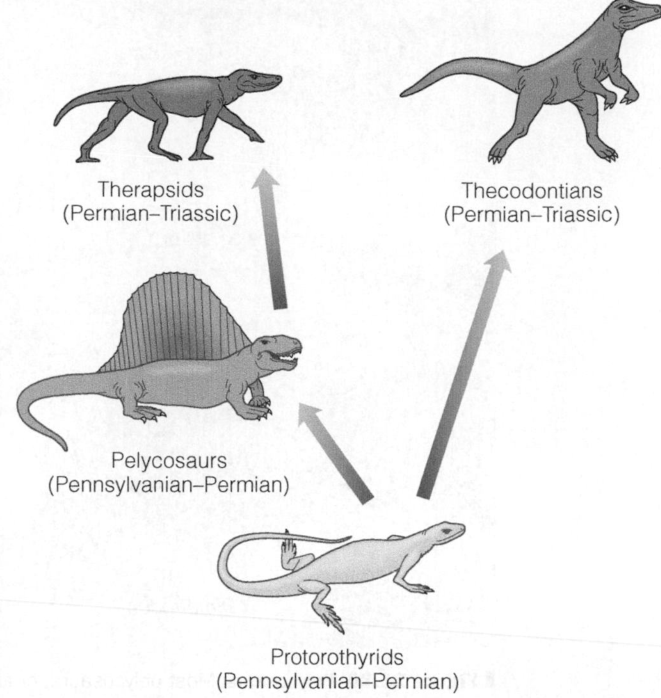

Figure 21.29 Evolutionary Relationship Among the Paleozoic Reptiles

Critical Thinking Question Although the transition from reptiles to mammals will not be discussed in detail until the next chapter, what is one characteristic in common between pelycosaurs and therapsids that indicates the pelycosaur to therapsid to mammal lineage is a correct evolutionary lineage?

evolved, it is interesting that they may have had some sort of body-temperature control.

The pelycosaurs became extinct during the Permian and were succeeded by the **therapsids**, mammal-like reptiles that evolved from the carnivorous pelycosaur lineage and rapidly diversified into herbivorous and carnivorous lineages (❚ Figure 21.31). Therapsids were small- to medium-size animals that displayed the beginnings of many mammalian features: fewer bones in the skull because many of the small skull bones were fused; enlarged lower jawbone; differentiation of teeth for various functions such as nipping, tearing, and chewing food; and more vertically placed legs for greater flexibility, as opposed to the way that the legs sprawled out to the side in primitive reptiles.

Furthermore, many paleontologists think that therapsids were *endothermic*, or warm-blooded, enabling them to maintain a constant internal body temperature. This characteristic would have let them expand into a variety of habitats, and, indeed, the Permian rocks in which their fossil remains are found are distributed not only in low latitudes but in middle and high latitudes as well.

As the Paleozoic Era came to an end, the therapsids constituted about 90% of the known reptile genera

Figure 21.30 Pelycosaurs Most pelycosaurs, or finback reptiles, have a characteristic sail on their back. One hypothesis explains the sail as a rudimentary thermoregulatory device. Other hypotheses are that it was a type of sexual display or a device to make the reptile look more intimidating. Shown here are the carnivore *Dimetrodon* and the herbivore *Edaphosaurus*.

Figure 21.31 Therapsids A Late Permian scene in southern Africa showing various therapsids, including *Dicynodon* and *Moschops*. Many paleontologists think therapsids were endothermic and may have had a covering of fur, as shown here.

and occupied a wide range of ecological niches. The mass extinctions that decimated the marine fauna at the close of the Paleozoic had an equally great effect on the terrestrial population. By the end of the Permian, about 96% of all marine invertebrate species were extinct, compared with more than two-thirds of all amphibians and reptiles. Plants, in contrast, apparently did not experience as great a turnover as animals.

21.10 Plant Evolution

When plants made the transition from water to land, they had to solve most of the same problems that animals did: desiccation, support, reproduction, and the effects of gravity. Plants did so by evolving a variety of structural adaptations that were fundamental to the subsequent radiations and diversification that occurred during the Silurian, Devonian, and later periods (Table 21.3). Most experts agree that the ancestors of land plants first evolved in a marine environment, then moved into a freshwater environment before finally moving onto land. In this way, the differences in osmotic pressures between saltwater and freshwater were overcome while the plant was still in the water.

The higher land plants are divided into two major groups: nonvascular and vascular. Most land plants are **vascular**, meaning that they have a tissue system of specialized cells for the movement of water and nutrients. The *nonvascular* plants, such as bryophytes (liverworts, hornworts, and mosses) and fungi, do not have these specialized cells and are typically small and usually live in low, moist areas.

The earliest land plants from the Middle to Late Ordovician were probably small and bryophyte-like in their overall organization (but not necessarily related to bryophytes). The evolution of vascular tissue in plants was an important step because it allowed for the transport of food and water.

Discoveries of probable vascular plant megafossils and characteristic spores indicate to many paleontologists that vascular plants evolved well before the Middle Silurian. Sheets of cuticle-like cells—that is, the cells that cover the surface of present-day land plants—and tetrahedral clusters that closely resemble the spore tetrahedrals of primitive land plants have been reported from Middle to Upper Ordovician rocks from western Libya and elsewhere.

The ancestor of terrestrial vascular plants was probably some type of green alga. Although no fossil record of the transition from green algae to terrestrial vascular plants has been found, comparison of their physiology reveals a strong link. Primitive *seedless vascular plants* (discussed later in this chapter), such as ferns, resemble green algae in their pigmentation, important metabolic enzymes, and type of reproductive cycle. Furthermore, green algae are one of the few plant groups to have made the transition from saltwater to freshwater.

The evolution of terrestrial vascular plants from an aquatic, probable green algal ancestry was accompanied by various modifications that let them occupy this new harsh environment. Besides the primary function of transporting

TABLE 21.3 Major Events in the Evolution of Land Plants. The Devonian Period was a time of rapid evolution for the land plants. Major events were the appearance of leaves, heterospory, secondary growth, and the emergence of seeds.

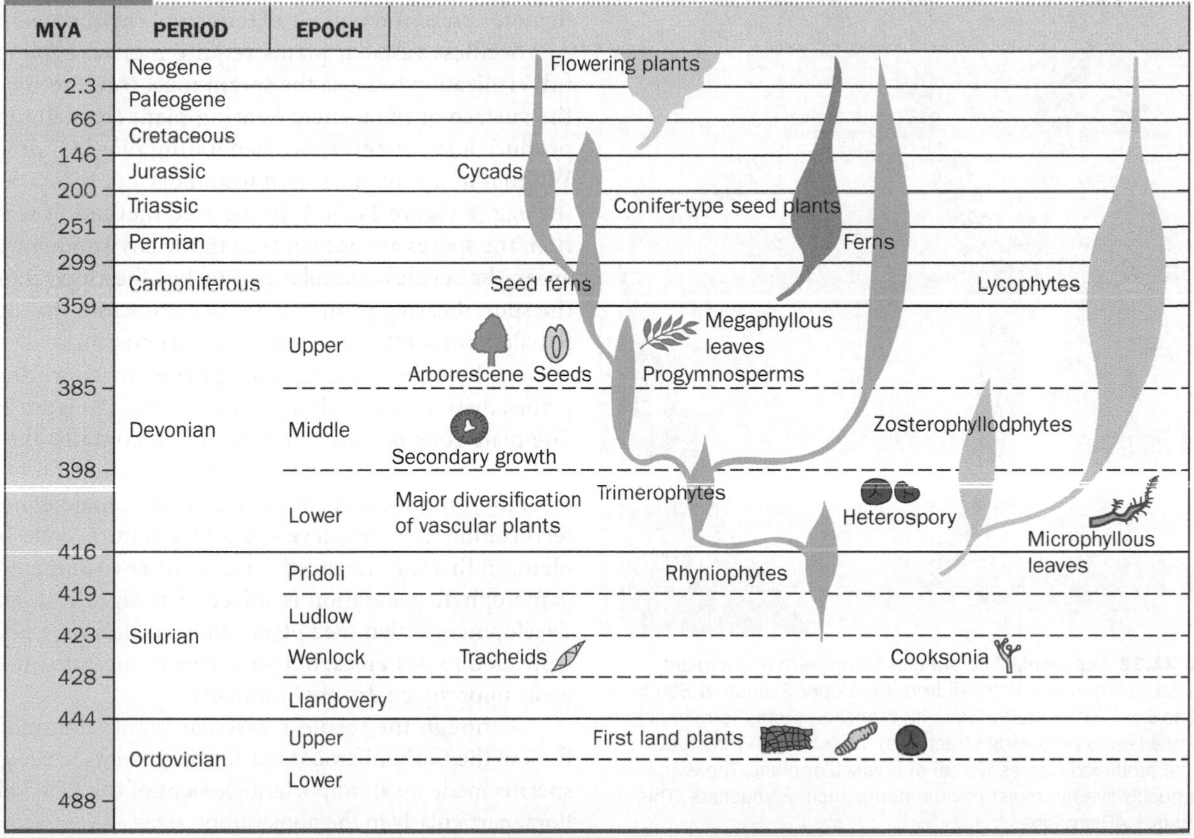

MYA	PERIOD	EPOCH	
	Neogene		Flowering plants
2.3	Paleogene		
66	Cretaceous		
146	Jurassic		Cycads
200	Triassic		Conifer-type seed plants
251	Permian		Ferns
299	Carboniferous		Seed ferns; Lycophytes
359	Devonian	Upper	Arborescence; Seeds; Megaphyllous leaves; Progymnosperms
385		Middle	Secondary growth; Zosterophyllodphytes
398		Lower	Major diversification of vascular plants; Trimerophytes; Heterospory; Microphyllous leaves
416	Silurian	Pridoli	Rhyniophytes
419		Ludlow	
423		Wenlock	Tracheids; Cooksonia
428		Llandovery	
444	Ordovician	Upper	First land plants
		Lower	
488			

water and nutrients throughout a plant, vascular tissue also provides some support for the plant body. Additional strength is derived from the organic compounds *lignin* and *cellulose*, found throughout a plant's walls.

The problem of desiccation was circumvented by the evolution of *cutin*, an organic compound found in the outer-wall layers of plants. Cutin also provides additional resistance to oxidation, the effects of UV light, and the entry of parasites.

Roots evolved in response to the need to collect water and nutrients from the soil and to help anchor the plant in the ground. The evolution of *leaves* from tiny outgrowths on the stem or from branch systems provided plants with an efficient light-gathering system for photosynthesis.

Silurian and Devonian Floras

The earliest known vascular land plants are small, Y-shaped stems assigned to the genus *Cooksonia* from the Middle Silurian of Wales and Ireland. Together with Upper Silurian and Lower Devonian species from Scotland, New York State, and the Czech Republic, these earliest plants were small, simple, leafless stalks with a spore-producing structure at the tip (▮ Figure 21.32); they are known as **seedless vascular plants** because they did not produce seeds. They also did not have a true root system. A *rhizome*, the underground part of the stem, transferred water from the soil to the plant and anchored the plant to the ground. The sedimentary rocks in which these plant fossils are found indicate that they lived in low, wet, marshy, freshwater environments.

▮ **Figure 21.32** *Cooksonia* The earliest known fertile land plant was *Cooksonia*, seen in this fossil from the Upper Silurian of South Wales. *Cooksonia* consisted of upright, branched stems terminating in sporangia (spore-producing structures). It also had a resistant cuticle and produced spores typical of a vascular plant. These plants probably lived in moist environments such as mudflats. This specimen is 1.49 cm long.

An interesting parallel can be seen between seedless vascular plants and amphibians. When they made the transition from water to land, both plants and animals had to overcome the problems that such a transition involved. Both groups, while successful, nevertheless required a source of water to reproduce. In the case of amphibians, their gelatinous egg had to remain moist, and the seedless vascular plants required water for the sperm to travel through to reach the egg.

From this simple beginning, the seedless vascular plants evolved many of the major structural features characteristic of modern plants, such as leaves, roots, and secondary growth. These features did not all evolve simultaneously, but rather at different times, a pattern known as *mosaic evolution*. This diversification and adaptive radiation took place during the Late Silurian and Early Devonian and resulted in a tremendous increase in diversity (▮ Figure 21.33). During the Devonian, the number of plant genera remained about the same, yet the composition of the flora changed. Whereas the Early Devonian landscape was dominated by relatively small, low-growing, bog-dwelling types of plants, the Late Devonian witnessed forests of large, tree-size plants up to 10 m tall.

In addition to the diverse seedless vascular plant flora of the Late Devonian, another significant floral event took place. The evolution of the seed at this time liberated land plants from their dependence on moist conditions and allowed them to spread throughout all terrestrial environments.

Seedless vascular plants require moisture for successful fertilization because the sperm must travel to the egg on the surface of the gamete-bearing plant (gametophyte) to produce a successful spore-generating plant (sporophyte). Without moisture, the sperm would dry out before reaching the egg (▮ Figure 21.34a). In the seed method of reproduction, the spores are not released to the environment, as they are in the seedless vascular plants, but they are retained on the spore-bearing plant, where they grow into the male and female forms of the gamete-bearing generation.

In the case of the **gymnosperms**, or flowerless seed plants, these are the male and female cones (▮ Figure 21.34b). The male cone produces pollen, which contains the sperm and has a waxy coating to prevent desiccation, and the egg, or embryonic seed, is contained in the female cone. After fertilization, the seed develops into a mature, cone-bearing plant. In this way, the need for a moist environment for the gametophyte generation is solved. The significance of this development is that seed plants, like reptiles, were no longer restricted to wet areas, but were free to migrate into previously unoccupied dry environments.

Although the seedless vascular plants dominated the flora of the Carboniferous coal-forming swamps, the gymnosperms made up an important element of the Late Paleozoic flora, particularly in the nonswampy areas.

Walter Myers/Photo Researchers, Inc.

Figure 21.33 Early Devonian Landscape Reconstruction of an Early Devonian landscape showing *Asteroxylon*, a common seedless vascular plant with curling branches that grew up to 4 m tall, and various low-growing seedless vascular plants.

Late Carboniferous and Permian Floras

As discussed earlier, the rocks of the Pennsylvanian Period (Late Carboniferous) are the major source of the world's coal. Coal results from the alteration of plant remains accumulating in low, swampy areas. The geologic and geographic conditions of the Pennsylvanian were ideal for the growth of seedless vascular plants, and consequently, these coal swamps had a very diverse flora (Figure 21.35).

ConnectionLink

To review the formation of coal and its distribution, see Chapter 7 and Chapter 20.

It is evident from the fossil record that whereas the Early Carboniferous (Mississippian Period) flora was similar to its Late Devonian counterpart, a great deal of evolutionary experimentation was taking place that would lead to the highly successful Late Paleozoic flora of the coal swamps and adjacent habitats. Among the seedless vascular plants, the *lycopsids* and *sphenopsids* were the most important coal-forming groups of the Pennsylvanian Period.

The lycopsids were present during the Devonian, chiefly as small plants, but by the Pennsylvanian, they were the dominant element of the coal swamps, achieving heights up to 30 m in such genera as *Lepidodendron* and *Sigillaria*. The Pennsylvanian lycopsid trees are interesting because they lacked branches except at their top, which had elongated leaves similar to the individual palm leaf of today. As the trees grew, the leaves were replaced from the top, leaving prominent and characteristic rows or spirals of scars on the trunk. Today, the lycopsids are represented by small, temperate-forest ground pines.

The sphenopsids, the other important coal-forming plant group, are characterized by being jointed and having horizontal underground stem-bearing roots. Many of these plants, such as *Calamites*, average 5 to 6 m tall. Small seedless vascular plants and seed ferns formed a thick undergrowth or ground cover beneath these treelike plants. Living sphenopsids include the horsetail (*Equisetum*) or scouring rushes (Figure 21.36).

Figure 21.34 Generalized Life History of a Seedless Vascular Plant and Gymnosperm Plant

Critical Thinking Question What is it about the life cycle of the gymnosperms that gave them an advantage over the seedless vascular plants in terms of the climatic changes that took place during the Permian Period, when the continents came together to form the supercontinent Pangaea, and the subsequent Triassic Period?

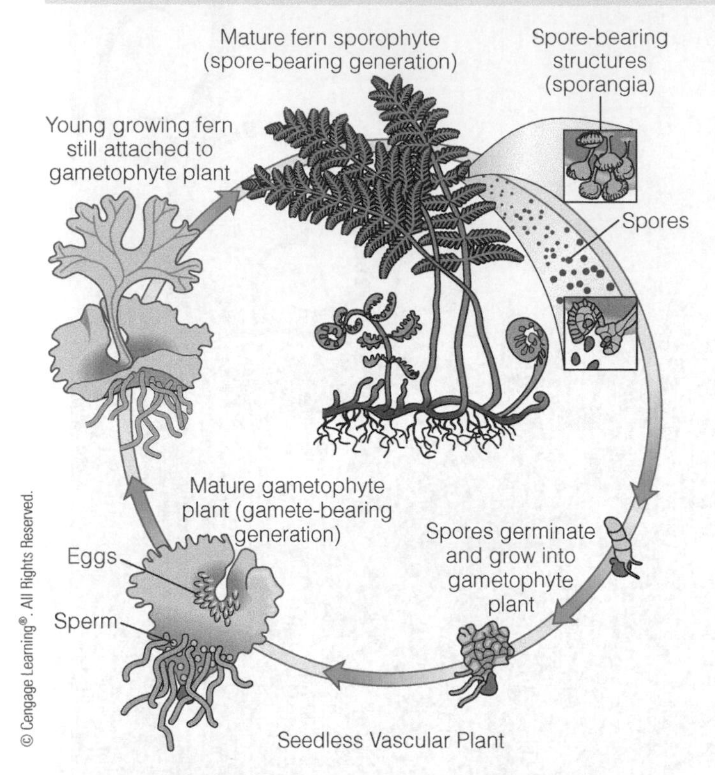

a Generalized life history of a seedless vascular plant. The mature sporophyte plant produces spores, which, upon germination, grow into small gametophyte plants that produce sperm and eggs. The fertilized eggs grow into the spore-producing mature plant, and the sporophyte–gametophyte life cycle begins again.

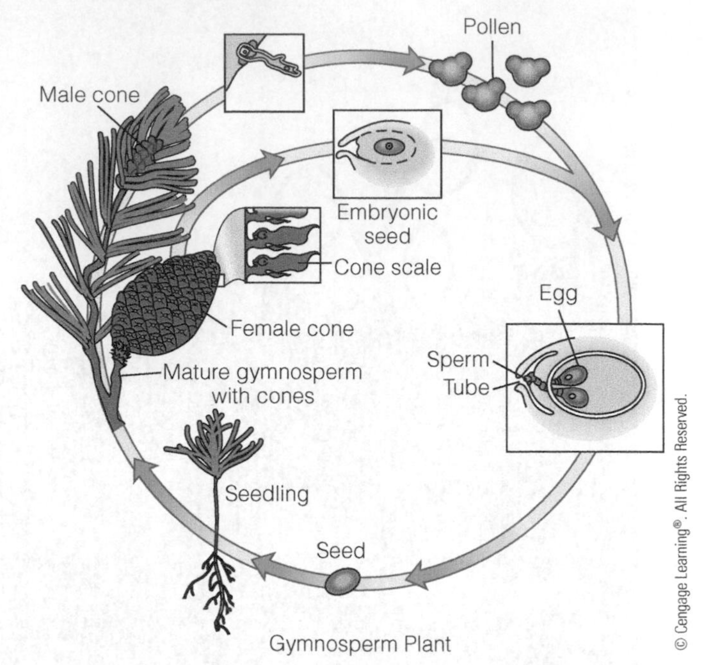

b Generalized life history of a gymnosperm plant. The mature plant bears both male cones that produce sperm-bearing pollen grains and female cones that contain embryonic seeds. Pollen grains are transported to the female cones by the wind. Fertilization occurs when the sperm moves through a moist tube growing from the pollen grain and unites with the embryonic seed, which then grows into a cone-bearing mature plant.

Figure 21.35 Pennsylvanian Coal Swamp Reconstruction of a Pennsylvanian coal swamp with its characteristic vegetation of seedless vascular plants.

Figure 21.36 Horsetail Rush (*Equisetum*) Living sphenopsids include the horsetail (*Equisetum*) or scouring rushes.

Figure 21.37 Late Carboniferous Cordaite Forest Cordaites were a group of gymnosperm trees that grew up to 50 m tall.

Not all plants were restricted to the coal-forming swamps. Among those plants that occupied higher and drier ground were some of the Cordaites, a group of tall gymnosperm trees that grew up to 50 m high and probably formed vast forests (Figure 21.37). Another important nonswamp dweller was *Glossopteris*, the famous plant so abundant in Gondwana (see Figure 2.1), whose distribution is cited as critical evidence that the continents have moved through time.

The floras that were abundant during the Pennsylvanian persisted into the Permian, but because of climatic and geologic changes resulting from tectonic events (see Chapter 20), they declined in abundance and importance. By the end of the Permian, the Cordaites became extinct, and the lycopsids and sphenopsids were reduced to mostly small, creeping forms. Gymnosperms with lifestyles more suited to the warmer and drier Permian climates diversified and came to dominate the Permian, Triassic, and Jurassic landscapes.

Key Concepts Review

- Table 21.4 summarizes the major evolutionary and geologic events of the Paleozoic Era and shows their relationships to one another.
- Multicelled organisms presumably had a long Precambrian history during which they lacked hard parts.
- Invertebrates with hard parts "suddenly" appeared during the Early Cambrian in what is called the "Cambrian explosion." Skeletons provided such advantages as protection against predators and support for muscles, enabling organisms to grow large and increase locomotor efficiency. Hard parts probably evolved as a result of various geologic and biologic factors rather than a single cause.
- Marine organisms are classified as plankton if they are floaters, nekton if they swim, and benthos if they live on or in the seafloor.
- Marine organisms are divided into four basic feeding groups: suspension feeders, which consume microscopic plants and animals as well as dissolved nutrients from water; herbivores, which are plant eaters; carnivore-scavengers, which are meat eaters; and sediment-deposit feeders, which ingest sediment and extract nutrients from it.
- The marine ecosystem consists of various trophic levels of food production and consumption. At the base are primary producers, on which all other organisms are dependent. Feeding on the primary producers are the primary consumers, which in turn are fed on by higher levels of consumers. The decomposers are bacteria that break down the complex organic compounds of dead organisms and recycle them within the ecosystem.

TABLE 21.4 Major Evolutionary and Geologic Events of the Paleozoic Era

Age (millions of years ago)	Geologic Period	Invertebrates	Vertebrates	Plants	Major Geologic Events
251	Permian	Largest mass extinction to affect the invertebrates	Acanthodians, placoderms, and pelycosaurs become extinct Therapsids and pelycosaurs are the most abundant reptiles	Gymnosperms diverse and abundant	Formation of Pangaea Alleghenian orogeny Hercynian orogeny
299	Carboniferous: Pennsylvanian	Fusulinids diversify	Amphibians abundant and diverse	Coal swamps with flora of seedless vascular plants and gymnosperms	Coal-forming swamps common Formation of Ancestral Rockies Continental glaciation in Gondwana
318	Carboniferous: Mississippian	Crinoids, lacy bryozoans, blastoids become abundant Renewed adaptive radiation following extinctions of many reef-builders	Reptiles evolve	Gymnosperms appear (may have evolved during Late Devonian)	Ouachita orogeny Widespread deposition of black shale
359	Devonian	Extinctions of many reef-building invertebrates near end of Devonian Reef building continues Eurypterids abundant	Amphibians evolve All major groups of fish present—Age of Fish	First seeds evolve Seedless vascular plants diversify	Widespread deposition of black shale Acadian orogeny Antler orogeny
416	Silurian	Major reef building Diversity of invertebrates remains high	Ostracoderms common Acanthodians, the first jawed fish evolve	Early land plants—seedless vascular plants	Caledonian orogeny Extensive barrier reefs and evaporites
444	Ordovician	Extinctions of a variety of marine invertebrates near end of Ordovician Major adaptive radiation of all invertebrate groups Suspension-feeders dominant	Ostracoderms diversify	Plants move to land?	Continental glaciation in Gondwana Taconic orogeny
488	Cambrian	Many trilobites become extinct near end of Cambrian Trilobites, brachiopods, and archaeocyathids are most abundant	Earliest vertebrates—jawless fish called ostracoderms		First Phanerozoic transgression (Sauk) onto North American craton
542					

- The Cambrian invertebrate community was dominated by three major groups: the trilobites, brachiopods, and archaeocyathids. Little specialization existed among the invertebrates, and most phyla were represented by only a few species.
- The Middle Cambrian Burgess Shale contains one of the finest examples of a well-preserved, soft-bodied biota in the world and provides us with an important glimpse of not only rarely preserved organisms but also the soft-part anatomy of many extinct groups.
- The Ordovician marine invertebrate community marked the beginning of the dominance of the shelly fauna and the start of large-scale reef building. The end of the Ordovician Period was a time of major extinctions of many invertebrate phyla.
- The Silurian and Devonian periods were times of diverse faunas dominated by reef-building animals, whereas the Carboniferous and Permian periods saw a great decline in invertebrate diversity.
- Chordates are characterized by a notochord, dorsal hollow nerve cord, and pharyngeal slits. The earliest chordates were soft-bodied organisms that were rarely fossilized. Vertebrates are a subphylum of the chordates.
- Fish are the earliest known vertebrates, with their first fossil occurrence in Cambrian rocks. They have had a long and varied history, including jawless and jawed armored forms (ostracoderms and placoderms), cartilaginous forms, and bony forms. It is from the lobe-finned bony fish group that amphibians evolved.
- The link between crossopterygian lobe-finned fish and the earliest amphibians is convincing and includes a close similarity of bone and tooth structures. New fossil discoveries, however, show that the transition between the two groups is more complicated than originally hypothesized and includes several intermediate forms.
- Amphibians evolved during the Late Devonian, with labyrinthodont amphibians becoming the dominant terrestrial vertebrate animals during the Carboniferous.
- The Late Mississippian marks the earliest fossil record of reptiles. The evolution of an amniote egg was the critical factor that allowed reptiles to completely colonize the land.
- Pelycosaurs were the dominant reptiles during the Early Permian, whereas therapsids dominated the landscape for the rest of the Permian Period.
- In making the transition from water to land, plants had to overcome the same basic problems as animals—namely, desiccation, reproduction, and gravity.
- The earliest fossil record of land plants is from Middle to Upper Ordovician rocks. These plants were probably small and bryophyte-like in their overall organization.
- The evolution of vascular tissue was an important event in plant evolution as it allowed nutrients and water to be transported throughout the plant and provided the plant with additional support.
- The ancestor of terrestrial vascular plants was probably some type of green algae based on such similarities as pigmentation, metabolic enzymes, and the same type of reproductive cycle.
- The earliest seedless vascular plants were small, leafless stalks, with spore-producing structures on their tips. From this simple beginning, plants evolved many of the major structural features characteristic of today's plants.
- By the end of the Devonian Period, forests with tree-size plants up to 10 m tall had evolved. The Late Devonian also witnessed the evolution of gymnosperms (flowerless seed plants) whose reproductive style freed them from having to stay near water.
- The Carboniferous Period was a time of vast coal swamps, where conditions were ideal for the seedless vascular plants. With the onset of more arid conditions during the Permian, the gymnosperms became the dominant element of the world's forests.
- A major extinction occurred at the end of the Paleozoic Era, affecting the invertebrates as well as the vertebrates. Its cause is still the subject of debate.

Important Terms

amniote egg (p. 561)
benthos (p. 542)
bony fish (p. 557)
carnivore-scavenger (p. 543)
cartilaginous fish (p. 557)
chordate (p. 553)
crossopterygian (p. 558)
gymnosperm (p. 566)
herbivore (p. 543)
labyrinthodont (p. 561)
nekton (p. 542)
ostracoderm (p. 555)
pelycosaur (p. 562)
placoderm (p. 557)
plankton (p. 542)
primary producer (p. 543)
protorothyrids (p. 562)
sediment-deposit feeder (p. 543)
seedless vascular plant (p. 566)
suspension feeder (p. 542)
therapsid (p. 563)
vascular (p. 565)
vertebrate (p. 553)

Review Questions

1. Which plant group first successfully invaded land?
   a. ___ seedless vascular.
   b. ___ gymnosperms.
   c. ___ naked seed bearing.
   d. ___ angiosperms.
   e. ___ flowering.
2. Which evolutionary innovation allowed reptiles to colonize all of the land?
   a. ___ tear ducts.
   b. ___ additional bones in the jaw.
   c. ___ the middle-ear bone.
   d. ___ an egg that contained food and waste sacs and an embryo surrounded by a fluid-filled sac.
   e. ___ limbs and a backbone capable of supporting the animals on land.
3. The discovery of *Tiktaalik roseae* is significant because it is
   a. ___ the ancestor of modern reptiles.
   b. ___ an intermediary between lobe-finned fish and amphibians.
   c. ___ the first vascular land plant.
   d. ___ a "missing link" between amphibians and reptiles.
   e. ___ the oldest known fish.
4. Which reptile group gave rise to the mammals?
   a. ___ labyrinthodonts.
   b. ___ acanthodians.
   c. ___ pelycosaurs.
   d. ___ protothyrids.
   e. ___ therapsids.
5. An organism must possess which of the following during at least part of its life cycle to be classified a chordate?
   a. ___ notochord, dorsal solid nerve cord, lungs.
   b. ___ vertebrae, dorsal hollow nerve cord, pharyngeal slits.
   c. ___ vertebrae, dorsal hollow nerve cord, lungs.
   d. ___ notochord, ventral solid nerve cord, lungs.
   e. ___ notochord, dorsal hollow nerve cord, pharyngeal slits.
6. Discuss how changing geologic conditions affected the evolution of invertebrate life during the Paleozoic Era.
7. Discuss how the incompleteness of the fossil record may play a role in such theories as the Cambrian explosion of life.
8. Based on what you know about Pennsylvanian geology (Chapter 20), why was this time period so advantageous to the evolution of both plants and amphibians?
9. If the Cambrian explosion was partly the result of filling unoccupied niches, why don't we see such rapid evolution following mass extinctions such as those that occurred at the end of the Permian and Cretaceous periods?

10. **Creative Thinking Visual Question:** Until the recently discovered 395-million-year-old fossilized footprints in the Holy Cross Mountains in Poland, these fossilized footprints, which are more than 385 million years old, and are part of a tetrapod trackway at Valentia Island, Ireland, represented evidence of some of the oldest land-dwelling tetrapods. What evidence in this photo (Figure 1) indicates that the sediments were deposited in a wet environment? How can you use the footprints to establish that the animal making them was a tetrapod? And lastly, how do you think scientists estimated the size of the tetrapod that made this trackway?

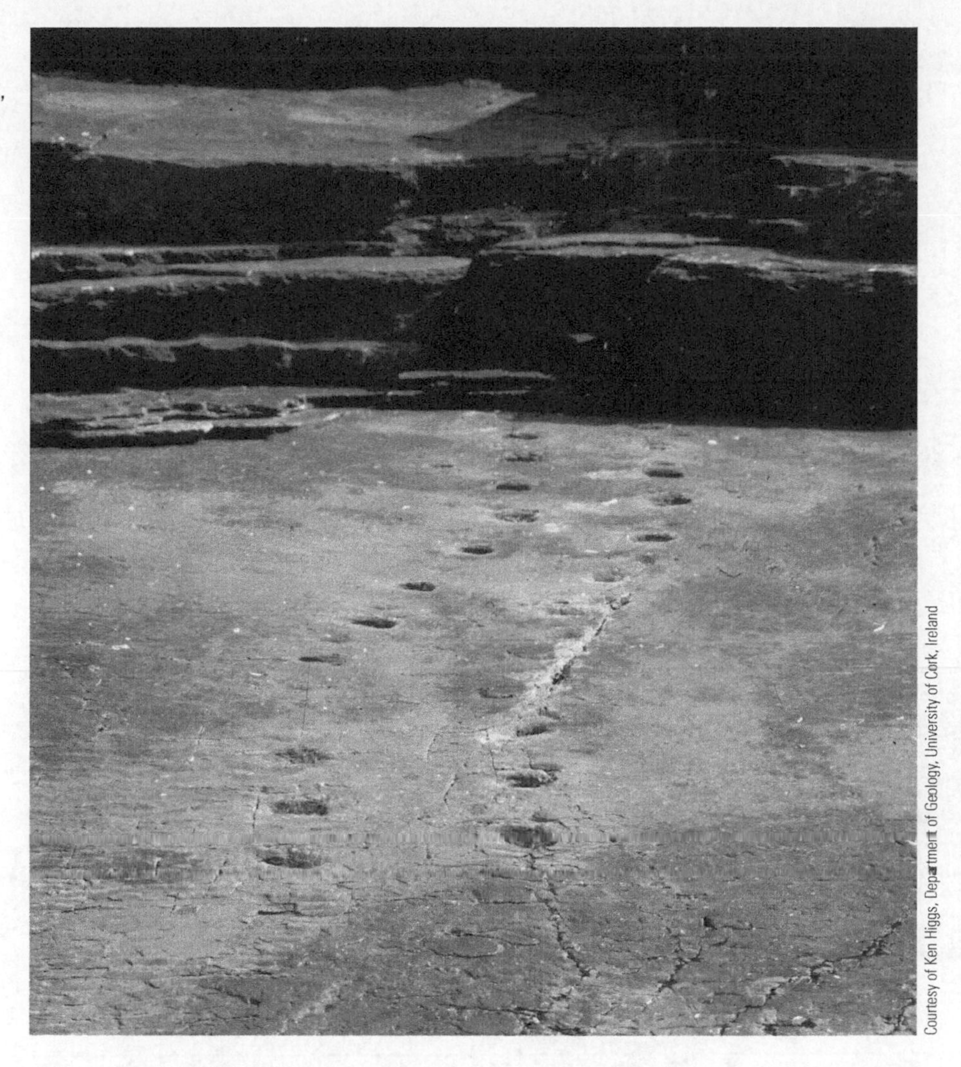

Courtesy of Ken Higgs, Department of Geology, University of Cork, Ireland

Figure 1 Devonian Fossilized Tetrapod Footprints These fossilized footprints, which are more than 385 million years old, are part of a tetrapod trackway at Valentia Island, Ireland.

Global GeoScience Watch Search for "mass extinctions" in the GREENR database, then click on "VIEW ALL" next to the "News" heading. There are a number of articles dated in the past several years warning of the effects of ocean acidification and its role in decreased biotic diversity. How does ocean acidification affect reefs and other marine ecosystems? How can increasing ocean acidification be linked to the Permian extinctions and specifically eruptions from the Siberian Traps?

Aerial view of the Sierra Nevada, showing Yosemite Valley and Half Dome, California. The rocks in this image are part of the Sierra Nevada batholith, a huge mass of granite and related rocks made up of many intrusive bodies. The massive plutons comprising the Sierra Nevada were emplaced during the Late Jurassic Nevadan orogeny and later exposed by uplift along faults on the east side of the present-day mountain range. Yosemite Valley and Half Dome were formed by valley glaciers during the Pleistocene.

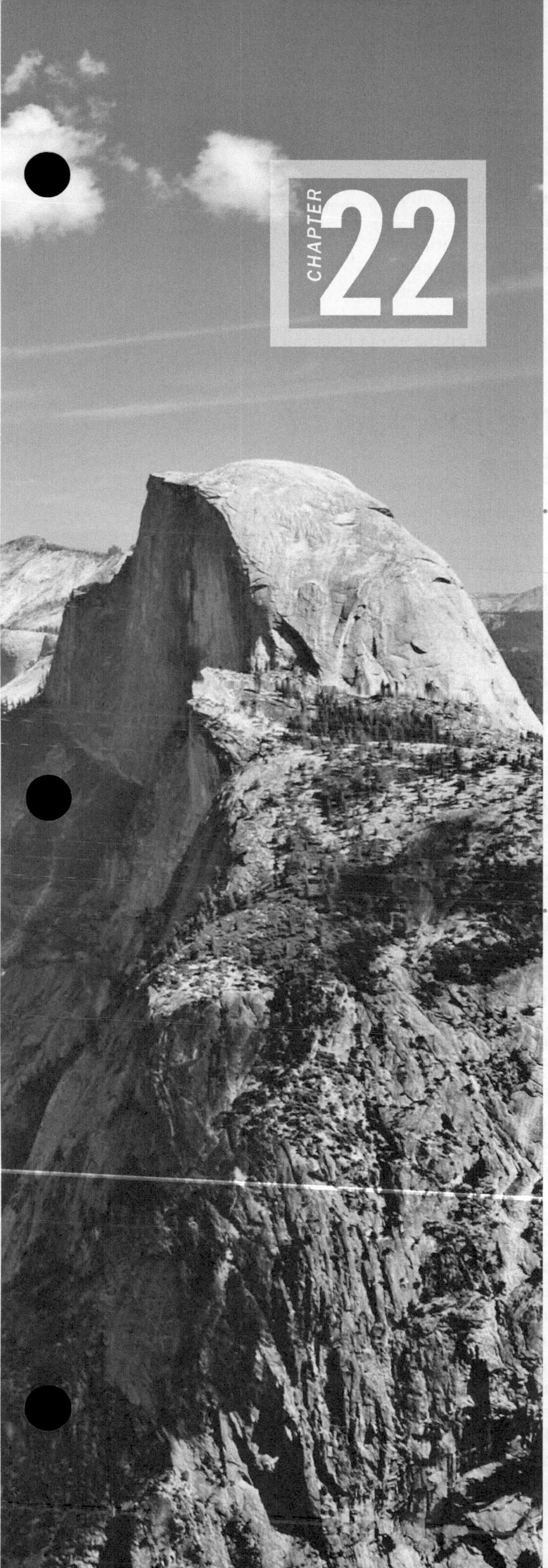

CHAPTER 22 Mesozoic Earth and Life History

OUTLINE

HAVE YOU EVER WONDERED?

- What caused the breakup of Pangaea and its effects on the geography, oceanic circulation patterns, climate, and distribution of natural resources during the Mesozoic Era?
- How land-plant communities changed during the Mesozoic Era?
- About the nature of dinosaurs and their relatives such as flying reptiles and marine reptiles?
- When birds first appeared in the fossil record?
- How the earliest mammals compare with those living today?
- What happened at the end of the Mesozoic to cause widespread extinctions?

22.1 Introduction

Approximately 150 to 210 million years after the emplacement of massive plutons created the Sierra Nevada (Nevadan orogeny), gold was discovered at Sutter's Mill on the South Fork of the American River at Coloma, California. On January 24, 1848, James Marshall, a carpenter building a sawmill for John Sutter, found bits of the glittering metal in the mill's tailrace. Soon settlements throughout the state were abandoned as word of the chance for instant riches spread throughout California.

Within a year after the news of the gold discovery reached the East Coast, the Sutter's Mill area was swarming with more than 80,000 prospectors, all hoping to make their fortune. At least 250,000 gold seekers prospected the Sutter's Mill area, and although most were Americans, prospectors came from all over the world, even as far away as China.

Although many prospectors searched for the source of the gold, or the mother lode, the gold that they recovered was mostly in the form of placer deposits (deposits of sand and gravel containing gold particles large enough to be recovered by panning). Unfortunately, most thought the gold was simply waiting to be taken and did not realize that prospecting was hard work. Life in the mining camps was extremely difficult and expensive, and the shop owners and traders frequently made more money than the prospectors. In reality, only a few prospectors ever hit it big or were even moderately successful. Most barely made a living wage working their claims before they finally abandoned their dream and went home. Nevertheless, during the five years from 1848 to 1853 that constituted the Gold Rush proper, more than $200 million in gold was extracted.

The Mesozoic Era (251 to 66 million years ago) was an important time in Earth's history. Just as the formation of Pangaea influenced geologic and biologic events during the Paleozoic Era, the breakup of this supercontinent had profound geologic and biologic effects during the Mesozoic. As a result of plate movements during the Mesozoic, the Atlantic Ocean basin formed, as well as the start of the Rocky Mountains. The accumulation of vast salt deposits eventually formed salt domes adjacent to which oil and natural gas were trapped, and the emplacement of huge batholiths account for the origin of various mineral resources, including the gold that fueled the California gold rush of the mid-1800s.

The Mesozoic Era is probably best known, however, as the time that dinosaurs roamed the land, flying reptiles (pterosaurs) ruled the skies, and various types of marine reptiles (ichthyosaurs, plesiosaurs, and mosasaurs) proliferated in the seas. Also present were turtles, lizards, snakes, and giant crocodiles. Birds also first appeared during the Jurassic Period, having evolved from small carnivorous dinosaurs. And mammals evolved from mammal-like reptiles during the Triassic. In fact, mammals were contemporaries of the dinosaurs, although they were not nearly as diverse and all were small.

A main emphasis of this book has been the systems approach to Earth and life history. Remember that the formation and movement of continents affects global climatic and oceanic regimes, as well as the distribution of Earth's biota. Populations become isolated, or are brought into contact with other populations, leading to evolutionary changes in the fauna and flora. So great was the effect of the breakup of Pangaea, that it forms the central theme of this chapter.

Connection Link

To review how plate movements affect the distribution of life, see Chapter 2.

22.2 The Breakup of Pangaea

Because of the magnetic anomalies preserved in the oceanic crust (see Figure 2.16), geologists have a very good record of the history of Pangaea's breakup, and the direction of movement of the various continents during the Mesozoic and Cenozoic eras. Pangaea's breakup began with rifting between Laurasia and Gondwana during the Triassic (Figure 22.1a). By the end of the Triassic, the newly formed and expanding Atlantic Ocean separated North America from Africa, and sometime during the Late Triassic and Early Jurassic, North America rifted from South America.

Separation of the continents allowed water from the Tethys Sea to flow into the expanding central Atlantic Ocean, whereas Pacific Ocean waters flowed into the newly formed Gulf of Mexico, which then was little more than a restricted bay (Figure 22.2). During that time, these areas were located in the low tropical latitudes, where high temperatures and high rates of evaporation were ideal for the formation of thick evaporite deposits.

The initial breakup of Gondwana took place during the Late Triassic and Jurassic periods. Antarctica and Australia, which remained sutured together, began separating from South America and Africa, whereas India began rifting from the Gondwana continent and moved northward.

South America and Africa began rifting apart during the Jurassic (Figure 22.1b) and the subsequent separation of these two continents formed a narrow basin where thick evaporite deposits accumulated from the evaporation of southern ocean waters (Figure 22.2). During this time, the eastern end of the Tethys Sea began closing as a result of the clockwise rotation of Laurasia and the northward movement of Africa. This narrow Late Jurassic and Cretaceous seaway between Africa and Europe was the forerunner of the present Mediterranean Sea.

By the end of the Cretaceous, Australia and Antarctica had detached from each other, and India had moved into the low southern latitudes and was nearly to the equator. South America and Africa were now widely separated, and Greenland was essentially an independent landmass with only a shallow sea between it and North America and Europe (Figure 22.1c).

A global rise in sea level during the Cretaceous resulted in worldwide transgressions onto the continents. Higher heat

Figure 22.1 Mesozoic Paleogeography Paleogeography of the world during the **(a)** Triassic Period, **(b)** Jurassic Period, and **(c)** Late Cretaceous Period.

Critical Thinking Question Why would increased seafloor spreading, resulting in higher heat flow and an increase and rapid expansion in oceanic ridges during the Cretaceous, result in worldwide transgression onto the continents?

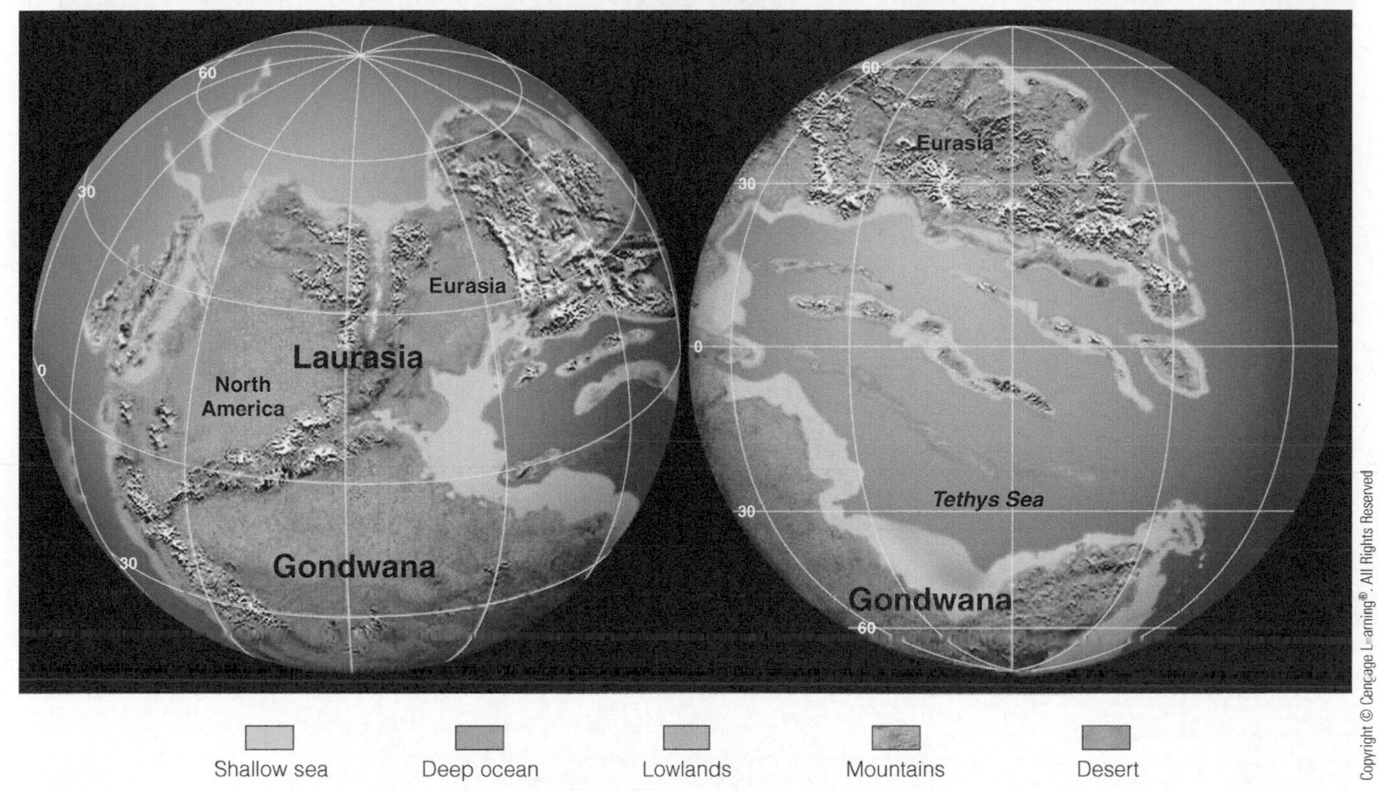

a Triassic Period.

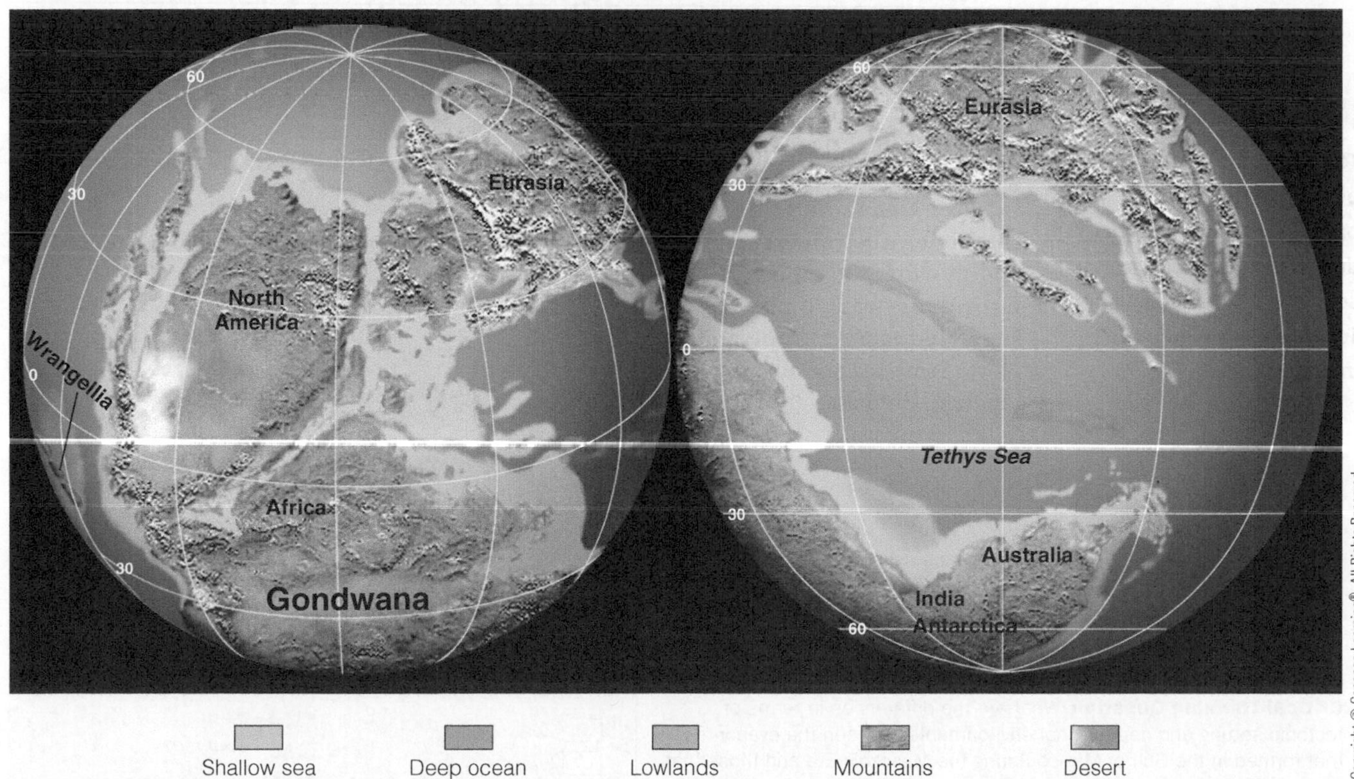

b Jurassic Period.

Figure 22.1 *(Continued)*

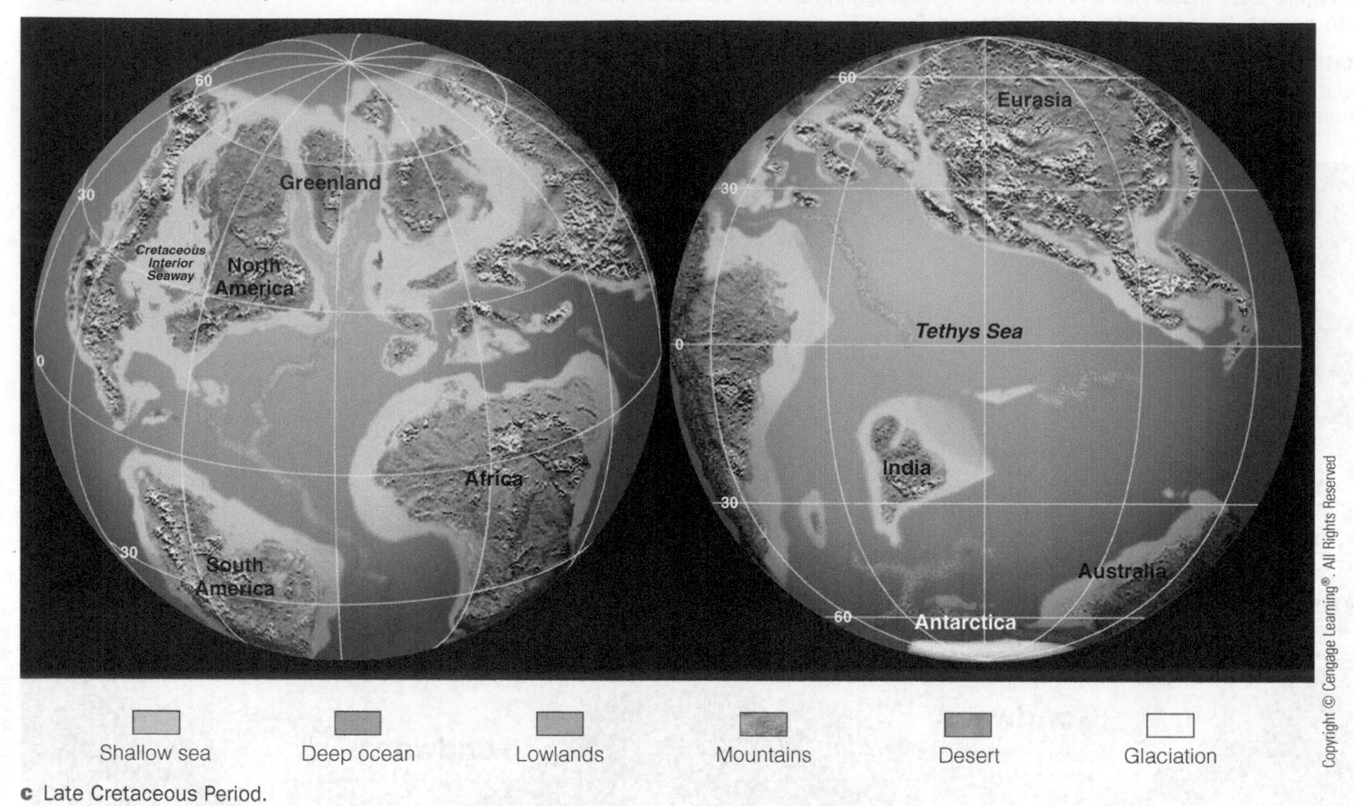

c Late Cretaceous Period.

flow and rapid expansion of oceanic ridges were responsible for these transgressions. By the Middle Cretaceous, sea level was probably as high as at any time since the Ordovician, and about one-third of the present land area was inundated by epeiric seas (Figure 22.1c).

ConnectionLink

To learn more about plate tectonics and the paleogeography of Earth during the Cenozoic, see section 23.2 Cenozoic Plate Tectonics and Figure 23.2 in Chapter 23.

The final stage in Pangaea's breakup occurred during the Cenozoic. During this time, Australia continued moving northward, and Greenland was completely separated from Europe and North America and formed a separate landmass.

The Effects of the Breakup of Pangaea on Global Climates and Ocean Circulation Patterns

By the end of the Permian Period, Pangaea extended from pole to pole, covered about one-fourth of Earth's surface, and was surrounded by Panthalassa, a global ocean that

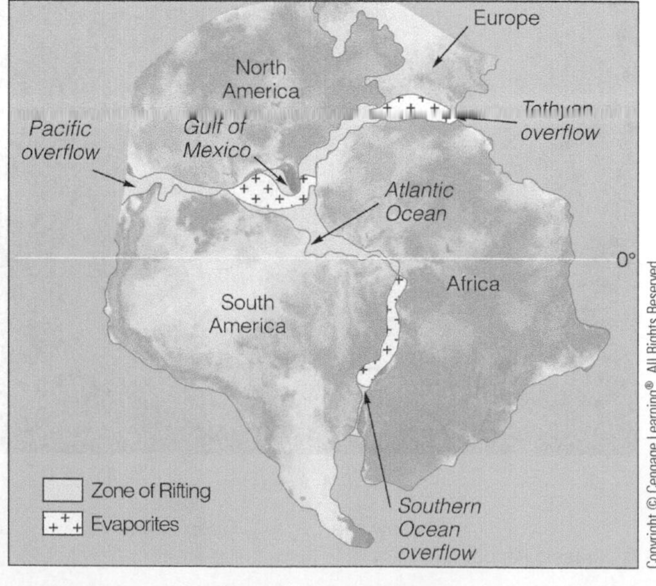

Figure 22.2 Gulf of Mexico Evaporites Evaporites accumulated in shallow basins as Pangaea broke apart during the Early Mesozoic. Water from the Tethys Sea flowed into the central Atlantic Ocean, and water from the Pacific Ocean flowed into the newly formed Gulf of Mexico. Marine water from the south flowed into the southern Atlantic Ocean.

Critical Thinking Question What are the differences in terms of tectonic setting and depositional environments between the evaporites that formed in the Gulf of Mexico during the Mesozoic Era and those that formed in North America during the Paleozoic Era?

ConnectionLink

To review the formation of Pangaea, see section 20.3 Paleozoic Paleogeography in Chapter 20.

encompassed approximately 300 degrees of longitude (see Figure 20.4b). Such a configuration exerted tremendous influence on the world's climate and resulted in generally arid conditions over large parts of Pangaea's interior.

The world's climates result from the complex interaction between wind and ocean currents and the location and topography of the continents. In general, dry climates occur on large landmasses in areas remote from sources of moisture and where barriers to moist air exist, such as mountain ranges. Wet climates occur near large bodies of water or where winds can carry moist air over land.

Past climatic conditions can be inferred from the distribution of climate-sensitive deposits. Evaporite deposits result when evaporation exceeds precipitation. Although sand dunes and red beds may form locally in humid regions, they are characteristic of arid regions. Coal forms in both warm and cool humid climates. Vegetation that is eventually converted into coal requires at least a good seasonal water supply; thus, coal deposits are indicative of humid conditions.

Widespread Triassic evaporites, red beds, and desert dunes in the low and middle latitudes of North and South America, Europe, and Africa indicate dry climates in those regions, whereas coal deposits are found mainly in the high latitudes, indicating humid conditions (Figure 22.1a). These high-latitude coals are analogous to today's Scottish peat bogs or Canadian muskeg. The lands bordering the Tethys Sea were probably dominated by seasonal monsoon rains resulting from the warm, moist winds and warm oceanic currents impinging against the east-facing coast of Pangaea.

The temperature gradient between the tropics and the poles also affects oceanic and atmospheric circulation. The greater the temperature difference between the tropics and the poles, the steeper the temperature gradient is, and thus, the faster the circulation of the oceans and atmosphere. Oceans absorb about 90% of the solar radiation they receive, whereas continents absorb only about 50%, even less if they are snow covered. The rest of the solar radiation is reflected back into space. Areas dominated by seas are warmer than those dominated by continents. By knowing the distribution of continents and ocean basins, geologists can generally estimate the average annual temperature for any region on Earth, as well as determine a temperature gradient.

The breakup of Pangaea during the Late Triassic caused the global temperature gradient to increase because the Northern Hemisphere continents moved farther northward, displacing higher-latitude ocean waters. Due to the steeper global temperature gradient, produced by a decrease in temperature in the high latitudes and the changing positions of the continents, oceanic and atmospheric circulation patterns greatly accelerated during the Mesozoic (Figure 22.3).

Although the temperature gradient and seasonality on land were increasing during the Jurassic and Cretaceous,

Figure 22.3 Mesozoic Oceanic Circulation Patterns Oceanic circulation evolved from **(a)** a simple pattern in a single ocean (Panthalassa) with a single continent (Pangaea) to **(b)** a more complex pattern in the newly formed oceans of the Cretaceous Period.

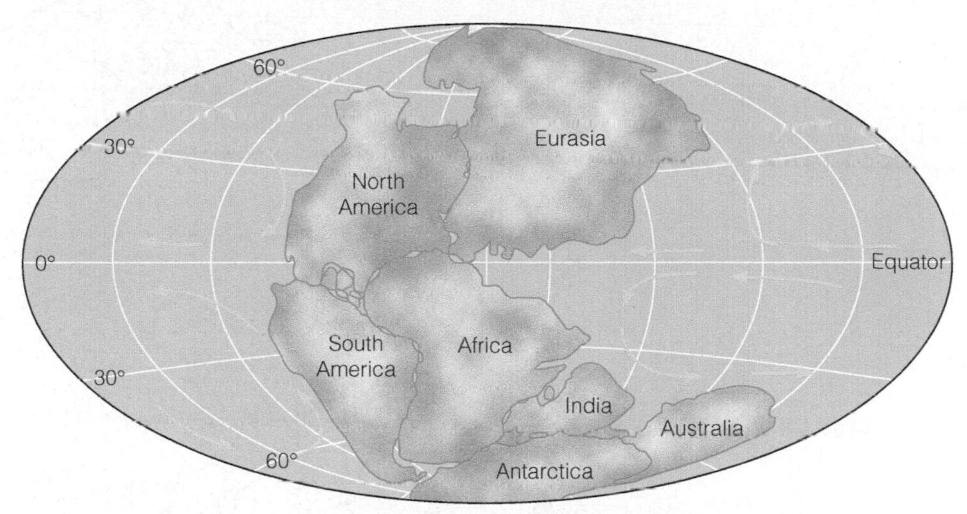

a Triassic Period.

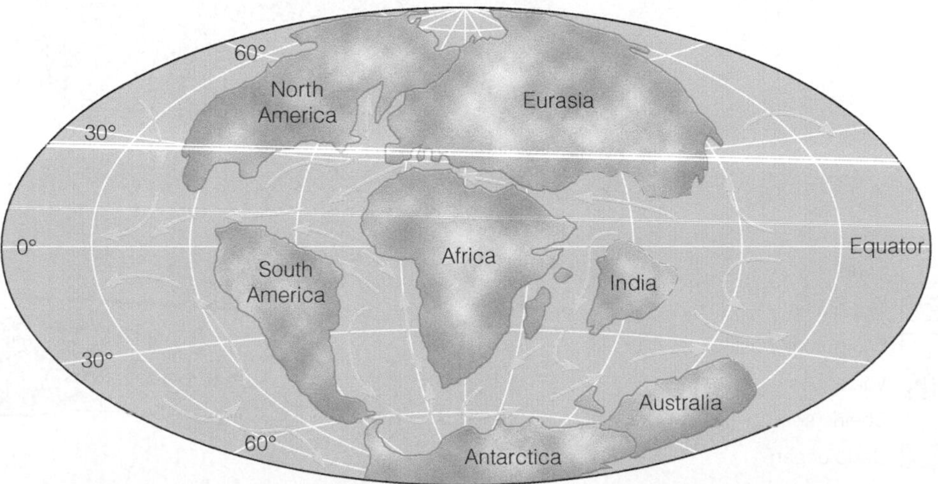

b Cretaceous Period.

the middle- and higher-latitude oceans were still quite warm because warm waters from the Tethys Sea were circulating to the higher latitudes. The result was a relatively equable worldwide climate through the end of the Cretaceous.

22.3 Mesozoic History of North America

The beginning of the Mesozoic Era was essentially the same in terms of tectonism and sedimentation as the preceding Permian Period in North America (see Figure 20.21). Terrestrial sedimentation continued over much of the craton, and block faulting and igneous activity began in the Appalachian region as North America and Africa began separating (▮ Figure 22.4). The newly forming Gulf of Mexico experienced extensive evaporite deposition during the Late Triassic and Jurassic as North America separated from South America (Figures 22.2 and ▮ 22.5).

A global rise in sea level during the Cretaceous resulted in worldwide transgressions onto the continents such that marine deposition was continuous over much of the North American Cordilleran (▮ Figure 22.6).

A volcanic island arc system that formed off the western edge of the craton during the Permian was sutured to North America sometime later during the Permian or Triassic. This event is referred to as the **Sonoma orogeny** and will be discussed later in the chapter. During the Jurassic, the entire Cordilleran area was involved in a series of major mountain-building

▮ **Figure 22.4** Triassic Paleogeography of North America

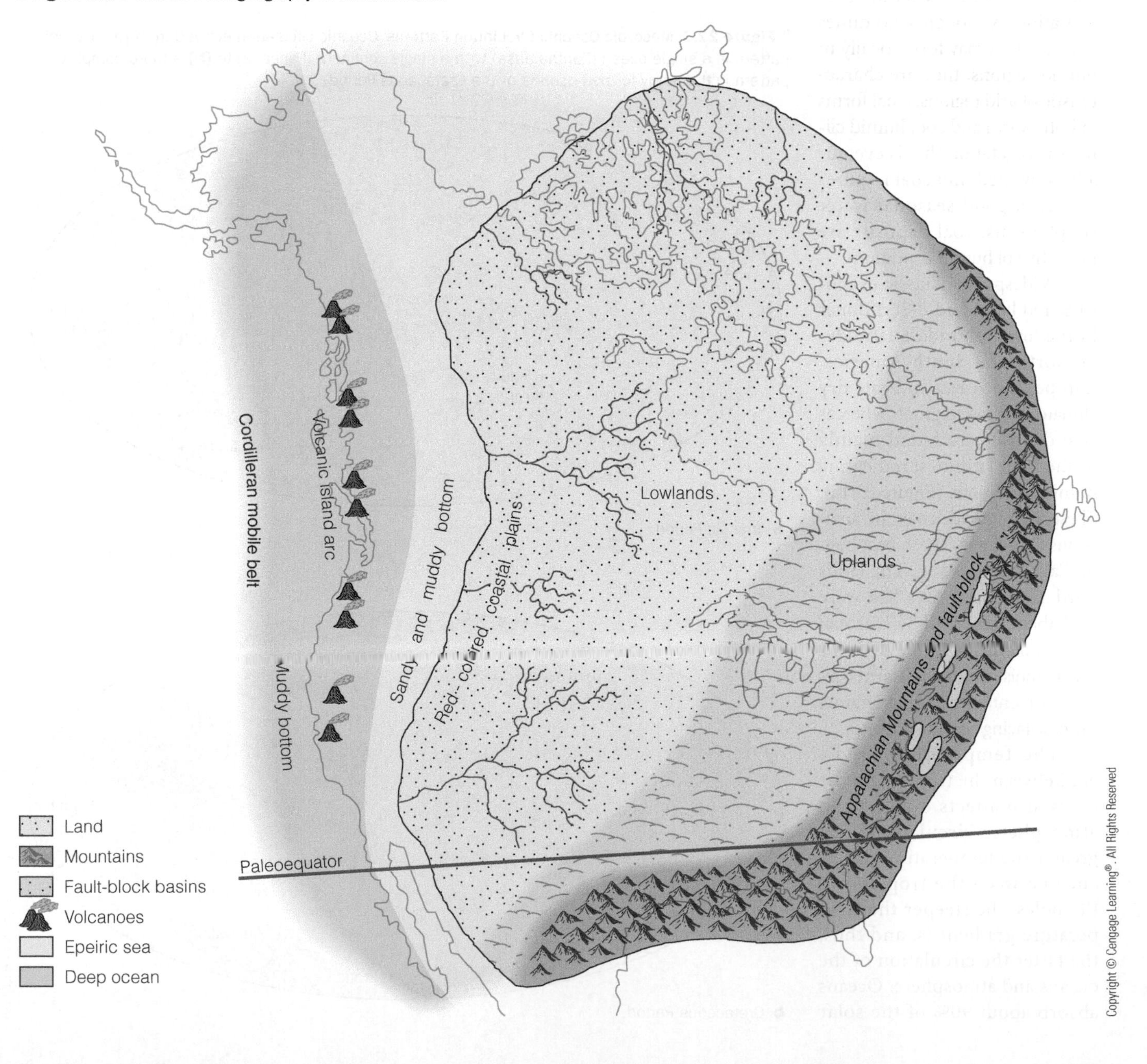

Figure 22.5 Jurassic Paleogeography of North America

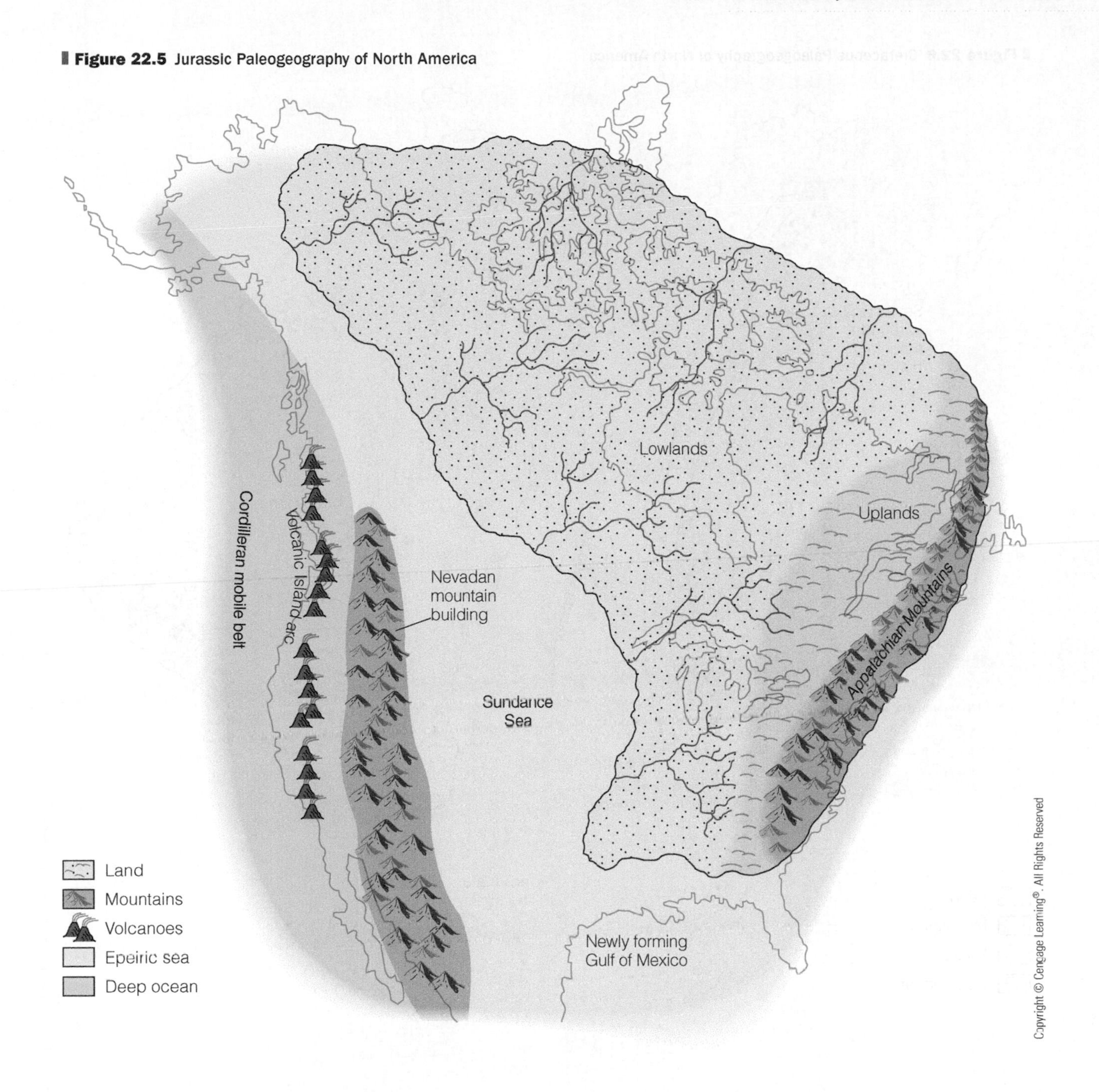

episodes resulting in the formation of the Sierra Nevada, the Rocky Mountains, and other lesser mountain ranges. Although each orogenic episode has its own name, the entire mountain-building event is simply called the *Cordilleran orogeny* (also discussed later in this chapter). With this simplified overview of the Mesozoic history of North America in mind, we will now examine the specific regions of the continent.

Continental Interior

Recall that the history of the North American craton is divided into unconformity-bound sequences reflecting advances and retreats of epeiric seas over the craton (see Figure 20.5). Although these transgressions and regressions played a major role in the Paleozoic geologic history of the continent, they were not as important during the Mesozoic.  Most of the continental interior during the Mesozoic was well above sea level and was not inundated by epeiric seas. Consequently, the two Mesozoic cratonic sequences, the *Absaroka Sequence* (Late Mississippian to Early Jurassic) and the *Zuni Sequence* (Early Jurassic to Early Paleocene) (see Figure 20.5) are not treated separately here; instead, we will examine the Mesozoic history of the three continental margin regions of North America.

Connection Link

To review cratonic sequences and sequence stratigraphy, see Figure 20.5 in Chapter 20.

Figure 22.6 Cretaceous Paleogeography of North America

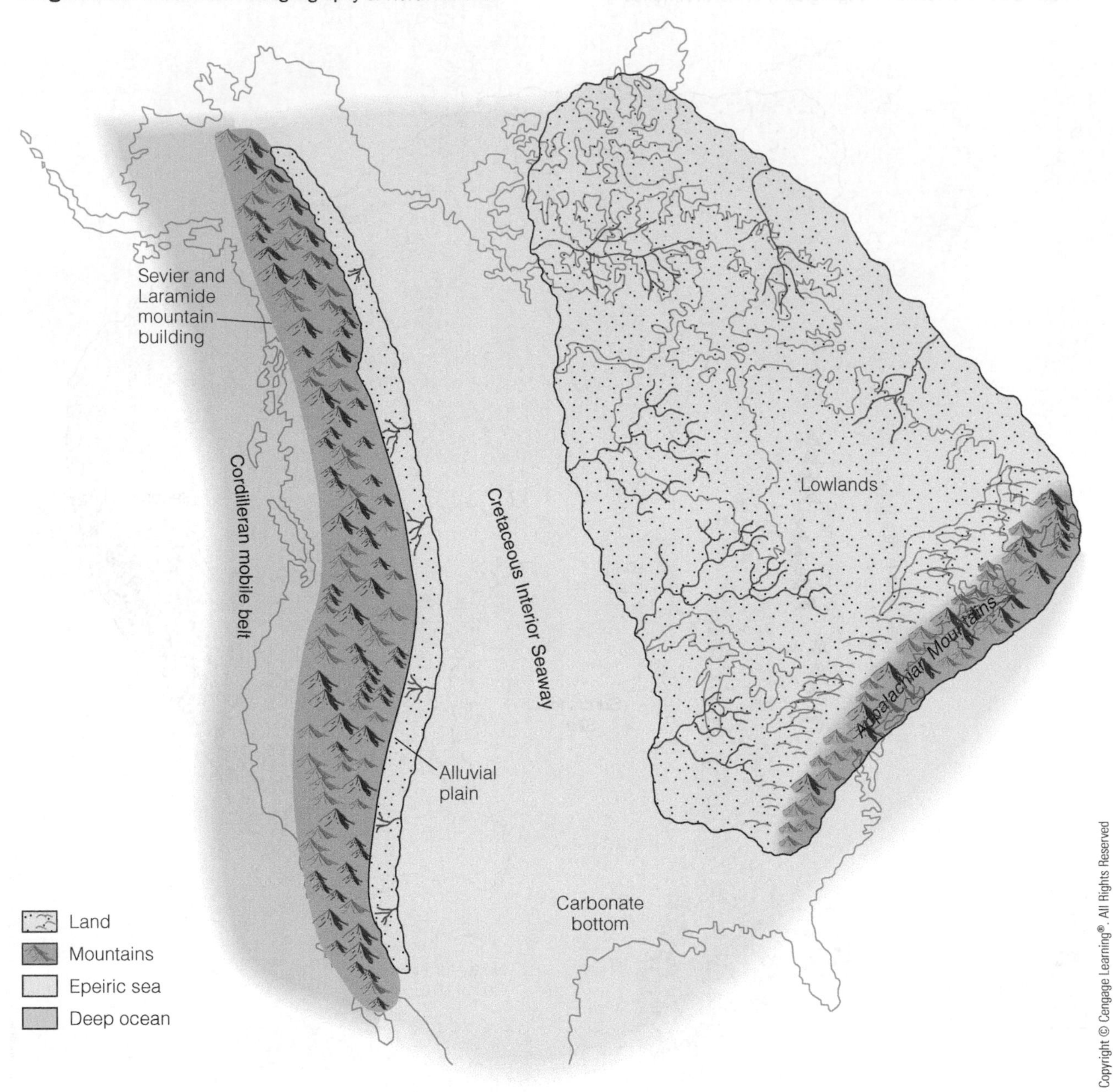

Eastern Coastal Region

During the Early and Middle Triassic, coarse detrital sediments derived from erosion of the recently uplifted Appalachians (Alleghenian orogeny) filled the various intermontane basins and spread over the surrounding areas. As weathering and erosion continued during the Mesozoic, this once lofty mountain system was reduced to a low-lying plain.

During the Late Triassic, the first stage in the breakup of Pangaea began with North America separating from Africa. Fault-block basins developed in response to upwelling magma beneath Pangaea in a zone stretching from present-day Nova Scotia to North Carolina (■ Figure 22.7). Erosion of the fault-block mountains filled the adjacent basins with great quantities (up to 6,000 m) of poorly sorted red nonmarine detrital sediments known as the *Newark Group*.

Reptiles roamed along the margins of the lakes and streams that formed in these basins, leaving their footprints and trackways in the soft sediments. Although the Newark rocks contain numerous dinosaur footprints, they are almost completely devoid of dinosaur bones. The Newark Group is mostly Late Triassic in age, but in some areas, deposition did not begin until the Early Jurassic.

Concurrent with sedimentation in the fault-block basins were extensive lava flows that blanketed the basin floors, as well as intrusions of numerous dikes and sills. The most famous intrusion is the prominent Palisades sill along the Hudson River in the New York–New Jersey area (Figure 22.7d).

Figure 22.7 North American Triassic Fault-Block Basins

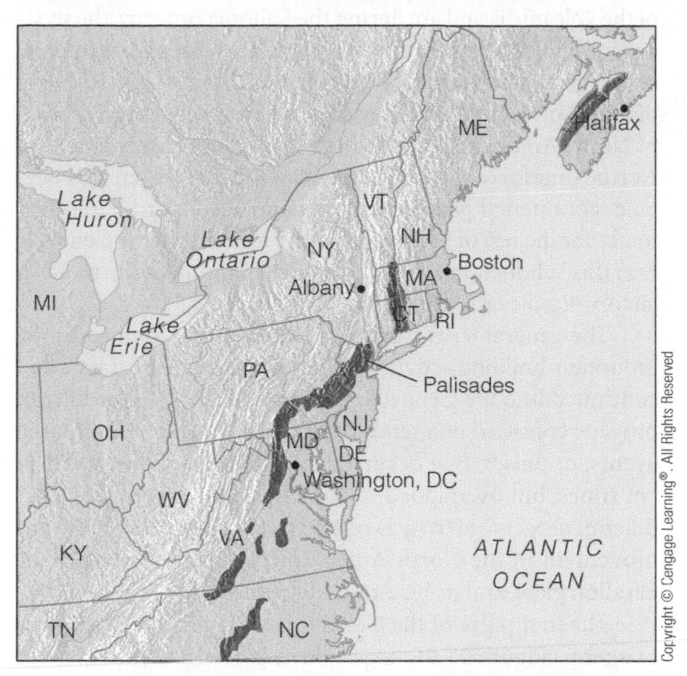

a Areas where Triassic fault-block basin deposits crop out in eastern North America.

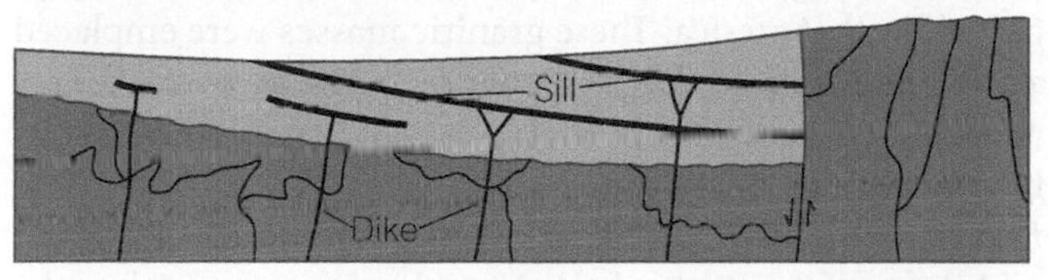

b After the Appalachians were eroded to a low-lying plain by the Middle Triassic, fault-block basins such as this one (shown in cross section) formed as a result of Late Triassic rifting between North America and Africa.

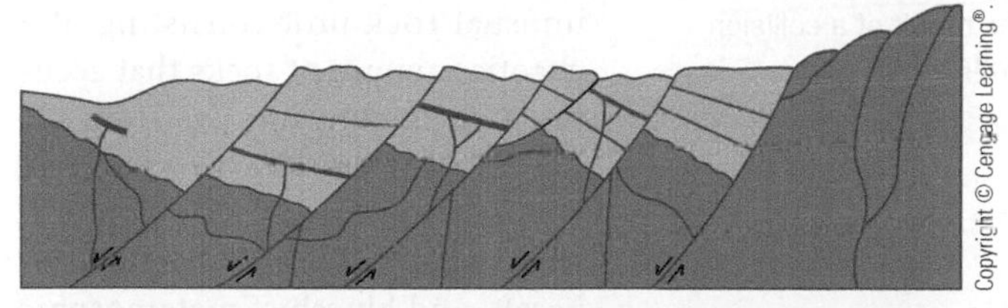

c These valleys accumulated tremendous thickness of sediments and were themselves broken by a complex of normal faults during rifting.

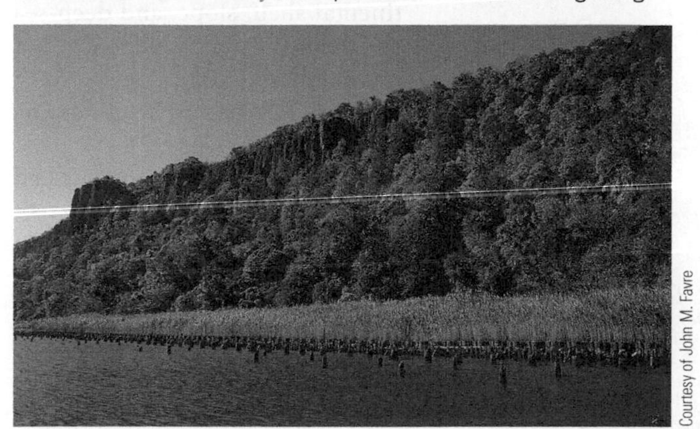

Courtesy of John M. Faure

d Palisades of the Hudson River. This sill was one of many intruded into the Newark sediments during the Late Triassic rifting that marked the separation of North America from Africa.

As the Atlantic Ocean grew, rifting ceased along the eastern margin of North America, and this once-active convergent plate margin became a passive, trailing continental margin. The fault-block mountains produced by this rifting continued to erode during the Jurassic and Early Cretaceous until all that was left was an area of low relief.

ConnectionLink

To review the difference between passive and active continental margins, see Figure 2.10 in Chapter 2.

The sediments produced by this erosion contributed to the growing eastern continental margin. During the Cretaceous Period, the Appalachian region was reelevated and once again shed sediments onto the continental shelf, forming a gently dipping, seaward-thickening wedge of rocks up to 3,000 m thick. These rocks are currently exposed in a belt extending from Long Island, New York, to Georgia.

Gulf Coastal Region

The Gulf Coastal region was above sea level until the Late Triassic (Figure 22.4). As North America separated from South America during the Late Triassic and Early Jurassic, the Gulf of Mexico began to form (Figure 22.5). With oceanic waters flowing into this newly formed, shallow, restricted basin, conditions were ideal for evaporite formation. More than 1,000 m of evaporites were precipitated at this time, and most geologists think that these Jurassic evaporites are the source for the Cenozoic salt domes found today in the Gulf of Mexico and southern Louisiana.

By the Late Jurassic, circulation in the Gulf of Mexico was less restricted, and evaporite deposition ended. Normal marine conditions returned to the area with alternating transgressing and regressing seas. The resulting sediments were covered and buried by thousands of meters of Cretaceous and Cenozoic sediments.

During the Cretaceous, the Gulf Coastal region, like the rest of the continental margin, was flooded by northward-transgressing seas (Figure 22.6). As a result, nearshore sandstones are overlain by finer sediments characteristic of deeper waters. Following an extensive regression at the end of the Early Cretaceous, a major transgression began during which a wide seaway extended from the Arctic Ocean to the Gulf of Mexico (Figure 22.6). Sediments deposited in the Gulf Coastal region during the Cretaceous formed a seaward-thickening wedge.

Reefs were also widespread in the Gulf Coast region during the Cretaceous. Because of their high porosity and permeability, reefs make excellent petroleum reservoirs. The facies patterns of these Cretaceous reefs are as complex as those in the major barrier-reef systems of the Paleozoic Era.

Western Region

Except for the fault-block basins along the East Coast of North America that developed in response to the separation between North America and Africa during the Late Triassic, tectonic activity and major mountain-building activity took place in the Cordilleran mobile during the Mesozoic. The western area of North America was also the site of two major seaways, one during the Jurassic (Figure 22.5) and the other, extending from the Arctic Ocean to the Gulf of Mexico, during the Cretaceous (Figure 22.6).

Mesozoic Tectonics The Mesozoic geologic history of the North American Cordilleran mobile belt is very complex, involving the eastward subduction of the oceanic Farallon plate under the continental North American plate. Activity along this oceanic–continental convergent plate boundary resulted in an eastward movement of deformation. This orogenic activity progressively affected the trench and continental slope, the continental shelf, and the cratonic margin, causing a thickening of the continental crust. In addition, the accretion of terranes and microplates along the western margin of North America also played a significant role in the Mesozoic tectonic history of this area.

Except for the Late Devonian–Early Mississippian Antler orogeny (see Figure 20.29), the Cordilleran region of North America experienced little tectonism during the Paleozoic. However, an island arc and ocean basin formed off the western North American craton during the Permian (Figure 22.4), followed by subduction of an oceanic plate beneath the island arc and the thrusting of oceanic and island arc rocks eastward against the craton margin (❚ Figure 22.8). This event, similar to the preceding Antler orogeny (see Figure 20.29), and known as the *Sonoma orogeny*, occurred at or near the Permian–Triassic boundary and resulted in the suturing of island-arc terranes along the western edge of North America.

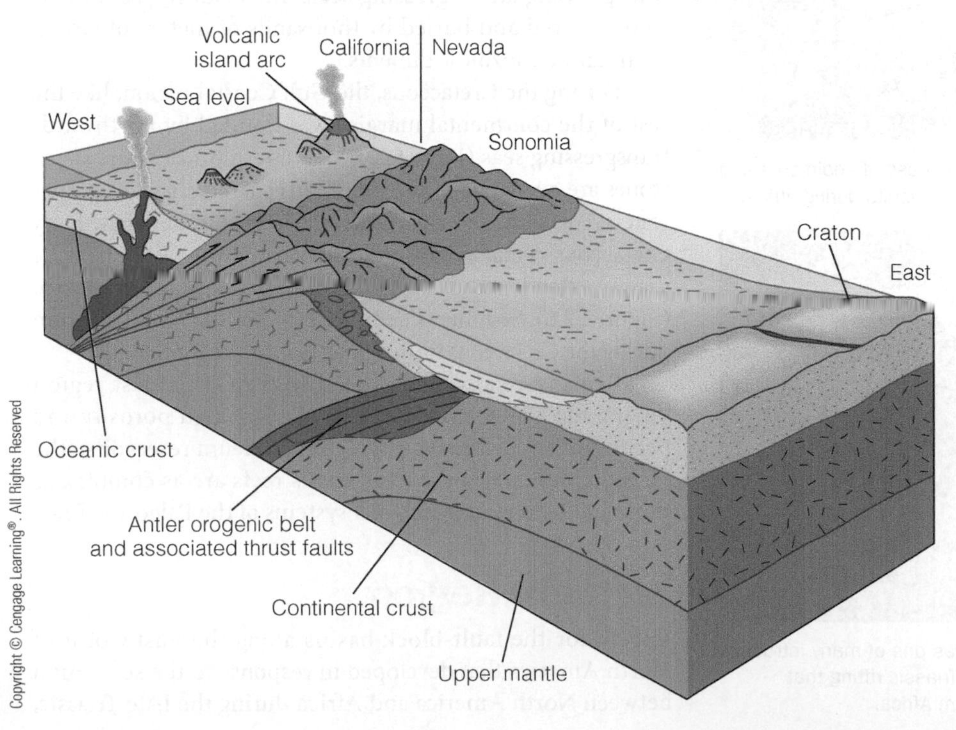

❚ **Figure 22.8 Sonoma Orogeny** Tectonic activity that culminated in the Permian–Triassic Sonoma orogeny in western North America. The Sonoma orogeny was the result of a collision between the southwestern margin of North America and an island arc system.

Following the Late Paleozoic–Early Mesozoic destruction of the volcanic island arc during the Sonoma orogeny, the western margin of North America became an oceanic–continental convergent plate boundary. During the Late Triassic, a steeply dipping subduction zone developed along the western margin of North America in response to the westward movement of North America over the Farallon plate. This newly created oceanic–continental plate boundary controlled Cordilleran tectonics for the rest of the Mesozoic and for most of the Cenozoic Era; this subduction zone marks the beginning of the modern circum-Pacific orogenic system (see Chapter 23).

The general term **Cordilleran orogeny** is applied to the mountain-building activity that began during the Jurassic and continued into the Cenozoic (❚ Figure 22.9). The Cordilleran orogeny consisted of a series of individual mountain-building events, or pulses, that occurred in different regions at different times, but overlapped to some extent. Most of this Cordilleran orogenic activity is related to the continued westward movement of the North American plate as it overrode the Farallon plate, and its history is highly complex.

The first pulse of the Cordilleran orogeny, the **Nevadan orogeny** (Figure 22.9), began during the Late Jurassic and continued into the Cretaceous as large volumes of granitic magma were generated at depth beneath the western margin of North America. These granitic masses were emplaced as huge batholiths that are now recognized as the Sierra Nevada, Southern California, Idaho, and Coast Range batholiths (❚ Figure 22.10). It was also during this time that the *Franciscan Complex* and *Great Valley Group* were deposited and deformed as part of the Nevadan orogeny within the Cordilleran mobile belt.

The Franciscan Complex, which is up to 7,000 m thick, is an unusual rock unit consisting of a chaotic mixture of rocks that accumulated during the Late Jurassic and Cretaceous. The various rock types—graywacke, volcanic breccia, siltstone, black shale, chert, pillow basalt, and blueshist metamorphic rocks—suggest that rocks of the continental shelf, slope, and deep-sea environments were brought together in a submarine trench when North America overrode the subducting Farallon plate (❚ Figure 22.11).

The Franciscan Complex, and the Great Valley Group that lies east of it, were both squeezed against the edge of the North American craton as a result of subduction of the Farallon plate beneath the North America plate. The Franciscan Complex and the Great Valley Group are currently separated from each other by a major thrust fault. The Great Valley Group consists of

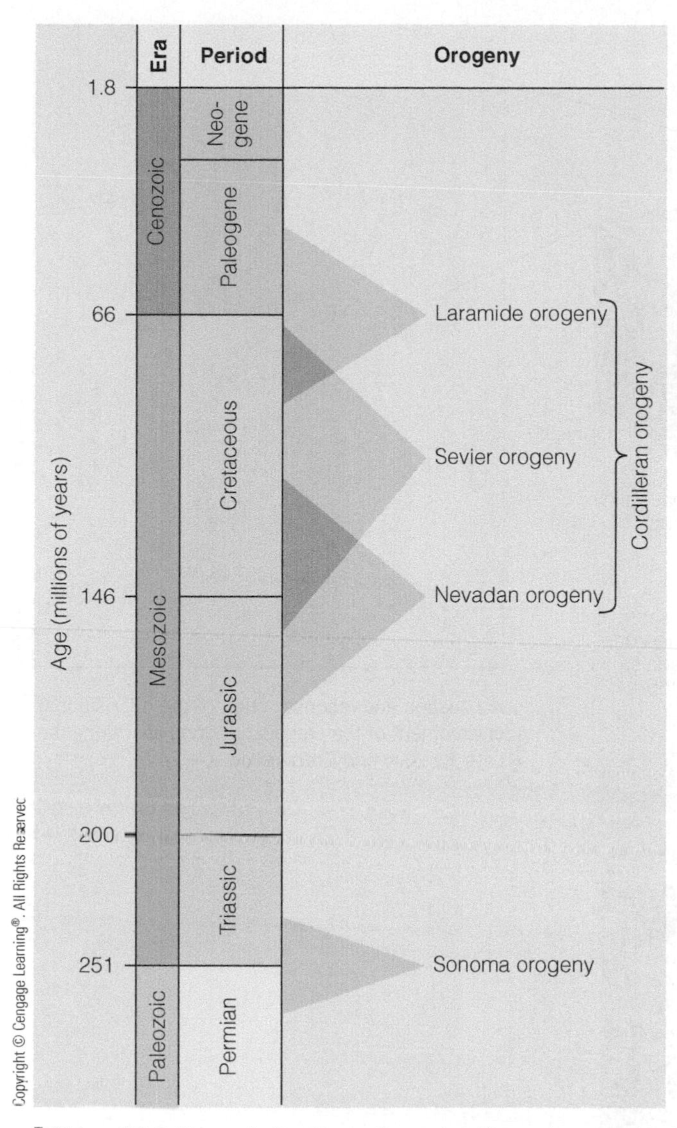

Figure 22.9 Mesozoic Cordilleran Orogenies Mesozoic orogenies occurring in the Cordilleran mobile belt.

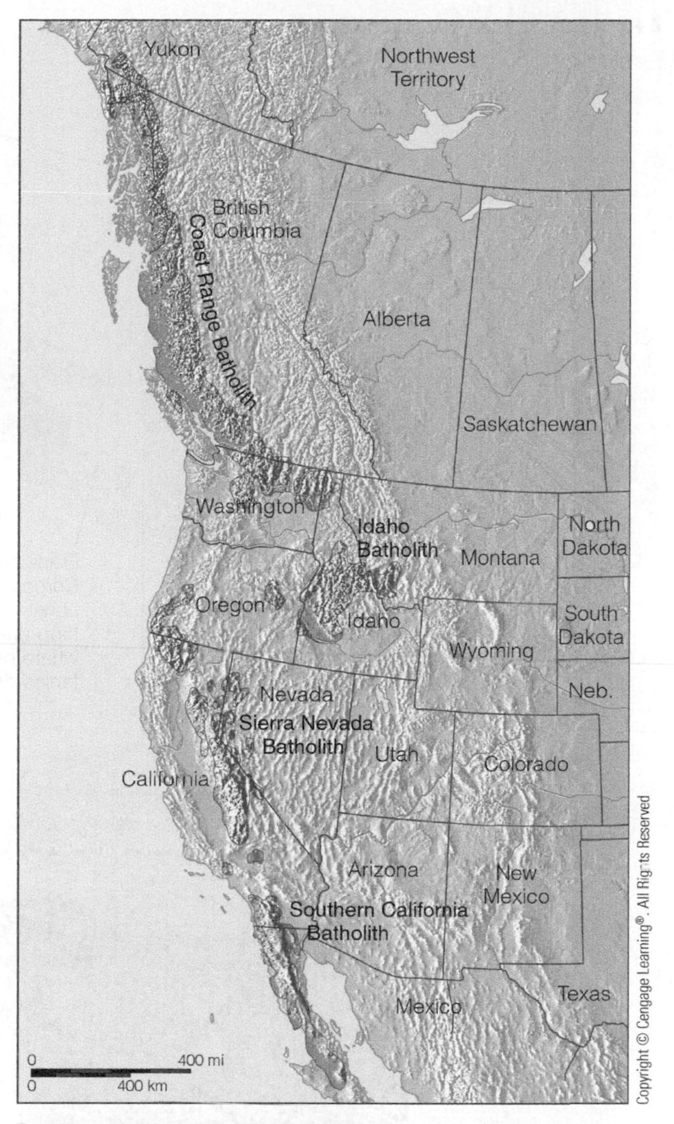

Figure 22.10 Cordilleran Batholiths Location of Jurassic and Cretaceous batholiths in western North America.

more than 16,000 m of conglomerates, sandstones, siltstones, and shales that were deposited on the continental shelf and slope at the same time that the Franciscan deposits were accumulating in the submarine trench (Figure 22.11).

By the Late Cretaceous, most of the volcanic and plutonic activity had migrated eastward into Nevada and Idaho. This migration was probably caused by a change from high-angle to low-angle subduction, resulting in the subducting oceanic plate reaching its melting depth farther east (Figure 22.12). Thrusting occurred progressively farther east, so that by the Late Cretaceous, it extended all the way to the Idaho–Washington border.

The second pulse of the Cordilleran orogeny, the **Sevier orogeny**, affected western North America from Alaska to Mexico and was mostly a Cretaceous event, even though it began in the Late Jurassic and is associated with the tectonic activity of the earlier Nevadan orogeny (Figure 22.9). Subduction of the Farallon plate beneath the North American plate during this time caused numerous overlapping, low-angle thrust faults in which blocks of older Paleozoic strata were thrust eastward on top of younger strata (Figure 22.13). This deformation resulted in crustal shortening and produced generally north–south-trending mountain ranges that stretch from Nevada to western Canada.

During the Late Cretaceous to Early Cenozoic, the final pulse of the Cordilleran orogeny took place (Figure 22.9). The **Laramide orogeny** developed east of the Sevier orogenic belt in what is now the Rocky Mountain areas of New Mexico, Colorado, and Wyoming. Most features of the present-day Rocky Mountains resulted from the Cenozoic phase of the Laramide orogeny, and for that reason, it will be discussed in Chapter 23.

Mesozoic Sedimentation Concurrent with the tectonism in the Cordilleran mobile belt, Early Triassic sedimentation on the western continental shelf consisted of shallow-water marine sandstones, shales, and limestones. During the Middle and Late Triassic, the western shallow

Figure 22.11 The Franciscan Complex

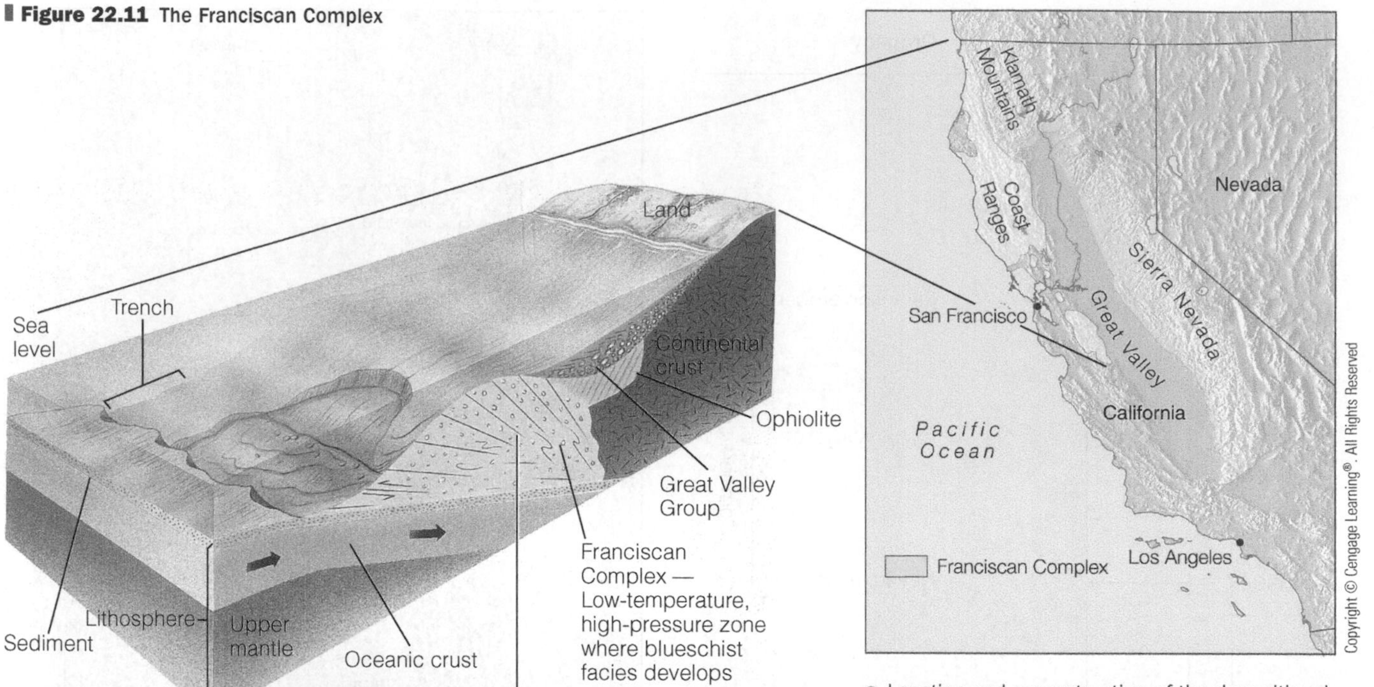

a Location and reconstruction of the depositional environment of the Franciscan Complex during the Late Jurassic and Cretaceous.

b Bedded chert exposed in Marin County, California. Most of the layers are about 5 cm thick.

James S. Monroe

seas regressed farther west, exposing large areas of former seafloor to erosion. Marginal marine and nonmarine Triassic rocks, particularly red beds, contribute to the spectacular and colorful scenery of the region.

These rocks represent a variety of continental depositional environments. The Upper Triassic *Chinle Formation,* for example, is widely exposed throughout the Colorado Plateau region and is probably most famous for its petrified wood, which is spectacularly exposed in Petrified Forest National Park, Arizona (Figure 22.14). This formation, as well as other Triassic formations in the Southwest, also contains the fossilized remains and tracks of various amphibians and reptiles.

Early Jurassic-age deposits covering a large part of the western region consist mostly of clean, cross-bedded sandstones indicative of wind-blown deposits. The thickest, best known, and most prominent of these is the *Navajo Sandstone,*

Figure 22.12 Cretaceous Subduction in the Cordilleran Mobile Belt A possible cause for the eastward migration of igneous activity in the Cordilleran region during the Cretaceous Period was a change from **(a)** high-angle to **(b)** low-angle subduction. As the subducting plate moved downward at a lower angle, the depth of melting moved farther east.

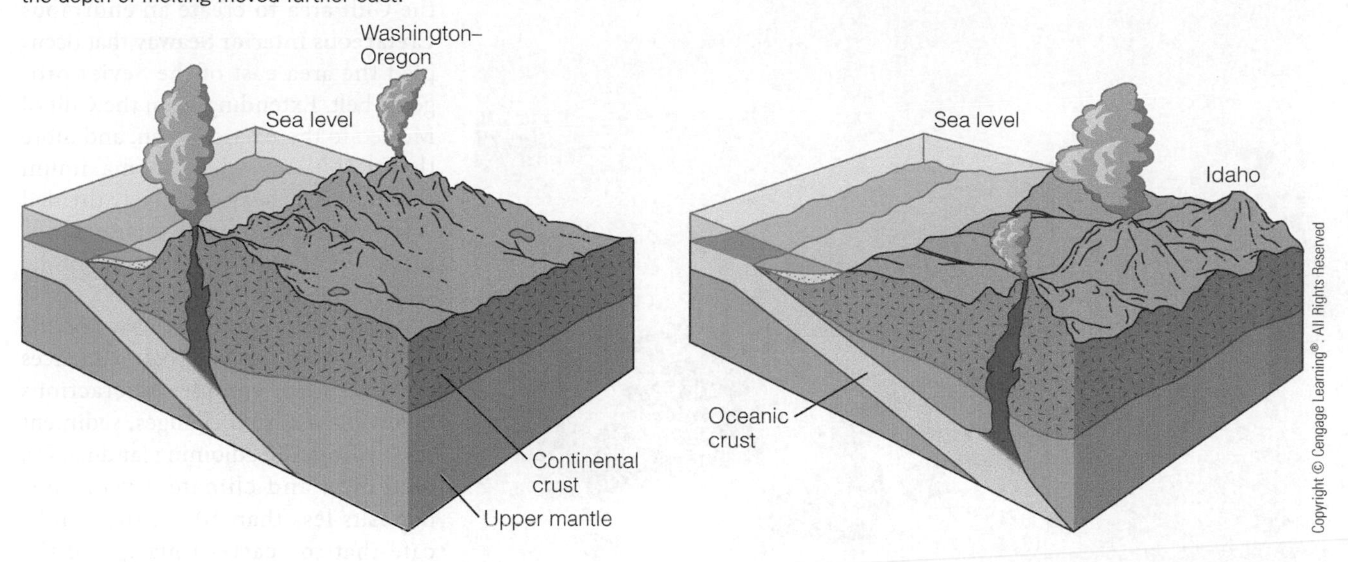

Figure 22.13 Sevier Orogeny

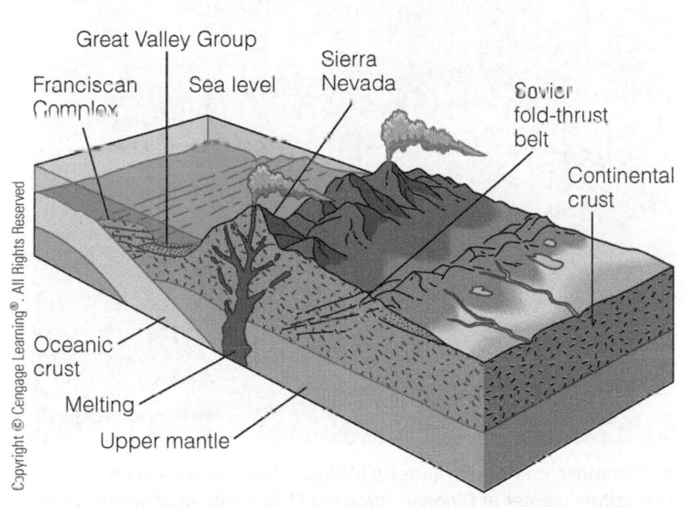

a Restoration showing the associated tectonic features of the Late Cretaceous Sevier orogeny caused by subduction of the Farallon plate under the North American plate.

b The Keystone thrust fault, a major fault in the Sevier overthrust belt, is exposed west of Las Vegas, Nevada. The sharp boundary between the light-colored Mesozoic rocks and the overlying dark-colored Paleozoic rocks marks the trace of the Keystone thrust fault.

a widespread cross-bedded sandstone that accumulated in a coastal dune environment along the southwestern margin of the craton. The sandstone's most distinguishing feature is its large-scale cross-beds, some of which are more than 25 m high (Figure 22.15).

Marine conditions returned to the region during the Middle Jurassic when a seaway called the **Sundance Sea** twice flooded the interior of western North America (Figure 22.5). The resulting deposits, the *Sundance Formation,* were produced from erosion of tectonic highlands to the west that paralleled the shoreline. These highlands resulted from intrusive igneous activity and associated volcanism that began during the Triassic.

During the Late Jurassic, a mountain chain formed in Nevada, Utah, and Idaho as a result of the deformation produced by the Nevadan orogeny. As the mountain chain grew and shed sediments eastward, the Sundance Sea began retreating northward. A large part of the area formerly occupied by the Sundance Sea was then covered by multicolored sandstones, mudstones, shales, and occasional lenses of conglomerates that constitute the world-famous *Morrison Formation* (Figure 22.16a).

The Morrison Formation contains the world's richest assemblages of Jurassic dinosaur remains (Figure 22.16b). Although most of the dinosaur skeletons are broken up, as many as fifty individuals have been found together in a

Figure 22.14 Petrified Forest National Park, Arizona All of the logs here are *Araucarioxylon*, the most abundant tree in the park. The petrified logs have been weathered from the Chinle Formation and are mostly in the position in which they were buried some 200 million years ago.

Helena Lovincic/iStockphoto.com

small area. Such a concentration indicates that the skeletons were brought together during times of flooding and deposited on sandbars in stream channels. Soils in the Morrison Formation indicate that the climate was seasonably dry.

Shortly before the end of the Early Cretaceous, Arctic waters spread southward over the craton, forming a large inland sea in the Cordilleran region. Mid-Cretaceous transgressions also occurred on other continents, and all were part of the global mid-Cretaceous rise in sea level that resulted from accelerated seafloor spreading as Pangaea continued to fragment. These Middle Cretaceous transgressions are marked by widespread black shale deposition within the oceanic areas, the shallow sea shelf areas, and the continental regions that were inundated by the transgressions.

Dr. Nicholas Short, Remote Sensing Tutorial, NASA

Figure 22.15 Navajo Sandstone Large cross-beds of the Jurassic Navajo Sandstone exposed in Zion National Park, Utah.

Critical Thinking Question From this photo, how can you tell that these cross-beds are the result of wind deposition?

By the beginning of the Late Cretaceous, this incursion joined the northward-transgressing waters from the Gulf area to create an enormous **Cretaceous Interior Seaway** that occupied the area east of the Sevier orogenic belt. Extending from the Gulf of Mexico to the Arctic Ocean, and more than 1,500 km wide at its maximum extent, this seaway effectively divided North America into two large landmasses until just before the end of the Late Cretaceous (Figure 22.6).

Deposition in this seaway and the resulting sedimentary rock sequences are a result of complex interactions involving sea-level changes, sediment supply from the adjoining landmasses, tectonics, and climate. Cretaceous deposits less than 100 m thick indicate that the eastern margin of the

Figure 22.16 Morrison Formation, Dinosaur National Monument, Utah

Reed Wicander

a Panoramic view of the Jurassic Morrison Formation as seen from the visitors' center at Dinosaur National Monument, near Vernal, Utah.

Reed Wicander

b North wall of visitors' center showing dinosaur bones in bas relief, just as they were deposited 140 million years ago.

Cretaceous Interior Seaway subsided slowly and received little sediment from the emergent, low-relief craton to the east. The western shoreline, however, shifted back and forth, primarily in response to fluctuations in the supply of sediment from the Cordilleran Sevier orogenic belt to the west. The facies relationships show lateral changes from conglomerate and coarse sandstone adjacent to the mountain belt through finer sandstones, siltstones, shales, and even limestones and chalks in the east. During times of particularly active mountain building, these coarse clastic wedges of gravel and sand prograded even farther east.

As the Mesozoic Era ended, the Cretaceous Interior Seaway withdrew from the craton. During this regression, marine waters retreated to the north and south, and marginal marine and continental deposition formed widespread coal-bearing deposits on the coastal plain.

22.4 What Role Did Accretion of Terranes Play in the Growth of Western North America?

In the preceding sections, we have discussed orogenies along convergent plate boundaries resulting in continental accretion. Much of the material accreted to continents during such events is simply eroded older continental crust; however, a significant amount of new material is added to continents as well, such as igneous rocks that formed as a consequence of subduction and partial melting. Although subduction is the predominant influence on the tectonic history in many regions of orogenesis, other processes are also involved in mountain building and continental accretion, especially the accretion of terranes.

Geologists now know that portions of many mountain systems are composed of small accreted lithospheric blocks that clearly formed elsewhere. These **terranes** differ completely in their fossil content, stratigraphy, structural trends, and paleomagnetic properties from the rocks of the surrounding mountain system and adjacent craton. In fact, these terranes are so different from adjacent rocks that most geologists think that they formed elsewhere and were carried great distances as parts of other plates until they collided with other terranes or continents.

Geologic evidence indicates that more than 25% of the entire Pacific Coast from Alaska to Baja California consists of accreted terranes. These accreted terranes are composed of volcanic island arcs, oceanic ridges, seamounts, volcanic plateaus, hot-spot tracks, and small fragments of continents that were scraped off and accreted to the continent's margin as the oceanic plate on which they were carried was subducted under the continent.

Geologists estimate that more than 100 different-size terranes have been added to the western margin of North America during the past 200 million years (▌Figure 22.17). The Wrangellian terranes (Figure 22.1b) are a good example of terranes that have been accreted to North America's western margin (Figure 22.17).

▌**Figure 22.17 Mesozoic Terranes** Some of the accreted lithospheric blocks, called terranes, that form the western margin of the North American craton. The dark brown blocks probably originated as terranes and were accreted to North America. The light green blocks are possibly displaced parts of North America. The North American craton is shown in dark green. See Figure 22.1b for the position of the Wrangellian terranes during the Jurassic.

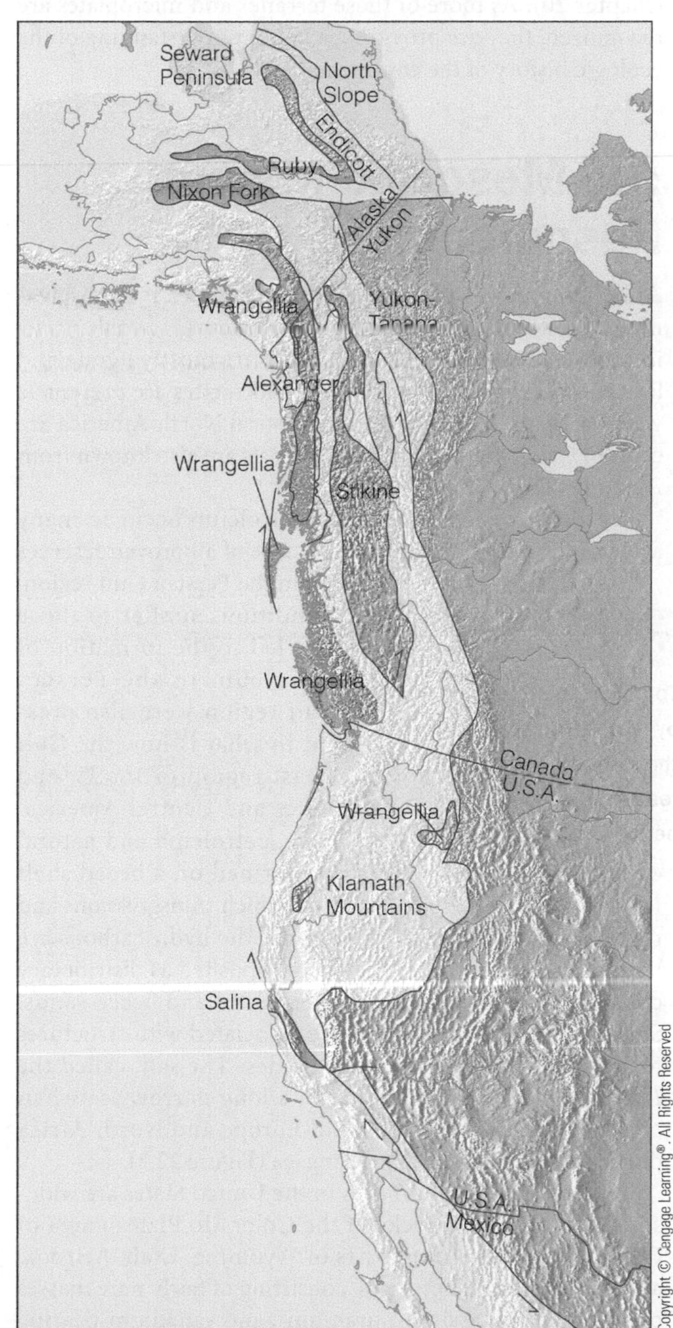

The basic plate tectonic reconstruction of orogenies and continental accretion remains unchanged; however, the details of such reconstructions are decidedly different in view of terrane tectonics. For example, growth along active continental margins is faster than along passive continental margins because of the accretion of terranes. Furthermore, these accreted terranes are often new additions to a continent rather than reworked older continental material.

So far, most Mesozoic terranes have been identified in mountains of the North American Pacific Coast region, but geologists suspect that a number of others are present in other mountain systems as well. They are more difficult to recognize in older mountain systems, such as the Appalachians, because of greater deformation and erosion (see Chapter 20). As more of these terranes and microplates are recognized, they are providing a better understanding of the geologic history of the continents.

22.5 Mesozoic Mineral Resources

Although much of the coal in North America is Pennsylvanian or Paleogene in age, important Mesozoic coals occur in the Rocky Mountain states. These are mostly lignite and bituminous coals, but some local anthracites are present as well. Particularly widespread in western North America are coals of Cretaceous age. Mesozoic coals are also known from Australia, Russia, and China.

Large concentrations of petroleum occur in many areas of the world, but more than 50% of all proven reserves are in the Persian Gulf region. Conditions similar to those that led to the formation of petroleum in the Persian Gulf region were also present in what is now the Gulf Coast region of the United States and Central America. Here, petroleum and natural gas formed on a broad shelf over which transgressions and regressions occurred. In this region, the hydrocarbons are largely in reservoir rocks that were deposited as distributary channels on deltas and as barrier-island and beach sands. Some of these hydrocarbons are associated with structures formed adjacent to rising salt domes. The salt, called the *Louann Salt*, initially formed in a long narrow sea when North America separated from Europe and North Africa during the fragmentation of Pangaea (Figure 22.2).

Connection Link

To review the role played by plate tectonics and the formation of natural resources such as petroleum, see Chapter 2.

The richest uranium ores in the United States are widespread in Mesozoic rocks of the Colorado Plateau area of Colorado and adjoining parts of Wyoming, Utah, Arizona, and New Mexico. These ores, consisting of fairly pure masses of a complex potassium-, uranium-, and vanadium-bearing mineral called *carnotite*, are associated with plant remains in sandstones that were deposited in ancient stream channels.

As noted in Chapter 19, Proterozoic banded iron formations are the main sources of iron ores. There are, however, some important exceptions. For example, the Jurassic-age "Minette" iron ores of Western Europe, composed of oolitic limonite and hematite, are important ores in France, Germany, Belgium, and Luxembourg. In Great Britain, low-grade iron ores of Jurassic age consist of oolitic siderite, which is an iron carbonate. And in Spain, Cretaceous rocks are the host rocks for iron minerals.

South Africa, the world's leading producer of gem-quality diamonds and among the leaders in industrial diamond production, mines these minerals from kimberlite pipes, conical igneous intrusions of dark gray or blue igneous rock. Diamonds, which form at great depth where pressure and temperature are high, are brought to the surface during the explosive volcanism that forms kimberlite pipes. Although kimberlite pipes have formed throughout geologic time, the most intense episode of such activity in South Africa and adjacent countries was during the Cretaceous Period. Emplacement of Triassic and Jurassic diamond-bearing kimberlites also occurred in Siberia.

In the Introduction, we noted that the mother lode or source for the placer deposits mined during the California gold rush is in Jurassic-age intrusive rocks of the Sierra Nevada. Gold placers are also known in Cretaceous-age conglomerates of the Klamath Mountains of California and Oregon.

Porphyry copper was originally named for copper deposits in the western United States mined from porphyritic granodiorite; however, the term now applies to large, low-grade copper deposits disseminated in a variety of rocks. These porphyry copper deposits are an excellent example of the relationship between convergent plate boundaries and the distribution, concentration, and exploitation of valuable metallic ores. Magma generated by partial melting of a subducting plate rises toward the surface, and as it cools, it precipitates and concentrates various metallic ores. The world's largest copper deposits were formed during the Mesozoic and Cenozoic in a belt along the western margins of North and South America (see Figure 2.29).

22.6 Life of the Mesozoic Era

Dinosaurs and other Mesozoic reptiles such as pterosaurs (flying reptiles) as well as ichthyosaurs, plesiosaurs, and mosasaurs (marine reptiles) are without a doubt the most popular prehistoric animals. They have been featured in numerous books, magazines, television specials, and movies, especially *Jurassic Park* (1993) and its two sequels, and several others. In fact, the Mesozoic Era is commonly referred to as the "Age of Reptiles," calling attention to the fact that reptiles predominated among land-dwelling vertebrate animals.

Oil Drilling in the National Parks?

Because of political events in the Middle East, the oil-producing nations of this region have reduced the amount of petroleum that they export, resulting in shortages in the United States. To alleviate United States' dependence on overseas oil, the major oil companies want Congress to let them explore for oil in and around many of our national parks. As director of the National Park system, you have been called to testify at the congressional hearing addressing this possibility. What arguments would you use to discourage such exploration?

Keep in mind, though, that the "Age of" designation simply reflects our preferences; there were far more species of Mesozoic fish and insects than reptiles.

Certainly, the Mesozoic diversification of reptiles was an important event in life history, but other equally important, though not as well known, events took place. Mammals made their appearance during the Triassic, having evolved from mammal-like reptiles, and birds evolved from small carnivorous dinosaurs by the Jurassic. And, of course, there were many other reptiles in addition to dinosaurs—flying reptiles and marine reptiles, as well as lizards, snakes, and turtles.

Important changes also took place in plants when the first flowering plants (angiosperms) evolved and soon became the most common plants on land. Even though the major land plants from the Paleozoic Era persisted, and many still exist, they now make up less than 10% of all land plants. Marine invertebrates such as clams, snails, and cephalopods made a remarkable resurgence following the Permian mass extinctions.

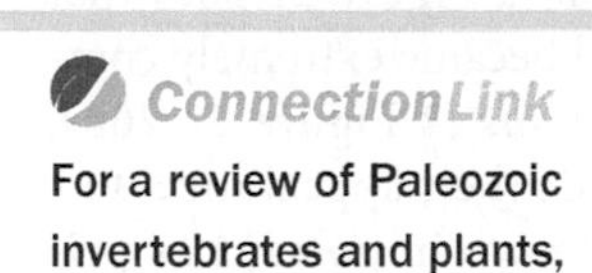

For a review of Paleozoic invertebrates and plants, see Chapter 21.

Although Pangaea began fragmenting during the Triassic (Figure 22.1), the proximity of continents and mild, Mesozoic climates made it possible for plants and animals to occupy extensive geographic ranges. But as the fragmentation continued, some continents became isolated, especially Australia and South America, and their faunas evolving in isolation became increasingly different from those elsewhere. Indeed, Australia has been a separate landmass since the Mesozoic and even now has a fauna unlike that of any other continent.

Marine Invertebrates and Phytoplankton

Following the Permian mass extinctions, the Mesozoic was a time when marine invertebrates repopulated the seas. The Early Triassic invertebrate fauna was not very diverse, but by the Late Triassic, the seas were once again swarming with invertebrates—from planktonic foraminifera to cephalopods. The brachiopods that had been so abundant during the Paleozoic never completely recovered from their near extinction, and although they still exist, the bivalves have largely taken over their ecologic niche.

Mollusks such as cephalopods, bivalves, and gastropods were the most important members of the Mesozoic marine invertebrate fauna. Their rapid evolution and the fact that many cephalopods were nektonic make them excellent guide fossils (Figure 22.18). The Ammonoidea, cephalopods with wrinkled sutures, constitute three groups: the goniatites, ceratites, and ammonites. The latter, though present during the entire Mesozoic, were most prolific during the Jurassic and Cretaceous. Most ammonites were coiled, some attaining diameters of 2 m, whereas others were uncoiled and led a near benthonic existence. Ammonites became extinct at the end of the Cretaceous, but two related groups of cephalopods survived into the Cenozoic: the *nautiloids,* including the living pearly nautilus, and the *coleoids,* represented by the extinct belemnoids (Figure 22.18), which are good Jurassic and Cretaceous guide fossils, and also by the living squid and octopus.

Mesozoic bivalves diversified to inhabit many epifaunal and infaunal niches. Oysters and clams (epifaunal suspension feeders) became particularly diverse and abundant, and despite a reduction in diversity at the end of the Cretaceous, they remain important animals in the marine fauna today.

Figure 22.18 Belemnoids Belemnoids are extinct squidlike cephalopods that were particularly abundant during the Cretaceous. They are excellent guide fossils for the Jurassic and Cretaceous. Shown here are several belemnoids swimming in a Cretaceous sea.

University of Michigan Exhibit Museum

As is true now, where shallow marine waters were warm and clear, coral reefs proliferated. However, reefs did not rebound from the Permian extinctions until the Middle Triassic. An important reef builder throughout the Mesozoic was a group of bivalves known as *rudists*. Rudists are important because they displaced corals as the main reef builders during the later Mesozoic and are good guide fossils for the Late Jurassic and Cretaceous.

A new and familiar type of coral also appeared during the Triassic, the *scleractinians*. Whether scleractinians evolved from rugose corals or from an as-yet-unknown soft-bodied ancestor with no known fossil record is still unresolved. In addition, another invertebrate group that prospered during the Mesozoic was the echinoids. Echinoids were exclusively epifaunal during the Paleozoic but branched out into the infaunal habitat during the Mesozoic.

A major difference between Paleozoic and Mesozoic marine invertebrate faunas was the increased abundance and diversity of burrowing organisms. With few exceptions, Paleozoic burrowers were soft-bodied animals such as worms. The bivalves and echinoids, which were epifaunal animals during the Paleozoic, evolved various means of entering infaunal habitats. This trend toward an infaunal existence may have been an adaptive response to increasing predation from the rapidly evolving fish, cephalopods, and marine reptiles (▌Figure 22.19). Bivalves, for instance, expanded into the infaunal niche during the Mesozoic, and by burrowing, they escaped predators.

The foraminifera (single-celled consumers) diversified rapidly during the Jurassic and Cretaceous and continued to be diverse and abundant to the present. The planktonic forms, in particular, diversified rapidly, but most genera became extinct at the end of the Cretaceous.

The primary producers in the Mesozoic seas were various types of microorganisms. *Coccolithophores* are an important group of calcareous phytoplankton (▌Figure 22.20a) that first evolved during the Jurassic and became extremely common during the Cretaceous. *Diatoms* (▌Figure 22.20b), which build skeletons of silica, made their appearance during the Cretaceous, but they are more important as primary producers during the Cenozoic. Diatoms are presently most abundant in cooler oceanic waters, and some species inhabit freshwater lakes. *Dinoflagellates*, which are organic-walled phytoplankton, were common during the Mesozoic and today are the major primary producers in warm water (▌Figure 22.20c).

In general terms, we can think of the Mesozoic as a time of increasing complexity among the marine invertebrate fauna. At the beginning of the Triassic, diversity was low and food chains were short. Near the end of the Cretaceous, though, the marine invertebrate fauna was highly complex, with interrelated food chains. This evolutionary history reflects changing geologic conditions influenced by plate tectonic activity, as discussed in Chapter 2.

▌Figure 22.20 Primary Producers Coccolithophores, diatoms, and dinoflagellates were primary producers in the Mesozoic and Cenozoic oceans. **(a)** A Miocene coccolith from the Gulf of Mexico (left); a Miocene-Pliocene coccolith from the Gulf of Mexico (right). Coccoliths are the calcareous plates covering the living coccolithosphere. **(b)** Upper Miocene diatoms from Java (left and right). **(c)** Miocene dinoflagellates from Sale, Morocco (left and right).

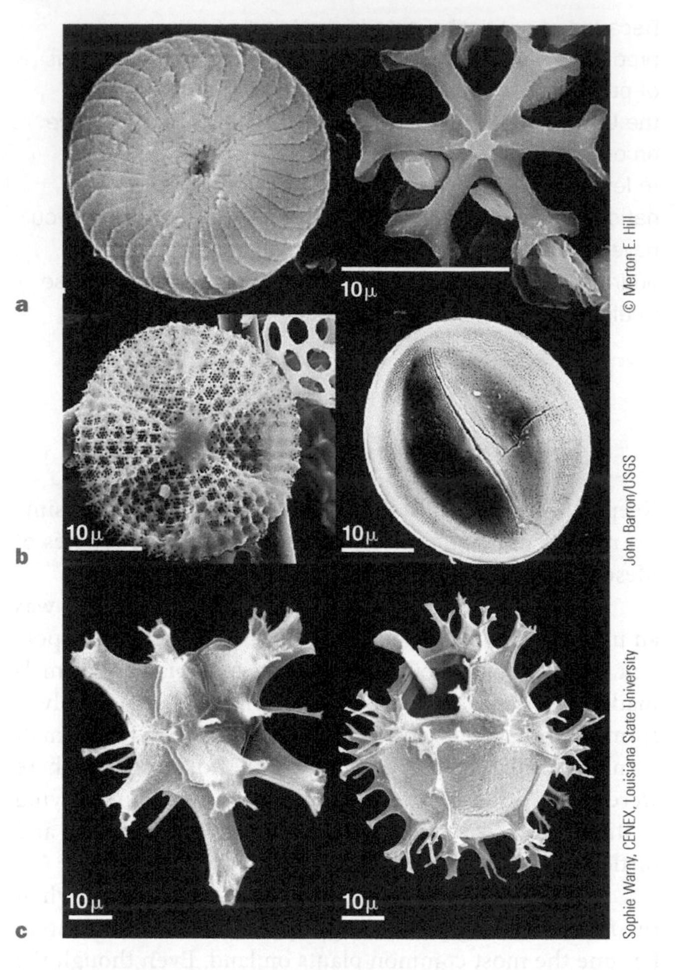

▌Figure 22.19 The Mesozoic Fish *Leedsichthys* (background) and the Short-Necked Plesiosaur *Liopluerodon* This fish, from the intermediate group of fishes (Holostei), was one of the largest ever, but because some of its spine is missing, its total length is not known; estimates range from 9 to 30 m. It was probably a plankton feeder.

Aquatic and Semiaquatic Vertebrates

Remember that fishes evolved by Cambrian time and then diversified, especially during the Devonian Period (see Figure 21.17). The sharks and their relatives became more abundant during the Mesozoic but never came close to matching the diversity of bony fish. A few lobe-finned fish were present at this time, but the ray-finned fish (see Figure 21.21) account for almost all bony fish. By Cretaceous time, the most advanced group of ray-finned fish, known as teleosts, were by far the most common fish in both fresh and saltwater habitats.

Amphibians were most abundant and diverse during the Pennsylvanian Period (see Figure 21.27), but a few labyrinthodonts survived until the end of the Triassic. Frogs and salamanders had evolved by the Mesozoic, although their fossil records are poor.

Plants—Primary Producers on Land

Just as during the Late Paleozoic, seedless vascular plants and gymnosperms dominated Triassic and Jurassic land-plant communities, and, in fact, representatives of both groups are still common. Among the gymnosperms, the large seed ferns became extinct by the end of the Triassic, but *ginkgos* remained abundant and still exist in isolated regions, and *conifers* continued to diversify and are now widespread in some terrestrial habitats, particularly at high elevations and high latitudes. A new group of palmlike gymnosperms called *cycads* made its appearance during the Triassic and became widespread in tropical and semitropical areas.

The long dominance of seedless plants and gymnosperms ended during the Early Cretaceous, perhaps the Late Jurassic, when many were replaced by **angiosperms,** or flowering plants (▌Figure 22.21). Studies of fossils and living gymnosperms show that they are closely related to angiosperms, but unfortunately, the early fossil record of angiosperms is sparse, so their precise ancestry remains obscure.

Since they first evolved, angiosperms have adapted to nearly every terrestrial habitat—from mountains to deserts—and some have even adapted to shallow coastal waters. Several factors account for their success, but chief among them is their method of reproduction (Figure 22.21a). Particularly important was the evolution of flowers, which attract animal pollinators, especially insects, and the evolution of enclosed seeds. With 250,000 to 300,000 species, angiosperms account for more than 90% of all land-plant species and occupy some habitats in which other land plants do poorly or cannot exist.

The Diversification of Reptiles

Late Paleozoic reptiles were the first animals to lay amniote eggs (see Figure 21.28), which was one feature that allowed them to occupy many environments unavailable to the

▌**Figure 22.21** The Angiosperms or Flowering Plants

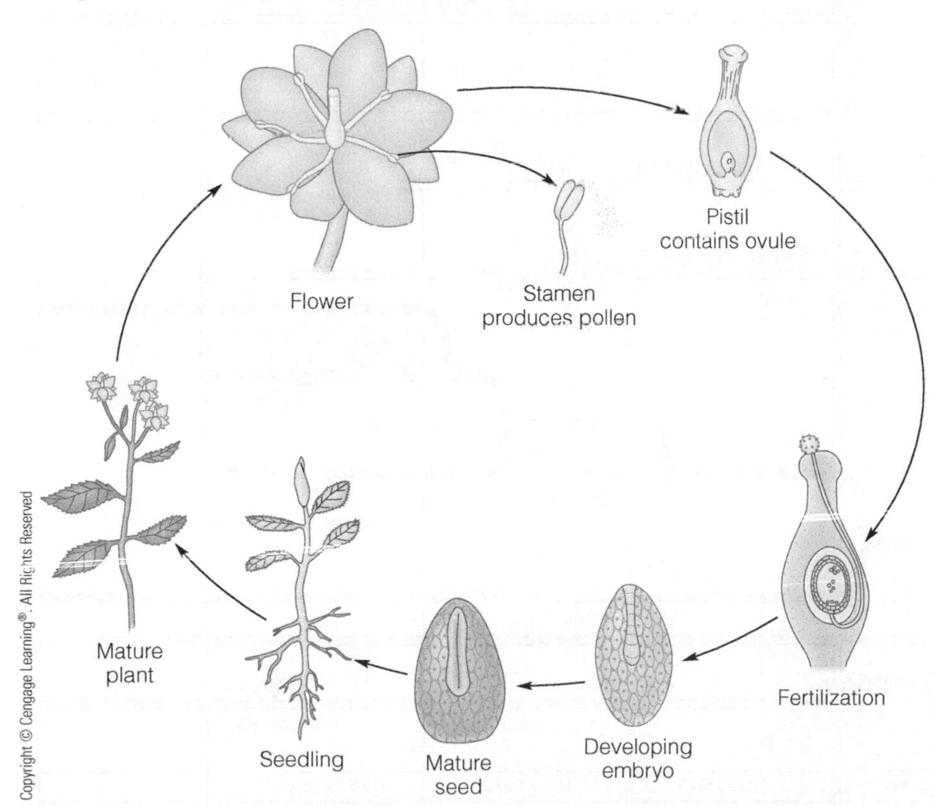

a The reproductive cycle in angiosperms.

David Dilcher and Ge Sun

b *Archaefructus sinensis* from Lower Cretaceous rocks in China is among the oldest known angiosperms.

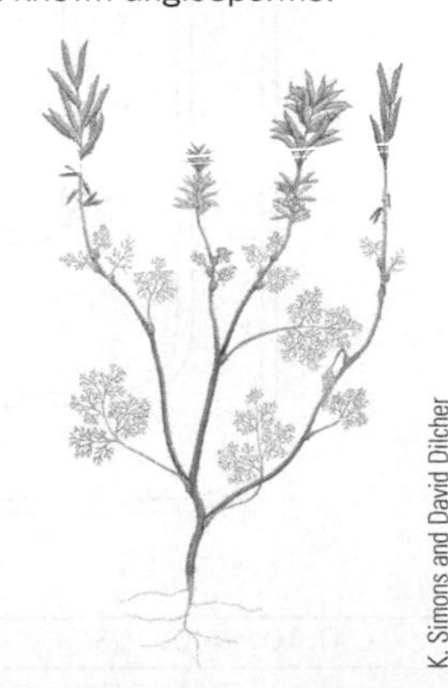

c Restoration of *Archaefructus sinensis.*

To learn more about the amniote egg in mammals, see Chapter 23.

amphibians. All reptiles lay amniote eggs and are accordingly called **amniotes**, but also included in this group are birds and mammals, even though the amniote egg has been highly modified in most mammals. From a basal stock of Late Paleozoic stem reptiles, two main branches emerged, one leading to the pelycosaurs and therapsids and eventually to mammals, and the other to all other reptiles and birds.

Dinosaurs are included among the reptiles, but they possess several characteristics that set them apart: teeth set in individual sockets, a ball-like head on the upper leg bone (femur), and a fully upright posture with the limbs directly beneath their bodies. Indeed, their upright posture and modifications in their limbs account for more efficient locomotion than in other land-dwelling reptiles, which may have been one factor in their incredible success.

Contrary to popular belief, there were no flying dinosaurs (although a few may have glided) or fully aquatic ones; there were other Mesozoic reptiles that occupied these niches (▌Figure 22.22). Nor were all dinosaurs large, although some certainly were. Further, dinosaurs lived only during the Mesozoic Era, unless we consider their evolutionary descendants, the birds.

Archosaurs and the Origin of Dinosaurs Reptiles known as **archosaurs** (*archo* meaning "ruling," and *sauros* meaning "lizard") include crocodiles, flying reptiles (pterosaurs), dinosaurs, and birds (Figure 22.22). The probable ancestors of dinosaurs were archosaurs much like those from Middle and Late Triassic-age rocks in Argentina (▌Figure 22.23). They were rather small (about 1 m long), long-legged carnivores that walked and ran on their hind limbs, so they were **bipedal,** as opposed to **quadrupedal** animals that moved on all four limbs. In any case, dinosaurs had evolved by Late Triassic time.

Sir Richard Owen proposed the term **dinosaur** in 1842 to mean "fearfully great lizard," although now "fearfully" has come to mean "terrible," thus the depiction of dinosaurs as "terrible lizards." Of course, we have no reason to

▌**Figure 22.22 Evolution of the Amniotes** All of the organisms shown here are amniotes—that is, animals that lay amniote eggs. The living amniotes are snakes, lizards, tuataras, birds, turtles, and mammals.

Critical Thinking Question Other than their pelvises, are there other anatomical features you can cite to differentiate the two main groups of dinosaurs?

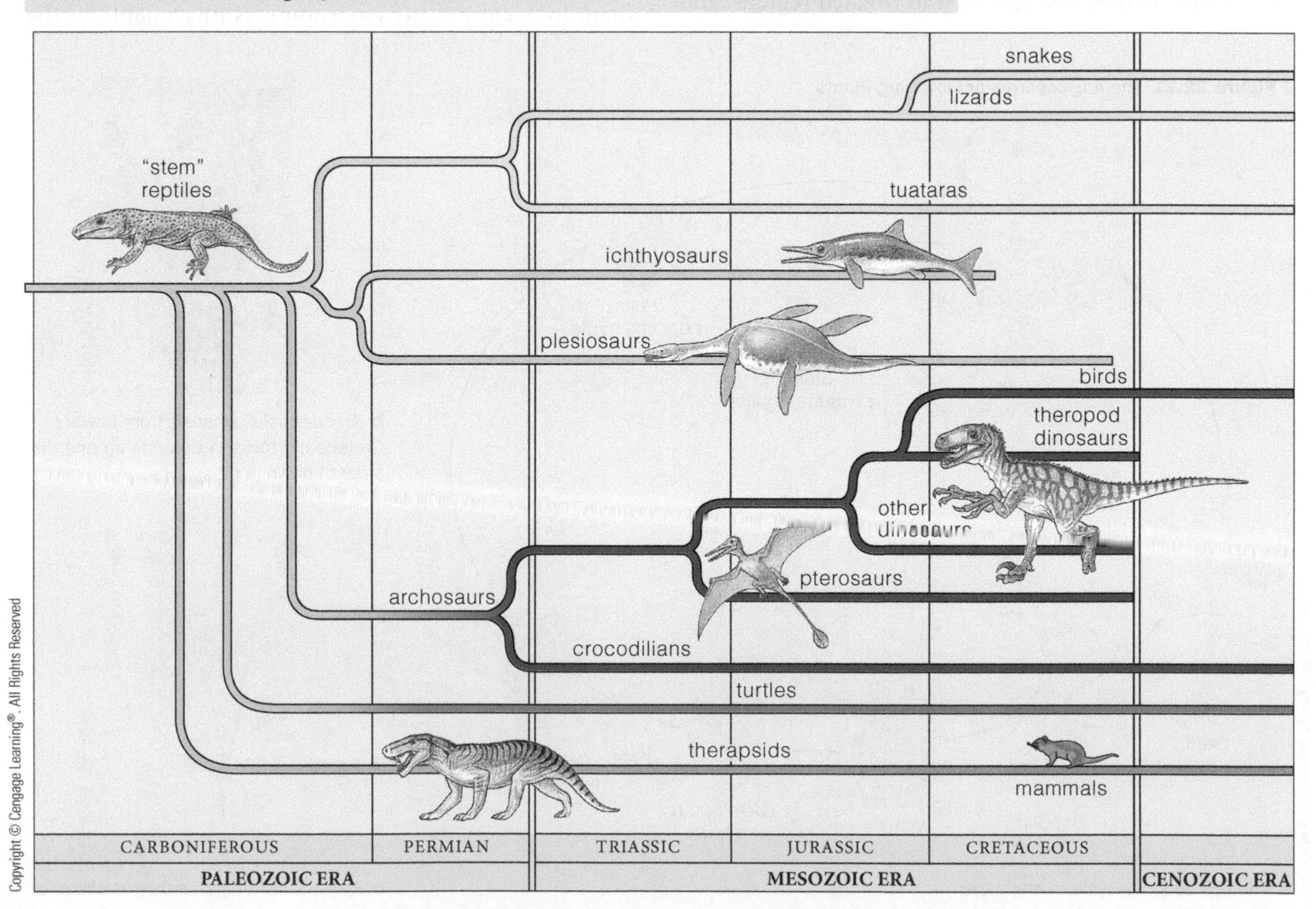

Figure 22.23 *Eoraptor* This small, carnivorous biped was one of the first dinosaurs. Populations of these animals, or ones very much like them, are the probable ancestor of all later dinosaurs. *Eoraptor* lived during the Late Triassic (about 230 million years ago) in what is now Argentina. It was only about a meter long and likely weighed no more than 10 kg.

think they were any more terrible than animals today, and they were not lizards. Nevertheless, these ideas persist and the popularization of dinosaurs in movies and television shows has often been inaccurate and has contributed to misunderstandings.

Although various media now portray dinosaurs as more active animals, the mistaken belief persists that they were dim-witted, lethargic beasts. Evidence available now indicates that some were more active than formerly thought, many, or perhaps all, may have been warm-blooded, and a few types cared for their young long after they hatched, a behavioral characteristic more typical of birds and mammals. Much remains to be learned about dinosaurs, but their fossils and the rocks containing them are revealing more and more about their evolution and behavior.

Dinosaurs are united by several shared features, yet they differ enough for paleontologists to recognize two distinct groups based mostly on their type of pelvis—the **Saurischia** and **Ornithischia.** Saurischian dinosaurs have a lizard-like pelvis, whereas ornithischians have a bird-like pelvis, and both groups are further divided into several distinct types (Figure 22.24). It is now clear that both saurischians and ornithischians had a common ancestor much like the Triassic archosaurs in Figure 22.23.

Saurischian Dinosaurs Saurischian dinosaurs have a lizard-like pelvis, hence their popularization as lizard-hipped dinosaurs, and include two subgroups, the *theropods* and *sauropods* (Figure 22.24). Theropods were bipedal carnivores that varied from tiny *Compsognathus* (2 or 3 kg) to giants such as *Tyrannosaurs* (5 metric tons) (Geo-Insight 22.1).* Beginning in 1966, Chinese scientists have found several types of theropods with feathers, including two juveniles and an adult of the genus *Yutyrannus*, which at 9 m long and 1.5 metric tons is the largest feathered dinosaur discovered so far. Feathered theropods have now been found in Germany and Canada.** No one doubts that these dinosaurs had feathers, and molecular evidence indicates that they were made of the same material as bird feathers.

One of the smaller theropods, the 1.8-m-long predator *Velociraptor,* was popularized in the movie *Jurassic Park.* This animal and its somewhat larger relative, *Deinonychus,* had large, sickle-like claws on the back feet that were probably used in a slashing type of attack. But despite what you might see on television or in movies, theropods, like predators today, no doubt avoided large, dangerous prey and went for the easy kill, preying on the young, old, or disabled, or they dined on carrion. And undoubtedly the larger predators simply chased smaller predators away from their kill.

As opposed to theropods, the sauropods were large, quadrupedal herbivores. Among the sauropods were the largest land-dwelling animals ever. *Brachiosaurus,* a giant even by sauropod standards, may have weighed 75 metric tons, and partial remains indicate that even larger sauropods existed. Evidence from trackways shows that sauropods moved in herds.

Sauropods were preceded in the fossil record by the smaller, Late Triassic to Early Jurassic prosauropods. In fact, there is a good record of small, bipedal prosauropods that were ancestors to the larger, quadrupedal sauropods. A recent discovery of at least 10 nests of the Early Jurassic-age prosauropod *Mossospondylus* in South Africa is the oldest known dinosaur nesting site. Some of the numerous eggs in the nests contained embryos. The true sauropods, represented by *Apatosaurus, Diplodocus,* and *Brachiosaurus,* were common during the Jurassic, but only a few types existed during the Cretaceous.

Ornithischian Dinosaurs In addition to their pelvis, the ornithischians, or so-called bird-hipped dinosaurs, differ from saurischians in other ways. For example, they had no teeth in the front of the mouth, whereas saurischians did, and they had ossified (bonelike) tendons in the back region. Although you may not know the names of the five subgroups of ornithischians, you surely have seen examples of *ornithopods, pachycephalosaurs, ankylosaurs, stegosaurs,* and *ceratopsians* (Figure 22.24) (Geo-Insight 22.1).

*One group of theropods, the poorly known therizinosaurs, reverted to a herbivorous diet.

**One feathered specimen purchased in Utah that was reported to have come from China turned out to be a forgery. It even appeared in *National Geographic* before scientists exposed it as a fraud.

GEO INSIGHT 22.1 Dinosaurs

Dinosaurs are the best-known extinct animals. Hundreds of dinosaur species existed from the Late Triassic through the Cretaceous. They varied from species that weighted no more than a few kilograms to giants weighing several tens of metric tons.

The United States Geologic Survey (USGS) estimates that paleontologists have named about 700 dinosaur species that are assigned to 300 genera. This is only a fraction of the species of living birds and mammals, so there must be many more dinosaurs as yet undiscovered.

▲ **1.** Lifelike restoration of the Early Cretaceous 3-m-long theropod *Deinonychus* "terrible claw" in its probable attack posture. Note the huge curved claws on the back feet and the well-developed forelimbs with sharp claws.

▲ **2.** *Gigantosaurus* was a huge theropod that lived during the Late Cretaceous in Argentina and Brazil. It was nearly 13 m long, weighed about 7.3 metric tons, and lived 25 million years before the more familiar *Tyrannosaurus rex*.

◄ **3.** This Mesozoic scene shows the 10-m-long *Baryonyx* and two large sauropods in the background.

► **4.** *Styracosaurus* was one of several ceratopsian dinosaurs that were common in North America during the Late Cretaceous. This dinosaur was about 5.5 m long and weighed 2.7 metric tons.

◄ **5.** *Sinosauruopteryx* was a 1-m-long theropod from Lower Cretaceous rocks in China. It was the first dinosaur with evidence of feathers. Recent studies of preserved structures called melanosomes, which partly control feather color in today's birds, indicate this animal was probably reddish brown and had stripes on its tail.

Pure Illustration

▲ **6.** A Late Cretaceous scene showing the heavily armored dinosaur *Ankylosaurus* defending itself from an attack.

Photo by De Agostini Picture Library/De Agostini/Getty Images

▲▼ **7–9.** The ornithopod dinosaurs were abundant and varied. *Maiasaura* (above) nested in colonies and tended its young long after hatching. It had no crest, but *Corythosaurus* (below left) and *Parasaurolophus* (below) had bony extensions of the skull.

Dave King/Graham High at Centaur Studios—modelmaker © Dorling Kindersley

▲ **8.** *Corythosaurus*.

Ann Winterbotham © Dorling Kindersley

▲ **9.** *Parasaurolophus*.

2011 by Karen Carr and Karen Carr Studio, Inc.

▲ **10.** *Stegosaurus* is noted for the rows of plates on its back and the bony spikes at the end of its tail. Stegosaurs died out by the end of the Jurassic Period.

Sergey Krasovskiy/Stocktrek Images/Getty Images

▲ **11.** The pachycephalosaurs, represented here by *Stegoceras*, had dome-shaped skulls from thickening of the bones. *Stegoceras* was small, measuring only 2 m long, but some of its relatives were up to 4.5 m long.

Figure 22.24 Cladogram Showing Relationships Among Dinosaurs Pelvises of ornithischian and saurischian dinosaurs are shown for comparison. All of the dinosaurs shown here were herbivores, except for theropods. Note that bipedal and quadrupedal dinosaurs are found in both ornithischians and saurischians.

In 1822, Gideon Mantell and his wife, Mary Ann, discovered some teeth from a dinosaur that he later named *Iguanadon* that proved to be an ornithopod. Another ornithopod, discovered in North America in 1858, was the first dinosaur to be assembled and displayed in a museum. All ornithopods were herbivores and primarily bipedal, but they had well-developed front limbs and could also walk in a quadrupedal manner. One group of ornithopods that was especially numerous during the Cretaceous is called duck-billed dinosaurs, or hadrosaurs. Some of these animals had crests on their heads that may have been used to amplify bellowing, for sexual display, or for species recognition. The Late Cretaceous ornithopod *Maiasaura* nested in colonies, used the same nesting area repeatedly, and apparently cared for its young long after hatching (Geo-Insight 22.1).

The pachycephalosaurs, or dome-headed dinosaurs, had thickened skulls that may have been used in butting contests for dominance or mates. Not all agree with this interpretation and claim that only juveniles had thick skulls. In any case, pachycephalosaurs were bipedal herbivores that ranged from 1.0 to 4.5 m long. Their fossils are known only from Late Cretaceous-age rocks.

Ankylosaurs were quadrupedal herbivores with heavy, bony armor that protected the back, flanks, and top of the head. The tail of some genera, such as *Ankylosaurus*, ended in a heavy, bony club that could probably deliver a crippling blow to an attacking predator. If the tail proved inadequate, the animal probably simply hunkered down, making it difficult even for *Tyrannosaurs* to flip over a 4.5-ton *Ankylosaurus*.

The most familiar genus of stegosaur is the medium-size, herbivorous quadruped known as *Stegosaurus* (Geo-Insight 22.1). This dinosaur had distinctive spikes on its tail, almost certainly used for defense, and bony plates on its back. The arrangement of these plates is not certainly known, although they are usually depicted in two rows with the plates on one side offset from those on the other. Most agree that the plates functioned to absorb and dissipate heat.

The fossil record of ceratopsians (horned dinosaurs) shows that small, bipedal, Early Cretaceous animals were ancestors of large, Late Cretaceous, herbivorous quadrupeds such as *Triceratops*. *Triceratops* and related genera had huge heads, a bony frill over the top of the neck, and a horn or horns on the skull. Fossil trackways show that these were herding animals. A bone bed in Canada with numerous fossils of a single species indicates that large numbers of ceratopsians perished quickly, probably during a river crossing.

We have every reason to think that dinosaurs behaved much as land animals do today. They had to acquire resources, avoid predators, and reproduce, and it seems likely that some of them had complex social structures. Certainly, some dinosaurs lived in herds and no doubt interacted by bellowing, snorting, grunting, and foot stomping in defense, territorial disputes, and attempts to attract mates. Furthermore, some dinosaur species nested in the same area repeatedly and tended their young long after they hatched.

Warm-Blooded Dinosaurs? Were dinosaurs **endotherms** (warm-blooded) like today's mammals and birds, or were they **ectotherms** (cold-blooded) like today's reptiles? Almost everyone now agrees that there is compelling evidence for dinosaur endothermy, but opinion is divided among (1) those holding that all dinosaurs were endotherms, (2) those who think only some were endotherms, and (3) those proposing that dinosaur metabolism, and thus the ability to regulate body temperature, changed as they matured.

Bones of endotherms typically have numerous passageways for blood vessels, but the bones of ectotherms have considerably fewer. Proponents of dinosaur endothermy note that dinosaur bones are more similar to the bones of living endotherms. Crocodiles and turtles have this so-called endothermic bone, but they are ectotherms, and some small mammals have bone more typical of ectotherms. Perhaps bone structure is related more to body size and growth patterns than to endothermy, so this evidence is not conclusive.

Endotherms must eat more than comparably sized ectotherms because their metabolic rates are so much higher. Consequently, endothermic predators require large prey populations and thus constitute a much smaller proportion of the total animal population than their prey, usually only a few percent. In contrast, the proportion of ectothermic predators to prey may be as high as 50%. Where data are sufficient to allow an estimate, dinosaur predators made up 3%–5% of the total population. Nevertheless, uncertainties in the data make this argument less than convincing for many paleontologists.

A large brain in comparison to body size requires a rather constant body temperature and thus implies endothermy. And some dinosaurs were indeed brainy, especially the small- and medium-size theropods, so brain size might be a convincing argument for these dinosaurs. Even more compelling evidence for theropod endothermy comes from their relationship to birds and from the discoveries in China of dinosaurs with feathers or a feather-like covering (Geo-Insight 22.1). Today, only endotherms have hair, fur, or feathers for insulation.

Some scientists point out that some duck-billed dinosaurs grew and reached maturity much more quickly than would be expected for ectotherms and conclude that they must have been warm-blooded. Furthermore, a fossil ornithopod discovered in 1993 has a preserved four-chambered heart much like that of living mammals and birds, convincing many scientists that this animal was an endotherm.

There are good arguments for endothermy for several types of dinosaurs, particularly theropods, although the large sauropods may not have been endothermic, but nevertheless were capable of maintaining a rather constant body temperature. Large animals heat up and cool down more slowly than smaller ones because they have a small surface area compared with their volume. With their comparatively smaller surface area for heat loss, sauropods probably retained heat more effectively than their smaller relatives.

Flying Reptiles Paleozoic insects were the first flying animals, but the first among vertebrates were reptiles called **pterosaurs,** which were common from the Late Triassic until their extinction during the Cretaceous (▌Figure 22.25). Many pterosaurs were no longer than today's sparrows, robins, and crows, but there were a few exceptions. One Cretaceous species from Texas had a wingspan of about 12 m, but even the largest species probably weighed no more than few tens of kilograms.

Pterosaur adaptations for flight include a wing membrane supported by an elongated fourth finger; light, hollow bones; and development of those parts of the brain associated with muscular coordination and sight. Experiments and studies of fossils of large pterosaurs, such as *Pteranodon* (Figure 22.25b), indicate that these animals likely relied on thermal updrafts to stay airborne but occasionally flapped their wings for maneuvering. In contrast, small pterosaurs must have flapped their wings vigorously, just as present-day small birds do.

Wing flapping requires a high metabolic rate and efficient respiratory and circulatory systems as in birds, so it seems likely that pterosaurs were warm-blooded. And, in fact, one pterosaur species had a coat of hair or hairlike feathers, and a fossil found in China in 2009 has hairlike fibers on its wings. In today's animals, only endotherms have hair or feathers.

Marine Reptiles Mesozoic turtles, some crocodiles, and the Triassic mollusk-crushing placodonts adapted to a marine environment, but here we concentrate on the ichthyosaurs, plesiosaurs, and the less-familiar mosasaurs (Figure 22.22). All were thoroughly aquatic marine predators, only the ichthyosaurs and plesiosaurs were closely related, and none was a dinosaur even though they are commonly depicted as such.

The streamlined, rather porpoise-like **ichthyosaurs** (Greek for "fish lizard") ranged from only 0.7 m to 15 m long (▌Figure 22.26a). They used their powerful tail for propulsion and maneuvered with their flipper-like forelimbs, which are simply modified forelimbs of their land-dwelling ancestors. They had numerous sharp teeth, and preserved stomach

■ **Figure 22.25** Flying Reptiles (Pterosaurs)

a *Pterodactylus* from the Late Jurassic is a good example of the earlier pterosaurs. It has a long tail, numerous teeth, and a wingspan of about 1.5 m.

b Later pterosaurs such as *Pteranodon* of the Late Cretaceous were much larger, were tailless, and had toothless beaks. *Pteranodon's* wingspan was about 6 m.

contents reveal a diet of fish, cephalopods, and other marine organisms. It is doubtful that ichthyosaurs could come onto land, so females must have retained the eggs in their bodies and given birth to live young. Fossils with small ichthyosaurs in the appropriate part of the body cavity support this interpretation. An interesting side note on ichthyosaurs is the story of Mary Anning (Geo-Focus 22.1), who, when she was only 11 years old, discovered and excavated a nearly complete ichthyosaur in southern England.

Plesiosaurs were also aquatic animals, but they probably came ashore to lay their eggs. It is unlikely that they were graceful on land, but then neither are walruses and seals today, but they certainly manage. Plesiosaurs were either long-necked or short-necked (■ Figure 22.26b) and most were modest size at 3.6–6.0 m long, although one species found in Antarctica measures 15 m. All plesiosaurs were predators.

Mosasaurs were Late Cretaceous marine lizards that measured from 2.5 to 9.0 m long (■ Figure 22.26c). All were predators, and preserved stomach contents shows that they ate fish, birds, smaller mosasaurs, and several invertebrate animals including amminoids. Mosasaur limbs resembled paddles and were used mostly for maneuvering, whereas their long tail provided propulsion. A possible mosasaur ancestor, found in Texas in 1989, has a mosasaur skeleton, but rather than paddle-like flippers, it retained complete limbs for walking on land.

The Origin and Evolution of Birds

Since the discovery of feathered dinosaurs beginning in the mid-1990s, it has become increasingly difficult to make a distinction between dinosaurs and birds. In fact, many paleontologists now refer to birds as *avian dinosaurs* and call all the others *nonavian dinosaurs*. Living birds do not resemble any of today's reptiles, and yet the best evidence indicates that they evolved from a group of small theropods. Both reptiles and birds lay shelled amniote eggs, they also share several skeletal features, such as the way the jaw attaches to the skull, and several fossils point to a close relationship between birds and theropods.

Since 1860, about 10 fossils of a bird-like animal called *Archaeopteryx* have been recovered from the Solnhofen Limestone in Germany (■ Figure 22.27). *Archaeopteryx* (from Greek *archaeo*, "ancient," and *pteryx*, "feather") had feathers and a wishbone and the fused clavicles so typical of birds, but most of its skeletal features, teeth, clawed wings, and a long tail, are much more like those of theropods (see Geo-Focus 18.1). *Archaeopteryx* is classified as a bird, but if feather impressions had not been found on some of these fossils, scientists would no doubt have called them theropods.

Feathers and a wishbone may seem like definitive features, but even these are no longer sufficient to distinguish birds from theropods, because several theropods had wishbones and paleontologists have discovered several theropods with feathers, especially in China, but also in Germany, and Canada. The few that oppose the idea that birds evolved from theropods point to minor skeletal differences and note that theropods are from Cretaceous-age rocks, whereas *Archaeopteryx* is Jurassic. However, one feathered theropod fossil in

Figure 22.26 Mesozoic Marine Reptiles

John Sibbick/The Natural History Museum/The Image Works

a Ichthyosaur.

Roger Harris/Science Source

b A long-necked plesiosaur.

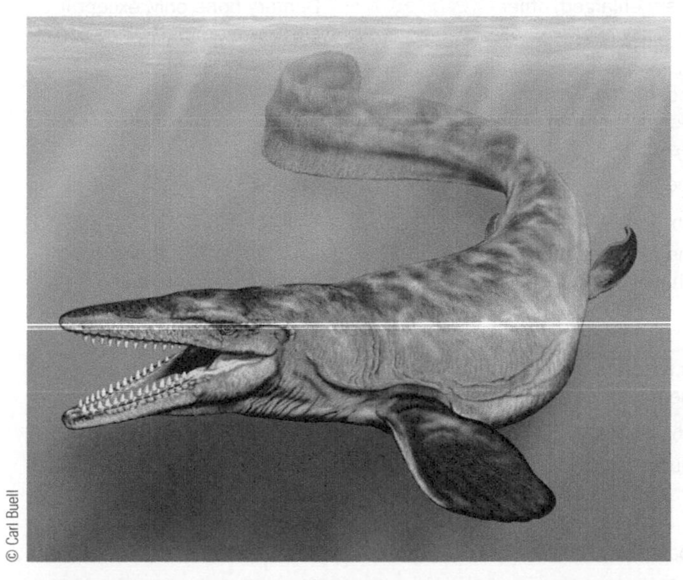

© Carl Buell

c Mosasaur.

Figure 22.27 Restoration of *Archaeopteryx* Several specimens of *Archaeopteryx* have been found in the Jurassic-age Solnhofen Limestone in Germany. This animal had feathers and a wishbone, so it is classified as a bird, but in most details of its anatomy, it more closely resembles small theropod dinosaurs—it had reptile-like teeth, a long tail, and claws on its wings.

De Agostini/Getty Images

China came from Late Jurassic-age rocks, and others are the same age as *Archaeopteryx*.

One recent fossil find in China is an *Archaeopteryx*-like fossil bird known as *Xiaotingia* and is from the Late Jurassic, and thus older than *Archaeopteryx*, so it may be the oldest known bird. Perhaps so, but the fact that it, as well as *Archaeopteryx,* show many features of theropods reinforces the idea that birds evolved from theropods. Another contender for the first bird is Late Triassic-age fossils known as *Protoavis* found in Texas. Unfortunately, the fossils are fragmentary and had no feather impressions, leading most paleontologists to conclude that they were small theropods.

One hypothesis for the origin of bird flight—*from the ground up*—holds that the ancestors of birds were bipedal, fleet-footed ground dwellers that used their wings to leap into the air, at least for short distances, to catch insects or to escape predators. The *from the trees down* hypothesis holds that bird ancestors were bipeds that climbed trees and used their wings for gliding or parachuting. A more recent hypothesis holds that birds may have taken both routes. Researchers know that many of today's bird chicks flap their rudimentary wings to run up steep slopes to avoid predators, and they use the same rudimentary wings to help them descend.

A recent fossil find from China reported in May 2013 might be the earliest birdlike creature known, and it predates *Archaeopteryx* by about 10 million years. However, no feather impressions were preserved, so some paleontologists think it is simply a small birdlike dinosaur.

Evolution of Mammals

Recall from Table 18.1 that one prediction of evolutionary theory is, "If today's organisms descended with modification from ones in the past, there should be fossils showing characteristics that connect orders, classes, and so on." *Archaeopteryx* certainly fills the bill, but so do the fossils showing the transition from ancient therapsids (mammal-like reptiles) to mammals. In fact, the transition is so well documented that some fossils are difficult to assign to one group or the other. Among the therapsids, the most mammal-like of all is a group called the **cynodonts,** which by Late Triassic time had given rise to true mammals.

Cynodonts and the Origin of Mammals We can easily recognize living mammals as warm-blooded vertebrate animals that have hair or fur and mammary glands and, except for the platypus and spiny anteater, give birth to live young. However, these criteria are not sufficient for recognizing fossil mammals; for them, we must rely on skeletal structure only. Several skeletal modifications took place during the transition from mammal-like reptiles to mammals, but distinctions between the two are based mostly on details of the middle ear, the lower jaw, and the teeth (Table 22.1).

Reptiles as well as typical cynodonts have one small bone in the middle ear (the stapes), whereas mammals have three: the incus, the malleus, and the stapes. Also, the lower jaw of a mammal is composed of a single bone called the dentary, but a reptile's jaw is composed of several bones (▌Figure 22.28). In addition, a reptile's jaw is hinged to the skull at a contact between the articular and quadrate bones, whereas in mammals the dentary contacts the squamosal bone of the skull (Figure 22.28).

During the transition from cynodonts to mammals, the quadrate and articular bones that had formed the joint between the jaw and skull in reptiles were modified into the incus and malleus of the mammalian middle ear (Figure 22.28, Table 22.1). Fossils document the progressive enlargement of the dentary until it became the only element in the mammalian jaw. Likewise, a progressive change from the reptile to mammal jaw joint is documented by fossil evidence. In fact, some of the most advanced cynodonts were truly transitional, because they had a compound jaw joint consisting of (1) the articular and quadrate bones typical of reptiles, and (2) the dentary and squamosal bones as in mammals (Table 22.1).

In Chapter 18, we noted that the study of embryos provides some of the evidence for evolution. Opossum embryos show that the middle-ear bones of mammals were originally part of the jaw. In fact, when opossums are born, the middle ear elements are still attached to the dentary, but as they develop further, these elements migrate to the middle ear, and a typical mammal jaw joint develops.

Several other aspects of cynodonts also indicate they were ancestors of mammals. Their teeth were becoming double rooted as they are in mammals, and the teeth were somewhat differentiated into distinct types that performed specific functions. In mammals, the teeth are fully differentiated into incisors, canines, and chewing teeth (premolars and molars), but typical reptiles do not have differentiated teeth (Figure 22.28a). In addition, mammals have only two sets of teeth during their lifetimes—a set of baby teeth and

TABLE 22.1 Summary Chart Showing Some Characteristics and How They Changed During the Transition from Reptiles to Mammals

Features	Typical Reptile	Cynodont	Mammal
Lower Jaw	Dentary and several other bones	Dentary enlarged, other bones reduced	Dentary bone only, except in earliest mammals
Jaw–Skull Joint	Articular quadrate	Articular-quadrate; some advanced cynodonts had both the reptile jaw–skull joint and the mammal jaw–skull joint	Dentary-squamosal
Middle-Ear Bones	Stapes	Stapes	Stapes, Incus, malleus
Secondary Palate	Absent	Partly developed	Well developed
Teeth	No differentiation; chewing teeth single rooted	Some differentiation; chewing teeth partly double rooted	Fully differentiated into incisors, canines, and chewing teeth; chewing teeth double rooted
Tooth Replacement	Teeth replaced continuously	Only two sets of teeth in some advanced cynodonts	Two sets of teeth
Occipital Condyle	Single	Partly divided	Double
Occlusion (chewing teeth meet surface to surface to allow grinding)	No occlusion	Occlusion in some advanced cynodonts	Occlusion
Endothermic vs. Ectothermic	Ectothermic	Probably endothermic	Endothermic
Body Covering	Scales	One fossil shows that it had skin similar to that of mammals	Skin with hair or fur

Figure 22.28 Evolution of the Mammal Jaw and Middle Ear (a) Cynodonts had four bones in the jaw, whereas **(b)** mammals have only one. Note that the cynodont jaw–skull joint is between the articular and the quadrate bones, but in mammals, it is between the dentary and squamosal bones. **(c)** Cynodonts had one middle-ear bone—the stapes—but **(d)** mammals have three: the malleus, the incus, and the stapes. The malleus and incus were derived from the cynodont articular and quadrate bones.

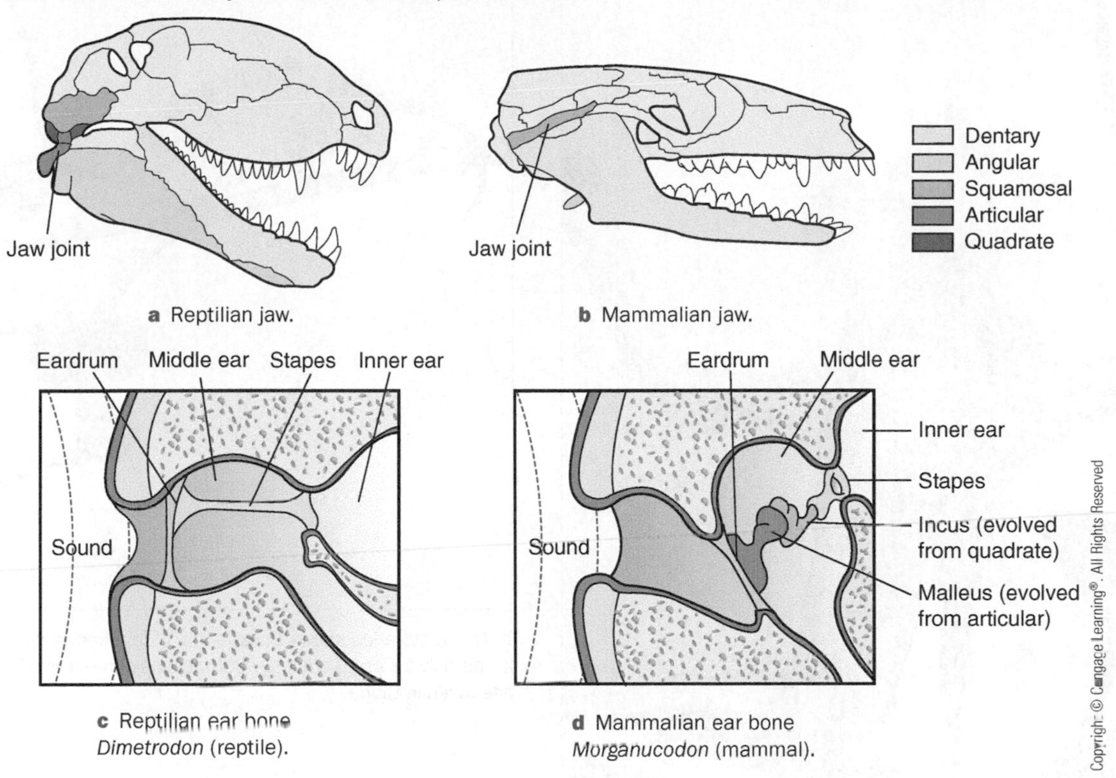

the permanent adult teeth. Reptiles have teeth replaced continuously throughout their lives, the notable exception being some cynodonts, that in mammal fashion had only two sets of teeth. Another important feature of mammal teeth is occlusion; that is, the chewing teeth meet surface to surface to allow grinding. Thus, mammals chew their food, but reptiles, amphibians, and fish do not. However, tooth occlusion is known in some advanced cynodonts (Table 22.1).

Reptiles and mammals have a bony protuberance from the skull that fits into a socket in the first vertebra—the atlas. This structure, called the *occipital condyle,* is a single feature in reptiles, but in cynodonts it is partly divided into a double structure typical of mammals. Another mammalian feature, the secondary palate, was partially developed in advanced cynodonts (Table 22.1). This bony shelf separating the nasal passages from the mouth cavity is an adaptation for eating and breathing at the same time, a necessary requirement for endotherms with their high demands for oxygen.

Mesozoic Mammals Mammals evolved during the Late Triassic not long after the first dinosaurs, but for the rest of the Mesozoic Era, most of them were small. There were, however, a few exceptions. One is a Middle Jurassic-age aquatic mammal found in China that measures about 50 cm long, which also has the distinction of being the oldest known fossil with fur. The other is an Early Cretaceous-age mammal called *Repenomamus giganticus,* also from China, that was about 1 m long, weighed 12 to 14 kg, and had the remains of a juvenile dinosaur in its stomach. Nevertheless, most other Mesozoic mammals were about the size of mice and rats, and they were not nearly as diverse as they were during the Cenozoic Era. Furthermore, they retained reptile characteristics but also had mammalian features. The Triassic triconodonts, for instance, had the fully differentiated teeth typical of mammals, but they too had both the reptile and the mammal types of jaw joints. In short, some mammal features appeared sooner than others (again, recall the concept of *mosaic evolution* from Chapter 18).

The early mammals diverged into two distinct branches, one leading to the **monotremes,** or egg-laying mammals such as the living spiny anteater and platypus of the Australian region. The other branch includes a group of mammals that gave rise to both **marsupial** (pouched) **mammals** and the **placental mammals** during the Early

For more information about the diversification of marsupial and placental mammals, see Chapter 23.

Figure 22.29 Mesozoic Mammals

Mark A. Klinger/Carnegie Museum of Natural History

a Restoration of the oldest known marsupial mammal, *Sinodelphym,* which measures 15 cm long.

Mark A. Klinger/Carnegie Museum of Natural History

b This restoration shows *Eomaia,* the oldest known placental mammal. It was only 12 or 13 cm long. Both fossils come from Early Cretaceous-age rocks in China.

Cretaceous (Figure 22.29). Most mammals that lived during the Cenozoic Era were placentals, although marsupials were successful in South America and even today make up most of the mammalian fauna of Australia.

22.7 Mesozoic Climates and Paleogeography

Fragmentation of the supercontinent Pangaea began by the Late Triassic and continues to the present, but during much of the Mesozoic, close connections existed between the various landmasses. The proximity of these landmasses, however, is not sufficient to explain Mesozoic biogeographic distributions because climates are also effective barriers to wide dispersal. During much of the Mesozoic, though, climates were more equable and lacked the strong north-south zonation characteristic of the present. In short, Mesozoic plants and animals had greater opportunities to occupy much more extensive geographic ranges.

Pangaea persisted as a single unit throughout most of the Triassic (Figure 22.1a), and the Triassic climate was warm-temperate to tropical, although some areas, such as the present southwestern United States, were arid. Mild temperatures extended 50 degrees north and south of the equator, and even the polar regions may have been temperate. The fauna had a truly worldwide distribution. Some dinosaurs had continuous ranges across Laurasia and Gondwana; the peculiar gliding lizards were in New Jersey and England, and reptiles known as phytosaurs lived in North America, Europe, and Madagascar.

By the Late Jurassic, Laurasia had become partly fragmented by the opening North Atlantic, but a connection still existed (Figure 22.1b). The South Atlantic had begun to open so that a long, narrow sea separated the southern parts of Africa and South America. Otherwise, the southern continents were still close together.

The mild Triassic climate persisted into the Jurassic. Ferns, whose living relatives are now restricted to the tropics of southeast Asia, are known from areas as far as 63 degrees south latitude and 75 degrees north latitude. Dinosaurs roamed widely across Laurasia and Gondwana. For example, the giant sauropod *Brachiosaurus* is known from western North America and eastern Africa. Stegosaurs and some families of carnivorous dinosaurs lived throughout Laurasia and in Africa.

GEO FOCUS 22.1 Mary Anning's Contributions to Paleontology

Paleontologists use fossils to study life of the past, so part of their efforts are spent in finding and collecting fossils. Western European men dominated the early history of this field, but this situation no longer prevails. Indeed, men and women from many countries are now making significant contributions. Perhaps the most notable early exception is Mary Anning (1799–1847), who began a remarkable career as a fossil collector when she was only 11 years old.

Mary Anning was born in Lyme Regis on England's southern coast. When only 15 months old, she survived a lightning strike that, according to one report, killed three girls and, according to another, killed a nurse tending her. In 1810, Mary's father, a cabinet maker who also sold fossils part time, died, leaving the family nearly destitute. Mary Anning (Figure 1) expanded the fossil business and became a professional fossil collector known to the paleontologists of her time, some of whom visited her shop to buy fossils or gather information. She collected fossils from the Dorset coast near Lyme Regis and is reported to have been the inspiration for the tongue twister,

She sells seashells on the seashore
The shells she sells are seashells, I'm sure
So if she sells seashells on the seashore
Then I'm sure she sells seashore shells.

Soon after her father's death, Mary Anning made her first important discovery—a nearly complete skeleton of a Jurassic ichthyosaur, which was described in 1814 by Sir Everard Home. The sale of this fossil specimen provided considerable financial relief for her family. In 1821, she made a second major discovery and excavated the remains of a plesiosaur. And in 1828, she found the first pterosaur in England, which was sent to the eminent geologist William Buckland at Oxford University.

Mary Anning (1799–1847) (oil on canvas), English School (19th century)/Private Collection/The Bridgeman Art Library International

Figure 1 Mary Anning, who lived in Lyme Regis on England's southern coast, began collecting and selling fossils when she was only 11 years old.

By 1830, Mary Anning's fortunes began declining as collectors and museums had fewer funds with which to buy fossils. In fact, she may once again have become destitute were it not for her geologist friend Henry Thomas de la Beche, also a resident of Lyme Regis. De la Beche drew a fanciful scene called *Duria antiquior*, meaning "An earlier Dorset," in which he brought to life the fossils Mary Anning had collected. The scene was made into a lithograph that was printed and sold widely, and its proceeds went directly to Mary Anning.

Mary Anning died of cancer in 1847, and although only 48 years old, she had a fossil-collecting career that spanned 37 years. Her contributions to paleontology are now widely recognized, but, unfortunately, soon after her death, she was mostly forgotten. Apparently, the people who purchased her fossils were credited with finding them.

> It didn't occur to them to credit a woman from the lower classes with such astonishing work. So an uneducated little girl, with a quick mind and an accurate eye, played a key role in setting the course of the 19th century geologic revolution. Then—we simply forgot about her.*

Following Mary Anning's death, her contributions to paleontology were mostly forgotten. Charles Dickens wrote about her life in 1865, but otherwise little notice was taken of her work. However, since 2002, the Palaeontological Association has presented the Mary Anning Award to someone who is "not professionally employed within palaeontology but who made an outstanding contribution to the subject**." She was further recognized in 2010 when the Royal Society of London named her as one of the 10 British women who have most influenced the history of science.

---

*John H. Lienhard, Professor, College of Engineering, University of Houston.

**The Palaeontological Association, http://www.palass.org/modules.php?name=palaeo&sec=Awards&page=122.

By the Late Cretaceous, the North Atlantic had opened further, and Africa and South America were completely separated (Figure 22.1c). South America remained an island continent until late in the Cenozoic, and its fauna, evolving in isolation, became increasingly different from faunas elsewhere. Marsupial mammals reached Australia from South America via Antarctica, but the South American connection was eventually severed. Placentals, other than bats and a few rodents, never reached Australia, thus explaining why marsupials continue to dominate the continent's fauna even today.

Cretaceous climates were more strongly zoned by latitude, but they remained warm and equable until the close of that period. Climates then became more seasonal and cooler, a trend that persisted into the Cenozoic. Dinosaur and mammal fossils demonstrate that interchange was still possible, especially between the landmasses that made up Laurasia.

22.8 Mass Extinctions—A Crisis in the History of Life

Though the greatest mass extinction took place at the end of the Paleozoic Era, the one at the close of the Mesozoic has attracted much more attention because among its victims were dinosaurs, flying reptiles, and marine reptiles. Several kinds of marine invertebrates also went extinct, including ammonites, which had been so abundant during the Mesozoic, rudistid bivalves, and some planktonic organisms.

Numerous hypotheses have been proposed for Mesozoic extinctions, but most have been dismissed as improbable, untestable, or inconsistent with the available data. A proposal that has become popular since 1980 is based on a discovery at the Cretaceous–Paleogene boundary in Italy—a 2.5-cm-thick clay layer with a remarkably high concentration of the platinum-group element iridium (❚ Figure 22.30a). High iridium concentrations have now been identified at many other Cretaceous–Paleogene boundary sites.

The significance of this **iridium anomaly** is that iridium is rare in crustal rocks but is found in much higher concentrations in some meteorites. Accordingly, some investigators propose a meteorite impact to explain the anomaly and further postulate that the impact of a meteorite, perhaps 10 km in diameter, set in motion a chain of events leading to extinctions. Some Cretaceous–Paleogene boundary sites also contain soot and shock-metamorphosed quartz grains, both of which are cited as additional evidence of an impact.

According to the impact hypothesis, approximately 60 times the mass of the meteorite was blasted from the crust high into the atmosphere, and the heat generated at impact started raging forest fires that added more particulate matter to the atmosphere. Sunlight was blocked for several months, temporarily halting photosynthesis; food chains collapsed; and extinctions followed. Furthermore, with sunlight greatly diminished, Earth's surface temperatures were drastically reduced, adding to the biologic stress. Another consequence of the impact was that vaporized rock and atmospheric gases produced sulfuric acid (H_2SO_4) and nitric acid (HNO_3). Both would have contributed to strongly acid rain that may have had devastating effects on vegetation and marine organisms.

Now some geologists point to a probable impact site centered on the town of Chicxulub on the Yucatán Peninsula of Mexico (❚ Figure 22.30b). The 180-km-diameter structure lies beneath layers of sedimentary rock and appears to be the right age. Evidence that supports the conclusion that the Chicxulub structure is an impact crater includes shocked quartz, the deposits of huge waves, and tektites—small pieces of rock melted during the impact and hurled into the atmosphere.

❚ **Figure 22.30** Mesozoic Extinctions

a Close-up view of the iridium-rich Cretaceous–Paleogene boundary clay in the Raton Basin, New Mexico.

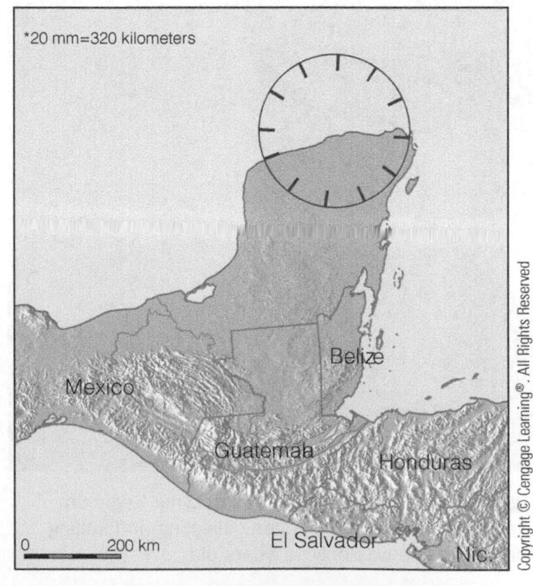

b Proposed meteorite impact crater centered on Chicxulub on the Yucatán Peninsula of Mexico.

Even if a meteorite did hit Earth, did it lead to these extinctions? If so, both terrestrial and marine extinctions must have occurred at the same time. To date, strict time equivalence between terrestrial and marine extinctions has not been demonstrated. The selective nature of the extinctions is also a problem. In the terrestrial realm, large animals were the most affected, but not all dinosaurs were large, and crocodiles, close relatives of dinosaurs, survived, although some species died out. Some paleontologists think that dinosaurs, some marine invertebrates, and many plants were already on the decline and headed for extinction before the end of the Cretaceous. A meteorite impact may have simply hastened the process.

In the final analysis, Mesozoic extinctions have not been explained to everyone's satisfaction. Most geologists now concede that a large meteorite impact occurred, but we also know that vast outpourings of lava were taking place in what is now India. Perhaps these brought about detrimental atmospheric changes. Furthermore, the vast shallow seas that covered large parts of the continents had mostly withdrawn by the end of the Cretaceous, and the mild equable Mesozoic climates became harsher and more seasonal by the end of that era. Nevertheless, these extinctions were very selective, and no single explanation accounts for all aspects of this crisis in life history.

Key Concepts Review

Tables 22.2 and 22.3 summarize many Mesozoic geologic and biologic events.

- We can summarize the breakup of Pangaea as follows:
  1. During the Late Triassic, North America began separating from Africa. This was followed by the rifting of North America from South America.
  2. During the Late Triassic and Jurassic periods, Antarctica and Australia—which remained sutured together—began separating from South America and Africa, and India began rifting from Gondwana.
  3. South America and Africa began separating during the Jurassic, and Europe and Africa began converging during this time.
  4. The final stage in Pangaea's breakup occurred during the Cenozoic, when Greenland completely separated from Europe and North America.
- The breakup of Pangaea influenced global climatic and atmospheric circulation patterns. Although the temperature gradient from the tropics to the poles gradually increased during the Mesozoic, overall global temperatures remained equable.
- An increased rate of seafloor spreading during the Cretaceous Period caused sea level to rise and transgressions to occur.
- Except for incursions along the continental margin and two major transgressions (the Sundance Sea and the Cretaceous Interior Seaway), the North American craton was above sea level during the Mesozoic Era.
- The Eastern Coastal region was the initial site of the separation of North America from Africa that began during the Late Triassic. During the Cretaceous Period, it was inundated by marine transgressions.
- The Gulf Coastal region was the site of major evaporite accumulation during the Jurassic as North America rifted from South America. During the Cretaceous, it was inundated by a transgressing sea, which, at its maximum, connected with a sea transgressing from the north to create the Cretaceous Interior Seaway.
- Mesozoic rocks of the western region of North America were deposited in a variety of continental and marine environments. One of the major controls of sediment distribution patterns was tectonism.
- Western North America was affected by four interrelated orogenies: the Sonoma, Nevadan, Sevier, and Laramide. Each involved igneous intrusions, as well as eastward thrust faulting and folding.
- The cause of the Nevadan, Sevier, and Laramide orogenies was the changing angle of subduction of the oceanic Farallon plate beneath the continental North American plate. The timing, rate, and, to some degree, the direction of plate movement were related to seafloor spreading and the opening of the Atlantic Ocean.
- Although the structural features of North America's western margin are associated with activity along an oceanic–continental convergent plate boundary, geologists think that more than 25% of the western margin originated from the accretion of terranes.
- Mesozoic rocks contain a variety of mineral resources, including coal, petroleum, uranium, gold, and copper.
- The marine invertebrate survivors of the Permian mass extinction diversified and gave rise to increasingly diverse Mesozoic marine invertebrate communities.
- Land plant communities of the Triassic and Jurassic consisted of seedless vascular plants and gymnosperms. The angiosperms, or flowering plants, evolved during the Early Cretaceous, diversified rapidly, and were soon the most abundant land plants.
- Dinosaurs evolved during the Late Triassic but were most abundant and diverse during the Jurassic and Cretaceous. The two distinct groups of dinosaurs,

based on pelvic structure, are Saurischia (lizard-hipped) and Ornithischia (bird-hipped).

- Bone structure, predator–prey relationships, and other features have been cited as evidence of dinosaur endothermy. Although there is still no solid consensus, many paleontologists think that at least some dinosaurs were endotherms.
- That some theropods had feathers indicates that they were warm-blooded and provides further evidence of their relationship to birds.
- Small pterosaurs were probably active, wing-flapping fliers, whereas large ones may have depended on soaring to stay aloft. At least two pterosaur species had hair or feathers, so they, and perhaps all pterosaurs, were probably endothermic.
- The fish-eating, porpoise-like ichthyosaurs were thoroughly adapted to an aquatic life, whereas the plesiosaurs with their paddle-like limbs could most likely come out of the water to lay their eggs. The marine reptiles known as mosasaurs are most closely related to lizards.

TABLE 22.2 **Mesozoic Geologic Events**

Age (millions of years)	Geologic Period	Sequence	Relative Changes in Sea Level (Rising / Falling)	Cordilleran Mobile Belt	Gulf Coast Region	Eastern Coastal Region	Global Plate Tectonic Events
66–146	Cretaceous	Zuni	Present sea level	Cordilleran orogeny; Laramide orogeny; Sevier orogeny; Jurassic and Cretaceous tectonism controlled by eastward subduction of the Pacific plate beneath North America and accretion of microplates	Major Late Cretaceous transgression; Reefs particularly abundant; Regression at end of Early Cretaceous; Early Cretaceous transgression and marine sedimentation	Appalachian region uplifted; Erosion of fault-block mountains that formed during the Late Triassic to Early Jurassic	South America and Africa are widely separated; Greenland begins separating from Europe
146–200	Jurassic	Zuni; Absaroka		Nevadan orogeny	Sandstones, shales, and limestones are deposited in transgressing and regressing seas; Thick evaporites are deposited in newly formed Gulf of Mexico	Fault-block mountains and basins develop in eastern North America	South America and Africa begin separating in the Late Jurassic
200–251	Triassic	Absaroka		Subduction zone develops as a result of westward movement of North America; Sonoma orogeny	Gulf of Mexico begins forming during Late Triassic	Deposition of Newark Group; lava flows, sills, and dikes	Breakup of Pangaea begins with rifting between Laurasia and Gondwana; Supercontinent Pangaea still in existence

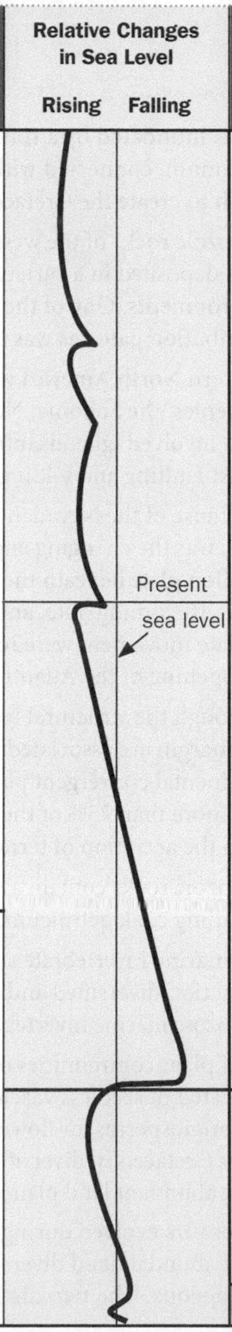

- Birds probably evolved from small theropod dinosaurs. The oldest known bird, *Archaeopteryx,* appeared during the Jurassic; however, few other Mesozoic birds are known.
- The earliest mammals evolved during the Late Triassic, but they are difficult to distinguish from advanced cynodonts. Details of the teeth, middle ear, and lower jaw are used to distinguish the two.
- Several types of Mesozoic mammals existed, but most were small and their diversity was low. Both marsupials and placentals evolved during the Cretaceous from a group known as eupantotheres.
- Because the continents were close together and climates were mild during much of the Mesozoic, animals and plants occupied much larger geographic ranges than they do now.
- Among the victims of the Mesozoic mass extinctions were dinosaurs, flying reptiles, marine reptiles, and several groups of marine invertebrates. A meteorite impact may have caused these extinctions, but some paleontologists think that other factors also contributed.

TABLE 22.3 Mesozoic Biologic Events

Age (millions of years)	Geologic Period	Invertebrates	Vertebrates	Plants	Climate
66–146	Cretaceous	Extinction of ammonites, rudists, and most planktonic foraminifera at end of Cretaceous Continued diversification of ammonites and belemnoids Rudists become major reef builders	Extinctions of dinosaurs, flying reptiles, marine reptiles, and some marine invertebrates Placental and marsupial mammals diverge	Angiosperms evolve and diversify rapidly Seedless plants and gymnosperms still common but less varied and abundant	Climate becomes more seasonal and cooler at end of Cretaceous North-south zonation of climates more marked but remains equable
146–200	Jurassic	Ammonites and belemnoid cephalopods increase in diversity Scleractinian coral reefs common Appearance of rudist bivalves	Greatest Diversity of Dinosaurs First birds (may have evolved in Late Triassic) Time of giant sauropod dinosaurs	Seedless vascular plants and gymnosperms only	Much like Triassic Ferns with living relatives restricted to tropics live at high latitudes, indicating mild climates
200–251	Triassic	The seas are repopulated by invertebrates that survived the Permian extinction event Bivalves and echinoids expand into the infaunal niche	Mammals evolve from cynodonts Cynodonts become extinct Ancestral archosaur gives rise to dinosaurs Flying reptiles and marine reptiles evolve	Land flora of seedless vascular plants and gymnosperms as in Late Paleozoic	Warm-temperate to tropical Mild temperatures extend to high latitudes; polar regions may have been temperate Local areas of aridity

Important Terms

amniote (p. 594)
angiosperm (p. 593)
archosaur (p. 594)
bipedal (p. 594)
Cordilleran orogeny (p. 584)
Cretaceous Interior Seaway (p. 588)
cynodont (p. 602)
dinosaur (p. 594)
ectotherm (p. 599)
endotherm (p. 599)
ichthyosaur (p. 599)
iridium anomaly (p. 606)
Laramide orogeny (p. 585)
marsupial mammal (p. 603)
monotreme (p. 603)
mosasaur (p. 600)
Nevadan orogeny (p. 584)
Ornithischia (p. 595)
placental mammal (p. 603)
plesiosaur (p. 600)
pterosaur (p. 599)
quadrupedal (p. 594)
Saurischia (p. 595)
Sevier orogeny (p. 585)
Sonoma orogeny (p. 580)
Sundance Sea (p. 587)
terranes (p. 589)

Review Questions

1. The jaw–skull joint in mammals is between the
   a. ______ atlas and mandible.
   b. ______ occlusal and zygomatic.
   c. ______ dentary and squamosal.
   d. ______ articular and maxillary.
   e. ______ occipital and quadrate.
2. The time of greatest post-Paleozoic inundation of the craton occurred during which geologic period?
   a. ______ Triassic.
   b. ______ Jurassic.
   c. ______ Cretaceous.
   d. ______ Paleogene.
   e. ______ Neogene.
3. Among the therapsids, the cynodonts were
   a. ______ the most mammal-like.
   b. ______ the descendants of birds.
   c. ______ the smallest dinosaurs.
   d. ______ the ancestors of pterosaurs.
   e. ______ extinct by the end of the Paleozoic Era
4. A possible cause for the eastward migration of igneous activity in the Cordilleran region during the Cretaceous was a change from
   a. ______ high-angle to low-angle subduction.
   b. ______ divergent plate margin activity to subduction.
   c. ______ subduction to divergent plate margin activity.
   d. ______ oceanic–oceanic convergence to oceanic–continental convergence.
   e. ______ divergent to convergent plate margin activity.
5. Which one of the following is not an amniote?
   a. ______ reptile.
   b. ______ mammal.
   c. ______ bird.
   d. ______ dinosaur.
   e. ______ amphibian.
6. What is the most popular hypothesis for the extinctions at the end of the Mesozoic, and what evidence supports this idea?
7. From a plate tectonic perspective, how does the orogenic activity that occurred in the Cordilleran mobile belt during the Mesozoic Era differ from that which took place in the Appalachian mobile belt during the Paleozoic Era?
8. Describe the changes that took place in the jaw, middle ear, and teeth during the transition from the mammal-like reptiles to the mammals.
9. Explain how plate positions and climate influenced the geographic distribution of Mesozoic plants and animals.

10. **Creative Thinking Visual Question:** This view (▌Figure 1) in Capitol Reef National Park, Utah, shows excellent exposures of various Triassic and Jurassic sedimentary rock formations.
    a. What accounts for the reddish color of some of the rock layers?
    b. The rocks visible on the skyline belong to the Navajo Formation, which is exposed over a large area in the southwest. It is composed of well-sorted, well-rounded quartz sandstone, it has tracks of land-dwelling animals, including dinosaurs, and the sandstone has cross-beds up to 30 m high. How do you think it was deposited?

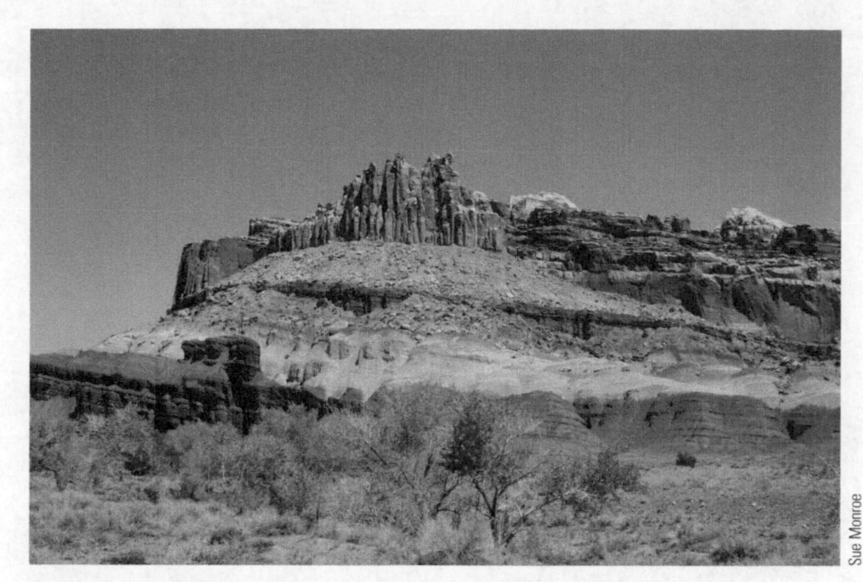

Sue Monroe

▌**Figure 1** This view in Capitol Reef National Park, Utah, shows excellent exposures of several sedimentary rock formations that were deposited during the Triassic and Jurassic periods.

Global GeoScience Watch Within the GREENR database, search for the term "Mammaliaform." This will take you to an article titled "A New Mammaliaform from the Early Jurassic and Evolution of Mammalian Characteristics." What features of this fossil indicate that it is a mammal rather than the mammals' closest reptile relatives (cynodonts), and what does it imply about the evolution of mammals?

Gastronis was a large, flightless, predatory bird that stood about 2 m tall and may have weighed 270 kg. It lived in North America and Europe during the Paleocene and Eocene epochs. Following the extinction of dinosaurs at the end of the Mesozoic Era, the large, land-dwelling predators during the Paleocene and into the Eocene were mostly flightless birds such as *Gastronis* and related genera.

James S. Monroe

CHAPTER 23

Cenozoic Earth and Life History

OUTLINE

HAVE YOU EVER WONDERED?

- How plate tectonics played a role in the distribution of Cenozoic continents?
- What events led to the origin of the present-day Rocky Mountains and Appalachian Mountains?
- Why western North America has a chain of active volcanoes?
- When and how the San Andreas Fault originated?
- What caused the extensive glaciers of the Ice Age?
- Why the Cenozoic Era is called the Age of Mammals?
- What caused the extinctions at the end of the Ice Age?
- What the evolutionary history of humans is (i.e., who our ancestors are, and how complete the fossil record is of our species and primates in general)?

23.1 Introduction

Remember the twenty-four-hour clock that represented the entire 4.6 billion years of geologic time (see Figure 19.1). On this hypothetical clock, the Cenozoic Era is only 20 minutes long, or simply 1.4% of all geologic time. Perhaps this is a "short" interval in the context of Earth's history, but by any other measure, it is far longer than we can even imagine. Certainly it was long enough for significant evolution of Earth and its biota. In fact, if we could somehow go back to the beginning of the Cenozoic Era, Earth's geography would be strange to us because Earth's tectonic plates were not where they are now; North America was similar in appearance but not the same as it is today; and some mountain systems, such as the Himalayas, had not yet formed. And even though we would likely recognize some of the land animals as mammals, most would not be at all familiar.

Until recently, the Cenozoic Era was divided into two periods, the Tertiary and the Quaternary, terms that you are likely to encounter in the geological literature. However, the International Commission on Stratigraphy now divides the Cenozoic into three geologic periods—the *Paleogene* (66 to 23 million years ago), the *Neogene* (23 to 2.6 million years ago), and the *Quaternary* (2.6 million years ago to the present). Each of these periods is further divided into epochs (see Figure 17.1). This usage is now recommended and is followed in this book.

Geologists know more about Cenozoic Earth and life history than they do about other intervals of geologic time because Cenozoic-age rocks, being the youngest, are the most accessible at or near Earth's surface, and they have been altered less by deformation and metamorphism. Furthermore, the fossil record for Cenozoic life, although not truly representative of all life of that time, is very good for some groups of organisms, especially some orders of mammals. In fact, even the fossil record of humans is better than most people realize.

We emphasize the evolution of mammals in this chapter, but other important events occurred. For instance, the angiosperms continued to diversify and now represent more than 90% of all land plant species. Following the mass extinctions at the end of the Mesozoic Era, marine invertebrates diversified once again, giving rise to today's familiar marine biota. The present-day families of birds made their appearance during the Paleogene and Neogene and are now the most numerous land-dwelling vertebrate animals.

The Cenozoic was also the time that Earth became more and more like it is today as Pangaea continued to fragment, which in turn had profound effects on oceanic circulation patterns and climate. And, of course, these changes also resulted in migrations of organisms or their isolation in areas such as Australia. The Appalachian Mountains began their long, complex evolution during the Proterozoic Eon, but their present-day topography developed mostly as a result of Cenozoic uplift and erosion. And likewise, the San Andreas Fault of California, the Himalayas of Asia, the Andes of South America, and the Grand Canyon in Arizona owe their distinctive features to Cenozoic events.

Previous chapters emphasized the dynamic nature of Earth as it has evolved from an accumulation of planetesimals to its present state. Recall also, that Earth's dynamic nature results from its internal heat, as well as its interaction with the atmosphere, biosphere, and hydrosphere. Furthermore, Earth and its biota are not finished products, but rather both are works in progress. Although scientists cannot predict Earth's far distant future, they can be certain that our planet will continue to evolve, but eventually it will lose its internal heat, both residual and that generated by radioactive decay, and heat-driven geologic activities (moving plates, volcanism, and seismic activity) will cease.

23.2 Cenozoic Plate Tectonics

The ongoing fragmentation of Pangaea accounts for the present distribution of Earth's landmasses. We noted in Chapter 22 that as the Americas separated from Europe and Africa, the Atlantic Ocean basin opened, first in the south and later in the north. The Mid-Atlantic Ridge and East Pacific Rise were established, as spreading ridges along which new oceanic crust formed and continues to form. However, the age of the oceanic crust in the Pacific is very asymmetric, because much of the crust in the eastern Pacific Ocean basin has been subducted beneath the westerly moving North and South American plates (see Figure 2.16).

Neogene rifting also began in East Africa, the Red Sea, and the Gulf of Aden (see Figure 2.19). Rifting in East Africa is in its early stages, because the continental crust has not yet stretched and thinned enough for new oceanic crust to form from below. In the Red Sea, rifting and the Late Pliocene origin of oceanic crust followed vast eruptions of basalt. In fact, ongoing volcanism formed a new island along the Red Sea rift in December 2011. And in the Gulf of Aden, Earth's crust had stretched and thinned enough by Late Miocene time for upwelling basaltic magma to form new oceanic crust. The Arabian plate is moving north, so it, too, causes some of the deformation taking place from the Mediterranean through India.

In the meantime, the North and South American plates continued their westerly movement as the Atlantic Ocean basin widened (▮ Figure 23.1). Subduction zones bounded both continents on their western margins, but the situation changed in North America as it moved over the northerly extension of the East Pacific Rise, and it now has a transform plate boundary, a topic we discuss more fully in a later section.

23.3 Cenozoic Orogenic Belts

Cenozoic orogenic activity took place in two major zones or belts: the *Alpine–Himalayan orogenic belt* and

For a review of orogenies, see Chapter 10.

Figure 23.1 Paleogeography of the World for the Eocene and Miocene Epochs

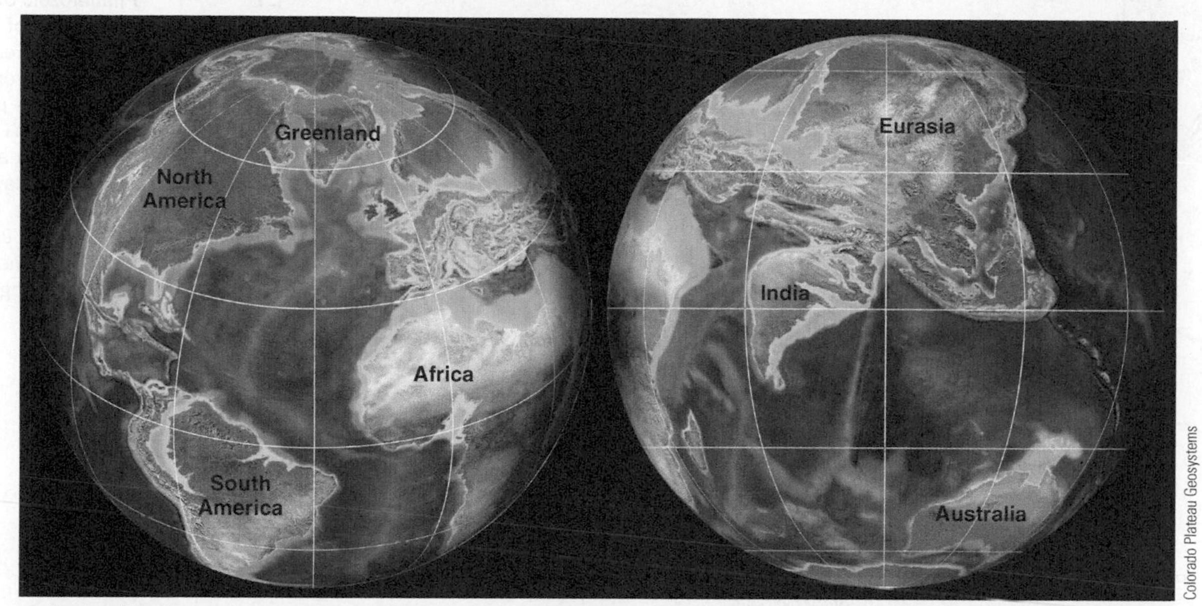

a Eocene Epoch.

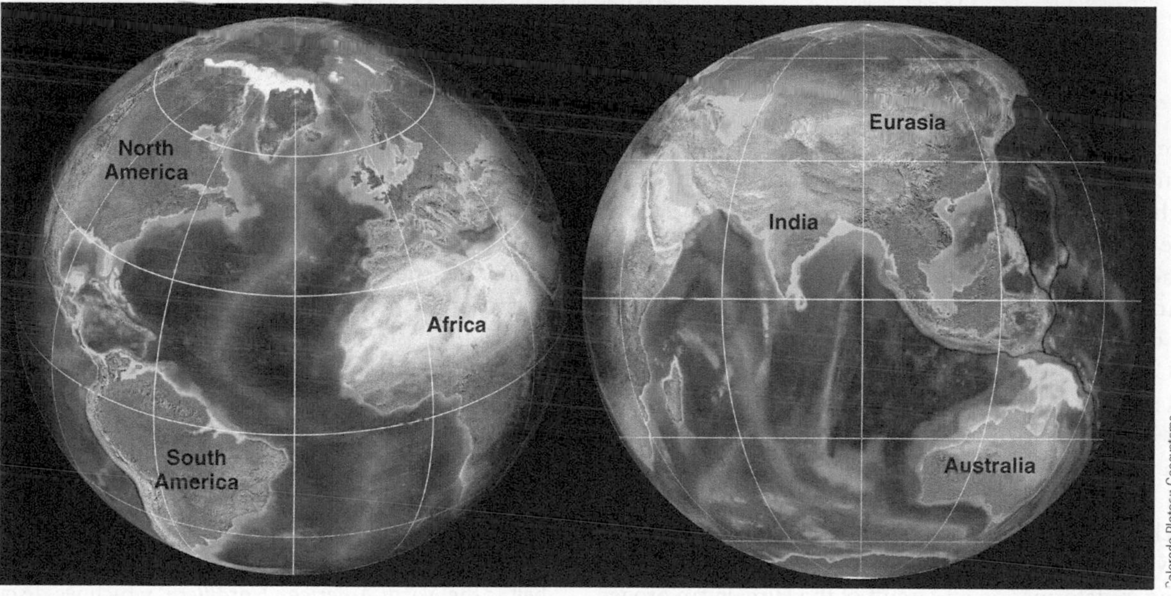

b Miocene Epoch.

the *circum-Pacific orogenic belt* (Figure 23.2). Both belts are made up of smaller segments known as **orogens,** each of which shows the features of an orogeny—deformation, metamorphism, emplacement of plutons, and thickening of Earth's crust.

The Alpine–Himalayan Orogenic Belt

The **Alpine–Himalayan orogenic belt** extends eastward from Spain through the Mediterranean region, the Middle East and India, and on into Southeast Asia (Figure 23.2). Ongoing deformation, volcanism, and seismicity remind us that it remains active. Remember that during the Mesozoic, the Tethys Sea separated much of Gondwana from Eurasia, but this sea closed during the Cenozoic when the African plate collided with the Eurasian plate to the north.

During the *Alpine orogeny,* deformation occurred in a linear zone extending from Spain eastward through Greece and Turkey. Mountain building yielded the Pyrenees between Spain and France and the Apennines of Italy, as well as the Alps of mainland Europe. Plate convergence also produced an almost totally isolated sea in the Mediterranean basin where Late Miocene deposition in an arid environment accounts for evaporite deposits up to 2 km thick.

The collision of the African plate with Eurasia also accounts for the origin of the Atlas Mountains of northwest

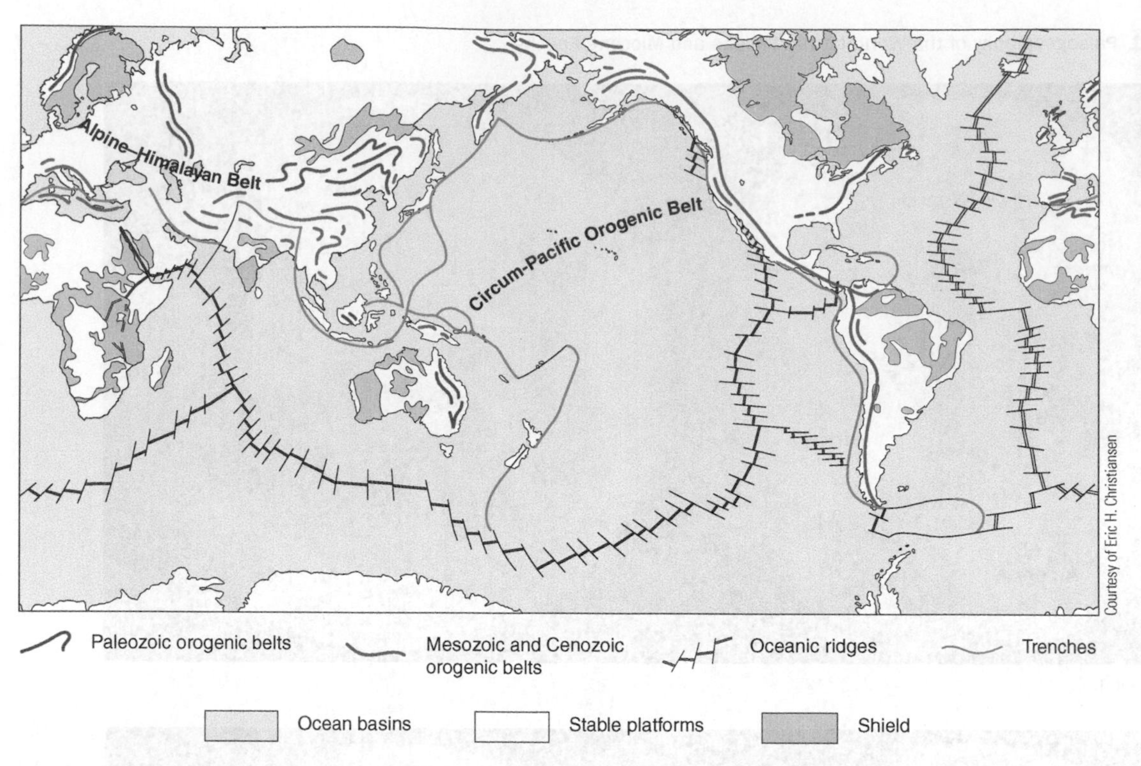

Figure 23.2 Phanerozoic Orogenic Belts Most of Earth's geologically recent and continuing orogenic activity takes place in the circum-Pacific orogenic belt and the Alpine–Himalayan orogenic belt. Each belt is made up of smaller segments known as orogens.

Africa, and farther to the east, Africa continues to force oceanic lithosphere northward beneath Greece and Turkey. Erupting volcanoes in Italy and Greece and seismic activity throughout the entire region indicate that southern Europe and the Middle East remain geologically active.

The *Himalayan orogen* also resulted from plate convergence, but in this case, two continental plates collided, giving rise to the loftiest mountains on Earth. During the Early Cretaceous, India broke away from Gondwana and moved north, and as it did so, oceanic lithosphere was consumed at a subduction zone along the southern margin of Asia (see Figure 10.21a). The descending plate partially melted, yielding magma that rose to form granitic plutons and a chain of volcanoes in what is now Tibet. The Indian plate eventually collided with and became sutured to Asia, resulting in deformation and uplift of the Himalayan orogen (see Figure 10.21b).

India's collision probably began about 40–50 million years ago when the northward movement of the Indian plate decreased from 15–20 cm/yr to about 5 cm/yr. This decrease probably marks the time of collision and India's resistance to subduction, but the Indian plate has been thrust about 2,000 km beneath Asia, causing crustal thickening. This underthrusting continues at a rate of about 5 cm/yr.

The Circum-Pacific Orogenic Belt

The **circum-Pacific orogenic belt** is an area of tectonism made up of several orogens along the western margins of South, Central, and North America, as well as the Aleutian Islands, the eastern margin of Asia, the islands north of Australia, and New Zealand (Figure 23.2). Subduction of oceanic lithosphere causes deformation and volcanism in the orogens in the western and northern Pacific. For example, the Pacific plate is subducted beneath Japan, the Philippines, and the Aleutian Islands, accounting for continuing tectonism in these areas.

In the eastern part of the Pacific, the Cocos and Nazca plates moved west from the East Pacific Rise only to be consumed at subduction zones along the west coasts of Central and South America (see Figure 10.20). This entire region also remains tectonically active. In South America, for instance, the Andes Mountains, with about 50 peaks higher than 6,000 m, also have many active volcanoes, and it is an area of seismic activity. Another important part of the circum-Pacific orogenic belt is the *North American Cordillera*, which we discuss in the next section.

23.4 Paleogene and Neogene Evolution of North America

We mentioned in the Introduction that many of Earth's features have long histories, but they owe their present-day distinctive characteristics to events that took place during the Cenozoic Era. Here we concentrate on the events of the Paleogene and Neogene periods, although we should note that Earth's evolution continued during the Quaternary Period, which we cover later in this chapter.

The North American Cordillera

The **North American Cordillera,** a complex mountainous region in the West, is a huge segment of the circum-Pacific orogenic belt extending from Alaska southward through Canada and the continental United States and into central Mexico. In the United States, it widens to 1,200 km, stretching east-west from the east flank of the Rocky Mountains to the Pacific Ocean (▮ Figure 23.3). Although the geologic evolution of this region began during the Proterozoic, a protracted episode of deformation known as the *Cordilleran orogeny* began during the Late Triassic as the Nevadan, Sevier, and Laramide orogenies progressively affected areas from west to east (see Figure 22.9). The first two of these orogenies were discussed in Chapter 22, so here we are concerned with the Laramide orogeny.

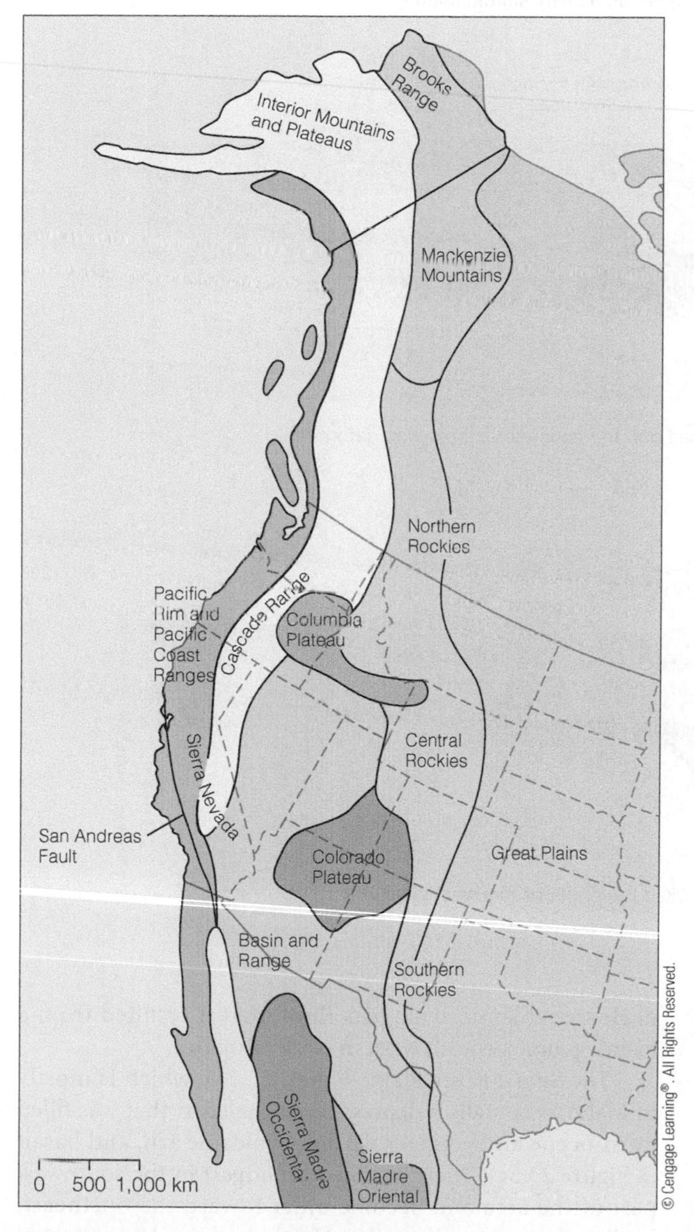

▮ **Figure 23.3 The North American Cordillera** The North American Cordillera is a complex mountainous area extending from the eastern margin of the Rocky Mountains to the Pacific coast.

The Late Cretaceous to Eocene **Laramide orogeny** was the last episode of mountain building that took place during the Cordilleran orogeny, but it differed in important ways from its two predecessors. For one thing, it took place much farther inland from a convergent plate boundary than the Nevadan and Sevier orogenies. In addition, there was little or no volcanism or the emplacement of plutons during this orogeny. And lastly, deformation was mostly in the form of vertical, fault-bounded uplifts with cores of Precambrian rocks, rather than the compression-induced folding and thrusting typical of orogenies at convergent plate boundaries.

To account for these differences, geologists have modified their model for orogenies at convergent plate boundaries. During the Nevadan and Sevier orogenies, the oceanic Farallon plate was subducted beneath North America at about a 50-degree angle, and volcanism and plutonism took place 150–200 km inland from the oceanic trench. Most geologists now agree that by Early Paleogene time, there was a change in the subduction angle from steep to gentle, and the Farallon plate moved nearly horizontally beneath the continent. However, they disagree on what may have caused this change in angle of subduction.

According to one hypothesis, a buoyant oceanic plateau that was part of the Farallon plate was carried beneath the continent and resulted in shallow subduction. Another hypothesis holds that North America overrode the Farallon plate beneath which was the deflected head of a mantle plume (▮ Figure 23.4a). The lithosphere above the plume was buoyed up, accounting for shallow subduction. As a result, igneous activity shifted farther inland and finally ceased because the descending plate no longer penetrated to the mantle (▮ Figure 23.4b).

The changing angle of subduction also resulted in a change in the style of deformation that involved large-scale buckling and fracturing, which yielded fault-bounded vertical uplifts. Erosion of the uplifted blocks yielded mountainous topography and erosion of the mountains supplied sediments to the intervening basins. Deformation also accounted for huge overthrust faults, especially in the Northern Rocky Mountains. By Middle Eocene time, Laramide deformation ceased and volcanism resumed when the mantle plume beneath the lithosphere disrupted the overlying oceanic plate (▮ Figure 23.4c). By the Neogene, the mountains had been deeply eroded and the basins filled with sediment; however, renewed uplift and erosion during the Late Neogene accounts for the present ranges.

Cordilleran Igneous Activity

The enormous batholiths in British Columbia, Canada, Idaho, and the Sierra Nevada of California were emplaced during the Mesozoic (see Chapter 22), but intrusive activity continued into the Paleogene Period. Numerous small plutons formed, including copper- and molybdenum-bearing stocks in Utah, Nevada, Arizona, and New Mexico.

Figure 23.4 Laramide Orogeny The Late Cretaceous to Eocene Laramide orogeny took place as the Farallon plate was subducted beneath North America.

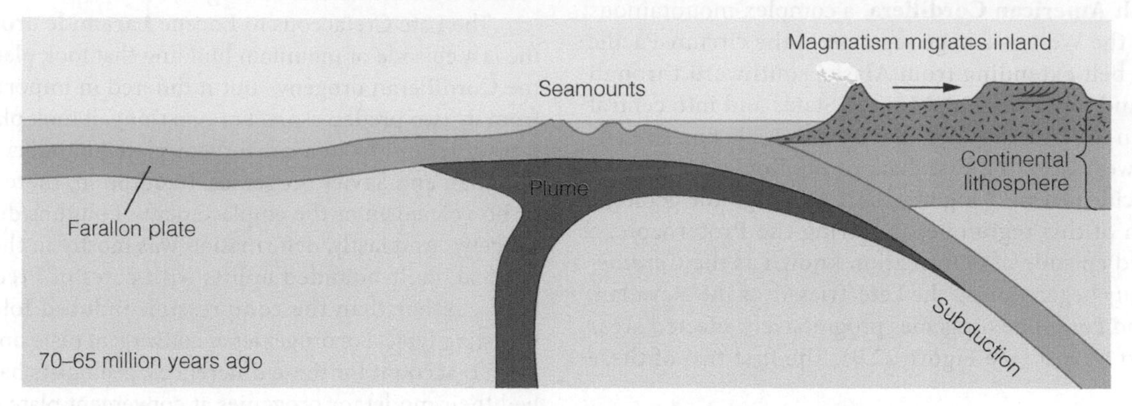

a As North America moved westward over the Farallon plate, beneath which was the deflected head of a mantle plume, the angle of subduction decreased, and the igneous activity shifted inland.

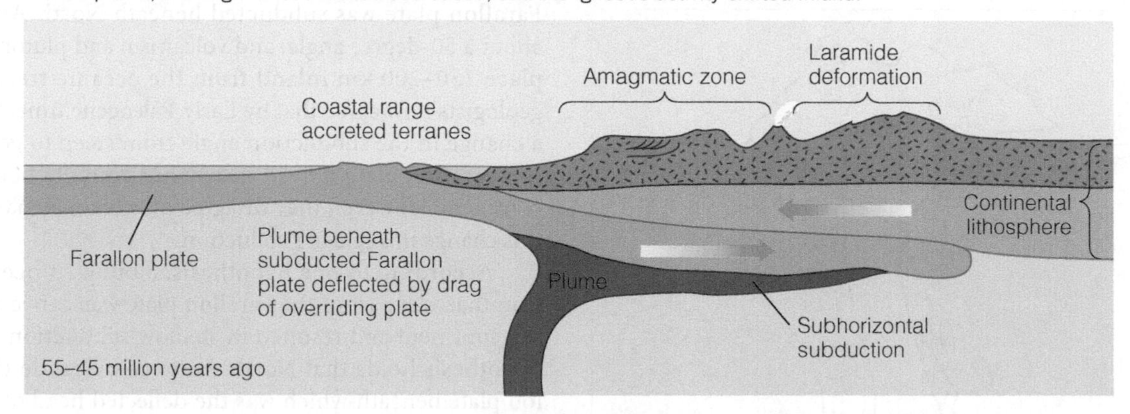

b With nearly horizontal subduction, igneous activity ceased and the continental crust was deformed, mostly by vertical uplift.

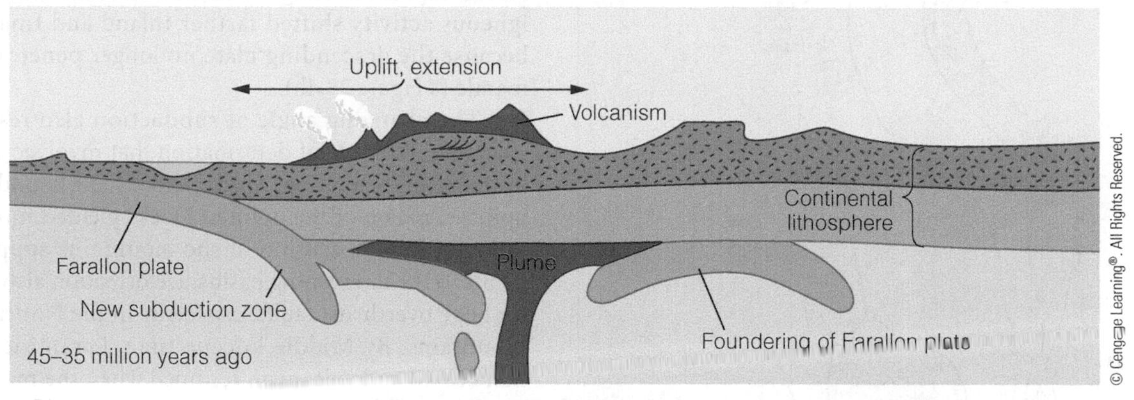

c Disruption of the oceanic plate by the mantle plume marked the onset of renewed igneous activity.

Volcanism was more or less continuous in the Cordillera, but it varied in location, intensity, and eruptive style, and it ceased temporarily in the area of the Laramide orogen (Figure 23.4b). In the Pacific Northwest, the Columbia Plateau (▮ Figure 23.5a) is underlain by 200,000 km^3 of Miocene lava flows called the Columbia River basalts that have an aggregate thickness of about 2,500 m (Geo-Insight 23.1). These vast lava flows are well exposed in the walls of the canyons eroded by the Columbia and other rivers (▮ Figure 23.5b). The relationship of this huge outpouring of lava to plate tectonics remains unclear, but some geologists think that it resulted from a mantle plume beneath western North America.

The Snake River Plain (Figure 23.5a), which is mostly in Idaho, is actually a depression in the crust that was filled by Miocene and younger rhyolite, volcanic ash, and basalt (▮ Figure 23.5c). These rocks are youngest in the southwest part of the area and become older toward the northeast, leading some to propose that North America has migrated over a mantle plume that now lies beneath Yellowstone National Park in Wyoming. Other geologists disagree,

Figure 23.5 Cenozoic Volcanism

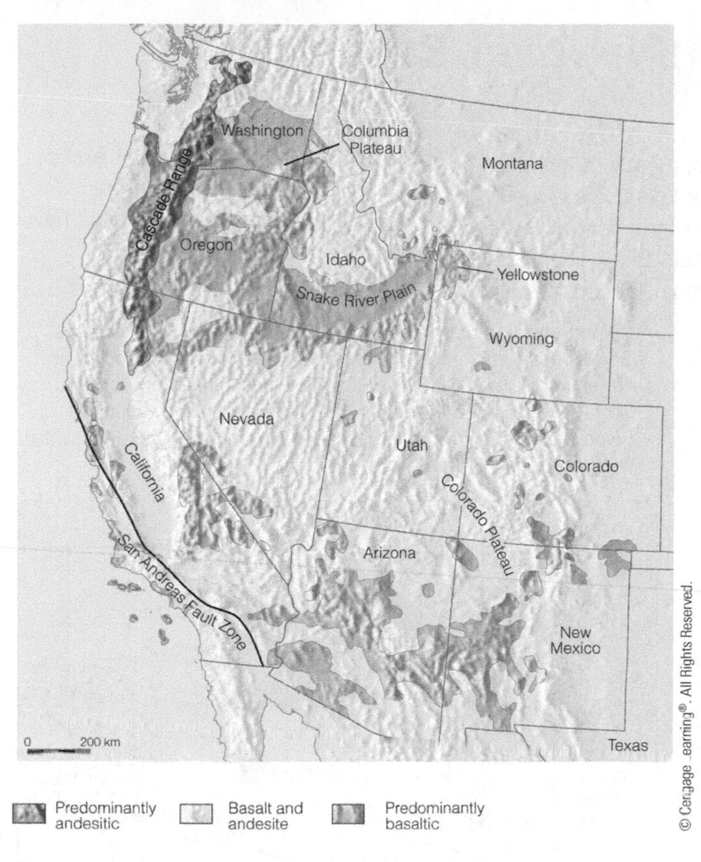

a Distribution of Cenozoic volcanic rocks in the western United States.

b Columbia River basalts are exposed in the walls of this canyon eroded by a tributary of the Columbia River in Oregon. Multnomah Falls plunge 189 m from a small tributary valley.

c Basalt lava flows of the Snake River Plain at Malad Gorge State Park, Idaho.

thinking that these volcanic rocks erupted along an intracontinental rift zone.

Bordering the Snake River Plain on the northeast is the Yellowstone Plateau (Figure 23.5a), an area of Pliocene and Pleistocene volcanism. Perhaps a mantle plume lies beneath the area, as just noted, that accounts for the ongoing hydrothermal activity there, but the heat may come from an intruded body of magma that has not yet completely cooled.

Some of the most majestic and highest mountains in the Cordillera are in the **Cascade Range** of northern California, Oregon, Washington, and southern British Columbia, Canada (Figure 23.5a). Thousands of volcanic vents are present, the most impressive of which are the dozen or so large composite volcanoes and Lassen Peak in California, the world's largest lava dome. Volcanism in this region is related to subduction of the Juan de Fuca plate beneath North America. Volcanism in the Cascade Range goes back at least to the Oligocene, but the most recent episode began during the Late Miocene or Early Pliocene and continues to the present.

Connection Link

For more on Cascade Range Volcanism, see Chapter 5.

Basin and Range Province

Earth's crust in the **Basin and Range Province** (Figure 23.6a)—an area centered on Nevada but extending into adjacent states and northern Mexico—has been stretched and thinned, yielding north-south-oriented mountain ranges with intervening valleys or basins. The ranges are bounded on one or both sides by steeply dipping normal faults that probably curve and dip less steeply with depth.

Before faulting began, the region was deformed during the Nevadan, Sevier, and Laramide orogenies. Then, during the Paleogene, the entire area was highlands undergoing extensive erosion, but Early Miocene eruptions of rhyolitic lava flows and pyroclastic materials covered large areas. By the Late Miocene, large-scale faulting had begun, forming the basins and ranges. Sediment derived from the ranges was transported into the adjacent basins and accumulated as alluvial fan and playa lake deposits.

At its western margin, the Basin and Range Province is bounded by normal faults along the east face of the Sierra Nevada (Figure 23.6b). Pliocene and Pleistocene uplift tilted the Sierra Nevada toward the west, and its crest now stands 3,000 m above the basins to the east. Before this uplift took place, the Basin and Range Province had a subtropical climate,

GEO INSIGHT 23.1 The Columbia River Basalts

The Columbia Plateau is a large area mostly in Washington State, but parts of it are in Idaho and Oregon. This vast area is underlain by at least 300 huge lava flows as well as many smaller ones, mostly of basaltic composition. These lava flows issued from fissures rather than central vents, and they were so fluid that volcanic cones did not develop. Indeed, the Columbia Plateau is rather flat except where rivers and the catastrophic floods from Glacial Lake Missoula have eroded deep canyons.

▼ **1.** Map showing the distribution of the Columbia River basalts. Some basalt lava flows of this region were of vast proportions. For example, the 30-m-thick Roza flow advanced along a front 100 km wide and covered 40,000 km^2. Notice the Columbia River, which is by far the largest river in terms of discharge in the western United States.

CANADA
WASHINGTON
Columbia River
Grand Coulee Dam
SEATTLE
Pacific Ocean
Mount Rainier
Clearwater River
Mount St. Helens
Columbia River
McNary
LEWISTON-CLARKSTON
PASCO-KENNEWICK-RICHLAND
PORTLAND
Mount Hood
Mount Jefferson
Willamette River
Snake River
OREGON
IDAHO
Columbia River Flood Basalts
Cascade Volvano
0 100 100
Km

http://vulcan.wr.usgs.gov/Volcanoes/ColumbiaPlateau/Maps/map_columbia_river_flood_basalts.html

James S. Monroe

▲ **2.** Much of the Columbia Plateau consists of rather flat-lying land and rolling hills except where eroded by rivers.

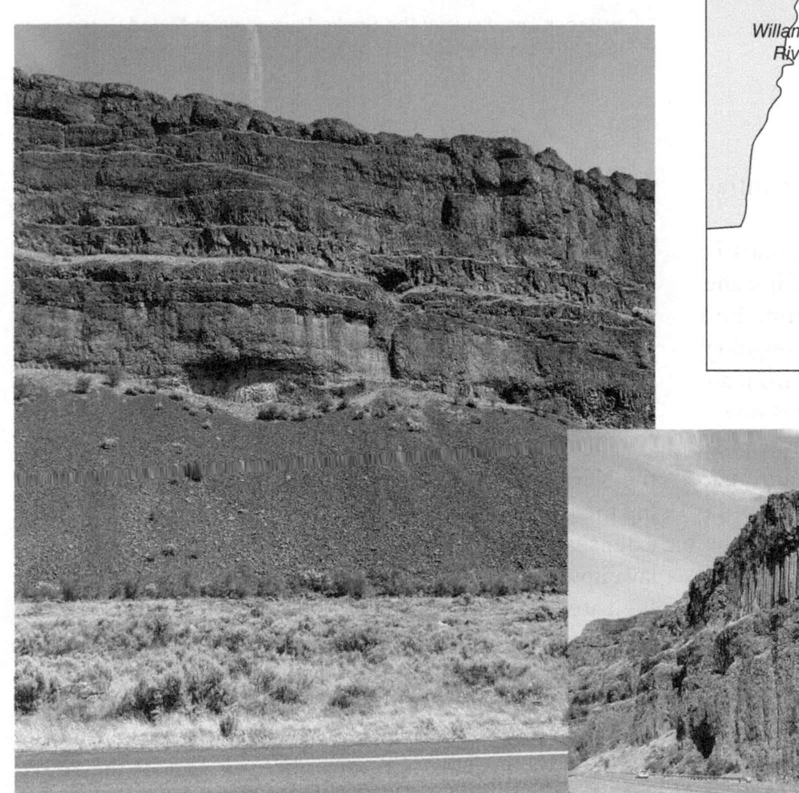

James S. Monroe

James S. Monroe

◄ **3. and 4.** Exposures of several lava flows of the Columbia River basalts in Moses Coulee, which was eroded in part by the catastrophic floods of Glacial Lake Missoula (see Figure 23.16a). Notice the well-developed columnar joints in 4.

▼ **5.** The Grand Coulee Dam on the Columbia River, built between 1933 and 1942, stands 168 m high and is 1,592 m across. The reservoir on the upstream side of the dam is Lake Roosevelt. The Grand Coulee Dam provides water for irrigation, recreation, and power generation. Notice in the lower left part of the image the three tube-like structures leading from the face of the dam. These are called penstocks, which conduct water from the reservoir to generators where electricity is generated.

James S. Monroe

James S. Monroe

▲ **6.** Notice the parallel lines on the side of this mountain. These are ancient shorelines of Glacial Lake Missoula at Mount Jumbo in Missoula, Montana. This huge lake drained catastrophically several times during the Pleistocene causing extensive erosion of the Columbia Plateau. A geologist at the United States Geological Survey estimates that these floods occurred many times, perhaps as many as 25, between 15,000 and 13,000 years ago.

▼ **7.** When the ice dam impounding Glacial Lake Missoula failed, the floodwaters rushed across this area known as Dry Falls State Park in Washington State. Dry Falls are about 3.5 km long and stand 120 m high.

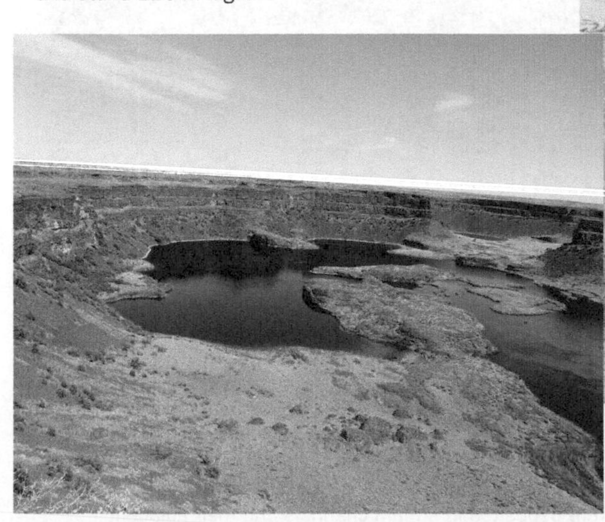

James S. Monroe

Karl Anders/http://drfumblefinger.com/

▲ **8.** Restoration of what Dry Falls was like during one of the floods from Glacial Lake Missoula (also called the Spokane floods). Although the results of these floods are widespread in Washington, their effects are also seen in western Montana, Idaho, and northern Oregon.

Figure 23.6 The Basin and Range Province

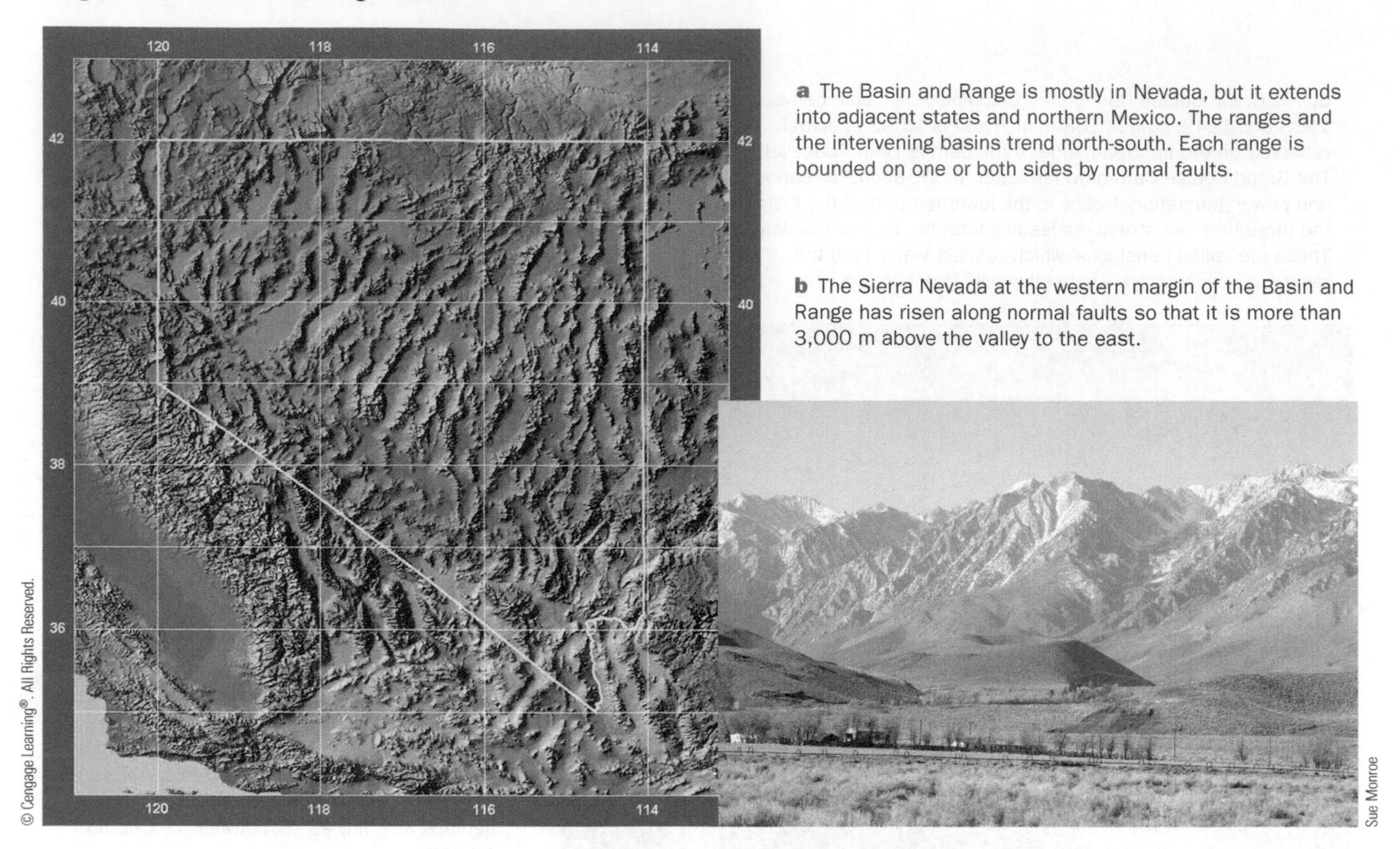

a The Basin and Range is mostly in Nevada, but it extends into adjacent states and northern Mexico. The ranges and the intervening basins trend north-south. Each range is bounded on one or both sides by normal faults.

b The Sierra Nevada at the western margin of the Basin and Range has risen along normal faults so that it is more than 3,000 m above the valley to the east.

Sue Monroe

but the rising mountains created a rain shadow, making the climate increasingly arid.

Colorado Plateau and Rio Grande Rift

The vast, elevated **Colorado Plateau** in Colorado, Utah, Arizona, and New Mexico, has volcanic mountains rising above it, brilliantly colored rocks, and deep canyons. In Chapters 20 and 22, we noted that during the Permian and Triassic, the Colorado Plateau region was the site of extensive red bed deposition; many of these rocks are now exposed in the uplifts and canyons (Figure 23.7).

Cretaceous-age marine sedimentary rocks indicate that the Colorado Plateau was below sea level then, but during the Paleogene Period, Laramide deformation yielded broad anticlines and arches and basins, and a number of large normal faults. However, deformation was much less intense than it was elsewhere in the Cordillera. Neogene uplift elevated the region from near sea level to the 1,200- to 1,800-m elevations seen today, and as uplift proceeded, streams and rivers began eroding deep canyons.

Geologists disagree on the details of how the deep canyons so typical of the region developed—such as the Grand Canyon. Some think that the streams were *antecedent*, meaning that they existed before the present topography developed, in which case they simply eroded downward as uplift proceeded. Others think that the streams were *superposed*, implying that younger strata covered the area on which streams

James S. Monroe

Figure 23.7 Cenozoic Rocks of the Colorado Plateau The rocks of the Colorado Plateau range from Proterozoic to Cenozoic. As uplift of the plateau took place, the area was deeply eroded, so that now we see deep canyons and the erosional remnants of these ancient rocks. This image shows the 31- to 34-million-year-old Claron Formation in Bryce Canyon National Park, Utah. These sedimentary rocks consists of conglomerate, sandstone, shale, and limestone that were deposited in stream channels and their adjacent floodplains and in lakes.

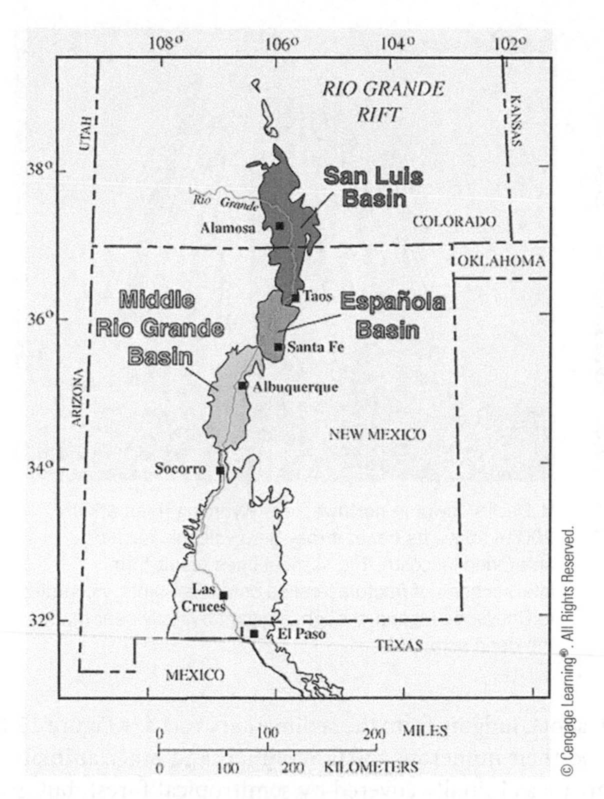

Figure 23.8 The Rio Grande Rift Location of the basins making up the Rio Grande rift. A complex of normal faults is present on both sides of the rift.

were established. During uplift, the streams stripped away these younger rocks and eroded down into the underlying strata. In either case, the landscape continues to evolve as erosion of the canyons and their tributaries deepens and widens them.

The **Rio Grande rift** extends north to south about 1,000 km from central Colorado through New Mexico and into northern Mexico (Figure 23.8) where Earth's crust has been stretched and thinned. It is bounded on both sides by normal faults, volcanoes and calderas are present in the rift, and seismic activity continues. Rifting began 29 million years ago and persisted for 10–12 million years; after a hiatus, renewed rifting began 17 million years ago and continues even now, but only at about 2 mm per year. As rifting took place, the rift was filled with huge amounts of sediments and volcanic rocks.

Pacific Coast

Before the Eocene, the entire Pacific Coast was a convergent plate boundary where the **Farallon plate** was consumed at a subduction zone that stretched from Mexico to Alaska. Now there are only two small remnants of the Farallon plate—the Juan de Fuca and Cocos plates (Figure 23.9). Continuing subduction of these small plates accounts for the present seismic activity and volcanism in the Pacific Northwest and Central America.

Another consequence of plate interactions in this region was the westward movement of the North American plate and its collision with the Pacific–Farallon ridge. Because the Pacific–Farallon ridge was at an angle to the margin of North America, the continent–ridge collision took place first during the Eocene in northern Canada and only later during the Oligocene in southern California (Figure 23.9). In southern California, two triple junctions formed, one at the intersection of the North American, Juan de Fuca, and Pacific plates, and the other at the intersection of the North American, Cocos, and Pacific plates. Continued westward movement of the North American plate over the Pacific plate caused the triple junctions to migrate, one to the north and the other to the south, giving rise to the **San Andreas transform Fault**

Figure 23.9 Origin of the San Andreas and Queen Charlotte Faults Three stages in the westward movement of North America and its collision with the Pacific–Farallon Ridge. As North America overrode the ridge, its margin became bounded by transform faults except in the Pacific Northwest.

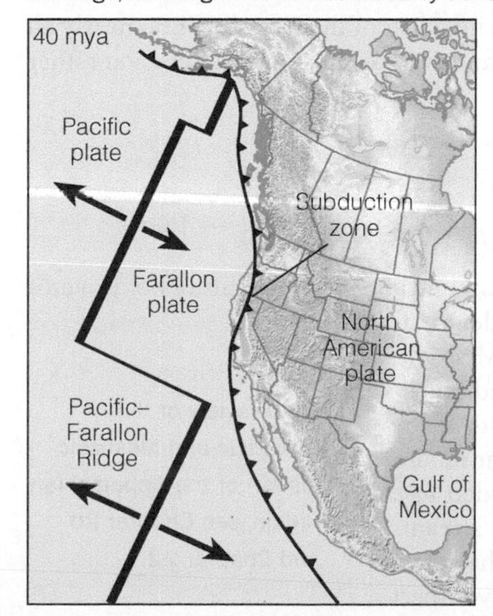

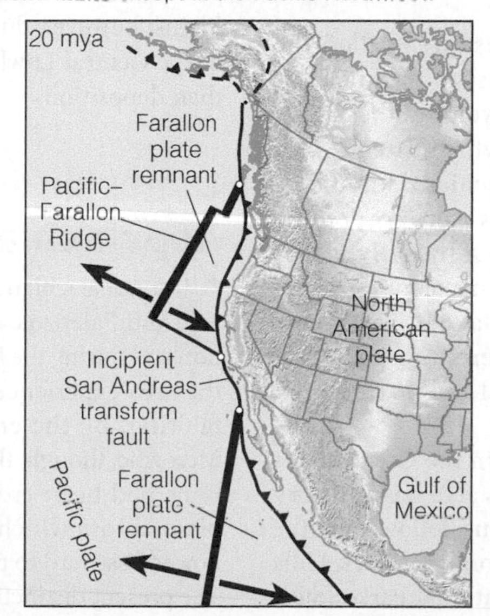

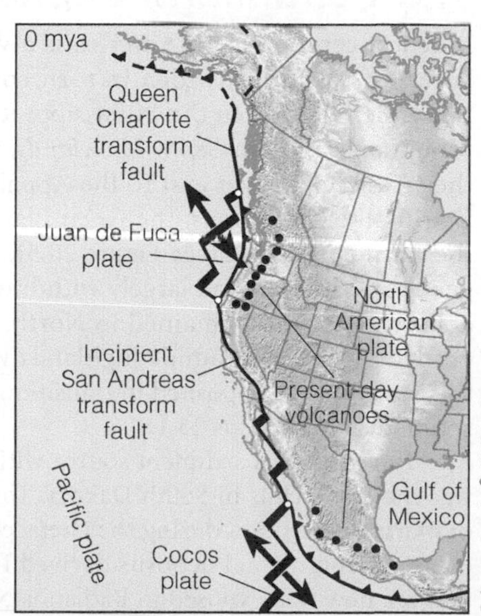

Figure 23.10 Cenozoic Sedimentary and Igneous Rocks in the Great Plains

Sue Monroe

a The Oligocene-age Brule Formation in Badlands National Park in South Dakota was deposited mostly in stream channels and on their floodplains.

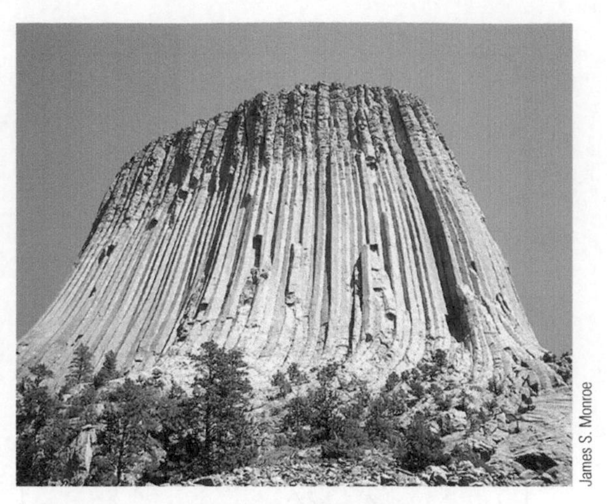

James S. Monroe

b Devil's Tower in northwestern Wyoming rises about 260 m above its base. It may be a volcanic neck or an eroded laccolith. The vertical lines result from intersections of fractures called columnar joints. According to Cheyenne legend, though, a gigantic grizzly bear made the deep scratches.

(Figure 23.9). A similar event along Canada's west coast produced the *Queen Charlotte transform Fault.*

Seismic activity on the San Andreas Fault results from continuing movements of the Pacific and North American plates along this complex zone of shattered rocks. Indeed, where the fault cuts through coastal California, it is actually a zone as much as several kilometers wide, and it has numerous branches. Movements on such complex fault systems subject blocks of rocks adjacent to and within the fault zone to extensional and compressive stresses, forming basins and elevated areas, the latter supplying sediments to the former. Many of the fault-bounded basins in the southern California area subsided below sea level and soon filled with turbidites and other deposits. A number of these basins are areas of prolific oil and gas production.

The Continental Interior

Much of central North America is a vast area called the **Interior Lowlands,** which in turn consists of the *Great Plains,* lying between the Mississippi River and the Rocky Mountains, and the *Central Lowlands,* which extend from the Mississippi River east to the Appalachian Mountains. During the Cretaceous, the Great Plains were covered by the **Zuni epeiric sea** (see Figure 20.5), but by Early Paleogene time, this sea had largely withdrawn except for a sizable remnant that remained in North Dakota. Sediments eroded from the Laramide highlands were transported to this sea and were deposited in transitional and marine environments (Geo-Focus 23.1).

The only local sediment source within the Great Plains was the Black Hills in South Dakota. This area has a history of marine deposition during the Cretaceous, followed by the origin of continental deposits derived from the Black Hills that are now well exposed in Badlands National Park, South Dakota. Judging from the sedimentary rocks (Figure 23.10a) and their numerous fossil mammals and other animals, the area was initially covered by semitropical forest, but grasslands replaced the forests as the climate became more arid.

Igneous activity was not widespread in the Interior Lowlands, but it was significant in some parts of the Great Plains. For instance, igneous activity in northeastern New Mexico was responsible for volcanoes and numerous lava flows, and several small plutons were emplaced in Colorado, Wyoming, Montana, South Dakota, and New Mexico. Indeed, one of the most widely recognized igneous bodies in the entire continent, Devils Tower in northeastern Wyoming, is probably an Eocene volcanic neck, although some geologists think that it is an eroded laccolith (Figure 23.10b). Another prominent volcanic feature is Shiprock in New Mexico, a volcanic neck that dates from the Late Oligocene.

Pleistocene glacial deposits are present in the northern part of the Central Lowlands, as well as in the northern Great Plains; however, during most of the Cenozoic Era, nearly all of the Central Lowlands was an area of active erosion rather than deposition.

Cenozoic History of the Appalachian Mountains

The Appalachian Mountain region has a history of Proterozoic and Paleozoic deformation and during the Mesozoic, the area experienced block-faulting. By the end of the Mesozoic, though, the mountains had been eroded to a plain across which streams flowed eastward to the ocean. The present distinctive aspect

ConnectionLink

For a review of the Paleozoic and Mesozoic history of the Appalachian region, see Chapter 20 and Chapter 22.

GEO FOCUS 23.1

Geology Along the Oregon Trail in Nebraska

The Oregon Trail flourished as a 3,200 km avenue for migration from Independence, Missouri, to Oregon, a journey that would take four to six months. When it was first established in the early 1830s, the trail crossed what would become Kansas, Nebraska, Wyoming, Idaho, and Oregon, and branches from the trail led to California and Utah. As part of the historical heritage of the United States, the trail is interesting in its own right; some estimate that 400,000 people migrated west along the trail from 1841 until 1869, when the Transcontinental Railroad finally linked east and west.

You can see many interesting geologic features along the Oregon Trail, but here we concentrate on a segment in Nebraska, particularly in the western part of the state. We should also note, though, that the Oregon Trail followed the Platte and North Platte rivers across Nebraska and into Wyoming. In Wyoming, the river is swift flowing, but when it reaches Nebraska, it is wide, shallow, and muddy, or, as it is more popularly described, "too thin to plow, too thick to drink."

Following the Laramide orogeny, which ended during the Eocene, the area we now call Nebraska and adjacent parts of the Great Plains was a site of detrital deposition in stream channels and their floodplains as well as lakes and swamps. Some of the sediment from the mountains to the west was carried across the Great Plains and eventually deposited on the Gulf Coastal Plain or in the Gulf of Mexico. Our interest here, however, is the Paleogene and Neogene sedimentary rocks at Chimney Rock National Historic Site and Scotts Bluff National Monument, Nebraska.

Chimney Rock is one of several landmarks along the Oregon Trail where it crossed the Great Plains, or what the immigrants called the Great American Desert. Indeed, Chimney Rock, an 87-m-high erosional remnant with a spire about 36 m high was, according to one traveler in 1832, visible from about 48 km (Figure 1). It is made up of rocks of the Eocene- to Oligocene-age Arikaree Group, which is mostly siltstone but also contains layers of sandstone, claystone, volcanic ash, and some limestone. In fact, hard layers of sandstone in Chimney Rock's spire protect the softer rocks below. During the time that wagon trains followed the Oregon Trail, many people climbed the lower slopes of Chimney Rock to carve their names on the spire.

Zack Frank/Shutterstock

Figure 1 Chimney Rock is a prominent landmark along the Oregon Trail in western Nebraska. This erosional remnant rises 87 m above the surrounding plain. It is made up of several types of Eocene to Oligocene sedimentary rocks.

John Brueske/Shutterstock.com

Figure 2 Scotts Bluff in Nebraska is made up mostly of Oligocene- to Miocene-age siltstone and sandstone, but it also has two prominent layers of volcanic ash. It is about 150 m high, and, like Chimney Rock, it was a landmark along the Oregon Trail.

After passing Chimney Rock, wagon trains heading west passed Scotts Bluff, another erosional remnant of Great Plains rocks that were once much more widespread (Figure 2). In fact, you can still see ruts made by wagons as they crossed Mitchell Pass, and the Visitors' Center at Scotts Bluff National Monument features displays of historical significance as well as literature on the monument and the Oregon Trail.

Scotts Bluff is an imposing edifice rising more than 150 m above the surrounding plain. It is composed of Oligocene- to Miocene-age sedimentary rocks, mostly siltstone and sandstone, as well as volcanic ash layers. Indeed, two prominent layers of volcanic ash are visible that must have come from volcanoes that were active far to the west. One formation at Scotts Bluff and in adjacent areas, the 30- to 40-million-year-old Brule Clay, has yielded many fossil mammals, including the now extinct piglike oreodonts and the rhinoceros-sized titanotheres, as well as three-toed horses, camels, dogs, cats, and rodents.

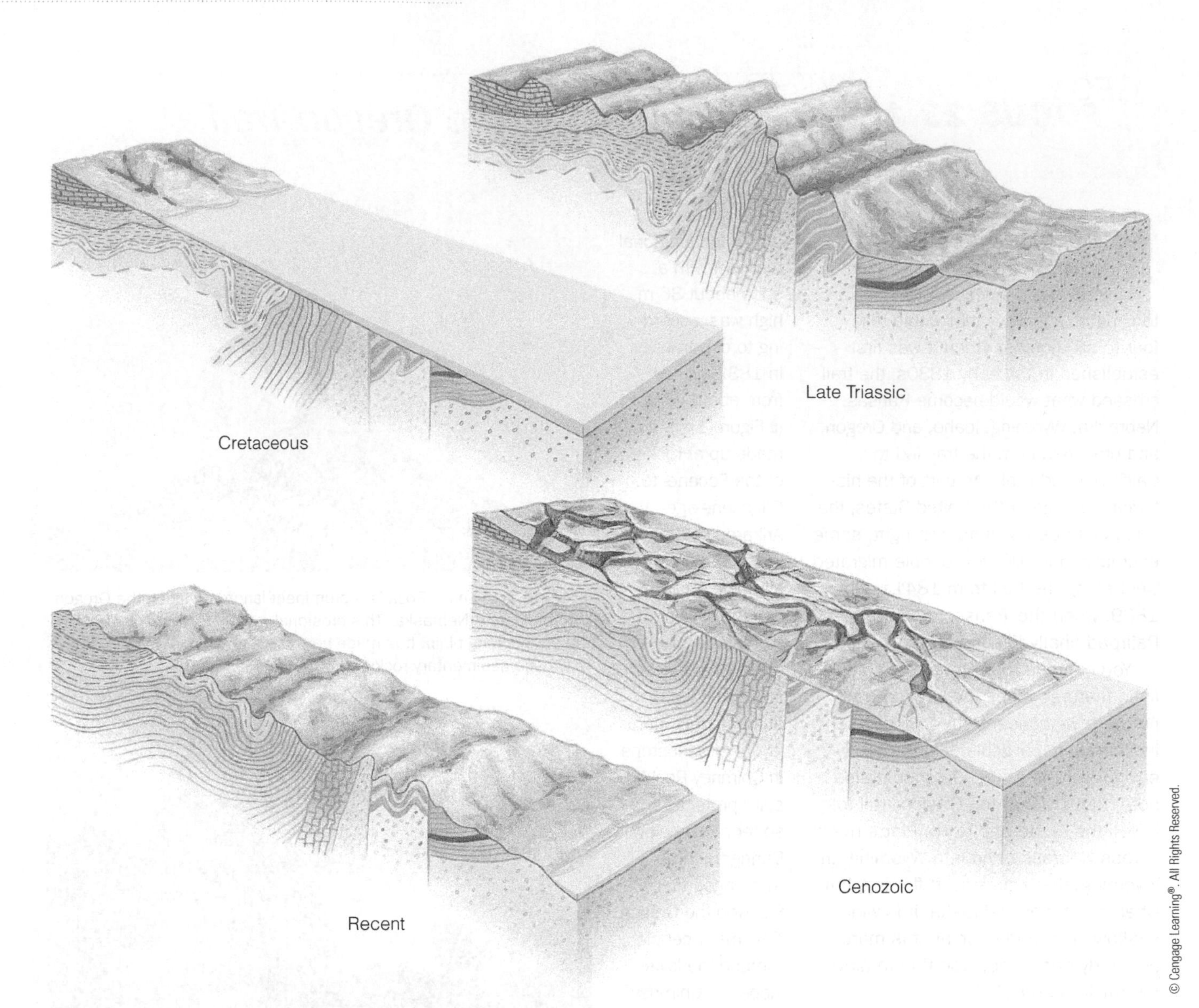

Figure 23.11 Evolution of the Present Topography of the Appalachian Mountains Although these mountains have a long history, their present topographic expression resulted mainly from Cenozoic uplift and erosion.

of the Appalachian Mountains developed as a result of Cenozoic uplift and erosion (Figure 23.11). As uplift proceeded, upturned resistant rocks formed northeast–southwest-trending ridges with intervening valleys eroding into less resistant rocks. The preexisting streams eroded downward while uplift took place and were superposed on resistant rocks, thereby cutting large canyons across the ridges, forming *water gaps*, deep passes through which streams flow, and *wind gaps*, that no longer contain streams.

Erosion surfaces at different elevations in the Appalachians are a source of continuing debate among geologists. Some are convinced that these more or less planar surfaces show evidence of uplift followed by extensive erosion and then renewed uplift and another cycle of erosion. Others think that each surface represents a differential response to weathering and erosion. According to this view, a low-elevation erosion surface developed on softer strata that eroded more or less uniformly, whereas higher surfaces represent weathering and erosion of more resistant rocks.

North America's Southern and Eastern Continental Margins

In a previous section, we mentioned that much of the Interior Lowlands eroded during the Cenozoic. Even in the Great Plains, where vast deposits of Cenozoic rocks are present, sediment was carried across the region and into the drainage systems that emptied into the Gulf of Mexico. Likewise, sediment eroded from the western margin of the Appalachian Mountains ended up in the Gulf, but these mountains also shed huge quantities of sediment eastward that was deposited along the Atlantic Coastal Plain. Notice in ▮ Figure 23.12 that the **Atlantic Coastal Plain** and the **Gulf Coastal Plain** form a continuous belt extending from the northeastern United States to Texas.

The Gulf Coastal Plain

After the withdrawal of the Cretaceous to Early Paleogene Zuni Sea, the Cenozoic **Tejas epeiric sea** made a brief appearance on the continent (see Figure 20.5). But even at its maximum extent, it was restricted to the Atlantic and Gulf Coastal plains and parts of coastal California. It also extended up the Mississippi River Valley, where it reached as far north as southern Illinois.

The overall Gulf Coast sedimentation pattern was established during the Jurassic and persisted throughout the Cenozoic. Sediments derived from the Cordillera, western Appalachians, and the Interior Lowlands were transported toward the Gulf of Mexico, where they were deposited in continental, transitional, and marine environments. The sediments form seaward-thickening wedges, grading from continental facies in the north to marine facies in the south (Figure 23.12).

Sedimentary facies development was controlled mostly by regression of the Tejas epeiric sea. After its maximum extent onto the continent during the Paleogene, this sea began its long withdrawal toward the Gulf of Mexico. Its regression, however, was periodically reversed by minor transgressions—eight transgressive–regressive episodes are recorded in Gulf Coastal Plain sedimentary rocks, accounting for the intertonguing among the various facies.

Geological Education

As the only resident of your community with any background in geology, you are asked by a curious person why western North America has volcanoes, earthquakes, huge mountain ranges, and small valley glaciers, whereas these same features are absent or nearly so in the eastern part of the continent. How would you explain this disparity? Can you think of how the situation might be reversed—that is, what kinds of geologic events would lead to these kinds of phenomena in the east?

Most of the Gulf Coastal Plain was dominated by detrital deposition, but in the Florida section of the region and the coast of Mexico, significant carbonate deposition took place. Florida was a carbonate platform during the Cretaceous and continued as an area of carbonate deposition into the Early Paleogene; carbonate deposition continues even now in Florida Bay and the Florida Keys.

The Atlantic Continental Margin

The East Coast of North America includes the Atlantic Coastal Plain and extends seaward across the continental shelf, slope, and rise (Figure 23.12). It is a classic example of a passive continental margin. When Pangaea began fragmenting during the Triassic, continental crust rifted, and a new ocean basin began to form. Remember that the North American plate moved westerly, so its eastern margin was within the plate, where a passive continental margin developed.

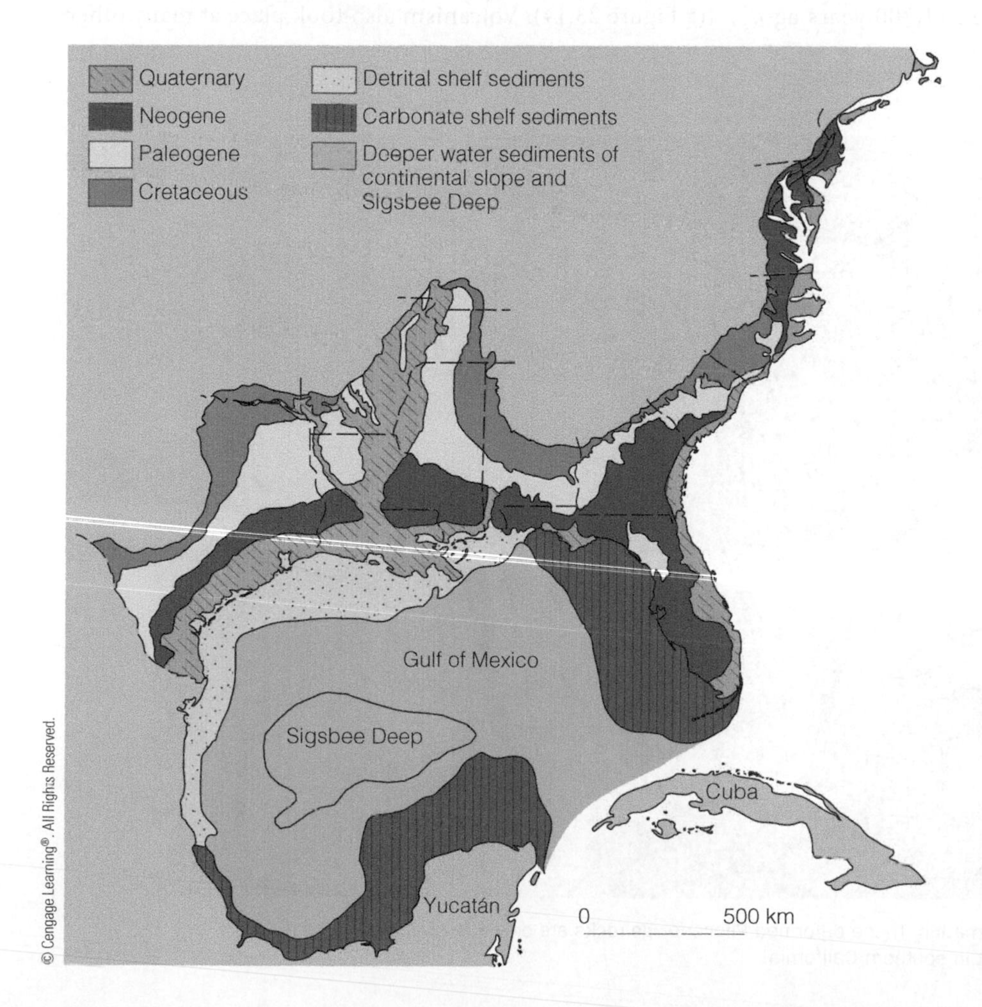

▮ **Figure 23.12 Cenozoic Deposition Along the Gulf and East Coasts** Depositional provinces and the surface geology of the Gulf and Atlantic coasts. Note the carbonate shelf deposits near Florida and the Yucatán Peninsula of Mexico.

The Atlantic continental margin has a number of Mesozoic and Cenozoic basins, formed as a result of rifting, in which sedimentation began by Jurassic time. Even though Jurassic-age rocks have been detected in only a few deep wells, geologists assume that they underlie the entire continental margin. The distribution of Cretaceous and Cenozoic rocks is better known, because both are exposed on the Atlantic Coastal Plain, and both have been penetrated by wells on the continental shelf.

Sedimentary rocks on the broad Atlantic Coastal Plain, as well as those underlying the continental shelf, slope, and rise, were derived from the Appalachian Mountains. Numerous rivers and streams transported sediments toward the east, where they were deposited in seaward-thickening wedges (up to 14 km thick) that grade from terrestrial deposits on the west to marine deposits farther east. For instance, the Calvert Cliffs in Maryland consist of rocks deposited in marginal marine environments.

23.5 The Pleistocene and Holocene Epochs

The *Pleistocene* and *Holocene* or *Recent* epochs make up the Quaternary Period, the most recent 2.6 million years of Earth history (see Figure 17.1). The Pleistocene, which began 2.6 million years ago and ended 11,700 years ago, is the focus of our discussion here because it was the time of extensive glaciation; hence, the Pleistocene is commonly called the *Ice Age.*

Pleistocene and Holocene Tectonism and Volcanism

The Pleistocene Epoch is best known for widespread glaciers, but it was also a time of continuing tectonic activity and volcanism. Orogeny continued in the Himalayas and the Andes, and deformation at convergent plate boundaries proceeded unabated in the Aleutian Islands, Japan, the Philippines, and elsewhere. Interactions between the North American and Pacific plates persisted along the San Andreas Fault (a transform plate boundary), yielding folds, faults, and a number of basins and uplifted areas (▮ Figure 23.13). For example, several east-west–trending mountain ranges in southern California owe their existence to stresses created along a bend in the San Andreas Fault.

Ongoing subduction of remnants of the Farallon plate beneath Central America and the Pacific Northwest accounts for Pleistocene and present-day volcanism in these two regions. Although the Cascade Range began evolving during the Paleogene, the large composite volcanoes such as Mount Shasta and Mount Rainier, as well as the lava dome known as Lassen Peak, are mostly Pleistocene and Recent (▮ Figure 23.14). Volcanism also took place at many other

Marli Bryant Miller

▮ **Figure 23.13 Pleistocene Deformation** These deformed Pliocene-age rocks are only a few hundred meters from the San Andreas Fault in southern California.

Figure 23.14 Pleistocene and Holocene Volcanism in the Cascade Range Volcanism has occurred in this area since the Oligocene Epoch, but the large volcanoes now present developed mostly during the Pleistocene and Holocene.

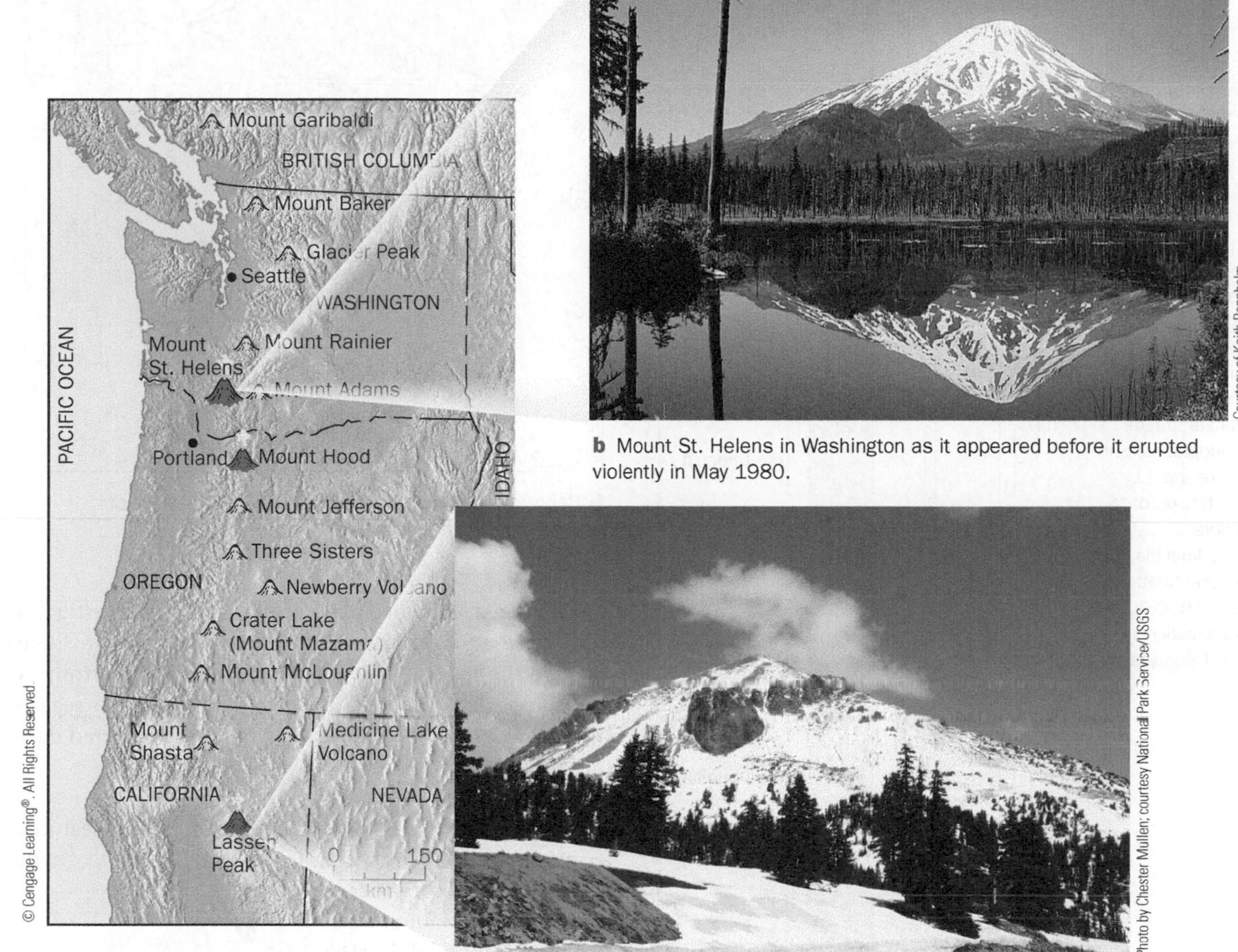

a The Cascade Range extends from northern California into southern British Columbia, Canada. This map shows the locations of the major volcanoes in the range. All are composite volcanoes with the exceptions of Lassen Peak (lava dome) and two shield volcanoes, Medicine Lake Volcano and Newberry Volcano.

b Mount St. Helens in Washington as it appeared before it erupted violently in May 1980.

c Lassen Peak is a lava dome that formed about 27,000 years ago on the northeast flank of a deeply eroded composite volcano. Lassen Peak erupted numerous times from 1914 to 1917.

locations in the western United States (Figure 23.14), and huge calderas formed in what is now Yellowstone National Park in Wyoming.

Pleistocene Glaciation

In 1837, the Swiss naturalist Louis Agassiz argued that large boulders (erratics), polished and striated bedrock, U-shaped valleys, and deposits of sand and gravel in parts of Europe resulted from huge glaciers moving over the land. Although the idea initially met with considerable resistance, scientists finally came to realize that Agassiz was correct and accepted the idea that an Ice Age had taken place in the recent geologic past.

The Distribution and Extent of Pleistocene Glaciers In the past, scientists identified four stages of Pleistocene glaciation in North America, but now we know that the most recent glaciation (the Wisconsinan) was preceded by 12 episodes of glacial advance and retreat (Figure 23.15a). In short, Pleistocene glaciation was more complex than formerly accepted. Glaciers of regional extent (continental glaciers) were widespread, especially on the Northern Hemisphere continents, and smaller valley glaciers were much more numerous than they are now.

As one would expect, the climatic effects responsible for Pleistocene glaciers were worldwide. Nevertheless, Earth was not as frigid as portrayed in movies and cartoons, nor was the onset of the climatic conditions leading to glaciation very

Figure 23.15 Pleistocene Glaciers

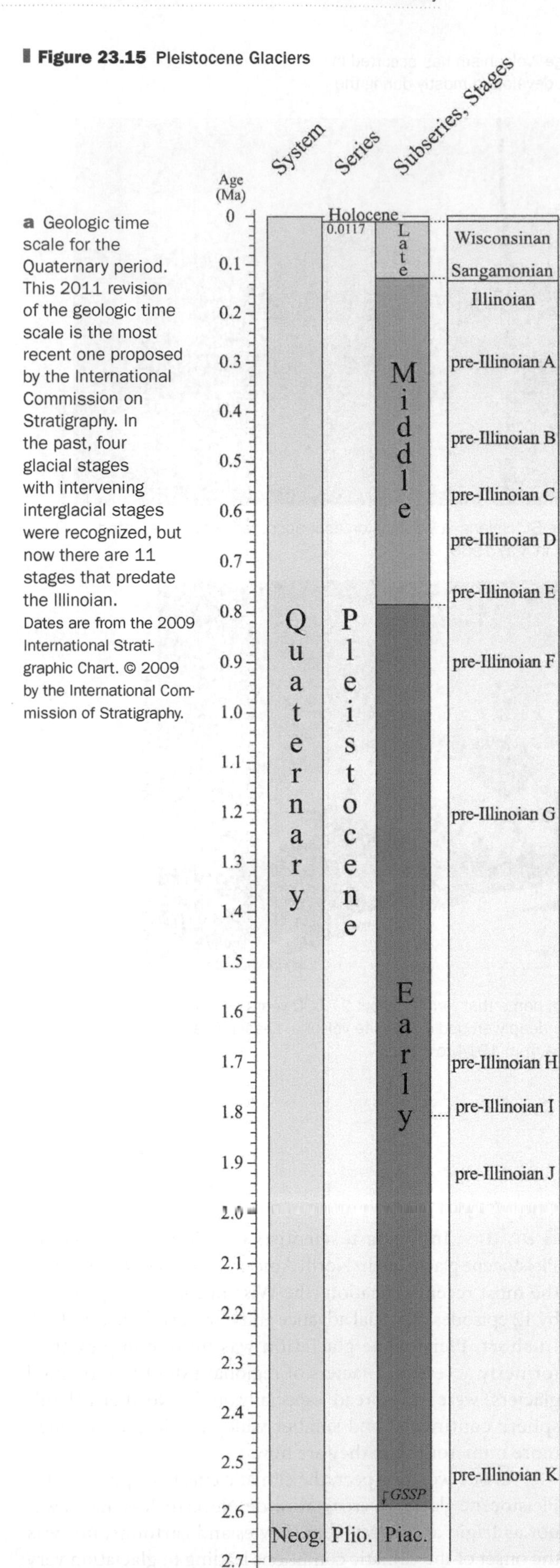

a Geologic time scale for the Quaternary period. This 2011 revision of the geologic time scale is the most recent one proposed by the International Commission on Stratigraphy. In the past, four glacial stages with intervening interglacial stages were recognized, but now there are 11 stages that predate the Illinoian. Dates are from the 2009 International Stratigraphic Chart. © 2009 by the International Commission of Stratigraphy.

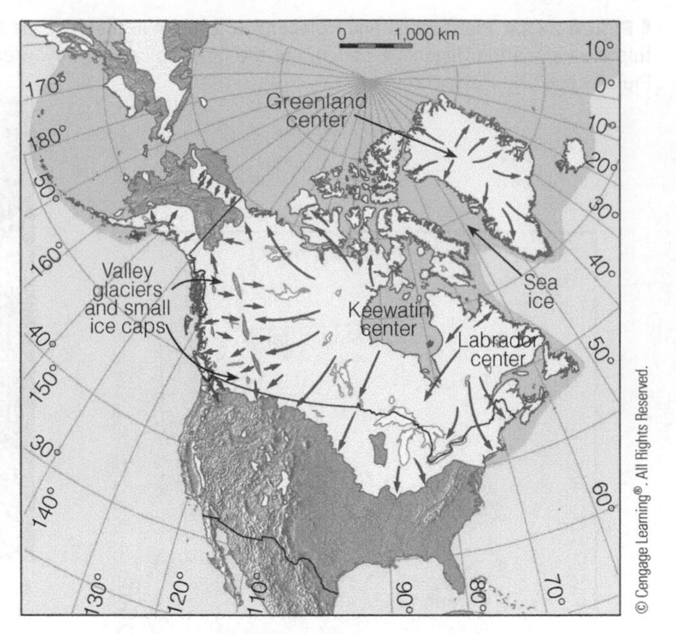

b Centers of ice accumulation and maximum extent of Pleistocene glaciers in North America.

rapid. Indeed, evidence from several types of investigations indicates that the climate gradually cooled from the Eocene through the Pleistocene. Furthermore, evidence from oxygen isotope data (the ratio of O^{18} to O^{16}) from deep-sea cores shows that 20 major warm–cold cycles have occurred during the last 2.6 million years.

The Effects of Glaciation From such glacial features as terminal moraines, erratics, and drumlins, it seems that at their greatest extent, Pleistocene glaciers covered about three times as much of Earth's surface as they do now (Figure 23.15b). That is, they covered more than 40 million km^2, and like the vast ice sheets now present in Greenland and Antarctica, they were probably as much as 3 km thick.

ConnectionLink

See Chapter 14 for a review of how glaciers form and move, and how they modify Earth's surface by erosion, transport, and deposition.

Glaciers moving over Earth's surface have produced distinctive landscapes in much of Canada, the northern tier of states, and the mountains of the West (see Chapter 14). Sea level has risen and fallen with the formation and melting of glaciers, and these changes in turn have affected the margins of continents. Glaciers have also altered the world's climate, causing cooler and wetter conditions in some areas that are arid to semiarid today.

More than 70 million km^3 of snow and ice blanketed the continents during the maximum glacial coverage of the Pleistocene. The storage of ocean waters in glaciers lowered sea level 130 m and exposed large areas of the present-day continental shelves, which were soon covered by vegetation. Lower sea level also affected the base level of rivers and

Figure 23.16 Pleistocene Lakes in the Western United States

Critical Thinking Question Why were there large lakes in places such as Death Valley, California, during the Pleistocene?

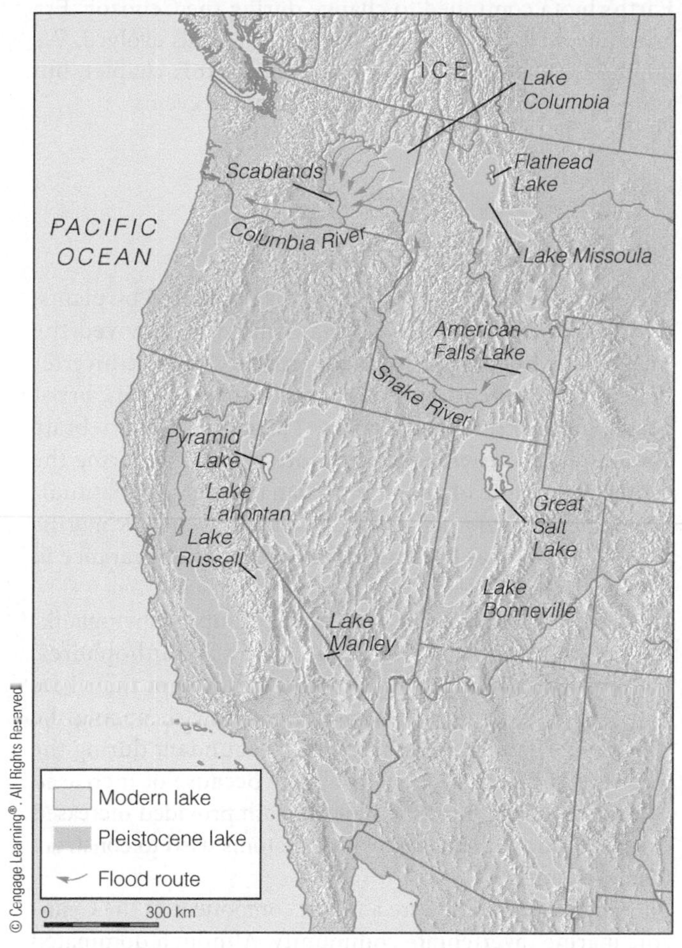

a Pyramid Lake and Great Salt Lake are shrunken remnants of much larger ancient lakes. Of all the lakes shown, only Lake Columbia and Lake Missoula were proglacial lakes. When the 600-m-high ice dam impounding Lake Missoula failed, it drained westward and scoured out the scablands of eastern Washington.

b The area in this image east of Fallon, Nevada, was covered by Lake Lahontan. In fact, the image was taken from Grimes Point Archaeological Site where Native Americans lived near the lakeshore.

streams. When sea level dropped, streams eroded downward as they sought to adjust to a lower base level.

During the Pleistocene, stream channels in coastal areas were extended and deepened along the emergent continental shelves. When sea level rose with the melting of the glaciers, the lower ends of stream valleys along the East Coast of North America were flooded and are now important harbors (see Figure 16.21a), whereas just off the West Coast, they form impressive submarine canyons. Great amounts of sediment eroded by the glaciers were transported by streams to the sea and thus contributed to the growth of submarine fans along the base of the continental slope.

We noted in Chapter 10 that as the Pleistocene ice sheets formed and increased in size, the weight of the ice caused the crust to slowly subside deeper into the mantle. In some places, Earth's surface was depressed as much as 300 m below the preglacial elevations. As the ice sheets retreated by melting, the downwarped areas gradually rebounded to their former positions.

During the Wisconsinan glacial stage, many large lakes formed in what are now dry basins in the southwestern United States (Figure 23.16a). These so-called *pluvial lakes* formed far from the glaciers because of the overall greater precipitation and cooler temperatures, which lowered the evaporation rate, and they fluctuated as the glaciers elsewhere expanded and contracted. The largest was Lake Bonneville, which attained a maximum size of 50,000 km^2 and was at least 335 m deep (Figure 23.16a). The Great Salt Lake that lies just west of Salt Lake City, Utah, is the shrunken remnant of this once-vast lake. Lake Lahontan, mostly in Nevada (Figure 23.16b), covered 22,000 km^2, and even Death Valley, California, which is now the driest place in North America, had a large lake.

In contrast to pluvial lakes, *proglacial lakes* form when meltwater accumulates along the margins of glaciers. Lake Agassiz was a large proglacial lake that covered about 250,000 km^2 of North Dakota, Manitoba, Saskatchewan, and Ontario. It persisted until the glacial ice along its northern margin melted and it drained northward into Hudson Bay. A proglacial lake in Montana called Lake Missoula (Figure 23.16a) was impounded by an ice dam in Idaho that periodically failed, and the lake drained catastrophically. Many depositional and erosion features in western Montana, Idaho, and eastern Washington formed during these gigantic floods (Geo-Focus 23.1).

23.6 Cenozoic Mineral Resources

The United States is the third leading oil producer in the world, and yet it imports just over half of its oil needs, mostly from Canada, Mexico, and Venezuela, but more than 20% comes from the Middle East. Much of the United States domestic production comes from Cenozoic reservoirs on the Gulf Coastal Plain and adjacent continental shelf, especially in Texas and Louisiana. On the Gulf Coastal Plain, most of the petroleum production is from structural traps, many of which formed adjacent to rising salt domes. Several Cenozoic-age basins in southern California are also important areas of oil production. The Eocene Green River Formation in Wyoming, Utah, and Colorado has huge deposits of oil shale, which at some time in the future may be processed for oil and combustible gases. No oil is derived from this formation now because it is cheaper to obtain oil from conventional sources.

Diatomite is a sedimentary rock composed of the microscopic shells of diatoms, which are single-celled marine and freshwater plants that secrete skeletons of silica (SiO_2). This rock, also called diatomaceous earth, is used chiefly in gas purification and to filter liquids such as molasses, fruit juices, water, and sewage. The United States is the leader in diatomite production, mostly from Cenozoic deposits in California, Oregon, and Washington.

Huge deposits of low-grade lignite and subbituminous coal in the northern Great Plains are becoming increasingly important resources. These coal deposits are Late Cretaceous to Early Paleogene in age and are most extensive in the Williston and Powder River basins of Montana, Wyoming, and North and South Dakota. In addition to having a low sulfur content, some of these coal beds are 30–60 m thick.

Gold production from the Pacific Coast, particularly California, comes mostly from Cenozoic gravels. The gold is found in placer deposits, which formed as concentrations of minerals separated from weathered debris by fluvial processes.

A variety of other mineral deposits is also important. For example, the United States must import almost all manganese used in the manufacture of steel. The largest manganese deposits are in Cenozoic rocks in Russia. One molybdenum deposit in Colorado accounts for much of the world production of this element. Cenozoic sand and gravel, as well as evaporites, building stone, and clay deposits, are quarried from areas around the world.

Sand and gravel deposits resulting from glacial activity are a valuable resource in many formerly glaciated areas. Most Pleistocene sands and gravels originated as floodplain deposits, outwash sediment, or esker deposits. The bulk of the sand and gravel in the United States and Canada is used in construction and as roadbase and fill for highway and railway construction.

The periodic evaporation of pluvial lakes in the Death Valley region of California during the Pleistocene led to the concentration of many evaporite minerals such as borax. During the 1880s, borax was transported from Death Valley by the famous 20-mule-team wagon trains.

Another Neogene resource is peat, a vast potential energy resource that has been developed in Canada and Ireland. Peatlands formed from plant assemblages as the result of particular climate conditions.

23.7 Paleogene and Neogene Life History

Earth's biota continued to change during the Cenozoic Era as more and more familiar plants and animals evolved. We emphasize the evolution of mammals in this chapter, but you should be aware of other important life events.

Marine Invertebrates and Phytoplankton

The Cenozoic marine ecosystem was populated by plants, animals, and single-celled organisms that survived the Mesozoic extinctions. Especially prolific Cenozoic invertebrate groups were foraminifera, radiolarians, corals, bryozoans, mollusks, and echinoids. The marine invertebrate community in general became more provincial during the Cenozoic because of changing ocean currents and latitudinal temperature gradients. In addition, the Cenozoic marine invertebrate faunas became more familiar in appearance to those of today.

Only a few species in each major group of phytoplankton survived into the Paleogene. The coccolithophores, diatoms, and dinoflagellates all recovered from their Late Cretaceous reduction in numbers to flourish during the Cenozoic. Diatoms were particularly abundant during the Miocene (▮ Figure 23.17a), probably because of increased volcanism during this time. Volcanic ash provided increased dissolved silica in seawater, which diatoms used to construct their skeletons.

The foraminifera were a major component of the Cenozoic marine invertebrate community. Although dominated by relatively small forms (▮ Figure 23.17b, c), it included some exceptionally large forms that lived in the warm waters of the Cenozoic Tethys Sea. Shells of these larger foraminifera accumulated to form thick limestones, some of which the ancient Egyptians used to construct the Sphinx and the Pyramids of Giza.

Having relinquished their reef-building role to rudists, which are mollusks, during the mid-Cretaceous, corals again became the dominant reef builders during the Cenozoic. Other suspension feeders such as bryozoans and crinoids were also abundant and successful during the Paleogene and Neogene. Perhaps the least important of the Cenozoic marine invertebrates were brachiopods, with fewer than 60 genera surviving today. Brachiopods never recovered from their reduction in diversity at the end of the Paleozoic.

Just as during the Mesozoic, bivalves and gastropods were two of the major groups of marine invertebrates during the Cenozoic, and they had a markedly modern appearance. Following the extinction of the ammonites and belemnites at the end of the Cretaceous, the Cenozoic cephalopod fauna consisted of nautiloids and shell-less cephalopods such as squids and octopuses.

Figure 23.17 Cenozoic Diatoms and Foraminifera

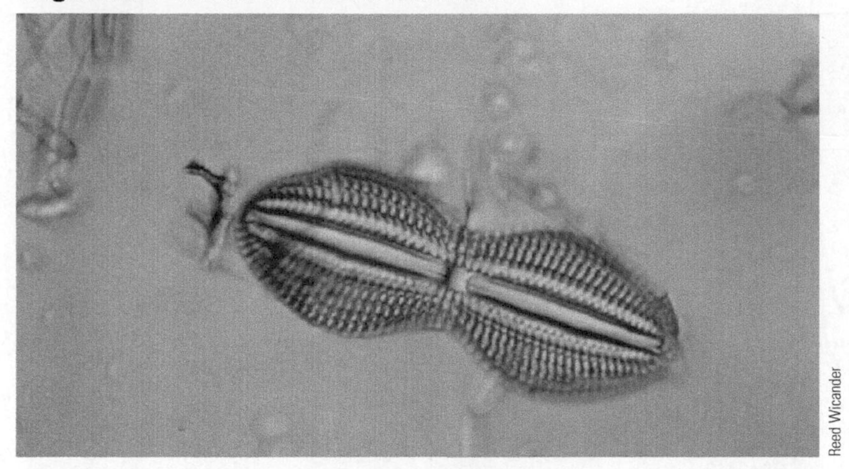

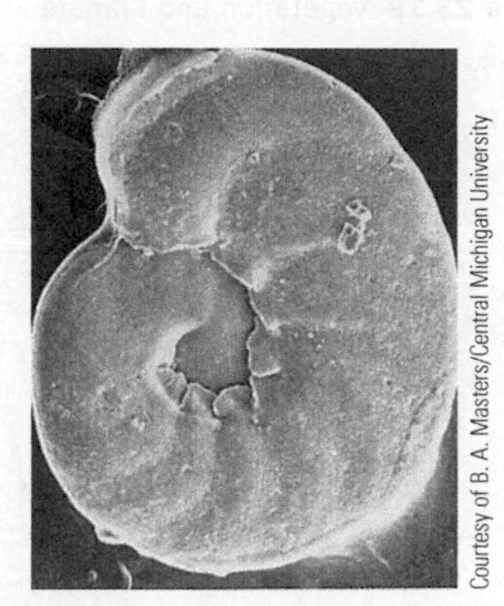

b *Cibicides americanus* from the Early Miocene of California is a benthic foraminifera.

Courtesy of B. A. Masters/Central Michigan University

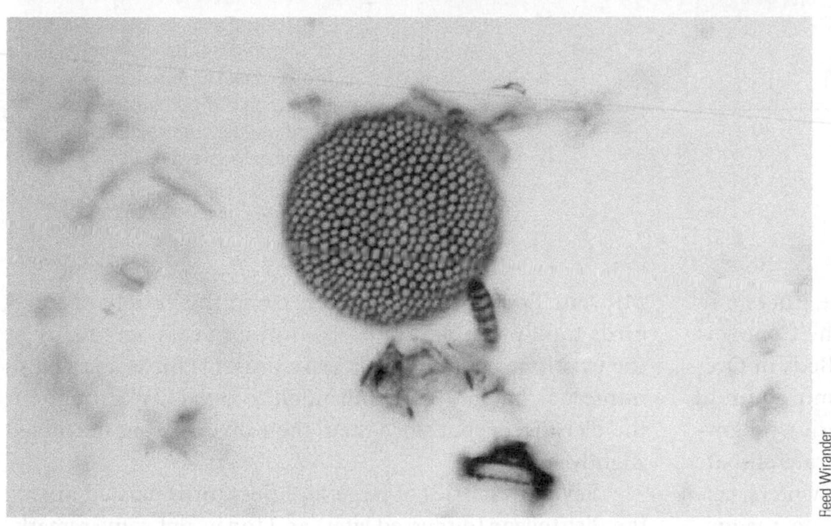

a Diatoms from the Miocene-age Monterey Formation of California.

Reed Wicander

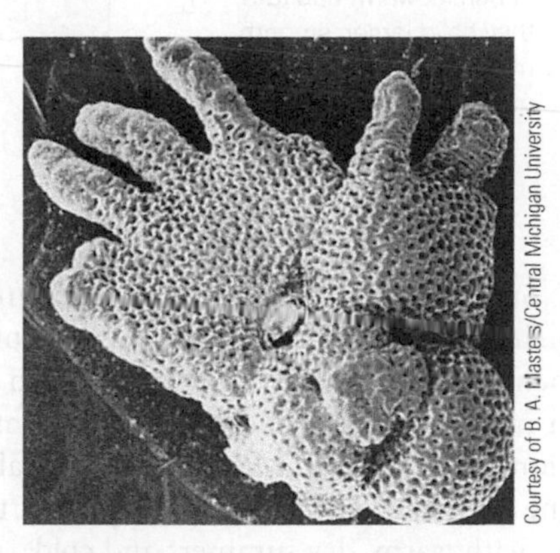

c *Globigerinoides fistulosus* is a Pleistocene planktonic foraminifera from the south Pacific Ocean.

Courtesy of B. A. Masters/Central Michigan University

The echinoids continued their expansion in the infaunal habitat and were very prolific during the Cenozoic. New forms such as sand dollars evolved during this time from biscuit-shaped ancestors.

Cenozoic Vegetation and Climate

During the Cenozoic, angiosperms continued to diversify, although seedless vascular plants and gymnosperms were also present in large numbers. In fact, many Paleogene plants would be familiar to us today, but their geographic distribution was not what it is now, because changing climatic conditions along with shifting plant distributions were occurring.

The makeup of ancient floras and the types of leaves found as fossils are good climatic indicators. Some plants today are confined to the tropics, whereas others have adapted to drier conditions, and we have every reason to think that climate was a strong control on plant distribution during the past. Leaves with entire or smooth margins, many with pointed drip-tips, are dominant in areas with abundant rainfall and high annual temperatures. Smaller leaves with incised margins are more typical of cooler, drier areas (Figure 23.18a).

In a recently discovered Paleocene flora in Colorado with about 100 species of trees, nearly 70% of the leaves are smooth margined, and many have drip-tips. The nature of the leaves, coupled with the diversity of plants, is much like that found in today's rain forests. In fact, the Early Oligocene fossil plants at Florissant Fossil Beds National Monument in Colorado, indicate that a warm, wet climate existed then.

Seafloor sediments and geochemical evidence indicate that about 55 million years ago an abrupt warming trend took place. During this **Paleocene–Eocene Thermal Maximum,** large-scale oceanic circulation was disrupted so that heat transfer from equatorial regions to the poles diminished or ceased. As a result, deep oceanic water became warmer, resulting in extinctions of many deepwater foraminifera. Some scientists think that this deep, warm oceanic water released methane from seafloor methane hydrates, contributing a greenhouse gas to the atmosphere that caused the temperature to increase at this time.

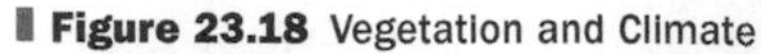

Figure 23.18 Vegetation and Climate

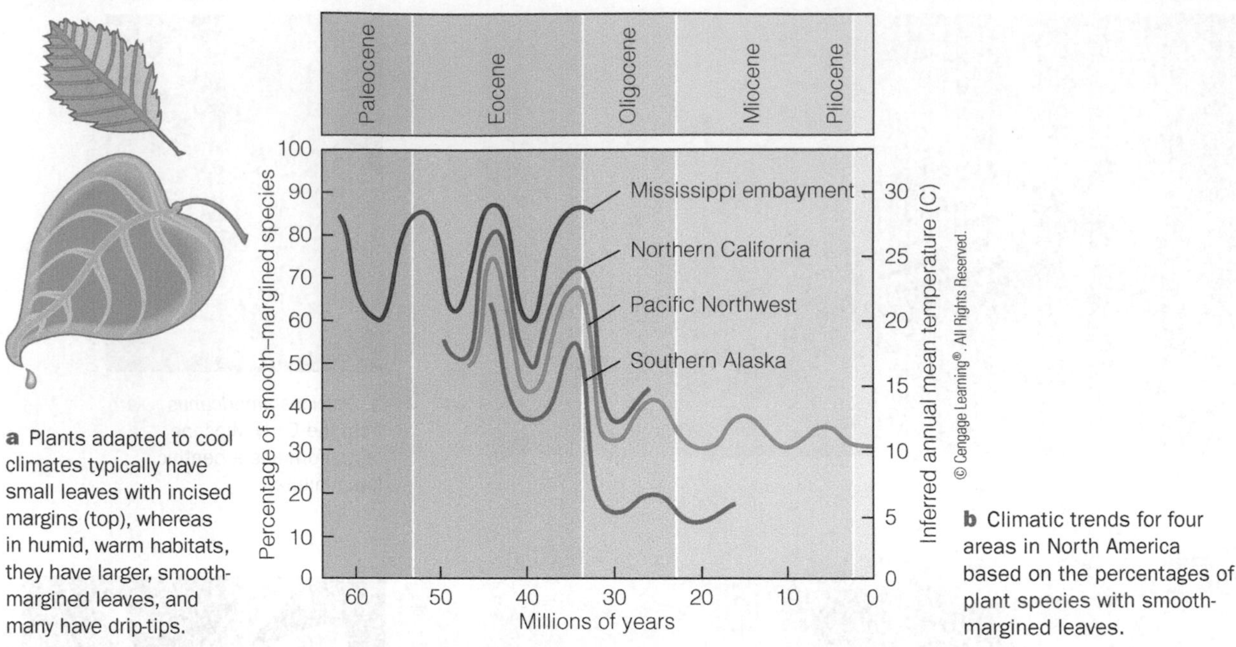

a Plants adapted to cool climates typically have small leaves with incised margins (top), whereas in humid, warm habitats, they have larger, smooth-margined leaves, and many have drip-tips.

b Climatic trends for four areas in North America based on the percentages of plant species with smooth-margined leaves.

Subtropical conditions persisted into the Eocene in North America, probably the warmest of all the Cenozoic epochs. Fossil plants in the Eocene John Day Beds in Oregon include species that today live in much more humid regions, as in parts of Mexico and Central America. Yellowstone National Park in Wyoming has a temperate climate now, with warm, dry summers and cold, snowy winters, certainly not an area where you would expect avocado, magnolia, and laurel trees to grow. Yet their presence there during the Eocene indicates that the area then had a considerably warmer climate than it does now.

A major climatic change took place at the end of the Eocene, when mean annual temperatures dropped as much as 7°C in about 3 million years (Figure 23.18b). Since the Oligocene, mean annual temperatures have varied somewhat worldwide, but overall, they have not changed much in the middle latitudes except during the Pleistocene Epoch.

A general decrease in precipitation during the last 25 million years took place in the mid-continent region of North America. As the climate became drier, the vast forests of the Oligocene gave way first to *savannah* conditions (grasslands with scattered trees) and finally to *steppe* environments (short-grass prairie of the desert margin). Many herbivorous mammals quickly adapted to these new conditions.

Paleogene and Neogene Birds

The first members of many of the living orders of birds, including owls, hawks, ducks, penguins, and vultures, evolved during the Early Paleogene. Beginning during the Miocene Epoch, a marked increase in the variety of songbirds took place, and by 5–10 million years ago, many of the existing genera of birds were present. Birds adapted to numerous habitats and continued to diversify throughout the Pleistocene, but since then, their diversity has decreased slightly.

Several varieties of large flightless birds existed during the Pleistocene (discussed later) and today, but some remarkable predatory birds lived during the Paleogene and Neogene. For instance, *Gastronis,* a heavily built, flightless bird that stood about 2 m high, lived during the Paleocene and Eocene in North America and Europe (see the chapter opening photo). About 25 species of flightless birds up to 3 m tall are known from South America, where they were the dominant predators until they were replaced by big cats and dogs that migrated from North America.

Large flightless birds are truly remarkable, but the success story belongs to the flying birds. Even though few skeletal modifications have occurred during the Cenozoic, a bewildering array of adaptive types evolved. In fact, with 9,000 to 10,000 living species, birds are the most numerous land-dwelling vertebrates.

Diversification of Mammals

Recall from Chapter 22 that reptiles, birds, and mammals are *amniotes*—that is, animals that have an amniote egg in their reproductive cycle, although it is highly modified in most mammals. The mammals known as **monotremes,** such as the platypus and spiny anteater, lay amniote eggs and thus are egg-laying mammals. **Marsupial mammals**

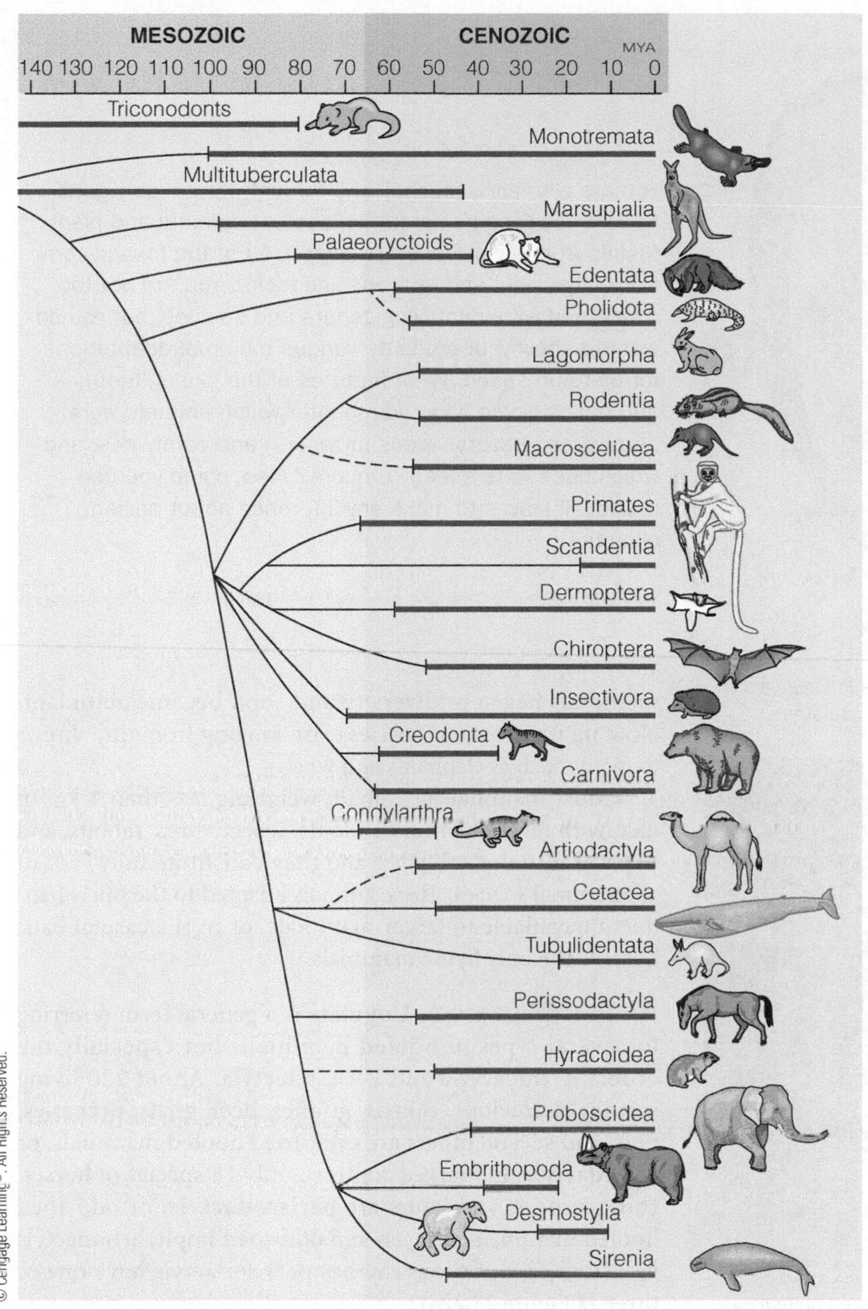

▌Figure 23.19 Diversification of Mammals Mammals existed during the Mesozoic, but most pacental mammals diversified during the Paleocene and Eocene epochs. Among the living orders of mammals, all are placentals except for the monotremes and marsupials. Several extinct orders are not shown. Bold lines indicate actual geologic ranges, whereas the thinner lines indicate the inferred branching of the groups.

and **placental mammals** give birth to live young, but in marsupials (pouched mammals) the young are born in an immature, almost embryonic, condition and then undergo further development in their mother's pouch. The amnion of the amniote egg is fused to the wall of the uterus in placental mammals. In fact, the membranes in the amniote egg form the placenta through which nutrients and oxygen are carried from the mother to the developing embryo, allowing the embryo to develop much more fully before birth.

Judging from the fossil record, monotremes have never been very common, and the only living ones are found in the Australian region. Marsupials have been more successful in terms of number of species and geographic distribution, but even they have been mostly restricted to South America and the Australian region. In marked contrast, the success of placental mammals is indicated by the fact that more than 90% of all mammal species, fossil and living, are placental.

Although mammals had evolved during the Triassic, a major diversification began during the Paleocene and continued throughout the Cenozoic (▌Figure 23.19). In fact, many of the mammals of the Paleocene are called archaic, calling attention to the fact that they were holdovers from the Mesozoic or they were mammals that did not give rise to any of today's mammals. For example, the rodent-like multituberculates persisted from the Jurassic until their extinction during the Oligocene (▌Figure 23.20a). Among the predators were creodonts that slightly resembled wolves and weasel-like "miacids," as well as the large, predatory birds we mentioned in the previous section. Among the herbivores were several groups that survived only until the Eocene.

Also among these Paleocene mammals were the first rodents, rabbits, primates, and hoofed mammals. They had not become fully differentiated from their ancestors, and differences between carnivores and herbivores were slight—even the carnivores were short-legged, flat-footed creatures that undoubtedly were not very speedy. Furthermore, most of these early mammals were small, although some fairly large ones were present by the Late Paleocene, but the first land-dwelling giant mammals did not appear until the Eocene (▌Figure 23.20b).

Diversification continued during the Eocene when several more types of mammals appeared, but if we could go back and visit that time, we would probably not recognize many of these animals. Some would be vaguely familiar, but the ancestors of horses, camels,

 **ConnectionLink**

You can review the origin and Mesozoic evolution of mammals by referring to Chapter 22.

Figure 23.20 Paleocene and Eocene Mammals

Nobumichi Tamura

a This squirrel-sized animal is the Paleocene genus *Ptilodus*, a member of a group of mammals known as multituberculates.

Field Museum Library/Getty Images

b Scene from the Eocene showing the rhinoceros-sized animal known as *Unitatherium,* one of the first giants among mammals. It had three pairs of horns and saber-like upper canine teeth.

rhinoceroses, and elephants would bear little resemblance to their living descendants. By Oligocene time, all the orders of existing mammals were present; however, diversification continued as more familiar families and genera appeared. Miocene and Pliocene mammals were mostly mammals that we could readily identify, although a few types were unusual (Figure 23.21).

Cenozoic Mammals

Mammals evolved from *cynodonts* by the Late Triassic, so two-thirds of their evolution was during the Mesozoic Era (see Chapter 22). However, following the Mesozoic extinctions,

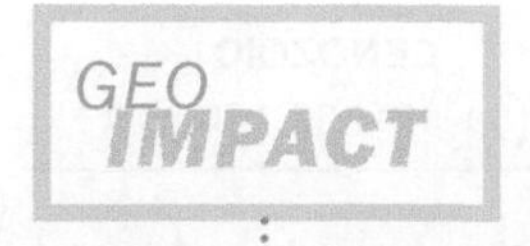

Fossils and Life History

You are a science teacher who through remarkably good fortune receives numerous unlabeled mammal and plant fossils from a generous benefactor. All of the fossils come from Oligocene- and Miocene-age rocks. You are not too concerned with identifying genera and species, but you do want to show your students various mammal adaptations for diet and speed. What features of the skulls, teeth, and bones would allow you to infer which animals were herbivores (grazers versus browsers) and carnivores, and which ones were speedy runners? Also, could you use the fossil leaves to make any inference about ancient climates?

mammals began to diversify and soon became abundant. Now, more than 4,000 species exist, ranging from tiny shrews to giants such as elephants and whales.

Most mammals are small, weighing less than 1 kg. In fact, with few exceptions, rodents, insectivores, rabbits, and bats fall into this category, and they constitute fully 75% of all mammal species. These animals adapted to the microhabitats unavailable to larger mammals, or in the case of bats, became the only flying mammals.

Hoofed Mammals **Ungulate** is a general term referring to several types of hoofed mammals, but especially the orders Artiodactyla and Perissodactyla. About 220 living species of antelope, camels, giraffes, deer, goats, peccaries, pigs, and several others are even-toed hoofed mammals, or **artiodactyls.** In marked contrast, only 18 species of horses, rhinoceroses, and tapirs are **perissodactyls,** or odd-toed hoofed mammals. As even and odd-toed imply, artiodactyls have two or four toes, whereas perissodactyls have one or three (Figure 23.22a).

Some ungulates are small and depend on concealment to avoid predators; others, such as rhinoceroses, are so large that size alone is enough to discourage predators, at least for adults. But many of the more modest-size ungulates are speedy runners. Adaptations for running include elongation of some of the limb bones, as well as reduction in the number of bony limb elements, especially toes. Accordingly, the limbs of speedy ungulates are long and slender (Figure 23.22b).

All ungulates are herbivores, but some are **grazers,** meaning that they feed on grasses, and others are **browsers,** feeding on the tender shoots, twigs, and leaves of trees and bushes. When grasses grow through soil, they pick up tiny pieces of sand that are quite abrasive to teeth, so the grazing ungulates developed high-crowned chewing teeth resistant to abrasion (Figure 23.22c). Browsers, on the other hand, never developed these kinds of chewing teeth.

Figure 23.21 Ashfall Fossil Beds State Historical Park, Nebraska Restoration of fossils found at this state park near Orchard, Nebraska, include **(1)** one-toed horses, **(2)** small camels, **(3)** turtles, **(4)** rhinoceroses, **(5)** cranes, **(6)** giraffe-like camels, and **(7)** three-toed horses. In the distance you can see some carnivores and mastodons. These animals were buried in volcanic ash during the Miocene, about 12 million years ago.

Figure 23.22 Evolutionary Trends in Hoofed Mammals

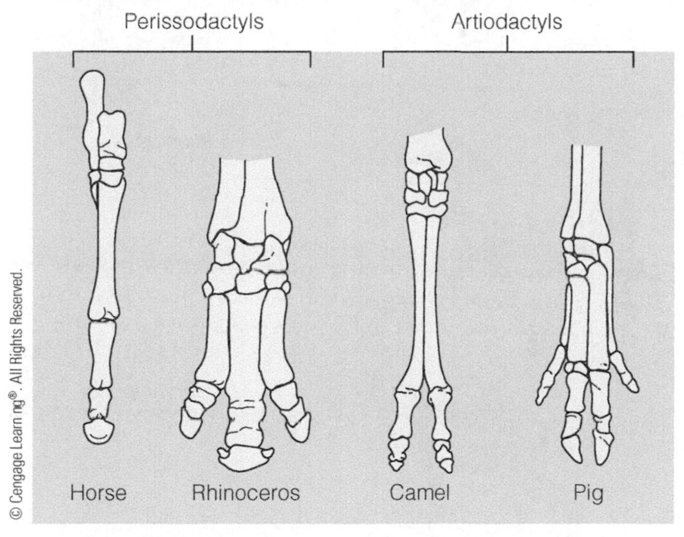

a Perissodactyls have one or three functional toes, whereas artiodactyls have two or four. Note that in perissodactyls the weight is borne on the third toe, but artiodactyls walk on toes three and four.

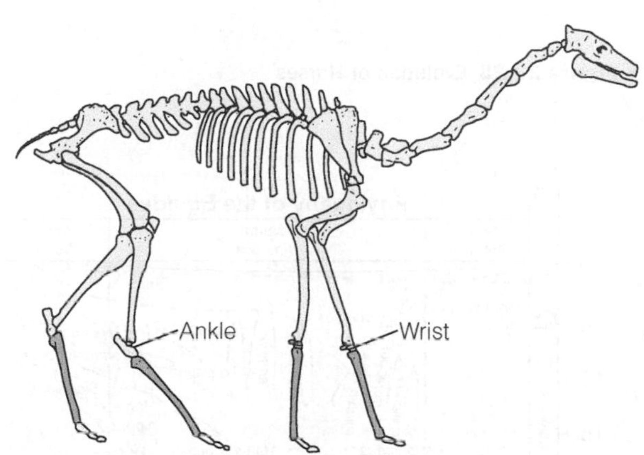

b In many hoofed mammals, long, slender limbs evolved as bones between the wrist and toes, and the ankle and toes, became longer.

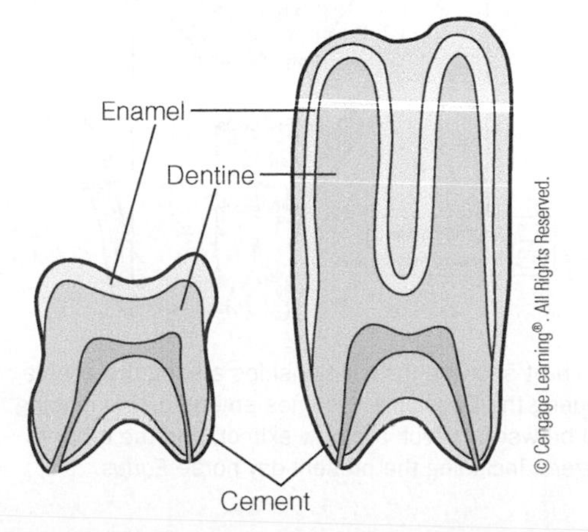

c High-crowned, cement covered chewing teeth (right) evolved in hoofed mammals that adapted to a diet of grass. Low-crowned teeth are found in many other mammals, including primates and pigs, both of which have a varied diet.

Rabbit-size ancestral artiodactyls of the Early Eocene differed little from their ancestors, but gave rise to numerous families, several of which are now extinct. Small, four-toed camels, for instance, appeared early in this diversification and were common in North America well into the Pleistocene. In fact, most of their evolution took place on this continent, and only during the Pliocene did they migrate to South America and Asia where they now exist.

The perissodactyls—horses, rhinoceroses, tapirs—and the extinct titanotheres and chalicotheres are united by several shared characteristics. Furthermore, the fossil record shows that they all evolved from a common ancestor during the Eocene (see Figure 18.17). Their diversity increased through the Oligocene, but since then, they have declined markedly and now constitute less than 10% of the world's hoofed mammal fauna.

Several groups of mammals such as rhinoceroses, whales, elephants, and camels have good fossil records, but certainly one of the best known is the fossil record of horses. Fossils indicate that present-day *Equus* evolved from a tiny Eocene ancestor called *Hyracotherium* (❙ Figure 23.23). Most of horse evolution took place in North America although now they survive in the wild only in Asia and Africa. Fossils of horses reveal several trends such as increased size, lengthening of the limbs, reduction of the toes to one, and development of high-crowned chewing teeth among others as horses became speedy grazing animals. There was, however, another branch in horse evolution that led to three-toed browsers, which became extinct during the Pleistocene (Figure 23.23a).

Other Mammals—Carnivores, Elephants, and Whales During the Paleocene, the *miacids,* small carnivorous mammals with short heavy limbs, made their appearance (❙ Figure 23.24a). These small creatures were ancestors to all later members of the order Carnivora, which includes, among others, today's dogs, cats, hyenas, bears, weasels, and seals. All carnivores have well-developed canine teeth for slashing and tearing, and most also developed a pair of large shearing teeth (carnassials) (❙ Figure 23.24b). Some of the better-known fossil carnivores are the saber-toothed cats, or what are more commonly called saber-toothed tigers.

❙ **Figure 23.23** Evolution of Horses

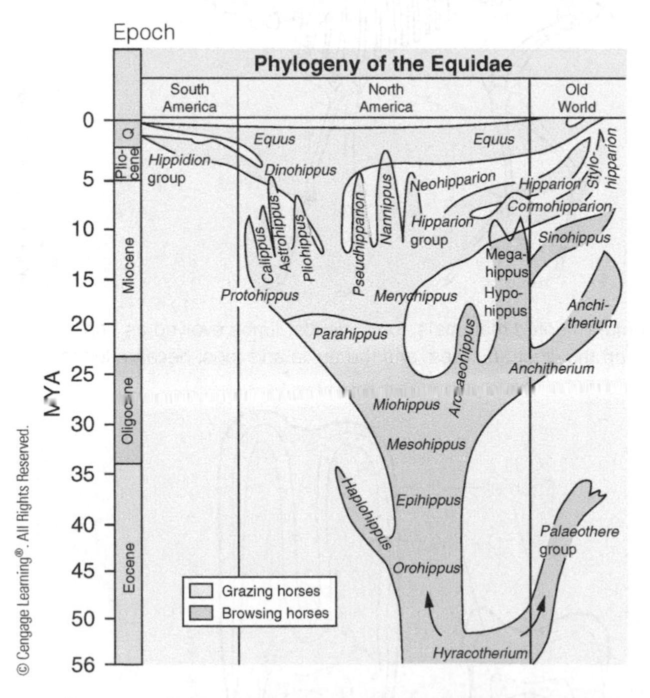

a Summary chart showing the relationships among the genera of horses. During the Oligocene two lines emerged, one leading to three-toed browsers, which are now extinct, and the other to one-toed grazers, including the present-day horse *Equus.*

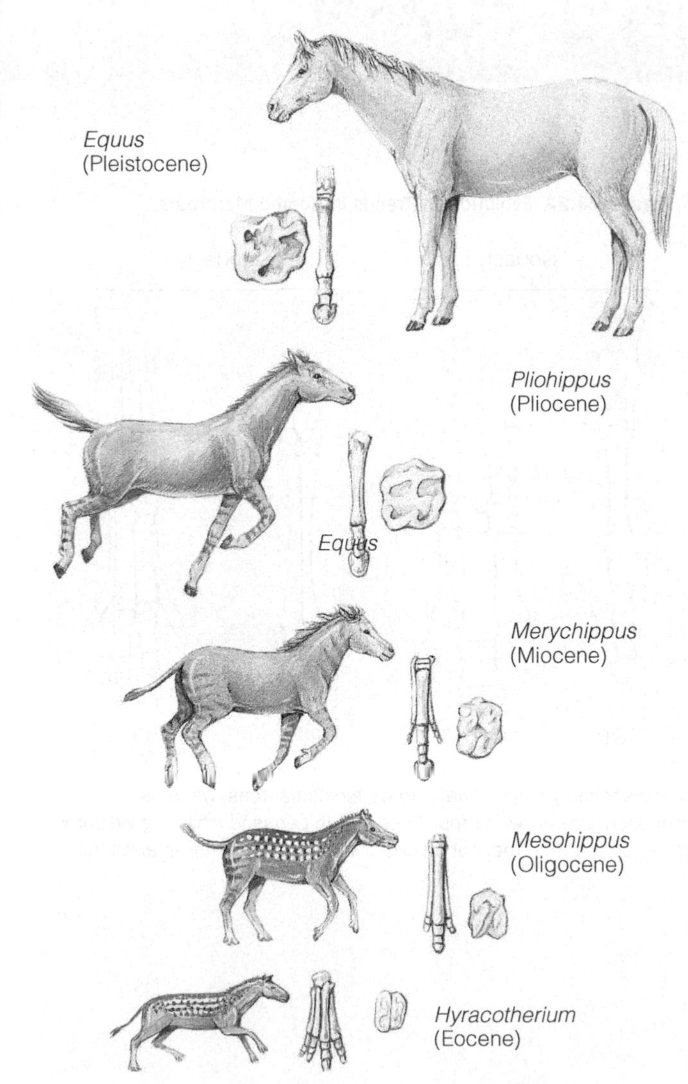

b Simplified diagram showing some trends in horse evolution. Trends include a size increase, a lengthening of the limbs and a reduction in the number of toes, and the development of high-crowned teeth with complex chewing surfaces.

Figure 23.24 Carnivorous Mammals

Laurie O'Keefe/Photo Researchers/Getty Images

a This miacid, known as *Topocyon*, was a coyote-sized animal that lived during the Eocene. All weasels, otters, skunks, badgers, martins, wolverines, seals, sea lions, walruses, bears, raccoons, dogs, hyenas, mongooses, and cats evolved from an ancestor very much like this creature.

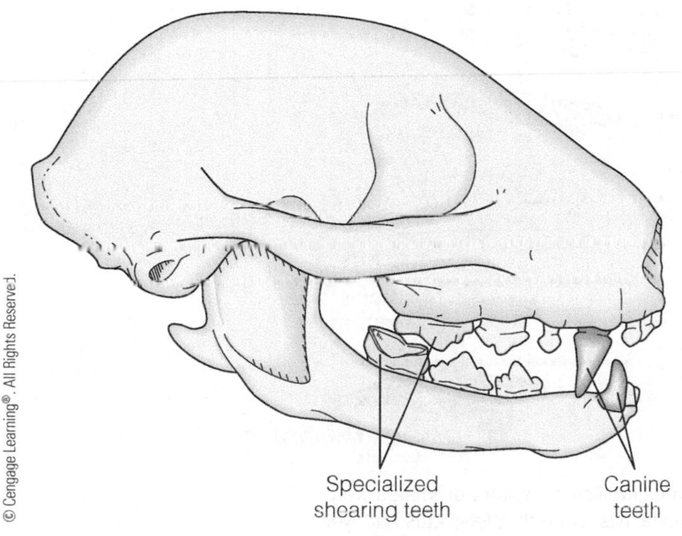

b This present-day skull and jaw of a large cat show the specialized sharp-crested shearing teeth, or carnassials, of carnivorous mammals. Note that carnivorous mammals also have well-developed canine teeth.

Elephants (order Proboscidea), the largest land mammals, evolved from pig-size ancestors during the Eocene. And by Oligocene time, they clearly showed the trend toward large size, a long snout (proboscis), and large tusks. Mastodons, with teeth adapted for browsing, were present by the Miocene, and during the Pliocene the present-day elephants and mammoths diverged from a common ancestor. During most of the Cenozoic, elephants were widespread on the northern continents, but now only two species exist in southern Asia and Africa.

Until recently, little was known about the transition of whales from land-dwelling ancestor to fully aquatic whales. Although a number of questions remain unanswered, the fossils now available show that whales appeared during the Early Eocene and by the Late Eocene had become diverse and widespread. Eocene whales still possessed vestigial rear limbs, their teeth resembled those of their land-dwelling ancestors, their nostrils (blowhole) were not on top of the head, and they were proportioned quite differently from living whales (❚ Figure 23.25). By Oligocene time, both groups of living whales—the toothed whales and the baleen whales—had evolved.

23.8 Pleistocene Faunas

We devote much of this section to the evolution of primates, particularly the hominids, which include present-day humans. Primates as an order evolved by Late Cretaceous time, but the ones of interest to us here date from the Pliocene and Pleistocene.

As for mammals other than primates, most of the present-day genera had evolved by Pleistocene time. We would have little difficulty recognizing most Pleistocene mammals; there were a few unusual types that persisted from earlier times, but they are now extinct. A good example is the chalicotheres, a group of horselike mammals with claws on their forefeet. Likewise, we would recognize most Pleistocene birds, but some large ground-dwelling species are now extinct.

Mammals and Birds

One of the most remarkable aspects of the Pleistocene fauna is that so many large species of mammals and birds existed. In North America, for example, there were mastodons and mammoths, giant bison, huge ground sloths, giant camels, and beavers nearly 2 m long. Kangaroos standing 3 m tall, wombats the size of rhinoceroses, leopard-size marsupial lions, and large platypuses characterized the Pleistocene fauna of Australia. Cave bears, elephants, and the giant deer, commonly called the Irish elk, with an antler spread of 3.35 m, lived in Europe and parts of Asia. The evolutionary trend toward large body size was perhaps an adaptation to the cooler temperatures of the Pleistocene. Large animals have proportionately less surface area compared with their volume and thus retain heat more effectively than do smaller animals.

In addition to mammals, some other Pleistocene vertebrate animals were of impressive proportions. The giant moas of New Zealand and the elephant birds of Madagascar were very large, and Australia had giant birds standing 3 m tall and weighing nearly 500 kg, and a lizard 6.4 m long and weighing 585 kg. Huge ground sloths and armored mammals known as glyptodonts have been found in Florida (❚ Figure 23.26a) The tar pits of Rancho La Brea in southern California contain the remains of at least 200 kinds of animals. Many of these are fossils of dire wolves, saber-toothed cats, and other mammals (❚ Figure 23.26b), but some are the remains of birds, especially birds of prey, and a giant vulture with a wingspan of 3.6 m.

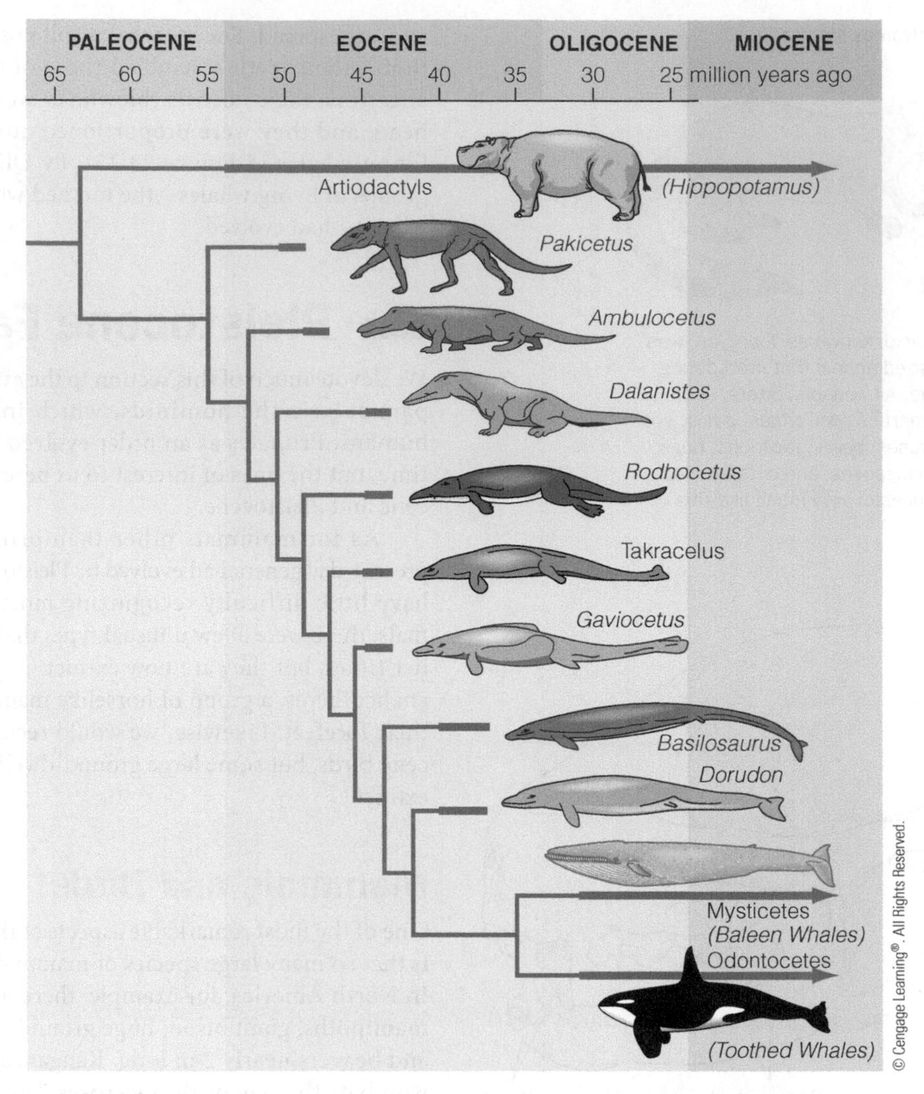

Figure 23.25 The Evolution of Whales The land-dwelling ancestors of whales were among the even-toed hoofed mammals, the artiodactyls. Note that *Pakicetus* had well-developed hind limbs, but some of the others, such as *Basilosaurus,* do not. Although *Basilosaurus* was a fully aquatic whale, it differed considerably from today's whales.

Primate Evolution

Primates are difficult to characterize as an order because they lack the strong specializations found in most other mammalian orders. We can, however, point to several trends in their evolution that help define primates and that are related to their *arboreal,* or tree-dwelling, ancestry. These trends include changes in the skeleton and mode of locomotion; an increase in brain size; a shift toward smaller, fewer, and less-specialized teeth; stereoscopic vision; and the evolution of a grasping hand with an opposable thumb. Not all of these trends took place in every primate group, and they did not evolve at the same rate in each group. In fact, some primates have retained certain primitive features, whereas others show all or most of these trends.

The order Primates is divided into two suborders, the Prosimii and Anthropoidea (Figure 23.27, Table 23.1). The *prosimians* include the lemurs, lorises, tarsiers, and tree shrews (Figure 23.27a). They are generally small, ranging from species that are the size of a mouse up to those as large as a house cat. They are arboreal, have five digits on each hand and foot with either claws or nails, and are typically omnivorous. They have large, forwardly directed eyes specialized for night vision—hence, most are nocturnal (Figure 23.27a). As their name implies (*pro* means "before," and *simian* means "ape"), they are the oldest primate lineage, with a fossil record extending back to the Paleocene.

During the Eocene, prosimians were abundant, diversified, and widespread in North America, Europe, and Asia. As the continents moved northward during the Cenozoic and the climate changed from warm-tropical to cooler mid-latitude conditions, the prosimian population decreased in both abundance and diversity. Presently, prosimians are

Figure 23.26 Pleistocene Fossils from Florida and California

Erika Simons/Florida Museum of Natural History

a Among the diverse Pliocene and Pleistocene mammals of Florida were 6-m-long ground sloths and armored glyptodonts that weighed more than 2 metric tons.

Tom Mchugh/Science Photo

b Restoration of a camel trapped in the sticky tar (asphalt) at the La Brea Tar Pits in Los Angeles, California. Vultures and a saber-toothed cat look on. Notice the giant ground sloths in the background.

found only in the tropical regions of Asia, India, Africa, and Madagascar.

Sometime during the Late Eocene, the *anthropoids,* primates that include monkeys, apes, and humans, evolved from a prosimian lineage. By the Oligocene, the anthropoids were a well-established group with both Old World monkeys (Africa, Asia) and New World monkeys (Central and South America) having evolved during this epoch.

Old World monkeys (superfamily Cercopithecoidea) include the macaque, baboon, and proboscis monkey and are characterized by close-set, downward-directed nostrils (like those of apes and humans), grasping hands, and a nonprehensile tail (Figure 23.27b). Present-day Old World monkeys are distributed throughout the tropical regions of Africa and Asia and are thought to have evolved from a primitive anthropoid ancestor sometime during the Oligocene.

New World monkeys (superfamily Ceboidea) are found only in Central and South America. They are characterized by a prehensile tail, flattish face, and widely separated

Figure 23.27 Primates Primates are divided into two suborders: the prosimians **(a)** and the anthropoids **(b)–(d)**, which are further subdivided into three superfamilies: Old World monkeys **(b)**, New World monkeys **(c)**, and hominoids, which include the apes **(d)** and humans.

© G. Austing/Photo Researchers Inc.

a Tarsier.

© Tony Camacho/Photo Researchers, Inc.

b Baboon.

© Renee Lynn/Photo Researchers

c Spider monkey.

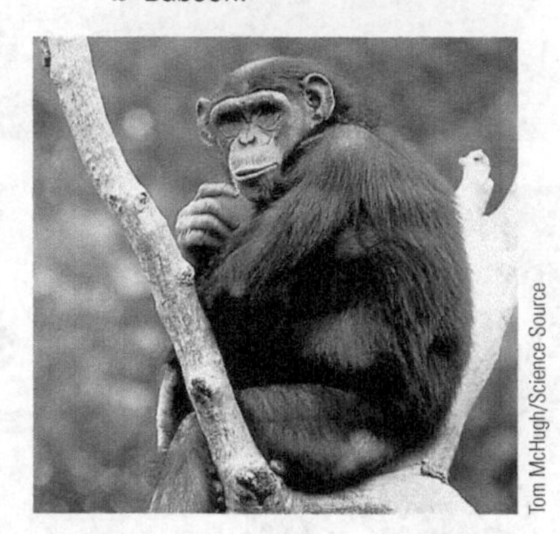

Tom McHugh/Science Source

d Chimpanzee.

nostrils and include the howler, spider, and squirrel monkeys (Figure 23.27c). They probably evolved from African monkeys that migrated across the widening Atlantic sometime during the Early Oligocene, and they have continued evolving in isolation to this day.

The *hominoids* (superfamily Hominoidea), the group containing apes (Figure 23.27d), humans, and their extinct ancestors, diverged from Old World monkeys sometime before the Miocene, but exactly when is still being debated. It is generally accepted, however, that hominoids evolved in Africa from an ancestral anthropoid group. Although there is still debate on the evolutionary relationships among the early hominoids, fossil evidence and molecular DNA similarities between present-day hominoid families is providing a clearer picture of the evolutionary pathways and relationships among the hominoids.

TABLE 23.1 Classification of the Primates

- Order Primates: Lemurs, lorises, tarsiers, tree shrews, monkeys, apes, humans
  - Suborder Prosimii: Lemurs, lorises, tarsiers, tree shrews
  - Suborder Anthropoidea: Monkeys, apes, humans
    - Superfamily Cercopithecoidea: Macaque, baboon, proboscis monkey (Old World monkeys)
    - Superfamily Ceboidea: Howler, spider, and squirrel monkeys (New World monkeys)
    - Superfamily Hominoidea: Apes, humans
      - Family Pongidae: Chimpanzees, orangutans, gorillas
      - Family Hylobatidae: Gibbons, siamangs
      - Family Hominidae: Humans and their extinct ancestors

Hominids

The **hominids** (family Hominidae), the primate family that includes present-day humans and their extinct ancestors (Table 23.1), have a fossil record extending back almost 7 million years. Several features distinguish the hominids from other hominoids. Hominids are bipedal; that is, they have an upright posture, which is indicated by several modifications in their skeleton (▮ Figure 23.28a). In addition, they show a trend toward a large and internally reorganized brain (▮ Figure 23.28b). Other features include a reduced face and reduced canine teeth, omnivorous feeding, increased manual dexterity, and the use of sophisticated tools.

Many anthropologists think that these hominid features evolved in response to major climatic changes that began during the Miocene and continued into the Pliocene. During this time, vast savannas replaced the African tropical rain forests where the prosimians and Old World monkeys had been so abundant. As the savannas and grasslands continued

▮ Figure 23.28 Comparison of Gorilla and Human Locomotion and Hominid Brain Size

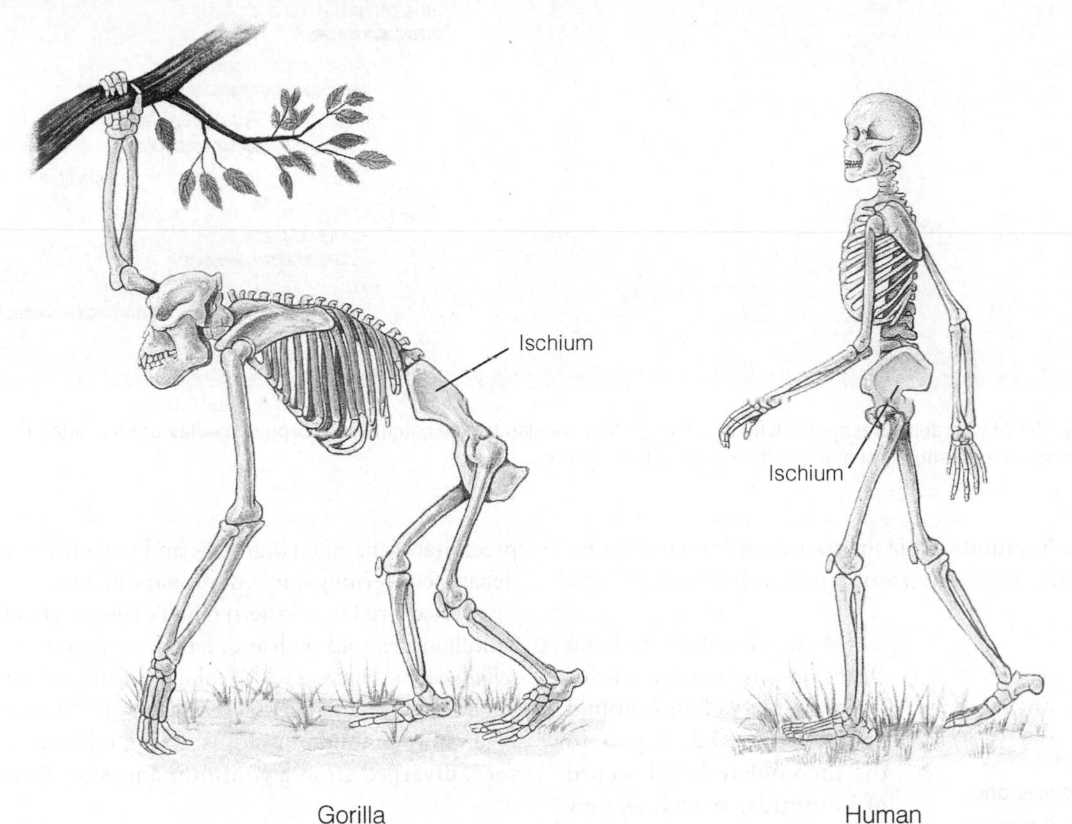

a In gorillas, the ischium bone is long, and the entire pelvis is tilted toward the horizontal. In humans, the ischium bone is much shorter, and the pelvis is vertical.

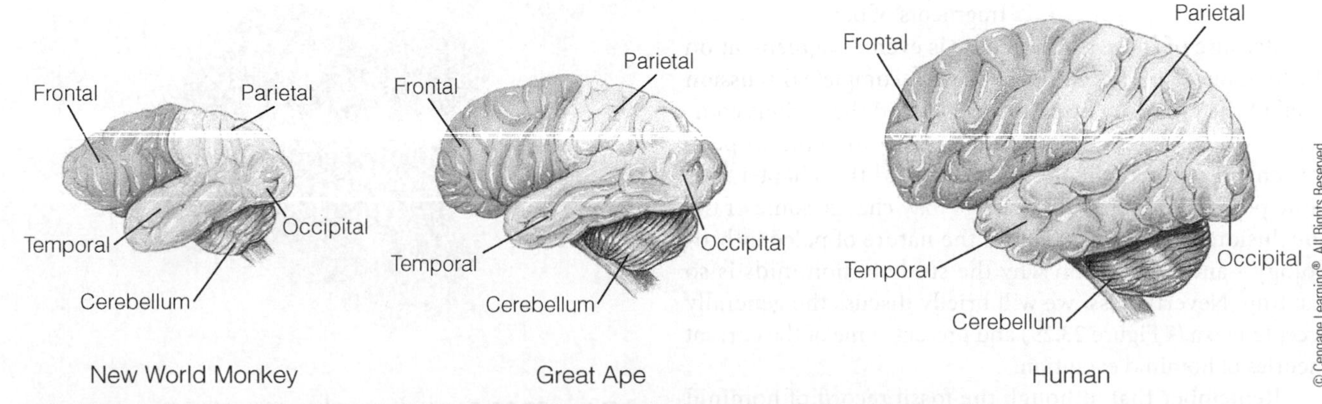

b An increase in brain size and organization is apparent in comparing the brains of a New World monkey (superfamily Ceboidea), a great ape (superfamily Hominoidea; family Pongidae), and a present-day human (superfamily Hominoidea; family Hominidae).

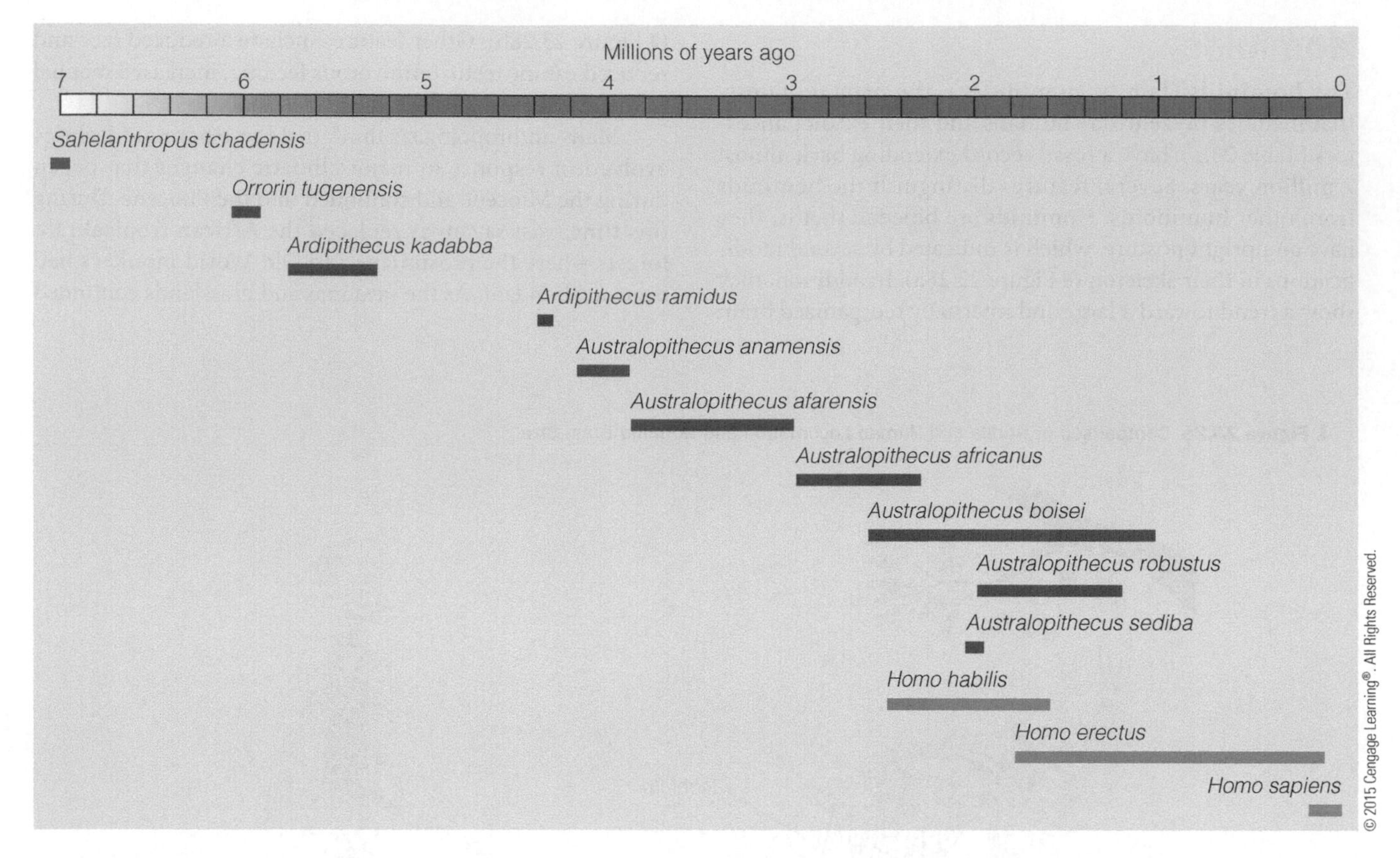

Figure 23.29 The Stratigraphic Record of Hominids The geologic ranges for the commonly accepted species of hominids (the branch of primates that includes present-day humans and their extinct ancestors).

to expand, the hominids made the transition from true forest dwelling to life in an environment of mixed forests and grasslands.

ConnectionLink

To review what you have learned about fossils and fossilization, see Chapter 7 and Chapter 18.

At present, there is no clear consensus on the evolutionary history of the hominid lineage. This is due, in part, to the incomplete fossil record of hominids, as well as new discoveries, and also because some species are known only from partial specimens or fragments of bone.

Because of these factors, there is even disagreement on the total number of hominid species. A complete discussion of all of the proposed hominid species and the various competing schemes of hominid evolution is beyond the scope of this chapter. In fact, by the time you read this chapter, it is quite possible that new discoveries may change some of the conclusions stated here. Such is the nature of paleoanthropology—and one reason why the study of hominids is so exciting. Nevertheless, we will briefly discuss the generally accepted taxa (Figure 23.29) and present some of the current theories of hominid evolution.

Remember that although the fossil record of hominid evolution is not complete, what exists is well documented. However, it is the interpretation of that fossil record that precipitates the often vigorous and sometimes acrimonious debates concerning our evolutionary history.

Discovered in northern Chad's Djurab Desert, the nearly 7-million-year-old skull and dental remains of *Sahelanthropus tchadensis* (Figure 23.30) make it the oldest known hominid yet unearthed and one that existed at or very near to the time when humans and our closest living relative, the chimpanzees, diverged from a common ancestor. Currently, most

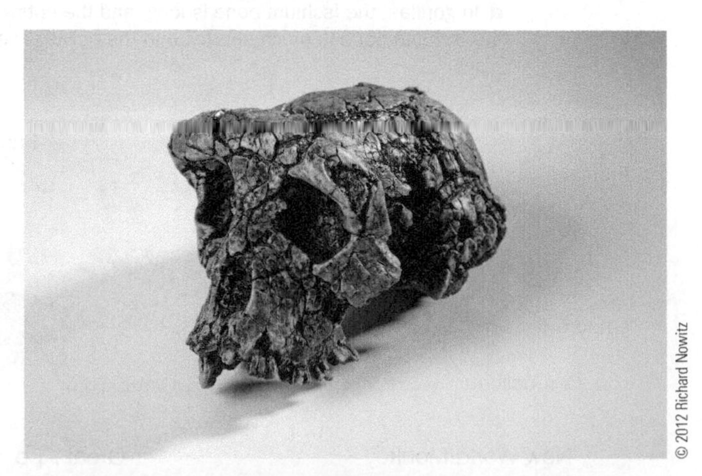

Figure 23.30 ***Sahelanthropus tchadensis*** Discovered in Chad in 2002, and dated at nearly 7 million years, this skull of *Sahelanthropus tchadensis* is presently the oldest known hominid.

paleoanthropologists accept that the human–chimpanzee stock separated from gorillas approximately 8 million years ago and that humans separated from chimpanzees about 5 million years ago.

The next oldest hominid is *Orrorin tugenensis,* whose fossils have been dated at 6 million years old and consist of bits of jaw, isolated teeth, and finger, arm, and partial upper leg bones. At this time, there is still debate as to exactly where *Orrorin tugenensis* fits in the hominid lineage.

Sometime between 5.8 and 5.2 million years ago, another hominid, *Ardipithecus kadabba,* was present in eastern Africa (Figure 23.29). The recent discovery of a second species of *Ardipithecus, A. ramidus,* has caused paleoanthropologists to once again rethink the human evolutionary lineage. Nicknamed "Ardi," this skeleton and skull of a female, 119 cm tall, and weighing approximately 50 kg, lived about 4.4 million years ago (▌Figure 23.31). Ardi displays an interesting mosaic of evolutionary characteristics that includes a dexterous hand for grasping and a foot with an opposable big toe, but lacking the flexibility of an ape's foot. Based on these and other features, scientists think that Ardi was able to walk upright on the ground, while also retaining the ability to climb and maneuver in trees.

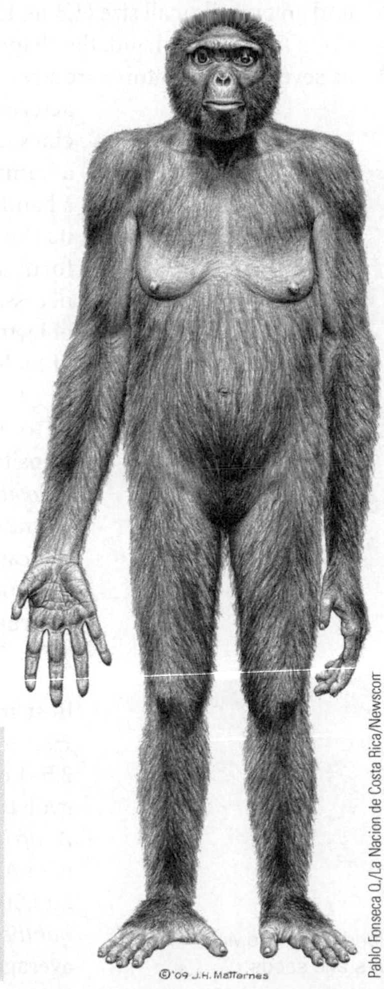

▌Figure 23.31 ***Ardipithecus ramidus*** Frontal view of a female *Ardipithecus ramidus*, known as "Ardi." *A. ramidus*, who lived approximately 4.4 million years ago in what is now Ethiopia shows an interesting variety of features, such as a dexterous hand for grasping, and a foot with an opposable big toe (clearly shown here) that allowed it to walk upright, yet also climb and maneuver in trees.

Critical Thinking Question
With features that allow for climbing in trees like a gorilla, and also walking upright on the ground like a hominid, should "Ardi" be considered a "missing link?" Why do you think "missing link" is not a good term to use in evolution?

Pablo Fonseca Q./La Nacion de Costa Rica/Newscom

Australopithecines **Australopithecine** is a collective term for all members of the genus *Australopithecus.* Currently, five species are generally recognized: *A. anamensis, A. afarensis, A. africanus, A. robustus,* and *A. boisei.* However, a recently discovered sixth species may result in a new interpretation of the origin of our own genus, *Homo*. Notwithstanding this new discovery, many paleoanthropologists accept the evolutionary scheme in which *A. anamensis,* the oldest known australopithecine, is ancestral to *A. afarensis,* who in turn is ancestral to *A. africanus* and the genus *Homo,* as well as the side branch of australopithecines represented by *A. robustus* and *A. boisei.*

The oldest known australopithecine is *Australopithecus anamensis.* Discovered at Kanapoi, a site near Lake Turkana, Kenya, by Meave Leakey of the National Museums of Kenya and her colleagues, this 4.2-million-year-old bipedal species has many features in common with its younger relative, *Australopithecus afarensis,* yet is more primitive in other characteristics, such as its teeth and skull. *A. anamensis* is estimated to have been between 1.3 and 1.5 m tall and weighed 33–50 kg.

A discovery of fossils of *Australopithecus anamensis* from the Middle Awash area in northeastern Ethiopia has shed light on the transition between *Ardipithecus* and *Australopithecus.* Prior to this discovery, the origin of *Australopithecus* had been hampered by a sparse fossil record. The discovery of *Ardipithecus* in the same region of Africa and at the same time as the earliest *Australopithecus* provides strong evidence that *Ardipithecus* evolved into *Australopithecus* and links these two genera in the evolutionary lineage leading to humans.

Australopithecus afarensis (▌Figure 23.32), who lived 3.9–3.0 million years ago, was fully bipedal and exhibited great variability in size and weight. Members of this species ranged from just over 1 m to about 1.5 m tall and weighed between 29 and 45 kg. They had a brain size of 380–450 cubic centimeters (cc), larger than the 300–400 cc of a chimpanzee, but much smaller than that of present-day humans (1,350 cc average).

The skull of *A. afarensis* retained many apelike features, including massive brow ridges and a forward-jutting jaw, but its teeth were intermediate between those of apes and humans. The heavily enameled molars were probably an adaptation to chewing fruits, seeds, and roots (▌Figure 23.33).

A. afarensis was stratigraphically succeeded by *Australopithecus africanus,* who lived 3.0–2.3 million years ago. The differences between the two species are relatively minor. They were both about the same size and weight, but *A. africanus* had a flatter face and a somewhat larger brain. Furthermore, it appears that the limbs of *A. africanus* may not have been as well adapted for bipedalism as those of *A. afarensis.*

Both *A. afarensis* and *A. africanus* differ markedly from the so-called robust species, *A. boisei* (2.6–1.0 million years ago) and *A. robustus* (2.0–1.2 million years ago). Both robust species were about the same size (1.1–1.4 m tall) and weight (32–49 kg) and had similar morphologic features. For example, *A. robustus* had a flat face, and the crown of its skull had an elevated bony crest that provided additional area for the

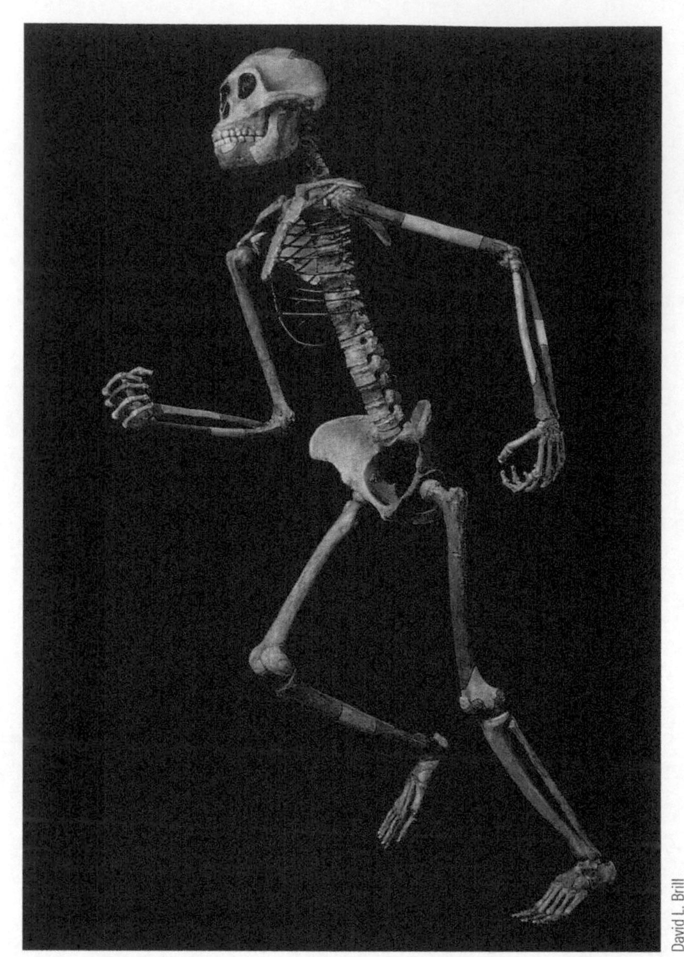

David L. Brill

Figure 23.32 Skeleton of Lucy (*Australopithecus afarensis*) A reconstruction of Lucy's skeleton by Owen Lovejoy and his students at Kent State University, Ohio. Lucy, whose fossil remains were discovered by Donald Johanson, is an approximately 3.5-million-year-old *Australopithecus afarensis* individual. This reconstruction illustrates how adaptations in Lucy's hip, leg, and foot allowed for a fully bipedal means of locomotion.

attachment of strong jaw muscles. In addition, its broad, flat molars indicated that it was a vegetarian. Most scientists accept the idea that the robust australopithecines form a separate lineage from the other australopithecines and went extinct around 1 million years ago.

At a site called Malapa, approximately 40 km northwest of Johannesburg, South Africa, fossils from an eroded limestone cave have yielded the bones of hominid individuals that might represent the oldest known species of the human lineage. Named *Australopithecus sediba* (Figure 23.34a), it appears to be an intermediate form between *Australopithecus africanus* and either *Homo habilis* or *Homo erectus*. What makes *A. sediba* such an exciting discovery is that it contains a combination of primitive and advanced features. Among the primitive features it shares with australopithicines is a small brain, comparable to that of other australopithicines, yet smaller than the larger brained *Homo habilis* or *H. erectus*. Other primitive features include long upper arms; long, high cheekbones; primitive molar cusps; a primitive heel bone; and an overall small size (1.3 m. tall).

On the other hand, the shape of *A. sediba*'s brain, as well as several facial features are advanced traits that are not characteristic of earlier australopithicines. Other advanced features are a humanlike pelvis, long legs, and a hand with the capability to bring its thumb and fingers together to form a precision grip, a feature necessary for the manipulation of objects and the making and using of tools (Figure 23.34b).

Currently, there is great debate as to whether *A. sediba* is a true transitional species between *Australopithecus* and either *H. habilis* or *H. erectus*, or simply a late southern African branch of *Australopithecus* that coexisted with members of the evolving *Homo* genus.

© Darwen and Vally Hennings 1990

Figure 23.33 African Pliocene Landscape Recreation of a Pliocene landscape showing members of *Australopithecus afarensis* gathering and eating various fruits and seeds.

The Human Lineage The earliest member of our own genus ***Homo*** is *Homo habilis,* who lived 2.5–1.6 million years ago. *H. habilis* evolved from the *A. afarensis* and *A. africanus* lineage and coexisted with *A. africanus* for approximately 200,000 years (Figure 23.29). *H. habilis* had a larger brain (700 cc average) than its australopithecine

▌Figure 23.34 *Australopithecus sediba*—The Ancestor of *Homo*?

AP Photo/Brett Eloff/Lee Berger/University of Witwatersrand/dapd

a The skull of *Australopithecus sediba* contains both primitive and advanced features. One of the primitive features is a small brain, comparable in size to that of other australopithicines. On the other hand, the skull displays such advanced features as relatively small premolars and molars, as well as facial features that are more similar to *Homo*.

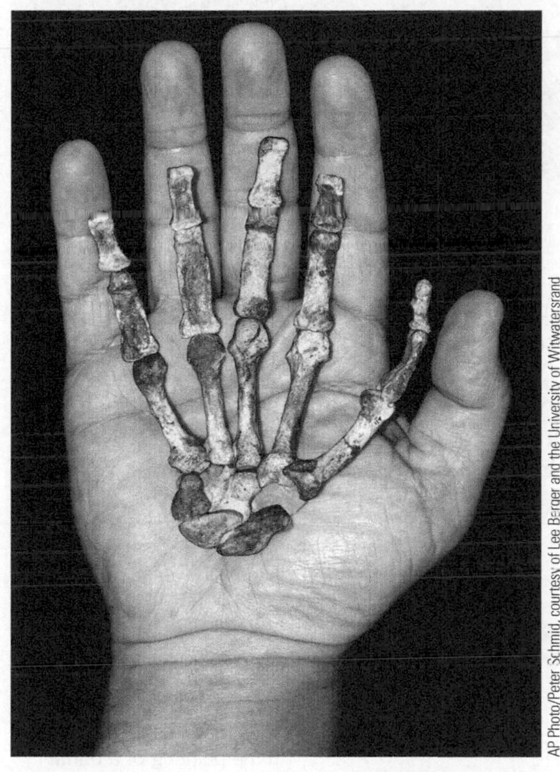

AP Photo/Peter Schmid, courtesy of Lee Berger and the University of Witwatersrand

b This image shows the skeleton of the right hand of *Australopithecus sediba* against a modern human hand. The hand lacks three wrist bones and four terminal phalanges (fingertip bones), but is otherwise complete. An important and advanced feature of the hand of *A. sediba* is that the thumb and fingers can be brought together to grip items, a feature necessary for the manipulation of objects.

ancestors, but smaller teeth. It was between 1.2–1.3 m tall, had disproportionately long arms compared with modern humans, and weighed only 32–37 kg.

The evolutionary transition from *H. habilis* to *H. erectus* appears to have occurred in a short period of time, between 1.8 and 1.6 million years ago. However, evidence indicating that *H. habilis* and *H. erectus* may have coexisted for approximately 500,000 years has led some scientists to suggest that *H. habilis* and *H. erectus* evolved from a common ancestor and represent separate lineages of *Homo,* rather than the traditional linear view of *H. erectus* evolving from *H. habilis.*

In contrast to the australopithecines and *H. habilis,* which are unknown outside of Africa, *Homo erectus* was a widely distributed species, having migrated from Africa during the Pleistocene. Specimens have been found not only in Africa, where it evolved 1.8 million years ago, but also in Europe (▌Figure 23.35), India, China ("Peking Man"), and Indonesia ("Java Man").

Although *H. erectus* developed regional variations in form, the species differed from modern humans in several ways. Its brain size of 800–1,300 cc, although much larger than that of *H. habilis,* was still less than the average for *Homo sapiens* (1,350 cc). The skull of *H. erectus* was thick walled, its face was massive, it had prominent brow ridges, and its teeth were slightly larger than those of present-day humans. *H. erectus* was comparable in size to modern humans, standing between 1.6 and 1.8 m tall and weighing between 53 and 63 kg.

The archaeological record indicates that *H. erectus* was a toolmaker. Furthermore, some sites show evidence that its members used fire and lived in caves, an advantage for those living in more northerly climates (Figure 23.35).

Debate still surrounds the transition from *Homo erectus* to our own species, *Homo sapiens.* Paleoanthropologists are split into two camps. On the one side are those who support the "out of Africa" view. According to this view, early modern humans evolved from a single woman in Africa, whose offspring then migrated from Africa, perhaps as recently as 100,000 years ago, and populated Europe and Asia, driving the earlier hominid population to extinction.

The alternative explanation, the "multiregional" view, maintains that early modern humans did not have an isolated origin in Africa, but, rather, that they established separate populations throughout Eurasia. Occasional contact and interbreeding between these populations enabled our species to maintain its overall cohesiveness while still preserving the regional differences in people that we see today. Regardless of which theory turns out to be correct, our species, *H. sapiens,* most certainly evolved from *H. erectus.*

Perhaps the most famous of all fossil humans are the **Neanderthals,** who inhabited Europe and the Near East from about 200,000 to 30,000 years ago and, according to the best estimates, never exceeded 15,000 individuals in western Europe (▌Figure 23.36). Some paleoanthropologists regard the Neanderthals as a variety or subspecies (*Homo sapiens neanderthalensis*), whereas others consider them a separate species (*Homo neanderthalensis*). In any case, their name comes from the first specimens found in 1856 in the Neander Valley near Düsseldorf, Germany.

The most notable difference between Neanderthals and present-day humans is in the skull. Neanderthal skulls were long and low, with heavy brow ridges, a projecting mouth,

Figure 23.35 ***Homo erectus*** Recreation of a Pleistocene setting showing members of *Homo erectus* using fire and stone tools at an encampment site.

Publiphoto/Photo Researchers, Inc.

and a weak, receding chin. Their brain was slightly larger, on average, than our own and somewhat differently shaped. The Neanderthal body was more massive and heavily muscled than ours, with a flaring rib cage and rather short lower limbs, much like those of other cold-adapted people of today. Neanderthal males averaged between 1.6 and 1.7 m tall and weighed about 83 kg.

Based on specimens from more than 100 sites, we now know that Neanderthals were not much different from us, only more robust. Europe's Neanderthals were the first humans to move into truly cold climates, enduring miserably long winters and short summers as they pushed north into tundra country. Their remains are found chiefly in caves and hut-like rock shelters, which also contain a variety of

Robert Harding World Imagery

Figure 23.36 Pleistocene Cave Setting with Neanderthals Archaeological evidence indicates that Neanderthals lived in caves and participated in ritual burials, as depicted in the painting of a burial ceremony such as occurred approximately 60,000 years ago at Shanidar Cave, Iraq.

specialized stone tools and weapons (Figure 23.36). Furthermore, archaeological evidence indicates that Neanderthals commonly took care of their injured and buried their dead, frequently with such grave items as tools, food, and perhaps even flowers.

As more fossil discoveries are made and increasingly sophisticated techniques of DNA extraction and analysis are carried out, our view of Neanderthals and their society as well as their place in human evolution will continue to change.

About 30,000 years ago, humans closely resembling modern Europeans moved into the region inhabited by Neanderthals and completely replaced them. **Cro-Magnons** (▌Figure 23.37), the name given to the successors of Neanderthals in France, lived from about 35,000 to 10,000 years ago. During this period, the development of art and technology far exceeded anything the world had seen before. Using paints made from manganese and iron oxides, Cro-Magnon people painted hundreds of scenes on the ceilings and walls of caves in France and Spain, where many of them are still preserved.

Cro-Magnons were also skilled nomadic hunters, following the herds in their seasonal migrations, and using a variety of specialized tools in their hunts, including perhaps the bow and arrow. They sought refuge in caves and rock shelters and formed living groups of various sizes.

With the appearance of Cro-Magnons, human evolution has become almost entirely cultural rather than biologic. Since the evolution of the Neanderthals approximately 200,000 years ago, humans have gone from a stone culture to a technology that has allowed us to visit other planets with space probes and land men on the Moon. It remains to be seen how we will use this technology in the future and whether we will continue as a species, evolve into another species, or become extinct as many groups have before us.

Pleistocene Extinctions

Extinctions are part of the evolutionary history of life, but during times of mass extinction, Earth's biotic diversity sharply declined, as at the end of the Paleozoic and Mesozoic eras (see Figure 21.13). In marked contrast, the Pleistocene extinctions were rather modest in comparison, but they did have a profound impact on large terrestrial mammals (those weighing more than 44 kg) and some large flightless birds. Particularly hard hit were the mammalian faunas of Australia and the Americas.

ConnectionLink

To review the mass extinctions of the Paleozoic and Mesozoic eras, see Chapter 21 and Chapter 22.

In Australia, 15 of the continent's 16 genera of large mammals died out. North America lost 33 of 45 genera of large mammals, and in South America, 46 of 58 large genera went extinct. In contrast, Europe had only 7 of 23 large genera go extinct, whereas Africa south of the Sahara lost only 2 of 44 genera.

▌**Figure 23.37 Pleistocene Cro-Magnon Camp in Europe** Cro-Magnons were highly skilled hunters who formed living groups of various sizes.

These facts lead to three questions, none of which has been answered completely: (1) What caused Pleistocene extinctions? (2) Why did these extinctions eliminate mostly large mammals? and (3) Why were extinctions most severe in Australia and the Americas? Scientists are currently debating two competing hypotheses for these extinctions. One, the *climatic change hypothesis*, holds that rapid changes in climate at the end of the Pleistocene caused extinctions, whereas proponents of *prehistoric-overkill*, contend that human hunters were responsible.

Rapid changes in climate and vegetation did occur over much of Earth's surface during the Late Pleistocene, as glaciers began retreating. The North American and northern Eurasian open-steppe tundras were replaced by conifer and broadleaf forests as warmer and wetter conditions prevailed. The Arctic region flora changed from a productive herbaceous one that supported a variety of large mammals, to a comparatively barren, water-logged tundra that supported a much sparser fauna. The southwestern United States region also changed from a moist area with numerous lakes, where saber-toothed cats, giant ground sloths, and mammoths roamed, to a semiarid environment unable to support a diverse fauna of large mammals.

Rapid changes in climate and vegetation can certainly affect animal populations, but there are several problems with the climate-change hypothesis. First, why didn't the large mammals migrate to more suitable habitats as the climate and vegetation changed? After all, many other animal species did. For example, reindeer and the Arctic fox lived in southern France during the last glaciation and migrated to the Arctic when the climate became warmer. The second argument against the climate-change hypothesis is the apparent lack of correlation between extinctions and earlier glacial advances and retreats throughout the Pleistocene Epoch. Previous changes in climate were not marked by episodes of mass extinction.

Proponents of the prehistoric-overkill hypothesis argue that the mass extinctions in North and South America and Australia coincided closely with the arrival of humans. Perhaps hunters had a tremendous impact on the faunas of North and South America about 11,000 years ago because the animals had no previous experience with humans. The same thing happened much earlier in Australia soon after people arrived about 40,000 years ago. No large-scale extinctions occurred in Africa and most of Europe because animals in those regions had long been familiar with humans.

One problem with the prehistoric-overkill hypothesis is that archaeological evidence indicates that early human inhabitants of North and South America, as well as Australia, probably lived in small, scattered communities, gathering food and hunting. How could a few hunters decimate so many species of large mammals? However, it is true that humans have caused extinctions on oceanic islands. For example, in a period of about 600 years after arriving in New Zealand, humans exterminated several species of the large, flightless birds called moas.

A second problem is that present-day hunters concentrate on smaller, abundant, and less dangerous animals. The remains of horses, reindeer, and other small animals are found in many prehistoric sites in Europe, whereas mammoth and woolly rhinoceros remains are scarce.

Finally, few human artifacts are found among the remains of extinct animals in North and South America, and there is usually little evidence that the animals were hunted. Countering this argument is the assertion that the impact on the previously unhunted fauna was so swift as to leave little evidence.

The reason for the extinctions of large Pleistocene mammals is still unresolved and probably will be for some time. It may turn out that the extinctions resulted from a combination of different circumstances. Populations that were already under stress from climatic changes were perhaps more vulnerable to hunting, especially if small females and young animals were the preferred targets.

Key Concepts Review

- Cenozoic tectonism was concentrated in the Alpine–Himalayan and circum-Pacific belts.
- The Cenozoic evolution of the North American Cordillera included deformation during the Laramide orogeny, extensional tectonics that yielded basin-and-range structures, extensive intrusive and extrusive igneous activity, and uplift and erosion.
- One model for the Laramide orogeny involves near-horizontal subduction of the Farallon plate beneath North America, resulting in fault-bounded uplifts in the area of the present-day Rocky Mountains.
- As the North American plate drifted westward, it collided with the Pacific–Farallon Ridge, at which time subduction ceased and the continent became bounded by large transform faults, except in the Pacific Northwest where subduction continues.
- Sediments eroded from Laramide uplifts were deposited in intermontane basins and in the Great Plains, whereas a wedge of sediments pierced by salt domes is found on the Gulf Coastal Plain.
- Cenozoic uplift and erosion were responsible for the present topography of the Appalachian Mountains. As the Appalachians eroded, much of the sediment was deposited on the Atlantic Coastal Plain.
- Vast glaciers covered about 30% of Earth's land surface during the Pleistocene. About 20 warm–cold Pleistocene climatic cycles are recognized from evidence found in deep-sea cores.

- Cenozoic mineral resources include sand and gravel, placer deposits of gold, some evaporite minerals such as borax, and oil and natural gas.
- Marine invertebrate groups that survived the extinctions at the end of the Mesozoic continued to diversify, giving rise to the present-day marine fauna.
- The Paleocene mammalian fauna was composed of Mesozoic holdovers and several new orders, and by Eocene time, most living mammal orders had evolved.
- Shrewlike placental mammals that evolved during the Late Cretaceous were the ancestors for the placental mammalian orders that evolved during the Cenozoic.
- Among the hoofed mammals (artiodactyls and perissodactyls), adaptations include modifications of the teeth for grinding vegetation and changes in their limbs for speed.
- The evolutionary history of horses is particularly well documented by fossils, but scientists also know much about the evolution of other hoofed mammals, as well as elephants, whales, and some carnivores.
- Primates, which evolved during the Paleocene, differ from other mammalian orders on the basis of overall skeletal structure and mode of locomotion, an increase in brain size, stereoscopic vision, and a grasping hand with opposable thumb.
- The primates are divided into two suborders: prosimians, which are the oldest primate lineage and include lemurs, lorises, tarsiers, and tree shrews, and anthropoids, which include the New and Old World monkeys, apes, and hominids (humans and their extinct ancestors).
- The hominid lineage begins at nearly 7 million years ago with *Sahelanthropus tchadensis,* followed by *Orrorin tugenensis* at 6 million years, and then two species of *Ardipithecus,* at 5.8 and 4.4 million years ago, respectively. These early hominids were succeeded by the australopithecines, a fully bipedal group that evolved in Africa 4.2 million years ago.
- The human lineage began about 2.5 million years ago in Africa with the evolution of *Homo habilis,* superseded by *Homo erectus,* sometime between 1.8 and 1.6 million years ago, which was the first hominid to migrate out of Africa, spreading to Europe, India, China, and Indonesia.
- Although the transition from *H. erectus* to *H. sapiens* is still unresolved, the most famous of all human fossils are the Neanderthals, who inhabited Europe and the Near East between 200,000 and 30,000 years ago and differed mainly from present-day humans in being more robust and having a long, low skull with heavy brow ridges.
- The Cro-Magnons, highly skilled hunters and cave painters, were the successors of the Neanderthals and lived from about 35,000 to 10,000 years ago. Modern humans succeeded the Cro-Magnons and have spread throughout the world, as well as having set foot on the Moon.
- Modest in comparison to the mass extinction events at the end of the Paleozoic and Mesozoic eras, the Pleistocene extinctions, nevertheless, had a profound effect on large terrestrial mammals. Two competing hypotheses have been advanced to explain these Pleistocene extinctions: climatic change and prehistoric-overkill.

Important Terms

Alpine–Himalayan orogenic belt (p. 615)
artiodactyl (p. 636)
Atlantic Coastal Plain (p. 627)
australopithecine (p. 645)
Basin and Range Province (p. 619)
browser (p. 636)
Cascade Range (p. 619)
circum-Pacific orogenic belt (p. 616)
Colorado Plateau (p. 622)
Cro-Magnon (p. 649)
Farallon plate (p. 623)
grazer (p. 636)
Gulf Coastal Plain (p. 627)
hominid (p. 643)
Homo (p. 646)
Interior Lowlands (p. 624)
Laramide orogeny (p. 617)
marsupial mammal (p. 634)
monotreme (p. 634)
Neanderthal (p. 647)
North American Cordillera (p. 617)
orogen (p. 615)
Paleocene–Eocene Thermal Maximum (p. 633)
perissodactyl (p. 636)
placental mammal (p. 635)
primate (p. 640)
Rio Grande rift (p. 623)
San Andreas transform Fault (p. 623)
Tejas epeiric sea (p. 627)
ungulate (p. 636)
Zuni epeiric sea (p. 624)

Review Questions

1. Which of the following evolutionary trends characterize primates?
   a. _____ increase in brain size.
   b. _____ change in overall skeletal structure.
   c. _____ steroscopic vision.
   d. _____ grasping hand with opposable thumb.
   e. _____ all of these.
2. The Himalayas formed when the _____ plate collided with the _____ plate.
   a. _____ Farallon/Pacific
   b. _____ Nazca/Cocos
   c. _____ Australian/South American
   d. _____ Indian/Asian
   e. _____ African/European
3. One of the defining features of the carnivorous mammals is
   a. _____ carnassials teeth.
   b. _____ an increase in the number of toes.
   c. _____ short, massive limbs.
   d. _____ hooves.
   e. _____ teeth adapted to grazing.
4. A complex part of the circum-Pacific orogenic belt in the United States is the
   a. _____ Tejas sedimentary sequence.
   b. _____ North American Cordillera.
   c. _____ Rio Grande rift.
   d. _____ Atlantic Coastal Plain.
   e. _____ Pacific–Farallon Ridge.
5. During the Pleistocene Epoch
   a. _____ warm, shallow seas covered most of North America.
   b. _____ the most common large predators on land were monotremes.
   c. _____ continental glaciers were present on the Northern Hemisphere continents.
   d. _____ the Laramide orogeny began and continues even now.
   e. _____ the vast lava flows of the Columbia River basalts formed.
6. Discuss the currently held view of the evolutionary history of hominids and how the discovery of *Aridopithecus ramidus* and *Australopithecus sediba* have changed what we formerly thought was the hominid phylogeny.
7. How would you explain the Cenozoic evolution of the North American Cordillera in terms that a nongeologist could readily understand?
8. Explain how the Laramide orogeny differed from other orogenies at convergent plate boundaries.
9. Briefly discuss the major mammal diversification that took place during the Cenozoic Era.

10. **Creative Thinking Visual Question:** The image below (▌Figure 1) shows Mount Shasta in California, which is one of a dozen or so large volcanoes in the Cascade Range. What kind of volcano is shown in this image? If you were to sample a lava flow from Mount Shasta, what kind of volcanic rocks would you expect to find? In addition to lava flows, what other materials are found in volcanoes such as this?

James S. Monroe

▌**Figure 1** Mount Shasta in California.

Global GeoScience Watch Search for "Grand Canyon" in the GREENR database. Next to the "News" heading, click on "VIEW ALL," then scroll to find the article titled "60-Million-Year Debate on Grand Canyon's Age." What are the ages under debate for the origin of the Grand Canyon? Do the scientists proposing opposing views cite any evidence to support their conclusions?

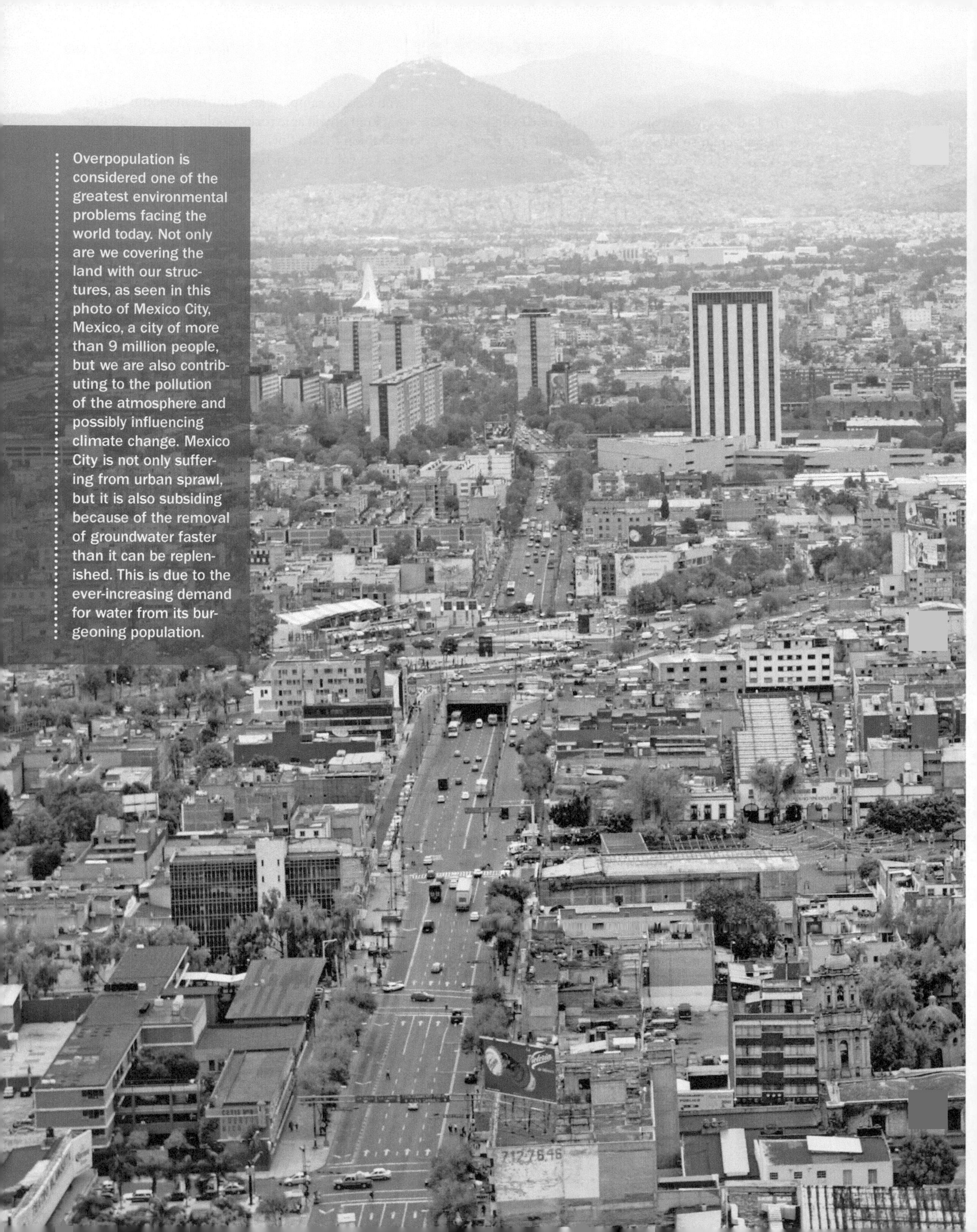

Overpopulation is considered one of the greatest environmental problems facing the world today. Not only are we covering the land with our structures, as seen in this photo of Mexico City, Mexico, a city of more than 9 million people, but we are also contributing to the pollution of the atmosphere and possibly influencing climate change. Mexico City is not only suffering from urban sprawl, but it is also subsiding because of the removal of groundwater faster than it can be replenished. This is due to the ever-increasing demand for water from its burgeoning population.

Wendy Connett/Robert Harding World Imagery/Getty Images

CHAPTER 24 Geology in Perspective

OUTLINE

24.1 Introduction

Throughout this book, we have emphasized that Earth is a complex, dynamic planet that has changed continuously since its origin some 4.6 billion years ago. These changes, and the present-day features we observe, are the result of interactions among the various interrelated internal and external Earth systems, subsystems, and cycles. In addition, these interactions have influenced the evolution of the biosphere, as is obvious from an examination of the fossil record. Perhaps the most important lesson to be learned from the study of geology is that Earth is a dynamic and changing planet in which continual change is taking place, although not always at a rate we can readily appreciate from a human lifetime or perspective.

The rock cycle (see Figure 1.15), with its recycling of Earth materials to form the three major rock groups (see Figure 1.16), illustrates the interrelationships between Earth's internal and external processes. The hydrologic cycle (see Figure 12.2) explains the continuous recycling of water from the oceans to the atmosphere to the land and eventually back to the oceans again. Changes within this cycle can have profound effects on Earth's topography as well as its biota. For example, a rise in global temperature will cause the ice caps to melt, contributing to rising sea level, which will greatly affect coastal areas where many of the world's large population centers are presently located. We have seen the effect of changing sea level on continents in the past, which resulted in large-scale transgressions and regressions (see Chapter 20). Some of these were caused by growing and shrinking continental ice caps when landmasses moved over the South Pole as a result of plate movements.

On a larger scale, the movement of plates has had a profound effect on the formation of landscapes, the distribution of mineral resources, and atmospheric and oceanic circulation patterns, as well as the evolution and diversification of life.

The launching in 1957 of *Sputnik 1*, the world's first artificial satellite, ushered in a new global consciousness in terms of how we view Earth and our place in the global ecosystem. Satellites and manned space missions have provided us with the ability to view not only the beauty of our planet but also the fragility of Earth's biosphere and the role humans play in shaping and modifying the environment. The pollution of the atmosphere, oceans, and many of our lakes and streams; the denudation of huge areas of tropical forests; the scars from strip mining; the depletion of the ozone layer are all visible in the satellite images beamed back from space and attest to the impact humans have had on the global ecosystem.

Accordingly, we must understand that the changes we make to the global ecosystem can have wide-ranging effects. For this reason, an understanding of geology, and science in general, is of paramount importance if we are to be good stewards of Earth. On the other hand, we must also remember that humans are part of the global ecosystem, and, like all other life-forms, our presence alone affects the ecosystem. We must therefore act in a responsible manner, based on sound scientific knowledge, so future generations will inherit a habitable environment.

One of the most important objectives of this book, as well as much of your secondary education, is to develop your critical thinking skills. As opposed to simple disagreement, critical thinking involves evaluating the supporting evidence for a particular point of view. Although your exposure to geology is probably still limited, having read this book, you now have the fundamental knowledge needed to appraise why geologists accept plate tectonic theory, why they think that Earth is 4.6 billion years old, and why scientists are convinced that the theory of evolution is well supported by evidence. Furthermore, your abilities as a critical thinker will help you more effectively evaluate the arguments about global warming, ozone depletion, groundwater contamination, resource management, energy independence, and many other geologic- and environmental-related issues.

When such environmental issues as acid rain (see Chapter 6), the depletion of the ozone layer, the greenhouse effect, and global warming (see Figure 1.4) are discussed and debated, it is important to remember that they are not isolated topics but rather are part of a larger system that involves the entire planet (see Figure 1.1). In addition, it is important to remember that Earth goes through cycles of much longer duration than the human perspective of time (see Chapter 17). Although they may have disastrous effects on the human species, global warming and cooling are part of a larger cycle that has resulted in numerous glacial advances and retreats during the past 2.6 million years (see Chapter 14).

In fact, geologists can make important contributions to the debate on global warming because of their geologic perspective. Long-term trends can be studied by analyzing deep-sea sediments, ice cores, changes in sea level during the geologic past, and the distribution of plants and animals through time. As we have seen throughout this book, such studies have been done, and the results and synthesis of that information can be used to make intelligent decisions about how humans can better manage the environment and the effect we are having in altering the environment.

24.2 Geology and the Environment

Geologists who specialize in environmental geology are involved in understanding the relationship between humans and their geologic environment. Put another way, environmental geologists are concerned with studying and resolving the problems that human occupation has caused to the environment. As we discussed in Chapter 1, many scientists consider overpopulation to be the greatest environmental problem facing the world today (▮ Figure 24.1). The world's population was slightly more than 7 billion at the end of 2012, and it is projected to reach 8 billion by 2025. Much of this population growth will be in areas that are already

Figure 24.1 Overpopulation Overpopulation is perhaps the greatest environmental problem facing the world today. Until the world's increasing population is brought under control, people will continue to strain Earth's limited resources. Shown here is the crush of people at the Sadar Bazaar in New Delhi, India, which currently has a population of nearly 14 million people.

at risk from earthquakes, tsunami, volcanic eruptions, and floods. In addition, adequate and clean water supplies, as well as additional energy resources, must be found and properly managed, and all with a minimal impact on the environment.

Interestingly, most of the major environmental issues currently being discussed and debated, such as acid rain, ozone depletion, and global warming, seemingly have a common denominator: humans and their contribution to these global problems. The role that humans play in many of these issues, as well as the degree to which they play that role, are being examined and debated not only by scientists but also by economists, sociologists, and politicians, to name a few. It is, therefore, more important than ever that citizens become versed in what science is and in what it can, and cannot do, as well as becoming critical thinkers in the arguments regarding the impact humans have on the global ecosystem.

We have already discussed in previous chapters such environmental problems and concerns as the greenhouse effect (Chapter 1), soil degradation (Chapter 6), acid rain (Chapter 6), flood control (Chapter 12), groundwater contamination (Chapter 13), and climate change (Chapters 14 and 17).

For example, you learned in Chapter 6 how soils form, how they are categorized, and the factors that control soil formation and degradation. Having learned the basic science of soils and their formation, it is important to remember that our future as a species is clearly linked to the wise management of our soil resources. Like other conservation efforts, using soils sustainably depends on a combination of scientific understanding and political will. Your generation will probably see new pressures on world soil resources and may potentially contribute new solutions to perennial soil problems.

Additionally, you have learned what acid rain is, how it forms, and why it is such a multifaceted problem that knows no national boundaries. Acid rain is just one of many environmental issues that are interconnected and are part of a dynamic and ever-changing Earth (▮ Figure 24.2).

Probably the major environmental issue currently being discussed is global warming, a topic we covered in Chapter 1 and continued to emphasize in subsequent chapters in various contexts. Climate change is not only important to humans but also is an excellent example of how Earth's systems, subsystems, and cycles are interrelated, and why it is important to look at the issue both from a human perspective

Figure 24.2 Acid Rain

Yoram Lehman/Photo Researchers (Science Source)

a Pollution from the smokestacks of steel mills in Pittsburgh, Pennsylvania, during the 1970s contributed to the formation of acid rain in the eastern United States and Canada.

Mary Terriberry/shutterstock.com

b The effects of acid rain can be seen in these dead trees in Grayson Highlands State Park, Virginia.

in terms of short-term changes as well as from a geologic perspective of time and Earth history (see Chapters 14, 17, 20, 22, and 23).

Recall that carbon dioxide is produced as a by-product of respiration and the burning of organic material. As such, it is a component of the global ecosystem and is constantly being recycled as part of the carbon cycle. The concern in recent years over the increase in atmospheric carbon dioxide has to do with its role in the greenhouse effect (see Figure 1.4). The issue is not whether we have a greenhouse effect, because we do, but to what degree human activity, such as the burning of fossil fuels, has contributed to global warming. It is therefore important to look at preanthropogenic changes in climate using the most precise dating techniques and high-resolution sampling that are available. The debate concerning human's role in climate change must be addressed using sound scientific methods, and not become politicized, as it seemingly has in the past few years.

Related to the degree to which humans are adding carbon dioxide to the atmosphere through the burning of fossil fuels is the issue of energy independence. The development of horizontal drilling and hydraulic fracturing is another example of competing interests pitting those who favor its benefits in reducing energy dependence and the economic benefits resulting from "fracking," against those who are concerned with its environmental impact in regards to groundwater pollution and possible increase in global warming (see Chapter 1, Geo-Impact, and Chapter 13, Geo-Focus 13.1).

The ability to exploit gas- and oil-rich shales has been a game changer in terms of redrawing the world's energy map. Instead of being a net importer, the United States may soon become an exporter of oil and natural gas due to improved drilling and extraction technology. Although the United States might become an exporter of oil and natural gas in the near future, it should be noted that its reserves are still far less than those in the Middle East. The availability of large supplies of natural gas is also affecting the economy, as many industries switch from coal or oil to natural gas. Although there are many benefits resulting from gas- and oil-rich shale production, there is still fervent opposition to "fracking," and the concerns of those opposed to it must be fully addressed before there is wide public acceptance of this technology.

Another theme basic to geology and one that affects all of us regardless of where we live is geologic hazards. Any geologic phenomenon that endangers property and people is a geologic hazard. Earthquakes, volcanic eruptions, landslides, and river and coastal flooding during hurricanes, such as Hurricane Sandy in 2012, immediately come to mind (▌Figure 24.3). These are spectacular geologic hazards that commonly cause property damage, injuries, and fatalities.

MC Images/Alamy

▌**Figure 24.3 Geologic Hazards—Coastal Flooding** Coastal flooding is just one of many geologic hazards that can cause loss of human life as well as significant property damage. Shown in this aerial view of the New Jersey coast taken on October 30, 2012, is just some of the destruction caused by Hurricane Sandy.

Some of the more subtle hazards, however, such as expansive soils and soil creep (see Chapters 6 and 11), cause more property damage, but rarely rate headlines in newspapers or on news reports.

During 2011 and 2012, we witnessed major earthquakes in New Zealand, Japan, and Iran, as well as a devastating tsunami that resulted from the 2011 Japanese earthquake (see Figure 9.17). Massive flooding devastated Brisbane and environs in Australia in 2011. The United States Midwest also suffered from massive flooding in 2012 (see Chapter 12, Geo-Insight 12.1).

Geologic hazards account for thousands of injuries and fatalities each year, as well as billions of dollars in property damages. You may read in the more sensational press that geologic hazardous events, especially earthquakes, have been increasing. The incidence of these events has not increased, but there has been an increase in the amount of property damages and the number of injuries and fatalities because more and more people live in disaster-prone areas. Population density and poor construction practices are obvious contributors to the devastation caused by flooding and earthquakes in developing countries (▮ Figure 24.4).

Of course, we cannot eliminate geologic hazards, but we can better understand these phenomena, enact prudent zoning and land-use regulations, and at the very least, decrease the amount of human suffering and monetary costs associated with disasters (see Chapter 9, Geo-Insight 9.1).

Unfortunately, geologic information that is readily available is often ignored. A case in point is the Turnagain Heights subdivision in Anchorage, Alaska, that was heavily damaged when the soil beneath it liquefied during the 1964 earthquake (see Figure 11.18). Not only were reports on soil stability ignored or overlooked before homes were built there but also, since 1964, many new homes have been built on part of the same site.

La Conchita, California, lies at the base of steep, unstable slopes that collapsed in 1995 and again in 2005, killing 10 people in the latter event (see Figures 1 and 2 in Chapter 11, Geo-Focus 11.1). There are many other examples of building subdivisions on known landslide masses in the Los Angeles area, and development in some coastal areas continues where erosion will surely destroy homes and other structures.

Although most of you are not going to become licensed geological engineers, you now have the knowledge to make

▮ Figure 24.4 Geologic Hazards Earthquakes are just one of the numerous geologic hazards that can cause loss of human life as well as significant property damage. One example is the May 12, 2008, 7.9-magnitude earthquake that devastated Sichuan province in China, killing approximately 69,000 people, and leaving millions homeless.

informed decisions on the suitability of the geology of the area where you might buy or build a house (see Chapter 11, Geo-Impact). Hopefully, with this knowledge, you will not make the same mistakes that many people made in the examples given throughout this book.

24.3 A Final Word

In conclusion, the most important lesson to be learned from the study of geology is that Earth is an extremely complex and ever-changing planet. Interactions between its various systems and subsystems have resulted in changes in the atmosphere, lithosphere, and biosphere through time. By studying how Earth has evolved in the past, we can better understand how the different Earth systems and subsystems work and interact with one another, and, more importantly, how our actions affect the delicate balance between these various components.

We tend to view our planet from the perspective of a human lifetime and commonly overlook the fact that Earth has changed markedly in the context of geologic time and that it continues to do so. Geology is not a static science, but one, like the dynamic Earth it seeks to understand, that constantly evolves as new information and methods of investigation become available. It is thus incumbent on all of us to apply the geologic lessons we have learned to make Earth a more sustainable planet for both current and future generations.

Global GeoScience Watch On the GREENR database home page, click on the "Environment and Ecology" heading on the left side of the page. Next, click on the "Climate Change: Science and Mitigation" portal. Start by reading the "Overview" for this portal, and then browse the "News" section and the "Magazines" section. There are hundreds of current articles to enrich your understanding of the debate concerning climate change and the effect humans might have in contributing to it. After reading some of the articles in the aforementioned sections, what do you think are the salient arguments concerning climate change, its causes, and its effect on the world's biota? Write a short report to summarize your observations.

English-Metric Conversion Chart

	English Unit	Conversion Factor	Metric Unit	Conversion Factor	English Unit
Length	Inches (in.)	2.54	Centimeters (cm)	0.39	Inches (in.)
	Feet (ft)	0.305	Meters (m)	3.28	Feet (ft)
	Miles (mi)	1.61	Kilometers (km)	0.62	Miles (mi)
Area	Square inches (in.^{2})	6.45	Square centimeters (cm^2)	0.16	Square inches (in.^{2})
	Square feet (ft^2)	0.093	Square meters (m^2)	10.8	Square feet (ft^2)
	Square miles (mi^2)	2.59	Square kilometers (km^2)	0.39	Square miles (mi^2)
Volume	Cubic inches (in.^{3})	16.4	Cubic centimeters (cm^3)	0.061	Cubic inches (in.^{3})
	Cubic feet (ft^3)	0.028	Cubic meters (m^3)	35.3	Cubic feet (ft^3)
	Cubic miles (mi^3)	4.17	Cubic kilometers (km^3)	0.24	Cubic miles (mi^3)
Weight	Ounces (oz)	28.3	Grams (g)	0.035	Ounces (oz)
	Pounds (lb)	0.45	Kilograms (kg)	2.20	Pounds (lb)
	Short tons (st)	0.91	Metric tons (t)	1.10	Short tons (st)
Temperature	Degrees Fahrenheit (°F)	−32° × 0.56	Degrees centrigrade (Celsius) (°C)	× 1.80 + 32°	Degrees Fahrenheit (°F)

Examples:

10 inches = 25.4 centimeters; 10 centimeters = 3.9 inches

100 square feet = 9.3 square meters; 100 square meters = 1080 square feet

50°F = 10.1°C; 50°C = 122°F

APPENDIX B

Topographic Maps

Nearly everyone has used a map of one kind or another and is aware that a map is a scaled-down version of the area depicted. For a map to be of any use, however, one must understand what is shown on a map and how to read it. A particularly useful type of map for geologists, and people in many other professions, is a *topographic map*, which shows the three-dimensional configuration of Earth's surface on a two-dimensional sheet of paper.

Maps showing relief—differences in elevation in adjacent areas—are actually models of Earth's surface. Such maps are available for some areas, but they are expensive, difficult to carry, and impossible to record data on. Thus, paper sheets that show relief by using lines of equal elevation known as *contours* are most commonly used. Topographic maps depict (1) relief, which includes hills, mountains, valleys, canyons, and plains; (2) bodies of water such as rivers, lakes, and swamps; (3) natural features such as forests, grasslands, and glaciers; and (4) various cultural features, including communities, highways, railroads, land boundaries, canals, and power transmission lines.

Topographic maps known as *quadrangles* are published by the United States Geological Survey (USGS). The area depicted on a topographic map is identified by referring to the map's name in the upper right and lower right corners, which is usually derived from some prominent geographic feature (Lincoln Creek Quadrangle, Idaho) or community (Mt. Pleasant Quadrangle, Michigan). In addition, most maps have a state outline map along the bottom margin, and shown within the outline is a small black rectangle indicating the part of the state represented by the topographic map.

Contours

Contour lines, or simply contours, are lines of equal elevation used to show topography. Think of contours as the lines formed where imaginary horizontal planes intersect Earth's surface at specific elevations. On maps, contours are brown, and every fifth contour, called an *index contour*, is darker than adjacent ones and labeled with its elevation (▮ Figure B1). Elevations on most USGS topographic maps are in feet, although a few use meters; in either case, the specified elevation is above or below mean sea level. Because contours are defined as lines of equal elevation, they cannot divide or cross one another, although they will converge and appear to join in areas with vertical or overhanging cliffs. Notice in Figure B1 that where contours cross a stream they form a V that points upstream toward higher elevations.

The vertical distance between contours is the *contour interval*. If an area has considerable relief, a large contour interval is used, perhaps 80 or 100 feet, whereas a small interval such as 5, 10, or 20 feet is used in areas with little relief. The values recorded on index contours are always multiples of the map's contour interval, shown at the bottom of the map. For instance, if a map has a contour interval of 10 feet, index contour values such as 3600, 3650, and 3700 feet might be shown (Figure B1). In addition to contours, specific elevations are shown at some places on maps and may be indicated by a small X, next to which is a number. A specific elevation might also be shown adjacent to the designation BM (benchmark), a place where the elevation and location are precisely known.

Contour spacing depends on slope, so in areas with steep slopes, contours are closely spaced because there is a considerable increase in elevation in a short horizontal distance. In contrast, if slopes are gentle, contours are widely spaced (Figure B1). Furthermore, if contour spacing is uniform, the slope angle remains constant, but if spacing changes, the slope angle changes. However, one must be careful in comparing slopes on maps with different contour intervals or different scales.

Topographic features such as hills, valleys, plains, and so on, are easily shown by contours. For instance, a hill is shown by a concentric pattern of contours with the highest elevation in the central part of the pattern. All contours must close on themselves, but they may do so beyond the confines of a particular map. A concentric contour pattern also might show a closed depression, but in this case, special contours with short bars perpendicular to the contour pointing toward the central part of the depression are used (Figure B1).

Map Scales

All maps are scaled-down versions of the areas shown, so to be of any use, they must have a scale. Highway maps, for example, commonly have a scale such as "1 inch equals 10 miles," by which one can readily determine distances. Two types of scales are used on topographic maps. The first and most easily understood is a graphic scale, which is simply a

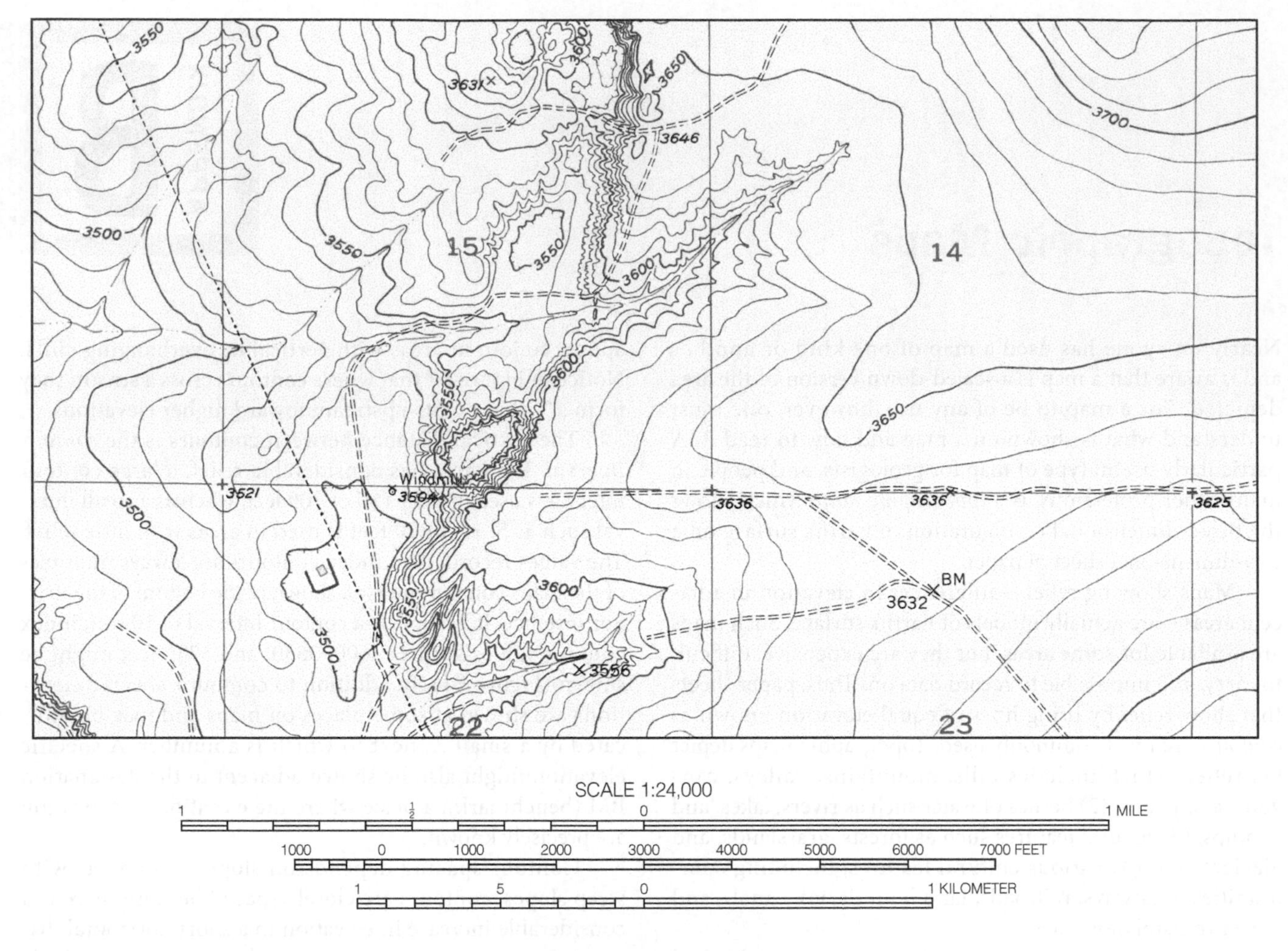

Figure B.1 Part of the Bottomless Lakes Quadrangle, New Mexico, which has a contour interval of 10 feet; every fifth contour is darker and labeled with its elevation. Notice that contours are widely spaced where slopes are gentle and more closely spaced where they are steeper, as in the central part of the map. Hills are shown by contours that close on themselves, whereas depressions are indicated by contours with hachure marks pointing toward the center of the depression. The dashed blue lines on the map represent intermittent steams; notice that where contours cross a stream's channel they form a V that points upstream.

bar subdivided into appropriate units of length (Figure B1). This scale appears at the bottom center of the map and may show miles, feet, kilometers, or meters. Indeed, graphic scales on USGS topographic maps generally show both English and metric distance units.

A ratio or fractional scale, which represents the degree of reduction of the area depicted, appears above the graphic scale. On a map with a ratio scale of 1:24,000, for instance, the area shown is 1/24,000th the size of the actual land area (Figure B1). Another way to express this relationship is to say that any unit of length on the map equals 24,000 of the same units on the ground. Thus, 1 inch on the map equals 24,000 inches on the ground, which is more meaningful if one converts inches to feet, making 1 inch equal to 2,000 feet. A few maps have scales of 1:63,360, which converts to 1 inch equals 5,280 feet, or 1 inch equals 1 mile.

USGS topographic maps are published in a variety of scales such as 1:50,000, 1:62,500, 1:125,000, and 1:250,000. One should also realize that large-scale maps cover less area than small-scale maps, and the former show much more detail than the latter. For example, a large-scale map (1:24,000) shows more surface features in greater detail than does a small-scale map (1:125,000) for the same area.

Map Locations

You can locate features on topographic maps in two ways. First, the borders of maps correspond to lines of latitude and longitude. Latitude is measured north and south of the equator in degrees, minutes, and seconds, whereas the same units are used to designate longitude east and west of the prime meridian, which passes through Greenwich, England. Maps depicting all areas within the United States are noted in north latitude and west longitude. Latitude and longitude are noted in degrees and minutes at the corners of maps, but usually only minutes and seconds are shown along the margins. Many USGS topographic maps cover 7 1/2 or 15 minutes of latitude and longitude and are thus referred to as 7 1/2- and 15-minute quadrangles.

Beginning in 1812, the General Land Office (now known as the Bureau of Land Management) developed a standardized method for accurately defining the location of property in the United States. This method, known as the General Land Office Grid System, has been used for all states except those along the eastern seaboard (except Florida), and parts of Ohio, Tennessee, Kentucky, West Virginia, and Texas.

As new land acquired by the United States was surveyed, the surveyors laid out north–south lines they called *principal meridians* and east–west lines known as *base lines*. These intersecting lines form a set of coordinates for locating specific pieces of property. The basic unit in the General Land Office Grid System is the *township*, an area measuring 6 miles on a side and thus covering 36 square miles (❚ Figure B2). Townships are numbered north and south of base lines and are designated as T.1N., T.1S., and so on. Rows of townships known as *ranges* are numbered east and west of principal meridians—R.2W and R.4E, for example. Note in Figure B2 that each township has a unique designation of township and range numbers.

Townships are subdivided into 36 1-square-mile (640-acre) *sections* numbered from 1 to 36. Because of surveying errors and the adjustments necessary to make a grid system conform to Earth's curved surface, not all sections are exactly 1 mile square. Nevertheless, each section can be further subdivided into half sections and quarter sections designated NE 1/4, NW 1/4, SE 1/4, and SW 1/4, and each quarter section can be further divided into quarter-quarter sections. To show the complete designation for an area, the smallest unit is noted first (quarter-quarter section) followed by quarter section, section number, township, and range. For example, the area shown in Figure B2 is the NW 1/4, SW 1/4, Sec. 34, T.2N., R.3W.

Because only a few principal meridians and base lines were established, they do not appear on most topographic maps. Nevertheless, township and range numbers are printed along the margins of 7 1/2- and 15-minute quadrangles, and a grid consisting of red land boundaries depicts sections. In addition, each section number is shown in red within the map. However, small-scale maps show only township and range.

Where to Obtain Topographic Maps

Many people find topographic maps useful. Land use planners, personnel in various local, state, and federal agencies, as well as engineers and real estate developers might use

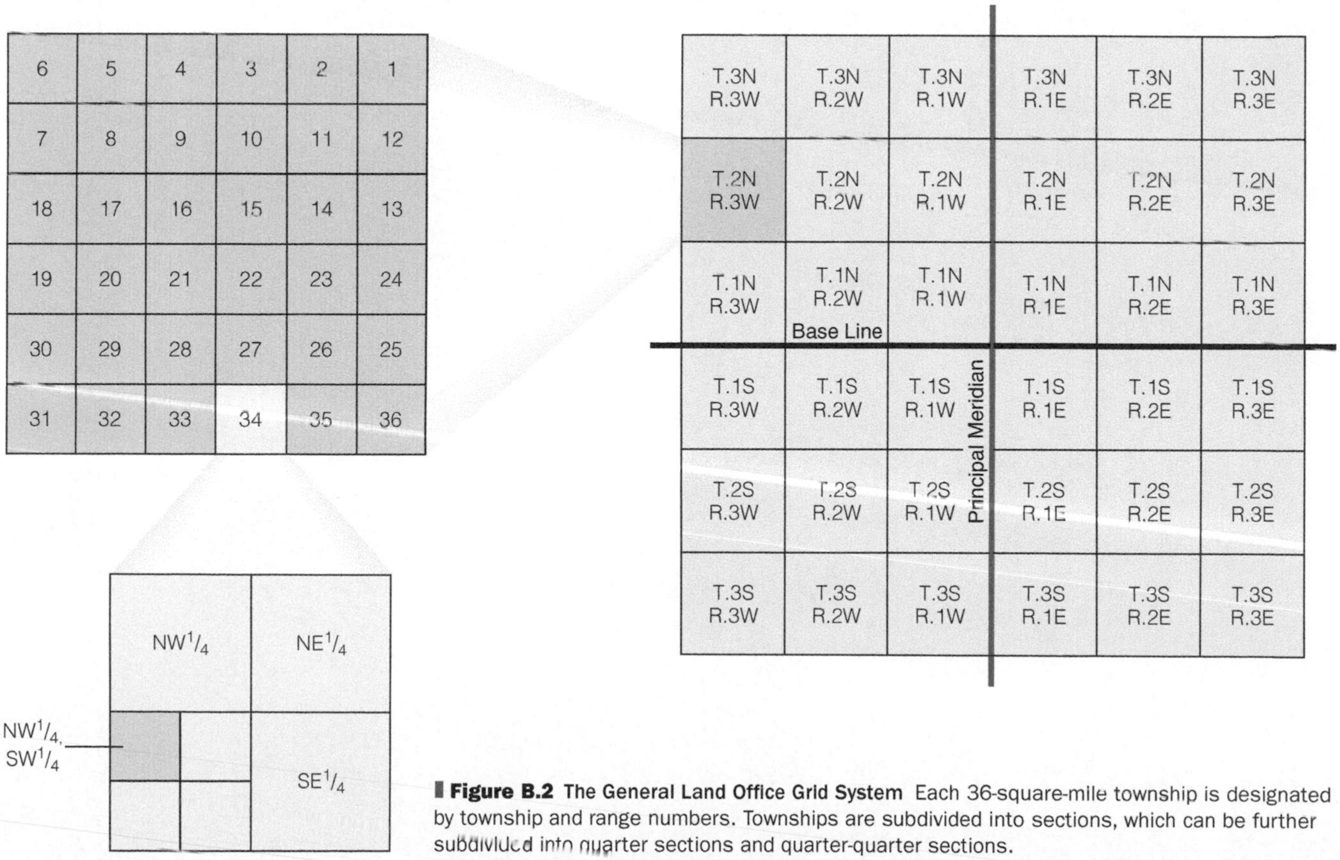

❚ **Figure B.2 The General Land Office Grid System** Each 36-square-mile township is designated by township and range numbers. Townships are subdivided into sections, which can be further subdivided into quarter sections and quarter-quarter sections.

these maps for a variety of reasons. In addition, hikers, backpackers, and others interested in exploring undeveloped areas commonly use topographic maps because trails are shown by black dashed lines. Furthermore, map users can readily determine their location by interpreting the topographic features depicted by contours, and they can anticipate the type of terrain they will encounter during off-road excursions.

Topographic maps for local areas are available at some sporting goods stores, at National Park Visitor Centers, and from some state geologic surveys. Free index maps showing the names and locations of all quadrangles for each state are available from the USGS to anyone uncertain of which specific map is needed. Any published topographic map can be obtained from the United States Geologic Survey (USGS) at http://topomaps.usgs.gov/ or from many commercial map dealers.

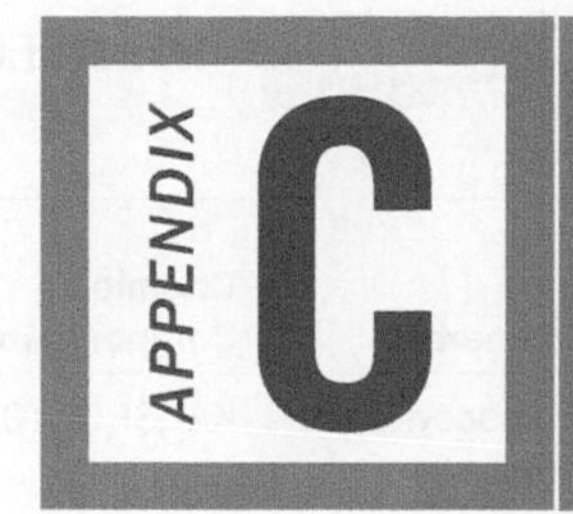

Mineral Identification

To identify most minerals, geologists use physical properties such as color, luster, crystal form, hardness, cleavage, specific gravity, and several others (see Chapter 3, Section 3.5). The mineral identification table (C1) is divided into two parts: The first part lists minerals with a metallic luster, and the second part has minerals with a nonmetallic luster. After determining luster, ascertain hardness and note that each part of the table is arranged with minerals with increasing hardness. Thus, if you have a nonmetallic mineral with a hardness of 6, it must be augite, hornblende, plagioclase, or one of the two potassium feldspars (orthoclase or microcline). If this hypothetical mineral is dark green or black, it must be augite or hornblende, and if it has 2 cleavage plane intersecting at nearly right angles it is augite.

TABLE C1 Mineral Identification Tables

Metallic Luster

Mineral	Chemical Composition	Color	Hardness / Specific Gravity	Other Features	Comments
Graphite	C	Black	1–2 2.09–2.33	Greasy feel; writes on paper; 1 direction of cleavage	Used for pencil "leads." Mostly in metamorphic rocks.
Galena	PbS	Lead gray	2½ 7.6	Cubic crystals; 3 cleavages at right angles	The ore of lead. Mostly in hydrothermal rocks.
Chalcopyrite	$CuFeS_2$	Brassy yellow	3½–4 4.1–4.3	Usually massive; greenish black streak; iridescent tarnish	The most common copper mineral. Mostly in hydrothermal rocks.
Magnetite	Fe_3O_4	Black	5½–6½ 5.2	Strong magnetism	An ore of iron. An accessory mineral in many rocks.
Hematite	Fe_2O_3	Red brown	6 4.8–5.3	Usually granular or massive; reddish brown streak	Important iron ore. An accessory mineral in many rocks.
Pyrite	FeS_2	Brassy yellow	6½ 5.0	Cubic and octahedral crystals	Found in some igneous and hydrothermal rocks and in sedimentary rocks associated with coal.

Nonmetallic Luster

Mineral	Chemical Composition	Color	Hardness / Specific Gravity	Other Features	Comments
Talc	$Mg_3Si_4O_{10}(OH)_2$	White, green	1 2.82	1 cleavage direction; usually in compact masses; soapy feel	Formed by the alteration of magnesium silicates. Mostly in metamorphic rocks.
Clay minerals	Varies	Gray, buff, white	1–2 2.5–2.9	Earthy masses; particles too small to observe properties	Found in soils, mudrocks, slate, phyllite.
Chlorite	$(Mg,Fe)_3(Si,Al)_4O_{10}(OH)_8$	Green	2 2.6–3.4	1 cleavage; occurs in scaly masses	Common in low-grade metamorphic rocks such as slate.
Gypsum	$CaSO_4 \cdot 2H_2O$	Colorless, white	2 2.32	Elongate crystals; fibrous and earthy masses	The most common sulfate mineral. Found mostly in evaporite deposits.

TABLE C1 **Mineral Identification Tables (*continued*)**

Nonmetallic Luster					
Mineral	**Chemical Composition**	**Color**	**Hardness Specific Gravity**	**Other Features**	**Comments**
Muscovite (Mica)	$KAl_2Si_3O_{10}(OH)_2$	Colorless	2–2½ 2.7–2.9	1 direction of cleavage; cleaves into thin sheets	Common in felsic igneous rocks, metamorphic rocks, and some sedimentary rocks.
Biotite (Mica)	$K(Mg,Fe)_3AlSi_3O_{10}(OH)_2$	Black, brown	2½ 2.9–3.4	1 cleavage direction; cleaves into thin sheets	Occurs in both felsic and mafic igneous rocks, in metamorphic rocks, and in some sedimentary rocks.
Calcite	$CaCO_3$	Colorless, white	3 2.71	3 cleavages at oblique angles; cleaves into rhombs; reacts with dilute HCl	The most common carbonate mineral. Main component of limestone and marble.
Anhydrite	$CaSO_4$	White, gray	3½ 2.9–3.0	Crystals with 2 cleavages; usually in granular masses	Found in limestones, evaporite deposits, and the cap rock of salt domes.
Halite	NaCl	Colorless, white	3–4 2.2	3 cleavages at right angles; cleaves into cubes; cubic crystals; salty taste	Occurs in evaporite deposits.
Dolomite	$CaMg(CO_3)_2$	White, yellow, gray, pink	3½–4 2.85	Cleavage as in calcite; reacts with dilute hydrochloric acid when powdered	The main constituent of dolostone. Also found associated with calcite in some limestones and marble.
Fluorite	CaF_2	Colorless, purple, green, brown	4 3.18	4 cleavage directions; cubic and octahedral crystals	Occurs mostly in hydrothermal rocks and in some limestones and dolostones.
Augite	$Ca(Mg,Fe,Al)(Al,Si)_2O_6$	Black, dark green	6 3.25–3.55	Short 8-sided crystals; 2 cleavages; cleavages nearly at right angles	The most common pyroxene mineral. Found mostly in mafic igneous rocks.
Hornblende	$NaCa_2(Mg,Fe,Al)_5(Si,Al)_8O_{22}(OH)_2$	Green, black	6 3.0–3.4	Elongate, 6-sided crystals; 2 cleavages intersecting at 56° and 124°	A common rock-forming amphibole mineral in igneous and metamorphic rocks.
Plagioclase feldspars	Varies from $CaAl_2Si_2O_8$ to $NaAlSi_3O_8$	White, gray, brown	6 2.56	2 cleavages at right angles	Common in igneous rocks and a variety of metamorphic rocks. Also in some arkoses.
Potassium Feldspars: Microcline	$KAlSi_3O_8$	White, pink, green	6 2.56	2 cleavages at right angles	Common in felsic igneous rocks, some metamorphic rocks, and arkoses.
Potassium Feldspars: Orthoclase	$KAlSi_3O_8$	White, pink	6 2.56	2 cleavages at right angles	
Olivine	$(Fe,Mg)_2SiO_4$	Olive green	6½ 3.3–3.6	Small mineral grains in granular masses; conchoidal fracture	Common in mafic igneous rocks.
Quartz	SiO_2	Colorless, white, gray, pink, green, black	7 2.67	6-sided crystals; no cleavage; conchoidal fracture	A common rock-forming mineral in all rock groups and hydrothermal rocks. Also occurs in varieties known as chert, flint, agate, and chalcedony.
Garnet	$Fe_3Al_2(SiO_4)_3$	Dark red, green	7–7½ 4.32	12-sided crystals common; uneven fracture	Found mostly in gneiss and schist.
Zircon	Zr_2SiO_4	Brown, gray	7½ 3.9–4.7	4-sided, elongate crystals	Most common as an accessory in granitic rocks.
Topaz	$Al_2SiO_4(OH,F)$	Colorless, white, yellow, blue	8 3.5–3.6	High specific gravity; 1 cleavage direction	Found in pegmatites, granites, and hydrothermal rocks.
Corundum	Al_2O_3	Gray, blue, pink, brown	9 4.0	6-sided crystals and great hardness are distinctive	An accessory mineral in some igneous and metamorphic rocks.

Multiple-Choice Questions

CHAPTER 1
1. e; 2. c; 3. c; 4. e; 5. c.

CHAPTER 2
1. a; 2. e; 3. b; 4. c; 5. c.

CHAPTER 3
1. a; 2. d; 3. b; 4. e; 5. b.

CHAPTER 4
1. d; 2. a; 3. e; 4. d; 5. b.

CHAPTER 5
1. b; 2. c; 3. d; 4. c; 5. e.

CHAPTER 6
1. b; 2. a; 3. e; 4. d; 5. b.

CHAPTER 7
1. a; 2. e; 3. d; 4. d; 5. a.

CHAPTER 8
1. e; 2. d; 3. e; 4. d; 5. a.

CHAPTER 9
1. e; 2. c; 3. d; 4. c; 5. d.

CHAPTER 10
1. e; 2. a; 3. b; 4. b; 5. a.

CHAPTER 11
1. d; 2. d; 3. e; 4. b; 5. e.

CHAPTER 12
1. c; 2. a; 3. d; 4. c; 5. e.

CHAPTER 13
1. b; 2. c; 3. c; 4. d; 5. e.

CHAPTER 14
1. c; 2. b; 3. a; 4. d; 5. d.

CHAPTER 15
1. d; 2. b; 3. b; 4. a; 5. d.

CHAPTER 16
1. b; 2. c; 3. d; 4. a; 5. c.

CHAPTER 17
1. b; 2. b; 3. d; 4. b; 5. a.

CHAPTER 18
1. e; 2. d; 3. a; 4. e; 5. b.

CHAPTER 19
1. c; 2. b; 3. e; 4. d; 5. a.

CHAPTER 20
1. b; 2. a; 3. c; 4. e; 5. c.

CHAPTER 21
1. a; 2. d; 3. b; 4. e; 5. e.

CHAPTER 22
1. c; 2. c; 3. a; 4. a; 5. e.

CHAPTER 23
1. e; 2. d; 3. a; 4. b; 5. c.

ANSWERS

Selected Short Answer Questions

CHAPTER 1

6. Plate tectonic theory provides a unifying explanation for many geologic features and events by relating many seemingly unrelated phenomena and providing a framework for interpreting Earth's composition, structure, and internal processes. It fits into a systems approach for the study of Earth because it shows how the continents and ocean basins are part of a lithosphere-atmosphere-hydrosphere system that evolved along with Earth's interior.
9. Interaction between plates determines, to some extent, which of the three rock groups will form. Heat within Earth's interior results in convection cells that power the movement of plates and magma, which forms intrusive and extrusive igneous rocks. The interaction among the atmosphere, hydrosphere, and biosphere contributes to the weathering of rocks exposed at Earth's surface, leading to sedimentary rocks. Plates descending into Earth's interior are subjected to increasing heat and pressure, resulting in metamorphic rocks, and if heated to their melting point, the generation of magma, which starts the cycle over again.

CHAPTER 2

6. The evidence convincing Wegener and others that continents have not remained in their present position, but rather have moved around, includes the fit of the shorelines of continents, the appearance of the same rock sequences and mountain ranges of the same age on continents now widely separated, the matching of glacial deposits and paleoclimatic zones, and the similarities among many extinct plant and animal groups, whose fossils are found today on widely separated continents.
8. Many metallic mineral deposits are related to igneous and associated hydrothermal activity. The magma and associated hydrothermal fluids generated by subducting plates along convergent plate boundaries contain minute amounts of metallic elements. As this magma rises toward the surface, it cools and precipitates and concentrates metallic ores. In the same way, rising magma along divergent plate boundaries precipitates metallic elements adjacent to hydrothermal vents.

CHAPTER 3

6. Some minerals have a range of chemical compositions because one element can substitute for another if they have the same charge and are of about the same size. In the plagioclase feldspars, the elements calcium (Ca) and sodium (Na) meet these criteria, so the composition of plagioclase ranges from calcium-rich ($CaAl_2Si_2O_8$) to sodium-rich ($NaAlSi_3O_8$) varieties.
9. Only about two dozen minerals are common because even though more than 90 naturally occurring elements are known, only 8 are abundant in Earth's crust. And even among these 8 elements, oxygen and silicon are by far the most common, making up more than 80% of all elements to make minerals. Thus, most common minerals are silicates that contain silicon and oxygen, and in most, one or more of the other common elements, such as iron, aluminum, calcium, sodium, potassium, and magnesium.

CHAPTER 4

7. The most important controls on the viscosity of a lava flow are temperature and composition, although the presence of gases, mineral crystals, and the surface over which it flows have some influence. In general, the hotter the lava the less viscous (more fluid) it is, but composition has a large affect, too. For example, silica-rich lava (>65% silica) has numerous chemical bonds between atoms that inhibit flow and thus the lava is viscous (not very fluid). In contrast, mafic lava with 45%–52% silica is much more fluid.
8. Specimen 2 is gabbro, so it would be darker and denser because its predominant minerals (olivine, pyroxene, and hornblende) are ferromagnesian silicates that contain iron or magnesium or both and tend to be dark colored. Specimen 1 is granite, which is made up mostly of nonferromagesian silicates.

CHAPTER 5

6. Geologists monitor several physical and chemical aspects of volcanoes to better anticipate eruptions. They monitor gas emissions, changes in hot spring activity, and temperature, as well as changes in the volcano itself as when it bulges as magma is injected into it. Other important criteria are changes in the local magnetic and electrical fields, and geologists are keenly aware of volcanic tremor and a volcano's eruptive history.
10. The volcano in the background is a cinder cone and the lava flow is an aa flow.

CHAPTER 6

8. Mechanical weathering involves the breakdown of Earth materials with no change in composition, as when granite breaks down into small pieces of granite and

individual minerals. Chemical weathering, in contrast, decomposes rocks and minerals by chemical changes in parent material. Mechanical weathering contributes to chemical weathering because chemical alteration takes place faster on small particles.

9. Any soil that expands by 6% or more is an expansive soil. As expansive soil expands and then contracts, it may cause stresses that damage overlying structures such as roadways, sidewalks, and foundations.

CHAPTER 7

7. Sedimentary rocks preserve evidence of how they were deposited in the form of textures, sedimentary structures, and fossils. The only requirement for geologists to interpret such ancient rocks is that the evidence for their origin has been preserved in the geologic record. If it has, they use their knowledge of present-day processes and sediment deposits to make inferences about ancient deposits.

9. If these figures are correct, the United States coal reserve should last 234 years. However, current rates of consumption are not likely to remain the same, and our calculation depends partly on the reserve base, which is probably not fully accurate. Furthermore, as any resource is depleted, it is more difficult and expensive to recover what is left. In short, we will never mine all of the coal in our reserve base.

CHAPTER 8

8. Metamorphism could still exist without plate tectonic movement. However, it would only be contact metamorphism related to rising magma, assuming there was enough internally generated heat to melt rocks in Earth's interior, and thus generate a magma, which would rise to Earth's surface.

9. Foliated texture results when rocks are subjected to heat and differential pressure causing the minerals to align in a parallel fashion. Nonfoliated texture shows a mosaic of roughly equidimensional minerals that result from contact metamorphism or regional metamorphism of rocks with no platy or elongate minerals.

CHAPTER 9

7. The inner core is thought to be solid because of the arrival of weak P-wave energy in the P-wave shadow zone, and the outer core is liquid because of the S-wave shadow zone, within which no S-waves are recorded. Hence, the outer core must be liquid or behave like a liquid.

   The core is largely composed of iron, with lesser amounts of sulfur, nickel, and possibly some silicon, oxygen, and potassium. The evidence for this is based primarily on the composition of meteorites.

8. Insurance companies use the Modified Mercalli Intensity Scale because it measures actual damage done to structures and areas. Therefore, insurance companies can determine the type of damage to expect from a given earthquake in a particular area based on the amount and type of damage that resulted from previous earthquakes in that area, and thus adjust their rates accordingly.

CHAPTER 10

6. Convert 200 km to millimeters (20 million) and divide by 5 million years, giving an average rate of movement of 4 mm per year. In many segments of faults, such as the San Andreas Fault, blocks on opposite sides of the fault do creep past one another slowly, but on occasion they also show movements of many meters during single earthquakes. Accordingly, the average does not adequately describe the history of movement on a fault.

7. Stress is the force applied to a given area of rock, usually expressed in kilograms per square centimeter (kg/cm^2). The force exerted by a person walking on an ice-covered pond is a function of the person's weight and the area beneath his or her feet. Strain is a measure of the amount of deformation (change in shape, volume, or both) as a result of stress. In our ice-covered pond example, the strain is the bending or fracturing of the ice by the stress.

CHAPTER 11

7. Climate plays an important role in the rate and type of weathering. Weathering disintegrates and decomposes material exposed at the surface, thus decreasing a slope's shear strength and increasing its susceptibility to mass wasting.

8. The factors that influence mass wasting are slope angle, weathering and climate, water content, vegetation, and overloading. These factors are interconnected by the fact that the amount of rain is related to climate, and rain can increase the water content of a slope's material, leading to an increase in weight, which might cause mass movement or decrease the friction between grains, leading to a loss of cohesion and sliding. Too little rain may cause plants to die, decreasing the vegetative root system, which helps bind soil particles together.

CHAPTER 12

6. A graded stream is one with a smooth profile of equilibrium and a delicate balance among all aspects of the stream, including gradient, discharge, flow velocity, sediment load, and channel characteristics, so that neither erosion nor deposition takes place within its channel. Such equilibrium conditions are rarely attained along the entire length of a stream, but along segments of a stream, erosion tends to eliminate places where the gradient is steep, whereas deposition takes place where the gradient is insufficient to transport sediment.

9. Point bars are bodies of mostly sand that are deposited on the gently sloping inner bank of a meandering stream. They form in that location because that is where flow velocity is least, resulting in deposition of some of the stream's sediment load.

CHAPTER 13

8. In a water well, groundwater rises to the top of the zone of saturation, or the water table, and from there, it must be pumped to the surface. In an artesian system, groundwater is confined to an aquifer by aquicludes above and below, the rock sequence is tilted so that it builds up high hydrostatic (fluid) pressure, and the aquifer is exposed at the surface, enabling it to be recharged.

   There is no intrinsic difference between water from a water well or from an artesian system. Any difference in quality is a function of dissolved minerals and any introduced substances.

9. In arid regions, such as Nevada, the water table is very low, in some cases tens and hundreds of meters below the surface. Furthermore, rainfall is scant, and what little rain does fall and percolate into the ground usually evaporates before it get very far. Thus, the likelihood of water reaching deeply buried canisters containing radioactive nuclear waste material where it could corrode the containers or become polluted by the radioactive material is virtually nonexistent for the foreseeable future.

   However, if the climate should change and the region become humid, say, in the next several thousand years, water from rainfall will percolate freely through the zone of aeration into the zone of saturation, and eventually it will cause the water table to rise, possibly to the level of the buried radioactive nuclear waste material, where the groundwater system could then become polluted by the radioactive material.

   The average rate of groundwater movement during the next 5,000 years would have to be 8 cm per year to reach radioactive waste canisters buried at a depth of 400 m (400 m = 40,000 cm [100 cm/m × 400 m = 40,000 cm]; 40,000 cm/5,000 years = 8 cm/year).

CHAPTER 14

7. You can recognize a mountain area that was once glaciated by landforms that resulted from erosion or deposition. For instance, erosion scours out U-shaped valleys, abrades bedrock (striations), and leaves other features, such as cirques, arêtes, and horns. The easily recognized deposits of former valley glaciers include lateral and medial moraines, and perhaps silt and clay deposits that formed in lakes.

10. The erosional landforms visible in this image include cirques (bowl-shaped depressions at the upper ends of glacial troughs), arêtes (the sharp ridges between glacial troughs), and several horns (pyramid-shaped peaks).

CHAPTER 15

8. Although not exclusive to deserts, evidence in the geologic record that might indicate former deserts would include ventifacts, desert pavement, large-scale cross bedding, evaporate minerals, and red beds.

9. The main line of evidence that indicates wind was an important agent in the formation of the Martian surface includes desert pavement, ventifacts, and dunes.

CHAPTER 16

7. Upwelling involves the vertical circulation of cold, nutrient-rich seawater from depth to the surface. It is important because it carries nutrients, especially phosphates and nitrates, into the photic zone that support large communities of organisms. Also, much of the sedimentary rock rich in phosphorous was deposited in areas of upwelling.

9. As a member of a planning commission, you might encourage the development of building codes and land-use regulations, such as how close to a shoreline is permissible for structures. Although expensive, you also may recommend the construction of seawalls, jetties, and groins to inhibit erosion in specific areas, or you might favor artificially replenishing eroding beaches with sand transported from elsewhere.

CHAPTER 17

6. In reconstructing the geologic history of an area, it is important to differentiate between a sill and a lava flow because they formed differently in relation to the rocks found in association with them.

   A sill is a tabular igneous intrusion and thus is younger than the rocks it intrudes. A lava flow moving over the land surface will be younger than the rocks that it is flowing over, and it will be older than any rocks deposited on it after it cools and solidifies.

   To identify a sill, one should look for baking (contact metamorphism) of the rocks immediately above and below the sill. A lava flow, however, will only show the effects of baking in the rocks below it, because the rocks above it were deposited on the solidified flow.

9. The K40/Ar40 and Rb87/Sr87 radiometric dating techniques could produce distinctly different ages for a volcanic ash fall due to the fact that if the volcanic ash was heated after deposition, some of the argon might have escaped, yielding an apparently younger age.

   To rectify the discrepancy, whole-rock analysis of the volcanic ash could be conducted, which might yield a more accurate potassium to argon ratio.

CHAPTER 18

6. Inheritance of acquired characteristics holds that organisms acquire features during their lifetimes that are then passed on to their descendants. According to natural selection, variations already exist in interbreeding populations, some of which are more favorable for survival, and that those individuals with these variations are more likely to survive long enough to pass on these variations. Both hypotheses are testable, but the former has failed all attempts to verify it whereas the latter has not.

10. The kinds of evidence from living organisms that might support this cladogram include biochemical, anatomical, and genetic investigations that show the degree of relationships indicated among these animals. In the fossil record, we would expect to find examples of animals that show the development of the features we associated with these animals.

CHAPTER 19

6. Given that 4,000 m (400,000 cm) of sediment was deposited over 550 million years, there was an average rate of 0.0007 cm/year. This computation is correct but of little value, because deposition in most environments is episodic. Long periods may pass with little or no deposition and then several centimeters, or even a meter or so, may be deposited in, for example, a single flood event, only to be followed by another long period of little or no deposition. Furthermore, we have no idea of how much sediment may have been lost to erosion. Remember, that uniformitarianism does not demand that processes such as sedimentation go on at an unchanging rate, only that deposition of sediment took place during the past by processes like those operating today.
7. Laurentia is a landmass made up mostly of North America and Greenland. Much of Laurentia assembled during the Proterozoic when smaller landmasses (minicontinents), as well as oceanic plateaus and oceanic islands, collided along its margins, a process called continental accretion. By the end of the Proterozoic, a succession of these events formed about 75% of present-day North America.

CHAPTER 20

7. Reconstructing paleogeography requires the synthesis of all pertinent paleoclimatic, paleomagnetic, paleontologic, paleomagnetic, sedimentologic, stratigraphic, and tectonic data available for the time period of interest. Such information as the distribution of plants and animals yields insights into the possible latitudes of landmasses and ocean basins. Sedimentary structures can tell something about depositional environments. Although most geology books and journals agree on the major features of paleogeography, some evidence preserved in the geologic record can be interpreted in different ways, leading to minor differences in paleogeographic reconstructions.
9. The sequence is the basic unit of sequence stratigraphy and represents a succession of rocks of related facies bounded by unconformities. These boundaries result from a relative drop in sea level, and hence represent local, regional, and global events. Thus, sedimentary rocks can be subdivided into related units on the basis of time-stratigraphic unconformable surfaces. Therefore, global correlations can be made on the basis of a global drop in sea level, such as occurred near the end of the Ordovician Period.

CHAPTER 21

7. It could be argued that the Cambrian explosion is nothing more than the fact that conditions were more favorable for the preservation of skeletonized animals at the beginning of the Cambrian than during the Proterozoic, thus giving the impression that skeletonized animals evolved rapidly at the Proterozoic–Cambrian boundary. No evidence in the geologic record, however, supports the notion that conditions in the marine realm during the Proterozoic were any different than they were at the beginning of the Cambrian Period.
8. The Pennsylvanian Period was a time of repetitive transgressive–regressive cycles that resulted in widespread coal-forming swamps. These swamps provided the perfect environment for amphibians and seedless vascular plants, both of which required moist conditions as part of their reproductive cycle.

CHAPTER 22

7. The orogenic activity that occurred in the Cordilleran mobile belt during the Mesozoic Era was the result of an oceanic–continental convergent plate boundary in which the oceanic Farallon plate was subducted beneath the continental North American plate. Changes in the angle of subduction of the Farallon plate resulted in the Nevadan, Sevier, and Laramide orogenies. This was similar to what took place in the Appalachian mobile belt during the Paleozoic Era in that this area was the site of an initial oceanic-continental convergent plate boundary in which the Iapetus plate was subducted beneath the continental North American plate. The main difference is that continued subduction in the Appalachian mobile belt resulted in the development of a continental–continental convergent plate boundary, as the supercontinent Pangaea formed by the end of the Paleozoic Era.
8. One trend during the transition from mammal-like reptiles to mammals was a reduction in the number of bones making up the lower jaw. Reptiles have several bones in their jaw but mammals have only one, the dentary. The reptile jaw-skull joint is between the articular bone of the jaw and the quadrate bone of the skull, but mammals have a joint between the dentary and the squomosal bone of the skull. The articular and quadrate bones of the reptile jaw-skull joint became the malleus and incus of the mammal middle ear.

CHAPTER 23

7. The Cenozoic is the most recent 66 million years of Earth history, and the North American Cordillera is a mountainous area made up of several component parts in western North America stretching from Alaska into Mexico. Plate convergence was an important element in the evolution of this region that helps account for mountain building, emplacement of intrusive igneous bodies, volcanism, and uplift of broad areas, such as the Colorado Plateau. Furthermore, a huge area centered

on Nevada was the site of uplift of numerous mountain ranges, each bounded by faults and separated from one another by board basins (valleys). Another important event was the collision of western North America with a spreading ridge, which gave rise to the San Andreas Fault.

8. The Laramide orogeny differed from other orogenies at convergent plate boundaries in several ways. First, it took place much farther inland than is typical; second, it was accompanied by little or no volcanism or emplacement of plutons; and, third, rather than the compression-induced folding and thrust faulting typical of convergent plate boundary orogenies, it involved the vertical uplift of huge blocks of rock along faults. Geologists think that the angle of subduction changed from steep to nearly horizontal to account for this orogeny.

GLOSSARY

aa Lava flow with a surface of rough, angular blocks and fragments.

abiogenesis The origin of life from nonliving matter.

abrasion The process whereby rock is worn smooth by the impact of sediment transported by running water, glaciers, waves, or wind.

Absaroka Sequence Widespread Upper Mississippian to Lower Jurassic sedimentary rocks bounded above and below by unconformities; deposited during a transgressive–regressive cycle of the Absaroka Sea.

absolute dating Using various radioactive decay dating techniques to assign ages in years before the present to rocks. Also known as numerical dating. (*See also* relative dating)

abyssal plain The flat surface covering vast areas of the seafloor. Abyssal plains are the flattest, most featureless areas on Earth, and their flatness is a result of sediment deposition covering the usually rugged topography of the seafloor.

Acadian orogeny An episode of Devonian deformation in the northern Appalachian mobile belt resulting from the collision of Baltica with Laurentia.

active continental margin A continental margin with volcanism and seismicity at the leading edge of a continental plate where oceanic lithosphere is subducted. (*See also* passive continental margin)

allele Alternative form of a single gene controlling different versions of the same trait.

allopatric speciation Model for the evolution of a new species from a small, geographically isolated part of a larger parent population.

alluvial fan A cone-shaped accumulation of mostly sand and gravel deposited where a stream flows from a mountain valley onto an adjacent lowland.

alluvium A collective term for all detrital sediment transported and deposited by running water.

Alpine–Himalayan orogenic belt An area of mountain building extending from Spain eastward to Southeast Asia.

amniote Any vertebrate animal that produces amniote eggs in which the embryo develops in a fluid-filled amnion; includes reptiles, birds, and mammals.

amniote egg Egg in which the embryo develops in a fluid-filled cavity called the amnion. The egg also contains a yolk sac and a waste sac.

anaerobic Refers to organisms that do not need oxygen for respiration.

analogous structure Body part, such as wings of insects and birds, that serves the same function but differs in structure and development. (*See also* homologous structure)

Ancestral Rockies Late Paleozoic uplift in the southwestern part of the North American craton.

angiosperm Vascular plants that have flowers and seeds; the flowering plants.

angular unconformity An unconformity below which older rocks dip at a different angle (usually steeper) than overlying strata. (*See also* disconformity and nonconformity)

anticline A convex upward fold in which the oldest exposed rocks coincide with the fold axis and all strata dip away from the axis.

Antler orogeny Late Devonian to Mississippian deformation that affected the Cordilleran mobile belt from Nevada to Alberta, Canada.

aphanitic texture A texture in igneous rocks in which individual mineral grains are too small to be seen without magnification; results from rapid cooling of magma and generally indicates an extrusive origin.

aphotic zone The depth in the ocean below which sunlight does not penetrate. (*See also* photic zone)

Appalachian mobile belt A mobile belt along the eastern margin of North America extending from Newfoundland to Georgia.

aquifer A permeable layer in which groundwater flows. From the Latin *aqua*, "water."

archosaur A term referring to the ruling reptiles—dinosaurs, flying reptiles (pterosaurs), crocodiles, and birds.

arête A narrow, serrated ridge between two glacial valleys or adjacent cirques.

artesian system A confined groundwater system with high hydrostatic pressure that causes water to rise above the level of the aquifer.

artificial selection The practice of selectively breeding plants and animals for desirable traits.

artiodactyl The even-toed hoofed mammals, such as goats, cattle, antelope, and swine; members of the order Artiodactyla.

assimilation A process whereby magma changes composition as it reacts with country rock.

asthenosphere The part of the mantle that lies below the lithosphere; it behaves plastically and flows slowly.

atavism An anomaly that looks like the reappearance of an ancestral trait in an organism, such as rear limbs in whales.

Atlantic Coastal Plain The broad, low-relief area of eastern North America extending from the Appalachian Mountains to the Atlantic shoreline.

atom The smallest unit of matter that retains the characteristics of an element.

atomic mass number The number of protons plus neutrons in the nucleus of an atom.

atomic number The number of protons in the nucleus of an atom.

aureole A zone surrounding a pluton in which contact metamorphism took place.

australopithecine A term referring to several extinct species of the genus *Australopithecus* that existed during the Pliocene and Pleistocene epochs.

autotrophic Describes organisms that synthesize organic molecules from inorganic raw materials, as in photosynthesis. (*See also* heterotrophic)

back-arc basin A marine basin, such as the Sea of Japan, that lies between a volcanic island arc and a continent.

Baltica One of six major Paleozoic continents; composed of Russia west of the Ural Mountains, Scandinavia, Poland, and northern Germany.

banded iron formation (BIF) Sedimentary rocks made up of thin alternating bands of silica (chert) and iron minerals (mostly hematite and magnetite).

barchan dune A crescent-shaped sand dune with its tips pointing downwind.

barrier island A long, narrow island of sand parallel to a shoreline but separated from the mainland by a lagoon.

basal slip Movement involving a glacier sliding over its underlying surface.

basalt plateau A plateau built up by horizontal or nearly horizontal overlapping lava flows that erupted from fissures.

base level The level below which a stream or river cannot erode; sea level is ultimate base level.

basin An oval to circular fold in which all strata dip inward toward a central point and the youngest exposed strata are in the center.

Basin and Range Province An area of Cenozoic block-faulting centered on Nevada, but extending into adjacent states and northern Mexico.

batholith An irregularly shaped, discordant pluton with at least 100 km^2 of exposed surface area.

baymouth bar A spit that has grown until it closes off a bay from the open sea or lake.

beach Any deposit of sediment extending landward from low tide to a change in topography or where permanent vegetation begins.

bed An individual layer of rock, especially sediment or sedimentary rock. (*See also* strata)

bed load That part of a stream's sediment load, mostly sand and gravel, transported along its bed.

benthos Those organisms that live on or in the bottom sediments of seas or lakes.

Big Bang A model for the evolution of the universe in which a dense, hot state was followed by expansion, cooling, and a less dense state.

biochemical sedimentary rock Any sedimentary rock produced by the chemical activities of organisms. (*See also* chemical sedimentary rock)

biogeography The study of the geographic distribution of organisms, both past and present.

biostratigraphic unit An association of sedimentary rocks defined by its fossil content.

bipedal Walking on two legs as a means of locomotion, as in humans, birds, and some dinosaurs. (*See also* quadrupedal)

body fossil The actual remains of an organism, such as bones, teeth, shells, and, rarely, the soft parts. (*See also* trace fossil)

bonding The process whereby atoms join to other atoms.

bony fish Fish with an internal skeleton of bone; the class Osteichthyes.

Bowen's reaction series A series of minerals that form in a specific sequence in cooling magma or lava; originally proposed to explain the origin of intermediate and felsic magma from mafic magma.

braided stream A stream with multiple dividing and rejoining channels.

breaker A wave that steepens as it enters shallow water until its crest plunges forward.

browser Any animal that eats tender shoots, twigs, and leaves. (*See also* grazer)

butte An isolated, steep-sided, pinnacle-like hill formed when resistant cap rock is breached, allowing erosion of less resistant underlying rocks.

caldera A large, steep-sided, oval to circular depression usually formed when a volcano's summit collapses into an underlying partially drained magma chamber.

Caledonian orogeny A Silurian-Devonian episode of deformation along the northwestern margin of Baltica as it collided with Laurentia.

Canadian shield The exposed part of the North American craton.

carbon-14 dating technique An absolute dating technique relying on the ratio of carbon 12 to carbon 14 in an organic substance; useful back to about 70,000 years ago.

carbonate mineral A mineral with the carbonate radical $(CO_3)^{-2}$, as in calcite ($CaCO_3$) and dolomite [$CaMg(CO_3)_2$].

carbonate rock Any rock, such as limestone and dolostone, made up mostly of carbonate minerals.

carnivore-scavenger Any animal that depends on other animals, living or dead, for nutrients.

cartilaginous fish Sharks, rays, and skates and their extinct relatives that have an internal skeleton of cartilage.

Cascade Range A mountain range with several active volcanoes in northern California, Oregon, Washington, and southern British Columbia, Canada.

Catskill Delta A Devonian clastic wedge deposited adjacent to highlands that formed during the Acadian orogeny.

cave A natural subsurface opening generally connected to the surface and large enough for a person to enter.

cementation The process whereby minerals crystallize in the pore spaces of sediment and bind the loose particles together.

chemical sedimentary rock Sedimentary rock made up of minerals that were dissolved during chemical weathering and later precipitated from seawater, more rarely lake water, or extracted from solution by organisms. (*See also* biochemical sedimentary rock)

chemical weathering The decomposition of rocks by chemical alteration of parent material.

China One of six major Paleozoic continents; composed of southeast Asia, including China, Indochina, part of Thailand, and the Malay Peninsula.

chordate Animals of the phylum Chordata; all have a notochord, dorsal hollow nerve cord, and pharyngeal pouches at some time during their life cycle.

chromosome Double-stranded, helical molecule of deoxyribonucleic acid (DNA); specific segments of chromosomes are genes.

cinder cone A small, steep-sided volcano made up of pyroclastic materials resembling cinders that accumulate around a vent.

circum-Pacific belt A zone of seismic and volcanic activity and mountain building that nearly encircles the Pacific Ocean basin.

circum-Pacific orogenic belt An area of mountain building along the west coasts of South, Central, and North America; the Aleutians Islands; Japan; the Philippines; and the islands north of Australia and New Zealand.

cirque A steep-walled, bowl-shaped depression on a mountainside at the upper end of a glacial valley.

cladistics A type of analysis of organisms in which they are grouped together on the basis of derived, as opposed to primitive, characteristics.

cladogram A diagram that shows the probable evolutionary relationships among members of a clade, a group of organisms including their most recent common ancestor.

clastic wedge An accumulation of detrital sediments eroded from and deposited adjacent to an uplifted area (e.g., the Catskill Delta).

cleavage The breaking or splitting of mineral crystals along planes of internal weakness.

Colorado Plateau An elevated area in Colorado, Utah, Arizona, and New Mexico with brilliantly colored rocks, deep canyons, and volcanic mountains raising above it.

columnar joint Columns in igneous rocks bounded by fractures that formed when lava or magma cooled and contracted.

compaction Reduction in the volume of a sedimentary deposit that results from its own weight and the weight of any additional sediment deposited on top of it.

complex movement A combination of different types of mass movements in which no single type is dominant; usually involves sliding and flowing.

composite volcano (stratovolcano) A volcano composed of lava flows and pyroclastic layers, typically of intermediate composition, and mudflows.

compound Any substance resulting from the bonding of two or more different elements (e.g., water, H_2O, and quartz, SiO_2).

compression Stress resulting when rocks are squeezed by external forces directed toward one another.

concordant pluton Intrusive igneous body whose boundaries parallel the layering in the country rock. (*See also* discordant pluton)

cone of depression A cone-shaped depression around a well where water is pumped from an aquifer faster than it can be replaced.

contact (thermal) metamorphism Metamorphism of country rock adjacent to a pluton and beneath a lava flow.

continental accretion The process in which continents grow by additions of Earth materials along their margins.

continental–continental plate boundary A convergent plate boundary along which two continental lithospheric plates collide.

continental drift The theory that the continents were joined into a single landmass that broke apart with the various fragments (continents) moving with respect to one another.

continental glacier A glacier that covers a vast area (at least 50,000 km^2) and is not confined by topography; also called an ice sheet.

continental margin The area separating the part of a continent above sea level from the deep seafloor.

continental red bed Sedimentary rock, mostly sandstone and shale, with its red color due to the presence of iron oxides.

continental rise The gently sloping part of the continental margin between the continental slope and the abyssal plain.

continental shelf The very gently sloping part of the continental margin between the shoreline and the continental slope.

continental slope The relatively steeply inclined part of the continental margin between the continental shelf and the continental rise or between the continental shelf and an oceanic trench.

convergent evolution The development of similar features in two or more distantly related groups of organisms resulting from adapting to a comparable lifestyle. (*See also* parallel evolution)

convergent plate boundary The boundary between two plates that move toward each other.

Cordilleran mobile belt A large region of deformation along the western margin of North America bounded on the west by the Pacific Ocean and on the east by the Great Plains.

Cordilleran orogeny An episode of deformation affecting the western margin of North America from Jurassic to Early Cenozoic time; divided into three separate phases called the Nevadan, Sevier, and Laramide orogenies.

core The interior part of Earth beginning at a depth of 2,900 km that probably consists mostly of iron and nickel.
Coriolis effect The apparent deflection of a moving object from its anticipated course because of Earth's rotation. Winds and oceanic currents are deflected clockwise in the Northern Hemisphere and counterclockwise in the Southern Hemisphere.
correlation Demonstration of the physical continuity of rock units or biostratigraphic units, or demonstration of time equivalence as in time-stratigraphic correlation.
country rock Any preexisting rock that has been intruded by a pluton or altered by metamorphism.
covalent bond A chemical bond formed by the sharing of electrons between atoms.
crater An oval to circular depression at the summit of a volcano resulting from the eruption of lava, pyroclastic materials, and gases.
craton The stable nucleus of a continent consisting of a shield of Precambrian rocks and a platform of buried ancient rocks.
cratonic sequence A widespread sequence of sedimentary rocks bounded above and below by unconformities that was deposited during a transgressive–regressive cycle of an epeiric sea.
creep A widespread type of mass wasting in which soil or rock moves slowly downslope.
Cretaceous Interior Seaway A Late Cretaceous arm of the sea that effectively divided North America into two large landmasses.
Cro-Magnon Humans that lived mostly in Europe from 35,000 to 10,000 years ago.
cross-bedding A type of bedding in which layers are deposited at an angle to the surface on which they accumulate, as in sand dunes.
crossopterygian A specific type of lobe-finned fish that had lungs; the ancestors of amphibians were among the crossopterygians.
crust Earth's outermost layer; the upper part of the lithosphere that is separated from the mantle by the Moho; divided into continental and oceanic crust.
crystal A naturally occurring solid of an element or compound with a specific internal structure that is manifested externally by planar faces, sharp corners, and straight edges.
crystal settling The physical separation and concentration of minerals in the lower part of a magma chamber or pluton by crystallization and gravitational settling.
crystalline solid A solid in which the constituent atoms are arranged in a regular, three-dimensional framework.
Curie point The temperature at which iron-bearing minerals in cooling magma or lava attain their magnetism.
cyclothem A sequence of cyclically repeating sedimentary rocks resulting from alternating periods of marine and nonmarine deposition; commonly contains a coal bed.
cynodont A type of carnivorous therapsid (advanced mammal-like reptile); ancestors of mammals are among cynodonts.
debris flow A type of mass wasting that involves a viscous mass of soil, rock fragments, and water that moves downslope; debris flows have larger particles than mudflows and contain less water.
deflation The removal of sediment and soil by wind.
deformation A general term for any change in shape or volume, or both, of rocks in response to stress; involves folding and fracturing.
delta An alluvial deposit formed where a stream or river flows into the sea or a lake.
density The mass of an object per unit volume; usually expressed in grams per cubic centimeter (g/cm^3).
deoxyribonucleic acid (DNA) The chemical substance of which chromosomes are composed; the genetic material in all organisms.
depositional environment Any site such as a floodplain or beach where physical, biologic, and chemical processes yield a distinctive kind of sedimentary deposit.
desert Any area that receives less than 25 cm of rain per year and that has a high evaporation rate.
desertification The expansion of deserts into formerly productive lands.
desert pavement A surface mosaic of close-fitting pebbles, cobbles, and boulders found in many dry regions; results from wind erosion of sand and smaller particles.
detrital sedimentary rock Sedimentary rock made up of the solid particles (detritus) of preexisting rocks.
differential pressure Pressure that is not applied equally to all sides of a rock body.
differential weathering Weathering that occurs at different rates on rocks, thereby yielding an uneven surface.
dike A tabular or sheetlike discordant pluton.
dinosaur Any of the Mesozoic reptiles belonging to the orders of Saurischia and Ornithischia.
dip A measure of the maximum angular deviation of an inclined plane from horizontal.
dip-slip fault A fault on which all movement is parallel with the dip of the fault plane. (*See also* normal fault and reverse fault)
discharge The volume of water in a stream or river moving past a specific point in a given interval of time; expressed in cubic meters per second (m^3/sec) or cubic feet per second (ft^3/sec).
disconformity An unconformity above and below which the rock layers are parallel. (*See also* angular unconformity and nonconformity)
discontinuity A boundary across which seismic wave velocity or direction of travel changes abruptly, such as the mantle–core boundary.
discordant pluton Pluton with boundaries that cut across the layering in the country rock. (*See also* concordant pluton)
dissolved load The part of a stream's load consisting of ions in solution.
divergent evolution The diversification of a species into two or more descendant species.

divergent plate boundary The boundary between two plates that are moving apart.
divide A topographically high area that separates adjacent drainage basins.
dome A rather circular geologic structure in which all rock layers dip away from a central point and the oldest exposed rocks are at the dome's center.
downwelling The slow transfer of ocean surface water to depth. (*See also* upwelling)
drainage basin The surface area drained by a stream or river and its tributaries.
drainage pattern The regional arrangement of channels in a drainage system.
drumlin An elongate hill of till formed by the movement of a continental glacier or by floods.
dune A mound or ridge of wind-deposited sand.
dynamic metamorphism Metamorphism in fault zones where rocks are subjected to high differential pressure.
earthflow A mass-wasting process involving the downslope movement of water-saturated soil.
earthquake Vibrations caused by the sudden release of energy, usually as a result of displacement of rocks along faults.
economic geology Any aspect of geology concerned with the search for minerals, rocks, or other commodities of economic value, such as oil and natural gas.
ectotherm Any of the cold-blooded vertebrate animals, such as amphibians and reptiles; an animal that depends on external heat to regulate body temperature. (*See also* endotherm)
Ediacaran faunas The name for all Late Proterozoic faunas with animal fossils similar to those of the Ediacara fauna of Australia.
elastic rebound theory An explanation for the sudden release of energy that causes earthquakes when deformed rocks fracture and rebound to their original undeformed condition.
elastic strain A type of deformation in which the material returns to its original shape when stress is relaxed.
electron A negatively charged particle of very little mass that encircles the nucleus of an atom.
electron shell Electrons orbit an atom's nucleus at specific distances in electron shells.
element A substance composed of atoms that all have the same properties; atoms of one element can change to atoms of another element by radioactive decay, but they cannot be changed by ordinary chemical means.
emergent coast A coast where the land has risen with respect to sea level.
end moraine A pile or ridge of rubble deposited at the terminus of a glacier. (*See also* terminal moraine and recessional moraine)
endosymbiosis A symbiotic relationship in which one symbiont lives permanently within another organism. Organelles in cells such as plastids and mitochondria are thought to have resulted from endosymbiosis.
endotherm Any of the warm-blooded vertebrate animals, such as birds and mammals, that maintain their body temperature within narrow limits by internal processes. (*See also* ectotherm)
epeiric sea A broad, shallow sea that covers part of a continent, such as the Sauk Sea.
epicenter The point on Earth's surface directly above the focus of an earthquake.
erosion The removal of weathered materials from their source area by running water, wind, glaciers, or waves.
esker A long, sinuous ridge of stratified drift deposited by running water in a tunnel beneath stagnant ice.
eukaryotic cell A cell with internal structures such as mitochondria and an internal membrane-bounded nucleus containing chromosomes. (*See also* prokaryotic cell)
expansive soil A soil in which the volume increases by at least 6% when water is present.
extrusive igneous rock An igneous rock formed when magma is extruded onto Earth's surface where it cools and crystallizes, or when pyroclastic materials become consolidated. (*See also* volcanic rock)
evaporite Any sedimentary rock, such as rock salt, formed by inorganic chemical precipitation of minerals from evaporating water.
Exclusive Economic Zone (EEZ) An area extending 370 km seaward from the coast of the United States and its possessions in which the United States clams rights to all resources.
exfoliation dome A large, rounded dome of rock resulting when concentric layers of rock are stripped from the surface of a rock mass.
Farallon plate A Late Mesozoic–Cenozoic oceanic plate that was largely subducted beneath North America; the Cocos and Juan de Fuca plates are remnants.
fault A fracture along which rocks on opposite sides of the fracture have moved parallel with the fracture surface.
fault plane A fault surface that is more or less planar.
felsic magma Magma with more than 65% silica and considerable sodium, potassium, and aluminum, but little calcium, iron, and magnesium. (*See also* intermediate magma, mafic magma, and ultramafic magma)
ferromagnesian silicate Any silicate mineral that contains iron, magnesium, or both. (*See also* nonferromagnesian silicate)
fetch The distance the wind blows over a continuous water surface.
fiord An arm of the sea extending into a glacial trough eroded below sea level.
firn Granular snow formed by partial melting and refreezing of snow; transitional material between snow and glacial ice.
fission-track dating The absolute dating process in which small linear tracks (fission tracks) resulting from alpha decay are counted in mineral crystals.
fissure eruption A volcanic eruption in which lava or pyroclastic materials issue from a long, narrow fissure (crack) or group of fissures.

floodplain A low-lying, flat area adjacent to a channel that is partly or completely water-covered when a stream or river overflows its banks.

fluid activity An agent of metamorphism in which water and carbon dioxide promote metamorphism by increasing the rate of chemical reactions.

focus The site within Earth where an earthquake originates and energy is released.

fold A type of geologic structure in which planar features in rock layers such as bedding and foliation have been bent.

foliated texture A texture in metamorphic rocks in which platy and elongate minerals are aligned in a parallel fashion.

footwall block The block of rock that lies beneath a fault plane. (*See also* hanging wall block)

fossil Remains or traces of prehistoric organisms preserved in rocks. (*See also* body fossil and trace fossil)

fracture A break in rock resulting from intense applied pressure.

frost action The disaggregation of rocks by repeated freezing and thawing of water in cracks and crevasses.

gene A specific segment of a chromosome constituting the basic unit of heredity. (*See also* allele)

geologic structure Any feature in rocks that results from deformation, such as folds, joints, and faults.

geologic time scale A chart that subdivides geologic time into a hierarchy of increasingly shorter time intervals, each of which has a specific name and duration.

geology The study of Earth, as well as the planets and moons in our solar system. It is generally divided into two broad areas—*physical geology*, which is the study of Earth's materials, such as minerals and rocks, as well as the processes operating within Earth and on its surface, and *historical geology*, which examines the origin and evolution of Earth, its continents, oceans, atmosphere, and life.

geothermal energy Energy that comes from steam and hot water trapped within Earth's crust.

geothermal gradient Earth's temperature increase with depth; it averages 25°C/km near the surface but varies from area to area.

geyser A hot spring that periodically ejects hot water and steam.

glacial budget The balance between expansion and contraction of a glacier in response to accumulation versus wastage.

glacial drift A collective term for all sediment deposited directly by glacial ice (till) and by meltwater streams (outwash).

glacial erratic A rock fragment carried some distance from its source by a glacier and usually deposited on bedrock of a different composition.

glacial ice Water in the solid state within a glacier; forms as snow partially melts and refreezes and compacts so that it is transformed first to firn and then to glacial ice.

glacial polish A smooth, glistening rock surface formed by the movement of sediment-laden ice over bedrock.

glacial striation A straight scratch rarely more than a few millimeters deep on a rock caused by the movement of sediment-laden glacial ice.

glacial surge A time of greatly accelerated flow in a glacier. Commonly results in displacement of the glacier's terminus by several kilometers.

glaciation Refers to all aspects of glaciers, including their origin, expansion, and retreat, and their impact on Earth's surface.

glacier A mass of ice on land that moves by plastic flow and basal slip.

***Glossopteris* flora** A Late Paleozoic association of plants found only on the Southern Hemisphere continents and India; named for its best-known genus, *Glossopteris*.

Gondwana A major Paleozoic continent composed of South America, Africa, Australia, India, and Antarctica.

graded bedding A sedimentary layer that shows a decrease in grain size from bottom to top.

graded stream A stream that has an equilibrium profile in which a delicate balance exists among gradient, discharge, flow velocity, channel characteristics, and sediment load so that neither significant deposition nor erosion takes place within its channel.

gradient The slope over which a stream or river flows; expressed in meters per kilometer (m/km) or feet per mile (ft/mi).

granite–gneiss complex The most common association of Archean age rocks.

gravity anomaly A departure from the expected force of gravity; anomalies may be positive or negative.

grazer An animal that eats low-growing vegetation, especially grasses. (*See also* browser)

Great Oxygenation Event An event that took place about 2.4 billion years ago when Earth's atmosphere became richer in free oxygen as a result of photochemical dissociation and, more importantly, photosynthesis.

greenstone belt A linear or podlike association of igneous and sedimentary rocks particularly common in Archean terrains.

Grenville orogeny An episode of deformation that took place in the eastern United States and Canada during the Neoproterozoic.

ground moraine The layer of sediment released from melting ice as a glacier's terminus retreats.

groundwater Underground water stored in the pore spaces of soil, sediment, and rock.

guide fossil Any easily identified fossil with an extensive geographic distribution and short geologic range useful for determining the relative ages of rocks in different areas.

Gulf Coastal Plain The broad, low-relief area along the Gulf Coast of the United States and Mexico.

gymnosperm A flowerless, seed-bearing plant.

gyre A system of ocean currents rotating clockwise in the Northern Hemisphere and counterclockwise in the Southern Hemisphere.

half-life The time necessary for half of the original number of radioactive atoms of an element to decay to a stable daughter product; for example, the half-life for potassium 40 is 1.3 billion years.

hanging valley A tributary glacial valley whose floor is at a higher level than that of the main glacial valley.

hanging wall block The block of rock that overlies a fault plane. (*See also* footwall block)

hardness A term used to express the resistance of a mineral to abrasion.

headland Part of a shoreline, commonly bounded by cliffs, that extends out into the sea or a lake.

heat An agent of metamorphism.

herbivore An animal that depends directly on plants as a source of nutrients.

Hercynian (Variscan)–Alleghenian orogeny Pennsylvanian to Permian orogenic event during which the Appalachian mobile belt in eastern North America and the Hercynian mobile belt of southern Europe were deformed.

heterotrophic Any organism that depends on preformed organic molecules from its environment for nutrients. (*See also* autotrophic)

hominid Abbreviated form of Hominidae, the family of bipedal primates that includes *Australophithecus* and *Homo*.

Homo The genus of hominids consisting of *Homo sapiens* and their ancestors *Homo erectus* and *Homo habilis*.

homologous structure Body part in different organisms that has a similar structure, similar relationship to other structures, and similar development, but does not necessarily serve the same function; for example, forelimbs in whales, dogs, and bats. (*See also* analogous structure)

horn A steep-walled, pyramid-shaped peak formed by the headward erosion of at least three cirques.

hot spot Localized zone of melting below the lithosphere that probably overlies a mantle plume.

hot spring A spring in which the water temperature is warmer than the temperature of the human body (37°C).

hydraulic action The removal of loose particles by the power of moving water.

hydrologic cycle The continuous recycling of water from the oceans, through the atmosphere, to the continents, and back to the oceans, or from the oceans, through the atmosphere, and back to the oceans.

hydrolysis The chemical reaction between hydrogen (H^+) ions and hydroxyl (OH^-) ions of water and a mineral's ions.

hydrothermal A term referring to hot water as in hot springs and geysers.

hypothesis A provisional explanation for observations that is subject to continual testing. If well supported by evidence, a hypothesis may be called a theory.

Iapetus Ocean A Paleozoic ocean between North America and Europe that closed when the continents collided during the Late Paleozoic.

ice cap A dome-shaped mass of glacial ice that covers less than 50,000 km^2.

ice-scoured plain A low relief bedrock surface with glacial striations and polish eroded by a glacier.

ichthyosaur Any of the porpoise-like, Mesozoic marine reptiles.

igneous rock Any rock formed by cooling and crystallization of magma or lava or the consolidation of pyroclastic materials.

incised meander A deep, meandering canyon cut into bedrock by a stream or river.

index mineral A mineral that forms within specific temperature and pressure ranges during metamorphism.

infiltration capacity The maximum rate at which soil or sediment absorbs water.

inheritance of acquired characteristics Jean-Baptiste de Lamarck's mechanism for evolution; holds that characteristics acquired during an individual's lifetime can be inherited by descendants.

intensity The subjective measure of the kind of damage done by an earthquake as well as people's reaction to it.

Interior Lowlands An area in North America made up of the Great Plains and the Central Lowlands.

intermediate magma Magma with a silica content between 53% and 65% and an overall composition intermediate between mafic and felsic magma.

intrusive igneous rock Igneous rock that formed from magma intruded into or formed in place within the crust. (*See also* plutonic rock)

ion An electrically charged atom produced by adding or removing electrons from the outermost electron shell.

ionic bond A chemical bond resulting from the attraction between positively and negatively charged ions.

iridium anomaly The occurrence of a higher-than-usual concentration of the element iridium at the Cretaceous–Paleogene boundary.

isostasy *See* principle of isostasy.

isostatic rebound The phenomenon in which unloading of the crust causes it to rise until it attains equilibrium with the underlying upper mantle.

joint A fracture along which no movement has occurred or where movement is perpendicular to the fracture surface.

Jovian planet Any of the four planets (Jupiter, Saturn, Uranus, and Neptune) that resemble Jupiter. All are large and have low mean densities, indicating that they are composed mostly of lightweight gases, such as hydrogen and helium, and frozen compounds, such as ammonia and methane.

kame A conical hill of stratified drift originally deposited in a depression on a glacier's surface.

karst topography Landscape consisting of numerous caves, sinkholes, and solution valleys formed by groundwater solution of rocks such as limestone and dolostone.

Kaskaskia Sequence A widespread sequence of Devonian and Mississippian sedimentary rocks bounded above and below by unconformities that was deposited during a transgressive–regressive cycle of the Kaskaskia Sea.

Kazakhstania One of six major Paleozoic continents; a triangular-shaped continent centered on Kazakhstan.

labyrinthodont Any of the Devonian to Triassic amphibians characterized by teeth with complexly folded enamel.

laccolith A concordant pluton with a mushroom-like geometry.

lahar A mudflow composed of pyroclastic materials such as ash.

Laramide orogeny Late Cretaceous to Early Cenozoic phase of the Cordilleran orogeny; responsible for many structural features of the present-day Rocky Mountains.

lateral moraine A ridge of sediment deposited along the margin of a valley glacier.

laterite A red soil, rich in iron or aluminum, or both, resulting from intense chemical weathering in the tropics.

Laurasia A Late Paleozoic Northern Hemisphere continent made up of North America, Greenland, Europe, and Asia.

Laurentia A Proterozoic continent composed of North America, Greenland, parts of Scotland, and perhaps part of the Baltic shield of Scandinavia.

lava Magma that reaches Earth's surface.

lava dome A bulbous, steep-sided volcano formed by viscous magma moving upward through a volcanic conduit.

lava tube A tunnel beneath the solidified surface of a lava flow through which lava moves; also, the hollow space left when the lava within a tube drains away.

lithification The process of converting sediment into sedimentary rock by compaction and cementation.

lithosphere Earth's outer, rigid part, consisting of the upper mantle, oceanic crust, and continental crust.

lithostatic pressure Pressure exerted on rocks by the weight of overlying rocks.

lithostratigraphic unit A body of rock, such as a formation, defined solely by its physical attributes.

Little Ice Age An interval from about 1500 to the mid- to late-1800s during which glaciers expanded to their greatest historic extent.

loess A wind-blown deposit of silt and clay.

longitudinal dune A long ridge of sand generally parallel to the direction of the prevailing wind.

longshore current A current resulting from wave refraction found between the breaker zone and a beach that flows parallel to the shoreline.

Love wave (L-wave) A surface wave in which the individual particles of material move only back and forth in a horizontal plane perpendicular to the direction of wave travel.

luster The appearance of a mineral in reflected light. Luster is metallic or nonmetallic, although the latter has several subcategories.

macroevolution Evolutionary changes that account for the origin of new species, genera, orders, and so on.

mafic magma Magma with between 45% and 52% silica and proportionately more calcium, iron, and magnesium than intermediate and felsic magma.

magma Molten rock material generated within Earth.

magma chamber A reservoir of magma within Earth's upper mantle or lower crust.

magma mixing The process whereby magmas of different composition mix together to yield a modified version of the parent magmas.

magnetic anomaly Any deviation, such as a change in average strength, in Earth's magnetic field.

magnetic field The area in which magnetic substances are affected by lines of magnetic force emanating from Earth.

magnetic reversal The phenomenon involving the complete reversal of the north and south magnetic poles.

magnetism A physical phenomenon resulting from moving electricity and the spin of electrons in some solids in which magnetic substances are attracted toward one another.

magnitude The total amount of energy released by an earthquake at its source. (*See also* Richter Magnitude Scale)

mantle The thick layer between Earth's crust and core.

mantle plume A cylindrical mass of magma rising from the mantle toward the surface; recognized at the surface by a hot spot, an area such as the Hawaiian Islands where volcanism takes place.

marine regression The withdrawal of the sea from a continent or coastal area, resulting in the emergence of the land as sea level falls or the land rises with respect to sea level.

marine terrace A wave-cut platform now above sea level.

marine transgression The invasion of a coastal area or a continent by the sea, resulting from a rise in sea level or subsidence of the land.

marsupial mammal Pouched mammals, such as kangaroos and wombats, that give birth to their young in a very immature state.

mass extinction Greatly accelerated extinction rate that results in a marked decrease in biodiversity, such as the mass extinction at the end of the Cretaceous Period.

mass wasting The downslope movement of Earth materials under the influence of gravity.

meandering stream A stream that has a single, sinuous channel with broadly looping curves.

mechanical weathering Disaggregation of rocks by physical processes that yields smaller pieces that retain the composition of the parent material.

medial moraine A moraine carried on the central surface of a glacier; formed where two lateral moraines merge.

Mediterranean belt A zone of seismic and volcanic activity extending through the Mediterranean region of southern Europe and eastward to Indonesia.

meiosis Cell division that yields sex cells, sperm, and eggs in animals and pollen and ovules in plants, in which the number of chromosomes is reduced by half. (*See also* mitosis)

mesa A broad, flat-topped erosional remnant bounded on all sides by steep slopes.

metamorphic facies A group of metamorphic rocks characterized by particular minerals that formed under the same broad temperature and pressure conditions.

metamorphic grade The degree to which parent rocks have undergone metamorphic change; the higher the grade the greater the change, as in low-, medium-, and high-grade.

metamorphic rock Any rock that has been changed from its original condition by heat, pressure, and the chemical activity of fluids, as in marble and slate.

metamorphic zone The region between lines of equal metamorphic intensity known as isograds.

metamorphism The phenomenon of changing rocks subjected to heat, pressure, and fluids so that they are in equilibrium with a new set of environmental conditions. Metamorphism takes place in the solid state.

microevolution Evolutionary changes within a species.

Midcontinent rift A Late Proterozoic rift in Laurentia in which volcanic and sedimentary rocks accumulated.

Milankovitch theory An explanation for the cyclic variations in climate and the onset of ice ages as a result of irregularities in Earth's rotation and orbit.

mineral A naturally occurring, inorganic, crystalline solid that has characteristic physical properties and a narrowly defined chemical composition.

mitosis Cell division that results in two cells that have the same number of chromosomes as the parent cell; takes place in all cells except sex cells. (*See also* meiosis)

mobile belt An elongated area of deformation generally along the margins of a craton, such as the Appalachian mobile belt.

modern synthesis A combination of ideas of scientists regarding evolution that includes the chromosome theory of inheritance, mutations as a source of variation, and gradualism, but rejects the idea of inheritance of acquired characteristics.

Modified Mercalli Intensity Scale A scale with values from I to XII used to characterize earthquakes based on damage.

Mohorovičić discontinuity (Moho) The boundary between Earth's crust and mantle.

monocline A bend or flexure in otherwise horizontal or uniformly dipping rock layers.

monomer A comparatively simple organic molecule, such as an amino acid, that can link with other monomers to form more complex polymers such as proteins.

monotreme The egg-laying mammals; includes only the platypus and spiny anteater of the Australian region.

mosaic evolution The concept that not all parts of an organism evolve at the same rate, thus yielding organisms with features retained from the ancestral condition, as well as more recently evolved features.

mosasaur A term referring to a group of Mesozoic marine lizards.

mud crack A crack in clay-rich sediment that forms in response to drying and shrinkage.

mudflow A flow consisting mostly of clay- and silt-size particles and up to 30% water that moves downslope under the influence of gravity.

multicelled organism Any organism made up of many cells, as opposed to a single cell; possesses cells specialized to perform specific functions.

mutation A change in the genes of organisms; yields some of the variation on which natural selection acts.

natural levee A ridge of sandy alluvium deposited along the margins of a channel during floods.

natural selection A mechanism that accounts for differential survival and reproduction among members of a species; the mechanism proposed by Darwin and Wallace to account for evolution.

Neanderthal A type of human who lived in the Near East and Europe from 200,000 to 30,000 years ago; may be a subspecies of *Homo (Homo sapiens neanderthalensis)* or a separate species *(Homo neanderthalensis)*.

nearshore sediment budget The balance between additions and losses of sediment in the nearshore zone.

nekton Actively swimming organisms, such as fish, whales, and squid. (*See also* plankton)

neutron An electrically neutral particle found in the nucleus of an atom.

Nevadan orogeny Late Jurassic to Cretaceous phase of the Cordilleran orogeny; most strongly affected the western part of the Cordilleran mobile belt.

nonconformity An unconformity in which stratified sedimentary rocks overlie an erosion surface cut into igneous or metamorphic rocks. (*See also* angular unconformity and disconformity)

nonferromagnesian silicate A silicate mineral that has no iron or magnesium. (*See also* ferromagnesian silicate)

nonfoliated texture A metamorphic texture in which there is no discernable preferred orientation of minerals.

normal fault A dip-slip fault on which the hanging wall block has moved downward relative to the footwall block. (*See also* reverse fault)

North American Cordillera A complex mountainous region in western North America extending from Alaska into central Mexico.

nucleus The central part of an atom consisting of protons and neutrons.

nuée ardente A fast-moving, dense cloud of hot pyroclastic materials and gases ejected from a volcano.

oblique-slip fault A fault showing both dip-slip and strike-slip movement.

oceanic–continental plate boundary A convergent plate boundary along which oceanic lithosphere is subducted beneath continental lithosphere.

oceanic–oceanic plate boundary A convergent plate boundary along which two oceanic plates collide and one is subducted beneath the other.

oceanic ridge A mostly submarine mountain system composed of basalt found in all ocean basins.

oceanic trench A long, narrow feature restricted to active continental margins and along which subduction occurs.

ooze Deep-sea sediment composed mostly of shells of marine animals and plants.

ophiolite A sequence of igneous rocks representing a fragment of oceanic lithosphere; composed of peridotite overlain successively by gabbro, sheeted basalt dikes, and pillow lava.

organic evolution See theory of evolution.

organic reef A wave-resistant limestone structure with a framework of animal skeletons.

Ornithischia One of the two orders of dinosaurs, characterized by a birdlike pelvis; includes ornithopods, stegosaurs, ankylosaurs, pachycephalosaurs, and ceratopsians. (*See also* Saurischia)

orogen A linear part of Earth's crust that was, or is, being deformed during an orogeny; part of an orogenic belt.

orogeny An episode of mountain building involving deformation, usually accompanied by igneous activity, metamorphism, and crustal thickening.

ostracoderm The "bony-skinned" fish characterized by bony armor but no jaws or teeth; appeared during the Late Cambrian, making them the oldest known vertebrates.

Ouachita mobile belt An area of deformation along the southern margin of the North American craton.

Ouachita orogeny A period of mountain building that took place in the Ouachita mobile belt during the Pennsylvanian Period.

outgassing The process whereby gases released from Earth's interior by volcanism formed an atmosphere and hydrosphere.

outwash plain The sediment deposited by meltwater discharging from a continental glacier's terminus.

oxbow lake A cutoff meander filled with water.

oxidation The reaction of oxygen with other atoms to form oxides or, if water is present, hydroxides.

pahoehoe A type of lava flow with a smooth ropy surface.

Paleocene–Eocene thermal maximum A warming trend that began abruptly about 55 million years ago.

paleomagnetism Residual magnetism in rocks, studied to determine the intensity and direction of Earth's past magnetic field.

paleontology The use of fossils to study life history and relationships among organisms.

Pangaea The name Alfred Wegener proposed for a supercontinent consisting of all of Earth's landmasses at the end of the Paleozoic Era.

Pannotia A supercontinent that existed during the Neoproterozoic.

Panthalassa Ocean A Late Paleozoic ocean that surrounded Pangaea.

parabolic dune A crescent-shaped dune with its tips pointing upwind.

parallel evolution The development of similar features in two or more closely related but separate lines of descent as a consequence of similar adaptations. (*See also* convergent evolution)

parent material The material that is chemically and mechanically weathered to yield sediment and soil.

passive continental margin A continental margin within a tectonic plate as in the East Coast of North America where little seismic activity and no volcanism occur; characterized by a broad continental shelf and a continental slope and rise.

pediment An erosion surface of low relief gently sloping away from the base of a mountain range.

pelagic clay Brown or red deep-sea sediment composed of clay-size particles.

pelycosaur Pennsylvanian to Permian reptiles that had some mammal characteristics; many species had large fins on the back.

perissodactyl The odd-toed hoofed mammals that includes present-day horses, rhinoceroses, and tapirs; members of the order Perissodactyla.

permafrost Ground that remains permanently frozen.

permeability A material's capacity to transmit fluids.

phaneritic texture Igneous rock texture in which minerals are easily visible without magnification.

photic zone The sunlit layer in the oceans where plants photosynthesize. (*See also* aphotic zone)

photochemical dissociation The process by which water molecules in the upper atmosphere are disrupted by ultraviolet radiation to yield free oxygen (O_2) and hydrogen (H_2).

photosynthesis The metabolic process in which organic molecules are synthesized from water and carbon dioxide, using the radiant energy of the Sun captured by chlorophyll-containing cells.

phyletic gradualism The concept that a species evolves gradually and continuously as it gives rise to new species. (*See also* punctuated equilibrium)

pillow lava Bulbous masses of basalt, resembling pillows, formed when lava is rapidly chilled under water.

placental mammal All mammals with a placenta to nourish the developing embryo, as opposed to egg-laying mammals (monotremes) and pouched mammals (marsupials).

placoderm Late Silurian through Permian "plate-skinned" fish with jaws and bony armor, especially in the head and shoulder area.

plankton Aquatic organisms that float passively, such as phytoplankton (plants) and zooplankton (animals). (*See also* nekton)

plastic flow The flow that takes place in response to pressure and causes deformation with no fracturing.

plastic strain Permanent deformation of a solid with no failure by fracturing.

plate An individual segment of the lithosphere that moves over the asthenosphere.

plate tectonic theory The theory holding that large segments of Earth's outer part (lithospheric plates) move relative to one another.

platform The part of a craton that lies buried beneath flat-lying or mildly deformed sedimentary rocks.

playa A dry lakebed found in deserts.

plesiosaur A type of Mesozoic marine reptile; short-necked and long-necked plesiosaurs existed.

pluton An intrusive igneous body that forms when magma cools and crystallizes within the crust, such as a batholith or sill.

plutonic rock Igneous rock that formed from magma intruded into or formed in place within the crust. (*See also* intrusive igneous rock)

point bar The sediment body deposited on the gently sloping side of a meander loop.

polymer A comparatively complex organic molecule, such as nucleic acids and proteins, formed by monomers linking together. (*See also* monomer)

porosity The percentage of a material's total volume that is pore space.

porphyritic texture An igneous texture with minerals of markedly different sizes; results from slow cooling of magma and generally indicates an intrusive origin.

pressure release A mechanical weathering process in which rocks that formed under pressure expand on being exposed at the surface.

primary producer Organism in a food chain, such as a bacterium or green plant, that manufactures its own organic molecules, and on which other members of the food chain depend for sustenance. (*See also* autotrophic)

primate Any of the mammals that belong to the order Primates; includes prosimians (lemurs and tarsiers), monkeys, apes, and humans.

principle of cross-cutting relationships A principle holding that an igneous intrusion or fault is younger than the rocks it intrudes or cuts across.

principle of fossil succession A principle holding that fossils, and especially groups or assemblages of fossils, succeed one another through time in a regular and predictable order.

principle of inclusions A principle holding that inclusions or fragments in a rock unit are older than the rock unit itself; for example, granite inclusions in sandstone are older than the sandstone.

principle of isostasy The theoretical concept of Earth's crust "floating" on a dense underlying layer. (*See also* isostatic rebound)

principle of lateral continuity A principle holding that rock layers extend outward in all directions until they terminate.

principle of original horizontality According to this principle, sediments are deposited in horizontal or nearly horizontal layers.

principle of superposition A principle holding that in a vertical sequence of undeformed sedimentary rocks, the relative ages of the rocks can be determined by their position in the sequence—oldest at the bottom followed by successively younger layers.

principle of uniformitarianism A principle holding that past events can be interpreted by understanding present-day processes; based on the idea that natural processes have always operated in the same way.

prokaryotic cell A cell that lacks a nucleus and organelles such as mitochondria and plastids; the cells of bacteria and cyanobacteria. (*See also* eukaryotic cell)

proton A positively charged particle found in the nucleus of an atom.

protorothyrid A loosely grouped category of small lizard-like reptiles that include the earliest reptiles.

pterosaur Any of the Mesozoic flying reptiles.

punctuated equilibrium A concept holding that a new species evolves rapidly, in perhaps a few thousand years, and then remains much the same during its several million years of existence. (*See also* phyletic gradualism)

P-wave A compressional, or push-pull, wave; the fastest seismic wave and one that can travel through solids, liquids, and gases; also called a primary wave.

P-wave shadow zone An area between 103 and 143 degrees from an earthquake focus where little P-wave energy is recorded by seismographs.

pyroclastic (fragmental) texture A fragmental texture characteristic of igneous rocks composed of pyroclastic materials.

pyroclastic materials Fragmental substances, such as ash, explosively ejected from a volcano.

pyroclastic sheet deposit Vast, sheetlike deposit of felsic pyroclastic materials erupted from fissures.

quadrupedal A term referring to locomotion on all four limbs, as in dogs and horses. (*See also* bipedal)

Queenston Delta A clastic wedge resulting from erosion of the highlands formed during the Taconic orogeny.

quick clay A clay deposit that spontaneously liquefies and flows when disturbed.

radioactive decay The spontaneous change of an atom to an atom of a different element by emission of a particle from its nucleus (alpha and beta decay) or by electron capture.

radon A colorless, odorless naturally occurring gas that results from the decay of uranium 238 to lead 206.

rain-shadow desert A desert found on the lee side of a mountain range because precipitation falls mostly on the windward side of the range.

rapid mass movement Any kind of mass wasting that involves a visible downslope displacement of material.

Rayleigh wave (R-wave) A surface wave in which individual particles of material move in an elliptical path within a vertical plane oriented in the direction of wave movement.

recessional moraine An end moraine that forms when a glacier's terminus retreats, then stabilizes, and a ridge or mound of till is deposited. (*See also* end moraine and terminal moraine)

reef A moundlike, wave-resistant structure composed of the skeletons of organisms.

regional metamorphism Metamorphism that occurs over a large area, resulting from high temperatures, tremendous pressure, and the chemical activity of fluids within the crust.

regolith The layer of unconsolidated rock and mineral fragments and soil that covers most of the land surface.

relative dating The process of determining the age of an event as compared to other events; involves placing geologic events in their correct chronological order, but does not involve consideration of when the events occurred in number of years ago. (*See also* absolute dating)

reserve The part of the resource base that can be extracted economically.

resource A concentration of naturally occurring solid, liquid, or gaseous material, in or on Earth's crust in such form and amount that economic extraction of a commodity from the concentration is currently or potentially feasible.

reverse fault A dip-slip fault on which the hanging wall block has moved upward relative to the footwall block. (*See also* normal fault)

Richter Magnitude Scale An open-ended scale that measures the amount of energy released during an earthquake.

Rio Grande rift A linear depression made up of interconnected basins extending from Colorado into Mexico.

rip current A narrow surface current that flows out to sea through the breaker zone.

ripple mark A wavelike (undulating) structure produced in granular sediment, especially sand, by unidirectional wind and water currents or by oscillating wave currents.

rock A solid aggregate of one or more minerals, as in limestone and granite, or a consolidated aggregate of rock fragments, as in conglomerate, or masses of rocklike materials, such as coal and obsidian.

rock cycle A pictorial representation of events leading to the origin, destruction, or changes, and reformation of rocks as a consequence of Earth's internal and surface processes. It also shows how the three major rock groups are interrelated and how any rock type can be derived from any other rock type.

rockfall A type of extremely fast mass wasting in which rocks fall through the air.

rock-forming mineral Any mineral common in rocks that is important in their identification and classification.

rock slide Rapid mass wasting in which rocks move downslope along a more or less planar surface.

Rodinia The name of a Neoproterozoic supercontinent.

rounding The process by which the sharp corners and edges of sedimentary particles are abraded and smoothed during transport.

runoff The surface flow in streams and rivers.

salinity A measure of the dissolved solids in seawater, commonly expressed in parts per thousand.

salt crystal growth A mechanical weathering process in which salt crystals growing in cracks and pores disaggregate rocks.

San Andreas transform Fault A transform fault extending from the Gulf of California in Mexico to its termination in the Pacific Ocean off the northern coast of California.

sandstone-carbonate-shale assemblage An association of sedimentary rocks typically found on passive continental margins during the Proterozoic Eon.

Sauk Sequence A widespread association of sedimentary rocks bounded above and below by unconformities that was deposited during the latest Proterozoic to Early Ordovician transgressive–regressive cycle of the Sauk Sea.

Saurischia An order of dinosaurs characterized by a lizard-like pelvis; includes theropods, prosauropods, and sauropods. (*See also* Ornithischia)

scientific method A logical, orderly approach that involves gathering data, formulating and testing hypotheses, and proposing theories.

seafloor spreading The theory that the seafloor moves away from spreading ridges and is eventually consumed at subduction zones.

sediment Loose aggregate of solids derived by weathering from preexisting rocks, or solids precipitated from solution by inorganic chemical processes or extracted from solution by organisms.

sediment-deposit feeder Animal that ingests sediment and extracts nutrients from it.

sedimentary facies Any aspect of a sedimentary rock unit that makes it recognizably different from adjacent sedimentary rocks of the same or approximately the same age.

sedimentary rock Any rock composed of sediment that forms at or near Earth's surface.

sedimentary structure Any feature in sedimentary rock that formed at or shortly after the time of deposition, such as cross-bedding, animal burrows, and mud cracks.

seedless vascular plant A plant with specialized tissues for transporting fluids and nutrients that reproduces by spores rather than seeds, as in ferns and horsetail rushes.

seismograph An instrument that detects, records, and measures the various waves produced by earthquakes.

seismology The study of earthquakes.

sequence stratigraphy The study of rock relationships within a time stratigraphic framework of related facies bounded by widespread unconformities.

Sevier orogeny Cretaceous phase of the Cordilleran orogeny that affected the continental shelf and slope areas of the Cordilleran mobile belt.

shear strength The resisting forces that help maintain a slope's stability.

shear stress The result of forces acting parallel to one another but in opposite directions; results in deformation by displacement of adjacent layers along closely spaced planes.

sheet joint A large fracture more or less parallel to a rock surface resulting from pressure release.

shield A vast area of exposed ancient rocks on a continent; the exposed part of a craton.

shield volcano A dome-shaped volcano with a low, rounded profile built up mostly by overlapping basalt lava flows.

shoreline The area between mean low tide and the highest level on land affected by storm waves.

Siberia One of six major Paleozoic continents; composed of Russia east of the Ural Mountains, and Asia north of Kazakhstan and south of Mongolia.

silica A compound of silicon and oxygen.

silica tetrahedron The basic building block of all silicate minerals; consists of one silicon atom and four oxygen atoms.

silicate A mineral that contains silica, such as quartz (SiO_2).

sill A tabular or sheetlike concordant pluton.

sinkhole A depression in the ground that forms by the solution of the underlying carbonate rocks or by the collapse of a cave roof.

slide Mass wasting involving movement of material along one or more surfaces of failure.

slow mass movement Mass movement that advances at an imperceptible rate and is usually detectable only by the effects of its movement.

slump Mass wasting that takes place along a curved surface of failure and results in the backward rotation of the slump mass.

soil Regolith consisting of weathered materials, water, air, and humus that can support vegetation.

soil degradation Any process leading to a loss of soil productivity; may involve erosion, chemical pollution, or compaction.

soil horizon A distinct soil layer that differs from other soil layers in texture, structure, composition, and color.

solar nebula theory A theory for the evolution of the solar system from a rotating cloud of gas.

solifluction Mass wasting involving the slow downslope movement of water-saturated surface materials.

solution A reaction in which the ions of a substance become dissociated in a liquid and the solid substance dissolves.

Sonoma orogeny A Permian–Triassic orogeny caused by the collision of an island arc with the southwestern margin of North America.

sorting A term referring to the degree to which all particles in a sedimentary deposit or sedimentary rock are about the same size.

species A population of similar individuals that in nature can interbreed and produce fertile offspring.

specific gravity The ratio of a substance's weight, especially a mineral, to an equal volume of water at 4°C.

spheroidal weathering A type of chemical weathering in which corners and sharp edges of rocks weather more rapidly than flat surfaces, thus yielding spherical shapes.

spit A fingerlike projection of a beach into a body of water such as a bay.

spring A place where groundwater flows or seeps out of the ground.

stock An irregularly shaped discordant pluton with a surface area smaller than 100 km^2.

stoping A process in which rising magma detaches and engulfs pieces of the country rock.

storm surge The surge of water onto a shoreline as a result of a bulge in the ocean's surface beneath the eye of a hurricane and wind-driven waves.

strain Deformation caused by stress.

strata (singular, stratum) Refers to layering in sedimentary rocks. (*See also* bed)

stratified drift Glacial deposits that show both stratification and sorting; deposited by streams that discharge from glaciers.

stream terrace An erosional remnant of a floodplain that formed when a stream was flowing at a higher level.

stress The force per unit area applied to a material such as rock.

strike The direction of a line formed by the intersection of an inclined plane and a horizontal plane.

strike-slip fault A fault involving horizontal movement of blocks of rock on opposite sides of a fault plane.

stromatolite A biogenic sedimentary structure, especially in limestone, produced by the entrapment of sediment on sticky mats of photosynthesizing bacteria.

submarine hydrothermal vent A crack or fissure in the seafloor through which superheated water issues.

submergent coast A coast along which sea level rises with respect to the land or the land subsides.

Sundance Sea A wide seaway that existed in western North America during the Middle Jurassic Period.

superposed stream A stream that once flowed on a higher surface and eroded downward into resistant rocks while maintaining its course.

suspended load The smallest particles (silt and clay) carried by running water that are kept suspended by fluid turbulence.

suspension feeder Animal that consumes microscopic plants and animals or dissolved nutrients from water.

S-wave A shear wave that moves material perpendicular to the direction of travel, thereby producing shear stresses in the material it moves through; also known as a secondary wave; S-waves travel only through solids.

S-wave shadow zone Those areas more than 103 degrees from an earthquake focus where no S-waves are recorded.

syncline A down-arched fold in which the youngest exposed rocks coincide with the fold axis and all strata dip toward the axis.

system A combination of related parts that interact in an organized fashion; Earth systems include the atmosphere, hydrosphere, biosphere, and solid Earth.

Taconic orogeny An Ordovician episode of mountain building that resulted in the deformation of the Appalachian mobile belt.

talus Accumulation of coarse, angular rock fragments at the base of a slope.

Tejas epeiric sea A Cenozoic sea largely restricted to the Gulf and Atlantic Coastal Plains, coastal California, and the Mississippi Valley.

tension A type of stress in which forces act in opposite directions but along the same line, thus tending to stretch an object.

terminal moraine An end moraine consisting of a ridge or mound of rubble marking the farthest extent of a glacier. (*See also* end moraine and recessional moraine)

terrane A small lithospheric block with characteristics quite different from those of surrounding rocks. Terranes probably consist of seamounts, oceanic rises, and other seafloor features accreted to continents during orogenies.

terrestrial planet Any of the four innermost planets (Mercury, Venus, Earth, and Mars). They are all small and have high mean densities, indicating that they are composed of rock and metallic elements.

theory An explanation for some natural phenomenon that has a large body of supporting evidence. To be scientific, a theory must be testable—for example, plate tectonic theory.

theory of evolution The theory holding that all organisms are related and that they descended with modification from organisms that lived during the past.

therapsid Permian to Triassic mammal-like reptiles; the ancestors of mammals are among one group of therapsids known as cynodonts.

thermal convection cell A type of circulation of material in the asthenosphere during which hot material rises, moves laterally, cools and sinks, and is reheated and continues the cycle.

thermal expansion and contraction A type of mechanical weathering in which the volume of rocks changes in response to heating and cooling.

thrust fault A type of reverse fault in which a fault plane dips less than 45 degrees.

tide The regular fluctuation of the sea's surface in response to the gravitational attraction of the Moon and Sun.

till All sediment deposited directly by glacial ice.

time-stratigraphic unit A body of strata that was deposited during a specific interval of geologic time; for example, the Devonian System (a time-stratigraphic unit) was deposited during the Devonian Period (a time unit).

time unit Any of the units such as eon, era, period, epoch, and age, used to refer to specific intervals of geologic time.

Tippecanoe Sequence A widespread body of sedimentary rocks bounded above and below by unconformities that was deposited during an Ordovician to Early Devonian transgressive–regressive cycle of the Tippecanoe Sea.

tombolo A type of spit that extends out from the shoreline and connects the mainland with an island.

trace fossil Any indication of prehistoric organic activity such as tracks, trails, and nests preserved in rocks. (*See also* body fossil)

Transcontinental Arch The area extending from Minnesota to New Mexico that stood above sea level as several large islands during the Cambrian transgression of the Sauk Sea.

transform fault A fault along which one type of motion is transformed into another; commonly displaces oceanic ridges; on land, recognized as a strike-slip fault, such as the San Andreas Fault.

transform plate boundary Plate boundary along which plates slide past one another and crust is neither produced nor destroyed.

transverse dune A ridge of sand with its long axis perpendicular to the wind direction.

tree-ring dating The process of determining the age of a tree or wood in a structure by counting the number of annual growth rings.

tsunami A large sea wave that is usually produced by an earthquake, but can also result from submarine landslides and volcanic eruptions.

ultramafic magma Magma with less than 45% silica. (*See also* mafic, intermediate, and felsic magma)

unconformity A break in the geologic record represented by an erosional surface separating younger rocks from older rocks. (*See also* nonconformity, disconformity, *and* angular unconformity)

ungulate An informal term referring to the hoofed mammals, especially the orders Artiodactyla and Perissodactyla.

upwelling The slow circulation of ocean water from depth to the surface. (*See also* downwelling)

U-shaped glacial trough A valley with steep or vertical walls and a broad, rather flat floor formed by the movement of a glacier through a stream valley.

valley A linear depression bounded by higher areas such as ridges or mountains.

valley glacier A glacier confined to a mountain valley or an interconnected system of mountain valleys.

valley train A long, narrow deposit of stratified drift confined within a glacial valley.

vascular The specialized tissue found in most land plants for transporting nutrients and fluids.

velocity A measure of distance traveled per unit of time, as in the flow velocity in a stream or river.

ventifact A stone with a surface polished, pitted, grooved, or faceted by wind abrasion.

vertebrate Any animal that has a segmented vertebral column, as in fish, amphibians, reptiles, birds, and mammals.

vesicle A small hole or cavity formed by gas trapped in cooling lava.

vestigial structure A structure in an organism that no longer serves any, or a limited, function, such as dewclaws in dogs and wisdom teeth in humans, or a structure that serves a completely different function, such as two of the middle ear bones in mammals.

viscosity A fluid's resistance to flow.

volcanic ash Pyroclastic materials that measure less than 2 mm.

volcanic explosivity index (VEI) A semiquantative scale for the size of a volcanic eruption based on evaluation of criteria such as the volume of material explosively erupted and the height of the eruption cloud.

volcanic neck An erosional remnant of the material that solidified in a volcanic pipe.

volcanic pipe The conduit connecting the crater of a volcano with an underlying magma chamber.

volcanic rock An igneous rock formed when magma is extruded onto Earth's surface where it cools and crystallizes, or when pyroclastic materials become consolidated. (*See also* extrusive igneous rock)

volcanic tremor Ground motion lasting from minutes to hours, resulting from magma moving beneath the surface, as opposed to the sudden jolts produced by most earthquakes.

volcanism The processes whereby magma and its associated gases rise through the crust and are extruded onto the surface or into the atmosphere.

volcano A hill or mountain formed around a vent as a result of the eruption of lava and pyroclastic materials.

water table The surface that separates the zone of aeration from the underlying zone of saturation.

water well A well made by digging or drilling into the zone of saturation.

wave An undulation on the surface of a body of water, resulting in the water surface rising and falling.

wave base The depth corresponding to about one-half wavelength, below which water in unaffected by surface waves.

wave-cut platform A beveled surface that slopes gently seaward; formed by the erosion and retreat of a sea cliff.

wave refraction The bending of waves so that they move nearly parallel to the shoreline.

weathering The physical breakdown and chemical alteration of rocks and minerals at or near Earth's surface.

Wilson cycle The opening and then closing of an ocean basin as a result of plate movements.

zone of accumulation The part of a glacier where additions exceed losses and the glacier's surface is perennially covered with snow. Also refers to horizon B in soil where soluble material leached from horizon A accumulates as irregular masses.

zone of aeration The zone above the water table that contains both air and water within the pore spaces of soil, sediment, or rock.

zone of saturation The area below the water table in which all pore spaces are filled with water.

zone of wastage The part of a glacier where losses from melting, sublimation, and calving of icebergs exceed the rate of accumulation.

Zuni epeiric sea A widespread sea present in North America mostly during the Cretaceous; however, it persisted into the Paleogene.

INDEX

Note: *Italicized* page numbers indicate illustrations.